Fourth E.C.
Photovoltaic Solar Energy Conference

The Photovoltaic Solar Energy Conferences
are organized on the initiative of the Solar Energy Division,
Directorate-General Science, Research and Development,
Commission of the European Communities, Brussels

Publication arrangements: D. NICOLAY
                              Directorate-General Information Market
                              and Innovation, Commission of the European Communities,
                              Luxembourg

Commission of the European Communities

Fourth E.C. Photovoltaic Solar Energy Conference

Proceedings of the International Conference, held at Stresa, Italy, 10-14 May, 1982

Edited by

W. H. BLOSS

University of Stuttgart

and

G. GRASSI

Commission of the European Communities, Brussels

D. REIDEL PUBLISHING COMPANY

DORDRECHT : HOLLAND / BOSTON : U.S.A.
LONDON : ENGLAND

Library of Congress Cataloging in Publication Data

Photovoltaic Solar Energy Conference (4th : 1982 : Stresa, Italy)
   Fourth E.C. Photovoltaic Solar Energy Conference.

   At head of title: Commission of the European Communities.
   Includes index.
   1. Photovoltaic power generation—Congresses. I. Bloss, W. H.
(Werner H.), 1930–   II. Grassi, G., 1929–   III. Commission
of the European Communities. IV. Title.
TK2960.P47   1982   333.79'23   82-9122
                                                    AACR2
ISBN-13:978-94-009-7900-0   e-ISBN-13:978-94-009-7898-0
DOI: 10.1007/978-94-009-7898-0

Organization of the conference by
Commission of the European Communities
Directorate-General Science, Research and Development, Brussels
in co-operation with the Italian Ministries of Research and Industry
and the Institute of Electrical and Electronic Engineers, New York

Publication arrangements by
Commission of the European Communities
Directorate-General Information Market and Innovation, Luxembourg

EUR 8042 EN
Copyright © 1982, ECSC, EEC, EAEC, Brussels and Luxembourg
Softcover reprint of the hardcover 1st edition 1982
LEGAL NOTICE
Neither the Commission of the European Communities nor any person acting on behalf of the
Commission is responsible for the use which might be made of the following information.

Published by D. Reidel Publishing Company
P.O. Box 17, 3300 AA Dordrecht, Holland

Sold and distributed in the U.S.A. and Canada
by Kluwer Boston Inc.,
190 Old Derby Street, Hingham, MA 02043, U.S.A.

In all other countries, sold and distributed
by Kluwer Academic Publishers Group,
P.O. Box 322, 3300 AH Dordrecht, Holland

D. Reidel Publishing Company is a member of the Kluwer Group

All Rights Reserved
No part of the material protected by this copyright notice may be reproduced or utilized
in any form or by any means, electronic or mechanical, including photocopying,
recording or by any informational storage and retrieval system,
without written permission from the copyright owner.

INTRODUCTION

by the General Chairman

Prof. W.H. BLOSS
Universität Stuttgart, Federal Republic of Germany

The 1982 Photovoltaic Solar Energy Conference, the fourth organized by the Commission of the European Communities, provided an international platform for the assessment of photovoltaic power generation, its present potential and its future developments.

Photovoltaic generators, which are now employed in a large number of demonstration and pilot projects, have received growing attention and commercial interest. Progress in cell and module production has led to improved economy in operation and brought us nearer the goal of large scale electricity production. New approaches in science and technology offer promising potential for the future.

The programme of the conference included invited lectures, presentations, poster sessions and an exhibition, which all reflected the scientific and technological state-of-the-art as well as the recent achievements of commercial products and their applications. Moreover it provided an ideal opportunity for individual discussions and exchange of information among experts in the field of photovoltaics.

The high reputation of the European Photovoltaic Conferences was clearly demonstrated by the large number of participants and by the enthusiastic response to the Call for Papers. Altogether 260 abstracts were received from 30 countries. However, because of time and other limitations, the reviewers had to select only 40 papers for oral presentation and 125 papers for the poster sessions. From the accepted papers, the Committee assembled an attractive conference programme which is published in this volume of proceedings.

Podium during the opening session (general view)

The audience at the opening session

CONFERENCE NOTEBOOK

F.C. TREBLE
Consultant to the Commission of the European Communities

The 1982 EC Photovoltaic Solar Energy Conference at Stresa, the fourth in a series of international gatherings which began at Luxembourg in 1977, can perhaps claim to have achieved the best balance so far between the interests of the 600 scientists, engineers, system designers, marketing executives, consultants, administrators and potential users who attended. Of the 186 papers presented, 85 were concerned with forward-looking aspects such as fundamental studies, advanced devices and thin-film solar cells, while 101 dealt with topics of more immediate interest such as crystalline silicon solar cell technology, components, systems and applications. Poster sessions, first introduced at the Berlin Conference in 1979, again proved their worth, enabling those interested in the more specialised papers to study them at leisure and discuss them with the authors. They also enabled the organisers to avoid parallel oral sessions, which tend to segregate the various disciplines instead of bringing them together to their mutual benefit.

In the opening session, Dr. W. Palz of the CEC reviewed the Commission's activities in photovoltaics. In the current programme, it was spending 6.5M ECU on research and development, 9.4M ECU (about one third of the total cost) on 16 pilot projects and 9M ECU on module qualification testing at the Joint Research Centre, Ispra. An additional 1.1M ECU had been spent on demonstration projects and 3M ECU on co-operative projects in Africa. It was hoped to start a new programme in July, 1983, when the current one comes to an end.

Dr. Farinelli of ENEA, the Italian national agency for the development of renewable energies, said that his Government's policy was based on the assumption of steady exponential growth from the present market for small-scale remote power applications, through an intermediate phase of implementation in developing countries to grid-connected applications in the year 2000 and beyond. He believed that the next few years would be crucial in the development of the market and said that his organisation intended to encourage both innovation and increased production to bring down costs. ENEA's expenditure on photovoltaics in 1982-4 would be $ 55M, about 20% of their total budget. Their plans included the construction of a 1.1 MWp photovoltaic generator.

Dr. M.B. Prince of the US Department of Energy revealed that, because of the success of their 1974-80 programme in stimulating the new photovoltaic industry, future Government aid, with a 1982 budget of $ 78M, would concentrate on basic and applied research, exploratory development and high risk technical development. In the latter category, he mentioned the Sacramento Municipal District (SMUD) 1 MWp photovoltaic plant, which will be built at an estimated cost of $ 12M. He advocated more international co-operation in the way of meetings, information transfer, standards and design review consultation.

A highlight of the Applications Session was an invited paper by Dr. M.R. Starr of Sir William Halcrow & Partners, Swindon, UK on the potential of photovoltaics in Europe. This was an extract from a recent independent study carried out under contract to the Commission. He pointed out that the growth of the market would depend not only on cost reduction but also on the escalation of electricity prices, discount rates, tax incentives, support by governments and international agencies, demonstration projects and the effective dissemination of information to the public at large. He foresaw that, by 1990, module prices would reach $1-2/Wp and system prices $3-6/Wp. On the assumption that the cost of electricity increases in real terms at an average rate of 5% per annum, he forecast that photovoltaic generators could become competitive with small diesels in southern Europe by the mid-1980s, with larger diesels and gas turbines by the late 1980s and with grid supplies by the mid-1990s. For central and northern Europe, the corresponding break-even dates would be some 3 to 5 years later. By the year 2000, he foresaw a market of at least 1000 MWp per annum.

Other interesting applications papers concerned water pumping, desalination, rural electrification, cooling, refrigeration and the use of solar power for houses, schools, commerce and industry.

Dr. K. Krebs of the CEC Joint Research Centre reported on the performance, qualification and durability tests which the Centre is carrying out on European modules as part of the Commission's pilot projects programme. He pointed out that long-term reliability was an important aspect of life-cycle cost reduction. The modules tested so far had stood up well but the variety of shapes and sizes indicated the need for some measure of standardisation.

The need for standardising array field designs was also stressed in a paper by Dr. E.L. Burgess of Sandia Laboratories, Albuquerque, NM. He called for more work on low-cost array structures, simplified wiring and cheaper power conditioning equipment.

Dr. E. Fabre of Photowatt International, Rueil Malmaison, France reviewed the state-of-the-art in flat plate silicon module technology. He said that, because of reductions in the costs of cell processing and module assembly, the silicon wafers now accounted for half the cost of the module, as compared with one third in 1977. This lent added urgency to the development of cheaper starting materials. He listed the various ingot and ribbon approaches and hazarded a guess that ribbon made from solar grade silicon may come into general use by 1986.

Dr. E. Sirtl of Wacker Heliotronic, Burghausen, Germany reported on the new silicon casting techniques being developed by his company. With Silso, their cast polycrystalline product, they had progressed to larger ingots and were now studying further cost reduction and the optimum charge size for large scale production. He described a continuous casting process for the manufacture of silicon ribbon, in which special cooling arrangements allowed faster growth. Another interesting paper in this session was that by Dr. H.A. Aulich of Siemens, Munich, describing how solar grade silicon may be cheaply produced by using high purity silica and carbon in an arc furnace. Pilot experiments with this method have yielded encouraging results. Progress in the development of the French RAD ribbon process was reported by Dr. C. Belouet of the Laboratoire de Marcoussis. Cells, 4 cm^2 in area, made from the latest batch of ribbon have shown an average conversion efficiency of 10%.

Papers on low-cost silicon cell processing employing automated screen printing techniques for junction diffusion, antireflective coating, back surface field formation and contact deposition were presented by Dr. K.-D. Rasch of AEG-Telefunken, Heilbronn and Dr.G. Cheek of the Catholic University of Leuven, Belgium. The University has manufactured several thousand cells using these processes and claimed to have achieved an average efficiency of 11.2%, with a 95% yield of cells over 10% in efficiency.

No less than four papers in the Fundamental Studies Session reported on studies of grain boundary effects and carrier lifetime in polycrystalline silicon, reflecting the intensive efforts being made to improve the performance of solar cells made with this material.

An overview of progress in the development of ion implantation in silicon cell manufacture was given in an invited paper by Dr. P. Siffert of CRN,

Strasbourg, France. He said that the annealing problem still remained to be solved but predicted that the technique would be in production in two years' time. Other news of progress in low-cost silicon and solar cell processing was presented in posters.

Despite the continued progress with silicon, it is generally agreed that cost reduction beyond the $1/Wp mark will depend on the successful development of an efficient, stable thin-film solar cell. Judging by this Conference amorphous silicon appears to have displaced cadmium sulphide/copper sulphide as the front runner in this field. None of the CdS/Cu_xS papers had any news to rival the announcement of a 10% cell at the Cannes Conference in 1981, although a team from the University of Delaware, which made the announcement, presented a progress report on their efforts to make such cells by a continuous deposition process. Runs producing up to 3000 cm^2 of 25 $micron_2$ CdS have been carried out at throughputs of 400 cm^2/hour. To date, 1 cm^2 cells processed from this material have shown efficiencies of over 7.5%.

The highlight of the amorphous silicon papers was the claim by Dr. Y. Tawada of Osaka University and Dr. Y. Kuwano of Sanyo that they had achieved an efficiency of 8% in 3.3 mm^2 p-i-n heterojunction cells. The cell structure was glass/ITO/p(aSiC)-i(aSi)-n(aSi). Dr. Kuwano described the technique of using separate reaction chambers for depositing the successive p, i and n layers, which reduces intermixing of the dopants. Sanyo have pushed on with the manufacture of 10 cm x 10 cm modules of 9 such cells series connected on a glass substrate, followed by 45 cm x 60 cm panels of 20 modules. They have also constructed a 2kW demonstration array on a house. The second phase of a production facility to manufacture amorphous silicon cells at the rate of 1.5MW/year is going ahead. Despite this confidence in Japan, some doubts remain regarding the stability of a-Si cells when continuously exposed to sunlight. This is being studied by a research team at Siemens, Munich, who presented a poster on the subject.

Among other thin-film approaches featured in the papers were CdS/CdTe, CdS/Cu ternary alloy, Zn_3P_2, InP/SnO_2-Sb, ITO/CdTe and ITO/CdSe. In an interesting review of advanced devices, Dr. D.L. Feucht of SERI, Golden, CO mentioned that an efficiency of 9.9% had been achieved in a $CdZnS/CuInSe_2$ cell, 8.5% in thin-film GaAs, 6.2% in CdTe and 8.7% in ITO/InP. With concentrator cells, 22% had been reached with a metal-interconnected AlGaAs/GaAs cascade cell at 103 suns.

Reviewing concentrator technology in the USA, Dr. E.C. Boes of Sandia disclosed that his laboratory has dropped all reflective concentrators from their programme of photovoltaic R&D. Development of point focusing and linear Fresnel lenses continues, however. Interest in fluorescent concentrators was also being maintained, although efficiencies were still very low.

A well organised exhibition supported by 45 companies and agencies from Europe and USA provided concrete evidence of the advances the industry has made in solar cells, modules and other photovoltaic products. One of the exhibitors, Arco Solar, a subsidiary of the Atlantic Richfield Oil Co., announced their plan to build a 1MWp flat-plate pv generator with 2-axis tracking on a 20 acre site in San Bernardino, CA. The plant, to be completed in December, 1982, will be owned and operated by Arco, who will sell the electricity to Southern California Edison, the local utility.

The general impression left by the Conference was one of continued steady progress in this important branch of solar technology. But it will also be remembered for the massed blooms of azalea, camelia and wistaria in the beautiful lakeland scenery of Stresa and the sunny weather that prevailed throughout the five days.

Dr. DINKESPILER during his opening speech

The conference committee from left to right :
MM. Chidester, Bucher, Strub, Fabre, Grassi, Koepke, Luque,
Van Overstraeten, Barnett, Mertens, Hamakawa, Nicolay, Makios, Bloss,
Schnell, Macomber, Pfisterer, Beresowski, Schock, Krebs, Brandhorst,
Treble, Kipperman, Broser, Palz, Sirtl

CONFERENCE COMMITTEE

General Chairman
Prof W.H Bloss
Univ of Stuttgart
Stuttgart, F R Germany

Technical Programme Chairmen
Dr K Krebs
JRC/CEC
Ispra, Italy

Prof R van Overstraeten
K U Leuven
Leuven, Belgium

Conference Secretariat
Dr G Grassi
CEC
Brussels, Belgium

Mr F Pfisterer
Univ of Stuttgart
Stuttgart, F R Germany

Mr W Schnell
CEC
Brussels, Belgium

Mr F C Treble
Consultant
Farnborough, Hants, UK

Local Organization
Dr M P Moretti
JRC/CEC, Ispra Establishment
Press & Public Relations
Ispra, Italy

Publications
Mr D Nicolay
CEC
Luxembourg

Posters
Mr H W Schock
Univ of Stuttgart
Stuttgart, F R Germany

Mr F C Treble
Consultant
Farnborough, Hants, UK

Exhibition
Mrs M P Moretti
JRC/CEC
Ispra, Italy

Mr W Schnell
CEC
Brussels, Belgium

Executive Board
Dr T Beresovski
UNESCO
Paris, France

Prof H Durand
COMES
Paris, France

Prof F Fittipaldi
ENEL
Rome, Italy

Dr H Klein
BMFT
Bonn, F R Germany

Dr L Magid
Solarex
Rockville, Ma, USA

Dr R M S Obaid
Nat Center for Sci, and Technol
Riyadh, Saudi Arabia

Dr W Palz
CEC
Brussels, Belgium

Dr A S Strub
CEC
Brussels, Belgium

Members
Prof C E Backus
Arizona State University
Tempe, Arizona, USA

Dr C J Bishop
SERI
Golden, Colorado, USA

Dr K Bogus
ESTEC
Noordwijk, Netherlands

Dr H W Brandhorst, Jr
NASA Lewis
Cleveland, Ohio, USA

Dr E L Burgess
Sandia Labs
Albuquerque, NM, USA

Mr L G Chidester
Lockheed
Sunnyvale, Ca, USA

Dr E DeMeo
EPRI
Palo Alto, Ca, USA

Mr A A Dollery
Royal Aircraft Est
Farnborough, Hants, UK

Dr E Fabre
Photowatt
Argenteuil, France

Prof H Gerischer
Fritz Haber Institut
Berlin, F R Germany

Mr E M Greby
Dansk Energi Teknik
Kopenhagen, Denmark

Prof Y Hamakawa
Osaka University
Osaka, Japan

Dr A H M Kipperman
University of Technology
Eindhoven, Netherlands

Dr R Koepke
KFA Julich
Julich, F R Germany

Dr K M Koliwad
JPL
Pasadena, Ca, USA

Mr M P Lequeux
CEC
Brussels, Belgium

Prof A Luque
Ciudad Universitaria
Madrid, Spain

Dr H Macomber
Monegon Ltd
Gaithersburg, Ma, USA

Prof V Makios
University of Patras
Patras, Greece

Prof D Nobili
CNR Lamel
Bologna, Italy

Prof E J Perez
Instituto Politecnico Nacional
Mexico, D F

Prof S Pizzini
Heliosil
Milano, Italy

Mr E Ralph
Spectrolab
Sylmar, Ca, USA

Prof F Rueda
Universidad Autonoma
Madrid, Spain

Ing G Simoni
AGIP Nucleare SpA
Rome, Italy

Prof E Sirtl
Wacker Heliotronic
Burghausen,
F R Germany

Dr A Taschini
ENEL
Milano, Italy

Dr G Wrixon
University College
Cork, Ireland

Guest Reviewers:
Dr A M Barnett
University of Delaware, USA

Prof I Broser
Technische Universitat Berlin, F R Germany

Prof E Bucher
Universitat Konstanz, F R Germany

Dr T F Ciszek
Solar Energy Research Institute, Golden, Co, USA

Dr G H Hewig
Universitat Stuttgart, F R Germany

Dr F L Hesse
Jet Propulsion Laboratory, USA

Dr Y Marfaing
CNRS, Meudon, France

Dr R Mertens
Katholieke Universiteit Leuven, Belgium

C O N T E N T S

Introduction
   Prof. W. BLOSS, University of Stuttgart, F.R. Germany   v

Conference notebook
   F.C. TREBLE, Consultant to the Commission of the European
   Communities   vii

Conference committee   xi

OPENING SESSION

Overview of the European Community's activities in photo-
voltaics
   W. PALZ, Directorate General for Science, Research and
   Development, Commission of the European Communities   3

The Italian programme in photovoltaic solar energy
   Prof. U. FARINELLI, ENEA, National Committee for Research
   and Development of Nuclear and Alternative Energies,
   Italy   9

Opening address - The ENEL's programmes in the field of
photovoltaic conversion
   Prof. F. FITTIPALDI, Member of the Board of Directors of
   the ENEL, Italy   13

Opening address - CNR programme for photovoltaics
   Prof. G. ELIAS, Director of "Progetto finalizzato energetica" - CNR, Italy   17

United States federal photovoltaic program status
   M.B. PRINCE and A.L. BARRETT, jr., Photovoltaic Energy
   Technology Division, U.S. Department of Energy, USA   20

SESSION I - APPLICATIONS

Application trends for photovoltaics
    Dr. H.L. MACOMBER, Vice President, MONEGON Ltd, USA   30

The potential for photovoltaics in Europe
    M.R. STARR, Sir William Halcrow and Partners, United
    Kingdom   40

The behaviour of large solar power stations in the Swiss
Alps
    M. REAL, Project Manager, Division for Solar Elec-
    tricity, Swiss Federal Institute for Reactor Research,
    Switzerland   51

Design, installation, and initial performance of 350-kW
photovoltaic power system for Saudi Arabian villages
    F. HURAIB, Dr. B. KHOSHAIM and A. AL-SANI, Midwest
    Research Institute, USA;
    M.S. IMAMURA and A.A. SALIM, Martin Marietta Corporation,
    USA   57

The photovoltaic-powered water desalination plant "SORO" -
Design, start up, operating experience
    G. NEUHäUSSER, J. MOHN, and G. PETERSEN, AEG-Telefunken,
    GKSS-Forschungszentrum Geesthacht GmbH, F.R. Germany   67

The UNDP World Bank solar pump project - preparing for phase II
    P.L. FRAENKEL, Intermediate Technology Power Ltd, United
    Kingdom;
    M.A.S. MALIK, Energy Department, World Bank, USA;
    D.E. WRIGHT, Sir William Halcrow & Partners, United
    Kingdom   74

Installation et évaluation d'un système simple de pompage
photovoltaique dans un village algérien
    A. MOUHOUB and M. BENMALEK, Laboratoire Cristaux et
    Couches Minces, Centre des Sciences et de la Technologie
    Nucléaires, Algérie   81

POSTER GROUP P1 - APPLICATIONS

- Intermediate

Advanced system design for solar power plants
    V. CORDES and K.H. KORUPP, AEG-Telefunken Neue Techno-
    logien, Raumfahrt, F.R. Germany   89

The Mississippi county community college large-scale demonstration project - a success story
   H.V. SMITH, President, Mississippi County Community College   94

TISO 15 - 15kW experimental photovoltaic solar power plant
   M. CAMANI, Dipartimento dell'ambiente, Bellinzona;
   D. BOZZOLO, O. DALDINI, R. PAMINI, G. SALVADE',
   F. SOLCA', C. SPINEDI and F. ZAMBONI, Laboratorio di fisica terrestre, Canobbio;
   T. CELIO, Ufficio d'ingegneria per l'elettroottica, Ambri;
   C. GIOVANNINI, Invertomatic, Locarno, Switzerland   97

- Residential

Photovoltaic retrofit feasibility in the United States
   J.L. JACKSON, PV System Definition Division 4723, Sandia National Laboratories, USA   101

Experiments on combined photovoltaic-aeolian electric generation in a residential stand-alone system
   P.U. CALZOLARI, Istituto di Elettronica, Facoltà di Ingegneria, Università di Bologna, Italy;
   G.C. CARDINALI, A. GARULLI, D. NOBILI and A. ZANI, CNR - Istituto LAMEL, Bologna, Italy   106

Simulations of the energy performance of a solar photovoltaic residence and hybrid electric automobile in Fresno, California
   J.S. REUYL and R.D. SCHUTT, JSR Associates, USA   111

Alicudi project
   V. ARCIDIACONO, S. CORSI, A. ILICETO, A. PREVI and A. TASCHINI, ENEL, Direzione Studi e Ricerche, Italy   115

A successful integration of solar energy in African life : Kolokani, Mali
   L. LEVHA, Elf Solaire, Développements énergétiques, France   120

Application and experience of photovoltaic pumps for irrigation in Pakistan
   R.G. PALLETT and T.E. BRABBEN, Overseas Development Unit, Hydraulics Research Station Limited, United Kingdom   125

Development and testing of a submersible motor pump driven by a solar cell generator
   Dr. E. PICMAUS and K.P. KIESSLING, KSB - Klein, Schanzlin & Becker AG, F.R. Germany   130

- Other applications

Solar generator performance with load matching to water electrolysis - longterm averages and range of instantaneous efficiencies
K. FREUDENBERG, Sektion Physik der Universität München, F.R. Germany ... 135

Photovoltaic power for walk-in coolers
L. SELLES and B. AUBERT, SERI Renault Ingénierie, France ... 139

Technical and economic requirements for small-scale photovoltaic refrigerators
M.R. STARR, Sir William Halcrow and Partners;
B. McNELIS, Intermediate Technology Power Ltd, United Kingdom ... 144

Residential applications of photovoltaics in the United States
S.J. STRONG, Solar Design Associates, USA ... 151

SESSION 2 - EXPERIENCE, PERFORMANCE, RELIABILITY, MONITORING

Performance, testing and module monitoring at the EC : necessary steps to develop cost-effective PV modules
K. KREBS, Commission of the European Communities, Joint Research Centre, Ispra Establishment ... 158

Reliability and performance experience with flat-plate photovoltaic modules
R.G. ROSS, jr., Jet Propulsion Laboratory, USA ... 169

Operational experience with intermediate flat-plate photovoltaic systems
V.V. RISSER and H.S. ZWIBEL, New Mexico Solar Energy Institute, New Mexico State University, USA ... 179

Operational experience with a 35-kWp concentrating photovoltaic system
R.M. SPENCER, Acurex Corporation, USA ... 184

Solar photovoltaic systems in the development of Papua New Guinea
G.H. KINNELL, Project Evaluation Engineer, Department of Minerals and Energy, Papua New Guinea ... 189

Field trial of rural solar photovoltaic system
P. BASU, K. MUKHOPADHYAY, T. BANERJEE, S. DAS and H. SAHA, Department of Physics, University of Kalyani, India ... 203

POSTER GROUP P2 - EXPERIENCE, PERFORMANCE, RELIABILITY, MONITORING

- Experience and Reliability

Study of a photovoltaic concentrating system like "Sophocle" in fluctuating mode
   B. LAURENT, PHAM VAN VUI and G. VIALARET, CNRS, Laboratoire d'Automatique et d'Analyse des Systèmes, France   212

Performance of 1kW peak concentrating photovoltaic array
   G. SALA and F. CHENLO, Instituto de Energia Solar, ETSIT, Universidad Politécnica Madrid, Spain   217

Photovoltaic concentrator module characterization
   H.J. GERWIN, Experimental Facilities Operations Division, Sandia National Laboratories, USA   222

- Performance and Monitoring

Results from a test facility for solar cells in Sweden
   J. HEDSTRÖM, B. KäLLBäCK and D. SIGURD, Institute of Microwave Technology, Sweden   227

A microprocessor-based instrument for automatic solar cell characterization
   G.C. CARDINALI, CNR, Istituto LAMEL, Bologna;
   E. FALDELLA, Istituto di Elettronica and Centro Interazione Operatore-Calcolatore, Università di Bologna, Italy   232

SESSION 3 - SYSTEMS, COMPONENTS AND ENGINEERING

Subsystem engineering and development of grid-connected photovoltaic systems
   E.L. BURGESS, H.N. POST and T.S. KEY, Sandia National Laboratories, USA   238

Array structures for fixed flat-plate photovoltaic power generators
   G. GRASSI, Commission of the European Communities, Directorate General for Research, Science and Development   248

Inverters for PV plants feeding the electrical network of small isolated communities
   V. ARCIDIACONO and A. TASCHINI, Direzione Studi e Ricerche, ENEL, Italy   258

POSTER GROUP P3 - SYSTEMS, COMPONENTS AND ENGINEERING

- Array design

A support structure for intermediate PV solar plant
    V. ALBERGAMO, Delphos Project, ENEA, Istituto Casaccia;
    P.L. BORLENGHI, Studio Moretti;
    F. CHELI and G. DIANA, Politecnico di Milano, Dipartimento di Meccanica;
    M. FALCO, Università di Catania, Istituto di Macchine, Italy   270

Low cost modular designs for photovoltaic array fields
    H.N. POST, Sandia National Laboratories, Albuquerque;
    D.C. CARMICHAEL, Battelle Laboratories, Columbus;
    J.A. CASTLE, Hughes Aircraft Co., Los Angeles, USA   275

The photovoltaic solar system, analysis and basic design rules
    A.H.M. KIPPERMAN, Eindhoven University of Technology, Eindhoven, The Netherlands   280

A detailed package of digital codes specially developed for the array system
    P. BULLO and M. GASBARRA, ENEL, Direzione Studi e Ricerche;
    G. EMANUELE, Phoebus, Catania, Italy   286

- Shadowing, Hot-Spots

Analysis of hot-spot-effects in encapsulated photovoltaic generators by laser scan and partial shadowing
    G.H. HEWIG and H.-P. HÜBNER, Institut für Physikalische Elektronik, Universität Stuttgart, F.R. Germany   291

Reverse bias power dissipation of shadowed or faulty cells in different array configurations
    P. SPIRITO, Istituto Elettrotecnico, University of Naples, Italy;
    V. ALBERGAMO, ENEA, Istituto Casaccia, Italy   296

Disequilibriums in series connected solar cells - an approach to the protection by parallel diodes
    J.A. ROGER, S. MASSAAD, J. POSBIC and J. PIVOT, Département de Physique des Matériaux, Université Lyon I, France   301

- Grid Connection and Inverters

Minimum cost of photovoltaic energy for a utility grid and general features of a generating plant using costless solar cells
    D. MADET, Ingénieur à la Direction des Etudes et Recherches, Electricité de France ... 307

A DC/AC modular interface for photovoltaic systems
    M. VAN GYSEL, IDE;
    N. LIMBOURG, ETCA, Belgium ... 312

Simple transformerless inverter with automatic grid-tracking and negligible harmonic content for utility interactive photovoltaic systems
    J. SCHMID and R. SCHäTZLE, Fraunhofer-Institut für Solare Energiesysteme, F.R. Germany ... 316

Regulated converter circuit for direct photovoltaic energy feedback into the power grid
    P. CEPPI, R. ULMI and G. GUEKOS, Institute of Applied Physics, Swiss Federal Institute of Technology, Switzerland ... 320

Power conditioning in solar photovoltaic array applications
    G.J. VACHTSEVANOS, C.K. KALAITZAKIS and E.J. GRIMBAS, School of Engineering, Democritos University of Thrace, Greece ... 325

- Battery Interface

Mismatch between batteries and two module types PV arrays Interest of DC-DC converters
    P. GUCHER, J.A. ROGER, S. MASSAAD, J. POSBIC and J. PIVOT, Service Electronique, Institut de Chimie et Physique Industrielles de Lyon - Département de Physique des Matériaux, Université de Lyon 1, France ... 330

Matching the characterstics of batteries with solar cell modules
    C.F. GAY, V.K. KAPUR, B. PYLE and J. RUMBURG, ARCO Solar, Inc., USA;
    A. MANFREDI, ARCO Solar Europe SpA, Italy ... 335

Calculation to improve power conversion efficiency in photovoltaic systems
    A. MAS, J. GARCIA, L. CLOSAS, M. INSAUSTI and L. CASTANER, ETSITB, Barcelona, Spain ... 340

SESSION 4 - FLAT PLATE MODULE TECHNOLOGY

Flat plate module technology - overview
   E. FABRE, Photowatt International S.A., France   346

AM/PM : The rating system for photovoltaic modules
   C.F. GAY, ARCO Solar, Inc., USA   353

POSTER GROUP P4 - FLAT PLATE MODULE TECHNOLOGY

Recent progress in terrestrial photovoltaic collector technology
   R.R. FERBER, Manager, Collector R&D, Photovoltaic TD&A Lead Center, Jet Propulsion Laboratory, USA   364

Low-cost solar array progress and plans
   W.T. CALLAGHAN, Manager, Flat-Plate Solar Array Project, Jet Propulsion Laboratory, USA   369

An all glass encapsulation of solar cells using silk-screening techniques
   M.C. MICHEL, Laboratoires de Marcoussis, Centre de Recherche de la Compagnie Générale d'Electricité, France   374

Laminated modules with new plastic material based on an ethylene-vinylacetate copolymer
   A. DESOMBRE and J. DONON, Photowatt International S.A.; Y. DE ZELICOURT and J.C. BOBO, Laboratoires de Marcoussis, Centre de recherches de la Compagnie Générale d'Electricité, France   377

Design and development of a module for the EEC PV pilot plants
   A. MULEO, F. NARDI and V. PARAGGIO, Solaris SpA - Pragma SpA, Italy   382

Module design for EC pilot projects
   J. DONON, J. ANGUET and A. DESOMBRE, Photowatt International S.A., Caen;
   P. COUREAU, Photowatt International S.A., Rueil Malmaison, France   387

Solar cells failure modes and improvement of reverse characteristics
   A.M. RICAUD, France-Photon, France   392

Fabrication of large area Cu_2S/CdS thin film solar modules
H. HUSCHKA, B. SCHURICH and J. WÖRNER, NUKEM GmbH, F.R. Germany   399

Front contacts for large area high efficiency Cu_2S-CdS solar cells
W. ARNDT and H.W. SCHOCK, Institut für Physikalische Elektronik, Universität Stuttgart, F.R. Germany   404

SESSION 5 - FUNDAMENTAL STUDIES

Solid solubility and precipitation of phosphorus and arsenic in silicon solar cells front layer
D. NOBILI, CNR, Istituto LAMEL, Italy   410

Grain boundaries and intragrain defects dependence of local and global electronic and photovoltaic properties of CGE polysilicon
J. OUALID, M. ZEHAF, H. AMZIL, J.P. CREST, G. MATHIAN, J. DUGAS, J. GERVAIS, F. MINARI, B. PICHAUD and S. MARTINUZZI, Université d'Aix-Marseille III, CGE; J. FALLY, Les Laboratoires de Marcoussis, Faculté des Sciences et Techniques de Saint-Jérôme, France   421

Gas immersion laser diffusion : A new method for making efficient Si solar cells
G.B. TURNER, D. TARRANT and D. ALDRICH, ARCO Solar Inc.; R. PRESSLEY and R. PRESS, XMR, Inc., USA   427

Validity of the effective lifetime concept in polycrystalline silicon
S.C. JAIN, Solidstate Physics Laboratory, India; R. JANSSENS, G. CHEEK, P. DE PAUW, R. MERTENS and R. VAN OVERSTRAETEN, Katholieke Universiteit Leuven, ESAT Laboratory, Belgium   432

Study of the grain boundary effects in n-p junction polycrystalline Si solar cells
N.C. HALDER, Department of Physics, University of South Florida, USA   437

Studies of the gap states density in undoped and doped amorphous hydrogenated silicon
J. KOCKA, M. VANECEK, J. STUCHLIK, O. STIKA, E. SIPEK, H.T. HA and A. TRISKA, Institute of Physics, Czechoslovak Academy of Sciences, Czechoslovakia   443

Influence of light exposure on the transport properties of a-Si:H films
D. HAUSCHILDT, W. FUHS, H. MELL and K. WEBER, Fachbereich Physik, Universität Marburg, F.R. Germany   448

Pulsed electron beam annealing of ion implanted germanium
for photovoltaic devices
   B. SAUTREUIL, A. LAUGIER and D. BARBIER, Laboratoire de
   Physique de la Matière, Institut National des Sciences
   Appliquées de Lyon;
   A. CACHARD, Département de Physique des Matériaux,
   Université Claude Bernard, Lyon I, France   453

Recent advances in ITO/InP and CdS/InP solar cells
   T.J. COUTTS and N.M. PEARSALL, School of Physics,
   Newcastle upon Tyne Polytechnic, United Kingdom   459

Intensity enhancement in textures optical sheets for solar
cells
   E. YABLONOVITCH, Exxon Research Center, USA   465

Spectrum shifting methods in photovoltaics : an evaluation
of model systems
   F. GALUZZI and E. SCAFE', Laboratori Ricerche ASSORENI,
   Italy   477

POSTER GROUP P5 - FUNDAMENTAL STUDIES

- Theoretical studies

Extrema of majority and minority carrier quasi Fermi levels
in p-n junction solar cells
   A. PIMPALE and P.T. LANDSBERG, Faculty of Mathematical
   Studies, The University, Southampton, United Kingdom   485

High-blocked heterojunction and Schottky barrier solar
cells
   K.W. BöER, University of Delaware and SES, Inc., USA   488

Computer-aided-characterization of the illuminated and dark
current voltage characteristics of solar cells
   P.H. NGUYEN, B. LEPLEY, C. BOUTRIT and S. RAVELET,
   Laboratoire d'Electronique et de Physique des Interfaces,
   Institut des Sciences de l'Ingénieur, France   492

Non linear model for shunt current in terrestrial silicon
solar cells
   J. CABESTANY, ETS Ingenieros de Telecomunicacion, Spain   498

Non-linear increase and decrease of open circuit voltage of
$n^+ p p^+$ silicon solar cells at high illumination level
   C.M. SINGAL and R.V. SINGH, Energy Centre, Department of
   Physics, University of Roorkee, India   503

Series resistance analysis of concentrator cells under high injection conditions
    J.M. RUIZ, M. CID, A. CUEVAS and A. LUQUE, Instituto de Energia Solar - ETSIT, Universidad Politécnica Madrid, Spain   511

- Silicon Solar Cells

The influence of grain-boundary recombination and grain size on the I(V)-characteristics of polycrystalline silicon solar cells
    M. BÖHM, R. KERN and H.G. WAGEMANN, Institut für Werkstoffe der Elektrotechnik, Technische Universität Berlin, F.R. Germany   516

Diffusion length of minority carriers in scanning electron beam annealed silicon
    H.J. SMITH and R. CILLIERS, CSP, Council for Scientific and Industrial Research, South Africa;
    A. BONTEMPS, DRF, Centre d'Etude Nucléaire de Grenoble, France   522

Pathology of solar cell contacts
    B. ROSS, Bernd Ross Associates, USA   527

Design of stable metal-insulator-semiconductor (MIS) solar cells by oxide thickness compensation
    G. RAJESWARAN and W.A. ANDERSON, Department of Electrical and Computer Engineering, State University of New York at Buffalo, USA   532

- Amorphous Silicon

Study of gap states in a-Si:H by transient current spectroscopy
    J. BEICHLER and H. MELL, Fachbereich Physik, Universität Marburg, F.R. Germany   537

Highly conductive boron doped Si-layers prepared by plasma decomposition of SiH_4
    H. SIMON, G. WINTERLING and G. MÜLLER, Messerschmitt Bölkow Blohm GmbH, F.R. Germany   542

A model for analysis of optical measurements carried on a-Si:H films for photovoltaic applications
    L. GUIMARAES, R. MARTINS, A.G. DIAS and R. BARRADAS, Centro de Fisica Molecular das Universidades de Lisboa, Portugal   546

- Other Materials

The assessment of thin film Cu_xS-CdS solar cells using cathodoluminescence techniques
    T.J. CUMBERBATCH and I.D. McINALLY, Thorn EMI plc, Central Research Laboratories, United Kingdom;
    W.K. KE, Peking Institute of Semiconductors, Chinese Academy of Sciences, China;
    B. HAMILTON, Solid State Electronics, UMIST, United Kingdom   551

Photovoltaic effect in SnTe/CdTe junctions
    M. KANE, G.W. COHEN-SOLAL and D. LAPLAZE, Laboratoire des semi-conducteurs et énergie solaire, Faculté des Sciences Dakar;
    G. COHEN-SOLAL, Laboratoire de physique des solides CNRS, France   557

Organic photovoltaic materials : polyacetylene
    J. KANICKI, S. BOUE' and E. VANDER DONCKT, Chimie Organique Physique, Université Libre de Bruxelles, Belgium;
    P. FEDORKO, Department of Physics, Electrotechnical Faculty, Slovak Technical University, Bratislava, Czechoslovakia   562

Compositional analysis of $CuInS_2$ chalcopyrite semiconductor
    H.L. HWANG, L.M. LIU, M.H. YANG, T.F. HUNG, P.Y. CHEN and J.R. CHEN, National Tsing Hua Univeristy, Taiwan, R.O.C.;
    C.Y. SUN, Industrial Technology Reserach Institute Hsin-chu, Taiwan, R.O.C.   568

Studies on CdS/n-InP PEC solar cells
    Y. RAMPRAKASH, S. BASU and D.N. BOSE, Materials Science Centre, Indian Institute of Technology, India   574

SESSION 6 - ADVANCED DEVICES AND CONCENTRATION

Advanced photovoltaic devices
    D.L. FEUCHT, Solar Energy Research Institute, USA   580

High efficiency GaAs solar cells for concentrator and flat plate arrays
    G. GUARINI, CISE SpA, Milan, Italy   591

Photovoltaic concentrator technology in the USA
    E.C. BOES and M.W. EDENBURN, Photovoltaic Concentrator Project, Sandia National Laboratories, USA   600

Luminescent solar concentrators (LSC) : technical and economic requirements for a residential system
    E. BERMAN and P.D. WILDES, ARCO Solar, Inc., USA   606

Recent progress in a residential solar energy system development
    E.L. JOHNSON and J.S. KILBY, Texas Instruments Incorporated, USA   611

High efficiency tandem type solar cells consisting of a-Si:H and a-SiGe:H
    G. NAKAMURA, K. SATO, H. KONDO, Y. YUKIMOTO and K. SHIRAHATA, LSI Research and Development Laboratory, Mitsubishi Electric Corp., Japan   616

High efficiency shallow p^+nn^+ cadmium telluride solar cells
    G. COHEN-SOLAL, D. LINCOT and M. BARBE', CNRS, Laboratoire de Physique des Solides, France   621

POSTER GROUP P6 - ADVANCED DEVICES AND CONCENTRATION

- Advanced Solar Cells

The photovoltaic advanced research and development program in the United States
    J.L. STONE, D.W. RITCHIE, T. SUREK and C.E. WITT, Solar Electric Conversion Research Division, The Solar Energy Research Institute, USA   628

Influence of plasma Si-nitride deposition on the dark I-V curves of MIS contacts for inversion layer solar cells
    R. SCHÖRNER and R. HEZEL, Institut für Werkstoffwissenschaften VI, Universität Erlangen-Nürnberg, F.R. Germany   638

Limitations of the open circuit voltage of induced junction silicon solar cells due to surface recombination
    R. GIRISCH, R.P. MERTENS and R. VAN OVERSTRAETEN, ESAT Laboratory, Katholieke Universiteit Leuven, Belgium   643

Some comments on sprayed ITO/semiconductor solar cells
    J.C. MANIFACIER, H. LUQUET, L. GOUSKOV, C. GRIL, A. OEMRY and A. CHAOUI, Centre d'Etudes d'Electronique des Solides, Université des Sciences et Techniques du Languedoc, France   648

Role of photoluminescence in the efficiency of a $Ga_{1-x}Al_xAs$-GaAs solar cell
    P. BARUCH and M. CUNIOT, Groupe de Physique des Solides de l'Ecole Normale Supérieure, Université Paris 7, France   654

Operating characteristics of thin thermophotovoltaic cells with minority carrier mirrors and optical mirrors using selective radiators of erbium and ytterbium oxides
    E.S. VERA, M. SPITZER, Department of Physics and J.J. LOFERSKI, Division of Engineering, Brown University; J. SEVERNS, Naval Research Laboratory, USA   659

- <u>Concentrators</u>

Lambertian analysis of mirrors and Fresnel lenses for solar concentration
    A. LUQUE and E. LORENZO, Instituto de Energia Solar, ETSIT, Universidad Politécnica Madrid, Spain   666

750 suns concentrator modules using GaAs solar cells
    E. FANETTI, C. FLORES, G. GUARINI and F. PALETTA, CISE SpA, Milan, Italy   671

A 500 W_{pk} photovoltaic concentrator using a glass laminated metal membrane reflector
    W. HAAF, Schlaich & Partner, Stuttgart; K. HAGENLOCHER, Zeppelin Metallwerke, Friedrichshafen, F.R. Germany   677

Fluorescent planar concentrator (FPC) : Monte-Carlo computer model - Limit efficiency and latest experimental results
    K. HEIDLER, A. GOETZBERGER and V. WITTWER, Fraunhofer-Institut für Solare Energiesysteme, F.R. Germany   682

SESSION 7 - THIN FILM SOLAR CELL TECHNOLOGY

Physical limitations of present thin film solar cells
    Y. MARFAING, Laboratoire de Physique des Solides, CNRS, France   688

8% efficiency a-SiC:H/a-Si:H heterojunction solar cells
    Y. TAWADA, K. TSUGE, M. KONDO, K. NISHIMURA, H. OKAMATO and Y. HAMAKAWA, Faculty of Engineering Science, Osaka University, Japan   698

Amorphous silicon solar cells produced by a consecutive, separated reaction chamber method
    Y. KUWANO, M. OHNISHI, S. NAKANO, T. FUKATSU,
    H. NISHIWAKI and S. TSUDA, Research Centre, SANYO
    Electric Co., Ltd., Japan   704

Charge collection in a-Si:H solar cells
    G. MÜLLER, G. MÜCK, M. SIMON and G. WINTERLING,
    Messerschmitt Bölkow Blohm GmbH, F.R. Germany   709

The effect of glow discharge excitation frequency on the performance of microcrystalline Si:H thin films and devices
    R.R. GAY, D.L. MOREL, D.P. TANNER, D. KANANI and
    H.S. ULLAL, ARCO Solar, Inc., USA   714

Electrodeposited CdS/CdTe heterojunction solar cells
    B.M. BASOL, R.L. ROD and E.S. TSENG, Monosolar Inc., USA   719

Thin film heterojunction CdS/Cu ternary alloys - solar cells with minority carrier mirrors
    M. KWIETNIAK, J.J. LOFERSKI, R. BEAULIEU, R.R. ARYA and
    E. VERA, Division of Engineering, Brown University;
    L. KAZMERSKI, Solar Energy Research Institute, USA   727

Large area CdS/Cu_xS thin film solar cells produced by electrophoretic deposition
    T.J. CUMBERBATCH, I.D. McINALLY, E.W. WILLIAMS,
    D.J. GIBBONS, M. CLAYBOURN, H. CLOW and P.M.G. DICKINSON,
    THORN EMI plc, Central Research Laboratories;
    R. HILL and N.M. PEARSALL, Newcastle Polytechnic;
    J. WOODS, G. RUSSELL and P.C. PANDE, Science Laboratories,
    Durham University, United Kingdom   732

Sprayed zinc-cadmium sulfide films for backwall $Cu_2S/(ZnCd)S$ cells
    V.P. SINGH, M.C. BOST, J.F. JORDAN and D.M. SPITZER, jr.,
    Photon Power, Inc., USA   737

POSTER GROUP P7 - THIN FILM SOLAR CELL TECHNOLOGY

- Amorphous Semiconductors

Large area and high efficiency a-Si:H solar cell
    Y. HIGAKI, M. KATO, M. AIGA and Y. YUKIMOTO, LSI R&D
    Laboratory, Mitsubishi Electric Corp., Japan   745

Post-hydrogenated CVD amorphous silicon p-i-n diodes for photovoltaic applications
    N. SZYDLO, E. CHARTIER, N. PROUST, J. MAGARINO and
    D. KAPLAN, Thomson-CSF, Laboratoire Central de Recherches,
    France   749

Stability of amorphous silicon solar cells with pin structure
W. KRÜHLER, M. MÖLLER, H. PFLEIDERER, R. PLÄTTNER
and B. RAUSCHER, Research Loaboratores of Siemens AG,
F.R. Germany   754

Carrier conduction in a-Si:H solar cells
M.K. HAN, P. SUNG, R. LAHRI and W.A. ANDERSON, Department
of Electrical and Computer Engineering, State University
of New York at Buffalo, USA   759

Optical optimization of amorphous silicon solar cells
W. DEN BOER and R.M. VAN STRIJP, Department of Electrical
Engineering, Delft University of Technology, The Netherlands   764

Large area hydrogenated amorphous silicon for photovoltaic
application
G.J. SMITH and W.I. MILNE, Cambridge University, Engineering
Department;
P. BLACKBOROW, Electrotech (ET Associates), United Kingdom   769

Novel plasma chemical methods for doping a-Si:H
G.H. BAUER and G. BILGER, Institut für Physikalische
Elektronik, Universität Stuttgart, F.R. Germany   773

Electronic properties of doped amorphous SiO_x
E. HOLZENKÄMPFER and J. STUKE, Fachbereich Physik,
Universität Marburg;
R. FISCHER, AEG-Telefunken, F.R. Germany   778

- Crystalline Silicon Based Cells

Antimony doping in vacuum deposited thin film silicon
photovoltaic cells
C. FELDMAN, F.G. SATKIEWICZ, N.A. BLUM and K.G. HOGGARTH,
The Johns Hopkins University, Applied Physics Laboratory,
USA   783

Photovoltaic performance of CdS heterojunctions on poly-
crystalline silicon
E. SCAFE', G. MALETTA, R. TOMACIELLO, P. ALESSANDRINI,
A. CAMANZI, L. DE ANGELIS and F. GALLUZZI, Laboratori
Ricerche ASSORENI, Italy   788

- II-IV Compound Solar Cells

Temperature dependence of the IV-characteristic of Cu_2S-CdS
thin film solar cells and related phenomena
G.H. HEWIG, F. PFISTERER and H.W. SCHOCK, Institut für
Physikalische Elektronik, Universität Stuttgart,
F.R. Germany   793

Continuous deposition of photovoltaic grade CdS sheet
at the unit operations scale
    R.E. ROCHELEAU, P.J. LUTZ, D.F. BRESTOVANSKY,
    B.N. BARON and T.W.F. RUSSEL, Institute of Energy Conversion, Department of Chemical Engineering, University
of Delaware, USA   798

$Cu_xS(p) - CdZnS(n) - CdS(n^+)$ evaporated thin film solar cells
    B. BOUCHIKHI, S. CHANDRASEKHAR, F. ZAPIEN NATAREN and
S. MARTINUZZI, Laboratoire de Photoélectricité des
Semi-conducteurs, University of Marseille, France   804

Thin film Cu_2S/CdS junctions produced by evaporation and
sputtering: effect of thermal treatments in vaccum
    E. ELIZALDE, M. LEON, F. RUEDA and F. ARJONA
Departamento de Fisica Applicada, Universidad Autonoma
de Madrid, Spain   809

Airless sprayed CdS solar cells
    J. VEDEL, B. THIEBAUT and M. LEVART, Laboratoire d'Electrochimie Analytique et Appliquée de l'ENSCP, Ecole
Nationale Supérieure de Chimie, France   818

Electrochemical preparation and conditioning of Cu_2S for
Cu_2S-CdS solar cells
    J. VEDEL, P. COWACHE and D. LINCOT, Laboratoire d'Electrochimie Analytique et Appliquée de l'ENSCP, Ecole
Nationale Supérieure de Chimie, France   822

Physico-chemical properties of Cu_xS
    H. RICKERT, H.-D. WIEMHOEFER, I.E. SCHMIDT, R. WAGNER
University of Dortmund, Dept. of Physical Chemistry I,
F.R. Germany   827

Photovoltaic behaviour of CdSe thin film solar cells
    E. RICKUS, Battelle Institute, F.R. Germany   831

All thin film n-CdTe/ITO Solar Cell
    C. MENEZES, F. SANCHEZ-SINENCIO, C. VAZQUEZ-LOPEZ
and A. SOUZA E., Departamento Fisica, CINVESTAV, Mexico;
R. BUBE, Department of Materials Science, Stanford
University, USA   836

Preparation of high purity II-VI compounds by laser
annealing
    L. BAUFAY, D. DISPA, A. PIGEOLET, M.-C. JOLIET and
L.D. LAUDE, IRIS, Faculté des Sciences, Université de
l'Etat à Mons, Belgium   839

- Other Thin Film Systems

Zn_3P_2 thin-film solar cells
    M. BHUSHAN, Institute of Energy Conversion, University
of Delaware, USA   844

Cadmium sulfide polyacetylene photovoltaic hererojunction
   M. CADENE and M. ROLLAND, Groupe de Dynamique des Phases
   Condensées;
   M. ALDISSI and M. ABADIE, Laboratoire de Chimie Macro-
   moléculaire, Université des Sciences et Techniques
   du Languedoc, France ... 848

Antimony sulphide thin films
   J.S. CURRAN and R. PHILIPPE, Laboratoire de Physicochimie
   des Interfaces, Ecole Centrale de Lyon, France ... 853

SESSION 8 - CRYSTALLINE SILICON SOLAR CELL TECHNOLOGY

Progress in unconventional crystallization of silicon
   E. SIRTL, Heliotronic GmbH, F.R. Germany ... 858

Production of solar grade silicon in an arc furnace using
high purity starting materials
   H.A. AULICH, K.-H. EISENRITH, Research Laboratories and
   W. DIETZE, J. SCHäFER, F.W. SCHULZE and H.-P. URBACH,
   Components Group, Siemens AG, F.R. Germany ... 868

Segregation of impurities at grain boundaries and other
compositional inhomogeneities in chill-casted silicon ingots
   S. PIZZINI, L. BRAICOVICH, L. CALLIARI, M. GASPARINI,
   C.M. MARI, F. REDEALLI and M. SANCROTTI,
   Istituto di Elettrochimica, Milan;
   Heliosil SpA, Milan;
   Istituto de Fisica, Politecnico di Milano;
   IRST, Trento, Italy ... 874

Advanced slicing techniques
   P.G. WERNER, Laboratory for Industrial Production Techniques,
   University of Bremen, F.R. Germany ... 883

Continuous growth of thin polysilicon sheets on a temporary
carbon shaper by the R.A.D. process
   C. TEXIER-HERVO, M. MAUTREF, C. BELOUET and E. KERRAND,
   Laboratoires de Maroussis, France ... 896

Possibilities of ion implantation in silicon solar cell
manufacturing
   P. SIFFERT, Laboratoire de Physique et Applications des
   Semiconducteurs, Centre de Recherches Nucléaires, France ... 901

Low cost processes for cast silicon solar cells
   K.-D. RASCH, K. ROY, W. SCHMIDT and G. WAHL, AEG-Telefunken,
   Electronic Components Division, F.R. Germany ... 919

Polycrystalline silicon solar cells utilizing an integral
screen printing technique
    G. CHEEK, R. JANSSENS, M. LEEMPOELS, L. FRISSON,
R. MERTENS and R. VAN OVERSTRAETEN, Katholieke Universiteit
Leuven, ESAT Laboratory, Belgium   926

Photovoltaic solar cell comparison methodology
    A.M. BARNETT, Electrical Engineering Department, University
of Delaware, USA   931

POSTER GROUP P8 - CRYSTALLINE SILICON SOLAR CELL TECHNOLOGY

- Silicon Material

Aluminothermic reduction of quartz sand
    J. DIETL, C. HOLM and E. SIRTL, Heliotronic GmbH, F.R. Germany   941

Solar grade floating-zone silicon
    A. LUDSTECK and H.J. FENZL, Siemens AG, F.R. Germany   946

Current aspects of the CGE semicrystalline silicon ingots
elaboration method
    J. FALLY and C. GUENEL, Laboratoires de Marcoussis, Centre
de Recherches de la CGE, France   955

An approach to solargrade silicon layers epitaxially grown
on mg silicon substrates
    V. SCHLOSSER, F. KUCHAR and K. SEEGER, Institut für Festkörper-
physik, Universität Wien, Austria   960

Method of raw material continuous feeding on silicon
ribbon growth
    N. MAKI, T. SAWADA, M. IIDA, T. MATSUI, K. TAMAI and
M. NAKAGAWA, Electron Device Engineering Lab., Electron
Tube & Device Div., Toshiba Corporation, Japan   965

Impurity incorporation in R.A.D. polysilicon layers and
consequences on their electrical properties
    G. REVEL and N. DESCHAMPS, Laboratoire Pierre Süe, CEN/SACLAY;
J.P. DEVILLE, Laboratoire d'Etude des Surfaces, Université
Louis Pasteur;
C. TEXIER-HERVO and C. BELOUET, Laboratoire de Marcoussis,
Centre de Recherches de la CGE, France   970

Fast silicon-sheet growth with the supported-web method
    J.G. GRABMAIER, H. FÖLL, B. FREIENSTEIN and K. GEIM,
Components Group, Discrete Semiconductors and
Research Laboratories, Siemens AG, F.R. Germany   976

Recent developments in multi-wire fixed abrasive slicing
technique (FAST)
    F. SCHMID, C.P. KHATTAK, M.B. SMITH and L.D. LYNCH,
    Crystal Systems Inc., USA   980

Critical technology limits to silicon material and
sheet production
    M.H. LEIPOLD, PV Components Research, California
    Institute of Technology, Jet Propulsion Laboratory, USA   985

Economic viability of the UCP semicrystalline silicon sheet
technology
    Z. PUTNEY, T. ROSENFIELD and C. WRIGLEY, Semix Inc., USA   990

- Ion Implantation, Cell Processing

Comparison between various ion beam doping procedures and
anneal techniques used in manufacturing silicon solar cells
    J.C. MÜLLER, A. MESLI and P. SIFFERT, Centre de Recherches
    Nucléaires, Groupe PHASE;
    J. COM-NOUGUE, C. TESSARI and J.P. DUMAS, Laboratoires de
    Marcoussis, Centre de Recherches de la Compagnie Générale
    d'Electricité, France   994

Status of ion-implanted silicon solar cells
    W. SCHMIDT and K.-D. RASCH, AEG-Telefunken, F.R. Germany   999

Optimization of pulsed electron beam annealing process
for silicon solar cells
    A. LAUGIER, D. BARBIER and G. CHEMISKY, Laboratoire de
    Physique de la Matière, Institut National des Sciences
    Appliquées de Lyon, France   1007

Silicon solar cells by ion implantation : E-beam and self
annealing
    G.F. CEMBALI, R. GALLONI, G. LULLI, A. MAZZONE, P.G. MERLI and
    R. NIPOTI, CNR, Istituto LAMEL;
    F. ZIGNANI, Istituto Chimico, Facoltà di Ingegneria,
    Università di Bologna, Italy   1013

An automated ion implant/pulse anneal machine for low cost
silicon cell production
    A.J. ARMINI, S.N. BUNKER and M.B. SPITZER, Spire Corpor-
    ation, USA   1018

Laser processing in the preparation of high efficiency polycrystalline silicon solar cells
    E. COURCELLE, E. FOGARASSY, J.C. MULLER and P. SIFFERT,
    Centre de Recherches Nucléaires, Groupe de Physique et
    Applications des Semiconducteurs (Phase), France   1023

The influence of surface texture and thermal treatment on
the performance of laser-annealed silicon solar cells
    W. SINKE, D. HOONHOUT and F.W. SARIS, FOM-Institute for
    Atomic and Molecular Physics, The Netherlands   1029

Implantation of boron and boron fluoride compounds into
silicon for production of solar cells
    A. NYLANDSTED LARSEN, F. NIELSEN and G. SØRENSEN,
    Institute of Physics, University of Aarhus, Denmark   1034

Grain boundary photocurrent enhancement in solar cells
made by laser diffusion
    G.B. TURNER, D. TARRANT and D. ALDRICH, Arco Solar Inc.;
    R. PRESSLEY and R. PRESS, XMR Inc., USA   1039

Dry process for economic cell manufacturing
    J. DONON, H. LAUVRAY, P. AUBRIL, G. DAVID and P. LOUBLY,
    Photowatt International SA, France   1044

LATE NEWS PAPERS

Low cost structures and optimization of support structures
    J. GLÖCKL, P. HELM and K. TRÄDER, WIP, F.R. Germany   1050

Screen printed SIS-type solar cells
    J.N. AVARITSIOTIS and C. CAROUBALOS, University of
    Athens, Department of Physics, Division of Electronics,
    Greece;
    D.S. CAMPBELL, University of Technology, Department of
    Electrical and Electronic Engineering, Loughborough,
    United Kingdom   1053

An improved derivation of solar cell parameters in terms of
transition probabilities
    P.T. LANDSBERG, The University, Southampton, United
    Kingdom   1056

Impurity diffusion in amorphous silicon and its implications
for solar cells
    S. KALBITZER, M. REINELT and W. STOLZ, Max-Planck-Institut für Kernphysik, Heidelberg, F.R. Germany   1059

The world's largest 12 Volt single string photovoltaic module
    P. LAUWERS and G.R. SMEKENS, Energies Nouvelles et
    Environnement SA, ENE, Belgium   1063

Light assisted pulsed annealing of photovoltaic silicon by
microwave energy
    P. CHENEVIER, J. COHEN and G. KAMARINOS, Laboratoire de
    Physique des Composants à Semiconducteurs, ERA CNRS No
    659 - ENSERG, France   1065

Microprogrammed coupling system for photovoltaic generators
with multiple receptors
   G. CHAUMAIN, M. BARLAUD, P. ROUAN and J.P. REQUIER, Laboratoire de microinformation et électronique appliquée aux énergies renouvelables, ENSUT de Dakar, Senegal   1068

LIST OF PARTICIPANTS   1071

INDEX OF AUTHORS   1099

OPENING SESSION

Overview of the European Community's activities in photovoltaics
    W. PALZ, Directorate General for Science, Research and Development, Commission of the European Communities

The Italian programme in photovoltaic solar energy
    Prof. U. FARINELLI, ENEA, National Committee for Research and Development of Nuclear and Alternative Energies, Italy

Opening address - The ENEL's programmes in the field of photovoltaic conversion
    Prof. F. FITTIPALDI, Member of the Board of Directors of the ENEL, Italy

Opening address - CNR programme for photovoltaics
    Prof. G. ELIAS, Director of "Progetto finalizzato energetica" - CNR, Italy

United States federal photovoltaic program status
    M.B. PRINCE and A.L. BARRETT, jr., Photovoltaic Energy Technology Division, U.S. Department of Energy, USA

Podium during the opening ceremony - from left to right :
Prof. Fittipaldi, Dr. Prince, Dr. Palz, Prof. Bloss,
Dr. Strub, Dr. Farinelli, Prof. Elias

OVERVIEW OF THE EUROPEAN COMMUNITY'S ACTIVITIES IN PHOTOVOLTAICS

W. PALZ

Directorate General (XII) for Science, Research and Development
Commission of the European Communities
200 rue de la Loi
B - 1049 Brussels

ABSTRACT

The Commission of the European Communities is currently implementing a comprehensive programme on photovoltaics. This programme includes basic R+D work, module testing, development of pilot projects, demonstration projects and applications in developing countries. Particular attention is being paid to the development of expertise and production capabilities in industry and to strengthening cooperation between the various European countries.

INTRODUCTION

The Commission of the European Communities has been implementing a photovoltaic research and development programme for the last six years. This programme was followed in recent years by activities for module testing, small demonstrations and early applications in some African countries.

Since its beginning, the R+D programme for photovoltaics was given high priority within the overall solar energy R+D programme. A clear strategy was followed with continuity, the basic options of which can be summarized as follows :

- formation of a European community of experts across national borders;
- involvement of European industry;
- integrated development of components and total systems;
- focus on cheap technologies in view of future application and markets;
- decreasing order of priorities for silicon cells and modules, thin film cells, basic research, concentrating devices.

It is important to note that more recently the commitment of European industry to photovoltaics has become very strong and its role in the Commission's programme developed favourably. There has also been a gratifying interest in the Commission's programme on the part of universities who have played an important role in developing alternative cells and processes.

At the beginning, there were not more than three manufacturers of cells and modules in Europe and technology was almost identical with that used for applications in space. Meanwhile many new companies all over Europe have decided to make major investments in this field. It is expected that within one or two years from now photovoltaic production will increase tremendously. The prospects for large photovoltaic markets in Europe and in other parts of the world will eventually result in thousands of new employment vacancies in European industry. Along with an accelerated business activity in photovoltaics now and in the next few years, considerable efforts will have to be deployed at considerable cost in order to establish the optimal technologies for cell and module production and to build up large-scale production facilities. Current European Communities' activities are well oriented towards this goal.

POTENTIAL OF PHOTOVOLTAICS

In order to establish the future potential of photovoltaics as seen from Europe, the Commission has initiated comprehensive work which includes an appraisal of the possible markets for photovoltaics in Europe and for export (1). The study which will be published within a few weeks from now confirms that there exists a large potential for photovoltaic applications for rural electrification and electrification in general.

Currently the photovoltaic market is still very similar to the one in the mid 1950's when the solar cell was first developed. This market is for toys and gadgets, telecommunications and various appliances. Quite a considerable part of today's photovoltaic market is new however, e.g. for small calculators where millions of pieces are being sold.

Obviously, in terms of energy production, current markets are totally insignificant and the question which arises is whether by the year 2000 there will be a real energy market for photovoltaics in Europe. In the assessment study mentioned above, a potential market of approx. 1 GW per year has been identified for the year 2000. This market lies for part in the housing area; there are currently a few hundred thousand houses in Europe which have not yet been provided with electricity and many "week-end houses" in the countryside and remote areas could well be provided with electricity by means of photovoltaics.

Once the cost of photovoltaic generators has dropped to the range of $ 3-5 per Watt, the market for grid-connected applications becomes also accessible i.e. for dwellings in general, industrial and commercial centres and for central power plants.

---

(1) "Photovoltaic Power for Europe, an Assessment Study", by Sir William Halcrow and Partners for the Commission of the European Communities, to be published by Reidel Publishing Company.

Outside Europe, there is considerable room for photovoltaic applications and the European Communities have started to cooperate with many overseas countries to set up photovoltaic generators for hospitals (cooling of vaccines, lighting) and for water pumping. Photovoltaic irrigation systems can be totally automated and provide for the optimal use of the scarce water resources. Eventually taking into account that there are millions of villages without electric power all over the world, photovoltaic applications have the potential for becoming implemented on a really large scale.

THE SITUATION IN EUROPEAN INDUSTRY

Involvement of industry in photovoltaics has steadily grown over the last few years. Last year, there were six companies in the European Communities each of them producing and selling approx. 150 kW or somewhat less. There are also some very small companies being active and about two or three new ones who plan to come into the market this year. Total production of photovoltaic modules in 1981 was nearly 800 kW. By next year this figure is expected to become three times as big. To this end, large production capacities have been built up namely in Germany, France, Italy, Belgium, the U.K. and the Netherlands. There has also been considerable progress for automation of production.

There has been quite a change in technology and the space technology which was used for many years has now been abandoned virtually everywhere. With the assistance of the European Community's photovoltaic programme, the national programmes and on the own initiative of industry, considerable progress has been made for the development of alternative technologies aimed at cheaper production costs.

European industry still has a leading role for low cost polycrystalline silicon sheets where production capacities are expanding dramatically. Bearing all this in mind, it can be considered that the way is largely open now towards solar cell production costs in the $ 2-4 per Watt range which should come on the market in the next 3-5 years.

Within its photovoltaic pilot programme, the Commission of the European Communities has tested at its Ispra establishment 11 different types of commercial modules produced by European industry. These tests all confirm that module technology in Europe is of a very high standard. European industry can now tackle the large markets of the future with considerable expertise in this field.

Concerning complete systems, European industry cooperates with a lot of enthusiasm within the Commission's photovoltaic pilot programme where know-how is being developed for low-cost structures, high efficiency power conditioning, batteries with long life etc.

ACTIVITIES OF THE COMMISSION OF THE EUROPEAN COMMUNITIES

Table 1 gives an overview of the Commission's activities and budgets in the photovoltaic area. The Commission's budgets normally extend over periods longer than just one year and we have tried to put together relevant budget figures for the period 1979-82.

Table 1

EC ACTIVITIES AND BUDGETS (1979 - 1982)		
Development of cell and module technology in industry, universities, research centres	DG XII, Brussels	6.5 MECU
Module qualification tests	DG XII, Ispra	~ 9.0 MECU
Pilot Projects	DG XII, Brussels	9.4 MECU
Demonstration projects	DG XVII, Brussels	1.1 MECU
Cooperation projects in Africa	DG VIII, Brussels	~ 3.0 MECU
Most projects are funded for approx. 40% of their cost by these budgets		

There is an important activity going on at the Ispra establishment which can be considered as an "in-house" activity of the Commission. The budget indicated here also includes personnel costs. Very large simulation facilities for photovoltaic modules have been built up at Ispra in the last few years which have been used very efficiently for the first time in the course of the Commission's photovoltaic pilot programme and have proved to perform satisfactorily.

The Commission also has a small programme for photovoltaic demonstrations where four projects are being supported; the most important being an irrigation system in the South of France.

In the frame of the European Development Fund, the Commission has cooperated with many overseas countries particularly with some African countries to set up a number of photovoltaic projects which perform very successfully. Most of these projects were for water pumping; however the biggest one which is now under preparation concerns the equipment of 750 rural hospitals in Zaïre.

It is obvious that the activities in cooperation with overseas countries could be expanded considerably to the benefit of the people living in rural areas for whom photovoltaics would really be the only technology to provide them with a minimum of electricity i.e. comfort.

The Commission's R+D activities are shown in Table 2.

Table 2

EUROPEAN COMMUNITIES' R+D PROGRAMME
16 pilot projects from 3 to 300 kW all over the EC to be completed between mid 1982 and mid 1983 built by industrial consortia. Design phase was completed in December 1981.
Component development (44 contracts) Polycrystalline silicon Screen printing contacts etc. Ion implantation Polymeric materials for encapsulation a-Si, CdS, CdSe cells Power conditioning Advanced concentration

They are part of the Commission's 2nd Energy R+D Programme which extends from July 1979 to June 1983. The programme is implemented on a cost-sharing basis with European industry, universities, research centres etc. with the participation of most national programmes of the member countries. It is split into two parts namely R+D and pilot plants. Comprehensive R+D work is being focused on new technologies for silicon solar cells and the development of amorphous silicon, CdS and CdSe solar cells. The effort on concentration systems was never very big and has been decreasing again with respect to the first programme. The programme activity has led so far to many interesting results in particular for the following :

- polycrystalline silicon material
- screen printing contacts
- ion implantation
- new annealing methods after doping
- better understanding of amorphous silicon solar cells and improvement of their efficiency
- new efficient methods to produce cheap CdS solar cells
- new polymeric material for encapsulation of cells.

A large part of the Commission's overall budget is devoted to the photovoltaic pilot programme which includes a budget of approx. ECU 30 million of which the Commission contributes for one-third from its own budget. Work started about one year ago and final design for all projects was completed in December last year; the first project will be completed in August this year. It is a 50 kW generator for powering the village of Aghia Roumeli on the Island of Crete in Greece. 14 other projects will be completed by June 1983. A complete list of these projects has been published earlier (2).

---

(2) Photovoltaic Outlook from European Community's Viewpoint, W. Palz, Proceedings of the 15th IEEE Photovoltaic Specialists Conference, p.17

The scope of the pilot programme is larger than just the development of photovoltaic power for intermediate loads. First of all, through this programme, cooperation within European industry could be strengthened and the general state of know-how increased everywhere. In a common effort of innovation, module technology has been brought to a high level and the basic technologies for supporting structures and cabling, power conditioning and batteries, monitoring and control have been implemented in industry. So far, in this field, industrial experience used to be relatively low. Furthermore, through the photovoltaic pilot programme, industry has the opportunity to familiarize itself with new applications of photovoltaic systems be it for electrification of islands and remote communities, large telecommunication systems, desalination, water disinfection, cold stores for food, hydrogen production etc.

It should also be mentioned that within the pilot programme, a small number of systems in the 5 kW range are being set up in Sicily. They are of interest in view of the future application of photovoltaic generators in houses which are of that very same size. In this perspective the Commission is also supporting the construction of one of the first photovoltaic solar houses in Europe which is now being built in Munich. The house is of a passive design and has the photovoltaic panels integrated in its roof.

CONCLUSION

The European Commission's photovoltaic programme has been successful in stimulating interest in photovoltaics and its financial support for R+D, testing, and advanced applications has proved to be effective. Considerable progress has been made over the last few years in developing the necessary technologies in view of the large scale production of cost-effective solar cells and modules. Investments in European industry were considerable and while sales have been small until now, it is expected that from this year onwards, they will spin off.

THE ITALIAN PROGRAMME IN PHOTOVOLTAIC SOLAR ENERGY

Ugo Farinelli

ENEA - National Committee for Research and Development
of Nuclear and Alternative Energies - Roma

Credit must be given to the Commission of the EEC for organizing this highly successful series of symposia on photovoltaic conversion. Indeed, even greater are the merits of the Commission for starting a programme on photovoltaics at a time in which very little research in this field was carried out in Europe, in promoting and supporting the development of these activities to a level that compares very well with what is done outside, as concerns both research and development, and the experimentation and demonstration of how photovoltaic conversion can successfully be applied.

And now let us turn to the situation in Italy.

Italy - as most parts of Europe - is densely pupulated and with an electric grid reaching nearly everywhere. Isolated houses for which an autonomous electricity generating system is less expensive than connecting them to the grid are few. A detailed study carried out by ENEL has shown there are a few thousands - a few peak MW of photovoltaic.
To these, some minor islands in the Mediterranean should be added, for a few more MW's. A law being approved by the Italian Parliament in these days gives a very consistent possibility of starting this operation, by allowing a government contribution of 80% of the capital cost for "photovoltaic systems (and for other renewable energy sources) for the production of electricity for rural buildings that are permanently inhabited by the farmer". A large scale application of photovoltaic systems, connected to the grid, even taking into account environmental and social incentives, is an objective for the year 2000 and beyond; at that point, however, even 1% of PV electricity generation in Italy would mean an installed peak capacity of several GW. How does one fill the

gap between the 80's and the 2000's, and between a few MW's and a few GW's? The answer must be sought in the market provided by the isolated communities and agricultural requirements in developing countries, where the local conditions on one side and the low maintenance and operation requirements of PV systems make them attractive in many cases already now and will make them even more desirable in 2-3 years' time, when the cost of the PV-generated KWh will be competitive in many isolated locations with that obtained by a diesel generator. An increasing larger fraction of funds devoted to supporting development in less developed countries can be expected to go to renewable energies and to photovoltaics in particular. The Italian strategy for the development of photovoltaic energy is based on a steadily exponential growth of production of PV devices with a market represented by three phases:

- *in the next few years, applications in Italy where this is economically reasonable, and demonstration of the realiability and life of PV systems (for a total of a few MW's);*
- *in the following years, up to the end of the century, essentially applications in developing countries, with rugged, simple, reliable systems totaling several tens to several hundreds of MW;*
- *finally, according to economics and other factors, introduction to grid connected applications in Italy on the GW scale.*

ENEA, the Italian Committee for Nuclear Energy and Alternative Energy, plays a key role in this strategy. ENEA is just over one month old; it derives from a transformation of CNEN (the Italian Nuclear Energy Commission) to include activities on non-nuclear, alternative energy sources and on energy conservation.

The budget available to ENEA for renewable energies and energy conservation in the three years 1982-84 is about 300 billion lire - 250 million dollars or 600 million DM.

A good fraction - something more than 20% - will be devoted to photovoltaics: about 70 billion lire (55 M$). This is a very consistent public effort, even when compared with countries like France and the USA. To this one has to add the efforts by ENEL, by CNR and the incentives included in the bill mentioned before. Industrial investments in this sector announced for the next years are also consistent. The next few years are thus crucial in determining whether photovoltaics will take off in Italy or not.

How are we ENEA going to get around it?

In addition to the stimulation of demand, provided by the incentives which are not our direct concern, there are three other types of action: one on the demand side - which is demonstration, and two on the offer side: industrial promotion and research. We are giving about 1/3 of our budget to each. We think that reductions in PV costs come from a combination of two factors: scale factors (depending on the market) and innovation. Both should be acted upon.

As concerns demonstration, we have already presented the DELPHOS project for a 1.1 MW plant in Apulia, which we are carrying on together with ENEL. We are participating in two other photovoltaic experimental plants (Giglio and Verona) to be built in Italy with EEC support. On the offer side, short range innovation (2-3 years) affecting the production lines, and based on improvements of present technologies, or on other technologies available and demonstrated, is encouraged by industrial promotion (i.e. supporting technically and financially research carried out at the industry with the aim of improving production methods). Research, as far as we are concerned, relates to alternative technologies and processes that may have commercial interest 5 years or more from now (up to 20). We are supporting, and will continue to support, research carried out in universities and in industrial laboratories in Italy, and we are building a new laboratory-CRIF - entirely devoted to photovoltaics near Naples;

this laboratory should start operating in 1984 and reach a staff of about 120 people.

Photovoltaic in English is often abbreviated as PV; the same initials in Italian are used for "prossimo venturo" - i.e., next to come. Let's take this as a good omen.

OPENING ADDRESS

THE ENEL'S (*) PROGRAMMES IN THE FIELD OF PHOTOVOLTAIC CONVERSION

Professor F. Fittipaldi
Member of the Board of Directors of the ENEL

It is a great pleasure for me to be able to address this conference today, particularly as I consider that the Community is making a major contribution to the development of solar energy and more specifically to photovoltaic energy conversion. I am also speaking in my capacity as member of the Consultative Committee for the Community Research Programme on Solar Energy.

Some of the biggest research projects we are now pursuing in Italy are being carried out within the framework of Community research programmes, and I should like to express the hope that the third four year programme due to begin in 1983 will also assign a major role to solar energy.

I should like to take this opportunity of giving you some information on the main programmes which the ENEL is now pursuing in the field of the photovoltaic conversion of solar energy.

However, let me say first and foremost that the ENEL's commitment towards developing solar energy is not restricted to photovoltaic conversion alone, but also covers thermodynamic conversion and low-temperature heat applications. I need only mention, where thermodynamic conversion is concerned, the Eurelios project, which is the first tower and mirror array power station to have been brought into operation in the world.

As you know, this installation, constructed as part of a European Community research programme by a group of Italian, French and German companies (one of which is the ENEL) has been connected up to the Italian electricity grid since April 1981.

The ENEL, which is co-owner of the plant together with the Community and responsible for the experimental phase of operation, will, once this test period is over, be in a position to assess in real terms the outlook for the development of this application of solar energy.

(*) ENEL = Italian National Electricity Board

As far as low-temperature heat applications are concerned, I should like to remind you that the ENEL has developed, and is on the point of launching a campaign for the distribution of, solar water heaters. Two of the measures which the ENEL intends to take in this connection are: first, the granting of financial support by advancing to users part of the cost of the solar installation (this advance will later be recovered by debiting the electricity bills) and secondly, the approval of types of solar panel eligible for use in this campaign.

The ENEL, through one of its subsidiaries, PHŒBUS & company, has already granted approval for 15 types of panel on the basis of specifications drawn up specifically for this campaign.

Having touched on these two measures, let us now return to the subject of this conference, which is photovoltaic conversion.

The ENEL is deeply committed in this sector through both its fundamental research activities and its practical demonstrations of this application through small, medium-sized and large-scale installations.

Where <u>fundamental research</u> is concerned, the ENEL, through the CISE, one of its research institutes, is at present developing <u>high-performance cells of the gallium and aluminium arsenide type</u> to be coupled to concentrators.

The first cells produced have given efficiency readings of approximately 21%.

Using cells of this sort, the CISE has built a prototype plant of 0.1 kW with concentrators and cell heat recovery.

In addition, within the framework of a European Community programme, the CISE is now completing work on a further prototype of 0.1 kW with coupled silicon and gallium arsenide cells and with solar spectrum separation.

<u>Amongst its practical demonstrations</u>, the ENEL is involved in the installation and testing of photovoltaic installations of various sizes and purposes.

The photovoltaic installations which have the greatest potential in the short and medium-term are <u>small plants for supplying power to individual consumers.</u> In Italy, users who are not yet linked up to the grid include a not inconsiderable number, approximately one thousand, for whom connection to the mains supply would be very expensive and this means that photovoltaic electricity supply, even at today's prices, can be a competitive solution. Similarly, because of their simplicity of design and operation, photovoltaic installations are fairly well suited to remote locations. The ENEL is currently carrying out a project called 'Case sparse' (Remote Homes). For this project a very exhaustive study was first carried out into the different characteristics (technical, energy consumption demographic, etc) of all potential users whose connection costs to the mains were higher than 20 million Lit.

The results of this study provided the basic data for selecting a preliminary group of individual users who could be supplied with power from photovoltaic installations.

It is the intention to build these installations around a basic modular unit which is capable of providing the user with a minimum annual power output of 750kWh.

The manufacturers were asked to develop not only the photovoltaic units but also the associated consumer equipment, (lighting units, radios, television sets and refrigerators with low energy consumption so that the electrical energy produced could be used economically.

By mid-1983 approximately 10 prototypes, consisting of a single module or two combined modules, will have been installed and tested. At a later stage, a small number of mini power stations (at least one hundred) will be built and installed in the vicinity of a number of remote houses. These will have varied power output ratings.

Another factor which aided in the preparation and launching of the 'Remote homes' programme was the testing of the 1 kW power station at Misterbianco (Catania) which has been operating for approximately one year, providing a power output similar to that required for low domestic consumption.

Another sector in which there might be suitable scope for photovoltaic conversion, even in the medium-term, is that of <u>photovoltaic installations for remote communities.</u> In practice this only involves in Italy the smaller islands.

In this sector the ENEL is participating in the 'Alicudi project'; it is planned to install and operate an 80 kW photovoltaic power plant on the island of Alicudi in order to provide power for the resident population at present supplied by the electricity board.

This installation is now at the design stage (it should come into operation in 1983) and will be partly financed by the European Community. This project, both because of its location and the social framework within which it will function, represents a typical example of applications through which photovoltaic systems could meet real needs in a useful and appropriate manner even in the short and medium-term.

The Alicudi experiment should therefore prove to be extremely significant.

The Alicudi project ought also to be considered as part of the wider-ranging 'Eolie Project' whose aim is to define a system for the production and use of energy based exclusively on locally available energy sources (solar, wind and geothermal power) and therefore without fossil fuel imports.

Up to now I have spoken of miniphotovoltaic power stations for supplying electricity to homes remote from the mains supply and to small islands. It is clear that the use of these installations can in no way be a major breakthrough in energy supply, given the restricted number of potential applications.

However, if we wish to consider the possibility that photovoltaic electricity production may in future make a significant contribution to electric power production, we must focus our attention on photovoltaic power stations linked up to the grid. Installations of this type can be medium or large, or even mini power stations for specific consumers, particularly industrial users.

The following are the ENEL's main projects in these two sectors:
- the Delphos project, the objective of which is to construct a photovoltaic power station for demonstration purposes, with an approximate power output of 1 000 kW, to be located in the Puglia region.

This project is being carried out by the ENEL and the ENEA (formerly CNEN) with the cooperation of industry and the Italian Universities.

The plant, which is at present in the preliminary design stage, will also be able to supply power to a large, remote community as an alternative to mains supply.

When it comes into operation (this is scheduled for 1984) the Delphos power station will be the largest photovoltaic plant in the world.

- the Adrano Photovoltaic project, the objective of which is to build in Adrano (near to the Eurelios power station) a miniphotovoltaic power station with a 30-kW output, made up of six 5-kW modules (two concentrating and four not).

This project is financed by the European Economic Community.

The siting of this plant will also make it possible to compare the various photovoltaic systems with a tower and mirror array system.

I am aware that I have only sketched the broad outlines of our programmes, but, given the short time at my disposal, my main aim was to give you some basic information on our major ones.

I should like to conclude my address by expressing my conviction that the ENEL's programmes will make a major contribution towards ensuring that the Italian photovoltaic cell industry has a production level high enough to provide adequate stimulus for research, technological progress and cost reduction, but on the other hand not so high as to necessitate investment and production facilities which are out of proportion to the state of the art and too far ahead of their time.

OPENING ADDRESS

CNR PROGRAMME FOR PHOTOVOLTAICS

Prof. G. ELIAS

Director of 'Progetto finalizzato energitica' - CNR

Ladies and Gentlemen,

In my short address, I would like to say a few words only about the Energy Project (PPE) of the National Research Council of Italy in general, and about its activity in the field of Properties and Technologies of Semiconductors.

As you know the Project was a 5 years research programme (1976-1981) concerning renewable energy resources and rational energy use.

The general Project lines have been the following :
1. utilization of the so-called 'minor' energy resources (especially solar and geothermal energy - the second one in collaboration with ENEL and ENI) and waste utilization; excluding research programmes of nuclear, hydraulic energy and oil products, fields in which the State Corporations are already working;
2. conservation studies and rational energy utilization in the field of road transport and building industry, besides other studies on innovation and improvement (energy storage, electro-chemical generation, heat pumps, enhanced engine efficiency etc.);
3. research on normative and juridical aspects to give the law-maker some help to choose possible incentives for public or private industry in the energy field, providing at the same time for the organization and up-dating of the energy flux in Italy.

The choice of these main lines depends on necessary interventions on technologically poor sectors and in fields of excessive oil consumption.

Nearly 2000 researchers from CNR, University, public and private industry, grouped in about 400 units have worked for the Project.

This represents the first experiment in Italy of a large range collaboration of people from public and private Bodies, working together and gaining a lot of experience on energy problems what will not fail to give fruitful results in the long run.

It is most important to stress that the CNR has a great experience in Italy concerning the research organization and planning, thanks to the management experience of its scientific and technological community in a large scale of purposes and problems, so different to those of the Energy Project.

As far as the special topic on 'Properties and Technologies of Semiconductors' is concerned I would like to mention that the research programme pursues two aims :
- the technological acquisition and development of the construction of silicon solar cells and complete photovoltaic panels with high conversion efficiency;
- the study of compound semiconductors and amorphous silicon cells as a

possible alternative to traditional cells will not be achieved in the near future; development of techniques for solar cell measurements to use in connection with photovoltaic cells.

With respect to the first aim the chemical and structural details have been defined to which the silicon solar grade has to comply. The endurable limit of the chemical impurity of various natures, penetrated by contamination of Czachralski monocrystals during the growing phase, has been defined by obtaining prototypes with an AM1 efficiency of 12 %.

Concerning the utilization of the ion implantation technique, the production process of solar cells has been developed at a working temperature not exceeding 750°C.

Besides the thermal annealing at low temperature, the technology of pulsed laser beams and pulsed electron beams have been tested.

The ion implantation process is suitable for the mass-production of solar cells and it has permitted the production of prototypes of 2 x 2 cm^2 cells with an AM1 efficiency around 14%. Photovoltaic demonstration panels have also been built; they are suitable for connection with 12 V batteries; an assembling technique for solar cells in double glazed panels, the air-space filled with inert gas, has been also developed.

The traditional technology to produce solar cells for an average concentration (50 suns) has been improved reaching an AM1 conversion efficiency of about 18%.

Basing on this technique, experimental photovoltaic panels, optimized for use in parabolic concentrators, have been built with a concentration ratio of 20 ÷ 25.

In the field of compound semiconductors solar cells of GaAs for high concentration have been set-up.

The thin-film photovoltaic cells represent a good long term alternative in respect of the devices presently used and therefore this way is also being explored.

A 10% conversion efficiency seems a reasonable aim to seek, and a limit beyond the thin-film cells could definitely represent the most adequate solution.

This is the very case for amorphous silicon, where it has recently happened that hydrogenation or flururation of the materials provokes a saturation of vacant bonds and improves the material properties as far as the use of p-n photovoltaic junction is concerned.

The radiofrequency glow discharge method has been used for preparing amorphous silicon films; this technique has been tested in a low pressure atmosphere of $SiH4$ (hydrogenated cells) and of $SiCl4$ (chlorurate cells).

Measurements have been performed on the mobility and lifetime as well as on the optical and structural properties of the materials for detecting the physical features of the samples. These measurements have allowed the establishment of the influence of defects and impurity centres on the cell efficiency.

The sputtering technology has been prepared to be used for the production of amorphous Si films. The thin-film technology has been used also for prototypes of solar cells by means of heterojunction of comound semiconductors and a new methodology has been developed for such a cell formation.

New ways have been explored as well for solar cells with electrolyte semiconductor junction, requiring a n-type semiconductor, normally functioning as photoanode, coupled with a cathodic transparent material by means of a redox couple in water solution.

The main advantage of this cell type to solid ones consists in their low construction cost. There exists the possibility of using polycristal photoelectrode materials, in some cases directly generated 'in situ' in the electrochemical cell for anodic effect. Photoelectrode mono- and poly-crystal semiconductors have been prepared and examined in redox-type electrolytes.

The following measurements have been carried out on sample cells :
- photoresponse
- current - voltage curves
- anode photodissolution.

The present prototypes have provided electrode stability for 1000 hours under simulated sunlight and an AM1 efficiency of 5 - 7% with a filling factor of about 60%.

A 'passive' cooling system for concentrator cells has been set up and the electrochemical battery features have been registered systematically in respect of the charge and discharge phases taking place in solar plants.

UNITED STATES FEDERAL PHOTOVOLTAIC PROGRAM STATUS

M. B. PRINCE AND A. L. BARRETT, JR.
Photovoltaic Energy Technology Division
U.S. Department of Energy

Summary

Since the 3rd EC Photovoltaic Solar Energy Conference, the strategy of the United States Federal Photovoltaic Program has been modified to stress long-range, high-risk, potentially high-payoff research. The level of Department of Energy funding for photovoltaics has been reduced to $74,000,000 for the present fiscal year. Most of the important accomplishments of the past 18 months will be covered by papers in the various sessions of this Conference. However, there are some accomplishments that are program-wide. There are now more than 15 cell suppliers in the United States with approximately 5 MWp of annual sales amounting to $75,000,000. Industry is investing approximately $100,000,000 in facilities and R&D, and approximately 2,700 experiments have been fielded with government funds.

1. INTRODUCTION

The United States Federal Photovoltaic Program has had, since the 3rd EC Photovoltaic Solar Energy Conference, a significant change in strategy. In the early years of the program (1975 - 1980), when the terrestrial photovoltaic industry was in its embryonic stage, the U.S. PV program stressed those activities that would encourage the growth of the industry by developing technologies that would reduce the cost of PV modules and systems and also field experiments to determine how these components and systems actually performed under various real conditions. Thus, both producers and users were able to evaluate the many PV products for determining their potential markets. That phase of the program turned out to be quite successful. From three or four small manufacturers, the U.S. industry has grown to more than a dozen manufacturers, some of which are moderate sized today and have significant financial backing to now grow as required. The terrestrial market sales have increased by more than an order of magnitude since 1975 both in quantity of devices and in dollar receipts. Today there exists a relatively mature industry which should now be able to invest in near-term development that will permit it to continue to grow rapidly. The markets that today's industry serves are the remote stand-alone applications in the U.S. and in countries overseas. An original purpose of the U.S. PV Program was to develop a renewable energy source that would complement fossil and nuclear electricity generating systems by the end of this century. We believe that the best way to meet this objective with the funds available today is to continue our activities in the long-term, high-risk, high-payoff research areas and allow industry to pursue on its own the near-term, lower-risk developments.

2. PROGRAM GOAL, OBJECTIVES, AND STRATEGY

The program goal is to support R&D that establishes the technology base needed for the private sector to develop and deploy advanced photovoltaic energy systems. In order to meet this goal, we have established two objectives: (1) to increase the efficiency of photovoltaic devices, and (2) to increase the stability of the device materials and the resulting devices. The strategy for pursuing these objectives is through two types of activities. The first is to be selective in research areas emphasizing advanced R&D on emerging materials (amorphous silicon and other thin film materials), conversion concepts (multi-bandgap cells and concentrator devices), theory (physics of amorphous materials and electrochemical mechanisms), and laboratory testing techniques (measurement of surface and sub-surface properties). The second type of activity relates to continuing the testing of existing experimental systems to obtain feedback to the R&D projects.

3. PROGRAM STRUCTURE AND FISCAL YEAR 1982 BUDGETS

The program structure is diagrammatically shown in Figure 1. Across the top are shown the six stages of product development used for planning purposes. In the vertical direction are the various program activities. In prior years the photovoltaic program supported development from basic and applied research through engineering development with demonstration and commercial production relegated to private industry. Now, the program supports work in the light area and has left to private industry those areas of the figure that are cross-hatched. In the material, cell and device R&D activities, support is given until laboratory technical feasibility (LTF) is reached. By LTF is meant the device or cell meets four criteria. These are: (1) minimum efficiency of 10 percent; (2) reproducibility; (3) stability; and (4) potential for being produced at low cost. In the high-risk collector R&D activities, support is given until technical feasibility (TF) is reached either at the $0.70/Wp for silicon subsystems or $0.40/Wp for non-silicon subsystems.

The Fiscal Year 1982 funding levels are shown in Figure 2 by program element. More details of these funding levels are shown in Figure 3 and 4 along with the responsible agency for overseeing the program element under the guidance from the Headquarters operation of the Department of Energy. It should be noted that the capital equipment budget is distributed in Figures 3 and 4. The three principle supporting laboratories are the Solar Energy Research Institute, the Jet Propulsion Laboratories, and the Sandia Laboratory. Many papers are being given by these organizations in various regular and poster sessions of this conference.

In Figures 3 and 4, several activities have been asterisked. These activities have significant cost-sharing with private industry. The actual amount of the cost-sharing is shown in Figure 5.

4. PROGRAM ACCOMPLISHMENTS

To indicate the significant general accomplishments of the program since its initiation in 1975, Figure 6 shows the position of several key indicators in 1975 and 1981. More specifically, Figure 7 shows how the efficiency of cells made from materials other than single-crystal silicon has improved during this period. Dr. D. L. Feucht (SERI) will describe these results in some detail in his talk on Thursday morning. Other key accomplishments will be presented by R. G. Ross (JPL) (Session 2), E. L. Burgess (Sandia) (Session 3), E. C. Boes (Sandia) (Session 6), R. R. Ferber (JPL) (Poster Group 4). W. T. Callaghan (JPL) (Poster Group 4) and J. L. Stone et al (SERI) (Poster Group 6).

As an outcome of the Federal Photovoltaic Program, a maturing industry presently exists in the United States. A summary of industrial type accomplishments is given in Figure 8. Finally, Figure 9 shows graphically the price reduction and production growth during the past 8 years that was stimulated by the photovoltaic program.

5. OPPORTUNITIES FOR COOPERATION WITH THE EUROPEAN COMMUNITY AND OTHER PROGRAMS

In the past the United States photovoltaic program has participated in many forms of cooperation with photovoltaic programs in many parts of the world. It is expected that this cooperation will continue in the future within the limitations of the U.S. program budget. The primary activities that offer significant opportunities for cooperation include information transfer, development of international performance criteria and standards, design review consultation, and the arrangement of formal visits of government and industry scientists and managers.

In the field of information transfer, quarterly and annual reports of progress can be exchanged in those areas of mutual programmatic interest. Also, mutual invitations to program and sub-program review meetings may be pursued. At those U.S. meetings where U.S. contractors give reports on their progress, foreign visitors should give reports of their progress at a similar level of detail. The same conditions should apply for visits by the U.S. side to foreign meetings.

In the development of international performance criteria and standards, the U.S. has recently funded the secretariat of the international group for this activity for 3 years since we believe that it is urgent that universal standards should be established early in the commercialization phase of such a potentially important energy industry.

In the field of design review consultation, the U.S. is working with the Italians on their 1 MWp utility experiment. In return the U.S. will receive detailed data of the system performance. Similar exchanges could be considered, subject to the availability of U.S. funds to do these additional tasks.

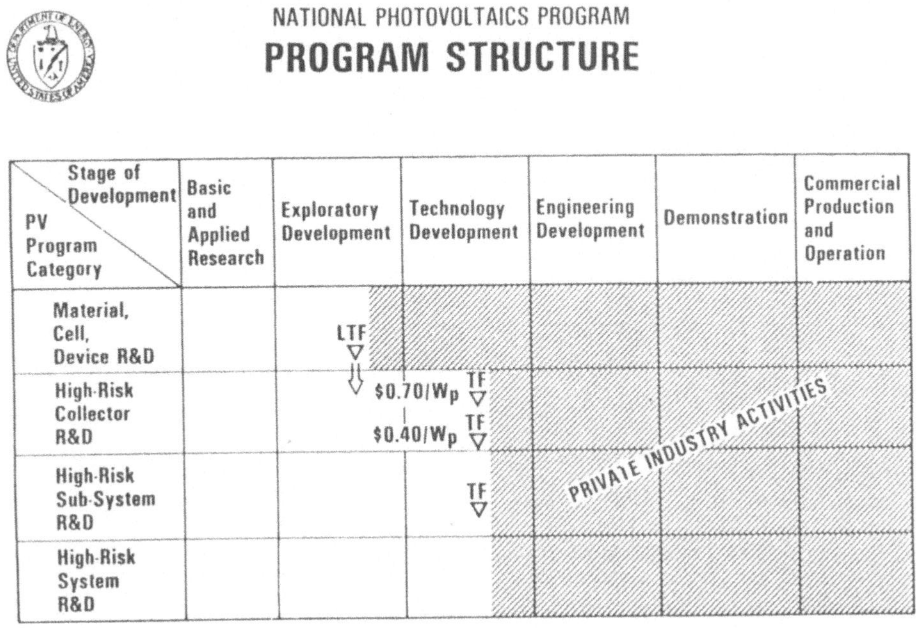

NATIONAL PHOTOVOLTAICS PROGRAM
PROGRAM STRUCTURE

Stage of Development / PV Program Category	Basic and Applied Research	Exploratory Development	Technology Development	Engineering Development	Demonstration	Commercial Production and Operation
Material, Cell, Device R&D		LTF ▽				
High-Risk Collector R&D		⇩ $0.70/W_p$ TF ▽				
		$0.40/W_p$ TF ▽				
High-Risk Sub-System R&D			TF ▽			
High-Risk System R&D						

(Shaded diagonal region labeled: PRIVATE INDUSTRY ACTIVITIES)

Figure 1. Program structure of the U.S. National Photovoltaic Program

PHOTOVOLTAIC ENERGY TECHNOLOGY PROGRAM
FUNDING LEVELS FY 1982
(MILLIONS OF DOLLARS)

PROGRAM ELEMENTS	FY 82 APPROPRIATIONS
MATERIALS RESEARCH	23.9
ADVANCED CONCEPTS	11.6
SUPPORTING RESEARCH	2.2
SYSTEMS RESEARCH	20.5*
TECHNOLOGY DEVELOPMENT	3.6
EXPERIMENTS	12.2**
CAPITAL EQUIPMENT	4.0
TOTAL	78.0

* INCLUDES THE SE RESIDENTIAL EXPERIMENTAL STATION
** INCLUDES COMPLETION OF NATIONAL EXEMPLAR PROJECT AND THE 1 MW SMUD PROJECT

Figure 2. Fiscal Year 1982 Funding Levels by Program Element

PHOTOVOLTAIC ENERGY TECHNOLOGY PROGRAM

PROGRAM ACTIVITIES

ACTIVITY	RESPONSIBLE AGENCY	FUNDING ($ MILLIONS)
MATERIALS RESEARCH		
AMORPHOUS MATERIALS*	SERI	3.8
STABILITY & EFFICIENCY OF THIN FILMS	SERI	4.1
HIGH EFFICIENCY DEVICE CONCEPTS	SERI	3.3
SILICON AND POLYCRYSTALLINE SHEET*	SERI	2.5
SILICON MATERIAL PURIFICATION*	JPL	2.9
RIBBON AND SHEET SILICON RESEARCH	JPL	6.8
ENVIRONMENTAL DEGRADATION RESEARCH	JPL	1.5
ADVANCED CONCEPTS		
ELECTROCHEMICAL MATERIALS AND CELLS*	SERI	2.0
CONCENTRATOR MATERIALS AND CELLS	SANDIA	2.2
POWER QUALITY AND CONTROL RESEARCH*	SANDIA	3.1
CELL AND MODULE FORMATION RESEARCH	JPL	4.7
SUPPORTING RESEARCH		
DIAGNOSTIC EQUIPMENT	SERI	2.2

Figure 3. Detailed Funding Levels with Organizational Responsibility

PHOTOVOLTAIC ENERGY TECHNOLOGY PROGRAM

PROGRAM ACTIVITIES (CONTINUED)

ACTIVITY	RESPONSIBLE AGENCY	FUNDING ($ MILLIONS)
SYSTEMS RESEARCH		
SYSTEMS RESEARCH	SANDIA	1.1
CRITICAL SUBSYSTEMS DEVELOPMENT	SANDIA	1.3
CONCENTRATOR RESEARCH AND TESTING	SANDIA	4.9
DATA COLLECTION OF EXPERIMENTS AND ANALYSIS	SANDIA	4.8
SYSTEMS EXPERIMENTS, OPERATIONS/ CLOSEOUT	VARIOUS	5.6
ENGINEERING SCIENCES RESEARCH	JPL	3.4
SE RES	SANDIA	2.0
TECHNICAL DEVELOPMENT		
SILICON TECHNOLOGY DEVELOPMENT	JPL	3.6
OTHER		
SMUD*	HQ	6.8
NATIONAL EXEMPLAR	ORO	5.4
	TOTAL	78.0

* ACTIVITIES WITH SIGNIFICANT COST SHARING

Figure 4. Detailed Funding Levels with Organizational Responsibility

PHOTOVOLTAIC ENERGY TECHNOLOGY PROGRAM

ACTIVITIES WITH SIGNIFICANT COST SHARING

ACTIVITIES	DOE SHARE*	PRIVATE SHARE
AMORPHOUS MATERIALS	$ .8 M	$ .2 M
SILICON AND POLYCRYSTALLINE SHEET	.6 M	.2 M
ELECTROCHEMICAL MATERIALS & CELLS	2.5 M	.7 M
SILICON MATERIAL PURIFICATION	1.5 M	8.5 M
CELL AND MODULE FORMATION RESEARCH	4.0 M	.9 M
SMUD PROJECT	6.8 M	5.2 M

* APPLIES TO COST-SHARED CONTRACTS ONLY

Figure 5. Program Cost-Sharing with Industry

PHOTOVOLTAIC ENERGY TECHNOLOGY PROGRAM

PROGRAM ACCOMPLISHMENTS

TECHNOLOGY ELEMENTS	1975	1981
o EFFICIENCY		
– SILICON	8%	20%
– THIN FILMS	1-3%	5-11%
o CELL COSTS	$50/Wp	$7/Wp
o MODULE LIFE (TERRESTRIAL)	1-2 YEARS	10 YEARS
o TERRESTRIAL SYSTEM EXPERIMENTS	FEW SMALL REMOTE USES	2,700 SMALL EXPERIMENTS (FPUP); 15 MAJOR PROJECTS (AS LARGE AS 350 KWp)
o INDUSTRIAL BASE	A FEW SMALL SPECIALITY COMPANIES (SEVERAL KWp SALES IN 1975)	RAPIDLY EXPANDING INDUSTRIAL BASE; MORE THAN 15 CELL SUPPLIERS (5 MWp SALES IN 1981)
o FABRICATION TECHNOLOGY	EXPENSIVE MANUAL OPERATION	SEMI-AUTOMATED PILOT PRODUCTION OF CELLS; RIBBON PROCESSES NEAR COMMERCIALIZATION

Figure 6. Major Accomplishments since Initiation of U.S. National PV Program

CELL EFFICIENCY VS. TIME

(THIN FILM TECHNOLOGIES)

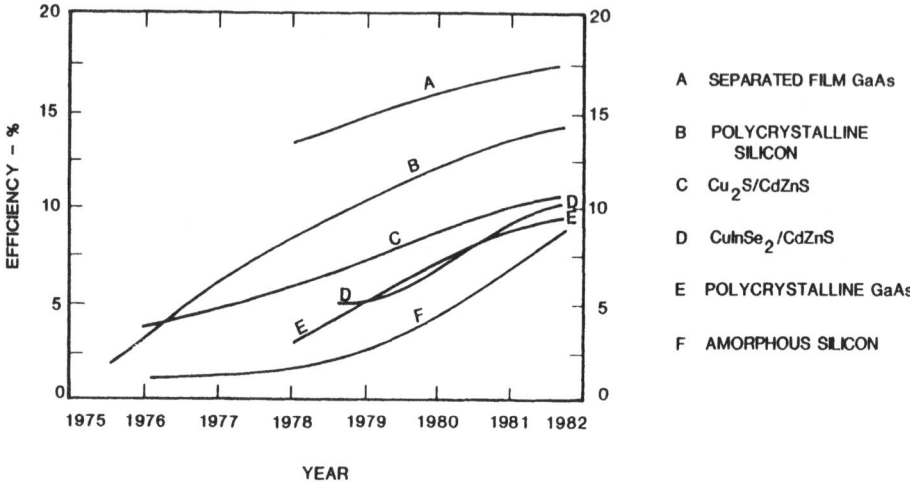

Figure 7. Cell Efficiency as a Function of Time for other than Single Crystal Silicon Devices

PHOTOVOLTAIC ENERGY TECHNOLOGY PROGRAM

SELECTED U.S. INDUSTRY ACCOMPLISHMENTS

(1) DEVELOPED HIGHLY-RELIABLE PRODUCTS TO SUPPLY REMOTE ELECTRICITY FOR COMMUNICATIONS, CORROSION CONTROL, NAVIGATIONAL AIDS, HOME AND FARM USES

(2) INVESTED IN INFRASTRUCTURE TO DELIVER AND SERVICE THESE PRODUCTS

(3) INCREASED SALES AND REVENUES AT A VERY HEALTHY RATE

	1979	1980	1981	1982 (EST)
SALES	1.4 MWP	3.2 MWP	5 MWP	7.5 MWP
REVENUE	$30 MILLION	$50 MILLION	$75 MILLION	$100 MILLION

(4) INVESTED APPROXIMATELY $40 MILLION IN 1981 INTO RESEARCH ON IMPROVED PHOTOVOLTAIC CELLS AND MODULES

Figure 8. State of U.S. Photovoltaic Industry

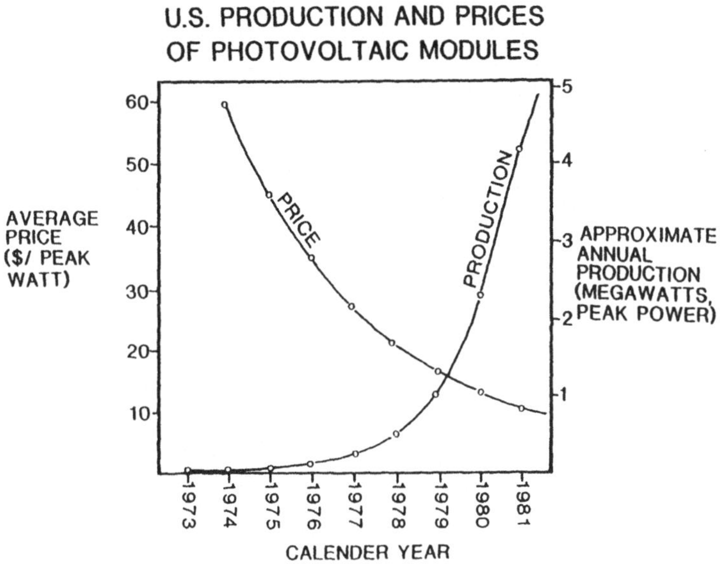

Figure 9. History of U.S. Prices and Production of PV Modules

SESSION I - APPLICATIONS

Chairmen : Prof. H. DURAND, Président, Commissariat à
l'Energie Solaire (COMES), Paris, France

Dr. H.W. BRANDHORST, jr., NASA Lewis,
Cleveland, Ohio, USA

Application trends for photovoltaics

The potential for photovoltaics in Europe

The behaviour of large solar power stations in the Swiss Alps

Design, installation, and initial performance of 350-kW photovoltaic power system for Saudi Arabian villages

The photovoltaic-powered water desalination plant "SORO" - Design, start up, operating experience

The UNDP World Bank solar pump project - preparing for phase II

Installation et évaluation d'un système simple de pompage photovoltaique dans un village algérien

APPLICATION TRENDS FOR PHOTOVOLTAICS

DR. H.L. MACOMBER
Vice President
MONEGON, LTD.
Gaithersburg, Maryland, U.S.A.

Summary

This paper presents the results of studies by MONEGON to develop forecasts of PV system application markets. These forecasts consider economic factors such as conventional energy costs now and in the future, the relationship between world economic conditions, as represented by each country's Gross National Product, and energy demand; the cost potential for PV systems technologies; and the application trends in the past and future. The application sectors analyzed are: remote, stand-alone systems; residential systems; service/commercial/industrial/institutional; and central utility systems. An overall market forecast is developed and this forecast is segmented into the four market sectors.

1. INTRODUCTION

Attempts to commercialize terrestrial photovoltaics (PV) began in the mid-1950s with Western Electric efforts to sell licenses for silicon photovoltaic technology. Markets were sought with applications such as photovoltaic-powered radios, hearing aids, highway construction warning flashers, and light controlled devices such as dollar-bill change machines and machines for decoding punched cards and tape. The costs of these devices, however, were too high to successfully compete with conventional electricity sources and the terrestrial markets were inadequate to sustain the two companies which opened production lines. The application of solar cells for spacecraft power opened a new market in 1958 and five companies entered the field.

Space applications dominated solar cell markets until pioneering entrepreneurs began a new terrestrial industry in 1973, shortly before the oil embargo. This industry began with consumer applications for solar cells, such as watches and toys, and small, remote applications such as communication repeater stations, environmental measurement instruments, navigational aids, highway call boxes and corrosion protection devices.

Various national governments in Europe, North America and the Far East began after 1973 to take photovoltaics seriously as a potential national source for electrical energy. Government programs for PV were begun in Japan, France, Germany, Canada and the U.S.A. The European Economic Community started a multi-national PV program and international aid institutions such as the World Bank, the United Nations, the European Development Fund and the U.S. Agency for International Development began to be interested in PV power systems for developing nations.

In addition to sponsoring research and technology development, the government programs have supported a large number of application demonstrations. Types of applications receiving government support range from current cost-effective stand-alone systems to residential applications, intermediate sized load center applications, and now, small central power applications.

The industry, meanwhile, has continued to grow strongly from its emergence in 1973. During the past nine years the industry has installed thousands of photovoltaic systems throughout the world. Industry sold PV applications have been primarily for relatively small and remote applications. Such applications include corrosion protection systems, navigation systems, pumping systems, communication systems, refrigeration systems, lighting systems, security systems, and, more recently, small power systems for remote residences. Emerging markets being developed by the industry include residential applications, manufacturing facility applications, and central power plants. Demand for these systems has continued to be sufficient to sustain a rapid growth in the photovoltaic industry along with steadily declining prices for PV power systems.

This paper briefly reviews the past and current applications, examines the more recent application areas in a framework of market penetration concepts and presents an assessment of the future major trends in PV applications based upon these considerations.

2. MARKET CONSIDERATIONS

In order to establish the basis for the application trends and market forecasts presented, it is necessary to briefly summarize some of the key influencing factors which are involved.

Oil and fossil fuel price increases have been one important factor in encouraging the photovoltaic power system industry development. Economic considerations such as the need to power remote equipment and to aid in the economic development of lesser developed countries also have been important factors. In the case of oil, however, the price of oil increased over 500 percent from 1970-1980. While there will be periodic slumps in oil prices, as is currently the case, and as occurred in the mid to late 1970s, the long-term trend will be for significant price increases, similar to the the trend from 1970-1980. Oil and other fossil fuels are finite resources; increasing world demand for energy will continue to increase the value of dwindling oil reserves. In addition, some fuels have traditionally received heavy government subsidies (in the U.S. these subsidies have approached $100 billion during the 1900s). Recently some of these subsidies have been relaxed which also has had the affect of increasing the final price of fossil fuels. As fossil fuels continue to become more expensive photovoltaic power systems will be competitive for a growing number of applications until virtually all major applications where electrical energy can be used will be cost-effective using photovoltaic power systems.

Another major factor affecting the attractiveness of photovoltaic power systems, has been the steady decline in system prices. Photovoltaic power system prices have been steadily falling since the production of the early photovoltaic panels for spacecraft applications at a cost in excess of $500 per peak watt. It is expected that continued price decreases will be experienced through the 1980s with a final leveling off in the early 1990s at a price of under $1.00 per peak watt for an installed system. It should be noted, however, that photovoltaic prices will be somewhat like energy prices in general with periods of price reductions interspersed with short periods of price leveling or even slight price increases. This is simply due to the introduction of new technologies, the writing off of investment of production costs, and the like. The long term trend will be in accordance with the historical trend: steadily declining prices. These price reductions will be achieved by: (1) the introduction of new technologies, (2) the adoption of mass production techniques, and (3) increased levels in production.

First, several new production technologies have been introduced and others are currently under development that offer the potential of significantly lower costs and improved performance for photovoltaic cells and panels. Some of the cost reductions experienced in photovoltaics technology have been through the development of technologies which permit the use of lower cost input materials which do not sacrifice the performance of photovoltaic devices. Certain manufacturing methods of producing polycrystalline silicon solar cells, for example, permit the use of a feedstock material that is 5-10 times less expensive than that used in conventional single crystal solar cell production. Advances in solar cell and panel technology will continue to be a major factor in reducing the cost of photovoltaic power systems.

Secondly, new manufacturing processes have introduced a higher degree of automation and are enabling increasingly higher-volume production levels. Increasing demand for photovoltaics power systems also is introducing greater efficiencies. Increases in demand also are requiring increases in production levels which is providing the means for achieving significantly lower cost photovoltaic systems.

In addition, total system cost reductions are coming from cost reductions in the "balance of systems" area. Items such as batteries for electrical storage, inverters, charge control units, and system structures, are beginning to be affected by the same economies of scale as for the photovoltaics area as demand increases. Furthermore, technology development is continuing in these areas to find lower cost materials, components and designs to meet the needs of the photovoltaics power system industry.

Another key factor to consider is the overall demand for energy throughout the world and, specifically the relationship between total energy requirements, electrical energy requirements and the use of photovoltaic power systems to meet these requirements.

Energy consumption increases as countries rise in their level of economic development. Various studies have shown that energy consumption is positively correlated with GNP per capita. It is interesting, however, to note the wide variation between oil importing and exporting countries. OPEC and oil producing countries enjoy a high GNP per capita relative to their energy consumption, while developing European countries, who import most of their oil, consume a significantly larger amount of energy relative to their GNP per capita.

Another factor to compare is the variance in energy consumption between developed regions of the world. This variance in consumption probably reflects how efficiently energy is used. Western Europe and Japan are commonly known to implement energy conservation measures and consume energy more efficiently than the U.S., although more recently there has been improvements in the U.S. use of energy conservation measures.

Figure 1 shows projections of total energy consumption and total electrical energy consumption for the United States and for the world. These projections are based upon historical trends and the effects of both increased energy conservation in the developed countries and increased uses of electricity worldwide

The 1980 energy consumption figures for the United States were 76.2 quads of total energy consumed of which about 32 percent, or 24.8 quads were consumed to produce electricity. This represents about 2.3 trillion kilowatt hours of electricity. Total world primary energy requirements for for 1982 are estimated at 363.4 quads or 5.4 times the U.S. requirements. Total world consumption of electricity in 1982 is estimated at 12.4 trillion kilowatt hours, also about 5.4 times the U.S. electricity consumption.

-32-

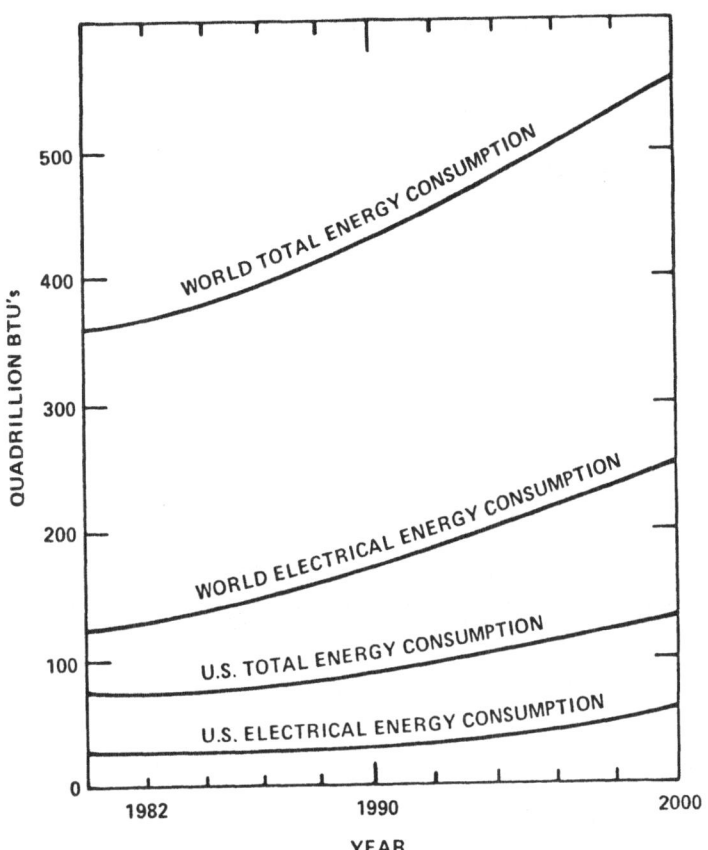

FIGURE 1

WORLD AND U.S. CONSUMPTION OF TOTAL
ENERGY VS. ELECTRICAL ENERGY

For the year 2000, U.S. energy consumption is expected to exceed 100 quads per year. These expectations, however, may not be realized. Until a year or two ago, estimates for the year 2000 ran in the 130 to 140 quad range. Energy conservation activity in the last few years, however, has far exceeded what most analysts believed possible, even in the most optimistic of scenarios. Of course, the current economic situation is depressing energy consumption to some extent, and a healthy economy will certainly increase energy demand. However, the important point to note is that the conservation reaction to increased prices for energy was quicker and more pervasive in the United States than was previously thought possible.

World energy consumption is expected to increase to about 550 quads. Energy growth projections on the world level are generally viewed as more accurate than for any given country. This is so because, while an individual country may react unpredictably within a given set of circumstances, on a global scale, the integration of individual country differences will tend to cancel each other. Moreover, while the United States or other developed countries may be able to cut back energy consumption through greater efficiencies or slower overall growth, developing countries have little if any waste to spare, and must increase consumption to meet growth needs.

It is also projected that electricity consumption, as a percentage of overall energy consumption, will show a marked increase. For the United States, electricity is projected by the year 2000 to account for up to 40 percent of all energy consumed. A similar consumption pattern, if somewhat lower, is projected for global electricity consumption. This is so primarily because, electricity, as seen by the end user, is the cleanest, most efficient, and easiest-to-use of available energy sources. It is also the energy source that lends itself best to either centralized production and distribution, or to distributed production and consumption.

3. <u>PHOTOVOLTAIC SYSTEM APPLICATIONS</u>

As photovoltaic power systems become cost-effective for increasing numbers and types of applications they can be expected to play a larger role in world electricity generation. During the early years of market penetration photovoltaic system generating capacity will represent only a small portion of the total electrical energy generation, continued growth in electricity requirements can be expected to provide opportunities for an increasing share of the market for photovoltaic power systems. In order to forecast where those market shares may occur we need to consider what are the current applications and where are the emerging opportunities for photovoltaic system applications.

Current Applications

Photovoltaic power systems have been installed for a wide variety of applications worldwide. While most of these applications can be classed as remote, stand-alone systems powering a single, non-diverse load, there have been a number of photovoltaic system installations which serve diverse loads. Most of the diverse load applications have been purchased or have been subsidized by government programs or agencies to serve as either demonstration applications or as engineering pilot plants.

The types of current applications have been widely publicized. They include: powering communication systems such as microwave repeaters, radiotelephones, television receivers, aircraft beacons and marine navigation aids; powering various instrumentation devices such as environmental monitoring instruments and various instruments used on weapon test ranges; powering corrosion protection devices for bridges, pipe lines and oil wells; powering pumps for potable water and irrigation systems; and generally acting as battery chargers for lighting systems, highway signs and security devices.

Over the past nine years the remote, stand-alone market has continued to grow. The annual market for these systems has grown from a few watts per year in the start-up years of 1973-1974 to nearly 250 kilowatts in 1976 and growing to an estimated 3.5 megawatts in 1981.

Emerging Applications

It is generally useful to classify system applications by either a system characteristic or by application sectors. In the case of photovoltaic power system applications one classification often used categorizes the system applications into the four areas of: remote/stand-alone, residential, service/commercial/industrial/institutional (or SCII, which is sometimes categorized by size and referred to as "Intermediate Load Centers"), and central utility. Applications for photovoltaic power systems are emerging in each of these four categories.

Remote, Stand-Alone

The remote, stand-alone market has been the major market for photovoltaic system applications as discussed above. Major markets for stand-alone systems still remain to be penetrated for certain applications in less developed countries, developing countries and in developed countries.

For the less developed countries the primary markets will be for agricultural applications such as water pumping, irrigation and refrigeration of food products. There also will be needs for electrical power for medical related applications such as remote clinics and medicine refrigeration and for powering educational devices including TV receivers.

For both less developed and developing countries village electrification will be a major need. Initial pilot plant demonstrations currently exist in Africa, the Middle East and parts of North America. In addition, the U.S. Agency for International Development and the Commission of the European Communities have pilot plant projects for photovoltaic system village power applications in the design or construction phase. Such pilot plant applications began as early as 1978 at an American Indian village. Total market for the pilot projects to date exceeds 300 kWp. The potential market for village electrification is very large. There are more than one billion people living in two million villages without electricity. Market estimates indicate that this market has a total potential of 20,000 MWp.

There are a few hundred pumps which are currently being powered by photovoltaic power systems. These systems are located in developing and developed countries. Markets for PV water pumps in remote areas of developed countries are emerging rapidly. Sales in the U.S. alone exceed 100 units today and more sales are being made constantly.

It has been estimated that the total small pump market will exceed 50 million units. Large pump applications for deep wells also is expected to be large. A few applications exist.

A few sales have been made for photovoltaic powered refrigerators for medical needs. And some prototype refrigerators for food storage have been built. The U.S. Agency for International Development has embarked on a major program to introduce PV refrigerators for medicine. It has been estimated that the medicine refrigerator application alone will exceed 5000 units per year.

Residential

Residential application markets exist for stand-alone systems for remote residences which are regularly occupied and remote recreational cottages. In both cases the cost to extend the grid to such homes is generally prohibitive. A few hundred sales of small (100-500 watt) photovoltaic systems for remote housing in developed countries have been made during the last 1-2 years. This market is expected to be significant and is beginning to grow at a rapid rate.

Grid connected residential applications have been a major focus of the U.S. government program. Applications representing about 150 kW have been installed.

Privately financed applications of PV systems for residential applications also have been initiated in the U.S. MONEGON has recently dedicated the first solar electric house in Potomac, Maryland, near Washington, D.C. This house, illustrated in the photograph below, also has passive design features and incorporates active solar thermal systems for domestic hot water and space conditioning.

The residential applications to date represent a total in excess of 200 kW. Analysis of both new and retrofit housing markets indicate that this market can reach 5000 MW per year. Figure 2 shows the residential forecast for the U.S. and Europe.

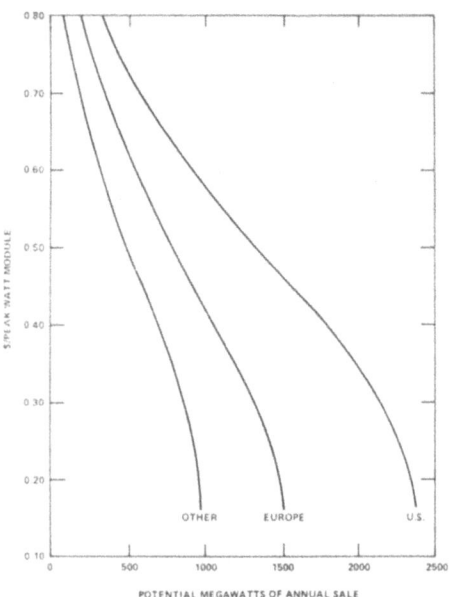

FIGURE 2
POTENTIAL INSTALLED PV SYSTEM CAPACITY FOR THE
RESIDENTIAL SECTORS IN U.S., EUROPE, AND OTHER COUNTRIES

Service/Commercial/Industrial/Institutional(SCII)

Although there have been a few small systems installed for SCII applications, the vast majority of such applications to date have been under government funded programs. The U.S. program has funded SCII applications of PV systems representing nearly 1.4 MW of power. The CEC PV pilot plant program includes more than 700 kW of such applications. It is forecast that this sector will be a major market reaching a level of 14,000 MW per year (See Figure 3).

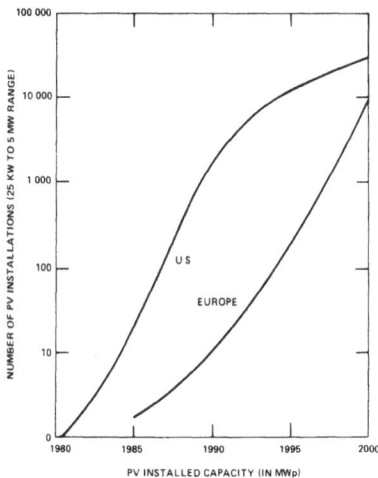

FIGURE 3

POTENTIAL PV INSTALLED CAPACITY FOR THE
SCII SECTORS IN EUROPE AND THE U.S.

The first privately funded industrial application is now under construction in Frederick, Maryland. This system is a 200 kW power plant for the Solarex production facility known as the "Solar Breeder."

Central Utility

Central utility applications have been perceived as a market area for the future when PV system costs are very low. Recently, however, there has been increasing interest in gaining experience with central power systems. There have been announcements of four 1 MW central power systems: two in the U.S., 1 in Japan and 1 in Italy. One of the U.S. systems is expected to be funded privately with the utility, Southern California Edison Company, buying the electricity generated at a negotiated rate. Other similar applications are being negotiated.

The capability to have privately financed systems while the PV system prices are still relatively high derives from U.S. tax laws. On the basis of this recent activity, market penetration for central power is expected to occur earlier than previously projected. This market is expected to grow to an annual rate of 110,000 MW per year.

5. FORECASTS

Analyses of past market trends and growth rates along with emerging application activities and electrical power needs have provided the basis for making market forecasts. Table 1 presents a forecast of the overall market growth together with the value in sales for these markets using forecasts of wholesale system prices. The total market presented in Table 1 has been analyzed for the market share by the four application sectors discussed. This estimate of the application sector markets is presented in Table 2.

MONEGON Estimate of the Worldwide Photovoltaics Market

Year	Annual New Installations GWp	Systems Cost per kWp (1980 dollars) (wholesale level)	Total Systems Sales (1980 billion $) (wholesale level)	Total Cumulative Installed Capacities GWp
Estimated Historic Data				
1976	0.0002	47,000	0.009	0.0002
1977	0.0005	41,000	0.021	0.0007
1978	0.0010	35,000	0.035	0.0017
1979	0.0015	29,000	0.044	0.0032
1980	0.003	23,000	0.070	0.006
1981	0.005	17,000	0.085	0.011
Forecast				
1982	0.010	12,000	0.120	0.025
1983	0.022	6,000	0.132	0.06
1984	0.053	4,100	0.217	0.16
1985	0.13	2,700	0.35	0.4
1986	0.3	2,100	0.7	0.9
1987	0.7	1,700	1.2	1.9
1988	1.7	1,500	2.5	4
1989	4	1,300	5.2	8
1990	8	1000	8	16
1991	15	900	14	31
1992	30	900	27	61
1993	46	900	41	107
1994	66	900	59	173
1995	88	900	79	260
1996	100	900	90	360
1997	120	900	108	480
1998	130	900	112	610
1999	130	900	112	740
2000	130	900	112	870

Table 1

Table 2
ESTIMATE OF WORLD
PV SYSTEM MARKET
BY APPLICATION SECTOR
(MWp)

Year	Remote Stand-Alone	Residential	Commercial Industrial/ Institutional	Central Utility	Annual Total	Cum. Total
1976	0.239	0.001	--	--	0.24	0.24
1977	0.449	0.001	--	--	0.45	0.69
1978	0.603	0.002	0.33	0.015	0.95	1.64
1979	0.900	0.005	0.58	0.015	1.50	3.14
1980	2.245	0.01	0.95	0.045	3.25	6.39
1981	3.52	0.25	1.55	0.08	5.40	11.8
1982	5.0	0.75	3.25	1.0	10	22
1982	10.0	3.5	5.5	3.0	22	44
1984	28.0	8.5	12.5	6.0	55	99
1985	70	20	25	15	130	229
1986	125	50	75	50	300	529
1987	200	125	250	125	700	1229
1988	320	430	600	350	1700	2929
1989	700	1000	1450	850	4000	6930
1990	900	1700	3000	2400	8000	14930
1991	1000	3000	5000	6000	15000	29930
1992	1000	4000	10000	15000	30000	59930
1993	1000	4500	10500	30000	46000	105930
1994	1000	5000	12000	48000	66000	172000
1995	1000	5000	12500	69500	88000	260000
1996	1000	5000	13000	81000	100000	360000
1997	1000	5000	13500	100500	120000	480000
1998	1000	5000	14000	110000	130000	610000
1999	1000	5000	14000	110000	130000	740000
2000	1000	5000	14000	110000	130000	870000

THE POTENTIAL FOR PHOTOVOLTAICS IN EUROPE

M R STARR
Sir William Halcrow & Partners
Burderop Park, Swindon SN4 0QD UK

Summary

There has been considerable investment worldwide in photovoltaic technology, by both private and public agencies. In Europe, there is a major photovoltaic research and development programme co-ordinated and financially supported by the European Communities. A large worldwide market is expected to develop, with important opportunities for European exports, but what are the prospects of this technology as an energy resource for Europe itself? Could a significant proportion of Europe's total electricity production one day be derived from photovoltaics? This paper addresses these questions. After a review of the prospects for cost reduction, the economics of electricity generation by photovoltaics in comparison with conventional alternatives are discussed. Break-even costs and the approximate dates when these could be achieved are presented. Market prospects are then discussed for various applications in Europe, such as stand-alone systems, grid connected residential and industrial systems and central generating plants. Given official support and full co-operation from the utilities the total market for applications in Europe could be 1000MWp per year by the year 2000 and continue to grow rapidly for many years thereafter. By 2025, the total installed capacity of photovoltaic systems in Europe could exceed 200 GWp, generating about 10% of total electricity production.

1. INTRODUCTION

In recent years, substantial investments have been made in all aspects of photovoltaic technology, from research and development through to production facilities and field installations. To date, it has been estimated that over US $1000 million has been committed to photovoltaics by public and private agencies. Much of the photovoltaic research and development work undertaken in Europe has been co-ordinated and supported by the Commission of the European Communities (CEC). From the beginning, solar energy has taken the major share of the funding available for the CEC energy R&D programme, and out of that share, the larger proportion has been spent on photovoltaics. Such a level of support reflects the high hopes placed in this technology. The question remains, can these hopes be realised? Will photovoltaics be technically and economically viable for widespread use in Europe, USA and in the developing countries within 10 years - less than the time it takes for a conventional or nuclear power plant to be planned, designed, constructed and commissioned? Even if a major worldwide market develops, could a significant proportion of Europe's total electricity production one day be derived from photovoltaics? If so, how much and when?

To attempt to answer these questions, it is first necessary to look at the economics of photovoltaics. What are the prospects for cost reduction? Then it is possible to review the potential market for systems of various types, from small stand-alone systems to large central generating systems feeding the grid.

The currency unit used in this paper is the European Currency Unit (ECU), which currently is equivalent to about 0.56 UK pounds or 1.01 US dollars. All prices for photovoltaic modules and systems, including projections, are given in 1980 currency values unless otherwise stated, since 1980 is the base year used by several authorities for comparing photovoltaic prices.

The introduction of photovoltaic power generation on a significant scale in Europe would have far-reaching social and environmental implications. The market projections presented here are based on the assumption that general support and encouragement will be forthcoming from national governments and utilities for the introduction of photovoltaic systems wherever they are economic or deemed desirable. Although beyond the scope of this present paper, this question of official suport and encouragement for photovoltaics is an important matter meriting further discussion and appropriate action at all levels.

2. PROSPECTS FOR COST REDUCTION

Photovoltaic research, development and demonstration work is proceeding worldwide on a broad front. A great variety of techniques and processes hold promise for low cost photovoltaic systems and whilst it is clear that progressive cost reductions have been made in the past, will the low cost targets be achieved in future and, if so, when?

From the time that major development work into photovoltaics for terrestrial applications began in the late 1960's and early 1970's, manufacturers and government organisations have been studying the prospects for reducing costs, through improved techniques and higher volume production, and various cost forecasts and targets have been produced. One such study, carried out by CNES in France in 1974, compared the module price forecasts of manufacturers in USA and France as part of a much wider exercise (1). Experience since 1974 has shown that these forecasts were remarkably accurate.

In 1980, the US Department of Energy published revised cost goals for photovoltaic modules and systems for the period 1975-1990 (2). Although current module prices are somewhat higher than the target, current system costs are within the target range. A useful survey of photovoltaic cost experience in the USA has been prepared by Burgess et al (3). Their conclusions are encouraging in that they lend support to the long term DOE cost targets for systems, without of course saying when these might be achieved.

Until recently, the emerging photovoltaic industry in the USA had been strongly supported by the US government. Cost targets had been consistently met each year, but the budget cut-backs by the Reagan administration will have a significant impact on the US photovoltaic research, development and demonstration programme. It is now generally accepted that the cost targets may in consequence be delayed by several years. Costs will no doubt continue to decrease in real terms, but not as quickly as previously hoped.

Although the cost targets for modules and complete systems are based on careful analyses of large scale production techniques, and may thus be considered as technically feasible given the assumptions, what is actually achieved and when will be strongly dependent on three main factors:

1. Technical progress

   Progress towards achieving cheaper materials and processes must continue as anticipated, and this is largely dependent on the level of public and private funding of research and development. In the search for low cost photovoltaic cells, five topics stand out as needing co-ordinated and concentrated efforts. These are: solar grade silicon production, ribbon silicon techniques, improved silicon wafer cutting techniques thin film cadmium sulphide cell processes and possibly the most promising for the long term, thin-film amorphous silicon cell processes. Development work is also needed for components and complete systems for various applications.

2. Industry build up

   Large integrated production plants for materials, cells, modules and complete sytems must be built, with annual capacity 50 to 100 MWp. This scale of operation is needed to reduce production costs to the minimum practicable.

3. Market development

   The output from these large manufacturing plants must find markets, in Europe and overseas. Successful demonstrations will help establish confidence in the technology, but official encouragement and appropriate tax and other incentives will be essential to provide the necessary inducement for private and public investment.

With the above discussion in mind, it has been possible to prepare Figure 1, which is an updated cost projection for photovoltaic modules and systems, based on the latest information from manufacturers and trends in the industry. A considerable degree of uncertainty has to be included in the cost projection due to the dependence on the three factors referred to above. The main features are as follows:

Year	Modules FOB ECU/Wp	Systems Installed ECU/Wp
1980	9	20 - 40
1985	3.00 - 3.50	8 - 14
1990	1.00 - 2.00	3 - 6
1995	0.70 - 2.00	1.8 - 3.8
2000	0.50 - 2.00	1.6 - 3.5

This cost projection can now be used to calculate the electricity unit cost from a photovoltaic generator.

3. ELECTRICITY UNIT PRICE

The key economic parameter when comparing alternative means of generating electricty for any specific application is the unit energy cost, the cost per kilowatt-hour. Assuming the full annual electrical output of the system is used, the unit energy cost is given by:

$$\frac{\text{annual capital charges} + \text{annual operating costs}}{\text{total annual energy output}}$$

Neglecting inflation, tax credits and other financial incentives, a simplified approach gives the relationship:

$$p = \frac{r.C + m.W_p}{\eta_{ave}.P_i}$$

where p = unit cost per kWh; r = interest plus amortization factor; C = total capital cost of system; m = specific operating cost, expressed as cost per peak Watt installed; W_p = peak power of photovoltaic array, η_{ave} = annual average conversion efficiency of system; and P_i = total annual solar energy incident on the array.

As an example, let us consider a 10kWp grid connected system built in southern Europe in the late 1980's when photovoltaic module costs are expected to have fallen to about ECU 2.00/Wp and total installed costs for a complete generator system including some battery storage to about ECU 5.00/Wp. The total capital cost would thus be ECU 50000. Assuming a 20 year amortization period and 5% rate of return, the factor r would be 0.0802 and thus the annual capital charges would be ECU 4012. The annual operating cost including maintenance and insurance, might be of the order of ECU 0.06/Wp and thus the total annual operating cost would be ECU 600. Taking the total area of a 10kWp array to be 100 square metres and assuming the installation was at a place where the total annual solar energy incident on the plane of the array was 1750kWh/m^2, Pi would thus be 175000kWh. Based on an annual average conversion efficiency of 7.5%, the total annual energy output would be 13125kWh. The unit energy cost would thus be:

$$p = \frac{4012 + 600}{13125} = \text{ECU } 0.35/\text{kWh}$$

This simplified approach to photovoltaic system economic analysis may be used to derive unit energy costs for different values of capital cost, interest rate, specific operating cost, total incident solar energy and average system conversion efficiency.

The next step is to consider the electricity unit cost given by alternative sources for comparison with the photovoltaic system, to see when the break-even points occur. For example, a typical price in Europe for grid electricity supplied to domestic consumers is ECU 0.08/kWh. A typical price for electricity generated by large (ie 2-10MW) diesel or gas turbine generators (GTs) for small networks is about ECU 0.20/kWh. For small diesel generators (10-100kW) in remote places, the unit cost can often exceed ECU 0.50/kWh. These unit prices may be expected to rise at a rate higher than the general inflation rate. In recent years, the differential inflation rate for commercial energy in most countries has been over 10% per annum, but it is generally considered unlikely that such a high rate will obtain in future. The probable range for the differential inflation rate for electricity supplied from conventional sources is between 5 and 10% in the long term.

Thus, whereas the price of photovoltaic systems, and hence the cost of the electricity generated by such systems, is expected to fall in real terms over the next 10-20 years, the cost of electricity from conventional generators is almost certain to rise in real terms over that period. The consequences are illustrated in Figure 2, which shows that photovoltaic systems in regions with high solar insolation (eg, southern Europe) could be competitive with small diesels in remote areas by the mid-1980's, with larger diesels and GTs by the late 1980's and with grid supplies by the mid-1990's. For places with less solar insolation (eg central and northern Europe), the corresponding break-even dates would be later by some 5 to 10 years.

Clearly, for specific systems and locations considerably more detailed economic analyses need to be made, taking into account other factors such as taxation and inflation, but the above simplified approach provides a good indication of the potential for photovoltaics to be competitive with conventional generating systems, including in time grid supplied electricity.

4. SMALL STAND-ALONE SYSTEMS

The market for small stand-alone photovoltaic systems (from a few watts to about 500W) is the one least subject to objections and difficulties posed by the utilities, by governments or from other members of the public. It is also the area where photovoltaics are first likely to be economic, in some applications even at today's prices.

Although it is known that most photovoltaic manufacturers have studied the market potential for stand-alone systems, their conclusions are rarely published. A recent study in France by COMES identified over 20 applications and listed the associated photovoltaic array power required (4). The applications of main interest in Europe may be divided into two categories: the consumer market and the professsional market. The consumer market consists mainly of systems that would be purchased as off-the-shelf items for remote farms, off-grid houses, holiday and mobile homes, such as lighting units, small pumps, refrigerators, battery chargers and security devices. The professional market for small stand-alone systems will also consist largely of packaged systems for remote locations. Again these would be mainly purchased as off-the-shelf items and typical applications include navigation aids on land and at sea, telecommunication systems, railway signalling, cathodic protection, automatic weather stations and lighting units for remote buildings.

There is no precise method of forecasting sales of these relatively small stand-alone photovolatic systems, but given large scale production and effective marketing, it seems probable that by 1990-1995, complete systems will be available for less than ECU 5/Wp. Considering the whole of the western Europe, with a total population of over 300 million, a large holiday industry and with many thousands of off-grid houses, particularly in southern Europe, the following mature market projection would seem reasonable:

(a) Consumer systems of all types - 100000 units per year by 1995. The average size of the photovoltaic array for each system is likely to be about 50Wp and thus the total power involved would be about 5MWp per year.

(b) Professional systems of all types - 50000 units per year by 1995. The average size of the photovoltaic array for ech system is likely to be about 150Wp and thus the total power involved would be about 7.5MWp per year.

It is anticipated that this level of market activity for small stand-alone photovoltaic systems, amounting to about 12MWp per year, would not increase significantly after 1995, even if system costs continued to decline, since the market will have become fully developed.

5. LARGER STAND-ALONE SYSTEMS

The market for stand-alone systems in the size range 500Wp to 150kWp is mainly that at present occupied by gasoline and diesel generators. Batteries or another form of energy storage will generally be required with these systems to provide continuous power capability. This has a significant effect on the economics of such systems and also on the applications where they could be used. In general they would not be suitable for use as portable generators due to the size of arrays and the great weight of batteries involved. The low use factors typical of this type of generator application will result in a high cost of power. There are however several situations where stationary generators operating at high use factor are needed and for which photovoltaic systems may be appropriate. The main application will be for remote houses and villages, but there is some scope for other applications such as radio and TV transmitters, radar installations and pumping stations.

There are over 70000 houses in Italy and probably at least this number in Greece, in Spain and in Portugal that are permanently occupied and are not grid connected. There are a further 20000 in France but elsewhere in western Europe the number of off-grid permanently occupied houses is considered to be relatively small. The total potential market for photovoltaic systems for off-grid houses could thus be of the order of 300000, with an average size of about 5kWp each. Based on the breakeven conditions indicated on Figure 2 and subject to appropriate finance being forthcoming, significant penetration could start in 1985, rise to about 3000 to 4000 systems a year by 1990 and level off at about 6000 to 8000 systems a year by 1995. The associated photovoltaic system market would thus be about 20MWp per year in 1990, rising to about 40MWp per year by 1995, with relatively little increase thereafter.

One other important application that is expected to become economic in the 1990's is large stand-alone systems to supply remote villages and island communities which at present either are served by diesel generators or have no electricity supplies. There are at least 500 such locations in southern Europe, for which the size of the photovoltaic system would need to be between 50 and 150kWp. Assuming on average some 50 of these per year will be equipped with photovoltaic generators throughout the 1990's, the associated market would be about 5MWp per year.

6. GRID-CONNECTED RESIDENTIAL SYSTEMS

The potential for grid-connected residential photovoltaic systems in Europe has not been studied in depth although organisations in France and Italy have carried out some work in this area (4, 5). In the USA such systems are considered by many to offer an early potential market and one which will grow rapidly.

A major point in favour of residential photovoltaic systems is that the load is close to the generation source and as a result there are minimal transmission and distribution losses. From the user's viewpoint, the reliability of his supply is improved, particularly if battery storage is incorporated. From the utility's viewpoint, the photovoltaic generator contributes to the total generating capacity of the system, thereby reducing the capacity that has to be provided by conventional generating plant. Studies by the Italian utility ENEL show that the capacity credit of a photovoltaic generator on their system would be about 30% of its rated power (6).

If installed as part of a new building during construction, the photovoltaic array may be incorporated into the roof structure and may, with some designs, replace part of the conventional waterproofing system, giving appreciable cost savings. Retrofit installations (ie, installing the array on existing houses) would be rather more expensive.

There would seem to be no unsurmountable technical barriers to prevent photovoltaic systems coming down to price levels that would enable them to generate electricity at costs comparable with grid selling prices in southern Europe by about 1990 and in northern Europe by about 2000 or soon after, given continued real inflation in the price of grid-supplied electricity. It would then be feasible, given the necessary finance and political will, for a large proportion of all new single family houses and a fair proportion of low-rise apartments within the break-even areas to be equipped with photovoltaic systems. A few years after systems for new houses started to be cost-effective, retrofit installations would also become cost effective, assuming continued real price inflation of grid supplied electricity plus possibly some further reduction in system prices.

By the year 2000, the market for grid connected residential systems could have reached 800 MWp/year and be growing rapidly. The proportion of new houses with photovoltaic generators would steadily increase and the retrofit market would grow. It is conceivable that the total market in Europe for grid connected residential systems could have grown to 2000 MWp/year by the year 2025, considering new houses, retrofit installations and replacements for early systems.

7. PHOTOVOLTAIC SYSTEMS FOR INDUSTRIAL SECTORS

The service, commerical, institutional and industrial (SCII) sectors of the economy encompass many business establishments with electrical loads in the range 25kW to 5MW. Many of the 16 European Community photovoltaic pilot plants come into this category. The loads are intermediate between those of single residences, which are generally considerably less than 25kW, and those of conventional central electricity generating stations, whose capacities range from 100MW to over 1000MW.

Detailed studies on the potential for photovoltaic systems for the SCII sectors have not been carried out for Europe and for present purposes only a preliminary assessment is possible. The first point to note is that there is a tremendous variety of possible applications leading to a great diversity of operating schedules, energy demands, space availability and access to solar energy.

It is also important to note that, for many electricity users in the SCII sectors, the consequence of failure in the grid electricity supply are very serious, and expensive stand-by diesel plant has to be maintained to meet essential loads in the event of power failure. Photovoltaic systems with energy storage could find a practical application as a second source of power, used to supplement grid power normally and to meet all emergency demands when necessary long before such systems become directly competitive with utility supplied electricity.

The total installed capacity of photovoltaic systems in the SCII sectors in Europe could reach as much as 1000MWp by the year 2000, assuming economic viability is achieved by 1990. Depending on general economic circumstances and on the tax and other incentives available, a possible development would be (excluding subsidised demonstration plants) 10 such installations in 1990, 20 in 1991, 40 in 1992 and so on, the number doubling each year. This would result in a cumulative total of about 10000 installations by the year 2000, requiring about 1000MWp of photovoltaics, on the assumption that the average size of an SCII instllation will be about 100kWp. The rate of installation in the year 2000 would be about 4000 systems per year, say 400MWp of photovoltaics. This scale of business would continue to grow until well into the next century.

8. PHOTOVOLTAIC CENTRAL GENERATING STATIONS

The rate at which photovoltaic systems are integrated into the electricity utility generating mix will depend upon several factors and not simply on the economics. National policies may require the utilities to provide a certain proportion of generating capacity from indigeneous sources to reduce dependence on imported fuels. The utility may find that a photovoltaic plant, being incremental, can be built up at a rate to suit a rising demand, whereas a major thermal installation would take much longer to bring on stream.

There is nevertheless no doubt that economic criteria will play the dominant part in determining the role for photovoltaic systems for central generation. The key factors will be (i) the cost savings that will accrue from displacement of fossil fuels whose prices will rise faster than general inflation; and (ii) the displacement of a significant proportion of conventional generating capacity (capacity credit).

There have been few studies in Europe on the potential for photovoltaic central generators, although there are plans for a 1MWp experimental plant in Italy (7). Some consideration has been given to the feasibility of building ground rectennas in Europe for large photovoltaic space power stations, but for various reasons this approach is not generally considered appropriate, at least for Europe. Others have proposed large photovoltaic plants in the Sahara desert feeding power to the European grid system through underwater cables across the Mediterranean. This approach is unlikely to be economic and has many associated problems.

In Europe, the availability of land for large photovoltaic plants is likely to be an important consideration, since competing interests are involved. It is unlikely that it will ever be acceptable for large areas of countryside to be used, bearing in mind transmission costs and environmental considerations. On the other hand, near most towns there are usually considerable areas of waste land, former factory sites, disused railway lines, obsolete thermal power stations and the like.

The first development will probably be for many existing power stations to have photovoltaic generators built within the same site boundary, to reduce costs for land, switchgear, transmission and operation. Later, the utilities may be expected to establish photovoltaic generating stations near towns, using under-utilised land areas. The size of these plants would be unlikely to exceed 200 MWp each, but the total contribution would be significant given many hundreds of such installations in each country, particularly in southern Europe.

The introduction of commercial photovoltaic central generating stations may possibly start in the late 1990's, but for economic reasons the main development in this important sector is unlikely to arise until later. Rapid growth in the first quarter of the next century could well occur, resulting in a total installed capacity in Europe for central generation of the order of 50GWp by 2025.

9. TOTAL EUROPEAN MARKET FOR PHOTOVOLTAICS

The above discussions on potential markets for photovoltaics in Europe are highly speculative and depend on many assumptions, particularly regarding future total system costs. Unless there is suitable support for the emerging photovoltaic industry, it is unlikely that the low system cost targets will be achieved within the time periods mentioned, in which case most of the markets for photovoltaics in Europe will not become economically viable until much later, when conventional energy prices have further risen.

Assuming however that full support is given to the industry and that appropriate financial and fiscal encouragements are offered to purchasers of systems, then the market projections for the various categories of system may be taken as an indication of the potential. The aggregated total sales projection for the various applications in Europe is shown in Figure 3. By the year 2000, the total European market could be between 650 and 1150MWp per year, implying average growth rates from a base of about 1MWp in 1983 of between 45% and 50% per year. A lower total sales line has ben included to indicate the situation that may obtain if grid-connected systems do not become viable within the time frame being considered. By the year 2025, the total installed capacity of photovoltaic systems in Europe could exceed 200GWp, generating about 10% of total electricity production.

It should be stressed that the market projections presented in this paper are for applications in Europe. The total world market will be several times larger and European manufacturers will have important opportunities to develop photovoltaic system exports. By the year 2000, the total value of photovoltaic business transacted in Europe could be of the order of ECU 5 to 10 billion.

10. CONCLUSION

Photovoltaic power is emerging from the research and development phase and the prospects are good for this technology, with its many attractive features, to become an energy resource of major significance for Europe as well as many other regions of the world. Continued public and private support and investment is thus justified to ensure technical progress towards the goal of low cost systems, to build up industrial capacity and develop the markets.

11. ACKNOWLEDGEMENT

This paper is largely based on the results of an assessment study carried out for the Commission of the European Communities under contract No ESC-P-049-81-UK(H) and is published with permission. The author is grateful for the help and advice of many organisations and individuals in the European Community and in the United States of America during the course of the assessment study.

REFERENCES

1   W Palz, A Simon and D Thieret : Les applications terrestres des generateurs solaires photovoltaiques; Centre National d'Etudes

Spatiales - CNES, March 1974 (CNES 109 2952 B).

2   P D Maycock : Overview - Cost Goals in the LSA Project; Proc 14th IEEE PV Specialists Conference, San Diego, 1980, p 6.

3   E L Burgess, K L Biringer and D G Schueler : Update of photovoltaics system cost experience for intermediate-sized applications; Proc 15th IEEE PV Specialists Conference Orlando, 1981, p 1453.

4   Y Chevalier, A Haentjens and B Meunier : French public photovoltaic market stand-alone applications; Proc 15th IEEE PV Specialists Conference, Orlando, 1981, p 201.

5   F Amman, E Antognazza, L Braicovich and G Panati : Energia Solare : Una Proposta Organica di Politica Industriale nei Comparti Fotovoltaico e Termodinamico; Instituto di Economia delle Fonti di Energia - IEFE, Milano, Italy, May 1980 (draft).

6   G P Pacati and A Taschini : Utility aspects of photovoltaics in Europe; Proc 3rd EC PV Solar Energy Conference, Cannes, 1980, p 46.

7   V Albergamo and P Bullo : The Delphos project, Proc 15th IEEE PV Specialists Conference, Orlando, 1981, p 1208.

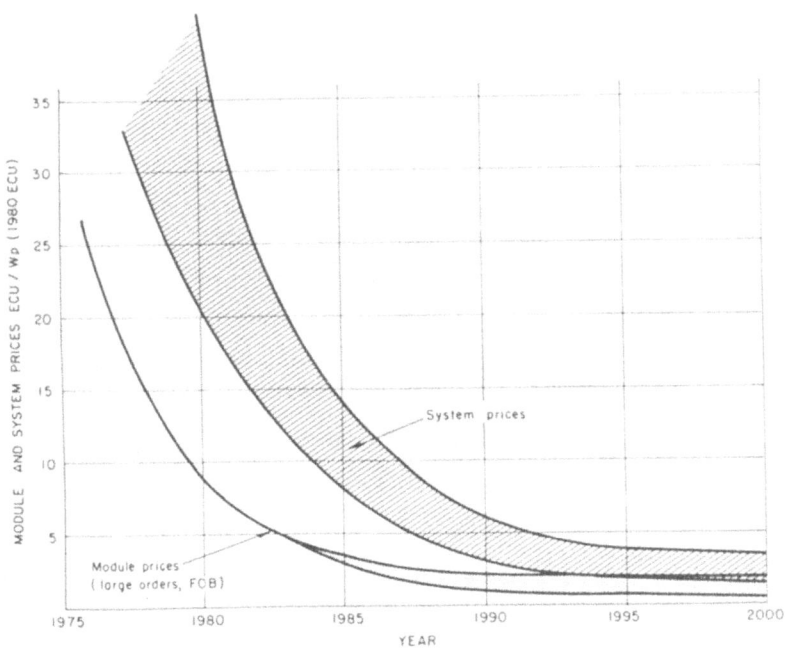

Fig. 1   Cost projection for modules and systems

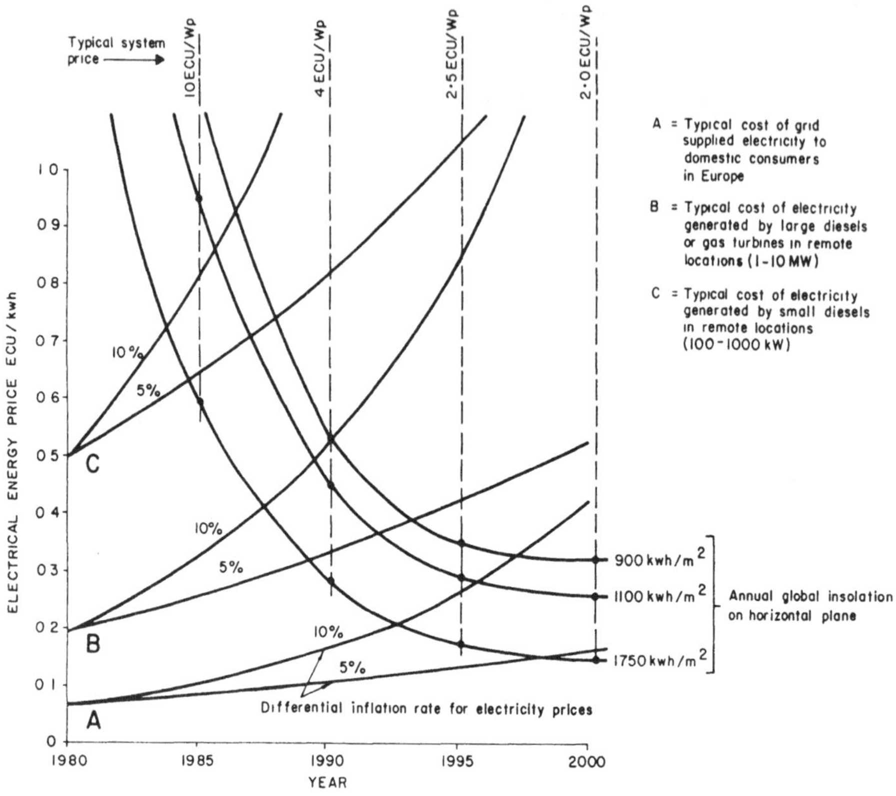

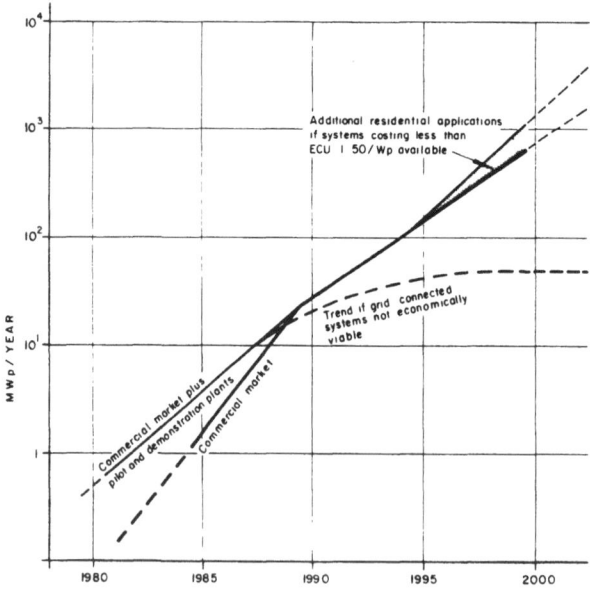

Fig. 2 - Photovoltaic system break-even points

Fig. 3 - Total market projection for photovoltaics in Europe

THE BEHAVIOUR OF LARGE SOLAR POWER STATIONS

IN THE SWISS ALPS

M. Real

Project Manager, Division for Solar Electricity
Swiss Federal Institute for Reactor Research

Summary

Electric utility acceptance of solar cell power stations as a reliable electricity generation alternative will require sufficient experience and know how to allow the technology to expand in Switzerland. This acceptance will require succesfull resolution of a number of key issues. Among the commonly well known problems related to solar electricity fed into the electrical grid, Switzerland is facing a somehow unique situation: for reason of land availability and higher solar insolation during the winter period, large solar power stations will have to be built in the higher regions of the swiss alps. The acceptance of solar electricity will therefore require the proof that large solar cell power plants are operational under the severe climatic conditions of the alps.
The objective of the test presented in this paper is to investigate this problem. Both full size test with a field of solar cells and scale model tests had shown, that at least at the chosen site during the test period of last winter, snow and ice accumulation will only cause small losses of the potential overall energy output of a photovoltaic power station.

1. INTRODUCTION

Electricity, generated by solar power plants in the swiss mountains, is suitable to be fed into the existing electric grid, considering the possible combination of the existing installations of hydro with solar power stations. The two main reasons to install those solar power plants in the swiss alps are:
- The expected energy output of photovoltaice power plants are much higher for installations in the mountains due to higher solar insolation. This is especially true for the winter season, when the electricity is needed most.
- There is much land available which is suitable for large scale installations. In fact, a study which about 12o km^2 of free land in the alps will meet the requirements of solar electric installations and the aspects which will arise from environmental constraints.

Obviously, solar plants installed on such sites will face very severe climatic conditions during the winter period.

To investigate this special issue, the Swiss Federal Institute for
Reactor Research started with experiments five years ago. Until last
year, this tests were related to problems of solarthermal power plants.
Within the very fast development of solar cells, however, the question
of a real comparison of both systems became more and more important.
Solarthermal versus photovoltaic power plants in the swiss alps - that
was the issue of a test initiated last year.
A solar cell field was therefore installed at the site of the heliostat
test facility above Davos, 2600 m sea level. Careful evaluation showed
that the minimal size for simulating a hole field of a large power
station requires three rows of cells, each row 15 meters in length.
The specific objectives of the test are:

- to investigate the operational behaviour of large solar cell arrays
  in the swiss mountains

- to compare actual snow drift accumulation problems with results of
  a scale model test

- to measure the actual energy produced by cells which are integrated
  in an array in order to predict the power output of larger systems

- to compare phtovoltaic versus solarthermal power systems.

2. Description of the test

Since the objective of the test is not to produce energy, most ob the
field lay-out had been simulated with wooden dummies.
In order to achive realistic donditions as they will occur in large
power plants, three rows of cells had been mounted. Each row is 15 meters heigh. Calculations showed, that a land use factor of four will
be required to minimize shadowing. This lead to a spacing of 4 meters
between each row. Each row is mounted on a support which keeps the lower
edge of the solar cells two meters above ground.
A tilt angle of the cells of $60°$ had been chosen to optimise energy output for the winter time and to facilitate accumulated snow to glide.
The field array, shown in Figure 1, includes 18 solar cell modules of 34
Watt peakpower output each. They are concentrated on four places within
the array, as indicated by the arrows in Figure 1.
These positions had been chosen to study the difference of the global
radiation, which is expected to vary for different places within the
array. Four modules are periodically scanned and the output power characteristics are recorded by a sophisticated Data Aquisition System,
shown in Figure 2. The load is simulated by a capacitor, a very simple
but efficient measurement technique. The idea had been adopted from
ingeneers from the MIT (1). In addition, a time-lapse 16 mm movie camera
has been mounted in front of the array to record snow-shedding rates and
snow drift profiles, built up between the rows.
The picture rate is set at one frame every ten minutes. Tests with the
entire field array leads to profiles of snow accumulation after each

season. Results of Modifications in the lay-out will produce only results after having waited another year. It was felt, however, that scale model tests should be carried out which give test results after a few hours test. The theory of scale model tests has been described by Ronald D. Tabler (2). Up to now, only a few tests had been carried out to verify the feasybility of scale model tests. Figure 3 shows the snow drift pattern of a outdoor scale model test. Steady state conditions were reached after one test period. The snow pattern is identically with that obtained in the field array, shown in Figure 4.

As a first result, it was found for example that the spacing after each group of two rows should be increased in order to allow the drifted snow to deposite. The presented snow pattern is the result of a snow drift direction normal to the rows. Profiles will drastically change with changing wind directions.

Most of the time, however, selected sites have dominant winds from the same direction. More serious concerns against scale-model tests is the fact, that on a mountain slope the air flow is turbulent.

This leads to an uneven snow profile, which diverges from one spot to another. However, this snow profile distribution is reproduced each year (3). Actually, this fact is used for designing avalanche barriers (4). It is felt that the result with both the field array and the scale model test will help to establish standards for designing large power fields, presupposed that the snow profile of sites in consideration is known.

3. Preliminary results and conclusion

The test with a solar cell field has shown that both snow and ice accumulation have only a small impact on the overall energy output of power station, mounted in the swiss alps. It is to mention, however, that in other places different problems may occur. The snow fall of last winter was about 30 % higher than normal and the accumulated data of daily snow fall measurements gives 13,5, equivalent to about 1100 mm ov water. Snow under the cell will be a higher problem than this on the cells, because for slope angle above $10°$ the snow cover starts creeping down hill. To avoid loads on the support frame, it is highly recommended to paint the support pedestals black. The absorbed light will heat the frame and trues melts the very surrounding snow cristalls. The result is a clear reduction of the forces of the creeping snow cover.

The chosen tilt angle of $60°$ is both optimal for solar energy collection during the winter period as well as to enable any snow and ice on the cell to glide downwards. Actually, there had been none recording, that snow will stay long on the cells when the sun is shining. Normal clearance time is about 2o minutes.

Since the Data Aquisition System (DAS) had not worked from the beginging of the test period, only few data are available and no statistical proven statements can be given. As the DAS records every 3o minutes the entire I - V - Curve of four specially selected modules, the peak power point is being calculated on line as well as two power outputs, es they would occur if the module would operate at a given voltage point.

The result should be an indication, what the losses would be if a power plant in cold climatic conditions will work without a Max Power tracer. The solar cell efficiency is decreasing for higher solar insolation. The reduction is about 15 % when the daily energy increases from 2,5 KWh per m^2 up to 8,5 KWh per m^2.

As mentioned, so far there are no statistical data available. The highest insolation during March was 1290W/m^2, averaged over half an hour of a sample of data which are recorded every 30 seconds. The daily energy insolation was 8.56 KWh/m^2 for a plane with a tilt angle of 60° (Maximum). The panel with nominal peak power of 36 Watt produced at that day 274.14 Watt, with sample every 30 seconds.

After having tested both a full size heliostat and a solar cell field, the following statements can be made:

- Both systems will be operational during the wintertime

- There was - for the short period considered - no maintainance required for the photovoltaic field. There was, however, some maintainance required for the heliostat.

- The time of operation will be much higher for solar plants with solar cells

- The solar cells will profit from the very high global radiation, although for nice days the direct solar radiation is very high in this altitude.

4. References

1. Warnes, T.H., Cox, C.H.; Commercial
   Photovoltaic Measurement Workshop, Vail, CO, 27-29 June 1981

2. Tablor, R.D.; Journal of Glaciology, Vol. 26, No. 94, 1980

3. Fruttiger, H.; Swiss Federal Institute for snow and avalanche Research, in house report No. 210
   February 1956

4. Report of the Swiss Federal Institute for snow and avalanche Research, Report No. 29, December 1968

5. Real, M., Kesselring, P.; Colloques internationaux du CNRS
   STS, No. 306-Systèmes solaires Thermodynamiques.

5. Acknowledgements

The autor acknowledge the very helpfull assistance of G. von Tobel and W. Naegeli, who had done most of the hardware work. The Swiss Federal Institute for snow and avalanche Research had supported the project with ideas and manpower support.

Figure 1 shows the entire test set up, consisting of three rows of solar cell dummies, each row 15 meters in length, with solar cells on four selected positions as indicated by the arrows. The test facility is 2600 meters above sea level. A time-lapse 16-mm movie camera is facing the field taking one frame every ten minutes.

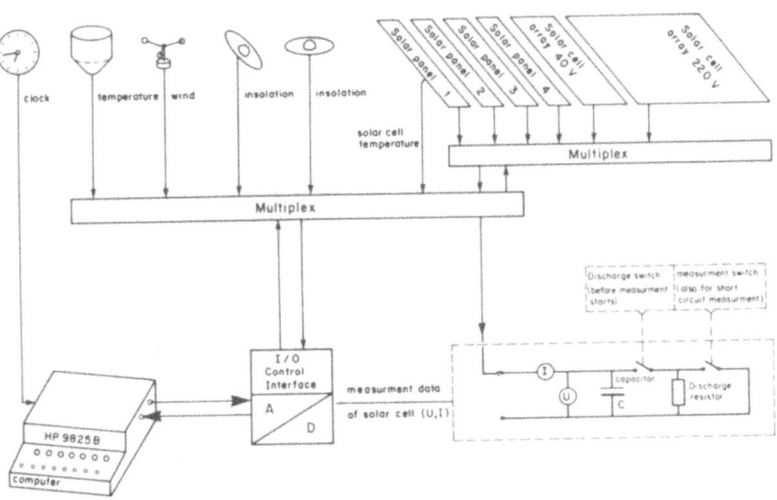

Figure 2 shows the diagram of the Data Aquisition System. The basic idea is to measure the entire I-V-curve of each solar module and calculate all relevant data needed; as for example Max Power Point, different Power Points for given voltages etc. The meteorological Data are recorded as well. Scanning rate is 3o seconds.

Figure 3 shows a result of a scale model test designed to study snow drift patterns for large solar fields. Steady state behaviour is achieved after a few hours of exposure. The results of this modelling are in good agreement with the observations made on the full size array.

Figure 4 snow drift pattern after four days of strong winds. Although snowfall was higher than normal (the accumulated snowfall during the winterperiod 1981/82 amounted to 13,5 meters), the losses of potential energy production due to snow and icing problems would have been less than 5 %.

DESIGN, INSTALLATION, AND INITIAL PERFORMANCE OF 350-KW PHOTOVOLTAIC POWER SYSTEM FOR SAUDI ARABIAN VILLAGES

Fahad Huraib, Dr. Bakr Khoshaim, and Ahmed Al-Sani
Midwest Research Institute, Kansas City, Missouri

Matthew S. Imamura and Abbas A. Salim
Martin Marietta Corporation, Denver, Colorado

Summary

This paper summarizes the design, fabrication, installation and initial performance of the world's largest photovoltaic (PV) power plant. The installation of this system was completed in August 1981, and a month later it began providing electrical power to three remote villages in Saudi Arabia. The facility includes a 350-kW photovoltaic array, 1-MW diesel powered generators, 1.6-MW-hr lead acid batteries, a 300-kVA inverter, dc and ac switchgear, 13.8-kV transformers, control and data acquisition subsystems, and solar and weather monitoring subsystem. The PV array is a point focusing concentrator type which uses an acrylic Fresnel lens with a geometric concentration ratio of 40. The system operates in stand-alone and cogeneration modes and is capable of completely automatic operation. The most significant aspects of the project to date are as follows: 1) the turn-key operation was completed within 19 months after contract go-ahead, 2) array dc power output exceeded 375-kW in February 1982 at an array field efficiency of 10.4%, 3) concentrator array module progressed from development stage to a commercial high production type within a year, and 4) it has the first operational array field to employ optical concentrators. With essentially trouble-free shipping, installation, and start-up operations, the project demonstrated that the PV system is an excellent candidate as a rapidly deployable dispersed power generator in remote areas with no utility transmission system.

1. INTRODUCTION

In late 1977, Saudi Arabia and the United States signed a project agreement in the field of solar energy under the auspices of the Saudi Arabia/United States joint commission on economic cooperation. As a result of this agreement, known as SOLERAS (Solar Energy Research American and Saudi), a five-year plan was established. One objective of this plan was to enhance the quality of rural life in Saudi Arabia by use of solar energy systems for domestic communal, agricultural, and local industrial applications.

This PV project is sponsored by the Saudi Arabia National Center for Science and Technology (SANCST) and the U.S. Department of Energy. Solar Energy Research Institute was responsible for implementing the project through installation and initial checkout. Martin Marietta Corporation in Denver, Colorado was selected to be the prime contractor from among 19 bidders for the overall design, fabrication, installation, and operation and maintenance. The project started in January 1980 and the PV facility installation and checkout was completed in September 1981.

The PV facility, shown in Figure 1, was installed near the villages of Al Jubaylah, Al Uyaynah, and Al-Hejrah, which are about 45 km northwest of Riyadh, the capital city of Saudi Arabia. They have a combined population of 4000. Previously, these villages had no reliable power sources and had no local transmission grid. Now, the PV power plant supplies all their electrical load needs. Figure 2 presents a simplified diagram of the PV system. The major elements in the power plant and their principal features are listed in Table I.

The array field is the most striking feature of the facility. Rectangular PV collectors, 160 of them, are neatly arranged on the smooth gravel field. Each collector has two wings mounted on a single concrete pedestal. The wings contain 256 circular silicon solar cells, 32 Fresnel lenses and module housing, sun tracking electronics and drive mechanism. Each of the 160 arrays can acquire and track the sun in two axes and be placed in the lens-down stow position at night and in the event of wind and dust storms. Figure 3 shows the arrangement of the PV system building that includes training, maintenance, and storage rooms. The PV facility site consists of a fenced-in area of about 230 m x 290 m on which a single-story building and a solar array field is situated. The building has an area of 670 m^2 and is of masonry and concrete constructions. An above-ground Diesel fuel tank and service water tank are located near the building. Previous publication (1) provides additional description of various PV system hardware.

Figure 1   PV Array Field

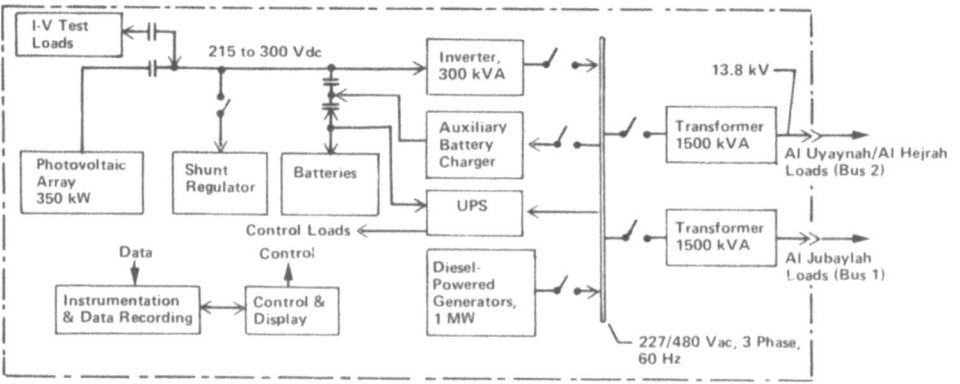

Figure 2 Simplified Block Diagram of PV System

Table I Major PV Power Plant Elements and Their Principal Features

Item	Features
PV System	
PV Array Field	160 Concentrator Arrays, 12.1 m x 2.7 m, 64 Parallel Strings of 640 Cells in Series, 40,960 Circular Silicon (Cz) Cells (5.7 cm-dia), 160 Sun Tracking Electronics and Drive Mechanisms; 5,120 Fresnel Lenses (Quad) and Plastic Housing, and Drive Mechanisms
Battery	4 Lead-Acid Batteries, 120 Cells in Series Each, 1.6 MW-hr Rated Capacity (1700 A-hr)
Battery Auxiliary Charger	60 kW, 300 Vdc, 200 A, for Off-line Maintenance
Inverter	300 kVA, 480-VAC, 3 Phase
Diesel Generators	1 MW (Four 250 kW Units)
Transformers	3 MVA (Two 1500 kVA Units), 480 to 13,800 VAC
Switch Gear	600 Vdc, 480 VAC, and 110 VAC
Control Equipment	Manual/Automatic Operation with HP 9845 Computer
Uninterruptable Power Supply	10 kVA, 110 VAC Inverters (2 Units) and 10 kW 300 Vdc Power Supply
Instrumentation & Data Recording Equipment	Magnetic Tape, HP 9845 Computer, and HP 3052 Data Acquisition System
Array Cleaning Equipment	Purified Water Spray (82°C at 1000 psi), 7.5 liters/min, Truck Mount
Facility	Building That Houses PV System Hardware; Training, Maintenance, and Storage Rooms, Diesel Fuel Storage Tanks

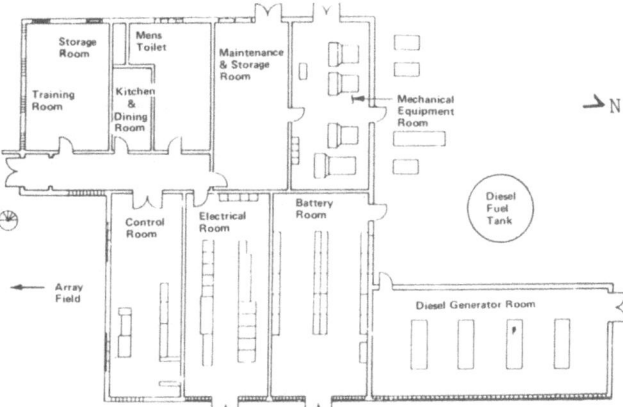

Figure 3 PV System Building Arrangement

2. SIGNIFICANT PROGRAM RESULTS

Martin Marietta was responsible for the overall project supervision and system design, and fabrication of the PV concentrator arrays. Ralph M. Parsons, International, performed architectural and construction engineering and other on-site tasks as a subcontractor to Martin Marietta. Their responsibilities consisted of site preparation, installation, and logistics support for shipping, personnel housing, and transportation. Other major subcontractors were: 1) Applied Solar Energy Corporation for over 41,000 5.72-cm diameter circular Silicon solar cells, 2) Helionetics for the 300 kVA inverter, 3) C&D for 480 1700 A-hr lead-acid battery cells, and 4) Stewart and Stevenson for the four 250-kW Diesel generators.

2.1 <u>Design and Fabrication</u> - No major problems were encountered during the design and fabrication phase. The two most significant areas were in the design of solar cell to ceramic substrate bonding method and fabrication of a large number of array modules.

A considerable amount of effort was spent in arriving at an acceptable method of attaching the cell P-contact to the metalized ceramic substrate. The problem was basically that of minimizing the excessive voids in the cell to ceramic solder interface. For the size of the cell involved (5.72-cm diameter), the best solution was to use copper paste rather than copper plating of the ceramic substrate.

Over 5,000 2.7-m long heat sink assemblies, each containing 8 solar cells interconnected by flat conductors, and about 10,500 lens/module plastic housing assemblies (30.5-cm x 122-cm) were required. Because of this large quantity, special tooling were designed and built to permit semi-automated assembly line. Once this assembly line was set up, the actual fabrication of some 5,500 array modules to satisfy the 350-kW field size took less than six weeks.

2.2 **Shipping and Installation** - Large number or bulky hardware like the support tubes, array modules, sun tracking electronics, and battery cells were shipped in sea vans via surface vessels from Houston to Dahran. Shipment by air was limited to computers, instrumentation and data recording racks, and junction boxes. All of these hardware arrived at the PV site on time without much difficulty.

Site preparation started in August 1980, and the construction and installation of the PV control building, underground cabling, and concrete pedestals were completed in May 1982. This was followed by the installation of the array modules on the support tube and the wiring and PV hardware in the control building. Figure 4 shows the array field construction in progress and the installed PV equipment, including batteries, control and display panel, instrumentation/data acquisition console, and control computer.

2.3 **Initial Checkout Phase** - The initial checkout phase was divided into three test categories: Electrical Isolation and Continuity Test, Subsystem Functional Tests, and System Functional Tests. System functional tests consisted of: 1) DC Power Test, Manual Mode, 2) Inverter AC Power Test, Manual Mode, 3) Diesel Generator AC Power Test, Manual Mode, 4) Power System Performance Tests, Auto Mode, 5) Array Subfield Current-Voltage (I-V) Performance Tests, 6) Battery Storage Capacity Test, and 7) System Acceptance Tests.

In general, equipment and wiring installation, and test sequences were arranged so that various subsystems could be tested in a logical sequence, to minimize the test repetitions and the risk of damaging the hardware. For example, all underground field cabling, both signal and power, were first installed, and individual arrays were checked out as they were mounted on the pedestal. Over 10,000 terminations in the interface wiring between subsystems were verified.

Both DC and AC "Power up" tests were first performed in the manual mode. Then system performance in the auto mode was verified. After the system performance was verified in the manual mode, the system was configured to operate in the auto mode where the control computer assumed the control of the complete system. All predefined modes of automatic operation were checked out. These operating modes are: 1) Diesel Generator Only Mode, 2) Cogeneration Mode (i.e., PV, Battery, Diesel Generation), 3) Stand Alone Mode (i.e., PV and Battery only), and 4) Stand-alone mode with PV supplying Bus 1 and Diesel generator Supplying Bus 2.

It should be noted that all control software were previously checked out in the U.S. using a PV computer simulation test bed. The final software verification on site in the actual PV system was essentially trouble free, perhaps due to thorough verification effort in the U.S. Various system emergency shutdown modes such as (1) disconnecting individual batteries and subfields from the DC bus, and (2) turning off the inverter and Diesel generators due to an abnormal operating condition were verified.

The solar array subfield I-V data were obtained automatically using the data acquisition computers. The I-V test arrangement is a permanent part of the system and utilized frequently to evaluate the performance of the array field. The test sequence is fully automatic. When an I-V data is requested through the data acquisition computer on a given array subfield, the control

Figure 4. Array Field Construction in Progress and Installed PV Equipment

computer disconnects that subfield from the DC bus and connects it to the I-V test bus. The rest of the I-V test is handled by the data acquisition computer. This computer controls the appropriate I-V load relays in binary steps to obtain the 256 I-V test points from open circuit voltage to short circuit current within 30 seconds. Solar insolation levels immediately preceeding and following the test are recorded. In addition to this, ambient air temperature and wind speed information is acquired and recorded at the time of the test. When the I-V test is completed, the control computer disconnects the subfield from the I-V test bus and connects it to the DC bus. The data acquisition computer transforms the measured I-V test data to the standard reporting conditions using a modified single exponential model (2), and displays the extrapolated I-V data on the CRT (cathode ray tube) for immediate evaluation purposes.

Battery storage capacity tests were performed to verify the specified energy availability predictions. The test consists of first charging the batteries to 100% state-of-charge, utilizing power from the solar array subfield, and then discharging them through the inverter to an AC load for a predetermined length of time or battery voltage drops to predefined value, whichever occurs first.

No major hardware and software problems were encountered at the subsystem and system level functional tests. This is considered very fortunate in view of remoteness of the site where quick availability of materials, parts and people with the proper skills was a problem.

2.4 <u>Acceptance Tests</u> - Following the system functional tests a formal system acceptance testing was conducted in September 1981. This test essentially repeated system level functional tests along with emergency field stow and a full day array tracking under clear and semi-clear sky conditions. This testing was witnessed by both U.S. and Saudi Arabian government scientists. The system operated very well and received high marks from these PV experts. Following the system acceptance tests, the 13.8 kV transmission line (installed by the Saudi Electricity Corporation) was energized and placed in operation in October 1981.

2.5 <u>Initial Performance Summary</u> - Since the installation of array subfields, the PV power system has operated successfully and well beyond expectations in several areas. The most significant ones are: 1) PV field efficiency of nearly 11% was often achieved, 2) the ease of array subfield installation and checkout, 3) relatively trouble-free installation and interfacing of various subsystem elements, and 4) inverter efficiency of over 94% at 300 kW outputs was achieved.

System DC power output has averaged between 280 and 320 kilowatts and reached as high as 375 kilowatts several times in February 1982 when the average ambient air temperature was between 8 to 15°C, wind speed between 5 and 10 meters per second, and direct normal insolation levels of 900 to 1000 w/m^2. Figure 5 shows the system block diagram and performance data displayed by the data acquisition subsystem computer when a 372 kW power output was achieved. It shows the amount of DC power generated in each subfield, dc power being supplied to batteries, battery state-of-charge, inverter AC output, power supplied by the Diesel generators and various system AC load status. Figure 6 shows the test I-V characteristic curves obtained from a typical subfield under varying insolation and temperature conditions during January thru March 1982. Maximum power point, the subfield photovoltaic efficiency, and fill factor are listed. The overall PV field efficiency including cable losses has consistently ranged between 9 and 10.5%. This is well over 6 to 7% typically obtained by other large flat plate systems. Figure 7 is a plot of system photovoltaic field efficiency versus

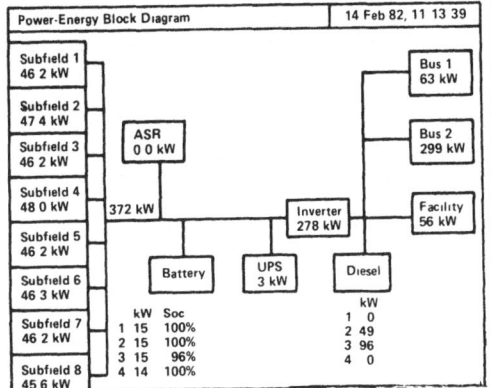

Figure 5   Typical CRT Display of Power Plant Performance

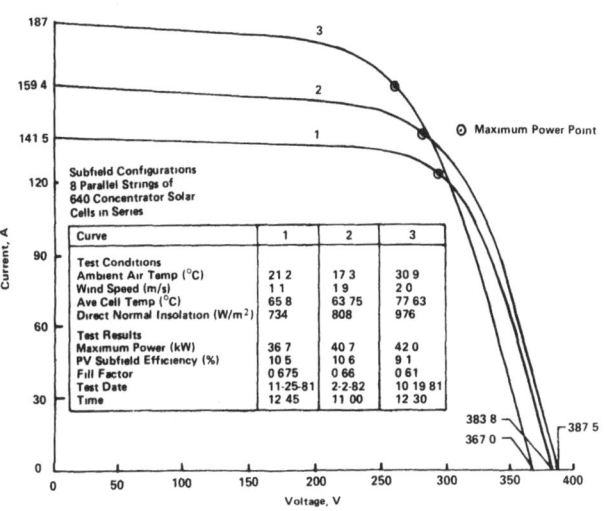

Figure 6   Typical Subfield I-V Characteristics

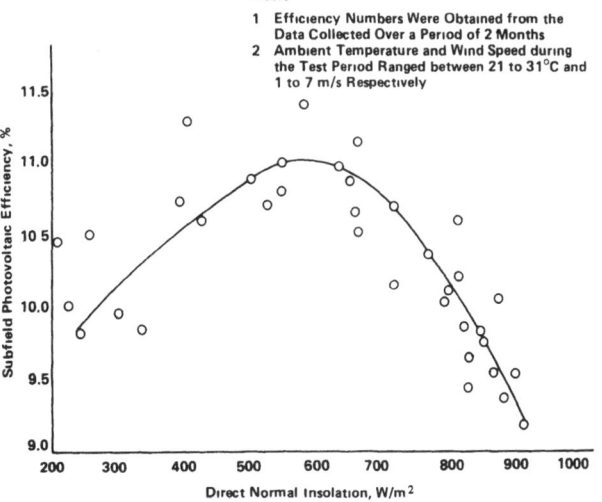

Figure 7   Effect of Insolation on Subfield Photovoltaic Efficiency

insolation at a fixed ambient temperature based on the data collected over a period of two months. The peak efficiency seen near 600 W/m^2 (see Figure 7) is due to non-uniform intensity profile at the plane of the solar cell. The expected linear relationship between a typical subfield short circuit current and direct normal insolation is shown in Figure 8. This information is based on performance data collected over a period of several weeks. The effect of variation in the solar cell temperature due to direct normal insolation is shown in Figure 8. Temperature and direct normal insolation data collected to date appear to indicate an average temperature rise of 40 to 44°C from ambient air to the solar cell in the insolation range of 700 to 900 W/m^2 and winds not exceeding 6 meters/second.

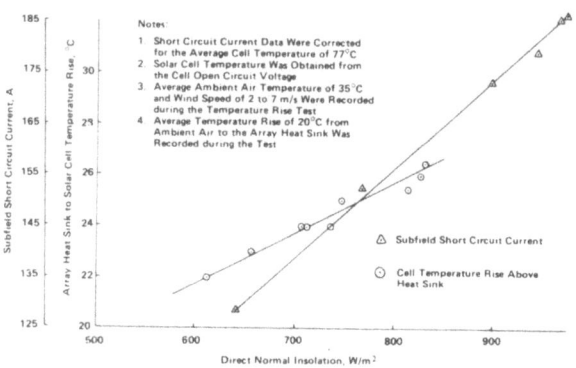

Figure 8 *Effect of Insolation on the Solar Cell Temperature and Subfield Short Circuit Current*

Typical performance of the system in various modes of operation is shown in Figures 9 and 10. Data shown were obtained in earlier phase of the system operation during November 1981. System software is being reviewed and improved continually to maximize the system solar power utilization based on the load demand and performance of the solar field. During the day light

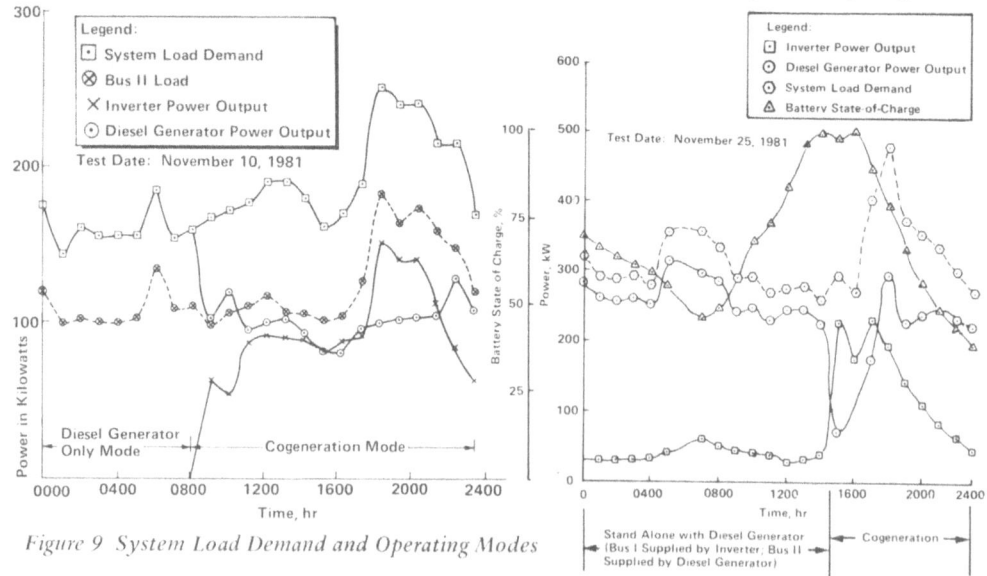

Figure 9 *System Load Demand and Operating Modes*

Figure 10 *System Performance in Various Operating Modes*

hours when the solar power is available, the control algorithm gives first preference to charging the batteries. When the battery state-of-charge approaches 60%, the computer turns on the inverter (thus placing the system in cogeneration mode), and programs it to dispatch power to the AC link. The amount of power dispatched is based on the state of charge of the batteries and the available solar power. The algorithm is designed to ensure 100% state-of-charge of batteries by the end of the day (or before sunset when the solar arrays are commanded to stow). During the night-time, the inverter dispatches power as a function of state-of-charge of batteries.

Figure 11 shows typical battery performance characteristics during cogeneration mode over a period of 24 hours. The control system software is so configured that the batteries are charged at the rate of 250 amperes per battery until their SOC reaches 90%. Beyond 90% SOC, batteries are charged at a trickle charge rate. The battery charge current and terminal voltage are controlled by both the system shunt regulator and the control computer. Array branches are shorted or power dispatch from the inverter is increased depending upon the availability of the excess solar power or the village load demand.

Inverter performance data collected to date have shown that the inverter average efficiency has never been lower than 90% for loads over 50 kW and has exceeded 94% for loads over 250 kW. These efficiency figures exceed the range specified by the inverter supplier. Typical power system load demand and power factor for a 24-hour period in the month of January 1982 is shown in Figure 12. Effort is being made on a continuous basis to review the system load demands and maximize the plant efficiency by properly identifying and matching the load demand with the power generated by the solar field, battery storage, and inverter power output capability.

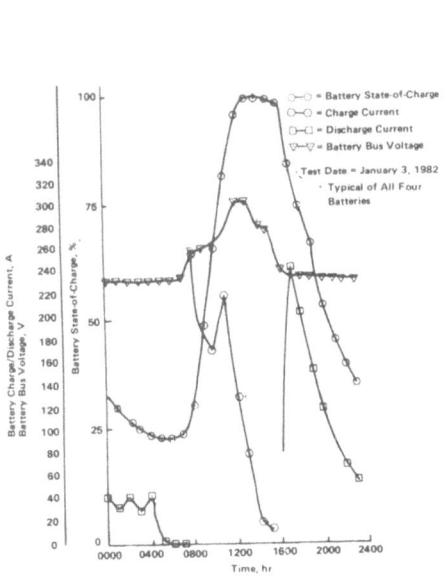

Figure 11 Battery Performance Data during Cogeneration Mode

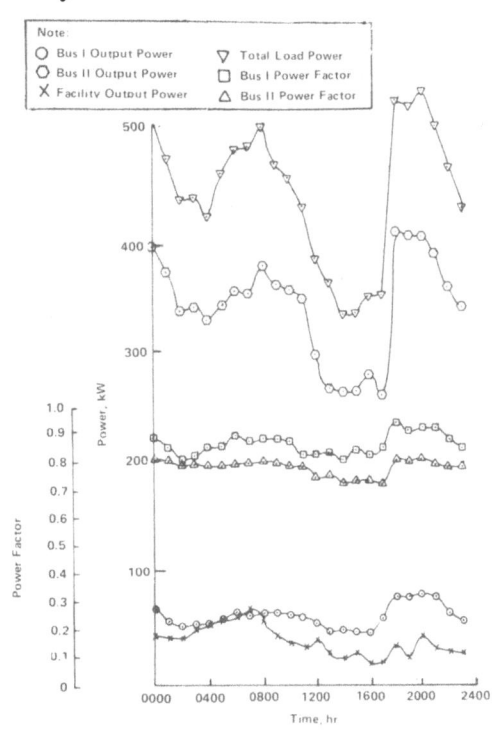

Figure 12 Typical Load Demand and Power Factor Profiles in January 1982

Limited space in this paper does not allow discussions of many other interesting operating aspects of the system. These will be presented along with the system performance in the future solar energy conferences.

3. FUTURE PLANS

Research activities planned in the future are intended to increase our knowledge of the technical and economic viability of the PV systems in the Saudi Arabian environment. These tasks include: 1) determination of the operational characteristics and limitations of the installed PV system, including array and battery performance degradation, maintenance methods, especially lens dry cleaning technique and cost, 2) identification of best operating strategy, 3) development of PV performance and economic analysis, and 4) assessment of PV applications for various sites which will involve analytical and empirical characteristics of solar irradiance and weather data at several regions in Saudi Arabia.

4. CONCLUDING REMARKS

The power plant described in this paper is the only high power PV system in Saudi Arabia. This facility has several unique and important features. It performs many functions typically incorporated in central power plant operation such as monitoring the load status and controlling the operation of power generation sources and dispatching of power into a totally captive high voltage grid.

This PV system uses many high technology equipment and is very complex compared to other PV systems deployed throughout the world. Along with the load management capability, it operates in stand-alone and cogeneration modes both manually and automatically. The system uses desk top computers for control, monitoring, and data acquisition/processing/ storage, and operation of the power plant. It has the largest PV array field in the world (over twice that of the next largest PV system). It is also the first PV system to utilize concentrating optics. The concentration of solar energy allowed the system to use significantly fewer PV solar cells than required typically by a flat plate array of comparable size (requires only 1/40th as many solar cells). These features help to make this installation one of the most interesting PV systems built to date.

This PV project is an excellent demonstration of the capability of the PV system to be deployed rapidly for a truly stand-alone high power application. The project was well organized and coordinated from its inception by the members of the Saudi Arabian government, the United States DOE, SERI/MRI, Martin Marietta, Parsons International Limited, and other key subcontractors. It is rather remarkable that a project of this magnitude and complexity can be implemented in such a short time. The system should provide an excellent data base to the solar energy community in the future.

5. REFERENCES

1. M. S. Imamura, F. Huraib, et. al., "470-kW Photovoltaic Power System for Saudi Arabia Villages," presented at the 1980 Intersociety Energy Conversion Engineering Conference, Seatle, Washington in August 1980.
2. A. A. Salim and M. S. Imamura, "A New Practical method for I-V Transformation for Concentrator Solar Arrays," Submitted for presentation in the Sixteenth IEEE Photovoltaic Specialist Conference, 1982.

THE PHOTOVOLTAIC-POWERED WATER DESALINATION PLANT 'SORO'
- DESIGN, START UP, OPERATING EXPERIENCE -

G. Neuhäusser, J. Mohn and G. Petersen
AEG-TELEFUNKEN, GKSS-Forschungszentrum Geesthacht GmbH
F. R. Germany

Summary

In the framework of a Mexican-German agreement for scientific and technical cooperation a photovoltaic powered water desalination plant is operated jointly as a "remote stand alone system" at Concepcion del Oro in the Mexican Federal State of Zacatecas. The solar cell array with a peak power of 2.4 kW and a total area of 30 m^2 has been designed and fabricated by AEG-TELEFUNKEN. The reverse osmosis desalination plant using one module of plate and frame configuration was developed by the GKSS research center at Geesthacht. The pretreatment system of the raw water, which is brackish water from a well at the test site, and the preparation of the site was done by the desalination agency of the Mexican Government DIGAASES (Direccion General de Aprovechamiento de Aguas Salinas y Energia Solar). The main objective of the plant with a fresh water output of approximately 1.5 m^3 per day is to demonstrate the reliability and the low maintenance requirements of a photovoltaic powered water desalination plant for the fresh water supply of a small remote community without connection to the public grid.

1. INTRODUCTION

The majority of plants presently producing fresh water from saline water are operating on the Multistage Flash (MSF) evaporation process. During the past decade, however, an increasing field of applications for the Reverse Osmosis (RO) process has developed, commencing with the desalination of brackish water and recently by the desalination of sea water. The advantage of the RO process over the MSF process is the lower energy requirement. While a MSF plant requires approximately 3 - 5 kWh of electrical energy plus about 60 - 80 kWh thermal energy per m^3 of distillate, independent of the salt content of the raw water, the electrical energy requirements for a RO plant are about 4.5 kWh per m^3 of product for raw water with a salt content of 2000 ppm and increase to about 12 kWh/m^3 for sea water with a salt content of 35000 ppm.
However, even with improvements in RO system efficiency, energy costs still account for nearly half the costs of water. Therefore, a natural means of reducing the impact of rising fuel costs is to power the desalination system with renewable sources of energy such as wind and solar (1, 2, 3). In case of Mexico many of the populations living in remote communities in arid and semi-arid regions experience saline water supplies. While these regions lack adequate electrical supplies for desalting water, they typically have an abundance of intensive sunlight. An attractive, potential alternative either to installing small fossil fueled power generators or to trucking fresh water supplies is to desalt the water by photovoltaic powered reverse osmosis.

2. DESIGN CONCEPT OF THE SORO PLANT

2.1 Desalination process

Water desalination in the reverse osmosis system is carried out in two steps, i.e.:
- the raw water pretreatment
- the desalination in the reverse osmosis system.

The plant systems required for these steps are described in the following and are schematically depicted in fig. 1.

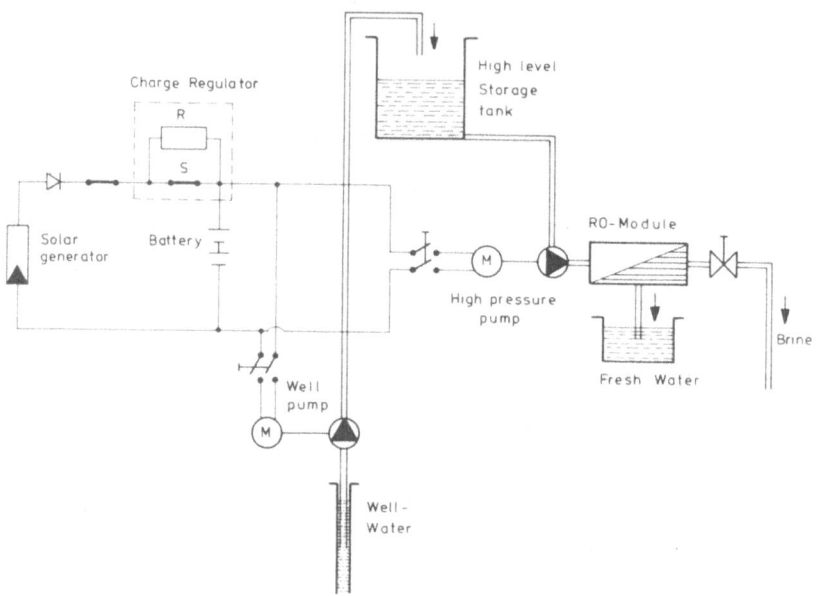

Fig. 1:   SORO-block diagram

The raw water with a salinity of approximately 2000 ppm is fed by the well pump into the high level storage tank. Before entering the RO-system the water is pretreated by adding different chemicals for the following purposes:
- to adjust and control the pH of the raw water
- to inhibit the formation of compounds which, when precipitated, will plug the water passages
- to desinfect the raw water and prevent slime growth or prevent contamination of the equipment.

The desalination of the raw water is carried out in a reverse osmosis module, depicted in fig. 2. Reverse osmosis is a separation phenomenon which separates fresh water from brackish water or sea water through semipermeable membranes. Being admitted with a hydrostatic pressure which overcompensates the osmotic pressure difference, desalinated water is separated from the concentrated phase through the membrane to the dilute phase. The module with a total membrane area of 8 m^2 is served by a high pressure piston pump of 500 l/h capacity at 30 bar. The module consists of a serial assembly of porous membrane backing plates alternating with special support plates which provide the necessary framing and contain the entire raw water channeling. The backing plates are discs of porous polyethylene which

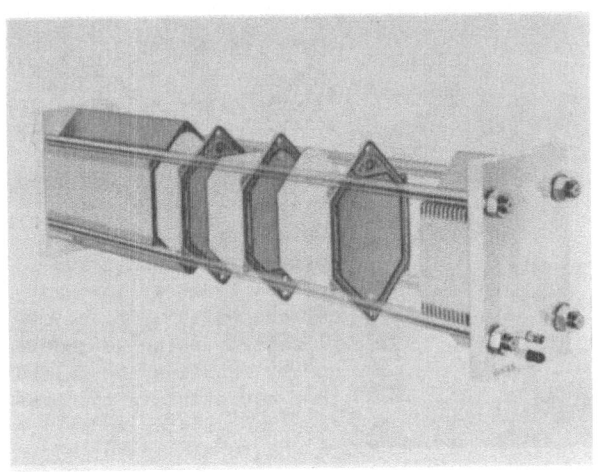

Fig. 2:   SORO-reverse osmosis module

are losely covered with sheet membranes on both sides. Each backing element is symmetrically framed by support plates, both faces of which carry identical raw water passages in the form of a circular recess of uniform cross sectional area each. The two raw water passages of each plate are connected into an U-shaped flow passage by series of bores contained in a peripheral channel at the lower end of the recessed area. The module is assembled between steel flanges by lining up the plates onto the tie rods of the support frame (4).

The product water, which passes the membranes with a flow rate of approximately 180 l/h is stored in a permeate tank and from there distributed to the fresh water consumers.

2.2 Solar power supply and control system

On days of normal automatic operation the desalination plant, represented in fig. 1 by its main electrical energy consumers high pressure pump and well pump, is started, when the solar generator produces enough energy for plant operation. During the time from sunrise to starting to RO-system the generated energy is used to charge the battery. With increasing insolation level after starting up the system the additional energy is also used to charge the battery until the gassing voltage is reached. Now the charge regulator opens the switch S, which means that the resistor R is connected in series to the battery. The resistor limits the solar generator current to the trickle charge current, so that even in case of disconnection of the load the battery is protected against overcharging. The drawback of this device is that in case of normal operation, when the resistor is switched on, the battery is discharged, although the solar generator could produce enough energy to operate the desalination plant. This means an effiency drop of the energy supply system particularly during the summer months with high insolation peaks around noon.

3. START UP AND OPERATING EXPERIENCE

The site for the SORO plant is Concepcion del Oro in the Mexican Federal State of Zacatecas. The self-contained desalination unit and the solar cell panel were prefabricated by GKSS and AEG-TELEFUNKEN, respectively. DIGAASES prepared the test site with respect to the foundations for the container and the solar cells. Design and construction of the raw water supply system consisting of a well and an elevated tank for the storage of the brackish well water was also done by DIGAASES. Fig. 3 shows the RO-container and a part of the solar cells at the test site after finishing the installation in May 1980.

After comprehensive tests with the raw water pretreatment system the plant was started up in August 1980. Fig. 4 and Fig. 5 show some test results obtained during the first year of operation from September 80 to August 81.

Fig. 3: SORO - plant at the test site Concepcion del Oro

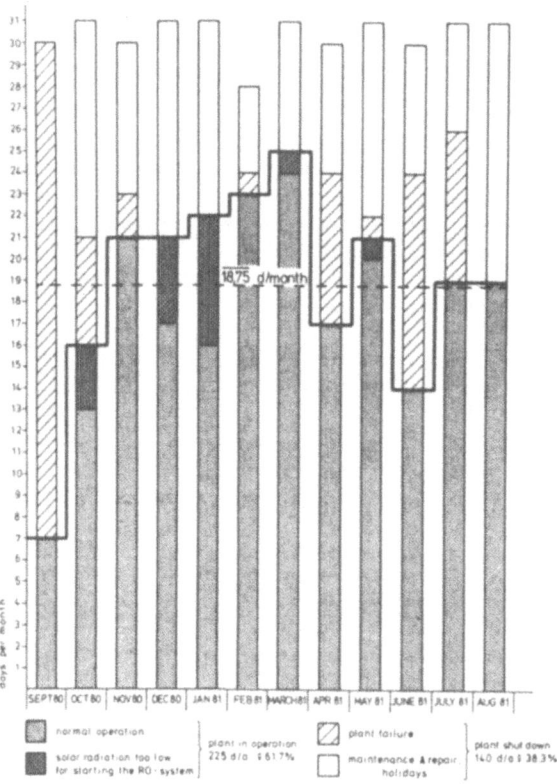

Fig. 4: SORO - availability of the plant September 80 - August 81

The graph in Fig. 4 shows the monthly availability of the plant. Plant failures were mainly caused by the raw water supply and pretreatment system. No failure was experienced with the energy supply system during the first annual operation period. The maintenance time was determined by membrane replacement in the RO-module and by changing the filtration equipment in the pretreatment system as well. It should be emphasized, that the plant was not operated in full automatic mode, but was shut down by hand at weekends and holidays when the operating personnel was not at the test site.

The two graphs in Fig. 5 show the monthly and annual means of the daily insolated energy and the corresponding fresh water output. The measured annual variation of the global radiation shows a fair agreement with other measurements at the same latitude (5). The fresh water production curve generally follows the insolation curve. Deviations as in October 80 and June 81 are due to membrane replacement and insufficient raw water supply, respectively.

A typical measured insolation profile for a day without clouds with the corresponding performance data of the plant is depicted in Fig. 6. An important fact is the already mentioned overcharge protection of the battery, which occurs for about 3 hours around noon. This effect is demonstrated in the figure by the jumps in the solar

- 70 -

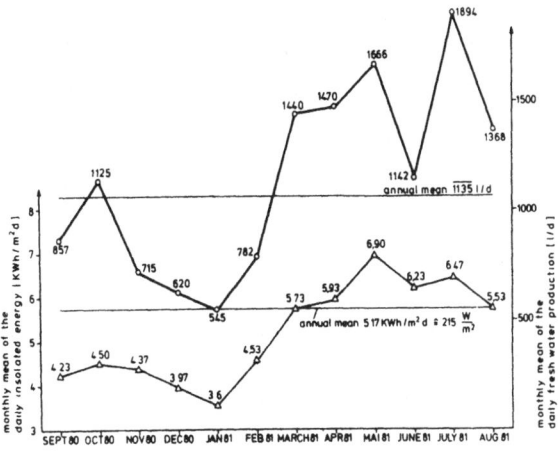

Fig. 5 : SORO - daily fresh water production and insolated energy

generator current curve from the maximum value to the trickle charge current of approximately 5 A. To prevent the corresponding losses in the electrical output of the energy supply system of almost 25 % a new well pump has been integrated into the system in autumn 81. The power characteristics of this pump effects a significant better adjustment of the total load to the insolation profile, because it does not operate continuously but only during the hours around noon.

4. CONCLUSIONS

The results obtained from the first year of SORO test operation indicate that a solar cell array is a very reliable energy supply system for a remote stand-alone desalination unit. With some refinements in the raw water supply and pretreatment it should be possible to achieve an increase of the plant performance in the range of the design capacity of 1.50 m^3 fresh water per day. It is planned to continue the test operation and the performance data collection in order to verify the design data and to investigate the economic potential of photovoltaic powered RO-plants on the basis of the operation results.

Reductions in the cost of solar cells, continued increases in the cost of fossil fuels and the recent rapid improvements in the quality and life of RO-membranes can be expected to increase the number of areas in Mexico and other countries where the combination of photovoltaic and reverse osmosis is the cost effective method for brackish water desalting. Recently published studies show that the size of the potential market for small solar powered desalination plants is substantial and that photovoltaic devices are economically competitive with solar distillation processes for raw water with salt contents up to 2000ppm(6,7).

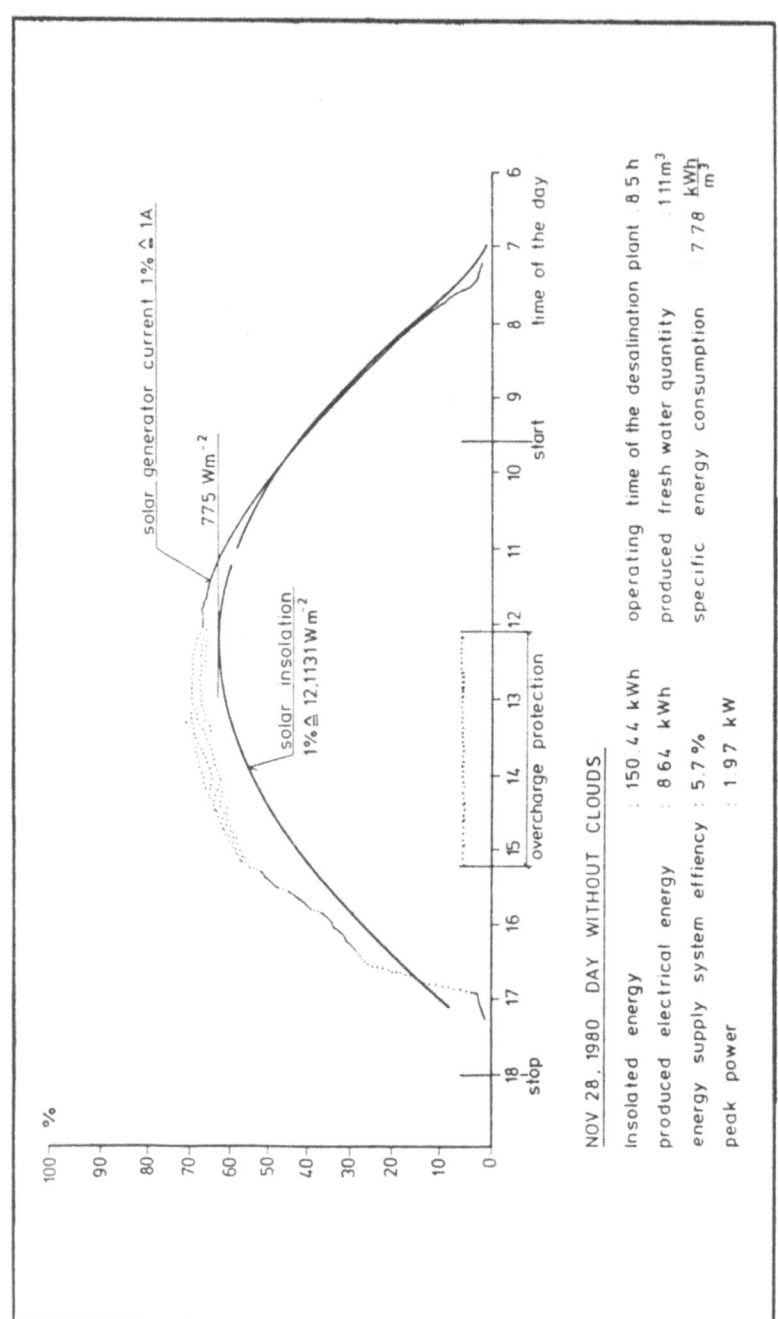

Fig. 7 : SORO - insolation profile November 28, 1980

5. REFERENCES

(1) The U.S.A. I.D. Desalination Manual. A Manual Prepared By CH2M Hill International Corporation, 7201 N.W. 11th Place, Gainsville, Florida 32 601 for the United States International Development Cooperation Agency, Office of Engineering, and for U.S. Agency for International Development, Washington, D.C. 20 523, USA, August 1980.

(2) G. Petersen, S. Fries, J. Mohn, A. Müller, Wind and solar powered reverse osmosis desalination units - Description of two demonstration projects. Proceedings of the International Congress on Desalination and Water Reuse, Nice, France, October 21 - 27, 1979, and DESALINATION, 31 (1979) 501 - 509.

(3) G. Petersen, S. Fries, J. Mohn, A. Müller, Wind and solar powered reverse osmosis desalination units - Design, start up, operating experience. Proceedings of the International Congress on Desalination and Water Reuse, Manama, State of Bahrein, November 29 - December 3, 1981, and DESALINATION, 39 (1981) 125 - 135.

(4) K.W. Böddeker, W. Hilgendorff, J. Kaschemekat, An alternative design in reverse osmosis desalination, GKSS 76/E/55.

(5) R. Schulze, Das Strahlenklima der Erde, Steinkopf Verlag, Darmstadt 1970.

(6) J.P. Gerofi, G.G. Fenton, Comparison of solar RO and solar thermal desalination systems. Proceedings of the International Congress on Desalination and Water Reuse, Manama, State of Bahrein, November 29 - December 3, 1981, and DESALINATION, 39 (1981) 95 - 107.

(7) Janet J. Turnage, Abdo A. Husseiny, An assessment of the market potential for small solar powered desalination plants. Proceedings of the International Congress on Desalination and Water Reuse, Manama, State of Bahrein, November 29 - December 3, 1981, and DESALINATION, 39 (1981) 43 - 52.

THE UNDP WORLD BANK SOLAR PUMP PROJECT - PREPARING FOR PHASE II
P L Fraenkel*, M A S Malik**, D E Wright***
*Intermediate Technology Power Ltd,Mortimer Hill,Mortimer, Reading,Berks,UK
**Energy Dept, World Bank, 1818 H St, Washington DC 20433, USA.
***Sir William Halcrow & Partners, Burderop Park, Swindon, Wilts,SN4 0QD,UK

Summary

Phase I of the UNDP financed World Bank executed Project was undertaken by Consulting Engineers, Sir William Halcrow & Partners in association with the Intermediate Technology Development Group Ltd between July 1979 and July 1981. Its purpose was to review proposals for the utilization of solar energy for pumping water for irrigation purposes and to advise on the way the technology should be developed. The paper briefly summarises the main findings of this part of the work.

Although no definite plans for Phase II can be made until funds for it have been approved by UNDP in 1983, the UNDP, World Bank and Consultants agreed there was useful work to be done in the interim period between the end of Phase I and the start of Phase II of the Project. A contract has therefore been placed by the World Bank under which the following actions are being progressed: continuation of Phase I field trials where appropriate; a study of the factors which have to be satisfied if solar pumps are to be viable for irrigation and water supply purposes; an assessment of ways in which prospective countries might participate in Phase II; the procurement and testing of PV commercial pumping systems; the evaluation of continuing development of thermal systems; further development of individual components in PV systems; and a study of manufacturing potential in selected developing countries. A brief description of these activities is given in the paper.

1. INTRODUCTION

Phase I of the UNDP financed World Bank executed Project undertaken by Sir William Halcrow & Partners in association with Intermediate Technology Development Group (the Consultants) has been extensively reported (e.g. references 1, 2 and 3) and reference to it will be made in this paper only to the extent necessary to set the Projects for Phase II Preparation and Phase II in context.

The basic purpose of Phase I of the Project was to advise the UNDP and World Bank on the way in which solar pumping technology should be developed to provide an appropriate technique for pumping water under the conditions which prevail on small farms in the developing world. To that end, in the period from July 1979 to July 1981 the Consultants undertook a state of art review (reported as reference 3), organised field trials and performance measurements on small-scale solar-powered pumping systems in Mali, Sudan and the Philippines, made laboratory tests on the performance of components of the pumping systems under full and part-load conditions, and studied performance of a number of feasible system options utilising mathematical modelling techniques. In all this work attention was directed towards systems delivering flows in the range from 1-5 litres per second through static heads of up to 7 metres. Pumps of this capacity have hydraulic power outputs in the range of 150-500 watts and can irrigate areas of between 0.5 and 1.0 hectares, depending on the crop and efficiency of water distribution.

2. MAIN CONCLUSIONS FROM PHASE I

It will be understood that in so short a paper it is impossible to do justice to the conclusions drawn from a two-year Project which culminated in major reports (references 2 and 3). In order to save space the principal points have been summarised in two Tables: Table 1 lists the mainly technical conclusions which have design significance, while Table 2 lists the factors which, if satisfied, will favour the application of solar pumps Since this paper is being presented at a Conference concerned in particular with photovoltaic sources of energy there is one general point which may usefully be made: this is that there is no market for PV cells by themselves but only as a part of systems which have technically appropriate applications and are economically viable.

3. PROSPECTS FOR PHASE II

The fact that Phase I of the Project was mounted presupposes an intention to proceed with Phase II at some point in the future. It is believed that the overall objective of Phase II should be to take the development of solar pumping systems to the point where they will be suitable for pilot manufacture in developing countries. It is thought that this will require four main activities, and these are:
o demonstrating by field trial and pilot use that selected solar powered pumping systems can provide a cost-effective, reliable and appropriate means of pumping water for agricultural and water supply purposes
o confirming the technical, economic, financial and social conditions under which the pumping systems will do this
o finalising designs for appropriate pumping systems
o arranging for the local assembly and/or manufacture of suitable systems and for their distribution, manufacture and financing in appropriate developing countries

It was felt however, that before this could be undertaken properly, certain preparatory work should be undertaken, hence the present Project entitled "Phase II Preparation". It should be made clear that there can be no definite commitment to Phase II until the UNDP Governing Council agree to a proposal at their Annual Meeting to be held in June 1983 and that the ideas sketched above are only in draft form and subject, therefore, to amendment.

4. PHASE II PREPARATION

The World Bank commissioned the Consultants (Sir William Halcrow & Partners now acting in association with Intermediate Technology Power Ltd, a new Company within the IT Group) to undertake the preparatory work necessary for Phase II in the period from May 1981 to December 1982.

Following on from the position reached at the end of Phase I it was agreed by the World Bank and the Consultants that the work should be divided into the following main Activities:
o Continuation of field trials started in Phase I in Mali, Sudan and Philippines and analysis of additional data (Activity 134)

- Study of the technical and economic factors which need to be satisfied if solar pumps are to be viable for agricultural and water supply purposes (Activity 136)
- Assessment of prospective countries for participation in Phase II (Activity 137)
- Procurement of improved commercial PV pumping systems and sub-systems and their testing in order to qualify them for use in Phase II (Activity 138)
- Evaluation of further developments of thermal pumping systems and the prospects for their commercialisation (Activity 140)
- Further developments of components in PV systems (Activity 141)
- Study of the potential for assembly and/or manufacture of systems in selected developing countries (Activity 142)
- Report to World Bank, including recommendations for Phase II (Activity 143)

Activities 136 and 138 are described more fully in sections 5 and 6 of the Paper while a brief note is given about the other Activities below:

Activity 134 - The field trials started under Phase I in Mali, Sudan and the Philippines are being continued and contact is being maintained with the participating Institutions. Visits are being made from time to time to check on the equipment and to encourage and advise on the collection of data and calibration of instruments. In addition to the three original countries, arrangements have also been made to monitor an ARCO system installed at the American University Cairo.

Activity 137 - Visits to six additional prospective Phase II countries have been made during this period to explain the Project and to explore the possibilities for their future involvement. Factors which have been assessed include: the pumping needs for irrigation or water supply which can be met by solar powered pumps; the presence of suitable solar energy resources and the absence of any more readily exploitable alternatives; the level of interest in solar pumping on the part of the Government; and the willingness and ability of host institutions to provide the necessary technical and logistical support for successful field trials and monitoring.

Activity 139 - On the agreement of the World Bank and UNDP to the participation of additional countries in Phase II, further visits will be arranged in order to select the sites to be used for field trials, the systems to be supplied and to agree on the administrative and technical support arrangements required. It is not feasible to do this in 1982 so this Activity has been planned as an option to take place early in 1983.

Activity 140 - During this Activity the Consultants are reviewing what further improvements might be made to photovoltaic systems by the additional development of their components, for example to see if it is possible to obtain improved efficiency at acceptable extra cost. Through this development it is hoped to move closer to the goal of developing second generation small scale PV systems which are suitable for pilot manufacture.

Activity 141 - Phase I has shown that the development of small scale thermal systems must start from a somewhat different point from that for PV systems. There is a marked lack of commercially available hardware and very little reliable performance data. After a check of developments worldwide, some basic studies will be made of the feasibility of the more likely possibilities: these might include, for example, small cheap Rankine engines and concentrating collectors.

Activity 142 - The manufacture (or at least the assembly) of solar pumping systems in developing countries is a very important factor in their widespread adoption in use. Ideally the systems should be wholly manufactured within the country in which they are to be used; the benefits of so doing include improvement of the balance of payments, improvement in local repair and maintenance capabilities, creation of jobs, upgrading of local technical skills and development of greater self reliance.

Activity 143 - A Report summarising the main conclusions to emerge from the Project and providing technical and economic justification for the recommendations to be made for Phase II itself will be prepared at the end of 1982.

5. REVIEW OF APPLICATIONS, SYSTEM SIZES & ECONOMICS

There are two main applications being examined in detail in Activity 136: these are irrigation, and water supply. Data are being collected from Thailand, Bangladesh, Kenya and Egypt to provide the data base for this part of the Project.

It is important to assess the pumping requirements of particular geographical regions in terms of head, flow and the power required for irrigation or water supply needs, so as to build up a picture of the market requirement for solar pumps. This enables the size requirements to be specified to match the end uses identified and so avoids the cost penalty of incorrect sizing of solar pumping systems.

It is important also to demonstrate that solar pumping technology is at least promising as alternative water lifting techniques such as human, animal, wind and engine powered pumps in those countries where it is hoped that it may eventually become viable. It would be impolitic to demonstrate solar pumping technology in a country where there are fundamental factors which mean it is unlikely ever to be technically or economically viable.

Attention must be given not only to the performance of the pumps themselves and the minimisation of their cost, but also to the manner and efficiency with which the water is stored, distributed and used in the field and the village.

Mathematical models will be constructed of the agricultural and water supply applications and these will be used to analyse the costs of water supplied in the baseline situations. The sort of factors which will be taken into account in the agricultural model include: solar irradiation; typical area commanded by the pumps; static lift; crops to be grown; the daily water requirement; the variation of water requirement from month to month; water losses in distribution; water storage; and the capital and recurrent costs of wells, boreholes, pumps, storage and distribution channels (or pipes). The factors which will be taken into account when examining the water supply application in the rural village model include for example; solar irradiation; typical village population; static lift; per capita consumption; the daily water requirement; the pattern of consumption distribution; location and depth of wells or boreholes; and capital and recurrent costs of wells, boreholes, pumps, storage and distribution networks. Sensitivity analyses will be done to show how these costs vary with changes in the baseline assumptions.

6. PROCUREMENT AND TESTING OF IMPROVED COMMERICAL PV PUMPING SYSTEMS

In this aspect of the work the more reliable and technically efficient of PV systems tested under Phase I, or other systems which have come on to the market since, have been reviewed. The Consultants prepared a Tender Document which called for the supply of three categories of pumping system:

A - to pump 60 m^3 per day through 2m design static head
B - to pump 60 m^3 per day through 7m design static head
C - to pump 20 m^3 per day through 20m design static head

The stated volumes are to be delivered under a global solar irradiation of 5kWh per square metre per day with a defined irradiance time distribution.

The Tender Document was issued to the suppliers or manufacturers in January 1982. Tenders were received from 26 separate companies: in all 18 systems were offered in category A, 25 systems in category B, and 21 systems in category C.

The Consultants received the Tenders at the end of February and after due consultation with their own advisors and with the World Bank have recently awarded contracts for the manufacture of the systems. Two complete pumping systems are being purchased in each of the three categories and in addition two subsystems are being purchased in categories A and B, and three subsystems in category C. All equipment is due for delivery in August 1982.

One of the lessons to come out of Phase I was the strong desirability of testing the systems under controlled conditions before being sent for field trial. To this end the Consultants are establishing a solar pump test facility at their office near Swindon, Wiltshire, UK. This will comprise the following basic facilities:
- 30m deep, 300m diameter watertight borehole
- 2m deep, 1500mm diameter sump
- 12m high tower for imposition of the necessary delivery heads
- pipework between borehole, sump, tower, flow meters and calibration tank
- PV array output simulator to provide a current and voltage input to the motors and pumps which matches that from arrays operating under sunshine
- instruments to record solar irradiance, current, voltage, total static head, total head across the pump, pump speed and flow rate
- data logging equipment

With this test facility the Consultants expect to be able to carry out the performance tests on the systems and subsystems to qualify them prior to the field tests in Phase II and to verify their performance under a range of solar and head conditions. The facility is due to be commissioned in July 1982 and tests on the complete systems will start in August. Tests on all the subsystems utilising the PV array simulator have been scheduled for September and October 1982.

REFERENCES

1. D E Wright - "The use of photovoltaic pumps for small scale irrigation in the developing world : a Progress Report on the UNDP/World Bank Project" Proc. 3rd E.C. Photovoltaic Solar Energy Conference, Cannes, France October 1980. p.117
2. Sir William Halcrow & Partners in association with Intermediate Technology Development Group Limited - "Small-Scale Solar-Powered Irrigation Pumping Systems: Phase I Project Report" World Bank July 1981.
3. Sir William Halcrow & Partners in association with Intermediate Technology Development Group - "Small-Scale Solar-Powered Irrigation Pumping Systems : Technical & Economic Review" World Bank September 1981

TABLE 1 - CONCLUSIONS FROM PHASE I ON DESIGN ASPECTS

System

o Choice of components governed by systems matching considerations for cost-effectiveness.

o Fail safe design important.

o Lives in excess of 10 000h operating time required before major replacement due.

o Maximum instantaneous overall system efficiency to exceed 4.5% (based on cell area) or 3.5% (on array area). Depends on good subsystem efficiencies.

o Daily overall system efficiency to exceed 3% (cell area). Depends on low start-up irradiance levels, high instantaneous and good part-load efficiencies.

PV Arrays

o Smaller modules (say 20W) better for systems optimisation, transport, and field replacement and operation.

o Attention to detailed design needed: electrical connectors; laminated glass/plastic covers; foolproof wiring; interconnect redundancy; dc voltage limit for safety.

o Array power and parallel/series cell connections to be optimised. Efficiency should exceed 11%.

o Electronic power conditioning feasible, but cost effectiveness to be confirmed. Manual switch to alter cell connections probably more cost effective.

o Occasional manual tracking of array very cost effective (for example, 3 times per day) - little point in automatic systems.

o Performance of arrays not entirely predictable and output should be verified. Degradation to be no more than 10% over life of system.

o Arrays preferably mounted well above ground level but requirements for security and demounting in case of very high winds to be considered.

Motors

o Permanent magnet dc motors have suitable characteristics with full load efficiency exceeding 87% and half-load 75%.

o Brushless motors attractive if electronic commutation can be made reliable

o Motors should be designed to be made fail safe to take account of conditions like; brushwear, overheating, high current or voltage transients.

o Electronic components must be robust and reliable and designed to work with any other electronic control elements (in power conditioner or pump).

o Motor/pump couplings to permit angular misalignment.

o Motors should be designed for submersion or if not, must be mounted above possible flood level.

Pumps and Pipework

o If not of submerged type, must be self priming.

o Single stage centrifugal pumps suitable for heads up to about 10-15m; for higher heads multistage turbine pumps or reciprocating pumps better.

o Head-flow characteristic (at constant speed) should slope for stable performance under varying irradiance.

o Full load efficiency should exceed 50% (efficiency at part load to be as high as possible).

o Suction head requirement to be moderate (preferably not more than 3m).

o Balance between low clearance and high efficiency and silt handling capability and lower efficiency to be determined for different applications.

o Pump to be designed to run dry (if not submersible), or given cut out protection.

o Pipework important part of system and diameter to be designed accordingly. Pipework to exclude air traps. For centrifugal pumps pipes to have rigid cross-section.

TABLE 2 - FACTORS FAVOURING APPLICATION OF SOLAR POWERED PUMPS

Factor	Comment
Good solar regime over major part of year, matching demand.	This is a fundamental prerequisite.
Volumes pumped per day to be moderate	Suited for irrigating areas of between 0.5-1.0 hectares, or villages with a population of 500 - 1 000.
Static water heads to be low	Up to 5 metres for irrigation, higher for water supply.
Small variation in static head (for contrifugal pumps)	Difficult to design centrifugal pumps whose efficiency remains high under a large variation in static head.
Demand to be present over a large part of the year	Utilisation factor must be high in order to give low cost per unit volumes water pumped.
Water to have a high value	The crop must have a high value and water used for domestic purposes must likewise be valued highly.
Low specific capital cost	Best 1982 costs were around US$2 per kJ per day, but this needs to drop further for economic viability.
System life to be in excess of 10 000 hrs between major replacements	This for economic life cycle costs.
Diesel fuel prices to be rising and supply erratic	Value is placed on a facility under local control and which does not depend upon long supply lines
No mains electricity	If available and reliable, this is an attractive alternative, provided its costs is subsidised.
Poor wind regime	Good wind regime power costs may be lower than solar.
Pigs are not part of the local culture	Methane generated from pig waste can provide low cost energy.
High opportunity cost for alternative applications for human labour	If no alternative uses for human labour exist, there is no real reason for not using hand pumps.
Suitability for assembly/manufacture in developing countries	This is essential to give repair/maintenance schemes credibility.
Farmers/villagers willing to adapt agricultural techniques and consumption to solar pump outputs	This probably needs to be associated with provision of storage.
Government willing to provide credit to meet high initial cost	Not seen as fundamental difficulty, once pumps have been shown to be economic.
Low discount rates	Favours capital intensive investment.

INSTALLATION ET EVALUATION D'UN SYSTEME
SIMPLE DE POMPAGE PHOTOVOLTAIQUE DANS UN VILLAGE ALGERIEN

A. MOUHOUB and M. BENMALEK

Laboratoire Cristaux et Couches Minces Centre des Sciences
et de la Technologie Nucléaires BP-1017 ALGER GARE ALGERIE

Summary

A study was made of a photovoltaic system for water pumping. This simple system was developped in the aim to need no maintenance. In this view we were interested in direct coupling including neither battery storage nor adaptation circuits. The photovoltaic generator consists of a 500 W array suppling an half kW D.C motor. The pump used is a single stage centrifugal model. The system was placed in a small community and was tested during ten months to set down the main characteristics such as the electrical parameters and the influence of external agents for a real remote site. Analysis and discussion of the obtained results are reported.

1. INTRODUCTION

L'Algérie par sa localisation géographique et son étendue se situe parmi les pays les mieux placés pour l'utilisation de l'énergie solaire |1| |2|.
Parmi les applications de l'énergie solaire, le pompage de l'eau à partir de systèmes photovoltaiques est certainement l'une **des plus adaptées** pour répondre aux besoins des sites isolés . En effet, il allie la simplicité d'emploi à l'autonomie, à condition de choisir un système répondant aux besoins et aux conditions particulières du lieu d'utilisation.
En Juillet 1981, notre laboratoire a démarré une expérimentation sur une installation de pompage solaire alimentée par un générateur photovoltaique. Le choix de cette opération test a été fait en tenant compte à la fois des conditions sociales, techniques et économiques. Il a conduit à l'implantation d'un groupe moto-pompe-générateur photovoltaique, couplé directement, d'une puissance nominale de l'ordre de 500 W. Les éléments de ce choix et les premiers résultats de l'exploitation sont reportés dans la suite.

2. DONNEES SUR LE SITE D'IMPLANTATION

Le système a été installé dans une communauté isolée située à la limite de l'Atlas Saharien, regroupant près de 200 personnes et environ 300 têtes de moutons. Le village se trouve au lieu dit Bouira Lahdeb, distant de 100 km de Djelfa et de 250 km d'Alger (altitude: 650 m, latitude: 31°N). Ce village dispose de 3 puits pour ses besoins en eau dont l'un est équipé d'un groupe diesel. Ceci permettra par la suite de faire éventuellement une étude comparative solaire-diesel. Il est à remarquer qu'aucune culture n'est exploitée dans le voisinage du site.

Du fait de l'isolement du site, il n'a pas été possible de recueillir ou d'effectuer des données climatiques suffisantes pour le calcul de l'installation. Aussi nous avons dû utiliser celles disponibles pour le Chef lieu de Wilaya (tableau I).

Tableau I
Moyennes des données d'ensoleillement da la région de Djelfa

Mois	Janvier	Avril	Juillet	Octobre
Durée d'ensoleillement en heures	6	8	11	7
Global orientation sud et suivant latitude du lieu en kWh/m^2/jour	4	6	6.5	5
Direct suivant la normal en kWh/m^2/jour	4	5	7	5

Les valeurs reportées dans ce tableau représentent des moyennes estimées à partir d'un programme de calcul et de données météorologiques fournies par Satellite |3|. La zone de localisation est très ventée (vent modéré) et on y a relevé des maxima et minima absolus de température respectivement de 43°C et -8°C.

L'installation a été placée sur un puits de un mètre de diamètre et huit mètres de profondeur. Le niveau statique de l'eau est situé à cinq mètres à partir du niveau du sol. Un réservoir d'eau d'une capacité voisine de 7 m3 se trouve à proximité (50 mètres).

3. CHOIX DU SYSTEME DE POMPAGE

Etant données les conditions sociales des populations isolées nous avons opté pour le système le plus simple possible qui élimine tout risque de panne provenant du dispositif d'adaptation ou de commutation (fig.1). Ceci est obtenu grâce à un couplage direct générateur-groupe moto-pompe. Le pompage étant effectué au fil du soleil et à la demande, un interrupteur permet la mise en marche du système, l'eau pompée est alors envoyée pour stockage dans le réservoir. Un tel système même s'il ne permet pas l'extration du maximum d'énergie répond mieux à la caractéristique de certains puits existants le long des parcours des populations nomades. Par suite de l'ensoleillement disponible le système a été dimensionné de manière à pouvoir fonctionner durant la majeure partie du jour soit près de 5 heures durant la période hivernale.

4. LE GENERATEUR PHOTOVOLTAIQUE

Il est formé de 18 modules de 33 W à 28°C chacun, disposés en deux panneaux fixés sur des armatures en aluminium. L'ensemble couvre une surface de 6,5 m^2. Les panneaux sont orientés vers le sud avec un angle d'inclinaison voisin de la latitude du lieu. Le couplage réalisé permet de délivrer une puissance optimale de 500 Watts.

5. LE GROUPE MOTO-POMPE

Un groupe compacte couplé en usine est utilisé. Il est constitué :

- d'un moteur à aimant permanent travaillant sous une tension de 90 V et une intensité de 5,5 A pour une vitesse de 2500 trs/mn, ce qui correspond à la puissance optimale des panneaux solaires sous un eclairement de 800 W/m^2.

- d'une pompe centrifuge à un étage. Ce type de pompe est utilisé généralement en surface pour les usages d'irrigation. Dans notre cas le groupe a été fixé à 0,5 mètre du niveau statique de l'eau. La pompe travaille en succion sur une hauteur maximale de 3 mètres.

6. ANALYSE DU FONCTIONNEMENT DU SYSTEME

La figure 2 donne la variation de l'ensoleillement sur le site en fonction de l'heure locale pour la journée du 4 Mars 1982. L'énergie reçue au niveau du site prend des valeurs significatives entre 9 et 16^{h}30. Durant cette période, la puissance délivrée par le générateur en fonction de l'heure, est reportée sur la figure 3. Les points a et b sur la courbe correspondante indiquent respectivement les points de démarrage et d'arrêt du groupe moto-pompe. Les résultats des mesures de débit effectuées pour différentes heures de cette même journée sont portés sur la figure 4. On voit que le maximum de débit est obtenue entre 11 et 13^h correspondant aux puissances maximales délivrées par le générateur pendant le même temps. Ces valeurs sont obtenue pour l'ensoleillement maximal de la journée soit 80 mW/cm^2. L'utilisation du système tel qu'il a été installé, convient parfaitement au mode de vie de la communauté. Le besoin en eau n'étant pas constant tout au long de la journée, une à deux heures d'exhaure, suffisent à satisfaire la demande, grâce au stockage dans le réservoir. Le dimensionnement a été fait de telle manière que même par temps courvert une extraction est possible. Il est à signaler que le système étant donné la structure du puits et sa régénération, ne peut fonctionner tout au long de la journée.

L'ensemble qui vient d'être décrit a été testé initialement au laboratoire pendant une durée de deux mois avant d'être installé sur le site actuel. Il totalise actuellement dix mois de fonctionnement et donne entière satisfaction à la population. Aucune modification des caractéristiques électrique et mécanique n'a été constatée durant cette période. La parfaite autonomie de l'installation par rapport au diesel fait qu'on envisage l'acquisition d'un groupe plus puissant pour le remplacement de cette dernière. Il est aussi envisagé dans le cadre de cette communauté une utilisation plus rationnelle de l'énergie disponible par l'acquisition d'un système de stockage électrique qui pourrait servir à alimenter une salle commune (T.V et conservateur de médicament).

7. CONCLUSION

Une installation simple de pompage photovoltaique a été placée dans un village isolé Algérien. Basée sur le principe du couplage direct groupe moto-pompe-générateur photovoltaique d'une puissance de 1/2 kW, elle permet de subvenir en partie aux besoins en eau d'une communauté de 200 personnes. Le mode d'utilisation simple est parfaitement adapté aux conditions sociales de cette population. L'acceptation sociale de cette station expérimentale laisse à envisager une extension de ce type dans d'autres régions plus reculées du pays.

BIBLIOGRAPHIE

1. J.F. DAY III, Third E.C Photovoltaic Solar Energy Conference CANNES 1980 p.124.
2. M. BENMALEK, 1ère Conférence Arabe sur l'Energie Solaire DAMAS 1981.
3. Données climatiques en Algérie non publiées.

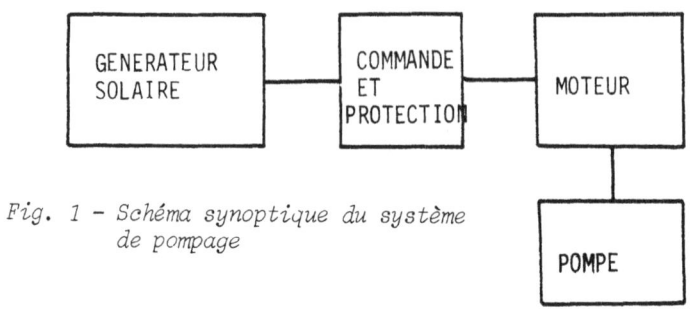

Fig. 1 - *Schéma synoptique du système de pompage*

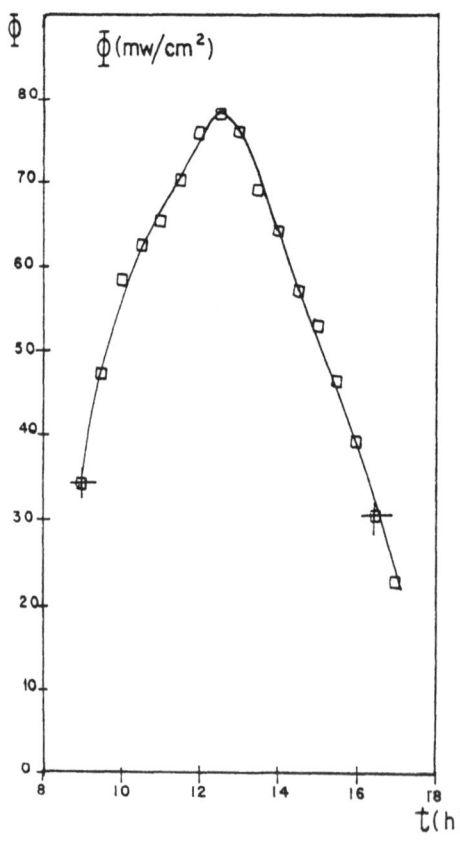

Fig. 2 - *Variation du flux lumineux sur le site pour la journée du 4 mars 1982*

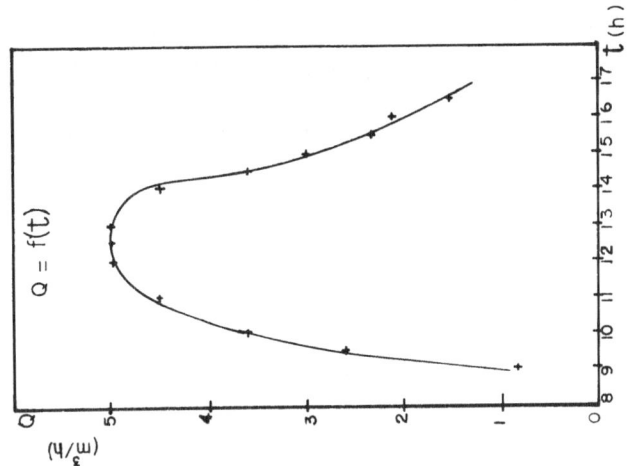

Fig. 4 — Relevé du débit correspondant aux données des fig. 2 et 3

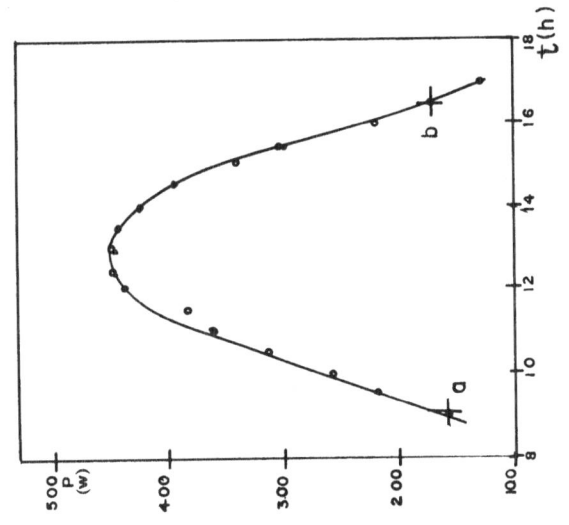

Fig. 3 — Puissance délivrée par le générateur photovoltaïque en fonction de l'heure locale

POSTER GROUP P1 - APPLICATIONS

- Intermediate

Advanced system design for solar power plants

The Mississippi county community college large-scale demonstration project - a success story

TISO 15 - 15kW experimental photovoltaic solar power plant

- Residential

Photovoltaic retrofit feasibility in the United States

Experiments on combined photovoltaic-aeolian electric generation in a residential stand-alone system

Simulations of the energy performance of a solar photovoltaic residence and hybrid electric automobile in Fresno, California

Alicudi project

A successful integration of solar energy in African life : Kolokani, Mali

Application and experience of photovoltaic pumps for irrigation in Pakistan

Development and testing of a submersible motor pump driven by a solar cell generator

- Other applications

Solar generator performance with load matching to water electrolysis - longterm averages and range of instantaneous efficiencies

Photovoltaic power for walk-in coolers

Technical and economic requirements for small-scale photovoltaic refrigerators

Residential applications of photovoltaics in the United States

In the poster hall

ADVANCED SYSTEM DESIGN FOR SOLAR POWER PLANTS

V. Cordes and K. H. Korupp
AEG-TELEFUNKEN Neue Technologien, Raumfahrt
Industriestraße 29, D-2000 Wedel/Holstein

SUMMARY

This report gives an overview on applied technology of Photovoltaic Systems as well as a view into the advanced subcomponent technology. It turns out more and more as a reliable source in the field of alternative energy. The now achieved standard of technology for complete systems as well as for subcomponents is already approved in various applications. In order to have the opportunity to build up power plants in the Megawatt range a continuous reduction of costs by intensive development work is necessary. In addition an approved wind generator technology would be a reasonable supplement for many applications and would decrease the storage part, which means a reduction of system costs.

1. INTRODUCTION

Photovoltaic power plants throughout the power range up to several Megawatt offer the advantage of a modular design. This is especially true for the solar generator, for the chargers and for the power conditioning system.

For large plants the relatively complex process control is realized with the help of micro-computers which enable an optimization of reliable operation and flexible adaptation in an economical way. For small and medium systems however, control is performed by analogous control units while still simpler systems use the chargers for regulation and monitoring.

All systems have in common that they will require a storage unit because of the insolation profile and the load profiles (yearly and daily period). Besides the pumping systems which store the medium water, photovoltaic plants use commonly lead batteries to store the electrical energy.

Fig. 1 shows a selection of photovoltaic systems and their power ranges. An impression of typical load profiles and of an idealized real solar generator characteristic is given by Fig. 2. The load profile of 21 fully electrified households is equivalent to complex loads (household or rural supplies) with strongly flucutating powers throughout the day, which represent high requirements to the power conditioning system with respect to high plant efficiencies.

2. SOLAR GENERATOR

In order to obtain higher system voltages and powers, an arbitrary number of solar generator modules (basic units) may be connected. The size of the plant determines the power and the reasonable voltage to be selected.

Fig. 3 shows the basic wiring of the modules as well as three different methods for their connections. The decoupling diodes for the decoupling of the strings connected in parallel and in addition for the suppression of the lead battery discharge via the solar generator during the night hours are installed in the distributing devices in case of larger plants and in the charger in case of smaller and medium systems

The shunt diodes which have been integrated in the module frame for the AEG-TELEFUNKEN modules PQ 10/20/0 have mainly two purposes. They have to prevent the destruction of the solar cells which are operated in the backward range of the characteristic, e.g. because of partial shading, and secondly they have to avoid the breakdown of a complete string as being caused by a defective module.

The integrated shunt diode simplifies the connection according to the methods of the bus bar and the plug connection techniques. The junction box technique and especially the plug connection technique represent the most economic solutions, but in the end the application of one of these methods is determined by the design of the solar generator. This holds true for small applications as well as for large solar power plants.

3. CHARGE REGULATORS

Any photovoltaic systems using lead batteries as a storage unit for the energy requires a regulation and monitoring of the battery charging current. For smaller and medium systems, this may be realized by separate chargers while large plants use the help of the plant control and of the switching devices. The power part of a charger may be fundamentally designed as shown in Fig. 4, applying one of four versions: as a series regulator, as a shunt regulator, as a switching regulator, or as a inverter for feeding the mains when a mains network is present. Switching regulators may be favourably applied in large plants, especially when they are integrated in the plant control.

Series regulators offer the advantage that the voltage drop across the regulator is increased after the gassing voltage threshold has been reached, which shifts the operating point on the solar generator characteristic towards a decreasing output power. Due to their permanent losses however they are suited for smaller plants only. For larger plants, the principle of the shunt regulators according to the short-circuit method or of the switching regulators may be used, for which only little heat must be carried of to the environment (Fig. 5) during the regulating operation.

There are two important additional functions which must be fulfilled by comfortable chargers: At first the automatic matching of the gassing voltage threshold according to the temperature of the lead battery to achieve optimum charging, and secondly a delay of the protective circuit against overdischarge. The latter prevents an unnecessary disconnection of the load in case of momentary high starting currents.

4. INVERTER TECHNOLOGY

In the total power range and for the most different applications (Fig. 6), static inverters are used in photovoltaic systems or will be provided for in future projects.

The most important applications are the driving technique in the power range up to several kilowatts, the supply of houses and villages and the feeding into another energy system (e.g. public supply mains) up to several hundred kilowatts.

The driving technique uses first of all transistorized inverters up to 10 kW; their power is matched to the motor in most cases. These inverters must be variable in their voltage and frequency in order to run the motor with a constant momentum in each mode of operation. As an example, the 5 kVA and the 10 kVA solarverter can be quoted (Fig. 7).

For more complex household or rural power supply systems, AEG-TELEFUNKEN apply stepped inverter systems: the individual inverters are synchronized and connected or disconnected, depending on the momentary load current. This ensures a high efficiency in the partial-load range also. These installations use inverters with silicon-controlled rectifiers, which can be overloaded momentarily and enable the release of automatic cutouts.

The feeding into a public supply mains or the operation in parallel with other energy converters may be performed with line-commutated inverters as well as with self-commutated inverters. Self-commutated inverters offer the advantage that an arbitrary amount of active and reactive power may be fed into the mains. Line-commutated inverters with six or higher pulsed circuits load the mains with a defined reactive power; but for economic reasons and due to their simplicity, they are better suited for many applications as compared with self-commutated inverters.

5. POWER PLANTS

The requirement for availability and reliability will determine the design of a solar power plant in any case, independent on its power. One of the solutions depicted in Fig. 8 may be applied as the basic structure for a power plant.

In many cases of application, it will be necessary to split up the solar generator: either for a better control or because of maintainability and modular design; in these cases, a switching device must be provided for. Moreover there may be the requirement of redundancy in the DC distribution and in storage systems, and a stepped inverter system may be necessary for load reasons. A split-up battery is favourable as it complies with the demand for an optimum battery maintenance; together with a sensitively switchable solar generator, it forms the base for a reasonable control possibility of the flow of energy.

Conception C shown in Fig. 8 is the base of the EEC project 300 kW Pellworm where central chargers should be dispensed with. For this large power plant, the additional task beside the self-sufficient supply of a recreation centre the energy feeding into a public supply mains will be realized with the help of a line-commutated inverter.

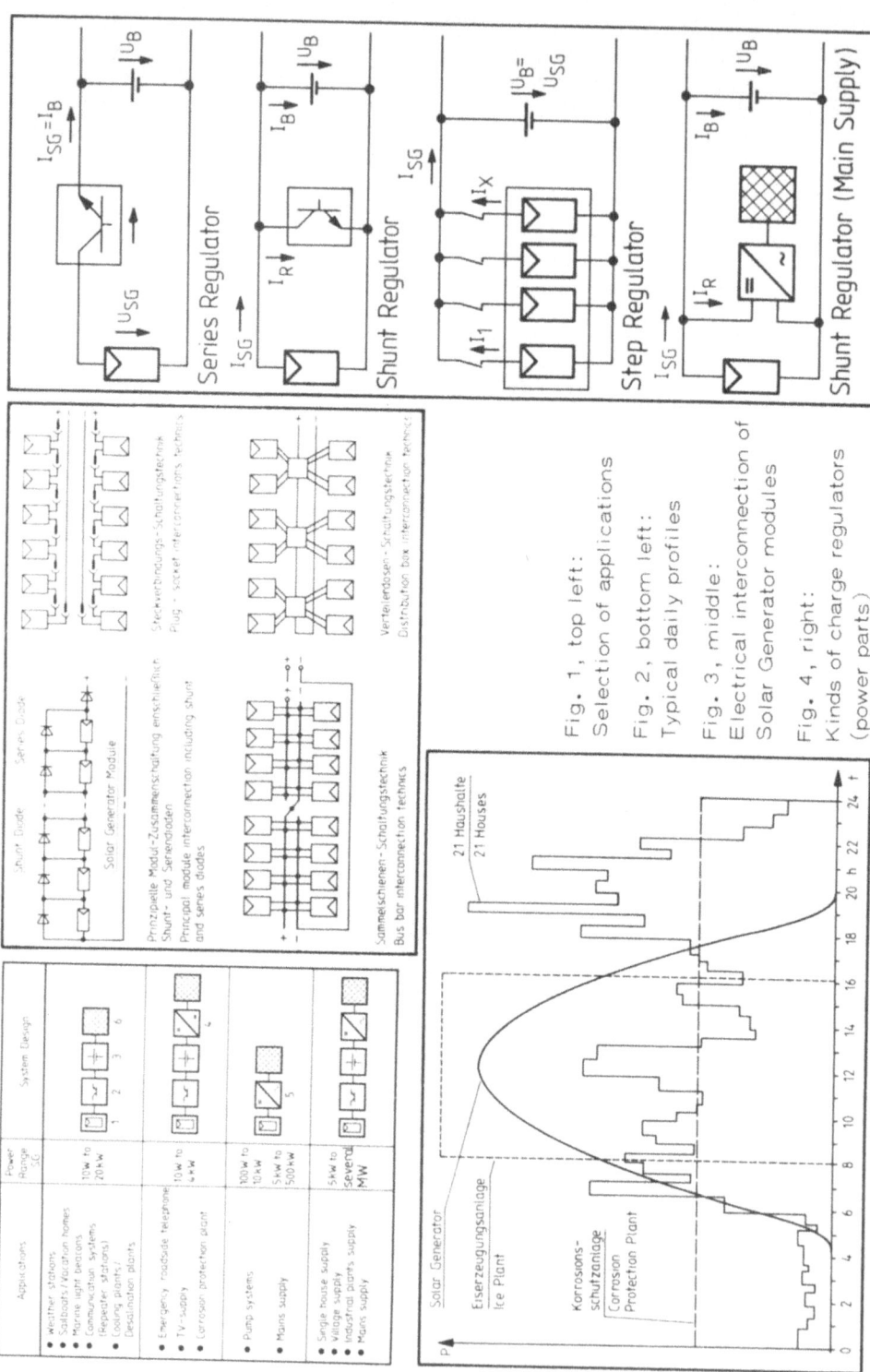

Fig. 1, top left:
Selection of applications

Fig. 2, bottom left:
Typical daily profiles

Fig. 3, middle:
Electrical interconnection of
Solar Generator modules

Fig. 4, right:
Kinds of charge regulators
(power parts)

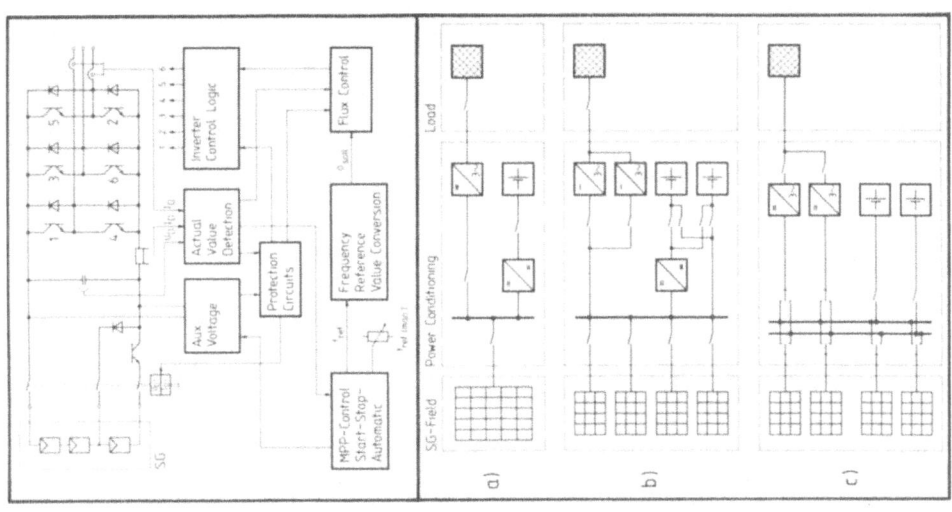
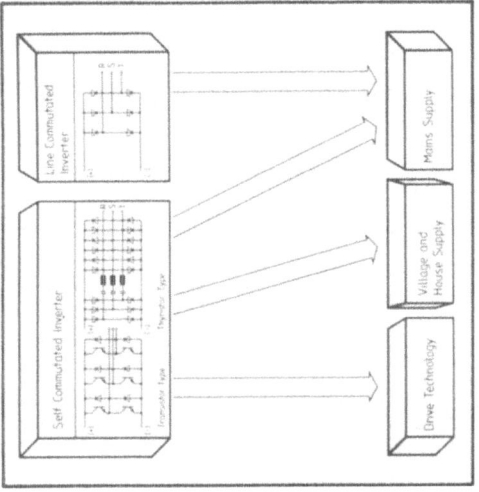

Fig. 5, left:
Power dissipation of the environment during operation

Fig. 6, middle:
Overview inverter technology

Fig. 7, top right:
Solarverter (drive technology inverter)

Fig. 8, bottom right:
Selected power plant design

THE MISSISSIPPI COUNTY COMMUNITY COLLEGE
LARGE-SCALE DEMONSTRATION PROJECT
A SUCCESS STORY

H.V. SMITH
President
Mississippi County Community College

Summary

The world's first large-scale photovoltaic concentrating system has now been completed and is operational. The project is located at Mississippi County Community College in Blytheville, Arkansas, U.S.A., and consists of a uniquely energy-conscious building with some 63,000 photovoltaic concentrator cells providing 320 kW (DOE standard) power. The system is actively cooled, single-axis tracking, and is designed for providing about 85 percent of the total energy needs of the 55,000 sq. ft. building.

In 1977 Mississippi County Community College signed a $6.8 million grant with the Department of Energy to build the first large-scale photovoltaic concentrating system and an exceptionally energy-conscious building. The college district made available from its capital construction fund some $2.6 million and work began in early 1978.

The 55,000 sq. ft. building, employing age-old passive solar and energy conservation concepts, was recently awarded first place for design by the Gulf State Region of the American Institute of Architecture. Not only is it exceptional in architectural integrity, but it is extremely efficient in its energy budget of less than 4.75 watts per square foot, unusual in typical institutions of higher education. Information concerning the design study and the various considerations for energy can be obtained from the architectural firm Cromwell, Truemper, Levy, Parker & Woodsmall of Little Rock, Arkansas.

The unique photovoltaic system was installed and became operational in late October of 1981. Solar Kinetics of Dallas, Texas, provided and

installed 270 single-axis tracking concentrating collectors. These are
parabolic troughs 20 feet by 7 feet and provide a concentration ratio of
about 28 suns optical and about 42 suns geometric. The reflective material
is a thin film (aluminized milar) of FEK 244. The collectors are ganged in
groups of six per row. Each row tracks east/west, driven by a hydraulic
system. The shadow band tracking mechanism is mounted on each row and is
activated by photo-diodes over a time interval of approximately 30 to 45
seconds.

The concentrator cells were provided by Solarex Corporation of Rockwell, Maryland, and are single crystal cells with minimum efficiency of 12 percent at 55° C. over the range of 1 W/cm^2 (10X) to 5 W/cm^2 (50X). Each cell is 2.5 x 5 cm.

The cells are laid down on a hollow V-shaped aluminum extrusion. Necessary cooling is provided by a
solution of ethylene glycol flowing through the hollow extrusion, and maximum temperature of 55° C. is maintained by the flow rate. When installed, each row contains 1396 cells and operates at about 400 volts d.c. Electrical interface between the solar system and the building is provided by a
d.c.-a.c. inverter built by Delta Electronic Control Corporation of Irvine, California, and provides peak power tracking as well as controlling the power mixture from the local utility and the photovoltaic system.

The final design of the utility interactive system required a special
relationship with the Arkansas Power & Light Company, an operating company

of the Middle South Utilities, Inc., the local electric company. This relationship is evident in a contract providing for a one-for-one electron exchange. In other words, if we generate more power than we need over the weekend, the utility will accept that on its power grid, and in essence, run the meter backward or return the power to us during the week when needed.

The byproduct of cooling the cells, thermal energy, is interfaced with the building's heating system through a standard heat exchanger and provides the primary heat source for winter heating. Included within the building heating loop are two 40,000-gallon, well-insulated and buried hot water storage tanks. This provides storage for some three to four days of heating capability.

Control of all aspects of the solar system is provided by a Hewlett-Packard desktop computer, Model 9845-B. This provides wake-up capabilities, emergency stow, and a multitude of redundant safety measures. A second Hewlett-Packard desktop computer gathers data for the National Data Bank located in Seattle, Washington. A complete weather station, as well as some 135 data gathering points, provides constant input to this system.

The design of the system calls for a peak electrical output of 320 kW (DOE standard) with a peak thermal output of 22 million Btu/day. Total cost of installed energy system (all components) is $9.70/watt electrical peak (1976 dollars). This does not take into account the value of the thermal energy, which could amount to about $25,000 per year. If our calculations are correct and if the system performs as designed, we should realize about 85 percent of our total yearly energy budget from the solar system.

Because the MCCC project is a large-scale actively-cooled concentrating system and the thermal output is essential for winter heating of the integrated buildings, it is an excellent source to study the effectiveness and long-term payback of such a system. The United States Department of Energy recognizes this potential and has agreed to support a maintenance, operation, and study effort over the next five years. We believe data gathered in this effort will have great influence on the future choice of such systems.

If you are a potential user of a photovoltaic system and your choice is for an actively-cooled concentrator system, the MCCC project should be carefully studied. There are many pitfalls to be avoided, particularly in the installation and troubleshooting stages.

The college is located in the rich farmland of the Mississippi River Delta at Blytheville, Arkansas, and is open for viewing.

TISO 15

15 kW Experimental Photovoltaic Solar Power Plant

M. Camani
Dipartimento dell'ambiente, CH-6500 Bellinzona

D. Bozzolo O. Daldini, R. Pamini, G. Salvadè
F. Solcà, C. Spinedi and F. Zamboni
Laboratorio di fisica terrestre, CH-6952 Canobbio

T. Celio
Ufficio d'ingegneria per l'elettroottica, CH-6775 Ambrì

C. Giovannini
Invertomatic, CH-6600 Locarno.

Summary

TISO 15 is an experimental, utility line interactive photovoltaic installation. It is intended for the collection of practical data related with the direct immission of photovoltaic electrical energy into a standard AC utility grid.
 A flat-plate array subfield provides the electrical power to a 10 kW inverter interacting with the local mains.
 A second array subfield (5 kW) with flat-concentrating modules and a separate inverter will be installed later on this year.

1. INTRODUCTION

In order to assess the relevance of photovoltaic energy in the general energy context clarification of the technical and economical problems related to the integration of photovoltaic power plants into the utility grid is utterly important.
 The project TISO 15 should help clarifying these questions. A 15 kW experimental photovoltaic solar power plant, including a flat-plate and a flat concentrating photovoltaic array, two inverters and a data acquisition system is being realised by the Dipartimento dell'ambiente (Environment Departement) of the Ticino State of Switzerland in collaboration with the Laboratorio di fisica terrestre at Lugano-Trevano.
 The size of the power plant should allow comparisons with the results of the photovoltaic programm of the European Communities.
 The main features of the solar plant and the research program are described below.

2. LAYOUT OF THE SOLAR PLANT

2.1 Solar Arrays

Two different arrays are used. The first one includes 288 flat-plate

Arco modules ASI-2300 with 37 W$_p$ nominal power each. They are lined up in three arrays of 8 (vertical) by 12 (horizontal) modules and installed on the roof of the Scuola Tecnica Superiore at Lugano-Trevano (Figure 1).

The tilt angle is 65° in order to maximize the power generation in winter: in the southern part of Switzerland 44% of the solar energy is available during the winter term, i.e. at the time of maximum energy consumption.

The distance between the blocks ist 10 m so that, in the worst case, only the lowest modules are shadowed for a short time.

The second array subfield consists of flat-concentrating modules providing electricity and heat. This new type of generators has been developed by Atlantis Energy Ltd., Bern, Switzerland, and is used for the first time in a solar plant. It will be installed at the end of 1982.

Figure 1

The 10 kW flat-plate array subfield installed on the roof of the Technical School at Lugano-Trevano

2.2 Inverter

The power of the flat-plate array subfield is fed directly into the utility grid by means of a 10 kW inverter (Abacus Controls Inc., N.J. USA, model 714-3-200). The control features include automatic start-up and shutdown at preset array powers, automatic synchronisation to mains frequency, maximum power tracking and selective fault shutdown and restart controls.

The measured efficiency at nominal power is 92 % and the first significant harmonic component the 23 rd.

A second inverter for power/conditioning of the concentrating array subfield, will be installed at the end of 1982.

2.3 Data Acquisition System

The data acquisition system comprises a front-end process computer (LSI 11) connected to a general purpose computer (PDP 11/34).

2.4 System Block Diagram

The block diagram of the flat-plate system is shown by figure 2.
12 modules on a horizontal line are connected in series (192 V / 2,3 A). The 24 resulting strings are conveyed separately to the control room where they are connected in parallel.

Further 12 modules will be placed on another building at about 80 m distance. This will allow simulation of dynamical effects due to non simultaneus shadowing of the arrays as it occurs in large scale solar plants.

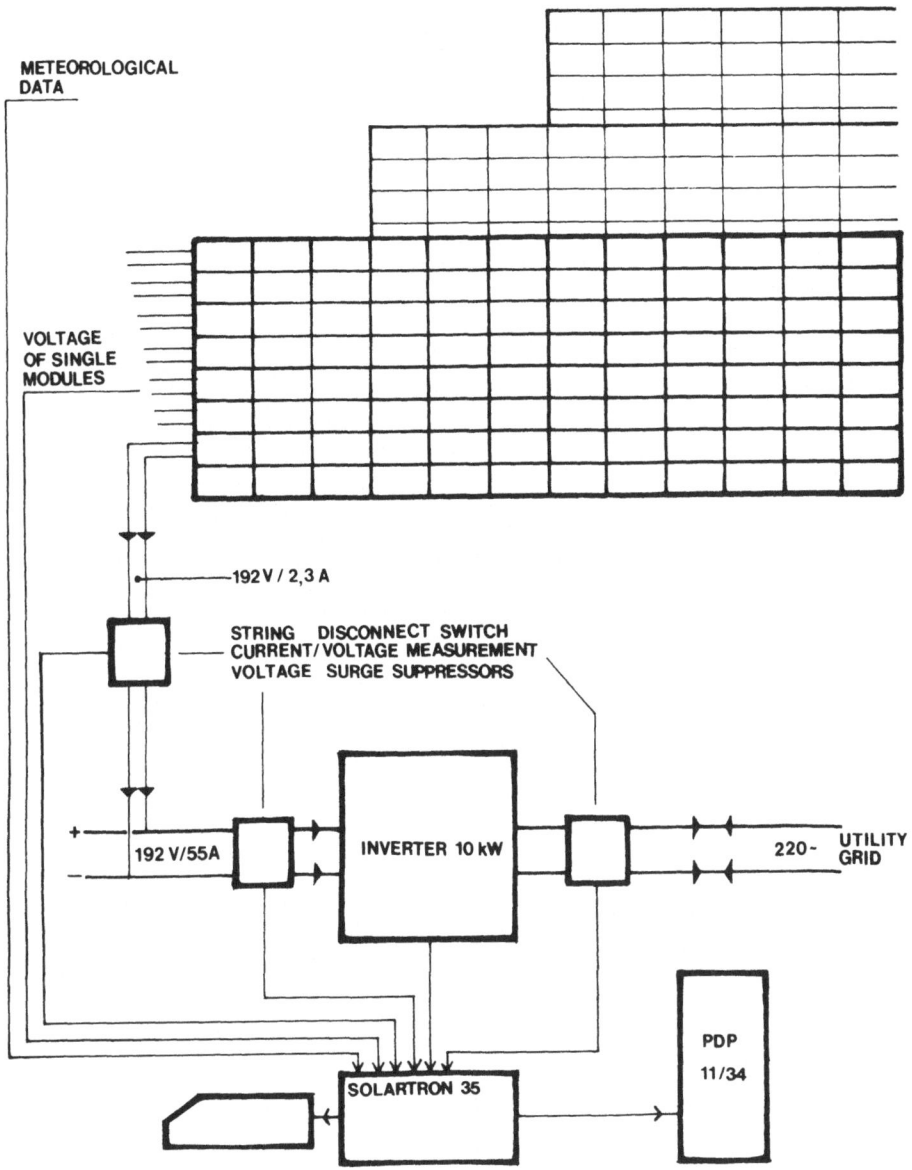

Figure 2 Block diagram of the flat-plate array subfield.
12 modules of a horizontal line are connected in series. The 24 resulting strings (only one is shown on the diagram) are conveyed separately to the control room where they are connected in parallel.

3. RESEARCH PROGRAM

The following problems will be studied:
- series or/and parallel connection of the modules (minimization of power losses caused by connections, shadowing and deterioration of single modules).
- magnitude of DC or AC voltage of the array subfields in large power plants
- effects of induced voltages (lightning protection)
- safety problems of grid connection
- level of AC-distortions (current harmonics)
- effects of partial shadowing of the arrays
- measurement of power and energy production of the flat-plate and flat-concentrating modules as a function of meteorological conditions
- electricity production of the flat-concentrating array as a function of the cooling water temperature
- module degradation.

Acknowledgement

The project is supported by Nationaler Energie-Forschungs- Fonds (NEFF), by Società Elettrica Sopracenerina (SES), by Migros-Genossenschafts-Bund and by other private institutions.

PHOTOVOLTAIC RETROFIT FEASIBILITY IN THE UNITED STATES†

J. L. JACKSON
PV System Definition Division 4723
Sandia National Laboratories
Albuquerque, NM 87185 U.S.A

Summary

Residential and commercial retrofits may represent a significant U.S. national market for photovoltaic (PV) systems. We discuss techniques for estimating this market and present conclusions about physical market size. We then briefly discuss possible PV retrofitting techniques for residential and commercial structures.

1. INTRODUCTION

Although retrofitted PV systems have been considered since the inception of the National PV program, the design of PV systems for residential and commercial buildings has focused on new construction. While new construction could represent an attractive market for PV systems, the existing building stock may present a potentially larger market, particularly in light of the slow turnover rate of existing buildings (less than 25 percent will be replaced by the year 2000)(1). Sandia National Laboratories has funded two studies to assess the technical and economic feasibility of widespread retrofits for residential and commercial buildings (2,3). We summarize the important results of these studies in this paper.

2. METHOD OF APPROACH

We first chose to study retrofit feasibility in the southwestern, northeastern, and southeastern regions of the country before proceeding on a nationwide basis. These regions, in the order given, have been found to be most economically suitable for PV applications (4,5).

The next step involves a general characterization of the existing building stock. Residential architectural styles vary from the suburban cape and victorian frame house of the Northeast to the tract and suburban ranch of the Southwest (2) (see Figure 1). Commercial buildings appear to be mostly industrial, retail or office structures. These three categories of commercial buildings comprise nearly 70 percent of the total surveyed for the three geographic regions. There is little regional architectural variation in commercial buildings with the vast majority having flat roofs (3). The next task is to estimate the percentage of existing buildings suitable for PV retrofits.

3. RESIDENTIAL MARKET SIZE ESTIMATE

Total Environmental Action (TEA) first used existing estimates of favorable solar orientation based on aerial photographs taken by the Jet Propulsion Laboratory (JPL) of residential areas in Hartford-New Haven, Miami and Los Angeles (6). These results were then crossed checked by TEA

---

†This work supported by Photovoltaic Energy Technology Division, United States Department of Energy.

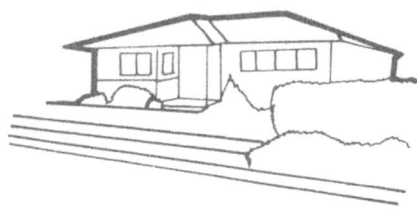
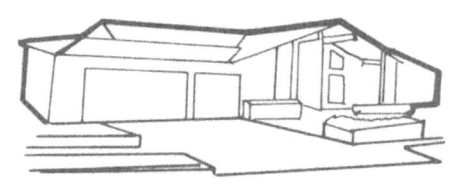

Suburban Cape

Victorian Two-Family Wood-Frame House

Post War Tract House

Shake-Roofed Suburban Ranch

Figure 1. Types of Residential Styles

using ground or "wind-shield" surveys. As a result, TEA determined that only 40-70 percent of those homes identified from the air as suitable for retrofitting were appropriate. The JPL/TEA results were then extrapolated on a nationwide basis by using Census Bureau estimates of total residential structures in the North Central, Northeast, South and West census regions (7). Table 1 summarizes this "discounting" process.

TABLE 1. RESIDENTIAL BUILDINGS SUITABLE FOR PV RETROFIT

	Northeast	North Central	South	West	U.S. Total
Number of Residential Buildings* x 10^6	11.5	16.4	21.0	11.1	60.0
Solar Orientation (JPL)	69%	69%	78%	71%	--
Number with Solar Orientation x 10^6	7.9	11.3	16.4	7.9	--
Unobstructed Roofs (TEA)	38%	38%	69%	52%	--
Number of Potential Retrofits x 10^6	3.0	4.3	11.3	4.1	21.0**
Residential Building Suitable for Retrofit	26%	26%	54%	37%	35%

*Data is for residential buildings having four dwelling units or less
**Total is less than the sum of the four census region numbers because areas where PV systems are not economic have been subtracted out. Average size of a residential system is around 4 kW_p.

This methodology produces an estimate of approximately 21 million total residences that are potentially suitable for PV retrofit, or 35 percent of the total U.S. residential building stock.

4. COMMERICAL MARKET SIZE ESTIMATE

Battelle Columbus Laboratories (BCL) first conducted a telephone survey of selected building appraisers, roofing and general contractors and compiled data on various physical characteristics. This information was then statistically analyzed for candidate building retrofits. Ranking and weighting factors were developed for evaluating the relative difficulty of accomplishing a retrofit installation. Scores were then totaled for each candidate application in each region to provide a relative ranking with regard to retrofit difficulty. Applications suitable for retrofit were then totaled for the three geographic regions and are presented in Table 2.

TABLE 2. COMMERCIAL BUILDINGS SUITABLE FOR PV RETROFIT

Type of PV System	PV System Size (kW_p)	NE	SE	SW	Final Totals
Flat Panel Roof Mounted	0-20	13,079	44,311	59,718	
	21-400	23,434	82,302	109,686	
	>400	9,236	23,716	19,649	
Subtotals		45,749	150,329	189,053	385,131
Flat Panel Ground Mounted	0-20	54,187	66,843	56,961	
	21-400	74,301	119,427	110,343	
	>400	16,117	30,658	24,498	
Subtotals		144,605	216,928	191,802	553,335
Concentrator Ground Mounted	0-20	60,340	72,166	59,592	
	21-400	79,623	132,425	114,835	
	>400	18,503	34,496	26,069	
Subtotals		158,466	239,087	200,496	598,049
Concentrator Roof Mounted	0-20	5,305	11,369	13,785	
	21-400	13,119	22,557	26,559	
	>400	5,552	8,995	5,785	
Subtotals		23,976	42,921	46,129	113,026

If we extrapolate to include the entire nation, an estimated inventory of 1.5 to 2.0 million feasible commercial retrofits results. This represents approximately 50-60 percent of the total U.S. commercial building stock.

5. RETROFITTING TECHNIQUES

TEA developed a total of 11 residential retrofit designs for typical houses in New Haven, Miami and Los Angeles. In addition to the usual engineering and architectural details and drawings, TEA prepared a sequence of drawings picturing the entire installation sequence from initial module assembly through installation on the roof. Figure 2 is an example of such a sequence.

BCL developed nine retrofit designs for typical commercial buildings in the northeastern, southeastern and southwestern regions of the country. Figure 3 is an example of such a commercial retrofit design which also illustrates a possible installation sequence for an actively cooled PV concentrator system.

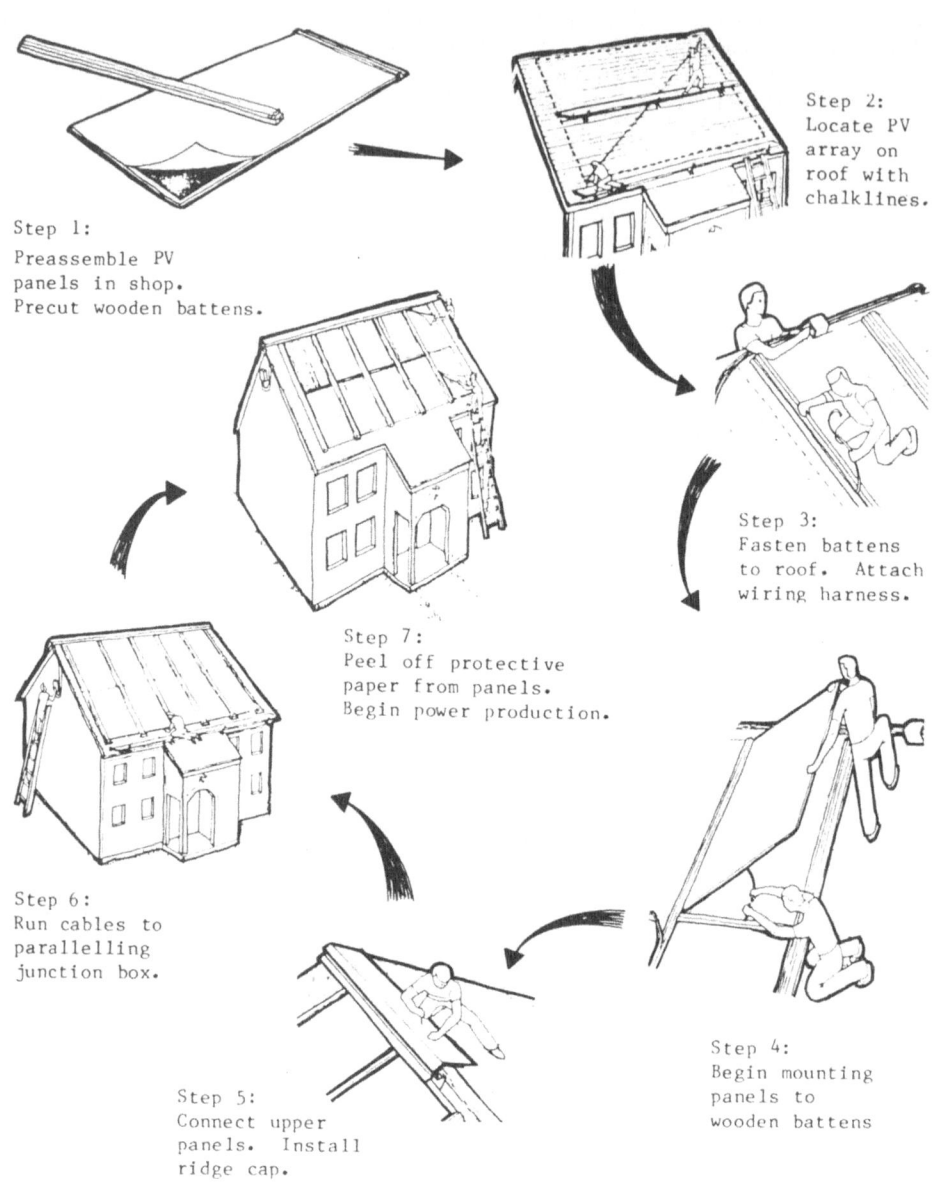

Figure 2. Residential Retrofit Sequence

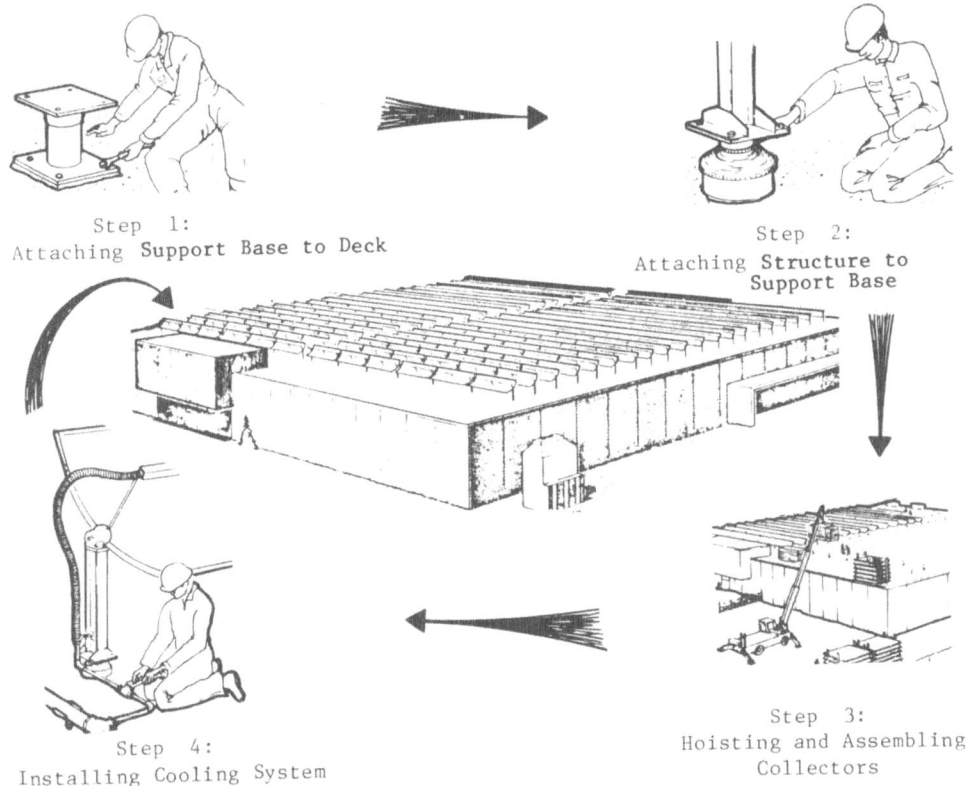

Figure 3. Commercial Retrofit Sequence

6. REFERENCES

(1) Department of Energy, Solar Energy Objectives, Calendar Year 1980, DOE/CS-0155, April 1980.
(2) D. Mahone, P. Temple, et al, Study of PV Residential Retrofits, SAND81-7019, (Albuquerque: Sandia National Laboratories, May 1982). Work performed by Total Environmental Action.
(3) G. Noel, J. Hagely, L. Stember, Design and Market Study of Retrofit PV Systems for Commercial Building Applications, SAND81-7179, (Albuquerque: Sandia National Laboratories, May 1982). Work performed by Battelle Columbus Laboratories.
(4) P. Pittman, et al, Regional Conceptual Design and Analysis Studies for Residential PV Systems, SAND78-1040, (Albuquerque: Sandia National Laboratories, September 1979). Work performed by Westinghouse R&D Center.
(5) E. J. Burger, et al, Regional Conceptual Design and Analysis Studies for Residential PV Systems, SAND78-7039, (Albuquerque: Sandia National Laboratories, January 1979). Work performed by General Electric.
(6) R. Fretz, et al, Metropolitan Roof Top Analysis for Three Cities: Los Angeles, Hartford-New Haven, and Miami, (in preparation), Earth Resources Application Image Processing Laboratory, Jet Propulsion Laboratory, Pasadena.
(7) Bureau of the Census, 1978 Annual Housing Survey.

EXPERIMENTS ON COMBINED PHOTOVOLTAIC-AEOLIAN ELECTRIC GENERATION IN A RESIDENTIAL STAND-ALONE SYSTEM

P.U.Calzolari
Ist. di Elettronica, Facoltà di Ingegneria, Università di Bologna
40136 Bologna (ITALY)

G.C.Cardinali, A. Garulli, D.Nobili, A. Zani
CNR - Istituto LAMEL
Via Castagnoli, 1 - 40126 Bologna (ITALY)

SUMMARY

The general design guidelines and the main features of an experimental photovoltaic-aeolian plant, with some preliminary observations on its working, are presented. The system has been operating for over a year and it meets the basic domestic needs of a remote farm situated in a hilly district of Northern Italy (Passo Mandrioli in Emilia Romagna). The plant was preliminarily conceived by the regional administration board which has supplied the financial support. This is the first project of a plan intended to collect information on alternate energy systems operating in a real-world environment, with a view to its diffusion between potential users and national manufacturers. To this end, the plan has been designed to accomodate a complete set for acquisition, recording and first elaboration of the main variables (both meteorological and directly relevant to the operation of the system).
Combined photovoltaic-aeolian generation was chosen on account of the expected complementary nature of the two sources during the year. Though extended through a limited period, the available data seem to confirm the above-mentioned characteristics and support the general design criteria which are based on them.

1. INTRODUCTION

This paper illustrates one of the first demonstrations of the combined aeolian-photovoltaic generation of electric energy for stand-alone applications. The plant is intended as an instrument in order to gain experience in the designing and in the maintenance problems of small isolated plants, a field where solar generation may be expected to compete with conventional systems in the near future. It is a well known fact that in order to reach the diesel on-site generation cost, the present PV system cost needs a reduction factor which largely depends on the extimation of the solar cell lifetime. Nevertheless, even in the less optimistic predictions, the gap is not too deep to be bridged in a few years. An effective contribution may be given by the combination of wind power to PV power which allows the reduction of energy storage requirements and hence the overall system cost (1).

This is a typical situation where public financial support through a far-sighted policy can favour market growth and help national industry. This is one of the purposes given to the Mandrioli Pass experimental plant by the local administration board (Comunità Montana dell'Appennino Cesenate). These experiments are part of a wider program which aims to provide with on-site generators the farmhouses not served by the national utility grid. The experimental nature of the plant is underlined by the installation of an acquisition system of all the relevant meteorological and operating data, which is necessary to investigate the optimization of controlling procedures and of power conditioning on a realistic basis; the installation of the data acquisition system has been completed in the last month with the financial support of the regional board for the economical development (ERVET), although the plant as been operating for about one and half years.

2. DESCRIPTION OF THE SYSTEM

The system structure is summarized in fig. 1, which shows the two power generators connected in parallel, the battery storage, the control circuit, the inverter and the supplementary generator which is automatically connected by the control circuit whenever it is required by the regulation strategy. The same figure shows all the environmental and electrical quantitaties which are continuously monitored and periodically recorded by the acquisition subsystem.

Switch SW1 disconnets the generation sub-system from the batteries thus protecting them against overcharging and making the excess power to flow into an auxiliary load. Battery voltage, and therefore the output voltage of the generators, has been fixed at 48V, which seems to be a reasonable trade-off between the inverter requirements and those of the module connection arrangement. In order to reduce inverter losses, lighting power is delivered to the load directly by battery through a small power, low cost line about 60 m long. This is the distance between the farm building and the low hill which has been considered as the most favorable site for the whole plant (see fig.2). The user operated switch SW2 connects the inverter to an AC line whenever AC power is needed, thus avoiding the no-load losses which are known to be the weakest point of the inverters.

Considering the reduced reliability of wind power predictions the introduction of a two-step priority scale in energy requirements is recommended. Accordingly, lighting and radio-TV have been classified as primary consumptions, which have been estimated at 1.7 KWh/day, and the PV surface has been chosen so as to meet this primary energy demand, by itself, even in the lowest insolation period. The remaining domestic appliances give raise to further consumption estimated at 2.8 KWh/day. Due to the numerous uncertainties of the problem, attempts to forcast seasonal load variation have been considered superfluous. Load requirements are superimposed in fig.3 on curves which show the estimated available energies of the PV and wind-generator. They have been calculated from the component characteristics summarized in table 1 and on information on insolation and wind from the nearest meteorological stations; those for the specific site not being available. Though collected in similar environments, as far as height and distance are concerned, wind information is by-far the most critical point, due to the strong dependence of the wind regime on the short-range morphological structure of the terrain. Accordingly, the system design has to account for the reduced reliability of these data.

Figure 3 clearly shows that the two energy sources join in a complementary way to meet the total power requirement during the year. This has made possible a substantial reduction of the storage capacity (see table 1) which guarantees an energy reserve for just 2-3 days; notwithstanding this, no energy shortage has yet occurred and hence no occasion has yet arisen to have recourse to the auxiliary conventional generator.

3. THE DATA ACQUISITION SYSTEM

In order to collect information on this alternate energy system, operating in a real-world environment, the plant has been designed to accomodate a complete set for the acquisition, the recording and first elaboration of the main variables. Both the meteorological variables (such as insolation, wind direction and speed, temperature, etc.) and those more directly connected to the system operation (such as the power available, the power consumption, the state of battery, the system configuration, etc.) are controlled; they are shown in the block diagram in fig. 1 and are specified in Table II. The number of these variables and above all their different nature, call for an acquisition system with considerable versatility. Besides, severe limitations must be respected expecially for energy consumption, which must be restricted so as to avoid significant interference with the main load. Therefore, a sensor set and a recording data system have been adopted, which are controlled by a microprocessor circuit, completely realized with CMOS technology. Signals produced by the different transducers are continuously detected, subjected to simple elaborations (most of them, for example, undergo an averaging operation) and finally recorded on a magnetic tape with a sampling period of 10', once they have been given a suitably compact digital form.

4. SOME EXPERIMENTAL RESULTS

The measuring and monitoring system has been operating for a few months; nevertheless, the first information collected already gives indication which are useful to the system designer. Two examples are given; the first referring to the primary data for the plant, that is the energy availability; the second more directly connected to the system arrangement. Fig. 4 shows the available solar and aeolian energies, the first directly measured by a solarimeter and the latter estimated by averaging the cube of the wind speed. These figures put into evidence the significant contribution of the wind source. In fact the available data, although limited, show that the actual wind energy was about 20% less of the espected one. As it is known the main problem of this source is the reliability in the short term. The following figure depicts a typical operating condition which may occur as a consequence of the internal resistance of the battery. In the night-time the wind-generator works and supplies current to the battery; as soon as the PV generator begins to be operative in the morning, the sum of currents rises the voltage drop on the battery series resistance untill the control circuit excludes the wind-generator. This takes place because the exclusion threshold voltage of the wind-generator has been fixed at a lower value than for the PV generator, according to the most convenient approach; of course, both generators would be excluded in the case of equal threshold values. This observation shows the need for a careful choise in the management of a multi-generator system which must be strictly linked to experimental information.

5. REFERENCES

1. J.A. Castle, et. al., 15th IEEE Photovoltaic Spec. Conf., 738, may 1981.

TABLE I

COMPONENT CHARACTERISTICS

Photovoltaic:
  72 modules in an array of 4 rows;
  36 cells Ø 3 in. series-connected
    in each module;
  Total PV area 11.45 m^2
  Peak power 1.4 KW

Aerogenerator:
  high speed twin blade Ø 3.4 m
  propeller with horizontal axis;
  pitch angle regulation through
  centrifugal masses;
  three phases alternator
  followed by rectifier;
  output voltage 48V;
  rated power 2.5 KVA

Battery:
  lead-acid series of 24 cells;
  250 Ah total capacity

Inverter:
  42-56 VDC input
  220 V, 50 Hz, 2 KVA output

TABLE II

MONITORED DATA

1) PV current
2) Aeolian current
3) Aux. Gen. current
4) DC current to user
5) DC current to inverter
6) Battery input current
7) Battery output current
8) Battery voltage
9) PV voltage
10) SW1 operating time
11) Irradiance horizontal
    plane with solarimeter
12) Irradiance 45° tilt plane
    with solar cell
13) Irradiance horizontal
    plane with solar cell
14) Propeller angular speed
15) Battery electrolyte
    temperature
16) Solar cells temperature
17) External temperature
18) Instruments temperature
19) Wind speed and direction

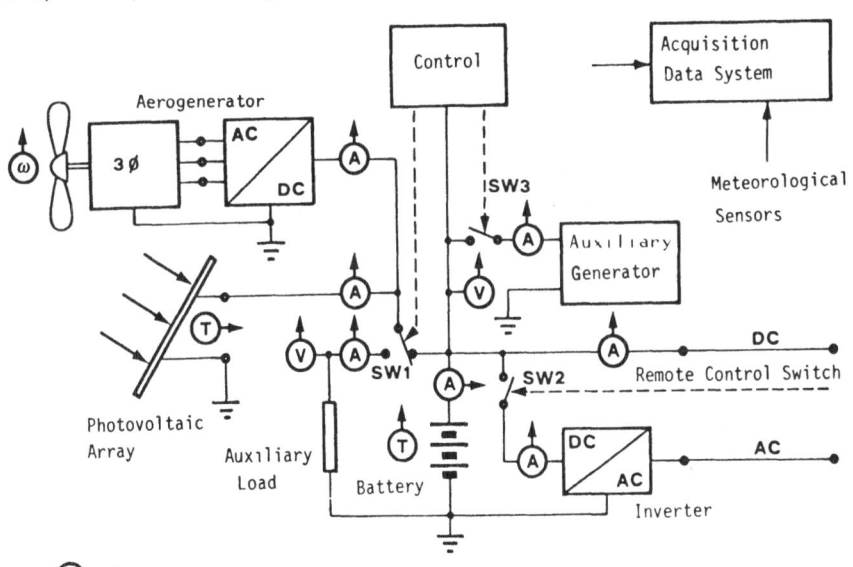

Fig. 1 - Block system diagram

Fig. 2 - View of the plant

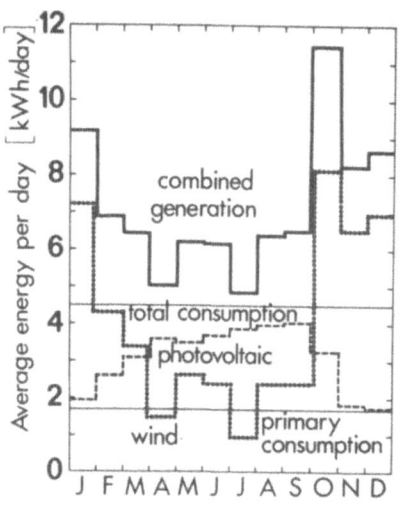

Fig. 3 - Expected energy availability

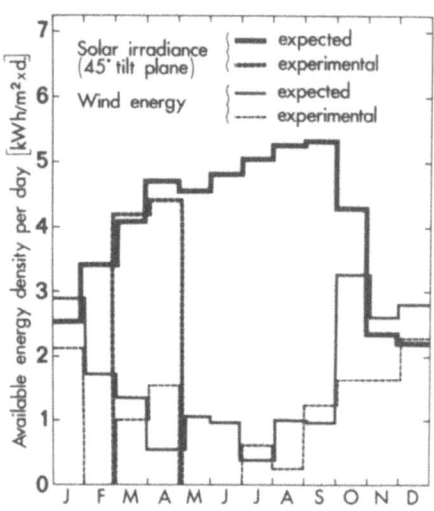

Fig. 4 - Available energy

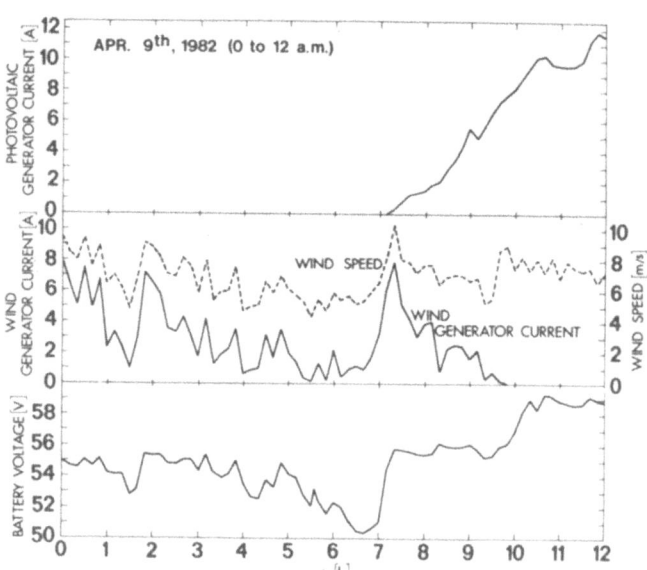

Fig. 5 - Typical operating condition

SIMULATIONS OF THE ENERGY PERFORMANCE OF A SOLAR PHOTOVOLTAIC RESIDENCE AND HYBRID ELECTRIC AUTOMOBILE IN FRESNO, CALIFORNIA

J.S. Reuyl and R.D. Schutt

JSR Associates
2280 Hanover Street
Palo Alto, California 94306   U.S.A

Summary

JSR Associates has designed an integrated system incorporating a solar photovoltaic residence and hybrid electric auto that (1) collects, converts, stores, and distributes incident solar energy on the residence to meet the time-varying needs for energy in both the residence and auto, (2) provides a source of supplemental energy for the residence and auto on cloudy days, and (3) enables the auto to travel extended distances whenever the owner so wishes. This study analyzed the hour-by-hour energy performance of a system that could be built with 1980s' technology. Thermal and electrical performance in both stand-alone and grid-connected configurations were simulated with computer models using actual hourly solar and weather data for Fresno, California. Results of the study (1) indicate that the integrated system works well as designed and uses a fraction of the energy required by a conventional residence and auto. A brief cost analysis indicates that the system will be competitive with grid electricity and conventional autos within this decade.

1. BASIC CONCEPT

On a sunny day, enough solar energy falls on a properly designed residence to power both the residence and an electric auto. Securing this energy is feasible, but it is not a trivial task. First, it is necessary to collect, convert, store, and distribute this energy so that the time-varying requirements for energy in both the residence and the auto are met. Second, one must provide a source of supplemental energy for the residence and auto during periods of extended cloudy weather and to allow longer trips by the auto.

To accomplish these objectives, one should begin with a properly designed residence -- one that requires a minimum of supplemental heat or electricity. Such residences typically are (1) energy conserving, to minimize winter heat loss and summer heat gain, and (2) passively heated and cooled, using the sun for winter heating and natural ventilation for summer cooling.

In the patented JSR design (2) the following features are then added:
- An array of photovoltaic (PV) modules cover the south facing roof of the residence. (The modules chosen for this study convert approximately 9% of the annual solar energy incident on the array directly to electricity).
- Batteries in the residence permit storage of electricity for later use.

- An inverter converts the direct current (DC) from the PV array and residence batteries to alternating current (AC) for AC loads.
- A battery-powered electric auto (with an electric motor for propulsion) uses electricity from the photovoltaic array for around town and commuter trips.
- A small liquid-fueled engine/generator in the auto provides supplemental electricity (hence, the name "hybrid electric") when there is insufficient stored photovoltaic electricity. In this way, the auto's range is limited only by the size of the fuel tank.
- The auto engine/generator also is used to recharge both the electrical and heat storage systems of the residence whenever there is too little solar energy to meet the residence needs.

Since the auto engine/generator serves as backup for both the auto and the residence, this integrated system requires no hookup to the electric utility grid. If the engine is a multi-fuel engine, then any liquid fuel (alcohol, gasoline, gasohol, diesel, kerosene, etc.) could be used, thereby rendering the system less dependent on any one fuel type. If alcohol is used, the system would consume only renewable energy; no conventional energy of any kind would be required. Although the system requires no utility hookup, there are instances when a grid connection is attractive to both residence owner and the utility. The analysis in this study addressed both applications.

2. SCOPE OF THE STUDY

The study analyzed the energy performance of a specially designed solar photovoltaic residence and hybrid electric auto that could be built with "off-the-shelf" hardware. The project was constrained to begin with a design for a photovoltaic residence developed by General Electric for sunbelt applications (e.g., Phoenix, Arizona) (3). This design was then modified by simulating natural ventilation and other passive solar features to improve its heating, ventilating, and air conditioning (HVAC) performance for California-type climates. A solar water heater was included (to reduce the demand upon the electric resistance backup heater); some reduction in the use of appliances (e.g., freezer and clothes dryer) was assumed.

The residence was then equipped with photovoltaic arrays, on-site battery storage, and a DC-AC inverter. This energy system was designed to serve as the primary source of electricity for both the residence and an electric automobile.

In addition, the auto was assumed to have an on-board liquid-fueled engine/generator that provides backup energy for the auto. In the stand-alone application (i.e., when the residence does not have a hook-up to the electric utility), this same engine/generator also serves as a backup source of electricity for the residence.

The benefits of utilizing the waste heat of the engine in the residence would have to be measured against the cost of the associated equipment. Initial study of the operation of the integrated system in California-type climates indicated that the need for supplemental space heating and supplemental hot water heating in the study residence would be quite low. Therefore, the use of the engine waste heat was not simulated in this study.

The study assessed the energy performance of the residence and auto on an hour-by-hour basis. Thermal performance of the residence was simulated with CALPAS, a computer model on-line at the Berkeley Solar Group (4) and first developed by California Polytechnic Institute/San Luis

Obispo for the California Energy Commission. Electrical performance of the residence was simulated with SOLCEL-II, a computer model developed by Sandia National Laboratories (5) and modified by JSR to simulate the hybrid electric automobile as both electrical load and a source of electricity. These programs were used to determine the optimal size of the various components.

Both programs require insolation (solar radiation) and weather data as input. For this study, the SOLMET Typical Meteorological Year (TMY) data for Fresno, California were used (6,7). Fresno is one of 26 sites around the United States for which a Typical Meteorological Year has been compiled from actual insolation and weather data. These TMY sites are increasingly used by researchers to permit easier comparisons among research studies.

California was chosen as the region to be simulated for this study because the study results would be broadly applicable to a large population. Fresno was chosen over Santa Maria (the only other TMY California site) because Fresno has heating and cooling seasons that are more extreme than Santa Maria -- in fact, more extreme than those encountered by most California residents. Hence, the performance results of this study would encompass the results that could be expected in most of the populated regions of the state.

The study simulated the performance of a stand-alone residence with an optimally-sized array of 110 square metres of GE PV shingle modules (3). This array produces 9.6 kilowatts of electricity (kW(e)) at standard operating conditions (SOC), i.e., insolation of 1 kilowatt per square metre, ambient (air) temperature of 20 degrees Celsius, wind speed of 1 metre/second, and nominal operating cell temperature (NOCT) of 64 degrees Celsius. A 10 kW(e) DC/AC inverter, 61.0 kilowatt-hours (kWh(e)) of on-site electrical storage, and "balance-of-system" (BOS) equipment (e.g., wiring) complete the system.

The hybrid electric auto was assumed to incorporate 30 kWh(e) (680 kilograms (kg) (1500 lb)) of lead-acid batteries and a 40-horsepower (hp) electric-drive motor. These assumptions result in an auto with a range capability of 100 kilometres (km) (62 miles) (on batteries alone and at a steady 65 km/hour), a top speed considerably in excess of legal limits in the U.S., and acceleration capability superior to many conventional autos.

The backup engine/generator assumed for the study is a state-of-the-art gasoline internal combustion design with a thermal efficiency of 25% and a fuel consumption rate of 4.5 litres/hour (1.2 gallons/hour). The engine/generator was sized at 25 hp/10 kW(e).

3. RESULTS

During sunny weather, all of the energy needs of the residence and auto are met from incident solar energy. The passive solar features of the residence meet the residence needs for space conditioning, and the solar thermal water heater supplies all the hot water. The 9.6 kW(e) PV arrays meet the daytime needs of the residence for electricity, and excess power is stored in the on-site batteries. At night, electricity from these batteries supplies the residence with all its needs for power and fully recharges the auto batteries. The auto travels 100 kilometres (62 miles) each day (36,500 km or 22,680 miles each year).

During cloudy weather, the PV array continues to produce some power (typically 5-10% of peak power even during storms), and the residence draws additional electricity from its batteries as needed. The auto engine/generator provides power for the auto while it is on the road.

When the auto returns at night, the auto engine/generator recharges the residence batteries. By operating 3 hours during the night, the auto engine/generator meets the 24-hour supplemental electricity needs of the residence during the most severe winter conditions of minimum insolation and maximum load.

For the Fresno Typical Meteorological Year, the PV arrays provide 100% of the electricity needed in the residence on 315 days. The auto engine/generator is used as backup on the other 50 days for that part of the residence load not met by the arrays. The PV arrays also provide all or part of the auto energy needs on 282 days. The array provides a total of 21,120 kWh(e) and the auto engine/generator provides 4,729 kWh(e). The auto engine/generator operates 102 hours in the backup mode to the residence and 371 hours while the auto is on the road. Total operating time is 473 hours, and total annual liquid fuel consumption is 2,128 litres (562 gallons). The operation of the system in the grid-connected mode was also simulated.

A cost analysis indicates that the integrated system would be cost-competitive with utility grid electricity and conventional autos within this decade. However, institutional barriers, some of which are identified in the final chapter, could inhibit the commercialization process. The report concludes with a discussion of a full-scale test of the system that could help to resolve these institutional issues as well as the engineering tradeoffs.

REFERENCES

1. Reuyl, John S. and Randall Schutt, Simulations of the Energy Performance of a Solar Photovoltaic Residence and Hybrid Electric Automobile in Fresno, California, JSR Associates for Sandia National Laboratories, Albuquerque, New Mexico, available from National Technical Information Service, U.S. Department of Commerce, 5285 Port Royal Road, Springfield, VA 22161, Report No. SAND81-7044, 130 pages, August 1981.

2. Reuyl, John S., "Integrated Residential and Automotive Energy System," U.S. Patent 4,182,960, Palo Alto, CA, January 8, 1980.

3. Kirpich, A. and E.M. Mehalick, et al., Regional Conceptual Design and Analysis Studies for Residential Photovoltaic Systems, Volume II, Technical Volume, General Electric Space Division, Valley Forge, PA for Sandia National Laboratories, Albuquerque, NM, Report No. 78-7039, 450 pages, January 1979.

4. CALPAS: A Passive Solar Simulation Model, Version 2.3, Berkeley Solar Group, Berkeley, CA, 32 pages, June 16, 1980.

5. Hoover, E.R., SOLCEL-II: An Improved Photovoltaic System Analysis Program, Sandia National Laboratories, Albuquerque, NM, 200 pages, February 1980.

6. SOLMET--Hourly Solar Radiation Surface Meteorological Observations, Volume 1--User's Manual, Report No. TD-9724, U.S. Department of Commerce, National Oceanic and Atmospheric Administration, Environmental Data and Information Service, National Climate Center, Asheville, NC for the Department of Energy, Division of Solar Technology, August 1978.

7. Typical Meteorological Year Data--User's Manual, Sandia National Laboratories, Albuquerque, NM, 7 pages, 1979.

ALICUDI PROJECT

V. Arcidiacono, S. Corsi, A. Iliceto, A. Previ, A. Taschini
ENEL - Direzione Studi e Ricerche

Summary

The chief aim of the Alicudi Project is to evaluate the technical and financial feasibility of supplying small isolated communities with electric power, using a photovoltaic solar energy system.
The experiment, conducted by ENEL, consists in the design, construction and operation, on a small island in the Aeolian Archipelago, of a plant for the production of electric power by means of a photovoltaic system.
The plant is one of a series of experimental plants whose construction is being promoted and partly financed by the Commission of the European Communities, R & D Department.

1. DESCRIPTION OF THE SYSTEM

The electrification of the island will be carried out by ENEL by means of a low-voltage system; two overhead lines supply a number of hamlets located at sea-level and up to an altitude of 300 ÷ 400 m a.s.l. situated on the south and south-eastern slopes of the island while the other slopes are completely uninhabited (Fig. 1 and 2).
The plant will consist of the following main sub-systems:
- A photovoltaic array with flat panels, subdivided into two 40-kWp sections (Fig. 3)
- A bank of electrochemical accumulators (3000 Ah, 220 V_{dc} rated)
- An inverter, to convert the direct current supplied by the bank of accumulators to three-phase alternating current (40 kVA, 380 V_{ac}, 50 Hz rated)
- A data acquisition system relating to plant operation.

The supply of users under emergency conditions is provided by a Diesel back-up group (60 kVA, 380 V_{ac} rated).

2. ENVIRONMENT AND STRUCTURAL PROBLEMS

The island is mountainous in appearance and cone-shaped, being, geologically speaking, a quiscent volcano (Fig. 4).
The vegetation of the island is typical of many areas of the Mediterranean coast with low rainfall: broom, prickly pear, cactus and wild olive.
Nearly the whole of the south side has a steep slope (between 30 and 50%) and consists of a series of terraces built by means of uncementated stone walls, in order to create patches of cultivable ground (Fig. 5).
The relative inaccessibility of the island and the absence of motorroads makes the plant assembly extremely awkward and costly.

The pv plant will be located on a small patch of ground with an average slope of 50% at an height of 150 m a.s.l. which is naturally suited to the tilt chosen for the photovoltaic modules (30°) thus making it possible to erect relatively low structures with a very compact array.

3. SYSTEM SIMULATION

The computer program PVSS, performed by ENEL, provides data characterizing the operation of a pv plant (power supplied from the array, battery voltage and state of charge, efficiency of the different sub-systems, etc.) on hourly basis. The picture (Fig. 6) shows the simulation of the operation of the Alicudi plant during July, using as input data the solar radiation of a typical year and a load estimated profile.

Fig. 1 - General view of Alicudi (South-East side)

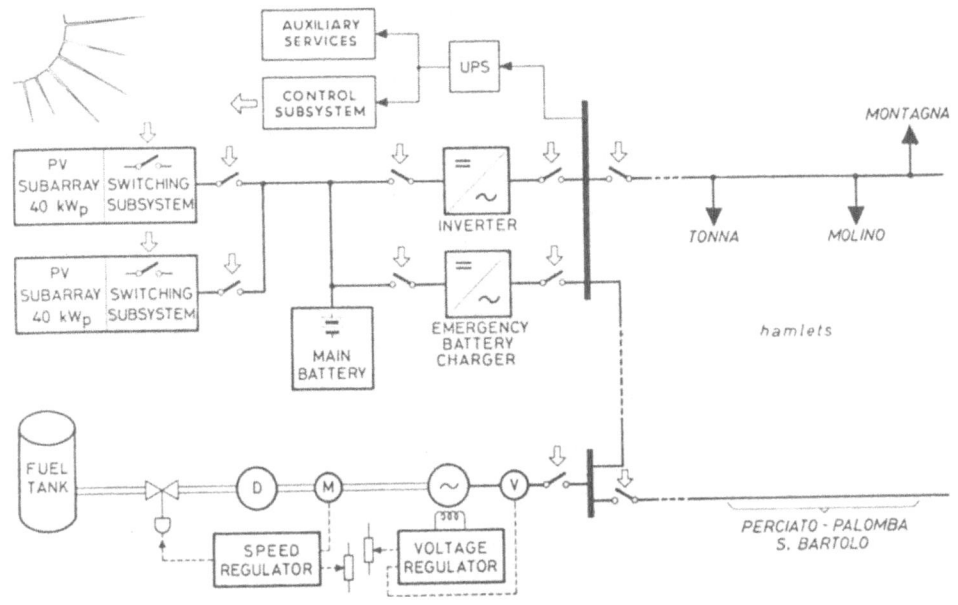

Fig. 2 - **Electric scheme of the distribution network**

Fig. 3 - **View of the pv plant** - The field is divided into two sub-arrays; a sub-array consists of 11 strings with 56 modules each. About 1500 m^2 of land are occupied by the whole array.

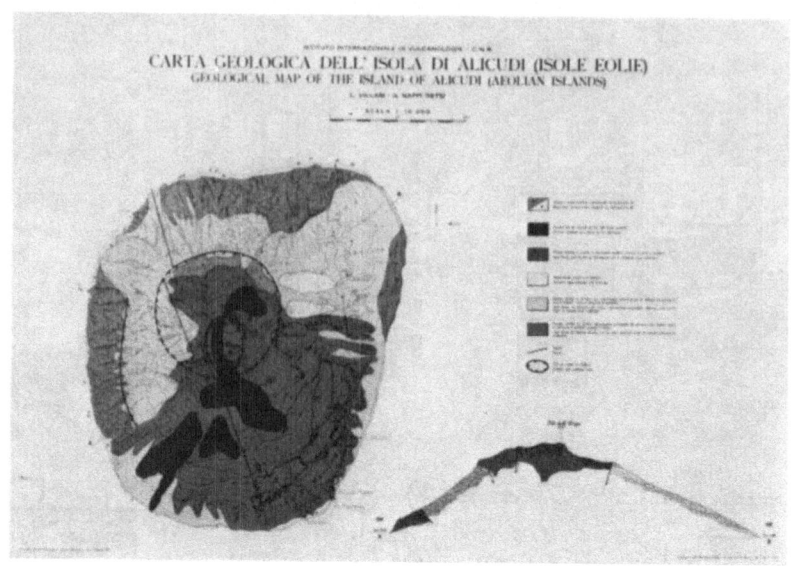

Fig. 4

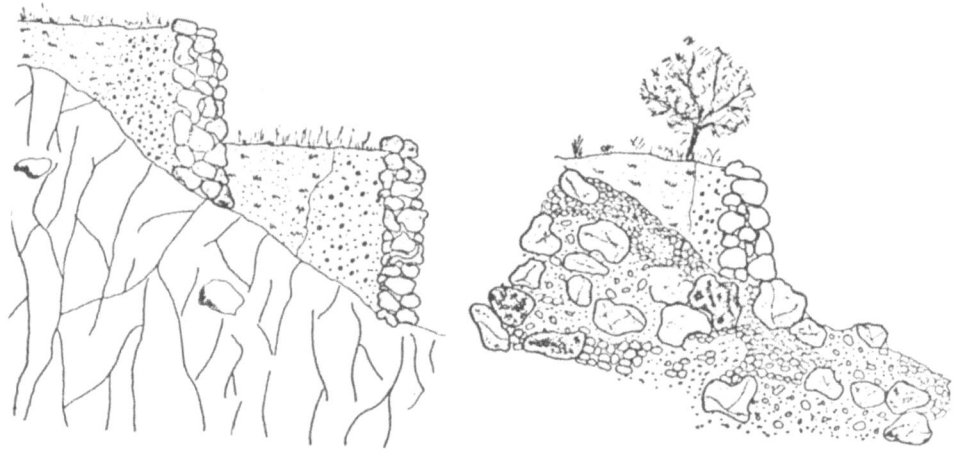

Fig. 5 - **Land profiles in the area interested by the pv plant.** On a lavic subsoil terraces with dry walls have been anciently realized for agricultural use.

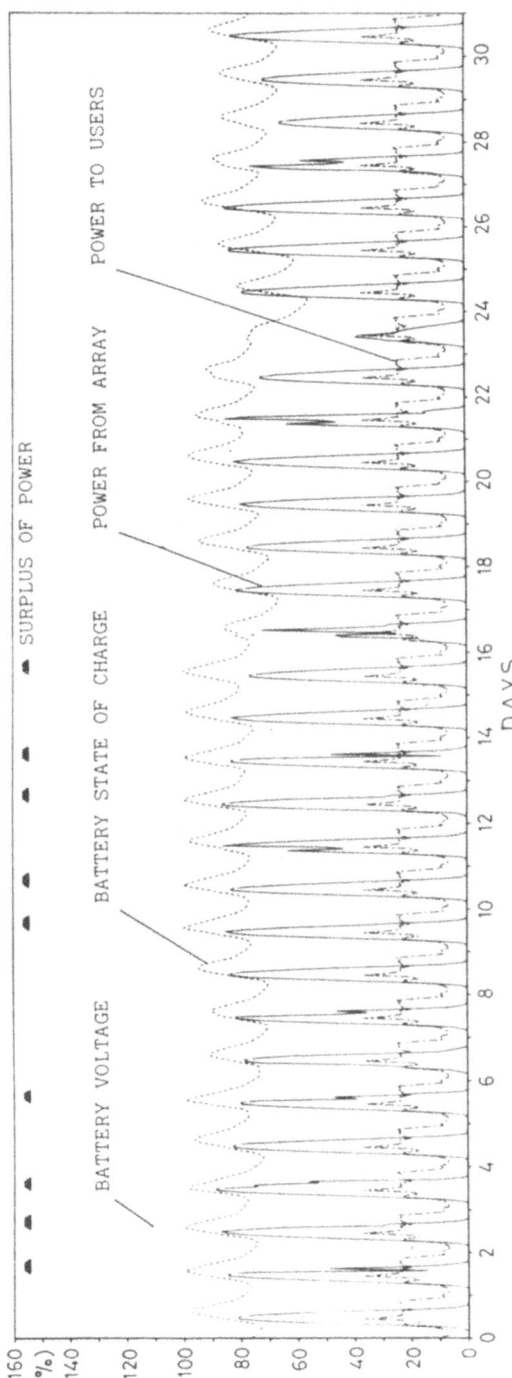

Fig. 6 - **Alicudi island** - Simulation of the operation of the plant during July (power, capacity and voltage are in p.u. of their rated values)

A SUCCESSFUL INTEGRATION OF SOLAR ENERGY IN AFRICAN LIFE : KOLOKANI, MALI

Lionel LEVHA
ELF SOLAIRE, Développements énergétiques

SUMMARY

Mali, in the trading heart of Western Africa, was at one time a sovereignty, and indeed the name 'Mali' comes from the Mandingues Dynasty, which followed the Soninkes Dynasty.

The country is traversed by two major methods of communication :
- The Niger River, flowing Koulikoro (south-west) to Gao;
- The Mabako-Mauritania axis, which has always been important for trade and cultural progress.

The Kolokani village was built on the Bamako-Mauritania axis, around a water well. ELF chose this village to supply its hospital with a solar generator.

The dispensary was built, but it could not ensure total medical care because of :
- No vaccines (no cooler to preserve them)
- No surgery (the surgery unit was not equipped for it).

The patients were transported to the hospital at Bamako, a trip of about 80 miles on mud strip, which helps to explain the very high death rate (70 %). The solar generator was a real necessity for Kolokani and its inhabitants.

From the beginning, the villagers were involved with the project, interested in the harnessing of solar rays to produce electricity, which until then, did not exist in Kolokani. Light was provided by petrol lamps and a very few people possessed generators. Many people doubted the 'toubab' project and its purpose (toubab in the Mali language means white man). It was difficult for them to believe that it was for their own use, and that we were not there only for our own interests. In fact we were there only to set up the installation and maintain it for an initial period. After a while, however, they realized that it was for their hospital and their attitude changed.

Within two monts, the hospital had a complete installation :
- Lighting
- Refrigeration
- Electric rectifiers (for medical purposes)
- A scialytic.

The system worked well, but it was strange for the villagers to see a fluorescent light outside the hospital, burning all through the nigt. It soon became a focal point for the whole village.

The energy management problems were predictable :
- At first, all the lights were left on all night, causing an overload and breakdown
- Then came a period when everyone was frightened to use the system, in case of another breakdown
- At last a rhythm of consumption developed, based on the centre's individual needs.

A surgeon was eventually assigned to the hospital, and now, after a year's use, the surgery copes with an operation every second day (excluding emergencies), and it will soon celebrate its 500th operation.

It is pleasing to see that the project has been accepted by the villagers of Kolokani, and also to know that they now intend to buy a solar generator for the maternity from their own resources.

Le Mali a connu des cycles d'expension et de retrait de peuples se taillant de puissants empires, avant de retourner aux frontières plus modestes du territoire ancestral. Aux SONINKES de GHANA ont succédé les MANDINGUES qui donnèrent au pays son nom : le MALI ... Mais, ce pays ne fut pas seulement une de ces places anonymes qui ne fournissent qu'un décor à l'histoire. Il a été longtemps au coeur de l'histoire africaine, source de vie, centre de civilisation rayonnante, et surtout un centre de commerce de l'Afrique de l'Ouest. Le coeur du vieux Mali a son artère : le fleuve Niger. Source de vie pour les uns, divinité pour les autres, de Koulikoro au Sud-Ouest jusqu'à Gao au Nord-Est, le fleuve a été de tout temps le moyen de communication d'une extrémité du territoire à une autre. Mais il y a d'autres voies de communication importantes et l'axe Bamako-Mauritanie a toujours été un chemin vital d'échanges tant commerciaux que des hommes et des idées. C'est sur cet axe, à peu près à 130 Kms du Nord de Bamako, que des paysans s'étaient rassemblés autour d'un point d'eau abondant et l'avaient aménagé pour faire un bon puits : Kolon Kagni (bon puits en Bambara). Ainsi était née Kolonkagni qui est devenue Kolokani au fil des époques.

Depuis toujours Kolokani a été une ville étape sur la route de la Mauritanie et du Sénégal. Elle a donc pris de l'importance et est devenue le chef-lieu du cercle de Kolokani (préfecture de département en France). A Kolokani la majorité des villageois sont des pays qui cultivent le mil et l'arachide, et c'est de façon très dure qu'ils ont ressentis les effets des dernières sécheresses. L'eau n'est plus si abondante que cela à Kolokani et même le puits qui a donné son nom au village aurait vu son niveau baissé.

C'est ce village qu'avait choisi ELF pour y installer un générateur solaire destiné à alimenter en électricité l'hôpital du cercle (assistance médicale), hôpital que l'on pourrait comparer à un grand dispensaire en France. Pour ce faire ELF avait engagé deux VSN (volontaires du service national) et c'est donc par un jour de Mars 1980 que je suis arrivé dans ce village où une maison m'attendait. Nous étions les seuls Toubabs (blancs en Bambara) à vivre en permanence à Kolokani et je dois dire que les premiers temps les relations furent très difficiles avec les gens du pays. Bien sûr il n'y avait aucune hostilité de leur part, les maliens sont des gens très souriants et très accueillants, mais plutôt une certaine méfiance envers nous. Aussi, en discutant beaucoup et longtemps avec eux, nous leur explicâmes le but de notre séjour à Kolokani : installer l'électricité dans l'hôpital et ce au moyen de l'énergie solaire. Mais ils restaient sceptiques car il faut préciser que l'électricité n'existe pas à Kolokani, seuls quelques privilégiés ont des groupes électrogènes et c'est à la lampe de pétrole que les villageois s'éclairent. Il était dur pour nous de leur expliquer que c'était un phénomène bien connu et que ni la magie ni la sorcellerie n'intervenaient. De plus les Bambaras, gens de la terre, ont le bon sens paysan et pour eux il faut voir pour croire que ce soit en matière de réalisations ou d'actes personnels. De plus, on comprend la méfiance des paysans maliens, dès qu'il s'agit de projets venant des toubabs. En effet, il ne se passe pas un mois sans qu'un expert de telle ou telle Organisation Internationale transite dans un village pour faire une préétude. Bien souvent ces projets restents lettres mortes sans qu'on se soucie d'en avertir les populations concernées.

Les quelques mois d'attente de la livraison du matériel nous ont permis de connaître ces gens, de voir leur misère, leur façon de vivre : c'est très dur d'être un enfant à Kolokani. Il faut d'abord savoir que 80 % des enfants qui naissent, meurent avant d'atteindre leur cinquième année et que pour les 20 % restants l'espérance de vie est de 38 ans. Nous nous sommes alors demandé si installer l'électricité dans l'hôpital de Kolokani était vraiment une priorité pour ces gens. L'argent nécessaire à une telle opération ne serait-il pas mieux utilisé à autre chose de plus concret pour eux. Nous étions donc saisis d'un doute concernant notre réelle utilité et nous nous demandions si nous n'allions pas installer un gadget de toubab de plus. L'avenir nous prouva le contraire.

Très vite, nous avons compris que pour qu'un minimum d'entretien, nécessaire au bon fonctionnement journalier de l'installation, puisse être effectué, il fallait dès le début associer le village à l'opération. Nous ne voulions pas que cette installation soit considérée comme une affaire de toubab. C'était leur hôpital et ce serait rapidement leur énergie et leur générateur. A ce stade, nous avons dû faire face à un problème de compréhension. En effet, il a été assez difficile de leur faire comprendre qu'ils étaient leur propre maître dans cette affaire, et que nous n'étions là que pour l'installation du matériel et pour veiller à son bon fonctionnement au démarrage. Le principal obstacle a été en fait un proverbe Bambara : " tout ce que tu fais tu le fais pour toi ". Si nous installions l'électricité dans leur hôpital c'est que nous y avions notre propre intérêt, donc ce que nous faisions était normal. Cette installation leur tombait du ciel, et bien ils l'utiliseraient comme telle. Les travaux commencèrent avec une participation locale. En un mois, nous avons passé plus d'un kilomètre de cables dans les murs de cet hôpital, nous avons posé une trentaine d'interrupteurs, une vingtaine de prises, quinze lampes, et installé le réfrigérateur solaire, premier du genre en Afrique. Au fur et à mesure de l'installation les villageois prennaient vraiment conscience que c'était leur hôpital qu'ils équipaient, sans jamais nous le montrer ouvertement. Ainsi en deux mois nous avions installé complètement le générateur solaire avec ses batteries et tous les bâtiments de l'hôpital disposaient d'une installation électrique digne de ce nom. Une chose restait à faire : relier le générateur aux huit bâtiments constituant l'hôpital. Nous avions décidé d'enterrer les cables de liaison par mesure de sécurité, et il fallait creuser un peu plus de 300 mètres de tranchée sur 50 centimètres de profondeur. Le travail se fit vite et bien grâce à l'efficacité des villageois. Ils avaient entendu notre message : c'était leur hôpital et nous pouvions être sûrs qu'ils feraient tout pour qu'il fonctionne bien et longtemps.

Enfin, l'installation était terminée et tout marchait à merveille. Ce fût une belle fête. Mais, pour les villageois cette lumière qui éclairait le devant de l'hôpital toute la nuit restait un mystère. La lumière sans bruit pour eux cela ne pouvait exister. Très vite, ce tube néon qui marchait toute la nuit devint un lieu de rencontre, et le devant le hôpital était le soir le lieu de rendez-vous de tout le milieu hospitalier de Kolokani et de tous leurs amis. Le plus difficile à leur faire comprendre fut la façon de gérer leur énergie. En effet, un générateur solaire photovoltaïque a ceci de particulier qu'il ne peut fournir qu'une certaine quantité d'énergie par jour et que donc on ne peut en consommer plus. Ayant expliqué ce phénomène au personnel de l'hôpital, leur premier reflexe a été de trop peu consommer. Mais là aussi, c'était gacher de l'énergie. Petit à petit, nous leur avons appris à atteindre l'optimum de consommation. L'installation avait été bien dimensionnée pour les besoins de l'hôpital et il fallait donc utiliser cette énergie au mieux. Ayant passé le cap de l'économie d'énergie il y eut une période de surconsommation et les batteries commençaient à se décharger. Il n'était pas rare de voir les lampes de l'Hôpital allumées toute la nuit, ou bien des salles fermées la nuit dont on avait oublié de commuter l'interrupteur. Là aussi, il fallut bien expliquer que sans se retenir de consommer de l'énergie, un minimum de discipline était nécessaire si l'on voulait que le générateur remplisse longtemps sa fonction. Et au bout de deux mois, on peut dire que la consommation énergétique de l'Hôpital était arrivée à un régime stabilisé optimal. L'électricité faisait partie intégrante de la vie de l'Hôpital.

Le bloc opératoire : un matin, nous vîmes arriver à Kolokani un nouveau médecin affecté à l'Hôpital, il était chirurgien et devait "ouvrir" le bloc opératoire. Celà faisait un an que les gens attendaient ce moment ; la moindre opération devait s'effectuer à Bamako, ce qui signifiait des frais de transport et d'hébergement élevés et surtout des risques énormes d'infection, les hôpitaux de Bamako étant trop surchargés pour pouvoir respecter correctement les règles d'asepsie et de propreté. De ce fait, une personne évacuée sur Bamako avait 40 % de chance de survivre à l'opération. On comprend alors l'utilité et l'espoir que représentait le bloc opératoire de Kolokani pour ces villageois. L'arrivée de ce chirurgien nous permit d'élargir les objectifs : nous avions dépassé le stade de la simple lumière dans l'Hôpital. En effet, nous nous sommes rapidement aperçus que le matériel médical n'était pas adapté à l'énergie solaire (faible tension, faible consommation) et il fallut donc se procurer un convertisseur 220 Volts et ensuite mettre au point, avec le chirurgien, un système de gestion d'énergie réalisé grâce à des horloges et des commutateurs. La vie énergétique de l'Hôpital fut entièrement régulée avec bien sûr une priorité au bloc opératoire. Nous prouvions ainsi qu'avec peu d'énergie, un minimum de discipline, et une connaissance des besoins médicaux, nous pouvions faire marcher un hôpital de brousse de façon aussi efficace que n'importe quel hôpital de Bamako.
Un seul appareil manquait : un scialytique. C'est une espèce de lampe très puissante, avec des réflecteurs qui concentrent la lumière sur la partie à opérer, son coût est de 8.000 FF représentant un trop gros investissement pour le village. On decida alors avec le chirurgien de le fabriquer nous-mêmes à l'aide des moyens locaux pour la somme de 1.000 FF. Le bloc était enfin complet : lumière, scialytique, ventilateur, matériel d'opération, chirurgien et anesthésiste. Dès la première opération, ce fut un succès complet. Il y avait beaucoup de gens qui attendaient à la sortie du bloc : la famille et les amis du malade.

Les quatre derniers mois que nous avons passé là-bas furent exaltants. Le bloc fonctionnait parfaitement avec deux opérations tous les deux jours, plus les urgences, il en était de même pour le réfrigérateur où de plus en plus de médicaments s'entassaient. Le personnel de l'Hôpital avait entièrement pris en charge l'entretien et le fonctionnement de l'installation électrique et du générateur solaire.
Parallèlement à celà, nous montions un nouveau type de générateur solaire à concentration, dans l'Hôpital, le SOPHOCLE, et cette fois, sans même la demander, je reçus beaucoup d'aide de la part des villageois. L'énergie solaire faisait désormais partie de leur vie et de nombreuses personnes venaient demander une explication sur le fonctionnement du générateur. C'était le sujet de conversation au café bar de Kolokani ... On peut dire que Kolokani a été conquis par l'énergie solaire, à un tel point que le village a commandé, pour la maternité, un petit générateur destiné à l'éclairage que les villageois paieront

eux-mêmes. Ils ont aussi demandé et obtenu de l'UNICEF un réfrigérateur solaire et deux pompes.

Investir une somme importante pour les villageois était un grand pari qu'ils ont relevé. Je crois que c'est la plus belle récompense que nous pouvions recevoir de leur part.

APPLICATION AND EXPERIENCE OF PHOTOVOLTAIC PUMPS FOR IRRIGATION IN PAKISTAN

R.G. PALLETT and T.E. BRABBEN
Overseas Development Unit, Hydraulics Research Station Limited, Wallingford, Oxon, U.K.

Summary

In recent years considerable interest has been expressed in the use of solar power to pump water for either irrigation or water supply, in remote regions of developing countries where grid electricity does not exist and diesel oil is difficult and expensive to obtain. There is a need for a suitably sized small scale pumping system capable of supplying the water requirements of small-holder farmers, farming less than 3 hectares.

Intermediate Technology Industrial Services (ITIS) of the UK purchased 20 solar powered micro-irrigation units (Sun pumps), of which 18 were supplied by Solar Electrical International (SEI) and 2 units, incorporating batteries, were supplied by Lucas Energy Systems of the UK.

Sun pumps were installed at research farms and private farms in Pakistan for a one year trial period with the objective of demonstrating, testing and evaluating the suitability of this technology to the needs of irrigated small holdings.

The trials have proven this technology to be acceptable to the farmers who readily appreciated the benefits of using sun pumps; the absence of running costs, the ease of use and the independence gained by controlling their water supply. It was also very evident that the effectiveness of this low flow irrigation system is more dependent, than conventional irrigation water supplies, on the plot layout, channel conveyance system and the farmer's water management technique.

INTRODUCTION

Pakistan is a most suitable country in which to carry out a project to field test sun pumps for a variety of reasons:
(a) an average of 8 sun hours per day, throughout the year,
(b) a long history of irrigation,
(c) over 8 million hectares of land has ground water, suitable for irrigation, within 6m of the surface,
(d) over 44% of farmers in Pakistan own less than 3 ha.

In Pakistan there are a number of sources of water for irrigation, namely:

(a) canal water supplied by either gravity or lift,
(b) pumped water supplied from diesel or electric powered tubewells with farmers owning their own tubewells or purchasing water from other tubewell owners,
(c) water from shallow open wells, lifted by an animal driven Persian wheel.

It is possible to use sun pumps either to replace the farmer's present method of irrigation or to supplement it, thereby allowing the farmer to change his cropping pattern to include cash crops requiring a perennial water supply or to allow irrigation of a larger portion of his land throughout the year.

Two organisations in Pakistan were involved in the trials, the Pakistan Agricultural Research Council (PARC) and the Agricultural Development Bank of Pakistan (ADBP). Four units including the two units incorporating batteries were installed at PARC research stations and the remaining units were installed at a number of farms throughout the country.

Of the two types of sun pump used in the trials data was collected from only the SEI units as neither of the Lucas units could be made to operate for an extended period of time. The SEI system as shown in Fig.1 produces 250 watts under insolation of 1000 mW/cm^2 at 30°C and is designed to pump 2.5 litres/sec at a head of 5 metres. The panels are mounted on two trolleys which can be easily manoeuvred to track the sun throughout the day. The system uses a submersible pump/motor set which is placed in a well as illustrated, and is suspended just below the surface of the water by a flotation buoy. A master power point tracker (MPPT) is incorporated, which matches the load of the pumpset to the varying power available from the solar array, thereby ensuring optimal use of the power available. The SEI design was well received by the farmers who in general encountered little difficulty in using this system.

FIELD TRIALS

During the initial trial period, units were allocated to 14 farmers in different regions throughout the country. This procedure was adopted for two reasons, firstly to promote and introduce solar pumping to a large number of farmers, and secondly to identify the regions where the introduction of solar pumping might be most beneficial. Following this initial period three regions, Peshawar, Muzaffargarh and Sukkur, were chosen as the most suitable because:

(a) The three regions illustrate the various cropping patterns practised in Pakistan.
(b) There were a large number of open surface wells in existence, permitting installation of the solar pumps without incurring additional costs of well construction.
(c) Groundwater was available within 3m of the surface throughout the year.
(d) A large percentage of the farmers used the Persian wheel to irrigate small, (less than 1.5 ha), plots.

At each of the farm sites, weekly data regarding hours of pump use, area of crops irrigated per day, and depth to the water table before and after pump use was collected by ADBP personnel. This information gives an idea of the effectiveness and capacity of the solar pumps as used by farmers throughout the year.

In Pakistan two distinct seasons exist, the Rabi (winter) from October to March and the Kharif (monsoon). The majority of the farmers carry out double cropping usually growing wheat in the winter and either rice, maize or cotton in the monsoon season. The farmers were asked to set aside whatever area of their land they thought the sun pump could command, and irrigate it exclusively with the solar pump throughout the year. The majority of the farmers selected carried out a subsistence type of cropping, growing sufficient quantities of staple foodstuffs such as wheat, rice and vegetables for their own consumption, whilst growing a small amount of a cash crop such as sugar cane to generate some income. This type of cropping pattern is most suited to the sun pump as the crop water requirement throughout the year is less variable when small areas of different crops are grown.

A sun pump was installed at each of two PARC research stations where daily records were made of insolation, sun hours, volume of water pumped, change in water level within the well and air temperature. This data defined the capacity of the sun pump, in terms of the volume of water pumped for the solar conditions experienced in Pakistan. This enabled an estimate of the volume of water pumped by the various farmers for irrigation to be made.

SYSTEM PERFORMANCE

This project was the first of its kind in field testing a large number of sun pumps, and the trials have illustrated several problems that exist with the equipment as well as with the irrigation practices of the small farmers using solar pumps.

In all, 9 out of the 18 SEI units became defective at some time during the one year of trials. The most common failure was found to be with the pump/motor unit. It appears that the pump was allowed to run whilst not immersed, causing the motor to overheat and burn out. Another problem was that the electrical leads and connectors between the two arrays became loose and required replacement. Two MPPT's were found to be defective whilst the units were being installed and had to be replaced, but apart from those two no further problems occurred with the MPPT's.

There does not appear to be any correlation between hours of pump use and pump breakdown. In fact, one pump that was used almost continuously by a farmer for over 3000 hours worked trouble-free throughout. It is thought likely that some degree of misuse on the farmers' part has caused the majority of system failures.

Fig 2 shows the maximum daily volume of water delivered from a depth of 3 metres by the SEI unit, in May and December. It is apparent that during May the pump is able to deliver more water than in December due to higher intensity radiation and longer daylight hours. However, evaporation rates also rise sharply during this period and Fig. 3, drawn using data collected from one farm, shows how the crop water requirement for a typical farm and the available water from the solar pump compare throughout the year.

During the winter period (October-March) the sun pump was able to irrigate 2 ha of mixed crops (wheat, fodder, sugar cane and vegetables), whilst in the monsoon season the pump was only able to irrigate 1.2 ha (rice, fodder, sugar cane). From Fig.3, it is evident that insufficient water was available for optimum crop growth of this smaller cropped area during the monsoon season. This problem is not so acute in Pakistan, as all areas of land with high groundwater levels receive limited amounts of irrigation water from Pakistan's extensive canal network during the monsoon season which could be used in conjunction with water from the sun pump to irrigate the whole 2 ha.

By observing the irrigation practices at many of the sites it is evident that farm efficiencies of water use, defined as the farm water requirement divided by the volume of water supplied to the farm, can be very low perhaps well below 50%. In several instances water was conveyed to various plots via channels designed to carry 28 litres/sec compared to approximately 3 litres/sec from the solar pump, resulting in large seepage losses as the channels are all unlined. Further, it is also apparent that the same channel had to be continually refilled each morning as water had seeped away during the night. An obvious solution would be to line a main feeder canal and alter the plot layout of the farm to minimise the length of channel required. However canal lining and altering the plot layout represents additional costs and loss of revenue during reconstruction for the farmer, which might prove unacceptable.

CONCLUSIONS

It is evident from these trials that a large potential exists within Pakistan for solar powered irrigation to supplement or replace existing sources of irrigation water provided that:
  (a) Sun pumps are available at a price level that farmers can afford. With the present size of system it is estimated that a six fold price reduction to say $1000 US might be required before farmers would purchase this size of system.
  (b) A range of different sizes of sun pump are available to suit different farm sizes and cropping patterns.
  (c) The reliability of the sun pumps could be improved. It is noted that the manufacturers are now producing motor/pumpsets including thermal switches which will eliminate what is thought to be the major cause of system failure.
  (d) Adequate arrangements can be made for the prompt replacement of any system components should they become faulty, so that crop failure would not occur due to lack of water over an extended period of time.

In general the large variation in crop water requirement and the low farm water use efficiencies noted, necessitate considerable oversizing of the sun pump to meet the peak crop water requirement. This required overcapacity obviously results in a higher cost of system, which could make sun pumps uneconomic for irrigation until a substantial price reduction occurs.

ACKNOWLEDGEMENTS

The help of the staff of ADBP and PARC is gratefully acknowledged. This work was carried out in the Hydraulics Research Station's Overseas Development Unit in co-operation with ITIS with funds provided by the Overseas Development Administration of the British Government. This paper is published with the permission of the Managing Director, Hydraulics Research Station Limited.

Fig.1 SEI 250 Sun Pump

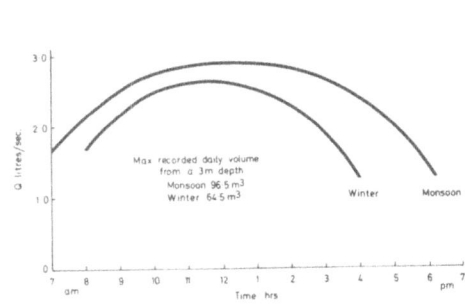

Fig.2 Flow Rate/time
for the SEI 250 Sunpump

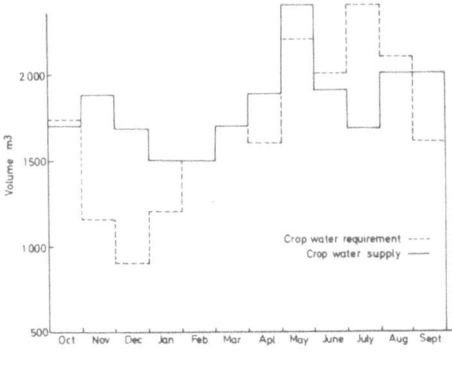

Fig.3 crop water requirement
and water supply

DEVELOPMENT AND TESTING OF A SUBMERSIBLE MOTOR PUMP DRIVEN BY A SOLAR CELL GENERATOR

Dr. Ernst PICMAUS
Mr. Karl P. KIESSLING
Dept. SVU

KSB - Klein, Schanzlin & Becker AG
Postfach 2 25
6710 Frankenthal

In applications of submersible motor pumps the total system-efficiency is most important. The pump- and motor-efficiency of a submersible motor-pump of series-standard could be improved for each 10 points, using an asysnchronous motor similar to series. The investigation of synchronous motors and brushless DC-motors proved no economical result.

To convert the DC-current delivered by solar cells into AC-current of variable frequency and voltage, a pulse-converter with MOS-power-transistors and a control-device considering the special demand of operation with solar cells was taken. It consists of on-off automatics and a maximum power point control-device. For testing the single components and the total system, a testing supply has been installed. As a result a total efficiency of 43 % at nominal load and of 33 % at 1/4 load was determined.

1. Economic possibilities for the application of solar cell generators for the supply of motor-pumps.

From a technical and economical point of view the development of solar cell generators has reached a level which enables their use as a source of energy for water pumps in areas highly exposed to sun light and where electricity is not generally available. Two fields of application are of special significance here:
- surface water
In areas where water is available only a few meters below ground, we require motor-pumps with high flow rates and small heads - somewhere in the range between 200 and 500 watts. The outer diameter of the units isn't very important here, but the floodability would be an advantage for operational reliability.
- deep wells
In areas where water is available in lower-lying strata, wells of which the cost mainly depends on their diameter, have to be drilled. The drilling depth lies between 25 m and 150 m in general. Here we have wells of 4" (100 mm $\emptyset$) or at the most 6" (150 mm $\emptyset$). For this we need special submersible motor pumps the power range of which, when supplied by solar cell generators of between 1 and 3 kW, is reasonable from an economic viewpoint.

2. Determination of production requirements for the present research project.

The present research project deals with the development and tryout of a submersible motor pump supplied by a solar cell generator. So, this project is based on a well with given diameter.

- environmental conditions

   The conditions prevailing at the place of installation make special demands on the unit to be developed. To be considered are the hot and often very humid climate, sites far away from traffic routes, the lack of skilled operating and maintenance personnel as well as the often grave consequences of a breakdown of the water supply.

- total efficiency

   What distinguishes these units from standard series units as well is the completely different evaluation of the total efficiency. As the solar screens are still very costly and the space requirement considerable, it is possible and necessary to spend more on motor, pump and electronic devices than in the case of series units, to achieve the best possible efficiency of the individual components.

- Function

   The function of the unit depends on the amount of energy made available by the sun. The best and cheapest method to store energy for water supply plants is to collect the delivered medium in a tank. The storage of energy by means of accumulators must normally be excluded, as these reduce the total efficiency, require regular maintenance and do not have a long lifetime.

   If the submersible motor pump is fed by the solar cell generator directly, special attention should be paid to the starting conditions at the onset of the incoming sunlight and the amount of energy changing with the weather conditions. The curve of total efficiency must be designed in such a way that we obtain an optimum effect under the given circumstances, i.e. that as much water as possible is pumped during a day.

- special features

   The motor shall not have sealing problems even if water is pumped from great depths. It should be possible to install and dismantle the unit quickly and easily.

   Parts subject to wear, such as brushes, collectors, waterair seals etc. as well as drive rods and the pertaining bearings must be avoided. Motors and pumps should be easily repairable which excludes complicated constructions, such as canned motors of casting resin.

3. Investigations and developments of submersible motor pumps

An available series of submersible motor pumps for 4 inch and 6 inch wells has been taken as a basis for our investigations.
For the application of photovoltaic methods a high efficiency is of utmost importance.
It is, therefore, our aim to improve the units so much that a best possible efficiency can be reached.

- Steps to improve the pump efficiency.

   In the paragraphs hereafter we describe the possibilites to improve the pump efficiency and examine their applicability for submersible motor pumps. The result is shown in Fig. 1.

Choice of pump construction.
The best efficiency is reached with volute casing pumps. But these have a relatively large diameter and are, therefore, not so well suited for installation in a narrow well.
Submersible pumps DN 100 and DN 150 are stage casing pumps for use in vertical wells.
Selection of specific speed nq.

$$nq = 16{,}65 \cdot n/100 \cdot Q^{1/2} \cdot H^{-3/4}$$

n (1/min)
Q (m³/h)
H (m)

For the given values of flow rate and head at the selected speed range, the best efficiencies are reached where nq is between 25 and 80. Pumps selected for photovoltaic plants have specific speeds of nq between 27 and 75. The exactly determined impellers and diffusers have been optimized in a series of tests.
Reduction of inlet and outlet losses.
In the pump inlet and at the point where the flow medium leaves the pump and enters the piping there are losses which, although relatively small, are still noticed here where the ratings employed are small. By carefully guiding the flow these losses may be reduced further.
Smooth surfaces.
Smooth surfaces reduce the losses at the circumcirculated surfaces. The best surfaces are obtained if impellers and diffusors are made of synthetic material.
Narrow sealing gap.
As, compared to standard fields of application, the number of stages of the pumps used here is small, the sealing gaps may be narrower than in the case of series pumps. But they must not fall below 0.1 mm, as the operational reliability would then no longer be guaranteed.
- Steps to improve the motor efficiency.
Also the motor efficiency may be improved - at relatively low expense - by carefully selecting the construction type of the motor as well as other suitable measures. The result is shown in Fig. 2.
Selection of the construction type of the motor.
We examined three types of motors.
The brushless D.C.-motor with permanent-magnetic rotor; the synchronous motor with permanent-magnetic rotor; the asynchronous motor with squirrel-cage rotor. Comparing the systems we found that the abt. 4 points higher efficiency of the two permanent exciter motors involves an additional price which medium-term can no longer be compensated by the resulting saving of solar cells.
Reduction of the friction losses.
The friction losses of the rotor in the motor fill can be kept at a minimum by choosing a streamline construction with smallest possible rotor diameter. By adjusting the other parts, such as shaft, radial bearing, thrust bearing and sealing elements, to the required power range, the friction losses of the water-filled motor can be reduced considerably.
Reduction of the copper losses.
Frequency start-up by means of a converter allowed the use of a squirrel cage in the rotor with larger copper section, which

reduced the rotor copper losses.
Reduction of the iron losses.
The use of higher-quality sheet metal for the laminations in combination with a not too high magnetic utilization resulted in a reduction of the iron losses.
Results of developmental research work.
By means of the steps described the efficiency referred to pump stage and the motor efficiency could be improved by 10 points each compared to the series pumps.

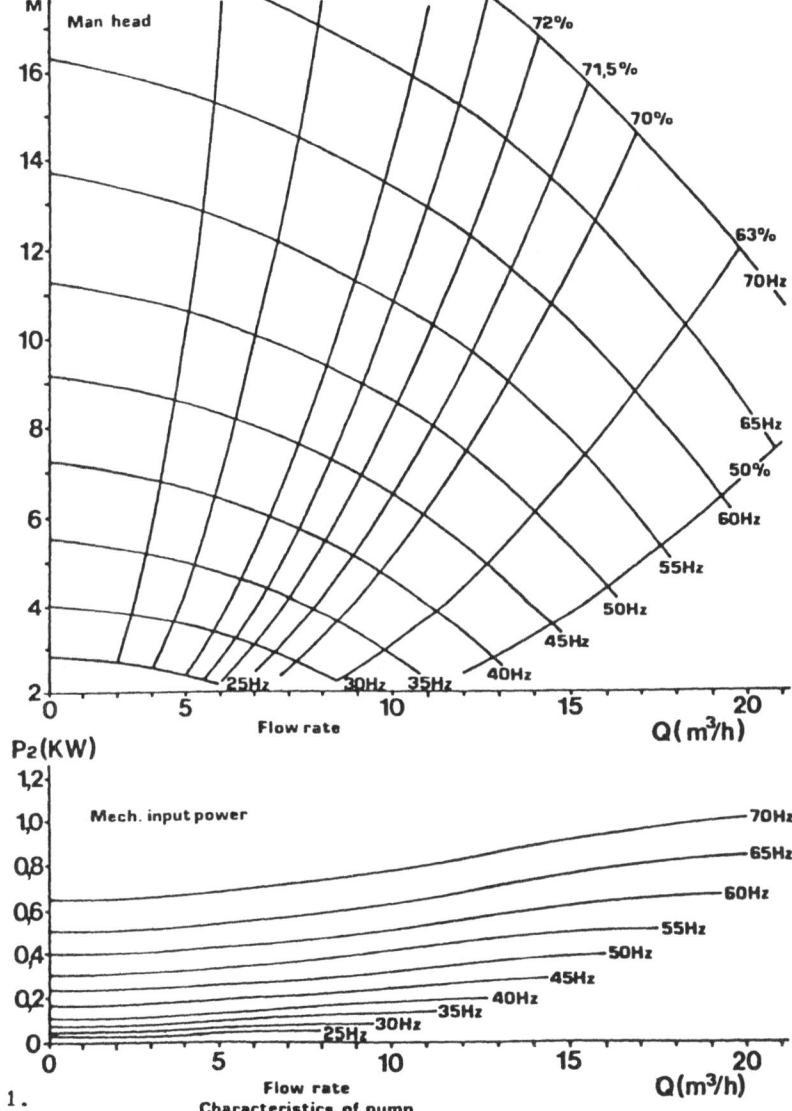

Fig. 1. Characteristics of pump
Manometric head and mechanical input power as function of flow rate
Frequency range 25 to 70 Hz.

K.S.B. A.G. Frankenthal.

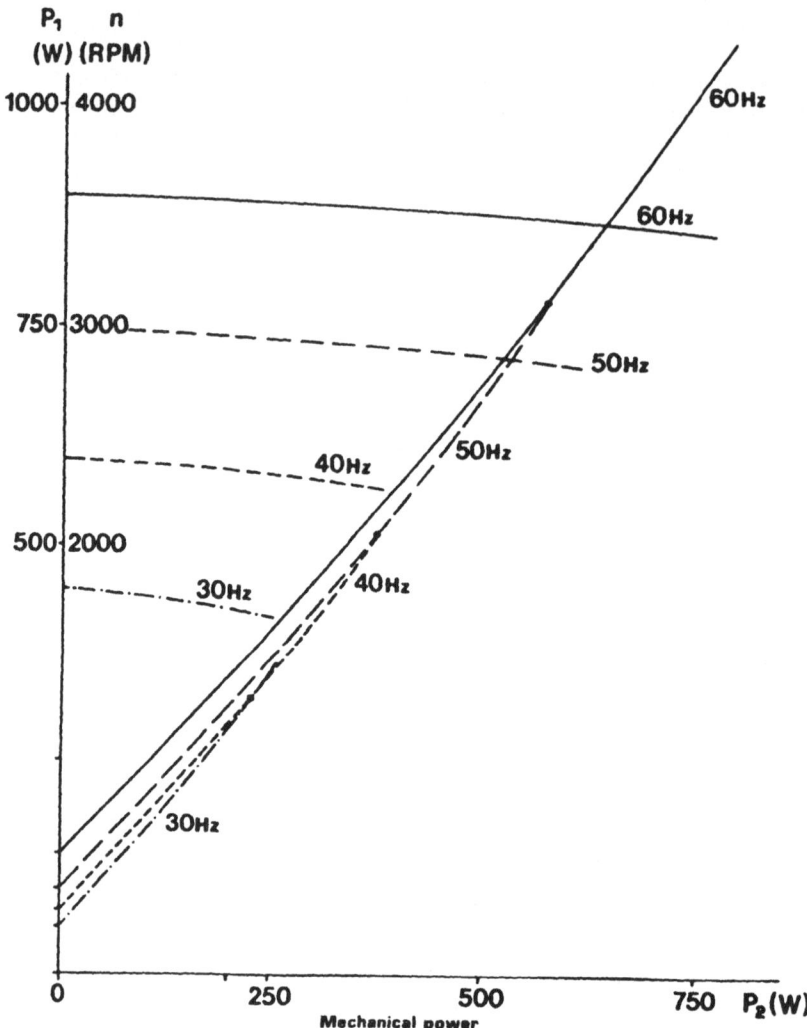

AC input power and rotational speed as function of mechanical power

K.S.B. AG. Frankenthal

Fig. 2
Characteristic curves of motor UMS 1508 for pump type
Corasol   10-9/1

SOLAR GENERATOR PERFORMANCE WITH LOAD MATCHING TO WATER ELECTROLYSIS
LONGTERM AVERAGES AND RANGE OF INSTANTANEOUS EFFICIENCIES

K.Freudenberg
Sektion Physik der Universität München
Lehrstuhl Prof.R.Sizmann

Summary

The efficiency of producing hydrogen by solar energy conversion via solar cells and water electrolysis is determined by the performance of the solar cells, the efficiency of the electrolyser and the quality of load matching. The strong variations of solar generator characteristics in the varying conditions of Central European climate necessitate longterm monitoring of system performance.
A solar generator is directly wired to an electronic simulation of an advanced water electrolyser. System performance is monitored from January 15 to July 7, 1980. Instantaneous generator efficiencies are capricious, but average generator efficiencies at specified solar irradiance and ambient temperature are well behaved. Load matching losses amount to only 8.9 % of the solar generators MPP output. Hydrogen production is calculated from the current through the load assuming 100 % current efficiency of the water electrolysis. Starting from 10.0 % generator efficiency an overall **energetic** efficiency of 8.0 % is achieved for hydrogen production. The exergetic efficiency is 9.4 %.

1. INTRODUCTION

The efficiency of solar energy systems must be analysed on the basis of longterm performance measurements in a natural environment. Overall efficiency is not just the product of component efficiencies, which are usually stated for optimal operating conditions. System configuration determines the operating conditions of the system components.They also depend strongly on environmental parameters. The efficiency of hydrogen production via photovoltaic solar energy conversion and water electrolysis can be analysed in terms of the MPP efficiency of the solar generator, the quality of load matching and the electrochemical efficiency of the water electrolysis.
Solar generator efficiency is determined by magnitude and spectral composition of irradiance and by the temperature of the solar cells. Performance predictions can only be made, if generator efficiency is related to meteorological data, such as global irradiance and ambient temperature. In a direct wiring configuration load matching losses are inevitable, since the water elctrolysis is not always drawing the generators current at the voltage appropriate for maximum power output. Load matching losses can be minimized by suitable dimensioning of the systems voltages. This study was conducted to gain insight into the electrical efficiency of hydrogen production via solar cells and water electrolysis in Central European climate.

2. SYSTEM CONFIGURATION

A solar generator based on a series x parallel (10 x 3) connection of monocristalline silicon solar cells is directly wired to an electronic simulation of the solid polymer electrolyte (SPE) water electrolyser developed by General Electric. The cell potential of the SPE electrolysis rises from 1.51 V to 1.83 V at current densities of 0.1 A/cm^2 and 1 A/cm^2 respectively, operating at 355 K. The current density in the electrolyser is essentially proportional to the solar irradiance received by the solar generator in a direct wiring configuration. Assuming a current density of 1 A/cm^2 in the electrolyser at 1000 W/m^2 solar irradiance the voltage of the SPE electrolyser is scaled by a factor of 1.85 to match the solar generators MPP voltage of 3.4 V at ambient temperature of 301 K. This is legitimate, since series connections of many electrochemical cells will be driven by high voltage solar generators in large scale solar hydrogen production. The voltage of the electronic load deviates less than 3 % from the current-voltage-characteristic of the SPE electrolyser within the current range determined by the solar generators output at irradiances between 80 and 1100 W/m^2.

The solar generator faces south with a tilt of 45 degrees. Solar irradiance is measured by an Eppley precision pyranometer in the plane of the generator. System performance is monitored on 143 days in the period from January 15 to July 7, 1980. Measurements included current and voltage at the load and at the point of maximum power output, solar irradiance in the plane of the generator and ambient temperature. The MPP measurements are performed by straightforward application of standard analog circuitry. Data sets are recorded once a minute, with solar irradiance and real time taken simultaneously with every data. The experimental setup and the data base are described in (1) in full detail.

3. GENERATOR PERFORMANCE

The average efficiency of the solar generator, as related to the area of the solar cells, is 10.0 %. The generator works at its average efficiency at 800 W/m^2 and 293 K ambient temperature, which has been defined as the Nominal Terrestrial Environment for photovoltaic power conversion in the United States. Instantaneous generator efficiencies ranged from 7.5 % to 11.5 % at ambient temperatures between 271 K and 307 K and solar irradiance above 50 W/m^2. The average generator efficiency at specified solar irradiance G (in W/m^2) and ambient temperature T_a (in K) is given to a precision of 0.2 % absolute by the relation:

$$\overline{\eta}_{MPP}(\%) = -7.5 + 3.3 \ln G - 0.0061 \, G - 0.062(T_a - 298) \quad (1)$$

Average generator efficiency is thus closely related to meteorological data.

Instantaneous generator efficiency is capricious and cannot be closely related to the magnitude of solar irradiance and ambient temperature. Differences in spectral composition of solar irradiance, angle of incidence wind velocity and the thermal capacity of the solar generator cause a large spread of instantaneous generator efficiencies at identical values of solar irradiance and ambient temperature. Figure 1 shows the frequency of instantaneous generator efficiencies recorded at ambient temperatures between 290 K and 292 K. Solar irradiance is subdivided into intervals of 50 W/m^2. The three pyranometer readings taken 5 seconds before, simultaneous with and 5 seconds after the MPP measurements shown in the figure agree within ten percent, excluding erraneous results caused by the thermal capacity of the pyranometer.

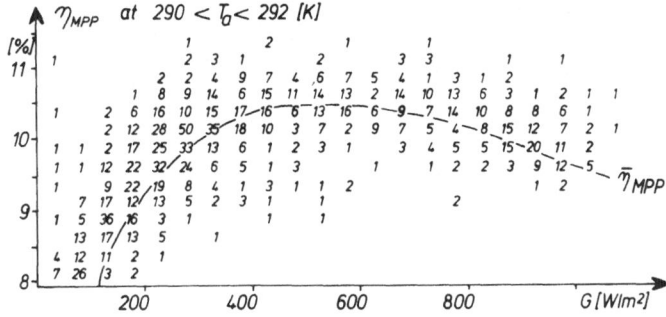

Figure 1: Frequency of instantaneous generator efficiencies

4. LOAD MATCHING

Load matching is greatly influenced by ambient temperature, since the generators voltage drops linearily with rising solar cell temperature. The dependence of the quality of load matching on ambient temperature is shown in figure 2 for solar irradiances of 300 W/m^2 and 900 W/m^2. The curves labeled A,B and C correspond to different configurations of the solar hydrogen system.

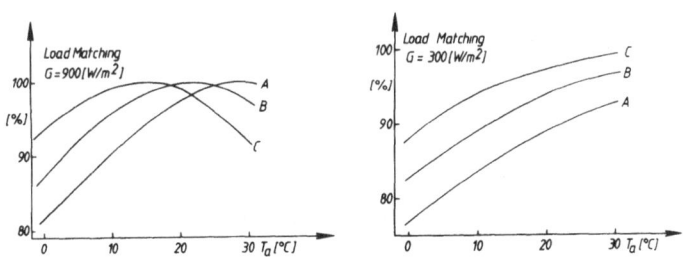

Figure 2: Quality of load matching

The ratio of solar cells to electrochemical cells determines the quality of load matching in a direct wiring series configuration. Configuration A, which achieves ideal load matching at high ambient temperature with high solar irradiance is realized in this study. Load matching losses exceed 20 % at ambient temperatures below 273 K, especially at low solar irradiance. Reducing the number of electrochemical cells by 10 % or 20 % respectively (configurations B and C) improves load matching at low ambient temperature and low solar irradiance at the cost of the high temperature-high irradiance range. Averaged over the half-year period the losses by non-ideal load matching amounted to only 8.9 % of the solar generators MPP energy. Direct wiring of water electrolysers to solar generators is thus very attractive, provided the Faraday efficiency of the water electrolysis does not degrade severely at low current densities. (2) and references therein may be consulted for further details.

5. EFFICIENCY OF SOLAR HYDROGEN PRODUCTION

The reversible cell voltage for water splitting is 1.23 V at standard conditions. An electrolyser working at this voltage would consume heat at ambient temperature, because the entropy of the products is higher than the entropy of the reactand. The thermoneutral voltage, corresponding to the enthalpy change ΔH^{298}= 285 kJ/mol of water splitting is 1.48 Volts. The ratio of the two voltages gives an exergetic content of 83.1 % for hydrogen energy. The SPE electrolysis has an almost ideal current efficiency at current densities around 1 A/cm^2. A discussion of the current efficiency at low current densities, say 0.1 A/cm^2, is not available in the literature. The quantity of hydrogen is therefor calculated from the current through the electronic load on the basis of Faradays Law.

Starting from 10.0 % solar generator efficiency, loosing 8.9 % of the electrical energy by non-ideal load matching with the water electrolysis, the energetic efficiency of solar hydrogen production is 8.0 %. Considering the fact, that solar irradiation has an exergy content of approximately 70 % (3), the exergetic efficiency is a respectable 9.4 %.

Acknowledgements

I would like to take this opportunity to thank Hubert Böck, Dieter Jung, Harald Ries, Wolfgang Schölkopf and Prof.Dr.Rudolf Sizmann for their contributions to this work.

References
1. K.Freudenberg: Diplomarbeit an der Sektion Physik der Universität München, Dezember 1980
2. K.Freudenberg: Applied Physics A, to be published
3. H.Ries: Journal of the Optical Society of America, Vol.72, No.3, March 1982

PHOTOVOLTAIC POWER FOR WALK-IN COOLERS

L. SELLES and B. AUBERT
SERI Renault Ingénierie

Summary

From a system analysis, the association of photovoltaic cells with motor compressors used in small industrial cold stores has been investigated by Séri Renault Ingénierie with financial support from the French Commissariat à l'Energie Solaire (COMES). With a view to assessing the market potential of photovoltaics for food conservation in hot climates and harsh environments stand-alone walk-in coolers have been studied and sized for:
- internal volumes ranging from 5 to 30 m^3
- temperature levels between 0°C and 7°C.
The alternative DC versus AC supply for the motor compressor has been analyzed with respect to specifications of the location site envisaged - very remote without technical staff available. To reduce routine maintenance a fully hermetic AC motor-compressor has been chosen in preference to a DC "open" compressor. Consequently a high efficiency (>90%) PWM 3 - phase 3 kVA inverter has been developed. A 10 m^3 capacity prototype powered by a 2.2 kW peak photovoltaic panel has been designed and is now being tested at Séri Renault's headquarters - Bois d'Arcy (France). By Fall' 82, prototype will be sent to an agricultural centre in Ouagadougou (Upper Volta) for field testing.

1. INTRODUCTION

The aim of this COMES R & D contract was the industrial analysis of "What is possible with photovoltaic refrigerating stores in the short term". Purpose-built for hot countries of various climates (equatorial, tropical, sahelian ...) these cold stores, in sizes ranging from a few cubic meters to 20 or 30 cubic meters, show good prospects in the near future to be commonly used for small-scale retailers scattered over an area situated at some distance from an utility network, they comply with the diffused nature of solar energy. Located in small agricultural villages, they could serve as links between villages' customers and important food storages in large conglomerations. The study concerns the following points:
- analysis of the cold balance sheets,
- analysis of the alternative DC versus AC supply,
- general design and dimensioning,
- development of a 10 m^3 prototype,
- test of the prototype in France,
- field testing at an agricultural centre in Ouagadougou (Upper Volta).
Today the system is being tested at Bois d'Arcy (Paris suburb) before it is shipped by Fall to Upper Volta.

2. METHOD OF APPROACH

System analysis has been performed following a multi-criteria approach taking into account quantitative and qualitative factors as well as according to identification of market.

<u>Market identification</u>: The selected temperature range is the "refrigeration band" between 0°C and +7°C. This band makes it possible to preserve milk and fish for a few days, meat for two or three weeks, fruit, vegetables and butter for more than a month.

<u>Cold balance sheets</u>: For each typical size selected, i.e. 2, 5, 10, 20 and 30 m^3, an analysis of the energy balance has been taken over the whole range of thermal levels and system designs. Taking into account the climatic conditions in two typical sites - Abidjan (tropical) and Niamey (sahelian) - the analysis shows the significant efficiency gain that can be achieved by using a "solar" design for cold stores. See tables I and II for both climatic types and both insulation coefficients ("standard" k = .40 W/m^2/°C and "solar" k = .14 W/m^2/°C). Table III points out the increase of the "cold used/cold produced" ratio with the size of the stores. It can be noticed that in the selected prototype (10 m^3) about 40 or 42% of the cold produced is actually used to cool the food stored while smaller sized coolers produce proportionately more expensive useable cold.

<u>DC vs AC supply</u>: The alternative DC versus AC supply for the motor-compressor has been analyzed with respect to specifications of envisaged sites (very remote and without technical staff available). A fully hermetic AC motor-compressor has been chosen in preference to a DC "open" compressor (i.e. no belt, no bearing, no brushes, no O-ring to be changed periodically). Therefore a high efficiency (>90%) PMW 3 phase 3 kVA inverter has been specially developed featuring a variable frequency drive to overcome the starting intensity of the motor-compressor (Figure 1).

<u>General design and dimensioning</u>: Since a reliable cold delivery by the frigorific machine is of prime importance (no cut-off of the "cold chain" allowed), a two storage system has been envisaged and then verified by a computer simulation. These two storages (a lead-acid battery and a set of eight eutectic plates) keep the temperature inside the store within the 0-7°C range for ca. a week without direct sunlight. System structure is given on Figure 2.

Energy management and dimensioning of subsystems (i.e. PV array, battery, inverter, condensing unit and eutectic cold storage) related to climatic data and operation's assumptions, have been achieved by the dynamic simulation (Figure 3). In particular, on the graph plotting (versus time) of the states of charge of battery storage and eutectic storage (Figure 4), their strong correlation can be noticed i.e. both storages act in a complementary way directed by the energy management cabinet (Figure 5).

3. DEVELOPMENT OF THE 10 m^3 PROTOTYPE

Prototype was designed as a compact package to minimize land occupation, civil engineering and associated costs. The package consists of:
- a modular walk-in insulated room fitted with slats. The extra thickness of insulation (18 cm of polyurethane) ensures a very low waste of cold (0.14 W/m^2/°C). A redundant internal polyethylene door limits the free convective loss when the door is opened,

- a roof-mounted photovoltaic array representing a 2.2 kW of installed peak power (at nominal conditions: 1 kW/m² and 25°C cells' temperature) (Figure 6),
- a specially designed 3 kVA DC-AC inverter. Reliability and efficiency were the key points for this development,
- a fully hermetic motor-compressor and condensing unit specially adapted for both high thermal efficiency and reliability. At normal operating conditions (380 Volts, 50 Hz, AC, 3 - phase) power consumption is 1560 W to produce 2.3 kW of cold at - 10°Celsius for a + 58.8°Celsius condensing unit,
- an electrochemical storage consisting of a 37 element battery. The 4 volts - 200 Amphours elements are pratically maintenance-free thanks to their tight casing (Figure 7),
- a cold storage consisting of 8 eutectic plates representing ca. 80MJ of cold (Figure 8).

The addition of a redundant conventional generating set is possible though it is not necessary in most cases of utilizations.

TABLE I : ENERGY BALANCE (KJ/24 HOURS) - CLIMATE : TROPICAL / TYPE : ABIDJAN

VOLUME (m3)		2		5		10		20		30	
LOAD (KG OF FOOD, 85 % WATER CONTENT)		250		700		1 600		3 200		4 800	
INPUT PER DAY (KG OF FOOD)		25		70		160		320		480	
C = CLASSICAL DESIGN S = SOLAR DESIGN		C	S	C	S	C	S	C	S	C	S
Qi$_s$ (KJ/DAY) REPAR- TITION	WALLS	84 000	42 000	15 000	7 500	22 000	11 000	35 000	19 500	56 000	28 000
	CONVECTION THRU DOOR	10 000	2 500	16 500	8 500	27 000	12 000	33 000	14 500	38 400	15 500
	FOOD COOLING	2 700		7 600		17 300		34 600		51 800	
	FOOD BREATHING	100		200		500		1 000		1 500	
	OTHER LOSSES	500		1 000		2 000		3 000		4 000	
TOTAL Σ Qi$_s$		21 700	10 000	40 300	24 800	68 800	42 800	106 600	72 600	151 700	100 800
GAIN (c - s) / c		54 %		39 %		37 %		35 %		34 %	

TABLE II: ENERGY BALANCE (KJ/24 HOURS) - CLIMATE : SAHELIEN / TYPE : NIAMEY

VOLUME (m3)		2		5		10		20		30	
LOAD (KG OF FOOD, 85 % WATER CONTENT)		250		700		1 600		3 200		4 800	
INPUT PER DAY (KG OF FOOD)		25		70		160		320		480	
C = CLASSICAL DESIGN S = SOLAR DESIGN		C	S	C	S	C	S	C	S	C	S
Qi$_s$ (KJ/DAY) REPAR- TITION	WALLS	11 200	5 600	19 000	9 500	26 800	13 400	45 800	22 900	60 400	30 200
	CONVECTION THRU DOOR	8 000	2 200	15 000	3 700	24 600	6 000	33 000	7 500	38 400	9 000
	FOOD COOLING	2 700		7 600		17 300		34 600		51 800	
	FOOD BREATHING	100		200		500		1 000		1 500	
	OTHER LOSSES	500		1 000		3 000		3 000		4 000	
TOTAL Σ Qi$_s$		22 500	11 100	42 800	22 000	72 200	39 200	117 400	69 000	156 100	96 500
GAIN (c - s) / c		51 %		49 %		46 %		41 %		38 %	

TABLE III "SOLAR" DESIGNED COLD STORES: COMPARISON OF THE COLD CONSUMPTION (IN kJ/24h) IN TWO TYPICAL SITES

SIZE OF THE COLD STORE (m3)		2	5	10	20	30
SAHELIAN TYPE (NIAMEY) Worst case : August	Total cold consumption Qi_s	11 000	26 500	44 800	75 000	102 000
	Cold effectively used $Qu_s + Q'u_s$	2 800	7 800	17 800	35 600	53 300
	Cold used/cold produced ratio $\frac{Qu_s + Q'u_s}{Qi}$	25 %	29 %	40 %	47 %	52 %
TROPICAL TYPE (ABIDJAN) Worst case : July	Total cold consumption Qi_t	10 000	24 800	42 800	72 600	100 000
	Cold effectively used $Qu_t + Q'u_t$	2 800	7 800	17 800	35 600	53 300
	Cold used/cold produced ratio $\frac{Qu_t + Q'u_t}{Qi_t}$	28 %	31 %	42 %	49 %	53 %
RELATIVE DIFFERENCE Δ BETWEEN BOTH SITES	$\frac{Qi_s - Qi_t}{Qi_s}$	10 %	6 %	4 %	3 %	2 %

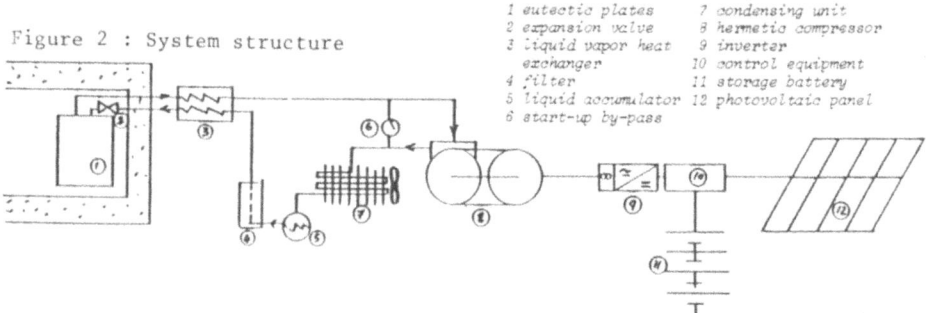

Figure 2 : System structure

1 eutectic plates
2 expansion valve
3 liquid vapor heat exchanger
4 filter
5 liquid accumulator
6 start-up by-pass
7 condensing unit
8 hermetic compressor
9 inverter
10 control equipment
11 storage battery
12 photovoltaic panel

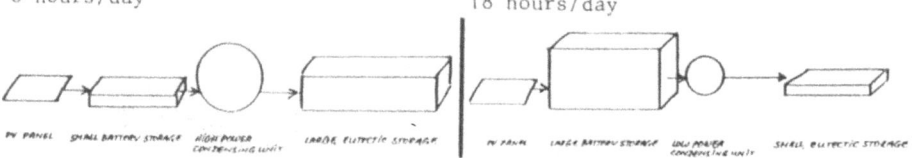

Figure 3 : Design optimization : Two examples of the compressor's operation
6 hours/day   18 hours/day

Figure 1 : PWM Inverter   Figure 4 : Dynamic simulation

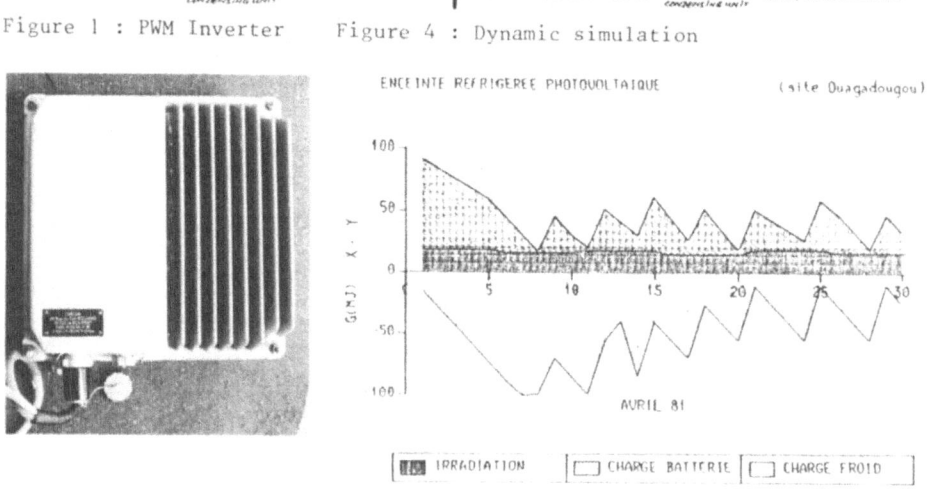

Figure 6 : general view of prototype

Figure 5 : control command cabinet

Figure 8 : battery storage

Figure 7 : eutectic storage

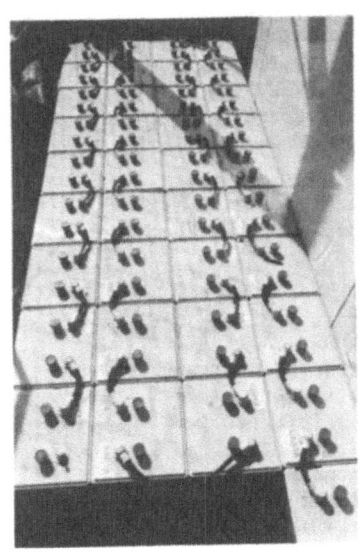

TECHNICAL AND ECONOMIC REQUIREMENTS FOR SMALL-SCALE PHOTOVOLTAIC REFRIGERATORS

M R STARR AND B McNELIS

Sir William Halcrow & Partners
Burderop Park
Swindon SN4 0QD, UK

Intermediate Technology Power Ltd
Mortimer Hill, Mortimer,
Reading RG7 3PG, UK

Summary

There is a great need for small refrigerators for hospitals and clinics in rural areas in developing countries, but conventionally powered units currently available have a number of disadvantages. Robust and reliable photovoltaic refrigerators could fulfil the need and the main technical requirements are described. The World Health Organisation and the US Government are currently engaged on testing programmes to identify suitable commercial products but it is clear that there is scope for considerable technical improvement. Based on discounted cash flow analyses, photovoltaic refrigerators are shown to be cost competitive with the main alternatives currently available. Large markets are foreseen when suitable systems become available.

1. INTRODUCTION

Hospitals and health centres need refrigerators for storing vaccines and sensitive medicines, yet often in developing countries there is no electricity supply in the rural areas where many of these hospitals and health centres are located or planned. A major constraint on the World Health Organisation's Expanded Programme on Immunization is the lack of suitable refrigerators for vaccine storage and transport. At present, absorption cycle refrigerators are often used, working on kerosene or bottled gas. These fuels are becoming increasingly expensive and it is often difficult to ensure regular supplies. Kerosene refrigerators are notoriously unreliable and sometimes are not capable of making ice needed for vaccine transport. Gas systems generally work well but the running cost is higher than for kerosene.

Often a small hospital in a rural area will have its own electricity supply from a diesel generator, in which case conventional electrically powered refrigerators may be used. Unfortunately, the diesel generator is rarely run on a 24 hour basis and furthermore there are periods when the generator is out of service waiting for maintenance or fuel, although a stand-by set is generally provided to reduce this problem. The insulation thickness on typical standard refrigerators is not sufficient to permit interruptions in the power supply of more than a few hours and consequently there is risk that the contents will be lost or deteriorate in quality, possibly without the medical staff realizing that there has been a critical temperature excursion.

There is thus a great need for robust and reliable small solar refrigerators, since they have the potential for good performance, better reliability and longer working life, with little maintenance and no fuel supply problems. A number of technical approaches have been investigated by the World Health Organisation (1), from which it was concluded that photovoltaic/vapour compression system showed the most promise, although it was recognised the solar thermal systems, if properly developed, could possibly be competitive.

2. TECHNICAL REQUIREMENTS

In 1981, the World Health Organisation issued an outline specification for photovoltaic refrigerators (2). The minimum requirements are summarized in Table I. Conventional mains powered

Net vaccine capacity	30-40 litres (top opening)
Ice-making performance:	Minimum 1 kg/24 hrs in +32°ambient
Refrigerating performance:	No part of the vaccine storage area to exceed +8°C or drop below -3°C in:
	a. +43°C ambient temp
	b. +32°C ambient temp
	c. +43°C day time/+15°C night time cycle
Hold-over time:	More than 6 hours below +10°C when power cut out in +43°C outside temp
External casing:	Non-corrodable
Minimum battery maintenance interval:	One year
Insulation:	Rigid polyurethane
1984 cost target:	Less than US$1500 complete

Table I   <u>WHO outline specification for photovoltaic refrigerator</u>

refrigerators could of course be used but there are disadvantages:
- this type of refrigerator needs an alternating current supply and thus an inverter must be interposed between the photovoltaic array and the motor:
- the overall efficiency is generally low because energy consumption has not been an overriding factor in the design of refrigerators and thus a large (and expensive) photovoltaic array would be necessary:
- a large capacity battery would be needed to power the refrigerator during periods of darkness or low insolation.

The complete system needs to be re-designed and optimised to achieve the most cost effective solution. A typical purpose-designed photovoltaic refrigerator consists of a photovoltaic array charging a battery with associated charge regulator, a dc motor directly coupled to a compressor, and a refrigeration cabinet with extra thickness insulation to reduce energy consumption plus cold storage in the form of ice or other eutectic material to extend the hold-over time when the compressor is not working. A thermostat controls the motor/compressor unit.

The battery is essential even though it adds to the capital cost, requires regular maintenance and has a life of only 2 or 3 years compared with a life of say 10 years for the refrigerator itself. (The photovoltaic array may be expected to last for 20 years or more.) The compressor has a high starting torque and the power demand for start up is some 4-6 times the running load. Without a battery buffer, the photovoltaic array would have to be grossly oversized.

All components need to be robust and corrosion resistant, suitable for long life in a tropical climate. The photovoltaic modules should meet recognised performance and environmental durability standards, such as for example those laid down in Specification 501 issued by the European Joint Research Centre, Ispra. The control system and the battery charge regulator must be fully automatic and fail safe.

3. TESTING PROGRAMMES

Independent laboratory and field testing is important to determine the suitability of different commercial systems. The main features to be investigated include:
- variation of internal temperature under normal operating conditions (eg, frequency of door opening and ambient temperature)
- energy consumption to maintain required temperatures
- the 'hold-over' time, with compressor not working
- the time taken to freeze a standard quantity of water
- general durability and reliability of the system.

An engineering assessment of the design as a whole is also needed to identify inappropriate features.

In the USA, the NASA-Lewis Research Center is evaluating small photovoltaic refrigerators on behalf of the US Center for Disease Control. Several manufacturers have supplied systems for laboratory testing and field trails are also being conducted in several countries. Co-ordinated with this programme, the World Health Organisation is currently undertaking laboratory testing of commercial and prototype systems at the Consumers Association Laboratory, Harpenden, UK (3). The World Health Organisation will later undertake field trials of promising systems. Although full results of these testing programmes are yet to be published, it is already clear that few commercial systems are sufficiently robust and reliable and there is scope for further improvement. It should be noted that problems so far experienced are in general related to the refrigerators, not the photovoltaic generators.

4. ECONOMIC ANALYSIS

There is little published data on the economics of small refrigerators, whether conventionally powered or with photovoltaic arrays. A comparative analysis of costs for three important types of small refrigerator has therefore been carried out, based on the following reference or base line assumptions:

i) <u>General</u>

Costs are compared on the basis of Present Value in United States dollars over a 20 year period for three systems:

System A	Photovoltaic/vapour compression
System B	Kerosene/thermal absorption
System C	Diesel generator/vapour compression

Discount rate	:	5%
General inflation rate	:	0
Fuel inflation rate	:	5%
Freeze compartment size (-7°C)	:	17 l
Cold compartment size (+3°C)	:	90 l

Hold-over capability : 2 days

ii) **System A**   Photovoltaic/vapour compression

    PV array : 120 Wp, 20 year life, $2400 FOB + $500 delivery and installation
    Battery : 30Ah, 3 year life, installed cost $100
    Refrigerator : 10 year life, installed cost $1500 including controls (high cost as current production very low)
    Maintenance : Allow 1% of total capital cost per year (average)

iii) **System B**   Kerosene/thermal absorption

    Fuel consumption : 1.25 l/day (average)
    Fuel cost : $0.80/l in year 1 (for remote site)
    System : 3 year life, installed cost $500
    Maintenance : Assume negligible expenditure

iv) **System C**   Diesel generator/vapour compression

Assume two 3 kVA diesel generators (one duty, one standby) are installed to provide power for a number of loads including the refrigerator. Allocate 25% of the total capital and running costs of the diesel generators to the refrigerator.

    Total capital cost of diesels : Each generator $3000 installed
    Running hours : 9 hours of one unit per day
    Fuel consumption : 0.5 l/hour average
    Fuel cost : $0.80/l in year 1 (for remote site)
    Maintenance : Assume negligible expenditure
    Total running cost : $1314 in year 1
    Generator life : One new generator every 2 years
    Refrigerator : 10 year life/installed cost $500

The cash flows for the three systems over 20 years are shown in Figure 1. The Present Values for the base-line assumptions are:

System A	Photovoltaic/vapour compression	$ 6350
System B	Kerosene/thermal absorption	$ 9650
System C	Diesel generator/vapour compression	$12860

The sensitivity of these Present Values to changes in the assumptions for each of the four key parameters of discount rate, capital costs, fuel cost and lifetime has also been investigated. The results are shown in Figure 2, from which it is clear that the photovoltaic system is cost competitive throughout the range of reasonable variation in the key parameters.

5. **CONCLUSION**

There is a great need for small photovoltaic refrigerators for small hospitals and clinics in rural areas of developing countries. Conventionally powered refrigerators are often unreliable, need fuel and maintenance and are expensive on a discounted cash flow basis over 20 years. Photovoltaic systems can be cost competitive even at today's prices and will become even more competitive given volume production of the refrigerators. The need now is for manufacturers to develop well-engineered, robust and reliable systems. When good systems have been identified on the basis field and laboratory testing, some 10000 units per year are likely to be needed for rural hospitals and clinics.

Substantial cost reduction for the photovoltaic system is anticipated given volume production of the refrigerator itself and lower cost photovoltaic arrays. Within a few years, it should be possible to buy systems for $1500 compared with the present $4500. Large markets for solar refrigerators for non-medical users such as schools, police posts and military camps may be expected to develop, followed by shops and domestic users. The world market in time be reckoned as several hundred thousand units per year.

REFERENCES

1. World Health Organisation : Solar Refrigerators for Vaccine Storage and Ice Making; EPI/CCIS/81.5 (1981)

2. World Health Organisation : Specification for Photovoltaic Refrigerator; EPI/9.81/CC (June 1981)

3. B McNelis and J S Lloyd: Evaluation of Solar Refrigerators for use in Vaccine Cold Chain; Proc of UK-ISES Conference C28 'Solar Energy for Developing Countries' January 1982, p 36.

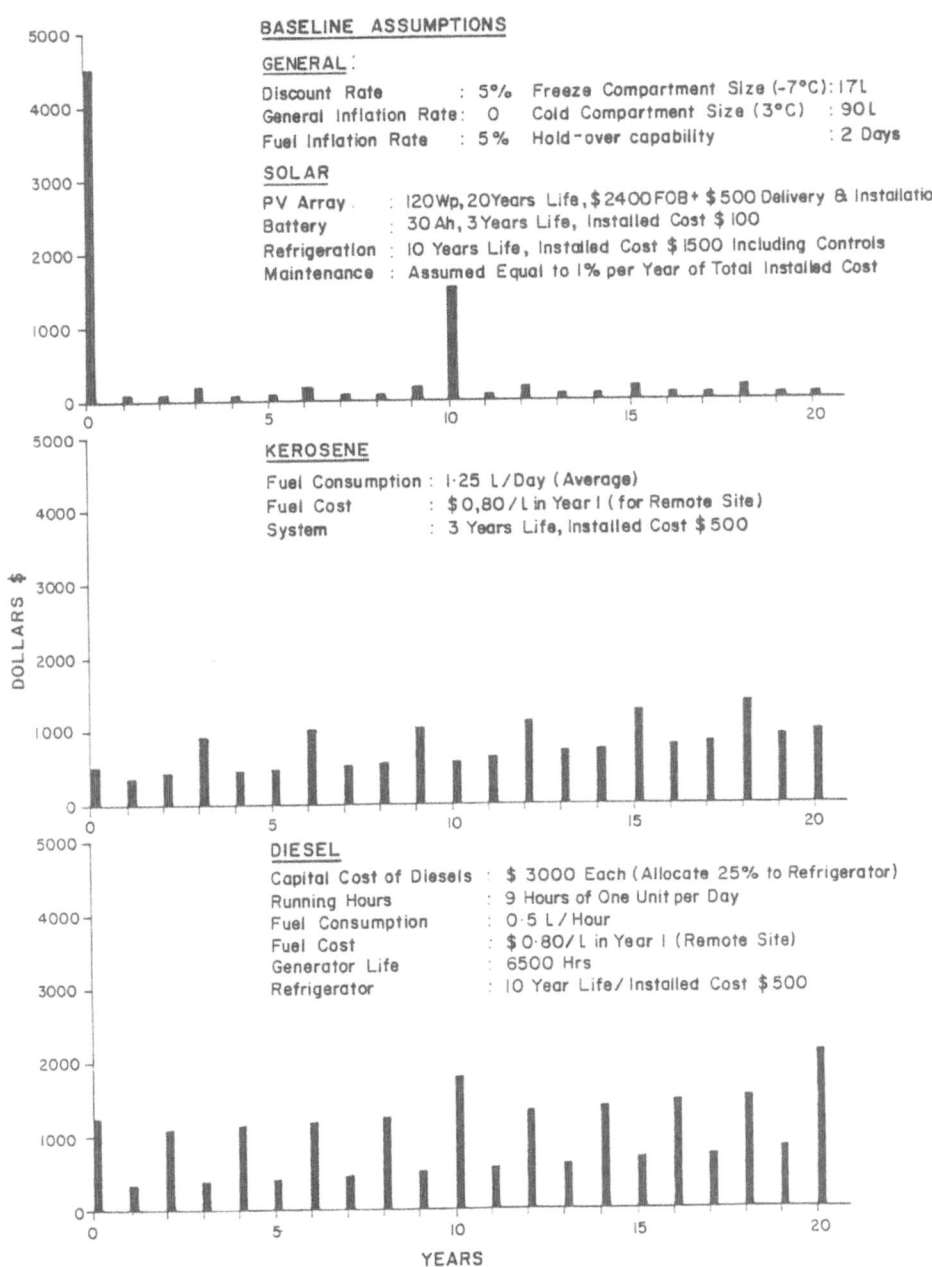

Fig. 1 Cash flows for three types of small refrigerator

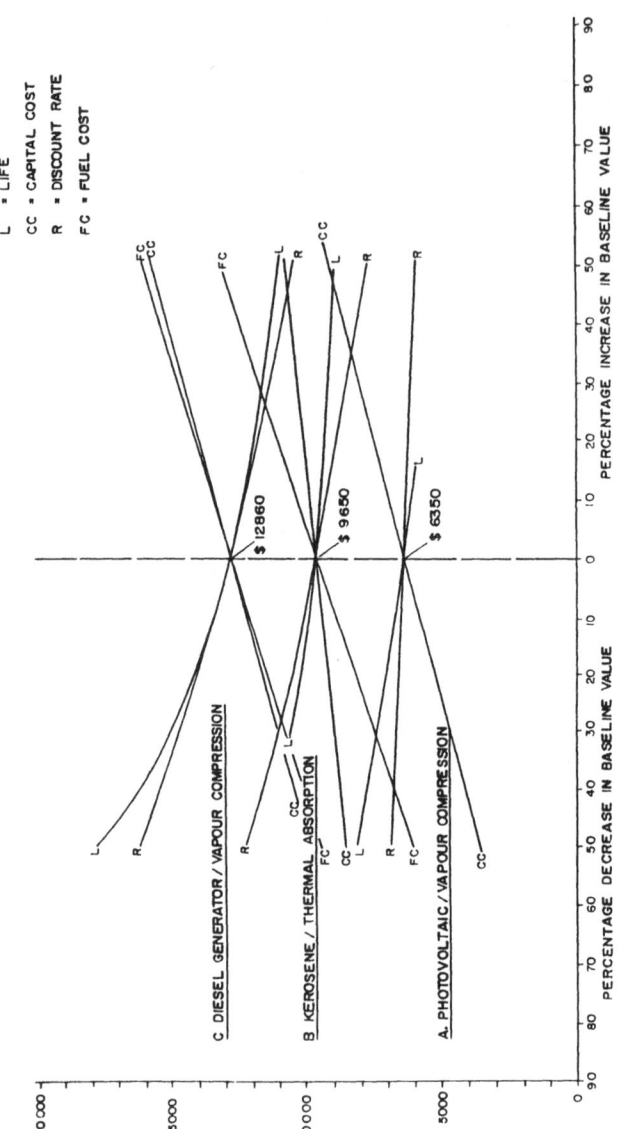

Fig. 2 Sensitivity Analysis

RESIDENTIAL APPLICATIONS OF PHOTOVOLTAICS IN THE UNITED STATES

S. J. Strong
Solar Design Associates
Lincoln, Massachusetts
01773   USA

Summary

The U.S. Department of Energy has predicted that residential photovoltaic applications will become cost-competitive with conventional utility-grid power during the 1980's - perhaps as early as 1986 in some regions of the country. This paper will present the results of our early work in the design, engineering and construction of photovoltaic residences, including the design trade-offs involved in the integration of PV systems with passive solar architecture, using our first lived-in PV residences as case studies.

1. INTRODUCTION

Solar Design Associates has completed the architectural design and engineering of six full-sized single-family passive solar residences with photovoltaic arrays for construction in various areas of the continental United States. These residences have been designed with state-of-the-art passive solar and energy conservation features.

The PV systems are sized to provide all or most of the residences' electrical requirements, depending on geographic location and the energy-use patterns of the occupants. Most of the PV arrays are roof-integrated, and all use high-density, high-efficiency flat plate PV modules. Five systems employ utility-interactive inverters with no on-site electrical storage and one is stand-alone with a battery bank.

Our first lived-in PV residence was completed in January 1981 in Carlisle, Massachusetts. Our second was completed in Milton, Massachusetts, in December 1981. Our third was completed during February 1982 in Santa Fe, New Mexico and, our fourth is now under construction in New York state. In addition, we have completed and tested prototype PV systems in the Northeastern and Southwestern regions of the United States to verify residential PV systems performance.

2. THE CARLISLE HOUSE

Located in Carlisle, Massachusetts, about 30 kilometers northwest of Boston, the Carlisle residence features 270 m^2 of net living area, super-insulation and direct passive solar gain. Appliances are all electric, including the high-efficiency heat pump that provides supplemental heating for space and domestic water. Primary domestic water heating is supplied by a 10-square-meter solar thermal array integrated with the south roof.

The Carlisle House faces south with the 98.4 m^2 PV array mounted on the 45-degree sloping south roof. 32.5 m^2 of floor-to-ceiling, double-glazed sliding glass doors allow passive solar gain to warm the living room, dining area and family room. A concrete slab finished with dark quarry tile and containing 11.5 m^3 of mass stores heat absorbed from the sun.

A large fireplace in the center of the living room offers woodburning heat with additional thermal mass in a massive central chimney. A woodburning stove in the family room also contributes to heating the house. The kitchen, a laundry, study and bath are also on the ground floor, with the master bedroom suite with dressing room, whirlpool tub room and bathroom, three more bedrooms and another bathroom on the second floor.

Double-stud outer walls have 20.5 cm of fibreglass batts plus 2.5 cm of rigid styrofoam under the siding (R-30); 30.5 cm of fibreglass insulation in the ceiling (R-40) and 20.5 cm under the floors (R-26). Triple-glazed windows on the east, west and north walls complete the low-heat-loss structure. Earth-berming and styrofoam insulation against the foundation outer walls also keep heat loss to a minimum. The predicted net heat loss for the house is 3.5 Btu/DD/ft^2, about two and one-half times better than conventional houses.

The Carlisle House

Much of the space heating comes from passive solar gain. Auxiliary space heating is provided by the air-to-air, dual-compressor, 38,500 Btu/hr, 11.3 kW, heat pump with a low-pressure air circulation system for the entire house. Supply registers are in the floor near the outer walls, and return air enters the chimney at the top of the cathedral ceiling and travels down a chase in the masonry core to the air handler, picking up some chimney heat along the way.

The low-speed fan may operate continuously to circulate the passively heated air throughout the house. South-facing clerestory windows along the ridge line provide natural lighting overhead and additional solar gain. In the summer, these windows, which are shaded by an overhang, may be opened for natural-convection ventilation and cooling. The heat pump can also be used to provide air conditioning if desired.

2.1 PV System

The PV system features a 7.3 kWp PV array using 126 63.5 cm x 122 cm glass-superstrate high-density PV modules with 100 mm square semicrystalline cells from Solarex Corporation. The south roof is divided into two sections for the PV array, one with 42 modules (6 rows of 7) and one with 84 modules (12 rows of 7). The 18 rows are connected 2-rows-in-series to provide 9 strings of 14 modules each to provide the 200 Vdc input voltage to the inverter. Each module is comprised of 72 semicrystalline cells (2 paralleled sets of 36 cells) and produces about 58 watts of power under full sunshine at nominal temperatures.

The 9 strings are paralleled and fed to the DC-to-AC inverter manufactured by Windworks Company. This utility-dependent inverter converts the DC current from the array to 240 volt single phase AC power to match the 60-cycle house electric service. Under full sunshine conditions at normal operating temperature, the PV array is rated at a nominal 7.3 kW peak; it produces somewhat more power in the winter cold and somewhat less in the summer heat.

A computer simulation of energy flow was performed for the house using a conservative load scenario and a typical meteorological year. On an annual basis, the house is expected to be virtually self-sufficient, with winter requiring considerable utility dependence, and summer providing a large surplus of electricity.

The Carlisle PV system has been in operation for more than eighteen months, and the results of field monitoring show that its output is a good match for the electrical demands of this large, energy-conserving residence, and that the house will likely generate a net surplus of energy annually.

3.0 THE MILTON HOUSE

Solar Design Associates was retained in early 1979 to design a solar-heated and cooled private residence for a professional man and his wife to be built on a large rural tract in Milton, Massachusetts, about 30 kM southeast of Boston.

Our clients saw this house as the place they wanted to spend the rest of their lives. Having a strong commitment to environmental issues and a good perception of economics and energy, they requested that this house be made as nearly energy-independent as possible.

Our response was to design a compact earth-sheltered house using masonry, concrete and steel construction to produce a low maintenance structure with high internal thermal mass. Full south floor-to-ceiling glass was employed for direct passive solar gain, with super insulation and earth covering on the roof and north, east, and west walls for energy conservation and temperature stability. We included a solar greenhouse and planned a large garden plot for food production, and installed a waste recycling system to complete the cycle.

The house has approximately 225 m^2 of living area with 3 bedrooms, 2 1/2 bathrooms, kitchen, open plan living and dining area, study, workshop, pantry, wine cellar, storage areas and two-car garage. Appliances are all electric with solar domestic water heating, solar space heating, occasional wood-fired cooking, and an old fashioned clothes line, resulting in a comparatively low diversified electrical demand.

The garage/workshop was to be unheated and therefore, it could be built above grade with conventional wood-frame construction. We had the south roof to work with and designed a 30 m^2 roof-integrated solar thermal collector array to provide domestic water heating and supplemental space heating.

The Milton House

3.1 PV System

To provide electricity for the house, we designed a 4.5 kWp PV array using 112 43.5 cm x 107 cm. glass-superstrate high-density PV modules from Mobil-Tyco Solar Energy Corporation. Each module is comprised of 80 silicon ribbon cells (2 paralleled sets of 40 cells) produced by Mobil's unique edge-defined film-fed ribbon growth (EFG) process, and generates about 40 watts of power under full sunshine.

The module strings are paralleled and fed to the DC-to-AC inverter manufactured by Windworks company which converts the 200 Vdc output from the array to 60 cycle single phase 240 Vac to run the household appliances and interface with the utility service.

Like the Carlisle House, the Milton House has no on-site electrical storage, although space has been allocated for the future installation of a battery bank when storage battery technology improves and costs are reduced.

When more electrical power is available than is needed, it is fed to the utility grid for use elsewhere and credited to the house account. When the power needs of the house exceed the output of the PV array, the utility provides the electricity required to make up the difference.

The PV array at the Milton house has been operational for over a half a year now. Performance estimates based on field monitored operational data have projected an annual electrical output in excess of 6000 kWh which could fully satisfy the home's yearly electrical appetite.

4.0 THE SANTA FE HOUSE

In early 1981, Solar Design Associates and Rational Alternatives, a solar builder/developer based in Santa Fe, New Mexico, decided that the time was right to design and construct the nation's first privately-funded speculative PV residence. Encouraged by a very positive reception from mortgage lenders, insurance underwriters, building officials and the local utility, we broke ground in the summer of 1981 at Eldorado at Santa Fe for the first private market effort at energy-independent housing.

The 220 m^2 single story home features 3 bedrooms, 2 baths, a library, an open plan living and dining area, kitchen, solar sunspace, an air-lock front entry foyer, utility and storage spaces, a 2 car garage, an expansive landscaped south patio, and a walled private garden terrace at the rear of the house which includes a solar-heated outdoor hot tub.

The home is designed in the characteristic pueblo-style architecture typical of the southwestern US. Exposed ceilings with vigas and pine decking, brick floors, and kiva fireplaces capture the charm of old Santa Fe. The house is earth bermed on the north side and features walls of masonry and adobe to provide internal thermal mass with south-facing glass for direct passive solar gain.

A roof-mounted solar thermal collector array was included to provide heating for domestic water and the outdoor hot tub. Appliances are all electric with the most energy efficient available on the market chosen. Auxiliary space heating comes from woodburning and electric resistance. Twenty homes of similar size and design built by Rational Alternatives were monitored by the local utility for a 12 month period during 1980 and 1981, and the results showed an average of twenty-three dollars per year in space heating costs.

4.1 PV System

The PV array for the Santa Fe house has an output of 3 kWp and is rack mounted above the flat roof. It features 78 121 cm. x 30.5 cm. Arco Solar glass-superstrate high-efficiency modules. Each module consists of 35 100 mm. diameter single crystal silicon cells in a single series string with redundant cell interconnects and produces 37 watts of power under full sunlight of 1000 watts per m^2 at 25°C.

Six strings of 13 modules are paralleled to provide the nominal 200 Vdc input for the utility-interactive DC to AC inverter manufactured by the Windworks Company. As with our first two PV houses, there is no on-site electrical storage at the Santa Fe house. All surplus power is sold to the utility and kilowatt hours are purchased from the utility to make up any shortfall.

The Santa Fe PV house was completed in mid-February of 1982 and sold the first day it was offered on the market. The PV system is expected to produce over 6500 kWh annually which, taken together with the other solar and energy conservation aspects of the home, will likely make it energy-independent.

5.0 THE PV/PASSIVE INTERFACE

Photovoltaic and passive systems complement one another well in energy terms, and in combination may produce a residence which can be truly energy-independent. In architectural terms, the combination requires careful design. One of the most important considerations is the competition for aperture area.

In general, at most U.S. and European latitudes, the passive aperture will tend to occupy the vertical south walls, and the PV system the roof. The optimum pitch for the PV array is usually the same as that for sunspace roof glazing and solar thermal collectors for domestic water heating, so the principal issues become the division of roof area and the avoidance of shading.

PV modules must be designed physically and electrically to give the architect/or building designer maximum flexibility. Key issues here are module voltage and string length.

Module operating voltage affects the number of modules required to be wired in series to form an electrical "string" to provide the optimum input voltage to the power conditioning equipment. A number of these series strings of modules are wired in parallel at the inverter and determine the power output of the PV array. Since each array must be sized electrically in increments of a full string of modules, the string becomes, after the single PV module, the second basic building block of the PV array.

String length has a strong influence on the number of options available to the designer when choosing a PV module and sizing a PV array, affecting both the physical size and shape of the array as well as the electrical characteristics. For example, a system which requires 13 modules in series to achieve operating voltage can be easily configured in only one way: rows or columns of 13 modules. A 14-module string can be arranged in two ways: 1x14 or 2x7. A 12-module string can be configured in many ways: 1x12, 2x6, 3x4, doglegs, to allow the array to be designed to fit the building and not the other way around.

Generally, the higher the module operating voltage, the fewer number of modules will be required to make up a string and the more flexible the array design can be, both in terms of physical configuration and power output options.

Since supporting structure for PV arrays is expensive, the design of large-area high-efficiency modules is considered an important priority. This is especially true if the supporting structure is a house. It is very hard to justify building a bigger house just to support a large low efficiency PV array. The task then, is to deliver the maximum power that is economically practical from a given array area. This keeps construction costs low and reduces the architectural impact of the array allowing the designer more freedom and making more roof aperture available for other solar applications.

When a designer must deal with the simultaneous requirements of passive space heating and the PV array, this kind of flexibility becomes critical to the generation of aesthetically pleasing buildings which people will enjoy.

6. CONCLUSION

Residences are currently being financed with mortgages that strech out over twenty-five or thirty years. During the financial lifetime of these homes, we will see major changes in the way the world produces, distributes and consumes energy. While no one is sure of the exact nature of these changes, one thing is certain: the era of abundant and inexpensive fossil fuels and electricity is over.

Perceptive clients and designers have begun to understand this, and the amount and type of energy a residence consumes and the prospect of energy independence are now receiving considerable attention from all sectors. The Carlisle House and our other early work in the design and construction of photovoltaic-powered utility-interactive residences demonstrates that energy independence on an annual basis is possible, without compromising the quality of life.

Our next generation of designs, such as the residence now under construction which we designed for a private client in New York state, will integrate passive solar design and active solar thermal with PV and other appropriate on-site electrical generating systems and on-site electrical storage to create energy-producing organisms... independent homes which provide the comfort and amenities consistent with our current standard of living without need of purchased energy.

SESSION 2

EXPERIENCE, PERFORMANCE, RELIABILITY, MONITORING

Chairmen : Dr. A.F. FORESTIERI, NASA/Lewis Research Center, Cleveland, Ohio, USA

F.C. TREBLE, Consultant, Farnborough, Hants, United Kingdom

Performance, testing and module monitoring at the EC : necessary steps to develop cost-effective PV modules

Reliability and performance experience with flat-plate photovoltaic modules

Operational experience with intermediate flat-plate photovoltaic systems

Operational experience with a 35-kWp concentrating photovoltaic system

Solar photovoltaic systems in the development of Papua New Guinea

Field trial of rural solar photovoltaic system

PERFORMANCE TESTING AND MODULE MONITORING AT THE EC:
NECESSARY STEPS TO DEVELOP COST-EFFECTIVE PV MODULES

K. Krebs
Commission of the European Communities
Joint Research Centre   Ispra Establishment
Ispra (Varese) Italy

Summary

Cost and reliability of modules are the most critical elements for any future large scale application of photovoltaics. The modular nature of photovoltaic generators allows to use fault-tolerant circuitry but economic solutions are only possible if cell and module failure rates are kept at very low levels. Testing is essential to reach this target but also to create confidence with users, research ministeries and financing bodies. During the last few years the EC has made substantial investments to support this view: facilities have been built, test procedures were developed and an extensive module test program for all EC pilot plants has been started. A short description of motivations and activities is presented. It is expected that these efforts, combined with a later centralized data collection from all field experiments could contribute in an essential way to the further development of reliable and cost-effective photovoltaic modules.

1. MOTIVATION AND NEED

Why are we interested at the EC in the testing of photovoltaic modules? Testing is often considered as a way of putting bureaucratic limits to industry or as a method to eliminate innovative concepts by formal rules. That there should be control during or after manufacturing is perhaps acceptable, but that this control should be enforced by public and even more by supranational organisations like ours may be less obvious.
It has also been stated that it is too early to issue strict regulations for testing, but this argument is probably more directed against a too hasty creation of legal standards. In our view testing is the experimental method, first, to verify on an objective basis that stated targets have been reached and that a device has been understood in all its aspects, second, it is considered to be a powerful tool to predict the behaviour of these devices with time. If the last item includes preventive failure analysis, it becomes then, in fact, an effective tool for the further improvement of

modular designs. The broader our information, the better we can fulfill these tasks: as European organisation we have, at least in principle, this wider access to data.

Let us regard for a moment the production of photovoltaic modules in its present situation. Compared to other energy relevant industries photovoltaics does not yet offer any substantial energy substitution potential. There is a market of the order of 10 MW/y, and this seems to be so interesting that already some dozen firms produce modules. In developed countries this market has a kind of luxury appeal: electricity for sailing-boats, for camping-cars or holiday-homes. Testing under this aspect means establishing guarantee conditions which could help to promote the sales volume. This should not be understood as a negative statement: photovoltaics will need this "cash flywheel" to sustain its momentum through the difficult phase of going from a novelty item to a widely accepted energy source. This transition period could be shortened by tax benefits: testing would then be needed to establish minimum requirements. Our intention at the EC is to create a basis for harmonization of these efforts within our member states. Solar energy arrives across all borders.

On the other hand there are the developing countries where the availability of pv power could signify a fundamental change in life conditions. Whereas in the first case testing stands for consumer protection, here testing would mean to guarantee that these in the majority of cases inexperienced and certainly not wealthy people are indeed receiving reliable and useful devices.

The first group could possibly manage without publicly supported testing, the second should not since in this case one does not work with private capital and public (for specific cases EC) control is therefore an inherent necessity.

A third field I did not yet mention, is the energy replacement option. If the other two fields could be characterized by the slogan: photovoltaics starts where the grid ends, it is our hope at the EC that one day there will be also photovoltaic electricity even where we use today our conventional systems. However, this task is by orders of magnitude more difficult than the two others, and only careful tests and dedicated continuing public support, including EC funding, can help to reach this ambitious goal: the naïve optimism of those self-fulfilling prophecies will get us nowhere. Public administrations, policy making bodies and even the public at large can in the end be convinced only by showing them facts. That is perhaps the most essential argument why we need - at least in the EC countries where we have not yet reached that famous point-of-no-return - unbiassed, neutral and critical testing allowing a fair comparison between all products on the market, independant by whom and in which European country they are produced. Testing at the EC is in this sense considered as a confidence producing measure serving both users and producers.

2. PROBLEMS, TEST PROCEDURES AND PROGRAMS

In photovoltaics one talks, perhaps too often, about prices or more recently about costs. It is true that photovoltaic facilities are systems, and that thus all components play a role, on the other hand it is also true that low-cost, long-life time modules are the basic elements of these systems. There are examples of modules with a 20 years life time but these are neither the modules of present design nor are these results, unfortunately, typical. In real field experiments between 1977 and 1981 one has e.g. observed (1) that of 14.000 modules about 500, i.e. more than 3 % for an average operation time of about 2 years, have failed. Designs have meanwhile changed and in a new installation one has obtained a failure rate of 0.6 % during the first year of its operation (2), which corresponds almost to the US target value of five modules per thousand per year (3). This is valid for the US, with respect to Europe not all producers, especially those who are new in this field, may have reached that level of experience or inside information to reach such low failure rates.

In high voltage applications the large number of series cells makes the system very sensitive to cell failure unless fault-tolerant circuitry is used. Appropriate strategies have been developed, like series-paralleling with an extensive use of bypass diodes and multiple cell interconnects. Details are discussed in recent JPL-publications (4,5). Another aspect concerns the optimisation of costs for the repair or replacement of modules (6). Present cell failure rates lie around 10^{-4} f/y but are expected to approach by known techniques 10^{-5} f/y (5). To demonstrate the dimension of this problem these numbers could be compared with 10^{-3} f/y for connectors or to 10^{-2} f/y for single battery cells.

The most appropriate safety factors, optimum design margins or good redundancy measures can in the end only be infered from field experience. In Table I we have listed the main defects one has seen so far in various demonstration plants. The large fraction of cracked cell and interconnect failures (63%) is remarkable. This was partially attributed to day/night thermal cycling, i.e. to an effect which is always present. In connection with another unavoidable effect, namely shadowing or soiling, cracked cells can cause major breakdowns as consequence of reverse bias conditions. It should be mentioned here that a failure was counted only when the electrical output of the module degraded by at least 25 %. Efficient testing means obviously that these defects should be eliminated by laboratory tests before field installation. In practice it means to establish a suitable list of test procedures. This has been done by several organizations as e.g. by JPL (7), by SERI (8) and also by the CEC (9,10).

The tests which are described in our EC Specification No. 501 are listed in Table II. There are at present 20 control tests: they cover inspection, verification and performance aspects (CT.1-8). The others should be useful to identify key environmental or operational factors and design features which could affect the attainment of a sufficiently long lifetime

(CT.9-20). Tests which are perhaps more critical in this list concern mechanical loading and hail impact, hot-spot heating, thermal cycling, humidity-freezing, damp heat (95 % RH at 40°, or possibly 90°C) and high temperature long exposure (95°C).

All tests are part of the contract conditions for the current Pilot Plant Programme of the EC and also for other projects which are financed by the Commission. The correspondance of these tests to failures observed with field experiments is shown in the last column of Table I.

In order to gain time we are at present using parallel test sequences (Fig.1), but for statistical reasons, this may be a questionable method. On the other hand one should realize that Fig. 1 already comprises 71 single tests for each module type.

So far 12 types were tested (Fig.2). The figure clearly demonstrates that we are still far from any standardization of dimensions. Some of the main characteristics of these modules are given in Table III. The table indicates that more efforts should and can be made to increase module efficiencies. Other comparisons will be made when more data will be available. All tests were executed at the ESTI facilities of the JRC in Ispra (Fig. 3 and Table IV). Fig. 3 is an artist's view of this installation which has been completed only recently. Table IV is a list of all facilities which are now available. There are 4 solar simulators, equipment for 11 qualification and endurance tests and 1 outdoor test field. Data acquisition is made through a number of station computers which are linked to a central computer for processing and storage.

A preliminary list of defects found during prototype testing is given in Table V. This list covers only the results of about 1/3 of all modules which will be tested for the Pilot Plant Program. The main defects observed were bubbles or voids in the encapsulation material. These defects fall into the so-called cosmetic category, a small percentage is permissible but the existence of these defects could also mean that further debonding may occur during the service life. In general it indicates that the lamination process should be improved. The other items in this list seem to be mainly related to poor workmanship, lack of sufficient quality control or to inexperience with a new production technology. But even if these defects could be called minor they reveal a certain lack of quality and create some doubts about the lifetime and the future maintenance costs of some of these devices. Looking at these first results one could also conclude that our test requirements are still too weak and that we should increase stress levels to more critical values. However, too severe test conditions could increase costs by a factor $1/(1-p)$, where p is the probability that a module fails in a certain test. A more interesting conclusion would be to reduce safety margins of present designs expecting that this would lead to a decrease in production costs.

It is obvious that our test statistics is not yet sufficient to go that far. It will therfore be one of the main objectives of the EC Pilot Plant Programme to increase the statistical basis for module failure analysis and to find out where present indoor test methods could be relaxed, where we need other tests or where an increase in stress level could lead to higher acceleration factors without losing the significance of the test. For these reasons the Commission is making a special effort for implementing efficient data acquisition systems at all EC test sites covering both aspects, energy performance and lifetime or failure rate of the main components. It is planned to centralize all data for further evaluation and for comparative studies which are one if not the most important objective of our testing efforts. The results from the pilot plants together with efficient indoor qualification testing should thus become a very essential part of the further module improvement process (Fig. 4).

3. CONCLUSION

This paper should have shown that EC-testing is not intended to be a bureaucratic way to control contract money or to keep committees busy in discussing formalistic aspects but rather as a means to arrive through neutral, comparative and innovative testing at better and less costly designs. Testing should thus help to convince the public that it is worthwhile to invest in a solar technology which is not based on educated guesses or on wishful thinking but rather on hard and "tested" facts.

REFERENCES

(1) L.N. Dumas and A. Shumka,
Proc. 15th IEEE Photovoltaic Specialists Conf. (1981), p. 1091

(2) J. Solman and B.L. Grossman,
Proc. 15th IEEE P.S. Conf. (1981), p.1231

(3) G. R. Mon,
Proc. 15th IEEE P.S. Conf. (1981), p.964

(4) C. Gonzalez and R. Weaver,
Proc. 14th IEEE P.S. Conf. (1980), p.528

(5) R. G. Ross,
Proc. 15th IEEE P.S. Conf. (1981), p. 1157

(6) R. G. Ross,
Proc. 14th IEEE P.S. Conf. (1980), p. 1126

(7) Jet Propulsion Laboratory Report
No. 5101-161 (1981)

(8) Solar Energy Research Institute Report SERI/TR-742-654 (1980)

(9) Specification No. 101 (Issue 2):
Standard Procedures for Terrestrial Photovoltaic
Performance Measurements
Report EUR 7078 EN (1981)

(10) Specification No. 501:
Photovoltaic Module Control Test Specifications
Report EUR 7545 EN (1981)

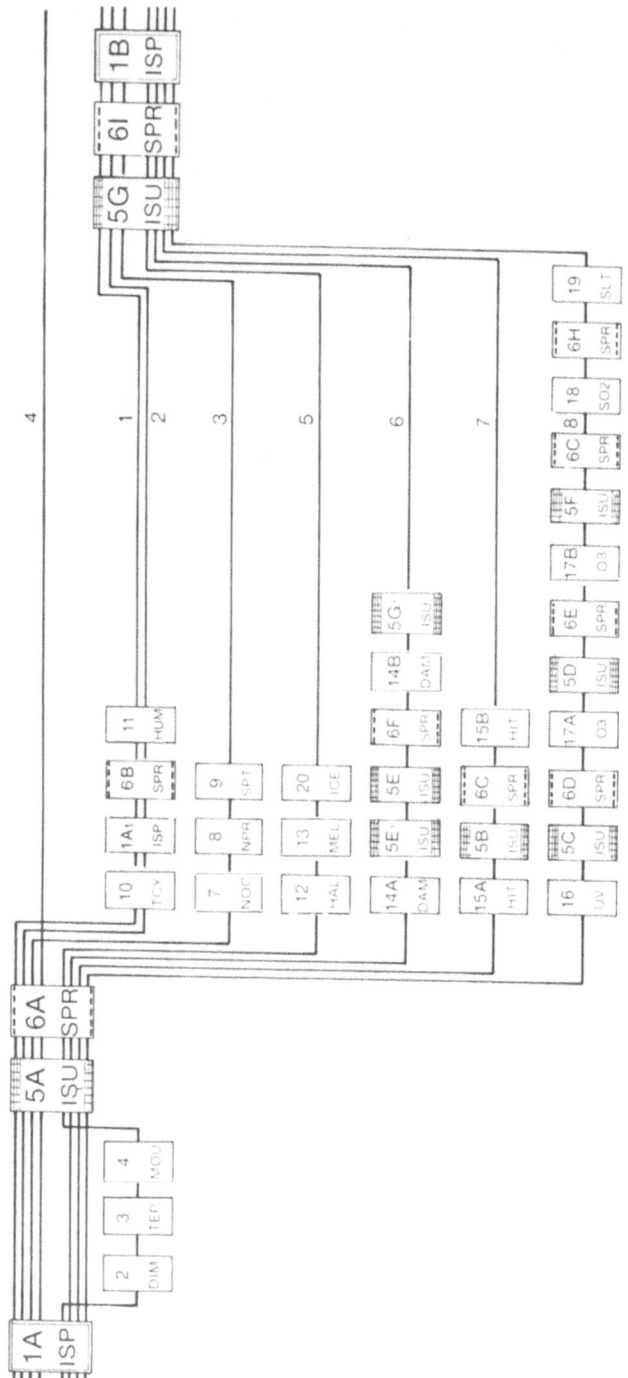

Figure 1 — Present test sequences for module testing at ESTI

Fig. 2 - Some of the modules tested at ESTI

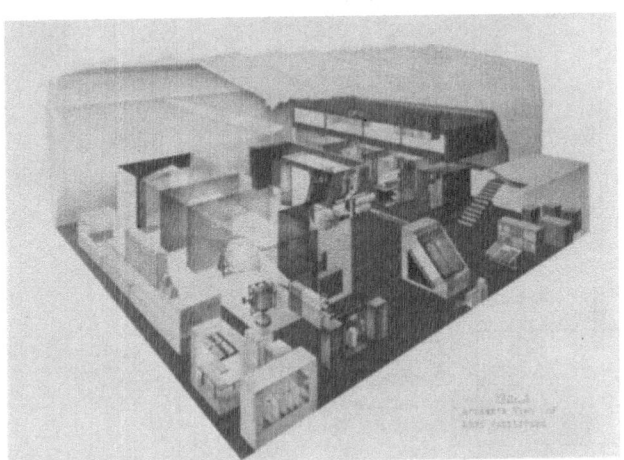

Fig. 3 - Artist's view of ESTI facilities

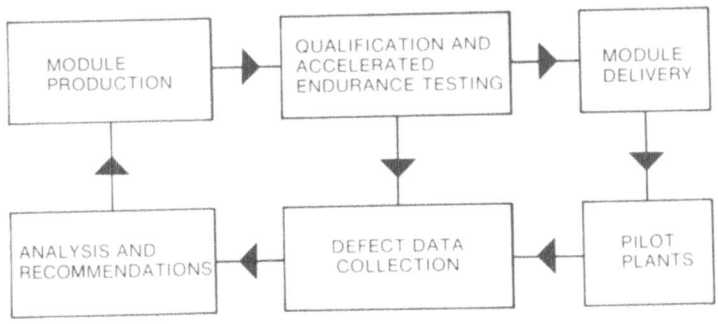

Fig. 4 - Integration of testing into module improvement

TABLE I   Field Experience and Corresponding Indoor Tests

Defects known from Field Experience	% of Failures	Probable Cause	Test No Acc to Spec 501
Cracked Cells	34	Day/Night Cycles, Hail, Mech Damage, Rev Bias Conditions	1, 9, 10, 11, 12, 13, 20
Interconnects (fractures, unsoldered, displaced or exposed contacts)	29	Day/Night Cycles, Diff Thermal Exp Coeff, Workmanship, Quality Control	1, 10, 11,
Delamination	22	Void Formation, Outgassing, Entraped Moisture, Curing, UV-Irradiation	1, 11, 14, 15, 16,
Insulation	9	Insulation Layers, Sharp Points	1, 5,
Corrosion (contacts, terminals, wires)	6	Reactive Solder, Defective Gaskets, Seals or Covers	1, 11, 14, 18, 19,

TABLE II	EUR-Specification No. 501: List of Test Procedures
CT 1	Visual Inspection
CT 2	Dimensions and Weight
CT 3	Robustness of Terminations
CT 4	Mounting Twist Test
CT 5	Insulation Test
CT 6	Electrical Performance at Standard Test Conditions (STC)
CT 7	Determination of Nominal Operating Cell Temperature (NOCT)
CT 8	Electrical Performance at Nominal Operating Cell temperature
CT 9	Hot-Spot Heating Test
CT 10	Temperature Cycling Test
CT 11	Humidity-Freezing Test
CT 12	Hail Resistance Test
CT 13	Mechanical Loading Test
CT 14	Damp Heat Long Exposure Test
CT 15	High Temperature Long Exposure Test
CT 16	UV Test
CT 17	O_3 Test
CT 18	SO_2 Test
CT 19	Salt Mist Test
CT 20	Ice Formation Test

TABLE III Characteristic Parameters of PV Modules Tested at ESTI

P_m at STC (W)	Module Aerea (m^2)	V_{oc} (V)	I_{sc} (A)	FF	η_{cell} (%)	η_{mod} (%)
65 3	0 98	41 7	2 41	0 65	11 6	6 7
60 9	0 82	41 9	2 02	0 72	10 8	7 4
56 6	0 82	41 6	2 00	0 68	10 0	6 9
54 7	0 85	6 8	13 00	0 63	9 7	6 4
34 0	0 44	22 4	2 23	0 68	11 4	7 7
33 3	0 37	19 3	2 40	0 72	12 8	8 9
31 3	0 44	7 1	6 48	0 68	11 1	7 1
30 7	0 45	21 2	2 01	0 72	10 9	6 8
18 2	0 26	11 2	2 35	0 69	9 2	7 1

TABLE V List of Defects Observed with some Prototype Modules

Defects Observed	% of Tested Modules
Discolorations	15
Bubbles or Voids	33
Interconnectors, Contacts	17
Chipped Cells	10
Broken or Cracked Cells	2
Defective Gaskets or Seals	21
Broken or Damaged Covers	6
Insulation Resistance	6
Electrical Degradation (>10%)	4

TABLE IV ESTI-Facilities at the JRC Ispra

SOLAR SIMULATORS

LS- 0	1 KW Xenon Simulator (20 × 20 cm²)	
LS- 1	Large Area Multiple Source Simulator (3.2 × 4.0 m²)	
LS- 2	25 KW Xenon Simulator (1.7 m diam)	
LS- 3	Large Area Pulsed Simulator (5 m diam)	

QUALIFICATION AND ENDURANGE TESTING

AT- 0	Temperature/Humidity/Freeze Cycle
AT- 1	Corrosion Test Chamber
AT- 2	UV - Irradiation
AT- 3	Pressure Loading
AT- 4	Hail Impact Facility
AT- 5	Overpressure and Leakage Test
AT- 6	Long Exposure Thermal Degradation
AT- 7	UV - Irradiation
AT- 8	High Humidity Chamber
AT- 9	High Temperature Chamber
AT-10	Salt Mist Chamber

OUTDOOR TESTING

TF- 1	PV System Test Facility

RELIABILITY AND PERFORMANCE EXPERIENCE WITH FLAT-PLATE PHOTOVOLTAIC MODULES*

R.G. Ross, Jr.**
Jet Propulsion Laboratory
Pasadena, California, 91109 U.S.A.

Summary

As part of the United States National Photovoltaics Program, the Jet Propulsion Laboratory's Flat-Plate Solar Array (FSA) Project has maintained a comprehensive engineering sciences activity addressed to understanding the reliability attributes of terrestrial flat-plate photovoltaic arrays and to deriving analysis and design tools useful for achieving the high levels of reliability necessary for future large-scale application. This paper builds upon the field-performance experience gained from several years of photovoltaic system-application experiments and the research results stemming from the development of reliability design and analysis techniques. The result is an overview of the array reliability problem in total, and available means of achieving high reliability at minimum cost.

1. INTRODUCTION

The reliability of photovoltaic solar arrays is probably second in importance only to cost in the list of factors influencing the market acceptance of this new technology. Because of their modular nature, photovoltaic arrays possess a higher than normal sensitivity to common-mode failures, but at the same time offer a wealth of redundancy options to increase reliability. Achieving the high reliability demanded by future large-scale application requires that these reliability design attributes be understood well and used effectively.

As a tool for managing module and array reliability development, the FSA Project has adopted a target for life-cycle-reliability costs equivalent to 20 years with no significant array-power degradation. Because small levels of degradation or replacement are economically justified, a typical array meeting the target life will last somewhat longer than 20 years to recapture the life-cycle costs associated with the gradual degradation expected (see Fig. 1).

A convenient means for quantifying the life-cycle costs associated with reliability attributes is based on computing the break-even photovoltaic energy cost over the expected life of the application (1,2):

---

* This paper presents the results of one phase of research conducted at the Jet Propulsion Laboratory, California Institute of Technology, for the U.S. Department of Energy through agreement with the National Aeronautics and Space Administration.

** Engineering Sciences Manager, Flat-Plate Solar Array Project, and Supervisor, Photovoltaic Engineering Group, Energy Technology Engineering Section.

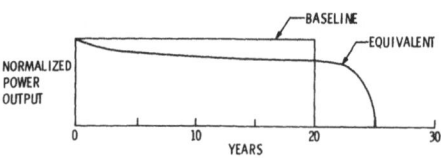

Figure 1. Reliability/Durability Target Relationship.

Table I. Economic Impact of Degradation Types.

TYPE OF DEGRADATION	UNITS	LEVEL CAUSING 10%* COST INCREASE	
		k = 0	k = 10
FIXED CELL FAILURE RATE**	FRACTION PER YEAR	0 0006	0 0008
FIXED MODULE FAILURE RATE	FRACTION PER YEAR	0 007	0 016
LINEAR DROP IN POWER	FRACTION PER YEAR	0 010	0 014
FIXED DROP IN POWER	FRACTION	0 10	0 10
DROP IN MODULE WEAROUT LIFE	YEARS	2 0	4 75

*10% INCREASE IN LIFECYCLE ENERGY COST, k = DISCOUNT RATE
**SOURCE CIRCUIT = 8 PARALLEL x 200 SERIES BLOCKS WITH DIODES

$$R = \frac{C_0 + \sum_{i=1}^{L} C_i M_i (1+k)^{-i}}{\sum_{i=1}^{L} E_i (1+k)^{-i}} \quad (1)$$

where: R = Cost (worth) of energy (startup-year $/kWh)
E_i = Energy generated in year i (kWh)
C_0 = Initial plant cost (startup-year $)
C_i = Cost per module replacement action (startup-year $/module)
M_i = Number of modules replaced in year i
k = Present-value discount rate
L = Plant lifetime (years)

Notice that the above expression explicitly includes the effects of array degradation versus time (E_i), module initial cost (C_0), and the cost of module replacement ($C_i M_i$).

In assessing the implications of the 20-year-equivalent-life target it is instructive to examine the life-cycle-cost impact of typical failure modes which are likely to exist in future low-cost modules and systems. Assuming a module cost of $0.70 per watt, Table I indicates the level of degradation, for each of five typical modes of array degradation, that will result in a 10% life-cycle system cost increase. Because these five failure modes may occur concurrently, the total cost impact must include the sum of these effects.

Table II carries the analysis a step further and suggests a possible "strawman" allocation of the allowable degradation among the five failure-mode categories to achieve a total performance consistent with the 20-year target. The distribution among the categories reflects this author's best judgment. The remainder of the paper addresses each category in light of the historical experience to date,

Table II. Strawman Degradation Allocations.

TYPE OF DEGRADATION	INCLUDED MECHANISMS	UNITS	DEGRADATION ALLOCATION
FIXED CELL FAILURE RATE	CELL CRACKING, INTERCONNECT FATIGUE	FRACTION PER YEAR	0 0001
FIXED MODULE FAILURE RATE	STRUCT FAILURE, INSUL BREAK	FRACTION PER YEAR	0 005
LINEAR DROP IN POWER	YELLOWING, AR COATING, CELL DEGRADATION	FRACTION PER YEAR	0 01
FIXED DROP IN POWER	SOILING	FRACTION	0 05
MODULE WEAROUT LIFE	OBSOLESCENCE, CORROSION	YEARS	25

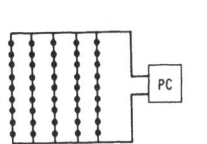

(CELL FAILURE RATE (R) = 0.0001 PER YEAR)

ARRAY VOLTAGE	SERIES CELLS (S)	POWER LOSS AT 5 YEARS
15	36	1.8%
150	360	16.5%
1500	3600	83.5%

POWER LOSS = $1 - [1 - (YEARS \times R)]^S$

Figure 2. Array degradation versus numbers of series cells.

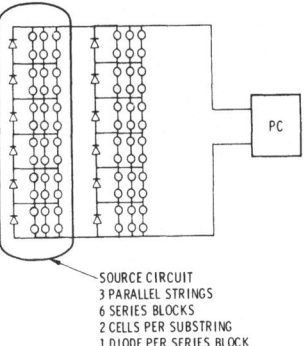

SOURCE CIRCUIT
3 PARALLEL STRINGS
6 SERIES BLOCKS
2 CELLS PER SUBSTRING
1 DIODE PER SERIES BLOCK

Figure 3. Series/Parallel Nomenclture.

and summarizes projected means of achieving the allocations listed in Table II.

2. CELL FAILURES

At the root of the high sensitivity to cell failures indicated in Table I is the need to interconnect electrically thousands of nearly identical solar cells in series and parallel to achieve the voltage and current levels of the intended application. For example, a 150-volt residential array will require 300 to 400 series cells, and a 1500-volt central-station application will require 3000 to 4000. This large number of series elements makes an array extremely sensitive to infrequent cell failures even when a high level of circuit redundancy is used.

Fig. 2 graphically illustrates this sensitivity by noting the effect of one cell failure per 10000 per year on various system configurations. To control this exaggerated sensitivity at high-voltage levels, extensive use of circuit redundancy techniques such as series/paralleling and bypass diodes is recommended (2,3).

Fault-Tolerant Circuit Design

Before the degradation allocation associated with solar-cell failures can be addressed further, the influencing effects of the available circuit redundancy solutions must be considered. The first step toward circuit redundancy is generally associated with dividing the large matrix of cells that makes up the array into a number of parallel solar-cell networks referred to as "branch circuits" or "source circuits". The circuits provide convenient points for monitoring array performance and provide an ability to isolate small areas of the total array for maintenance and repair.

As shown in Fig. 3, each source circuit may contain a single string of series solar cells or a number of parallel strings interconnected periodically by cross ties. The cross ties divide each source circuit into a number of series blocks. One or more series blocks may also be bridged by a bypass diode which is designed to carry the source-circuit current in the event that local failures constrict the current flow to the point of voltage reversal and power dissipation.

A key problem in assessing the impact of cell failures has been in quantifying the influence of specific series/parallel and bypass diode arrangements on array degradation. This problem has been solved in recent years by the development of an extensive parametric analysis based on the statistical distribution of failed substrings due to random cell open-circuit failures (4,5). Reference 5 contains a large number of parametric plots, an example of which is shown in Fig. 4, which allows rapid computation of the effects of cell failures and circuit redundancy on array power loss.

Using these techniques, together with typical array and balance-of-system costs and efficiencies per (2), and a cell failure rate of 0.0001 per year, we can calculate the break-even life-cycle energy costs for various redundancy and replacement options using Equation 1. Fig. 5 displays the calculated life-cycle energy costs for two replacement strategies as a function of the number of series blocks in branch circuits composed of 8-parallel x 2448 series cells. In the first strategy no module replacement is allowed and it can be seen that the life-cycle costs increase sharply with low numbers of series blocks. This reflects the rapid array degradation exhibited in Fig. 2 for series-string circuits without bypass diodes. For the second strategy (dashed curve) in Fig. 5, modules are replaced each time a solar cell fails during the 20-year life of the plant. This results in no power degradation, but does cause a substantial module replacement-cost contribution. This cost also varies with the number of series blocks due to improvements in module yield that occur when module series/paralleling achieves 8 parallel x 2 or more series blocks. This degree of module series/paralleling is only reached in this example when 272 or more series blocks are used per branch circuit.

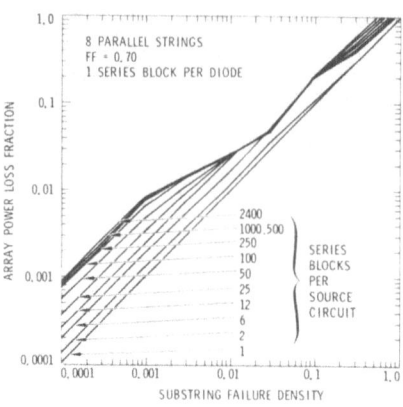

Figure 4. Example Plot for Power Loss Determination.

At this point it is important to note that the economic impact of cell failures presented in Table I assumes the high level of redundancy associated with the minimum life-cycle costs in Fig. 5. The critical question is therefore shifted to the feasibility of achieving the low cell-failure rates indicated as being necessary.

For purposes of assessing this feasibility a cell failure is

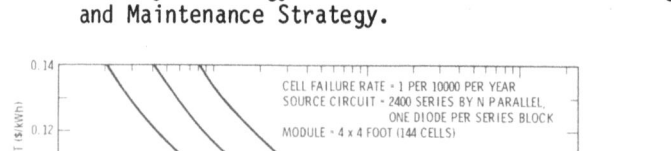

Figure 5. Life-cycle Energy Cost versus Series/Paralleling and Maintenance Strategy.

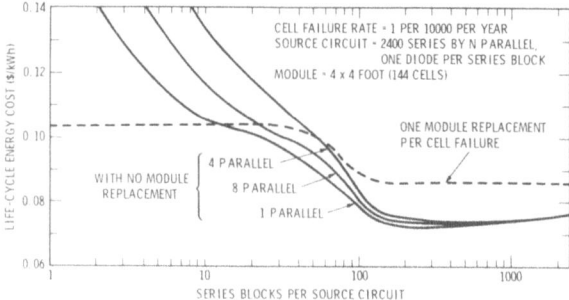

considered as more than 25% degradation in short-circuit-current compared with the average cell. This is generally sufficient to cause reverse biasing of the local bypass diode and therefore loss of power from the series block containing the failed cell.

Typical causes of cell failure include cell cracking, local shadowing and open-circuit cell interconnects.

Cell Cracking

Of the cell failures seen in the field, cell cracking is by far the most prevalent and is occurring at a rate of about 1% per year. The saving finding is that only 0.1 to 0.01 of these cracked cells have resulted in electrically-failed cells. The remainder remain electrically operative, quite often because the cell metallization bridges the break and holds the cell together.

Quantification of the cell failure rates has required extensive and expensive auditing of the actual field performance of several large multikilowatt photovoltaic application experiments. The three primary causes of cell cracking appear to be differential expansion between the cell and its support, impact loading by hailstones, and reduced strength due to cell damage occurring during cell processing and module assembly. Qualitative design techniques exist that address the first two causes (6,7), and cell proof-testing techniques can be utilized to screen out the one-out-of-a-thousand weak or damaged cell (8). Another cost-effective cell-failure solution is the use of multiple electrical attachment to each solar cell.

Cell Interconnects

Cell interconnects are both an important tool for reliability improvement and a source of failures. Given that a cell has cracked or otherwise degraded in a local area, the extent of module or array degradation can be substantially lessened by electrically attaching to the cell at more than one location. One means of assessing the degree of improvement possible is to consider analytically a large number of randomly oriented potential cracks and then to determine the fraction that would lead to open-circuiting or significant cell degradation ($\geq 10\%$ area loss). Many of the latest module designs are taking advantage of multiple interconnect attachment points and are expected to have substantially lowered failure rates. With present cell failure rates at about 0.0001 per year, it is expected that the improved redundancy will lead to values approaching 0.00001 per year.

The above optimistic projection of course assumes that the interconnects themselves don't fail, and of course they do. Interconnect open circuiting due to mechanical fatigue is a historical photovoltaic array failure mode and has even recently taken its toll on some modern installations. Like cell breakage, it is primarily caused by thermal and humidity expansion differences between the cell and its supporting substrate or superstrate. Also like cell breakage, interconnect fatigue is not easily predicted by available analytical models until the level of failure reaches major proportions. Mon, Moore and Ross (9), for example, have empirically characterized the probability of failure of a variety of photovoltaic interconnects versus number-of-cycles and developed a fatigue curve that treats probability-of-failure as a parameter (Fig. 6). As can be seen from Fig. 6, even identically loaded interconnects can be expected to fail over a broad range of cycles, the weakest failing one hundred times sooner than the average.

As with cell cracking, the solution is to design for a manageable number of failures (maybe 10% during the array design life) and then to incorporate interconnect redundancy to control power losses associated with those that fail. For example, if three interconnects were used to connect each cell, 10% failed interconnects would lead to a cell failure density of one per thousand at the array design life.

Future Cell-Failure Levels

With the conscientious application of available design and qualification techniques, this author predicts that cell failure levels well below 0.0001 per year will be achieved. The available techniques include:

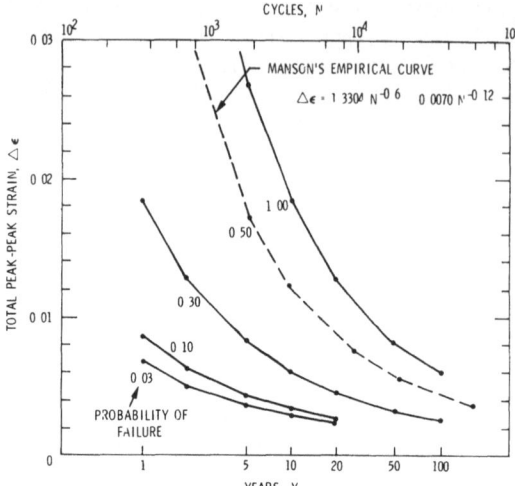

Figure 6. Fatigue curve for Copper Cell Interconnects with Probability of Failure as a Parameter

(1) Cell proof-testing to cull out weak or damaged cells before incorporation into modules.
(2) Redundant cell interconnects with multiple cell attachment to provide fault tolerance to cracked cells and broken interconnects.
(3) Proper materials and process selection and testing to limit differential expansion and process-related stressing of the cells.
(4) Adequate hail-impact protection.
(5) Bypass diodes and cell paralleling to limit array losses due to cell failures.

3. MODULE FAILURES

In addition to failure modes that are best treated at the cell level, a number of failures are more appropriately considered at the module level. These include glass breakage, electrical insulation breakdown and various types of major encapsulant failure such as delamination. Like cell failures, these failures are also flaw-related and must be treated statistically when considering quantities of modules in a large array. Moore (10) and Mon (11) in work with the author have developed empirical/analytical tools to design for given statistical levels of glass breakage and electrical breakdown, respectively.

When designing for appropriate levels of module failures, it is important to note that a module failure is likely to cause an electrical hazard or major power loss and will therefore require immediate maintenance or replacement. As a result, module failure rates are traded off against life-cycle maintenance costs as opposed to redundancy and life-cycle energy loss, which are associated with cell failures.

Although the data base on module-level failures is very poorly defined at this time, this author judges that it might total 2% per year from the following contributing mechanisms:

(1) Glass superstrates broken by projectiles or thermal-expansion loads.
(2) Electrical shorts from cell circuitry to frame ground.
(3) Major encapsulant delamination due to weathering or hot-spot cell heating.
(4) Failed (shorted or reversed) bypass diodes.

It is projected that future modules should be able to achieve module-failure levels consistent with the 0.005 per year indicated in Table II.

4. GRADUAL DROP IN POWER

In addition to the statistical failure mechanisms discussed previously, a variety of observed mechanisms typically lead to gradual degradation or loss of power over the life of a photovoltaic array. These generally fall into two categories: optical losses and cell power degradation. Both of these degradation modes tend to be generic as opposed to being statistical, i.e., the majority of modules and cells of the same type degrade at the same rate with little statistical scatter. Most systems must be designed to accommodate this gradual power decrease, since the only correction techniques involve incremental addition of array area or total module replacement. Aside from external surface soiling, discussed in the next section, "Fixed Drop in Power," the observed optical degradation mechanisms include:

(1) Encapsulant transmission loss (uniform yellowing) due to UV and temperature-induced self-degradation.
(2) Encapsulant transmission loss (local yellowing) due to foreign matter diffusing into the encapsulant from such things as edge seals, mounting hardware, and electrical-terminal hardware.
(3) Deterioration of the anti-reflective coating on the solar-cell irradiated surface due to leaching or contamination from plating or corrosion products.

Substantial research has been done on these various degradation mechanisms over the past two years and a variety of highly stable materials and coatings have been identified (7, 12).

The second important category of gradual-degradation modes includes mechanisms related to solar-cell power degradation due to increased series resistance or junction shunting. Increased series resistance is often associated with a gradual deterioration of the adherence between the cell metallization and the cell bulk material due to corrosion-related processes. Junction shunting may be caused by the diffusion or migration of metallization elements into the cell junction or over the external surfaces of the cell.

As with the optical degradation mechanisms, research has identified a variety of highly stable cell metallization systems that promise degradation rates well below 1% per year (12).

For both degradation categories the most effective techniques for quantifying expected levels of degradation involve accelerated temperature/humidity testing (possibly with ultraviolet exposure) together with Arrhenius and other means of relating the data to long-term use conditions (12, 13). Application of these techniques to present commercial modules suggests that the 1% per year drop of power assumed in Table II should be readily achievable with future low-cost concepts.

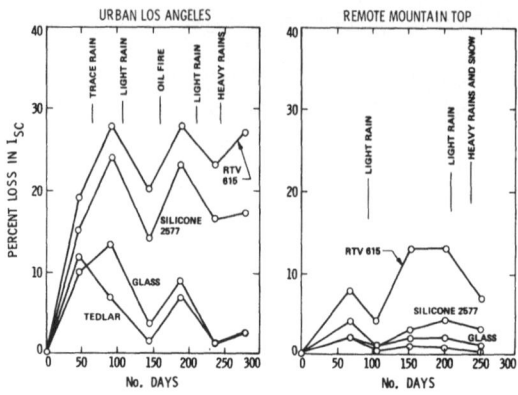

Figure 7. Array Performance Loss Due to Soiling.

5. FIXED DROP IN POWER

Although similar in effect to other optical-loss mechanisms, experimental data indicate that optical surface soiling due to dust and atmospheric contaminants reaches equilibrium levels in a few weeks and then fluctuates somewhat with natural cleaning mechanisms such as rain. The net result is most easily modeled as a fixed loss in array current and power over the life of the array.

Fig. 7 illustrates this soiling behavior for a variety of module surface materials in two site environments--one urban, the other remote. These and other data gathered by JPL over the past four years at a variety of sites in the United States indicate that average soiling levels below 5% should be easily achievable with glass or Tedlar-like optical-surface materials without washing (14). Very dusty remote sites and heavily polluted urban sites will, of course, exceed these levels, and may require periodic washing.

In response to these data Table II suggests a 5% allocation for fixed soiling-related losses.

6. MODULE WEAROUT LIFE

The last degradation category is the most difficult to quantify technically or to even define through known failure mechanisms. Nearly all of the known failure mechanisms have been studied and found to be gradual or statistical in nature and not associated with a wear-out end-of-life such as might be associated with automobile tires or light bulbs.

Mechanical fatigue of cell interconnects is a classic example of a wearout mechanism (Fig. 6). However, to achieve the desired low rate of random interconnect failures during the early life of the array, the wearout-life associated with 50% failures will typically be more than 100 years.

Gradual depletion of UV absorbers, or gradual embrittlement of the module encapsulant system are additional mechanisms that could lead to accelerated degradation in later years due to optical or mechanical-stress mechanisms.

The most likely mechanisms leading to array end-of-life may be technical obsolescences, decreased reliability, or the useful life of the application itself.

In consideration of the lack of known wearout mechanisms, an end-of-life in excess of 25 years seems quite likely for mature module designs of the future.

7. SUMMARY AND CONCLUSIONS

As part of the United States National Photovoltaics Program, the Jet Propulsion Laboratory Flat-Plate Solar Array Project has carried out a comprehensive array engineering activity addressed to understanding the reliability attributes of terrestrial flat-plate photovoltaic arrays and to deriving analysis and design tools useful for array optimization and cost reduction. Known array failure and degradation mechanisms have been carefully studied and grouped, for the purpose of discussion, into five categories. A target reliability allocation has been developed for each degradation category based on the life-cycle-cost requirements of future large-scale applications and the technical realities of available photovoltaic materials and processes. Comparison of these future requirements with present performance and design alternatives suggests that lives in excess of 20 years are very possible for module designs of the near future.

8. REFERENCES

1. Ross, R.G., Jr., "Photovoltaic Design Optimization for Terrestrial Applications," *Proceedings of the 13th IEEE Photovoltaics Specialists Conference*, Washington, D.C., June 5-8, 1978, pp. 1067-1073.

2. Ross, R.G., Jr., "Flat Plate Photovoltaic Array Design Optimization," *Proceedings of the 14th IEEE Photovoltaic Specialist Conference*, San Diego, California, January 7-10, 1980, pp. 1126-1132.

3. Ross, R.G., Jr., "Photovoltaic Module and Array Reliability," *Proceedings of the 15th IEEE Photovoltaic Specialists Conference*, Orlando, Florida, May 12-15, 1981, pp. 1157-1163.

4. Gonzalez, C., and Weaver, R., "Circuit Design Considerations for Photovoltaic Modules and Systems," *Proceedings of the 14th IEEE Photovoltaics Specialists Conference*, San Diego, California, January 7-10, 1980, pp. 528-535.

5. *Flat-Plate Photovoltaic Module and Array Circuit Design Optimization Workshop Proceedings*, JPL Document No. 5101-170, Jet Propulsion Laboratory, Pasadena, California, May 19-20, 1980.

6. Moore, D., and Wilson, A., *Photovoltaic Solar Panel Resistance to Simulated Hail*, JPL Document No. 5101-62, Jet Propulsion Laboratory, Pasadena, California, October 15, 1978.

7. Carroll, W. et al, *Photovoltaic Module Encapsulation Design and Materials Selection*, JPL Document No. 5101-177, Jet Propulsion Laboratory, Pasadena, California, November 1, 1981.

8. Chen, C.P., Fracture Strength of Silicon Solar Cells, JPL Document No. 5101-137, Jet Propulsion Laboratory, Pasadena, California, October 15, 1979.

9. Mon, G.R., Moore, D.M., and Ross, R.G., Jr., Interconnect Fatigue Design for Terrestrial Photovoltaic Modules, JPL Document No. 5101-173, (DOE/JPL-1012-62), Jet Propulsion Laboratory, Pasadena, California, March 1, 1982.

10. Moore, D.M., Proposed Method for Determining the Thickness of Glass in Solar Collectors, JPL Document No. 5101-148, Jet Propulsion Laboratory, Pasadena, California, March 1, 1980.

11. Mon, G.R., "Defect Design of Insulation Systems for Photovoltaic Modules," Proceedings of the 15th IEEE Photovoltaics Specialists Conference, Orlando, Florida, May 11-15, 1981, pp. 964-971.

12. Lathrop, J.W., et al, Investigation of Reliability Attributes and Accelerated Stress Factors on Terrestrial Solar Cells, 1st, 2nd, 3rd Annual reports, DOE/JPL-954929-79/4, 80/7 and 81/8, Clemson University, Clemson, South Carolina, May 1979, April 1980, January 1981.

13. Desombre, A., "Methodology for a Reliability Study on Photovoltaic Modules," Proceedings of Third E.C. Photovoltaic Solar Energy Conference, Cannes, France, 27-31 October, 1980, pp 741-745.

14. Hoffman, A.R., and Maag, C.R., Photovoltaic Module Soiling Studies, May 1978-October 1980, JPL Document No. 5101-131, Jet Propulsion Laboratory, Pasadena, California, November 1, 1980.

OPERATIONAL EXPERIENCE WITH INTERMEDIATE FLAT-PLATE PHOTOVOLTAIC SYSTEMS

V. V. RISSER and H. S. ZWIBEL
New Mexico Solar Energy Institute
New Mexico State University

Summary

The New Mexico Solar Energy Institute has been involved in the design, construction, test, operation, and data analysis of two intermediate flat-plate PV systems since 1978. These projects are funded by the United States Department of Energy. The 20-kilowatt peak system at El Paso, Texas, provides direct current power to an uninterruptible power supply. Operational since January 27, 1981, it has produced 37,300 kilowatt hours through March 31, 1982. The 100-kilowatt peak utility-grid-connected system at Lovington, New Mexico, has been operational since March 17, 1981, having produced 199,900 kilowatt hours direct current and 174,000 kilowatt hours alternating current through March 31, 1982. Primary user of this energy is a shopping center located approximately 1 kilometer from the system. Both systems operate automatically and unattended. The reliability of prime systems components has been quite high. NMSEI is collecting and analyzing data from both systems. The results are presented in this paper.

1. INTRODUCTION

The New Mexico Solar Energy Institute (NMSEI) has been evaluating system performance of two intermediate flat-plate photovoltaic (PV) systems since early 1981. The 20-kilowatt peak system in El Paso, Texas, (latitude 32.28°) and the 100-kilowatt peak system in Lovington, New Mexico, (latitude 32.95°) were designed and constructed as part of the United States Department of Energy (USDOE) program to encourage the development of PV systems technology. The program includes system performance monitoring and data analysis. Each site is equipped with identical computer controlled data acquisitions systems (DAS). The DAS scans weather sensors and system performance sensors at a one-minute rate, averages 10 readings, and records the average on magnetic tape. String currents are sampled and recorded each half hour. Currently, 84 parameters including 64 string currents are recorded at El Paso and 85 parameters including 42 string currents are recorded at Lovington. The DAS also provides for remotely accessing this recorded data via telephone line. NMSEI collects data daily and performs analysis in many areas.

2. SYSTEMS

El Paso 20-Kilowatt Peak System

The El Paso system generates power for an uninterruptible power supply (UPS) at the El Paso Electric Company generating station in El Paso, Texas. The UPS provides power for critical direct durrent (dc) loads and two inverters that supply power for a computer controlling a 197-megawatt combined cycle generator. The load on the UPS is a near constant 24 kilowatts. The voltage is fixed at 134 volts dc to maintain full charge on three battery banks, which provide backup in case of power failure. The system block diagram is given in Figure 1. Since the output of the PV system is always less than the load, the UPS voltage determines the operating voltage of the PV system.

The system operates automatically and unattended. If the insolation is sufficient, the array is connected to the load. Otherwise, the array is disconnected, shorted, and grounded. The array is also disconnected if the bus voltage exceeds preset limits. There have been no interface problems between the array and the UPS.

Panel structures at El Paso are made of treated pine. This material is easy to work and has been used successfully for a number of years in the arid climate of the Southwest. After one year, some panels (about 15 percent) exhibited excessive droop and required additional braces. However, no degradation of the module support is apparent.

Pertinent statistics on the El Paso system PV array are
  °278-square-meter panel area
  °168-square-meter active silicon area
  °fixed 26° tilt--due south
  °64 parallel-connected strings
  °9 series-connected modules per string
  °fixed voltage operation--134 volts
  °150 amperes maximum

The El Paso system produced 32 megawatt hours during the first 12 months of operation, Figure 2. The average system efficiency was 6.1 percent. There were 13 days when the system provided no power: 9 of these were due to maintenance on the UPS (non-PV related); 3, to insufficient solar insolation; and 1, to hardware failure (controller). The overall capacity factor for the first year was 24 percent.

Lovington 100-Kilowatt Peak System

This flat-plate system also operates automatically and unattended. Continuous monitoring of the array output current provides the information used to control the state of the system. When sufficient solar insolation is available, power is transferred to the utility. During periods of low incident solar radiation, the array is shorted and grounded.

The system consists of two equal subfields, each providing power to and controlled by a 60-kilowatt rated power conditioning unit (PCU). As shown in Figure 3, the PCU provides peak power tracking, inverts the nominal 240 volt dc to 480 volt 3Φ, 60 hertz alternating current (ac) power, and controls the power transfer to the utility grid. Loss of utility power causes automatic shutdown of the PCU. The first and primary user of the PV power is Lovington Square Shopping Center, which obtains approximately 8 percent of its annual energy needs from the PV system.

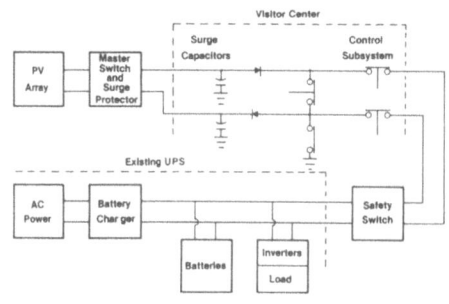

Figure 1. El Paso System Block Diagram

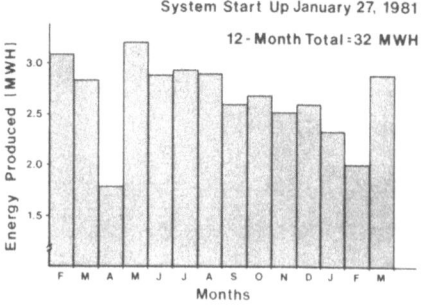

Figure 2. El Paso Energy Production (dc)

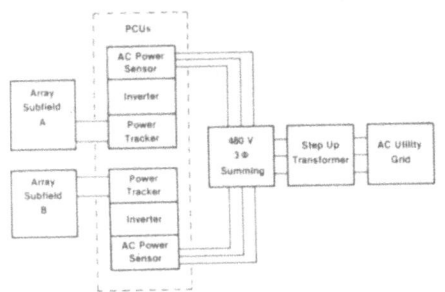
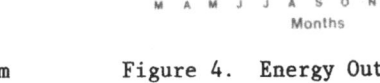

Figure 3. Lovington System Block Diagram

Figure 4. Energy Output (dc) at Lovington

The PV array faces true south; the galvanized steel frames may be manually adjusted to 10°, 30°, or 40° from horizontal. The pertinent statistics for each of the two subfields are

- 842-square-meter module area
- 507-square-meter active silicon area
- 21 subarrays in parallel
    80 modules
    5 parallel x 16 series electrical connection
- 220--260 volts dc
- 250 amperes maximum

This system has been operating successfully since March 17, 1981. During the first year, 192,000 kilowatt hours of dc energy were produced at an average efficiency of 6.4 percent, Figure 4. The PCU efficiency has averaged 87 percent, with peak efficiency of 91 percent at full power. The average system capacity factor for the first year was 25 percent.

3. SYSTEM STUDIES

Array Washing

Since both PV systems are located in semiarid high desert climates where blowing dust is a common phenomenon, the economic feasibility of washing the PV array has been investigated. Of particular concern at the El Paso site was the effect of the precipitation from plant cooling towers located 150 yards due east of the PV array. Although prevailing winds are from the west, there are times, particularly in the mornings, when an east breeze will blow clouds of mist generated by the cooling towers over the array. These clouds quickly dissipate in the dry atmosphere, but under some conditions, precipitation can occur.

During the design phase, several test panes of glass were placed on the proposed array site to determine the severity of the problem. The test frames consisted of common glass panels with one-half exposed and one-half covered. After 12 months of exposure, the glass was tested and the transmissivity of the exposed side was only 1.5 percent less than the covered side. One test panel remains in place to allow further study.

The opportunity to gather soiling data from the operating El Paso array happened almost accidentally in November 1981. The system operator, El Paso Electric Company, was doing some sand blasting operations 100 yards southeast of the solar array, and just south of the plant cooling towers. This sandblasting was done for a two-week period and, caused a thick layer of dirt to adhere to the PV modules. When the sandblasting operation ended, arrangements were made to wash all but three panels in the PV array. The operation of the panels that were not washed had been previously matched with other panels to provide a test basis. Figure 5 shows the deviation in string currents for two panels, taken under widely varying solar conditions. During the two weeks of sandblasting, the deviation averaged less than 0.5 percent for the three sets of test panels. After washing the array, the current difference between the matched washed and unwashed panels jumped to 6 percent on two sets of test panels and 7.5 percent on the third set. After a very light all-day rain (0.68 cm) on November 27, 1981, the dirty panels were cleaned well enough that the current comparison on the test panels fell to about 1 percent. This implies that the expense of array washing may not be economically justified for the soiling likely to occur at the site of these PV plants.

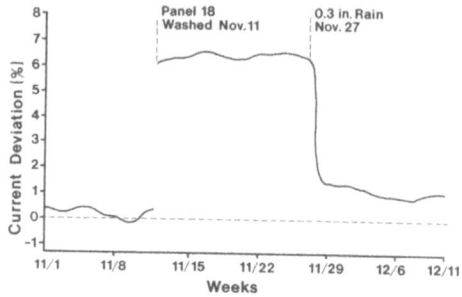

Figure 5. String Current Deviation for Two Test Panels

Array Tilt

The system at Lovington can be adjusted to three different tilt angles from horizontal--10°, 30°, and 40°. The array was installed in December 1980 at a 40° tilt angle. A plan was developed to compare the actual system output of the two 50-kilowatt subfields operating at different tilt angles. This plan called for setting one subfield at 30° in August 1981 for a comparison of 30° and 40° operation, and then setting the 40° field to 10° in late April 1982 to compare 10° and 30°. Figure 6 shows the actual dc energy production for the 30° and 40° tilt subfields from August 1981 through March 1982. The subfield at 40° produced 3,600 more kilowatt hours.

The labor used to lower one subfield of 168 panels from 40° to 30° was 36 man-hours at an estimated $8.50 per hour, for a total cost of $306, or approximately 10¢ for each additional ac kilowatt hour produced by the change. The testing will continue as planned.

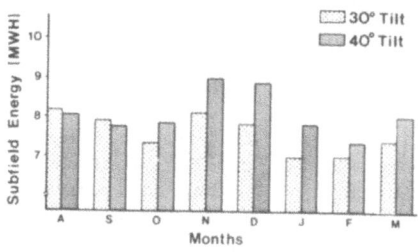

Figure 6. Subfield Energy Production (dc)

4. FAULT ISOLATION AND MAINTENANCE

Both of these PV systems are operated and maintained by electric utilities. Because their personnel are not trained in PV system maintenance, NMSEI undertook development of a fault location analysis that would isolate suspected failures to an individual string. The primary objective of the program is to locate any module with one of the bypass diodes conducting.

On the El Paso system, the string consists of nine series-connected modules; at Lovington the string has 16 series groups of five modules in parallel. Each of the groups of five modules also has a bypass diode. At El Paso, where the string voltage is fixed, the conduction of one bypass diode forces the other 8½ modules to operate at a higher voltage. Since the operating point is near the peak power voltage, any voltage increase causes a noticeable decrease in string current. At Lovington, the mechanism differs but the result is the same. If a failure causes conduction of a bypass diode in the module, the limited current-producing capability will cause the unaffected modules in the group to operate at higher current levels. This may cause one or more modules to become reverse biased and the group bypass diode will conduct. This causes a power loss in the string (1 of 16 groups shorted) and also causes the unaffected groups to operate at a higher voltage. This further reduces string current and allows detection of module failures in the string.

Both these systems have been monitored extensively and any slight decrease in performance has caused maintenance to be initiated. Maintenance activity required because of actual component failure has been quite low on both systems. Therefore, to present a detailed maintenance evaluation, all occurrences that required intervention in the automatic operation of the systems have been recorded, Table I. Examples of included activity would be the recycling of a PCU after a ground fault error or resetting the DAS computer after an error had caused a software time out. Through January 31, 1982, the total maintenance man-hours, exclusive of the daily scheduled walk-through, have been 267 at Lovington and 78 at El Paso. These figures do not include special test activity such as changing tilt angles or array washing. Travel time has been included in the Lovington figures when the action required a special trip from the utility facilities located two miles away. The system operators are pleased that most of the maintenance activity, particularly on the PV array itself, can be scheduled and performed on a nonemergency basis.

Table I. Maintenance Activity through January 31, 1982

Subsystem	Occurrences		Man-hours	
	LOV	ELP	LOV	ELP
PV Array	9	14	30	34
PCU	56	--	88	--
Control System	--	6	--	16
DAS	81	27	140	26
Other	3	2	9	2

5. CONCLUSION

These intermediate-size systems were funded by USDOE, with a major objective being the collection and analysis of operating data. Both systems have performed well and the operational and maintenance experience gained has been substantial and significant. Specifically--
° There are no technological barriers to prohibit operation of flat-plate PV systems.
° These systems can be installed by small businesses with no prior PV experience.
° Fixed voltage operation causes less than 3 percent loss in annual energy production.
° Array washing is normally not necessary for the two sites.
° For tilt angle adjustment to be economic, a labor saving method of panel support adjustment must be incorporated into the design.
° These types of systems can be controlled automatically and can operate unattended.
° System maintenance (preventive maintenance) can be scheduled and is seldom required on an emergency basis.

NMSEI plans to monitor system performance at both sites for at least 14 more months. With the complete performance documentation system in place, the data being obtained will provide a firm foundation for future system performance predictions.

OPERATIONAL EXPERIENCE WITH A 35-KWP CONCENTRATING PHOTOVOLTAIC SYSTEM

R. M. Spencer
Acurex Corporation

Summary

Acurex Corporation completed the installation and startup of a 35-kWp concentrating photovoltaic power system in October 1981. This paper describes the operation and performance of the field during the first 6 months of operation. The system has performed at a lower level than predicted, partly due to the unusual climatic conditions experienced at the site. The operational experience indicates three problem areas in the system design. First, some receiver degradation has ocurred due to manufacturing defects which have been corrected. Second, the power conditioning unit has demonstrated low reliability in the Hawaiian environment. Third, corrosion problems associated with the site have accounted for a major portion of the required maintenance. Overall, the system has operated reliably, except for the power conditioning unit, and most system components do not show signs of degradation.

1. SYSTEM HISTORY

In October of 1981, Acurex Corporation completed construction and startup of a 35-kWp concentrating photovoltaic system on the island of Kauai, Hawaii. The system was jointly sponsored by the U.S. Department of Energy and the State of Hawaii, Research Corporation of the University of Hawaii. The system incorporates 445m^2 of parabolic trough reflectors to concentrate sunlight onto photovoltaic receivers. Concentrating photovoltaic cells manufactured by Applied Solar Energy Corporation are mounted on two faces of a hollow aluminum extrusion to optimize the collection of the concentrated energy. Water, circulated through the receivers to cool the photovoltaic cells, is heated to 80°C. The heat is stored in a 12,000ℓ insulated tank to provide hot water for use in an adjacent hospital. The electrical energy is converted to 480 VAC and fed into the electrical distribution network for the island.

The system, the first actively cooled concentrating photovoltaic system to be commissioned in the United States, is shown in figure 1. Several unique features are incorporated in its design.

The receivers are designed to insure that they can be reworked inexpensively and simply onsite without any specialized equipment. They incorporate modular cell assemblies which are designed for mass production and offer an optimized series-parallel wiring of the cells for single-axis tracking.

The wiring in the field was accomplished with premanufactured, direct-burial cable harnesses. Installation was accomlished by digging the trenches, dropping the wiring in place, and back filling the trenches. Connections to the arrays and instruments were made with environmentally sealed connectors. This greatly simplifies the extensive amount of wiring required in photovoltaic applications.

The inverter system, used to transform the DC photovoltaic output into AC compatible with the local distribution network, is unique to this

Figure 1. Photovoltaic Energy System Located at Wilcox Hospital, Kauai, Hawaii

system. Built by Westinghouse Corporation, it incorporates a microprocessor-controlled, 12-step bridge which eliminates all harmonics lower than the 19th and greatly improves efficiency.

An extensive instrumentation and data acquisition system is employed to gather operational data. A microcomputer data acquisition system manufactured by Hewlett Packard gathers electrical, thermal, and meteorological data on system performance which is stored on magnetic tape and analyzed by Boeing computer services.

2. SYSTEM PERFORMANCE

The concentrating photovoltaic system commenced service in late October 1981 and operational data has been collected from the first of November through March 1982. These are the lowest insolation months of the year, as well as the maximum declination of the sun. The data in table I illustrates this fact. The prediction of performance is based on analysis of the Hawaiian site. The monthly and daily values are based on clear sky calculations as indicated by the number of days specified for system operations. The actual data incorporates both cloudiness and system downtime. The system was down for repairs throughout the month of December due to a component failure in the power conditiong unit.

The principal cause of nonoperation in the other months has been rain. This has been a record precipitation year in the islands and not a good one for solar energy collection. Several days during which the system was classified as operating have contained less than 1 hr of sufficient direct insolation to operate the field. Despite these facts, it appears that the performance prediction is optimistic and a more detailed analysis of the expected performance will be developed using actual insolation data.

The instantaneous performance of the solar field provides a much better analysis tool by removing the variable of weather. Single collectors were studied at length before the construction of this system and system performance can be extrapolated from those data. The system performance is dependent on the coolant temperature, insolation level, and the solar declination angle. The declination contributes not only to a cosine loss but also to shadowing of sections of the receivers. The effect of this shadowing can be partially alleviated by the installation of bypass diodes to shunt the shadowed cells.

The results of comparing performance predictions with actual performance of the field on an instantaneous basis are shown in table II. The initial comparisons in November and January indicated that the system was performing substantially as predicted. The more recent comparisons in February and March indicate poorer correspondence. This could imply two things. First, the performance prediction could be failing to adequately represent some parameter. Second, the field could be degrading in output. No degradation in performance has been specifically determined at this point, although several degradation paths have been identified. More detailed studies are underway to assess the cause of the discrepancy.

3. OPERATIONAL EXPERIENCE

Hawaii is an excellent site for testing of solar fields because it offers an accelerated aging environment compared to the more traditional desert site. The solar system on Kauai is exposed to salt spray due to its proximity to the ocean and steady shore breezes. The frequent partialy cloudy/rainy weather pattern causes the operational components to cycle in temperature and operation 10 to 100 times more often than a normally clear desert environment.

The rapid cycling has greatly accelerated the normal aging of the controls system. The hermetically sealed reed relays in the control system have a test cycle life of 100,000. During the 6 months of startup and operation some have cycled 20,000 times. There has been one relay failure in that time. Otherwise the control system has been entirely reliable. In particular, the performance of the tracking systems for the collectors has been flawless. There have been no problems with "false tracking" of bright cloud surfaces.

The photovoltaic receivers have demonstrated three failure modes that represent a serious problem for the system, which is composed of 1,530 photovoltaic modules.

Ground fault -- We have experienced two ground fault related failures, wherein the voltage insulation between the cell string and the grounded aluminum extrusion failed. The result is an electrical arc which deposits the cell metalization onto the extrusion. The heat generated in this process shatters the glass superstrate and cell and burns the PVB encapsulant. In all cases the receivers have been repaired onsite and are back in service. Sufficient data has not been gathered to predict a frequency for future failures.

Mechanical failure -- The glass superstrate over the cell modules is a lightly heat-tempered, low iron, soda-lime glass. We have experienced eight cracked glass modules. The cracks originate at cutting-induced flaws at a glass edge and propagate through the module surface until they can terminate on another edge. Because the crack allows moisture to penetrate the module, cracked modules must be replaced. All broken glass modules have been replaced in the field. Glass breakage of this sort is a statistical event and the current failure rate for the last 3 months indicates 0.5 percent per year anticipated failures.

Thermal failure -- The predominate receiver degradation in early operation was thermal/ultraviolet degradation of the PVB encapsulant. The degradation takes the form of bubbling and then brown discoloration of the water-clear PVB. The two phenomena have been linked to residual catalyst and vinyl alcohol interactions. The vinyl alcohol reacts to form a carbonile and causes the discoloration. The receivers manufactured for qualification testing did not demonstrate this effect in extensive

Table I. Monthly Performance Data

Month	Prediction			Actual		
	Daily kWh/day	Monthly kWh/month	Days of Operations	Daily Average kWh/day	Monthly kWh/month	Days of Operation
Jan	37.8	1,170	31	35.4	673.5	19
Feb	58.8	1,650	28	56.7	1,021.5	18
Mar	50.4	1,560	31	47.7	382.5	8
Apr	54.6	1,640	30			
May	75.6	2,340	31			
Jun	79.8	2,390	30	------ No Data ------		
July	75.6	2,340	31			
Aug	92.4	2,860	31			
Sept	79.8	2,390	30			
Oct	58.8	1,820	31			
Nov	33.6	1,010	30	42.0	294.0	7
Dec	22.7	704	31	13.5	13.5	1

Table II. Instantaneous Performance

Date	Number of Samples	Prediction	Actual
Nov 20	17	8.5	8.0
Jan 13	7	9.5	8.8
Jan 14	15	10.5	12.1
Jan 22	12	11.1	10.6
Feb 6	17	13.5	6.9
Feb 26	8	19.9	11.0
Mar 7	11	18.7	12.2
Mar 25	7	28.8	13.9

testing. Those manufactured for field installation were mass produced and the inadequate distribution of thermally conductive grease caused locally poor heat removal from the cell modules. As a result, 42 percent of the modules currently exhibit bubbling and 8 percent have some discoloration. We are approximately 50 percent complete in rebuilding the receivers to add thermally conductive grease and improve the efficiency of heat removal. The bubbling and discoloration do not noticeably degrade module performance. Receivers which have been rebuilt show no further indication of degradation.

The operational data acquisition system (ODAS) has had 11 failures in 6 months of operation. Most failures are related to the high humidity in Hawaii. The failure of the ODAS does not affect system performance.

The power conditioning unit has been the major source of downtime for the system. The erratic nature of the Kauai electric distribution network has proven to be a difficult operational environment for the Westinghouse inverter. In addition, the rapid fluctuation in insolation level causes the unit to drop offline frequently. The unit has been shown to cycle as often as 250 times in a single day. The result has been an extensive number of field modifications to cause the unit to function more reliably. The personnel from Westinghouse have been very helpful and supportive in bringing their unit to within specifications. Nonetheless, the unit has operated acceptably only 4 months out of 6 and it remains to be determined whether this is representative of its reliability.

The corrosion of system elements due to the salt spray environment has been the other significant problem with the field. An effort was made to protect the system components using galvanized metal for the collector components and painting all other exposed surfaces. The galvanizing is still holding up well. The painted surfaces have shown considerable failure to date. When the corrosion is observed, the area is stripped and repainted. This treatment appears to be effective in dealing with the problem but consumes a great deal of the operator's time.

Galvanic corrosion and pitting corrosion within the fluid loop have been controlled by using the combination of a sacrificial anode (zinc) and an anticorrosive additive (Calgon BTZ2000). No significant corrosion is noted in the aluminum receivers.

The reflective surfaces of the concentrating collectors are aluminized mylar (Scotch FEK-244) on anodized aluminum substrate. There has been no observable degradation in reflectivity and washing of the collectors is not required due to the frequent rains which adequately cleanse the reflective surface.

4. CONCLUSIONS

The first actively cooled concentrating photovoltaic system in the United States has been operating for 6 months. The performance is below predicted levels which may be coupled to unusual climatic patterns. The principal operational problems have come from the most advanced components: the receivers and the power conditioning unit. Both are operational and the problems are understood. The unusual environment at the Hawaiian site is proving to be a severe test of the system components.

SOLAR PHOTOVOLTAIC SYSTEMS IN THE DEVELOPMENT OF PAPUA NEW GUINEA

G.H. KINNELL MIE AUST.
Project Evaluation Engineer
Energy Planning Unit, Department of Minerals and Energy,
P.O. Box 2352, Konedobu, Papua New Guinea

Summary

The extreme tropical climate in Papua New Guinea, site remotness and cultural factors all present a challenge to designers and manufacturers of electrical and electronic equipment and to photovoltaic systems in particular.

This challenge has been accepted in order to provide rural and village people in this country with reliable and economical lighting, water pumping and medical refrigeration systems which incorporate solar photovoltaic stand-alone equipment.

Such a system, in addition to having low maintenance and operating costs, will reduce the dependence of Papua New Guinea's rural areas upon expensive and cost increasing non-renewable diesel fuel oil and kerosene used for electricity supply, lighting and water pumping at present.

Our experience, together with equipment performance safety, reliability, economics and policy considerations are presented.

In order to overcome the many problems encountered, particularly early failures, a firm policy has been adopted, based on sound scientific and engineering evaluation, knowledge and experience, to accept only good quality, safe, reliable and economic 'TROPICALIZED' photovoltaic and associated equipment for the implementation of the various projects undertaken.

Overseas manufacturers are fully advised on our requirements to enable them to consider all factors involved.

The simple solution for rural lighting has been to use standardized small low voltage DC lighting kits, on the basis of one per home, and not large PV AC systems supplying a complete village or patrol post.

Village water supplies can be provided by photovoltaic pumping systems for as low as one fifth of the capital cost of complex water reticulation schemes. Details are discussed.

The effectiveness of fluorescent lighting is compared with kerosene lighting and the results of simple photometric measurements are given. Battery types are discussed and reasons for selecting specific types are given.

Insolation measurements and records are discussed and comparisons given between the various provinces, particularly coastal and highlands regions.

Finally, the future is presented with projections of the potential total number of systems required together with the possible cultural, social and economic impact when such a level is reached.

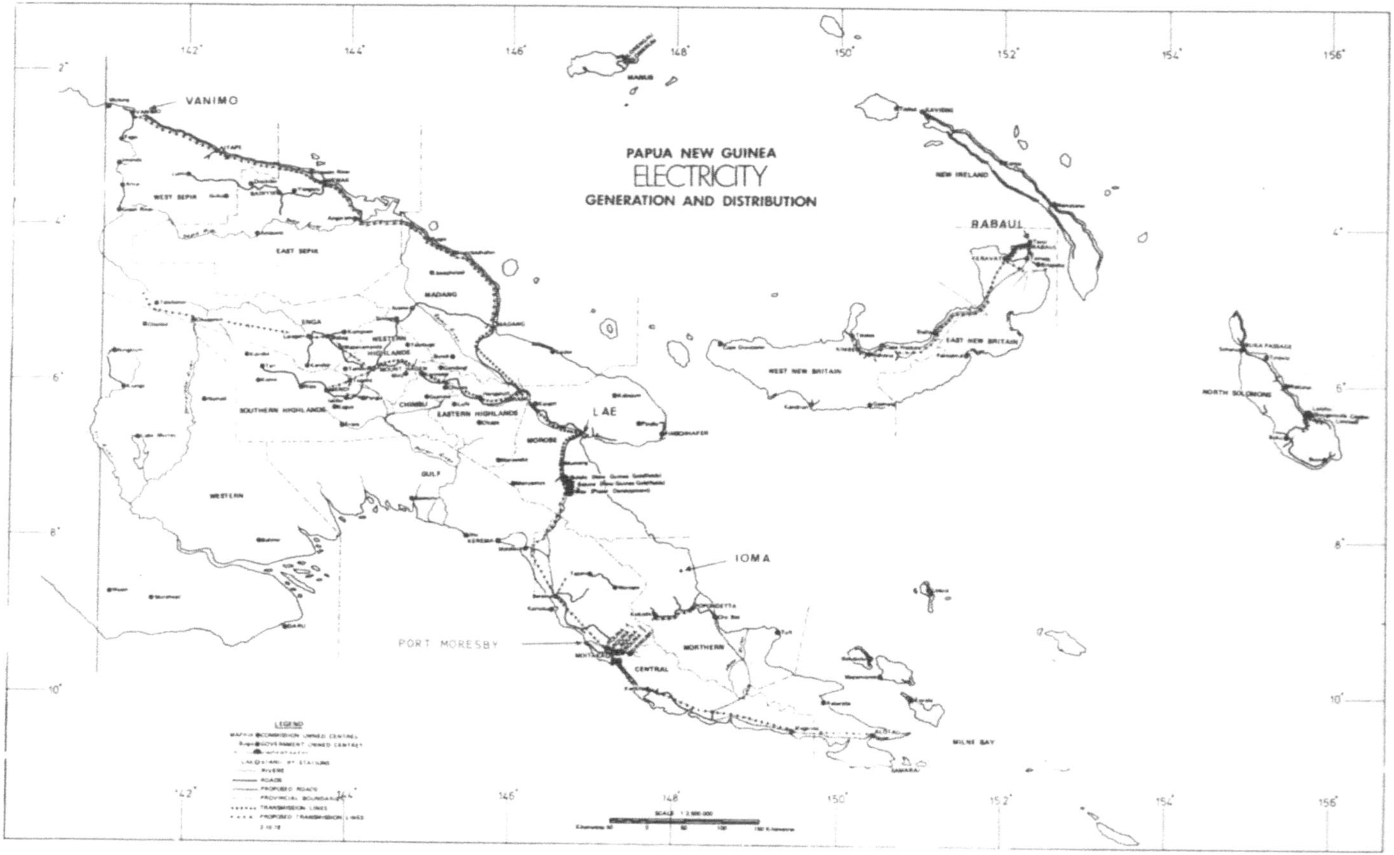
PAPUA NEW GUINEA
ELECTRICITY
GENERATION AND DISTRIBUTION
VANIMO
LAE
RABAUL
IOMA
PORT MORESBY
WEST SEPIK
EAST SEPIK
MADANG
ENGA
WESTERN HIGHLANDS
SOUTHERN HIGHLANDS
CHIMBU
EASTERN HIGHLANDS
MOROBE
GULF
WESTERN
NORTHERN
CENTRAL
MILNE BAY
NEW IRELAND
EAST NEW BRITAIN
WEST NEW BRITAIN
NORTH SOLOMONS
MANUS
LEGEND
142°
144°
146°
148°
150°
152°
154°
156°
2°
4°
6°
8°
10°

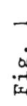

Fig. 1

INTRODUCTION

This paper is written in the belief that people are important and that equipment is to serve the needs of the people and therefore should be designed to meet their specific needs and environment. This is particularly important in the case of a developing country when a professional engineer accepts the responsibility to formulate policies evaluate equipment implement projects and train national people.

1. Government, geography and climate

Papua New Guinea, an independent and self governing state since 1975, is located directly North of Australia above the North Eastern State of Queensland.

The country extends from 141° east longitude, at the border with Indonesia (Irian Jaya) to 160° east longitude and between latitudes 1° and 12° south (see figure 1).

Papua New Guinea is a parliamentary democracy, with a single legislature known as the National Parliament (1). The State is divided into 19 provinces plus the National Capital District (Port Moresby) with decentralized Government established in each province. Before independence the country comprised the Australian territory of Papua in the southern regions and the United Nations Trust Territory of New Guinea in the North (1).

Land area is 462,840 square kilometres This includes the mainland, the three large islands of New Britain, New Ireland and Bougainville plus 600 small islands and archipelagos. Approximate direct distances from the capital city of Port Moresby to some of the other centres are : Vanimo 990 km, Rabaul 800 km, Arawa 990 km and Lorengau 825 km.

The elevation of the land varies from sea level on coastal areas to 4 706 metres at the peak of Mt. Wilhelm, the highest mountain.

Very rugged mountain ranges are a dominant feature of the mainland and extend in a north westerly direction from the eastern tip to the western border. Similar but lower mountain ranges are on the three largest and some of the other islands.

Two large valley systems are located in the North of the mainland along the Markham and Ramu rivers. These together with smaller valleys and plains in the highlands are utilized for agriculture.

Over 75 percent of the land is covered with tropical rain forest including some of the largest swamp systems in the world and in both the lowlands and the highlands areas of man-made grassland are present.

Rainfall : Annual rainfall throughout the whole country varies considerably as demonstrated in the following examples : Port Moresby 1200 mm (marked wet and dry season), Lae 4617 mm, Boku, on the west coast of New Britain, 8350 mm, areas along the coast of New Britain and the Gulf of Papua, over 9000 mm (1).

Temperatures : Temperatures in the lowlands and coastal areas are generally high, ranging from 30°C in the afternoon to 22°C at night. In the highlands lower temperatures are experienced an example being Goroka, elevation 1565 metres and co-ordinates : 06°04' latitude, 145°24' longitude, in the eastern highlands. Extreme temperatures for Port Moresby are 36°C and 14°C, Lae 39°C and 18°C, Mt. Hagen 31°C and 2°C (1).

Relative humidity : Relative humidity is generally high, being in the range of 75 - 90 percent in the lowlands and 65 - 80 % in the highlands (1). Extremes of relative humidity can be 100 % up to 25°C and 95% above 25°C (2).

Seismic activity can be expected as a line of active volcanoes extends along the north coast of the mainland through the islands of New Britain.

Zone A, equivalent to the worlds most active region includes East New Britain, southern part of New Ireland and western areas of Bougainville Island.

Lightning occurs throughout the country and requires consideration in power stations connected to overhead reticulation lines and in small stand-alone systems on exposed elevated locations.

2. The people and their way of life

The people of Papua New Guinea are Melanesian with their own distinct culture and social customs. They are generally friendly and helpful in villages and the author has found much co-operation and interest when going to villages and rural centres to discuss, demonstrate and sometimes install small photovoltaic lighting systems.

Results of the 1980 census shows that the population was 3 006 799 persons at that time with 86.9 % of the people living in rural areas.

The diverse culture is demonstrated by the fact that there are over 700 vernacular languages in Papua New Guinea. However, two common official languages are used : Pidgin and Motu and these are taught in the schools in addition to English with Pidgin being the main universal language.

The way of life of the people varies widely. Many rural village people depend upon subsistence farming for basic needs and then upon relatives working in the Government or private sectors for cash to purchase goods : e.g. manufactured items, radios, outboard motors, etc. or processed foods : rice, sugar etc.

Cash cropping is relatively extensive ranging from small holdings to large community or business undertakings. These include copra, cocoa, coffee, fish etc.

The national and provincial public services employ most of the educated national people with a smaller proportion being employed by the private sector.

Primary education is available with between 80 - 90 % of eligible children attending national primary school. For higher education, national and international high school are provided to full secondary standard.

A technical college for trades courses is located in Port Moresby and one for sub professional engineering courses at Lae. Professional engineering courses are undertaken at Lae University of Technology. Studies for professions in arts, science, law, medicine etc. are provided by the University of Papua New Guinea in Port Moresby.

3. Transport, telecommunications, radio, television

3.1 Transport : Because of the rugged nature of the country, roads are limited to an interlinked system in the northern area and the highlands. A much smaller one exists in the Central Province around Port Moresby. There is no road link through the mountains between these two systems and many remote centres and villages are only accessible by air or sea.

Thus Papua New Guinea relies heavily upon air transport for travel and movement of goods. This is demonstrated by the number of aerodromes which total 427, comprising 66 Government aerodromes including 6 international, 216 commercial and 145 restricted. Many of the aerodromes are landing strips in the jungle or mountain areas. The highest landing strip is at May River - Aume village (4°23' latitude, 141°48' longitude) with an elevation of 3 072 metres.

The national airline Air Niugine provides air transport and cargo between major centres using F 27 and F28 Fokker aircraft and third level

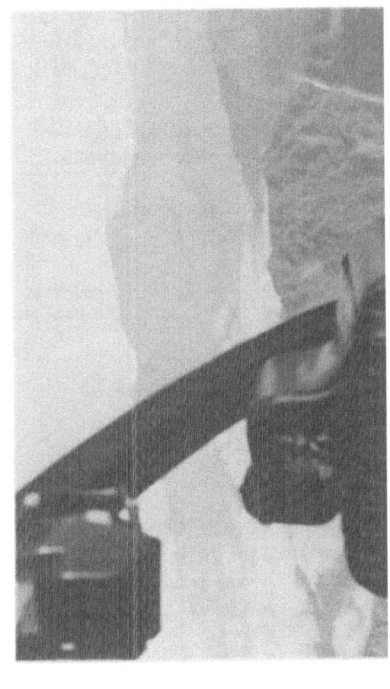

Approaching Mt. Yule by helicopter to visit P & T microwave repeater

PVC installation on top of Mt. Yule, 10400 ft. - P & T microwave repeater

Mountain landing strip near Mt. Yule

airlines provide the balance in a range of smaller aircraft (single and twin engine). Photovoltaic systems are used to power ground installations for radio beacons and navigational aids.

3.2 Telecommunications : The telecommunications network is installed through the country and is still being expanded. Posts and Telegraphs (P & T) are the responsible department and micro-wave repeater links are utilized with many of the repeaters located on mountain tops where electrical energy is supplied by photovoltaic systems. VHF and HF systems are also utilized.

3.3 Radio and television : The National Broadcasting Commission is responsible for broadcast radio and has studios in Port Moresby and other provincial locations.

A Government decision has been made to introduce broadcast television and consultants reports are at present being studied. A date has not yet been given for official commencement.

Photovoltaic systems can provide power supplies for remote area community television receivers combined with video audio visual education methods.

4. Energy

The Minister for Minerals and Energy is responsible for the activities of the Department of Minerals and Energy (DME) and the Papua New Guinea Electricity Commission (ELCOM).

4.1 ELCOM : Electricity generation and supply is mainly the responsibility of the Commission. However, the Government finances certain power stations in small and remote centres whilst rural missions and private industry i.e. timber mills, plantations etc. can have their own generating plant.

The official power stations are divided into Category A - Major Centres, Category B - Minor Centres and Category C - Outposts.

ELCOM is responsible for Category A and B stations plus high voltage transmission and distribution plus low voltage reticulation (415/240 Volt).

The Government owns and finances the Category C stations but equipment and low voltage reticulation is installed and maintained by ELCOM. Government Institution stations are installed and maintained by the Department of Works and Supply.

The Cat. A stations include the Rouna Hydro complex + 2 gas turbines (20 & 16 MW) at Port Moresby which feeds the National Capital District + the Ramu Hydro Station in the highlands which supplies certain northern centres and various highland centres. Cat. A stations connected to the Ramu Hydro system still maintain large diesel power stations for stand-by, part-time or full time generation depending upon conditions at Ramu. Three tariffs now apply :
Port Moresby Ramu Kreta/Arawa : all kWh - 11.5 t/kWh
Diesel category I : First 50 kWh - 11.5 t/kWh; Balance - 15.7 t/kWh
Diesel category II : First 50 kWh - 11.5 t/kWh; Balance - 26 t/kWh
(Note : PNG currency Kina and toea with 100 toea per kina and 1 kina = 10 French francs.

4.2 DME and alternative energy : The Department of Minerals and Energy is sub-divided into various divisions, namely : Policy and Planning, Mines, Geographical Survey, National Weather Service, Water Resources and Management Services.

The Energy Planning Unit (EPU) is a section of the Policy and Planning Division and is responsible for planning alternative energy projects through the Secretary of the DME, the National Energy Council

(NEC), ELCOM, National Planning Office (NPO), Department of Finance and other relevant Government Departments.

Each member of the EPU specializes in various areas of alternative energy with the author's responsibility being solar energy and wind energy, with particular accent on photovoltaic systems for rural and village development.

5. Photovoltaic systems

All systems throughout the country are summarized including all Government Departments, private sector and EPU projects.

It has been estimated that approximately 50 kWh of photovoltaic modules have been installed in Papua New Guinea to date.

The largest users have been :
- Posts and Telegraphs (P & T) : 22 - 23 kW in PVC/battery power supplies for micro-wave repeater systems and base stations;
- Civil Aviation Agency (CAA) : 15 kW, for air navigation aids;
- Mission stations and other users : 10.5 kW, for radio and lighting supplies.

Smaller users :
- ELCOM : 1.2 kW, for VHF vehicle communications. One large unit at approximately 0.9 kW and other smaller units totalling 3 kW;
- The Department of Works and Supply are now purchasing some two module (2 x 35 W) system for lighting in remote area construction camps.
- Water Resources and Geological Survey use small units as energy supplies for instrumentation.
- Boer village water pumping : 0.56 kW
- The Energy Planning Unit has arranged installation of approximately 1.4 kW of PVC lighting systems at 10 mA Patrol Post, 0.56 kW at Konedobu, office lighting and ventilation unit, 0.56 various village lighting kits, a total of 2.5 kW.

However, the potential for the next 10 years is estimated as :
- Village house lighting kits for the whole of PNG is approximately 500,000 : 17,500 kW
- Village street and area lighting up to 350 kW
- Government medical supply refrigerators - 350 kW
- Rural and village water pumping
- Government patrol posts
- Schools and churches
- Village business - trade stores, boat building, fish and crop processing
- Energy supplies to circulating pumps etc. in remote solar absorption air conditioning and fishing refrigeration units.

6. Rural lighting systems

6.1 Brief history : Early in 1980 commercially available 12 Volt DC fluorescent light units for use in PVC village lighting systems were evaluated electrically and also by simple photometric tests (Note: author was then Distribution Training Engineer at ELCOM).

Both 20 Watt and 40 Watt units were available but only 20 Watt will be discussed.

The 20 Watt unit consumed 24 Watts but the lamp gave approximately half its rated lumen output.

This was established by using a good quality light meter and comparing surface illumination one metre below the same tube mounted alternatively in a 12 Volt fitting and a 240 Volt AC fitting. Values obtained : 12 Volt - 50 lux, 240 Volt - 100 lux.

By fitting an industrial reflector the above values were increased to 90 lux and 180 lux respectively.

This data was fed back to the manufacturer via the local supplier with the result that 20 Watt 12 Volt DC fitting became available with efficient inverters with the lamps giving the same output as for the 240 Volt AC operation.

In addition the charge controllers were evaluated and found to be extremely complex in circuit design and operation and unreliable.

The first controllers have upper charge limiting but no discharge safeguard.

Both of the problems were fed back to the manufacturer but the following controllers, even though designed with discharge limiting, were unreliable.

From past highly reliable electronic design experience it was deduced that full tropicalized design and testing was required and this was discussed with the manufacturer's design engineers.

In 1980 two new type 'tropicalized' charge controllers became available from the same manufacturer and they have proved reliable during 12 months field operation.

6.2 Policy : The policy recommended and adopted for village and government patrol post lighting is to use individual 12 Volt DC PVC lighting systems for each house or building.

This overcomes problems of electrical safety in remote villages where bush material houses are often rebuilt every 5 to 10 years and allows environmental installation for village lighting, i.e. the business people are expected to be the first to install lighting, then other people, as they become aware of the advantages of improved lighting.

6.3 Lighting standards : Both British and Australian lighting standard give the following recommended values for illumination on a surface : office desks : 400 lux; home reading : 300 lux.

As shown by the measurements mentioned in 6.1 above, a 20 Watt fluorescent (white W 43 38 mm) lamp gives the following values on a surface one metre below the lamp (walls 2 metres away and ceiling 1 metre above fitting) : bare batten holder - 100 lux; with industrial reflector - 180 lux.

By using an adjustable height suspension system lowering the lamp can give the recommended standard values stated above.

Measurements carried out on a kerosene pressure lamp showed that such a device mounted with the mantle one metre above a desk gave a value of 12 lux on the desk surface (measured to one side away from shadow of tank).

Kerosene pressure lamps are thus a very inefficient light source for illuminating horizontal surfaces and are a high glare light source as most of the light radiates horizontally.

6.4 Village lighting kit : Various lighting kits are at present being evaluated and basic requirements are :
- one 35 Watt module
- one charge controller with upper and lower charge limiting plus overload protector (no fuses)
- one 12 Volt 100 A/h sealed and vented lead acid battery and lockable battery box
- plus fluorescent light fittings, wiring, switches, etc.

Note 1 : An alternative to the above is a self regulating system. The number and size of fluorescent lights will depend upon solar radiation at the site and the operating hours. Therefore the following could be used: one 20 Watt, one 13 Watt plus one 20 Watt or one 20 Watt plus one 20 Watt.
Note 2 : Sealed batteries are preferred to prevent contamination when topped-up.

6.5 High efficiency light sources : New 26 mm high efficiency lamps are being evaluated which can provide energy saving and thus increased operating hours or lighting. The normal range of savings is 10 %, however, the new Sylvania Octron 26 mm lamp is specified to give up to 25 % saving for 240 Volt AC operation and this is being investigated for use in 12 Volt fittings where special design might be required. Thus while development is required to improve PV module efficiency a large improvement can be gained in system efficiency with new improved light sources.

7. Lighting projects

7.1 Ioma patrol post - Oro Province : This patrol post was equipped with PVC 12 Volt DC lighting systems with individual system per building (23 total) (see fig. 2 for aerial view).

The buildings consist of : 16 small houses, 4 large houses, administration office block, medical centre and small school dormitory block (fig. 3).
The installation has been operating apparently reliably for 12 months and a follow-up visit will be made shortly.
Unfortunately the original funding did not include instrumentation, so evaluation depends upon verbal feed-back. Instrumentation is being purchased and installed within the next three months.

Fig. 2 : Ioma patrol post - aerial view

Fig. 3 : PVC lighting installation - school dormitory - Ioma

7.2 Basic lighting kits have been installed in remote areas and are apparently reliable. Here again we have to depend upon verbal feed-back. Note : Early equipment purchased and installed in 1980 before the author joined the Energy Planning were unreliable. The main basic problem being charge controller failures. The charge controllers used were those mentioned in section 6.1.

Because of these early failures confidence in PVC by the people and the Government was adversely affected and an important part of the author's responsibilities in restoring confidence has been to locate all previous PVC lighting units installed by the EPU, visit the locations and replace faulty equipment with new generation charge controllers provided under warranty.

7.3 PVC and wind generator lighting systems are to be installed this year at the Youth Development Centre near Port Moresby, complete with instrumentation. This installation will have the advantage of being only 20 minutes drive from the EPU Office which means it can be regularly visited for evaluation.

7.4 Batteries : The batteries preferred for village applications are sealed and vented lead acid types. The only battery to meet our requirements to date is the Delco 2000 105 A 12 Volt unit.

Parallel operation of batteries is predictably hazardous in our experience. In two installations parallel units were connected in series during installation with consequent damage to the equipment connected.

Battery manufacturers should ensure polarized connections if parallel operation is recommended.

The lead acid batteries used to date have been limited to a daily drain of 10 - 15 % of their capacity to ensure long cycle life according to manufacturers data.

8. Water pumping

8.1 Boer Village : One PVC pumping system has been installed at Boer Village situated about 30 kilometres from Port Moresby. The EPU was not directly involved in this project as it was initiated by the women of the village raising some funds, then applying to the Central Provincial Government for the balance of funding. The local supplier provided and installed the equipment.

The pumping system is apparently providing up to 27 kilolitres per day of fresh water from an underground supply to approximately 30 kilolitres of overhead storage.

The pumping system utilizes 16 - 35 Watt panels in series parallel connected to a vertically mounted 60 Volt 8 Amp. DC motor directly driving a centrifugal pump.

8.2 Future pumping projects : Various pumping and irrigation projects are being discussed with Provincial Governments and village people at present.

A pilot project in the Manus Province will be joint funded this year between the Manus Provincial Government and the EPU. They will provide a collection area of suitable size for rainwater which will funnel the water to underground storage. The EPU will provide the PVC pumping system to transfer water to an overhead tank which they will install and monitor.

As Manus has rainfall all year with no wet and dry season, maximum storage required will be 7 days.

Based on the results of this project, the costs of a convential water vehiculation system with a water supply from a creek storage pond in the mountains above a particular village will be compared with a PVC system.

From initial considerations based on limited data, the cost of the PVC unit could have been as low as one tenth of the conventional system. However, with more data available, this may be close to one fifth and this will depend upon the cost of the PVC pumping system and that of constructing the collection area and underground storage. Limited time with a large work load has prevented full investigation which would have given results for this paper and this project will be the result of a later report.

9. Solar radiation data

9.1 <u>Solar radiation maps</u> : A survey was carried out by Kalma (4) of data from three monitoring stations : Port Moresby, Rabaul in East New Britain and Djajapura-Sentai in Irian Jaya, cloud cover data and bright sunshine records. Based on this data the solar radiation distribution was calculated and maps prepared for January, April, July and October and for the year as a whole.

For Papua New Guinea values range from $4.25 - 5.75$ kWh/m^2 in January to $3.50 - 5.00$ kWh/m^2 in July for mean daily global radiation.

This report was published in 1972 and details are given in the references.

9.2 <u>Port Moresby</u> : The Department of Physics, University of Papua New Guinea (UPNG) has recorded global daily radiation on a horizontal surface every day since 1969 and published annual reports for years 1969 - 1979 inclusive. The data for 1980 and 1981 has not yet been published. In addition hours of bright sunlight each day have been recorded and published. An example of the range of radiation for 1979 follows - 6.09 kWh/m^2 - June.

Note : A cover of thick air pollution consisting of wood smoke, dust and vehicle emissions and water vapour can exist above this station for long periods during the year and the effects of this are to be investigated.

9.3 <u>Lae</u> : Ahmad has published a report based on recorded solar radiation data at the Papua New Guinea University of Technology and details are given in reference (5). The report published in 1979 uses data for the three years 1975 - 1977. The results show that daily radiation is in the range of 5.7 kWh - 6.2 kWh/m^2.

9.4 <u>EPU and other organizations</u> : We have funding this year for solar radiation recording equipment and ampere hour integrators.

At present suitable equipment for tropical operation is being sought and investigation and co-operation between the EPU, P & T, ELCOM and UPNG should result in selection and use of standard type of equipment by each organization with free exchange of recorded data.

All the above solar insolation recording stations are at coastal centres near sea level.

The only data available for the highlands is cloud cover and some sunshine recorders. Investigation is continuing using this data and we expect to instal a number of insolation recording stations in the highlands over the next two years with the first during the latter half of this year 1982.

The results of investigation to date show that generally lower value of daily solar energy can be expected in the highlands as compared to coastal areas because of greater cloud cover (7).

CONCLUSION

The experience in a developing country like Papua New Guinea with a good communication system shows that the loss in confidence in a new technology used directly by the people can lead to withholding of funds by the Government, Loan and Aid Organizations and village people.
(Note : the loss of confidence had been caused by unreliability).

It takes much time and effort (2 years) to reverse the process by all concerned in that new technology to ensure reliable equipment is evaluated, purchased, installed and proven in order to restore that confidence and hence funding.

However, now that has been achieved the application of photovoltaics for lighting, water pumping and irrigation can assist greatly in improving the standard of living of the people in rural areas and at the same time provide markets for reputable manufacturers and local agents prepared to put concerted effort into this area.

The importance of direct communication between PNG Government Engineers in the PVC area and manufacturers design engineers and research organizations in other countries is stressed.

With high costst of kerosene for lighting in remote areas (0.75 - 1.2 Kina/litre, i.e. 7.5 - 12 french francs/litre) pay back times of two to four years can be expected for simple PVC lighting kits (7).

The potential number of village house lighting kits is 500,000 during the next 10 years together with 5 - 10,000 street and area lighting kits.

The cultural and social impact of PVC lighting systems will be incremental with business people to be the first to install such units. As other people in the village realize the advantages, they will commence to also purchase at possibly an accelerated rate as time passes.

People will utilize their evening hours more for reading and study with the greatly improved lighting and business activities such as trade stores, boat building and crop processing can be improved.

PVC water pumping will have a more dramatic effect on village life as the long hard hours of water carrying each day will become available as a step function for other useful activities.

The economic impact will be to obtain improved lighting at less long term cost for the village people and the dependence of the country as a whole on imported kerosene will be progressively reduced.

We look forward to a good and expanding future in the application of photovoltaic system in Papua New Guinea and welcome the co-operation of research organizations and manufacturers.

ACKNOWLEDGEMENTS

Thanks are expressed to COMES, France, for making it possible for me to be in Europe at this time.

REFERENCES
1. Papua New Guinea Government Office of Information, 1981 'This is Papua New Guinea' - official book.
2. Papua New Guinea National Weather Services, official records
3. The University of Papua New Guinea - Department of Physics - Meteorological Bulletins.
4. KALMA J.D., 1972 'Solar Radiation over New Guinea and Adjacent Islands', Australian Meteorological Magazine, Vol. 20, June 1972.
5. AHMAD Q.A., 1979, 'Solar Radiation at Lae, Papua New Guinea', Papua New Guinea University of Technology, Dept. of Mechanical Engineering.
6. KINNELL G.H., 1981, 'Standards in the manufacturing industry - Energy

Conservation' Proceedings of National Standards Council Seminar, Lae, 13 - 15 April 1981.
7. KINNELL G.H., 1982 'Energy Conservation using new high efficiency light sources', Energy Planning Unit Report, 82/6.

A village photovoltaic demonstration

Large bush material house, Lou Island, Manus
PVC lighting installed

Delivering PVC lighting kit to Lou Island, Manu, by sea-going canoe

Village house near Vanimo

FIELD TRIAL OF RURAL SOLAR PHOTOVOLTAIC SYSTEM

P. BASU, K. MUKHOPADHYAY, T. BANERJEE, S. DAS, AND H. SAHA
Department of Physics, University of Kalyani, W. Bengal, INDIA 741235

Summary

A small stand-alone solar photovoltaic system using silicon solar cells has been set up in a village near Kalyani, India to meet two typical rural applications like a) village community cum adult literacy centre, b) irrigation pumping. A television set of 65 watts and three fluorescent lamps of total 100 watts and a d.c pump set of about 300 watts are being used as load. Power conditioning units include high frequency dc-dc convertor for TV set, dc-ac invertor for fluorescent lamps. Separate storage batteries have been used for the nighttime and daytime load. The performance of solar pump has been investigated for different options like with or without storage battery with a view to determine the cost effective option for stand-alone solar photovoltaic irrigation pumping. The balance of system (BOS) costs of this system has been estimated and compared with that reported in the U.S.A. The electrical and thermal performances of the solar cell modules have been investigated. The social impact of this system in the village has also been studied. These observations lead to some interesting features of a typical SPS in rural conditions in India.

1. INTRODUCTION

In the developing countries like India, where less than 20% of the small and remote villages are so far electrified, micro-irrigation and village community centre etc. have good future for large scale utilization of solar energy. It is high time to study the technical feasibility, economical viability, realiability and social impact of a photovoltaic system operating in real rural environments. The purpose of this work is to collect a first hand experience with such a system set up at a village Charsarati near Kalyani under the sponsorship of the Department of Science & Technology, Govt. of India. Initially a village community cum adult literacy centre (Fig.1) and irrigation pump set (Fig.2), both powered by solar photovoltaic modules, have been set up and the experience gathered are summarised in this paper. Similar installations have been carried out by developed countries in many parts of the world for the last few years but an important point to note is that the entire system reported here is built, installed and operated by locally available expertise in India.

2. SYSTEM DESCRIPTION

The Charsarati System is at present energizing two applications namely a) village community cum adult literacy centre and b) irrigation pumping. The functional block diagram of the system is shown below. This is a stand-

system having no back-up supply or any auxiliary load. The number of panels, array size, battery capacity etc. are calculated with the help of theoretical work done earlier (1,2).

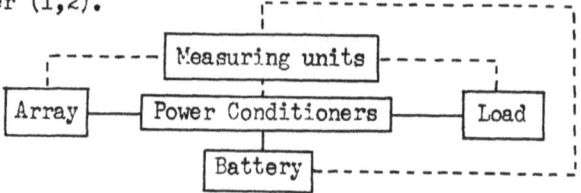

The detailed characteristics of these two applications are summarised in table - I.

TABLE - I

SITE	Location - Charsarati (near Kalyani), West Bengal, India Latitude - 23.5° N; Longitude - 89° E; Altitude - 9.75 m above sea level	
PV MODULE	Each module consists of 36 silicon single crystal solar cells each of 76 mm diameter connected in series and to generate about 15 peak watt at AM1 intensity at 28°C. These are supplied by M/s Central Electronics Ltd. India	
Date of installation	COMMUNITY CENTRE December, 1980	IRRIGATION PUMP October, 1981
Photovoltaic Array	Electrical characteristics Peak power - 200 watts; No. of Modules - 14 (7 rows in parallel and two modules in series in one row); Voltage - 24 volts (approx.); Substrate Al, Gl, FRP. Array structure Locally fabricated low cost fixed roof-top structure made of painted mild steel plate (1½ x ½ inch dimentions). The roof-top is specially designed to have an angle of 23.5° N, the latitude of the place.	Electrical characteristics Peak power - 300 watts; No. of Modules - 20 (4 in series in one row and 5 such rows are connected in parallel); Voltage - 48 volts. Substrate - Al. Array structure An angle from prefabricated structure which can be oriented about earth's N-S axis. The angle of the array with the horizontal can be changed monthly to account for the change in the declination of the sun.

Load	a) A commercial television set of 65 watt with 51 cm screen.
	b) Three fluorescent lamps
	   i) One 20 watts
	   ii) Two 40 watts
		Monoblock motor pump assembly of 300 watts rating
		Pumping head - 3-5 meters
		Motor - DC series motor.
		Pump - centrifugal.
		Manufactured in India.

	a) One DC-DC convertor (24-110V) for TV set	a) Change over electronic switch for battery to PV system and vice versa.

Power Conditioner	b) One 24 V DC - 150 V AC invertor for 20 W lamp. c) Two 24 V DC - 150 V AC high frequency invertors for two 40 W lamps.	b) Maximum power point tracker c) Waterhead measuring electronic arrangements.
Storage	Two 12 V, 120 Ah lead-Acid batteries of 20 hour rate. Specially designed for low loss and long life.	Four 12 V, 60 Ah lead-Acid Automobile batteries of 20 hour rate.
Data Recording	Data recorded of the different parameters like a) open circuit voltage, short circuit current of panels. b) Charging voltage and current, specific gravity of battery fluid. c) Insolation and ambient temperature. d) Degradation, NOCT of Modules etc. e) Load currents and total Ah recording during charging & discharging.	Data recorded of the different parameters like a) open circuit voltage and short circuit current of panels. b) Motor currents & voltages. c) Charging currents & voltages, specific gravity of battery fluid. d) Water flow rate and integrated flow. e) Static head and draw down of water level.

3. DATA COLLECTION AND ANALYSIS

The input output data of the entire system is being monitored regularly. The performance of the PV system, storage battery, load and the power conditioners etc. are also studied. It is observed that the performance of the system follows the well known Cosine law of insolation. Some of the important observations are reported here.

3.1 Nominal operating cell temperature (NOCT)

NOCT of the modules provide a convenient means of qualifying a modules thermal design and providing a meaningful reference temperature for rating output (3). An attempt has been made to measure this at nominal terrestrial environment (NTE) which is defined at insolation - 80 mw / Cm^2, Air temp - 20°C, wind average velocity - 1 m/s. NOCT has been measured for three different modules having different substrate materials. All modules under test were placed on the fixed roof-top array structure. The typical data for this measurement has been recorded in table II.

3.2 Dirt accumulation

As the system is operated near an open field in the village, it is very dust prone and the accumulation of dust is rather heavy. In table - II the effect of dirt accumulation of Isc & Voc of the modules are distinctly shown. It is recommended that the modules should be cleaned at least once a week.

3.3 Degradation of modules

The individual performance of the modules have been studied. It is interesting to note from Table - III that the modules show the degradation in the range of 7-12% since its installation in Dec., 1980. Apart from the well known causes of degradation (6,7) the most notable mode of degradation in these modules are the appreciable loss of transperancy of the glass cover. This loss occurs due to formation of a thin oily substance spread over the inner surface of the glass cover. Presumably, the elastomer encapsulant which have been used in these modules has softened under the hot humid condition and given rise to some organic vapours which has condensed on the glass cover. It is observed that it accounts for a 10% loss in Isc. But the details of this findings are being examined thoroughly and yet to be reported. The modules which has been used for pumping system are yet to show any appreciable degradation since its installation in October, 1981.

3.4 Failure of modules

Out of 14 panels used in community centre four have been nonfunctioning after Sept., 1981. Interestingly all the four panels have Al-substrate. Dismantling of these modules in the laboratory shows that the failure is due to hair line crack in the interconnection between the cells resulting into discontinuity. This may be due to lack of adequate strain relief, faulty soldering and manufacturing problem. But no such failure is observed in the modules of the pumping system till now.

3.5 Wind and Hail Impact loading

The modules and the array structure of both the applications are capable of withstanding the Nor'wester having velocity in the range of 80-100 km/hr. and severe hail storm which have been tested several times since their installation.

3.6 Battery performance

The performance of the batteries are quite satisfactory and it is working without any trouble since installation. The main feature of these batteries are low maintenance. The interval between two successive topping up is more than six months. No overcharging or voltage regulation are used in this systems. We feel that provided the batteries are properly sized and suitably designed, there is no need of overcharging or deep discharge protection circuits for such applications. This increases the reliability of the system.

3.7 Power conditioners

The usual commercially available 51 cm TV set requires about 65 watts of power and 220 volts Ac supply. With the minor modifications in the power supply unit of the TV set we have reduced the power consumption to 36 watts only. This needs a 24 V - 110 V DC-DC convertors. For the fluorescent lamps, 24 V DC - 150 V. AC invertors (3 to 4 K Hz Sq. wave output) have been used.

4. OPTIONS FOR PUMPING SYSTEM

As the photovoltaic pumping systems are to be used in remote areas, the systems should be as simple as possible. But due to the fluctuating nature of available solar energy and mismatching between generator and load there is a scope of opting into different systems with storage battery and maximum power point tracker. Some of the options are given in the following block diagram:

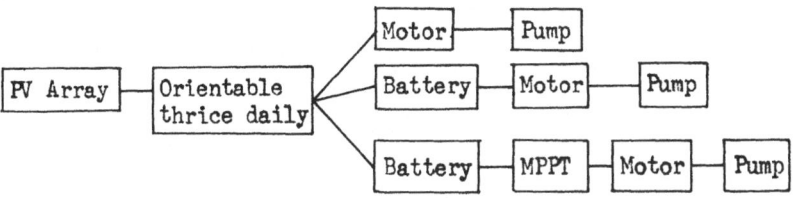

We have studied the first two options i.e. with and without storage battery in some details and the results are summarized in table IV which reflects that the use of battery for pumping is more promising. It increases the water output, pump motor efficiency, overall efficiency of the system and lower the specific capital cost to some extent but before making any conclusive remarks more critical investigation is called for. The third option i.e., the use of MPPT may also give an acceptable options for pumping work in this direction are on the process and to be reported shortly.

5. BALANCE OF SYSTEM COST

The BOS costs have been calculated for both the systems. From the Fig. 5 it is clear that the nature of variation of the BOS costs per peak watt for different applications with the size of installation in the developing countries like India is similar with that reported by NASA-LeRe, but the absolute BOS cost/peak watt is considerably less under Indian condition than that in USA (8). This is mainly due to the cheap labour cost involved in area-related component of BOS cost which is also reflected in the cost of some of the appropriate materials in the countries like India. The cheaper BOS cost in the developing countries like India allows the use of relatively low efficient ($\sim 5\%$) solar cells. An interesting outcome of low BOS costs is that it makes the allowable module cost almost independent of efficiency indicating that no bonus is achieved by developing high efficiency cells where BOS costs is relatively low. The fig.5 shows the effect of low area-related BOS costs on the allowable module costs for microirrigation system calculated from the actual expenses incurred in photovoltaic system (8).

6. SOCIAL IMPACT

The photovoltaic system has created a tremendous enthusiasm not only in the villagers of Charsarati but in the neighbouring villages also. A huge crowd gathers the centre to enjoy the TV programme which is the only means of entertainment in the dull life of the villagers in such villages. The fluorescent lamps are of special interest in the Adult Education Classes which runs regularly.

The villagers are interested in photovoltaic pumping. They are fully convinced by the fact that such simple almost maintenance free system can solve their irrigation problems to a large extent. The villagers also take care of the centre regarding its security, maintenance and operation on a co-operative basis.

TABLE - II Effect of dirt accumulation and NOCT for modules of different substrates.

Substrate	Performance of the modules before cleaning i.e., after dirt accumulation for 25 days		Performance of the modules immediately after cleaning		Degradation in short circuit current (I_{sc})	NOCT (°C)
	V_{oc} (V)	I_{sc} (mA)	V_{oc} (V)	I_{sc} (mA)		
Alluminium	18.1	710	18.5	748	5.1%	45°C
F.R.P.	18.4	710	18.6	755	5.9%	43°C
Glass	18.6	680	18.9	175	4.9%	43°C

TABLE - III Change in cell performances of six panels with three different substrates during the period March, 1981 to Nov., 1981.

Panel No.	Substrate	Performance measured on 9.11.81 at midnoon insolation 80 mw/cm² temp. 32°C.		Performance measured on 5.3.81 at midnoon insolation 80 mw/cm² temp. 37°C		Degradation short circuit current (I_{sc})
		V_{oc} (V)	I_{sc} (mA)	V_{oc} (V)	I_{sc} (mA)	
1	Al.	18.1	748	19.9	823	7.2%
2	Al.	18.0	-	19.8	877	Failure
3	Al.	-	-	19.9	862	Failure
4	FRP	18.6	755	29.01	842	8.5%
5	FRP	18.0	729	19.85	844	11.8%
6	Gl	18.9	715	20.0	815	12%

TABLE - IV Performance of irrigation pumping with and without battery for two typical days (occasional clouds)

System	Total water output	Motor-pump efficiency	Overall efficiency	Specific capital cost($ KJ)	Time of operation
Without battery	24 m³	22%	1.48%	4.48	7-30 to 16-30
With battery	27 m³	33%	2.0%	4.0	7-30 to 16-30

ACKNOWLEDGEMENTS - Authors are thankful to Mr. C. Neogi, University of Kalyani for his technical help throughout this work. The work has been carried out under the department of Science and Technology, Govt. of India research grant for the project on "Photovoltaic Electric Power Systems for Rural developments". Thanks are due to M/s Chloride India Ltd. for providing specially designed batteries free of costs.

REFERENCES

1. H. Saha, Solar Energy **27** (1981) 103.
2. H. Saha, P. Basu, and K. Mukhopadhyay. Proc. 3rd EC Photovoltaic Solar Energy Conf. (27-31 Oct 1980) Cannes 542.
3. J.W. Stultz. "Thermal and other tests of photovoltaic modules performed in natural sunlight" LSA task report 5101-76, July 31 1978.
4. D. Dignet and G. Anguet. Proc. 3rd EC Photovoltaic Solar Energy Conf. (27-31 Oct 1981) Cannes 726.
5. UNDP Project GLO/78/004 on "Small Scale Solar Powered Irrigation Pumping system" July 1981.
6. R.G. Ross Jr - 15th Photo Spec Conf Orlando Floria (May 12-15, 1981).
7. L.M. Dumas and A. Sbumka - 15th Photo Spec. Conf. Orlando Florida (May 12 - 15, 1981).
8. H. Saha, P.Basu & K.Mukhopadhyay - Impact of BOS cost on Solar Photovoltaic System" (In Press with "SOLAR CELLS").

FIG. 1

FIG. 2

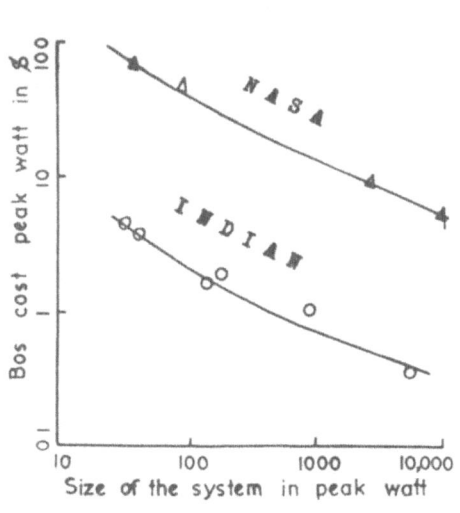

FIG. 3

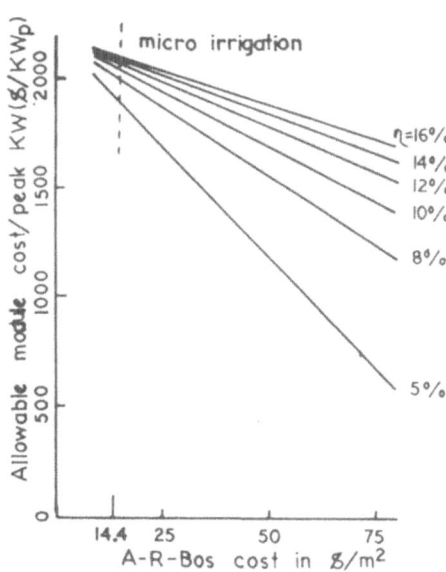

FIG. 4

POSTER GROUP P2

EXPERIENCE, PERFORMANCE, RELIABILITY, MONITORING

- Experience and Reliability

Study of a photovoltaic concentrating system like "Sophocle" in fluctuating mode

Performance of 1kW peak concentrating photovoltaic array

Photovoltaic concentrator module characterization

- Performance and Monitoring

Results from a test facility for solar cells in Sweden

A microprocessor-based instrument for automatic solar cell characterization

STUDY OF A PHOTOVOLTAIC CONCENTRATING SYSTEM LIKE "SOPHOCLE" IN FLUCTUATING MODE

B. LAURENT - PHAM VAN VUI - G. VIALARET

C.N.R.S. - Laboratoire d'Automatique et d'Analyse des Systèmes
7, avenue du Colonel Roche - 31400 TOULOUSE

Summary

SOPHOCLE is a Fresnel lenses concentration photovoltaic generator using silicon cells. It is tested under various climatic environment, by means of a measurement and registration system. The recorded data are treated at L.A.A.S. (Laboratoire d'Automatique et d'Analyse des Systèmes, Toulouse).
Different parameters are studied to define the generator performances and the dispersion of solar characteristics at different places. Their analysis by Monte-Carlo method has allowed to create a mathematical model, particularly for time constants, efficiency and storage.
The purpose is to understand at any time the whole behaviour of the system by means of statistics on real data and to detect any eventual defects.

1. INTRODUCTION

The L.A.A.S. has conceived and developped an electric generator named SO.PHO.C.L.E. (SOlaire PHOtovoltaique à Concentration Limitée d'Energie), in collaboration with the industrial groupment SOTEREM-SNIAS-SNEA.

This system is characterized by the use of a heliostat with an altazimuth mounting and by the choice of a medium concentration (C = 45) by Fresnel Lenses on Silicon Cells [1]. The lenses are integrated six by six in sealed modules provided also passive cooling.

The SOPHOCLE's Programme began in 1977 and is today in its test period in various climatic environnements (Algeria, Bresil, France, Gabon, Greece, Malaysia, Mexico, India, Saudi-Arabia, Senegal, Thailand).

We are presenting here a characterization of SOPHOCLE by using data, which were recorded in 1981 on measuring system.

2. THEORETICAL CONSIDERATIONS

In order to follow the behaviour of SO.PHO.C.L.E. to detect and to ensure a certain fiability in its working, a certain number of characteristics from the data are put into evidence. The latter of random nature is treated by the Monte-Carlo method which consists in the establishement of a distribution law of different physical parameters $\emptyset$ and in its association with a distribution of a pseudo random quantity r.

Thus $$\int_0^{\emptyset} p(\emptyset') d\emptyset' = \int_0^{r} p(r') dr'$$

For a uniform distribution, we have $p(r) = 1$. So $r = \int_0^{\emptyset} p(\emptyset')d\emptyset'$. Knowing the probability law $p(\emptyset)$, $\emptyset$ can be deduced in function of r (r is physically realised).

3. EXPERIMENTAL RESULTS AND INTERPRETATIONS

3.1. Data recorded

The parameters recorded on the magnetic tape are :
- ambient temperature T_{amb}
- radiation temperature : a few degrees close to the solar cell temperature T_{cell}
- total solar power received W_{GH}
- electric power delivered W_{el} integrated every 12 mn
- direct solar power received W_d
- rate of crossing over of the limit of 350 W/sq.m
- calculated efficiency : $\eta = W_{el}/Su \times Wd$
  where Su : collector area utilised

The daily curves (2, 3, 4, 1) drawn out from the data show the perfect and correlated relation ships between the direct power received and the electric power produced (covariance $(W_d, W_{el}) > 0,99$). On the efficiency curve, these is a certain variation during the day.

Moreover the phase shift exists between the sun's apparition (total energy) and that of the direct sunshine measured on the SOPHOCLE ; it is in the order of 2 hours and depends on the adjustment of the limit of the tracking sensor.

The phase shift exists equally in the evening. In average, the generator works out at 60% of the sunny day.

The classical statistics in the data enable to deduce especially :

- the average monthly efficiency = $\dfrac{\text{monthly energy produced}}{\text{monthly direct energy received}}$

for example : July $\eta_{Libreville}$ = 9,05 %
             July η_{Alger} = 8,5 %

- the value $\dfrac{\text{directed energy}}{\text{total energy}}$ which influences directly on the choice of the concentration.

for example : Libreville r_m = 40 %   watt-plane
               Alger r_m = 130 %   concentration

3.2. Variations in the generator performances

The curves for Alger (5), (5 bis) show that the electric efficiency depends on the cooler temperature (in fact of the ambiant temperature and of the wind speed) and on the direct sunshine.

The heating of the solar cell is bad for its working. The relationship is of the form :

$$\eta(T) = \eta(T_o)(1 - a(T - T_o))$$

The efficiency in function of the concentration C (that is in function of the incident power) is written as :

$$\eta(c) = \eta(1)\left[1 + \frac{nkT}{q\,V_{oc}(1)} \log c\right] \frac{CF(C)}{CF(1)}$$

n : type factor (ideal)
V_{oc} : open circuit voltage
CF : curve factor

As long as the curve factor CF remains constant, the efficiency increases as the logarithme of the sunshine. Above a certain limit to be determined (W_{ds}) the curve factor CF decreases strongly, so also the efficiency [3].

The experimental results (data) are utilised to determine α, β, δ, W_{ds} so that :

$$W_d \leqslant W_{ds} \quad \eta(T, W_d) = \eta(T_o, W_{ds}) + \delta \log W_d/W_{ds} - \alpha(T - T_s)$$
$$W_d \geqslant W_{ds} \quad \eta(T, W_d) = \eta(T_o, W_{ds}) - \beta(W_d - W_{ds}) - \alpha(T - T_s)$$

The statistical study of the measurements enables to conclude that :
W_{ds} = 550 W/sqm ; α = 0,0006°C^{-1} ; β = 225 x 10^{-7} (W/sqm)$^{-1}$; δ = 0,014
The Figure(7) represents the theoretical simulation thus defined the variations of the efficiency in function of T and W_d.

3.3. The behaviour of the generator SOPHOCLE

3.3.1. Time constant

The coefficient characterising the variation of the power of module per unit time (minute) is called "Time constant". The following two constants are utilised C_1 for the power rise, C_2 for the power descent.

Due to the values recorded every 12 mn, the frequenly curve representing the distribution of constants C_1 and C_2, figure (6) can be determined.

3.3.2. Thermal resistance of the generator

The study of the fluctuating values of the temperature difference between the ambiant and the cooler temperature, makes possible to show that:
- during the night, the radiator painted black, is cooler than the ambiant temperature (exchanges by radiation with the sky),
- practically one hour is to be waited in order to get the permanent working (conduction phenomenon wich the mass of the radiator).
The values of the thermal resistances calculated from the measurements are between 0.01 and 0.025°C/W/sqm with the mean value about 0.017°C/W/sqm.
This value is confirmed by the theoretical calculation (program ELLPACK)[2]

3.3.3. Storage Capacity

During fourty days in the year 1981, at Libreville and at Alger for different limits of electric energy, the number frequency of times f (and their duration d) when the energy is higher than those threshold E, cf. figures (8,9), is determined.

The storage capacity should neither be too small (all the produced energy is not utilisable) nor too big (storage rarely full and costly). The research of the maximum of the product E x d x f (E : threshold of energy, d : duration, f : frequency) enables to know the capacity (C = Ed) the most useful for a given site.

Moreover, the functionning point, closest to the demand (E_d, d_d, f_d) on minimising $(E - E_d)^2 + (d - d_d)^2 + (f - f_d)^2$ can be found.

4. CONCLUSIONS

The analysis of the results reveals two contradictory characteristics :
- Alger is a site more propicious for concentration for receiving more energy in the form of direct than fixed total,
- The sunshine values being greater, the efficiency is lower but the tracking is better.

This problem can be resolved in utilising the solar cells for which the curve factor falls for higher concentration values.

The distribution of different parameters are characterized. Thus we can follow their evolution. Due to these characteristics, we hope to ensure a certain fiability to the functionning of SOPHOCLE.

REFERENCES :
(1) Thèse D. FOLLEA "Conception, réalisation et expérimentation d'un générateur photovoltaïque à concentration - prototype Sophocle 1000", LAAS 1979
(2) Thèse J.F. PORTA "Etude de différents procédés de refroidissement des photopiles dans les centrales photovoltaïques à concentration", LAAS 1981.
(3) Thèse O. SOUMAORO "Etude des cellules solaires au Si au GaAs et de leur couplage optique au moyen d'un miroir dichroïque", LAAS 1982.

FIGURES : Place : Libreville - Date : 24.7.81

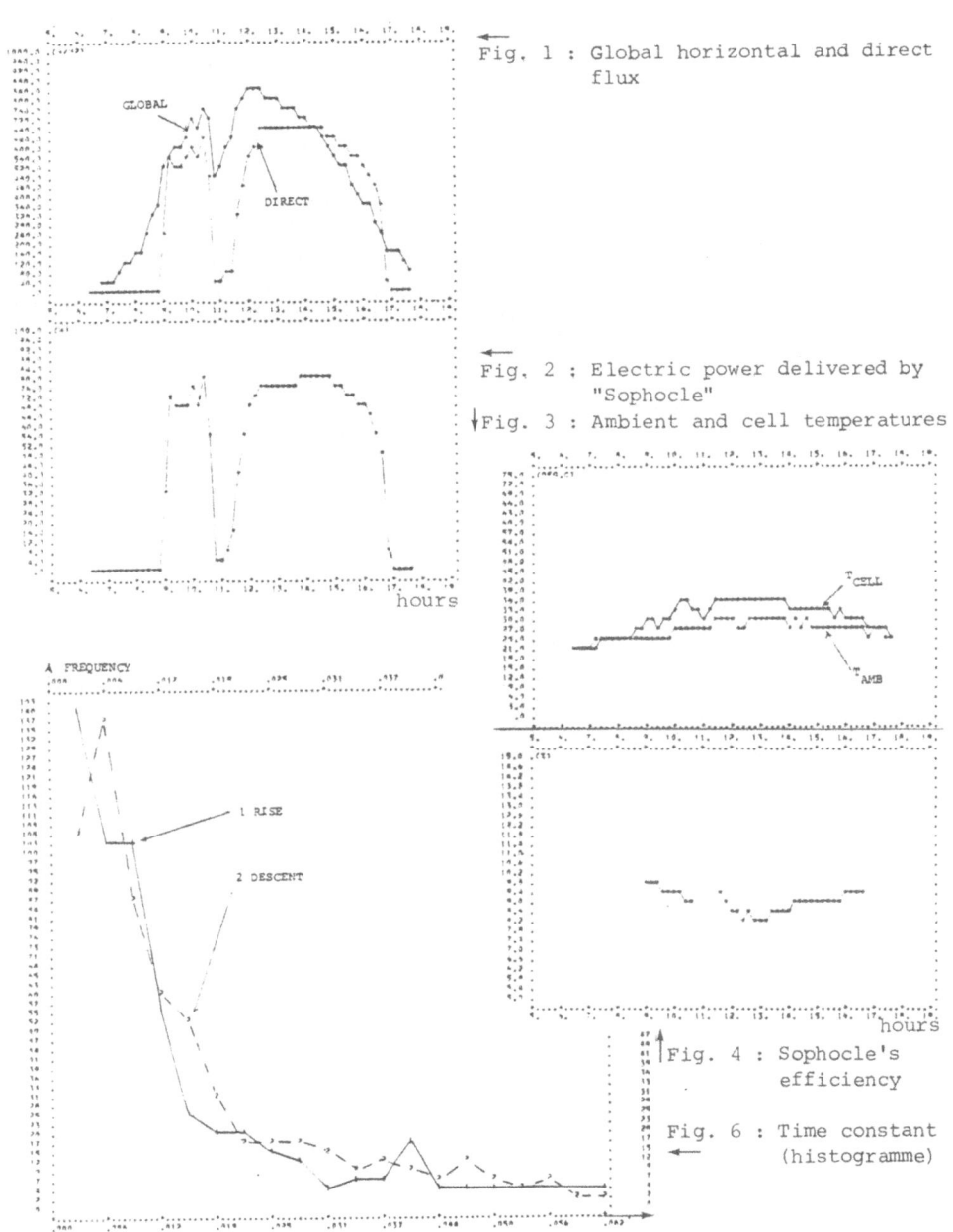

Fig. 1 : Global horizontal and direct flux
Fig. 2 : Electric power delivered by "Sophocle"
Fig. 3 : Ambient and cell temperatures
Fig. 4 : Sophocle's efficiency
Fig. 6 : Time constant (histogramme)

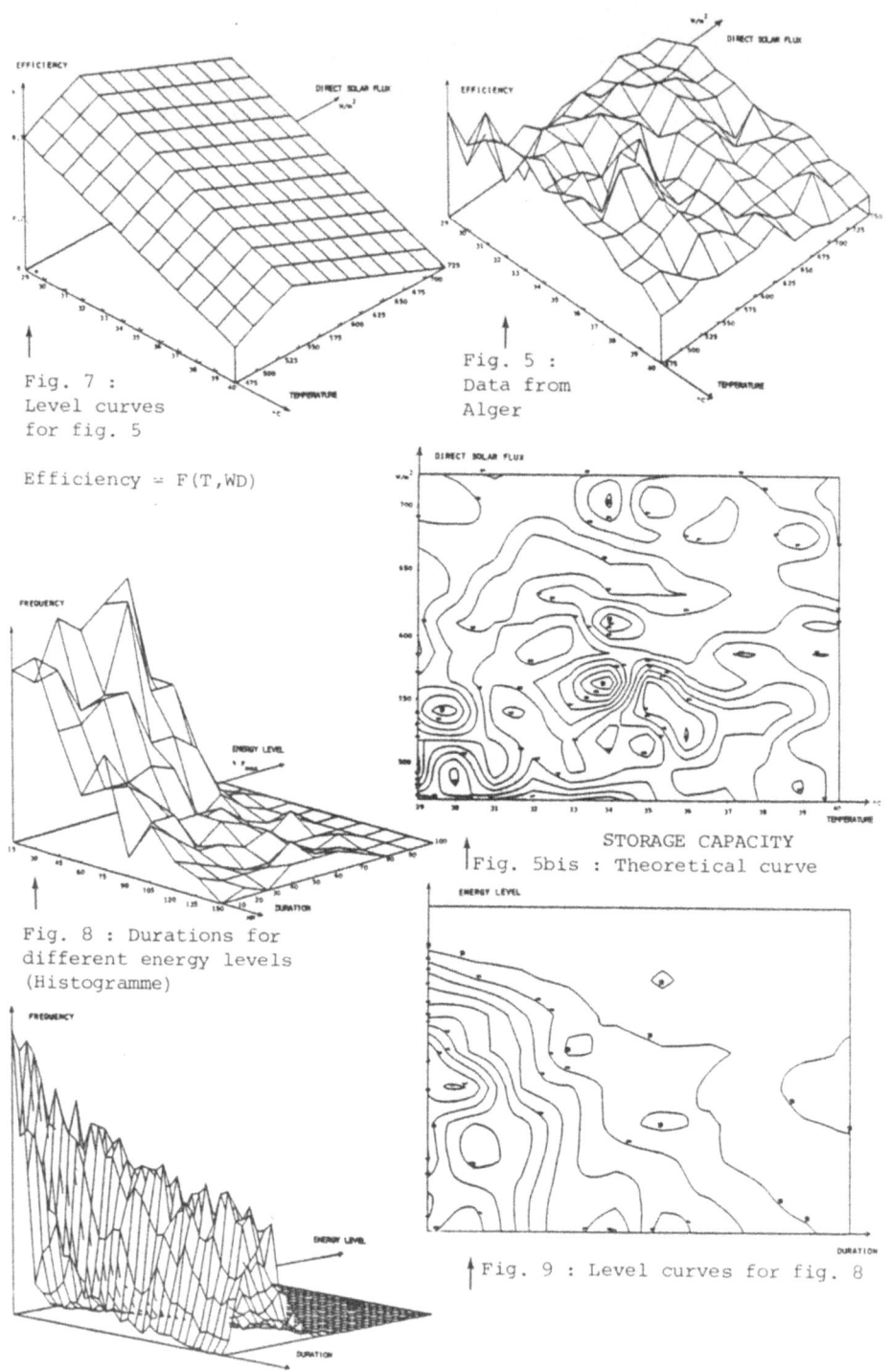

Fig. 7 : Level curves for fig. 5

Efficiency = F(T,WD)

Fig. 5 : Data from Alger

Fig. 5bis : Theoretical curve

Fig. 8 : Durations for different energy levels (Histogramme)

Fig. 9 : Level curves for fig. 8

Fig. 10 : Simulation by Monte-Carlo method.

PERFORMANCE OF 1 KW PEAK CONCENTRATING PHOTOVOLTAIC ARRAY

G. Sala and F. Chenlo

Instituto de Energía Solar - E.T.S.I.T. - Universidad Politécnica Madrid

Summary

The R.A. concentrating photovoltaic array has been measured during its first year of operation. Electrical, optical and mechanical performances are presented in this paper. The overall efficiency is 4%. This low value is due to non uniform illumination and to dispersion of the solar cell characteristics.

1. INTRODUCTION

The R.A. concentrating array has been developed and constructed in Madrid under the sponsorship of the R. Areces Foundation. The project began in April 1978 (1)(2) and the array was fully operative by May 1981. Characterization of the array performance has been carried out during the last 12 months, financed mainly by the Ministerio de Industria y Energía. Output power (of the cells, modules and arrays), optical characteristics of concentrators, cooling efficiency, series-parallel connection effects, tracking losses and light non-uniformity effects have been measured. The acquired experience is now being devoted to design a second generation concentrating array of 3 KW peak (array MINER).

2. DESCRIPTION OF THE ARRAY

The R.A. array consists of 36 modules with an effective area of 20.80 m^2. Each module includes 4 point focus hybrid Fresnel lenses 38x38 cm^2. Solar cells are made of silicon wafers 50 mm. in diameter. The geometrical concentration is 72 X, and the cooling of the cells is passive, by means of finned aluminum beams. The structure is of pedestal type, having azimuth and elevation drive mechanisms. Sun tracking is achieved with a linear sensor and associated driving motors. The 36 modules are series connected. The four cells of each module are connected in parallel. Fig. 1 shows a view of the array and a picture of the encapsulated solar cell.

3. PERFORMANCE OF THE "LENS-SOLAR CELL UNIT"

The unit consists of an encapsulated solar cell and a Fresnel lens mounted on a tracking mechanism. The solar cell power generation is highly dependent on the distribution of the incident irradiance on it. (3). Fig. 2 shows the V-I curves of a cell under two different illumination profiles, which are also plotted in Fig. 3. We can see that the fill factor is reduced from 0.53 to 0.39, decreasing the efficiency as well. The sharp-pointed illumination profile corresponds to a hybrid glass-silicone Fresnel lens type wich has been mounted on the R.A. array. Such a profile has its origin in the use of a master mould designed for acrylic but used for silicone, which has a different refractive index. The most uniform profile in Fig. 3 corresponds to an acrylic lens. Effective concentration of hybrid Fresnel lenses, defined as I_{sc} (concentrated) I_{sc} (1 sun direct radiation) is 53: then the optical efficiency is 74% because geometrical concentration is 72. The Histogram of the efficiency of the fabricated solar cells (Fig. 4) at 1 sun shows a great dispersion which has contributed., together with the non uniform illumination profile, to the power output reduction of the array. The efficiency at 36 suns, 55°C, of a solar cell is 11.6%. Fig. 5 shows the variations of fill factor, Vmpp and generated power versus the pointing error for the uniform acrylic lens. It is remarkable that the fill factor remains constant in spite of the large variation of I_{sc}. This effect shows that the equivalent series resistance of the cells increases when they are not uniformely illuminated.

4. PERFORMANCE OF THE MODULES

The modules consist of a finned aluminum beam which supports the lenses and the encapsulated solar cells. The interface between the capsules and the aluminum beam has been filled with epoxy resin containing aluminum powder to improve the thermal contact. Fig. 6 shows the temperature variations on the beam and on the solar cells surface.

The recorded anomalous high temperature of cell n° 2 is due to a failure of the epoxy-aluminum paste. The existence of voids in the interface cell-capsule of some cells has caused the burning in some points, of the protective transparent silicone.

One year and a half after installation of the array there were no lenses or cells broken (excluding the burnt ones), and they didn't show corrosion or variations of the VI curve.

The 4 cells of the module are parallel-connected. A previous paper (4) proposed that connection in parallel of solar cells having different I_{sc}'s and F.F's but not very different V_{oc}, should have very low or negligible connection losses. This effect has been experimentally verified in our modules as shown in Fig. 7.

The effects of the misorientation of module is shown in Figs. 8 and 9 using the I_{sc} as parameter. In Fig. 9 we can see that the angular aperture of the module (between 90%'s) is $\pm$ 23 mrad. The r.m.s. value of I_{sc} due to $\pm$ 9 tracking angular correction is only 0.6% of the I_{sc} average.

The efficiency of a module equipped with hybrid Fresnel lenses and solar cells chosen at random is 6.6% at a 43°C (operating temperature of the cell) which is equivalent to 7.0% at 25°C.

With conventional but selected solar cells and uniform acrylic lenses the overall efficiency of the module is 10.2% at 40°C (quivalent to 10.7% at 25°C).

5. PERFORMANCE OF THE ARRAY

The connection of very different modules, in series, promotes a significant loss of efficiency as Fig. 10 shows. In fact. efficiency of quarter B of the array is only 5% at 25°C. The characteristic VI curve of the array is shown in Fig. 11. Shunt diodes, one for each 3 modules, have saved all cells from hot spots, at any load condition.

The tracking mechanisms have failed only once at the "end swich" level, but the protective sliding coupling saved the array, as provided.

6. CONCLUSIONS

After the array construction and its evaluation we conclude that it is not difficult to reach efficiencies of 10% if uniform illumination lenses are used and if solar cells are carefully selected in series connection. We estimate that the cost of this type of array is 6 $/Wp installed.

REFERENCES

1 A. Luque et al. Project of the Ramon Areces Concentrated Photovoltaic Power station. Rec. of the 13th IEEE Photovoltaic Specialists Conf. IEEE, New York, 1978, p. 1139.

2 G. Sala et al. The Ramon Areces Concentrating photovoltaic array. XXe Conf. Int. Comples, Rabat, 1981.

3 E. Lorenzo et al. Experimental verification of the illumination profile influence on the series resistance concentrator solar cells. J. Appl. Phys. 52, 1, Enero 1981.

4 A. Luque et al. Connection Losses in Photovoltaic Arrays. Solar Energy 25, pp. 171-178.

FIG.1.- View of the array and the concentrating solar cell

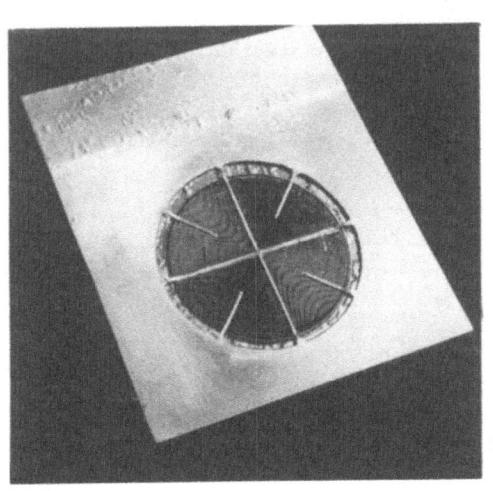

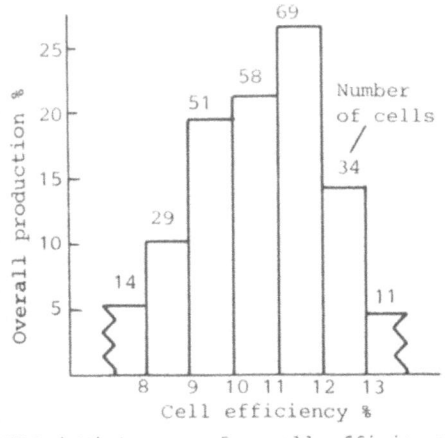

FIG.4. Histogram of s.cell efficiency

FIG.3. Illumination profiles of two Fresnel lenses acrilic (left) and hibrid (right)

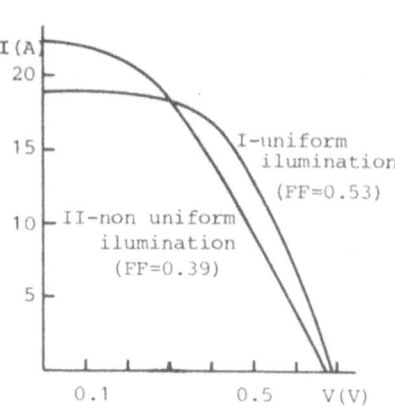

FIG.2.-**Effect** of concentracted light **uniformi**ty on V-I curve

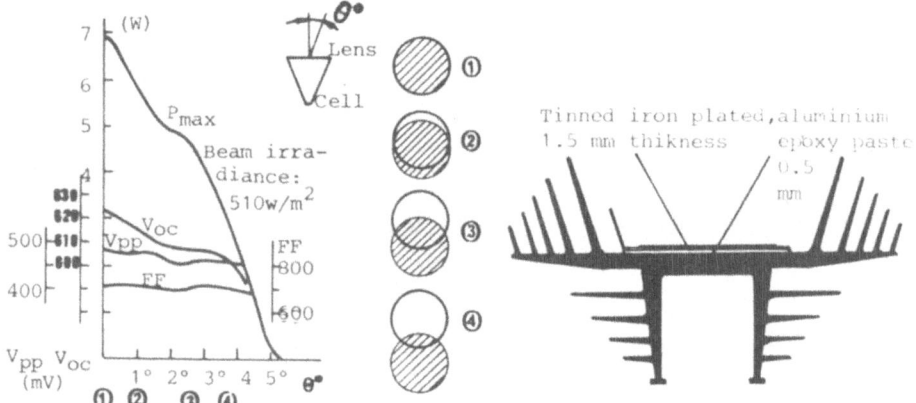

FIG.5. V-I curve parameters vs. pointing error (acrilic lens)

FIG.6a. Section of the aluminium cooling beam

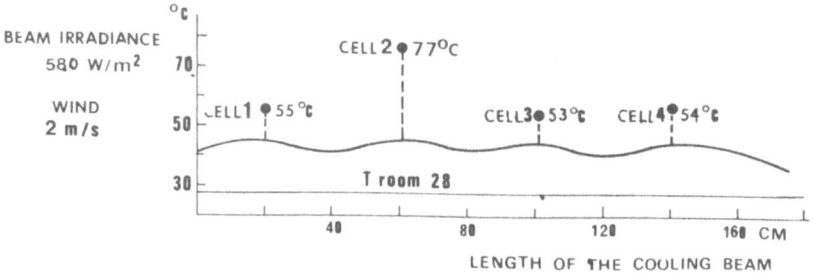

FIG.6b. Temperature along the beam and the cells of a module

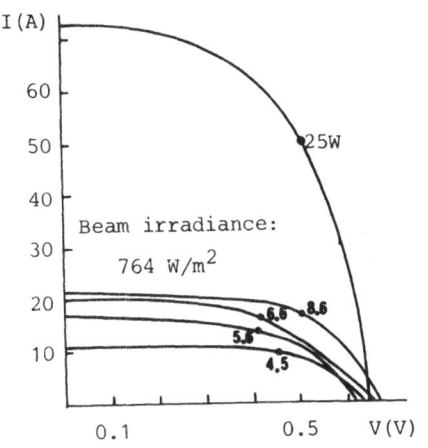

FIG.7. Connection of 4 concentrator solar cells in parallel shows very low matching losses.

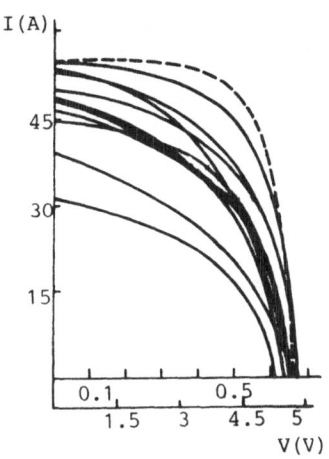

—— module of parallel cells
━━ 9 modules in series
----- module under uniform illumination

FIG.10.- Connection of 9 modules in series. V-I curve of Quarter B.

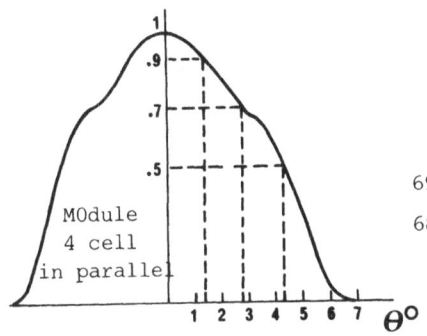

FIG. 9. Module Isc versus misorientacion angle.

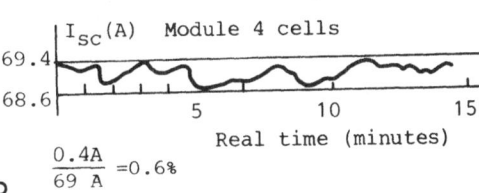

FIG. 8. Module Isc versus time in tracking mode.

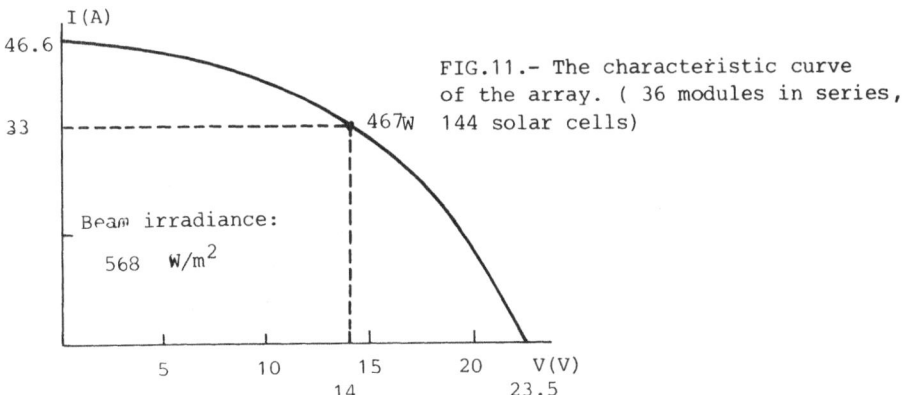

FIG.11.- The characteristic curve of the array. (36 modules in series, 144 solar cells)

PHOTOVOLTAIC CONCENTRATOR MODULE CHARACTERIZATION†

H. J. GERWIN
Experimental Facilities Operations Division
Sandia National Laboratories
Albuquerque, NM 87185 U.S.A.

Summary

A number of photovoltaic (PV) concentrator module designs have been tested under similar conditions, and the results are presented to show meaningful performance characteristics. The concentrator designs include actively cooled linear parabolic troughs and linear Fresnel lenses, as well as passively cooled point-focus Fresnel lenses. These modules were produced within the past three years and represent significant improvements over earlier concentrator designs.

All testing was performed with natural sunlight in the Sandia National Laboratories test facility in Albuquerque, New Mexico. Each module was run through a series of standard tests to determine performance at normal operating and off-design conditions. These tests included maximum power tracking under a range of insolation values and temperature variations.

Complete current-voltage scans were made at frequent intervals throughout the test period for each module. Using a multiple linear regression analysis technique, maximum power efficiencies were determined by normalizing the operating points for direct normal insolation (800 W/m^2) and average cell temperature (28°C).

The test facility is shown below, and several of the more interesting modules/arrays and their characteristics are shown in subsequent figures.

Sandia Test Facility

†This work was supported by the U.S. Department of Energy, Division of Photovoltaic Technology.

Acurex Actively-Cooled Line Focus Reflector (1980)
500 W Peak $\eta_{Electrical} = 10.0\%$ $\eta_{Thermal} = 60\%$
5.3 m² Aperture Area 108 Silicon Cells Application: Hawaii

BDM Actively-Cooled Line Focus Reflector (1981)
700 W Peak $\eta_{Electrical} = 9.7\%$ $\eta_{Thermal} = 65\%$
11.3 m² Aperture Area 166 Silicon Cells Application: New Mexico

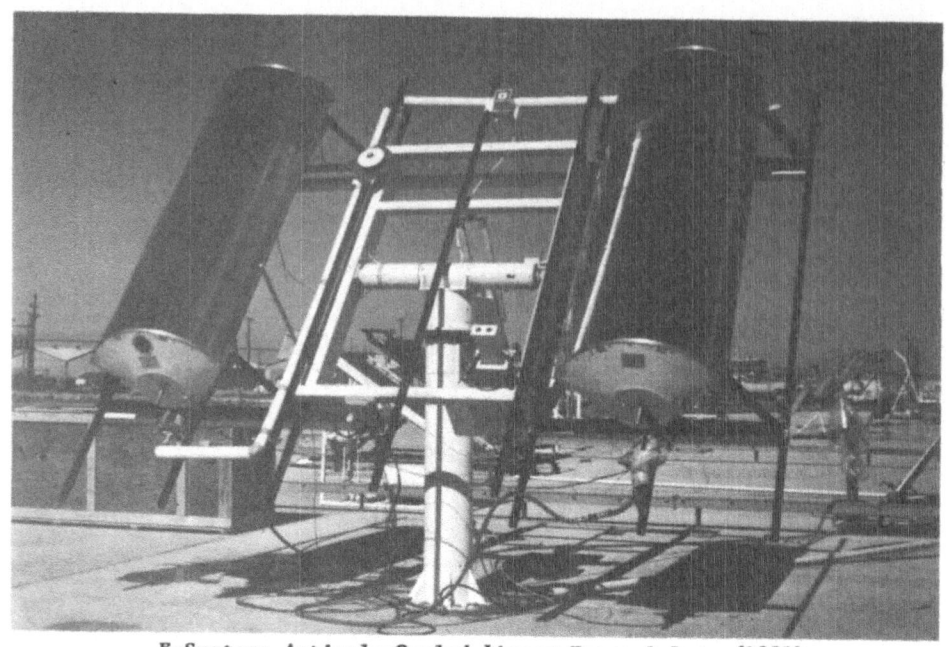

E-Systems Actively-Cooled Linear Fresnel Lens (1981)

225 W Peak $\eta_{Electrical}$ = 12.7% $\eta_{Thermal}$ = 55%
2 m² Aperture Area 94 Silicon Cells Application: Texas

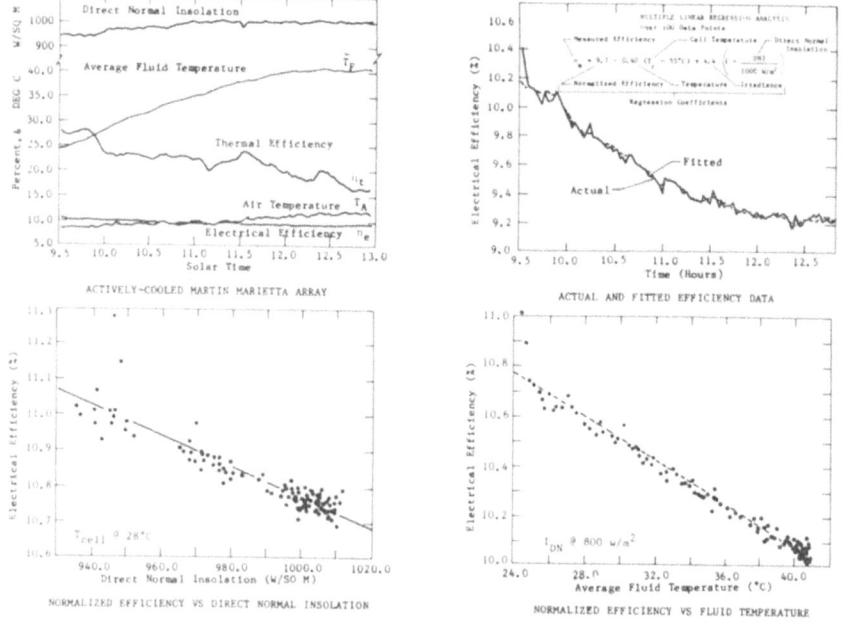

Martin Marietta Actively-Cooled Point Focus Fresnel Lens (1980)

2330 W Peak $\eta_{Electrical}$ = 11.5% $\eta_{Thermal}$ = 42%
25.3 m² Aperture Area 272 Silicon Cells

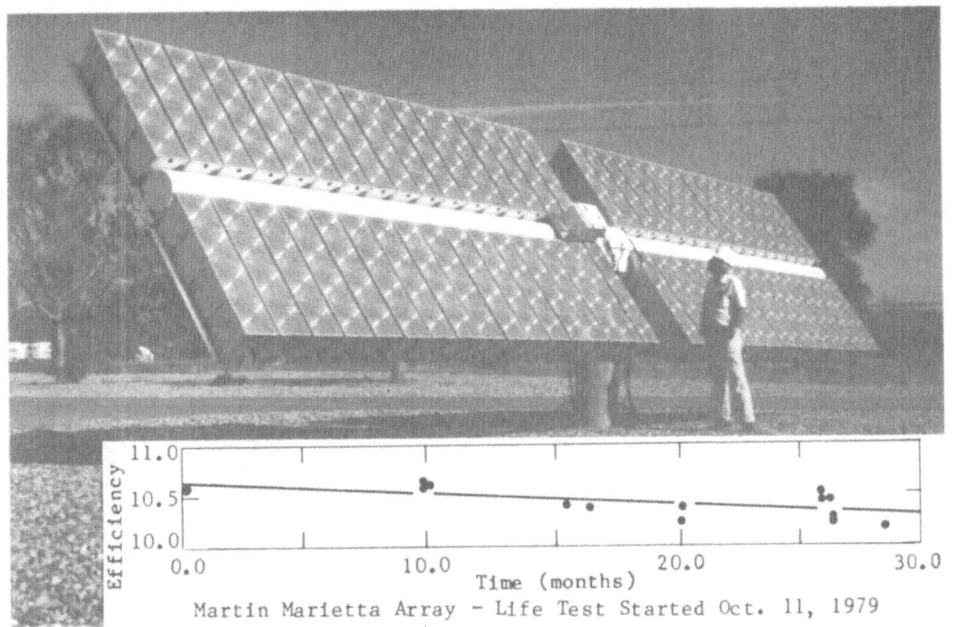

Martin Marietta Passively-Cooled Point Focus Fresnel Lens (1979)
2040 W Peak $\eta_{Electrical} = 10.6\%$ 25.3 m^2 Aperture Area
272 Silicon Cells Applications: Saudi Arabia and Arizona

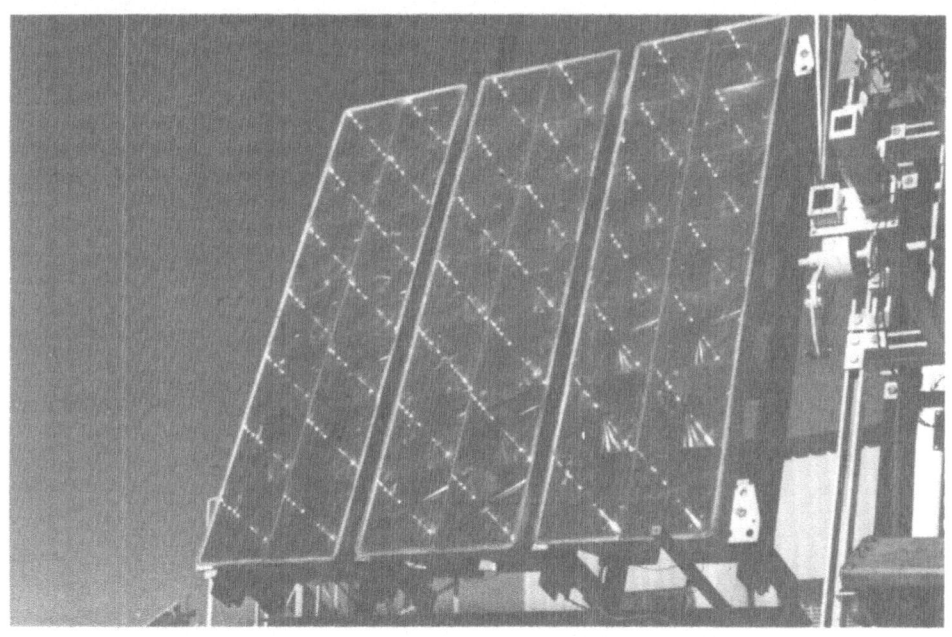

Martin Marietta Passively-Cooled Point Fresnel Lens (1981)
70 W Peak $\eta_{Electrical} = 14.0\%$
0.6 m^2 Aperture Area 14 Silicon Cells

Sandia Passively-Cooled Point Focus Fresnel Lens (1980)

80 W Peak

0.7 m² Aperture Area

η Electrical = 13.3%

25 Silicon Cells

Varian Actively-Cooled Point Focus Fresnel Lens (1980)

40 W Peak

0.3 m² Aperture Area

10 AlGaAs Cells

$\eta_{Electrical}$ = 19.0%

10 Silicon Cells

RESULTS FROM A TEST FACILITY FOR SOLAR CELLS IN SWEDEN

J. Hedström, B. Källbäck and D. Sigurd
Institute of Microwave Technology
S-100 44 Stockholm, Sweden

Summary

A 1 kW photovoltaic test facility has been built about 20 km north of Stockholm. The test station which has been in operation for nearly 3 years, is equipped with 5 different types of commercial solar cell modules, chosen to be representative for different kind of cell encapsulation methods. The main objective of this work has been to study how solar cell modules perform in the Swedish climate, both with respect to power output and degradation. The most important results are that one 1 kW_p photovoltaics generate 1000 kWh per year, effects of dust accumulation and snow coverage included, and that no degradation of the different modules is seen, except in one case where a module has a broken interconnection. Calculations based on the measurements show that if a solar cell is loaded at an optimum constant voltage the energy generated per year would be 96% of what is achieved with peak power tracking.

1. INTRODUCTION

The objective of the work reported here has been to get experience of how solar cells perform in the Swedish climate. For this purpose a test station has been built 20 km north of Stockholm, (59°N, 18°E), with solar cell modules from different manufactures.

Interesting parameters to study are energy produced, insolation, daily and annual variation of power output, cell efficiency variations with cell temperature, effects of dust and snow coverage and degradation. Data are collected with an automatic data acquisition system and stored on flexible disks.

All modules were measured in a solar simulator prior to installation and they have also been tested once a year in the simulator to detect any degradation of their performance.

2. DESCRIPTION OF THE TEST STATION AND THE DATA ACQUISITION SYSTEM

The test station is equipped with solar cell modules from 5 different manufacturers, with peak power output ranging from 10 to 35 W. The total number of modules is 56, distributed among different types to obtain about 200 W_p from each manufacture. The modules are mechanically mounted in eight panels and electrically wired to 6 groups. Each group comprises only modules from one manufacturer. The panels are 1.8 m high and are mounted one meter above ground to avoid snow disturbances. All panels are oriented to the south with an inclination of 58°. A front row with a free horizon contains all panels except one which is placed 3.6 m behind the front row. The purpose of this is to study effects of shadowing on the rear panel by comparing its output with an identical panel in the front row. A picture of the test station is shown in figure 1.

Figure 1. Overview of the test station.

All 6 module groups are loaded at the peak power point by means of peak power tracking circuits. Current, voltage, temperature on the rear side of the panels, ambient temperature, wind vector and insolation both in the horizontal plane and in the plane of the panels are measured continously and the collected data is stored on flexible disks. Figure 2 shows a block diagram of the test station.

Insolation is monitored with pyranometers of the Eppley PSP type. The temperature on the rear side of selected modules is recorded with Pt-100 foil elements. From these temperature measurements actual cell temperature is calculated using elementary heat flow calculation with geometry and marial parameters specific of each module of interest.

3. RESULTS

The test station has been operative for nearly 3 years. Data collection efficiency over this period is about 80%. Loss of data is mostly due to break downs in the data acquisition system. The results presented here are corrected for such data losses. The six module groups are numbered from one to six with number six as the shadowed one and number five as the identical front row group to number six.

In table I the yearly output, normalized to 1 kW_p, from the six groups are shown together with their peak power output measured in the solar simulator. Annual variation of insolation and power output is shown in figure 3.

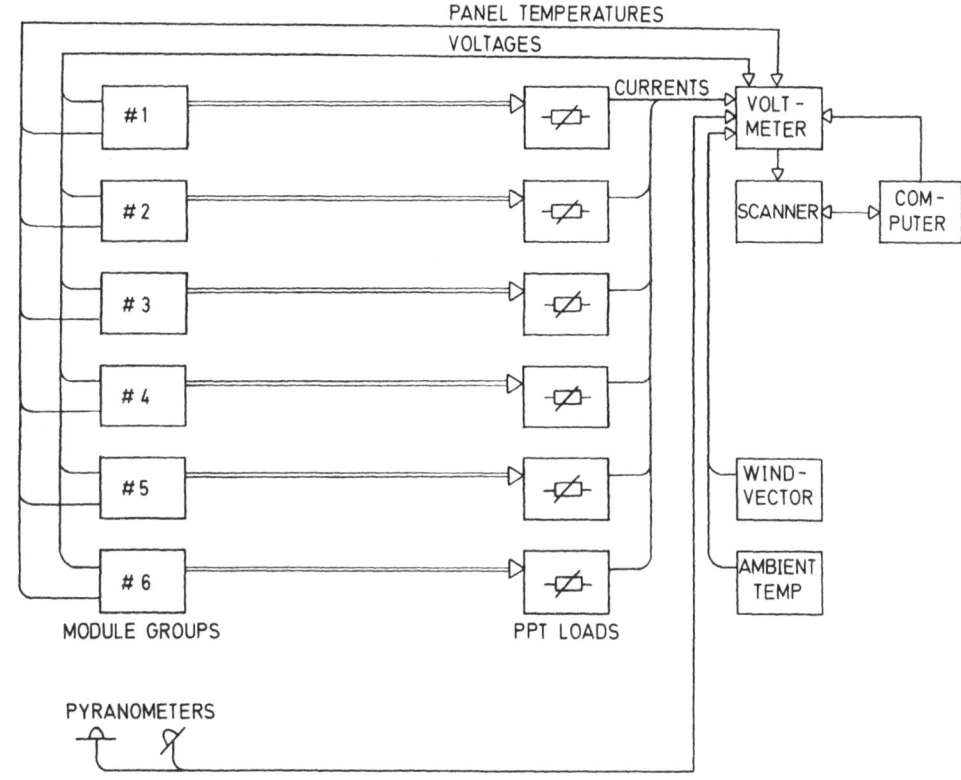

Figure 2. Block diagram of the test station.

TABLE I

Peak power and yearly output from the six groups. The yearly outputs are normalized to 1 kW_p. Insolation on a horizontal surface and in the plane of the panels.

Group number	W_p (W)	Year 1* (kWh)	Year 2* (kWh)
1	185	1060	1040
2	170	1000	1010
3	199	970	1000
4	176	1060	1090
5	89	1030	1090
6	89	910	930

<u>Insolation</u> (kWh/m^2)

Horizontal		930	950
58^0		1080	1100

* Year 1 is 79-06-01 to 80-05-31 and year 2 is 80-06-01 to 81-05-31

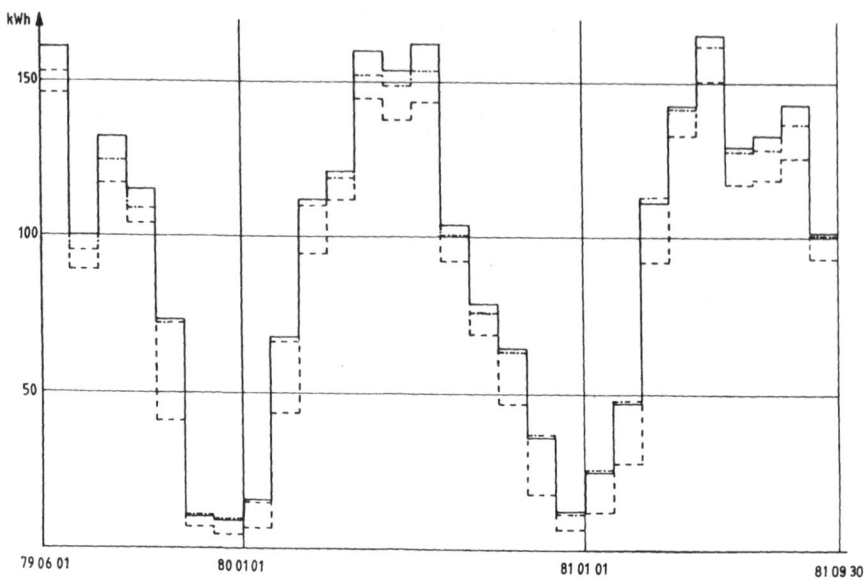

Figure 3. Annual variation of insolation per m² (——)(kWh/month) and generated energy (kWh/month) from module group nr 5 (-·-) and nr 6 (---), normalized to 1 kW$_p$. Module group nr 6 is the shadowed group.

The solar cell modules are also analyzed with respect to degradation and effects of dust and snow coverage. The modules have once a year been measured, before and after cleaning, in a solar simulator. No degradation has been detected except for one module which has a broken interconnection. The panels have during winter time been completely covered with snow about 30 days. Since the insolation during the winter season is very low, this has only a minor effect on total yearly generated energy.

The result can be summarized as follows:

- Insolation per year: 1100 kWh/m² in the plane of the panels
- Insolation per year: 940 kWh/m² in a horizontal surface
- Generated energy: 1 kWh/W$_p$ per year
- Generated energy: 0,9 kWh/W$_p$ per year for the cells in the back row
- Loss of power due to dust coverage: 4 - 5%
- Loss of energy due to snow coverage: 3 - 4%

4. CONSTANT VOLTAGE LOAD

To obtain maximum power output from a photovoltaic cell a peak power tracking load is required. In certain applications it may be easier and less costly to use a more simple type of load, for example a constant voltage. In this study the data collected have been used to calculate power output from a solar cell held at a constant voltage. Using the simple model for a solar cell, shown i figure 4, cell current and voltage are given by:

$$I = I_g - I_0(e^{(U+I \cdot R_s)/KT} - 1) - (U+I \cdot R_s)/R_{sh}$$

$$I_g = I_g(S,T)$$

$$I_0 = I_0(T)$$

Where S is insolation, T is cell temperature and I_0 is the reverse saturation current. With values for R_s, R_{sh}, I_G and I_0 typical for silicon solar cells together with collected data on insolation and cell temperature from the test station, the yearly output from a solar cell held at a constant voltage has been calculated. With an optimal choice of voltage the energy generated was found to be 96% of what would be achieved with peak power tracking. The optimal constant voltage found in this calculation is 70% of V_{oc} at 25°C cell temperature and an insolation of 1000 W/m². The reason that a constant voltage load only gives a few percent less produced energy over a year than a peak power tracking load would give is that a higher insolation is associated with higher cell temperatures giving a decreasing optimum voltage. In fact the optimum voltage is almost constant with constant ambient temperature.

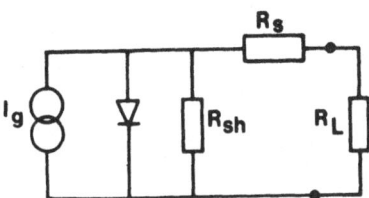

Figure 4. Solar cell model used in the calculations.

Acknowledgement

This project was financially supported by the National Swedish Board for Energy Source Development.

A MICROPROCESSOR-BASED INSTRUMENT FOR AUTOMATIC SOLAR CELL CHARACTERIZATION

G.C. CARDINALI*, E. FALDELLA**

*CNR - Istituto LAMEL - Via Castagnoli, 1 - Bologna - Italy
**Istituto di Elettronica and Centro Interazione Operatore-Calcolatore
Università di Bologna - Bologna - Italy

SUMMARY

This paper describes a microprocessor-controlled instrument for automatic characterization of photovoltaic cells. Accuracy, speed and flexibility, especially as regards run-time selection of measurement and data processing modalities, represent the main characteristics of the proposed system. As output the instrument provides plots of the current-voltage and power-voltage curves, together with information related to the measured data, such as cell fill factor and efficiency. The system turned out to be particularly useful for automatic classification and selection of photovoltaic cells for solar panels.

1. INTRODUCTION

The performance and reliability of photovoltaic modules consisting of several cells, are greatly influenced by the characteristics of the separate elements, according to a well-known mechanism which advises against interconnection of solar cells having inhomogeneous performance. Even though appropriate choice of the electrical connection scheme may reduce this problem, the need for photovoltaic cell characterization and selection arises. In general, automation would be desirable and certainly mandatory when solar panels are extensively produced.

With this aim in mind, different solutions have been developed in recent years: almost all involve either quite sophisticated and expensive minicomputers or cheap, but fairly efficient microcomputer-based systems (1),(2),(3). Applicability of the latter, however, has been mainly oriented to specific goals as the study of some technological aspects related to the implementation of solar cells.

The microprocessor-based system proposed in this work has been developed to find an economically viable and generally applicable solution to the problem of automatic measurement of solar cell characteristics. It has been envisaged to support experimental and development studies within laboratories, as well as automatic classification and performance evaluation within extensive production environments.

The firts part of the paper points out the problems related to digital evaluation of solar cell characteristics as regards measurement accuracy and resolution. System design criteria are then discussed, with particular emphasis on the instrument's microcomputer-based architecture and the overall software organization. The paper ends with some pertinent remarks on system performance and flexibility.

2. DIGITAL APPROACH TO AUTOMATIC CHARACTERIZATION OF SOLAR CELLS

Evaluation of the electrical performances of a photovoltaic cell generally involves measurement of its current-voltage characteristic. To this end, the cell is connected to a load circuit whose equivalent diagram is shown in Fig. 1-a. Most of the measurement methods involve sweeping of the bias voltage, E_o, so as to vary the cell output voltage from zero to the open circuit value, with the load resistance, R_o, set to a constant value, independently of the electrical characteristics of the cell under test. This solution, which is certainly simple and inexpensive as far as the load circuit configuration is concerned, may well satisfy the requirements involved in an analog measurement process, but it is not suitable for digital acquisition of the characteristic. In fact, in order to guarantee that the required number of samples is uniformly distributed along the I-V curve, it is necessary to vary the sweeping voltage according to a law which greatly depends on the non-linear relationship existing between I and V. Moreover, severe constraints may be imposed on E_o resolution, due to its small variations required by the scanning process near the open circuit operating region.

These problems may be overcome provided the value of the load resistance may be suitably varied according to a criterium which takes into account the specific electrical characteristics of the cell to be tested. A generally-applicable and easy-to-implement method is shown in Fig. 1-b, where R_o is set to the value given by the ratio between the open circuit voltage, V_{oc}, and the short circuit current, I_{sc}, of the cell, and E_o is varied from $-V_{oc}$ to $+V_{oc}$, with constant increments dependent only on the resolution required. This method involves a more sophisticated load circuit, but permits uniform distribution of the samples along the I-V curve for any type of cell or composite photovoltaic module.

3. SYSTEM ARCHITECTURE

The architecture of the µP-based instrument developed for photovoltaic cell automatic characterization, is schematically shown in the block-diagram of Fig. 2. The control system consists of a digital section, which performs the tasks of data acquisition and processing, as well as printout and recording of information related to the measured data, plus a D/A and A/D section interfacing the µP with a specially designed electronic load circuit used for I-V characteristic scanning.

The digital section is implemented with 3 in-house developed boards.
a) CPU board. This board contains the central processing unit (an INTEL 8085A µP running at 6.144 MHz), a Programmable Communication Interface (PCI) based on an 8251A USART, an 8253-5 Programmable Interval Timer (PIT), and the logic necessary for buffering and driving the system bus. The PCI provides the designer with a powerful tool for handling the full-duplex serial exchange of information between the µP and the operator console. The PIT makes available three independent 16-bit counters. One of them is used as the programmable baud rate generator for the communication interface, a second as the real-time clock generator for proper timing of the measurement process.

b) <u>Memory board</u>. This board contains four 2716 EPROM (for a total program storage capacity of 8 Kbytes), and for eight 8185 RAM (for a total data storage capacity of 8 Kbytes), together with the associated address decoding logic and bus drivers/receivers.

c) <u>I/O board</u>. This board contains four 8255A-5 progammable peripheral interface chips, thus making available up to 96 progammable input/output lines. A subset of these lines is used for driving an HP 7015B XY-recorder, whereas the other lines provide the digital interface with the analog section implementing the load circuit of the cell under test.

The analog section of the controller (Fig. 3) is built around a control circuit which supports programmable sweeping of the I-V characteristic according to the basic scheme of Fig. 1-a.

The cell bias voltage, E_o, is derived from the output of a digital-to-analog conversion unit based on a AD561 DAC. Provisions have been made at the design stage for three program-selectable output voltage ranges so as to allow sweeping of composite module characteristics.

The equivalent resistance, R_o, of the cell load circuit may be controlled by suitably selecting the input resistance R_2 of amplifier A_2. It turns out in fact from analysis of the circuit that $V = R (R_3/R_2) I + E_o$, and, therefore, $R_o = R (R_3/R_2)$.

Switches S_6 and S_7 allows the µP to establish, respectively, the short circuit and open circuit load operating conditions on the cell under test. In this way the controller can immediately select the value of R_o which guarantees uniform distribution of the sampled points along the I-V characteristic. Each sample involves measurement of the two parameters I and V. Their analog-to-digital conversion is obtained with a single ADC by properly multiplexing the current and voltage signals through S_9.

4. SYSTEM OPERATION AND PERFORMANCE

Performance of the instrument as regards automatic characterization of photovoltaic cells may be brought out by considering the tasks performed by the µP handling the system overall operation.

After executing the power-on hardware initialization procedure, the µP establishes a link with the operator console in order to get information about some configuration parameters involved in the measurement process. Basically 3 sets of parameters have to be defined by the operator:
- Measurement modalities: number of points to be sampled on the current-voltage characteristic, sampling repetition rate, software filtering specifications (if required, a measure may be iterated and the relevant data averaged so as to minimize noise effects).
- Output information selection: list of the measured data, plots of the current-voltage and power-voltage curves (the µP automatically chooses the widest scale compatible with data to be traced, unless otherwise specified by the operator), printout of maximum power, fill factor, efficiency and other information related to the measured characteristics.
- Cell selection criteria.

Following input parameter definition, the µP waits for a start-operation command which triggers the automatic measurement process. The measurement process involves first of all acquisition of the open circuit voltage and the short circuit current of the tested cell, hence evaluation and selection of the optimum value of load circuit resistance. Samples of the I-V characteristic are then obtained and processed so as to produce the required output information.

Software for the tasks at higher level has been designed according to structured programming principles in PLM/80 language. I/O drivers, on the contrary, have been written in assembler, in order to limit processing time. Storage requirements involve 8 Kbytes of program memory and about 6 Kbytes of read/write memory.

5. CONCLUSIONS

Extensive experimental tests carried out with the prototype system have emphasized the characteristics of accuracy and speed of the measuring instrument and the validity of the solution adopted for its design. What is particularly remarkable is the flexibility afforded by the system as a consequence of its µP-based implementation, especially as regards the possibility of defining the measurement modalities, the values of some basic parameters involved in the computations, and the cell selection criteria.

At present a feasibility study is being undertaken in order to evaluate the convenience of linking the instrument with a remote computer for off-line statistical elaborations and automatic archive purposes. The possibility of direct control over the cell light source is also investigated.

- REFERENCES

1. L. Castaner et Al.,'µP-based Solar Cell Measurement System', IEEE Transactions on Instrumentation and Measurement, IM-27, pp.152-156 (1978)
2. K. Emery, J. DuBow,'Automated Electronic Analysis of Solar Cells', IEEE Photovoltaic Specialist Conference, CH1508, pp. 506-510 (1980)
3. C.R. Saylor et Al.,'Short Interval Testing of Solar Cells', IEEE Photovoltaic Specialist Conference, CH1644, pp. 534-538 (1981)

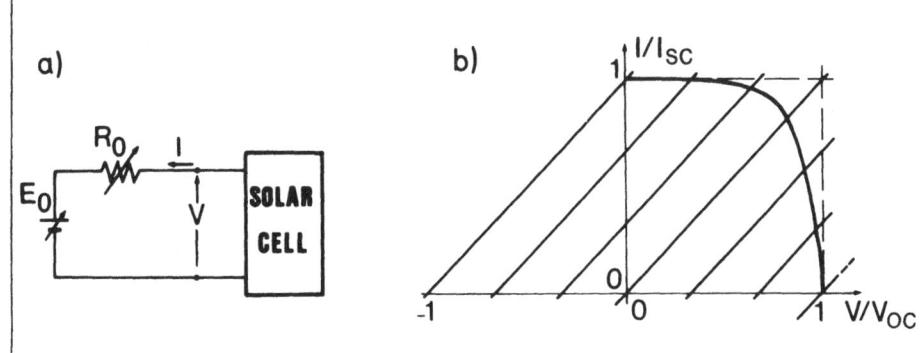

Fig. 1: Measurement of a solar cell characteristic
 a) equivalent load circuit diagram
 b) characteristic scanning process

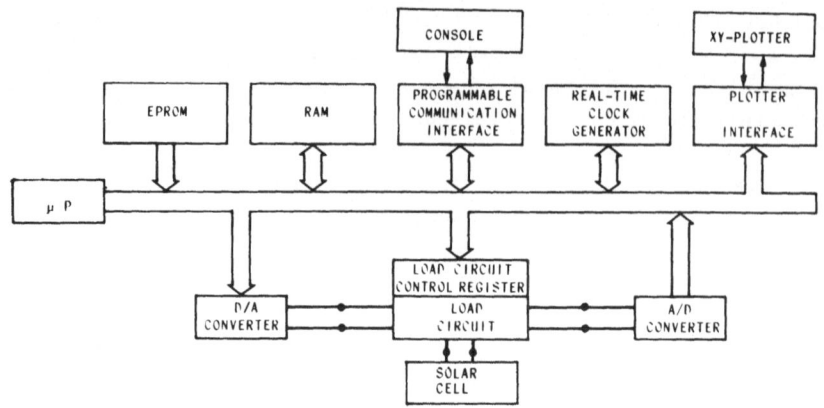

Fig. 2: Schematic block-diagram of the system architecture.

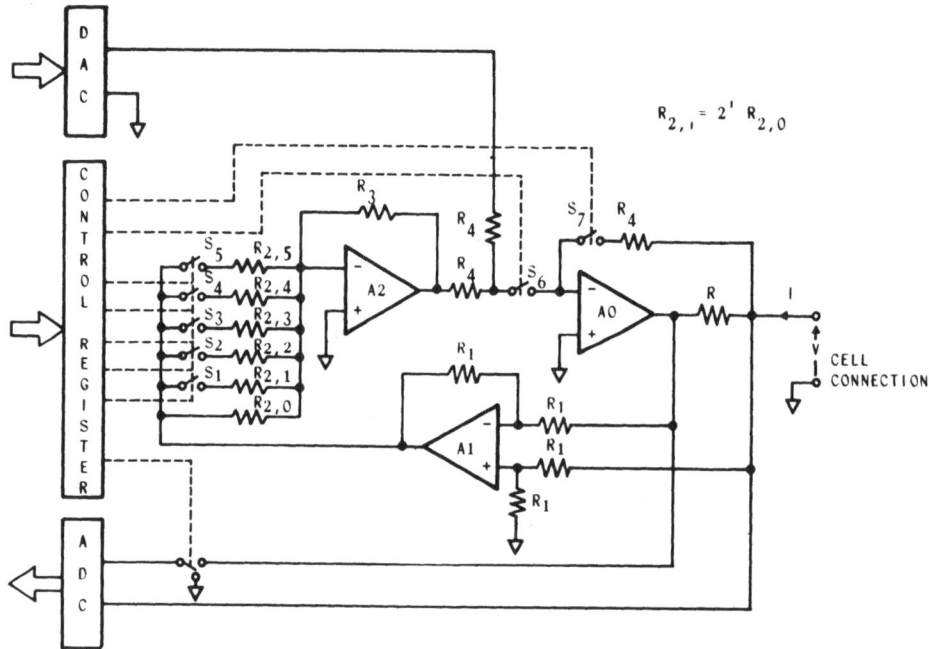

Fig. 3: Schematic circuit diagram of the controller analog section.

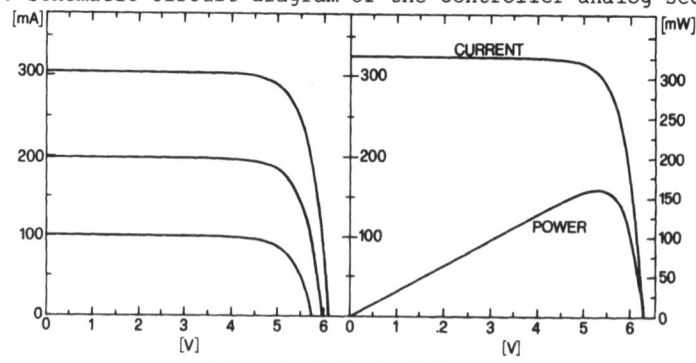

Fig. 4: Output plots.

SESSION 3

SYSTEMS, COMPONENTS AND ENGINEERING

Chairmen : Prof. A. LUQUE, University of Madrid, Spain

Ing. G. SIMONI, AGIP Nucleare SpA, Rome, Italy

Subsystem engineering and development of grid-connected photovoltaic systems

Array structures for fixed flat-plate photovoltaic power generators

Inverters for PV plants feeding the electrical network of small isolated communities

SUBSYSTEM ENGINEERING AND DEVELOPMENT OF GRID-CONNECTED
PHOTOVOLTAIC SYSTEMS*

E.L. BURGESS, H.N. POST, and T.S. KEY
Sandia National Laboratories
Albuquerque, New Mexico, USA

Summary

The experience gained in fielding residential and intermediate-sized photovoltaic application experiments is summarized. This experience is used to guide the engineering and development of array and power conditioning subsystems for grid-connected photovoltaic systems. A major consideration in this development effort is cost. Through innovative engineering, using a modular building block approach for the array subsystem, it is now possible to construct array fields, in moderate quantities, for about $52/m^2 excluding the photovoltaic modules. Similarly, results of power conditioning subsystem development indicate a projected cost of about $0.25/W_p$ for advanced units with conversion efficiencies in excess of 90%.

1. INTRODUCTION

The major emphasis of the U.S. Department of Energy's photovoltaic (PV) program has been on the development of grid-connected systems. The worth of a PV system, in this scenario, is measured by how well the levelized energy cost competes with the cost of conventional grid-supplied energy. For this reason, effort has been placed on reducing cost, particularly for the PV modules. In the last couple of years, increasing attention has been given to cost and performance considerations of the balance-of-system (BOS) element, i.e., the non-photovoltaic part of the system.

The major system elements are the array and power conditioning subsystems. The array subsystem incorporates all hardware necessary to deliver unregulated dc power to the power conditioning subsystem and typically includes, in addition to the PV modules, array field area-related items such as land, structures, foundations, field wiring, electrical subsystem protection, and site preparation. The power conditioning subsystem includes the equipment necessary to regulate dc power, convert it to ac power, and deliver it to a distribution grid. This involves all the hardware needed to interface with the array and the utility, as well as power conversion and control.

Considerable experience has been gained with these subsystems through the fielding of full-scale PV system experiments. Detailed performance and cost analyses of the various system components have been performed and the results used to guide subsystem engineering and development efforts. This paper summarizes this experience and the resultant advances in subsystem performance and cost.

---

*This work was supported by the U.S. Department of Energy, Division of Photovoltaic Energy Systems Technology.

2. PHOTOVOLTAIC SYSTEM EXPERIMENTS

The U.S. PV program has fielded a number of intermediate-sized systems over the past five years. The oldest of these is an irrigation application in Mead, Nebraska. That system, a 25 kW$_p$ flat-plate array, became operational in July 1977. It has operated reasonably trouble-free but is beginning to experience some PV module degradation due to cell interconnect failures. More recently, other intermediate-sized experiments have been planned and constructed or are in the process of being implemented. These experiments are listed in Table I along with characteristics of each system. More details of several of these systems are given elsewhere (1).

TABLE I. U.S. INTERMEDIATE-SIZED PV SYSTEMS

APPLICATION	LOCATION	SIZE (kW)	ARRAY TYPE	INITIAL OPERATION	COST ($/W)
OFFICE BLDG.	ALBUQUERQUE	47	PARABOLIC TR.	MAY 82	30
HIGH SCHOOL	BEVERLY, MA	96	FLAT-PLATE	JAN 81	30
AIRPORT	DALLAS, TX	27	LIN. FRESNEL	JUN 82	42
HOSPITAL	KAUAI, HI	35	PARABOLIC TR.	JAN 82	51
SHOPPING CENTER	LOV'TON, NM	100	FLAT-PLATE	MAR 81	29
UPS INVERTER	EL PASO, TX	16	FLAT-PLATE	JAN 81	30
LIGHT MFG.	SAN BERNARDINO	35	FLAT-PLATE	JAN 82	23
MUSEUM	OKLA. CITY	135	FLAT-PLATE	FEB 82	21
AIRPORT	PHOENIX, AZ	225	PT. FRESNEL	JUN 82	18
IRRIGATION	MEAD, NB	25	FLAT-PLATE	JUL 77	28
NAT. PARK	UTAH	100	FLAT-PLATE	JUN 80	31
RADIO STATION	BRYAN, OH	15	FLAT-PLATE	AUG 79	20
RADAR STATION	MT. LAG., CA	60	FLAT-PLATE	JUN 79	22
COMMUNITY COLLEGE	B'VILLE, AR	240	PARABOLIC TR.	MAY 81	30
UNIVERSITY	WASH., DC	300	FLAT-PLATE	SEP 84	21
UTILITY	SACRAMENTO	1000	TBD	JUN 84	12

An application receiving increased attention within the last two years is private residences. Because of concerns about safety, much of the experimentation in the U.S. has been in residential experiment stations (RES). The concept of a RES is to implement a number of prototype PV systems on simulated houses in a controlled-access test facility. Two such RES's have been implemented and a third will be implemented this year. Not all experimentation with these small roof-mounted systems has been restricted to the RES's. Seven other systems have been constructed. The U.S. PV program residential activity is summarized in Table II.

TABLE II. U.S. RESIDENTIAL PV SYSTEMS

PROJECT	LOCATION	NUMBER SYSTEMS	SYSTEM SIZE (kW)	INITIAL OPERATION	COST ($/W)
NE RES	CONCORD, MA	5	5.7	NOV 80	25
SW RES	LAS CRUCES, NM	8	5.4	MAR 81	23
CARLISLE	CARLISLE, MA	1	7.8	FEB 81	14
FSEC	CAPE CAN., FL	1	4.0	AUG 80	-
HI HOUSES	HAWAII	3	3.3	MAY 81	29
LONG HOUSE	PHOENIX, AZ	1	6.6	MAY 80	39
UNIV. TEX.	ARLINGTON, TX	1	8.0	OCT 78	21
SE RES	TBD	TBD	TBD	83	-

These two large experimental sets have provided a very significant data base from which to assess system performance and to highlight areas which dominate system cost. In the remainder of this section a summary is presented of the lessons learned which deal with the two major subsystems, i.e., the array and power conditioning.

Systems incorporating concentrator PV arrays are just now becoming operational. For this reason, most of the discussion here will deal with the flat-plate systems. The costs for six flat-plate, intermediate-sized systems have been discussed in detail in a paper by Burgess and Biringer (2). The data in Table III are taken from that paper.

TABLE III. PV SYSTEM COST BY COST CATEGORY

PROJECT	UPS		SHOP. CTR.		HIGH SCH.		MUSEUM	
COST CATEGORY	$/m^2	$/W	$/m^2	$/W	$/m^2	$/W	$/m^2	$/W
ENGINEERING	593	10.56	560	8.41	495	7.42	373	3.06
PV MODULES	576	10.23	753	11.32	725	10.86	1435	11.77
STRUCTURE	90	1.60	44	0.66	46	0.69	277	2.27
FOUNDATION	18	0.32	70	1.05	121	1.81	84	0.69
CIVIL WORK	54	0.96	137	2.05	178	2.67	195	1.60
ELECTRICAL	219	3.91	173	2.59	231	3.46	104	0.85
PCS	108	1.91	180	2.70	159	2.38	56	0.46
BUILDINGS	54	0.96	38	0.57	20	0.29	–	–
TOTAL	1712	30.45	1955	29.35	1975	29.58	2524	20.70

As shown, the costs for design can run as high as 30-35% of the total project, the PV modules represent about one-third of the cost, and the remaining cost is for the BOS elements. Analysis of this data and the system design details has highlighted the following reasons for the high costs in the design and BOS categories.

- All the designs were developed essentially independent of each other, yet much similarity exists. This suggests that standardized designs are applicable and that considerable cost savings are possible if they were available.

- Structures and foundations were designed according to standard A & E practices which incorporate large "margins-of-safety". The resultant costs of $2-4/$W_p$ are not acceptable. More innovative approaches are necessary and some margins-of-safety will have to be sacrificed. Because of the conservativeness of the present designs (e.g., 2-3 times over-design on wind loading), this is possible without sacrificing personnel safety.

- Site preparation has tended to be overdone in order to obtain perfectly level sites and aesthetically pleasing results. This has been the practice because of the experimental nature of the projects and the large number of visitors expected at the sites. In actual applications, "favorable" sites (i.e., reasonably level, no rock outcroppings, and minimum vegetation) should be selected and very little site preparation performed.

- Electrical wiring costs have frequently exceeded the total allowed system cost. As in the case of structures and foundations, standard A & E practices have been applied in electrical wiring design.

Numerous junction boxes, interconnections, and conduit have been incorporated. Innovative approaches to electrical wiring are required, eliminating essentially all of the junction boxes.

The next section discusses how many of these features and lessons learned have been incorporated into standardized, modular building block array subsystems for ground-mounted arrays.

Four types of mounting techniques have been examined for small roof-mounted systems: standoff, direct, integral and rack mounts. The evaluation of these mounting techniques is still in progress; however, some preliminary results are available. These include:

- Standoff mounts are adaptable to retrofits and the array operates at an acceptable temperature.

- Direct mounts are also adaptable to retrofits but the array operates at a temperature 10-20°C higher than arrays in the other mounting schemes. This is due to cooling from one side of the module only.

- Integral mounts are potentially less expensive since the array is the major roof element and roofing credit can be applied to the array cost. Operating temperature is not a problem since innovative schemes for cooling the underside have been developed.

- The rack mount is simple but adaptable only to flat-roof houses.

More details of new array mounting techniques are discussed by Ross (3).

The lessons learned regarding intermediate-size power conditioning subsystems can be classified into two major categories: 1) excessive cost for the adapted uninterruptible power supply inverters and 2) problems with both the array and utility interfaces. These are discussed in a later section.

3. ARRAY SUBSYSTEM DEVELOPMENT

Array subsystems have been the focal point of extensive development activities within the U.S. PV program, primarily as a result of the lessons learned through system experiments and other system definition activities. Recent studies have examined cost reduction in array subsystems for ground-mounted systems through component integration, array field design optimization, and detailed installation analyses (4). Similar activites have emphasized the development of integrated and optimized roof-mounted arrays for residential systems (5). This latter work is detailed by Ross (3) and and will not be covered here. The remaining discussions on array subsystem development will concentrate on medium- and large-size systems, primarily commercial/industrial type applications, and will highlight the development of modular, low-cost array field designs.

3.1 Cost Reduction

The ideal approach toward low-cost array field subsystem development is to examine actual costs and installation experience for operating systems and to identify areas where cost reductions can be made. This has been done for the Lea County Electric Cooperative application experiment, typical of a medium-size, ground-mounted, flat-panel system. This system (identified as a shopping center application in Tables I and III) has a nominal power output of 100 kW and includes 1503 m^2 of collector area. As shown in Table IV, specific items were examined for potential cost reduction. The proposed activities which were identified are based on detailed evaluation of the major cost drivers for each subsystem.

TABLE IV. MAJOR ARRAY FIELD ITEMS REQUIRING COST REDUCTION

ITEM	LEA CO. SHOPPING CENTER COST	COST REDUCTION ACTIVITY
ARRAY FIELD ENG. DESIGN	$2.58/W	DEVELOP STANDARDIZED ARRAY FIELD DESIGNS AND BUILDING BLOCK HARDWARE; MINIMIZE SITE SPECIFIC DESIGN WORK
SITE PREPARATION	$137/m^2	SELECT SITES WITH FAVORABLE CONDITIONS (MINIMAL CLEARING AND EXCAVATION WORK)
STRUCTURAL SUBSYS.	$114/m^2	USE LOW-COST FOUNDATIONS; DEVELOP INTEGRATED PANEL/STRUCTURE HARDWARE; USE SIMPLIFIED MECHANICAL ATTACHMENT; MAXIMIZE FACTORY PANEL ASSEMBLY
ELECTRICAL SUBSYS.	$173/m^2	DEVELOP INTEGRATED FIELD WIRING; MINIMIZE SUBARRAY JUNCTION BOX INSTALLATION; USE DIRECT-BURIAL AND LOW-COST MODULE WIRING CONNECTORS; MINIMIZE INSTRUMENTATION

A key element for cost reduction is the development of standardized array field designs and component hardware. Estimates of the site specific costs for the application experiment systems noted earlier range from 30 to 35 percent of the total project costs. Standardized designs should significantly reduce the site specific or custom design work as well as reduce the engineering costs (system design, A & E fees, construction management) associated with future installations.

The poured concrete slab foundations represent 60 percent of the Lea County project structural costs. Integrated panel/structure hardware coupled with low-cost foundations were identified as cost-effective options. Similarly, the design approach for these structures should adopt a philosophy of being operational at reasonable environmental loads and remain safe, although allowing structural yielding and deformation, at survival condition loads.

Site preparation costs for the Lea County project were driven by excavation and backfill operations, representing 33 percent of the total costs, and the use of stone ground cover to control dust and weeds at a cost fraction of 23 percent. To reduce these costs, site selection should be restricted to locations which require minimal clearing and excavation. Leveling requirements should be minimized, through innovative foundation design, and lower-cost options, such as gravel instead of stone, should be incorporated in site plans.

Subarray conduit and junction box costs dominate the Lea County electrical subsystem costs, representing 45 percent of the total cost. Integrated field wiring designs incorporating direct-burial cabling, few if any junction boxes, low-cost module interconnection, and minimal instrumentation were identified for cost reduction options. Similarly, lightning protection should consist of terminal protection devices (to limit transients at the array and power conditioning equipment) and protective shielding for long cable runs. Direct strike protection should be considered only for areas with high thunderstorm activity where favorable risk/cost benefits so warrant.

3.2 Modular Array Field Designs

To significantly reduce array field costs, Sandia National Laboratories through two contractors, Battelle-Columbus Laboratories and Hughes Aircraft Company, has designed and developed totally integrated, modular array field designs for flat-plate systems which implement the cost reduction options noted previously (6,7). These designs, which are proposed for system sizes ranging from 20 to 500 kW, are presented in detail elsewhere (8). Based on estimated array field BOS costs (1980$) including site preparation, structural subsystems, and electrical subsystems, the Battelle design and Hughes design are projected at $51.93/m^2 and $62.92/m^2, respectively, for low production quantities and favorable installation sites. Complete engineering drawing packages, including detailed construction specifications, are available for each design.

The development of modular array field designs incorporating concentrator arrays should be completed by June of this year. These designs utilize two state-of-the-art concentrating collector arrays. The first is a point-focus, fresnel-lens array using a center-post mount. The second is a line-focus, fresnel-lens module mounted in an end-supported frame. Both systems are passively cooled and use two-axis tracking. These designs will be detailed in engineering drawing packages with example layouts for for 20-,100-, and 500-kW array fields.

3.3 Other Development Activities

Other key activities for medium- and large-size systems include the following:

- Electrical Subsystem Optimization
- Integrated Structure Design
- Automated Installation Methods
- System Grounding and Fault Protection Study

The subsystem optimization work examined the effects of array characteristics, system power, and voltage levels, as well as other design and application specific factors on first and operating costs for large (0.5 to 1000 MW_p) PV power systems (9). The complete electrical subsystem including dc wiring, ac wiring, grounding, and transient surge protection was addressed. A key result of this work is that array subfields with size and voltage ratings of 5 MW and 2000 Vdc, respectively, appear optimum for the combined costs of dc wiring and power conditioning equipment.

Bechtel Group, Inc., through the integrated structure design work, has developed an optimized design incorporating PV modules into a structurally integrated panel assembly. The projected cost (1980$) for the installed structural subsystem is $31/m^2 of collector area. This structural design was included in the automated installation methods work conducted by Burt Hill Kosar Rittelmann Associates (10). Installation scenarios, both conventional and automated, were developed for large photovoltaic array fields as part of a comprehensive study of installation methods. Results for a 10-MW flat-panel array field, installed using conventional methods at a southwestern U.S. location, show a projected installation cost (including overhead and profit) of $28/m^2 of collector area. This projection includes all the site preparation (roads, fencing, drainage), electrical subsystems, and foundation costs. Adding in the cost of the panel developed by Bechtel gives a total installed cost (1980$) of approximately $50/m^2. It is estimated that installation costs can be reduced by approximately $11/m^2 by using an appropriate mix of automated and conventional installation methods. Preliminary designs for the required robotic equipment, using currently available component hardware, have been prepared.

The last activity, system grounding and fault protection study, is focused on a design manual to be published during 1983.

4. POWER CONDITIONING DEVELOPMENT

The role of the power conditioner in a grid-connected PV system goes beyond dc to ac inversion. Other critical functions include interfacing the array to the inverter and the inverter to the utility line as well as providing overall system control and protection. As depicted in Fig. 1, there are four major functions of a power conditioning subsystem (PCS).

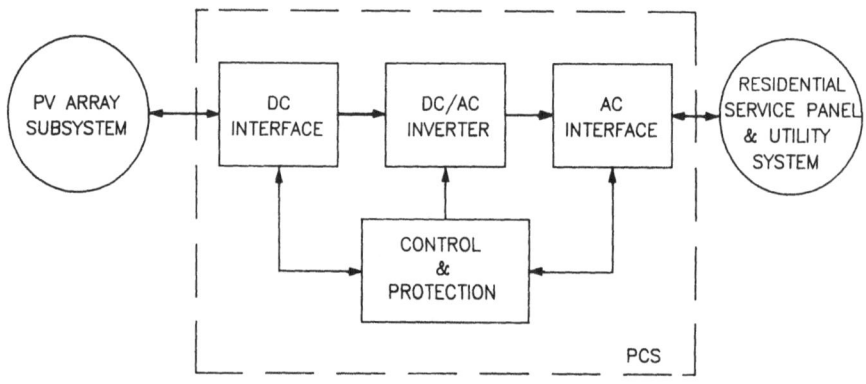

FIG. 1 - ROLE OF PCS IN GRID-CONNECTED PV SYSTEM

In most PV system application experiments the major PCS performance problems have been related to interfacing and control functions. On the dc-side, PV unique source characteristics such as high array V_{oc} and a limited I_{sc} have created early morning and late afternoon start-up problems. Periods of low array power output have lead to excessive on/off cycling of some inverters. As a synchronous ac power source on the utility-grid, the PCS has been exposed to line-voltage fluctuations and temporal phase shifting. Such transients have required special attention in the design of PCS power flow controls and protection schemes.

Another design constraint experienced by the grid-connected PCS is the power quality requirement imposed by the local utility company. At the present time, there is no technical precedence for connecting a large number of inverter sources into the utility grid. Therefore, requirements have varied from site to site.

4.1 Current Activities

Field experience has pointed out several complex issues in the PCS interfacing and controlling functions. A summary of these issues and other technical concerns surrounding the PCS is provided in Table V.

TABLE V. TECHNICAL ISSUES SURROUNDING THE GRID-CONNECTED PCS

ARRAY INTERFACE	PCS HARDWARE DEVELOPMENT	UTILITY INTERFACE
• POWER TRACKING	• PERFORMANCE	• POWER FACTOR
• Vdc MATCHING	• EFFICIENCY	• HARMONICS
• POWER LIMIT	• RELIABILITY	• STABILITY
• Vdc TOLERANCE	• MAINTAINABILITY	• LINESMAN SAFETY
• PROTECTION	• SURVIVABILITY	• PROTECTION
• USER SAFETY	• EMI & ACOUSTICAL COMPATIBILITY	• CONTROL
• ON/OFF CYCLING		• ISOLATION X'FORMER

REQUIREMENTS FOR THE PCS IN A PV SYSTEM

- IN-SITU TESTING AND EVALUATION
- CODES AND STANDARDS
- GROUNDING
- DESIGN CRITERIA/PCS SIZING

Using typical cell models, weather tapes, and a computer applied algorithm, the array interface has been characterized from the point-of-view of the PCS. These data give the potential PCS manufacturer an estimated percent of annual array energy loss for several power-tracking approaches from fixed voltage to a full-range input voltage tracking window. Also provided are array voltage and current levels under extremes of operation and a plot of the annual power vs hours profile of array output. These array data allow the PCS manufacturer to make array interface design "tradeoffs" that maximize the ac energy output of the PV system.

Research in the area of optimum power conversion has suggested abandoning very small (100 W) distributed parallel inverter schemes. The major drawbacks were inefficient use of magnetic materials, high total part count and low efficiency projections. An approach at the 4 kW level, which is considered to be close to optimum, shapes half-sine waves in a single-quadrant, push-pull buck converter with alternate halves inverted with a transistor bridge. This prototype has performed well in a breadboard stage.

Consensus utility interface requirements currently used for a residential PCS are: low harmonic distortion (5% current THD into a linear load) and nearly unity power factor (.95 lagging to .95 leading from 1/8 to full load). Also an isolation transformer is required between the utility and the PV system. Examples of on-going research intended to determine long-term utility connection requirements are; dynamic simulations of dispersed PV at varying penetration levels, power factor and harmonic effects on the distribution feeder, and identification of requirements for transformerless systems.

With respect to future safety codes, wiring and grounding standards, two documents have been drafted. One sets forth safety standards for residential power conditioning units (11). The other covers special grounding and wiring requirements of PV systems (12).

PCS specifications are being developed to improve technical communication with industry. For example, a residential baseline specification prepared for single-phase, 2-10 kW power conditioners (13) includes:

- Photovoltaic array output characterizations for flat-plate modules
- Protection, efficiency, and other performance requirements
- Latest available consensus utility interface requirements
- Recommended test procedures

This information allows the designer to optimize ac energy output of the PCS within known performance and interface constraints.

Small power conditioners are presently available that will meet the specifications in (13). Table VI provides cost and efficiency details for inverters that have operated in the U.S. DOE residential experiments.

TABLE VI. PCS CANDIDATES EXPECTED TO MEET OR EXCEED THE RESIDENTIAL BASELINE SPECIFICATION

AVAILABLE MFG./MODEL	OUTPUT RATING	EFFICIENCY (100%)	OPERATING PRINCIPLE	POWER TRACKER	PRICE ($/W)
WINDWORKS/GEMNI WITH FILTER	4kW 8kW	92* 92*	LINE COMM., CUR. SOURCED, SCR BRIDGE (120 Hz) WITH WAVESHAPING FILTER	N/A	— 0.94*
ABACUS CONTROLS/ SUNVERTER	4kW 6kW 10kW	91* 88 92*	SELF OSCILLATING, VOLTAGE SOURCED, X-SISTER BRIDGE (1.44 kHz) WITH PRE PROGRAMMED SINE WAVE IN PROM	PILOT CELL	1.59* 2.04 0.98*
HELIONETICS INC. (DECC)/61289 SERIES (2-12kW)	5kW 6kW 8kW	91* 91 91	SELF OSCILLATING OR LINE SYNC, VOLTAGE SOURCED X-SISTER BRIDGE WITH PULSE DURATION MODULATION CLASS "D" AMP (28 kHz)	PERTURB/ OBSERVE	1.00* 1.63 1.50
AMERICAN POWER CONVERSION CORP. SUNSINE VI-2000 & 4000	2kW 4kW	92 92*	SELF OSC. (LINE SYNC) CURRENT SOURCED PWM WAVESHAPING HF FRONT-END DOWN CONV. (13-20 kHz)	PERTURB/ OBSERVE	1.99 1.25*

* PROJECTED BASED ON DETAILED ANALYSES OR BREADBOARD TESTS OF PREPRODUCTION PROTOTYPE

Several 3-phase PCS units (20-200 kW) have also been built for the intermediate PV experiments. Most use a conventional modified Uninterruptible Power System technology (e.g. self-oscillating, voltage sourced, 6 or 12 step wave synthesis with output harmonic filtering). Another design used is a 3-phase line-commutated unit with appropriate output filtering and power factor correction. Neither of these approaches are considered promising in the long run. Price experience in this size range has been $0.50-2.50/W.

4.2 Future Plans

Seven studies were recently completed (14) that identify promising conceptual designs in the 2-10 kW and 20-200 kW size range. Single and 3-phase advanced prototype hardware will be developed during the next 18 months using the baseline performance requirements as a guide. Anticipated results include greater than 91% efficiency and future selling prices in the $0.20-0.30/W range for production levels of 1000 units/yr.

Another development activity in the near future will optimize a single-phase transformerless design based on anticipated future grounding, safety and utility interface requirements for residential PV systems. A 2% increase in efficiency over the 60 Hz isolation transformer approach is contemplated. Three promising advanced conceptual designs that have been identified for a 3-phase, 20-200 kW PCS include; a self-oscillating parallel transistor bridge (with advanced switching module), a line-commutated 6-pulse SCR step-wave synthesis (with complimentary parallel VAR inverter), and a hybrid, line/forced commutated 6-pulse multipattern programmed wave (with GTO commutated SCR for low-cost and high-efficiency). The projected efficiency range for these approaches is 93-96%.

5. CONCLUSIONS

Subsystem development activities have been implemented which lead to low-cost BOS. Array subsystem costs are now projected to be approximately $52/m^2 for modular array field designs, a factor of 5-8 below current field experience. Similarly, the PCS is projected at $0.25/$W_p$ for medium-sized units in large production quantities. For a system with 10% conversion efficiency, these projections result in a BOS contribution of $0.77/$W_p$ toward the overall system cost.

6. REFERENCES

1. E. L. Burgess and E. A. Walker, editors, Summary of Photovoltaic Application Experiments Designs, ALO-71 (Albuquerque: U.S. Department of Energy, October 1979).
2. E. L. Burgess and K. L. Biringer, "Update of Photovoltaic System Cost Experience for Intermediate-Sized Applications," Proceedings of the Fifteenth IEEE Photovoltaic Specialists Conference, June 1981.
3. R. G. Ross, Jr., "Photovoltaic Array Technology for Residential Roof-Top Applications" presented at the Fourth EC Photovoltaic Solar Energy Conference, Stresa, Italy, May 10-14, 1982.
4. H. N. Post and D. G. Schueler, "Array Subsystem Development for Photovoltaic Array Fields," Proceedings of the Fifteenth IEEE Photovoltaic Specialists Conference, June 1981.
5. N. F. Shepard, Integrated Residential Photovoltaic Array Development, DOE/JPL 955894-3 (Pasadena: Jet Propulsion Laboratory, August 1981). Work performed by General Electric Company.
6. D. C. Carmichael, et al, Development of a Standard Modular Design for Low Cost Flat-Panel Photovoltaic Array Fields, SAND81-7183 (Albuquerque: Sandia National Laboratories, to be published). Work performed by Battelle-Columbus Laboratories.
7. J. A. Castle, et al, Photovoltaic Array Field Optimization and Modularity Study, SAND81-7193 (Albuquerque: Sandia National Laboratories, to be published). Work performed by Hughes Aircraft Company.
8. H. N. Post, D. C. Carmichael, J. A. Castle, "Low Cost Modular Designs for Photovoltaic Array Fields" presented at the Fourth EC Photovoltaic Solar Energy Conference, Stresa, Italy, May 10-14, 1982.
9. W. J. Stolte, Photovoltaic Subsystem Optimization and Design Tradeoff Study, SAND81-7013 (Albuquerque: Sandia National Laboratories, March 1982). Work performed by Bechtel Group, Inc.
10. J. R. Oster, Automated Installation Methods for Photovoltaic Array Fields, SAND81-7192 (Albuquerque: Sandia National Laboratories, June 1982). Work performed by Burt Hill Kosar Rittelmann Associates.
11. Draft "Safety Standard for Power Conditioning Units for use in Residential PV Systems," Underwriters Laboratories, April 1982.
12. "Proposed 1984 National Electric Code Article on Solar Photovoltaic Systems," National Fire Protection Association, NFPA 70.
13. T. S. Key, Baseline Specification of a Utility-Interactive, Power Conditioning Subsystem for Residential Photovoltaic System Applications, SAND82-0290 (Albuquerque: Sandia National Laboratories, January 1982).
14. Investigations of a Family of Power Conditioners Integrated into the Utility Grid, Sandia National Laboratories, Albuquerque, NM, June 1981-January 1982: SAND81-7016 and 7041, United Technology Corp; SAND81-7030 and 7043, Westinghouse R&D Center; SAND81-7031 and 7042, General Electric R&D; SAND81-7032 AiResearch Manufacturing Co.

ARRAY STRUCTURES FOR FIXED FLAT-PLATE
PHOTOVOLTAIC POWER GENERATORS

G. GRASSI
Commission of the European Communities, DG XII
Brussels, Belgium

Summary

Support structures for photovoltaic arrays are minor elements of the photovoltaic systems but, in the near future, low cost solutions must be developed to obtain a better and satisfactory cost balance between the various system components if a large penetration of photovoltaic power generators is to be encouraged.

As perhaps you know, a programme of 15 photovoltaic pilot projects is now under construction in many member states of the European Community. The total cost of the programme is 30 million ECU[*], the total peak power is approximately 1100 KW, the power range being between 30 and 300 KWp.

In this paper, after a review of different solutions proposed for the photovoltaic pilot projects of the Commission, we present a new type of light weight flexible and economical array support structure, made of natural or treated timber and steel wire, which is presently being experimentally investigated on a field of about 3 ha. The utilisation of such materials will certainly help to minimise the visual impact on the environment, this being one of the objectives of the pilot plants sponsored by the Commission of the European Communities. Preliminary forecasts suggest an installed cost of about 0,2 ECU/Wp to compare with the cost of about 0,5-1,7 ECU/Wp relating to several solutions proposed in the framework of the photovoltaic pilot projects programme of the Commission.

1. INTRODUCTION

Photovoltaic solar electricity, with the further cost reductions foreseen, has the potential to become a major energy resource for many countries. To obtain a significant contribution and a rapid penetration of this new electro-technology, we first of all need higher cell efficiency and lower cell cost. Furthermore, it is important to develop and optimize all the "balance-of-system" components (that is all those other than the photovoltaic modules) often involving more familiar engineering technology.

Up to now, module cost has been high and consequently there has been little incentive for designers to develop cheap structures. In some pilot projects the structure accounts for about 10% of total costs. As the cost of modules continues to drop, it will become increasingly important to minimize the structure contribution to the total investment; i.e. the installed cost of structures proposed in the frame of photovoltaic pilot projects programme of the Commission is in the range

[*] ECU (European Counting Unit) = ~1 $ (April 1982)

of 50 – 170 ECU*/m2 (about 0,5 – 1,7 ECU/Wp). Today, this represents a marginal contribution to the total cost but in, say, 10 years time when module cost may be as low as 1 ECU/Wp, it would represent a significant and too high proportion of the total cost of the installed generator.

For this reason, important research efforts are currently being made to define low-cost solutions for durable array support structures. In this paper, an overview is given of several structure concepts, adopted for the pilot projects of the Commission. Furthermore, the utilization of timber poles and galvanized steel wire is proposed for a new type of light and flexible array structure. Preliminary information shows a cost in the range of 0,10 + 0,2 ECU/Wp depending on soil nature, remoteness of siting, wood type for poles.

2. REVIEW OF ALTERNATIVE SOLUTIONS OF SUPPORT STRUCTURE PROPOSED FOR THE PHOTOVOLTAIC PILOT PROJECTS PROGRAMME OF THE COMMISSION

To begin, we have shown in table I some reference values concerning such pilot projects.

In figure 1 and figure 2 we have presented some alternative solutions and characteristics of the proposed support structures.

The large variety of solutions clearly shows that the construction of PV support structures is a new issue for which no standard solution or rules are yet available.

The predominant construction material is galvanized steel, but aluminium, reinforced concrete or even wood have been chosen as well.

Although the foundations belong to the most cost-intensive elements of the support structure, in most cases the proposed solutions can be termed conventional. The foundation holes are either excavated or drilled whereas the foundations are made of concrete or precast concrete slabs. The support beams are either encased in concrete or screwed to the foundation with base plates.

Assembling the components of the structure in most cases is effected with screws or by welding. In some instances assembling is made through bracketing.

In figure 3, we have presented the preliminary data of the total cost of support structures for several E.C. pilot projects and the portion for site preparation, support structure construction, PV module assembling and wiring.

Figure n°1

Project	Alicudi (80 kWp)	Marchwood (80 kWp)	Tremiti (65 kWp)	Chevetogne (63 kWp)	Mont Bouquet (50 kWp)	Fota Island (50 kWp)
Structure scheme				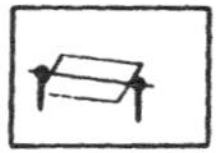	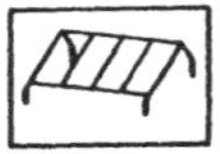	
Reference nr. of project	4	5	7	8	9 [*]	11
Measurem. Height Span	max. 5.8 m appr. 6.7 m	3.0 m 2.2 m	1.1 m 3.0 m	1.1 m 3.8 m	1.3 m 3.2 m	13.0 m 53.0 m
Modules Number Area Weight Tilt angle	1232 0.85 m2 15.0 kg 27°	2496 0.51 m2 n.a. 55°	378 648 1.47m2/0.44m 32.5kg/10.2kg 22° 22°	1920 0.42 m2 11.0 kg 30°	760 0.98 m2 16.7 kg 60°	2664 0.26 m2 3.9 kg 45°
Foundation type	n.a.	dowel screws in an existing concrete pad	concrete single footing	concrete single footing	concrete slabs	concrete
Construction material	galvanized steel, guy wires	galvanized steel	precast concrete leg, galvanized steel beams	painted steel	aluminium	steel with enclosed block walls
Components inter-conn.	n.a.	screws	screws	screws	aluminium nuts end bolts	n.a.

[*] Project 10 also has the same type of structure

Figure n°2

	Giglio (45 kWp)	Terschelling (50 kWp)	Aghia Roumeli (50 kWp)	Rondulinu (42 kWp)	Hoboken (30 kWp)	Pellworm (300 kWp)	Kythnos (100 kWp)
	14	12	13 *	15	17	1	2 - 3
	1.5 m 1.4 m	1.5 m 4.0 m	1.4 m 3.9 m	1.4 m 1.9 m	appr. 0.8 m appr. 2.8 m	2.6 m 2.8 m	0.9 m 5.0 m
	723 0.86 m2 15 kg 25°	2610 0.26 m2 3.9 kg 40°	768 0.83 m2 14.0 kg 25°	648 0.83 m2 14.0 kg 60°	912 0.44 m2 11.0 kg 30°	15792 0.26 m2 3.9 kg 40°	800 1.47 m2 32,5 kg 35°
	concrete kerb + plinths	concrete slabs	concrete foundation	concrete benches parallel to panels row	- - - -	continuous concrete footing	concrete in drilled foundation holes
	galvanized steel (L section)	galvanized steel cross sections	galvanized steel cross sections	galvanized steel with aluminium bars to inter-connect the modules	galvanized steel	galvanized steel or bongossi wood	galvanized steel
	screwed	welded	screwed	screwed	welded or screwed	stainless steel screws	U-shaped clamps

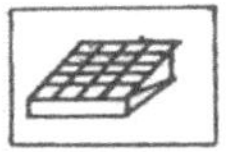

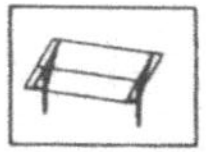

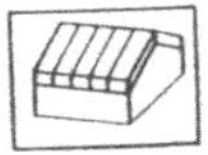

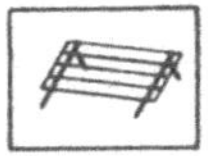

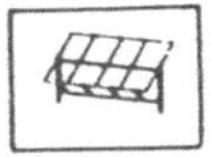

* Project nr. 16 also has the same type of structure

- 251 -

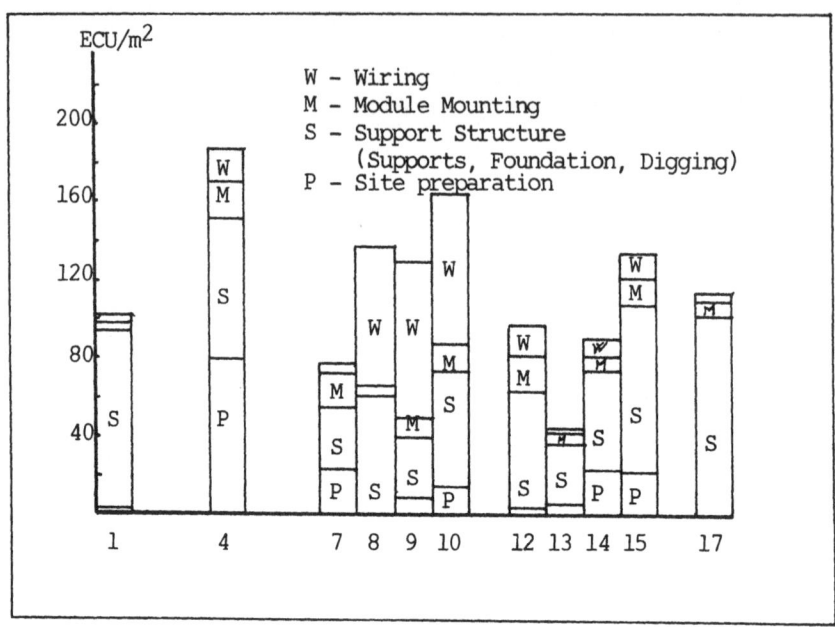

Figure n° 3 - Total cost of support structures

TABLE n° 1

Reference n°	Project	Power output kW	Panel area m^2	Total nr. of panels	Total weight of panel (kg)
1	Pellworm	300	4,081	15,792	61,589
3	Kythnos Isl.	100	1,180	800	26,000
4	Alicudi Isl.	80	1,052	1,232	18,480
5	Marchwood	80	1,264	2,496	
7	Tremiti Isl.	65	849	648	6,610
8	Chevetogne	63	813	1,920	21,120
9	Mont Bouq.	50	747	760	12,692
10	Nice	50	747	760	12,692
11	Fota Isl.	50	743	2,664	10,390
12	Terschelling Isl.	50	675	2,616	10,179
13	Aghia Roumeli	50	693	768	10,752
14	Giglio Isl.	45	622	723	10,845
15	Rondulinu	44	533	648	9,072
16	French Guy.	35	519	625	8,750
17	Hoboken	30	403	912	10,032
	T O T A L	1,092	14,921	33,364	229,203

List of the EC Photovoltaic Pilot Projects

3. NEW TYPE OF LOW COST SUPPORT STRUCTURE

Generally structure costs total 80 ECU/m2 and above.

Nevertheless a substantial cost reduction is possible, adopting a type of construction, now under investigation, suited for sites between some tens of m2 and ha with a maximum slope of about 10%.

3.1 Concept

The suggested structure (figure 4) is a complex, light, continuous reticular system, utilizing wood and steel wire as main structural materials in elements which are customary for modern vineyard support structures. The manufacturing assembling and management technologies involved are very common and elementary.

Here below we have shown a sketch of this new type of array support structure.

Fig. 4 - Sketch of new type of structure

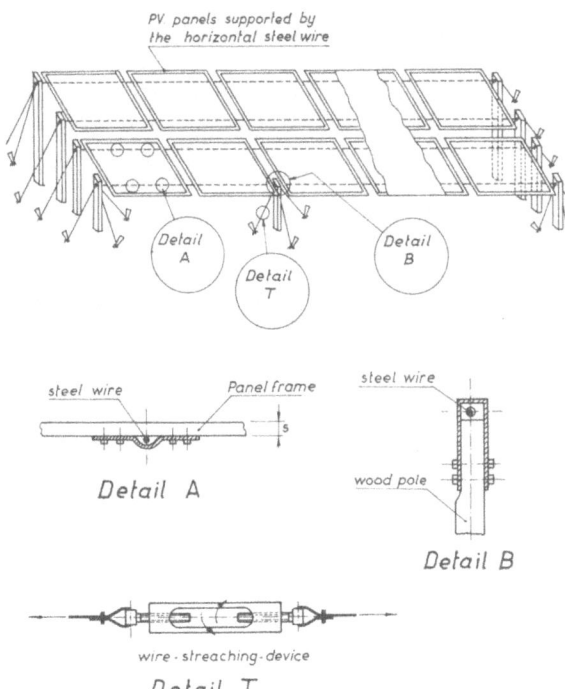

The wood elements may be available on the site. Metal elements : panel connection devices (C) and wire-bearing devices (A), (A'), (B), (D) may be easily removed for eventual panel and structure element maintenance.

Following experimental results, which will be available in the near future, a basic support structure of this kin will be further developed and defined.

3.2 Wood as construction material for support structures

Wood is an important construction material and has been used for structures since the beginning of civilisation because of its favourable strength, working characteristics, cost and worldwide availability. Strength depends on wood species, density, grade, temperature and moisture conditions. The desired in-service duration (not less than 30 years) can be obtained by proper choice of wood species and by adequate preservative treatment.

The most important factors influencing wood characteristics and life are :

- fungi (like Coriolus versicolor, Coniophora cerebella, Leuzites trabea, etc.) found in all parts of the world and by far the most important of the living enemies of wood. The essential requirements for fungal attack are 1) adequate moisture, 2) a favourable temperature, and 3) at least a small supply of oxygen. For this reason, wood parts in contact with soil are mostly exposed to fungal decay, because of moisture absorption, especially near the ground.

- Insects : termites are the most dangerous, mainly but not exclusively in tropical and subtropical regions. Powder past beetles and carpenter ants should also be considered.

- Weathering (i.e. U.V. radiation, changes in temperature and humidity...) which should be considered as a minor problem.

The two main features of a preservative treatment are :

a) the type of preservative used : coal-tar, creosote, solutions of pentachlorophenol, salts of copper, Cr, As, B ... Coal-tar creosote appears to be the most appropriate for the treatment of support structures; it afford protection against fungi and insects, retards checking and weathering, tends to inhibit corrosion of metals in contact with wood.

b) the methods of application : the **high pressure, low pressure** and double diffusion method. The most effective of these is the double diffusion method by which partially seasoned timber is held successively in contact for several days with solutions of two chemicals that react with each other within the wood to deposit a preservative compound. (The equipment may be just tanks). The hot-and-cold process is a most important method for treating wood to be used in gound contact. It consists of immersing wood for several hours in hot preservative, which is then allowed to cool down. This requires limited facilities, namely tanks and heating devices.

	Bending strength	Module of elasticity	Material loss by fungi	Average life	
				Natural	Treated
"Chestnut" wood	1100 kg/cm2	116.000 kg/cm2	6%	15	25
"Azobé" wood	2250 kg/cm2	170.000 kg/cm2	2%	25	40

In table II, we have presented the main characteristic of pole materials which may be used for this new type of array support structure.

As a rough indication of cost, the price in Italy of treated wood poles for agricultural uses, $\geq$ 10cm diameter at top, ranges somewhere around Lit. 3,500 per meter of length (length available up to about 3 meters).

3.3 **Realisation of the structure** can be subdivided into :

a) <u>Site selection</u>
Well-exposed position, gentle degree of slope to reduce severe soil erosion by water, sufficient depth of stoneless soil, are the most suitable characteristics.

The mechanical composition of soil is another important aspect. Loamy texture is better than a very sandy or clayey one. A very fine textured soil can present during the dry season (Mediterranean region) wide and deep cracks. Drainage of soil is very important. The site selected for our study, facing South-South-East, is situated at Castellina in Chianti, near Siena (I). The elevation is 450m, geological formation is a marly linestone, the soil is a brown forest soil with clay-loamy texture.

b) <u>Site preparation</u>
First of all we have to classify the site, from the point of view of difficulty in preparation, utilizing e.g. the following classification :

<u>1st category</u> (low difficulty soil) : 10% (in volume) of stones ($\emptyset$ = 30-40cm, max.)in a soil of a thickness of 100cm;
<u>2nd category</u> (medium difficulty soil) : 30% (in volume) of stones (max. $\emptyset$ = 150cm);
<u>3rd category</u> (high difficulty soil) : 50% (in volume) of stones and 3% (max.) of hard rocks to be removed by explosive;
<u>4th category</u> (extremely high difficulty soil) : 100% of hard rock

Our study on structure is carried out in a 2nd category soil.

The operations sequence for the site preparation before installing the array structure is the following (data between brackets refer to our experimental investigation on a site classified as 2nd category which is carried out by the E. Colpizzi Co. of Florence (I) with the scientific support of the University of Florence) :

- removal of natural vegetation by roller machine (130 H.P.);
- first site levelling by a high power operating machine (Caterpillar 275 H.P. - total volume displacement 1100 m3/ha);
- first deep soil ploughing (1.3 m) by special operating tool for stone removal (about 2000 m3 stones/ha having dimensions of 30 ÷ 150 cm were removed by a Caterpillar 275 H.P.);
- transportation of stones to the site borders by a roller machine (130 H.P. - average distance : 70m) and a special large-toothed tool;
- second deep soil ploughing, and levelling;
- achievement of a series of draining ducts (see sketch) by an excavator to avoid water stagnation in the soil (long ditches, 140 cm deep and partially filled by stones for a thickness of 40 cm; total length about 500 m/ha)

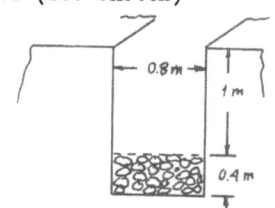

c) <u>Structure assembling</u>
- Insertion of wood or reinforced concrete poles into the soil :
    - by means of a special operating machine effecting the penetration of poles up to a depth of 1 m into a soil free of big stones ($\emptyset \leq 20$cm),
    - by an excavator equipped with an hydraulic device for making holes in a stony soil.

    Poles are then manually introduced into holes and firmly anchored by compressing soil material using a pneumatic hammer.

- Settlement of the galvanized steel wire by a special stretching machine. The wire is stretched by means of special devices;
- Mounting the mechanical devices in order to enable attachment of photovoltaic modules on the structure and correct orientation towards the sun;
- Assembly of electrical cables and terminal boxes.

3.4 Economic evaluation of the new type of structure

In summary, the estimated average cost for a medium difficulty site (2nd category) is the following:

- site preparation	10.000 ECU/ha
- squaring and picketing	500 ECU/ha
- materials	10.500 ECU/ha
- structure assembling	17.000 ECU/ha
Total	38.000 ECU/ha

So, the average specific cost for this type of structure should be approximately 0,1 ECU/Wp (minimum cost) and in any case not higher than 0,2 ECU/Wp. Such cost should be reduced by 30% for a site larger than 10 ha. On the other hand, such cost should be increased by about 30% for distances above 200 -300 km from the main operating centers of the company. Presently the estimated cost for disassembling modules and dismantling a PV array support system of this new type is apprximately 0,01 ECU/Wp.

4. CONCLUSIONS

- Present cost of array support structures is too high and unacceptable in a near future.

- Definition of general rules and guidelines for support structures is necessary for the PV power generators' programme.

- New type of support structure (made of wood and steel wire), now under investigation, may considerably reduce such cost to the level of 0,2 ECU/Wp or less.

REFERENCES

- Critical Comparative Analysis of Support Structure Cost (WIP - 8000 München, FRG ....) March 1982

- Photovoltaic Power Generation (Proceeding of the Final Design Review Meeting on EC PV Pilot Projects (30 Nov. - 2 Dec. 1981)- D. Reidel Publishing Co. (1982)

- Photovoltaic Power for Europe - An assessment study made for the Commission of the E.C. (to be published).

ACKNOWLEDGEMENTS

The author would like to express his thanks to the E. Colpizzi Co. of Florence and Prof. Caparrini, Prof. Selleri, Prof. Mancini, Prof. Uzielli of the University of Florence (I) and Mr. P. Helm of Wirtschaft und Infrastruktur GmbH & Co, München (RFA) for their support.

INVERTERS FOR PV PLANTS FEEDING THE ELECTRICAL NETWORK OF
SMALL ISOLATED COMMUNITIES

V. Arcidiacono and A. Taschini
Direzione Studi e Ricerche, ENEL, Italy

Summary

This report, after having reviewed the main circuit schemes and the working principles of the self-commutated inverters, examines the factors that have to be taken into account in choosing the type of inverter for PV plants feeding the electrical networks of small isolated communities.

1. INTRODUCTION

For the electric utilities, one of the more promising applications of photovoltaic conversion in the medium term is the supply of small isolated communities that cannot be connected to the national electricity grid for technical or financial reasons.

The load of small isolated communities is usually limited to appliances for which the use of electric power may be considered compulsory. The total load results from domestic loads (mainly lightning, TV, and refrigerators), loads of small commercial and public premises (schools, churches, etc.), and public lighting outlets, which normally, taken singly, represent but a small fraction of the total load. Sometimes, however, there may be a predominant load that forms a large share of the overall power requirement (e.g., desalination plants, refrigeration plants of agricultural or fishing co-operatives, etc.).

Within the normal power range envisaged (i.e., less than 100 kW_p), if the PV plant and the back-up diesel are not too far away from the load centre, distribution of electric energy will be provided with low voltage, three-phase lines (400 V). Any predominant loads can, if necessary for technical reasons, be served by dedicated lines.

2. INVERTER TECHNOLOGY

The technology of an apparatus is always conditioned by the characteristics of the materials and components available at the time, and is therefore subject to change when technical development and the market make available new materials with more advanced functional characteristics. All this applies to all electronic equipment, especially inverters, given the fact that the development of power semi-conductors has been, and continues to be, very rapid. For this very reason, a given phase in the development of the technology, the manufacturing characteristics, and the performance of an electronic apparatus may be called a "generation".

In the case of solid-state inverters, the first-generation type can be associated with SCR's (thyristors) that switch only once every half-cycle, while the second-generation type are associated with SCR's that switch several times every half-cycle (usually five-ten times), and third-generation inverters are associated with power transistors that have relatively few commutations every half-cycle. Lastly, fourth-generation inverters are associated with the new power transistors (characterized by a high breakdown voltage), which switch at a very high frequency compared with that of the a.c. output voltage.

To evaluate the advantages offered by the most advanced fourth-generation transistor inverters, compared with classic SCR inverters, it should first be mentioned that, in any type of inverter, the power semi-conductors, whether they be SCR's or transistors, simply perform the function of commutating devices. But while the transistor can be turned off very easily (by means of a control signal), the SCR once triggered, cannot be turned off until the current flowing through it drops below the holding current: that is, in practice, until the current goes to zero. Moreover, for the SCR to return to the off-state, a certain time has to pass after the instant at which the current falls to zero. This time, known as turn-off time, is a characteristic of the various types of SCR's, and may range from 20 to 100 μ sec and over, depending on the SCR rated current. However, special series of fast SCR's are available, for which the turn-off time is between 20 and 40 μ sec, even for the highest rated currents.

As a consequence of what above said, in the case of SCR inverters, it is necessary to make use of appropriate forced commutation circuits, given the fact that the current in the SCR's would not naturally go to zero, the supply being d.c.

These forced commutation circuits have a common characteristic, i.e. the presence of heavy ancillary power components, such as inductances and capacitors (or even ancillary SCR's), whose job it is to hold the load current for a short time, thus leaving the power SCR previously carrying the current the time to turn off.

The presence of power circuits to force commutation, apart from representing a high proportion of the cost and bulk of the inverter, has a strongly adverse effect on conversion efficiency.

By way of an example of the operation of a classic SCR inverter, see Fig. 1a, relating to a single-phase inverter. Let us assume that at the beginning the current flows through Th_1 and Th_2. Then capacitors C_1 and C_2 are short-circuited, while C_3 and C_4 are charged. If Th_3 is triggered, C_3 discharges through Th_3 and half of inductance L_2, inducing in the second half of the same inductance a voltage that make Th_2 switch off. Th_1 is turned off as well since there is no path for the current following through it. At this stage, if no other control signals are given, Th_3 switches off as well, once the discharge current of C_3 stops flowing. Current flow can start again with a simultaneous trigger signal to Th_3 and Th_4.

The circuit in Fig. 1a) allows a pause time between the two half-wave of opposite sign. This pause time may be used to control the r.m.s. value of the output voltage and to achieve a shape closer to the sinusoidal wave than the square wave (Fig. 1b).

The operation of a SCR three-phase inverter is the same as that previously described for the single-phase inverter. In some solutions, transformers are used to combine two three-phase systems with a phase-shift of 30°. The set of voltages thus obtained has a step wave closer in shape to the sinusoidal wave. The combination of two three-phase systems also allows a certain regulation range of the output voltage, by varying the phase-shift between the two systems around the optimal value (30°), which minimizes the harmonic distortion of the waveshape.

So far we have been speaking of classic SCR inverters, in which each SCR performs only one commutation every half-cycle. But there are also SCR inverters (2nd generation) in which each SCR, being of the fast type, performs several commutations in a period.

Referring again to the scheme in Fig. 1a, it is possible, by means of a suitable sequence of triggering impulses, to obtain a digitally-synthesized sinusoidal waveshape (see Fig. 2). The advantage of this method is that, by optimizing the commutation sequence (which can simply be memorized by means of digital integrated circuits), we have high-amplitude harmonics only in the relatively high frequency ranges, which therefore lessens the filtering problems. However, it should be pointed out that it is in any case impossible to use a particularly high commutation frequency, bearing in mind the relatively high turn-off time of the SCR's (even if they are of the fast type) and the presence of ancillary components to force commutation. In this case too, output voltage regulation can be carried out by combining two systems by means of transformers and controlling the phase-shift between them.

If we pass on to consider transistor inverters, we find that the methods used in the third-generation type to produce the output sinusoidal voltage are the same as those already described for SCR inverters. In this case, too, the output waveshape is of the step, or the digital-synthesis type. The advantages of using transistors are limited, in this case, to the elimination of costly capacitors and inductances to force commutation.

Even greater advantages, on the other hand, are offered by the modern technology of fourth-generation transistor inverters. In these inverters, the sinusoidal waveshape is obtained by means of pulse width modulation (PWM), operating at a very high frequency compared with that of the output sinusoidal wave. Working at a high frequency, does indeed make it possible to use a lighter output filter, which is a great advantage, as regards efficiency problems. A further advantage offered by the PWM technique is that it enables a very rapid control to be made of the instantaneous value of the a.c. output voltage, thus ensuring lower harmonic distortion, excellent dynamic regulation performance, and also the possibility of very fast limitation (less than one cycle) of the output current.

The operating principle of these inverters is illustrated, for the single-phase type, in Fig. 3. As can be seen, there is still a commutation bridge in which each of the four breakers consists of a certain number of power transistors, connected in parallel, and of a fast by-pass diode.

The output sinusoidal voltage is produced in the following way. In the positive half cycle, T_{r4} is keept conductive and T_{r2} and T_{r3} are switched on and off. It is therefore performed a high frequency[1] sharing of the d.c. voltage, where the duty cycle is modulated according to a sinusoidal wave. Similarly, in the negative half-cycle, T_{r1} is kept conductive and d.c. voltage partialization is still carried out by means of T_{r2} and T_{r3}.

3. QUALITY OF A.C. ELECTRIC POWER

a) <u>Voltage Range of Variation</u>. On the assumption that an inverter without voltage control is being used, the voltage at the far end of the distribution lines, as we pass from maximum electrical load and minimum battery charge to no-load operating conditions and maximum battery charge, will undergo a voltage variation equal to about $\Delta U + \Delta V$ (%)[2]. The voltage at the customer's site will therefore vary within the $\pm \frac{\Delta U + \Delta V}{2}$ (%) range around the rated voltage V_n of the distribution network (Fig. 4).

In low-voltage distribution networks connected to national grids, the standards laid down in contracts provide that the voltage at the customer's site be kept within $\pm$ 10% of the rated value. Should it be used an inverter without voltage control, except in special cases, both the electrical storage and the conductor of the distribution network would be unjustifiably overdimensioned. Indeed, it should be remembered that only $\frac{\Delta U}{2}$ (%) is, for many types of battery and for the depth of discharge used, greater than $\pm$ 15%.

Let us now assume that ΔV_o is the maximum voltage drop along the distribution line, in the event of the line conductor being chosen in such a way as to optimize line cost including the cost of Joule losses. Given the high cost of losses and the type of line ($R \gg X$), $\frac{\Delta V_o}{2}$ will be less than the $\pm$ 10% allowed for voltage variation at the customer's site.

Going on this assumption, when the following normal stationary conditions occur (Fig. 4):
- Variation of the inverter d.c. input voltage over the $\Delta U = U_M - U_m$ range (voltage window)
- Variation of the power output from 0 to 100% of the rated value
- Variation of the cos φ of the load in the specified range

The inverter a.c. output voltage must present a voltage variation in percent of the rated value V_n not higher than $\pm$ (10% - $\frac{\Delta V_o}{2}$), so as not to exceed, at any point of the network, deviations from V_n greater than $\pm$ 10% during normal operation of the system.

Should it be possible to achieve a voltage static stability of the inverter more accurate than $\pm$ (10% - $\frac{\Delta V_o}{2}$), then it will be acceptable a reduction of the inverter voltage window (ΔU) by the same percentage

---

(1) - In the range of 3 ÷ 10 kHz.

(2) - Maximum voltage variation of the battery voltage (ΔU) and maximum voltage drop along the distribution line (ΔV), expressed as a percentage of the rated battery voltage (U_n) and of the rated voltage of the distribution network (V_n), respectively.

value.

If it is desired to give network users a more uniform voltage value, by regulating the voltage at an intermediate point in the line (current compound) (Mode 2 in Fig. 4), it will, on the other hand, be necessary to have a larger voltage window for the inverter. In this case, the window will be equal, as a percentage, to the sum of $\Delta U + \frac{\Delta V o}{2}$ plus the voltage static stability of the inverter, also expressed as a percentage.

b) <u>Frequency Accuracy</u>. The frequency accuracy offered by most of the inverters (0.1% - 0.5%) may be considered largely acceptable for small isolated communities also in relation with the type of customer to be supplied.

c) <u>Admissable Harmonics Distortion</u>. At present, there are no international Standards laying down limits as to harmonic content of a network voltage. Based on what has so far been published on the subject, however, it is possible to suggest limits on which there would seem to be a large measure of international agreement.

Two limits may be stated, the first of which relates to individual harmonic distortion, and the second of which concerns the total harmonic distortion of the voltage wave

$$(1) \qquad D = \frac{\sqrt{\sum_{i=2}^{\infty} U_i^2}}{U_1} \cdot 100$$

in which U_1 is the (fundamental) 50 Hz voltage component, and U_i are the harmonic components.

These limits - at least for low voltage (400 V) - might be as follows:
- Odd harmonics: individual harmonic distortion not exceeding 5 - 6% up to the 7th - 9th harmonic, and then gradually decreasing up to the 19th - 21st harmonic.
- Even harmonics: individual harmonic distortion not exceeding 0.5 - 1%
- Total harmonic distortion: 5%.

The lack of special equipment requiring a low harmonic pollution in the type of load fed by the solar plant (e.g. x-ray equipment), as well as the acceptance of a lower accuracy in the measurement of the electric energy supplied to customers allow utilities to double the limits of the harmonic content given above in the networks of small isolated communities.

4. THE PERFORMANCE OF INVERTERS AND OF NETWORK PROTECTION SYSTEM IN DEALING WITH A.C.-SIDE DISTURBANCES

a) <u>A.c.-side Short Circuits</u>. It is considered that selective clearing of line faults is not needed due to the low probability of permanent faults on the lines, especially if overhead cables on wooden poles are used. On the other hand, it is necessary to achieve satisfactory coordination between equipment protecting the customer circuits against short-circuits and protection system of the inverter.

Inverters, as regards the possibility of carrying out selective protection against network faults, have the serious limitation that they cannot supply current values significantly greater than I_n for any length of time.

In the specific case of small isolated communities selective protection against faults on the customer circuits is somewhat simplified by the fact that the power of individual electric loads is small as compared with inverter rated power, except for possible predominant loads that can be dealt with by a dedicated line and special protection systems. Furthermore, in the case in question, the occurrence of several simultaneous faults in customer's plants may be disregarded.

The first approach is to protect consumer circuits by means of fuses compatible with the inverter protection. Inductances in series with the fuses may be necessary to avoid operation of the inverter protection especially for faults at customers close to the PV plant. The electric utilities, however, would prefer to use electromagnetic breakers rather than fuses, to prevent the customer using unsuitable fuses or none at all when replacement is required.

In the event of the inverter being able to limit the current, and with special reference to fourth-generation inverters, it may be expected that a regulation philosophy may be used that can keep the a.c. inverter voltage constant until the current reaches a value ($k_1 I_n$) acceptable for a time of the order of tenths of seconds, afterwards switching to a current regulation (from $k_1 I_n$ to $k_2 I_n$) for times of the order of seconds, before the inverter is taken out of service (Fig. 5). The values of k_1 and k_2 may be considered of the order of $1.25 \div 1.50$.

The current that the inverter injects into the fault will therefore be relatively slight, of the order of $k_2 I_n$ less the current drained by the loads, and will also depend on the way in which the current of the loads varies as a function of the a.c. bus-bar voltage of the inverter working in the current regulation mode (from $k_1 I_n$ to $k_2 I_n$). In general, the current of the loads should diminish as a result of the reduction of the a.c. inverter voltage.

With the type of inverter current regulation described above, and on the assumption regarding loads and faults previously made (small loads as a percentage of the rated power of the inverter, negligible significance of events such as simultaneous faults on several customer circuits), the use by the customer of electromagnetic breakers of the type now installed is possible, although the times required to clear the faults are fairly long. This naturally calls for careful checking of the customer's ground mat, and possibly the installation of a differential residual current device, to ensure the safety of the customer in the event of faults to ground.

Disturbances in the distribution network, in the event of the inverter having current-limiting circuits, are larger than in a standard low voltage network, because the current control system amplifies the voltage drops that naturally occur in the network when there are faults. However, taking into account the number of short circuits in customer's plants to be expected in the elctric power systems of small communities (of the order of 100 per year), it is felt that the greater amount of disturbance is entirely acceptable in the case in question.

b) Overloads. The most serious case of overload of inverters may occur upon restoration of the service after a certain period of interruption. In this case, many customer appliances (refrigerators, pumps with minimum-level signal, etc.) start up at the same time. For some types of load, there is therefore a high inrush current with a diversity factor near to unity. It may thus be necessary to energize the different lines separately. Should this not be sufficient to reduce the current during network energization to values acceptable for the inverter, it will be necessary to split the lines into several sections and reclose the various sections in succession, one after the other, with a certain delay between the closure of two successive sections. On the other hand, it must be considered that the inverter energizes the line with a ramp voltage inherently.

5. EFFICIENCY OF D.C.-A.C. CONVERSION

The load duration curve of the electrical load in the case of small isolated communities shows ratios between max. load and min. load of the order of 6-10 times (Fig. 6). Taking into account the high cost - even in the future - of energy produced by photovoltaic conversion, it is almost imperative to have inverters with a high efficiency within a wide range of output power. We may consider two technical alternatives. The first consists of an inverter with a very flat efficiency curve, so that efficiency exceeds about 90%, even for loads of the order of 0.1-0.2 times inverter rated power. The second technical alternative would be the use of several inverters in parallel, in operation or not, depending on the load, so that the inverters would always work within the high-efficiency range. At least in the case of isolated communities, the first solution would be preferred by electric utilities, due to the simplicity and reliability of the scheme.

6. RELIABILITY OF COMPONENTS AND OF THE WHOLE CONVERSION SYSTEM

UPS that have been in service for a long time make use of SCR apparatus. The reliability of these apparatus seems acceptable to electric utilities. As regards transistor apparatus, however, less evidence of reliable service is available. One point that deserves attention is the greater exposure of inverters for small communities to electromagnetic disturbances from low-voltage lines, as compared with the disturbances affecting the inverters used in conventional UPS.

Obviously, therefore, even if the voltage levels made it possible to omit the transformer between the inverter and the network, this would not be acceptable due to problems of electromagnetic compatibility, even though this involves a small loss in conversion efficiency. Furthermore, the transformer should be such as to limit any overvoltages that might be transferred from the network to the inverter in order not to exceed the surge withstand ability of power conditioning equipment.

7. CONCLUSIONS

Transistors are components extremely suitable for the switching function required in inverters. Their application is at present limited to low and medium power sizes, due to their voltage and current limits. However, the rapid technical development of these components has already shifted their limit of application from 30 kVA, as it was only a few years ago, to the present 100 kVA. The programmes and efforts involved in developing power transistors give grounds, moreover, for expecting that very soon single transistor inverter units of up to 500 kVA will become available. This rating will largely cover the need of small isolated communities.

The advantages offered by the modern technology of fourth-generation inverters can be summed up as follows:
- Transistor inverters are more compact and lighter than the SCR type, which therefore makes transportation, installation, etc. easier.
- Transistor inverters have a conversion efficiency considerably higher than that of SCR inverters, due to the absence of ancillary power commutation circuits. Their efficiency remains, moreover, extremely high even at low output power, given the fact that transistor losses (connected with conduction and commutation) are roughly proportional to the output current. Transistor inverters working at a high commutation frequency generate an a.c. output voltage with very low harmonic distortion and only require a light output filter, which has a further positive effect on efficiency.
- The high working frequency also makes for excellent dynamic regulation performance and offers new possibilities for selective clearing of faults in the network besides meeting any foreseeable requirement for voltage regulation at the bus-bars of the PV plant.

Drawbacks of transistor inverters compared to SCR inverters are on the contrary: lower thermal capability and lower experience as regards reliability.

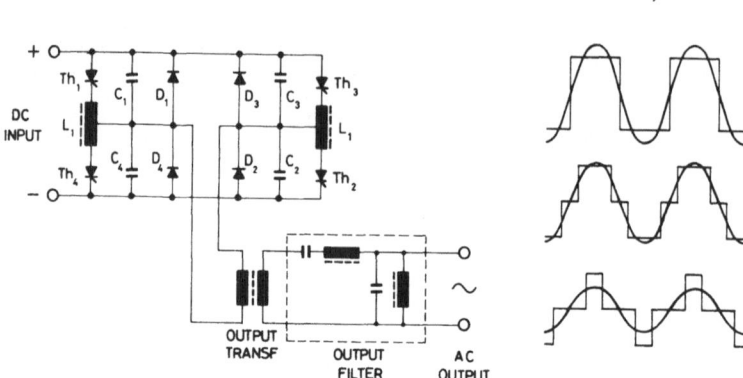

Fig. 1 - Forced-commutated single-phase SCR inverter:
   a) Basic scheme
   b) Voltage regulation

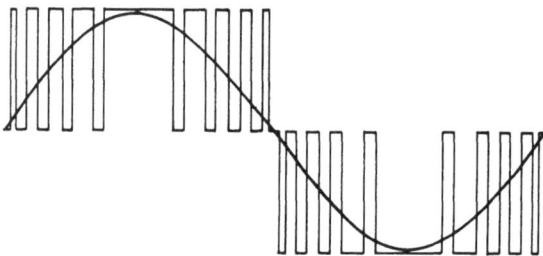

Fig. 2 - Digitally synthesized sinusoidal wave shape

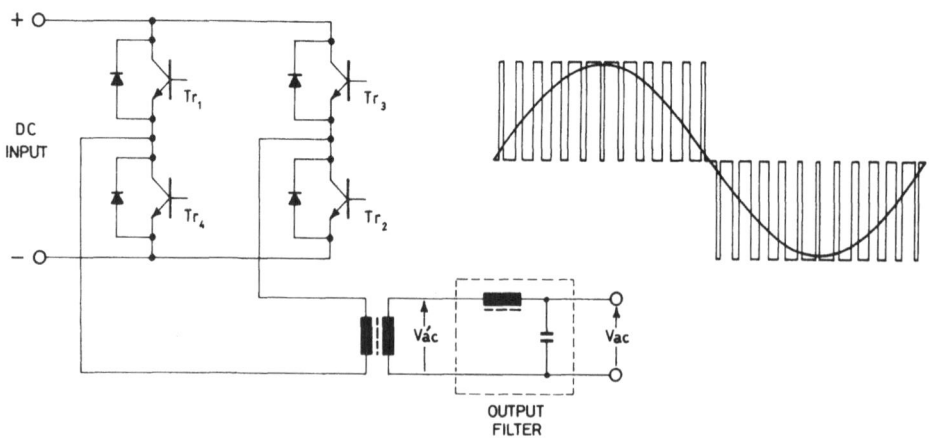

Fig. 3 - PWM transistor inverter

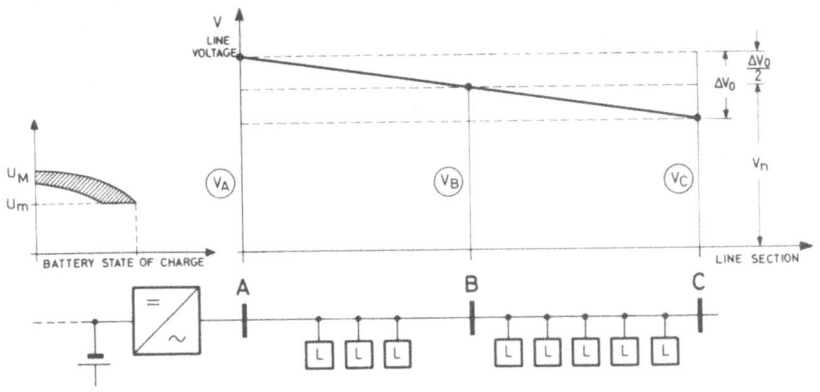

MODE 1: $V_A = V_n + \dfrac{\Delta V_o}{2} = $ constant $V_B = V_n \div (V_n + \dfrac{\Delta V_o}{2})$ $V_C = (V_n - \dfrac{\Delta V_o}{2}) \div (V_n + \dfrac{\Delta V_o}{2})$

MODE 2: $V_A = V_n + (V_n + \dfrac{\Delta V_o}{2})$ $V_B = V_n = $ constant $V_C = (V_n - \dfrac{\Delta V_o}{2}) + V_n$

Fig. 4 - Voltage profile along the line

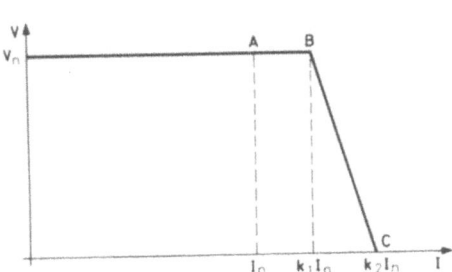

Fig. 5 - Possible voltage-current characteristic of PWM inverters

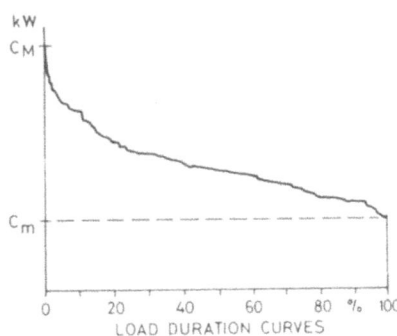

Fig. 6 - Example of load duration curve of small isolated communities

POSTER GROUP P3
SYSTEMS, COMPONENTS AND ENGINEERING

- Array design

A support structure for intermediate PV solar plant

Low cost modular designs for photovoltaic array fields

The photovoltaic solar system, analysis and basic design rules

A detailed package of digital codes specially developed for the array system

- Shadowing, Hot-Spots

Analysis of hot-spot-effects in encapsulated photovoltaic generators by laser scan and partial shadowing

Reverse bias power dissipation of shadowed or faulty cells in different array configurations

Disequilibriums in series connected solar cells - an approach to the protection by parallel diodes

- Grid Connection and Inverters

Minimum cost of photovoltaic energy for a utility grid and general features of a generating plant using costless solar cells

A DC/AC modular interface for photovoltaic systems

Simple transformerless inverter with automatic grid-tracking and negligible harmonic content for utility interactive photovoltaic systems

Regulated converter circuit for direct photovoltaic energy feedback into the power grid

Power conditioning in solar photovoltaic array applications

- Battery Interface

Mismatch between batteries and two module types PV arrays Interest of DC-DC converters

Matching the characterstics of batteries with solar cell modules

Calculation to improve power conversion efficiency in photovoltaic systems

A SUPPORT STRUCTURE FOR INTERMEDIATE PV SOLAR PLANT

V.Albergamo, Delphos Project, ENEA, Istituto Casaccia, Roma
P.L.Borlenghi, Studio Moretti, Roma
F.Cheli e G.Diana, Politecnico di Milano, Dipartimento di Meccanica
M.Falco, Università di Catania, Istituto di Macchine

SUMMARY

In this paper, a new concept of support structure for intermediate PV plant is described.
The structure is realized with parallel wire-ropes, anchored to the ground and supported by A-type poles. Flat plate PV row of panels are suspended between parallel ropes.
This concept of structure offers numerous advantages: modularity, extremely high PV field packing factor, lightness, easiness of trucking and mounting in remote regions, easiness of maintenance and cable routing, very low ground levelling and foundation work.
On the other hand, as a result of aerodynamic forces due to wind acting on the PV panels, instability phenomena may occur, as structure flutter and buffeting. Therefore theoretical and experimental analysis are necessary and are now in progress. First results are shown.

* This work has been performed in the frame of the ENEA-ENEL Delphos Project.

1. PREMESSA

Negli impianti fotovoltaici finora realizzati, i pannelli, inclinati con un opportuno angolo, vengono disposti su molte file parallele nella direzione Est-Ovest, distanziate fra di loro in modo da evitare che una fila ombreggi l'altra. Questa sistemazione del campo fotovoltaico richiede un'area di suolo da due a tre volte superiore rispetto all'area dei pannelli. Il suolo tra i pannelli inoltre è del tutto inutilizzabile, anzi deve essere reso sterile per evitare che la vegetazione copra i pannelli solari.
  Per superare questo problema, nel corso della progettazione dell'impianto Delphos (1,2) sono stati analizzati, accanto a strutture adatte per sistemazioni del campo a file di pannelli, anche concetti di strutture sopraelevate, che permettono di disporre tutti i pannelli su di un'unica falda, riducendo così la quantità di suolo richiesta e realizzando impianti molto più compatti (3).
  La sistemazione a falda unica elimina il problema dell'ombreggiamento sistematico, anche con sole molto prossimo all'orizzonte, semplifica i cablaggi dei pannelli (che possono essere eseguiti dal di sotto), riduce drasticamente la lunghezza dei cavi di collegamento.
  Tra i vari tipi di struttura che permettono la sistemazione dei pannelli a falda unica, la tenso-struttura a funi parallele, che costituisce l'oggetto della presente memoria, presenta alcuni aspetti particolarmente interessanti per le applicazioni in aree remote.
  Nella versione presentata nelle figure, la struttura portante è costituita da una orditura principale di funi parallele, poste ad un interesse di circa dieci metri tra loro. Le funi sono sostenute da puntoni. I puntoni sono costituita da cavalletti metallici a V rovesciata, in modo da garantire la stabilità trasversale della struttura. Essi sono realizzati con profilati metallici saldati fuori opera e sono incernierati al piede in modo da poter essere posizionati ed ancorati tenendoli adagiati sul terreno e

b - analisi preliminare su modello elastico in scala, per la determinazione dei modi di vibrare della struttura e per la validazione del modello analitico;
c - analisi aero-elastica su un modello completo dell'impianto, in scala 1:60 da eseguire nella galleria del vento della FIAT di Orbassano (TO), con vento in regime sia laminare che turbolento
d - realizzazione di una sezione della struttura, di dimensioni pari a circa 50x10 m, presso l'Istituto Casaccia dell'ENEA, per la verifica dei problemi tecnologici e di montaggio.

Il programma è in avanzato corso di attuazione.

3. PRIMI RISULTATI

Tra i primi risultati, si vuole innanzitutto dire che è stato messo a punto un modello analitico tridimensionale ad elementi finiti, che permette di definire la deformata statica ed i modi principali di vibrare della struttura. Una descrizione particolareggiata di tale modello analitico sarà fornita in una memoria di prossima pubblicazione (4). In linea generale, la determinazione della deformata statica comporta la soluzione di un problema geometricamente non lineare; infatti le grandi deformazioni a cui è soggetta la struttura, sottoposta ai carichi esterni, rendono necessario riferire le equazioni di equilibrio ai nodi, non alla geometria iniziale, ma a quella modificata per effetto degli spostamenti dei nodi stessi. L'equazione di equilibrio statico non lineare non può essere risolta in forma esplicita, ma solo mediante un processo iterativo.

I primi risultati ottenuti con il modello analitico (fig. 3 e 4) sono stati essenzialmente finalizzati al confronto con i dati sperimentali ottenuti su un modello fisico in scala, che riproduce una campata della struttura.

I risultati si riferiscono alla sola struttura portante, con i suoi carichi statici, senza alcuna struttura stabilizzante.

Le frequenze proprie di oscillazione della struttura reale senza struttura stabilizzante sono state confrontate (fig. 5) con le frequenze proprie della pulsazione della vena. In accordo con la letteratura (5), la descrizione dello spettro del vento è stata fatta con il diagramma adimensionale della densità di potenza spettrale, in funzione del parametro adimensionale $n\delta/u$.

Come di vede, il campo delle frequenze proprie della struttura si colloca in una zona con densità di potenza spettrale relativamente bassa. C'è però da tener conto dell'influenza del vento, che gioca a seconda della sua direzione, nel senso di aumentare o diminuire le frequenze proprie, perchè modifica i tiri della struttura e la sua geometria.

In caso di riduzione del tiro (vento proveniente da nord) è possibile che le frequenze proprie della struttura vadano a sovrapporsi alla zona di massima energia, con conseguente fenomeno di vibrazioni dovute alla pulsazione della vena (buffeting).

Questi primi dati forniscono perciò una indicazione ad aumentare i valori delle frequenze proprie, con l'adozione di un sistema di stralli. In questo modo è possibile risolvere contemporaneamente sia il problema statico (spinte dal vento verso l'alto) che quello del buffeting.

4. CONSIDERAZIONI ECONOMICHE

Le possibilità di successo commerciale dei sistemi PV a pannelli piani fissi sono legate principalmente all'elevata affidabilità e alla semplicità di esercizio e di manutenzione di questi impianti. Queste caratteristiche rendono questi sistemi particolarmente adatti ad operare in loca-

quindi sollevati nella giusta posizione dopo aver effettuato le operazioni di ancoraggio delle funi e dei tiranti.

Una orditura di travetti leggeri, posti ad un interasse di circa 2,2 m, costituisce la struttura secondaria sulla quale sono montati i pannelli PV. In questo modo il campo PV assume la forma di una superficie parabolica. Per l'impianto Delphos, le dimensioni previste per l'intero campo sono di circa 350x50 m.

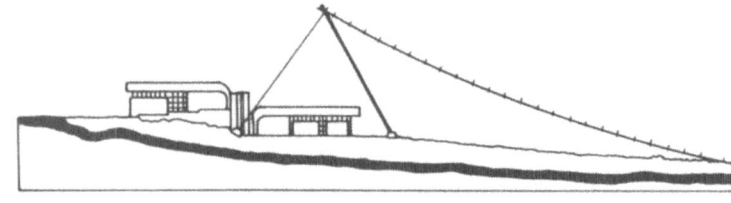

Fig. 1
EAST-SIDE
VIEW

Fig. 2
NORTH-SIDE
VIEW

2. PROGRAMMA DI SVILUPPO DELLA TENSO-STRUTTURA PIANA

Qualsiasi campo PV a pannelli piani è caratterizzato dall'avere un peso estremamente ridotto rispetto alla sua superficie.

Questa caratteristica, molto favorevole nelle applicazioni spaziali, comporta dei problemi nelle applicazioni terrestri, a causa dell'interazione con il vento.

In particolare, per la struttura leggera a grande falda da noi proposta, oltre a problemi di carattere statico, è possibile che insorgano eccessive oscillazioni dovute alla turbolenza o al distacco di vortici, ed al limite fenomeni di instabilità aeroelastica, con l'innesco di pericolose oscillazioni espansive nel tempo.

E' perciò attualmente in corso un programma teorico-sperimentale finalizzato a definire il comportamento aero-elastico della tenso-struttura, e a verificare sperimentalmente l'efficacia degli elementi strutturali stabilizzanti, sia per i fenomeni statici che dinamici.

Il programma di sviluppo prevede:
a - messa a punto di un modello analitico per l'analisi statica e dinamica della struttura;

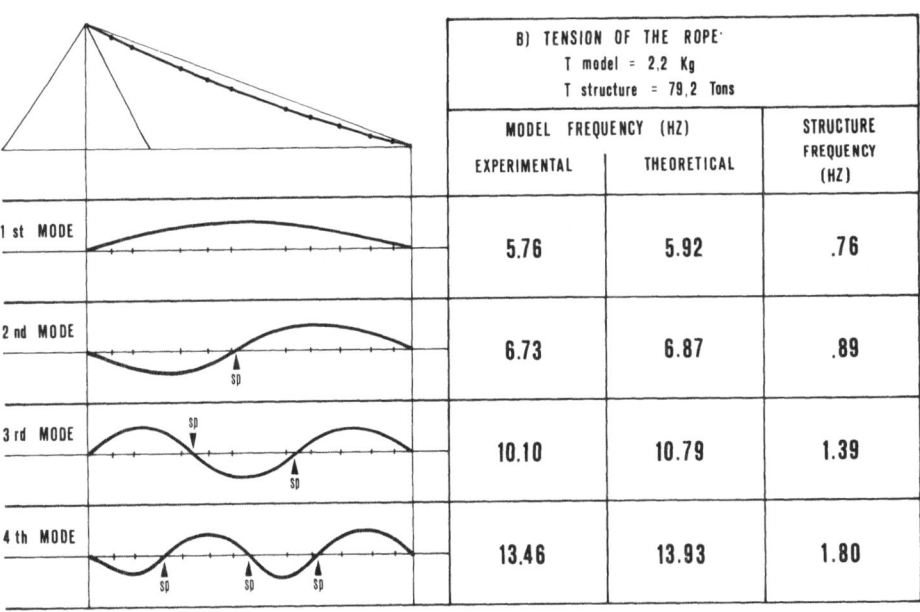

Figs. 3 and 4: Natural vibration modes in the vertical plane for different tension of the rope.

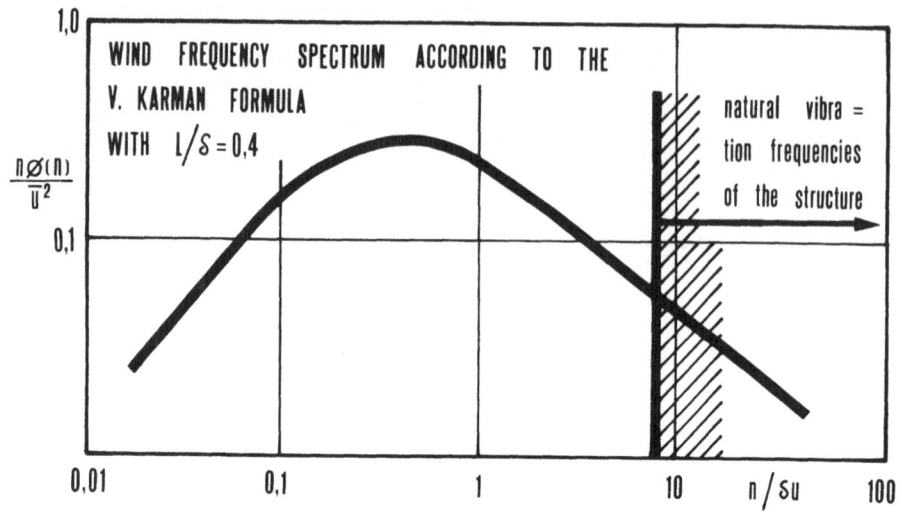

Fig. 5: Comparison of natural vibration frequency of structure and wind spectrum.

lità isolate, dove in generale il costo dell'approvvigionamento elettrico è molto elevato, ma dove la disponibilità di energia elettrica può svolgere un ruolo decisivo per la valorizzazione di risorse oggi scarsamente utilizzate (6), e per lo sviluppo economico di un'ampia fascia di paesi tropicali e sub-tropicali.

La giustificazione economica della struttura a funi deve ricercarsi in una ipotesi di realizzazione in località remota, tenendo conto di tutti i costi, inclusi quelli di preparazione del terreno, cablaggio, trasporto e montaggio, nonchè i costi indiretti di cantiere durante la realizzazione.

BIBLIOGRAFIA

1) A.A.W. Studio preliminare sulla fattibilità di un impianto fotovoltaico Dimostrativo - CNEN-RT/FARE SDI (80) 1, Roma Maggio 1980
2) V. ALBERGAMO, U. FARINELLI: Delphos, Una centrale fotovoltaica da 1000 KW - Notiziario CNEN Anno 27, n.9, 1981
3) A. RAITHEL, G. NICOLOSI: Strutture di sostegno per impianti fotovoltaici a pannalli piani - CNEN-RT/FARE SDI(81)1
4) F. CHELI ed AL.: Frequenze proprie e modi di vibrare di una tensostruttura di sostegno per un impianto fotovoltaico a pannelli piani. Indagine analitica e sperimentale. Memoria da presentare al VI Congresso AIMETA - Genova
5) Davenport, A.G. The Dependence of Wind on Meteorological Parameters. Proc. of the Int.Seminar on Wind Effects on Building and Structures, National Research Council, Ottawa, Sept. 1967.
6) G. LANZAVECCHIA: Growth of the rural areas and energy system. Symposium on "Rural change and public management" Paris, 19-22 October 1981

LOW COST MODULAR DESIGNS FOR PHOTOVOLTAIC ARRAY FIELDS[†]

H. N. POST
Sandia National Laboratories
Albuquerque, NM

D. C. CARMICHAEL
Battelle Laboratories
Columbus, OH

J. A. CASTLE
Hughes Aircraft Co.
Los Angeles, CA

Summary

This paper describes the design and development of optimized, modular array fields for photovoltaic (PV) systems. As a part of this activity, design criteria and performance requirements have been defined and evaluated for specific array subsystems. These subsystems include support structures, foundations, intermodule connection, field wiring, lightning protection, system grounding, site preparation, and monitoring and control. Fully integrated flat-panel array-field designs, optimized for lowest life-cycle costs, have been developed for systems ranging in size from 20 to 500 kW_p. These designs are applicable for near-term implementation (1982-1983) and reduce the array-field balance-of-system (BOS) costs to a fraction of previous costs. Key features, subsystem requirements, and projected costs are presented and discussed.

1. INTRODUCTION

Experience with application experiment systems and other PV field installations has demonstrated that array-field BOS and engineering costs must be substantially reduced before PV power systems can be cost-effective for grid-connected applications. Actual cost data for a representative application experiment (a 100-kW flat-panel, ground-mounted installation) gives the array-field BOS costs (site preparation, structural subsystem, electrical subsystem) as $424/$m^2$; engineering costs (system design, A&E fees, construction management) were $560/$m^2$ of collector area (1). These costs are far too great to encourage widespread commercial interest, regardless of the cost of PV modules.

To reduce these costs significantly, Sandia National Laboratories, as manager of the U.S. Department of Energy PV Systems Definition Project, is engaged in a program to optimize and integrate array-field subsystems and to develop standardized, modular building blocks which can be used to construct PV array fields ranging from 20 to 500 kWp. Two contractors, Battelle-Columbus Laboratories and Hughes Aircraft Company, were independently tasked with carrying out the objectives of this program for flat-panel systems.

2. DESIGN APPROACH

Actual reported costs and installation experience for operating systems were examined to identify areas in which costs could be reduced. In the area of array-field engineering, the development of standardized array-field designs and component hardware to minimize site-specific design work was identified as a key element for cost reduction. The use

---

[†]This work was supported by the U.S. Department of Energy, Division of Photovoltaic Technology.

of low-cost foundations and integrated panel/structure hardware, coupled with simplified mechanical attachments, was emphasized for the structural subsystem. To avoid the very expensive site preparation associated with many of the array-field installations, sites entailing minimal clearing and excavation work were recommended. Also recommended was an integrated field wiring design to incorporate minimal subarray junction box installation and instrumentation. These cost-saving features were determined through detailed examination of the major costs for each subsystem. Each of these features has been incorporated in the designs developed by Hughes and Battelle.

The following design requirements for the array field were specified.

- Field operating voltage of approximately 400 Vdc
- Compliance with applicable codes and safety requirements
- Suitability for installation at favorable sites
- Module efficiency (for comparison purposes) of 10 percent at NOC conditions (80 mW/cm^2, AM 1.5, NOCT)
- Suitability for near-term (1982-1983) use.

Note that development work on power conditioning equipment was not included in the design activity; however, the requirements imposed by this equipment on the array-field design were considered.

3. ARRAY-FIELD DESIGNS

A summary of key features of each design is presented in Table I.

Table I. Characteristics of Modular Array Field Building Block Designs

	Battelle Design	Hughes Design
PV Module	- 0.6x1.2 m (2x4 ft) - 5-Vdc output	- 0.6x1.2 m (2x4 ft) - 5-Vdc output
Foundation	- Steel hat-section stakes - Treated wood beam - Support spacing at 1.2-m (4-ft) intervals	- Steel footing in front; concrete curb in back - Support spacing at 1.2-m (4-ft) intervals
Structure	- Lightweight steel - Two-module panel	- Lightweight steel angle - Four-module panel
Intermodule Wiring	- Crimp-spliced pigtails with insulating pad	- Folded daisy chain with quick-disconnect connector
Field Wiring	- Direct-burial cabling - Buried shield conductor	- Direct-burial cabling - Power collection center
Building Block	- 10-kW peak output at NOC - 400 Vdc - Two rows of structures - 50 m (165 ft) east-west by 5.5 m (18 ft) north-south with aisles	- 10-kW peak output at NOC - $\pm$200 Vdc bipolar - One row of structures - 53.9 m (177 ft) east-west by 2.4 m (8 ft) north-south

Both of the designs use commercially available PV module hardware of the same size and output voltage. Both of the designs can be adapted to 0.3 x 1.2 m (1 x 4 ft) module sizes with minor modifications. Similarly, both designs, although developed independently, incorporate many of the same design features.

The modular array-field building block developed by Battelle is a 400-Vdc monopolar unit with a nominal output power rating of 10 kW. As shown in Fig. 1, the building block corresponds to a branch circuit of two modules wired in parallel and 82 pairs wired in series and includes one junction box per building block. The high degree of paralleling in the circuit design minimizes the effect of cell failures and allows for 20-year operation with a maximum estimated power degradation of 15 percent, assuming normal cell failure rates and no module replacement. The design incorporates an innovative support structure using galvanized steel stakes and panel support members. Treated wood beams permit simple fastening of the structural members using lag screws and provide flexibility in alignment during installation. All field wiring is direct-burial cabling, and the continuous metal support structure is attached to a bare copper wire counterpoise network to provide a uniform site ground. Details of the design are described in an engineering drawing package and construction specifications document which includes example layouts for 20-, 100- and 500-kW array fields (2). As an illustration, the 100-kW array field shown in Fig. 1 consists of ten building blocks which occupy on overall fenced area of approximately 56 x 59 m (185 x 195 ft).

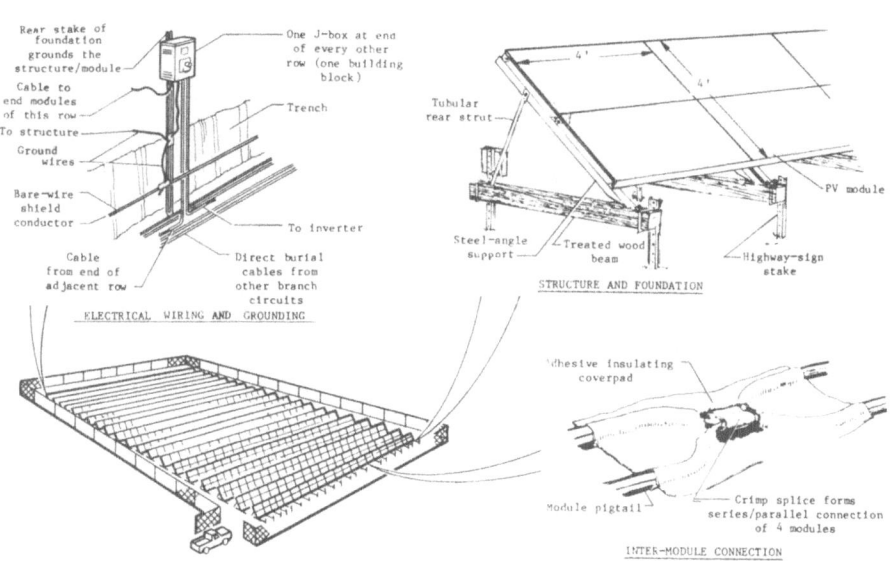

Fig. 1. 100 kW Array Field Layout Using Battelle's Building Block Design

The modular building block developed by Hughes is a ±200-Vdc bipolar unit with a nominal output power rating of 10 kW. The building block, which is a branch circuit, consists of two 200-Vdc monopolar subarrays positioned in an east-west-oriented row. Each subarray structure accommodates two parallel circuits of 40 series-connected 5-Vdc PV modules. As shown in Fig. 2, this design uses a galvanized steel support structure which incorporates a hybrid foundation. The front foundation is a buried metal foot which is an integral part of the grounding network; the rear foundation is a concrete curb. This design also incorporates a high level of parallelism in the module circuitry to enhance reliability and reduce lifetime power loss. A daisy chain module-to-module wiring scheme is used: the circuits run in horizontal rows, fold back on themselves, and all input-output leads terminate in a common junction box at one end of the subarray structure. The modules are connected by a quick-disconnect cable assembly manufactured by AMP, Inc. Power from each building block is routed from a subarray junction box to a power collection center (PCC) via direct-burial cable. The PCC contains a standard circuit breaker switch and bus panel, fault detection sensing, and a power control module (PCM) for every two building blocks. The PCM contains transient protection devices (MOVs), blocking diodes, and snubbers. Thus, the control of the array field is sectionalized in 20-kW increments. Design details are included in an engineering drawing package and construction specifications document (3). As an illustration, the 100-kW array field shown in Fig. 2 consists of ten building blocks which occupy a fenced area of approximately 66 x 74 m (217 x 242 ft).

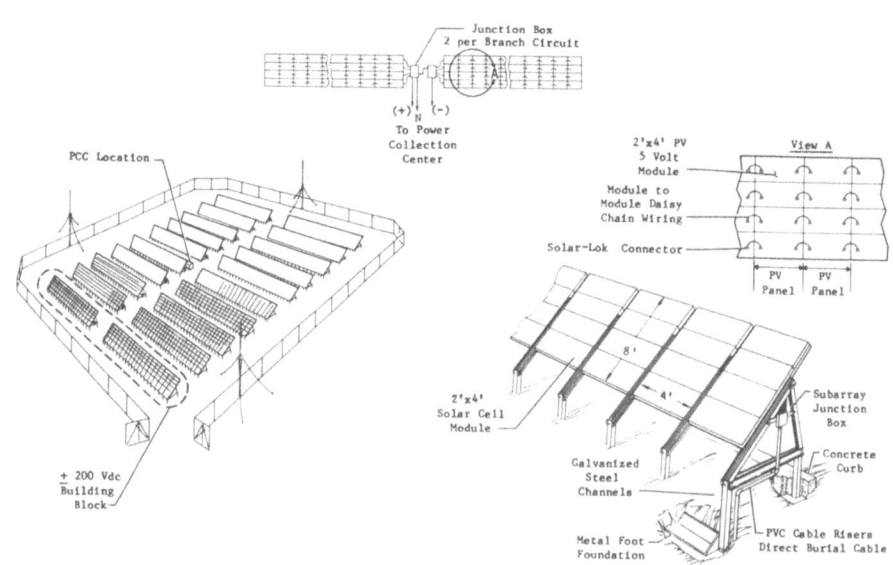

Fig. 2. 100 kW Array Field Layout Using Hughes' Building Block Design

4. COST PROJECTIONS

Table II presents a cost comparison of a typical (as-built) application experiment system with the modular array-field designs. The cost projections for the Hughes and Battelle designs are estimated prices derived primarily through cost quotations from several general and electrical contractors and component suppliers. The estimates assume a field size of 100 kW and a business volume of 1 MW/yr and 10 MW/yr, respectively, for the Battelle and Hughes designs.

Table II. Comparison of Array-Field BOS Costs for a Typical Application Experiment System and Price Estimates for the 100-kW Modular Array Field Designs

	1980 $/m^2		
	Application Experiment System	Battelle Design	Hughes Design
Site Preparation	137	8.07	9.46
Structural Subsystem	114	27.82	34.70
Electrical Subsystem	173	16.04	18.76
TOTAL	424	51.93	62.92

5. CONCLUSIONS

Modular array-field designs, optimized for lowest life-cycle energy cost, have been developed for PV systems ranging in size from 20 to 500 kW_p. These designs are applicable for near-term (1982-1983) implementation and reduce the array-field BOS costs to a fraction of previous costs. The designs are also more reliable than comparable existing installations and minimize maintenance requirements and costs. From estimated array-field BOS costs (1980$), the Battelle and Hughes designs are projected at $51.93/m^2 and $62.92/m^2, respectively, for low-production quantities and favorable installation sites. Similarly, the standardized designs and components should significantly reduce project engineering costs. Complete engineering drawing packages, including detailed construction specifications, are available for each design.

6. REFERENCES

(1) E. L. Burgess, K. L. Biringer, and D. G. Schueler, "Update of Photovoltaic Cost Experience for Intermediate Sized Applications," *Proceedings of the 15th IEEE Photovoltaic Specialists Conference*, Orlando, FL, June 1981.

(2) D. C. Carmichael, et al, *Development of a Standard Modular Design for Low Cost Flat-Panel Photovoltaic Array Fields*, SAND81-7183 (Albuquerque: Sandia National Laboratories, to be published). Work performed by Battelle-Columbus Laboratories.

(3) J. A. Castle, et al, *Photovoltaic Array Field Optimization and Modularity Study*, SAND81-7193 (Albuquerque: Sandia National Laboratories, to be published). Work performed by Hughes Aircraft Company.

THE PHOTOVOLTAIC SOLAR SYSTEM, ANALYSIS AND BASIC DESIGN RULES [1]

A.H.M. KIPPERMAN
Eindhoven University of Technology, Eindhoven, the Netherlands

SUMMARY

The basic structure of a photovoltaic system includes as a feed-back the state of charge of the battery and as a feed-forward signal the consumption pattern of the user, the time of the day, the season and the actual insolation. This knowledge is necessary to assure the coverage of user's basic energy needs. The system design is based on the minimum array area to cover the yearly basic needs and the minimum storage based on the probable number of consecutive days with low insolation in a critical supply period. Between these two boundaries the system can be optimized in terms of cost. As an example the same consumption of 1kWh/day is used to calculate the optimum system for Den Helder (NL) and for Trapani(I) resulting in a clear minimum in cost for the latter. In the summer period is energy available above the the basic need for comfort of the user; the average price in Trapani is $2/kWh and in Den Helder $4.5/kWh.

1. INTRODUCTION

In this paper the main boundary conditions of a photovoltaic system design will be presented starting from the requirement that in a system the mismatch between the generation and the consumption has to be bridged. Usually the size of the array is adapted to a period of low insolation and an estimate of the required storage is made. An average cost figure is calculated without any differentiation between the different needs of the consumer, wich can be labelled as basic or vital needs and energy for comfort. The value of the delivered energy covering these basic needs is the higher one. It can be estimated by using the 'value of the non-delivered kWh' which is e.g. the cost of buying water in case of a water-pumpig system, the cost of primary batteries at that particular location or the replacement of food or medicins in case of a cold store.

The basic steps of a system design following these pattern will be given and illustrated by the results for a constant consumption pattern at two locations in Europe: Den Helder (NL) and Trapani (I, Sicily).

2. BASIC STRUCTURE OF A SYSTEM

In a photovoltaic system, in general, the energy flow from the generator has to be directed to the user as far as his basic needs have to be covered and to the storage to assure the basic energy for the coming night or for periods of low insolation. This means that in the energy managment system the knowledge of the consumption pattern of the user, the time of the day, the season and the actual irradiance has to be combined with the state of charge of the battery. These feed-forward and feed-back lines are indicated in figure 1. In very large systems or in case the consumption pattern is complicated it can be necessary to follow also the insolation over a longer period to adapt statistical fluctuations. In case this indicates a coming shortage of energy, the managment system can reduce the delivery to the user in order to prevent an interruption of the basic energy supply. In simple systems a warning signal can be given instead of an automatic control action.

[1] This work is partilly supported by Holec, the Netherlands

3. SYSTEM DESIGN

The design procedure of the system follows a number of steps to determine the boundaries in between which the optimization has to be carried out. These boundaries are fixed by the energy consumption during the night at one hand and the minimum energy to generate with the available yearly insolation on the other hand.

With respect to the appropriate storage time, one has to realize that it is only possible to transfer energy from a period with a relative excess to a critical supply period. The latter being a period in which the consumption relative to the insolation is high.

3.1 BASIC NEED: Identify the required kWh's in the daytime and in the night (have to be stored; storage efficiency), if relevant take into account the kWh's/week and /weekend, and kWh's/season; this makes up the total kWh's/y to be generated. The efficiency of the storage depends on the depth of discharge (d.o.d) and the rate of charge or discharge (the live of a battery or the number of cycles depends on this d.o.d. e.g. 60% d.o.d. results in 70% efficiency and the equivalent of 800 cycles).

3.2 MINIMUM ARRAY AREA: this area follows directly from the average insolation and number of kWh's to be geberated given the array efficiency. This calculation can be made taking the whole year or an appropriate supply period, still considering the energy transfer from an excess to a critical supply period.

THE CORRESPONDING STORAGE CAPACITY: The shortage of energy in the critical supply period (given the minimum array area) over the allowed d.o.d. gives the required storage capacity.

3.4 MINIMUM STORAGE CAPACITY: The required number of consecutive days with a low insolation during a critical supply period, that should be supplied with the basic energy, have to be defined initially. The probability of the occurence of such a situation should be deduced from statistical meteorological data and gives the probability of a shortage in the delivery of the basic energy. Supposing the energy to be stored for the consumption in the night is not larger, this gives with the allowed d.o.d. the minimum storage capacity.

THE CORRESPONDING ARRAY AREA: As the storage is not capable of transfering large amounts of energy over a long period, the average insolation during the critical supply period has to provide the basic energy need by a suiable array area. This area is larger than the minimum area by the ratio of yearly insolation and the corresponding insolation in the critical supply period .

3.5 DESIGN FLEXIBILITY: The system design can only be varied between these two boundaries of array area and storage capacity. Optimizing the system interms of cost can now be done by introducing prices af array and battery, expected lifetime and interest.

4 CASE STUDIES: Two examples at different locations in Europe have been calculated in detail as an illustration of the design procedure lined out in chapter 3.
* Trapani, Italy Sicily; $37°.55'N$; average insolation 5 kWh/m^{2}day on a tilted surface; average insolation Dec. and Jan. 3 kWh/m^{2}day; 5 days of low insolation 1.5 kWh/m^{2}day.
* Den Helder, the Netherlands, $52°.55'N$, average insolation 3 kWh/m^{2}day on a tilted surface, average insolation Dec. and Jan. 1.5 kWh/m^{2}day; 5 days of low insolation 0.1 kWh/m^{2}day.
* consumption: basic need 1 kWh/day, 20% in daytime.
* battery: 60% depth of discharge, 70% efficiency, 4 years life $200/kWh

investment cost.
* array: 10% efficiency, 20 years life, $2000/$m^2$ investment cost.

From these figures it follows that 490 kWh/year have to be generated by the array.

4.1 DEN HELDER (NL):
* minimum array area = 4.5 m^2 + 52 kWh battery capacity
* 5 consecutive days of low insolation + 1 night: 9.7 kWh minimum battery capacity;
  1.34 kWh/day to be generated in Dec. and Jan.: 9.1 m^2 corresponding array area.

4.2 TRAPANI (I):
* minimum array area = 2.7 m^2 + 40 kWh battery capacity
* 5 consecutive days of low insolation + 1 night: 4.7 kWh minimum battery capacity;
  1.34 kWh/day to be generated in Dec. and Jan.: 4.47 m^2 corresponding array area.
* only day + night storage: 9 m^2 array + 1.9 kWh battery capacity (this is the absolute limit for the storage capacity).

4.3 SIZE OF THE SYSTEM:
Between the two boundaries, the size of the array has been varied and the corresponding battery capacity calculated, see fig. 2. It can be seen that due to the different average insolations and the different ratio's between average and minimum insolation the curves differ both in position and slope. Moreover, in the case of Trapani there exist a transion region indicating that an optimum can be found around 4.5 - 5 m^2 array area.

In figures 3.1 and 3.2 the performance of the system in terms of generated energy per month is plotted with the array area as a parameter. Note that the curves are not symmetric around June, July because at the end of the summer, the excess supply period, the storing of energy has to be started for the winter, the critical supply period. The excess energy in the summer is not necessary for the coverage of the basic needs, so available as an extra for comfort.

5. COST FIGURES
Introducing the investment costs of the array and of the storage figure 4 is obtained, lines of constant investment are indicated as well. These lines show clearly that in the case of Den Helder no clear optimum design is present in contrast to Trapani where the optimum at $9000 is significant.

Due to the different lifetimes of array and battery also the cost/year have to be considered for optimization, in figure 5 both investment and cost/year are plotted as a function of the array area. It shows that in the case of Den Helder a large array and a minimum storage capacity gives the lowest cost; for Trapani the choice between 4.5 and 5 m^2 array depends only on a marginal higher investment, the best design is clear.

The cost in $/kWh for the two array areas in each location is plotted in figure 6.1 and 6.2 related to the total number of kWh's available for the user. The minimum array's do not give any additional energy for comfort and have, moreover, a higher cost. Somewhat larger area's provide an interesting amount of energy for comfort and on the average at a lower price; note the reduction of the energy available for comfort at the end of the summer.

CONCLUDING REMARKS
Although the details of the optimisation of the system in the two cases depends on the exact price and lifetime of the components, it illustrates

that, with a limited amount of calculation, the optimum system size can be determined.

The approach of storing part of the energy at the end of the summer for the basic needs in wintertime implies an active energy management system that reduces the energy flow to the consumer.

The identification of critical supply period allows a large variety of users such as remote dwellings, water pumping and telecommunication systems.

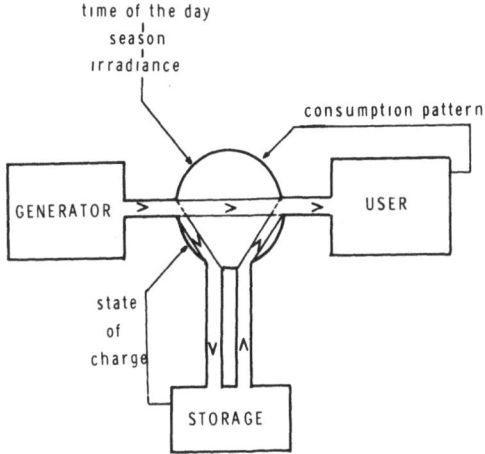

Fig. 1 - Basic system lay-out including the necessary feed-back and feed-forward lines, allowing the Energy Management (EM) to assure the supply of the basic need to the user.

Fig. 2 - The necessary storage capacity versus the array area between the two boundaries of minimum area and minimum storage for the two locations Den Helder (NL) and Trapani (I).

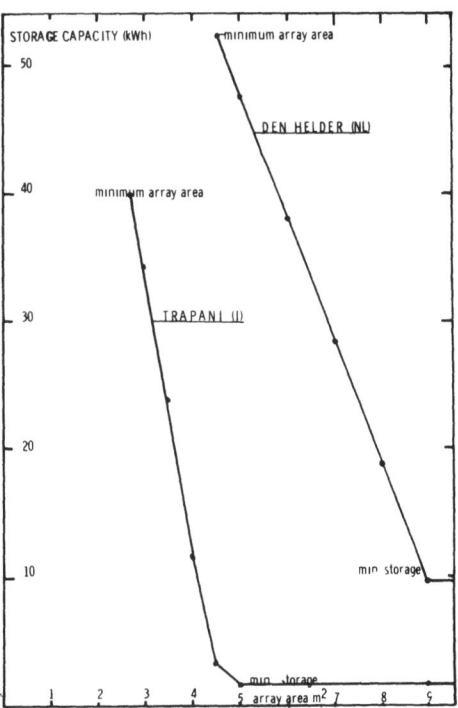

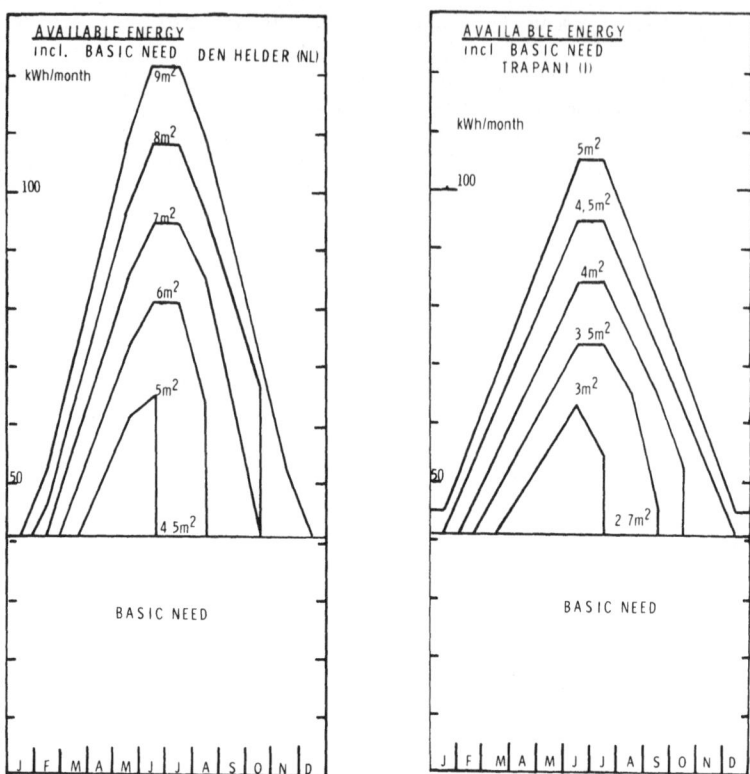

Figure 3.1 and 3.2
The available energy per month with the array area as a parameter; the energy above the basic need is available for comfort of the user. The curves are non-symetric due to the storage

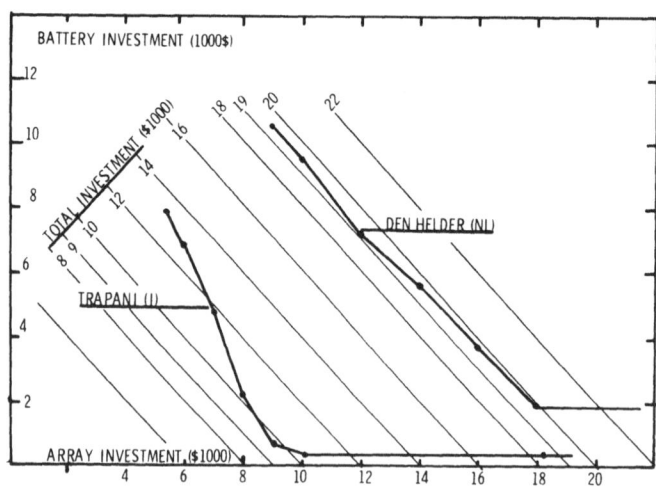

Figure 4
Investment cost for the system at the two locations showing a significant minimum in case of Trapani but only a very shallow one for Den Helder

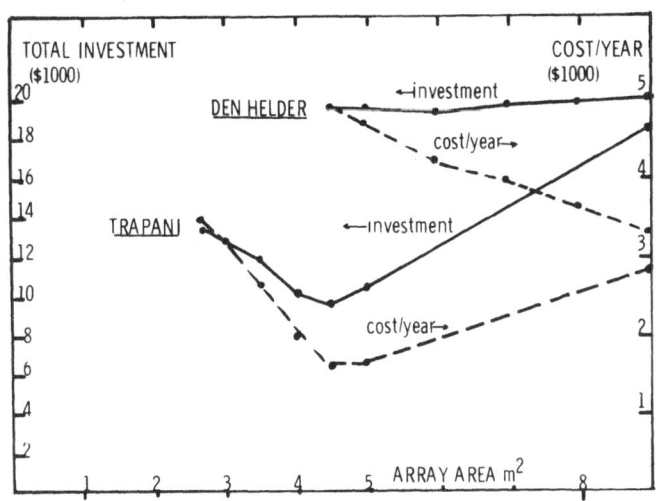

Figure 5
Comparison of an optimization in terms of minimum investment and/or cost per year

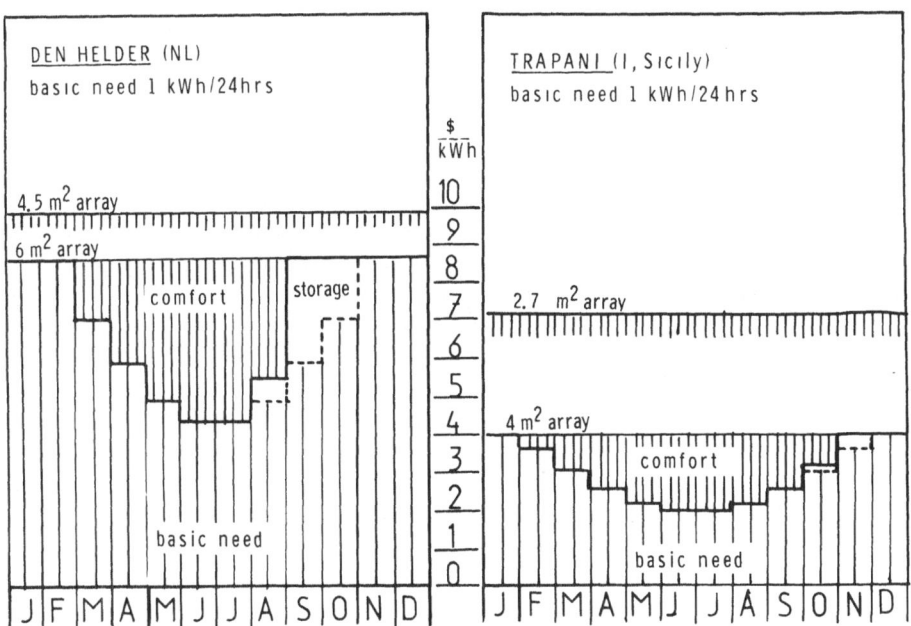

Figure 6
The cost per kWh for the minimum array area and a more optimized system; the cost reductions are significant in particular in Trapani

- 285 -

A DETAILED PACKAGE OF DIGITAL CODES SPECIALLY DEVELOPED FOR THE ARRAY DESIGN

P. Bullo, M. Gasbarra
ENEL - Direzione Studi e Ricerche, Milan
G. Emanuele
Phoebus - Catania

Summary

In the present paper a complete modular set of digital codes is described, specially developed for a sensitivity analysis of the different technical parameters affecting the design and construction of a PV generating unit, in particular of a large-scale type. The scope is the evaluation of the plant energy deficit on an hourly, daily and monthly basis, in respect of a base-case where optimum values of the panel tilt angles, azimuth and the N-S step between rows are chosen.

Assuming a lower ceiling in the B.O.S. cost of the PV power station, based on a foreseeable reduction of the panel costs, an indication is given of the region of tilt and azimuth loci which makes it possible to accept a moderate annual energy deficit and consequently low cost of the PV energy produced by the PV generating unit.

1. INTRODUCTION

This paper deals with the problem involved in the construction of PV plants characterized by support structures consisting of extremely simple, low-cost panels.

Particular attention has been paid to the comparison between the annual energy produced by a conventional plant (i.e. built with fixed panels in optimal arrangement), and the energy produced by hypothetical plant consisting of panels that follow the ground profile. For this and other reasons, computer programs have been developed that will make it possible to acquire sensitivity on various technical parameters inherent in the project and in the construction of photovoltaic generating plants.

Based on the results obtained, technical and economic conclusions are drawn that appear to be of some interest for the development both of PV plants for the supply of isolated communities in the medium term and for large plants to be connected to the electric national grid in the long-term.

2. DESCRIPTION OF THE COMPUTER PROGRAMS

The SOL-3 package is a collection of programs, written in FORTRAN, developed for purposes of supplying as far as possible accurate and complete information concerning solar energy, in respect of any site where data on the subject is available. The data resulting from the processing done with the package may be the point of departure for codes for the dynamic simulation of photovoltaic plants already developed by us that will be the subject of other publications [1].

The SOL-3 package consists mainly of three programs that operate in close co-ordination (see Fig.1):

1. SEPRA

This program operates the separation of global solar radiation on a horizontal plane - this item is commonly available - in its direct and diffuse components. If necessary, the program uses a considerable series of methods, which are open to choice and may be found in recent literature on the subject. The program contains an interface routine with module DIST of the package, so as to be able to accept as input the output files of the latter.

2. DIST

This program operates the simulation of the distribution of solar radiation during the hours of daylight. An original characteristic of the code is that of being able to obtain curves that simulate sufficiently closely the trend of solar radiation, even on partly cloudy days. This program also contains an interface routine with module SEPRA of the package, so that the latter's output files can be accepted as input.

3. INCLI

This program operates the computation of the solar radiation (direct and diffuse) from the horizontal plane to a tilted plane (i.e., fixed or with one/two-axis tracking. The program supplies, as output, a file of data on radiation intensity (direct, diffuse, and albedo) on a preselected plane. This program also accepts, as input, the output files of the aforementioned modules of the package.

With the help of this package, it has been easy to plot the maps (which we have called "iso-energetic") shown in Fig. 2. These maps make it possible to visualize, for a given site, the ratio between the annual solar energy striking a surface tilted at any angle and any azimuth and the energy striking the same surface with optimal tilt and azimuth arrangement.

To be able to evaluate the energy deficit due to the different mounting of the panels, the planimetry (scale 1 : 500) was taken of a real site on which medium-size PV plant might be constructed.

The following work stages were performed, based on digital codes which apply the finite elements method [2] :

- Recording and plotting of the contour lines by means of a digitizer.
- Approximate calculation, by means of digital programs, of the orographic data, complete with drawing (Fig. 3).
- Plotting of a hypothetic PV plant consisting of panels arranged in such a way as to follow the contour lines mentioned above (Fig. 4).

After completing the above three stages, the first step was to check that the site was a suitable one from the orographic point of view, by evaluating the dispersion of the direction cosines of the planes of the panels. It was found that the tilt varies between 2.2° and 7.7°, while the azimuth shows a wider range, between - 40° and + 40°. However, it should be remembered that this latter spread has little influence, due to the low tilt value.

Using the iso-energetic maps and the orographic data described above, an evaluation was made of the annual solar energy striking the whole PV field, comparing it with the value for the energy striking a field with the same number of panels, but arranged in such a way to obtain the maximum annual energy yield. The result (6.3% of deficit) confirms what it had already been possible to deduce from the iso-energetic maps: that is, that the annual energy striking a surface does not vary much for tilts of between 0 and 35° and azimuths of between - 45° and + 45°.

Nevertheless, since the designer is more interested in the electric energy produced than in the solar energy collected, the last step was to simulate the electrical behaviour. It was assumed that 50 panels (whose characteristics are given in Table I) were connected in an East-West direction. The individual panel was simulated by means of a single-diode model (ignoring the shunt and series resistances), and it was also assumed that each panel was protected by a by-pass diode.

The mathematical model that was developed presupposes extraction of the maximum power of the PV field, exactly as though a real MPPT were present. The results obtained from this simulation indicate a deficit of only 11% of the annual electric energy produced, as compared with a conventional PV field with E-W oriented rows of panels with optimum tilt and azimuth values.

This result would seem to have been dictated not only by the low tilt values in respect of the site chosen, but also by the fact that the methods used for separating direct and diffuse radiation operate with an average diffuse fraction of the order of 0.3.

3. CONCLUSIONS

At the present time, taking into consideration the cost of panels, the cost breakdown of a PV system is approximately as follows [3, 4]:
- Modules and their protections: 75%
- Support structures and ground preparation: 25%

taking the unit cost of modules as about $ 10/W_p, and neglecting the land cost. However, it is generally accepted that, by using production lines of the order of a few MW_p/year, the cost might drop asymptotically to about $ 2/82 per W_p. This reduction might therefore produce the following cost breakdown for the PV field:
- Modules 38%
- Support structures 62%

As may be seen, as compared with the previous case, the cost of the support structures is greater, these being difficult to cut down on, except by a radical change in technology. It is therefore obvious how desirable it is to reduce the share of the total cost of a PV field represented by the cost of the structures, even if this involves an energy deficit.

For example, in the case described in the report, and taking a situation of low panels costs, assuming that arranging the modules in accordance with the contours of the ground makes it possible to cut the costs of preparing the site and the structures by 50% - which would be reasonable [5]- the cost of the PV field, even taking into account the greater number of modules required ($\sim$10%), would be reduced by about 25%.

These simple financial considerations are being carefully considered in Italy. In particular, an examination is beginning to be made, together with Italian manufacturers, of low-cost mechanical solutions that would accord with the theoretical studies described in this report.

ACKNOWLEDGEMENTS

The authors would like to extend their warmest thanks to Mr. G. Fenini of ENEL-CRIS for making available to them the computational codes and equipment without which their work could not have been completed in time.

REFERENCES

[1] G. Emanuele, M. Gasbarra, A. Iliceto: Simulation Digital Code for the Analysis of Dynamic Behaviour of a PV Plant, paper under press.

[2] G. Fenini: Animazione grafica mediante "computer output microfilm", Quaderni di informatica, n. 1 - 1979 pag. 41-47.

[3] E.L. Burgess, K.L. Biringer: Update of PV System Cost Experience for Intermediate-sized Application, XV Spec. Conference, Orlando, May 81 pag. 1453-1457.

[4] H. Durand: A keynote Speech, Solar Energy for Development, proceedings of the International Conference held at Varese, Italy, March 26-29, 1979, by the CEC, pp. 59-66.

[5] G.J. Jones, D.G. Schueler: Status of the DOE Photovoltaic Systems Engineering and Analysis Project, XIII IEEE Spec. Conference, Washington, June 1978, pag. 1160-1165.

Tab. I - Module Electric Characteristics at 1 kW/m^2

Open circuit voltage	6,7 V
Short circuit current	40 A
Max power voltage	5,3 V
Max power current	36,3 A

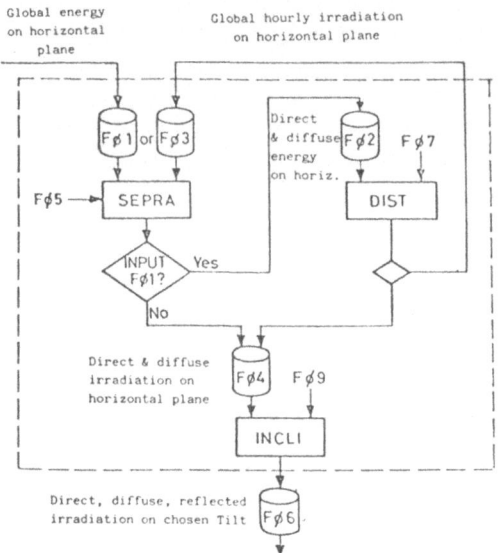

Fig. 1 - Flow-chart of SOL-3 package

Fig. 2 - Isoenergetic curves of yearly solar global energy as a function of tilt and azimuth (values are normalized to the max power)

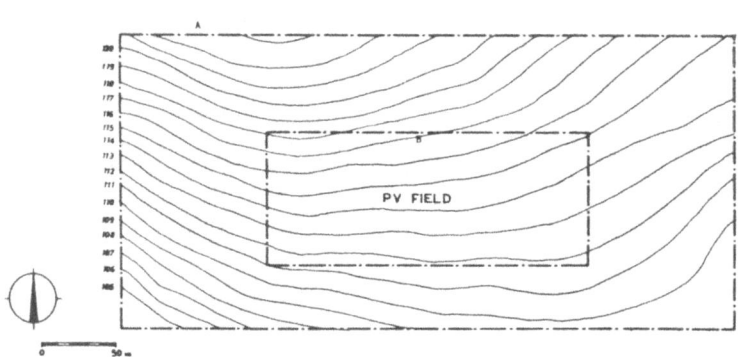

Fig. 3 - Site investigated by the study: A Fences; B Territory occupied by the PV field. Level lines expressed in meters.

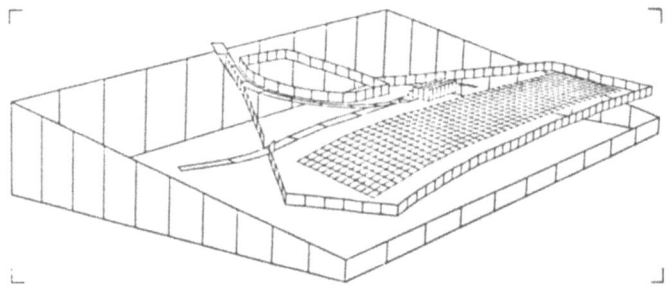

Fig. 4 - A finite elements artist's view of a central PV plant whose panels follow the contour lines of the site of fig. 3.

ANALYSIS OF HOT-SPOT-EFFECTS IN ENCAPSULATED PHOTOVOLTAIC GENERATORS BY LASER SCAN AND PARTIAL SHADOWING

G.H. HEWIG and H.-P. HÜBNER
Institut fuer Physikalische Elektronik
Universitaet Stuttgart
F.R. Germany

SUMMARY

The connection of several solar cells in a generator implicates some problems unknown to a single solar cell. Due to variations in the short circuit current of the different solar cells, caused either by partial shadowing or variations during the production process, some cells may act as a load, thus resulting in hot-spot-effects. The operating condition of a solar cell in a photovoltaic generator, with the single cells connected either in series or parallel, is analysed.

To investigate encapsulated modules the methods of laser scan and partial shadowing are employed. The combination of both methods yields the short circuit current and the reverse IV-characteristic of each solar cell, which both are important to classify the different cells according to the possible appearance of hot-spots. Hot-spot - effects in Si-solar cell generators are quite possible due to their high shunt resistance, thus withstanding quite high reverse voltages, which may result in a high power consumption of that solar cell, if their short circuit current is lower than the average current of the generator.

The possibility of the appearance of hot-spots in a Cu_2S-CdS solar cell generator is much lower due to fact that thin film solar cells in general show a breakthrough in their reverse characteristic at much lower reverse voltages.

1. INTRODUCTION

Fundamental research and development on a broad basis have led to high-efficient and stable solar cells. Both, the Si and the Cu_2S-CdS solar cells are fabricated in pilot-line productions and connected to modules and generators. As the fabrication process implies small variations in the IV-characteristics of different solar cells, they usually have to be matched before they are assembled in a generator. Nevertheless, there are still some variations in the short circuit current and the reverse characteristic which may cause serious problems if a large number of solar cells is connected in a generator.

2. OPERATING CONDITIONS OF SINGLE CELLS IN A GENERATOR

If N solar cells are connected in series in a generator, the IV-characteristic of the generator can be obtained from the IV-characteristics of the single cells by adding the voltage of all cells at a given current. Fig. 1 shows the IV-characteristic of a solar cell generator consisting of solar cells connected in series assuming that one solar cell whose IV-curve is also shown, has a lower short circuit current than the others.

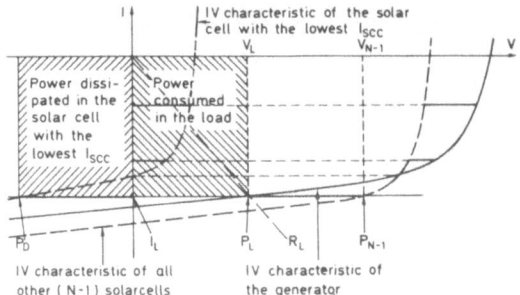

Figure 1: Distribution of the generated power in a solar cell generator (all cells connected in series)

The construction of the IV-curve of the (N-1) solar cells is demonstrated in the figure and assuming a constant load R_L, which may have been optimized at higher levels of illumination, the operating point P_L at the IV-curve of the generator, the operating point P_{N-1} at the IV-curve of the (N-1) solar cells and the operating point of the single solar cell, which acts as an additional load, is shown. For all cells connected in parallel the construction of the IV-curve of the generator can be obtained from the IV-curves of the single cells by adding the different currents of the solar cells at a given voltage. In this case the power is consumed in the load and in the solar cell with the lowest V_{OC}. The sum of both powers is equal to the power generated in the (N-1) other solar cells. The IV-characteristics of all types of generators can be composed in the way described above

3. DETERMINATION OF THE SHUNT RESISTANCE AND SHORT CIRCUIT CURRENT OF THE DIFFERENT SOLAR CELLS

To be able to predict, which of the different solar cells may act as a load in the generator, one needs to know the shunt resistance and the reverse dark characteristic of each solar cell in the generator. Employing the method of complete shadowing of one solar cell of the illuminated generator, as proposed by Hoffmann and Miller /1/, the dark reverse IV-curve of the shadowed solar cell can be obtained as is shown in Fig. 2.

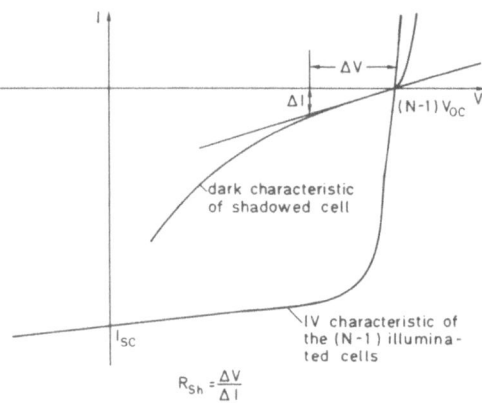

Figure 2: Measurement of the shunt resistance and breakthrough behaviour by shadowing of single cells

The slope $V/I = R_{Sh}$ and the breakthrough behavior, which is also obtained, can be used for further classification of the type of solar cell /2/.

To get the short circuit current of a single solar cell in an encapsulated generator the method of laser scan is used.

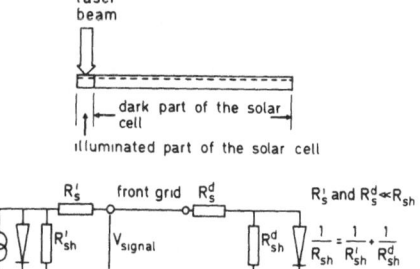 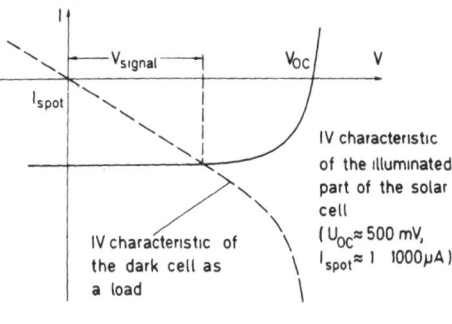

Figure 3: Analog circuit of a partially illuminated solar cell

Figure 4: Operating point on the IV-characteristic of a solar cell illuminated with a focused laser

Fig. 3 shows schematically the laser scanning of a single cell and the analog circuit of that partially illuminated cell. That part of the solar cell which is illuminated by the laser beam is connected with the remaining dark solar cell due to the common front grid and back contact. Assuming that the illuminated area of the solar cell is small compared to the total cell area and that $R_S \ll R_{Sh}^d$, then the shunt resistance of the dark cell, R_{Sh}^d, acts as the load of the illuminated part of the solar cell as is shown in Fig. 4. The voltage V_{signal} between front grid and back contact is therefore equal to

$$V_{signal} = R_{Sh}^d \cdot I_{spot}$$

where I_{spot} is the average short circuit current created in the solar cell during the scanning process. The voltage V_{signal} can be measured at the two contacts of the generator.

The intensity of the laser beam has to be low enough so that the influence of the series resistance can be neglected. Furthermore, the voltage V_{signal} has to be small compared to V_{OC}, which is usually the case, if the illuminated area of the cell is small in comparison to the total cell area.

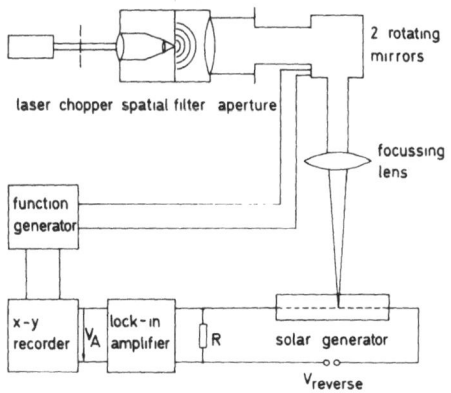

Figure 5: Blockdiagramm of a laser scan apparatus

Fig. 5 shows the blockdiagram of the laser scan apparatus /3/. The beam of a He-Ne laser is chopped and passed through a spatial filter, an aperture, a system consisting of two deflecting mirrors and a focussing lens before it hits the solar cell. To avoid some difficulties with the curvature of the IV-curve near the zero voltage point, a reverse bias voltage is applied to the generator. Employing a lock-in amplifier, the signal of the chopped laser beam can be detected and plotted on an x-y recorder which is controlled by the same function generator as the deflecting mirrors.

4. RESULTS

Employing the methods described above a solar cell generator consisting of 36 Si solar cells connected in series was investigated. The results are listed in table 1. In the first column the output voltage V_A of the lock-in amplifier is listet which is, without regard to a constant factor for all solar cells, equal to the product of R_{Sh} and $I_{SCC_{spot}}$. In the second column the shunt resistances R_{Sh} of the different solar cells are listed. In the last column the average values of the short circuit currents of the different cells created by the scanning laser beam are specified. This generator has been operated under short circuit conditions in full sunshine and a heating of some cells f.e. no. 3, has been observed, but due to the relatively large variations in short circuit current and R_{Sh} of the different cells, the dissipated power under these operating conditions was disseminated among various cells thus preventing a real hot-spot, whereas in other Si solar cell generators hot-spots under short circuit conditions have been observed and quantitatively analysed using an AGA-camera-system /4/. Local temperatures of the solar cells up to 170° C have been observed.

A Cu_2S-CdS thin film solar cell generator which was built at the University of Stuttgart and which consisted of 18 solar cells in series also has been investigated. The results look quite similar to those listed in table 1 with the exception that the shunt resistance is an order of magnitude lower than for Si solar cells due to the fact that the polycrystalline solar cells show a breakthrough already at relatively low voltages.

5. CONCLUSION

Completely encapsulated solar cell generators can be investigated in terms of their sensitivity to the appearance of hot-spots employing the methods of laser scanning and partial shadowing techniques. In Si solar cell generators the solar cells with the highest values of the shunt resistance in connection with a reduced short circuit current are the most hazardous cells for hot-spot-effects. To prevent internal dissipation of power at lower light levels, the generators always should be operated with an electronic load. Cu_2S-CdS solar cell generators are much less sensitive to hot-spot-effects due to the fact that the breakthrough occurs at relatively low voltages which limits the maximum power dissipated in a single solar cell of the generator.

6. ACKNOWLEDGEMENT

This work has been supported by the "Bundesministerium für Forschung und Technologie", F.R. Germany.

cell	V_A(V)	R_{Sh}(KΩ)	$I_{SCC_{spot}}$ (mA)	cell	V_A(V)	R_{Sh}(KΩ)	$I_{SCC_{spot}}$ (mA)
1	2,23	3,1	0,584	19	2,48	3,5	0,575
2	1,31	2,2	0,483	20	2,70	3,5	0,626
3	3,56	5,1	0,567	21	2,52	3,1	0,660
4	2,81	4,2	0,543	22	2,98	4,65	0,520
5	4,52	5,4	0,679	23	3,74	6,0	0,506
6	3,78	5,4	0,568	24	3,94	4,9	0,653
7	4,30	5,4	0,646	25	2,06	2,85	0,587
8	2,64	4,5	0,512	26	3,16	4,2	0,611
9	2,90	3,85	0,611	27	3,55	4,8	0,600
10	3,42	4,4	0,631	28	4,8	5,9	0,660
11	3,25	5,2	0,507	29	4,32	5,6	0,626
12	3,44	3,9	0,716	30	4,10	5,6	0,594
13	2,10	2,95	0,578	31	3,27	4,55	0,583
14	3,09	4,1	0,612	32	4,60	6,45	0,579
15	3,77	5,15	0,594	33	4,58	5,75	0,646
16	4,12	5,53	0,605	34	2,35	3,9	0,489
17	2,55	3,2	0,647	35	2,61	3,9	0,543
18	2,86	3,85	0,603	36	4,25	5,75	0,600

Table 1: Evaluation of I_{SCC} and R_{Sh} of all single solar cells of an encapsulated Si solar cell generator

7. REFERENCES

/1/ A.R. Hoffman and E.L. Miller,
IPL international document 5101-170,
Pasadena, California, October 15, 1978

/2/ J.C. Arnett and C.C. Gonzalez,
Conf. Proc. 15th IEEE Photovoltaic Specialists Conference,
Orlando, Florida, 1981

/3/ W. Arndt,
Diplomarbeit, Universität Stuttgart, 1976

/4/ H.-P. Hübner,
2. Semesterarbeit, Universität Stuttgart, 1981

REVERSE BIAS POWER DISSIPATION OF SHADOWED OR FAULTY CELLS IN DIFFERENT ARRAY CONFIGURATIONS †

P. SPIRITO[+] and V. ALBERGAMO[*]
+ Istituto Elettrotecnico, University of Naples, Italy
* ENEA, Istituto Casaccia, Rome, Italy

Summary

An analysis has been made of the maximum power dissipation in reverse biased shadowed (or faulty) cells. The results of the analysis are presented for different series-parallel configurations; for each case the bias point of the shadowed cell is determined as a function of the "shadowing degree", defined as the ratio between the I_{sc} of the shadowed cell and the I_{sc} of the illuminated ones. The maximum power dissipation allowable for the cell in a given module is computed by using the NOCT values of the module itself.
The results show that a large paralleling of cells can lead to severe stresses if more than one cell is shadowed, and to a greater sensitivity to the "hot spot" phenomenon.

1. INTRODUCTION

In recent years the reliability of PV fields has been one major point in system design, especially in large-scale applications, due to its relevance on the system performance and on the overall cost of the energy supplied.
Usually the reliability studies are concerned with the power reduction of the array caused by a given failure rate of the subcomponents (cells, interconnects, etc)(1). However the cell failure rate can be substantially increased if, as a result of a mismatch of the short-circuit currents of different cells, some of these become reverse biased. This mismatch may be due to different reasons, such as cracked cells, partial shadowing, encapsulant deterioration, or technological mismatches between cells.
The field experience indicates (2) that the overheating due to the reverse biasing of such faulty cells can lead to a fault propagation through the module, or to a complete cell failure, that will cause a larger power degradation in the array itself than the one predicted by a constant cell failure rate.
In this paper the influence of different array configurations on the reverse bias power dissipation of a faulty cell is analyzed, taking into account the thermal characteristics of the module and the different cell behaviours in reverse bias.

2. MAXIMUM POWER DISSIPATION OF A CELL IN A MODULE

The maximum temperature that the cell can safely reach depends mainly on the module encapsulant, and it may be taken as 100°C (3), because at that temperature outgassing of the encapsulant begins, that may cause cell bending and cracking. The cell temperature T_c is related to the power dissipated into it through the termal conductivity K_T of the module, defined as:

$$K_T = P_D/(T_c - T_a) \tag{1}$$

where T_a is the ambient temperature. Due to the variety of module designs, it is no meaningful to refer to a "mean" value of K_T; instead it can be evaluated by referring to the "Nominal Operating Cell Temperature" (NOCT), a parameter that is normally given for commercial modules. Using the definition of NOCT (4), the value of K_T can be obtained from the following expression:

$$K_T = \frac{800 (1 - \varepsilon)}{NOCT - 20°C} \quad (W/m^2 \, °C) \tag{2}$$

where ε is the amount of input radiation reflected by cover, inglobant and cell surface, usually taken as 10%.

Using eq. 1 and 2, the maximum power that can be applied to a cell in a given module is obtained as:

$$P_{DMAX} = K_T A (100°C - T_a - (NOCT - 20°C)) \tag{3}$$

where A is the area of the cell, and it has been assumed that the cell is exposed to the solar radiation of 800 W/m^2, corresponding to the Nominal Thermal Environment (NTE) value.

In fig. 1 it is plotted the maximum power dissipation allowable for a cell as a function of the NOCT of the module, for two ambient temperatures. It is seen that, for usual values of NOCT of commercial modules, the maximum power dissipation for 4 inch cells ranges from 8 W to 17 W.

3. REVERSE BIAS CHARACTERISTICS OF THE CELL

The power dissipation by the cell in reverse bias depends on:
a) the shape of reverse bias I-V curve of the faulty cell
b) the degree of mismatch between the faulty cell and the other cells
c) the array configuration

Regarding the first point, the cells can be roughly classified in two groups, the first one (curve (a) in fig. 2) with a reverse bias curve dominated by the avalanche multiplication, and the second one (curve (b) in fig. 2) with an almost linear shape of the I-V curve, due to a low shunt resistance.

For type (a) cells, the I-V curve can be described by the expression:

$$I = (I_{sc} - I_o (\exp \frac{V}{AV_T} - 1)) M(V) \tag{4}$$

where M(V) is the multiplication factor, that can be expressed by the Miller formula $M = 1/(1 - (|V|/V_{br})^a)$, with V_{br} the breakdown voltage.

For type (b) cells, the I-V curve can be described as:

$$I = I_{sc} - I_o (\exp \frac{V}{AV_T} - 1) - \frac{V}{R_{sh}} \tag{5}$$

The effect of different mismatches between the faulty cell and the other cells, either if the cell is partially shadowed or cracked, can be taken into account through the "shadowing degree" $\bar{I}$, defined as the ratio between the reduced I_{sc}^* of the faulty cell and the standard I_{sc} of the "good" cells.

4. INTERCONNECTIONS CONSIDERATIONS

For each array configuration, the worst-case condition arises if the array output is short-circuited (5). In this case if there are some faulty cells in a diode loop, the diode will be brought into conduction.

With reference to the parallel-series configuration of fig. 3a, if it is supposed that m' cells are shadowed (or cracked), the operating point of the system can be found by equating the I-V curve of the n-1 parallels with the I-V curve in reverse bias of the nth parallel.

For type (a) cells, recalling eq. 4 written for V<0 and neglecting the exp term in reverse bias, one has:

$$(\frac{m-m'}{m} + \frac{m'}{m}\overline{I}) = (1 - \frac{I_o}{I_{sc}}(\exp\frac{V}{(n-1)AV_T} - 1))/M(V-1) \qquad (6)$$

The power dissipated respectively by each of the m' faulty cells, and each of the "good" cells of the same parallel, is:

$$P_{D1} = (V-1)\,\overline{I}\,I_{sc}\,M(V-1) \qquad P_{D2} = (V-1)\,I_{sc}\,M(V-1) \qquad (7)$$

where it has been assumed a voltage drop of 1 volt across the diode in conduction.

Eq. 6 and 7 allow one to evaluate for each $\overline{I}$ value the operating point of the faulty parallel and the power dissipated both in the "good" and in the "bad" cells of the parallel.

For type (b) cells, eq. 6,7 become respectively:

$$(\frac{m-m'}{m} + \frac{m'}{m}\overline{I}) = 1 - \frac{I_o}{I_{sc}}(\exp\frac{V}{(n-1)AV_T} - 1) - \frac{V-1}{R_{sh}I_{sc}} \qquad (8)$$

$$P_{D1} = (V-1)\,\overline{I}\,I_{sc} + \frac{(V-1)^2}{R_{sh}} \qquad P_{D2} = (V-1)\,I_{sc} + \frac{(V-1)^2}{R_{sh}} \qquad (9)$$

a) Series-parallel connection

In the case of fig. 3b, the diode in conduction decouples the m strings from each other, so one can refer to a single series string with a bypass diode, and use eq. 6 or 8 by putting m=m'=1.

In fig 4a there is plotted the power dissipation for a 4 inch cell of type (a), as a function of the "shadowing degree" $\overline{I}$, for different n values. One can note that:

a) the maximum of P_D increases with the number n of cells in series; for commercial modules a maximum safe value of n=18÷20 is obtained

b) the complete shadowing of a cell ($\overline{I} = 0$) does not give the maximum of P_D; instead the maximum of P_D is reached at a value of $\overline{I}$ depending on the V_{br} of the cells, but the value of P_{DMAX} is independent from V_{br}.

Fig. 4b refers to the case of type (b) cells. Again one may note that the maximum of P_D shifts toward lower $\overline{I}$ values at lower R_{sh} values. The maximum however has the same value unless R_{sh} is lower than 5 ohms; this value of R_{sh} is unacceptable for practical cells because it reduces the efficiency of more than one point.

b) Parallel-series connection

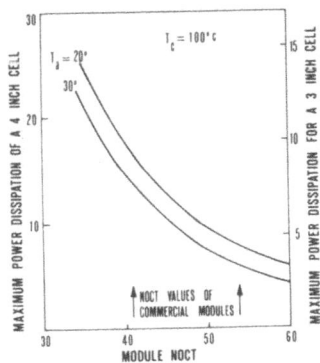

Fig. 1 - Power dissipation of a cel as a function of module NOCT

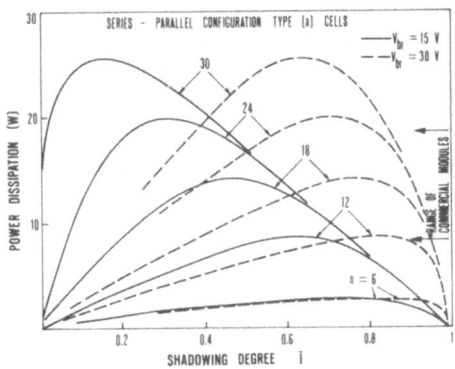

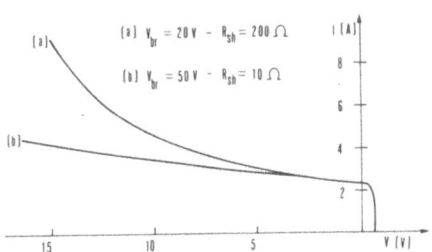

Fig. 2 - Reverse bias I-V curves of the cells

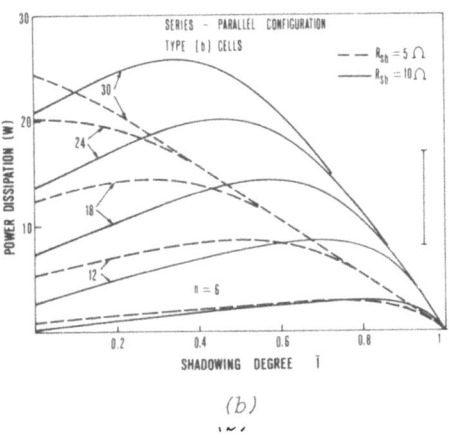

Fig. 4 a and b - Power dissipation of the shadowed cell in a series-parallel configuration

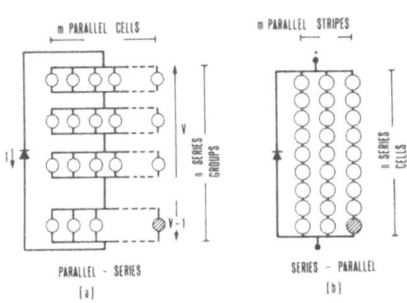

Fig. 3

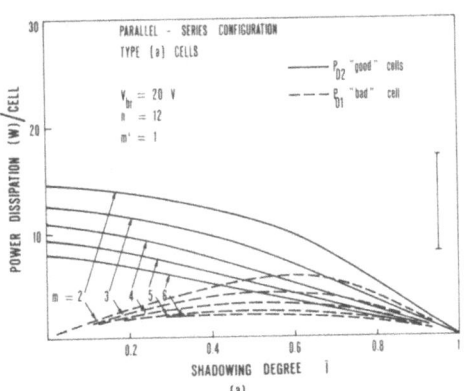

Fig. 5 - Power dissipation of the "bad" and "good" cells of the same parallel, for the parallel-series configuration

In this case (fig. 3a) eq. 6÷9 hold; in fig. 5 there are reported the results for type (a) cells. The results for type (b) cells are omitted because the plots are quite the same for the two cases; in each case the power dissipation is larger in the m-1 "good" cells than in the "bad" cell, and the maximum of P_D for the "good" cells is always obtained for $\bar{I} = 0$ (complete shadowing). Moreover this maximum of P_D drops significatively if m is increased, also if higher values of n are choosen; reasonable values of P_D are obtained for n = 24, m = 6.

However the worst case in parallel-series configurations arises if more than one cell is completely shadowed (or failed). In this case the "good" cells in the same parallel must now carry a larger current at reverse bias. These situations are summarized in Tab. I, where the P_{DMAX} values for the "good" cells are reported for an increasing number m' of shadowed cells, respectively for type (a) or (b) cells.

Tab. I

		n = 12 m = 6			n = 18 m = 6			
	m'	1	3	5	1	3	5	
V_{br}	= 15 V	6.7	15.6	18.1	7.1	28.0	43.4	
	20 V	8.2	14.6	16.8	8.9	27.6	34.9	type (a) cells
	30 V	10.1	12.6	12.7	16.8	21.4	23.6	
R_{sh}	= 5 Ω	5.1	13.2	16.0	5.3	18.5	28.2	
	10 Ω	8.4	13.0	14.0	9.9	22.1	25.9	type (b) cells
	20 Ω	9.8	12.3	13.4	15.5	21.4	23.3	

5. CONCLUSIONS

The results of the above analysis show that:
a) for the series-parallel connection, a single shadowed (or faulty) cell represents the worst-case condition, and there is no fault propagation because the power is dissipated only by the faulty cell; moreover the diode decouples the series string from each other.
b) for the parallel-series connection, a fault propagation is likely to occour, because the higher power dissipation is now experienced by the "good" cells of the faulty parallel. This power dissipation increases as more cells are shadowed in the same parallel, giving a worst-case for m'= m-1. In this case even a series of 18 cells leads to an unacceptable value of P_{DMAX}.

† Work performed in the mainframe of the ENEA-ENEL Delphos Project.

6. REFERENCES

(1) R.G. Ross, 15th IEEE Phot. Spec. Conf., 1157, Orlando, 1981
(2) S.E. Farman, M.P. Themelis, 14th IEEE Phot. Spec. Conf., 1284, S.Diego, 1980
(3) J.A. Arnett, C.C.Gonzales, 15th IEEE Phot. Spec. Conf., 1099, Orlando, 1981
(4) J.W. Stultz, J. Energy, 3, n.6, 363, 1979
(5) J.C. Larue, E. du Trieu, 3rd E.C. Phot.Sol.En.Conf., 490, Cannes, 1980

DISEQUILIBRIUMS IN SERIES CONNECTED SOLAR CELLS

An approach to the protection by parallel diodes

J.A. ROGER, S. MASSAAD, J. POSBIC and J. PIVOT

Département de Physique des Matériaux (Associé au C.N.R.S.)
43 Bd du 11 Nov. 1918 . 69622 . Villeurbanne (FRANCE)
Université Lyon 1

Summary
This paper presents the first systematic comparison between experiments and theory leading to a better understanding of the "hot spot" problem in PV arrays, and to a better determination of the major parameters influencing these disequilibriums. Measurements are performed on a single string of 408 series connected solar cells, and currents and dissipated power are recorded as a function of the number of cells that are protected by a given by-pass diode. Experimental data are in a very good agreement with a theoretical approach, summarized here, that allows a more complete investigation : the cell's parameters, especially the shunt resistance, are found to govern the "hot spot" formation and influence the aptitude of the diode to protect the cells. Various loading conditions are considered, and numerical values are given. This paper describes also the data acquisition system designed in our laboratory, that will be extended to analyze in summer 1982 the mismatch effects and the possible disequilibriums in the 5 kW array of the Refuge des Evettes in France.

1. INTRODUCTION
High power PV arrays installed today are designed to operate under high voltages to avoid large current densities. A great number of individual cells have then to be connected in series, and as a consequence disequilibriums in such strings are more likely to occur and are more and more critical. They are due either to disparities between cells or to accidental shading of a given surface of the modules. The result is an inversion of the voltage on one or several cells. If a given cell is to operate as a receiver, this junction has to dissipate the incoming solar energy plus the electrical energy contributed by the other parts of the array. Depending on the encapsulation properties and outside conditions, an equilibrium temperature is reached {1} that can be below or above the maximum value supported by the connections or the encapsulation material . By-pass diodes are put as a protection to derive the string current, but such diodes are expensive as well as their connection to the module. The main questions to answer are the following : how many cells is it possible to protect by one by-pass diode, and what is the efficiency of this protection as a function of the cells and diodes characteristics ? The two extreme cases are : (i) no diode and (ii) one diode per cell. This last possibility has been studied under the hypothesis of an integration of a diode directly into the silicon wafer for each cell {2}. It must be noted that such a technique is far to be controled yet and is subjected to the corresponding process complexity and cost.

2. OPERATING PROCEDURE AND DATA ACQUISITION SYSTEM

The string configuration is presented on Fig.1 : NCPROT is the number of cells protected by one diode (total voltage VBLOC) and NS is the total number of cells in the string. In this study three kinds of cells must be considered :
- 1 cell (n°2 in fig.1) is the shaded cell,
- (NCPROT-1) cells (such as 3) are correct cells inside the disequilibrated block,
- (NS-NCPROT) correct cells (such as 1) pass the total current I which is the sum of I_B and I_D.

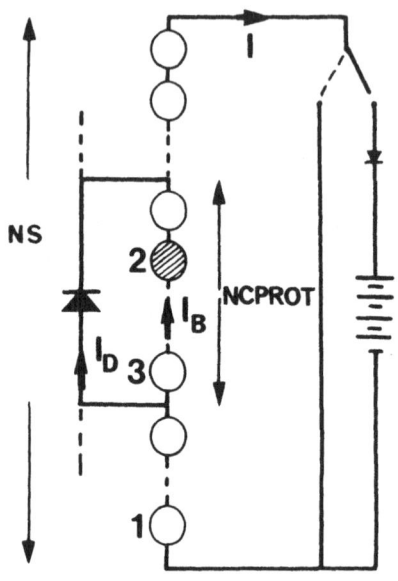

The experimental set up consists in 12 series connected modules located on the roof of the laboratory. One of these modules has been modified such as to allow electrical contacts on the terminals of one individual cell (n°2) which is gradually shaded from the sun. The illumination coefficient η is varied from 1 to 1/6. The in-time evolution of the resulting disequilibrium is analyzed by measuring the following quantities : the light flux ∅, the load current I, currents I_B and I_D, the temperatures, the voltage on the diode VBLOC and the voltage on the shaded cell VOLTC(2). The total power to be dissipated by cell n°2,PDIS(2), can then be computed.

Fig. 1

Signals are recorded on a small computer using a sequential multiplexing of several channels. This data acquisition system corresponds to the necessity of a <u>simultaneous</u> recording of the different quantities. All the data are taken within an adjustable very short period and stored on a magnetic tape, through an analog digital 8 bit converter. All incoming data are transformed into differential voltages using suitable interfaces. 16 measures can be accepted by the computer in 1/10 of a second. The tape storage capacity is about 20,000 values and this system has been extended up to 50 simultaneous measurements for the instrumentation of the 5 kW PV array of the Refuge des Evettes in France. The recorded quantities will permit a precise analysis of the losses due to mismatch effects : between modules on one side, and between the modules and the batteries on the other side, as a function of the light flux , the temperature and the state of charge of the batteries. The reduced prototype has been successfully tested in the present experiment.

3. EXPERIMENTAL DATA AND COMPARISON WITH THEORY

Experimental values have been compared with a simulation which has been developped in a previous paper {1}: <u>actual</u> characteristics of the individual cells making the array are determined as well as the thermal properties of the encapsulation. By taking into account the feed-back that exists between electrical and thermal effects, the equilibrium values for temperature and currents can be precisely determined if the resulting array characteristic is <u>built up</u> from the individual characteristics of all the cells. It must be noted at that point that statistical analysis found in the litterature are unsuited to describe the present effects.

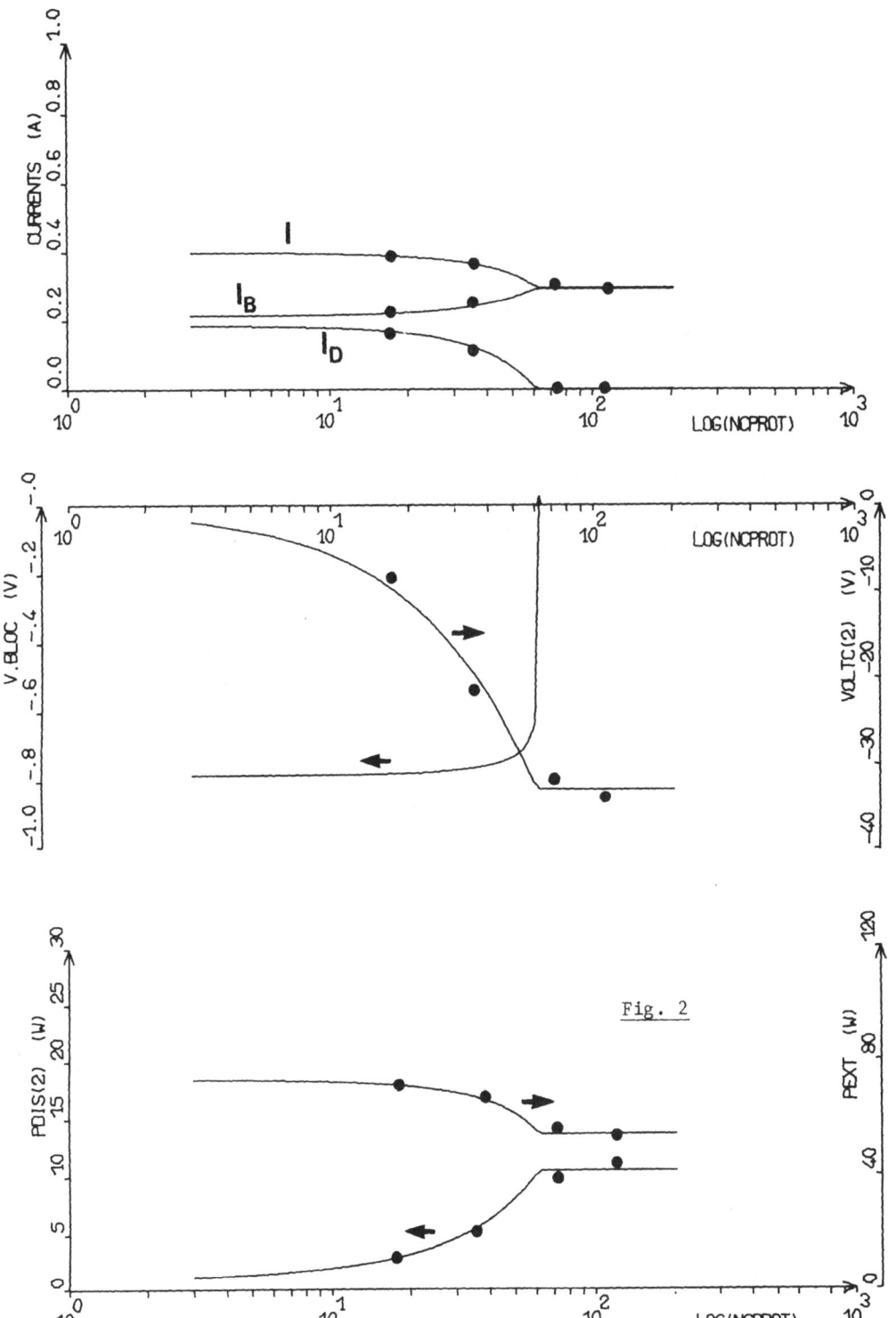

Fig. 2

Figure 2 shows that a very good agreement is found between theory and experiments (lines and dots respectively). Only one case, among many others, can be presented here : $\emptyset = 600$ W/m², load = battery, NS=408 x 57 mm cells, $\eta = 3/6$ for the shaded cell. The variable on the horizontal axis is the number of cells protected by one by-pass diode NCPROT. As precised above, the theoretical curves of Fig.2 have been obtained by introducing the actual characteristics, and in the present study the shunt resistance of these cells is very high with a negligible effect on the I-V curves.

Figure 2 indicates that, in the choosen conditions : above 60 cells for one diode, the by-pass diode is no more forward biased (VBLOC > 0), the total current delivered to the load is lower, the voltage on cell 2 saturates near -35V, the extracted power PEXT is decreased and the power dissipated by the shaded cell PDIS(2) is about 10 W.

Numerous similar experimental diagrams have been obtained for various values of $\emptyset$, η and NS and for other loading conditions. They will be published in a further paper. As an example, under short circuit conditions, it can be demonstrated that the diode would be always forward biased whatever the value of NCPROT, but the power dissipated by cell 2 can be as high as 23 W which leads to an unacceptable high temperature {1}.

Using the presented computation, the data obtained from the above experiments can be extrapolated to PV strings made up of other types of cells . The influence of the cell's parameters on the disequilibrium effects has so been studied, and the shunt resistance has been found to play a very important role.

In fig.3 is presented the variation of a string current (string closed on a battery) as a function of the illumination coefficient η of the shaded cell if no by-pass diodes are used . This decrease in current is determined (in %) with regards to the current delivered by the string if one cell was protected by one diode (optimum case). If the shunt resistance is high (noted $R_{sh\infty}$) a noticeable decrease of the current is found and is a function

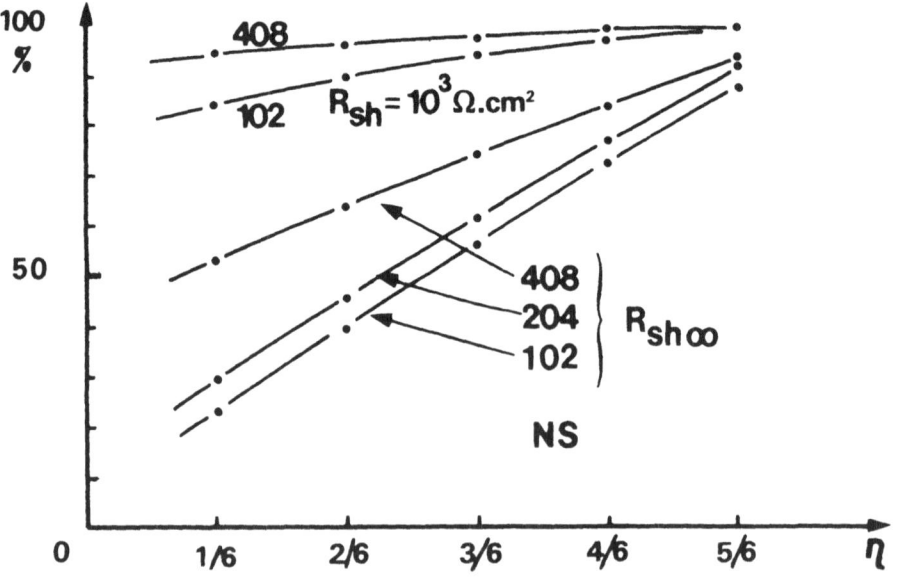

Fig. 3

of the number of cells in series as shown on the diagram. If junctions with low shunt resistance are concerned, the decrease in current, which also corresponds to the decrease in extracted power if closed on batteries, is minimized. The shunt resistance considered in this case (10^3 $\Omega.cm^2$) is still above actual values for many new 100 mm cells (300 $\Omega.cm^2$). It must be noted that such low values of R_{sh} do not affect too much the cell efficiency.

One major result of this short presentation is contained in fig.4 that presents the power dissipated by the shaded cell as a function of η, with the shunt resistance as a parameter. The 408 cells under test connected to batteries with no protective diodes gives curve a : the power dissipated by the shaded cell is very high as well as the limit temperature, and they both culminate for $\eta = 3/6$. If one diode per cell is used, curve b is obtained : no degradation occurs. If cells have a low shunt resistance, the hot spot effect is strongly decreased (curve c), the dissipated power being lower than 0.2 watts per cm^2.

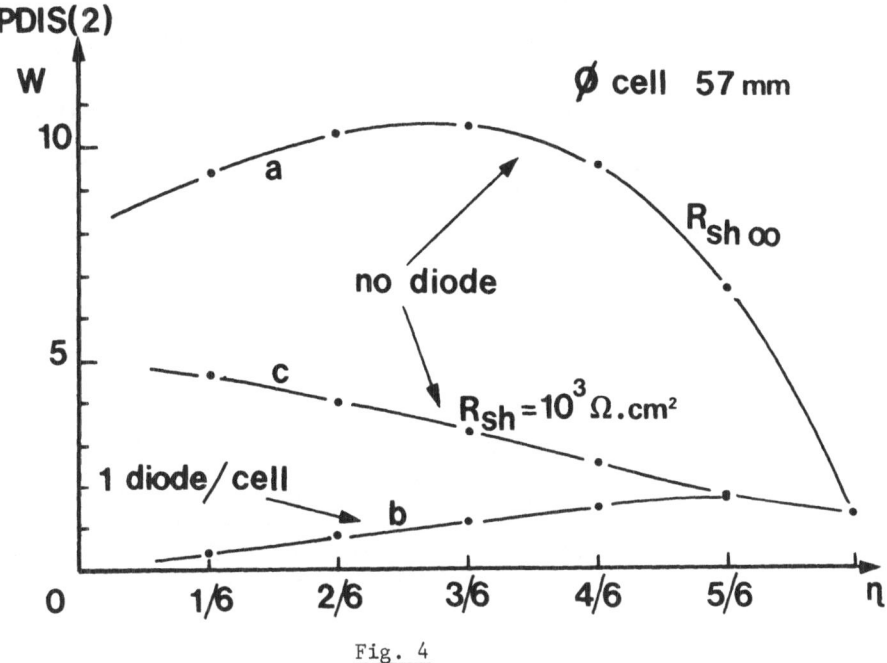

Fig. 4

As a consequence, a compromise would have to be found between the very small loss in power associated to a small shunt resistance, and the possibility of PV strings necessitating no by-pass diodes, the shunt resistance acting as a channel limiting the power dissipated in the shaded cell as numerically demonstrated in this paper.

Figure 5 has been obtained from set of curves such as 2, and gives, as a function of η, the value of NCPROT above which the by-pass diode is no more active ($I_D = 0$) : as an example, for a series of 204 cells (R_{sh} infin.) on batteries, a diode put on more than 40 cells will never pass any current whatever the shading conditions. The influence of low values of R_{sh} is again pointed out.

- 305 -

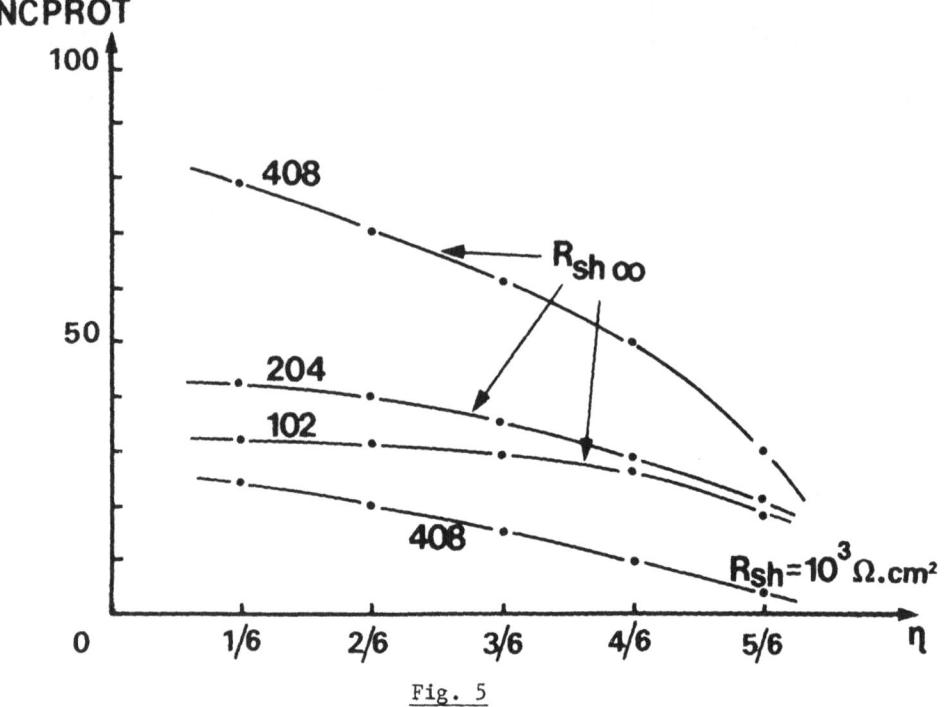

Fig. 5

In the low shunt resistance case, it can even be demonstrated that in some configurations, no current will pass through parallel diodes whatever the shading and loading conditions.

Finally it must be noted that the present research is confined strictly to panel made up of a single string of series connected solar cells. Similar experiments and the corresponding computation are now developped to describe series-parallel arrangements in high power PV arrays. A comparison between theory and the results that will be taken from the 5 kW PV array of the Refuge des Evettes in France, will permit a better understanding of the losses and failures in large systems. For instance, the net positive or negative role of equilibrium lines will be analyzed.

REFERENCES

1. J.A. ROGER and C. MAGUIN, Accepted for publication in the Jal of Solar Energy
2. CURLCANO, 14th IEEE PV Conference, San Diego 1981

MINIMUM COST OF PHOTOVOLTAIC ENERGY FOR A UTILITY GRID AND GENERAL FEATURES OF A GENERATING PLANT USING COSTLESS SOLAR CELLS.

Daniel MADET
Ingénieur à la Direction des Etudes et Recherches.
Electricité de France.

Summary

The purpose of this work is to evaluate the minimum long term cost of electricity produced by future photovoltaic plants connected to a utility grid. As the cost of photovoltaic cells is supposed to drop dramatically, our basic assumption is that, in the long run, it will be near zero.

Our approach consisted in imagining a very simple kind of photovoltaic plant based on this assumption, then in estimating the lowest value of the total cost per peak kW of the plant and finally the lowest cost of generated electricity.

A computer program has been worked out to optimize the various electrical, physical and economic characteristics of the plant.

According to various assumptions concerning the cell life time under such conditions, their efficiency and the improvement of inverters, the capital cost of each peak kW was found to range from 2200 FF to 5600 FF.

Therefore the hypothetical and ultimate kWh price would vary from 0,15 FF to 0,40 FF for the weather conditions typical of the South of France, and a little bit more than half those values in very sunny countries.

1. MAIN FEATURES OF A SOLAR PLANT WITH COSTLESS CELLS.

First let us stress the fact that if practical problems arise concerning the feasibility of the plant we designed, it will only mean that the evaluated cost of that plant was been underestimated. Therefore, the study would anyway provide a good evaluation of the lower bound for the cost of delivered energy.

The choice of the adequate size of the plant is a compromise between two antagonistic tendencies. On the one hand, the insertion in the grid as near as possible to the end-users is desirable since it eliminates the need for investments in lines and transformers. On the other hand, economies of scale are great, particularly as regards converters but they become much smaller beyond a peak power of one or several megawatts. Indeed, this represents the adequate size of a plant. It roughly corresponds to medium voltage feeders in the grid.

On the assumption that solar cells are costless, the costs of all the other parameters (land, civil works, encapsulation, supporting devices, cables and wires, inverters) as well as operating and maintenance costs must be as low as possible.

Many studies showed that the costs of supporting devices, even when the latter are fixed, were rather high. In order to avoid these costs, we decided to lay the modules flat on the ground.

Furthermore, with this solution the land is well used because of the absence of shade. On the other hand, the annual output of each module may be reduced, although our studies showed that if an appropriate site - such as a 10 to 15 % North South oriented slope -is used, this drawback disappears.

In order to reduce installation (and replacement) costs we imagined a special type of module and factory made arrays (including cables and wires), as shown in figure 1. As can be noticed, spaces between modules will enable the elimination of water and other materials and facilitate the adjustment to the ground.

Taken separately, the modules are rigid to avoid shade and strong wind-induced vibrations but the arrays are flexible. A cheap installation-method for the arrays is shown in Figure 2. Once installed, the arrays are fixed to the ground by steel ropes stretched between concrete blocks. Maintenance is limited to the replacement of defective arrays.

As far as current photovoltaic plants are concerned, the voltage level of modules (and of arrays, since modules are connected in parallel) is a compromise between the cost of wires, heat losses and insulating materials (and sometimes the converter cost). On the other hand, our future plant, with arrays laid flat on the ground, must have a low voltage level for obvious safety reasons. Thus the encapsulation materials will anyway be thick enough to provide adequate insulation.

Instead, attention must be paid to the progressive decrease in power of the modules, when an increasing number of cells run down while the module still works, as the array on the whole is in good condition (the higher the voltage level in the module, the higher the relative value of this decrease). A rough preliminary study showed that the voltage of each array should be a few tens of volts.

Thus, several arrays must be connected in series to form a group of arrays to obtain the adequate voltage level (Figure 3). Finally, numerous groups of arrays will be connected in parallel, thanks to two double buses, to form the plant. Power conditioning and control devices will be placed in a building at the junction of the two double buses. The general plant lay-out is given in Figure 4.

2. PROGRAM TO OPTIMIZE ELECTRICAL, PHYSICAL AND ECONOMIC CHARACTERISTICS

A computer program has been worked out to optimize the capital cost per each kWh produced. It first evaluates the total cost of a 1 MW peaking power plant : land, land preparation, cells, encapsulation, cables and wires (in modules, arrays and buses), manufacturing of factory-made arrays, installation, buses, civil works, conversion and control... This evaluation includes the levelized replacement costs of the short-lived parts of the plant (mainly arrays).

Then the program evaluates the energy output of the plant, taking account of heat losses in modules, arrays and buses, losses in converters and progressive decrease in power of the modules (before the entire array is replaced). Of course, the levelized cost and the amount of energy produced depend on the way the cells are arranged in the modules, on the number of modules in an array, of arrays in a group and of groups of arrays in the plant, on the size of the modules and on the sections of connecting wires and buses. Finally the computer can calculate the values of all these variables which minimize the cost of the energy produced.

3. APPLICATION OF THE OPTIMIZATION PROGRAM TO VARIOUS CASES

The results of this optimization program are dependant upon external economic parameters (land cost, copper cost) and technical parameters (efficiency of the inventer, life time of the modules...). About one hundred sets of parameter values were successively used in the program. Since the optimization process would not work with strictly costless cells, a minimum value of 30 FF/m2 was used, which corresponds to the minimum cost for depositing a few microns of costless materials with widespread technologies. The results roughly showed that technical parameters influence technical results whereas economic parameters influence economic results.

Figure 5 gives the capital cost per peak kW for several cell efficiencies (8 to 14 %) and considering two sets of values (optimistic and more realistic) for other parameters. The same figure shows how the various losses increase the cost per kW. On the righ hand side, the cost of one kWh of output is shown, assuming an average insolation of 1500 kWh/m2/y and using EDF's economic computation rules.

Figure 6 shows the sensitivity of the total capital cost to changes in the cost of modules : according to what we previously thought, we found out that if we raise the cost of the cells, the cost of the other components rises too.

4. EVALUATION OF THE MINIMUM LONG-TERM COST OF PHOTOVOLTAIC ENERGY IN A UTILITY GRID

The main characteristics of a photovoltaic plant equipped with very cheap cells were listed above :
- modules are laid flat on the ground
- low voltages are used
- defective cells and modules are not repaired.

The results of the optimization study of the electric and geometric values of that ideal plant, give a good, though very general, idea of the lowest value of the capital cost per peak kW. Even with very optimistic assumptions, this lowest value would exceed 2000 FF/kW.

The corresponding evaluation for average energy costs based on EDF's economic computation rules and the insolation of the South of France reveals that, even with costless cells, the energy produced by such a plant will not be free (more than 15 cF/kWh anyway). Therefore large scale production for the grid is likely to be uneconomical in France, though solar cells are useful for specific uses in remote places not connected to the grid.

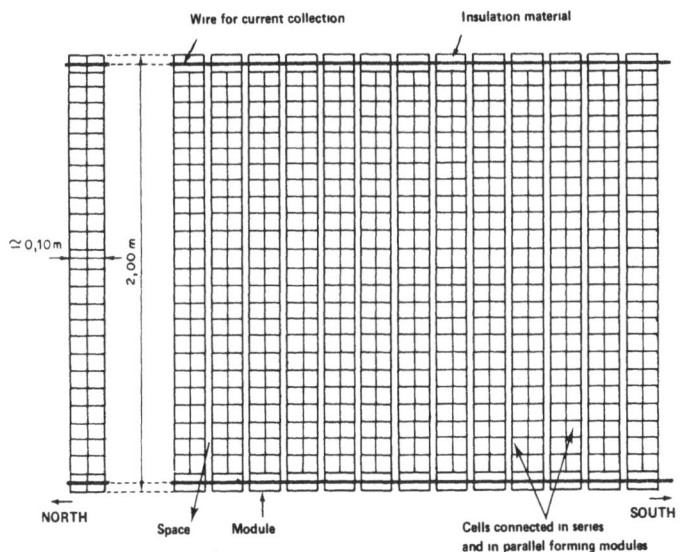

Fig 1   Partial view of an array several dozen meters long

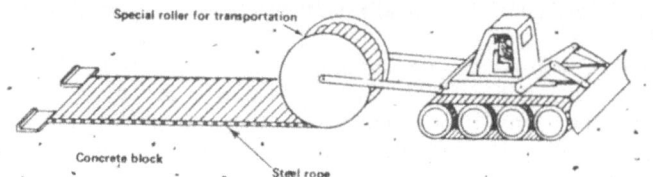

Fig 2 Possible way to install an array

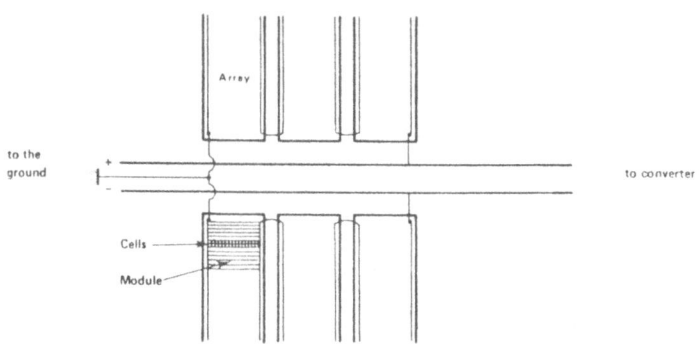

Fig 3 Several arrays, connected in series, form a group of arrays

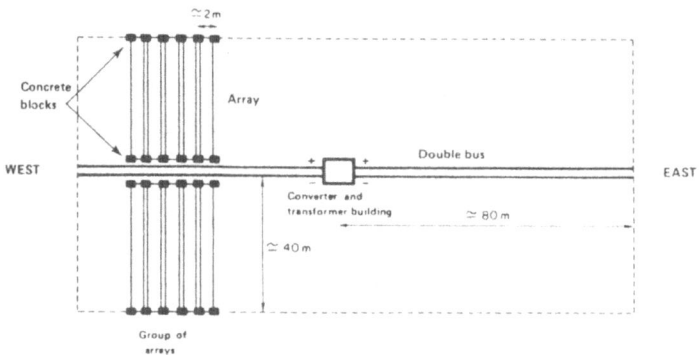

Fig 4 Many groups of arrays, connected in parallel, form the plant

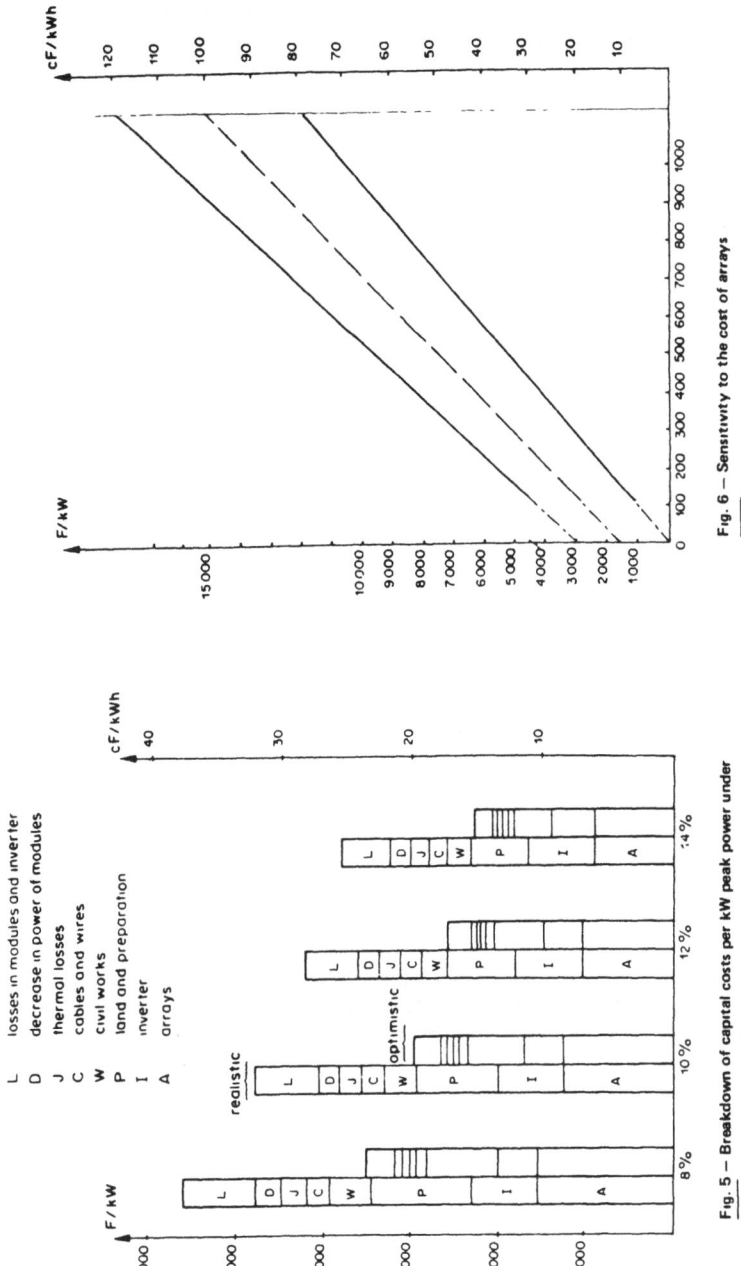

Fig. 6 — Sensitivity to the cost of arrays

Fig. 5 — Breakdown of capital costs per kW peak power under various assumptions

L losses in modules and inverter
D decrease in power of modules
J thermal losses
C cables and wires
W civil works
P land and preparation
I inverter
A arrays

A DC/AC MODULAR INTERFACE FOR PHOTOVOLTAIC SYSTEMS

M. VAN GYSEL - I.D.E. - and N. LIMBOURG - E.T.C.A. - BELGIUM

Summary

This development was supported by the Ministère de la Région Wallonne and the C.E.C. as an R & D contract.
It aims to the industrial design and prototype testing of a power conditioning interface specific for PV systems. The techniques chosen include modular design, microprocessor control, PWM transistor inverter with stored wave. The interface is able to convert AC/DC, synchronize with a grid, control a battery, and perform maximum power tracking. The module is 10 KW (15 KVA), 220 V DC input, regulated 220-380 3 phase output, and has an efficiency of 90 to 94% in the 10-100% load range, with a total harmonic distorsion less than 3%.

1. INTRODUCTION

The production of AC power from photovoltaic intermediate size power stations requires the development of specific interfaces.
The requirements are :
- high efficiency
- modularity so as to accomodate various sizes of application, and eventually allow up-scaling
- ability to parallel with an existing grid (including meeting the grid quality requirements)
- ability to be combined with batteries.

From those requirements, I.D.E. and E.T.C.A. worked out a development program, which was funded by the Ministère de la Région Wallonne, and the Commission of the European Communities.

2. FEATURES OF THE PRODUCT

Functions :
- DC/AC inverter
- Modular. Modules can be connected in parallel.
- Modes : a stand alone with battery
          b grid connected with battery
          c grid connected without battery
- Maximum power point tracking in mode c.
- Battery management (over-and undercharge protection)
- Synchronization with grid and other modules.
- Self-protection.

Technique :
- Power Wave Modulation.
- Self commutated.
- Hexfet transistors.

- Microprocessor drive (stored wave).

Characteristics :
- Input 220 V nominal (380 V open circuit).
- Output 220 V/380 V three-phase.
- Nominal power 10 KW (15 KVA)
- 50 Hz frequency.
- Voltage variation $\pm$ 1%
- Frequency variation $\pm$ 2‰
- Total harmonic distorsion < 3% (see fig. 1).
- Cos ∅ min 0,67.
- Tolerates 30 KVA transients.
- Efficiency : see fig. 2.

3. DESIGN

Figure 3 shows the general block diagram.
The hardware comprises the control, interface and power modules, the auxiliary power supply and the transformer.

Control module :
The operation is controled by two microprocessors.
The first microprocessor performs the data acquisition by sampling the output voltages and currents on each phase. It compares the ratio of the output voltage to the required voltage and sends this information to the second microprocessor. It determines the amplitude and phase with respect to the network so as to inject power.
It performs maximum power point tracking and battery management.

The second microprocessor generates modulated pulses to drive the power cells, so as to synthetise a sinusoidal wave.

Interface modules :
The power modules are driven through opto-couplers, to avoid problems due to high voltage differences between the driver circuits and to the high speed of commutation.
They include a watchdog facility which allows the processor to detect a failure and turn off the power.

Power modules :
Figure 4 shows a power module (one phase). The transistors used are of the hexfet technology.

Auxiliary power modules :
Each interface and power module needs a seperate, floating power module.
A multiple output flyback converter was chosen, with two hexfet high voltage transistors.
The mean consumption is 10 W for all those modules.

Transformer :
The transformer size is 15 KVA and it has an efficiency of 98%.

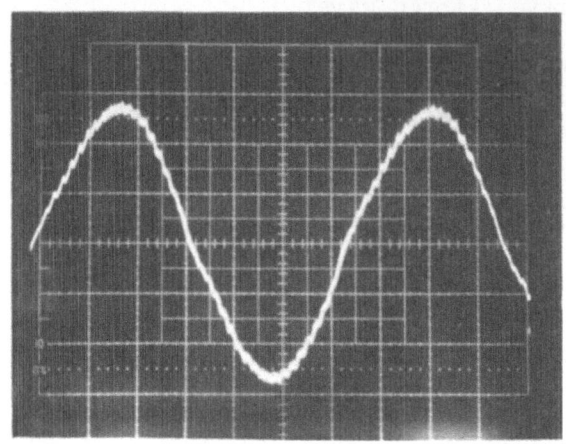

Figure 1 OUTPUT WAVEFORM

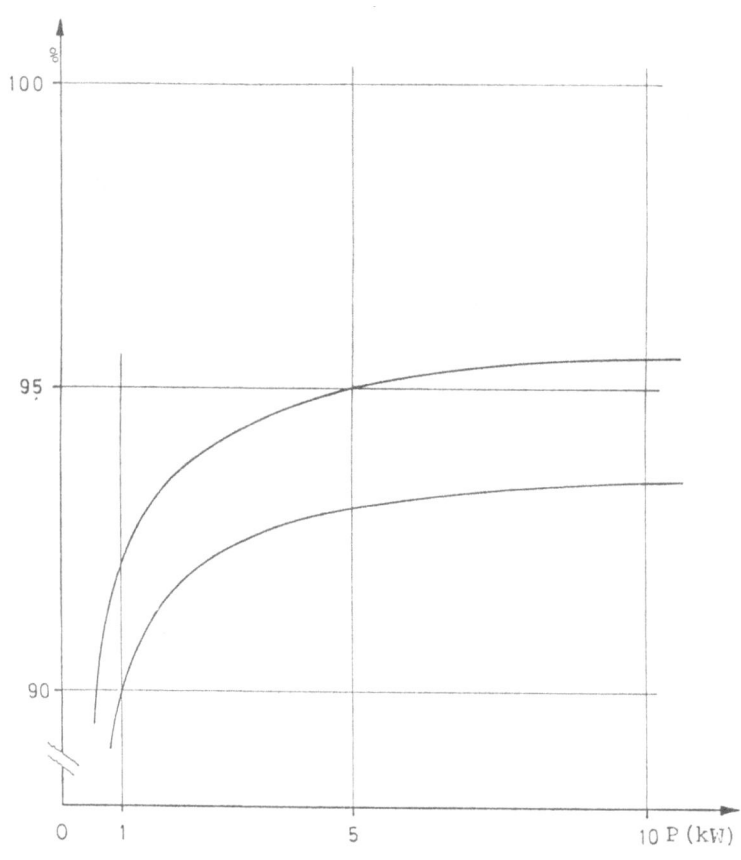

Figure 2 EFFICIENCY WITHCUT TRANSFORMER (98%)

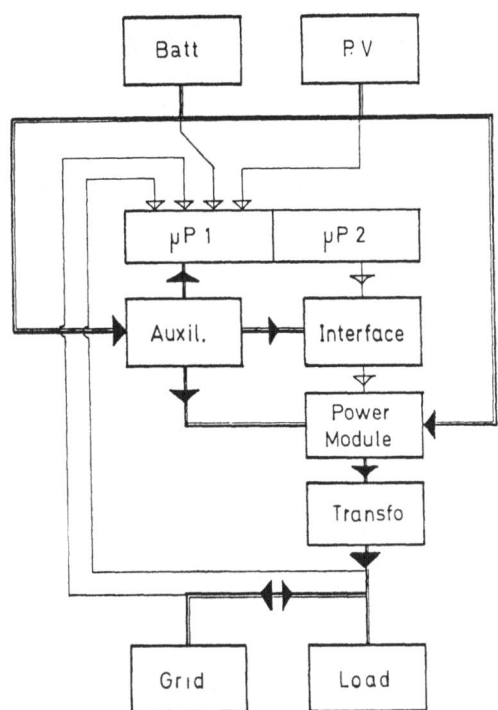

Figure 3   GENERAL BLOCK DIAGRAM

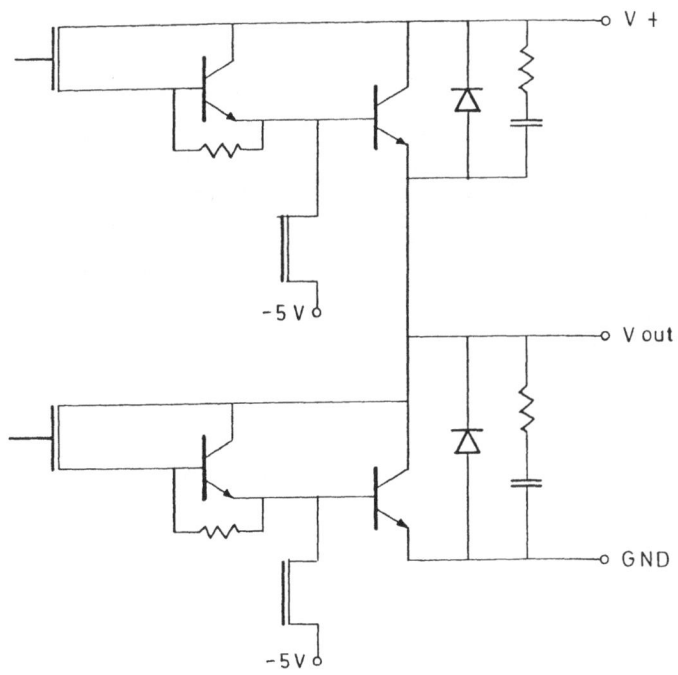

Figure 4   POWER MODULE

SIMPLE TRANSFORMERLESS INVERTER WITH AUTOMATIC GRID-TRACKING AND NEGLIGIBLE HARMONIC CONTENT FOR UTILITY INTERACTIVE PHOTOVOLTAIC SYSTEMS

J. Schmid and R. Schätzle
Fraunhofer-Institut für Solare Energiesysteme
Oltmannsstr. 22, D-7800 Freiburg, W-Germany

Summary

For commercially competitive photovoltaic residential systems the costs of all components have to be reduced significantly. One of the most costly components is the DC-AC-Converter satisfying all the necessary safety requirements. Under these boundary conditions, the FhG developed a converter which is especially designed for photovoltaic application. The new converter has an excellent efficiency over a wide power range and it satisfies all necessary safety requirements without additional components. Because it contains only electronic components, further cost reduction can be expected.

1. INTRODUCTION

If solar cell prices decrease further, the cost of the additional photovoltaic-system components becames more important for the economics of the whole plant. In utility interactive residential systems, it is mainly the cost of the DC-AC-Converter which must be reduced drastically. On the other hand, high efficiency and high reliability are expected. Cost reduction of components like transformers or choke coils cannot be expected.
Functional description: With these conditions in mind the FhG developed a new type of a DC-AC-Converter by using fully electronic components. Its basic principle can be described as follows:
Different solar cell arrays are switched in series in such a way, that the sinusoidal AC-voltage is approximated by fine steps. To keep the number of transistor switches low, the voltages of the solar cell-arrays are different. For the 200 Watt prototype, four solar cells are used with voltages of 22, 42, 84, and 168 Volts resp.. They are switched by the four corresponding transistor stages in a binary mode (see fig. 1). In this way the amplitude of the AC-output can be devided into 16 steps (4 bits). If further refinement is necessary, a subdivision into 32 steps can be performed by one further stage. The binary control for the switching transistors is carried out in an extremely simple way by using a commercially available Analog-Digital-Converter with binary output. This ADC converts the measured line voltage to corresponding binary signals, which are used by transistor switches to give a corresponding AC-output voltage. This simple combination performs the synchronization of the converter output to the grid automatically, without any time delay. Fig.2 shows how the stepped wave output is synchronized to the grid's voltage.

2. SAFETY FEATURES

All necessary safety functions are included without any additional components (see fig. 3). The under-voltage relay function is performed by the diode between the transistor switches and the bridge. If the solar cell-voltage is lower than the grid-voltage, no power can flow back into the converter.

The over-voltage relay is not necessary because the maximum voltage is determined by the solar cell voltage, which lies in the accepted range. Inverters which contain coils can produce significant overvoltages when not protected.

The overcurrent relay is used when an overvoltage at inverter output or an undervoltage (short circuit) in the utility line occurs. The overvoltage is not possible by definition, and in the case of utility undervoltage, the converter's ADC will sense the lower voltage and its output voltage will follow these changes without any production of overcurrent.

Probably the most important safety requirement is the self shut-off capability, in case the utility line shuts off for maintenance reasons. Otherwise the maintenance personell could be endangered by further power input of running converters. This requirement is fulfilled by the design in such a way that the ADC will sense a Zero-Voltage - or at least a lower voltage - and reduce its output to zero.

3. EFFICIENCY

The no-load loss of the 200 Watt prototype is 1 Watt. The efficiency at 200 Watt has been measured to be 94%.

4. SOLAR ARRAYS

The four solar cell arrays differ in voltage as well as in current because of the sinusoidal current-output. The solar cells have to be connected in parallel in such a way that the current ratio
$I_1 : I_2 : I_3 : I_4 : = 6 : 7 : 8 : 9$ is obtained. Because of the large number of solar cells used in this application, this subdivision is not a problem. To gain good efficiency, the power produced between discharge pulses has to be stored. For this purpose, capacitors can be used as well as batteries. If batteries are used, the pulse load does not decrease its lifetime because this short-term storage is performed by its double-layer capacity, which is in the order of some hundred Farad! Batteries are of advantage for the grid-voltage quality, when short-term clouding is changing the photovoltaic plant's output.

5. CONCLUSIONS

A DC-AC-Converter has been developed, which uses the modular features of solar cells for a simple and reliable DC-AC-conversion. All required safety features are contained in the basic system without additional components. Further cost reductions of the fully electronic components can be expected.

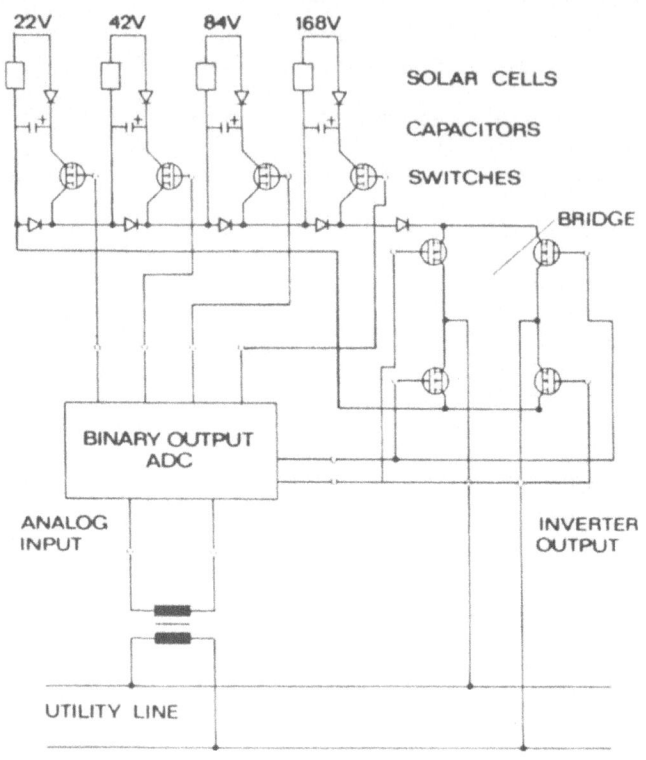

Fig. 1 Simplified circuit diagram of AC-DC-Converter

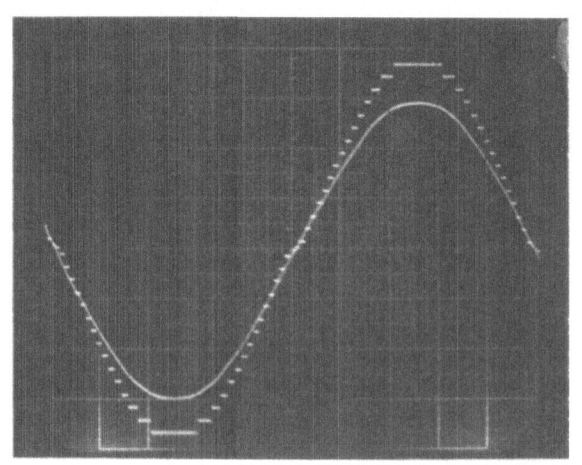

Fig. 2
Stepped sine-wave at inverter output compared to utility-voltage (smooth line)

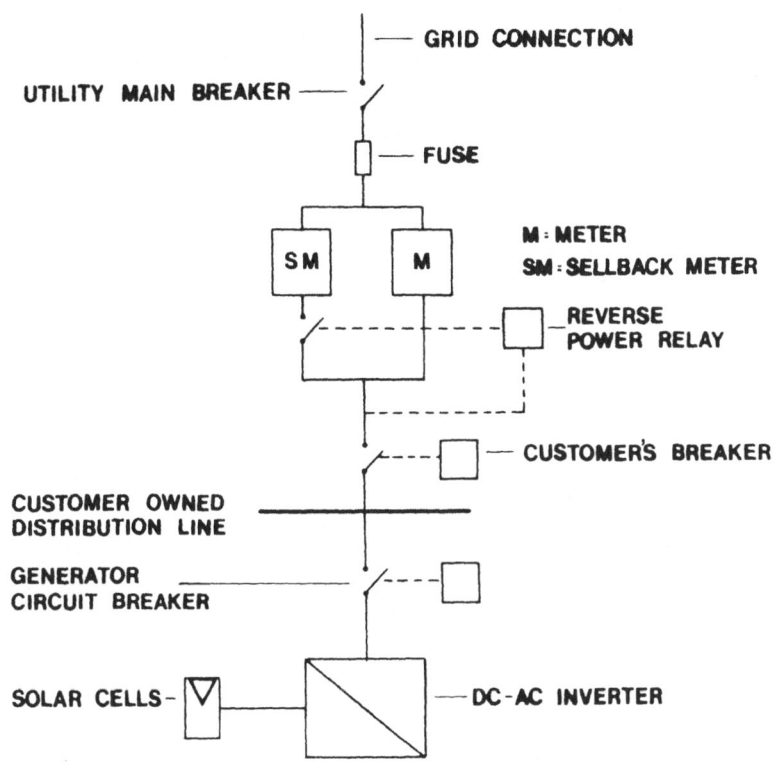

Fig. 3 Integration of the FhG-Inverter in a residential photovoltaic power plant

REGULATED CONVERTER CIRCUIT FOR DIRECT PHOTOVOLTAIC
ENERGY FEEDBACK INTO THE POWER GRID

P. CEPPI, R. ULMI, G. GUEKOS
Institute of Applied Physics
Swiss Federal Institute of Technology, CH-8093 Zurich, Switzerland

Summary

A new regulated utility interactive (UI) inverter for direct photovoltaic energy feedback to the mains has been realized. The inverter acts as a current source, its output is a sinewave with negligible harmonic distortion and has a projected efficiency of over 90%. A novel maximum power tracking (MPT) strategy based on a simple current peak error detection is described.

1. INTRODUCTION

Utility interactive photovoltaic systems have the advantage, with regard to the systems using batteries, of overcoming the problem of energy storage by injecting the photovoltaic power into the utility grid. The feedback of the power into the grid is accomplished by means of an electronic inverter which matches the electrical characteristics of the solar array to those of the grid [1,2]. An important prerequisite for the implementation of such an inverter is the generation of sinusoidal current waveform with low harmonic content [3,4,5]. This can be achieved through various means, such as digital waveform synthesis [6] and multiple step sinewave approximation [7,8]. Another possibility is the application of pulse width modulation (PWM) [9]. We present a new type of utility interactive inverter using pulse width modulation and a novel method for the tracking of the maximum array power. This method is based on the continuous control of the waveform of the output current and allows the minimization of the harmonic distortion.

2. UTILITY INTERACTIVE INVERTER

Since the inverter is an interface between solar array and mains, it should meet the requirements of both energy source and load.

In utility interactive (UI) inverters the load (utility line) determines the inverter output voltage. The utility connection generally represents a very low impedance. Therefore the inverter should have a current source like output (high impedance). This can be realized by suitable converter topologies allowing high output impedances and with additional feedback loops.

A straightforward method for the control of the power flow into the power grid consists of varying the amplitude of the inverter output current. Moreover, the phase between voltage and current at the output terminals regulates the reactive component of the power vector.

Low harmonic distortion of the injected current is very important for UI inverters.

Power conditioning units (PCU) for UI photovoltaic systems must further meet the requirements of the solar array, such as variable operating voltage and available power.

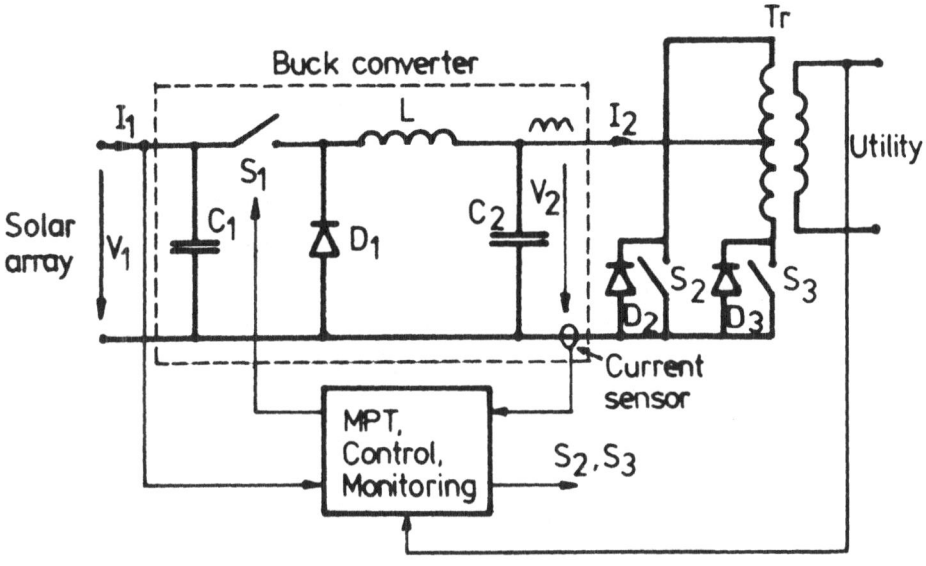

Figure 1: Schematic circuit of the utility interactive inverter

The inverter circuit shown in Fig.1 has been designed to meet all the aforementioned requirements. It consists of a high frequency (20KHz) chopped buck converter which feeds a current I_2 to a center tapped transformer. The amplitude of I_2 varies proportionally to the maximum power available from the photovoltaic array.

The self commutated half bridge circuit (transformer Tr and switches S_2, S_3) inverts the full wave rectified sinusoidal current I_2 and supplies it to the utility. Transformer coupling to the line was chosen for safety reasons. The voltage ratio of the transformer is determined from the minimum working voltage at the inverter input. Current waveshaping was accomplished through pulse width modulation at switch S_1. The buffer inductor L and the current control loop result to a high output impedance of the inverter.

Capacitor C_2 is used to minimize high-frequency ripple on the output current. However, it should be kept as small as possible to prevent low frequency harmonic distortion and unacceptable power dissipation due to reactive current.

The circuit works at unity power factor. Its nominal power is 1.5KW. S_1 was realized with power field effect transistors (FET) allowing fast switching times and low driving power consumption. Bipolar transistors were used for the switches S_2 and S_3.

Preliminary laboratory measurements, which were performed at 10% full load, show an overall inverter efficiency of 85%. The projected efficiency at nominal power is higher than 90%, the transformer being responsible for the main part of the losses.

3. MAXIMUM POWER TRACKING

Maximum power tracking (MPT) was realized with the simple current control loop shown in Fig.2.

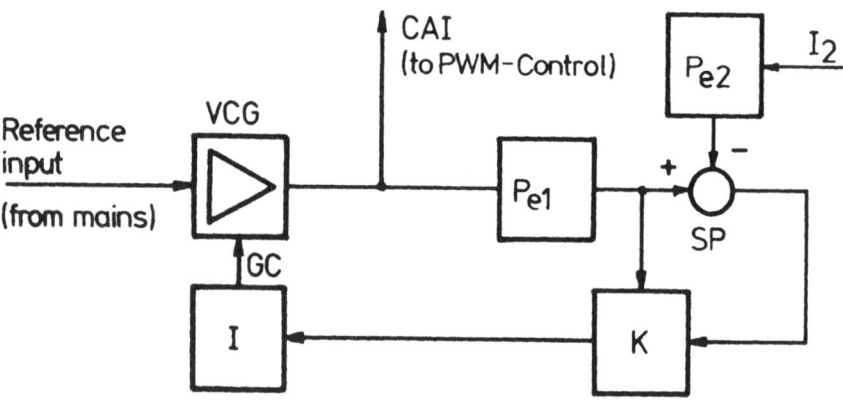

Figure 2: Schematic diagram of the MPT-Control

VCG: Voltage Controlled Gain-Amplifier, GC: Gain Control,
I: Integrator, CAI: Current Amplitude Input.

The MPT circuit has the following functions:
1) variation of the current amplitude I_2
2) detection and tracking of the maximum power to be delivered to the mains.

Function 1 is performed by a simple voltage-controlled-gain amplifier (VCG). Detection of the maximum power is achieved with the peak detectors Pe1 and Pe2, the summation point SP and the voltage comparator K (see Fig.2).

4. DESCRIPTION OF OPERATION

The output signal of the VCG-amplifier is increased as long as the amplitude difference ΔI between CAI and I_2 remains below some prescribed threshold value.

Once the maximum power of the solar array is reached, further increase of CAI causes ΔI to cross the threshold value thus resulting in a distortion of the output current waveform at the region of its maximum amplitude (Fig.3). CAI must now be reduced until ΔI decreases below that threshold.

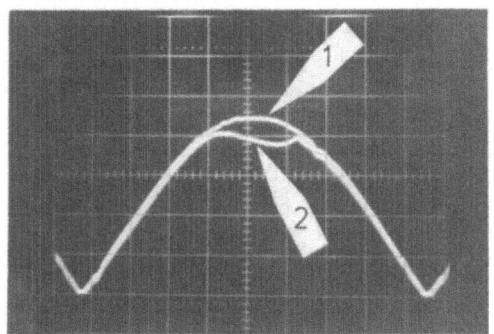

Figure 3: Waveform of CAI and I_2 signals
Excessive increase of CAI (1) causes distortion of the output current waveform (2).

Continuous control of ΔI allows the detection and tracking of the maximum available array power, independently of environmental changes such as irradiation and array temperature.

5. CONCLUSION

We have demonstrated a new regulated utility interactive converter for photovoltaic power systems in the kW range. The inverter acts as a current source, uses pulse width modulation and tracks the maximum power of the photovoltaic array. Maximum power tracking (MPT) is accomplished through a feedback loop which increases continuously the power flowing from the array into the line until a distortion in the sinusoidal shape of the output current is detected. This method enables automatically a minimization of the harmonic distortion of the current injected into the grid. Preliminary laboratory tests of an inverter for 1.5kW power showed an 85% overall efficiency at 10% load, the projected efficiency at full load being well over 90%.

ACKNOWLEDGMENTS

The authors would like to thank Prof. H. Melchior of the ETH-Zurich for helpfull suggestions and encouragement.

Support by the Swiss National Energy Research Foundation (NEFF) is acknowledged.

REFERENCES

[1] P.Wood: "DC-AC inversion for utility power systems", Int.Semicond. Power Conversion, 1977, pp.453-460.
[2] R.Ulmi, P.Ceppi, G.Guekos: "Methoden zur Uebertragung der photovoltaischen Solarenergie zum Verbraucher, inklusive 220V-Netz", Informationstagung über photovoltaische Energieumwandlung, 15.April 1982, ETH-Zürich, pp.35-46.
[3] E.E.Landsman: "Static inverters for use in PV power systems", Intersociety Energy Conversion Conference, 1979, pp.239-244.
[4] J.Jalade et al.: "New dc/ac high power cell structure improve performances for sine generator", IEEE Power Electronic Specialists Conference, 1980, pp.326-331.
[5] M.F.Schlecht: "A line interfaced inverter with active control of the output current waveform", IEEE Power Electronic Specialists Conference, 1980, pp.234-241.
[6] R.L.Pickrell et al.: "An inverter/controller subsystem optimized for photovoltaic applications", IEEE Photovoltaic Specialists Conference, 1978, pp.984-991.
[7] D.J.Roesler: "A 60kW solar cell power system with peak power tracking and utility interface", IEEE Photovoltaic Specialists Conference, 1978, pp.978-983.
[8] L.R.Suelzle et al.: "Power conditioning for large photovoltaic systems with battery storage", DECC Technical note DECC-79-101, May 1979.
[9] R.Ruble: "A new technique for sine wave synthesis inverter design", 6th nat.Solid State Conv.Conference, 1979, pp.F4.1-F4.7.

POWER CONDITIONING IN SOLAR
PHOTOVOLTAIC ARRAY APPLICATIONS

G. J. VACHTSEVANOS, C.K. KALAITZAKIS AND E.J.GRIMBAS
School of Engineering
Democritus University of Thrace
Xanthi, Greece

Summary

The design and implementation of appropriate power conditioning apparatus for the interconnected operation of PV arrays with the utility grid are presented. Problems of maximum power transfer from the PV array to the utility grid, reliability of the interconnected system operation and component protection are addressed. Because of the intermittent nature of the available solar energy, the need for conversion of dc electrical power to ac and the fluctuating loads of the utility system, optimization methodologies and associated equipment are developed for converting the maximum available solar power to electrical form and transfering it to the grid. A dual methodology is adopted containing a theoretical - system simulation - study of the performance of the interconnected system and a parallel experimental facility for the implementation of the system design. Results indicate higher energy transfer efficiencies in combination with an increased quality and reliability of performance of the PV array - utility grid system.

1. INTRODUCTION

In recent years there has been growing interest in utilizing PV arrays to provide some of the electrical demand of small autonomous power utility systems where the cost of electricity production using conventional fuels is becoming prohibitive. Such systems may be interfaced with the existing power grid for " fuel displacement " purposes as well as for earning some " capacity credit ".

Grid interface problems associated with the parallel operation of PV arrays with the utility grid have been studied recently [1,2]. The technoeconomic requirements for the optimum and reliable operation of PV arrays interconnected with the power grid include (a) a high degree of efficiency for the PV array system for the whole region of its power output operation; (b) a grid capability of absorbing the maximum power produced by the array: (c) the electrical power must be introduced into the grid without distortion, and (d) the operation of the PV array system should be reliable with adequate protection and minimum possible maintenance requirements.

The interconnected operation requires the development of appropriate interfacing equipment for purposes not only of quality of operation but also of transfering maximum power from the PV generator to the load or grid (3). For these reasons, an interface - matching methodology and associated implementation devices are proposed utilizing inexpensive components and addressing a wide range of operational requirements depending upon the specific application under consideration. Basic design goals included maximum economy in system components as well as the possibility of operation under variable input - output conditions.

2. SYSTEM DESCRIPTION

A PV array may be considered as a dc source of power with strongly nonlinear voltage - current characteristics. Such typical characteristic curves are shown in Figure 1(a).

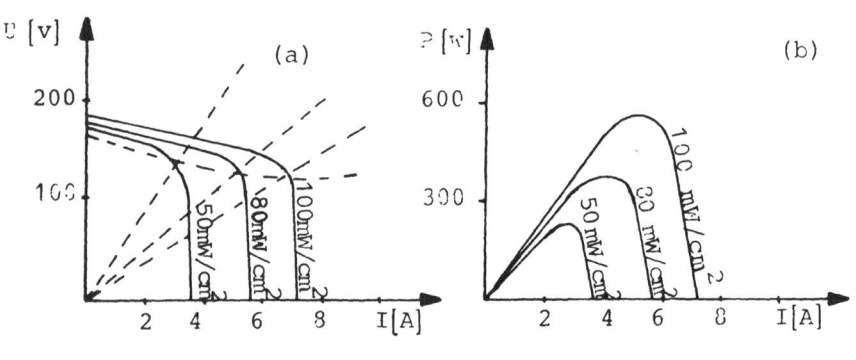

FIGURE 1

It is observed that the array voltage drops almost linearly as the current increases up to a certain value. From that point on, an abrupt change in the characteristics occurs. It is also noteworthy that the general shape of the v-i characteristics remains essentially unchanged for various values of incident solar radiation, H. Figure 1(b) shows that the array power output, P, as a function of the current, attains a maximum value near the knee of the characteristic curve. For varying solar intensity, the maximum output power is delivered to a load whose value places the operating point in the region of the knee of the characteristic curve. Since this point is constantly changing as a function of the solar intensity, a matching circuit must be interposed between the PV array and the grid which will guarantee maximum power to be transfered from the generator to the power lines, at each instant of time. Moreover, the constant (dc) form of the voltage waveform at the array output terminals imposes an additional requirement upon the interconnecting mechanism: The power output must be synchronized to the ac constant frequency characteristics of the utility mains. This goal is achieved with a new type synchronous inverter (4); it is based upon a four thyristor bridge designed so that an external voltage signal may be used to vary the thyristor firing angle and, therefore, the amount of electrical power being transfered from the PV generator to the utility grid. Variation of the firing angle accomplishes the dynamic matching required for maximum power transfer between the input characteristics of the inverter (dc voltage) and its output characteristics (power grid). Figure 2 shows a conceptual block diagram of the proposed scheme.

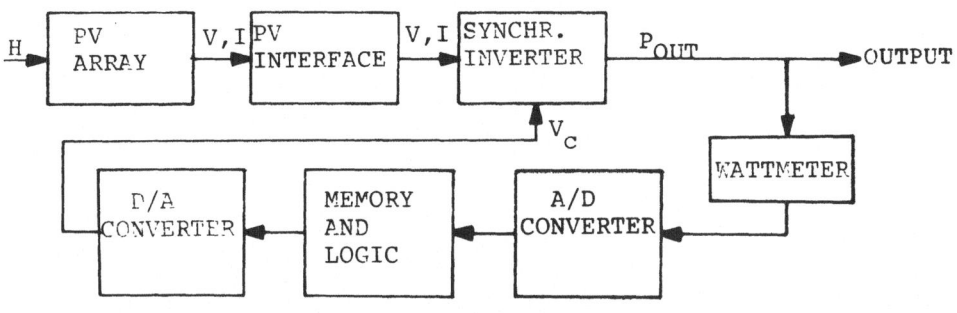

FIGURE 2

For any solar intensity H, the array output feeds into the inverter via a PV interface circuit which assists in improving signal quality while maintaining an approximate "static" matching between the generator and load impedance characteristics. The values of the L,C components are design parameters of the optimization procedure.

The feedback circuit includes the following subsystems:

1. An electronic wattmeter that continuously measures the power level at the utility grid terminals and provides a signal output proportional to actual power.

2. An A/D converter which converts the analog signal into digital form.

3. The "Memory and Logic" circuit is effectively a digital power level gradient detector which samples the wattmeter signal output and holds it for comparison with the next sample. The detector includes a comparator that works in combination with a logic circuit to determine if a given sample represents a power level that is greater or smaller than the previous sample. The logic circuit changes state whenever a new sample is smaller than the preceeding one, but remains in the same state if the new sample is larger than the preceeding, thus representing an increase in power level.

4. A D/A converter provides a constantly changing analog output whose direction of change is increasing for one state of the logic circuit and decreasing for the other. This control signal is used to fire the inverter thyristors, thus controlling the power level transfered to the grid.

The design approach used has both a theoretical and an experimental component. The theoretical study consists mainly of computer simulations of the steady - state behavior of the interconnected system while the experimental method involves testing of a particular PV conversion system interfaced to the power grid. Experimental results are compared, on a step by step basis, to those obtained from the simulation studies. This dual nature of the approach allows for an optimum design of system components and verification of theoretical results.

The photovoltaic array characteristics are modelled using a series approximation of the form:

$$V = c_1 I - c_2 I^{c_3} \tag{1}$$

For any given grid impedance, $Z \angle \varphi$, at the point of interconnection, the mathematical model for the filter - inverter system relates the inverter average power output, P_{out}, to the dc input voltage level, V, and the control signal V_c. This relation may be represented, functionally, by

$$P_{out} = g(V, V_c, Z \angle \varphi) \tag{2}$$

The power grid is viewed, at the point of interconnection, as a voltage source of fixed amplitude in series with an impedance $Z \varphi$. Protective devices are employed for the safe operation of the PV array - grid interconnected system. The nature of the overall system design prevents a bidirectional power flow thus elliminating the possible negative consequences of such an operation. Their mathematical representation employs simple magnitude constraints on certain variables which are easily implementable during the simulation runs. Finally, the model describing the operation of the maximum power tracker is represented by a set of switching state equations describing the state of the control variable V_c, at each instant of time, depending upon the gradient of the power at the grid terminals, P_{out}.

The computer simulation establishes initially an equilibrium state between the inverter input - output characteristics, for a step change in solar radiation intensity, assuming an arbitrary initial condition for the control variable V_c and a constant value for the grid impedance $Z \angle \varphi$. Next, the control signal V_c is given a new value and the equilibrium procedure is repeated. The two consequtive values for the power output, P_{out}, are used as inputs to the maximum power tracker model whose goal is to set a new value for the control voltage V_c which increases the power output. The previous computational steps are repeated until the power output, P_{out}, oscillates about a maximum value at which instant the program is terminated.

3. RESULTS

Figure 3 shows a typical power output vs. solar radiation intensity characteristic with and without the maximum power tracker in operation. A typical photovoltaic array with characteristics as shown in Figure 1(a) has been used and the experimental results are in agreement with those obtained from the computer simulation runs. Design parameters were chosen to optimize system performance with the assistance of sensitivity analysis techniques using the computer model. It is observed that the power transfer efficiency improves by as much as 33% with the maximum power tracking mechanism in operation.

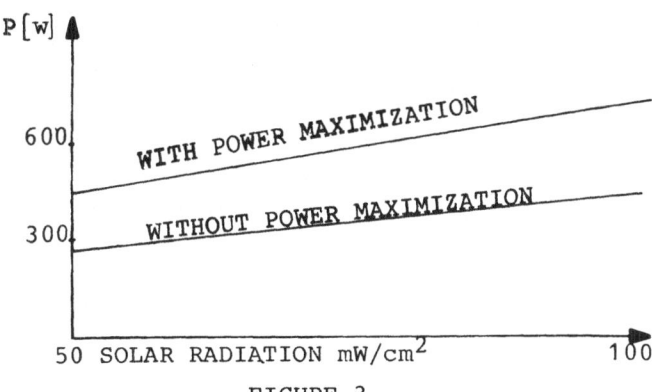

FIGURE 3

Total harmonic distortion introduced by the inverter operation is of the order of 1.5%. The safety equipment employed protect effectively both the PV array and the power grid from various fault conditions.

Both impedance matching and power tracking result in maximizing the energy transfered from the PV array to the grid. Thus, an optimum operation, from a technical and economic standpoint, is achieved for a large range of solar radiation variations.

Finally, the low cost of components (inverter - tracker) and maintenance, adds to the economic attractiveness of the proposed scheme.

4. REFERENCES

1. R. Corbefin and G. Vacelet, "Photovoltaic Power Generation System Interconnected with Mains", Proc. of Third E.C. Photovoltaic Solar Energy Conf., pp. 1038-1040, October 1980.

2. G. Beghin and V.T. Nguyen Phouoc, "Power Conditioning Unit for Photovoltaic Power Systems", Proc. of Third E.C. Photovoltaic Solar Energy Conf., pp. 1041-1045, October 1980.

3. Z. Zinger and A. Braunstein, "Dynamic Matching of a Solar - Electrical (Photovoltaic) System An Estimation of the Minimum Requirements on the Matching System", IEEE Trans. on Power Apparatus and Systems, Vol. PAS -100, No.3, pp. 1189-1192 March 1981.

4. G.J. Vachtsevanos and K.C. Kalaitzakis, "A Synchronous Inverter for Wind Energy Conversion Systems", Proc. of MELECON '81 Conf., 5.4.3 pp. 1-3, 1981.

MISMATCH BETWEEN BATTERIES AND TWO MODULE TYPES P.V. ARRAYS

Interest of D.C.-D.C. Converters

P. GUCHER, J.A. ROGER, S. MASSAAD, J. POSBIC, J. PIVOT

Service Electronique - Institut de Chimie et Physique Industrielles de Lyon, 31 place Bellecour. 69288 LYON Cedex 02 ;

Département de Physique des Matériaux (Associé au C.N.R.S.), Université de Lyon 1, 43 bd du 11 nov., 69622 VILLEURBANNE Cedex.

Summary

This paper concerns the study of mismatch effects between P.V. arrays and batteries. Due to the modifications of the experimental conditions (depending on the batteries state of charge, the external temperature and light flux), it is not possible to extract at each time the maximum available power from a P.V. array. Similar problems result from the parallel use in the same array of different modules, mismatch effects being enhenced by the fact that different encapsulation lead to different equilibrium for the junctions. The relative power losses in both cases have been analyzed on a small system made of two kinds of modules. Extrapolation to high power P.V. arrays has been done and a D.C.-D.C. converter for 5 kW P.V. arrays is presented. The interest of converters in the case of large generators coupled to batteries is discussed and some applications are proposed.

1 - INTRODUCTION

When designing P.V. arrays connected to batteries, the junction temperature evolution, which depends on the outside conditions must be taken into account : in order to obtain good conditions for charging the batteries whatever the respective temperatures, light flux, and state of charge, it is necessary to optimize the system for the highest expected working temperature, which involves losses for the other operating conditions. As an example, in the case of the French Alpine Refuge "Les Evettes" (5 kw P.V. arrays (1)) the junction temperature varies from 0°C to 40°C. The maximum power point for the array shifts then from 48 V to 85V, since the batteries stay between 48V and 56V. As a consequence, strong mismatch effects exists between the modules and the batteries. Today large P.V. arrays (10-300 kw) are being installed with modules from different origins. These modules are connected in parallel on the same storage in a given installation. The differences in cells process and in the encapsulation materials induce different I-V characteristics as well as different operating cell temperatures. For security the choice of the storage elements number is imposed by the modules presenting the highest temperature. Power losses on the other parts of the generator result.

2 - COUPLING BETWEEN P.V. ARRAYS AND BATTERIES

Experiments have been performed using two kinds of modules with different encapsulation structures : type A is a glass -glass encapsulation and type B uses an organic material rear face. Furthermore, the cell's processing being different, the I.V characteristic are not similar concerning the short circuit currents for the same temperature and light flux.

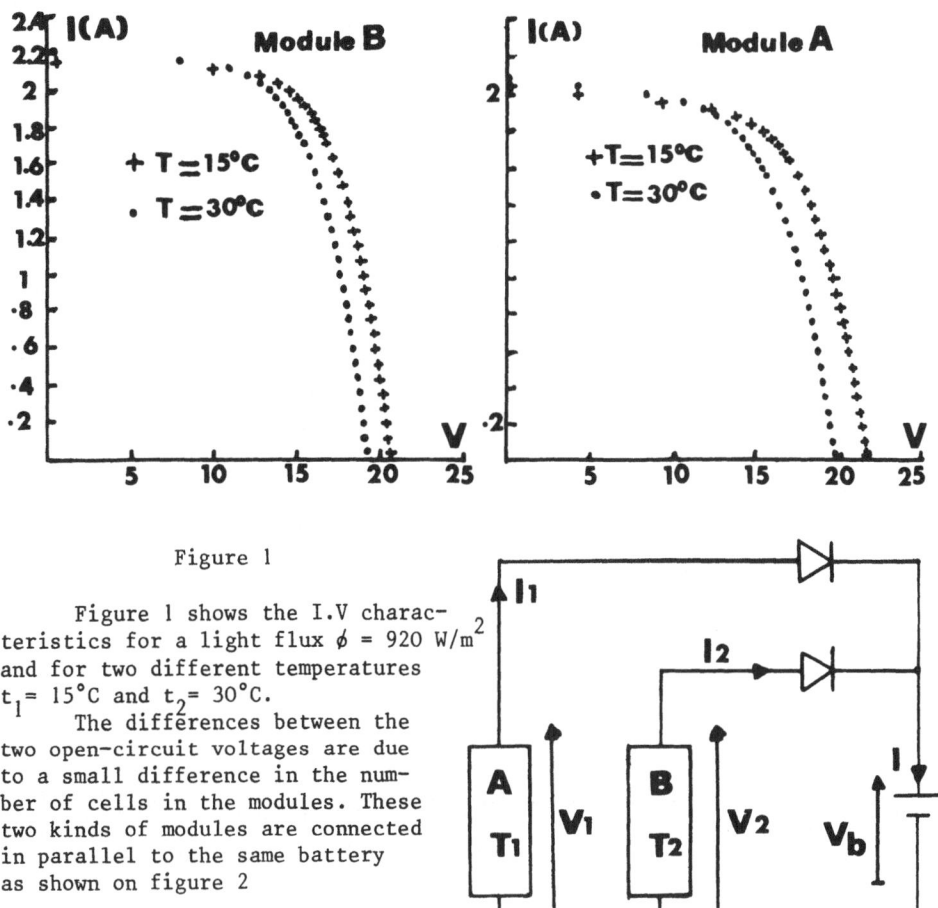

Figure 1

Figure 1 shows the I.V characteristics for a light flux $\phi = 920$ W/m^2 and for two different temperatures $t_1 = 15°C$ and $t_2 = 30°C$.
The differences between the two open-circuit voltages are due to a small difference in the number of cells in the modules. These two kinds of modules are connected in parallel to the same battery as shown on figure 2

Figure 2

The following quantities are automatically recorded as a function of time , a given day : the battery voltage V_B, the temperatures T_1 and T_2 of the modules A and B, and the outside temperature T_a. The I.V characteristics of the modules can be taken simultaneously at any time through the data acquisition system. Figures 3 and 4 present the results obtained along a day. Vopt is the voltage at the maximum power point at each time. On figure 5, the different relative power losses have been computed : curve A refers to type A P.V. array and curve B to type B. Losses range between some percent up to 18% in the morning when the batteries are discharged and when the modules temperature is low. Dots correspond to the relative power losses in the two blocking diodes for comparison.

An extrapolation of the results obtained on this small power installation shows that for higher power the use of a specially studied DC.-DC. converter can, in some cases, increase the overall efficiency of the system by minimizing the mismatch losses between the array and the batteries. This is especially true for systems installed in areas presenting large temperature variations.

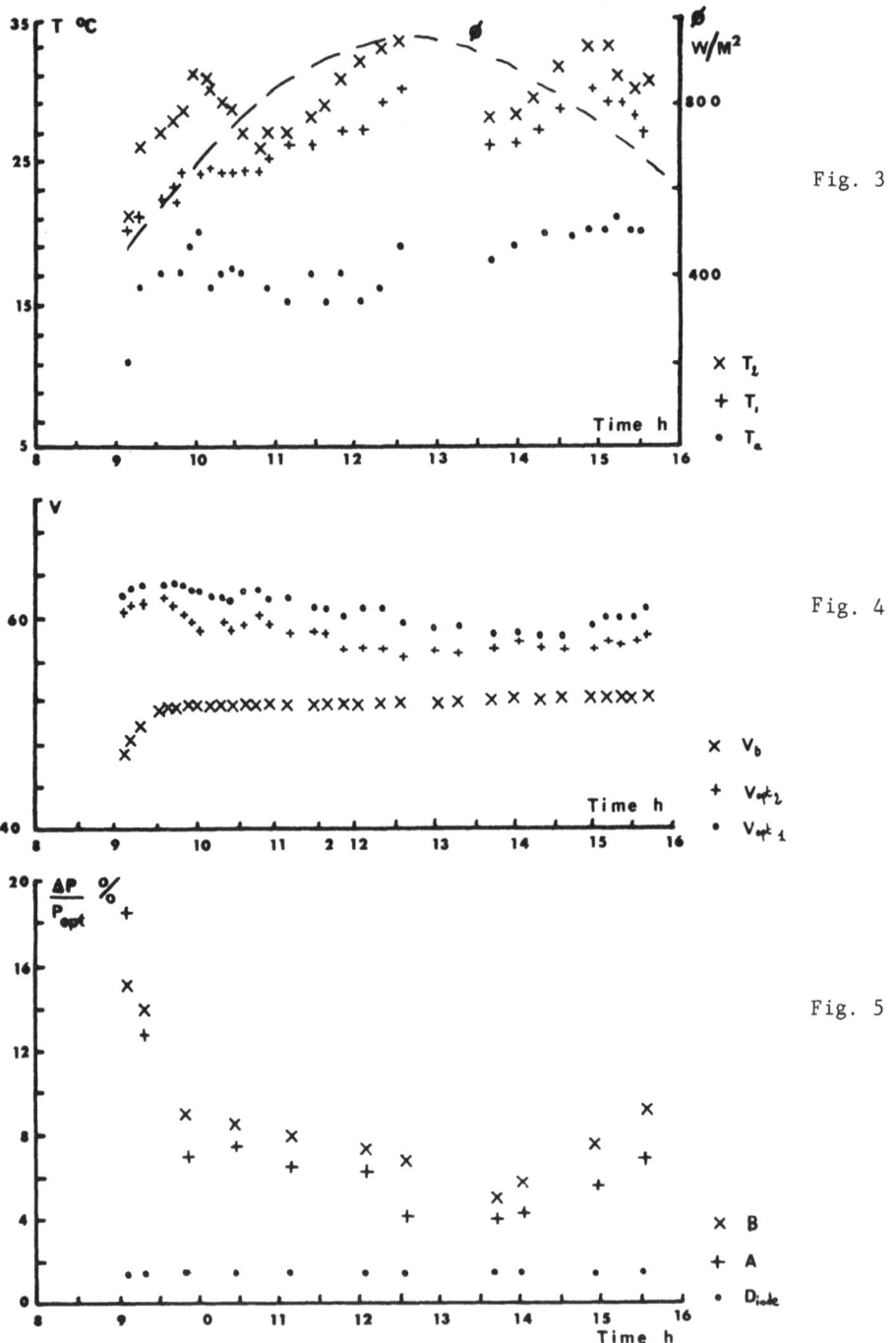

Fig. 3

Fig. 4

Fig. 5

3 - USE OF A DC.-DC. CONVERTER

D.C.-D.C. converters to be introduced between P.V. arrays and batteries are designed on the same principles than the good efficiency converters, used on motors or resistive loads, that were developped in our laboratories (2)(3)(4). However in the peculiar case of batteries, some new problems appear, mainly for the smoothing of the current delivered to the load. In the case of batteries, the present load adaptator lowers the voltage delivered by the P.V. array. The intermediate storage of energy to be transfered is done by means of a capacitance. The trigger of the chopper fixes for the array a voltage ajustable in a narrow window centered on the optimum voltage and takes into account the shift due to the cell's temperature evolution.

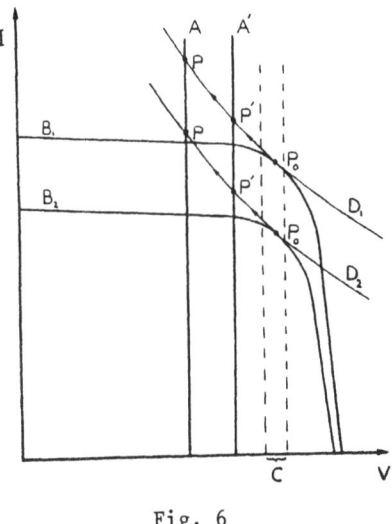

Fig. 6

The maximum power available is thus transfered to the batteries as shown on figure 6 :

. A and A' : I-V curves of the battery shifting with the state of charge.

. D_1 and D_2 : Constant power curves

. Zone C : Operating points on the P.V. arrays.

. P_o : maximum power point depending on the light flux and temperature.

. P and P' : operating points on the battery.

As a battery is not an inductive load, the chopper output current would be discontinuous without any coil. It is compulsary to deliver to such a load a continuous current in order not to decrease the battery life time. Thus the output current must be smoothed by a self-inductance coil associated to a fly-wheel diode.

As the period of the on-off ratio of the chopper current will vary, it is necessary to optimize the value of the coil in order to decrease its volume for a nearly continuous current whatever the experimental condition.

The power switch has been built in a MOS technology. These transistors can easily be put in parallel, which allows high intensity currents in the power circuit controled by a very simple driving circuit. Even for high currents, switching speeds are high enough to operate under frequencies from some kHz to some 10 kHz without lowering the efficiency. This drecreases the values of the input capacitance and the output self-inductance ; thus the cost and the volume of the converter are decreased while total efficiency increases.

Figure 7 presents a summarized electronic circuit of the converter, and figure 8 shows the corresponding efficiency. This converter has been built to be used on the 5 kW. P.V. array settled in the French Alpine refuge "Les Evettes"(1).

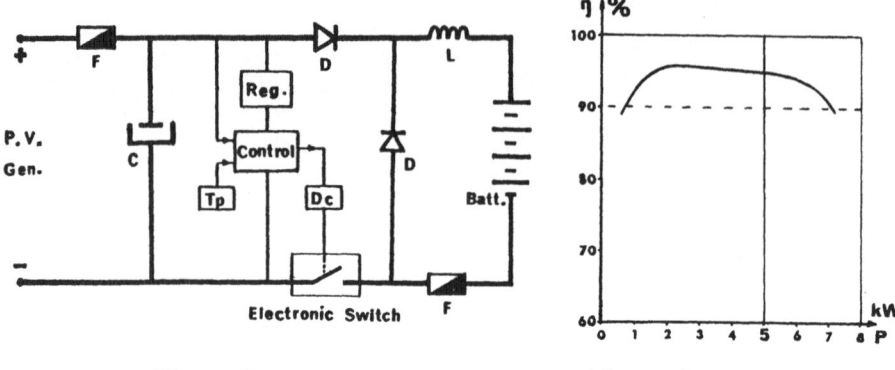

Figure 7   Figure 8

4 - CONCLUSION

From our experiments we can conclude that the use of converters coupled on batteries is inadequate in the case of low power systems : their price is to high as regards to the energy gain, and moreover, due to constant losses, their efficiency decreases at low power. However the use to D.C.-D.C. converters becomes of interest for power higher than 1 kW where their efficiency becomes very good and thus the maximum power can allways be extracted from the P.V. array whatever the light flux, the junction temperature variations and the state of charge of the batteries.

Some applications of D.C.-D.C. converters are as follow :

. High power stations can operate under high voltage module groupings and a much lower voltage on the batteries, decreasing the ohmic losses in the connection wires.

. If power arrays are constituted by different kinds of modules, each kind can be closed to the same battery through different converters. Such is the case if an increase in power of our old array is obtained by addition of new different modules.

These experiments derive from works sponsored by D.G.R.S.T. Contract 772011 and COMES contracts 7803192 and 7803292.

REFERENCES

(1) L. SELLES - 3rd E.C. Photovoltaïc Solar Energy Conference (1980) p.1028 (D.REIDEL - Publishing Company).

(2) J.A. ROGER - J. PIVOT - P. GUCHER - Revue de Physique Appliquée volume 15, p.603 (1980)

(3) J. PIVOT - J.A. ROGER - P. GUCHER - B. CANUT - E. DESCOURS 3rd E.C. Photovoltaïc Solar Energy Conference (1980) p. 1033 (D. Reidel - Publishing Company).

(4) J. PIVOT - G. BARREAULT - Compte-rendu des contrats COMES n° 7803192 and n° 7803292 (1980)

MATCHING THE CHARACTERISTICS OF BATTERIES WITH SOLAR CELL MODULES

C. F. GAY, V. K. KAPUR, B. PYLE and J. RUMBURG
ARCO Solar, Inc.
20554 Plummer Street
Chatsworth, California 91311

A. MANFREDI
ARCO Solar Europe - SpA
Viale Milanofioro, fabbr. E
20090 ASSAGO (Milano), Italia

Summary

This paper puts together the characteristics of the storage battery commonly used in worldwide photovoltaic applications with those of the photovoltaic generator and of the solar source. The natural matching of these characteristics may be exploited in meeting the typical requirements of the user. If the module is properly designed, and if due care is taken in battery installation for passive temperature control, there is no need for complex regulation or periodic adjustment. A well-designed photovoltaic module can function as both generator and controller, providing enhanced system reliability, lower system cost, and simpler operation for the user. The critical design parameter, the charging voltage delivered to the battery by the module, is determined by the number of cells in the series string.

1. INTRODUCTION

The typical photovoltaic (PV) power system in world application at present consists of one to a few PV modules of 32 to 40 cells each, coupled to one or more six-cell (12-volt) lead-acid batteries through a voltage regulator or charge controller. For this analysis, it is assumed that the daily energy demand is constant through the year.

Problems in sizing and designing PV systems of this kind, i.e., matching the system components to one another and matching energy supply to demand, are fairly common. The system may be put together by the user or a vendor who selects modules, batteries, and controller from different sources; in most cases these components have not been adequately optimized for the application. Reliability depends on different factors for the different system elements: the generator is a solid-state device, the battery is a liquid electrochemical device, and the controller is usually a mechanical device. Finally, in many cases, the best available data on site environmental parameters are very rough estimates, and the only analytical algorithm may be a rule of thumb.

New and emerging component design developments, accumulated field experience and environmental data, and improvements in analytical tools based on system experiments (1) make possible a new and more systematic approach to this task and indicate possibilities for simplifying the system design itself.

2. INPUT CONDITIONS FOR DESIGN OPTIMIZATION

The basic system has one module and one battery (Fig. 1).

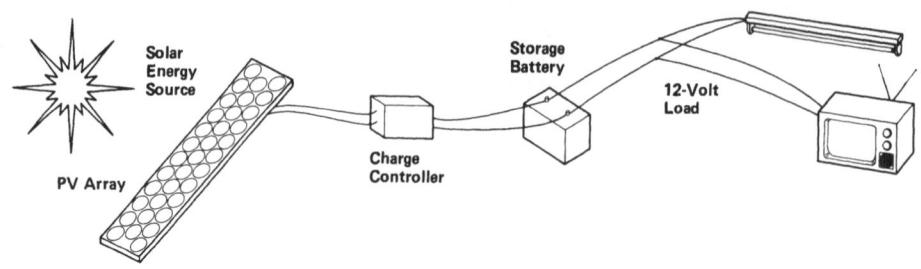

Fig. 1. The Basic System

System inputs for a PV power system are dominated by insolation and site characteristics. These are analyzed in another paper in this conference (2) and include daily profiles of insolation, air mass, and ambient temperature. Yearly, effectively global (-60 to +60 degrees latitude) averages of these profiles are derived in Ref. 2 and are presented in Fig. 2. Variations of this Standard Solar Day, also derived in Ref. 2, are summarized in Table I.

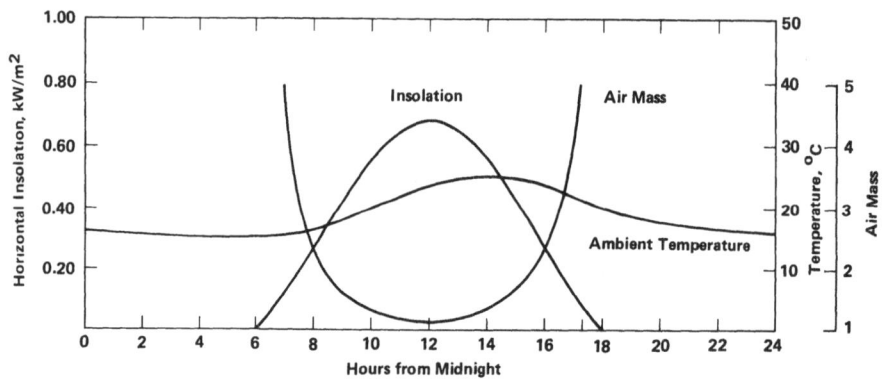

Fig. 2. Standard Solar Day

Table I. The Standard Solar Day and Typical Variations

Condition	24 - Hour Total Horizontal Langleys	Module - Normal Total kWh/m^2	Temp. °C Min to Max	Daylight Hours
Extreme Summer	722	7.4	25 to 40	13 1/2
Average Summer	567	6.0	20 to 35	13
Standard	413 (200 W/m^2)	5.3	15 to 25	12
Average Winter	206	3.6	5 to 15	10
Extreme Winter	21	0.63	-5 to 0	5 1/2

Fundamental characteristics of the battery include nominal output (12 volts) and capacity (100 amp-hours). A design for a module to match the battery must consider the relationship between charge rate and required charging voltage as a function of the state-of-charge. Most battery manufacturers recommend that state-of-charge be kept above 50% and battery temperature between -5 and 35°C to maximize battery life. State-of-charge is maintained by proper power-system management and temperature by appropriate installation which may, in some climates, require underground battery housing.

Figure 3 represents the charging characteristic for a 12-volt, 100 amp-hour battery. It is derived from laboratory study of available batteries, which include pure lead, lead-antimony, and lead-calcium plate types. A continuous range of charging-rate curves, from capacity/200 or 0.5 amp on the left to capacity/50 or 2 amp on the right, is shown. The 50-80% state-of-charge region exhibits very little voltage sensitivity to charge rate, but as the battery approaches full charge, the high-rate charging becomes voltage-controlled. If this is maintained against rising voltage, electrolysis begins in the battery and is the reason a regulator or charge controller is normally used.

PV module performance is sensitive to insolation and temperature, which vary with time and location as indicated above and discussed in Ref. 2. Figure 4 shows I-V curves of an idealized single-crystal silicon solar cell module for a range of insolation and temperature conditions. Note that the maximum-power region (knee of the curve) remains relatively stable as to voltage. Figure 5 gives daily voltage profiles of this module for the Standard Solar Day and four variations defined in Table I. Voltage is inversely dependent on temperature but, except for the extreme winter case, clusters over a wide range of conditions for most of the daylight hours.

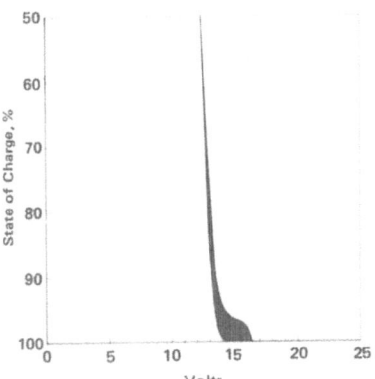

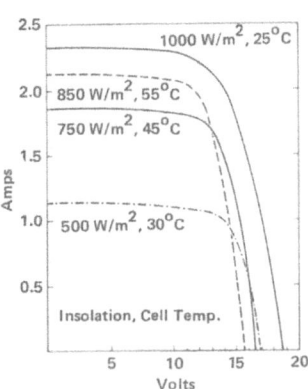

Fig. 3 (above, left). Charging Characteristics of Battery

Fig. 4 (above, right). I-V Curves of Idealized Module

Fig. 5 (right). Maximum Power Voltage Profile for Idealized Module

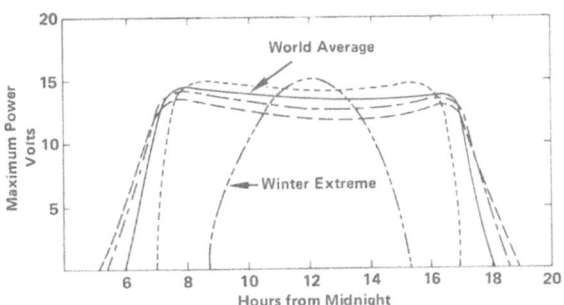

3. THE PRINCIPLE OF SELF-REGULATION

The desired charging profile for a storage battery of the type under consideration is to maintain a moderate charge rate until a relatively high state-of-charge is achieved, and then to taper the rate towards a low rate, filling the battery completely without shifting into electrolysis and undesirable heating. To achieve this, an inverse relation between voltage and current is needed, such as that provided in a sophisticated controller, or, as shown in Fig. 4, a PV module. The module must be sized so that its voltage-sensitive region coincides with the battery's critical region, as shown in Fig. 6.

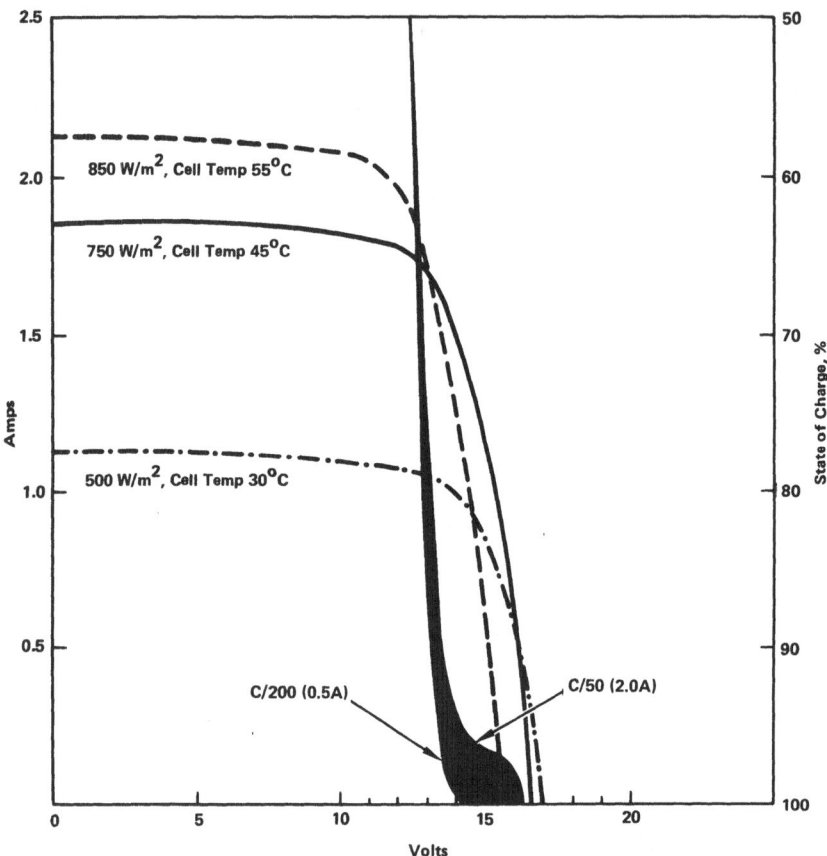

Fig. 6. Matching Module Voltage to Battery Characteristics

With an appropriately sized module, the battery's rising voltage demand in the critical region (95-100% state-of-charge) drives the module off the knee of the I-V curve to a lower current or charge rate, and the desired tapering-off of charging rate is automatically maintained. The only requirement is to design the module to match the battery's critical voltage region, with an allowance for wiring resistance.

4. RESULTS

ARCO Solar has designed a module to meet the requirements developed above. Using a single-crystal silicon cell whose V_{oc} is 0.59 volts and I_{sc} is 2.5 amps under test conditions, the module contains 30 cells. It has a superstrate configuration with tempered-glass front, PVB pottant, and Tedlar/polyester/aluminum/Tedlar back to provide environmental isolation and temperature moderation. This module provided the characteristics used to generate the "ideal" module simulations given in Figs. 4 and 5.

An annual state-of-charge simulation was run for this module charging the defined battery, assuming location at 30 degree N latitude where the average day is that described by Fig. 2 with seasonal excursions; installation is at 40-degree tilt to maximize winter input. Daily use is 10% of battery capacity or 10 amp-hours. As shown in Fig. 7, the battery reaches a 50% state-of-charge in midwinter and recovers to the 90-100% range in early March. Data supporting this plot and those in Figs. 4 and 5 were obtained using ARCO Solar's PV system simulation program, to be described in a forthcoming paper.

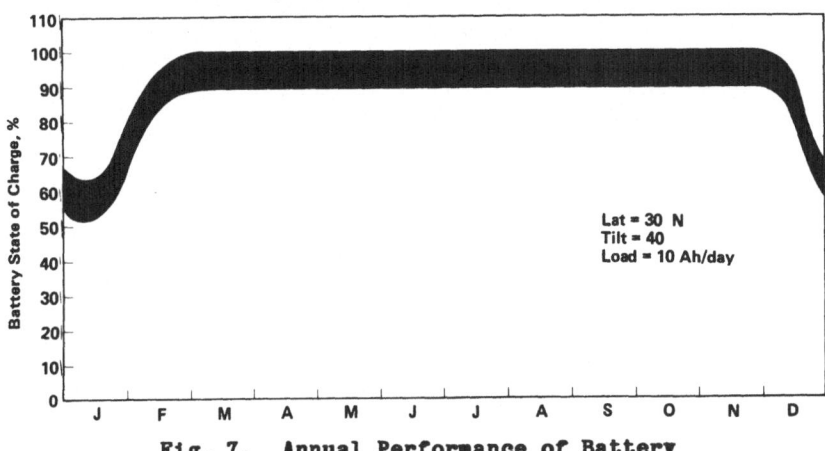

Fig. 7. Annual Performance of Battery

5. CONCLUSIONS

On the basis of this analysis, supported by the experiments (1), and by both in-house laboratory evaluation and manufacturers' data on batteries, it is clearly possible to design a photovoltaic generator to match appropriate characteristics of the battery so well as to maintain self-regulation in practical field operations. Eliminating the regulator will remove one parasitic, adjustable, failure-vulnerable component from small remote PV system. If proper installation of module and battery are carried out, maintenance should be minimized in the self-regulating system, and long working life assured.

REFERENCES

(1) J. E. Avery, E. Berman, and B. Pyle, "Photovoltaic Powered Village Lighting Systems: A System Design Experiment," Solar World Forum, Proceedings of the International Solar Energy Society Congress, Brighton, England, Aug. 23-28, 1981, Vol. 4, Pergamon Press.

(2) C. F. Gay, "AM/PM: The Rating System for Photovoltaic Modules," this conference.

CALCULATION TO IMPROVE POWER CONVERSION EFFICIENCY
IN PHOTOVOLTAIC SYSTEMS

Authors: A.Mas, J.García, L.Closas, M.Insausti and L.Castañer.
E.T.S.I.T.B. Aptdo.30002. Barcelona   SPAIN.
Tel. 2046551

Summary

This paper deals with several photovoltaic system configurations in order to improve the adaptation between a photovoltaic panel and an AC load, in low power applications. Panel to battery and battery to AC load interfaces are considered and their efficiencies evaluated for each configuration. Several values of the system main parameters as battery voltages, nominal output power and others are also considered. Experimental results indicated that the panel to battery interface efficiency is substantially improved by inserting an adaptative battery interface (ABI). The behaviour of the battery to load interface for the configurations proposed is analised.

1. INTRODUCTION

A photovoltaic system for residential or rural applications under 1 kW(220v rms) peak power consumption has several sources of efficiency limitation. First of all the panel to battery interface is non optimal in most of the cases where a simple diode (p-n junction or Shottky) is used, because the working point of the panel is far from the maximum power point mainly when the cell temperature is low. We have calculated the loss efficiency for a typical photovoltaic connection system.

The second source of efficiency limitation of the photovoltaic system concerns the electronic system in itself, that means the dc to ac conversion to supply 220v, 50 Hz. Generally the output of the accumulator is converted through a bridge dc to ac converter that supplies square wave power. In our case we will consider a sinusoidal output and two step conversion process: dc to square wave and further filtering to 50 Hz sinusoidal wave.

We have made a calculation to assess the efficiency improvement if an alternative configuration of the system is used. The alternative consists on a switching dc-dc converter to rise the voltage value and then a dc to dc conversion as in the previous scheme.

2. PANEL TO ACCUMULATOR INTERFACE

We have considered three kinds of interface, with a p-n junction diode, a Shottky diode and a relais to compare with an adaptative battery interface(ABI) that would bias the panel to its maximum power point under any condition of illumination and temperature. Experimental details of this ABI have been given elsewhere (1).

A typical 12v accumulator and a standard 30W peak panel are used in our experimental test. The results we have obtained are shown in figure 1. Inserting our 90% efficiency ABI instead of the previous diodes and relais the interface efficiency is substantially improved specially in the range of lower temperatures. An additional point of interest of this system is the possibility to integrate in the same electronic circuit the temperature control of the working point and the accumulator protection against overcharge. Efficiency improvement up to 20% can be obtained in the worst case occuring with the accumulator in its lower charge state.

3. ACCUMULATOR TO LOAD INTERFACE

We will consider first of all the two dc to ac conversion configurations that are shown in figure 2. The configuration number 2 uses immediately after the accumulator a dc-dc booster converter reducing the bridge losses and avoiding the use of output transformer(50Hz). A calculation of the two systems efficiency have been made assuming bipolar transistors as theconmutation elements in all the converters. Let us see the basic building blocks.

a) n_B, bridge dc-ac converter. The following sources of power loss have been taken into account: transistors saturation losses(P_{s1}), transistor conmutation losses(P_{cm1}), emitter junction losses(P_{e1}), and constant losses(P_{ct1}) which are independent of the load current and mainly due to the control circuit consumption.

b) n_T, the losses of the 50 Hz output transformer have been taken from commercial data for several nominal output power.

n_F, the filter efficiency we have taken is 81% that is the theoretical maximum for a filter output without armonics.

c) The losses in the dc-dc switching converter are mainly introduced by the losses in the saturation(P_{s2}),conmutation (P_{cm2}),emitter junction(P_{e2}) of the transistor,the losses in the coil L_0(P_r) mainly due to the resistive wire since the ripple current in the coil is small,the losses in the transformer (P_{T2}),the diode losses(P_D), and constant losses(P_{ct2}). It must be pointed out that in this alternative we can use,in the dc-dc converter,frecuencies in the range of 20 to 50kHz reducing the size of the components; these two frecuencies have been included in the calculation.

We have calculated the overall efficiency of the two systems for several output power and battery voltage values as well as for two frecuencies of the dc-dc converter (2) (3).

The efficiency of the systems 1 and 2 can be expressed as:

$$\eta_2 = \frac{P_L}{\dfrac{P_L}{\eta_B \eta_F} + P_{s2} + P_{cm2} + P_{e2} + P_{ct2} + P_{T2} + P_D}$$

$$\eta_1 = \frac{P_L}{\dfrac{P_L}{\eta_T \eta_F} + P_{s1} + P_{cm1} + P_{e1} + P_{ct1}}$$

We have plotted these results in the figures 3 and 4 for 150W and 600W and so we can compare the efficiency of the two systems. Figure 5 shows the evolution of the efficiency of the two systems versus the nominal output power at a fixed battery voltage of 48v.

There is an important source of losses in the filter to get sinusoidal output. There could be also a great improvement in the efficiency if the generated waveform has lower armonic contents. Some experimental systems use mechanical conmutation of different accumulator combinations to preshape this waveform (4).

In our case we modify the dc-dc converter in such a way that generates an square wave of 100Hz and appropiate amplitude over a DC level. This system 3 is shown in figure 6. We have calculated the system efficiency for the same parameters than previous systems 1 and 2; the results are plotted in figure 7.

In this case we observe that the efficiency of the system 3 can even be greater than that of system 1 if V_{BB}= 246v (for this battery voltage the system 1 does not need the 50 Hz transformer and its efficiency is quite high, about 80%).

For high battery voltages and rising conmutation frecuencies in the dc-dc converter the transistor conmutation losses rises dramatically. We can reduce these lossesusing VMOS devices instead of bipolar transistors. The VMOS devices have lower conmutation times, need lower input power to be driven, and can be easily connected in a shunt configuration.

REFERENCES

(1) L.Closas et al. Pilot instalation for small power photovoltaic systems. 3rd. E.C. Ph.Solar Energy Conference (1980)

(2) E.R.Hnatek. Design of Solid-State Power Supplies. Van Nostrand Reinhold (1981)

(3) B.D.Bedford, R.G.Hoft. Principles of inverter circuits. John Wiley (1964).

(4) G.J.Naaijer. New type of transformerless high efficiency inverter. 3rd. E.C. Ph.Solar Energy Conference (1980)

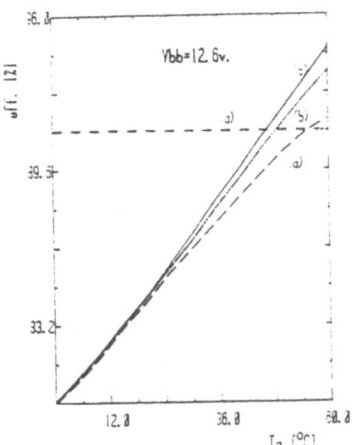

Fig. 1 - *Adaptation efficiency vs. ambient temperature for (a) junction diode, (b) Schottky diode, (c) relais, (d)ABI.*

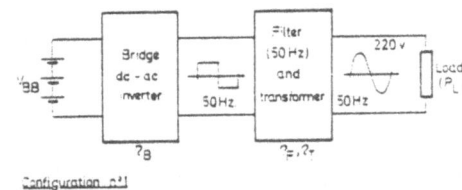

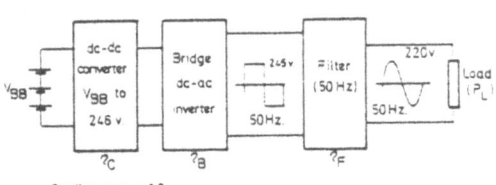

Configuration n° 2

Fig. 2

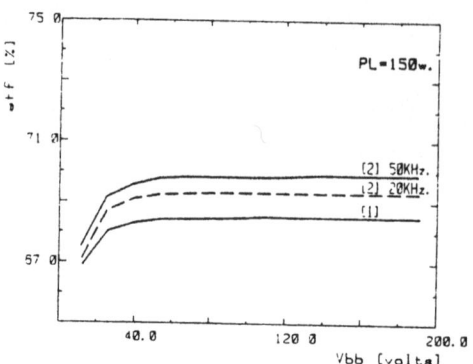

Fig. 3 - Efficiency vs. battery voltage, for systems (1) and (2), at 150 watts.

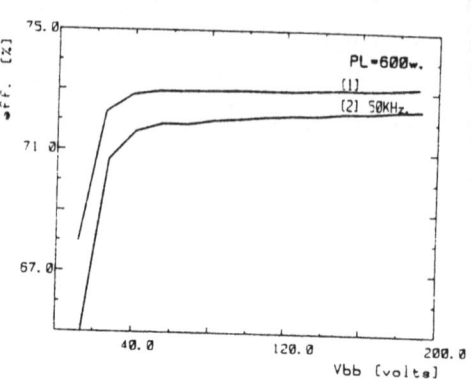

Fig. 4 - Efficiency vs. battery voltage, for systems (1) and (2), at 600 watts.

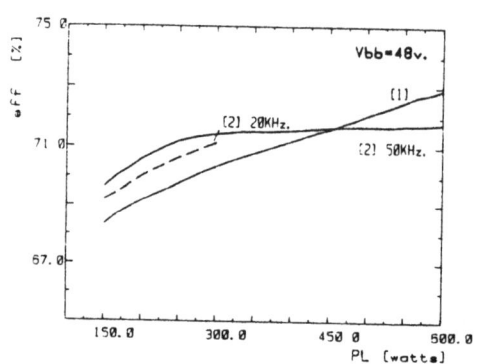

Fig. 5 - Efficiency vs. output power, for systems (1) and (2), at 48 volts battery voltage.

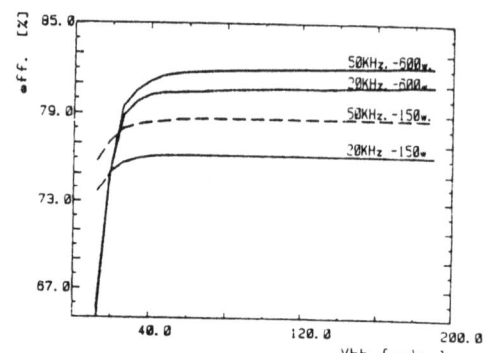

Fig. 7 - Efficiency vs. battery voltage for system (3).

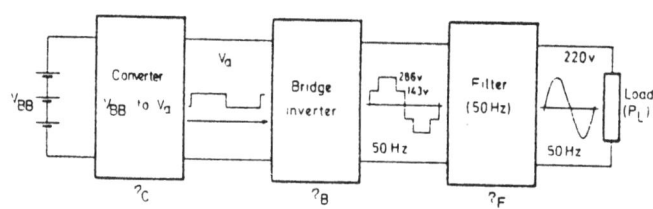

Configuration n° 3

Fig. 6

SESSION 4

FLAT PLATE MODULE TECHNOLOGY

Chairmen : Dr. K. KREBS, Commission of the European
Communities, Joint Research Centre, Ispra
Establishment

Dr. R. KOEPKE, Kernforschungsanlage Jülich,
Federal Republic of Germany

Flat plate module technology - overview

AM/PM : The rating system for photovoltaic modules

FLAT PLATE MODULE TECHNOLOGY - OVERVIEW

E . FABRE
PHOTOWATT INTERNATIONAL S.A.
131, route de l'Empereur
F-92500 - Rueil Malmaison

Summary

The paper presents a synthetic overview of the different technologies which can be used for manufacturing a photovoltaic module. The emphasis is stressed on the cost aspect.
The whole process is divided in three steps :
. material preparation and shaping
. cell processing
. cell interconnection and encapsulation
The present state of the art and the general technical trends for the next few years are presented and discussed for each step, in the frame of the silicon technology. The respective contributions to the total cost are also discussed, in view of defining the general orientations for R + D which have to lead to drastic cost reduction. Once all these steps have been individually reviewed, it is much important to investigate the technical compatibility between all the different possible technologies in order to build up an economic and consistent process.

1. INTRODUCTION

The photovoltaic industry is now an actual reality. Several megawatts of solar cells have been produced and commercialized throughout the world using the single crystal silicon technology. The manufacturing cost of the photovoltaic modules has steadily decreased following some learning curve, due to an increase of the production throughput and changes in the technological process. It is expected that the crystalline silicon solar cells will further be produced on a very large scale during the next decade, whereas their cost will substantially be reduced due to important changes and / or breakthrough in the material preparation.
The purpose of this paper is to give a general survey of all the present and potential steps in the whole module manufacturing process, and to review the critical paths which can lead to cost reduction, in the frame of the crystalline silicon technology.

2. PRESENT SITUATION

A general flow-chart of the present process is described in figure 1, from sand to the final utilization of the photovoltaic module.

Sand is reduced into metallurgical grade silicon, the purity of which is in the 98 - 99 % range. This material is far too impure to be used for any electronic application, and it has to be purified to the so-called "semi-conductor grade". The photovoltaic industry uses that material as feedstock, which has been developed mostly for IC's applications : its specifications are not optimized for the use in solar cells, and its cost is much too high.

The semi-conductor grade silicon is molten and pulled as single crystalline ingots using the Czochralski technique. The ingot is cut into wafers which are etched in order to remove the damages induced during the slicing operation.

The wafers are then ready to be processed, i.e. provided with a junction, presently achieved by diffusion, and with contacts. An anti-reflective coating can eventually be deposited to increase the number of absorbed photons.

The cells need to be protected against corrosion and to be interconnected in series-parallel association in order to get a useful voltage : these operations are realized within the module assembly process.

The photovoltaic module can then be used, but for most applications some power conditioning is required.

Figure 1 shows how the whole process can be divided into four technologically different parts :
- silicon material
- cell process
- module assembly
- system and applications

This paper is restricted to the first three parts, but is investigating all the different crystalline silicon approaches besides the present single crystal silicon technology.

3. ALTERNATIVES

3.1. SILICON MATERIAL

Several ways are presently looked at for getting a silicon flat plate product ready to be processed, and at a much lower cost as compared to the single crystal wafer. Figure 2 explains the different approaches which are the most investigated throughout the world at the moment.

First of all, a very important effort is devoted to the definition and the preparation of the so-called "solar-grade" material for replacing the expensive "semi-conductor grade". The target is more an economical one than a technical one : how to get a feedstock material for less than $ 20 / kg, compatible with the use in efficient solar cells. Several companies are facing this challenge, and it is generally agreed that such a material could come up in 1985 - 1986 on an industrial scale.

This material has to be "shaped", and two kinds of methods are investigated : bulk growth, leading to ingots which have to be cut into wafers with optimized slicing techniques, or ribbon pulling for which all the limitations bound to slicing are definitely overcome.

There is of course an economic advantage of using ribbons against ingots, but the ribbon pulling techniques are much less mature as compared to ingot growth methods and most of them still have to be proved viable before being industrialized. On the contrary most of the bulk growth techniques are now already being operated under pilot or even industrial production, starting with a semi-conductor grade feedstock.

But if the solar grade material does not appear in 1986 as expected, due to unforeseen technological difficulties, then the advantage of ribbon pulling from semi-conductor grade silicon will be predominant over the ingot growth : the silicon consumption is between 2 to 8 times smaller in ribbons as compared to sliced wafers, depending upon the different processes.

3.2. CELL PROCESS (figure 3)

The cell process most often starts with some surface preparation : damage removal after slicing, texturization for decreasing losses due to reflected photons in the case of (100) oriented single crystal wafers... Whereas the damage removal will always remain necessary with the use of polycrystalline wafers cut from ingots, the texturization will not be effective any more in that case. And the ribbons will most probably be ready for processing just as-grown, without any surface preparation.

Two ways will soon become available for achieving the collecting junction : either the High Temperature Diffusion (HTD) or the Cold Junction Processing (CJP).

HTD is a very well known process which is presently used for solar cell production everywhere, whereas CJP is still in the development stage : the doping species are introduced by ion or molecular implantation, or evaporated, or spray-deposited onto the surface of the semi-conductor and driven in by a pulsed annealing (laser or e-beam) : only a very thin layer is molten just underneath the surface and the bulk is kept at low temperature, preventing the minority carrier lifetime to be degraded. Such a lifetime decrease can be very important during HTD in the case of a lower grade (solar grade ?) polysilicon. The main advantages of CJP over HTD are :
- lower energy consumption,
- process amenable to full automation,
- bulk kept at low temperature,
- lower overall cost.

Three techniques are being used for depositing the metallic contacts. Evaporation still leads to the most reliable contacts, but it is far too expensive and has been abandoned by all the manufacturers. The two alternative methods are plating and screen-printing. Both techniques should lead to similar cost, provided one can get rid of noble metals. The reliability under hot and humid conditions should be carefully looked at, especially in the case of screen-printing, but new ink compositions are now available so that this problem can now be solved.

The cell process may be ended with the deposition of an antireflective coating, and this layer can either be evaporated, or sputtered, or chemically sprayed. The interest of depositing this layer results from an economic trade-off, depending mostly upon the nature of the starting material and the surface preparation.

3.3. MODULE ASSEMBLY

The main functions of the module assembly can be summarised as :
- interconnection of solar cells for series - parallel association in order to get useful voltage and current characteristics for given applications,
- protection of the solar cells, their contacts and their interconnections,
- supporting structure.

A key feature of the encapsulation is to provide a high reliability to the photovoltaic module, especially with regard to the resistance against humidity.

Two methods of module assembly have been or are now widely used : cast resin or lamination. Each one has advantages and disadvantages, depending upon the nature of the protection which is required, the kind of material used (glass, aliminium, plastics, resins...). Lamination is more amenable to automation. It also provides a more hermetic sealing and therefore can tolerate less reliable contacts. But on the other hand cast resin technique requires a smaller capital investment.

4. TRENDS FOR COST REDUCTION

Table I gives a tentative cost breakdown of the photovoltaic module based on crystalline silicon technology. A basis of 100 has been assumed for the present time (1982). The cost is broken down according to the three parts that are considered in this paper : material, cell process, module assembly. But for discussing the anticipated cost reduction, it would seem more interesting to use another cost breakdown, according to silicon, other materials and manpower (table II).

We can observe than the absolute silicon contribution has slightly decreased between 1977 and 1982, from 80 to 50, due to the increase of the size of the Czochralski ingots, from 2- 3" up to 4" ; but in the same time the relative silicon contribution has increased from 35 % up to 50 % because the other components of the cost have decreased considerably more :
i) "other materials", from 90 down to 25, mainly due to the replacement of evaporation by plating or screen-printing,
ii) manpower, from 70 down to 25, due to the increase of the size of Crochralski ingots and a beginning of automation.

Altogether, the cost has dropped by a factor 2.5 from 1977 to 1982, roughly.

A more drastic decrease can be expected from 1982 to 1986 - 87 because of important changes in material preparation :
- silicon is anticipated to decrease down to 8 - 12, depending upon different parameters : availability of solar grade silicon, ability to use an optimized slicing, readiness of ribbon technology...
- other materials : noble materials will be replaced by cheaper ones, material consumption will be optimized at each step of the process... The spread 7 - 10 is mostly related to the different kinds of encapsulation which will be used : glass, self-supporting structure, etc...
- manpower will be greatly reduced, from 25 down to 5 - 8, due to automation, the degree of which will depend upon the production capacity, i.e. depending upon the opening of the market.

An overall cost reduction by a factor 3 to 5 can then be expected within the next five years, the large spread still coming from uncertainty in the silicon shaping technology. But it is obvious that the relative contribution of silicon to the total cost is presently the highest ever reached and will be reduced by the use of polycristalline ingot growth and the availability of solar grade feedstock and / or ribbon technologies. However, this contribution will always remain in the 40 % range, as long as the crystalline silicon technology is involved (figure 4).

5. CONCLUSIONS

The photovoltaic industry will only survive and grow if the cost of modules can be substantially decreased : this is a prerequisite. A tremendous effort has already been made in the past few years for achieving actual cost reduction : but such an effort would have no sense if it were not continued on an even greater scale. The past cost reduction has been obtained within the single crystal silicon technology by continuous changes : further cost reduction will imply more fundamental changes or even breakthrough, as well as a learning curve assuming large increase in production capacity. The high level of R + D which is necessary, together with the introduction of new technologies and the increasing throughput require an important capital investment.

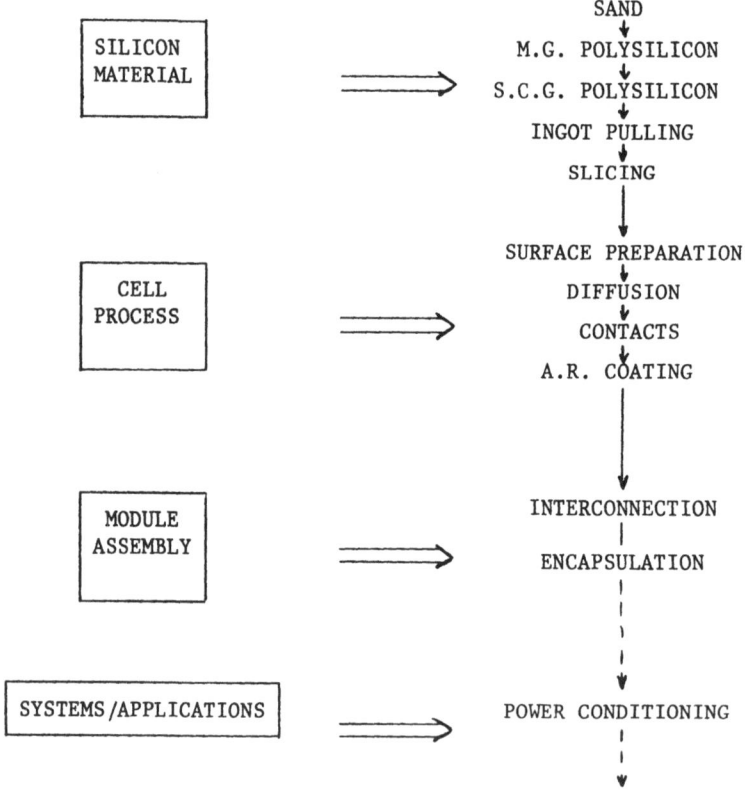

Fig. 1 : general flow-chart of the present process

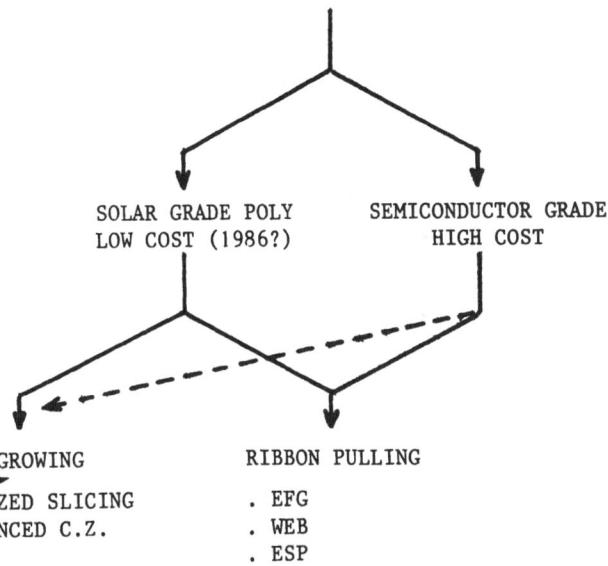

Fig. 2 : Silicon material - preparation and shaping

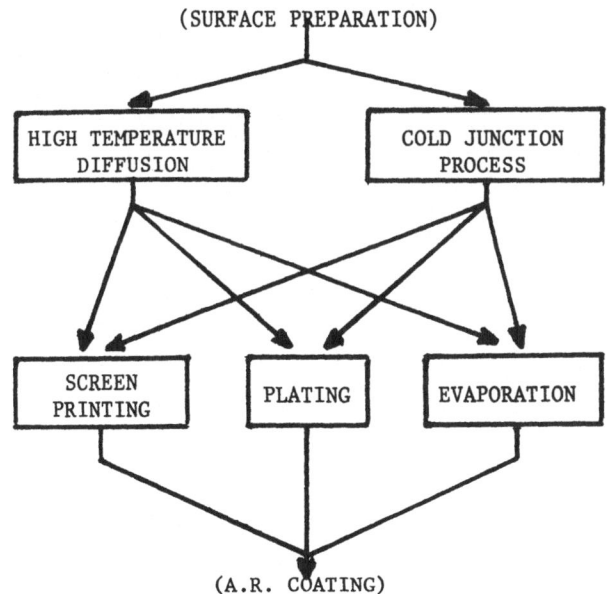

Fig. 3 : Cell process

SILICON CONTRIBUTION IN COST OF Wp

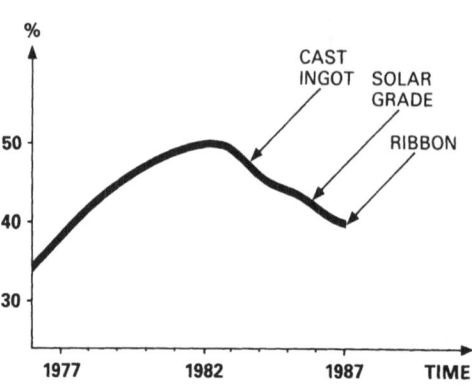

Fig. 4 : Relative contribution of shaped silicon in the total cost of Wp.

	1977	1982
SILICON	80	50
CELL PROCESS	80	30
MODULE ASSEMBLY	80	20
	240	100

Table I : Cost breakdown, according to the different steps of the process.

	1977	1982	1986-87
SILICON	80	50	8 - 12
OTHER MATERIALS	90	25	7 - 10
MANPOWER	70	25	5 - 8
	240	100	20 - 30

Table II : Cost breakdown, according to silicon, other materials and manpower.

AM/PM: THE RATING SYSTEM FOR PHOTOVOLTAIC MODULES

C. F. Gay
ARCO Solar, Inc.
20554 Plummer Street
Chatsworth, California 91311

Summary

This paper defines an approach to straightforward measures and ratings for photovoltaic modules which match the energy user's needs. The rating measurement proposed, that of electric energy delivered (Wh/day), is based on a Standard Solar Day averaged over the year and the world, indexable according to climatological zones. This Standard Day sees a 24-hour average irradiance of 200 W/m^2, average temperature of 20°C. The required module measurement involves the typical peak-watt power output and sensitivity to temperature, light level, and spectral content. It then becomes possible for the energy user to estimate the cost per kilowatt-hour of electrical energy received over the life of the system. The paper illustrates the concept of and need for energy-based ratings by demonstrating various case calculations for the standard day. One commercial module type rated at 40 peak watts has an AM/PM rating of 194 Wh/day.

1. INTRODUCTION

At the present time solar photovoltaic modules for terrestrial use are most commonly rated (and even nominally priced) from a measurement of maximum power under laboratory conditions of relatively high insolation, favorable air mass, and low cell temperature (the peak-watt rating). These conditions are rarely duplicated in the field. A somewhat more realistic rating is also used in which, under slightly reduced insolation, the module is allowed to come to equilibrium with a specified ambient temperature, so that maximum power is measured at a Nominal Operating Cell Temperature (the NOCT rating), resulting in a lower numerical watt value than the peak-watt rating. Not accounted for are the effects of (a) the load on the operating point, (b) local-site and time-of-day variations from the test conditions, and (c) the non-linear response of modules to changing conditions. This is a natural consequence of the fact that for nearly the first two decades of its existence the major use of photovoltaics was in outer space. It was reasonable that the initial rating methods were consistent with space conditions.

It is true that most electric generators are rated at peak instantaneous power, in watts or horsepower. The small diesel generator set, a present-day competitor of photovoltaics, is notable for this. But it operates at peak, converting fuel oil into electricity at very near its rated output from the time it is started until it stops. Hydro and nuclear plants are intended to run continuously at rated value unless shut down for maintenance or repairs. Fossil-fueled steam turbine generators are similar, but more likely to be shut down in load management. With all three of these classes of electricity supply, the long-term or short-term quantity of energy can be predicted, planned, and promised from the generator rating, given adequate fuel supply and system reliability.

By their very nature, solar and especially photovoltaic generators differ from all others. Photovoltaics directly converts a natural, dispersed energy source which fluctuates both regularly and irregularly. The fuel of photovoltaics is both free of charge and free of controls: we must take it as it is. Yet we must compete for customers with these other systems. We compete in the small and medium remote applications with batteries and diesel generators; in water pumping with animal power, grid electricity, and diesel engines; in grid-connected applications with utility peaking generators. We must meet these competitors on their own terms, in terms that the user understands, in energy units.

2. THE PROBLEM

The brutal fact is that the peak-watt rating and its improvement, the NOCT rating, so effective in communicating within and among government programs, technology developers, manufacturers, and demonstration operators, are virtually useless to the practical user.

What is needed, then, is a new way of rating our generators which will bear something like the same relation to meeting the user's practical needs as does the rating of any other generator. We are not attempting to compare equipment costs and fuel costs, but are merely measuring the ability to do work with electric energy. What we need is a reliable prediction of the energy per unit product for an average day or year that will be delivered from an average site. The average day's output rating must take into account the solar array's response to varying conditions, which is cell-specific as well as module-specific.

This is what we need in order to compete as a mature industry against other mature industries supplying electric generators.

3. THE SOLUTION

I propose the adoption of a rating standard which will meet these requirements. The product testing and calculations called for are within the capability of any photovoltaic manufacturer. I call it the AM/PM standard, because it is based on the whole day rather than just the "peak" sunshine hours, or solar noon, or a millisecond in a laboratory simulator.

It is based on the description of a standard solar global-average day, or a practical global average, in terms of insolation, ambient temperature, and air mass.

The survey which leads to this definition also puts limits on the degree of variation of specific sites, too few of which have been adequately measured thus far.

It is also based on photovoltaic module performance modelling studies, which in turn are based on field and laboratory measurement of many types of cells and modules, and a theory of performance prediction expressed as a simulation program built from several models.

3.1 The Standard Solar Day

It should be noted that the accuracy of our standard day is not of first-order importance, but only that its parameters be standardized by international agreement. For though the amount of sunlight reaching the Earth's surface typically varies by a factor of three from point to point, local temperature tracks this variation and, surprisingly, damps the effect of wide excursions in received insolation down to about ± 30% in photovoltaic output in today's silicon cells and modules.

Our planet intercepts 1.7×10^{14} kW of solar power continuously: one-third of this is reflected to space, one-fifth is absorbed in the atmosphere, and the remainder, a little less than half, reaches the surface. Averaged over the whole surface of the globe, lit and unlit, land and sea, of 5.1×10^{14} m^2, this amounts to about 155 W/m^2 continous equivalent. This value is distorted by inclusion of the polar regions, which receive a very small share of this power (evidenced by sustained low temperatures). If we segregate out the sectors beyond 60 degrees N and S latitude, which offer relatively dim prospects for year-round photovoltaic applications in any case, the average over the remaining Solar Zone rises to about 200 W/m^2. Measured annual averages within the Zone include, for example, about 100 W/m^2 in the United Kingdom and 300 W/m^2 in the Red Sea area. The Zone-wide average value accounts for diurnal and seasonal fluctuations, and is consistent with considerations of the solar cycle developed by, e.g., Rose (1). The 24-hour solar energy input is 413 langleys or 4.8 kWh/m^2 across the zone.

Two other world-characteristic parameters besides insolation affect the energy output of photovoltaic modules: temperature and air mass. The mean ambient temperature in the Solar Zone is about 20°C; the daytime high is about 5° higher, the nighttime minimum 5° lower. Regionally, the equatorial mean is about 27°C, while near the 60-degree limits it drops rapidly to about -3°C. Naturally, near the equator seasonal variations are suppressed and near the 60-degree limits they are accentuated.

As noted above, average local and instantaneous temperature tend generally to track insolation, and since cell temperature and conversion efficiency are inversely related for today's crystalline silicon products, there is considerable damping of excursions. There is usually a further decrease in variation caused by rain cleaning of modules in winter months versus dust accumulation in summer.

A representative plot of the annual cycle of total radiation and average temperature range for a 24-hour day each month is given in Figure 1. This is intended to represent a "Zone average" cycle which has been selected as 30 degrees N latitude. Inverting the seasons shifts it to 30 degrees S. The summer period represents average equatorial conditions, and the winter period approximates the annual averages at the 60-degree limits of the zone, when the sky is generally clear. The supplementary parameter, air mass, varies considerably with geography (altitude and latitude) and exhibits a daily variation on the basis of path length through the atmosphere to the surface. Air mass has two components, sunlight intensity, which is already covered in the insolation data, and spectral distribution. The spectral effect is neglected in the present analysis, because crystalline silicon solar cells are relatively insensitive to spectral variations within a reasonable range. However this will not necessarily be the case with newer solar cell materials to be introduced in the future.

The distribution of insolation into direct and diffuse radiation has been incorporated into this analysis using the method developed by Liu and Jordan (2).

The standard solar day may now be described. As indicated above, it may be described in terms of solar energy received per day as 4.8 kWh/m^2 (200 W/m^2), 20°C average temperature, air mass 1.5, and 30 degrees latitude. As argued above, and to be demonstrated in a subsequent section, this description proves inadequate even at present because of the variation in photovoltaic module response curves, primarily to varying insolation, to some degree to temperature, with future materials to spectral distribution, and possibly also to diffuse-radiation distribution. Seasonal variations were indicated in Figure 1; daily cycles for the Standard Day are given in Figure 2.

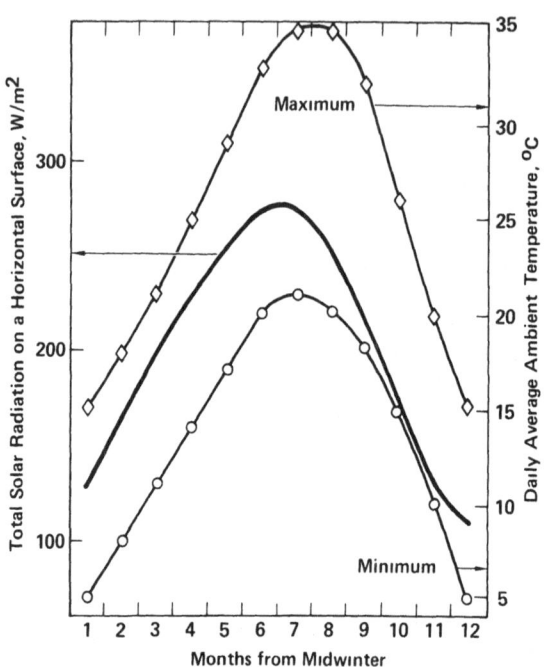

Figure 1. Solar Zone Average Insolation and Temperature Ranges, Yearly Cycle

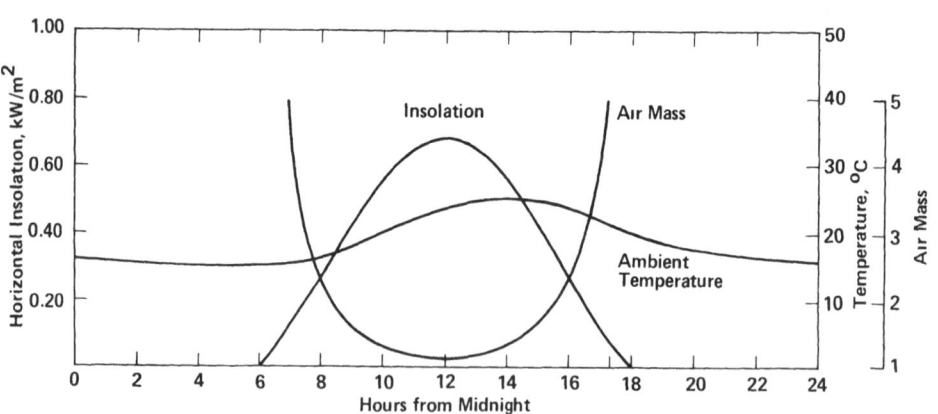

Figure 2. Standard Solar Day Insolation, Air Mass, and Temperature Cycles

3.2 The Range of Local Variations

A number of studies of local insolation and temperature, using historical observational data and various interpolation models, have been compiled, among them Smith (3). The greatest part of the work of which I am aware has concentrated on the continental United States. The various studies to date do not yet provide the comprehensive, global, and multiparameter data base needed to compile the necessary site-indexing gazetteer which will localize the standard accurately. Still, it is a feature of this standard approach that it is progressively perfectible.

At present it is possible to approximate some levels of variation from the Standard Solar Day baseline, given in Table I. The association between

Table I. Standard Solar Day (24 Hours) and Typical Variations

Condition	Total Horizontal Langleys	Module-Normal Total kWh/m^2	Temp. °C Min to Max	Daylight Hours	Latitude	Ratio of Diffuse/Total Horizontal
Extreme Summer	722	7.4	25 to 40	13 1/2	25	0.164
Average Summer	567	6.0	20 to 35	13	20	0.286
STANDARD	413 (200 W/m^2)	5.3	15 to 25	12	30	0.344
Average Winter	206	3.6	5 to 15	10	40	0.375
Extreme Winter	21	0.63	-5 to 0	5 1/2	60	0.450

seasonal variations within the Standard Day's year and variations in annual average due to latitude was discussed in connection with Figure 1 above. It appears again in this Table.

The data in this Table could be used (with the appropriate computer models) to characterize present-day photovoltaic modules for service in the five classes of conditions indicated. The necessary global data-acquisition efforts should be undertaken to support and perfect the use of this standard in the development of indices for specific sites.

Comparative daily plots of horizontal insolation and ambient temperature, as given in the Table, are shown in Figures 3 and 4.

3.3 Module Characteristics

The fundamental photovoltaic module datum is the illuminated I-V curve. A single curve can be used to indicate, for example, the way in which electrical load will drive the module or system away from the maximum-power operating point. A family of curves can indicate the module's response to changes in application-site parameters.

To illustrate the effect of varying response, a comparison between two nominally equal modules has been made. Both are designed for 12-volt battery charging, rated at 40 watts peak (at AM 1.5, 25°C, 100 mW/cm^2) and have NOCT's of 41°C. Type A has the characteristics of ARCO Solar's M51 module; Type B is a synthetic creation which exhibits characteristics which have been measured on some current market products.

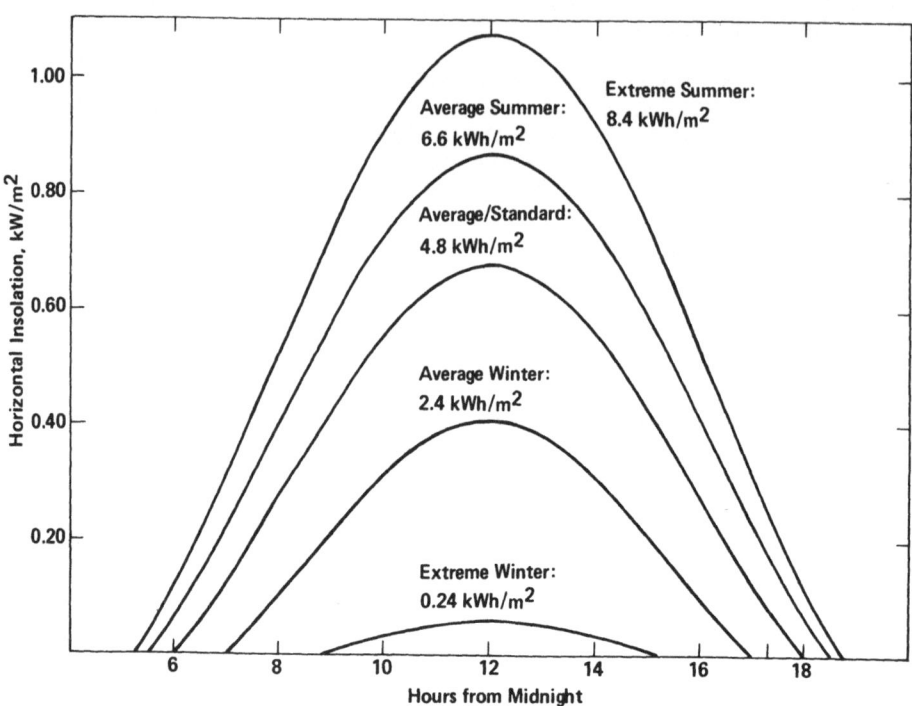

Figure 3. Variation of Insolation Daily Cycles

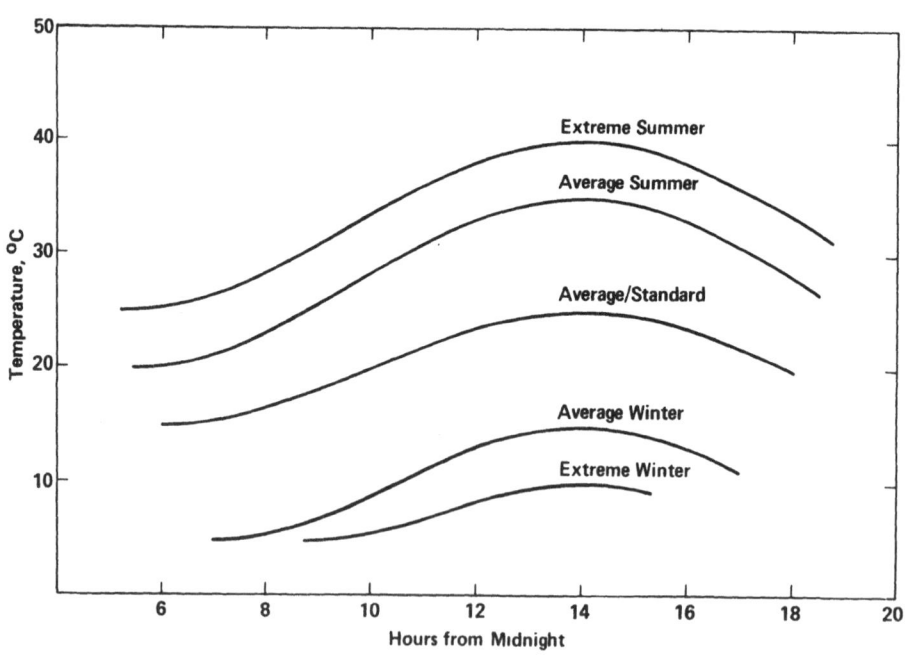

Figure 4. Variation of Ambient Temperature Daily Cycles

The runs were made using ARCO Solar's developmental performance prediction program, which has shown a predictive accuracy of ± 2% in outdoor testing in California. It includes an insolation model, cell temperature model, and module I-V and installation models with additional validation (4).

Responses of Type A and Type B to varying insolation and ambient temperature are shown in Figure 5. Performance of these two modules during the Standard and four variant days described in Table I is given in Table II (assuming the associated latitudes).

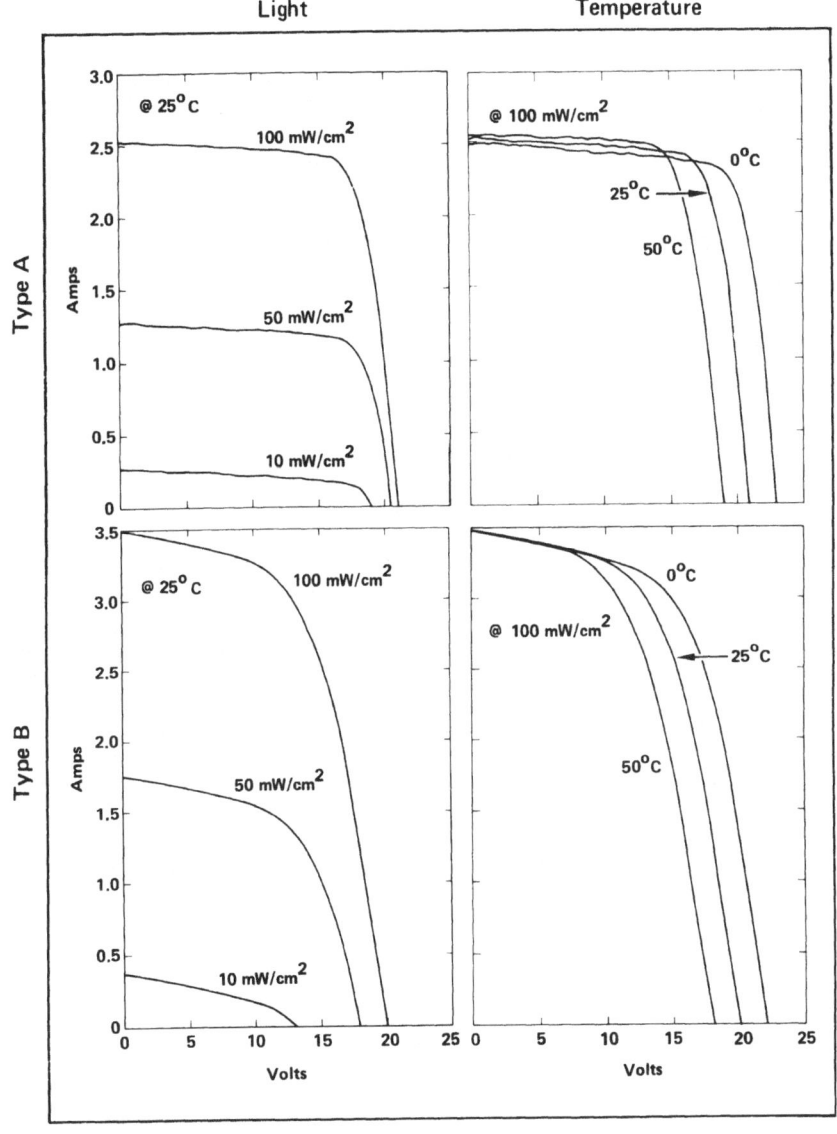

Figure 5. Variation of Module Response to Insolation and Temperature

Table II. Performance of Type A and Type B Modules During the Standard Solar Day and Defined Variations

	Fixed Tilt Angle, Deg.	Peak Cell Temp., °C	Total Energy, W-h/day	
			Type A	Type B
Extreme Summer	25	75	251.7	205.4
Average Summer	20	55	214.8	173.5
STANDARD	30	45	194.0	160.8
Average Winter	40	30	136.0	111.4
Extreme Winter	60	5	18.52	15.86

A new installation approach, two-axis tracking of flat-plate modules, has been given considerable study at ARCO Solar as a low-cost means of increasing daily energy delivery by eliminating substantial cosine loss. Applying this condition to the simulation gives an average of 43% greater energy output. The effect in increasing NOCT by 10°C reduces this output by only 3.5% in the summer and 1.5% in the winter.

4. CONCLUSIONS

The two total energy figures for the Standard Day, as shown in Table II, represent nameplate ratings derived from the AM/PM Photovoltaic Energy Standard. They are, respectively, 194 watt-hours per day for the Type A module and 160.8 watt-hours per day for the Type B module.

These ratings may be generated by a process nearly as simple as taking an I-V and determining max power; indeed it is simpler than measuring NOCT. The conditions of the Standard Day are clearly defined. Various validated performance prediction programs, with one of which the AM/PM values given above were derived, already exist. A standard rating program may quickly be adopted, to use appropriate multiple measurement to rate modules for their Standard Day energy delivery.

An even simpler method has been developed which requires a single maximum-power measurement and hand calculation. This method, performable without or prior to adoption of the standard rating program, gives a close approximation of standard daily energy, the proposed AM/PM rating, and could serve as an interim and "practical" standard. Its rationale is as follows: Horizontal daily total insolation (langleys or watt-hours/square meter) is converted to module-normal insolation, as was done in Table I; the module-normal Standard Solar Day value is 5.3 kWh/m^2. This represents the integral of a sinusoidal curve segment over 12 hours, the average daily photoperiod. The same module input conditions are represented by a steady insolation rate of 440 watts at a cell temperature of 45°C over the 12 hours, and these conditions may be easily reproduced in widely available module test equipment. Measuring module maximum power under these conditions, and multiplying it by the 12-hour photoperiod, gives the daily energy. This simplification has been partially validated with the ARCO Solar prediction program, and is compatible with the use of a simple conversion factor from Standard to site-specific conditions. Independent validation with a new program of field experiments is underway.

The next step is implementation. First in importance is international review, discussion, and adoption of the principle and the measurement and calculation methods. This would constitute partial implementation, since Standard Day and approved variation performances could immediately be used and would be useful.

ACKNOWLEDGEMENT

The help of Jeff Rumburg and Elliot Berman in developing this concept is gratefully acknowledged.

REFERENCES

1. Albert Rose, "Solar Energy: A Global View," *Chemtech* 11 (9), p. 566, September 1981.

2. B. V. H. Liu and R. C. Jordan, "The Interrelationship and Characteristic Distribution of Direct, Diffused, and Total Solar Radiation," *Solar Energy* 4, pp. 1-19, 1960, and subsequent papers.

3. Jeff H. Smith, *Handbook of Solar Energy Data for South-Facing Surfaces in the United States*, JPL Publication 79-103, DOE/JPL 1012-25, January 15, 1980, 3 volumes.

4. J. Avery, B. Pyle, and J. Rumburg, private communications.

POSTER GROUP P4

FLAT PLATE MODULE TECHNOLOGY

Recent progress in terrestrial photovoltaic collector technology

Low-cost solar array progress and plans

An all glass encapsulation of solar cells using silk-screening techniques

Laminated modules with new plastic material based on an ethylene-vinylacetate copolymer

Design and development of a module for the EEC PV pilot plants

Module design for EC pilot projects

Solar cells failure modes and improvement of reverse characteristics

Fabrication of large area Cu_2S/CdS thin film solar modules

Front contacts for large area high efficiency Cu_2S-CdS solar cells

RECENT PROGRESS IN TERRESTRIAL PHOTOVOLTAIC COLLECTOR TECHNOLOGY*

R. R. FERBER
Manager, Collector R&D, Photovoltaics TD&A Lead Center
Jet Propulsion Laboratory, Pasadena, California, USA

Summary

The U.S. Department of Energy's Photovoltaics Program has as a goal, the development of economic photovoltaic systems for terrestrial applications. A key element of a photovoltaic system is the collector, which converts the solar energy to electrical energy. The targeted cost for the collector is $0.70/$W_p$ (1980$). Two collector concepts are being developed to meet this cost target: (1) flat-plate by the Jet Propulsion Laboratory Flat-Plate Solar Array Project, and (2) concentrator by the Sandia Laboratories. The achievement of the cost target is dependent upon the Technical Feasibility demonstration of a number of different R&D areas in coordination with industry. The current Administration R&D policy changes have impacted the Program direction; however, the advances in both flat-plate and concentrator technology support the conclusions that collector R&D objectives are attainable.

1. INTRODUCTION

The U.S. Photovoltaics R&D Program has functioned since 1975. Its specific objective is to develop the technology necessary to foster widespread grid-competitive electric power generation by the late 1980s. The flat-plate and concentrator collector R&D activities form the nucleus of the Program. Flat-plate collector development is the responsibility of the Jet Propulsion Laboratory Flat-Plate Solar Array (JPL/FSA) Project. The Project includes the refining of silicon, silicon sheet production, solar cell processing and fabrication, encapsulation materials development, and collector design and production. Concentrator collector development is the responsibility of the Sandia Laboratories Concentrator Project. The Project includes the development of solar cells and devices for high concentration, highly efficient concentrating optical components, and concentrator collector arrays.

2. FLAT-PLATE COLLECTOR TECHNOLOGY

The primary objective of the JPL/FSA Project is to develop the technology for efficient, low-cost ($0.70/$W_p$), and long-life photovoltaic flat-plate collectors. This objective will be accomplished by performing a number of research and technology development activities which will be discussed on the following page. The completion of these activities will assist and encourage industry to complete the development and commercialization of photovoltaics technology.

---

*The work described in this paper was carried out or coordinated by the Jet Propulsion Laboratory, California Institute of Technology, and was sponsored by the U.S. Department of Energy through an agreement with the National Aeronautics and Space Administration.

A. <u>Silicon Material Task</u>. The objective of the Silicon Material Task is to establish technical feasibility of processes to refine silicon with a cost target of less than $14/kg (1980$). Two processes have shown considerable progress: (1) The process being developed by Union Carbide; hydrochlorination of metallurgical grade silicon (Si) to yield trichlorosilane ($SiHCl_3$), disproportionation to convert to silane (SiH_4), and the pyrolysis of SiH_4 to produce silicon. An experimental process system development unit (EPSDU) of 100 Mt/yr is now being built and is expected to be completed with Union Carbide funds. The latest economic analysis shows that the Union Carbide process with the fluidized bed reactor is capable of producing silicon at $11.65/kg (1980$ and 10% ROI) by using a 1000-Mt/yr production plant. (2) A modified Siemens process[15] being developed at Hemlock Semiconductor using dichlorosilane (SiH_2Cl_2) feedstock. The process had been demonstrated with the start of an integrated process development unit (PDU) using an intermediate-scale production-type reactor.

B. <u>Large-Area Silicon Sheet Task</u>. The objectives of the Large-Area Silicon Sheet Task are to develop and demonstrate the feasibility of several methods for producing large area silicon sheet material suitable for fabricating low-cost, high-efficiency solar cells. Progress is as follows: (1) Kayex Corp. and Siltec Corp. have been developing advanced Czochralski experimental sheet growth units (ESGU). The Kayex ESGU has been completed and has been operational for about 4 months. Kayex Corp. has been successful in growing 150 kg of ingots from a single crucible. Siltec Corp. has successfully grown a 100-kg ingot from one crucible with a process based on continuous melt replenishment. The Siltec process work has been terminated because of budget cutbacks. (2) The Crystal Systems Inc. heat exchanger method (HEM) has been successful with the growth of a 35-kg ingot within the specification requirements. (3) Semix Corp. is progressing with the development of a semicrystalline casting process. DOE funding reduction has significantly reduced the scope, and the Semix Corp. is continuing the development using corporate funds. (4) Wafering of silicon ingots continues to be a problem area for meeting the cost goals, despite advancements in silicon material utilization. Kerf losses still remain high, and cutting rates remain low. (5) Ribbon growth technology development work continues to show progress. The dendritic web ribbon process by Westinghouse Electric Corp. has successfully demonstrated continuous width control. Width control was the last goal to be achieved for around-the-clock continuous production. A ribbon almost 5 m long has been produced, with width held uniform to +0.1 mm. The edge-defined, film-fed growth (EFG) technique by Mobil Tyco continues to show promise. Three ribbons, 10 cm wide, have been grown simultaneously at 3.3 cm/min.

C. <u>Encapsulation Task</u>. The objective of the Encapsulation Task is to develop and qualify one or more solar array module encapsulation systems that have demonstrated high reliability for 20-year lifetime expectancy in terrestrial environments. The progress achieved is as follows: (1) A non-sticking ethylene vinyl acetate (EVA) pottant has been developed by Springborn Laboratories for module encapsulation. EVA is now widely used in flat-plate module production. Another pottant, ethylene/propylene rubber (EPR), has recently been developed as an alternate to EVA. (2) Spire Corp. has developed a novel, integral electrostatic, bonded-glass encapsulation technique. This process seems to have real promise for future module use. (3) An Encapsulation System Design and Material Specification document (JPL 5101-177) has been published describing the various functional elements of PV module encapsulation in terms of materials properties, performance, life, and cost. (4) Degradation mechanisms and anticipated property changes caused by environmental aging have been defined and are being evaluated.

D. **Process Development Area.** The objectives of this task are to develop techniques, processes, equipment, and assembly methods for low-cost production of flat-plate collectors. Progress is as follows: (1) Spire Corp. has developed a junction formation processing system using a high-throughput, continuous-feed ion implanter, followed by a pulsed electron beam annealer. (2) MB Associates has developed an automated assembly sequence using robotics. A prototype machine has been built using vacuum cell pickup and induction heating for interconnect bonding. (3) Kulicke and Soffa Industries has developed a cell stringing and string application machine that is flexible enough to handle a variety of cell and assembly strings. (4) Illinois Tool Works (ITW) has demonstrated production solar cells having front and back metallization and antireflective (AR) coatings deposited by gasless ion plating. (5) Two module process system development unit (MEPSDU) activities are progressing at a slower rate than originally planned because of reduced funding. Westinghouse is using the dendritic web sheet; Solarex is using the Semix semicrystalline process for producing cells. The rescoped efforts are directed to the development of solar cell process steps and interactions.

E. **Engineering Area.** The objective of the Engineering Area is to develop the design criteria, test methods, analysis tools, and trade-off data that provide the engineering technology base required for technical feasibility for flat-plate modules and arrays. Recent progress includes: (1) Significant advances in module and array design, including new design and test methods for hotspot endurance, interconnect fatigue, temperature-humidity endurance, array wind-loading endurance, and module and array electrical safety. (2) Development has been completed of an installed array structure that is estimated to cost less than half of the Project's allocated cost of $50/m^2. The structure uses a planar frame made of members formed from light-gauge galvanized steel sheet and is supported in the field by treated wood trusses using trenched and back-filled soil to carry uplift wind loads.

F. **Module Performance and Failure Analysis Area.** The objective of the Module Performance and Failure Analysis (MPFA) Area is to evaluate the reliability and durability of modules that are constructed using improved materials and techniques. Recent progress is as follows: (1) Block IV module procurement for residential and intermediate-load applications has been conducted in two phases: Phase I, Design, and Phase II, Production. Most contractors have now successfully completed their production contract activities. (2) Seven contractors have participated in Block V module design contracts. Block V module designs for residential and intermediate load applications are expected to incorporate improvements to increase reliability and enhance the performance over Block IV modules.

3. **CONCENTRATOR TECHNOLOGY**

The Sandia Concentrator Technology objectives are to demonstrate: (1) the technical feasibility for baseline concentrator designs to achieve installed array prices of $150 to $250/m^2 of collector aperture with annual efficiencies of 15%, and (2) the technical feasibility for advanced concentrator designs to achieve annual efficiency levels of 25 to 30%.

A. **Concentrator Cell Research Task.** The objectives of this task are to develop: (1) high-efficiency, long-life single crystal silicon solar cells for concentrations in the range of 30 to 100, and (2) advanced devices operable for concentrations between 100 and 1000. A very important item to be achieved with silicon concentrator cells is the reduction of series resistance so that the ohmic drop does not excessively lower the output voltage and the cell efficiency. A concept to improve the aspect ratio of the metallization involves vacuum-deposited TiPdAg on the cell which was previously delineated by a lift-off process. Subsequently, silver is electroplated through a thick photoresist mask. Another process uses both liquid and dry-film resist to increase the aspect ratio of the grid lines. An active cell area as large as 93% of total area was achieved by this metallization technique while maintaining low series resistance. The cell (n_+pp_+), made using low resistivity material, achieved an efficiency of 20% at 100X. An advanced concept based on a grooved cover glass is also being evaluated for a potential increase of the effective area.

Advanced cell types such as the etched multiple vertical junction (EMVJ) cell, and the interdigitated back contact (IBC) cell are also being investigated for potential reduction in contact resistance without increasing the shadowed area by the grid. The EMVJ cell has vertically etched grooves that provide large areas for current-collecting grids without reducing the solar absorbing area. Cell efficiency as high as 18.5% has been measured for concentrations between 400 and 1000. One variation of the IBC cell has both p and n contacts on the back to eliminate the conventional front-grid area entirely.

Because GaAs has high-temperature capability, it is suitable for high (>100) concentrations and for collectors where both electricity and heat are utilized. The GaAs cell absorbs the short wavelengths of the visible solar spectrum more effectively than silicon, so it is being considered for multi-cell beam-splitter or stacked cell modules. The longer wavelength component will be absorbed by low bandgap cells such as silicon in the above concepts. The average efficiency of AlGaAs/GaAs cells fabricated by OM-CVD process is in a range between 21 and 24% at 400X-500X.

To evaluate the integrity of fabricated point focus cell assemblies, circular cells having a back metallization of TiPdAg are directly soldered to the copper metallized alumina substrate which are then mounted on a heat sink. For linear focus cell assemblies, especially for actively cooled receivers, the cells are placed on a cooling channel via a bonding substance with a good thermal conductance. Sandia Laboratories has developed an acoustic means to detect voids between the cell and heat sink.

B. <u>Concentrator Module Research Task</u>. The objectives are to: (1) design and fabricate solar concentrators that are compatible with point-focus and linear-focus concentrator modules, and (2) develop measurement and analysis capabilities for effective evaluation of concentrators. Refractive concentrators use point-focus and linear-focus Fresnel lenses. Typically, point-focus Fresnel lenses are fabricated by a molding process. The Fresnel lenses are subsequently bonded to an appropriate superstrate before integration into a module. Typical linear Fresnel lenses are produced in a lens film form. These lenses are bonded to a relatively thin superstrate, which is arched to provide mechanical strength and to increase optical quality.

The maximum optical efficiency of various lenses developed in this task is 87%. Sandia Laboratories recently developed a computerized lens analyzer, which automatically generates a three-dimensional plot of the concentrated solar flux for an effective display of lens performance. The results obtained from the analyzer provide important data for lens design.

C. <u>Concentrator Array Task</u>. There are approximately 900 kW of concentrator array systems installed throughout the world at this time. These installations include three parabolic trough arrays, one linear-focus Fresnel array, and two point-focus Fresnel arrays.

The Soleras array system (350 kW), which is installed in Saudi Arabia, uses a point-focus Fresnel array, with each unit rated at 2.4 kW. The array is mounted on a single hollow beam, which is supported on a single ground-installed post to which a full two-axis tracker is attached. The array system, which is the world's largest, is designed for 20-year life in a harsh environment. Also, a 240-kW parabolic trough system is installed and operating at Mississippi County Community College. Four other array systems, including a BDM linear-focus trough (50 kW) in Arizona; Wilcox Hospital, linear-focus trough (35 kW) in Hawaii; Dallas-Fort Worth Airport, linear-focus Fresnel (25 kW) in Texas; and Sky Harbor Airport, point-focus Fresnel (225 kW) in Arizona, are now in the process of performance verification checkout. These arrays will be excellent vehicles for accumulating operation experience and reliability data.

4. FUTURE COLLECTOR R&D PROGRAM ACTIVITIES

The FY 81 budget rescissions, the FY 82 budget allocation and the present U.S. administration guidelines have made a significant impact on the flat-plate and concentrator collector R&D effort. The original Photovoltaics Program plans called for the R&D activities to be continued up to and through the Technology Readiness demonstration of collectors at the cost targets identified. Technology Readiness definitions included the successful subscale demonstration of each individual step in a production process that would yield an economically competitive and reliable product if produced in sufficient quantity. Prototypes of the collectors must also have successfully undergone intensive performance and reliability analysis. The Photovoltaics Program has been restructured to address high-risk, high-payoff PV research and advanced development with limited technology development. For the Photovoltaics Program to be of maximum value to industry, it must contain balanced activities spanning from the high-risk, high-payoff technological challenges in advanced concepts and components to Technical Feasibility demonstrations for transferring technology to industry. The collector R&D program activity has been redirected from Technology Readiness demonstration to address component technical feasibility. The definition includes the successful subscale demonstration of the most critical (high-risk) aspects of component technology. Beginning in 1983, the collector R&D activities will encompass, in addition to the current development of silicon technology, the initiation of module exploratory development for advanced thin-film materials such as amorphous silicon and $CuInSe_2$.

5. CONCLUSIONS

Advances in both flat-plate and concentrator collector technology have shown progress toward meeting the objectives to reduce cost, improve performance, and extend the lifetime expectancy. Although the imposed budget reductions have heavily impacted Program direction, the collector cost target of $0.70/$W_p$ is expected to be met with increased industry participation and cost sharing. As a result, a variety of economic flat-plate and concentrator collectors will become commercially available for grid-connected applications.

LOW-COST SOLAR ARRAY PROGRESS AND PLANS
W. T. Callaghan
Manager, Flat-plate Solar Array Project
Jet Propulsion Laboratory, California Institute of Technology
4800 Oak Grove Drive, Pasadena, California

SUMMARY

The Flat-Plate Solar Array Project (FSA), sponsored by the U.S. Department of Energy (DOE) and managed by the Jet Propulsion Laboratory, has achieved considerable technical progress since that reported at the Third European Communities Conference. The basic technological direction has also changed since then. Progress in the areas of silicon feedstock refinement; large-area sheet formation; flat-plate module process sequence development, and engineering and environmental testing are discussed here in the context of a revised thrust developed by DOE toward higher-risk, longer-term-payoff work that emphasizes the assumption of technological risks that industrial firms are not likely to finance in the effort to develop reliable, low-cost photovoltaic modules.

1. INTRODUCTION

Significant redirection in the U.S. Department of Energy (DOE) Photovoltaics Program, and thus in the Flat-Plate Solar Array Project (FSA), has occurred since the 3rd European Communities Conference. Emphasis has been shifted from price goals and commercial development to longer-term, higher-risk, higher-potential-payoff opportunities. Many FSA goals and contractual activities have therefore shifted significantly. Goals, status and plans of FSA in this changed environment are discussed below.

2. SILICON REFINEMENT RESEARCH TASK

Formerly the Silicon Materials Task, previously dedicated to the design, construction and operation of pilot-plant silicon-refinement operations, its objective is now more generally to sponsor theoretical and experimental research on silicon (Si) material refinement technology suitable for photovoltaic (PV) flat-plate solar arrays.

2.1 STATUS

Experimental Process System Development Unit (Union Carbide Corp):

-- Process technology feasibility established; design and procurement completed; all major equipment fabricated and delivered.

-- Mechanical and electrical installation package completed.

-- East Chicago, Indiana, facility dismantled and moved to Washougal, Washington, at UCC expense.

Hydrochlorination Reaction (Massachusetts Institute of Technology and Solarelectronics, Inc.):

-- Process proof of concept completed through definition of reaction kinetics, catalyst and reaction characteristics.

Dichlorosilane (DCS) Chemical Vapor Deposition Process (Hemlock Semiconductor Corp.):

-- Completed preliminary design of experimental process system development unit with capacity of 100 to 200 MT/yr of Si.

-- Completed preliminary process design and economic analysis of 100-MT/yr plant for producing Si. Results indicate product price of $19.23/kg Si and plant capital cost of $22.8 million.

-- Operation of experimental DCS unit integrated with reactors being continued to obtain data with larger reactors. Goals: reduce Si deposition on reactor walls; decrease reactor power consumption.

2.2 PLANS

-- Research new reactor concepts that enable significant increases in Si deposition rates using chlorosilane and silane precursors.

-- Research new concepts in fluidized-bed reactor technology for chlorosilane or silane chemical systems.

-- Solve the critical technical problems remaining in the silane-to-Si and the DCS-to-Si processes.

3. SILICON-SHEET FORMATION RESEARCH AREA

Formerly the Large-Area Silicon Sheet Task, previously dedicated to the pilot demonstration of quality, large-area sheet throughput rates at prices consistent with module price goals, its objective is now more generally to research the critical elements of Si sheet growth to achieve a silicon-sheet technology compatible with future solar-cell requirements.

3.1 STATUS

Edge-Defined Film-Fed Growth (Mobil Tyco Solar Energy Corp.):

-- Ribbon widths of 10 cm and 5-h growth periods have been demonstrated simultaneously with 55% ribbon yields and AM1 cell efficiencies of 9.5% (unencapsulated 5 x 10-cm cell). Remaining technical barriers are growth rate, material quality, ribbon thickness and flatness, die lifetime.

Dendritic Web Growth (Westinghouse Electric Corp.):

-- Ribbon widths of 3 cm and 8-h growth periods have been demonstrated simultaneously with 60% yields and unencapsulated 2 x 2-cm cell, AM1 efficiencies of 15.0%. Key remaining problems: barriers are growth rate, ribbon flatness.

Advanced Czochralski Growth (Kayex Corp.):

-- Crucible outputs of 150-kg, 15-cm ingots have yielded 90% usable ingot, unencapsulated 2 x 2 cm cell, AM1 efficiencies of 15.5%. Key remaining problems: crystallization rate, material quality, crucible/melt interaction, wafering.

Semicrystalline Casting Technology (Semix Corp.):

-- Cast ingots of 14 kg, ingot yield 83%, have been demonstrated simultaneously with unencapsulated 10 x 10-cm cell, AM1 efficiencies of 11% to 12%. Remaining problems: material quality, wafering.

Advanced Wafering (Crystal Systems, Inc.; P.R. Hoffman Co., and Silicon Technology Corp.):

-- Simultaneous demonstrations with 15-cm dia, 12-mil wafers, yielded 17 wafers/cm, sliced at 0.25 wafers/min, 90% yield. Remaining technical barriers are kerf loss and wafer thickness reduction, slicing rate, saw-induced surface damage and expendable materials.

3.2 PLANS

-- Research silicon growth crystallization rate limits.

-- Do theoretical and experimental research on thermal stresses generated in the growth of wide and thin Si ribbons.

-- Research the influence of growth ambient atmosphere chemistry on the crystallization process and Si material quality.

-- Continue research on the basic mechanisms of cutting Si and the interaction of Si surfaces with experimental parameters.

-- Continue characterization of Si sheet with innovative techniques.

4. ENVIRONMENTAL ISOLATION RESEARCH

Formerly the Encapsulation Task, its objectives continue as finding solutions to the problems of encapsulation of modules. Research will be sponsored on aging-degradation characteristics and their influence upon module durability and reliability.

4.1 STATUS

-- Definition and specification of encapsulant functional requirements completed.

-- Analysis methods developed and being verified that relate materials and configuration to module performance.

Materials Development:

-- Solar-grade ethylene vinyl acetate (EVA) developed and available (Springborn Laboratories, Du Pont Co.).

-- Primer systems formulated and available (Dow-Corning).

-- Cover films identified, evaluated and available (3M Corp., Du Pont Co.).

-- Interlaminate spacer and air release film evaluated and available (Craneglas).

Process Development:

-- Cover surface treatments being evaluated for antireflection (AR) coating, soiling resistance, abrasion resistance.

-- Gasless ion plating for AR coating and metallization feasibility demonstrated with good cost potential.

Materials Durability and Life Potential:

-- Photothermal stability data base compiled and reported.

-- Material degradation mechanisms modeled and being validated.

4.2 PLANS

-- Research long-term photothermal degradation mechanisms in polymers: establish models and validate.

-- Investigate effects of bonding, dissimilar materials, and operational environments on encapsulant-interface stability criteria.

- Research corrosion mechanisms in module internal circuit elements; verify degradation rates and control criteria.
- Investigate operating-temperature limitations imposed by module design and mounting and by hot-spot sensitivity.
- Investigate and apply accelerated and durability testing techniques and life-prediction methods.

5. CELL AND MODULE FORMATION TASK

Formerly the Production Process and Equipment area, and previously dedicated to pilot demonstrations of module process sequence development, the Cell and Module Formation Task is now more generally committed to sponsor research on advanced cell and module formation technologies.

5.1 STATUS

<u>Sheet Surface Preparation</u>: Key recent accomplishment was in hot-sprayed AR coating; remaining problems are process reliability and efficiency improvement through surface passivation, and optical coupling.

<u>Junction Formation</u>: Key accomplishments are a demonstrated pulsed electron-beam annealer; a non-mass-analyzed ion implantation breadboard, and demonstrated large-area laser annealing. Problems and concerns: back-surface field improvement, large-area uniformity, polycrystalline Si effects.

<u>Metallization</u>: Key accomplishments are evaporated and/or plated copper cell collector grids; copper paste successful for back surface; initiated research on fired-through metal on AR coat; successful wave soldering; ion milling of excess metal, and Midfilm process developed for Ag and for MoSn. Problems and concerns: cell shunting, series resistance, reliability, formation of barrier to base-metal diffusion.

<u>Assembly</u>: Key accomplishments are: programmable robot; ultrasonic bonding (rolling spot); visual feedback and force sensing; light trapping background analysis. Problems and concerns: module reliability; handling breakage of emerging Si sheet forms.

5.2 PLANS

- Research the formation and characterization of electrically conductive silicides.
- Research the influence of polycrystalline grain boundaries upon junction formation and metallization.
- Research the physics of surface-field formation and of corrosion reactions at metallic interfaces.
- Continue to research non-mass-analyzed ion-implantation techniques, metallization and cell interconnection systems, and module assembly techniques.

6. ENGINEERING SCIENCES AREA

The objectives of this Area are to continue sponsoring advanced module and array engineering science activities that will lead to high-performance, safe, reliable, long-life designs.

Interface and environment research includes residential and commercial building codes; central power-station practices; module and array safety requirements and liability; hail-impact and cloudy-day probability statistics (60% to 70% complete). Array concepts research includes residential, roof-mounted, intermediate-load and central-station work (40% to 50% complete). Electric-circuit analysis research includes circuit-reliability modeling, hot-spot heating, and bypass diode studies, (about 80% complete except the diode work, about 30% complete). Safety research includes electric-insulation breakdown work; arcing, module flammability and system safety research (about 60% complete). Durability research includes interconnect fatigue work; glass and cell fracture mechanics; cell and module temperature and humidity endurance; module hail-impact resistance; optical surface soiling (70% to 80% complete).

6.1 PLANS

-- Continue research to characterize and define safe, reliable module and array design concepts and associated technology.

-- Continue to evolve methods of evaluating modules and arrays incorporating experience gained by FSA and DOE activities.

7. MODULE PERFORMANCE AND FAILURE ANALYSIS AREA

Formerly the Operations Area, previously dedicated to a wider, in-depth involvement with photovoltaics manufacturers and in the field testing of modules, this Area is now dedicated to the evaluation of reliability and durability of modules developed in various activities of FSA.

7.1 STATUS

Block IV Modules: The design objectives were to improve power-output specifications; improve fault tolerances; address larger applications; improve array efficiency; emphasize process-control-related quality assurance programs, and encourage innovative design approaches. The Block IV procurement resulted in modules with many firsts: hail-resistant and residential designs, close-packed cells, semicrystalline cells, ion-implanted junctions, back-surface fields, integral by-pass diodes, machine-soldered interconnects, 14% AM1 encapsulated-cell efficiencies, use of EVA encapsulant, polymer/metal-foil-laminate moisture barriers.

Qualification Test Evolution: An increasingly sophisticated test activity has developed over the years as a result of PV module improvements. The test elements now include thermal and humidity cycling, mechanical loading, twist, hail impact, electrical isolation, ground continuity and hot-spot endurance.

7.2 PLANS

-- Procure module samples constructed using innovative concepts.

-- Improve performance characteristics measurements.

-- Perform a broad program of laboratory environmental testing.
-- Place modules in field sites for endurance testing.

-- Correlate field and laboratory testing results to evaluate the environmental testing program.

-- Perform diagnostic analyses of module problems or failures.

AN ALL GLASS ENCAPSULATION OF SOLAR CELLS USING SILK-SCREENING TECHNIQUES

M.C. MICHEL
Laboratoires de Marcoussis, Centre de Recherche de la
Compagnie Générale d'Electricité, 91460-Marcoussis-France

Summary

The interest and originality of this process is threefold :

. glass only is used as an encapsulating material, which gives lower cost and higher reliability,
. the making of the front contacts and the sealing of the cells are simultaneous, which reduces the number of operations,
. the process involves a major use of screening techniques which are cheap and can be automated.

The main part of our work has dealt with the bonding of the cells to the glass sheet. We made modules of one or six cells.
This study involves :

. the study of a sealing glass with mechanical, chemical, optical properties fitted to the all glass encapsulation and the fabrication of the screenable paste using this glass,
. the definition of the process which involves the screening of the sealing glass, the electrical contacts and the electrical connections between the cells, and which makes these three parts compatible.

Properties of modules were measured before and after accelerated aging tests : the modules behaved satisfactorily after thermal cycles between - 40°C and + 90°C and humid conditioning cycles. Measurements of the electrical properties of encapsulated cells still show some scatter, nevertheless yields up to 12 % have been achieved.

INTRODUCTION

We present the work progress of the studies made on the glass providing the bonding between the solar cells and the front glass sheet, and the definition of the process for encapsulation and the properties of the encapsulated cells.

THE SEALING GLASS

The fabrication of the glass paste involves different steps :
- the composition of the glass with properties fitted to the encapsulation of the solar cells : an expansion coefficient fitted to the silicium and the front glass, a low melting-point, a good optical transmission (no devitrifying, no coloration), a chemical durability (with UV radiation and humidity).
As a general rule, a low expansion coefficient and a low melting point are inconsistent properties.

- the grinding of the glass. Difficulties about a polluted and a too coarse powder were solved.
- the composition of the organic part was defined as the process for mixing which plays a great part in the rheology of the paste.

The expansion coefficient of the used glass is $8.7 \times 10^{-7} {}^\circ C^{-1}$ and the glass paste is fired at 650°C for 15 minutes.

THE FRONT GLASS

The front glass is a pyrex glass chosen for its expansion coefficient near the one of the silicium, its good mechanical and optical properties. A same cost as the one of the glass used for laminated modules is achieved with a "glossy" quality (a rather rough surface) which requires a lower viscosity of the glass when the pressure is applied.

THE ENCAPSULATION PROCESS

The all glass encapsulation process of solar cells involves the following steps :

- layout on a glass sheet (the front side of the module) of sealing sites for each cell, by deposit of a glass paste screened through a mask complementary of the collecting grid pattern,
- firing of the sealing glass,
- layout of the grid pattern by the screening of a conductive paste through a second mask complementary of the first "sealing" mask,
- firing of the conductive paste,
- positioning on the sealing sites of slices of silicon previously fitted with their screened back contacts,
- heat treatment under pressure to secure the sealing and provide the front contacts. The highest temperature reached during this heat treatment is about 600°C,
- screening and firing of a conductive paste to establish the connections between cells,
- back encapsulation by hot spraying of a powder glass.

RESULTS

Samples with one and six cells were made. The samples have undergone thermal cycles between - 40°C and + 90°C without damage. The properties of the encapsulated cells were not degraded after aging in a weatherometer (Θ = 70°C, UV radiation, 80 % RH). Electrical properties of the cells have been measured under AM 1. The values are scattered and often too low, nevertheless yields of 12 % have been achieved. We have shown that the thermal treatment during the sealing cycle increases the series resistance of the cell, related to the oxidation of the contacts : the encapsulation process has been modified and the sealing is made under nitrogen. Then, we have shown that the interactions between the sealing glass and the silicium are harmful for the cell, so we direct our work towards the layout of coatings able to limit these interactions.

CONCLUSION

This all glass encapsulation process has three main interests :
- glass only is used as an encapsulating material, which gives lower cost and higher reliability,
- the making of the front contacts and the sealing of the cells are simultaneous, which reduces the number of operations,
- the process involves a major use of screening techniques which are cheap and can be automated.

The properties of the modules with one and six cells are attractive, nevertheless we had to solve problems concerning the interactions between the sealing glass and the silicium which are responsible for the too low electrical properties of the encapsulated cells.

LAMINATED MODULES WITH NEW PLASTIC MATERIAL
BASED ON AN ETHYLENE-VINYLACETATE COPOLYMER

A. DESOMBRE, J. DONON
PHOTOWATT International S.A.

Y. DE ZELICOURT, JC. BOBO
Laboratoires de Marcoussis
Centre de recherches de la C.G.E.

Summary

A large scale commercial production of solar modules is based on a sheet lamination process. The polymer predominantly used is a polyvinylbutyral film (PVB).

We have made a new formulation with a copolymer of ethylene and vinylacetate including several additives. This product is more interesting than PVB for the following reasons :

- low cost,
- better thermal stability due to the crosslinking,
- better UV resistance with an appropriate stabilization system,
- good adhesion to glass cover with a silane type internally blended primer,
- very high fluidity of the basic resin allowing a shorter degassing period.

The basic steps of encapsulation process involve :

- assembling module components in sandwich form,
- placing in a vacuum bag frame,
- evacuating the assembly,
- heating to fuse and cure pottant,
- cooling,
- removing completed module.

We have made some experimental modules which are now under long term ageing tests.

1. INTRODUCTION

PVB is a thermoplastic material which is very sensitive to temperature and humidity and which needs to be processed in an air conditionned room. Adapted formulations of EVA can easily replace PVB with many advantages such as cost reduction per kilogramme of material (EVA is four times less expensive than PVB), flexibility of adaptation to the intended use by using differents grades of EVA and by incorporating additives, possibility to have a cross linked material and to incorporate silane to improve the adhesion to the different components of the modules.

2. IMPROVEMENT OF NEW EVA FORMULATIONS

EVA is a copolymer of 'ethylene' and 'vinyl acetate', various grades of EVA can be used which are different in vinyl acetate percentage and in molecular weight (table I).

TABLE I
Properties of ELVAX (DUPONT)

	Vinyl acétate (%)	Molecular weight	Melt index (g/10 min.)	Hardness shore A	Elasticity modules (MPa)	Melting point (°C)
ELVAX 40 P	40.5		57	40	2.1	104
ELVAX 150	33		43	65	4.8	110
ELVAX 210	28		402	62	4.1	82
ELVAX 220	28		150	69	5.9	88
ELVAX 230	28		110			
ELVAX 240	28		43	73	7.6	110
ELVAX 250	28		25	75	9.0	127
ELVAX 260	28		6	80	11	154
ELVAX 265	28	increase ↓	3	83	14	171
ELVAX 310	25		402	70	7.6	88
ELVAX 350	25		19	80	12	132
ELVAX 360	25	increase ↓	2.0	85	18	188
ELVAX 410	18		502	80	14	88
ELVAX 420	18		150	84	19	99
ELVAX 450	18		8.0			
ELVAX 460	18		2.5	90	24	199
ELVAX 470	18	increase ↓	0.7			
ELVAX 550	15		8	92		
ELVAX 560	15		2.5			
ELVAX 565	15	increase ↓	1.5			
ELVAX 650	12		8	93		
ELVAX 660	12		2.5			
ELVAX 670	12	increase ↓	0.3			
ELVAX 750	9.0		7	95		
ELVAX 760	9.3		2			
ELVAX 770	9.5	increase ↓	0.8			
ELVAX 3120	7.5					

The variation of the percentage of vinyl acetate changes the mechanical properties of the resins : if the percentage decreases, the hardness and the elasticity modulus increase. With a same percentage of vinyl acetate, when the molecular weight increases, the hardness, the elasticity modulus viscosity and melting point increase.

To adapt the EVA for photovoltaic use, it is possible to mix the former resin to obtain intermediate properties. Optical transmission decreases when the vinyl acetate percentage decreases. Elvax 150 transmission is higher than Elvax 420 transmission, but for the last one the transmission is increased by crosslinking the resin.

Different formulations were tested and for example the three following: standard formulation A, laboratoires de Marcoussis (LDM) formulations B and C :

	Standard formulations ELVAX 150 (A)	LDM formulations ELVAX 150 (B)	ELVAX 420 (C)
EVA	100 g	100 g	100 g
péroxyde	1,5 g	1,5 g	1,5 g
stabilizer	0,1 g	0,5 g	0,5 g
UV absorber	0,25 g	0,9 g	0,9 g
anti oxydant	0,2 g	0,2 g	0,2 g

The ageing results under weather - O - meter show that the UV and thermal stabilities increase from formulation A to B and that the UV stability increases from B to C. From B to C the material becomes less sticky and is easy to handle.
These results show the importance of the stabilizer, the UV absorber and of the grade of ELVAX used. An optimization is under way to know the best mix to use.

3. ADHESION PROPERTIES OF EVA FORMULATIONS

Adhesion of the resin on the different components of a module (glass, cells back sheet) is a very important factor for the encapsulation itself.
The tests done with formulations A, B and C without primer give too low adhesion for use in a module. Two solutions to increase the adhesion are under investigation : components coated with primer , a silane type primer incorporated into the resin :

TEST	PEEL STRENGTH
1. Formulation A coated with primer	2,8 kg/cm
2. Formulation C coated with primer	4 samples destroyed for 6 tested
3. Formulation A with 1 g. of primer incorporated	5 destroyed for 5 tested
4. Formulation C with 1 g. of primer incorporated	6 destroyed for 6 tested
5. (1) and (2) after 630 h 95°C 100% RH	0 kg/cm
6. (3) and (4) after 630 h 95°C 100% RH	2,94 kg/cm

The interest of the primer being incorporated into the resin is to avoid the operation of   putting the primer on and the results show better adhesion after humidity tests.

4. EXTRUSION OF EVA

The rheological behaviour of EVA is very important for the extrusion of film and for the encapsulation cycle.
For extrusion the best method is to use a resin with low melt index, but for encapsulation and to reduce the duration of the cycle, the best method is to have a high melt index. This shows the importance of the optimization of the grade of Elvax from both points of view : stability and utilization.

The following resins were extruded with the optimized extrusion temperature :

ELVAX 150 formulation A .... 80°C
ELVAX 420 formulation C + primer incorporated .... 75°C
(extrusion temperature)

ELVAX 420 is more difficult to extrude than the ELVAX 150, this is due to the difference in melt index of the two resins.

5. ENCAPSULATION PROCESS AND RELIABILITY INVESTIGATIONS

Two series of samples were manufactured :
- formulation A and C were used to do samples of 4 cells with electroless nickel and solder contacts with a primer deposited on the components,
- formulation A without primer and C with primer incorporated were used to do sample of 10 cells with silver printed contacts.

The encapsulation cycle is very similar to the one used with PVB and is not very critical and has three main phases :

a) vacuum in and out of the module + heat from 20 to 100°C in 10 minutes (vacuum outside of the module is broken after 5 minutes),
b) heat from 100 to 130°C in 10 minutes,
c) heat from 130 to 160°C in 10 minutes.

The different overstress tests used for the technological investigations on these modules are :
  humidity test : 85°C 85% RH (for very accelerated tests 95°C 100% RH)
  high temperature storage : 100°C (for very accelerated tests 150°C)
  thermal cycle : -40°C + 85°C

Some of the accelerated tests are only valid for particular technological investigations and we only consider the intended specific point (in these tests we can observe secondary effects without relation with use).

The results obtained with the first series of samples show a modification of the output power of 8% after 3000 h of humidity test at 95°C and 100% RH for both formulations and of 9% for formulation A and 3% for formulation C after 1000 h at 150°C and no degradation at 100°C.
From the visual point of view the formulation C after the tests looks better than the formulation A.

The results obtained with the second series of samples are not yet significant. After 500 hours there is no degradation for formulation C. Formulation A is not under test because of the too low adhesion of the resin to the glass

6. CONCLUSION

EVA is very promising material. Flexibilities to modify the starting material and to introduce additives make it possible to obtain a very stable resin for lamination.

The adhesion of EVA on all the components of the module can be done by the incorporation of a primer in the resin itself.

All the ageing tests show that EVA withstands very well under various conditions.

REFERENCE :

Investigation of test methods, material properties and processes for solar cell encapsulants

SPRINGBORN LABORATORY INC.
JPL CONTRACT 254 527

This study is partly supported by CEC and COMES.

DESIGN AND DEVELOPMENT OF A MODULE FOR THE EEC PV PILOT PLANTS

A. MULEO, F. NARDI and V. PARAGGIO

Solaris S.p.A., Florence, Italy - Pragma S.p.A., Rome, Italy

Summary

The modules for intermediate load applications ought to be particularly designed in accordance with higher standards and ad-hoc criteria. The module developed in this work accepts 72 cells either round, 100mm in diameter, or square, 100 mm by 100 mm. The cells are located in a 6x12 matrix and the connections form 6 parallel strings and 4 series blocks with 3 cells per substring. This particular module delivers a current that is six times the current and a voltage that is twelve times the voltage of a single cell. Parallel interconnections (total of five) run across every substring of three cells in series. There are two main advantages of this module. The first is that it is a high current-low voltage module, better suited for large systems since mismatching problems due to the use of modules in parallel are reduced. The second is the high reliability.

In this paper we discuss reliability recording the behavior of the module to shadowing, short circuiting of a cell, opening of interconnections, etc. Also the design of the module has been improved by closely matching the antireflection coating to the combination of low iron glass and of encapsulant. Spectral response of the cells prior to encapsulation and after paneling are shown.

1. INTRODUCTION

Modules for intermediate and large load applications have some particular boundary conditions to satisfy. One of the first requirements is to have a large-size module thus reducing the number of modules of the PV field. In any case, sizes larger than $1m^2$ (corresponding to weights in excess of 20 Kg) are very difficult to handle both in production and during installation, causing either higher production or installation costs. Also, the bigger the size, the higher the risk of breakage, the lower the production yield of the modules and the higher the cost per peakwatt. Have choosen about $1m^2$ will mean that each module will deliver about 100 peaw watts. This will mean that for a 10 KWp plant approximately 100 of those modules will be needed. This number (100 modules) is sufficient so that the system designer has a wide range of I-V characteristics of the PV field by interconnecting this modules in series-parallel, without being limited by the I-V characteristics of a single module.

Thus, the internal interconnection of the module will have a unique goal: the reliability of the module. The module that we have developed accepts 72 cells either round, 100 mm in diameter, or square, 100 mm by 100 mm.

2. THE CELLS INTERCONNECTIONS

It is well known that in a module, the cells can be connect all in series, all in parallel and in parallel-series mixed configurations. When the cells are series connected, if one cell fails (usual failure is breakage of either the cell or the interconnections or, in general, an opening of the cell circuit) or does not generate power (shadowing, etc.) the entire module will not produce power. If there is no protection at the system level, the module can also be damaged. If the cells are all parallel-interconnected, in the event of a failure or shadowing, the power loss of the module itself will be approximately equal to the power delivered by the single cell. In this case, the failure of one cell in the module is very difficult to detect and this module will be a very low-voltage and very high-current module causing high cost of interconnection inthe PV field, due to critical series resistence problems in the cable connections, junction boxes, etc. In our modules the cells are located in a 6x12 matrix, as seen in figure 1. The module itself is rectangular about 1.30x.65 meters, in size, with a weight of about 15 kg, and is very easy to handle and install.

The connections between the cells form 6 parallel strings and 4 series blocks with 3 cells per substring. This particular module delivers a current that is six times the current and a voltage that is twelve times the voltage of a single cell. A total of five parallel interconnections run across every substring of three cells in series. With this mixed series-parallel interconnection scheme, the failure of a cell in a module will give a power loss of less than 5%. Assuming an acceptable power deviation from average of 10% and a module cell failure rate of 1/10000, the module manufacturing yield is above 99,9% (1,2).

3. CONSTRUCTION CRITERIA

We have used a high-transmission, low-iron, tempered glass with a texturized surface to reduce the reflection and the limit angle effects. In figure 2, the transmittance of such glass, 3mm thick, is shown.

Infigure 3, we show : a) spectral responce of a cell in air, b) the response of the same cell under the glass with air in between and c) the response of the same cell under the glass with a thin layer of silicone in between. It can be observed that since the antireflection coating of the solar cell is matched to the index of refraction of the silicone encapsulant, so the response of the cell with a layer of silicone is at least 10% better than the response of the cell in air, over all the interesting wavelengths. See also figure 4. Therefore in a structure such as that of our modules, as shown in figure 5, the reduction in response due to absorption or reflection of the glass is less than the increase of response due to the presence of the silicone layer over all the wavelengths of interest.

Figure 6 shows the double redundant interconnections between the cells with a stress relief interconnection loop.

Figures 7,8,and 9 show the I-V characteristics of these modules under various conditions. It has to be noted that these modules have satisfactorily passed the tests of the EUR Specification No.501 of the ESTI facility at the Ispra CEE/JRC. In particular, they have satisfactorily passed :

-Temperature Cycling Test
-Humidity-Freezing Test
-Damp Heat Long Exposure Test
-Insulation Test
-Hail Resistence Test
-Mechanical Loading Test

4.CONCLUSION

The module designed is ideal medium and large photovoltaic generation plants.With dimensions smaller than one square meter,it is well handled both in production and during installation.The connections between the cells have been studied to achieve the highest reliability with respect to both failure of the cells and shadowing.

Finally,the encapsulation is such that the spectral response of the cell is not penalized by the combination of the glass and the silicone encapsulant over all the wavelengths of interest.

5.REFERENCES

1. Ross R.G. : 14th IEEE Photovoltaic Specialist Conference, San Diego California,Jan. 1980
2. Gonzales C.- Weaver R. : 14th IEEE Photovoltaic Specialist Conference,San Diego California,Jan 1980

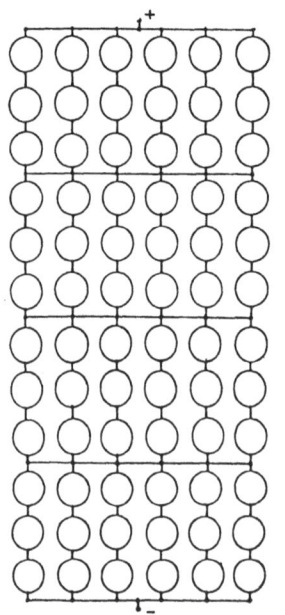

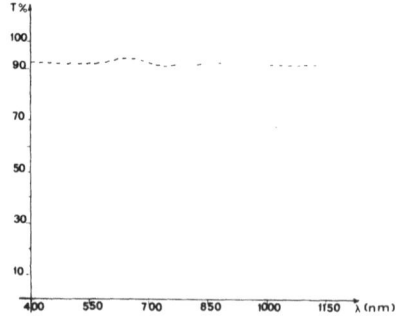

Fig. 1 - *Electrical interconnections of cells in the module*

Fig. 2 - *Transmission in % versus λ for the glass used.*

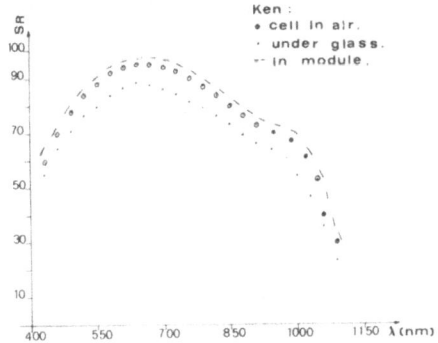

Fig. 3 - Spectral response versus λ for a typical cell in air, under glass, and in module

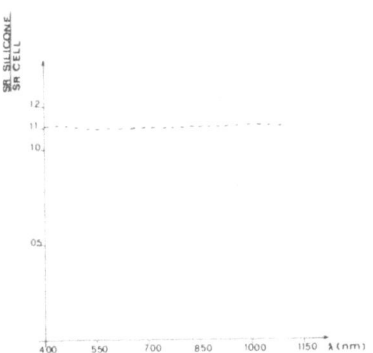

Fig. 4 - Improvement of cell response with silicone layer. Ratio of SR with silicone to SR without silicone is plotted versus λ

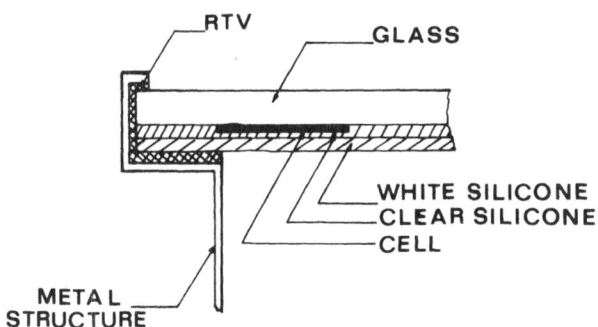

Fig. 5 - Cross section of the module

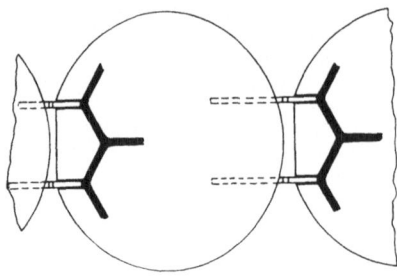

Fig. 6a - Cell interconnections

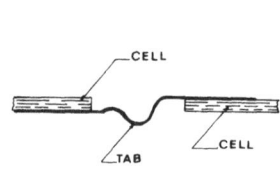

Fig. 6b - Cross section of a single interconnection

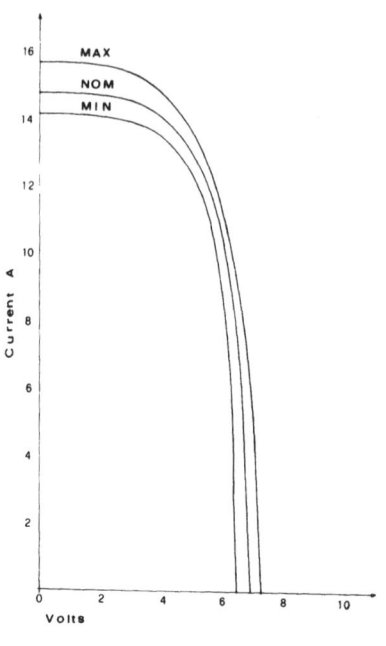

Fig. 7 - Typical range in I-V characteristics of the module

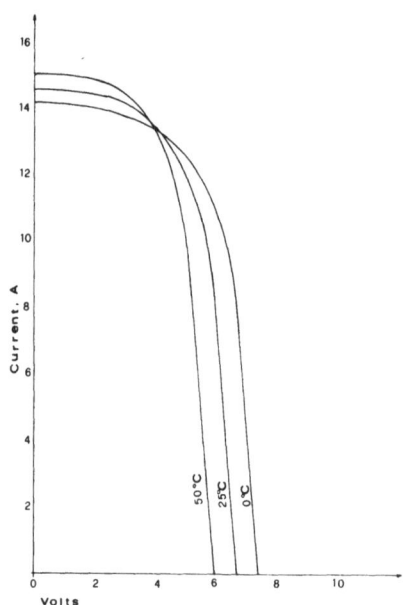

Fig. 8 - I-V characteristics of the module at different temperatures

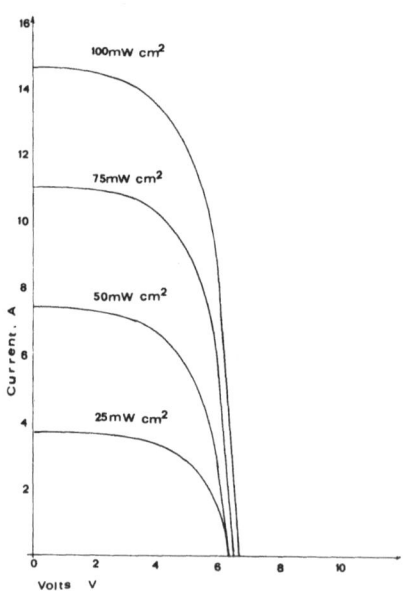

Fig. 9 - I-V characteristics of the module at different illumination levels.

MODULE DESIGN FOR EC PILOT PROJECTS

J. DONON, J. ANGUET, A. DESOMBRE
PHOTOWATT INTERNATIONAL S. A.
6, rue de la Girafe
14001 - CAEN

P. COUREAU
PHOTOWATT INTERNATIONAL S. A.
131, route de l'Empereur
92500 RUEIL MALMAISON

Summary

A new module was designed of the CEC pilot projects of NICE AIRPORT and MONT BOUQUET (50 kW each).
It is a 73 W module made of 72 100 mm diameter single crystalline silicon cells.

A cost reduction is achieved in encapsulation, frame and wiring. The technology of the encapsulation is a laminated structure including a low iron content tempered glass (1455 x 610 mm) polyvinyl butyral and a tedlar / aluminium tedlar sheet for the back.

A connection box allows an easy wiring in parallel or in series under 12 or 24 Volts.

The frame around the module is self-supporting, allowing "very simplified" on-site mounting structures.
According to the results of an investigation of the necessary protection against the hot spots, each string of 12 cells in series is protected by a diode.

The encapsulation structure allows to insert the 6 protecting diodes in the connection box. The whole technology has been selected to ensure a high reliability in severe environmental conditions and to allow a fast on-site mounting of the power plant.

A reliability programme has been set up to verify the good fit to the intended use in EUROPE and the possibility of also using this module under other kinds of climate, especially tropical ones.

1. INTRODUCTION

PHOTOWATT International is one of the contractors for the CEC pilot projects and is involved in building the two projects of NICE AIRPORT and MONT BOUQUET (50 kW each).
A new module was especially developped to be used in these projects from an economical and technical point of view. The optimisation was done not only for the module itself but also for the whole system : frame supporting structure, wiring, utilisation, protection...
This led PHOTOWATT International to develop the new, now commercially available, 73 W module (PW-P 800).

2. ENCAPSULATION DESIGN

The following components were selected to do the encapsulation :

- low iron content tempered glass superstrate,

- PVB encapsulation material,

- back sheet of tedlar Aluminium - tedlar PVB was chosen instead
  of EVA because the EVA film commercially available shows to low
  adhesion to the components of the module.

The three layers - glass, PVB and back sheet are laminated together
during the encapsulation cycle.

The other technological choices are :

- 72 100 mm diameter high efficiency monocrystalline solar cells
  with highly reliable silver printed contacts.

- high redundancy of the interconnections between cells.

- geometric arrangement in a six wide and twelwe long configuration.

- one by pass diode for each row of twelve cells to avoid all the
  hot spots problem.

- connection box   allowing an easy wiring in parallel or
  in series under 12 or 24 Volts and with the six protecting diodes
  incorporated in this box.

- connection box not positioned behind a cell to have the same
  heat dissipation for each cell (a connection box behind a cell
  led to an excessive heating of the cell and can give hot spot or
  bad electrical adaptation problems).

- self supporting frame around the module allowing very simplified
  and cheap on site mounting structures. This frame can support a
  wind load of 200 km/h with the module supported only by the two
  small sides.

3. ELECTRICAL PERFORMANCES AND GEOMETRICAL DIMENSIONS

The over all dimensions of module are 640 mm x 1535 mm x 40 mm
with a weight of 16,7 kg (Figure 1).

The electrical performances are the following :

T	P_{max}	I_{sc}	V_{oc}	NOCT	$\Delta V / \Delta T$	$\Delta I / \Delta T$
25°	73 W	2,4 A	42,4 V	45°C	2 mV/°C/ cell in series	1,5 mA/°C/ cell in parallel

4. STUDY CONCERNING THE PROTECTIONS OF THE MODULE

A computer programme has been set up to optimize the protections for
the module.

Different cases were studied :

n cells in series with one in reverse
n cells in series with one in reverse and having a Rsh
n cells in series with one partially shadowed

This programme takes into account the power to be dissipated by a cell $P_1 = E_1 \frac{\pi D^2}{4} a \, r$, where E_1 represents the incident power in W/cm2, D diameter in cm, a coefficient of transmission of the glass and $r = 1 - \eta$ with η conversion efficiency of the cell.

the total power that a cell can dissipate, this power is dependant upon the maximum temperature that the cell can support without degradation $P_2 = \frac{T_{max} - T_a}{T_j - T_a} P_1$ T_j junction temperature
   T_a ambient temperature

the electrical power delivered by the module $P_j = I_j V_j$ with I_j and V_j defined as the current and the voltage at the temperature T_j

For n cells with one in reverse the balance of the powers are

$$P_2 = \frac{P_1}{r} + (n - 1) P_j$$

By solving this equation and for a Tmax choosen at 100°C which is a good value to have no degradation of the cell and of the encapsulant, with a Ta = 35°C at 1KW/m2 and with $T_j = T_a + \Delta TE$

ΔTE	n
15°C	30
20°C	21
25°C	15

With the same computer programme and considering that the cells have a shunt resistance

$$\frac{Pe}{100} = \frac{Icc - IM}{Icc}$$

At 1 KW/m2 and Tmax = 100°C

$\Delta TE = 25°C$		$\Delta TE = 20°C$	
$\frac{Pe}{100}$	n	$\frac{Pe}{100}$	n
0 - 3 %	15	0 - 3 %	21
5 %	18	5 %	24
10 %	19	10 %	25
15 %	21	15 %	27

Cells with low Rsh gives self-protection of the module but also exhibit bad fill factor. For the module PW-P 800 the slope is between 5 and 10 % which shows that 18 cells in series have to be protected by a by-pass diode. In fact with the selected design 12 cells are protected giving more reliability for the module.

The same calculations with a cell partially shadowed gives the following results.

For a $T_j = 60°C$ and 1 KW/m2 the temperature of the cell partially shadowed are :

Number cells in serie	% of the area shadowed										
	0	10	20	30	40	50	60	70	80	90	100
11	61	64	67	70	73	76	79	82	85	88	91
12	64	66	69	72	75	78	81	84	87	90	93
13	66	69	72	75	78	81	84	87	89	92	95
14	68	71	74	77	80	83	86	89	92	95	98
15	70	73	76	79	82	85	88	91	94	97	100
16	72	75	78	81	84	87	90	93	96	99	102
17	74	77	80	83	86	89	92	95	98	101	104
18	77	80	83	86	89	92	95	97	100	103	106
19	79	82	85	88	91	94	97	100	103	106	109
20	81	84	87	90	93	96	99	102	105	108	111
21	83	86	89	92	95	98	101	104	107	110	113
22	85	88	91	94	97	100	103	106	109	112	115
23	88	91	94	97	100	103	105	108	111	114	117
24	90	93	96	99	102	105	108	111	114	117	120

4. RELIABILITY RESULTS

This module is approved by CEE and is to be used in the photovoltaic pilot project after test at JRC - ISPRA (Spec 501).

The following results were obtained from the PHOTOWATT International qualification programme :

High temperature storage 100° C	> 4000 h
Humidity storage 85°C 85 % RH	> 4000 h
Thermal cycling (-40 + 85°C)	> 500 cycles
Wind loading (module fixed only on the two small sides)	> 180 Km/h
Twist test deformation angle	$1,2°$

Hot spot heating test behaviour guaranteed by the 6 by pass diodes

Hail test ice ball of 25 mm diameter at 23m/s

insulation test > 2000V DC

5. CONCLUSION

The new PW-P 800 PHOTOWATT module is now approved by CEC for the pilot project. During the whole study of this module we learned how to protect the module efficiently. This module can be very interesting for the system designers and installators from the economical and technical point of view due to the high power, new design (frame, box, protection).

Reference : Photovoltaic module hot spot durability design and test methods
J. C. ARNETT, C. C. GONZALES
15th IEEE Photovoltaic Specialist conference 1981

FIGURE 1. View of the modules

SOLAR CELLS FAILURE MODES AND IMPROVEMENT OF REVERSE CHARACTERISTICS

A.M. RICAUD

FRANCE - PHOTON

16015 ANGOULEME CEDEX

Abstract

The growing number of large photovoltaic installations calls attention to the possible development of "hot spots" leading to a loss of power in the best case. Catastrophic consequences on the whole array can be avoided by particular precautions taken at the cell, panel and system design level. Hot spots problems arise in modules when one or several cells become back biased and operate in the negative voltage quadrant, as a result of short-circuit current mismatch, cell cracking or shadowing. Optimization of system designs with diodes protections has already been demonstrated by France-Photon in a previous paper(1). Further to the industrialised matching of the cells in a same module, an important result of this study is that we can improve the capability of dissipating reverse power by decreasing the shunt resistance with negligible sacrifice of the direct peak power. After a complete study of the dark direct current, we show for our basic technology that the main parameter which governs the slope of the reverse characteristic is the shunt resistance and we indicate trends to understand its fondamental causes and their relative weight for single and semicrystalline cells.

INTRODUCTION

Industrial solar cells are usually classified in terms of forward voltage characteristics by measuring their current at a given voltage in the vicinity of the maximum power point, but an understanding of their failure modes under reverse voltage is very important for the design of reliable large scale systems.

A cell is forced into reverse bias operation when it is shaded, cracked or electrically poorly matched with other cells in its string. Shading must be considered a normal event in photovoltaic system operating, occuring for example as a result of foliage, dirt accumulation, birds drops or washing.

In reverse operation a cell works as a receptor, dissipating supplementary power as heat. This power has to be low enough to be dissipated by natural convection depending upon encapsulant material.

Our experience has shown as many as 36 cells in series can easily be protected by one by-pass diode limiting the maximum reverse voltage to a value below that which would damage a shaded cell (18 V $_{at}$ 60°C). That prudent practice showed very efficient to protect a single cell against hot spots, but at the same time, the direct power loss of the module supporting the defective cell must be reduced to the minimum so as to be more tolerant of shading in real life condition.

So we have tried to define an optimum value of the shunt resistance compatible with the two following conditions :
- the direct power of a normal operating cell and its reliability is not much affected by the leaky junction.
- a partially shaded cell induces the minimum power loss at the module level.

In order to manipulate the shunt resistance in the manufacturing process with no trouble for long term reliability we present in this paper what we think to be the fundamental causes of the shunt for our basic technology.

A) Statistical study of reverse characteristics :

Three groups of cells have been tested :
Technology A on monocrystalline substrates (Type 1)
Technology B on monocrystalline substrates (type II)
Technology B on semicrystalline substrates (type III)

One finds that even in the same group (type1) reverse voltage responses of cells can vary widely. For example, junction breakdown can occur at 25 mA/cm2 and 40 V for one cell and 150 mA/cm2 and 10 V for another one of the same group. Experimental results are summarized in Fig 1 for cells of type 1 in the darkness. So we have determined a safe region delimited by a power hyperbola where junction breakdown do not occur even in a long exposure to reverse voltage at high junction temperature (60°C)

However, cells of type I had a shunt resistance higher than 50 Ω for most of them. It is the reason why 26% of them reach the breakdown limit for current less than 1,5 A which is normally their short circuit current under AM1. In that limit, breakdown occurs as a classical diode avalanche and seems to be mostly related to a field effet at the junction level.

Fig 2 shows that the three curves representative of each type of cell present the same general shape with three typical inflexion points measured at -1V,-18V and -30 V respectively for type II. Although semicrystallines cells behave the same way, their breakdown knee is generally very soft and extends towards high currents with a resistive comportment of about half the value of their shunt resistance.

The classical theory of a reverse diode characteristic which recognizes three main components for silicon (surface effects, generation current in the space-charge region and diffusion current) cannot explain the established shape of the curve. Moreover, the main parameter that governs the general curve orientation is the shunt resistance which value can be accurately appreciated in the short circuit vicinity.

B) Shunt effect on direct and reverse bias characteristics.

1 - Dark current measurements :

Using the two expomentials model for dark current, one can write the general expression of an illuminated cell current as :

$$I = I_L - I_{go}(e^{\frac{q(V+R_sI)}{2k_T}} - 1) - I_{do}(e^{\frac{q(V+R_sI)}{k_T}} - 1) - \frac{V+R_sI}{R_{SH}}$$

Where I_{go} is the generation component in the space-charge region for V=o and I_{do} is the diffusion component for V = o

Fig 3 shows a semi-log plot of the direct dark current of a type II cell and gives excellent agreement with the theorical model when the current conducted through the shunt resistance is substrated from the current values of the diode forward characteristics. Then, the curve is matched by the here-above equation with $I_{go} = 8.10^8$ A/cm2 and $I_{do} = 2.10^{12}$ A/cm2.

These values are compatible with those calculated by several authors

- 393 -

in the case of N+P silicon structures and prove the quality of the diffused junction. As they are almost independant of the shunt resistance values, we must investigate other trends to detect the shunting reasons.

2) Where does the shunt come from ?

The mean values of shunt resistance for each type of technology were respectively 63 Ω, 14 Ω and 6 Ω

An elaborate investigation over the cells of type II and III (single and semicrystalline cells of technology B) has revealed the following sources of shunting.

a) Printed insulation annulus :

A sectionned annulus shorting front and back of the cell is the most evident source of shunt. Values as low as 1Ω(80ΩXcm2) are possible. Normal shunt resistance as high as 50Ω(4000 Xcm2) can be restored by grinding the edge of the cell. An other cause of shunt can result from a porous ring because the oxide is not well densified. That sort of shunt cannot be removed. In that case, the quality of the chemicals must be questionned.

b) Unalloyed regions of aluminium (Ref 2)

Unalloyed regions in the back surface field alloy layer tends to decrease the shunt resistance but this defect in the back structure can be detected on the dark-current I-V curve because the extra PN+ diode is a source of recombination for minority carriers within a diffusion length of the unalloyed area. So, the incompleteness of the BSF can be evaluated by checking a soft knee on the direct I-V curve further to the shunt measurement.

That sort of shunt is highly undesirable because it is a major cause of poor fill factors. So it has to be totally eliminated in the manufacturing process.

c) Jagged edges and sawing damages :

A small percentage of as cut silicon wafers presents fracture beginnings at their edge as the result of deep sawing damages. These jagged edges can propagate inside the whole single crystal wafer with a high concentration of constraints inducing localised dislocations. Further investigations are needed to understand as these dislocations are sources of shunting.

We have also discovered on semicrystalline materials (SEMI X) a very high density of dislocations within sub-grains of slight disinclinations. These sub-grains are always related to a straight forward short circuit. Wolfgang Schröter and all(Ref 3) have shown that edge dislocations can be associated in Silicon and Germanium to the presence of half or partially filled bands within the band gap and act as conducting channels across the whole thickness of the wafers.

d) Grain boundaries in semicrystalline materials(Ref 4)

Since the grain boundaries extend from the front to the back cell surface, a tendency for cell shunting has always been noted. Control of this problem presents the major technical challenge in implementing semicrystalline materials at an industrial level.

We have noticed that a pre-gettering with phosphorous diffusion followed by sodium hydroxyde etch can improve the shunt resistance from 6 Ω to 30 Ω on a series of SIL60 material. Further work is required to improve the compromise between the enhancement of shunt and the decrease of short-circuit current due to longer exposure at high temperature in the diffusion furnace.

e) Front metallization micro-shunting :

Alkakine bath of electroless plating Nickel have a tendency to etch silicon before the start of autocatalytic deposition of Nickel (ref 5). This is of catastrophic effect on very uneven shallow diffusions since the junction can be shorted here and there in the thinnest parts of the diffusion.

This shunting effect does not appear normally on single crystal cells but is very wide spread on the grain boundaries of semi-crystals especially for the samples which have suffered a long alkaline etch exposure.

We have also noticed for single as for semi-crystalline wafers, temperature non uniformities on a given cell exposed to a high reverse voltage probably due to local micro shunts under the metallizations.

At breakdown, the solder metallization always fuses under a pattern showing local weakness. So, a good cosmetic aspect of the front pattern often correlates a hard resistance to breakdown.

3 - Optimization of the shunt resistance value :

Some manufacturers have maximized the peak power of their cells by increasing the shunt resistance and correspondingly have flat reverse curves. When shading occurs, the shaded cell moves quickly into a high reverse voltage inducing serious failures if not protected or in the best case an important power loss at the module level.

Fig 4 shows the relative influence of shunt resistance on the fill factor and photovoltaic efficiency of 100 mm diameter single crystal solar cells. These calculated efficiencies are based upon an ideal fill factor of 0,792 obtained from dark current experimental measurements, a short-circuit current of 32,4 mA/cm2 and an open circuit voltage of 600 mV also measured as the average values of that series of samples.
The series and shunt resistance are classical values of our basic technology. It is obvious, especially for these large diameter cells that the series resistance has a very high effect compared to the one of shunt resistance.

When R_{SH} varies from 50 to 10 Ω , the direct power loss is 2% and when R_{SH} varies from 10 to 5 Ω the loss is less than 3%. Comparatively, same power loss are expected with series resistance variations as small as 5mΩ. For semi-crystalline cells (SEMIX 100 X 100 mm) with 6Ω shunt resistance, the power loss is only 5% with the additional advantage that, when reverse biased, the heat is dissipated over a large percentage of the cell surface area. It has been demonstrated that a small fraction of the reverse power can be converted to light which is measured during a prolonged exposure photograph. The light emission patterns on a semicrystalline cell undergoing extreme reverse bias, are identical to the cell heat generation patterns with dissipation of heat over junction boundaries which are many times the area of a single crystal cell perimeter. As an additional advantage, most of the grain boundaries of the semicrystalline cell are surrounded by silicon, an excellent thermal conductor. The boudaries of a single crystall cell are surrounded by encapsulant, a poor thermal conductor (Ref 4).

Fig 5 shows the available direct power on a 34 cells (100 mm$\emptyset$) module when a cell suffers complete occultation for different values of its shunt resistance at constant charge (resistive load) : 8Ω at 25°C
The power dissipated into the occulted cell is also shown.

Obviously, we see that a shunt resistance of 10Ω is highly preferable than 50 Ω when a dead leaf occults a cell for example.

CONCLUSION

This paper has attempted to explain in basic terms how reverse bias heating occurs and how it can be minimized at the cell level. We have indicated the main sources of shunting on our technology, which of them are dangerous for long term reliability and those which are benefic from heat dissipation and power loss points of view.

We think that the continuous effort on photovoltaic efficiency improvements has neglected the real-life conditions of cells and modules. A small sacrifice of peak power can be made at the shunt level in order to have a safer reverse characteristic. It is worth improving cells efficiency by the way of other parameters such as :
dark current, short-circuit current, series resistance and open circuit voltage to win both challenges of efficiency and long term reliability. This study is supported by COMES contract n° 81 110 016 3311

REFERENCES :(1) A.M. RICAUD and all in 15 th I.E.E.E. photovoltaic specialists conference- Orlando 1981 p. 1117.
(2) J. CULIK and S. KATZ "Method to monitor the quality of screen-printed aluminum paste back surface fields" in 15h I.E.E.E. p.512
(3) Wolfgang SCHROTER and all, Journal of microscopy Vol 118 Pt 1 January 1980 p.23.
(4) J.LINDMAYER "characteristics of semicrystalline silicon solar cells" in 13 th I.E.E.E. photovoltaic specialists conference Washington 1978 p. 1096
(5) R. PETERSEN and A. MULES "Silicon solar cells with nickel/Solder metallization" Third EC photovoltaic solar energy conference, Cannes 1980 p. 684.

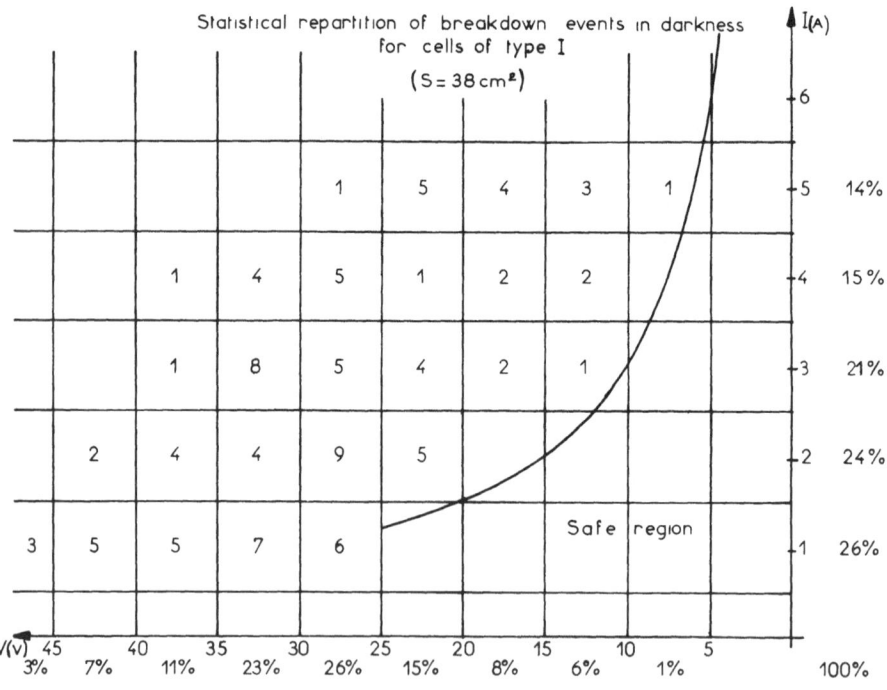

Figure 1

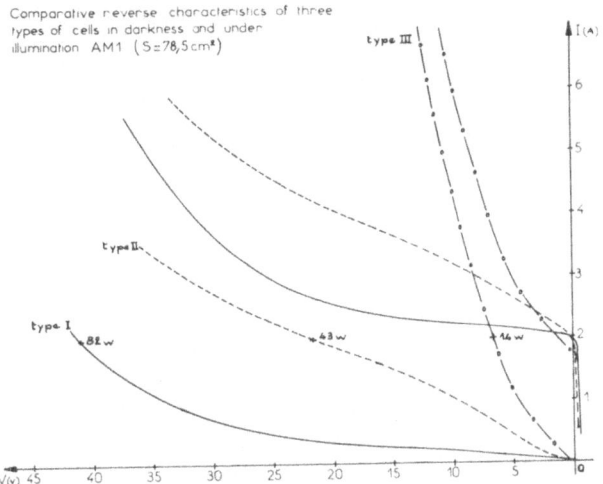

Figure 2

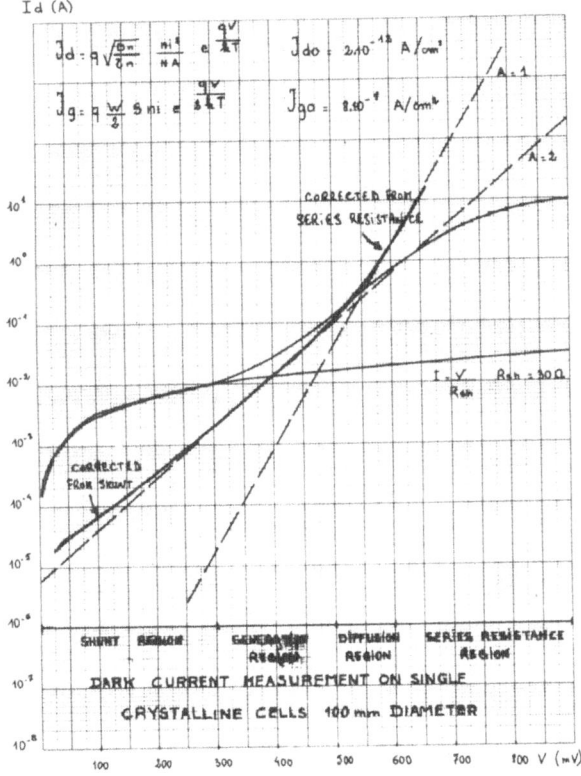

Figure 3

Rs Ω	Rsh Ω	FF	η%
0,015	5	0,70	13,6
	10	0,72	14
	50	0,74	14,4
0,020	5	0,69	13,4
	10	0,71	13,7
	50	0,72	14
0,025	5	0,68	13,1
	10	0,69	13,4
	50	0,70	13,7

COMPARATIVE INFLUENCE OF THE SHUNT AND SERIES RESISTANCE ON THE FILL FACTOR AND PHOTOVOLTAIC EFFICIENCY FOR 100 mm DIAMETER SINGLE CRYSTALLINE CELLS OF TECHNOLOGY B

$$J_{sc} = 32,4 \, mA/cm^2 \qquad V_{co} = 600 \, mV$$

$$FI = 0,792$$

$$FF = FI \left(1 - \frac{V_{co}}{R_{sh} I_{sc}} - \frac{I_{sc} R_s}{V_{co}} + \frac{R_s}{R_{sh}} \right)$$

Figure 4

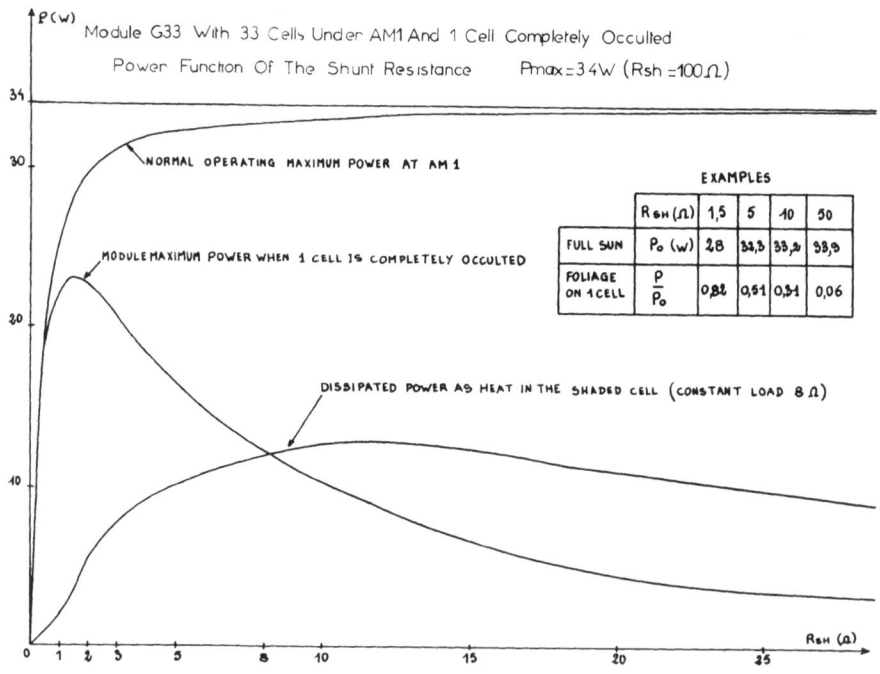

Figure 5

-398-

FABRICATION OF LARGE AREA Cu_2S/CdS THIN FILM SOLAR MODULES

H. HUSCHKA, B. SCHURICH, J. WÖRNER

NUKEM GmbH, D 6450 Hanau 11, Rodenbacher Chaussee 6

Summary

Thin film photovoltaic generators based on the heterojunction system Cu_2S/CdS are the most developed alternative to Si solar cells. The results and experience obtained prove the Cu_2S/CdS cell to be a stable and highly reliable element with good conversion efficiencies. To demonstrate the feasibility of a large scale production a pilot plant was built by NUKEM. Most conventional methods and techniques are used for this line.
It is expected that such a generator could reach the price goal which was defined for economic operations.

1. INTRODUCTION

Thin film photovoltaic generators based on a heterojunction system have been considered as an interesting alternative to Si solar cell systems for many years.

Thin film generators promise an economic fabrication due to inherent system features: they require a minimum of material and energy input and scale up and mass production is possible.

So far the Cu_2S/CdS cell is the most investigated photovoltaic thin film system and a high technological standard and quality was achieved.

Due to the high absorption coefficient of Cu_2S only a very thin absorber thickness of 0,2 µm is necessary. 90 % of the light is absorbed in the Cu_2S, the remaining 10 % in the 20 µm CdS layer.

The most effective way of preparing the CdS-layer is the evaporation technique because of its high deposition rate ($\sim$ 1 µm/min), large grain size of the CdS-crystals, a defined crystal structure and the absence of any reactions by additives.

Due to the topotaxial preparation of the p, n-junction, the Cu_2S/CdS conversion efficiencies are close to 10 % and overall efficiencies can be as high as 9 %. These figures were demonstrated lately.

Long time performance tests showed that Cu_2S/CdS generators work reliable. Efficiency reduction was due to failures of the encapsulation or electrical contacts, but not inherent to characteristics of either the semiconductor or the heterojunction.

2. PILOT PLANT

In 1981 NUKEM built up a pilot plant for the large scale manufacture of Cu_2S/CdS frontwall generators.

The used technology is based on experiences gained in the Institute für Physikalische Elektronik (IPE) at the University of Stuttgart

(Professor Bloss and his colleagues). The pilot plant has to demonstrate the reproducibility and economic feasibility of the fabrication process.

The back contact of the cell is deposited on a clean glass substrate by the well known sputter technique. The CdS layer is evaporated by a newly designed large area evaporator. In order to reduce the time consuming steps for evacuation a semi continous process is used for the sputtering and evaporating stages. In the first approach, up to 30 large substrates can be coated continuously in one cycle (fig. 1).

Table 1 shows the design criteria for the pilot plant.

glass panel size	500 x 500 mm^2
rate of feed for glass substrate	5 - 7 cm/min
deposition area	2 x 500 x 500 mm^2
quantity of CdS	100 g/m^2
capacity of the plant 8 hour shift	6 m^2 ≃ 350 W

Table 1: Pilot plant design criteria

The components for the wet chemical process of texturing the CdS surface and topotaxially preparing the Cu_2S layer are based upon technologies used in the industry for galvanotechnique. Special attention is spent to the control of the process parameters in order to obtain reproducible results for optical and electrical properties.

In order to form the heterojunction and to stabilize the cell a post-dipping treatment is followed consisting of a reducing hydrogen glow discharge and oxidizing heat treatment. For these processes conventional equipments have been installed.

The frontgrid technology of the press contact as done by IPE is presently transferred to the NUKEM process. For its production a copper foil is laminated on a glass plate where a hot-melt adhesive film was placed homogeneously. The following fabrication of the grid itself is made by the industry for printed circuits.

The technology of multilayer glass production is used for the final production stages of encapsulating and sealing the substrate and the cover glass (fig. 2).

The conception of the pilot plant allows to vary the size of the cells and the modules in a wide range. In figure 3 a 3 Wp module is shown. This module consits of 16 cell units of the size 7 x 7 cm^2.

3. QUALITY CONTROL

To ensure a proper quality of the modules, the physical and chemical properties of the starting materials, the cell units and of the integrated modules, are controlled and tested after the specific production stage.

Particular the following tests are relevant

- chemical analysis, DTA, DTG, X-ray diffractometry, B.E.T. for the starting materials Cadmium Sulfide, adhesives, copper compounds
- structure of the semiconductor layers by SEM (fig. 4)
- conductivity of the semiconductor layers by the 4-point method

- film thickness of various materials by β-back-scattering
- I, V characteristic for the determination of Isc, Uoc, FF, Pmax, Rsh and Rs at different stages of the production line
- Coulombmetric titration for the determination of the Cu_2O/Cu_{2-x}-stoichiometry and its thickness
- Recording the spectral response of the modules by measuring the photocurrent as function of the wavelength
- Laser scanning of the module surface shows the photoelectrical homogeneity

An example of a scan is shown in figure 5. A failure in a cell area can be recognized clearly.

4. QUALIFICATION TESTS

Tests for qualifying photovoltaic modules and arrays are of special interest to the user and producer of the devices.
Accellerated age tests of the modules show the optical and electrical stability as well as the mechanical stability under extreme conditions. To prove the reliability of these tests on operation an outdoor field test is performed.

5. CONCLUSIONS

NUKEM has built up a pilot plant for Cu_2S/CdS thin film solar cells which has to show the technical and economic feasibility. Based on the well established technology of the IPE and the first results of the described plant, NUKEM feels confident that stable and efficient solar generators will be produced at a cost range of 15 - 25 DM/Wpeak within the next two years. By scale-up of this pilot plant NUKEM expects to reach a cost goal of 2 - 5 DM/Wpeak by 1986. The cost might be further reduced by additional improvements of the production technology and the replacement of some expensive materials.

ACKNOWLEDGMENT

This work has been supported by the Bundesministerium für Forschung und Technologie, contract No. ET 4445 A and by the helpful collaboration of the Institut für Physikalische Elektronik at the university of Stuttgart (Professor Bloss and his colleagues).

Figure 1:
CdS evaporation plant

Figure 2:
Isostatic encapsulation plant

Figure 3:
3 Wp module

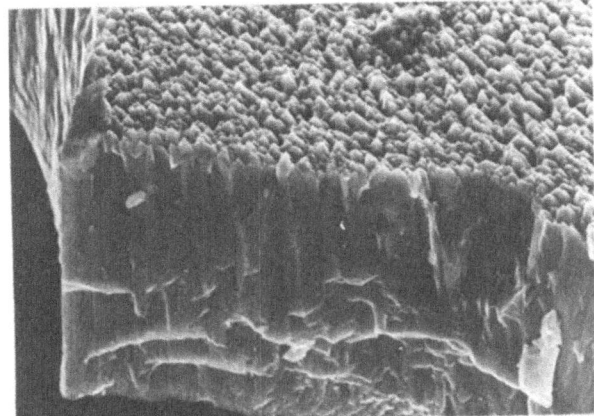

Figure 4:

SEM picture of a typical textured CdS-layer

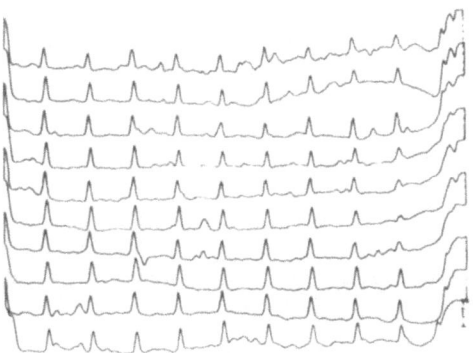

Figure 5:

Partial inhomogeneity of photo current by laser scan

FRONT CONTACTS FOR LARGE AREA HIGH EFFICIENCY Cu_2S-CdS SOLAR CELLS

W. ARNDT and H.W. SCHOCK
Institut fuer Physikalische Elektronik
Universitaet Stuttgart
F.R. Germany

SUMMARY

Cu_2S-CdS thin film solar cells have been fabricated by evaporating CdS onto metallized glass substrates and by forming the Cu_2S layer topotactially by a wet dip process in a solution of CuCl. The front contact is formed on the cover glass by silk screen printing, etching and electroplating techniques. The Cu_2S is contacted and encapsulated by sticking the front contact onto the Cu_2S and sealing it at high pressure and elevated temperature. Two different techniques have been used to produce the front contact:
1.) A Cu foil is laminated to the cover glass by means of a hot sealing adhesive. Employing silk screen printing, etching and electroplating, the complete grid is fabricated.
2.) Contact lines are directly vacuum deposited onto the Cu_2S-film. The lines can be obtained employing either an aperture mask or lift-off photolithography. Bus bars are provided by the cover glass, where they have been produced according to 1.).

I. INTRODUCTION

Cu_2S-CdS cells are now beeing investigated since many years. Various structures of the cells adapted to the technologies for fabrication have been developed. The transparent electrodes for large area devices cannot be realized easily because of the high sheet resistance of the semiconducting films either CdS or Cu_2S.
Cells with chemically sprayed CdS-films are in general made on ITO or SnO_2 films as transparent electrodes to the CdS whereas a Cu layer forms the contact to the Cu_2S /1,2/. Cells based on vacuum evaporated CdS are usually deposited on metallic substrate, hence the Cu_2S has to be contacted by a transparent electrode /3/. In this frontwall cell structure, the high energy part of the spectrum is absorbed near the surface of the cell.

2. TRANSPARENT ELECTRODES FOR THE Cu_2S LAYER

The special features of the Cu_2S layer require a carefully designed contact. These features are:
- high work function materials for ohmic contacts are needed
- the surface is sensitive to chemical reactions and the formation of alloys, this may be the reason for the poor stability of Ni contacts
- the thin film structure usually exhibits pinholes
- the surface roughness of high efficiency textured cells is ~ 2 μm according to the SEM micrograph Fig. 1

- high efficiency cells have a sheet resistance of the Cu_2S of about 5 k$\Omega_\square$ as shown in Fig. 2

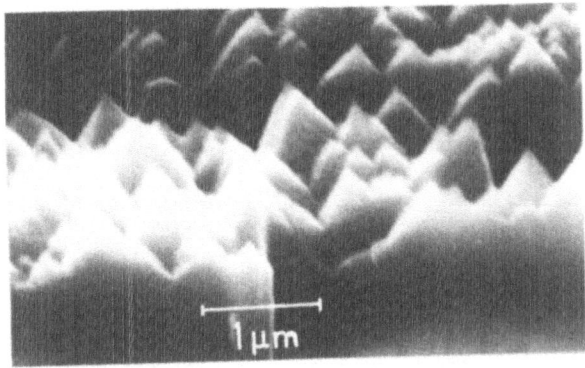

Fig. 1 SEM picture of the surface topography of a textured Cu_2S-CdS cell.

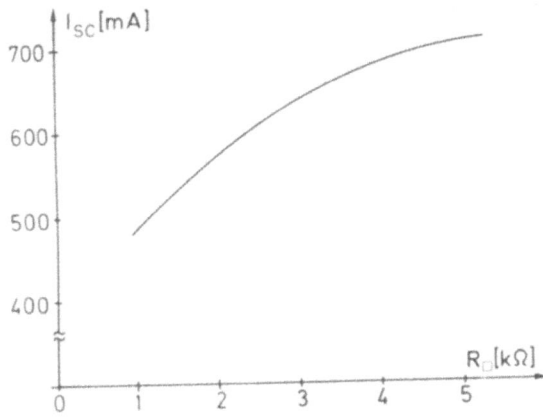

Fig. 2 Dependence of the short circuit current on the sheet resistance of the Cu_2S.

These features require special contacting technologies. Conductive oxide films are not compatible with the surface of the Cu_2S, furthermore the deposition of continuous conductive films increase the problem of pinholes. The most promising candidates for contact materials appear to be Au,Ni,Pd,and C. In view to these features and to economical reasons mainly two techniques and their combination have been applied.

3. GRIDDING TECHNIQUES

The first techniqe is described in previous papers in detail /4/. A copper foil is laminated to a glass sheet using a hot sealing adhesive. The grid structure is printed onto the foil by screen printing. After electroplating the grid with a thin Au layer the glass is laminated to the cell, hereby the grid is pressed to the surface of the Cu_2S. This kind of contact proofed to be very reliable as indicated by accelerated life tests /5/and long term rooftop tests as they are displayed in figure 3. The main advantage of this technique is the combination of gridding and encapsulation in one process-step. Furthermore it is very insensitive to pinholes, which can hardly be avoided in large area thin films. However, the minimum width of

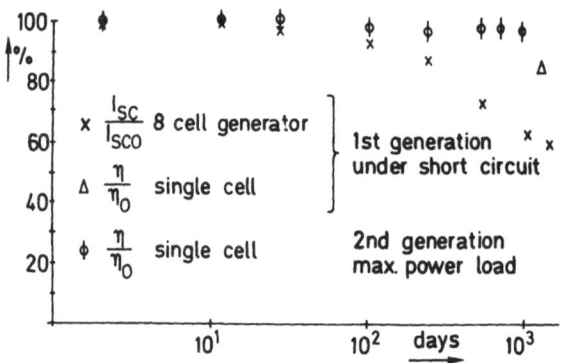

Fig. 3 Results of outdoor-tests.

the grid lines is limited to about 100 µm. This limitation is due to the large area screen printing and due to the kinetics during the hot-sealing process where the adhesive has to penetrate the open spaces of the grid in order to stick to the Cu_2S layer.

Much finer grids can be realized using "lift-off" photolithography based on dry photoresist which is laminated onto the surface of the Cu_2S layer. A line width of 20 µm can be achieved easily also on large areas. The grid lines are prepared by depositing a conductive layer in vacuum onto the photolithographic mask and by stripping the unexposed parts by means of an organic solvent.

The deposition of Cu directly onto the Cu_2S layer has to be avoided. Otherwise similar effects as observed during the post treatment (drastic decrease of the short circuit current) could occur /6/. Multilayer contacts have been fabricated in lift-off technique using very thin rare metal or carbon films as contact to the Cu_2S and conductive films of Cu or Al for the transport of the current. In some cases an additional layer is necessary in order to prevent interdiffusion of the materials. The combinations Au-Ni-Cu, Au-Al, and C-Cu have been investigated. They all show good stability during accelerated testing with thermal cycles (-20 to +80 °C) under illumination.

Thus a substantial reduction of material costs is possible especially for the combination C-Cu. The final grid structure is achieved by providing cross bars in the screen printing and hot sealing technique as described above. Cross bars with a separation of 10 mm require a thickness of the contact lines of .5 to 1 µm. The schematic structure of a solar cell gridded and encapsulated as described above is displayed in Fig. 4. The IV-characteristic of such a cell with optimized gridding technique which has 8% efficiency is demonstrated in Fig. 5.

4. OPTIMIZATION OF THE GRID

To minimize the losses due to shading by the contact grid and due to ohmic losses in the Cu_2S layer, the optimal grid layout as a function of sheet-resistance and current-density was calculated /7/.

$$\frac{P}{P_0} = 1 - \frac{nw}{a} - \frac{njR_\square}{12au} \left(\frac{a^3}{n^3} - 3\frac{a^2w}{n^2} + 3\frac{aw^2}{n} - 2w^3 \right)$$

This formula expresses the output power of a square shaped cell related to its output power without losses where is: a length of the cell, j current density, n number of grid lines, $R_\square$ square resistance of Cu_2S, u voltage in maximum power point.

- 406 -

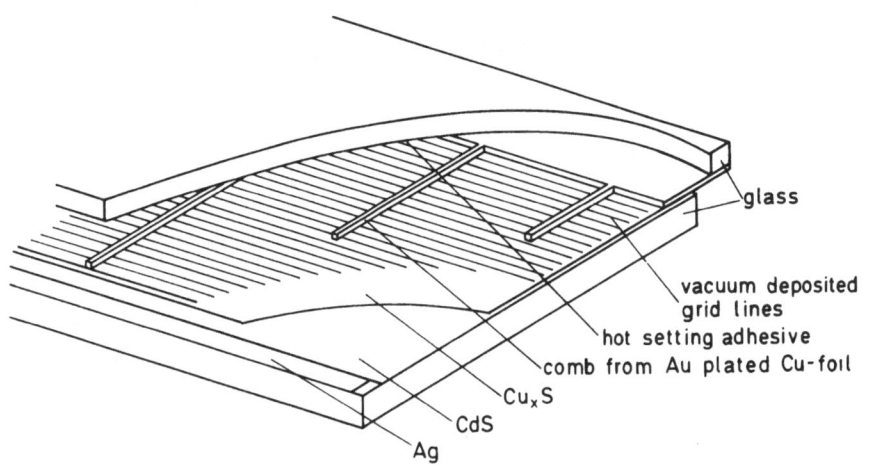

Fig. 4 Schematic structure of encapsulation.

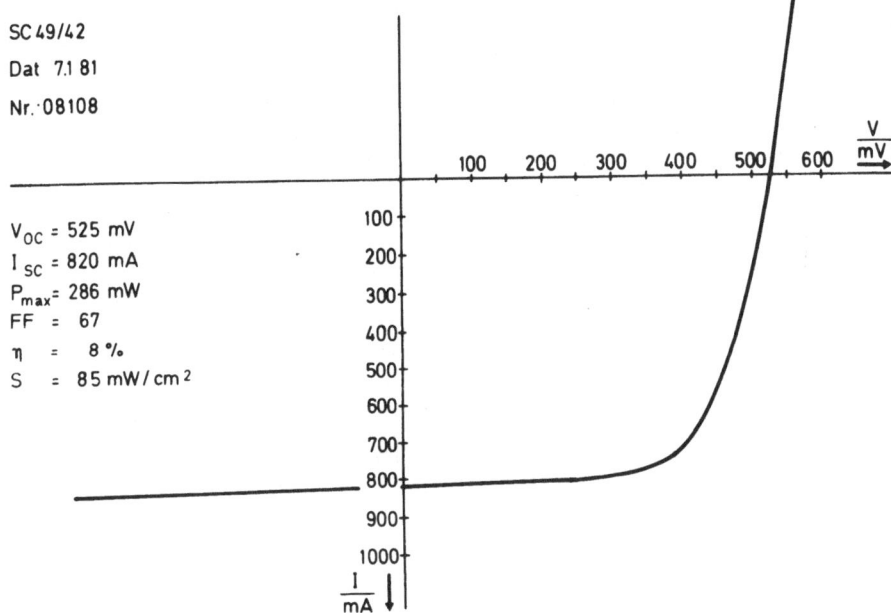

Fig. 5 IV-characteristic of a Cu_2S-CdS cell.

In Fig. 6 curves 1 show for both techniques the efficiency related to the efficiency of a cell with ideal frontcontact as a function of the sheet resistance of the Cu_2S. Switching from the screen printed pressure contact to the evaporated lift-off contact reduces the front contact losses by more than a factor of 2 as curve A1 and B1 demonstrate. At higher current densities (concentrating devices) the advantage of the lift-off technique increases. Curves 2 show the additional losses by a grid-layout optimized for 4 kΩ if the resistance of the Cu_2S deviates from this value. As Fig. 2 indicates, the square-resistance of the Cu_2S varies from about 1 k$\Omega_\square$ to 5 k$\Omega_\square$ with a tendency to higher values. The additional losses caused by a grid

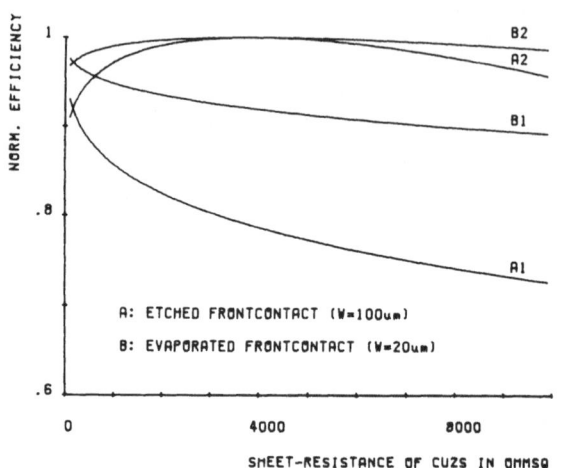

Fig. 6 Normalized efficiency for optimized front contacts versus sheet resistance of Cu_2S.

structure not optimized with respect to the sheet resistance are below 5 % between 1 k$\Omega_\square$ and 10 k$\Omega_\square$. Hence it is sufficient to optimize the grid structure for a mean resistance.

5. CONCLUSIONS

The techniques considered here for contacting the thin film solar cells were selected not only to yield high efficiency but also to be applicable in economical large scale productions. The described techniques are used in large scale productions e.g. in the printed circuit technology. By using different materials for contact and for conductor in the front contact, the amount of rare metal can be reduced or can be replaced by graphite. With the lift-off technique, the fill-factor and thus the efficiency can be improved considerably.

Large area Cu_2S-CdS cells (42 cm²) with more than 8 % efficiency and good stability could be realized by the optimized gridding technology.

6. ACKNOWLEDGEMENT

This work has been supported by the Bundesministerium für Forschung und Technologie, F.R. Germany, Contract No ET 4045 B.

7. REFERENCES

/1/ C.M. Lampkin, Proc. 2nd E.C. Photov. Solar Energy Conf., Berlin, 1979, (D. Reidel Publ. Comp., Dordrecht 1979) p. 396.
/2/ G. Bordure, C. Llinares, J. Bougnot, M. Perotin, H. Luquet, S. Belgacem, M. Savelli, Proc. 3rd E.C. Photov. Solar Energy Conf., Cannes, 1980, (D. Reidel Publ. Comp. Dordrecht 1980) p. 335.
/3/ W.H. Bloss, F. Pfisterer, and H.W. Schock, ref.2, p. 340.
/4/ W. Arndt, G. Bilger, F. Pfisterer, H.W. Schock, J. Wörner, and W.H. Bloss, ref.2, p. 798.
/5/ Final Report, Contract ET 4045 B, BMFT, to be published.
/6/ F. Pfisterer, H.W. Schock, and W.H. Bloss, Conf. Rec. XIIth IEEE Photov. Spec. Conf. Baton Rouge, 1976 (IEEE, New York, 1976), p. 502.
/7/ N.C. Wyeth, Solid-State Electronics, 1977, Vol. 20, pp. 629 - 634.

SESSION 5

FUNDAMENTAL STUDIES

Chairmen : Dr. A.H.M. KIPPERMAN, University of Technology, Eindhoven, The Netherlands

Prof. R. VAN OVERSTRAETEN, Katholieke Universiteit Leuven, ESAT Laboratory, Belgium

Solid solubility and precipitation of phosphorus and arsenic in silicon solar cells front layer

Grain boundaries and intragrain defects dependence of local and global electronic and photovoltaic properties of CGE polysilicon

Gas immersion laser diffusion : A new method for making efficient Si solar cells

Validity of the effective lifetime concept in polycrystalline silicon

Study of the grain boundary effects in n-p junction polycrystalline Si solar cells

Studies of the gap states density in undoped and doped amorphous hydrogenated silicon

Influence of light exposure on the transport properties of a-Si:H films

Pulsed electron beam annealing of ion implanted germanium for photovoltaic devices

Recent advances in ITO/InP and CdS/InP solar cells

Intensity enhancement in textures optical sheets for solar cells

Spectrum shifting methods in photovoltaics : an evaluation of model systems

SOLID SOLUBILITY AND PRECIPITATION OF PHOSPHORUS AND ARSENIC IN SILICON SOLAR CELLS FRONT LAYER

D. NOBILI

C.N.R. - Istituto LAMEL - Via Castagnoli, 1 - 40126 Bologna (Italy)

Summary

The physical nature of the electrically inactive P and As in silicon is discussed, and the results of experiments performed on specimens doped in a wide range of concentration obtained by ion implantation and laser annealing are reported. Thermal treatments of these alloys provided solid solubility values which correspond to the carrier density in equilibrium with excess dopant. Additional confirmation that the electrically inactive phosphorus and arsenic are a precipitated phase was obtained by the kinetics of the annealing process. The occurence of precipitation in an amount consistent with that of the inactive phosphorus was confirmed by TEM examinations using the weak beam dark field (WBDF) technique. The precipitation of arsenic was detected by small angle X ray scattering (SAXS) which provided also details on the size and shape of the particles. These results clearly contradict the well known models developed to account for the difference between carrier and chemical concentration profiles of P and As. These models hypothesize the formation of a high equilibrium concentration of complex point defects which compensate or make the excess dopant electrically inactive. As a consequence the models for the diffusion of P and As based on the above mentioned hypothesis are also contradicted. The influence of these precipitates on the minority carrier lifetime in the n^+ emitter region and on the conversion efficiency of silicon solar cells is considered, with reference to experimental procedures which reduce the amount of inactive dopant.

1. INTRODUCTION

The solid solubility of dopants in Silicon is still a poorly known property. Reference is often made to Trumbore data, which reflect the status of knowledge existing over 20 years ago (1).
A more precise information of the solubility phosphorus and arsenic is important in order to clarify the nature of the electrically inactive state of these dopants. Moreover it can contribute to the comprehension of their diffusion mechanisms and to the determination of the factors affecting the minority carrier lifetime in heavily doped layers.
The carrier concentration of silicon crystals heavily doped with phosphorus by thermal diffusion shows an upper limit which increases with temperature attaining $\sim 4 \times 10^{20}$ cm^{-3} at 1100°C (2). The same behaviour is shown by arsenic doped silicon; in this case carrier concentrations result sligtly lower and attain upper values of the order of 3×10^{20} cm^{-3} (3,4).
The excess phosphorus or arsenic is electrically inactive. A commonly accepted hypothesis advanced to explain this behaviour is the formation of high equilibrium concentrations of complex point defects which compensate or make the excess dopant electrically inactive (5-8).

This hypothesis, which is the basis of well known models of the diffusivity of these solutes, is in contrast with the results of investigations on the thermal predeposition of phosphorus (2,9). Moreover it is clearly contradicted by recently reported results concerning the solubility and the precipitation of this dopant (10,11).

These items are reviewed here together with new results concerning the solubility and the precipitation of arsenic. These latter findings will be reported in detail in a next paper.

For sake of clarity, although the behaviour of phosphorus and arsenic is similar, it will be discussed separately.

2. INACTIVE PHOSPHORUS

A typical phosphorus diffusion profile is reported in Fig. 1, which

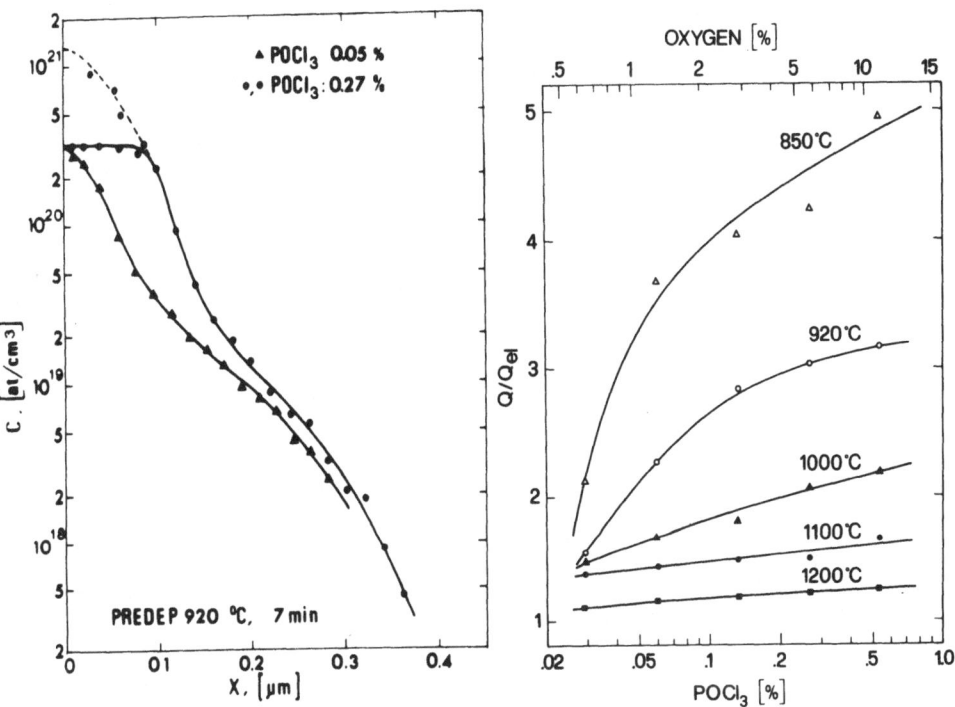

Fig. 1 - Chemical (open symbols) and carrier concentration profiles of predeposited specimens. Flux composition is $POCl_3$, 5,7% O_2 and N_2.

Fig. 2 - Ratio Q/Q_{el} between total and electrically active phosphorus vs oxygen and $POCl_3$ conc. increased by keeping constant their ratio (21.5). Predep.time is constant(27min)

shows the characteristic plateau of carrier density and the amount of inactive phosphorus. Reducing the concentration of $POCl_3$ in the flux, leads chemical and electrical profiles to coincide. In fact the amount of inactive dopant increases with the partial pressure of $POCl_3$. The amplitude of this effect as it is shown in Fig. 2, decreases with increasing predeposition temperature.

The flat region of the profile, associated to the presence of the electrically inactive dopant was termed 'dead layer' by Lindmayer and Allison (12) as it is characterized by very short minority carrier lifetime. This phenomenon was further analyzed by Kendall et al (13) who advanced the hypothesis that the so called inactive phosphorus had merely suffered a form of self compensation. On their model, at very high P concentrations the positively charged P^+ donor was severely compensated (5) by the phosphorus vacancy pair, the E center defect, (14) in a negatively charged $P^+V^=$ state.

By an analysis of the corresponding equilibria they concluded that the dead layer near the solar cell surface is a consequence of the extremely high E center density in this region.

The well known model of phosphorus diffusivity in Silicon developed by Fair and Tsai (6) also starts with considering the electrically inactive P anomaly and takes into account the reaction
$$V^- + P^+ + e^- \rightleftharpoons P^+V^=$$
and the corresponding equilibria. Moreover these A. report an empirical equation relating the electron concentration n to total donor concentration: $C_T = n + 2{,}04 \; 10^{-41} \, n^3$

The alternative hypothesis of precipitation, first advanced by Tannembaum (15) to account for the inactive dopant, was supported by investigations on the kinetics of thermal predeposition (9,16,2). These observations put into evidence that in conditions typical of device technology the chemical potential of phosphorus at the PSG-Silicon interface and in the adjacent Si matrix can overcame the value corresponding to the solubility limit. This was confirmed by the detection of orthorombic SiP precipitates grown concomitant to the predeposition process. The amount of these precipitates however couldnt account for the electrically inactive dopant.

The hypothesis of precipitation is in contrast with well known solubility data which reported values largely exceeding the electrically active equilibrium concentration (17,1). Actually these values were determined in conditions which do not comply the correct requirements: Kooi (17) used neutron activation analysis after predeposition from a P_2O_5 source and subsequent annealing and equilibration with the PSG layer. This procedure, is open to criticism (18). In fact the doped oxide film which is formed on the slice is a ternary system of silicon, oxygen and the dopant. Therefore the chemical potential of the dopant cannot correspond to that at the solubility limit in silicon. This limit results from equilibrium, or in other words competition, between the solid solution and its conjugate phase (SiP in this case). This criticism is further strengthened by the observation of cubic $SiO_2 \cdot P_2O_5$ at the PSG/silicon interface of heavily doped slices (19).

Trumbore (1) referred to unpublished results of Mackintosh obtained by sheet resistance and junction depth measurements after thermal diffusion. The approximation involved in order to calculate the surface concentration, a complementary error function distribution of the dopant, is surely not justified, as it was also pointed out by Mackintosh himself in a further work (20), considering the actual shape of the profile (see Fig. 1). These values are therefore unreliable.

Lower solubility values in satisfying agreement with the electrically active concentrations were determined by Abrikosov et al. by microhardness measurements on bulk doped specimens (21). This procedure is correct (22); although the technique seems poorly sensitive.

Another obstacle encountered by the precipitation model is connected to the slight slope in the concentration profile of the electrically active dopant in the plateau. A very high diffusivity is necessary in this region to account for the mass transport.

To verify the hypothesis of an equilibrium between point defects leading to self-compensation, determinations were performed on ion implanted and laser annealed specimens doped in a very broad range of composition.

Laser annealing is capable of completely activating very high implanted concentrations of dopants in Silicon. Values attaining $\sim 5 \times 10^{21}$ cm^{-3} were reported for phosphorus and $\sim 4 \times 10^{21}$ cm^{-3} for arsenic, corresponding to electrical resistivities of 110 and 195 μ ohm cm respectively. Structural examinations showed that up to this limit the crystal is free of observable defects and that the solute occupies substitutional lattice positions (23,10).

Thermal stability decreases markedly with supersaturation leading to a reduction of the carrier density. The electron mobilities of these alloys are reported in Fig. 3. These determinatios showed, in contrast with the compensation model, that the electrically inactive phosphorus is not a charged point defect, because, even when present at high concentrations, it does not affect the electron mobility, which depends only on carrier density (10).

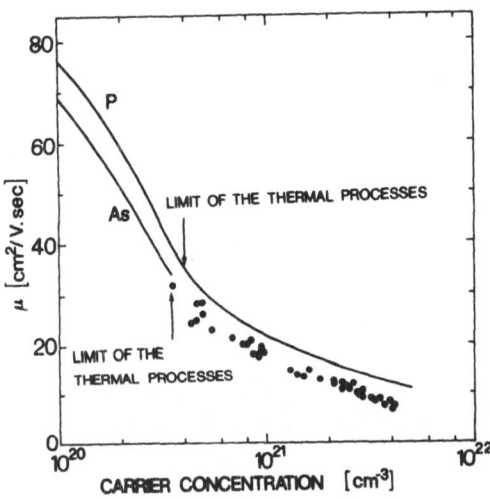

Fig. 3 - Electron mobility vs carrier density. Concentrations exceeding the limits of the thermal process were obtained by ion implantation and laser annealing. Experimental points clearly evidence the reproducibility and the trend of As doped compositions.

Moreover thermal equilibration of the alloys doped with phosphorus, showed that the electrically active concentration depends only on temperature and is insensitive to the excess dopant (11). An example is reported in Fig. 4. Similar results were previously obtained on diffused specimens (2) and the electrically active equilibrium concentrations resulted the same in both type of experiments.

A constant equilibrium concentration of P cannot result by a point defects reaction, in a single phase system. Only a two phase equilibrium, that is a precipitation process, can result in a constant chemical potential, independent on the total concentration of the solute.

Therefore these results unambiguously show that a two phase equilibrium takes place and that the electrically active concentration of phosphorus corresponds to the solubility.

This was confirmed by TEM examinations with the weak beam dark field (W.B.D.F.) technique which detected a high density of a very small coherent precipitates. This method allowed to observe particles of the same kind even on specimens predeposited in conditions typical of device technology (see Fig. 5) in an amount consistent with the electrically inactive dopant (11).

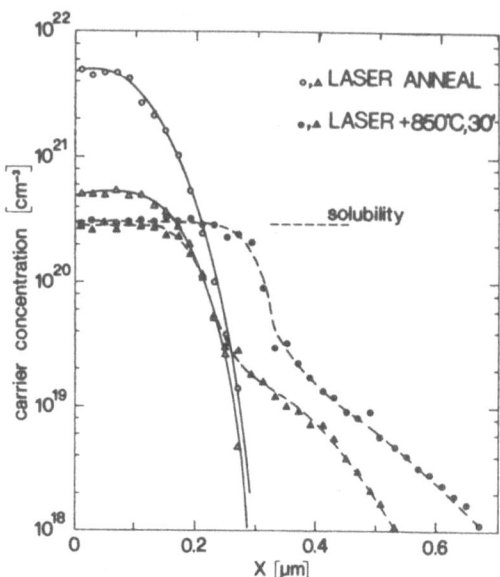

Fig. 4 - Carrier profiles of ion implanted and laser annealed samples doped with very different concentrations of phosphorus. Open symbols refere to the initial conditions (all dopant is electrically active), and black symbols to those after thermal equilibration (30min at 850°C). Maximum electrically active concentrations, in equilibrium with inactive dopant, turn out to be equal and correspond to the value (solubility) determined on thermally diffused specimens (ref. 2).

Due to the very small size of these precipitates and to their high density they could hardly be observed by conventional TEM techniques; moreover due to their coherent structure with the silicon matrix only a fraction of the precipitated phosphorus could be detected by Rutherford back scattering determinations (24).

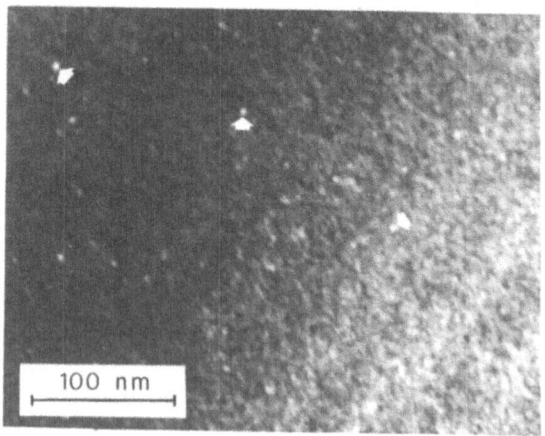

Fig. 5 - (g,3g) WBDF TEM micrograph (g=220) taken on a specimen predeposited 27 min at 920°C showing a very high density of small precipitates (about 15-20 Å in radius) a few of which are arrowed.

In addition it was verified, by using ion implanted laser annealed specimens as those in Fig. 4 that precipitation is associated to a very high enhancement of the diffusivity of phosphorus; this increase depends on the precipitation rate and can attain orders of magnitude (11). This phenomenon can account for the plateau region of the carrier profiles after thermal predeposition.

The precipitation process in heavily doped specimens, was found to be accompanied by the formation of extrinsic stacking faults. This indicates supersaturation of interstitials.

It can be concluded that the hypthesis of the formation of an high concentration of complex point defects, which compensate or make the excess phosphorus electrically inactive thus accounting for the difference between chemical and carrier concentration, is incorrect.

As a consequence the models for the diffusion of phosphorus in Silicon based on the above mentioned hypothesis have not a real physical significance and should be considered as an empirical fit to observations. In particular it must be pointed out that a Boltzmann-Matano analysis of the diffusivity from a concentration profile is invalid for any process involving precipitation or generation of non equilibrium defect concentrations.

3. INACTIVE ARSENIC

This problem is in several respects similar to that of phosphorus. Dopant concentration exceeding the electrically active value did not give rise, at high temperature, to clearly identified precipitation.

The formation of complex point defects in thermal equilibrium is the commonly accepted cause of the discrepancy between the number of carriers and that of arsenic atoms at high dopant concentrations. A small dependence of the carrier density on the chemical concentration, occuring in heavily doped layers, was in fact reported, although the values determined by different A. differ widely (8). The variety of models proposed reflects this discrepancy. As an example the equilibrium:

$$3As^+ + e^- \rightleftharpoons As_3^{++} \xrightarrow{25°C} As_3$$

was proposed by Tsai et al. (7), and the formation of an arsenic vacancy complex:

$$2As^+ + V + 2e^- \rightleftharpoons As_2V$$

by Fair (8,25).

Accurate R.B.S. measurements showed that excess arsenic still keeps the substitutional lattice site. The limited displacement from this position (~ 0.15 A) was attributed to the formation of the neutral complex defect As_2V (26).

On the other side, based on equilibrium experiments performed at 900°C on ion implanted laser annealed layers, Leitola et al. (3) concluded that the electrically active concentration depends solely on the annealing temperature. These A. confirmed that most of the excess arsenic maintains the substitutional lattice position.

Equilibrium experiments performed at 800, 900 and 1000°C on ion implanted and laser annealed specimens, doped in a very broad range of composition, confirmed that the electrically active arsenic concentration depends only on temperature. An example is reported in Fig. 6. These results strongly support the occurrence of precipitation.

This conclusion was further confirmed by the phenomenon of reversion, taking place when specimens that had been partially annealed at the temperature T_1, were further annealed at $T_2 > T_1$. As it is shown in Fig. 7, were T_1 and T_2 are 650°C and 750°C respectively, at the start of the higher temperature anneal the active arsenic concentration was larger than the equilibrium one at the same temperature. Nevertheless it initially increases and than decreases toward the equilibrium value.

Fig. 6 - Carrier profile of ion implanted laser annealed samples doped with very different concentration of arsenic. Open symbols refer to the initial condition (all dopant is electrically active) and black symbols to those after thermal equilibration (10 min at 1000°C) Maximum electrically active concentrations turn out to be equal and hence correspond to the solubility value.

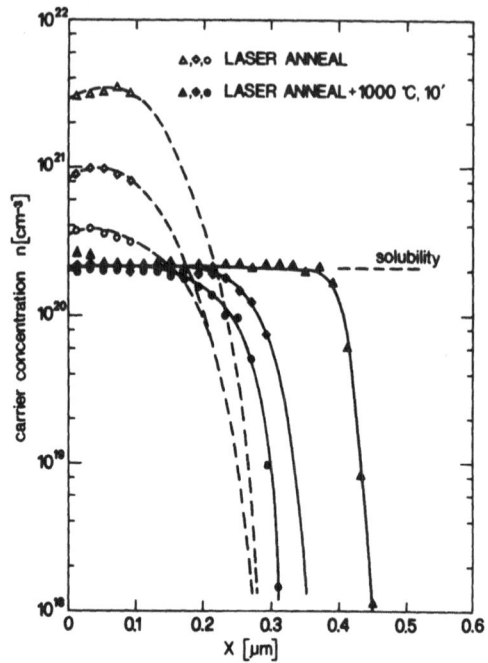

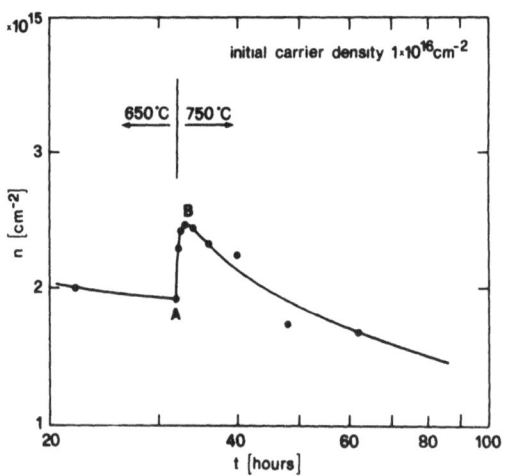

Fig. 7 - Carrier concentration per unit surface vs heating time in an As implanted and laser annealed specimen (maximum concentration 4×10^{21} cm^{-3}). Isothermal treatments are performed at 650°C up to 32h (point A) and subsequently at 750°C. Point B shows reverse annealing after 1 hour heating at the higher temperature.

The only model which can account for the reversion process is based on precipitated particles.

If the population consists of precipitates of different free energy, the higher energy particles redissolve when the temperature is raised enough, while the remaining more stable population, still in conditions of supersaturation, gives rise to further precipitation (27,28).

A similar effect was observed in phosphorus doped specimens containing a high density of small precipitates (11).

The occurrence of precipitation is compatible with the back scattering results previously referred (3,25) if the precipitates are coherent, i.e. they keep the lattice structure of silicon.

The observation of precipitates of this type by TEM techniques should be difficult, as the tetrahedral radius of arsenic (1.13 - 1.15 A) (29) is very similar to that of silicon (1.17 A).

On the other side, due to the marked difference of atomic number, 33 and 14 respectively, the occurrence of precipitation can be determined by small angle X ray scattering (SAXS) techniques (30,31). SAXS analysis of ion implanted and laser annealed specimens, subsequently heated for 30 min at 900°C, confirmed that the thermal treatment gives rise to extensive precipitation. The particles are coherent and have the shape of thin disks. The distribution curve of the particle size of a specimen with an arsenic concentration of about 4×10^{21} cm^{-3} is reported in Fig.8.

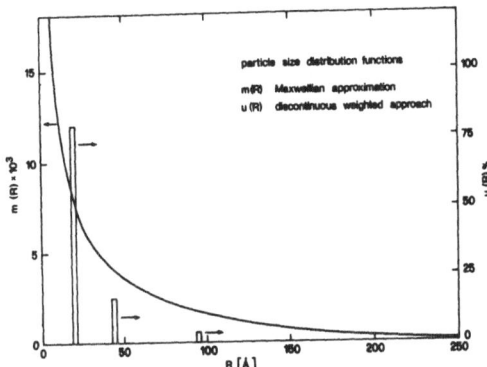

Fig. 8 - Statistical distribution of precipitate size obtained by SAXS determinations performed on As doped Si annealed at 900°C (see text for details). The curve m(R) represents the maxwellian continuous distribution, while the bars refer to the discrete fractions obtained from the tangents method. Evaluations were performed by considering the bidimensional slope of the particles.

The hypothesis of point defects, which make the excess dopant electrically inactive is clearly contradicted by this experiments. It can be concluded that arsenic solubility is equal to the electrically active concentration in equilibrium with excess dopant, and that the latter is a precipitated phase.

4. SOLID SOLUBILITY VALUES OF PHOSPHORUS AND ARSENIC IN SILICON

Values of the electrically active concentration in equilibrium with coherent precipitates are reported in Fig. 9. In both cases they correspond to equilibrium with a phase which has a higher molar free enthalpy than the conjugate one, hence the reported values exceed, at least in principle, the 'true' solubility.

The value at 450°C on the phosphorus curve was reported by Ekstrom et al. (32). These A. in fact determined at 550°C and 700°C solubility values which coincide with those in Fig. 9.

The values reported for arsenic, although lower, are in satisfying agreement with those determined by Leitola et al. (3).

5. DOPANT PRECIPITATION AND MINORITY CARRIER LIFETIME

A minority carrier diffusion length, comparable to the junction depth of conventional cells is expected by considering the experimental data obtained on bulk specimens n-doped up to about 3×10^{19} cm^{-3} (33). The same conclusion can be deduced by theoretical evaluations recently reported (34). On the other side measurements of minority carrier lifetime in heavily phosphorus doped (up to 3×10^{20} cm^{-3}) layers, indicated that the value is about 3×10^{-11} sec, an order of magnitude smaller that expected, and is process dependent (35). The A. attribute this effect to lattice strain or deformation.

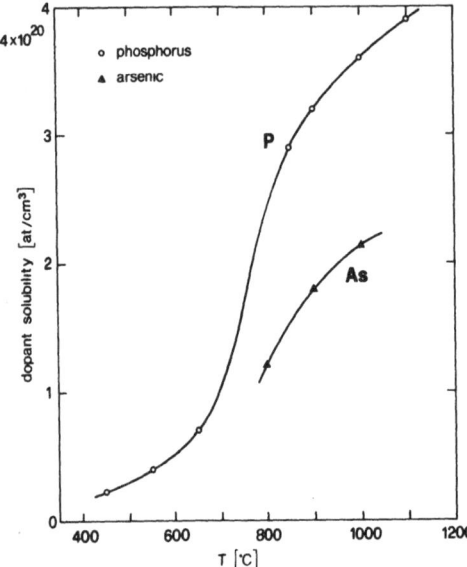

Fig. 9 - Solubility vs temperature of phosphorus and arsenic in silicon.

Consideration of the very small radius (15-20 A) of the phosphide precipitates in a typical thermal diffusion, and of their very high density, puts into evidence that the area of the interface between phases can attain values which exceed that of the corresponding cell surface. The lifetime of the photogenerated carriers could therefore be significantly affected, depending on the density and properties of the surface states of these internal surfaces. A further contribution is expected by the excess of interstitials which was put into evidence by the observation of extrinsic stacking-faults (11).

Technologies leading to a reduction of the amount of precipitation could therefore improve the conversion efficiency of the cell. The 'violet' cell was obtained by reducing the junction depth at about 1500 A in order to reduce the amplitude of the region were the carrier profile is nearly flat (12). This region consisted, in the opinion of the A., of a mixture of substitutional an interstitial phosphorus.

Moreover it was reported that laser irradiation of n^+/p diffused cells, which leads to partial activation of the excess dopant, and hence to dissolution of the corresponding precipitates, resulted in an increased of I_{sc} and V_{oc} (36).

As a further piece of information the relative spectral responses of two distinct types of n^+/p cells are reported in Fig. 10. The concentration profile of the cell which presents the lower conversion efficiency corresponds to the more heavily doped one, reported in Fig. 1. A measurable improvement is obtained with a flux composition which prevents extensive precipitation. A significantly higher quantum efficiency is observed at very short wavelength. Furthermore the V_{oc} increased of about 10 mV.

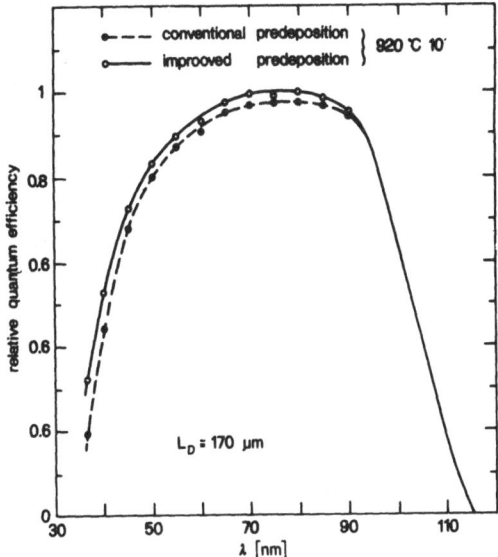

Fig. 10 - Relative quantum efficiency vs wavelength for two distinct phosphorus doped n^+/p cells, evidencing influence of the precipitates. L_D refers to the diffusion length in the base.

These results are in keeping with the conclusion that the minority carrier lifetime in the n^+ emitter is process dependent and suggest that dopant precipitation can significantly contribute to this phenomenon.

ACKNOWLEDGMENTS

Thanks are due to A. Armigliato, G.C. Celotti, P. Ostoja and S. Solmi for useful discussion and cooperation.

REFERENCES

1) F.A. Trumbore, Bell Syst. Technical J. <u>105</u>, 205 (1960)
2) G. Masetti, D. Nobili, S. Solmi in "Semiconductor Silicon 1977" R.H. Huff and E. Sirtl Editors The Electrochem. Soc. Softbound Symposium Series Princeton N.Y. (1977) p. 648
3) A. Leitola, J.F. Gibbons, T.W. Sigmon Appl. Phys. Lett. <u>36</u>, 765 (1980)
4) R.B. Fair, C.C. Tsai J. Electrochem. Soc. <u>122</u>, 1689 (1975)
5) D.L. Kendall, D.B. De Vries In Semiconductors Silicon (R.R. Haberect and E.L. Kern Eds.) The Electrochemical Society Inc. New York (1969) p. 358
6) R.B. Fair, J.C.C. Tsai J. Electrochem. Soc. <u>124</u>, 1102 (1977); R.B. Fair J. Appl. Phys. <u>50</u>, 862 (1979)
7) M.Y. Tsai, F.F. Morehead, J.E.E. Begin, A.E. Michel J. Appl. Phys. <u>51</u>, 3230 (1980)
8) R.B. Fair in "Semiconductor Silicon 1981" H.R. Huff, R.J. Kriegler, Y. Takeishi Editors. The Electrochem. Soc. Softbound Symposium Series Minneapolis (1981) p. 963

9) P. Negrini, D. Nobili, S. Solmi   J. Electrochem. Soc. $\underline{122}$, 1254, (1975)
10) M. Finetti, P. Negrini, S. Solmi, D. Nobili, J. Electrochem. Soc. $\underline{128}$, 1313 (1981)
11) D. Nobili, A. Armigliato, M. Finetti and S. Solmi J. Appl. Phys. $\underline{53}$, 1484, (1982)
12) J. Lindmayer, J.F. Allison   Conf. Record IEEE Photovoltaic Spec. Conf. 9th Silver Spring p. 83 and Comsat Techn. $\underline{3}$, 1, (1973)
13) D.L. Kendall, R. Carpio, Conf. Proceedings of IEEE Sectional Meeting, Monterrey N.L. Mexico (1977) p. 1, and Conf. Proceedings of the Electrochem. Soc. Meeting, Pittsburg, Pennsylvania (1978)
14) G.D. Watkins, J.W. Corbett   Phys. Rev. $\underline{134}$ A 1359 (1964); G.D. Watkins. In Radiation Damage in semiconductors, Dunod (1964) p. 97
15) E. Tannenbaum   Solid State Electron 2, 123 (1961)
16) A. Armigliato, D. Nobili, M. Servidori, S. Solmi   J. Appl. Phys. $\underline{47}$, 5489 (1977)
17) E. Kooi, J. Electrochem. Soc. $\underline{111}$, 1383 (1964)
18) V.M. Glazov, V.S. Zemskov, Physicochemical principles of semiconductor doping. Izdatel'stvo'Nauka' Moskva (1967)
19) S. Solmi, G. Celotti, D. Nobili, P. Negrini, J. Electrochem. Soc. $\underline{123}$, 654 (1976)
20) J.M. Mackintosh, J. Electrochem. Soc. $\underline{109}$, 392 (1961)
21) N.K. Abrikosov, V.M. Glazov, Liu Chen-Yuan - Russ. J. Inorg. Chem. $\underline{7}$, 429 (1962)
22) F.A. Shunk 'Constitution of binary alloys' II Suppl. Mc Graw Hill Co, 1969, p. 598
23) N. Natsuaki, M. Tamura, T. Tokuyama. J. Appl. Phys. $\underline{51}$, 3373 (1980)
24) E. Fogarassy, R. Stuck, J.C. Muller, A. Grob, J.J. Crob, P. Siffert, J. Electron Material $\underline{9}$, 1977 (1980)
25) R.B. Fair, G.R. Weber   J. Appl. Phys. $\underline{44}$, 273 (1973)
26) W.K. Chu, B.J. Masters in "Laser Solid interactions and Laser processing" AIP Cons. Proc. 50 (Am. Inst. of Phys.) N.Y. (1978) p. 305
27) J.W. Christian - The theory of transformations in metals and alloys, Pergamon Press (1975)
28) A. Kelly, R.B. Nicholson - Prog. Mater. Sci. $\underline{10}$, 149 (1963)
29) M. Servidori - Unpublished Results
30) V. Gerold, in "Small Angle X-ray Scattering" H. Brumberger (ed.) Gordon Breach. N.Y. (1967) p. 277
31) V. Gerold, K. Kostorz, J. Appl. Cryst. $\underline{11}$, 376 (1978)
32) L. Ekstrom, J.P. Dismukes J. Phys. Solids $\underline{27}$, 857, (1966)
33) J. Van Meerbergen, J. Nijs., R. Mertens, R. Van Overstraeten. Proc. 13th Photov. Specialist Conf. (1978) p. 66
34) R.N. Hall. Solid State Electronics - $\underline{24}$, 595 (1981)
35) D.J. Roulston, N.D. Arora, S.G. Chamberlain IEEE Trans. Electron Dev. $\underline{29}$, 284 (1982)
36) J.C. Muller, E. Fogarassy, D. Salles, R. Stuck, P.M. Siffert, IEEE Trans. Electron Dev. $\underline{4}$, 815 (1980)

GRAIN BOUNDARIES AND INTRAGRAIN DEFECTS DEPENDENCE OF LOCAL AND GLOBAL ELECTRONIC AND PHOTOVOLTAIC PROPERTIES OF CGE POLYSILICON.

J. OUALID, M. ZEHAF, H. AMZIL, J.P. CREST, G. MATHIAN, J. DUGAS,
J. GERVAIS, F. MINARI, B. PICHAUD, S. MARTINUZZI
Université d'Aix-Marseille III
Faculté des Sciences et Techniques de Saint-Jérome
F.13397 MARSEILLE Cedex 13

J. FALLY
Les Laboratoires de Marcoussis CGE
Route de Nozay, F.91460 MARCOUSSIS

Summary

Minority carrier properties (L_n, τ_n, μ_n) and photovoltaic parameters (I_{sc}, V_{oc}) have been measured in small MESA diodes revealed in N^+P photocells cut out from different regions of polysilicon ingots (top, middle, bottom or periphery, centre). Their dependence on the physical properties (number of grains N_G per diode, dislocations, impurities) has been analysed. It has been shown that the electronic and photovoltaic properties decrease strongly when the number of grains per diode N_G increases. This influence of the grain size is often screened by intragrain defects. The dispersion of the results can be partially resolved by taking into account the dislocation distribution. Local measurements of photocurrents I_{ph} and diffusion lengths L_n performed with a 5 µm light spot confirm the influence of dislocations on the grain response.

1. INTRODUCTION

In polycrystalline silicon, the defects are very different in nature (grain boundaries, dislocations, impurities ...). These defects degrade the electronic properties (1) (mobility μ, lifetime τ, diffusion length L) and so the photovoltaic parameters I_{sc}, V_{oc}, fill factor FF and efficiency η). As they are randomly distributed, the characterization of a given material must be achieved on a great number of small samples or must be made locally.

The purpose of this work is to determine the respective influence of grain size, grain boundaries (g.b.) and intragrains defects on the electronic and photovoltaic properties of p-type CGE polysilicon.

The investigated samples were conventionnal phosphorus diffused photocells made by PHOTOWATT Int.S.A. (2) on different ingots of C.G.E. polysilicon prepared by a fast variant of Bridgman's method using electronic grade silicon doped with boron (3). Three typical cells made with wafers cut out from the top (T), the middle (M) and the bottom (B) of each ingot have been studied. Arrays of twelve Mesa diodes (1,5 x 1,5 mm) have been revealed by photolythographic methods on samples (2 x 1 cm) cut out from the centre (C) or the periphery of each cell (Fig.1). Three ingots were investigated. Each ingot presents a resistivity which increases from the bottom to the top and from the centre to the periphery. Impurities, especially iron, are accumulated at the top of each ingot by the Bridgman process.

The density of dislocations revealed by Sirtl etch is generally higher at the periphery than at the centre of the cells and increases from the top to thebottom of each ingot. The etch pits show that the grains are generally (80 %) oriented in the direction (1 1 1).

The three ingots present different resistivities, dislocation densities and grain sizes.

2. MINORITY CARRIER PROPERTIES L_n, τ_n AND μ_n.

. Diffusion length L_n.

The diffusion length L_n is measured by means of a variant of the well known surface photovoltage method (4,5). Fig.2 gives the dependence of L_n on the number of grains per diode N_G. When the densities of dislocations or impurities are low as for the middle of each ingot, L_n decreases strongly with N_G (curve 1, Fig.2) as reported by Sopori (1).

Generally for a same grain size, the diffusion length is higher at the centrethan at the periphery of the cells, because the dislocation density is higher at the periphery. When the dislocation or impurity density is high as it is the case of samples accrued from the periphery of the bottom or from the top of the ingots, L_n is practically independent of grain size (curve 2, Fig.2). Fig.2 shows that L_n has a limiting value of about 20 µm which has been found also by Sopori (1) for R.T.R. Ribbon and for Wacker Silso.

. Lifetime τ.

Generally, lifetimes are deduced from the values of L_n, using the minority carrier mobility of monocrystalline silicon. In the present work τ_n and L_n were measured separately in the same MESA diodes and τ_n is determined by the transient photocurrent method (6).

τ_n decreases abruptly when N_G increases for low intragrain defect density as shown by curve 1 of Fig.3. Like L_n, τ_n smoothly varies with N_G (curve 2, Fig.3) at the bottom or at the top of the ingots where the dislocations and impurities are respectively concentrated. The decrease of τ saturates to about 0.3 µs as shown Figure 3. The saturation values of L_n and τ_n give a minimum of the minority carrier mobility : $\mu_n \simeq 500$ cm²/V.s.

We have tried to correlate the lifetime with the grain size assuming square grains of size $\bar{g}$ given by $\bar{g} = (A/N_G)^{1/2}$ where A is the area of MESA diodes (A = 2.25 mm²). The inset of Fig.3 shows that the variation versus $\bar{g}$ of the mean value of $\bar{\tau}_n$ (curve 1) is linear for $\bar{g}$ < 0.6 mm. This is in agreement with the variation of the effective lifetime due to the recombination via the grain boundary interface states as proposed by H.C. Card and J.S.Yang (7). We have found, when $\bar{g}$ < 0.6 mm, the very simple empirical law : $\bar{\tau}_n$ (µs) = $\bar{g}$ (mm). Gosh et al (8) have proposed an estimated similar law : τ_n (µs) = 0.5 g (mm).

. Minority carrier mobility.

Minority carrier mobilities μ_n can be deduced from the slope of the graph given in Figure 4 which represents the variation of L_n^2 versus τ_n. The lack of linearity of this graph suggests that μ_n is also dependent on all the defects which degrade the lifetime. The L_n^2 variation is well given by the empirical law : L^2 (µm) = 2500 $\tau^{1.3}$(µs) and μ_n varies as μ_n(cm²/V.s = 950 $\tau^{.3}$(s). The mean value of μ_n is about 750 cm² V⁻¹.s⁻¹.

Fig.1

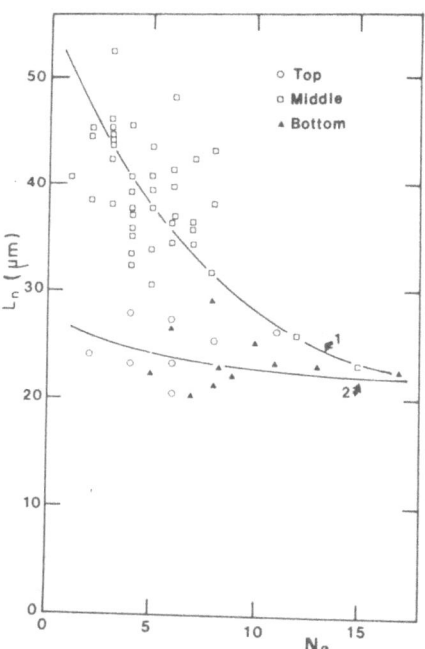

Fig.2

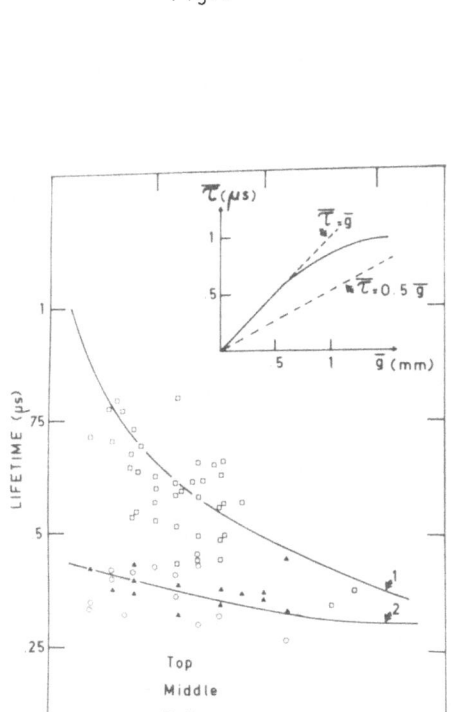

Fig.3

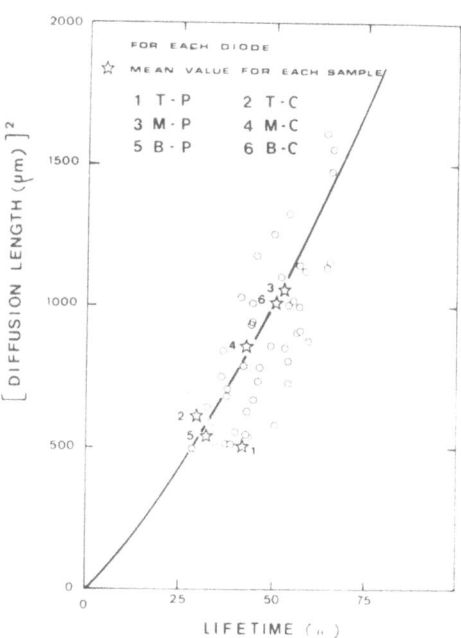

Fig.4

3. SHORT CIRCUIT PHOTOCURRENT I_{sc}.

I_{sc} is measured by means of a quartz-iode lamp giving an illumination of 100 mW/cm^2. Figure 5 represents the variation of the photocurrents of each MESA diode versus N_G. For the same N_G this figure confirms that the photocurrent is higher at the middle than at the top or than at the bottom of ingots where deep impurities or dislocations are concentrated. Curves 1, 2 and 3 give respectively the variation of the average photocurrents when the densities of dislocations or impurities are low, when the dislocation density N_{dis} is near 10^5 cm^{-2} and when $N_{dis} \gg 10^6$ cm^{-2}. Figure 5 exhibits the decreaseof the photocurrent with N_G down to a saturation value very similar tothat given by Sopori (1).

The variation of L_n, τ and I_{sc} against N_G present evidently a high dispersion on account of the complexe distribution of intragrain defects. To resolve this dispersion we have revealed the dislocations of MESA diodes belonging to lot 121. Fig.6 shows the variations of I_{sc}, L_n and τ_n versus the average of the density dislocation N_{dis} for the MESA diodes which present a similar total length of grain boundaries per unit area $\ell_{gb} \simeq 20$ cm^{-1} ± 10 cm^{-1}.

To determine this average value we have estimated the dislocation density and measured the area of each grain in a MESA diode. We have verified by LBIC (9) that the photoelectric responses and diffusion lengths of each grain depend onthe dislocation density.

It is clear that the influence of dislocations explains the major part of the dispersion of results presented in figures 2, 3 and 5. As it was expected, the variation of τ_n against the dislocation density is more important than the variations of L_n and I_{sc}. We have verified((13)F.10),that there is a fairly good correlation between I_{sc} and L_n. We can observe that the mean values of I_{sc} for a given L_n are comparable with the corresponding values given by Sopori (1) and obtained with Wacker or RTR polysilicon. This verifies the general character of this curve. Nevertheless, we can note that the mean values of I_{sc} for a given L_n are higher at the periphery (curve 1) than at the centre (curve 2) which confirms the observation already discussed. The grains being smaller at the centre, it appears consequently that I_{sc} is more sensitive to g.b. than L_n.

4. LOCAL INVESTIGATIONS.

Grain responses and g.b. activity have been determined by means of a monochromatic light spot (size : 5 µm). L_n and also the g.b. interfacial recombination velocity s were measured (10,11). The photocurrent attenuation observed at a g.b. suffices to get a suitable estimation of s when the wavelength is sufficiently high, provided the spot size is smaller than L_n. Three cells of ingot 104 were investigated, which the respective resistivities ρ are 1, 0.5 and 0.25 Ω.cm. Typical photoelectric lines at λ = 0.98 µm have been reproduced in Fig.7.

These wafers have been chosen because of their high density N_{dis} of dislocations ($N_{dis} > 5.10^5$ cm^{-2}) which is probably due to the presence of a great concentration of impurities. This explain the small and relatively homogeneous response of the cells except for some grains (A,B) where the dislocation density measured are lower ($N_{dis} = 5.10^4$ cm^{-2}). The g.b. are indicated by arrows and other dips shown by the photoelectric line are due to sub-grain boundaries which have been revealed by SIRTL etch. The g.b. attenuation ΔI/I of the photocurrent decreases as the resistivity does : ΔI/I ≃ 40 %, 25 % , 18 % for ρ = 1 Ωcm, 0.5 Ωcm, 0.25 Ωcm. This proves the

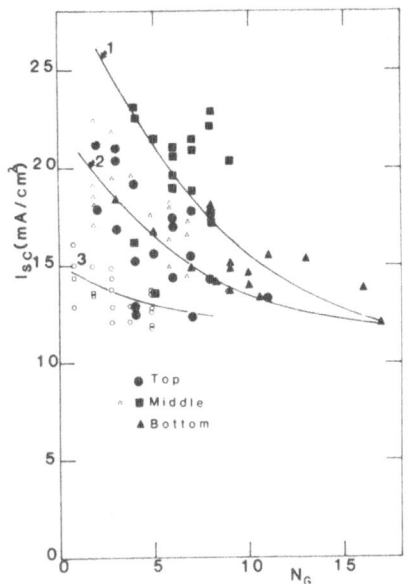
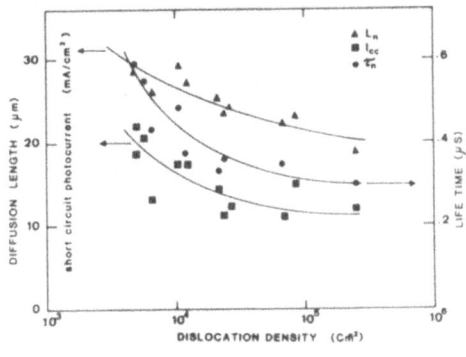

Fig.6

Fig.5

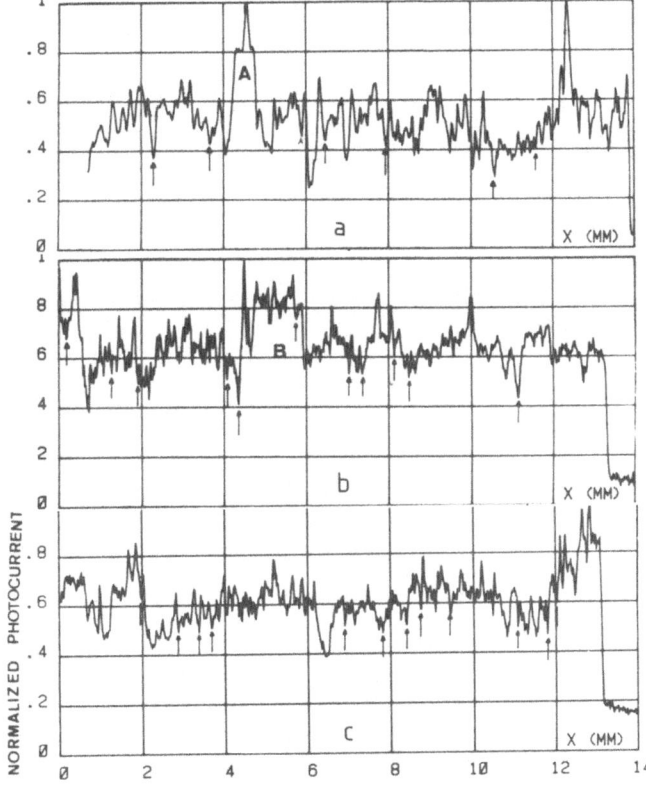

Fig.7

decrease of s with the doping concentration. The influence of ρ on s may be due to the variation of g.b. activity which is caracterized by an effective s depending exponentially on the g.b. barrier height E_B (7). When the interfacial state density is about 5.10^{11}cm^{-2}, we have shown (12) that E_B decreases for $N_A > 10^{16}$ cm^{-3} ($\rho \simeq 1$ Ωcm), and consequently the g.b. activity is drastically reduced. This explains the weak variation of the electronic properties with N_G observed in the lot 104.

5. CONCLUSION

It has been verified in this work that minority carrier properties (L_n, τ_n, μ_n) and photovoltaïc properties (I_{sc}) of cast polysilicon are dependent on the grain size up to 1 mm. This reveals the importance of scattering and recombinations at grain boundaries. As it was expected τ presents the most important variation with the grain size. This influence of grain boundaries is often screened by intragrains defects, especially the dislocations and the impurities. This explains why, generally, L_n and τ_n are shorter at the bottom or at the periphery of ingots where the dislocation density is often very high, and at the top of ingots where the impurities are concentrated by the Bridgman process.

Local investigations show that the difference between the photoelectric responses of grains is essentially due to variations of dislocation density. The g.b. activity decreases with the doping level when ρ is smaller than 1 Ω.cm.

ACKNOWLEDGMENTS

We would like to thank J. DONON and D. SARTI (PHOTOWATT Int. S.A.) for photocell supplying.

This study has been partially sponsored by COMES (contracts n° 79099 and 81.11.013.3413) and PIRDES (contract n° 2005).

REFERENCES

(1) B.L. SOPORI - SPIE 248, 8 (1980).
(2) S. CHITRE and J. DONON - Proc. 3rd Photovoltaïc Solar Energy Conference, Cannes (1980)p.608.
(3) J. FALLY and C. GUENEL - Proc. 3rd Photovoltaïc Solar Energy Conference, Cannes (1980)p.598.
(4) M.M. MANDURAL, K.C. SARASWAT, C.B. HELMS and T.I. KAMINS - J.Appl.Phys. 51, 5755 (1980).
(5) E.D. STOKES and T.L. CHU - Applied Phys.Lett. 30, 425 (1977).
(6) D.M. BIELLE-DASPET and G.D. GASSET - Solid State Electronics, 21, 1219 (1978).
(7) H.C. CARD and E.S. YANG - IEEE Trans. Electron. Dev. ED.24, 387 (1977).
(8) A.K. GOSH, C. FISHMAN, T. FENG - J.Appl.Phys. 51, 446 (1980).
(9) J.P. CREST, H. AMZIL, M. ZEHAF and J. OUALID - 161st Meeting of the Electrochemical Society - Montreal (1982).
(10) J.D. ZOOK - 3rd Photovoltaïc Solar Energy Conference, Cannes, 569 (1980).
(11) J. OUALID, M. BONFILS, J.P. CREST, G. MATHIAN, H. AMZIL, J. DUGAS, M. ZEHAF and S. MARTINUZZI - Rev.Phys.Appl. 17, 119 (1982).
(12) J. DUGAS, J.P. CREST and J. OUALID - 161st Meeting of the Electrochemical Society - Montreal (1982).
(13) M. ZEHAF, H. AMZIL, J.P. CREST - 161st Meeting of the Electrochemical Society - Montreal (1982).

GAS IMMERSION LASER DIFFUSION: A NEW METHOD FOR MAKING EFFICIENT Si SOLAR CELLS

G. B. TURNER, D. TARRANT and D. ALDRICH
ARCO Solar, Inc.
20554 Plummer Street
Chatsworth, California 91311

R. PRESSLEY and R. PRESS
XMR, Inc.
3350 Scott Boulevard #57
Santa Clara, California 95051

Summary

A new method for making both p^+/n and n^+/p junction solar cells is described. A Si wafer is immersed in a transparent dopant gas and irradiated with a pulsed alexandrite laser (0.73 μm). The surface of the wafer melts for less than a microsecond and dopant from the gas is dissolved in it before epitaxial regrowth. There is no implant or deposition step and no photolysis is required so a UV laser is not needed. The technique is termed GILD, an acronym for Gas Immersion Laser Diffusion. Using PH_3 or B_2H_6 dopant, junctions were formed in 3 x 3 mm squares by single 200 ns, 2 J cm^{-2} pulses. By repeating the pulses at a 10 Hz rate and stepping the substrate between pulses, large areas were treated and 2 x 2 cm solar cells cut from them. Efficiencies of 9.2% for both p^+/n and n^+/p cells have been achieved without antireflection coating or back surface fields. The process is simple and rapid and offers significant advantages for cell production, including elimination of back etch and masking. Substrates are not heated except at the surface.

1. INTRODUCTION

The use of a laser to diffuse dopants from a gas into Si has recently been described. A pulsed UV laser was first used to photochemically dissociate $Al(CH_3)_3$ or $B(CH_3)_3$ gas, as well as to melt the surface of the Si substrate (1). The dopant was dissolved and diffused in the liquid state before epitaxial regrowth. Subsequently, lasers of longer wavelength were used to melt Si surfaces immersed in BCl_3 or PH_3 (2,3). In these cases, the dopant was dissociated pyrolytically by the molten Si. We call the process Gas Immersion Laser Diffusion (GILD). This paper describes the application of the GILD process to both p^+/n and n^+/p Si solar cells. The process has a number of advantages for cell production in addition to its inherent simplicity. Since the doping occurs only where the wafer is irradiated, no back etch or masking is needed. Projection optics could be used to make interdigitated cells or other designs requiring pattern definition without photoresist. Substrates are not heated except at the surface, so grain boundary recombination is not activated in polycrystalline substrates (4).

2. GILD PROCESS

Cells were fabricated on chemically-mechanically polished Czochralski wafers, about 330 μm thick. Boron doped p-type wafers were 1 Ω-cm, 103 mm diameter and phosphorus doped n-type wafers were 2 Ω-cm, 75 mm diameter. The wafers were treated in dilute HF, rinsed in deionized water and propanol, and blown dry with N_2. The substrates were then placed horizontally beneath a quartz window in a stainless steel chamber and held in place with clips. The chamber and its associated optics are shown schematically in Fig. 1.

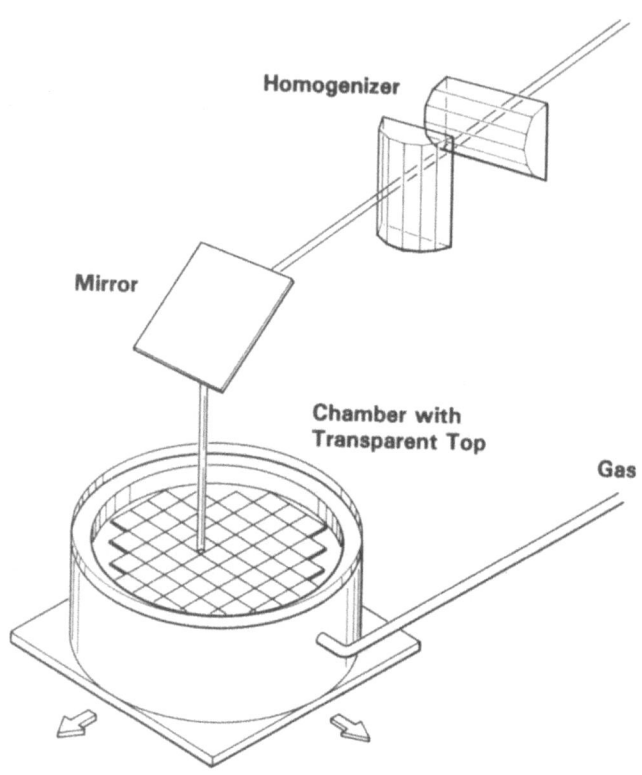

Fig. 1. Schematic diagram of chamber and associated optics.

The chamber is normally evacuated, purged, and evacuated again before filling with dopant-containing gas. For n^+/p cells the gas was usually 10% PH_3 in Ar, while for p^+/n cells, 10% B_2H_6 in He was used. For most experiments the pressure was 500 Torr for safety reasons. The laser was Q-switched alexandrite operating at 0.73 μm. The beam was expanded to 25 mm diameter and passed through a homogenizer to uniformly illuminate a 3 x 3 mm square area. The energy density was about 2 J/cm^2 with a duration of about 200 nsec, pulsed 10 times per second. Although junctions resulted from single pulses, the chamber was moved horizontally 1 mm between pulses by an automatic positioner in order to uniformly treat a large area. Repeating this process, each area was melted by 18-36 overlapping pulses, probably more than necessary.

3. RESULTS

A typical carrier concentration profile resulting from this treatment using PH_3 is shown in Fig. 2 (profiles using B_2H_6 are similar). This

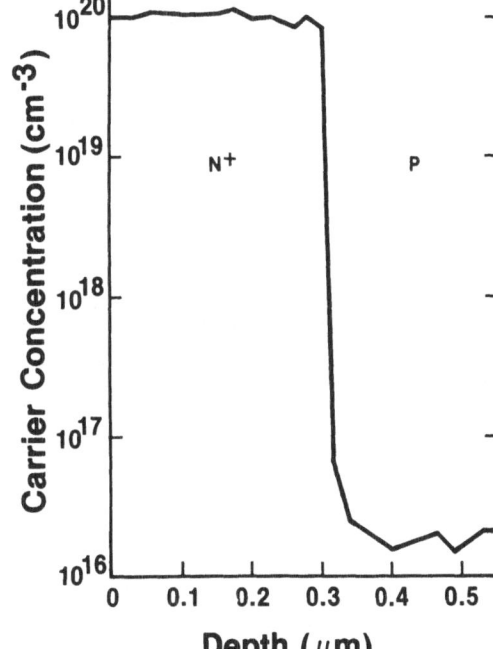

Fig. 2. Typical carrier concentration from spreading resistance.

profile was obtained by spreading resistance which indicates only carrier concentration and hence active dopant concentration. A profile obtained by secondary ion mass spectroscopy (3), which indicates all dopant, showed similar concentrations and the measured sheet resistivities are consistent with all, or almost all, of the dopant being active. Melt duration has been measured by optical reflection of 633 nm light and found to be somewhat longer than the laser pulse but less than 300 ns at 2 J cm^{-2}. This corresponds to a total molten time of about 10 μsec for the 36-exposure sample shown in Fig. 2. During that time about 1.8×10^{17} PH_3 molecules/cm^2 hit the surface, or about 60 times as many as the number of P atoms dissolved in the Si. Adsorption, however, plays a significant and perhaps major role. Deutsch et al. have reported (2) evidence of an adsorbed layer from exposure to BCl_3 gas prior to evacuation and laser melting in vacuum. Our own preliminary experiments seem to confirm this effect for PH_3.

The treated wafers were diced into 20 x 20 mm cells and Cr-Pd-Ag contacts applied by evaporation. The back contact was annealed for 10 s at < 500°C. The front grid contact, covering about 8.5% of the surface, was applied after this treatment. Dark I-V curves for one of each type of cell are shown in Fig. 3. The difference between p$^+$/n and n$^+$/p cells are small and within the cell-to-cell variation. Short circuit current of each cell was measured in sunlight near 100 mW/cm² and corrected to that value. An incandescent (ELC lamp) simulator was adjusted to give the corrected short circuit current for each cell and light I-V curves shown in Fig. 4 were obtained. Performance of the two types was almost identical. Antireflection

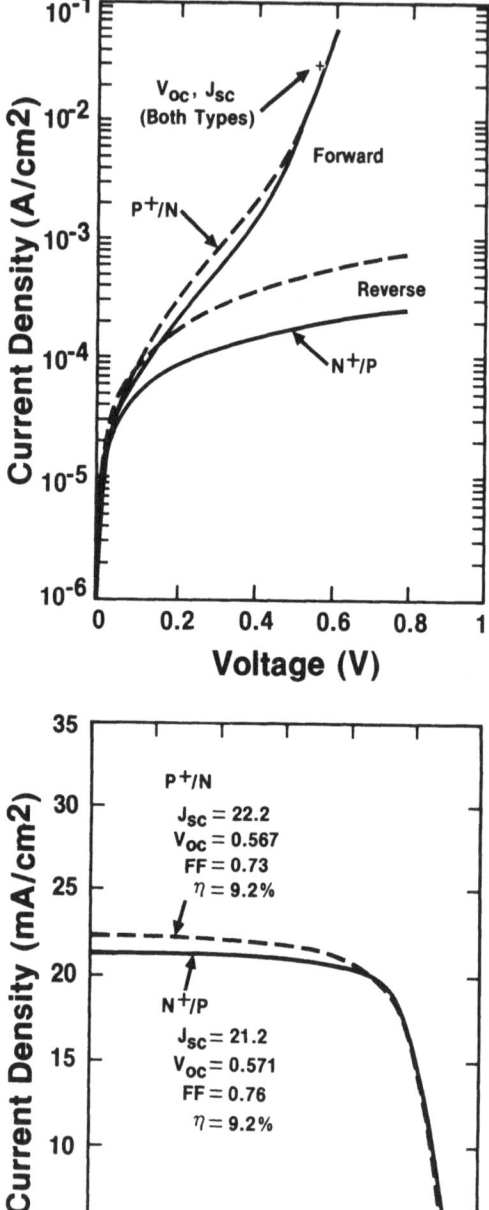

Fig. 3. Junction characteristics of cells in the dark. The cross represents the V_{oc}, I_{sc} points from Fig. 4.

Fig. 4. Photocurrent density vs. voltage curves for cells under 100 mW/cm² light. No antireflection coating was employed.

coatings commonly improve efficiency by a factor of 1.4 (e.g., from 9.2% to 12.9%). Spectral quantum efficiency is shown in Fig. 5. The p+/n cells have slightly higher response at short wavelengths.

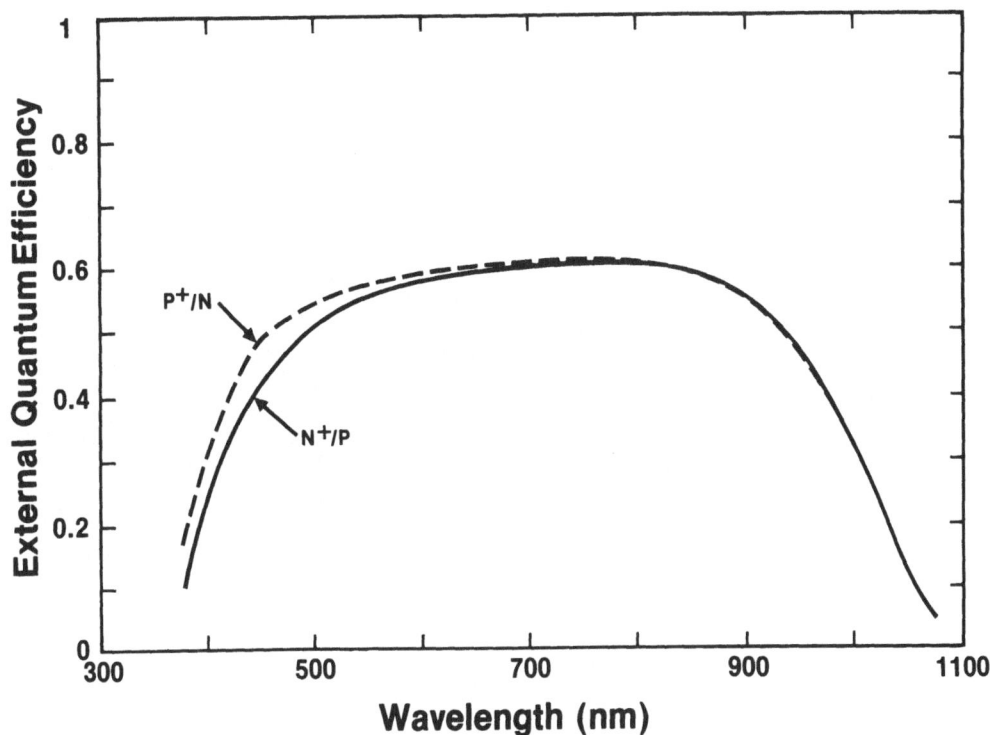

Fig. 5. External quantum efficiency measured under approximately one sun bias light. Result without bias is similar. No antireflection coating was employed.

We wish to thank Gary Brown, Twyla Byers, Kim Eskenas, Dorothy Houk and Ron Merkord for valuable technical assistance.

5. REFERENCES

(1) Deutsch, T. F., Fan, J. C. C., Turner, G. W., Chapman, R. L., Ehrlich, D. J. and Osgood, R. M. Jr., Appl. Phys. Lett. 38, 144 (1981).

(2) Deutsch, T. F., Ehrlich, D. J., Rathman, D. D., Silversmith, D. J., and Osgood, R. M. Jr., Appl. Phys. Lett. 39, 825 (1981).

(3) Turner, G. B., Tarrant, D., Pollack, G., Pressley, R., and Press, R., Appl. Phys. Lett. 39, 967 (1981).

(4) Turner, G. B., Tarrant, D., Aldrich, D., Pressley, R., and Press, R., "Grain Boundary Photocurrent Enhancement in Solar Cells Made by Laser Diffusion," this conference.

VALIDITY OF THE EFFECTIVE LIFETIME CONCEPT
IN POLYCRYSTALLINE SILICON

S.C.Jain
Solidstate Physics Laboratory, Lucknow Road
Delhi, 110007, India

R.Janssens, G.Cheek, P.DePauw, R.Mertens and R.Van Overstraeten
Katholieke Universiteit Leuven, E.S.A.T. Laboratory
Kardinaal Mercierlaan 94, B-3030 Heverlee, Belgium

Summary

Polycrystalline silicon currently has much technological importance for potential photovoltaic applications. In this paper the effective lifetime concept of minority carriers, often used for solar cell calculations, is examined. The effective grain boundary surface recombination velocity, S_{eff}^{GB}, is not linear in n, the minority carrier concentration. The problem can be made linear if S_{eff}^{GB} is very large such that the diffusion current to the grain boundary is limited by the diffusion velocity. However, in general a large S_{eff}^{GB} makes the series $n(x,y,z) = \Sigma \Sigma n_{i,j}(x,y,z)$ slowly convergent and the cell cannot be described by one τ_{eff}. It is found that for calculating Isc, the concept of τ_{eff} works as a special case. The error in the calculation of the dark saturation current becomes large, the principle of superposition becomes invalid and Voc cannot be calculated using the concept of τ_{eff}.

1.0 Introduction

The effect of loss of minority carriers at surfaces is described by a recombination velocity, S, and the boundary condition

$$D_n \frac{\delta n}{\delta x} = S n (x,y,z) \quad (1)$$

has to be used at the surfaces (1). Shockley (1) calculated the rate of recombination of the minority carriers in a thin semiconductor filament and showed that under certain conditions the surface can be ignored and the recombination rate in the filament can be described by one single constant effective lifetime, τ_{eff}, given by

$$\frac{1}{\tau_{eff}} = \frac{1}{\tau_B} + \frac{1}{\tau_S} \quad (2)$$

where τ_S is a function of S and of the dimensions of the filament. Since recombinations at the grain boundaries can also be described by a recombination velocity, S_{eff}^{GB}, the concept of using τ_{eff}, given by equation (2), to take into account the effect of grain boundaries is very attractive (2,3,4). The effective lifetime approach is interesting because with this approach, existing equations for Isc, Voc etc. can be used for a polycrystalline silicon solar cell simply by replacing τ_S by τ_{eff} in these equations.

It is the purpose of the present paper to determine to what extent the effective lifetime concept can be used to calculate the photovoltaic properties of the polycrystalline silicon.

2.0 The Grain Boundary Model

In recent years, several papers on grain boundary modelling have been published (2-5); today it is generally assumed that as a result of grain boundary traps, a band bending exists at a grain boundary creating a space charge layer. The existance of this band bending causes the grain boundary recombination to be a non-linear process (4,5). As a result of this non-linearity, the effective grain boundary recombination velocity, S_{eff}^{GB}, is not constant and will depend on the excitation given by the difference between the two quasi-Fermi levels E_{fn} and E_{fp} (Figures 1 and 2). It is important to note that two grain boundary recombination velocities can be defined (Figure 2): one designated as $S(0)$ at the grain boundary itself and another, (S_{eff}^{GB}), at the edge of the space charge region surrounding the grain boundary. For device calculations,

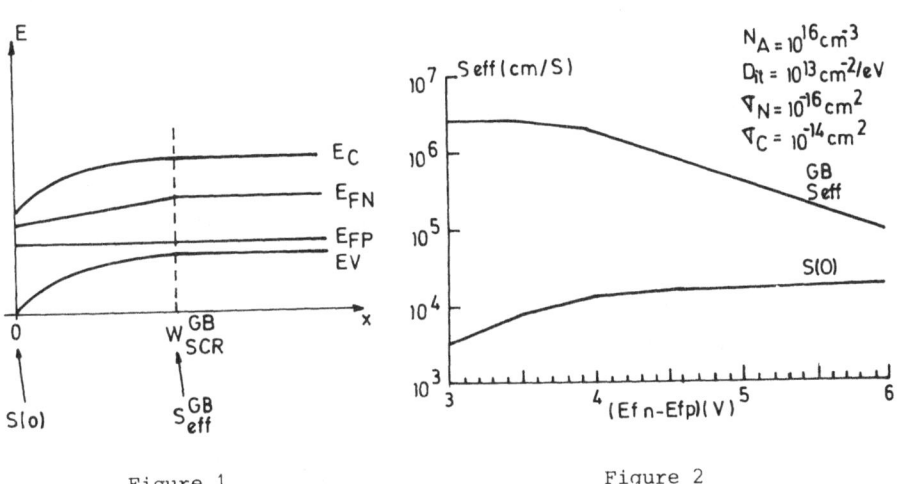

Figure 1

Figure 2

Figure 1: The band diagram near the grain boundary.
Figure 2: The calculated values of S_{eff}^{GB} and $S(0)$ are given for different values of $E_{fn}-E_{fp}$. The values of parameters used in the calculations are listed on the figure.

the second one defined at the depletion region edge, is the important quantity. It also can be seen that for the parameter values listed in Figure 2, which may be considered to be typical, the effective recombination velocity defined at the space charge region edge is high: typically in the 10^4 to 10^5 range. The most important conclusion, however, is that S_{eff}^{GB} is not constant.

3.0 The Solution of the Transport Equation

We consider a solar cell made of polycrystalline silicon with columnar grains. Let each grain have a cross section W X W and let the length of the grain in the z direction be much longer than the diffusion length. Also consider the n^+p junction in one grain, with the junction plane being perpendicular to the z axis. If the diffusion equation is non-linear in n, it cannot be easily solved. However in this case the nonlinearity comes in through the dependence of S on n(0) in the boundary condition. For a fixed n(0), S is constant and the diffusion equation can be solved. The minority carrier density in the base of the cell can be written as (6)

$$n(x,y,z) = \sum_{i=0}^{\infty} \sum_{j=0}^{\infty} n_{i,j}(x,y,z) \qquad (3)$$

and

$$\tau_{i,j} = f(i,j,\tau_B,W,D_n,S_{eff}^{GB}) \qquad (4)$$

The values $n_{i,j}(x,y,z)$ and the rapidity with which the series converges depend on the boundary conditions as well as on the distribution profile of the carriers. If the series in (3) is rapidly convergent i.e. if only the first mode dominates, we can write

$$n(x,y,z) \cong n_{0,0}(x,y,z) \qquad (5a)$$

and

$$\frac{1}{\tau_{eff}} \cong \frac{1}{\tau_{0,0}} \cong \frac{1}{\tau_B} + \frac{1}{\tau_S} \qquad (5b)$$

In addition to the series being rapidly convergent, τ_S must be independent of the injected density of minority carriers so that the boundary condition (1) becomes linear in n and τ_{eff} remains constant. It is only under these conditions that τ_{eff} is a physically meaningful parameter. We examine the conditions under which this can be accomplished in the next section.

4.0 The Validity of the Effective Lifetime Concept

Since τ_S involves S_{eff}^{GB}, which varies with the injection level, the concept of constant effective lifetime to describe the polycrystalline silicon solar cell is not valid in all cases. Since S_{eff}^{GB} is not constant, (Figure 2), τ_{eff} will be constant only if the recombination current flowing towards the grain boundaries is not dependent on the value of S_{eff}^{GB}. This can happen if S_{eff}^{GB} is sufficiently large. In other words, the grain boundary recombination current becomes independent of the value of S_{eff}^{GB} only when the grain boundary recombination velocity, S_{eff}^{GB}, is much larger than the average diffusion velocity, $2D_n/W$, of the carriers flowing from the bulk towards the grain boundary. In this case, the minority carrier current is diffusion limited and is independent of S_{eff}^{GB}. For example, in the case of W=100 μm, one finds $2D_n/W$ is 5×10^3 cm/s. Since S_{eff}^{GB} is considerably larger than this value (Figure 2), the current is diffusion limited.

To summarize the results of this discussion, we can say that will be independent of the excitation and the boundary condition will be linear if the condition

$$\frac{S_{eff}^{GB}}{2D_n/W} \gg 1 \qquad (6)$$

is satisfied.

We see that for equation 5b to be valid with a constant τ_{eff}, two conditions have to be satisfied. Firstly, series (3) must be rapidly convergent and secondly equation (6) must hold to make the problem linear in n and τ_{eff} independent of n. Shockley has shown that in general the series is rapidly convergent if

$$\frac{S_{eff}^{GB}}{2Dn/W} \ll 1 \tag{7}$$

We see that in general, equations (6) and (7) cannot be satisfied simultaneously and the τ_{eff} concept is not valid. We discuss two special cases in the next section.

5.0 Particular Cases: Voc and Isc

We consider a forward biased n^+ p junction in the grain. The low injection Shockley boundary condition requires a carrier profile $n(x,y,z)$ near the junction in the base given in Figure 3a if S_{eff}^{GB} is large. The profile corresponding to the dominant mode i.e. the first term in series (3) is given in Figure 3b. The first term poorly represents the profile

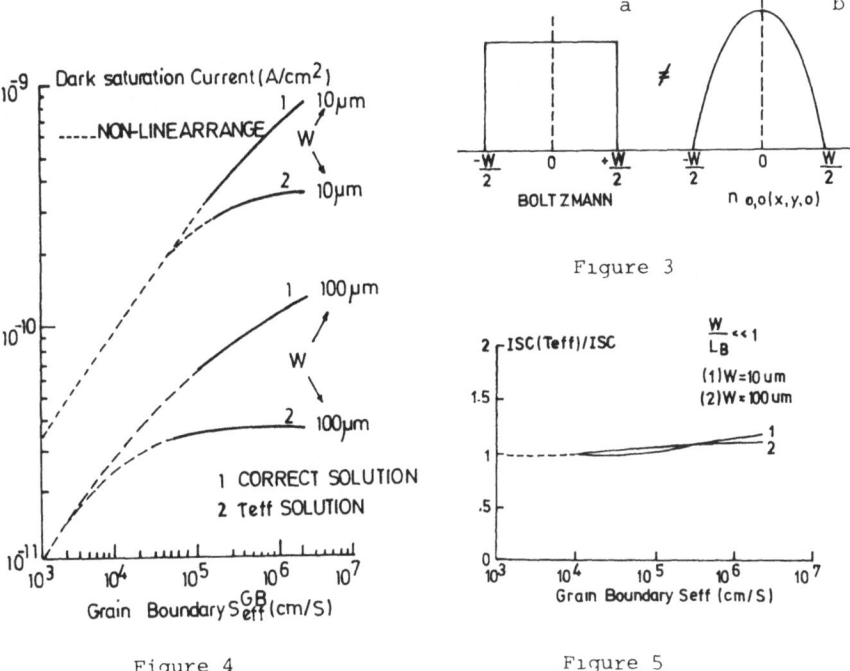

Figure 3

Figure 4

Figure 5

Figure 3a: The initial carrier profile in the base of a forward biased diode in a thin grain.
Figure 3b: The carrier profile required for the first mode in (3) to be dominant.
Figure 4: Dark saturation current as a function of S_{eff}^{GB}. Note that the dashed lines correspond to S values where τ_{eff} is a function of excitation.
Figure 5: The ratio of the correct short circuit current values and those using the τ_{eff} concept as a function of S_{eff}^{GB}.

and as implied by condition (7), the series is slowly convergent. The approximation of a single τ_{eff} is not valid in this case. Figure 4 indicates that the dark saturation current or forward diode current calculated using single τ_{eff} concept are, in fact, in large error for large values of S_{eff}^{GB}. For small values of S_{eff}^{GB} the problem is non-linear and τ_{eff} as well as S_{eff}^{GB} depend on excitation. In the short circuit configuration however, the initial profile in a solar cell does not have to satisfy the Shockley boundary condition at the junction, it corresponds more nearly to Figure 3b and the series becomes rapidly convergent even though S_{eff}^{GB} is large. This is a special case when the first mode is the dominent mode even though condition (7) given by Shockley for the general case is not valid. Since we have now identified the conditions when (6) is valid and the series is rapidly convergent, the effective lifetime concept should work in this case. The calculated values for Isc for a large S_{eff}^{GB}, in Figure 5 indicate that τ_{eff} concept is in fact valid in this case. Note that the largest error in Isc for $S_{eff}^{GB} > 10^6$ cm/sec and for $W = 10\,\mu m$ is less than 25%.

6.0 Conclusions

We have shown that for the concept of a single, constant τ_{eff} to be valid to describe the physics of polycrystalline silicon solar cells, in general two conditions, (6) and (7) must be satisfied. These conditions are mutually inconsistent, and the τ_{eff} concept in general does not work. However, in the short circuit configuration, the series (3) is shown to be rapidly convergent even for large S_{eff}^{GB}. In this case, equation (6) is satisfied and equation (7) is not necessary; the concept of τ_{eff} becomes valid. The τ_{eff} concept does not work, however, for calculating dark saturation current, the superposition principle cannot be used and the Voc can not be calculated for a polycrystalline silicon cell using the simple equations applicable to the case of single crystal cells.

7.0 Acknowledgement

This paper is part of a report of the Belgian National R-D Energy Program (Prime Ministers Office of Scientific Research, Wetenschapstraat 8, Brussels, Belgium). R.Mertens is supported by the N.F.W.O. One of us (SCJ) is grateful to professors R.Van Overstraeten and R.Mertens for making his visit to K.U.Leuven / E.S.A.T. possible which resulted in this collaboration.

8.0 References

1) W. Shockley, Bell Sys. Tech. J. 28, 435 (1949)
2) S.I.Soclof and P.A.Iles Proc. of the 11th IEEE Photovoltaic Specialists Conference May 1975, p.56
3) R.A.Smith "Semiconductors", p.286-290, Cambridge University Press 1978
4) J.G.Fossum and F.A.Lindholm IEEE Trans. on Electron Devices, ED-27 692, (1980)
5) P.Panayotatos and H.C.Card IEEE Electron Device Letters Vol. EDL-1 No.12 December 1980
6) H.S.Carslaw and J.C.Jaeger, "Conduction of Heat in Solids", Chapt.6 p.176, Clarendon Press, 1967

STUDY OF THE GRAIN BOUNDARY EFFECTS IN N-P JUNCTION POLYCRYSTALLINE SI SOLAR CELLS

N. C. HALDER
Department of Physics, University of South Florida
Tampa, FL 33620, U. S. A.

Summary

The effects of grain boundary, such as recombination velocity, grain size, and grain orientations have been studied on the efficiency of n-p junction polycrystalline Si solar cells. Numerical calculations have been made for the photocurrent, dark current and fill factor utilizing our previously developed theory in three dimension. The results of this study indicate that it is possible to produce an efficiency as high as 15% with fibrously grained samples of grain size 100μm, recombination velocity 100m/sec, and thickness 100μm.

Symbols

AM1	solar spectrum at earth's surface with sun at zenith; ideal conditions; Air Mass One
D_p	diffusion coefficient of holes in n-type material; cm^2/sec
F	solar irradiance in photons/cm^2/sec
H	cell thickness
L_p	diffusion length of holes in n-type material; $L_p^2 = D_p \tau_p$
N_p	hole concentration different from n_p; $N_p = S_p \gamma_p / D_p$
FF	fill factor
p_n	hole carrier density in n-type material
p_{no}	hole density in thermal equilibrium
q	electron charge
R	reflectance
S_g	grain boundary recombination velocity when $S_{ng} = S_{pg}$
W	width of depletion region
X_g	grain size ($X_g = 2a$)
z_j	junction depth
α	absorption coefficient
γ_p	effective polycrystalline diffusion length for holes in n-type material
η	efficiency

1. INTRODUCTION

In polycrystalline Si solar cells the primary loss mechanism has been shown to be the recombination of hole-electron pairs at the grain boundaries. Card and Yang (1) have discussed the electronic processes taking place at grain boundaries in polycrystalline semiconductors and noted that recombination velocities at these locations range from 10^2 to 10^6 cm/sec depending on the interface state density and the doping concentration of the semiconductor. Ghosh et al. (2) made calculations for the currents and other cell parameters based on an effective lifetime assumed evenly distributed throughout the grain and directly dependent on the grain size. Recently, Lindholm and Fossum (3) have suggested an approximate theory where some grain boundary effects have been included. However, there is no three dimensional (4) theory for polycrystalline samples with all the grain boundary effects.

Recently, however, we have solved (5) the three dimensional diffusion equation by the Green's function method for the photocurrent and the dark current which include the grain boundary effect in polycrystalline Si solar cells. The lifetime is assumed to be that which would normally be associated with the doping density of the particular semiconductor used in a fibrously grained n-p junction model. The fact that this lifetime is effectively reduced due to grain boundary recombination becomes apparent in the analysis. The result is that a grain boundary effect factor appears (which consists of grain orientation, grain size and recombination velocity) in the current density expressions suggested previously for the equivalent single crystal model. This factor allows a determination of the photocurrent and the dark current at any position across the grain. The three dimensional theory, which is an improvement over the previous one dimensional theory (3,4), points out correctly the limiting factors of the solar cell. It will be shown here that the I-V characteristics and efficiency of a polycrystalline Si cell approach the respective values of the corresponding single crystal Si cell in the zero grain boundary limit.

2. THEORETICAL CONSIDERATION

Theory for single crystal Si solar cell is well understood (6). For polycrystal Si cells, while there have been many attempts, the three dimensional diffusion equation has not been solved (3) fully. This is due to the fact that the partial differential equations are nonlinear and presence of many (at least eight) boundary conditions. For this reason, in the past, simpler approach, i.e., single crystal theory in one dimension with some modifications for grain boundary effects has been proposed.

Actually a polycrystalline sample has many grains with random orientations as shown in Fig. 1(a). In the present paper, however, we will consider a fibrously oriented rectangular grains as described in Fig. 1(b) for the sake of mathematical simplicity and standard geometry. The present solar cell is essentially an n-p junction polycrystalline Si with vertical grain boundary which are not difficult to make with present day technology and material processing. An exact solution of the partial differential equation of the type

$$\nabla^2(p_n - p_{no}) - \frac{(p_n - p_{no})}{L_p^2} = -\frac{\alpha F(1-R)\exp(-\alpha z)}{D_p} \qquad (1)$$

has already been suggested by the Green's function method in our previous publication (5) for the appropriate boundary conditions. Therefore, we

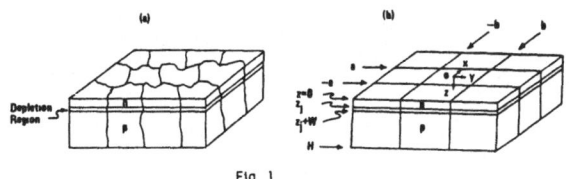

Fig. 1

will only use the main results of this paper to study the efficiency of a polycrystalline Si cell with reference to various cell parameters.

(a) <u>Photocurrent Density</u>

The hole current density per unit band width for the n-region at the junction edge is

$$J_p = 4q\alpha F(1-R) \sum_{m,n}^{\infty} \frac{\gamma_p M_a^2 N_b^2 \sin(ma)\sin(nb)}{mn(\alpha^2 \gamma_p^2 - 1)} \cos(mx)\cos(ny)$$

$$\times \left[\frac{(N_p + \alpha\gamma_p) - \exp(-\alpha z_j)\left(N_p \cosh\frac{z_j}{\gamma_p} + \sinh\frac{z_j}{\gamma_p}\right)}{N_p \sinh(z_j/\gamma_p) + \cosh(z_j/\gamma_p)} \right.$$

$$\left. - \alpha\gamma_p \exp(-\alpha z_j) \right]. \tag{2}$$

and electron current density in the p-region at the same junction edge is

$$J_n = 4q\alpha F(1-R)\exp[-\alpha(z_j+W)] \sum_{k,\ell}^{\infty} \frac{\gamma_n K_a^2 L_b^2 \sin(ka)\sin(\ell b)}{k\ell(\alpha^2\gamma_n^2 - 1)}$$

$$\cos(kx)\cos(\ell y)$$

$$\left[\alpha\gamma_n - \frac{N_n\left(\cosh\frac{H'}{\gamma_n} - \exp(-\alpha H')\right) + \sinh\frac{H'}{\gamma_n} + \alpha\gamma_n\exp(-\alpha H')}{N_n \sinh(H'/\gamma_n) + \cosh(H'/\gamma_n)} \right] \tag{3}$$

The expression for the photocurrent in the depletion region, however, is assumed to be approximately the same as in the corresponding single crystal

$$J_{dr} = qF(1-R)\exp(-\alpha z_j)[1 - \exp(-\alpha W)] \tag{4}$$

Thus the total photocurrent density is obtained as

$$J_{ph} = J_p + J_n + J_{dr} \tag{5}$$

(b) <u>Dark Current Density</u>

Similarly the dark current densities for holes and electrons respectively are

$$J_p = -4qD_p \sum_{m,n}^{\infty} \frac{M_a^2 N_b^2}{mn\gamma_p} \sin(ma)\sin(nb)\cos(mx)\cos(ny)$$

$$\times \left[\frac{N_p \cosh \frac{z_j}{\gamma_p} + \sinh \frac{z_j}{\gamma_p}}{N_p \sinh \frac{z_j}{\gamma_p} + \cosh \frac{z_j}{\gamma_p}} \right] p_{no}[\exp(qV_j/kT) - 1] \quad (6)$$

and

$$J_n = -4qD_n \sum_{k,\ell}^{\infty} \frac{K_a^2 L_b^2 \sin(ka)\sin(\ell b)}{k\ell \gamma_n} n_{po}[\exp(qV_j/kT) - 1]$$

$$\cos(kx)\cos(\ell y) \left[\frac{N_n \cosh \frac{H'}{\gamma_n} + \sinh \frac{H'}{\gamma_n}}{N_n \sinh \frac{H'}{\gamma_n} + \cosh \frac{H'}{\gamma_n}} \right] \quad (7)$$

which gives the total dark current density:

$$J_{dark} = J_p + J_n = J_o[\exp(qV_j/kT) - 1] \quad (8)$$

where

$$J_o = \frac{4qD_p n_i^2}{N_d} \sum_{m,n}^{\infty} \frac{M_a^2 N_b^2}{mn\gamma_p} \sin(ma)\sin(nb)\cos(mx)\cos(ny)$$

$$\times \left[\frac{N_p \cosh \frac{z_j}{\gamma_p} + \sinh \frac{z_j}{\gamma_p}}{N_p \sinh \frac{z_j}{\gamma_p} + \cosh \frac{z_j}{\gamma_p}} \right] + \frac{4qD_n n_i^2}{N_a} \sum_{k,\ell}^{\infty} \frac{K_a^2 L_b^2}{k\ell \gamma_n} \sin(ka)\sin(\ell b)\cos(kx)$$

$$\cos(\ell y) \times \left[\frac{N_n \cosh \frac{H'}{\gamma_n} + \sinh \frac{H'}{\gamma_n}}{N_n \sinh \frac{H'}{\gamma_n} + \cosh \frac{H'}{\gamma_n}} \right]. \quad (9)$$

3. RESULTS AND CELL BEHAVIOR

We shall now investigate the behavior of Si solar cells with respect to the above equations. We shall particularly concentrate on three parameters: (i) cell thickness, (ii) grain size and (iii) recombination velocity. The data used in this study are given below (7).

	S_g	D	L	τ	N
p	100 m/s	1.295×10^{-4} m²/s	6.233×10^{-7} m	3×10^{-9} s	1.5×10^{22}/m³
n	∞	2.70×10^{-3}	1.64×10^{-4}	10×10^{-5}	5×10^{25}

Figs. 2-5 show the variation of I_{sc}, I_{dark}, V_{oc}, and FF with grain size for various cell thicknesses. Based on these results we have finally obtained efficiencies of polycrystalline Si cells. These are shown in Figs. 6-9. These calculations have been made for AM1 solar irradiance and above solar cell parameters.

4. DISCUSSIONS AND CONCLUSIONS

Several important results emerge by this investigation. First of all we find from Figs. 2-5 that (i) I_{dark} decreased, but I_{SC}, V_{OC} and FF increased with the increase of grain size X_g for a fixed cell thickness H, and (ii) I_{dark} decreased, but I_{SC}, V_{OC} and FF increased with the increase of H for a fixed X_g. In addition to this, we find from Figs. 6-9 that (i) η decreases with the position of axis x (at y = 0) both for a fixed grain boundary velocity S_g and X_g, (ii) η decreases with S_g for a fixed X_g, and (iii) η increases with X_g for a fixed H.

These results are in general agreement with the previous one dimensional calculations (1-3). We have, however, obtained for the first time exact limiting factors for an n-p junction polycrystalline Si solar cell with vertical grain boundaries. In summary, for any cell size greater than 100μm there is no improvement in efficiency regardless of the grain boundary recombination velocity or thickness; an efficiency greater than 10% can be achieved with polycrystalline cells as thin as 20μm over the entire range (10 ~ 10^4 m/sec) of grain boundary recombination velocity. Finally, this study suggests that an efficiency of about 15% is possible for a cell (operating on this model) with grain size of 100μm, thickness of 100μm, and grain boundary recombination velocity of 100 m/sec.

Figs. 2-5

5. REFERENCES

1. H. C. Card and E. S. Yang, IEEE Trans. Electron Devices ED-24, 397, (1977).
2. A. K. Ghosh, C. Fisherman and T. Feng, J. Appl. Phys. 51, 446 (1980).
3. F. A. Lindholm and J. G. Fossum, Fifteenth IEEE Photovoltaic Specialist Conf., p. 422, May 22, 1981, Orlando, Florida.
4. H. J. Hovel, Semiconductors and Semimetals Vol. II, Solar Cells, Academic Press, New York, New York, 1955.
5. N. C. Halder and T. R. Williams - to be published.
6. J. J. Loferski, J. Appl. Phys. 27, 777 (1956).
7. R. B. Hilborn, Jr. and T. Lin, Proc. of Nat. Workshop on Low Cost Polycrystalline Silicon Solar Cell, p. 246, May 18, 1976, Dallas, Texas.

STUDIES OF THE GAP STATES DENSITY IN UNDOPED AND DOPED
AMORPHOUS HYDROGENATED SILICON

J. KOČKA, M. VANĚČEK, J. STUCHLÍK, O. ŠTIKA, E. ŠÍPEK, H.T. HA, A. TŘÍSKA
Institute of Physics, Czechoslovak Academy of Sciences
180 40 Prague 8, Czechoslovakia

Summary

Density of the gap states in undoped and phosphorus and antimony doped n-type a-Si:H has been determined from the absorption coefficient measurement. A constant photocurrent method was used to determine the value of optical absorption in a low absorption region.
A quantitative model has been suggested to determine the density of states (DOS) within a gap below the Fermi level from the spectral dependence of the absorption coefficient. The model is based on a Gaussian shaped maximum, connected with the dangling bonds, between the exponential valence and conduction band tails. The doping rises this maximum in the DOS.

1. INTRODUCTION

Amorphous hydrogenated silicon has received much attention as a material for low cost photovoltaic solar cells. Value of the optical gap and the density of states (DOS) within the gap are of great interest since they determine the optical and electrical properties of material and so the efficiency of a-Si:H solar cells. To rise the efficiency from present 8% above "10% economical barrier" the following suggestions have been made:
- suppression of the density of the midgap states down to the $10^{14} cm^{-3} eV^{-1}$ range (1)
- improvement of the doping efficiency of the n^+ and p^+ layers (2)
- modification of the n^+ and p^+ layers (a-SiC:H) to get the heterojunction window effect (3).

The optical absorption coefficient measurement in the 10^{-2} to $10^5 cm^{-1}$ range is important in view of above proposals as will be shown further.

2. EXPERIMENTAL METHOD

Photoconductivity related measurements and photothermal spectroscopy (PAS, PDS) are used to complement the direct optical absorption measurement in the low absorption region (below $10^3 \div 10^2 cm^{-1}$). In our recent paper (4) we have discussed some photoconductivity techniques for determination of the absorption coefficient in the low absorption region and suggested to use the constant photocurrent method (CPM) (5) for the measurement of $\alpha(E)$ as a function of photon energy E. In the CPM the inverse of the number of photons, which are necessary to keep the photocurrent constant, is plotted versus the photon energy. This spectrum follows directly the spectral dependence of $\alpha(E)$ (4,6) and the spectrum is set up to absolute scale ($\alpha(E)/cm^{-1}$/scale) by matching the results of the transmission measurements (Fig. 1). Details are given in ref. (6). We measure in the region of homogeneous absorption an average bulk properties of the material.

3. EXPERIMENTAL RESULTS

The samples have been prepared by the glow discharge decomposition of pure silane or mixture of silane with doping gas in capacitavely coupled rf system. The quartz, glass or metal substrates were held at 260°C. SiH_4 flow rate was about 5 sccm, pressure during deposition 40 Pa. Other experimental conditions are described elsewhere (6).

Measurements were made on samples with coplanar configuration of electrodes. The same results were obtained on forward biased Schottky diodes. All spectra exhibit interference fringes.

Value of the optical gap E_g^O was determined from the interception of $\alpha^{1/2} \cdot E^{1/2}$ with the E axis and $E_g^O = 1.8$ eV was found. After a rearrangement of reactor chamber and a change of the sample potential with respect to plasma, "optimized material" with $E_g^O = 1.7$ eV was received (sample C 143).

All measured spectra exhibit a shoulder for photon energies below approximately 1.4 eV (Figs. 1,2). The value of α at this shoulder is sensitive to the preparation condition and doping, being below $1 cm^{-1}$ for "optimized" undoped samples. We can see the rise of this shoulder with increasing Sb doping. (Fig. 1). For comparison a measurement on a sample doped with P (1% PH_3 in SiH_4) is presented in Fig. 1.

Room temperature dark conductivity changes from 10^{-10} to $10^{-3} \Omega^{-1} cm^{-1}$ and the activation energy E_A decreases from 0.8 eV for undoped a-Si:H to 0.25 eV due to doping. On the same samples we have measured a "threshold energy" E_{TH} for the onset of absorption spectra (from the intersection of $\alpha^{1/2} E^{1/2}$ versus E in the lowest absorption region). For all samples E_{TH} falls into 0.6 ÷ 0.8 eV interval.

4. DISCUSSION

Our CPM spectra reflect the optical emission of electrons from the valence band (v.b.) or filled gap states to the conduction band (extended states) or to the localized tail states within the depth D(T) which are in thermal equilibrium with the conduction band (c.b.). At our low illumination level (and low absorption) a quasi-Fermi level for electrons is near the dark Fermi level E_F. Simple estimate of D(T) gives the depth less than 0.7 eV at 300 K (or 0.2 eV at 80 K) (7).

To see the connection of our experimental data $\alpha(E)$ with density of gap states N(E) we shall start with a conventional model based on field effect measurements (8). We are particularly interested in the shoulder in absorption spectra below 1.4 eV. There are four possible absorption processes leading to the shoulder in the photocurrent spectra. Process 1-excitation from the v.b. to the tail states under the c.b. which are in thermal equilibrium with the c.b.-is not a good candidate for the observed shoulder, as it directly follows from the temperature dependence of the absorption (CPM) spectra in Fig. 2. At 163 K process 1 can start at 1.4 eV but the shoulder starts at about 0.7 eV. The similar arguments can rule out process 2 - localized to localized state transition with subsequent thermal emission to the c.b.

Process 3 - excitation from the occupied gap states just below the E_F to the states deeper and deeper in the c.b. - could explain the observed shoulder for undoped a-Si:H, with E_F about 0.7 to 0.8 eV below c.b., by simple fit based on constant DOS in the gap between the tails and with a square root rising DOS in the c.b. (7). But it is not true for n-doped

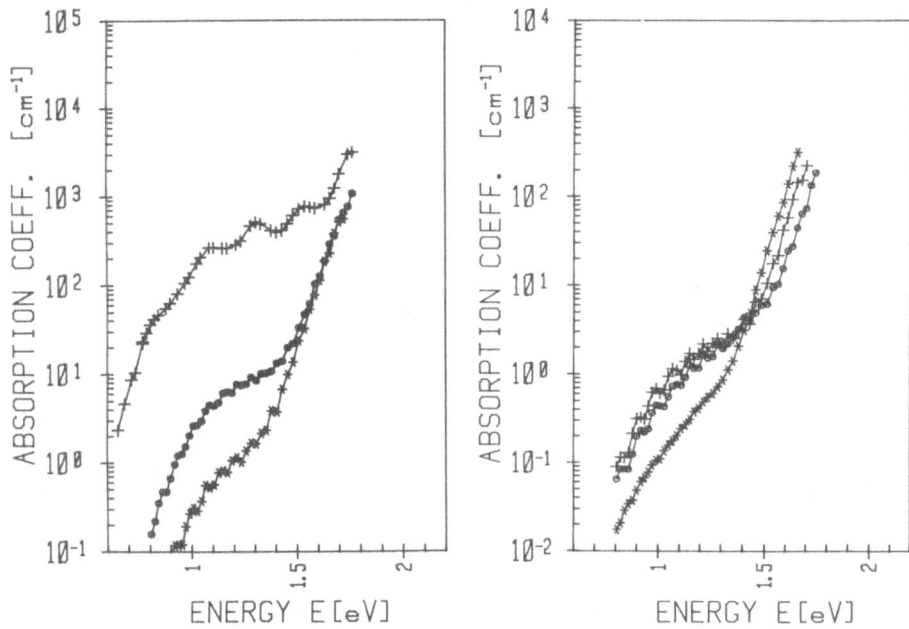

Fig. 1 Spectral dependence of the absorption coefficient $\alpha/cm^{-1}/$ for the samples doped with antimony (A 102 ∗∗ , A 103 oo) and with phosphorus (F 110 ++), 1%PH_3 in SiH_4.

Fig. 2 Temperature dependence of $\alpha(E)$ for undoped sample C98 (298K ++, 163K oo) and $\alpha(E)$ for undoped "optimized" sample C 143 ∗∗(room temperature).

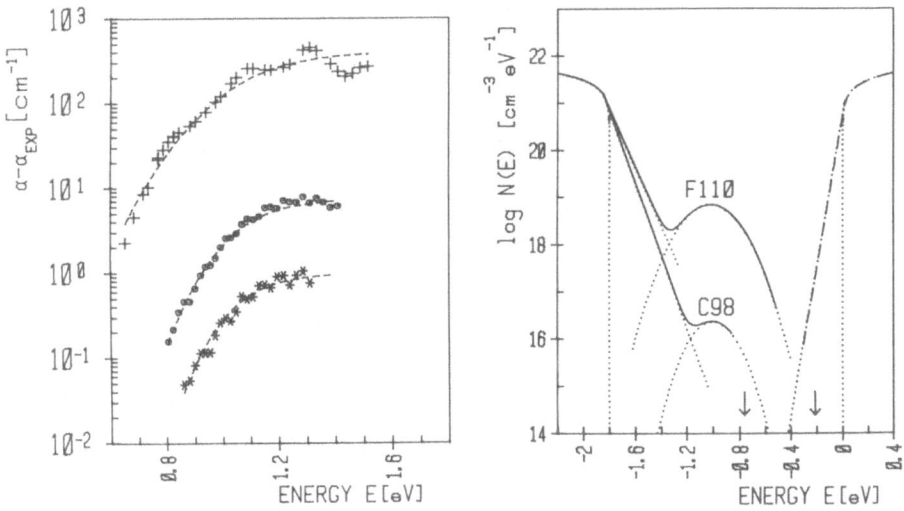

Fig. 3 Spectral dependence of $\alpha(E) - \alpha_{EXP}(E)$ where $\alpha(E)$ is taken from Fig.1 and $\alpha_{EXP}(E)$ is given by eq./1/. Dashed curves are theoretical fits according to eq./3/ with parameters given in Table I.

Fig. 4 Model of DOS within the gap for undoped sample C98 and strongly phosphorus doped sample F110. Full curve is determined from our model. Dashed curve is taken over from (1). Position of E_F is marked by arrow.

samples. On phosphorus doped sample (Fig. 1) with activation energy E_A=0.25 eV the process 3 cannot explain the data because the threshold energy E_{TH} of this shoulder equals 0.62 eV.

So the excitation from a deep gap states to the c.b. (process 4) appears as a likely candidate for the observed dependence of $\alpha(E)$ for all doped and undoped samples. We use it for explanation of our experimental data.

We should not omit another possible process which does not lead to conducting state (process 5) - localized to localized state transition with subsequent recombination. This process can cause underestimation of $\alpha(E)$. Definite answer about the role of this process can be given by comparison of results of the CPM and the PAS. Process 5 gives a signal in photoacoustic spectroscopy but not in the CPM. Photoacoustic measurements of Yamasaki et al (9) on a quite analogous samples of a-Si:H (undoped and doped with 1% PH_3) are in good quantitative agreement with our results (Figs. 1,2). Nevertheless, we have to mention that the PAS is about hundred times less sensitive than the CPM.

The part of the region of exponential absorption tail and the region of fundamental absorption above optical gap are accessible by direct transmission and reflection measurement and they will not be in detail discussed here. The value of the slope E_O of exponential absorption tail

$$\alpha_{EXP}(E) = \alpha_o \exp(E/E_o) \qquad /1/$$

is an important parameter directly determined from the CPM spectra. E_O=0.055 eV for the "optimized" undoped sample C 143. We assume that the exponential absorption tail is given by the transitions from localized states in the exponential valence band tail to extended states in the c.b. This follows from our measurements and from the observation of other authors (7, 10) because both optical and photoelectrical measurements give the same results and the photocurrent is carried by electrons in the c.b. (8).

5. MODEL FOR THE DENSITY OF GAP STATES

The absorption coefficient $\alpha(E)$ for the transitions from the extended or localized states with the density of states $N(\varepsilon)$ to the extended states with the density of states $g(\varepsilon)$ can be written as (11,12)

$$\alpha(E) = \frac{const.}{E} \int N(\varepsilon) \, g(\varepsilon + E) \, d\varepsilon \qquad /2/$$

where integration is over all pairs of the initial and final states separated by an energy E. We ignore the variation of the refraction index n with E and suppose matrix element of the optical transition to be independent of E. The simple arguments for this approximation are given in (12). We use eq. /2/ for the valence to the conduction band transition, taking for both densities of states the free electron value (12), to get the value of the constant which includes the unknown transition matrix element (6).

It follows from Figs. 1,2 that a model of transition from a broad exponential tail to conduction (valence) band (11) does not correspond to our experimental data. We do not observe exponential shape of $\alpha(E)$ below 1.4 eV.

In Fig. 3 we replotted our data as $\alpha(E) - \alpha_{EXP}(E)$, where α_{EXP} is given by eq. /1/ and α_o and E_o are computed to fit the exponential part of experimental spectra. On the basis of these replotted data we suppose excitation from a broad Gaussian shaped level, located E_I below the c.b., to the square root shaped density of states above the optical gap E_g^o. For simplicity we take E_g^o and the mobility gap value approximately equal and suppose excitation above mobility gap. A similar model was also suggested by the Marburg group for the interpretation of conductivity, thermopower and spin resonance data (13).

We fit the experimental data (Fig. 3) by means of our model and eq. /2/ by the expression

$$\alpha(E) = N(E_I) \frac{const.}{E} 6.7 \times 10^{21} \int_a^b \exp\left[-\frac{(\varepsilon + E_I - E)^2}{2W^2}\right] \varepsilon^{1/2} d\varepsilon \quad /3/$$

where the interval (a,b) is from 0 to some reasonable limit, e.g. 2 eV and $N(E_I)$ is the height of the Gaussian level with halfwidth 2W.

We calculated the values of the parameters E_I, W, $N(E_I)$ on an IBM 370 computer. Results are given in Table I.

Sample Characteristic		$N(E_I)/cm^{-3}eV^{-1}/$	$E_I/eV/$	$2W/eV/$
C 98	undoped	2.3×10^{16}	1.00	0.26
A 102	light SbF_5 doping	7.3×10^{15}	1.00	0.20
A 103	moderate SbF_5 doping	5.9×10^{16}	1.00	0.22
F 110	1% PH_3/SiH_4 doping	6.9×10^{18}	1.02	0.32
C 143	"optimized" undoped	3×10^{15}		

Table I: Optimized parameters $N(E_I)$, E_I and 2 W of our model for samples from Figs. 1,2. $N(E_I)$ is the peak value of the Gaussian shaped DOS with the halfwidth 2 W, E_I below the conduction band.

The model (Fig. 4) fits the experimental data well, as can be seen from Fig. 3. The main result is directly seen from Table I: Position of a Gaussian shaped maxima in the DOS is the same within the experimental error for all the doped and undoped samples prepared under the same experimental conditions. It points to correlation of these maxima in DOS with some structural defect (e.g. dangling bond) or with some persistent impurity (oxygen, nitrogen, carbon). But the latter is not consistent with our quantitative data and it was also ruled out by SIMS analysis (14) in favor of the first explanation.

In this way our optical absorption measurement can be interpreted as a determination of a position of the dangling bond level (peak in DOS) within the pseudogap in amorphous hydrogenated silicon. This peak in DOS rises with rising n-type doping. The density of midgap states for "optimized" undoped samples is in the order $10^{15} cm^{-3} eV^{-1}$.

REFERENCES

(1) B. Abeles, G.D. Cody, Y. Goldstein, T. Tiedje, C.R. Wronski, to be published in Thin Solid Films
(2) G.A. Swartz, J. Appl. Phys. 53, 712 (1982)
(3) Y. Tawada, H. Okamoto, Y. Hamakawa, Appl. Phys. Lett. 39, 237 (1981)
(4) M. Vaněček, J. Kočka, J. Stuchlík, A. Tříska, Sol.St. Comm. 39, 1199 (81)
(5) H.G. Grimmeiss, L.A. Ledebo, J. Appl. Phys. 46, 2155 (1975)
(6) M. Vaněček et al., to be published in Solar Energy Mat.
(7) G. Moddel, D.A. Anderson, W. Paul, Phys. Rev. B22, 1918 (1980)
(8) D.A. Anderson, W.E. Spear, Phil. Mag. 36, 695 (1977)
(9) S. Yamasaki et al., Japan. J. Appl. Phys. 20, L655 (1981)
(10) B. Abeles, C.R. Wronski, T. Tiedje, G.D. Cody, Solid St. Comm. 36, 537 (80)
(11) D.L. Wood, J. Tauc, Phys. Rev. B 5, 3144 (1972)
(12) G.A.N. Connell, in: M.H. Brodsky, ed., "Amorphous Semicond.", (Berlin 79)
(13) H. Overhof, W. Beyer, phys. stat. sol. (b) 107, 207 (1981)
(14) B. Abeles, C.R. Wronski, Y. Goldstein, G.D. Cody, to be published in Solid State Commun.

INFLUENCE OF LIGHT EXPOSURE ON THE TRANSPORT PROPERTIES OF a-Si:H FILMS

D. HAUSCHILDT, W. FUHS, H. MELL and K. WEBER
Fachbereich Physik, Universität Marburg,
F.R. Germany

Summary

The influence of strong illumination on the conductivity σ, thermopower S and field effect in glow discharge deposited a-Si:H-films is studied. The activation energies E_σ and E_S increase appreciably by exposure to light. Simultaneously the difference, $E_\sigma - E_S$, rises from $\sim$ 0.1 to 0.21 eV. These changes anneal completely at temperatures above 170°C. The field effect data suggest that mainly defect levels near midgap are created, whereas no marked changes occur in the band tail region. Our results cannot be explained by models which assume that the samples are homogeneous. It is argued that transport takes place in extended states and that light exposure leads to an enhancement of spatial fluctuations of the mobility edge. These may arise from a strongly inhomogeneous distribution of charged defects.

1. INTRODUCTION

The reversible conductivity changes in a-Si:H induced by light exposure are of general interest both for fundamental study and for device applications. Staebler and Wronski [1] first discovered that strong illumination decreases both the photo- and the dark conductivity. Annealing above 150°C completely restored the initial state. This reversible effect has been attributed to changes in the density or occupation of deep gap states. Dersch et al. [2] observed an increase of the dark ESR-signal after light exposure and Pankove et al. [3] reported an enhancement of the 0.8 eV luminescence band. Both results suggest that dangling bond type defects are generated. Recently an increase of the density of states near midgap has been observed in field effect [4,5] and current transient studies [6].
   In this paper we report on a study of the influence of light exposure on field effect, conductivity σ and thermoelectric power S. Whereas from field effect data changes in the density of gap states are deduced, the concomitant investigation of σ and S as a function of temperature provides information on the properties of the current path.

2. EXPERIMENTAL DETAILS

The a-Si:H-films were prepared by decomposition of silane in a capacitively coupled glow discharge system using the following parameters: deposition rate 1.5 to 2 Å/s, pressure 0.4 mbar, flow rate 6 sccm, rf-power < 5 W, substrate temperature 280°C, film thickness 0.5 to 1 µm. For the field effect measurements the films were deposited onto SiO_2 grown thermally on n-type Si-wafers. Details of the experimental technique and the method of evaluation are described elsewhere [5]. For measurements of σ and S sapphire or fused quartz substrates were used. Four Cr-stripes were evaporated on top of the film as electrodes. A temperature gradient $\Delta T < 5°C$ was established across the inner two stripes which were 7 mm apart. Both ΔT and the thermoelectric voltage ΔV_T were measured by means of two NiCr-CuNi thermocouples, 50 μm thick, which were attached to the inner Cr stripes. Each value of S was determined by a least square fit using about 10 pairs of ΔV_T and ΔT values, all measured at a given average temperature.

Light exposure was provided by means of a tungsten-iodine lamp. The light was filtered to pass only uniformly absorbed radiation with photon energies ranging from 1.2 to 2 eV. The incident power was ∿ 100 mW/cm^2.

3. RESULTS AND DISCUSSION

The field effect data clearly reveal an increase in the density of gap states upon light exposure (Fig. 1). Curve 1 was obtained after the film had been annealed for 2 hours at 170°C. After strong illumination the conductivity at $V_F = 0$ has decreased by an order of magnitude. By annealing at 170°C the original curve is quite well reproduced. Similar characteristics have been reported by Powell [7] and Goodman [4].

The form of these I-V_F-curves clearly indicates that the Staebler-Wronski-effect is a bulk effect and does not arise from a decrease of an accumulation layer at the surface or film-substrate interface [8]. In the latter case, namely, one would expect to find the same values for the current at negative gate voltages when the accumulation layer is removed by either external voltage (curve 1) or strong illumination (curve 2). On the other hand, if there were an accumulation layer at the free surface of the film before illumination, the onset of the steep current increase would be at positive values of V_F. Then a reduction of the accumulation layer would shift this point to the left and lead to identical values of the current at high positive gate voltages. However, just the opposite behaviour is observed.

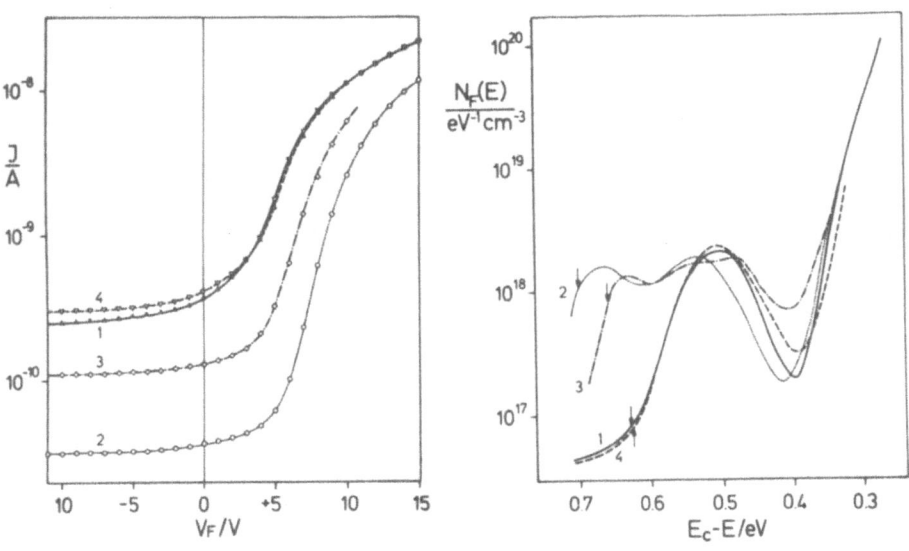

Fig. 1: Field effect in an undoped a-Si:H-film: a) current I vs. gate voltage V_F and b) calculated density of states distribution $N_F(E)$. (1) annealed at 170°C, (2) after strong illumination, (3) after 0.5 hour anneal at 170°C, (4) after a 4 hours anneal at 170°C. The arrows indicate the position of the Fermi level.

The general features of the density of states distribution (Fig. 1) before illumination are consistent with other reports in the literature. After illumination we find that the density of states is raised near midgap to as much as $10^{18} cm^{-3} eV^{-1}$ (curve 2). No significant changes are observed in the region of the tail states in accordance with measurements of the drift mobility on annealed and light exposed samples [9]. For the calculation of $N_F(E)$ it was assumed that there is no initial band bending at the film-substrate interface i.e. the flat-band-voltage V_{FB} is zero, independent of light exposure. In principle, the decrease of the slope of the I-V_F-curves near $V_F = 0$ after illumination can be caused by an increased density of gap states or alternatively by a shift of V_{FB} to positive values. Such a shift could arise from the creation of a depletion layer at the film-substrate interface by trapping of negative charge. For example a flat band voltage of 3 V would correspond to a depletion layer type band bending of about 0.35 eV. Such a large value seems to be highly unreasonable. Even in this case, however, we find an increase in the density of gap states to $4 \cdot 10^{17} cm^{-3} eV^{-1}$. We believe, therefore, that the enhancement of $N_F(E)$ upon light exposure is not an artefact caused by an unknown shift of V_{FB}, but a real change in the bulk density of gap states.

Light exposure also has a strong influence on the properties of the current path in a-Si:H films. This can be concluded from the changes of the conductivity σ and thermoelectric power S we have found on n-type films deposited from pure or PH_3-doped SiH_4. Figs. 2 and 3 show typical results obtained for a film which was doped with 100 ppm PH_3. A 60-hour optical exposure results in a pronounced increase of the activation energies of σ and S. In addition the difference of both quantities $E_\sigma - E_S$ has increased from 0.09 to 0.21 eV. This result strongly suggests that optical exposure does not merely shift the Fermi level towards midgap [1], but also affects the current path in the a-Si:H-film. By annealing at temperatures $T_A > 80°C$ the conductivity and thermopower gradually approach their original values.

Beyer and Overhof [10] have shown that it is advantageous to analyse such data in terms of the quantity

$$Q = \ln \sigma + \left| \frac{e}{k} S \right| \quad (1)$$

For unipolar conduction this quantity is independent of the (temperature dependent) position of the Fermi level E_F and, consequently, is only determined by the properties of the current path. The "activation energy of Q", i.e. its slope in a 1/T plot is

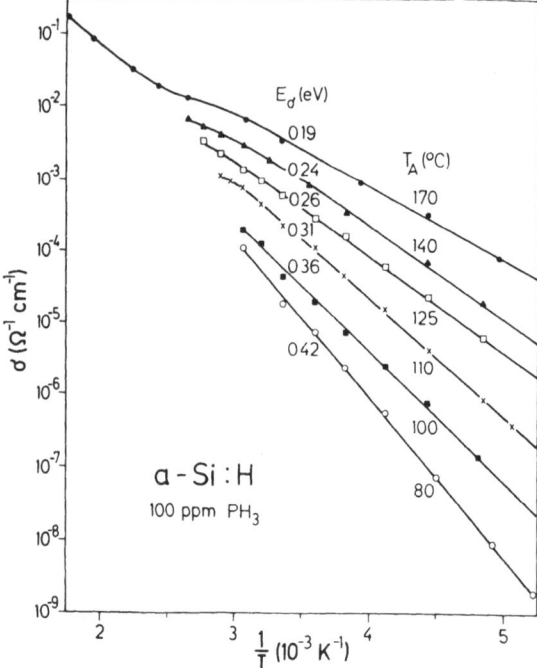

Fig. 2:
Temperature dependence of the conductivity σ. The curves were measured after a 60-hour optical exposure and subsequent annealing at various temperatures T_A given in the figure.

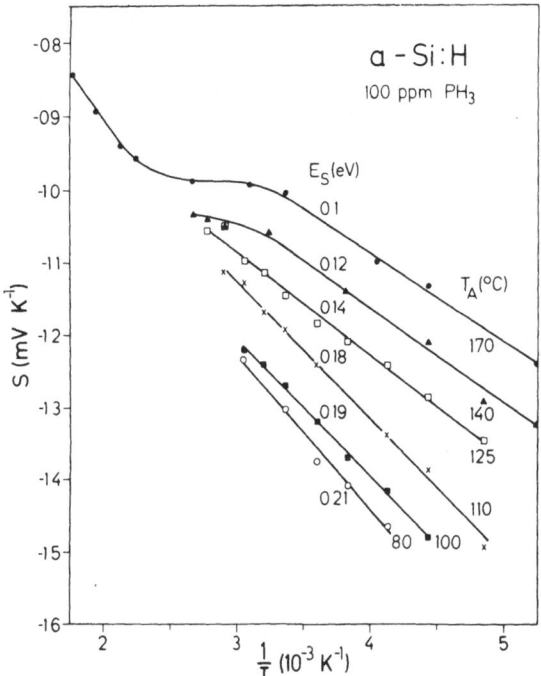

Fig. 3: Temperature dependence of the thermoelectric power S for the sample in Fig. 2.

equal to $E_\sigma - E_S$. The magnitude of E_Q depends sensitively on the transport mechanism. As an example, for conduction in extended states above a spatially constant mobility edge one expects $E_Q \lesssim kT$ [10]. Experimental investigations have revealed E_Q values ranging from 50 to $\sim$ 200 meV [10, 11]. Such values can formally be described as a mobility activation energy which in principle could be associated with hopping transport in localized states or with conduction by small polarons. The dependence of E_Q on the preparation conditions and on the dopant content [10,11], on the other hand, speaks in favor of alternative interpretations. The important result of the present study is that E_Q also depends on light exposure and on annealing temperature (Fig. 4). In the initial state ($T_A = 280°C$) $E_Q = 0.09$ eV, whereas after light exposure E_Q increases to 0.21 eV. By annealing the value of E_Q decreases continuously with increasing T_A similar to the activation energy E_σ of the conductivity (Fig. 2).

Various models have been put forward to explain large E_Q-values. It was proposed that the states responsible for charge transport are widely distributed in energy e.g. in a band tail [12,13]. $E_Q > kT$ then can arise from a strong temperature shift of the energy of maximum conduction [13]. This model does not explain the present results satisfactorily because it suggests a pronounced flattening of the bandtail by light exposure. The field effect and drift mobility data [9], however, indicate that illumination does not significantly affect the tail state distribution. Overhof and Beyer [14] have proposed that charge transport takes place in the extended states above the mobility edge E_C and attribute the result $E_Q > kT$ to spatial fluctuations of E_C. The origin of these fluctuations might be density fluctuations, strain fields and/or potential fluctuations caused by charged centers. Assuming a random distribution of charged centers it is found that their concentration must be appreciably larger than the concentration of deep gap states. This model, therefore, implicates that a-Si:H-films contain relatively large and almost equal concentrations of ionized donors and acceptors. For instance a value of $E_Q = 0.2$ eV can be explained by a concentration of charged centers of $\sim 10^{19} cm^{-3}$ and a density of states at the Fermi level near $10^{17} cm^{-3} eV^{-1}$. It is hard to believe that light exposure creates, in addition to some 10^{17} dangling bond type defects, about $10^{19} cm^{-3}$ self-compensation donors and acceptors without stronger changes in the tail state distribution.

Although the model of Overhof and Beyer cannot quantitatively explain the present results we propose that the increase of E_Q is caused by potential fluctuations. The quantitative difficulties obviously result from the

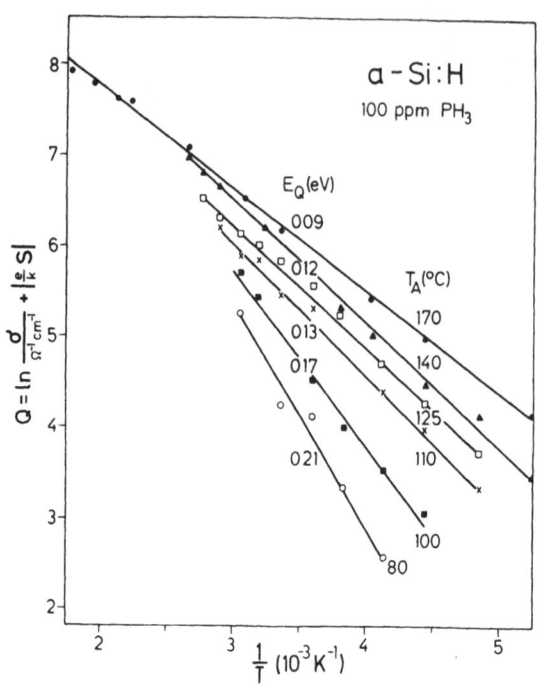

Fig. 4: Plot of the data in Figs. 2 and 3 in terms of the quantity $Q = \ln \sigma + |\frac{e}{k} S|$

assumption, that the charged centers are randomly distributed. However, recent NMR data [15] have suggested that the distribution of hydrogen in a-Si:H-films is strongly inhomogeneous. It is, therefore, very likely that also the defect centers are inhomogeneously distributed. Possibly light exposure causes the breaking of weak Si-Si bonds [2,3]. These can be saturated by hydrogen from nearby Si-H sites and the resulting dangling bonds can move further apart via hydrogen exchange. The light-induced defects then should be located predominantly in or close to the hydrogen rich regions. If part of them are charged they will give rise to much stronger potential fluctuations than those caused by randomly distributed centers.

ACKNOWLEDGEMENT

The authors gratefully acknowledge financial support by the Bundesminister für Forschung und Technologie (BMFT).

REFERENCES

[1] D.L. Staebler and C.R. Wronski, J.Appl.Phys. 51, 3262 (1980)
[2] H. Dersch, J. Stuke and J. Beichler, Appl.Phys.Lett. 38, 455 (1981)
[3] J.I. Pankove and J.E. Berkeyheiser, Appl.Phys.Lett. 37, 705 (1980)
[4] N. Goodman, J. de Physique 42, C4-375 (1981)
[5] K. Weber, M. Grünewald, W. Fuhs and P. Thomas, J. de Physique 42, C4-523 (1981); phys.stat.sol. (b) 110 (1982)
[6] J. Beichler and H. Mell, this conference
[7] M.J. Powell, B.C. Easton and D.H. Nicholls, J. de Physique 42, C4-379 (1981)
[8] I. Solomon, T. Dietl and D. Kaplan, J. de Physique 39, 1241 (1978)
[9] W. Fuhs, M. Milleville and J. Stuke, phys.stat.sol. (b) 89, 495 (1978)
[10] W. Beyer and H. Overhof, Solid State Comm. 31, 1 (1979)
[11] W. Beyer and H. Overhof, Solid State Comm. 38, 891 (1981)
[12] G.H. Döhler, Phys.Rev. B 19, 2083 (1979)
[13] M. Grünewald and P. Thomas, phys.stat.sol. (b) 94, 125 (1979)
[14] H. Overhof and W. Beyer, Phil.Mag. B 43, 433 (1981)
[15] J.A. Reimer, R.W. Vaughan and J.C. Knights, Phys.Rev.Lett. 44, 193 (1980)

PULSED ELECTRON BEAM ANNEALING OF ION IMPLANTED GERMANIUM
FOR PHOTOVOLTAIC DEVICES

B. SAUTREUIL, A. LAUGIER, D. BARBIER
Laboratoire de Physique de la Matière
Institut National des Sciences Appliquées de Lyon
20, Avenue Albert Einstein - 69621 VILLEURBANNE CEDEX (France)

A. CACHARD
Département de Physique des Matériaux
Université Claude Bernard - Lyon I
43, Boulevard du 11 Novembre 1918 - 69621 VILLEURBANNE CEDEX (France)

SUMMARY

This paper deals with the application of short (50 ns) high energy, pulsed electron beam for shallow p-n junction formation in implanted Germanium. From time-resolved spectroscopy achieved by means of analysis of diode current and voltage, energy deposition profiles are obtained on the basis of a Monte-Carlo simulation. Temperature-depth profiles are computed. Structure of the annealed samples are studied by R.B.S. plus channeling and S.E.M. A good annealing is obtained for energy densities above 0.45 J/cm^2 for 12-15 keV mean electron energy. The use of the Ge photovoltaic devices achieved by pulsed electron beam annealing is discussed in the case of Ge-GaAs bicolor systems.

1. INTRODUCTION

Much recent work have been devoted to the use of laser and electron beam to add energy to semiconductor materials in order to anneal ion implantation damage (1,2). Pulsed electron beam annealing (PEBA) was previously reported for high throughput production of Silicon solar cells (2,3). Previous work (4) reports regrowth of implanted Germanium using a Q-switched laser. This work is a study of the application of PEBA to Ge.

2. PULSED ELECTRON BEAM ANNEALING OF GERMANIUM

2.1. Electron beam-Germanium interaction

A Monte-Carlo calculation has been used to determine the energy deposition profile in Ge (5). Values of the Bethe's range R_B are obtained versus the incident electronic energy E_o by integration of the Beth's Law. The projected path (Gruen's range) is given by :

$$R_G = 0.5 \ R_B = 7.6 \ 10^{-3} \ E_o^{1.76}$$

with E_o in keV and R_G in microns. Typical values are $R_G = 0.45$ microns for $E_o = 10$ keV and 1.5 for 15 keV respectively. These values correspond to the mean energy range of electrons used in this work. The normalized electronic energy losses are independant of E_o in the 5-50 keV range but depend of the incidence angle. A detailed analysis can be found in ref. 5. With the choosen gun parameters the average angle is 45 degrees. So the reflexion coefficient is 0.27.

2.2. Pulsed electron beam

The pulsed electron beam is obtained by discharge of a low inductance capacitance C in a field emission diode with planar geometry (2,7). The cathode material is graphite. The storage voltage V_o is in the 30-100 keV range. A Tungsten mesh is used as anode above the Germanium sample.

A polykinetic beam is obtained and the spectral distribution for the electrons varies during the pulse. The parameters that control the beam characteristics are, for a given mmachine, the storage voltage, the cathode-diameter and the anode-cathode gap. Time-resolved spectroscopy of the electron beam pulse has been achieved by means of analysis of diode current and voltage waveforms.

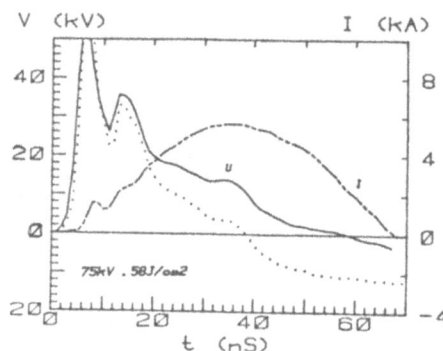

Figure 1 - Diode current and voltage waveforms. The dotted line is the as-measured voltage. The solid line U is the corrected voltage taking into account the characteristic of the voltage monitor.

Fig. 1 gives a typical recording for a 0.58 J/cm^2 shot. In this work, a 19 mm diameter cathode is used with a 1.3 mm gap. The anode-sample gap is 7 mm. From the current I(T) and the corrected voltage U(T) curves we get the spectral distribution of the electrons vs time.

Fig. 2 shows this distribution at the end of the shot. Then, using the results of ref. 5, the normalized dose-depth profile and its evolution during the pulse are computed.

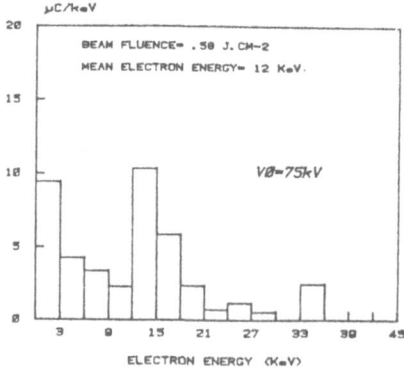

Figure 2 - Spectral distribution.

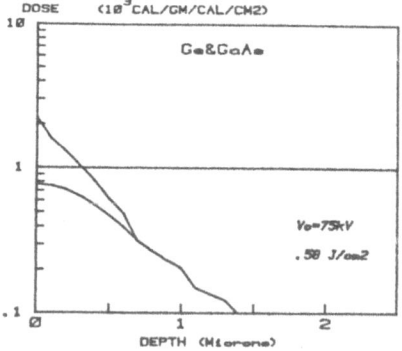

Figure 3 - Normalized dose profiles at 30 and 60 ns.

2.3. Analysis of the energy deposition-profile curves in Germanium

A set of shot has been performed with a storage voltage Vo from 50 keV to 100 keV. The corresponding fluences (measured with a build in calorimeter) are in the range F = 0.18 - 1.3 J/cm^2 (table 1).

Fluence (J.cm^{-2})	Vo (kV)	Effect	min χ
0.18	50		
0.36	64	not annealed	
0.40	68	annealed	9.1 %
0.80	85	annealed	5.1 %
1.00	92	annealed	6.1 %

Table 1 - Fluence and mean electronic energy vs storage voltage and effect of PEBA on In implanted Ge.

Fluence measurements reproducibility is $\pm$ 10 %. A typical dose profile is given in fig. 3 (shot of fig. 1). With the selected diode parameters the profiles are very similar and do not very significantly with Vo, below 80 kV.

So, Vo controls directly F and the thermal treatment. In order to illustrate this effect, the temperature-profile is deduced from the dose-profile. This deduction is more direct using the adiabatic hypothesis : during the first 40 ns the deposited energy is essentially used to heat the solid, to melt a part of the solid and, eventually, to heat the liquid. Conduction heat flow is assumed negligible. The validity of this hypothesis has been verified by comparison with a computer simulation solving the one-dimensional time dependent heat flow equation. The used thermal date are listed table 2.

T_M : 937 C   L = 508 J/g (crystal)   L = 300 J/g (amorphous, 0.04 micron)

Solid heat capacity C_p : linear variation from 0.32 (300 K) to 0.38 (T_M)
(J/g.degree)

Liquid heat capacity C_p : 0.38 J/g.degree

Thermal conductivity : data of Maycok (ref. 7)

Table 2 - Ge thermal data used for the T(X) profiles.

Then, it is easy to determine two minimum fluence thresholds. Using the value for the surface dose given fig. 3, 0.17 J/cm^2 are necessary to reach the melting point of Ge and 0.42 J/cm^2 for complete melting. Fig. 4,5 show the T(X) profiles for F = 0.18, 0.45 respectively.

The time step is 5 nanoseconds. For F = 0.18, only 0.1 micron are partly melted during only 5 ns. Below 0.4 J/cm^2 one can expect that the dose is not sufficient for good annealing. For F = 0.45, a first 0.1 micron thick layer is melted and the liquid is heated up to 1600°C. The computer simulation taking into account the conduction heat flow gives for this pulse the following results : liquid temperature maximum : 1400°C, duration of the liquid phase 150 ns, thickness of molten zone, 0.35 micron, recrystallization growth : 3 m/s.

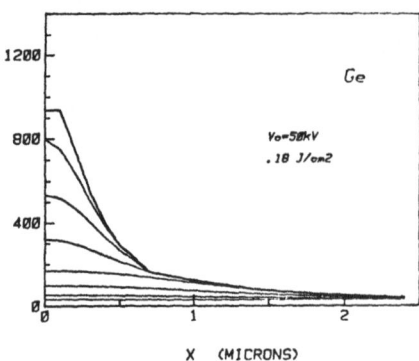
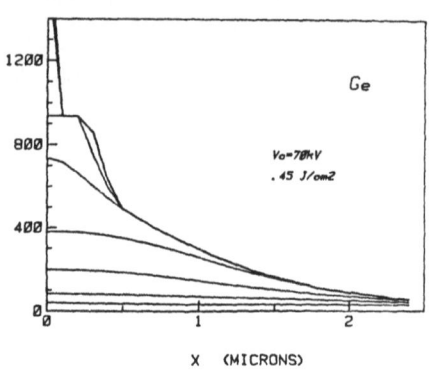

Figure 4 - Temperature profile during the pulse. Fluence : 0.18 J/cm^2. Time step 5 nanoseconds. (adiabatic hypothesis).

Figure 5 - Temperature profile during the pulse with a 0.45 J/cm^2 fluence. Time step 5 nanoseconds (adiabatic hypothesis).

3. EXPERIMENTAL RESULTS

Germanium samples were 150 keV In implanted at room temperature with two doses $1.25 \; 10^{13}$ at.cm^{-2} and $1.25 \; 10^{14}$ at.cm^{-2} (tilt angle : about 6° with respect to the 111 axe). With such implantation conditions, the Germanium surface appears slighty bluish.

3.1. Pulsed electron beam annealing

We use Ge samples implanted with $1.2 \; 10^{14}$ In cm^{-2}. We empirically obtain a fluence threshold necessary to anneal such samples by determining the smallest fluence for which the sample coulour, above mentioned, disappear.

Table 1 shows that a fluence threshold is observed for these samples. This fluence threshold has been determined on a sample previously submitted to a 0.36 J/cm^{-2} fluence with no visible change in aspect, as mentioned in table 1.

3.2. Rutherford backscattering study

The effect of PEBA has been studied by the Rutherford backscattering technique. A 1 MeV He$^+$ beam strikes the sample under normal incidence. We measure the energy distribution of the backscattered particles at 160°. The detector resolution of 300 Å for a 1 MeV incident beam in Ge. Fig.6 gives spectra obtained with samples of table 1. The width at half maximum of surface peaks of spectra a) and b) gives an "amorphous" layer of 900 Å and 630 Å for doses of $1.2 \; 10^{14}$ at.cm^{-2} and $1.2 \; 10^{13}$ at.cm^{-2} respectively. Spectra c), d) and e) correspond to annealed samples. Higher fluences than the threshold fluence give lower backscattering yield (see table 1), but further experiments are needed to study the effect of the annealing fluence on the backscattering yield of Ge samples, for various implantation conditions. The indium spectrum do not appear on spectra of fig.6. Taking into account the implantation conditions and assuming that the indium implantation profile is predicted by the theoretical profile for implantation in amorphous substrate, we obtain a maximum concentration of implanted of 2.10^{19} at.cm^{-3} or an atomic fraction of $4.5 \; 10^{-4}$ of In in Ge. This is below the detection limit of RBS for In in Ge (around 6.10^{-3}).

3.3. Pseudo Kikuchi lines

Pseudo Kikuchi lines are observed with a scanning electronic microscope using 30 keV electrons. Two In implanted samples with a dose of $1.2\,10^{13}\,at.cm^{-2}$ have been studied. No pseudo Kikuchi lines are observed for as implanted sample. Photograph of fig.7 shows an annealed sample with $F = 1.2\,J.cm^{-2}$. The average direction of the beam is nearly parallel to the 100 axe. So, different strips are visible and their presence indicate that monocrystallinity is restored in the implanted layer.

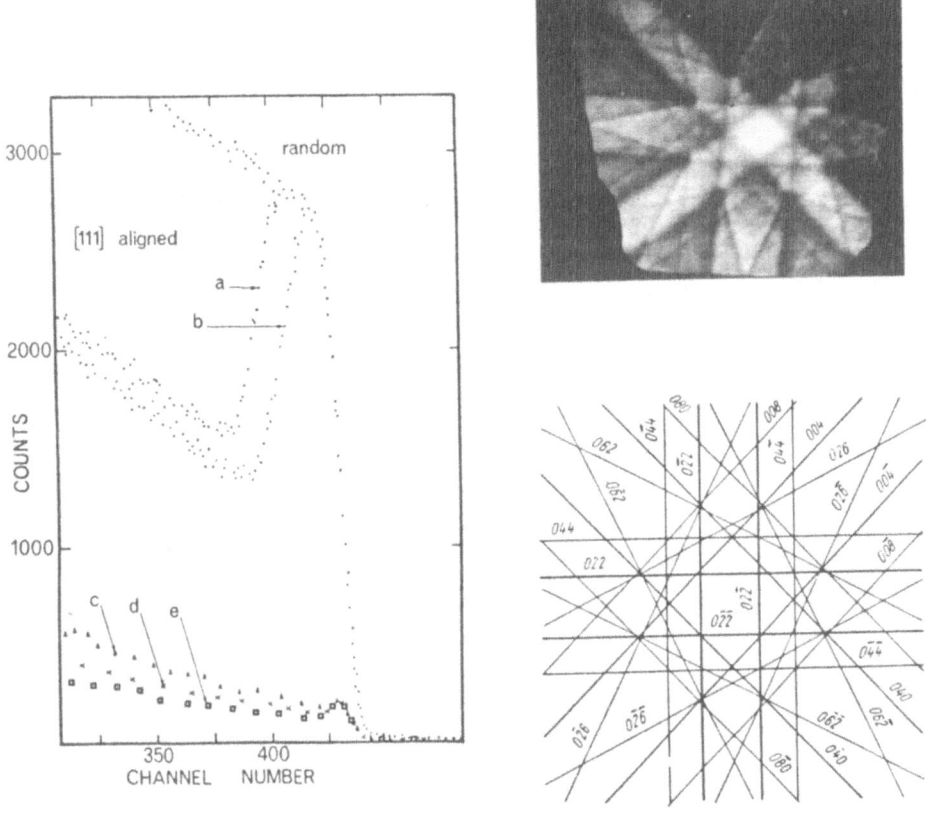

Figure 6 - R.B.S. spectra

Figure 7 - Pseudo Kikuchi lines on In-implanted Ge annealed at 1.2 J/cm^2 and indexation

4. CONCLUSION

We have determined the threshold of annealing for In implanted Ge : 0.40 J/cm^2. This value is smaller than the previously reported one for laser annealing (4) : 1 J/cm^2 for a 15 ns pulse Q-switched ruby laser. This difference can be due to a more superficial dose-profile and to strong energy losses by reflection. So we precise the conditions to realize Ge solar cells with abrupt p-n junctions by PEBA. Their development is justified by their potential use in a GaAs-Ge two cell system designed for solar concentration.

In a previous work (8), using the abrupt junction model, we derived the optimum doping level for the front (P type) and the base (N type) and showed that, provided that the junction depth was lower than 1 µm, the junction depth and the front surface recombination velocity had nearly no effect on the Ge cell performances, used in a GaAs-Ge system. A limit efficiency of 5.8 % (300 AM1.5, 300 K) for the Ge cell is computed. This, with the theoretical efficiency of 25 % for the GaAs cell (300 AM1.5) fixes the upper limit of the GaAs-Ge efficiency around 30 %.

BIBLIOGRAPHY

(1) A. GAT, J.F. GIBBONS, T.J. MAGEE, J. PENG, V.R. DELINE, P. WILLIAMS, C.A. EVANS : Appl.Phys.Lett. 32, 276, 1978
(2) A.C. GREENWALD, A.R. KIRKPATRICK, R.G. LITTLE, J.A. MINUCCI : J. Appl.Phys. 50, 783, 1979
(3) J. MICHEL, D. BARBIER, A. LAUGIER : Proceed. of 5thIEEE Phot.Volt. Spec.Conf.Orlando, 1007, 1981.
(4) G. DELLA MEA, G. FOTI, G. MAJNI : in Laser-Solid Interactions and Laser processing, MRS 1978, AIP 317, 1979
(5) D. BARBIER, M. BAGHDADI, A. LAUGIER, E. VICARIO : J.Microsc.Spectros. Electr. 6, 513, 1981
(6) SPI-300 T.M. from SPIRE Corp.
(7) P.D. MAYCOK : Sol.St.Electr. 10, 161, 1967
(8) B. SAUTREUIL, A. LAUGIER : Solar Energy Mat. 5, 21. 1981

RECENT ADVANCES IN ITO/InP AND CdS/InP SOLAR CELLS

T.J. COUTTS and N.M. PEARSALL
School of Physics,
Newcastle upon Tyne Polytechnic,
England.

Summary

Solar cells based on single crystal p-type indium phosphide (InP) have been fabricated by R.F. sputter deposition of either indium tin oxide (ITO) or cadmium sulphide (CdS). In an effort to determine the nature of these devices, the J-V characteristics have been measured and analysed under various conditions of illumination and temperature. Values of the reverse saturation current, Jo, and its activation energy have been obtained. These suggest that there may be a fundamental mechanism which limits the efficiency of ITO/InP heterojunctions to less than 10%. However, there is great potential for the CdS/InP cells. The variation of the quantum efficiency of the cells with wavelength has also been measured. The analyses clearly show that the substrate preparation procedure dominates cell behaviour.

1. INTRODUCTION

If solar cells based on InP have any real contribution to make in electrical power generation then they would almost certainly be as concentrator or thin film devices. The cost of the material ensures that this will remain the case. If devices based on single crystal InP can be manufactured using a low cost technique such as R.F. sputtering for the production of the n-type film, rather than fundamentally more expensive techniques such as MBE or LPE which are used with GaAs concentrator cells, then they could still have a cost advantage. Thin film InP is presently very little studied due to fundamental grain boundary problems and the, now, severe competition from other thin film systems. Thus, our studies of ITO/InP and CdS/InP cells are mainly related to understanding the physics of what should be, at least in the latter case, a nearly ideal heterojunction and which also has potential in concentrator systems. Much of our recent progress has been due to improved technology and analytical facilities coupled with an understanding of the detail of the fabrication procedure.

Fundamental interest in InP as a solar cell absorber stems from its advantageous physical properties, particularly when used in conjunction with n-type CdS window layers [1]. Several groups [2] have reported efficiencies of around 14% using p-type single crystal InP substrates and window layers of both CdS and ITO. It has been pointed out that there are no special reasons for believing that the ITO/InP combination would lead to such an efficient device and this has led to an interesting discussion in the literature in recent years [3,4,5,6] regarding the nature of this cell. The concensus of opinion is that the device operates as a buried InP homojunction although there is still an unresolved debate about whether this is due to plasma induced surface damage of the InP (and associated type conversion) or in-diffusion of tin. As we shall show in this paper, the former mechanism is not a serious possibility for our cells.

ITO is generally deposited by R.F. sputtering or spraying. The former technique can cause elevated substrate surface temperatures due to secondary electron bombardment, whilst the latter actually requires a heated substrate to effect the necessary chemical reaction for ITO formation. This heating could provide the driving force for tin diffusion. In addition, as we have previously shown [7], heating InP at temperatures as low as 200°C can cause the formation of a thin polycrystalline surface layer and it is quite feasible that this would further aid the in-diffusion of tin, which acts as a donor, and could thus cause the formation of a deeper buried homojunction. However, it would seem less likely that this would take place on InP substrates which are free of the polycrystalline layer prior to ITO deposition and, in this case, one would expect better behaved devices, perhaps even exhibiting heterojunction-like characteristics. Thus, to determine the nature of the junction in advance, one must start with a well defined surface and to this extent we believe that the pre-deposition treatment of the InP substrate is at least as important as the method of deposition itself. We shall show that it is this pre-deposition treatment which determines whether or not the ITO/InP cells are homo- or heterojunctions. It is significant that the sputtered CdS on InP devices have never behaved like buried homojunctions and this leads us to reject the sputter damage model. These findings may also be significant to devices other than solar cells based on InP.

2. DEVICE FABRICATION

The fabrication procedure has been discussed previously [6] and we shall only mention those aspects which are important to our current understanding of the cell behaviour. Single crystal p-type InP substrates, doped with either Cd or Zn to a level of 7 or 4×10^{17} cm^{-3} respectively and (111) oriented, were used. These were of 1 cm^2 in area. Having found that heating of the chips during contact sintering caused the formation of a polycrystalline surface layer, the front surface of the substrate was re-etched before deposition of the n-type film. The chips were supported against gravity by a mask around a 0.5 mm strip at their edge and only the central 9.0 x 9.0 mm was coated with the n-type film. It has been found that the method of chip support can also have a considerable effect on cell behaviour. Due to limitations of space, this aspect will be discussed in a subsequent publication.

Recently fabricated ITO/InP cells utilise a film of ITO which is only 7-800 Å thick. At this thickness the ITO forms its own anti-reflection coating but its sheet resistance is of the order of 200 Ω/□. Its transmittance is greater than 85% for 0.3 - 1.5 μm. Thus very little of the incident light fails to reach the InP and so one would expect high short circuit currents to be produced, although paying the price of a low fill factor. The indium doped, sputter deposited CdS has a much higher sheet resistance but, nevertheless, reasonable values of Jsc can be produced. We can now obtain chromium on glass grids, produced to our design using electron beam lithography. These can be used as masters to define the cell grid pattern photolithographically by positive contact printing. The grids are formed in 1 μm thick copper, which is R.F.sputter deposited immediately after the n-type layer. After grid fabrication, the cells are mounted on the dural support plate using a silver loaded epoxy resin which can withstand the extremes of temperature used in the characterisation procedure.

3. DEVICE MEASUREMENT

For the purposes of assessing the reproducibility of our device fabrication procedure, and its sensitivity to specific details, we have chosen to concentrate our recent efforts on the variation of the reverse saturation current with temperature and on the wavelength dependence of the quantum efficiency. The reverse saturation current, Jo, is estimated from the y intercept on a graph of $\ln Jsc \sim Voc$ (these quantities being varied by changing the level of illumination using neutral density filters). The devices are measured under vacuum and their temperature can be controlled to ± 0.1°C. An automatic system is used to acquire the very large volume of data for each cell (eight illumination

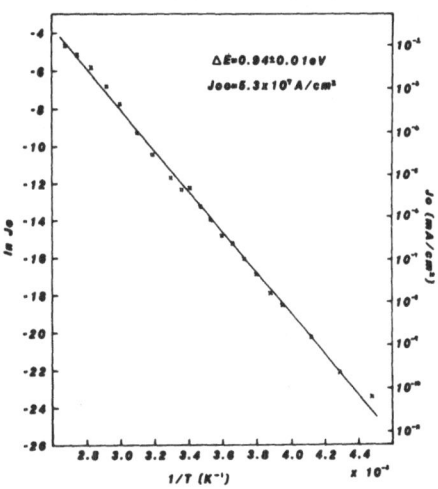

Figure 1

levels and approximately twenty temperatures over the range -50°C to + 100°C are used). The absolute spectral response of the short circuit current is obtained by comparison with a calibrated silicon cell. No correction has been made for reflectance losses since preliminary measurements have indicated that these are small. Measurements of the variation of depletion capacitance with voltage will be reported at a later date.

4. RESULTS

Figure 1 shows an Arrhenius plot for one of the second generation ITO/InP cells (ISC5), with the relevant, statistically calculated, parameters indicated and Figure 2 shows the quantum efficiency variation with wavelength for the same cell. This can be compared with Figures 3 and 4, which show the same plot for a first generation ITO/InP and a CdS/InP cell respectively.

Figure 2

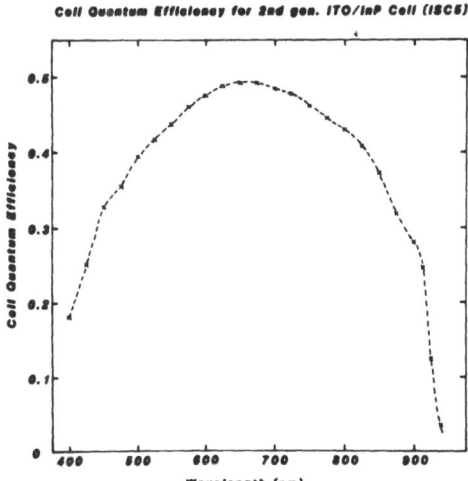

5. DISCUSSION

The $\ln Jsc \sim Voc$ plots indicate that there are two forward current mechanisms. This has been reported previously for most other classes of solar cells. Practically we are interested only in the high voltage region and so the discussion here concerns that region. A least squares fit was made to obtain the value of Jo for each temperature; these values then being used to produce an Arrhenius plot like that in Fig. 1. The most significant feature of these plots is the striking linearity of the data (over nearly <u>nine</u>

orders of current magnitude in the case of cell ISC5). Values of activation energy and Joo for typical cells are shown in Table 1. These quantities determine the magnitude and are indicative of the nature of the forward current mechanism. The values of ΔE for the two second generation ITO/InP cells are in reasonable agreement and are considerably higher than we have previously reported [6]. Their values of Joo are close enough to be encouraging considering the range of $1/T$ over which one must extrapolate to obtain Joo. We take this as a positive indication that our substrate preparation technique is reproducible. The corresponding values for the

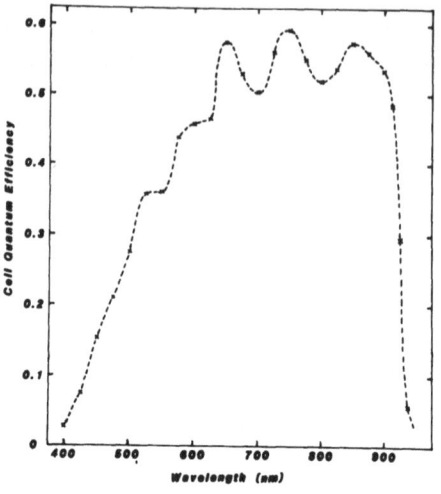

Figure 3

CdS/InP cell reported here are substantially lower. We postulate that this is due to the different properties of the interface, due to the different materials combination, since the substrate processing and deposition procedure were identical to those used for ITO/InP cells. The relatively low values of Voc for the ITO/InP cells appear to be due to high values of Joo, whereas the limiting factor in the case of the CdS/InP cell is the low activation energy. Further investigations are required to elucidate these matters.

The quantum efficiency curves for ISC5 and CSC2 indicate that these cells are behaving as a heterojunctions, rather than as homojunctions as suggested elsewhere [4]. We show, for comparison, the quantum efficiency of a device (I5) which was made without the second substrate etch and which would therefore have had the polycrystalline surface layer referred to earlier. Although this cell has a good quantum efficiency, it clearly exhibits a more triangular shape, typical of a homojunction. Therefore, we conclude that the present InP devices are behaving as

Figure 4

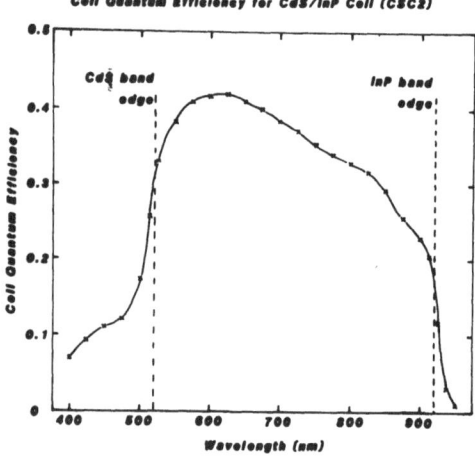

Table 1

Typical Activation Energies for Jo

Cell	ΔE (eV)	Joo (A/cm²)	Type
I5	0.465	3×10^4	1st gen. ITO/InP
ISC5	0.940	5×10^7	2nd gen. ITO/InP
ISC6	0.813	5×10^6	"
CSC2	0.650	2×10^4	CdS/InP

heterojunctions, although homojunctions can readily be produced by a different processing schedule. Further experiments using heat treatment of the heterojunctions may prove informative.

The efficiencies of the ITO/InP heterojunctions are modest despite their having high values of Jsc. This is due to the high series resistance and relatively low Voc. Whilst the former could readily be overcome by using better gridding or a thicker layer of ITO, it seems unlikely that Voc could be substantially increased due to the large values of Joo. On the other hand, the CdS/InP cells have a much poorer Jsc at present. This can only be due to the adverse properties of the CdS since it is known that the InP is capable of producing a large photocurrent. The lower Jo activation energy supports this, whilst the low value of Joo indicates that improvements should be realised if better quality CdS films can be deposited. It is this factor, rather than surface damage, which is the present limitation of R.F. sputter deposition.

Finally, preliminary measurement of the ITO/InP cells under moderate concentration (approx. 5 suns) has been undertaken. Whilst the high series resistance made it impractical to carry out extensive studies, the results were encouraging and further investigations will be undertaken in the near future.

6. CONCLUSIONS

Our main conclusions from this latest phase of our work are as follows:-

a) for our cells substrate preparation alone governs whether homo- or heterojunctions are produced;
b) in the case of the ITO/InP cells, the high Jo activation energies indicate that any effect of electron affinity mismatch must be small;
c) Voc for the ITO/InP cells is limited by a large value of Joo; this may prove a fundamental limitation on the efficiency of heterojunctions;
d) Jsc for the CdS/InP cells is presently limited by the poor quality of the sputtered CdS;
e) the relatively poor Voc of the CdS/InP cells is due to a low Jo activation energy, although the value of Joo is very low.

Having demonstrated that we can now reproducibly manufacture heterojunction cells of ITO/InP, our next goal is to produce heterojunctions using good quality CdS on InP, which theoretically should have high efficiencies as opposed to ITO/InP which theoretically should not and, in practice, do not.

ACKNOWLEDGEMENTS

The authors would like to thank all their colleagues within the Opto-Electronics Group at Newcastle Polytechnic (both permanent and visiting) who have assisted in the acquisition of data presented in this paper and the staff of the SERC facility at the Rutherford and Appleton Laboratories, who supplied the chromium-on-glass grids.

REFERENCES

1. M. Bettini, K.J. Bachmann and J.L. Shay, J. Appl. Phys., 49, (1978), pp. 865-870.

2. T.J. Coutts, Thin Solid Films, 90, (1982), to be published.

3. K.J. Bachmann, H. Schreiber, Jnr., W.R. Sinclair, P.H. Schmidt, E.G. Spencer, G. Pasteur, W.L. Feldmann and K. Sree Harsha, J. Appl. Phys., 50, (1979), pp. 4331-4346.

4. R.H. Bube, F.G. Courreges, A.L. Fahrenbruch and M.J. Tsai, Proc. of the 2nd E.C. Photovoltaic Solar Energy Conference, Berlin, 1979, Reidel, Dordrecht, Holland, pp. 432-439.

5. L. Gouskov, H. Luquet, J. Esta and C. Gril, Solar Cells, 5, (1981), pp. 51-66.

6. T.J. Coutts, N.M. Pearsall, R. Nottenburg, P.J. Ireland and L.L. Kazmerski, Proc. of the 15th IEEE Photovoltaic Specialists Conference, Orlando, Fla., 1981, [IEEE; New York, 1981], pp. 1077-1082.

7. N.M. Pearsall, T.J. Coutts, R. Hill, G.J. Russell and K.J. Lawson, Thin Solid Films, 80, (1981), pp. 177-182.

INTENSITY ENHANCEMENT IN TEXTURED OPTICAL SHEETS FOR SOLAR CELLS

E. Yablonovitch
Exxon Research Center

Summary

We adopt a statistical mechanical approach toward the optics of textured and inhomogeneous optical sheets. As a general rule, the local light intensity in such a medium will tend to be $2n^2(x)$ times greater than the externally incident light intensity, where $n(x)$ is the local index of refraction in the sheet. This enhancement can contribute toward a $4n^2(x)$ increase in the effective absorption of indirect-gap semiconductors like crystalline silicon. Also it may lead to a voltage increase equal to $KT\log 4n^2$.

1. Introduction

In the past decade, there have been a number of suggestions for the use of light trapping by total internal reflection to increase the effective absorption in the indirect-gap semiconductor, crystalline silicon. The original suggestions (1), (2) were motivated by the prospect of increasing the response speed of silicon photodiodes while maintaining high quantum efficiency in the near-infrared.

Subsequently, it was suggested (3) that light trapping would have important benefits for solar cells as well. High efficiency could be maintained while reducing the thickness of semiconductor material required. Additionally, the constraints on the quality of the silicon could be relaxed since the diffusion length of minority carriers could be reduced proportionate to the degree of intensity enhancement. With such important advantages, interest in this approach has continued, but progress in this field has been hindered because there was no method available to calculate the degree of enhancement to be expected.

For example, St. John (1) mentions that total internal reflection will result in two or more passes of the light rays with a proportionate intensity enhancement. On the other hand, Redfield (3) regards the number of light passes, or degree of enhancement, as an adjustable parameter which could vary anywhere between 1 and 100 and he plots the collection efficiency as a function of this parameter. In calculating the ideal efficiency of silicon solar cells, Loferski et al. (4) seemed to imply that perfect light trapping might be possible, which corresponds to an infinite degree of enhancement.

It is the purpose of this paper to show that the degree of intensity enhancement to be expected due to total internal reflection is $2n^2$ which for silicon is ~ 25. In other words, a light ray in silicon may be expected to make 25 passes on average before escaping. With a proper angular average of the longer path length of oblique rays, the "effective" absorption enhancement factor is $4n^2$ over the case of single pass normally incident rays. Also, we will show that the geometrical details of the silicon shape are relatively unimportant. Whether the silicon is simply roughened on one side (1), (2) or whether precisely angled grooves are etched into the surface (3), (4), the overriding tendency will be for an enhancement factor of $4n^2$ provided only that the surface is sufficiently strongly textured. Finally, we will show that in an inhomogeneous sheet, for example a composite, the enhancement will be given by the same formula employing the local index of refraction $n(x)$.

This intensity enhancement can translate into either an increase in the short circuit current or in a properly designed solar cell it can lead to a voltage improvement of $KT\log 4n^2$.

Two distinct derivations will be presented in the following two sections. In Chapter 2, we will give a derivation based on statistical mechanics. Such an approach is very powerful and it can be generalized to situations where geometrical optics is inapplicable (though we will not attempt such a generalization in this paper). In Chapter 3, a geometrical optics derivation will be presented. Its simplicity will permit us to better recognize some of the prerequisites and limitations of this type of intensity enhancement. In Chapter 4, we will show how these considerations are modified in the presence of absorption. Finally, in Chapter 5, we indicate how these effects can lead to a voltage enhancement in addition to a short circuit current improvement.

2. Statistical Mechanical Derivation

Consider an inhomogeneous optical slab with position-dependent index of refraction $n(x)$ as illustrated in Figure 1. Let the index of refraction vary sufficiently slowly in space, so that a density of electromagnetic modes may be defined, at least locally inside the sheet.

Now place the optical medium into a region of space which is filled with black-body radiation in a frequency band $d\omega$ at a temperature T. When the electromagnetic radiation inside the medium approaches equilibrium with the external black-body radiation, the electromagnetic energy denisty (5) is

$$U = \frac{\hbar\omega}{\exp\{\hbar\omega/KT\} - 1} \frac{2 d\Omega \, k^2 \, dk}{(2\pi)^3} \quad (1)$$

This is the standard Planck formula for black-body radiation in a vacuum, but as Landau and Lifshitz show (5), it can be adapted to any optical medium by making $k = n\omega/c$. In addition, the energy density may be changed to an intensity I (power per unit area), by multiplying (1) by the group velocity $v_g = d\omega/dk$. Making both changes in (1) we obtain

$$I \equiv U v_g = \frac{\hbar\omega}{\exp\{\hbar\omega/KT\} - 1} \frac{2 d\Omega n^2 \, \omega^2}{(2\pi)^3 \, c^2} d\omega \quad (2)$$

This differs from the vacuum black-body intensity simply by the factor n^2. Therefore, the intensity of light in a medium which is in equilibrium with external black-body (bb) radiation is n^2 times greater

$$I_{int}(\omega,x) = n^2(\omega,x) I^{bb}_{ext}(\omega) \quad (3)$$

This factor comes about simply due to the fact that the density of states in such a medium is proportional to n^2 and the equipartition theorem guarantees equal occupation of the states, internal as well as external.

Now let us decide whether or to what degree the situation changes when an arbitrary external radiation field replaces the black-body radiation. Since we are considering a transparent medium, inelastic events such as absorption and reemission at another frequency are not permitted. Therefore, each spectral component may be considered individually. In this circumstance, departure of the external field from an exact black-body frequency distribution will not affect Equation (3), which will remain valid separately at each frequency ω.

A much more serious question, which will be the main focus of this paper, is: What happens when the external radiation field departs from the isotropic distribution of black-body radiation? This situation is illustrated in Figure 2(a), where the external light is shown to be collimated. If the surface of the optical sheet is quite irregular in shape, then the light rays, upon entering the medium, will lose all memory of the external incident angle after the first, or at most the second, scattering from a surface. In other words, all correlation with the external angle will be lost almost immediately upon refraction or total internal reflection, especially when averaged over the illuminated surface of the sheet. If this condition is satisfied, then a collimated incident beam of intensity I_{ext} will produce, inside the optical sheet, a random angular distribution of light, no matter which direction the beam happens to be coming from. Therefore, a collimated beam, when subdivided so that it illuminates the optical medium equally from all directions produces identical internal light distributions under those respective conditions. The condition of isotropic illumination is, however, equivalent to that of black-body illumination. Therefore, Equation (3) remains valid whether the external field is isotropic as in the black-body case or whether it is collimated.

$$I_{int}(\omega,x) = n^2(\omega,x) I_{ext}(\omega) \qquad (4)$$

Equation (4) will be corrected for certain surface transmission factors in Section 3, but it is a key formula in this paper. It rests on the assumption that all correlation of the internal rays with the external angle of incidence is lost almost immediately upon entering the medium and/or upon averaging over the illuminated surface. Even optical sheets with ordered surface textures will show the type of randomization we are discussing here. The reasons are as follows:

    a) If light randomization does not occur upon the entering refraction, it can occur on the first internal reflection.
    b) If not on the first reflection, then on the second.
    c) If not then, it can still be the result of a spatial average over the illuminated surface area.
    d) If not even then, it can still result from angular averaging due to motion of the source, like the sun moving through the sky.

In other words, there is a rather overwhelming tendency toward randomization in the angular distribution of light and toward the validity of Equation (4), but it is not always satisfied.

Consider the simple plane-parallel slab shown in Figure 2(b). Clearly, there is no intensity enhancement in that case.[1] To distinguish between the class of geometries for which Equation (4) is valid and the class of geometries in which it is invalid is a problem in ergodic theory and in measure theory. We will not attempt in this paper to distinguish between these two classes mathematically in the measure theory sense. Instead, we will assume that the statistical approach is valid except in those few geometries where a cursory inspection shows that randomization cannot occur under any of the circumstances listed earlier, a) to d). Equation (4) will be valid provided any of those prerequisites, a) to d), is satisfied.

Now consider the situation shown in Figure 3. In that geometry, the light is confined to a half space by the presence of a white reflective plane. In effect, the light intensity external to the optical sheet has been doubled by virtue of reflection from the white surface. The total intensity enhancement will then be given by

$$I_{int}(\omega,x) = 2n^2(\omega,x) I_{inc}(\omega) \tag{5}$$

where $I_{inc}(\omega)$ is the incident light intensity.

One of the key implicit assumptions that we have not emphasized earlier is that there be no absorption, either within the volume or on the surface of the medium. Optical absorption effects will be treated in Section 4.

The main conclusion of this section is that a statistical mechanical approach results in the intensity enhancement given in Equation (5). The factor $2n^2$ can be quite substantial; i.e., approximately 25 for silicon and approximately 15 for TiO_2. Even conventional glass with an index of 1.5 has an enhancement factor equal to 4.5. In view of the importance of this result, we give an alternative derivation based on geometrical optics in the next section.

3. Geometrical Optics Derivation

Our approach is based on "Detailed Balancing" of the light which is incident on a small area element dA, and the light in the loss cone which escapes from it. Consider the geometry shown in Figure 4. Let I_{inc} be the incident radiation power per area element dA. A fraction $T_{inc}(\phi)$ of this light will be transmitted through the incoming interface, where ϕ is the angle of incidence. This must be balanced by the internal radiation which escapes. Let us assume that the internal radiation is isotropic due to the randomizing influence of refraction and reflection from the textured interfaces as discussed in the previous section. Let B_{int} be the internal intensity per unit internal solid angle. The internal intensity I_{int} on both sides of an area element dA is given by

$$I_{int} = \int B_{int} \cos\theta \, d\Omega$$

---

[1] Another situation in which Eq. (4) is not valid is an optically thick turbid sheet illuminated from only one side. In this case, angular averaging is no problem, but spatial averaging will be incomplete. The side of the sheet aways from the source of illumination will be dark.

where $\cos\theta$ is the reduction of intensity on the area element due to oblique incidence. In this paper we will follow the convention that internal intensity I_{int} is bidirectional while the incident intensity I_{inc} is unidirectional. Therefore

$$I_{int} = 2 \times 2\pi \int_0^\pi B_{int} \cos\theta \sin\theta \, d\theta$$

$$I_{int} = 2\pi B_{int}$$

Only a small fraction of this power per unit area will escape, since the loss cone solid angle is much less than 4π steradians. The intensity which escapes is

$$I_{esc} = 2\pi \int_0^{\theta_c} \frac{I_{int}}{2\pi} T_{esc}(\theta) \cos(\theta) \sin\theta \, d\theta \qquad (6)$$

where $n \sin\theta_c = 1$ and θ is the internal angle of incidence. If we substitute a weighted average transmission factor $\overline{T_{esc}}$ for the angle-dependent surface transmission factor $T_{esc}(\theta)$, then the integral in Equation (6) may be easily computed.

$$I_{esc} = I_{int} \frac{\overline{T_{esc}}}{2n^2}$$

If we now apply the principle of "Detailed Balancing", the entering intensity is made equal to the escaping intensity

$$T_{inc}(\phi) I_{inc} = I_{int} \times \frac{\overline{T_{esc}}}{2n^2}$$

Therefore

$$I_{int} = 2n^2 \times \frac{T_{inc}(\phi)}{\overline{T_{esc}}} \times I_{inc} \qquad (7)$$

As Equation (7) shows, the enhancement may be increased beyond $2n^2$ if $T_{inc}(\phi)$, the transmission factor into the medium, is greater than $\overline{T_{esc}}$ the average transmission factor out of the medium. Of course, time-reversal invariance guarantees that $T_{inc}(\phi) = T_{esc}(\theta)$. If the incident radiation is isotropic then the ratio $\overline{T_{inc}}/\overline{T_{esc}} = 1$ would appear in Equation (7), ensuring that the enhancement factor is $2n^2$ as it must be for black-body radiation. For collimated radiation, an additional small enhancement $T_{inc}(\phi)/\overline{T_{esc}}$ is possible, but this comes at the expense of angular selectivity. This is fully consistent with the ordinary "Brightness Theorem" of geometric optics (6), which states that intensity increases must come at the expense of angular selectivity. Because of the tendencies toward angular averaging described in Section II, the factor $T_{inc}(\phi)/\overline{T_{esc}}$ will be approximated as unity and, for most purposes, Equation (7) can be rewritten

$$I_{int} = 2n^2 \times I_{inc} \qquad (8)$$

It is interesting that Equation (8) itself could also have been derived directly from the "Brightness Theorem" (6) by taking note of the fact that the brightness defined in a medium differs from that in a vacuum simply by the factor n^2.

The intensity decreases discussed thus far in this paper do not necessarily translate directly into absorption enhancements. This will be the subject of the following section.

4. Absorption Enhancement

Two types of absorption can modify the results we have presented this far; volume absorption in the textured optical sheet and surface absorption. In general, both types may be expected to be present. For example, in a semiconductor solar cell material, there would be absorption in the semiconductor itself and also at the surfaces due to absorption in the "transparent" electrodes, and due to imperfect reflectors at the rear surface. In this section, we will model the intensity-enhancement effects allowing for absorption. First we will set up a general method. Then we will model a specific geometry that might be of interest for solar cells.

The approach we will follow is to balance the input of light from external sources with the loss of light from the optical medium by absorption and refraction through the escape cone. The light input is $A_{inc}I_{inc}T_{inc}$, where A_{inc} is the surface area on which light is incident and the other symbols have the same meaning as before. To estimate the loss of light, we will proceed along the same lines as in Section 3. We will assume that the light internal to the medium is isotropic due to the randomizing influence of refraction and reflection. There will be three contributions to light which is lost:

1) Light will escape through the escape cone at the rate

$$\frac{A_{esc}I_{int}\overline{T_{esc}}}{2n^2}$$

where A_{esc} is the surface area from which light can escape, which is not necessarily equal to A_{inc}, and the other symbols have the same meaning as before.

2) Light may be absorbed due to imperfect reflection from the boundaries

$$\int_0^{\pi/2} \eta A_{refl} I_{int} \cos\theta \sin\theta \, d\theta = \frac{\eta A_{refl} I_{int}}{2}$$

where η is the fractional absorption due to imperfect reflection at the boundaries and A_{refl} is the surface area of imperfect reflection.

3) Finally, there may be absorption within the bulk

$$\int \frac{\alpha I_{int}}{2\pi} \, dV \, d\Omega \approx \alpha \ell I_{int} A_{inc} \int_0^\pi \sin\phi \, d\phi = 2\alpha \ell I_{int} A_{inc} \qquad (9)$$

where dV is a volume element in the bulk, α is the absorption coefficient, and Equation (9) may be regarded as a definition of the effective thickness ℓ. This will be approximately the mean thickness of the sheet. Equation (9) implicitly assumes that the bulk absorption is sufficiently weak that I_{int} is uniform throughout the volume.

Equating the light gained to the light lost

$$A_{inc}T_{inc}I_{inc} = \left\{ \frac{A_{esc}\overline{T_{esc}}}{2n^2} + \frac{nA_{refl}}{2} + 2\alpha\ell A_{inc} \right\} I_{int} \tag{10}$$

Regarding I_{int} as the unknown, the expression may be rewritten

$$I_{int} = \frac{T_{inc}I_{inc}}{\{(A_{esc}/A_{inc})(\overline{T_{esc}}/2n^2)+(n/2)(A_{refl}/A_{inc})+2\alpha\ell\}} \tag{11}$$

Although Equation (11) is much more complex than Equation (7) and Equation (8), there are many realistic situations where the simpler expressions are adequate approximations.

One of the main questions we have in this paper is the extent to which the effects we have been discussing will act to enhance volume absorption. Using Equation (11) the volume absorption may be written

$$2\alpha\ell A_{inc}I_{int}$$

$$= \frac{2\alpha\ell A_{inc}T_{inc}I_{inc}}{\{(A_{esc}/A_{inc})(\overline{T_{esc}}/2n^2)+(n/2)(A_{refl}/A_{inc})+2\alpha\ell\}}$$

The fraction of the incoming light which is absorbed in the volume is

$$f_{vol} \equiv \frac{2\alpha\ell A_{inc}I_{int}}{A_{inc}I_{inc}}$$

$$= \frac{2\alpha\ell T_{inc}}{\{(A_{esc}/A_{inc})(\overline{T_{esc}}/2n^2)+(n/2)(A_{refl}/A_{inc})+2\alpha\ell\}} \tag{12}$$

This reduces simply to the transmission factor T_{inc} of the incoming light in the limit of very high absorption coefficient α.

A corresponding expression may be written for the total fraction absorbed, including absorption due to imperfect reflection at the surfaces

$$f_{tot} = \frac{2\alpha\ell+(n/2)(A_{refl}/A_{inc})}{(A_{esc}/A_{inc})(\overline{T_{esc}}/2n^2)+(n/2)(A_{refl}/A_{inc})+2\alpha\ell} T_{inc} \tag{13}$$

The absorption enhancement in Equation (12) and Equation (13) is of direct interest for weakly absorbing indirect-gap semiconductors like crystalline silicon. As Equation (12) shows, volume absorption can be very substantial even when $\alpha\ell$ is only $1/4n^2$ which is 1/50 for silicon. The absorption enhancement factor is twice the intensity enhancement factor due to angle averaging effects. The use of these formulas is best illustrated by some specific examples.

Consider the reflectivity of a silicon wafer as measured on an integrating sphere, which is shown in Figure 5. For these measurements, a white reflective medium was placed behind the wafer. The front surface of the wafer was polished. The main idea is to compare the overall reflectivity when the rear surface is either ground rough or polished smooth. The comparison is made in Figure 6. With both surfaces polished, we have a plane parallel plate, the situation described in Figure 2(b) where angular randomization within the silicon does not occur. The light simply makes a round trip in the wafer.

On the other hand, if the rear surface of the silicon is ground rough, internal angular randomization does occur. We may apply Equation (12) and Equation (13) to this situation. The Fresnel transmission of the silicon front surface T_{inc} is about 0.68. The areas A_{esc} and A_{inc} are the same and equal to the front surface area. The rear surface was covered with $MgCO_3$ an almost perfect white reflector, which is frequently used as a reference of whiteness. In the geometry of Figure 5, the edges of the silicon wafer are actually external to the intergrating sphere. Some of the internally trapped light, therefore, escapes across the cylindrical surface defined by the periphery of the round opening in the integrating sphere. This cylindrical surface in the silicon can be regarded as an imperfect reflector of area $A_{refl} = 2\pi r\ell$ where r is the radius of the opening in the integrating sphere. Therefore

$$A_{refl}/A_{inc} = 2\pi r\ell/\pi r^2 = 2\ell/r$$

The parameters in this experiment were r=1 cm and ℓ = 0.025 cm. The quantity η which represents the departure from unit reflectivity at this edge is difficult to estimate a priori, since it depends on the details of the roughness. The value η = 0.82 describes well the wavelength-independent backscattered light in Figure 6 in the transparent region between 1.2 and 1.35 µm. With these values for parameters and the known wavelength-dependent (7) absorption coefficient, a fairly good fit is obtained between $(1 - f_{tot})$ from Equation (13) and experiment through the band-edge transition wavelengths (dashed and smooth lines, respectively, in Figure 6).

The geometry described in Figures 5 and 6 is a very favorable one for solar cells and was first described (1), (2) some time ago. Figure 6 shows clearly the shift of the effective absorption edge toward the infrared for the light-trapping case.

5. Open Circuit Voltage Enhancement

It is obvious that any improvement in the "effective" absorption of a semiconductor sheet would result in an increased infrared photoresponse by a solar cell. This would produce an improvement in the short circuit current. It is important to point out that improved current is not the only way in which these intensity enhancement effects can manifest themselves. If the thickness of the semiconductor layer were reduced proprotionate to the improved effective absorption, then the short circuit current would be unchanged, but the open circuit voltage could be increased.

Such a voltage improvement is not automatic by any means. But if the voltage is limited by recombination in the active region, then a reduction in thickness of the active region could result in a voltage improvement. If, on the other hand diffusion current limits the voltage, as is often the case, then thickness reduction would not have a significant effect on the open circuit voltage.

The open circuit voltage of a solar cell is given by

$$V_{oc} = KT \log (J_{sc}/J_{rec})$$

where J_{sc} is the short circuit and J_{rec} is the recombination current in the dark. This may be due to nonradiative recombination at defects in the semiconductor, or more fundamentally due to unavoidable radiative recombination. Shockley and Quiesser (8) took the semiconductor layer to be surrounded non-absorbing material. They included only that recombination radiation which was in the loss cone, and only the light escaping from one of the two surfaces. These are the most favourable assumptions, as is appropriate for calculating an upper limit. They obtained:

$$J_{rec} = \frac{2\pi e}{c^2} \int_{\nu_g}^{\infty} [\exp(h\nu/KT)-1]^{-1} \nu^2 d\nu \qquad (14)$$

where ν_g is the band gap frequency. If on the other hand, one or both surfaces of the semiconductor are absorbing, then all internally emitted light counts as nonradiative recombination. If in addition the semiconductor is optically thin to its own luminescence then Equation (14) must be re-written as:

$$J_{rec} = \frac{8\pi n^2 e}{c^2} \int_0^{\infty} [\exp(h\nu/KT)-1]^{-1} \nu^2 \alpha(\nu)\ell d\nu \qquad (15)$$

An optically enhanced solar cell satisfies the requirements of Equation (14) and is in principle capable of an increase in $V_{oc} = KT\log 4n^2$ or about 100 mV for the index of crystalline silicon.

An optically unenhanced solar cell, such as one with absorbing electrodes, potentially pays a 100 mV penalty. Of course, in today's solar cells radiative recombination in the volume of the cell is not the limit on voltage performance. Whether the recombination is radiative or not, as long as it is determined by the volume of the active material, a thin solar cell will have a $KT\log 4n^2$ voltage advantage over a thick cell. If on the other hand diffusion current to the electrodes limits the voltage, then a thin solar cell would not necessarily present any advantage at all.

6. Conclusion

In this paper we have shown the utility of a statistical mechanical approach toward the optics of textured and inhomogeneous sheets. This work was motivated mainly by its applicability toward solar cells and other types of solar collectors. The basic enhancement factor for intensity of $2n^2$ becomes $4n^2$ for bulk absorption and n^2 for surface absorption, due to angle averaging effects. The open circuit voltage can be increased by $KT\log 4n^2$. It is because many semiconductors tend to have large indices of refraction n, that these effects are particularly important in those materials.

References

(1) E. St. John, "Multiple internal reflection structure in a silicon detector which is obtained by sandblasting," U.S. Patent 3 487 223, 1969.

(2) O. Krumpholz and S. Maslowski, "Schnelle photodioden mit Wellenlangen unabhangigen Demodulationseigenschaften," Z. Angew. Phys., vol. 25, p. 156, 1968.

(3) D. Redfield, "Multiple-pass thin-film silicon solar cell," Appl. Phys. Lett., vol. 25, p. 647, 1974.

(4) M. Spitzer, J. Shewchun, E. S. Vera, and J. J. Loferski, "Ultra-high efficiency thin silicon p-n junction solar cells using reflective surfaces," in Proc. 14th IEEE Photovoltaic Specialists Conf. (San Diego, CA, 1980), p. 375.

(5) L.D. Landau and E. M. Lishitz, "Electrodynamics of Continuous Media". New York: Pergamon, 1969.

(6) M. Borne and E. Wolf, "Principles of Optics". New York: Macmillan, 1964.

(7) M. Neuberger and S. J. Welles, "Silicon". Springfield, VA: Clearinghouse for Federal Scientific and Technical Information, 1969, p. 113.

(8) W. Shockley and H. J. Queisser, J. Appl. Phys. 32, 510 (1961).

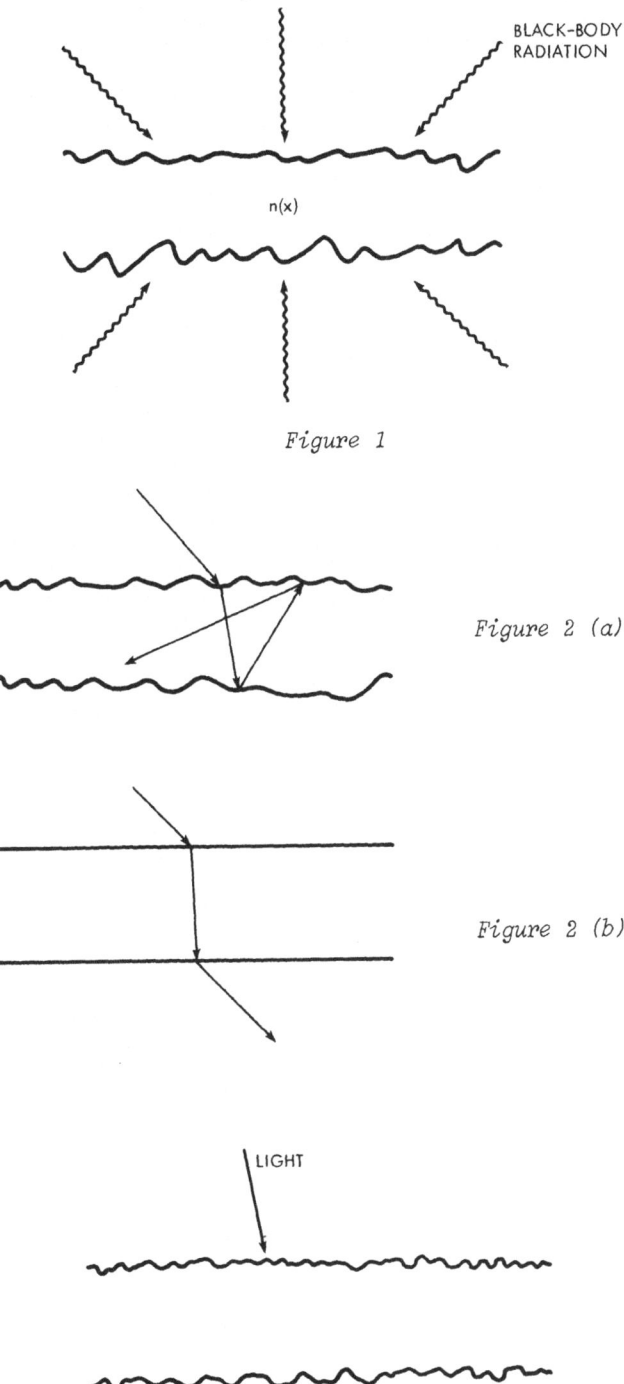

Figure 1

Figure 2 (a)

Figure 2 (b)

Figure 3

-475-

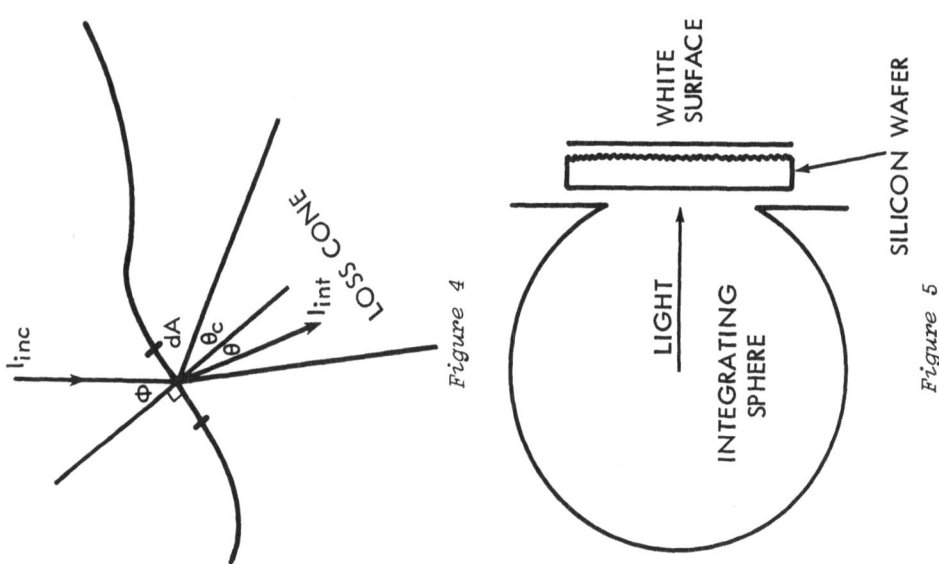

Figure 4

Figure 5

Figure 6

SPECTRUM SHIFTING METHODS IN PHOTOVOLTAICS:
AN EVALUATION OF MODEL SYSTEMS

F. GALLUZZI and E. SCAFE'
Laboratori Ricerche ASSORENI, Monterotondo (Rome), Italy

Summary

Photovoltaic methods using spectral distributions different from the solar spectrum are analyzed according to a simple unified approach. Introducing a "spectral transfer function", effective spectral responses and photocurrents are calculated for a few representative situations. In particolar thermophotovoltaic converters, planar luminescent concentrators and flat modules covered with fluorescent covers are analyzed and discussed.

1. INTRODUCTION

Photovoltaic (PV) cells are generally exposed to direct sunlight. However new methods have been proposed where the spectral distribution of the light incident on a cell is very different from the solar spectrum: this is the case - for instance - both in thermophotovoltaic systems and with planar fluorescent concentrators. Such systems, containing not only photovoltaic cells but also devices capable to shift and transform the solar spectrum, can be called simply spectrum shifting (SS) systems. Here we present a theoretical evaluation of a few SS systems using an unified approach.

2. ANALYTICAL MODEL

For treating in a general way spectrum shifting effects on photovoltaic devices we introduce a "spectral transfer function" $T(\lambda, \lambda')$ which describes the coupling between the solar spectrum and the cell photoresponse and allows a "conversion" from the wavelength λ of the sun radiation to the wavelength λ' of the actual radiation incident on the cell. Using this function the total photocurrent of the PV system is written

$$J_L = qC \int d\lambda \, d\lambda' \, S(\lambda) T(\lambda, \lambda') \phi(\lambda') \qquad (1)$$

where q is the electron charge, C is the concentration factor, $S(\lambda)$ is the solar spectral distribution (photons/cm^2/um), $\phi(\lambda)$ is the external quantum yield of the cell.

Equation (1) can be put in a form completely equivalent to that which holds for direct solar illumination, if we define also an effective quantum yield of the PV system

$$\phi_{syst}(\lambda) = \int d\lambda' T(\lambda, \lambda')\phi(\lambda') \qquad (2)$$

Indeed using this relationship the total photocurrent reads:

$$J_L = qC \int d\lambda\, S(\lambda)\, \phi_{syst}(\lambda) \qquad (3)$$

For a PV system directly illuminated by sunlight, for instance a transparent flat-plate module, the spectral transfer function is simply

$$T(\lambda, \lambda') = T_{cover}(\lambda)\delta(\lambda - \lambda') \qquad (4)$$

where $T_{cover}(\lambda)$ is the optical transmission of the flat-plate cover and $\delta(\lambda - \lambda')$ is a Dirac delta function.

On the contrary, when a spectrum shifting (SS) system is employed, the optical transfer function is factorized in two wavelength dependent terms: the former represents the absorption process of sunlight, the latter the emission of absorbed light to the photocell. Explicity we can write

$$T(\lambda, \lambda') = \eta_t A(\lambda)e(\lambda') \qquad (5)$$

where $A(\lambda)$ (≤ 1) is the absorptance of the system, $e(\lambda')$ is the normalized spectral distribution of emitted photons and η_t (≤ 1) is the overall efficiency of the absorption-emission process.

Particularly important for the final efficiency of the SS system is the overlap integral between emission spectrum and cell photoresponse

$$G = \int d\lambda'\, e(\lambda')\phi(\lambda') \qquad (6)$$

which essentially gives the mean value of the cell quantum yield at wavelenght where the SS system re-emits radiation. Both the effective quantum yield and the photocurrent are proportional to G and we have respectively

$$\phi_{syst}(\lambda) = \eta_t G A(\lambda) \qquad (7)$$

$$J_L = qC\, \eta_t G \int d\lambda\, S(\lambda)A(\lambda) \qquad (8)$$

The former equation says that spectral dependences of the system quantum yield and of the system absorbance are identical, while in the latter equation the integral simply represents the absorbed solar photon flux.

3. THERMOPHOTOVOLTAIC CONVERTERS

In thermophotovoltaic systems concentrated sunlight is absorbed by a black-body receiver and the high-temperature black-body emission is received by a photovoltaic converter [1].

For an ideal black-body an "ultimate" photovoltaic conversion efficiency can be evaluated using the simple equation

$$\eta_{ult}(T, \lambda_{gap}) = .93\; G(T, \lambda_{gap}) / \lambda_{gap}\,(\mu m) \qquad (9)$$

where we assumed $\eta_t = 1$, $Voc \sim 1/\lambda_{gap}$ and FF=1 and where G is the overlap integral depending only on the black-body temperature and the semiconductor gap.

As shown in fig. 1, when emission temperature grows up the photovoltaic efficiency increases (η_{ult}^{max} = 13% at 2000°K and 38% at 5000°K) and the optimum band gap shifts to shorter wavelengths (λ_{gap}^{max} = 3 µm at 2000°K and 1.3 µm at 5000°K). At realistic temperatures (≃ 3000°K) ultimate efficiencies are lower than 23% and hence practical efficiencies can be expected to be lower than 16%.

Owing to its band gap germanium appears a very attractive material for thermophotovoltaic systems [2], showing between 1500 and 3000°K (emission temperature) a ultimate efficiency ranging between 5 and 22%, while silicon cells possess in principle lower potentialities with efficiencies between 1% and 12% in the same temperature range.

4. FLUORESCENT CONCENTRATORS

The concept and the promise of planar fluorescent concentrators recently provoked a large interest [3]. Althought a dye system able to absorb a large position of the solar spectrum and to fluoresce in a narrow spectral interval is at present unavailable, it may be useful to evaluate the expected performance for such an "ideal" system.

In our model the luminescent plate absorbs completely light at wavelengths shorter than λ_A (absorption edge) and fluorescence occurs in a single line at λ_E. Then the photocurrents reads simply

$$J_L = qC \, \eta_t \phi(\lambda_E) \int_0^{\lambda_A} d\lambda \, S(\lambda) \tag{10}$$

The last equation has been applied to planar concentrators connected to silicon photocells, assuming η_t = .80.

Typical results are presented in fig. 2, where the dependence of the maximum photocurrent on the emission wavelength of the fluorescent plate is shown for different values of silicon diffusion lengths. In the visible region photocurrent increases rapidly with the emission wavelength owing to the increased portion of the absorbed solar spectrum, reaches a maximum in the range between 800 (short diffusion length Si) and 900 nm (high diffusion length Si) and then decreases sharply at longer emission wavelengths following the decrease of Si quantum yield. Maximum photocurrents for unit concentration are less than 20 mA/cm^2 and therefore always lower than those in direct sunlight concentrating systems (about 30-35 mA/cm^2). In other words, the effective concentration of luminescent plates can be at best about .5-.6 times the geometrical concentration.

5. FLAT MODULES WITH FLUORESCENT COVERS

In this case the usual transparent cover (glass or plastic) of a flat PV module is substituted by a partially absorbing fluorescent cover, in order to enhance and/or extend the spectral response of encapsulated cells [4]. Therefore both direct sunlight and fluorescent light are "seen" by the cell and the optical transfer function is obtained by summing eqs. (5) and (6):

$$T(\lambda, \lambda') = T_{cov}(\lambda)\delta(\lambda-\lambda') + A_{cov}(\lambda) e(\lambda') \qquad (11)$$

Neglecting for semplicity the reflectivity of the cover, the total quantum yield can be written in the form

$$\phi_{syst}(\lambda) = \phi(\lambda) + A_{cov}(\lambda)\left[\eta_t G - \phi(\lambda)\right] \qquad (12)$$

We see that, at wavelengths where fluorescent dyes are absorbing, we can have an enhancement of the cell spectral response if $\eta_t G > \phi(\lambda)$ (or even an extension of the photoresponse if the "bare" cell efficiency is zero), while, if $\eta_o G < \phi(\lambda)$, we have to suffer a decrease of the cell photoresponse.

Quantitatively the net photocurrent gain is given by

$$J_L = q\int d\lambda\, S(\lambda)\left[\phi_{syst}(\lambda) - \phi(\lambda)\right] = q\int d\lambda\, S(\lambda)\left[\eta_t G - \phi(\lambda)\right] A_{cov}(\lambda) \qquad (13)$$

Eqs. (12) and (13) have been applied to fluorescent covers on Si cells (fig. 3) and heterojunction cells with CdS "window" (fig. 4), assuming gaussian absorption and emission shape for dyes. Results on heterojunctions are particularly interesting. Spectral response extensions up to about 350-400 nm with quantum yields of about 40-50% could be obtained using dyes with absorption peak between 440 and 490 nm and Stokes shift of about 100 nm (assuming realistic transfer efficiencies $\eta_t \cong .50-.60$). The position of the dye absorption band is however very critical around the band gap of the "window" material and strong losses can occur in cell photoresponses cancelling any short-wavelength gain. For instance, in the system fluorescent cover/CdS, photocurrent gains up to 3 mA/cm^2 can be obtained for dye absorption peaks about 460 nm, but for dye absorbing at slightly longer wavelengths any gain disappears and losses higher than 2 mA/cm^2 can easily occur, depending on the transfer efficiency.

REFERENCES

(1) M.W. Edenburn, Solar Energy 24, 367 (1980).
(2) E.S. Vera, J.J. Loferski, M. Spitzer and J. Schewchun, Proceedings Third E.C. Photovoltaic Solar Energy Cong., W. Palz ed., Reidel, Dordrecht (1981), p. 911.
(3) A. Goetzberger and V. Wittwer, Advances in Solid State Physics XIX, 427 (1979); Solar Cells 4, 3 (1981).
(4) H.J. Hovel, R.T. Hodgson and J.M. Woodall, Solar Energy Mat. 2, 19 (1979); D. Sarti, F. La Foull and Ph. Gravisse; Solar Cells 4, 25 (1981).

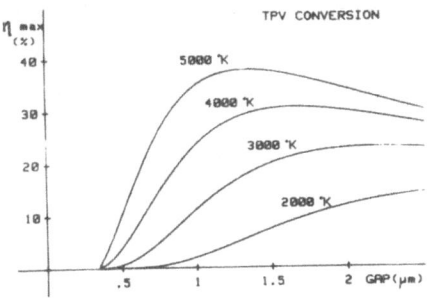

Fig.1: Ultimate efficiency in thermo photovoltaic conversion vs semiconductor band gap at different temperatures.

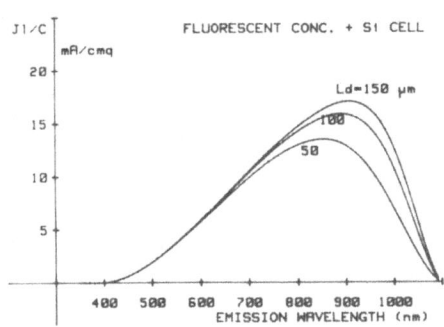

Fig.2: Maximum photocurrent of a Silicon cell coupled to "ideal" fluorescent concentrator vs concentrator emission wavelength for different Si diffusion lengths.

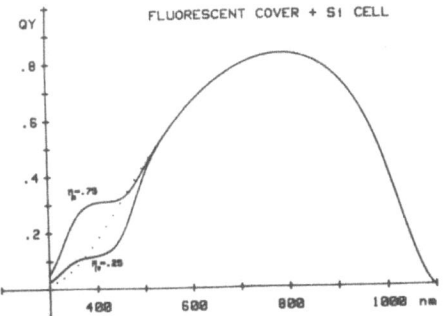

Fig.3: Effective spectral responses of a silicon cell (with poor blue response) covered by a single-dye fluorescent plate with $\lambda_A = .4$ um for two transfer efficiencies (.75, .25).

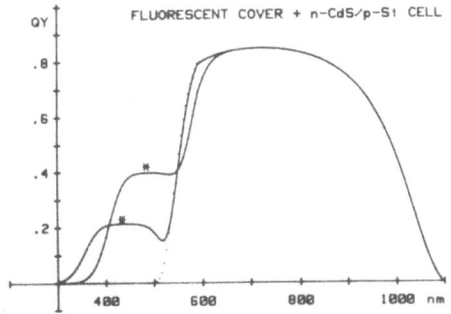

Fig.4: Effective spectral responses of a CdS/Si heterojunction cell covered by single-dye fluorescent plate with $\eta_t = .6$ for two absorption peak wavelengths (.44 and .48 um).

POSTER GROUP P5

FUNDAMENTAL STUDIES

- Theoretical studies

Extrema of majority and minority carrier quasi Fermi levels in p-n junction solar cells

High-blocked heterojunction and Schottky barrier solar cells

Computer-aided-characterization of the illuminated and dark current voltage characteristics of solar cells

Non linear model for shunt current in terrestrial silicon solar cells

Non-linear increase and decrease of open circuit voltage of $n^+ p^- p^+$ silicon solar cells at high illumination level

Series resistance analysis of concentrator cells under high injection conditions

- Silicon Solar Cells

The influence of grain-boundary recombination and grain size on the I(V)-characteristics of polycrystalline silicon solar cells

Diffusion length of minority carriers in scanning electron beam annealed silicon

Pathology of solar cell contacts

Design of stable metal-insulator-semiconductor (MIS) solar cells by oxide thickness compensation

- Amorphous Silicon

Study of gap states in a-Si:H by transient current spectroscopy

Highly conductive boron doped Si-layers prepared by plasma decomposition of SiH_4

A model for analysis of optical measurements carried on a-Si:H films for photovoltaic applications

- Other Materials

The assessment of thin film Cu_xS-CdS solar cells using cathodoluminescence techniques

Photovoltaic effect in SnTe/CdTe junctions

Organic photovoltaic materials : polyacetylene

Compositional analysis of $CuInS_2$ chalcopyrite semiconductor

Studies on CdS/n-InP PEC solar cells

EXTREMA OF MAJORITY AND MINORITY CARRIER QUASI FERMI LEVELS IN IN p-n JUNCTION SOLAR CELLS

A. Pimpale and P.T. Landsberg

Faculty of Mathematical Studies,
The University, Southampton SO9 5NH, England.

Summary

The movement of quasi-Fermi level extrema with applied voltage in p-n junction solar cells is discussed in a preliminary way.

Remarks containing results obtained so far.

Recent work on MIS - Schottky barrier solar cell has shown that the electron and hole quasi-Fermi levels have shallow extrema as a function of position in the cell[1]. From general considerations we expect a similar behaviour in a p-n junction solar cell. To study this we have developed systematic analytic approximations for the system of electron and hole continuity equations and Poisson equation for illuminated p-n junction in steady state. These equations are non-linear and coupled through through the recombination rate U and the electric field:

$$U = (np - n_i^2) \left[F + GHN/\{G(n + n_1) + H(p + p_1)\} \right],$$

where F involves band-band recombination coefficients, G and H involves band-trap recombination coefficients and N is the trap concentration.

In the zeroth approximation the system is linear and not coupled and only the minority carrier Fermi levels show shallow extrema: a maximum for electrons in the p-region and a minimum for holes in the n-region. These extrema persist all the way from short-circuit to open-circuit situation and, as voltage is increased from zero at short-circuit to the open-circuit voltage, these extrema move towards the transition region. Although the shapes of the quasi-Fermi levels near the front and back surfaces are dependent on the surface recombination velocities of the minority carriers, the positions and magnitudes of the extrema are not very sensitive to them. In this approximation no direct information about the majority carrier quasi-Fermi levels is available.

In the next approximation we maintain the depletion approximation but consider the electron and hole coupling through U and also introduce additional boundary conditions on the majority carriers. The majority carrier quasi-Fermi levels also now exhibit extrema which are at the same positions as the corresponding extrema in minority carrier quasi-Fermi levels for the open-circuit situation (some shapes are given in Fig. 1).

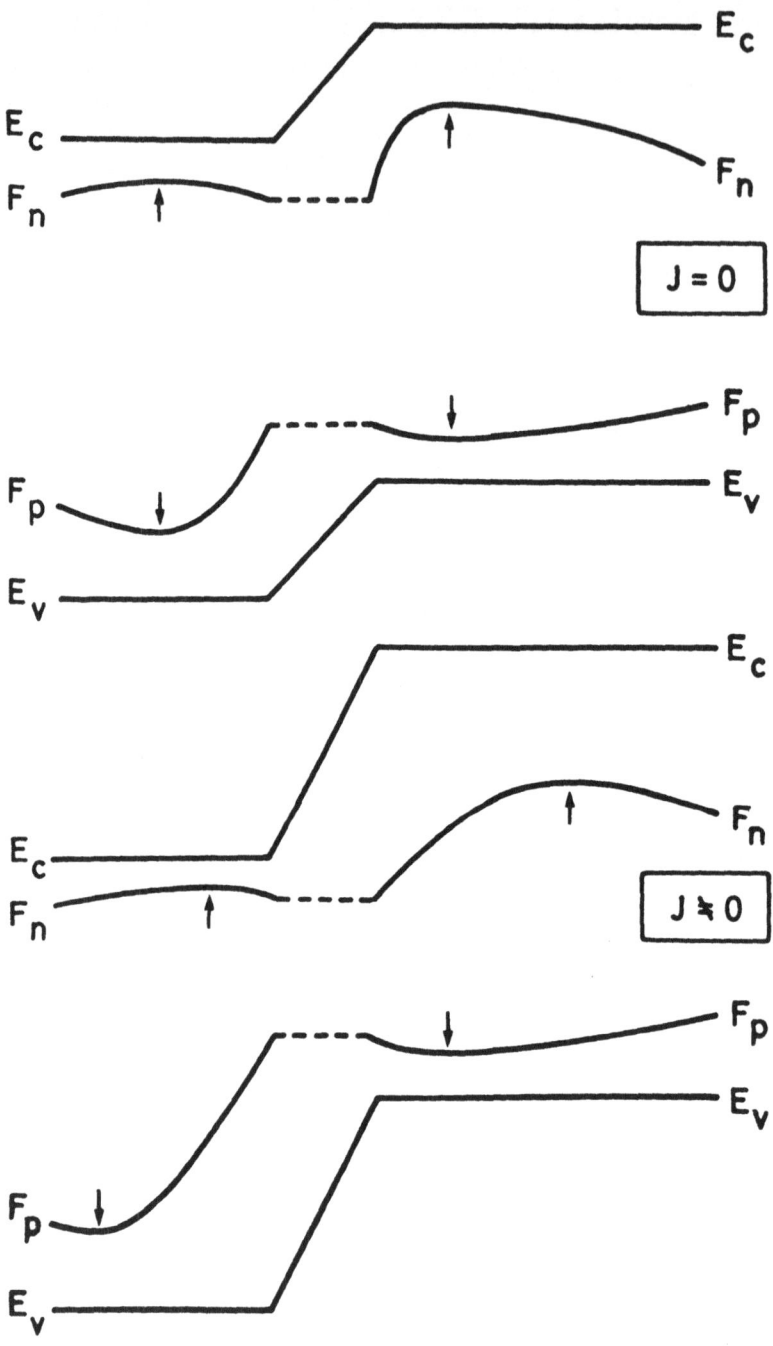

Fig. 1: Qualitative shapes of quasi-Fermi levels. The extrema are exaggerated for clarity and marked by arrow.

As non-zero current is drawn from the cell these extrema move away from each other.

The numerical calculations in this work pertain to silicon n-on-p cells with material parameters from Hovel[2] and treating the sun as a black body at a temperature of $6000°K$.

References

[1] C.M.H. Klimpke and P.T. Landsberg, Solid State Electr. <u>24</u> (1981) 401

[2] H.J. Hovel, Semiconductors and semimetals vol. 11: Solar cells (1975: Academic Press), p.26.

HIGH-BLOCKED HETEROJUNCTION AND SCHOTTKY BARRIER SOLAR CELLS

K. W. Bøer

University of Delaware and SES, Inc.

Newark DE. 19711

ABSTRACT

A simple formula is given for the current voltage characteristics of abrupt heterojunctions with a blocking barrier for the majority carriers in the highly doped region:

$$j = \frac{j_L[\exp(e(V - V_{oc})/(kT)) - 1]}{1 + s_j/(\mu_n F_j)}.$$

The meaning of the denominator, which acts as a shape factor of the characteristic, is discussed. This description replaces the one related to the exponential AkT-factor and can be directly interpreted as function of physical junction parameters, such as s_j, μ_n and space-charge-, e.g. donor- density.

This model applies to the CdS/Cu_2S solar cell where there is substantial experimental evidence that (except for series resistances and lateral inhomogeneities) it represents the major features of the shape of the observed characteristics.

1. SCHOTTKY BARRIERS

The developement of a heterojunction solar cell equation can be most easily understood by following the derivation of a Schottky barrier equation with one carrier only, using (for electrons here)

$$j_n = e\mu_n nF + \mu_n kT dn/dx \quad (1a)$$

$$dF/dx = \rho/(\varepsilon\varepsilon_0) \quad (1b)$$

$$d\psi/dx = -F. \quad (1c)$$

For a barrier with shallow donors in most of the barrier $\rho = eN_d$, hence the field is linear here:

$$F = F_c + eN_d x/(\varepsilon\varepsilon_0) \quad (2)$$

When this solution of eq. 1b is introduced in eq. 1a it yields a linear differential equation which can be easily integrated, resulting in

$$n(x) = n_c \exp[e(\psi_D - \psi(x))/(kT)]$$
$$+ [j_n/(e\mu_n F_c)](e\psi_D/(kT))^{0.5} D(e\psi(x)/(kT))^{0.5} \quad (3)$$

with ϕ_D the diffusion potential and $D(x)$ the Dawson's integral. Eq. 3 can be approximated with high accuracy (since $D(x) \simeq 1/(2x)$ for $x > 4$) for $|\phi(x)| > 4kT/e$ by

$$n(x) = n_c \exp[e(\phi_D - \phi(x))/(kT)] + j/(e\mu_n F_j) \qquad (3a)$$

For a <u>classical</u> <u>Schottky</u> <u>barrier</u> one obtains with

$$n(x = 0) = n_j = n_c = N_c \exp(-e\phi_{MS}/(kT)), \qquad (4)$$

when introduced into eq. 3a, the well-known <u>drift-limited diode equation</u> (and with $\phi_D - \phi(x = 0) = V$):

$$j = j_{o1}(\exp(eV/(kT)) - 1) \qquad (5a)$$

with

$$j_{o1} = e\mu_n n_c F_j. \qquad (5b)$$

When considering that the current at the maximum of the barrier, where $d\phi/dx = 0$, is given by diffusion only, and limited by Richardson – Dushman emission, one obtains instead of eq. 4 the following expression:

$$j = ev_n^*(n_j - n_c) \qquad (6)$$

with $v_n^* = v_n/(6\pi)^{0.5}$. When n_j from eq. 6 is introduced into eq. 3a for $x = 0$ one obtains the <u>modified diode equation</u>

$$j = \frac{j_{o2}[\exp(eV/(kT)) - 1]}{1 + v_n^*/(\mu_n F_j)}, \qquad (7a)$$

with

$$j_{o2} = en_c v_n^*. \qquad (7b)$$

For small barrier fields eq. 7a converts into eq. 5a, for large fields it converts to:

$$j = en_c v_n^*[\exp(eV/(kT)) - 1], \qquad (8)$$

the <u>diffusion limited diode equation</u>. Both, eqs. 5 and 8, are listed in the classical literature [1].

2. HIGH-BLOCKED HETEROJUNCTIONS

The same type of solution (eq. 3) applies to heterojunctions in which the highly doped part has a barrier at its interface which prevents leaking-out of its majority carriers into the lowly doped part. We will call such solar cells <u>high-blocked heterojunctions</u>. A typical example is the CdS/Cu_2S solar cell where a barrier (~1 eV) in the valence band prevents holes from the Cu_2S to enter the CdS. Such a barrier permits a single carrier approximation in each part of the junction (unlike in a homojunction).

In the low conductivity region again the approximation of shallow donor depletion may be used, yielding eN_d for the space-charge and consequently eq. 3a for the electron density distribution.

In solving the diffusion equation for the highly doped emitter, one obtains for $n(x = 0) = n_j$ a very similar equation to eq. 6, namely*

$$j = ev_D^*(n_j - n_{10}) \qquad (9)$$

with $n_{10} = g_0 \tau_n$ and v_D^* a modified diffusion velocity [2]

$v_D^* = (L_n/\tau_n)\tanh(x_m/L_n)$, with x_m the position of the electron density maximum in the emitter.

The resulting <u>high blocked heterojunction solar cell equation</u> is consequently very similar to eq. 7:

$$j = \frac{j_L[\exp(e(V - V_{oc})/(kT)) - 1]}{1 + v_D^*/(\mu_n F_j)} \qquad (10a)$$

with

$$j_L = en_{10}v_D^* \qquad (10b)$$

(after proper conversion of the electrostatical potential (ϕ) into the applied voltage, resulting in a shift of the characteristic from the non-illuminated diode by V_{oc}).

However, since for most solar cells $10^3 < v_D^* < 10^5$ cm/s, hence for reverse currents $v_D^* \ll \mu_n F_j$, the denominator in eq. 10 can be neglected, and eqs. 10 yield the well-known <u>ideal solar cell equation</u>:

$$j = j_L[\exp(e(V - V_{oc})/(kT)) - 1]. \qquad (11)$$

With non-negligible interface recombination ($s_j > v_D^*$), however, eq. 9 is modified to

$$j = en_j(v_D^* + s_j) - en_{10}v_D^* \qquad (9a)$$

and the <u>non-ideal heterojunction solar cell equation</u> results

$$j = \frac{j_L[\exp(e(V - V_{oc})/(kT)) - 1]}{1 + (v_D^* + s_j)/(\mu_n F_j)}. \qquad (12)$$

The denominator in eqs. 7a and 12 has the result of influencing the shape of the current voltage characteristic, hence shall be called the <u>shape factor</u>. For a solar cell characteristic it reduces the

---

*For simplicity here a constant generation rate $g(x) = g_0$ is assumed.

fill-factor, the more so, the higher s_j, and the lower μ_n and F_j is. Since F_j is a function of N_d:

$$F_j = [2eN_D(\phi_D + V_{oc} - V)/(\varepsilon\varepsilon_o)]^{0.5}, \quad (13)$$

the fill-factor also decreases with decreasing N_d (as long as $s_j \geq \mu_n F_j$). This behavior is demonstrated in Fig. 1 with $s_j/(\mu_n(N_d)^{0.5})$ as family parameter.

When more than one donor level is depleted with increasing reverse bias, the space-charge has to be evaluated more carefully:

$$\rho = e(p_{d1} + p_{d2} - n) \quad (14a)$$

which may be approximated with

$$\rho = e(N_{d1} + N_{d2}/[1 + \exp((E_F - E_{d2})/(kT))]) \quad (14b)$$

yielding the same non-ideal heterojunction solar cell equation, however with

$$F_j = [2e(p_{d1}(\phi_D + V_{oc} - \Delta E_{d2}) + (p_{d1} + p_{d2})(\Delta E_{d2} - V))/(\varepsilon\varepsilon_o)]^{0.5}. \quad (15)$$

This substitution may cause the characteristic to have an inflection point when a donor with sufficient density can be depleted while the bias is lowered (experimentally occasionally observed [3], and by some authors referred to as double diode characteristics).

3. ACKNOWLEDGEMENT

This work is supported by a grant from SERI. The assistance of M. Garcia in preparing the computations is greately appreciated.

4. LITERATURE

1. E. Spenke, Electronic Semiconductors, McGraw Hill Book Company, Inc., New York, Toronto, London (1958)
2. K.W. Böer, phys. stat. sol. 40, 355 (1977)
3. K.W. Böer, phys. stat. sol. 66, 11 (1981)

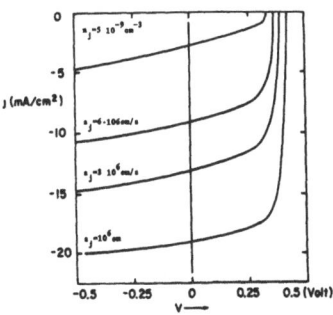

Fig. 1: Current-voltage characteristic for high-blocked heterojunction with $j_L = 24$ mA/cm^2, $N_d = 10^{16}$ cm^{-3}, $\varepsilon = 10$ and $\mu_n = 100$ cm^2/Vs. s_j is given as family parameter.
The first curve, however, is obtained for a constant $n_j = 5 \cdot 10^{-9}$ cm^{-3}, resulting in a characteristic similar to the Schottky barrier diode with constant n_c.

COMPUTER-AIDED-CHARACTERIZATION OF THE ILLUMINATED AND DARK CURRENT VOLTAGE CHARACTERISTICS OF SOLAR CELLS

P.H. Nguyen, B. Lepley, C. Boutrit and S. Ravelet
Laboratoire d'Electronique et de Physique des Interfaces
Institut des Sciences de l'Ingénieur
Parc Robert Bentz, 54500 Vandoeuvre, France.

Summary

The purpose of the present paper is the application of the computer-aided-characterization (C.A.C.) techniques to analyse the illuminated (IV^*) and dark forward (IV) current-voltage characteristics of solar cells. Reduced non linear least mean squares fits are performed on experimental IV^* characteristics to determine the parameters I_s, B (=q/nkT), R_s, G_{sh} and I_{ph} of the one diode equivalent circuit.

Under the dark conditions, the one diode model or the multiple diode equivalent circuit are investigated to determine the usual parameters I_{s1}, B_1 (=q/n_1kT), I_{s2}, B_2 (=q/n_2kT), R_s and G_{sh}.

1. INTRODUCTION

The application of optimization techniques in the study of semiconductor devices has been widely used since the development of the computer [1,2]. In the area of data analysis of solar cells a number of authors [3-7] have investigated the current-voltage characteristics using various iterative methods and different assumptions (i.e. $G_{sh} = 0$, $n_1 = 1$ and/or $n_2 = 2$ ...).

We present here a systematic study of these numerical models using a C.A.C. method based on a reduced non linear least mean squares fitting. In fact, the current-voltage characteristic of a solar cell can conveniently be described by the equation

$$I = -I_{ph} + I_{s1}[\exp\{B_1(V-R_sI)\}-1] + I_{s2}[\exp\{B_2(V-R_sI)\}-1] + G_{sh}(V-R_sI)$$

If B_1, B_2 and R_s are determined by a suitable minimization subroutine the other parameters (I_{ph}, I_{s1}, I_{s2} and G_{sh}) can be obtained now by a linear least mean squares fitting i.e. by a direct optimization technique. This might result in more rapid convergence [8] of the fitting program.

2. METHOD OF APPROACH

The multi-stage procedures of the C.A.C. techniques have been adopted as follows:
- The sums of the squares of the relative or absolute errors are the different objective functions to be minimized.
- The simplex minimization method is used for finding local minima, it does

not require any first or second derivatives of the objective function.
- The theoretical characteristics of ideal structures where the assigned values are allocated to each of the previous parameters, have been generated.
- The computer programs written in Fortran IV language which are available from the authors are used for both theoretical and experimental characteristics. Various different starting points can be fixed in order to determine the global minimum.
- A Quality Factor of Fit (QFF) is introduced to evaluate the "misfitting" of the theoretical model. One of the possible definitions is

$$QFF\ (\%) = 100\sqrt{1 - \frac{\Sigma\ (y_i - \hat{y}_i)^2}{\Sigma\ (y_i - \bar{y}_i)^2}}$$

where $\bar{y}_i = (\Sigma y_i)/N$, y_i is the measured value, $\hat{y}_i$ the computed value of the studied model and N the number of data points. The magnitude of the QFF provides information regarding the validity of the specified model. For example it is used to distinguish the conduction mode (one diode model or multiple diode equivalent circuit) under the dark conditions. The first necessary, although not sufficient, test for adequacy of the interpretation has to be the QFF of the theoretical relationship to the experimental data.
- The obtained minima are compared and tested for physical reality.
- The last stage of the procedure is the graphical comparison between the calculated current and the experimentally observed one. The mean standard deviation is also computed for each data point.

Table 1 shows different cases with assigned values to the parameters n and G_{sh} for approaching the IV* characteristics of solar cells. At the lower voltage levels (usually < 0.4 V) the photogenerated current is generally greater than the second component of current and at larger bias voltages the first diode current generally dominates. Thus, in most cases the second component effect can be disregarded under illumination. In the general case (IV2*5), the relative error function and the absolute criterion are defined respectively by

$$E_r^* = \Sigma\ ((I_j + I_{ph} - I_s A_j - G_{sh} V_1)/I_j)^2$$
$$E_a^* = \Sigma\ (I_j + I_{ph} - I_s A_j - G_{sh} V_1)^2$$

where $A_j = \exp(BV_1) - 1$ and $V_1 = V_j - R_s I_j$.

If the parameters B and R_s are known the remaining ones can be directly obtained by solving the system of linear algebraic equations:

$$\partial E_{a,r}^*/\partial I_{ph} = 0,\quad \partial E_{a,r}^*/\partial I_s = 0,\quad \partial E_{a,r}^*/\partial G_{sh} = 0.$$

Table 2 shows different numerical cases for fitting the dark IV characteristics of solar cells or of other semiconductor devices. In the general case the objective function is given by

$$E_r = \Sigma\ ((I_j - I_{s1} A_{j1} - I_{s2} A_{j2} - G_{sh} V_1)/I_j)^2$$

where $A_{ji} = \exp(B_i V_1) - 1$ [i=1,2] . The use of the relative criterion is due to the exponential nature of the relationship between the current and voltage. In the same manner, at each step one part of the parameters can be obtained from a straightforward computation by solving the system

$$\partial E_r/\partial I_{s1}=0, \quad \partial E_r/\partial I_{s2}=0, \quad \partial E_r/\partial G_{sh}=0.$$

3. RESULTS

The optimization programs find the unknown parameters from ideal IV^* and ideal IV characteristics with a good degree of accuracy and the QFF coefficients are very near 100% (99.999%) as the theoretical co-ordinates are given with three significant digits.

The values of the series resistance deduced from the cases IV1*3 and IV1*4, where the ideality factor n is taken to be equal to unity, are the over-estimated ones. They are in the same order as the graphical slope (dV/dI) for I=0. The results deduced from the two cases IV2*4 and IV2*5 are in agreement for all diodes with negligible shunt conductance ($G_{sh}<0.001$ mho). The salient feature is that the computed values B, I_s, R_s, G_{sh} and I_{ph} of the IV2*4 and IV2*5 models do not require any knowledge of the junction temperature during the IV measurements.

The programming techniques were used to study two types of solar cells: (a) a single crystal silicon solar cell, square, with an area of 4cm^2, supplied by SAT, France; (b) a series of cheap commercial solar cells, round, with an of area of 25.8 cm^2, from RTC. The real characteristics have been deduced from standard IV measurements which are not described here. Tables 3 and 4 show typical results obtained on ideal diodes and on real solar cells. Quality factors of fit near 99.5 percent generally indicate some caution in the representation of a data set by a given numerical model, while those of 99 % or less should give rise to severe doubt regarding the appropriateness of the empirical representation. The best fit of the illuminated IV data is the IV2*5 case with the absolute criterion since it generally corresponds to the highest QFF. The best fit for the dark current transport of the SAT cell is obtained by the two exponential equations. Figures 1 and 2 present respectively the illuminated and dark forward characteristics of the SAT cell. The values of the series resistance deduced from these characteristics are in agreement for the solar cell tested.

4. CONCLUSION

The illuminated solar cell is considered in this work as a generator and the usual parameters of its one diode equivalent circuit are assumed to be voltage independent within the range studied. The numerical analysis makes use of experimental data points from the characteristics.

In application, the computer programs can be implanted in a micro-computer which performs both an automatic acquisition of the IV^* data using suitable electronic interfaces [9] and an "on-line" characterization of the previous parameters using the IV2*5 model with the absolute criterion to study the performance of different devices on a solar cell production line.

ACKNOWLEDGEMENTS

The authors would like to thank Dr. Nguyen D. Thuoc (SAT, Paris) for supplying the single crystal silicon solar cell.

REFERENCES

1. M.J. Howes, D.V. Morgan and K.D. Al-Baidhawi, IEEE Trans. Electron Devices, 26 (1979) 1262.
2. P.H. Nguyen, C. Boutrit, B. Lepley and S. Ravelet, phys. stat. sol. (a), 54 (1979) 421.
3. R.T. Otterbein, D.L. Evans and W.A. Facinelli, Proc. 13th Photovoltaic Specialists' Conf., Washington, IEEE New York, (1978) pp 1074-1079.
4. R. Pommier, Thèse de Docteur-Ingénieur, USTL, Montpellier, France, (1981).
5. J.P. Charles, M. Abdelkrim, Y.H. Muoy and P. Mialhe, Solar Cells, 4 (1981) 169.
6. M. Wolf, G.T. Noel and R.J. Stirn, Proc. 12th Photovoltaic Specialists' Conf., Baton Rouge, IEEE New York, (1976) pp 44-52.
7. G.L. Araujo, E. Sanchez and M. Marti, Solar Cells, 5 (1982) 199.
8. W. Tantraporn, IEEE Trans. Electron Devices, 19 (1972) 331.
9. R. Schultz, A.A. Meilus, S.C. Hu and C. Goradia, IEEE Trans. Instrumentation and Measurement, 26 (1977) 295.

TABLE-1

Label \ Parameters	I_{ph}	n	I_s	R_s	G_{sh}
IV1*3	d.o.	1.	d.o.	sub	0.
IV2*4	d.o.	sub	d.o.	sub	0.
IV1*4	d.o.	1.	d.o.	sub	d.o.
IV2*5 (general case)	d.o.	sub	d.o.	sub	d.o.

Table 1 Numerical models of the one diode equivalent circuit for fitting the illuminated characteristics of solar cells

TABLE-2

Different Cases \ Parameters	n_1	I_1	R_s	G_{sh}	n_2	I_2	Remarks
IV2P4	sub	d.o.	sub	d.o.	-	0.	(one diode model)
IV2PO4	1.	d.o.	sub	0.	sub	d.o.	different
IV1QG4	1.	d.o.	sub	d.o.	2.	d.o.	cases
IV3PO5	sub	d.o.	sub	0.	sub	d.o.	of the multiple
IV2PG5	1.	d.o.	sub	d.o.	sub	d.o.	diode
IV2QG5	sub	d.o.	sub	d.o.	2.	d.o.	equivalent
IV3PG6 (general case)	sub	d.o.	sub	d.o.	sub	d.o.	circuit.

NOTATION sub : deduced from the minimization subroutine
d.o.: obtained by the direct optimization at each step

Table 2 Numerical models for fitting the dark forward IV characteristics of solar cells.

TABLE 3

IV^* data	I_{ph} (A)	B (V^{-1})	I_s (A)	R_s (Ω)	G_{sh} (Ω^{-1})	QFF / Best Fit (%)
IV*01	0.06	34.043	$1.3 \; 10^{-10}$	0.300	10^{-4}	(assigned values)
	0.059998	34.056	1.2910^{-10}	0.3005	$0.94 \; 10^{-4}$	99.999/IV2*5
IV*02	$2.1 \; 10^{-3}$	15.773	$9.7 \; 10^{-7}$	21.00	10^{-6}	(assigned values)
	$2.1 \; 10^{-3}$	15.745	$9.84 \; 10^{-7}$	20.92	0.	99.999/IV2*5
SAT1*04	0.0857	25.754	$8.83 \; 10^{-8}$	0.232	$2.4 \; 10^{-3}$	99.987/IV2*5 Abs. criterion.
	I_{sc}= 0.0856 A (measured value)					
RTC04*2	0.1379	19.215	$6.61 \; 10^{-6}$	0.262	$8.4 \; 10^{-3}$	99.992/IV2*5 Abs. criterion.
	I_{sc}= 0.137 A (measured value)					
RTC02*2	0.134	18.517	$8.91 \; 10^{-6}$	0.128	$9.22 \; 10^{-3}$	99.987/IV2*5 Abs. criterion.
	I_{sc}= 0.1335 A (measured value)					
	R_s (graphical method [5]) = 0.131 Ω					

Table 3 Optimized values found by computer for different IV^* characteristics

TABLE 4

IV data (Model)	B_1 V^{-1}	I_{s1} A	R_s Ω	G_{sh} Ω^{-1}	B_2 V^{-1}	I_{s2} A	QFF %
IVSCON	36.85	10^{-11}	0.053	10^{-4}	19.062	10^{-6}	(assigned values)
(IV2P4)	20.91	$5 \; 10^{-7}$	0.0485	$1.2 \; 10^{-4}$	-	I_{s2}= 0.	99.929
(IV3PG6)	36.79	$1.04 \; 10^{-11}$	0.0529	10^{-4}	19.058	$1.001 \; 10^{-6}$	99.9999
SATIV	[R_s (from IV^* data) = 0.232 Ω T = 295 K]						
(IV2P4)	31.99	$1 \; 10^{-9}$	0.262	$1.12 \; 10^{-4}$	-	I_{s2}= 0.	99.351
(IV2QG5)	39.03	$1.4 \; 10^{-11}$	0.265	$5.7 \; 10^{-6}$	19.68	$7.2 \; 10^{-6}$	99.9977
(IV3PG6)	38.54	$1.9 \; 10^{-11}$	0.263	$6.3 \; 10^{-7}$	12.45	$7.7 \; 10^{-6}$	99.9981 / Best Fit.

Table 4 Optimized values found by computer for different IV characteristics

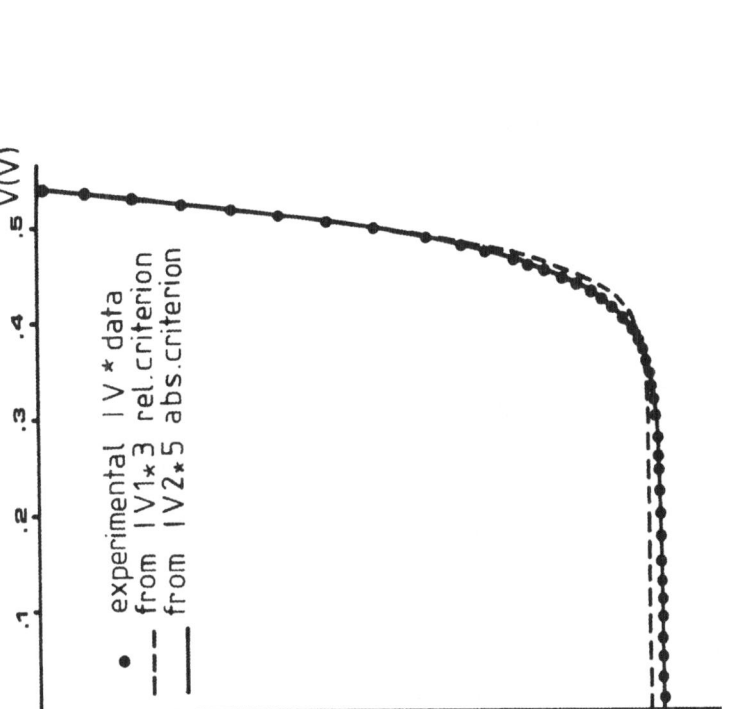

Fig. 2 - Dark forward IV characteristic of the SAT cell

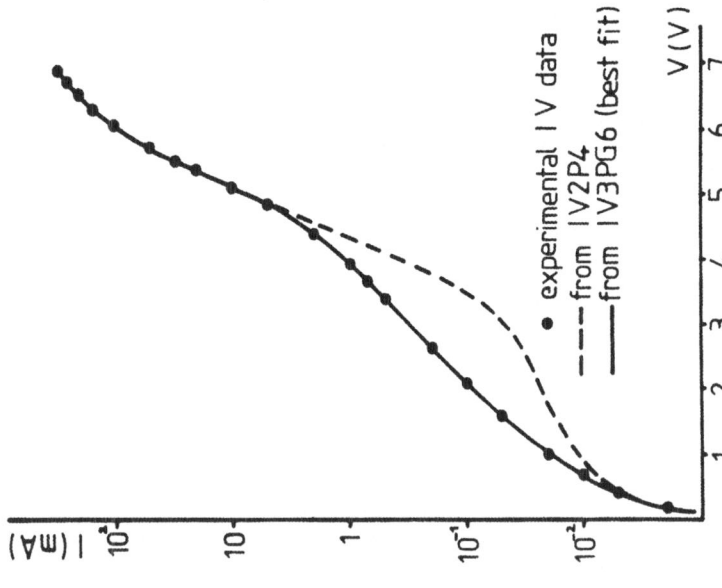

Fig. 1 - Illuminated IV* characteristic of the SAT cell

NON LINEAR MODEL FOR SHUNT CURRENT IN
TERRESTRIAL SILICON SOLAR CELLS

J. Cabestany
E.T.S. Ingenieros de Telecomunicación
C/Jordi Girona Salgado, s/n. Barcelona. Spain.

Summary

The circuit models commonly used for the I(v) characteristics in order to determine the equivalent circuit parameters of a solar cell fail when terrestrial textured surface solar cells are concerned with high percentage of metal covered area. Two alternatives are proposed, first is an empirical approach in which a parameter will be considered non constant and the second alternative is to include a new exponential term in the I(v) characteristic.

1. INTRODUCTION

The equivalent circuit parameters of a solar cell are usually related with the underliyng physics of the device and in some cases are empirical parameters representing several phenomena.
The double exponential model (1) has been the most widely used model: it has five parameters and can be written as follows:

$$I = I_{01} \left| \exp\left(\frac{q(V-IR_s)}{nkT}\right) - 1 \right| + I_{02} \left| \exp\left(\frac{q(V-IR_s)}{kT}\right) - 1 \right| + \frac{V-IR_s}{R_{sh}} \qquad |1|$$

The first term is the current contribution of the recombination in the space charge region, the second is the diffusion term and the third is the shunt current, here represented by an empyrical parameter called, shunt resistance, and accounting for the leakage currents in the very low voltage region. The shunt resistance has been usually taken as a constant parameter and accounts for an assembly of phenomena such as punctual junction short circuit or technological manipulation.
I_{01} and I_{02} are the reverse saturation currents of the two exponential terms and R_s is the series resistance that explains the behaviour of the I(v) characteristics in the upper voltage region.
The parameter n is the ideality factor and is related

with the recombination mechanism in the space charge region. All the other parameters of equation (1) have their usual meaning.

2. PARAMETER DETERMINATION

It is the purpose of the paper to discuss the validity of the model previously used. Recently (2) a systematic application of the non linear algorithm to double exponential model parameter determination has been reported. The method is based on the definition of a function accounting for the difference the analytical model (1) and an experimental I(v) characteristic.

The optimization algorithm applied to this function calculates the value of the parameters in order to obtain a minimum of the function. Two criteria have been used to define the object function: <u>least mean square</u> and <u>minimax</u>.

The error is evaluated in terms of the calculation of the standard deviation between the experimental and theoretical characteristics.

3. INFLUENCE OF THE SHUNT TERM

When the numerical method aforementioned is applied to several kinds of solar cells an interesting behaviour is observered. For space polished surface solar cells the proposed model of equation (1) gives enough accuracy in the parameters evaluation and a standard deviation under 10% is always found.

In terrestrial commercial silicon cells, textured surface and with high percentage of covered area, the proposed routine does not give appropiate accuracy. The differences in the theoretical and experimental characteristics is present in the lower region of the voltage.

Figure 1 shows such a behaviour for two cells.

4. PROPOSED REVISIONS

The inaccuracy of the model in a certain number of cases suggest the shunt term in equation (1) is not good enough to represent the shunt current through the device. Two alternatives have then been considered.

4(a) <u>Variable shunt resistance</u>.

This is an empirical alternative in the sense that a variable shunt resistance is difficult to relate to a physical phenomena in the device. If this hypothesis is applied to the sample of figure 1 we found the results that are shown in figure 2.

The values of shunt resistance resulting are obtained, prior to computer routine application, by graphic decomposition of the I(v) curve.

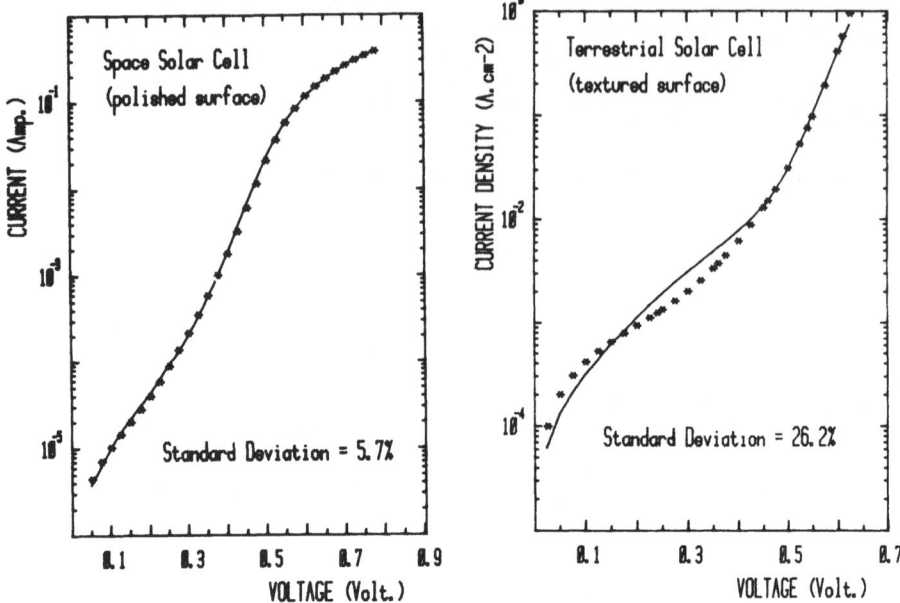

Figure 1.- Results obtained with the double exponential terms for (a) polished surface and (b) textured surface cell.

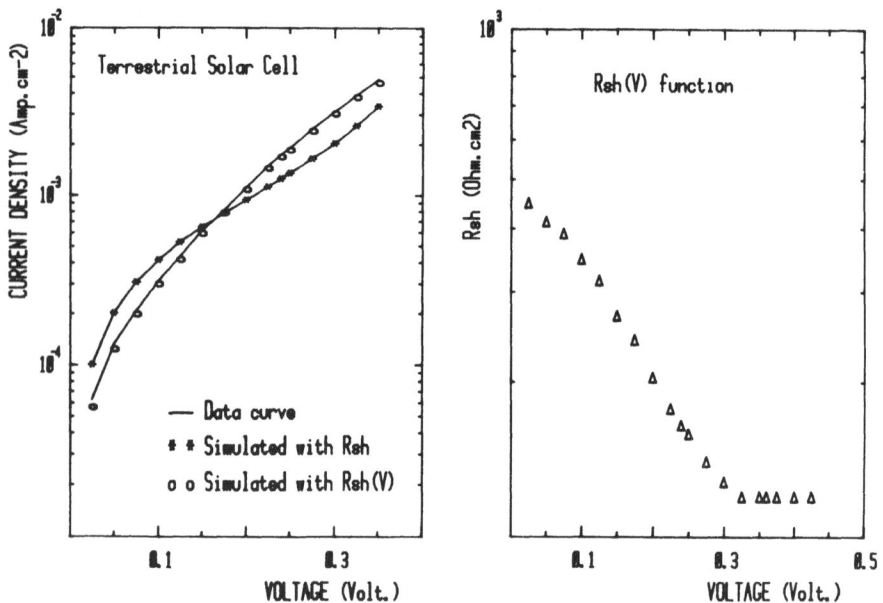

Figure 2.- Variable shunt resistance (a) adjust obtained and (b) shunt resistance values resulting.

4(b) Model modification.

If the origin of shunt current is assumed to be small and punctual short circuit through the junction became of the technological manipulations it is possible (3) to remodelize this shunt phenomena with a thirs exponential term for the I(v) characteristics

$$I = \frac{V-IR_s}{R_{sh}} + \sum_{i=1}^{3} I_{0i} \left| \exp \left| \frac{q(V-IR_s)}{n_i kT} \right| -1 \right| \qquad |2|$$

Table I shows comparative the results obtained by the three methods.

Table I

Sample	Equation |1|	With variable R_{sh}	Three exponential model
1	26'2%	8'6%	4'8%
2	40'8%	11'3%	8'6%

Table II

	Samples	THREE EXPONENTIAL MODEL					TWO EXPONENTIAL MODEL		
		Recombination term		New term			Recombination term		
		$I_{01}(\mu A)$	n	$I_{03}(\mu A)$	n	$R_{sh}(k\Omega)$	$I_{01}(\mu A)$	n	$R_{sh}(k\Omega)$
unpolished	1	6'7	2'29	272	5'4	37	4'2	2'27	0'25
	2	3'78	1'86	131	5	high	1'64	2'04	0'408
polished	3	0'46	1'96	0'0015	1'8	12'9	0'46	1'96	13'7
	4	0'73	1'93	5'9	51	12'5	0'79	1'99	11'23

Table II shows some results obtained for 4 samples, two of them presenting a clear influence of the new experimental term and the two others, space cells, very small influence of this terms is observered.

In the form cases of Table II the SD was lower than 10%.

It can be seen from Table II that effectively in samples 1 and 2 the relative wheigh of the new exponential term (I_{03} and n) is very high in the lower voltage region whereas for samples 3 and 4 the shunt resistance term keeps its influence and its value is not modified with respect to the results obtained with the two exponential model.

5. CONCLUSIONS

The poor adjust obtained in comparison of experimental and theoretical I(v) characteristics in terrestrial silicon cells leads to the conclusion that the two exponential model is not good enough for these cells. The solution of including a third exponential term in the characteristics gives the best results and can account for punctual junction short circuits.

REFERENCES

|1| Wolf M. et al, 1977, Investigation of the double exponential in the I(v) characteristics of Si. solar cells, IEEE Trans. on Electron Devices, vol. ED-24 (4), 419-428.

|2| Cabestany J. et al, 1982, Non linear algorithms application to irradiated solar cell parameters evaluation, Proc. "Third European Symp. Photovoltaic Gen. in Space". Bath. England, 4-6 May.

|3| Faith T.J. et al, 1978, High temperature contacts for Si. solar cells, Proc. 13th Phot. Spec. Conf, Washington. June 5-8.

NON-LINEAR INCREASE AND DECREASE OF OPEN CIRCUIT VOLTAGE OF n^+-p-p^+ SILICON SOLAR CELLS AT HIGH ILLUMINATION LEVEL

C.M. SINGAL and R.V. SINGH
Energy Centre, Department of Physics,
University of Roorkee, Roorkee-247672, India

1. INTRODUCTION

The photovoltaic power conversion efficiency as well as the total power generation from a silicon solar cell can be increased by operating it under concentrated sunlight [1-4]. The increase of the efficiency results from an increase in the open circuit voltage of the cell under concentrated sunlight. Additional enhancement of this voltage is predicted[5] through low high p-p^+ junction in a solar cell having an n^+-p-p^+ configuration. The evaluation of the voltages across the n^+-p junction (V_j) and p-p^+ junction (V_{lh}) have been done in a self-consistent manner since the carrier concentrations and their flow in the n^+, p and p^+ regions are coupled, to one another. Further as has been recently proposed by Lanyon and Tuft [6] that there is an energy band gap shrinkage which is directly dependent upon the concentration of mobile carriers, and not on the doping concentration alone, we expect significant modifications in the open-circuit voltage of the cell at high illumination levels.

The present analysis has been carried out incorporating the above requirements, and the developed voltages V_j, V_{lh} and Dember potential (V_D) as well as their total voltage V_{oc}, under open-circuit conditions, have been determined for illumination levels ranging from 10^{-10} to 10^8 suns for the purpose of investigating the onset, linearity, saturation and decrease of these voltages under different illumination conditions.

2. THEORY

We considered an n^+-p-p^+ solar cell structure as illustrated in Fig. 1(a). Under equilibrium situation, in the absence of any illumination or external bias voltage, electrostatic potential barriers are formed at the n^+-p and p-p^+ junctions. The height of these barriers in terms of the electron energy as shown by the energy band diagram in Fig. 1(b), are $-q\varphi_0$ and $-q\Psi_0$ for n^+-p and p-p^+ junctions respectively. At high illumination levels the charge carrier continuity equation [7], with space charge neutrality (i.e. $p' = p-p_0 = n' = n-n_0$), is given by:

$$\frac{dn'}{dt} = g - \frac{n'}{\tau^*} + \mu^* E \frac{dn'}{dx} + D^* \frac{d^2 n'}{dx^2} \qquad \ldots (1)$$

where D^*, μ^* and τ^* are the ambipolar diffusivity, mobility, life time and are given by

$$D^* = \frac{p+n}{p/D_n + n/D_p} \qquad \ldots (2)$$

$$\mu^* = \frac{p - n}{p/\mu_n + n/\mu_p} \qquad \ldots (3)$$

$$\tau^* = \tau_o \frac{(p + n + 2n_i)}{p + n_o} \qquad \ldots (4)$$

For the sake of simplicity we have assumed a uniform doping as well as uniform carrier generation rate g. Moreover, at high levels of injection, n and p become comparable to each other and there reduce μ^* to zero. As a result of this, the electric field E dependent drift term in equation (1) approaches zero. For the steady state equation (1) becomes:

$$D^* \frac{d^2 n'}{dx^2} - \frac{n'}{\tau^*} = -g \qquad \ldots (5)$$

For the purpose of solving this equation following approximations are incorporated. Generation and recombination of carriers in the space charge region are taken as negligibly small; thickness of the space charge region is considered as negligible compared to the thicknesses of n^+, p and p^+ regions; and the quasi-charge neutrality condition is applied in the cell outside the space charge region. On solving equation (5) for n^+, p and p^+ regions in one dimension subject to the above assumptions and suitable boundary conditions, the following expressions for the current density, and current continuity condition at the low-high junction are obtained:

$$J = -\frac{J_o + J_o' \exp[-q(\varphi_o - V_j)/kT]}{1 - \exp[-2q(\varphi_o - V_j)/kT]} \times [\exp(qV_j/kT) - 1]$$

$$+ \frac{J_o'' + J_o''' \exp[-q(\psi_o - V_{\varphi h})/kT]}{1 - \exp[-2q(\psi_o - V_{\varphi h})/kT]} \times [\exp(qV_{\varphi h}/kT) - 1] + J_I$$

$$\ldots (6)$$

and

$$J = J_L' + \frac{J_{W_p} + J_{W_p}' \exp[-q(\varphi_o - V_j)/kT]}{1 - \exp[-2q(\varphi_o - V_j)/kT]} \times [\exp(qV_j/kT) - 1]$$

$$- \frac{J_{W_p}'' + J_{W_p}''' \exp[-q(\psi_o - V_{\varphi h})/kT]}{1 - \exp[-2q(\psi_o - V_{\varphi h})/kT]} \times [\exp(qV_{\varphi h}/kT) - 1]$$

$$\ldots (7)$$

where,

$$J_o = \frac{q D_{n+}^*}{L_{n+}^*} \cdot \gamma \cdot p_{n+o} + \frac{q D_p^*}{L_{n+}^*} n_{po} \coth(W_p/L_p^*) \qquad \ldots (8)$$

$$J_o' = \frac{qD_{n+}^*}{L_{n+}^*} \gamma n_{po} + \frac{qD_p^*}{L_p^*} p_{n+o} \coth(W_p/L_p^*) \quad \ldots (9)$$

$$J_o'' = \frac{qD_p^*}{L_p^*} p_{po} \operatorname{cosech}(W_p/L_p^*) \quad \ldots (10)$$

$$J_o''' = \frac{qD_p^*}{L_p^*} n_{p+o} \operatorname{cosech}(W_p/L_p^*) \quad \ldots (11)$$

$$J_L = g_{n+} \frac{qD_{n+}^*}{L_{n+}^*} \tau_{n+}^* (\gamma-\theta) + g_p \frac{qD_p^*}{L_p^*} \tau_p^* [\coth(W_p/L_p^*)$$
$$- \operatorname{cosech}(W_p/L_p^*)] \quad \ldots (12)$$

$$\gamma = \frac{S_{n+} \cosh(W_{n+}/L_{n+}^*) + (D_{n+}^*/L_{n+}^*)\sinh(W_{n+}/L_{n+}^*)}{S_{n+} \sinh(W_{n+}/L_{n+}^*) + (D_{n+}^*/L_{n+}^*) \cosh(W_{n+}/L_{n+}^*)} \quad \ldots (13)$$

$$\theta = \frac{S_{n+}}{S_{n+} \sinh(W_{n+}/L_{n+}^*) + (D_{n+}^*/L_{n+}^*) \cosh(W_{n+}/L_{n+}^*)}$$
$$\ldots (14)$$

$$J_{W_p} = \frac{qD_p^*}{L_p^*} n_{po} \operatorname{cosech}(W_p/L_p^*) \quad \ldots (15)$$

$$J_{W_p}' = \frac{qD_p^*}{L_p^*} p_{n+o} \operatorname{cosech}(W_p/L_p^*) \quad \ldots (16)$$

$$J_{W_p}'' = \frac{qD_p^*}{L_p^*} p_{po} \coth(W_p/L_p^*) + \frac{qD_{p+}^*}{L_{p+}^*} n_{p+o} \coth(W_{p+}/L_{p+}^*)$$
$$\ldots (17)$$

$$J_{W_p}''' = \frac{qD_p^*}{L_p^*} n_{p+o} [\coth(W_p/L_p^*)] + \frac{qD_{p+}^*}{L_{p+}^*} p_{po}[\coth(W_{p+}/L_{p+}^*)]$$
$$\ldots (18)$$

$$J_L' = g_{n+} \frac{qD_{p+}^*}{L_{p+}^*} \tau_{p+}^* [\coth(W_{p+}/L_{p+}^*) - \operatorname{cosech}(W_{p+}/L_{p+}^*)]$$

$$+ g_p \frac{qD_p^*}{L_p^*} \tau_p^* [\coth(W_p/L_p^*) - \operatorname{cosech}(W_p/L_p^*)] \quad \ldots (19)$$

and L^*, S and W are the ambipolar diffusion length, surface recombination velocity and thickness with suffixes n^+, p and p^+ for the n^+, p and p^+ regions respectively.

2.1 Bandgap Shrinkage Effect

So far, the various researchers have considered bandgap shrinkage due only to the high levels of doping in the semiconductor. However, Lanyon and Tuft [6] showed theoretically that the shrinkage in energy gap is a result of the screening of the minority carrier charge by majority carriers, regardless of their origin. We included this improved theory of band gap narrowing for the present study. The value of ΔE_g at room temperature is given as:

$$\Delta E_g = 0.0225 \, (N/10^{18})^{1/2} \text{ eV} \quad \text{for non-degenerate semiconductor} \quad \ldots (20)$$

$$\Delta E_g = 0.162 (N/10^{20})^{1/6} \text{ eV} \quad \text{for degenerate semiconductor} \quad \ldots (21)$$

where N is the majority carrier concentration. As a result of this band gap narrowing the intrinsic carrier concentration for n^+, p and p^+ regions and the electrostatic potential barriers get modified.

Apart from V_j and $V_{\ell h}$, the value of Dember potential V_D is also calculated using the formula of Sabnis [8]:

$$V_D = -\frac{kT}{q} \left(\frac{b-1}{b+1}\right) \ln \left[\frac{N_A + (b+1) n_p'(x=0)}{N_A + (b+1) n_p'(x=W_p)}\right] \quad \ldots (22)$$

where b is the ratio of electron to hole mobilities, N_A is the doping in p-type base region and n_p' is excess carrier concentration at n^+-p and p-p^+ junctions. Now the total open circuit voltage across the cell terminals is:

$$V_{oc} = V_j + V_{\ell h} + V_D$$

3. RESULTS, DISCUSSION AND CONCLUSIONS

For the purpose of numerical analysis the values of dopings of n^+, p and p^+ regions are taken as 10^{18}, 10^{15} and 10^{19} cm^{-3} respectively. The minority carrier life times are taken as 1 µsec, 5 µsec and 1µsec respectively for the three regions and the thicknesses for these three regions are assumed equal to their ambipolar diffusion lengths. Front surface recombination velocity is also equated to diffusion velocity (D_{n+}^x / L_{n+}^x).

3.1 Open Circuit Voltage Linearity and Saturation

Fig. 2 shows the variation of V_j, $V_{\ell h}$ and V_{oc} with light generated current J_L for different levels of illumination. The values of V_D come out to be negligible (nearly 5mV) for this choice of solar cell parameters and is not shown in the figure. The potential V_j above $J_L \simeq 10^{-10}$ amp/cm^2 increases linearly upto $J_L \simeq 10$ amp/cm^2 and then it starts saturating. The p-p$^+$ junction potential $V_{\ell h}$ is negligibly small upto $J_L = 0.1$ amp/cm^2 and increases nearly linearly upto 100 amp/cm^2 and finally saturates at $J_L \simeq 10^4$ amp/cm^2. The combined effect V_j, $V_{\ell h}$ is reflected on V_{oc}, i.e. V_{oc} also saturates at $J_L \simeq 10^4$ amp/cm^2. In the first region of linearity ($J_L = 10^{-10}$ to 10^{-2} amp/cm^2) V_{oc} increases by 60 mV for every tenfold increase of J_L. In the second range of linearity ($J_L = 0.1$ to 100 amp/cm^2 it increases by 110 mV per tenfold increase in J_L value. From a practical stand point, these results provide us with a method of ascertaining the presence of p-p$^+$ junction at the back surface of the cell. In Fig. 3 the open circuit voltages are plotted against J_L for thin n^+ and p^+ regions ($W_{n+} = 0.25$ µm, $W_p = 150$ µm, $W_{p+} = 0.5$ µm). The effect of thinness of diffused region is such that the value of Dember potential now becomes measurable (~ 50 mV).

3.2 Energy Band Gap Shrinkage Effect

So far we have indicated that the open circuit voltages saturate at very high illumination levels. However when we include the carrier concentration dependent energy band gap narrowing, V_j and V_{oc} start decreasing instead of saturating while $V_{\ell h}$ starts increasing at high illumination levels as is shown in Fig. 4. There is no appreciable change in V_D. The solar cell parameters in this case are the same as in Fig. 3 except that the front region doping is taken as 2×10^{20} cm^{-3}. The sets of curves labelled 1 and 2 are without and with band gap shrinkage effect respectively. The minority carrier life times have been taken as fixed. The curves labelled 3 include the effect of band gap narrowing and the effect of illumination on the excess carrier life times. A comparison of curves 1 and 2 shows that the effect of band gap shrinkage at low injection levels is very small because at these levels this effect occurs only in diffused heavily doped regions. At very high levels of illumination, this effect begins to play its part and this turns to be play a dominant role in that it affects all the regions of the cell. Looking back at equation (6) for open circuit condition ($J=0$) and combining the fact that the second term in this equation involving p-p$^+$ junction potentials is not large, we see that a rapid enhancement of J_0 and J_0' due to band gap shrinkage results in a rapid decrease of V_j from saturation behaviour. Similarly V_{ph} should increase due to enhancement in J_{Wp} and J_{Lp}^+ as given by equation (7) under open circuit condition ($J=0$), at high levels of illumination, which is shown in Fig. 4. The combined effect of V_j, V_{ph} and V_D is reflected on V_{oc} as shown by curve 2, for V_{oc}. The maximum value of V_{oc} reached for the case under study is thus only 0.880 volt $[\ll (\varphi_0 + \psi_0)]$ at $J_L = 10^4$ amp/cm^2 (= 10^5 watt/cm^2). Similar behaviour of V_{oc} was observed experimentally by Blinov et al. [9] for n$^+$-p cell. They found the highest value of V_{oc} to be much below the diffusion potential φ_0 at intensities higher than 10^5 watt/cm^2. Thus our theoretical results are in very good agreement to their experimental values. Therefore, we believe that the phenomenon of V_{oc} decrease at very high intensities is a very real one and it results from the carrier concentration dependent energy band gap shrinkage. The effect of increased value of τ^* above τ_0 causes increased value of carrier concentration and band gap shrinkage, thus giving further lowering value of V_j, raising V_{ph} and keeping V_D essentially unchanged which is very much clear from curves labelled 2 and 3.

ACKNOWLEDGEMENT

The financial assitance of Indian National Science Academy and University Grants Commission, New Delhi for carring out this study is gratefully acknowledged.

REFERENCES

1. J. Lindmayer and C. Wringley, Photovoltage in Concentrated Cells, Record of 13th IEEE Photovoltaic Spec. Conf., p. 782, 1978.

2. J. Mandelkorn and J.H. Lamneck, Design, Fabrication and Characterisation of New Types of Back Surface Field Cell, Record of 10th IEEE Photovoltaic Spec. Conf., Pato Alto, CA, p. 207, 1973.

3. J.G. Fossum, E.L. Burgees and P.A. Lindholm, Silicon Solar Cell Designs Based on Physical Behaviour in Concentrated Sunlight, Solid State Electronics, Vol. 21, p. 729, 1978.

4. Ching-Yuan Wu and Wen-Zen Shen, The Open Circuit Voltage of Back Surface Field (BSF) p-n Solar Cells in Concentrated Sunlight, Solid State Electronics, Vol. 23, p. 209, 1980.

5. M.P. Godlewski, C.L. Baraona and H.W. Brandhorst, Jr., Low High Junction Theory Applied to Solar Cells, Record of 10th IEEE Photovoltaic Spec. Conf., Palo Alto, CA, p. 40, 1973.

6. H.P.D. Lanyon and R.A. Tuft, Bandgap Narrowing in Moderately to Heavily Dopend Silicon, IEEE Trans. Electron Devices, Vol. ED-26, p. 1014, 1979.

7. W. Van-Roosbroeck, Theory of Current Carrier Transport and Photo-conducting Semiconductors With Traping, Bell Syst. Tech. J., Vol. 39, p. 515, 1960.

8. A.G. Sabnis, Junction Potentials of Strongly Illuminated n^+-p-p^+, Solar Cells, Solid State Electronics, Vol. 21, p. 581, 1978.

9. L.M. Blinov, V.S. Vavilov and G.N. Galkin, Photo-EMF of p-n Junction in A Strongly Excited Semiconductor (Si), JETP Letters, Vol. 3, p. 234, 1966, and B.M. Vul, V-S. Vavilov, L.M. Blinov and G.N. Galkin, Photovoltaic EMF of An Intensly Illuminated p-n Junction in Silicon, Rev.Phys. Appl. (France), Vol. 1, p. 209, 1966.

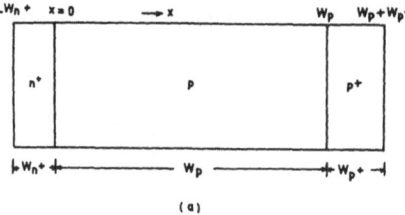

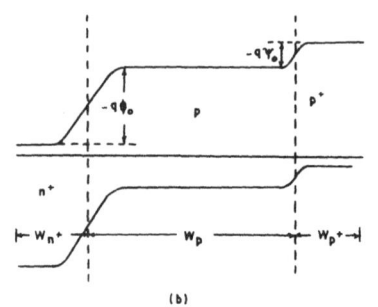

FIG 1 a-TYPICAL n^+-p-p^+ SOLAR CELL STRUCTURE
b-ENERGY BAND DIAGRAM OF n^+-p-p^+ SOLAR CELL

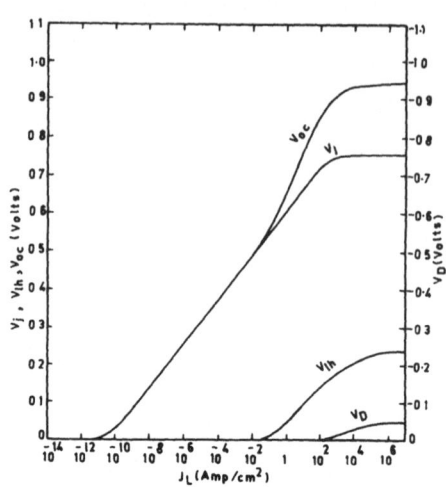

FIG 3-OPEN CIRCUIT VOLTAGES Vs LIGHT GENERATED CURRENT

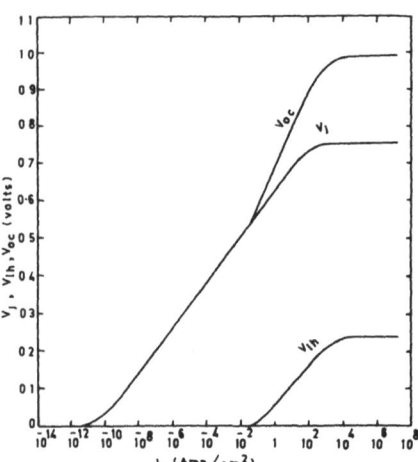

FIG 2-OPEN CIRCUIT VOLTAGES Vs LIGHT GENERATED CURRENT

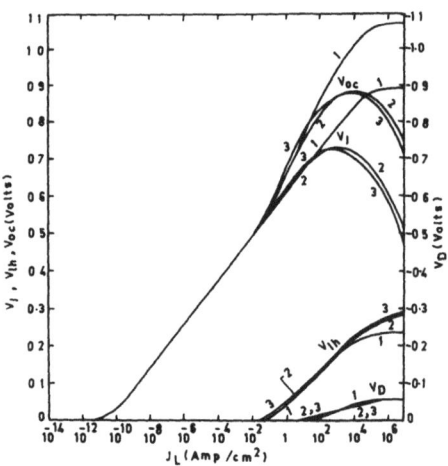

FIG 4-OPEN CIRCUIT VOLTAGES Vs LIGHT GENERATED CURRENT

SERIES RESISTANCE ANALYSIS OF CONCENTRATOR CELLS UNDER HIGH INJECTION CONDITIONS

J.M. Ruiz, M. Cid, A. Cuevas and A. Luque
Instituto de Energía Solar - E.T.S.I.T. - Universidad Politécnica Madrid

Summary

Two analytical models for low and high-injection levels, respectively, and an exact model for variable injection level are used to analyze the performance of silicon solar cells with either low or high base resistivity under concentration.
The ranges of aplicability of the two analytical models are determined by comparison with the exact model. Specially interesting is the very good agreement observed between the exact model and the high-injection one, because both are new and constitute a simplification of previous theories.
The theoretical results are that a high resistivity of the base region does not degrade the performance of the cell even at high concentration levels and that a low resistivity cell and a high resistivity cell have similar efficiencies.
Experimental results are included that agree qualitatively with the theoretical predictions.

1. INTRODUCTION

Most concentrator cells are presently done with low resistivity bases, among other reasons because of the high series resistance that is assumed to entail the use of low-doped bases. This assumption however, is based on conventional low injection models and a constant (not modulated) series resistance, and fails at sufficiently high concentration factors.

The purpose of this paper is to analyze the feasibility of high base resistivity cells, for concentration, paying particular attention to those effects related to high injection conditions such as conductivity modulation and reduction of recombination current components. For that, both theoretical and experimental results are presented.

2. THEORETICAL MODELS

For comparison purposes, the theoretical performance of concentrator cells with different base resistivities cannot be analyzed by means of just one simple model. Low base resistivity cells at moderate concentration factors are expected to work under low injection conditions specially near both the short-circuit and maximum power points of their I-V characteristics. On the contrary, near open-circuit conditions, high concentration factors or high base resistivities lead, in most cases, to high injection levels and (for B.S.F. cells) near p-i-n behaviour. Between those two extremes, for wich analytical, quite simple models can be used, a variety of intermediate situations nead to be considered.

Three different models are used in this paper for the theoretical analysis. One of them is based on the well known conventional low injection theory of p-n junctions and BSF effects. Corrections for a constant series resistance due to ohmic drops through the base are also introduced. A p-i-n high injection model (1) for high base resistivity cells leads, like the low injection one, to a linear carrier diffusion equation with constant coefficients. Modulated series resistance effects are calculated after the carrier profile is known for every particular case. The third model takes into account the possibility of variable injection conditions.

This variable-injection model, recently developed by one of us and still unpublished, is quasi-analytical, more exact but more complicated than the two preceeding ones. It considers lifetime, ambipolar diffusion coefficient, and ohmic field as injection-dependent parameters leading to a quasi-linear diffusion-drift equation for minority carriers. In the two extreme situations it coincides with the low or high injection models. For intermediate cases, it explains some recently published experimental results such as concentration enhancement of the internal spectral response and superlinearity of the uncompensated photocurrent with irradiance (2)(3) shunt resistance-like effects due to a lack of conductivity modulation at high irradiance levels near shortcircuit conditions (4), etc. The latter leads to an infralinearity of the short-circuit current at very high irradiances.

Since the purpose of this work is to analyze the base region of the cell, no attempt has been made to calculate the contribution of the highly doped emitter regions. Regarding the photogenerated current, the front emitter has been considered of negligible width. The contribution to the dark recombination current of both emitters has been included through the corresponding dark saturation currents J_{of} and J_{ob} rather than through the common effective surface recombination velocity wich, as it is well known is an injection-dependent concept.

3. THEORETICAL RESULTS

To do the calculations, we have chosen $J_{of} = 2\times 10^{-13}$ A cm^2 and $J_{ob} = 2.1\times 10^{-12}$ A cm^{-2}, based on previous experimental data (5). The values for the electron and hole mobilities have been taken from ref. 6. The electron and hole lifetimes have been assumed to be equal and to attain the fundamental limits calculated by Fossum et al. (7). The results of the three models are plotted in Figs. 1 to 4.

The low-injection model has been applied to the analysis of a cell with 0.3 Ωxcm base resistivity, and it proves to be correct up to a concentration of 30 suns. For higher irradiances the exact model predicts a super linearity of the short-circuit current (see Fig. 1) and an improvement in the fill factor because of resistivity modulation (see Fig. 3). Both factors lead to an increase in efficiency, and to a shift of the optimum concentration to 300 suns, as shown in Fig. 4.

To analyze high base resistivity cells the variable injection or the high injection models has to be used. The calculations for a 10 Ω.cm cells are also shown in Figs. 1 to 4. It can be seen that the high injection model is close to being correct for irradiances greater than 5 suns, at which the base region begins to be in high-injection level even for the maximum-power point. Although there is a small discrepancy in J_{sc} and FF, both models predict the same tendencies.

From an engineering viewpoint, the important result is that, for the particular values of the parameters used in the calculations, there is not a significant influence of the resistivity of the base in the performance of solar cells (see Fig. 4), although the lower resistivity is slightly preferable in the range of concentrations of practical interest. Looking at Figs. 1, 2 and 3 we see that the 0.3 Ω.cm cell has a lower short-circuit current, a similar open-circuit voltage and a significantly higher fill factor than the 10 Ω.cm cell.

4. EXPERIMENTAL RESULTS

To study the effect of the base resistivity on the efficiency, p^+nn^+ cells were made with FZ, n type silicon wafers of 2, 10 and 100 Ω.cm.
The electrical characteristics of some of these cells (simulated AM1.5 25°C conditions) are: J_{sc} = 30 mA cm^{-2}, V_{oc} = 616 mV, FF = 0.81 and η = 14.9% for a 2 Ω.cm cell and J_{sc} = 31 mA cm^{-2}, V_{oc} = 612 mV, FF = 0.77 and η = 14.7% for a 100 Ω.cm cell. A maximum efficiency of 16.5% has been measured at 25 AM1.5 suns and 25°C in a 2 Ω.cm cell; at 60AM1 suns the efficiency is still 16.3%.

The variation of the efficiency with the illumination level measured for a 10 Ω.cm cell has been plotted in Fig. 5 besides the theoretical predictions made with the high-injection model. To do the calculations, we have used J_{of} = 7.5x10^{-13} A cm^{-2} and J_{ob} = 1.2x10^{-12} A.cm^{-2} to obtain a good fit to the measured open-circuit voltages; the other parameters being the same as in section 3.

A preliminary calculation that does not include the contribution of the emitter and of the metal grid to the series resistance predicts an efficiency higher than the experimental one, showing that the fall in efficiency is not due to series resistance effects in the base region. In a second calculation the contribution of the emitter and the metal grid were included and a more reasonable fit to the experimental values was obtained. However, there is still a discrepancy that can be attributed to technological problems that are present in this particular cell and produce a degradation of the fill factor. In fact we have measured higher fill factors at 1 sun in other specimens with high-resistivity base (0.80 for a 100 Ω.cm cell).

To show the limits imposed by series resistance effects in the front emitter and metal grid, we have also plotted in Fig. 5 the variation of efficiency with concentration for one of our best cells that, however, has a base resistivity of 2 Ω.cm and is not in high injection conditions. Even for this good cell, the efficiency falls more steeply than the theoretical prediction, although such a fall can be due to undesired effects during the experimental measurements such as non-homogeneous illumination and/or differences between the measured temperature and the actual cell's temperature.

As a general conclusion, the experimental results show that the resistivity of the base has not a great influence on the behaviour of the cells, at least in the range of 2 to 100 Ω.cm, although there is a tendency to higher open-circuit voltages and fill factors as the base resistivity decreases.

To further demonstrate that high-resistivity base material can be used to make cells for concentrating systems, we have made several cells with n type, 600 Ω.cm silicon. The cells are small (1 cm^2) and have a very intrincate metallization grid pattern to obtain a low contribution to the series resistance from the emitter region and from the metal grid.

The experimental variation of the efficiency with the illumination level has been plotted in Fig. 6 besides the theoretical predictions, which have been done now with J_{of} = 5.7x10^{-13} A cm^{-2} and J_{ob} = 2.3x10^{-12} A cm^{-2} using the high-injection model.

Because of the ohmic drops in the base region, the model predicts a maximum efficiency at 150 suns for this particular cell. At this concentration level the ohmic drops in the emitter and in the grid are small (see curve) and it can be said that the behaviour of the cell is governed by high-injection effects.

There is again a misfit between theory and experiments. In the low-concentration range this misfit is due to shunt-resistance effects (see a higher experimental point measured at 1 sun before encapsulation) and other possible degrading effects associated with the small area of the cell. In the high-concentration range the efficiency has been probably lowered by non-homogeneous illumination and temperature-variations effects during the measurement.

In spite of the above mentioned misfit, the experiment is relevant to demonstrate qualitatively that high-resistivity cells can have a high efficiency at irradiances as high as 500 suns.

5. CONCLUSIONS

High-resistivity cells do not behave exactly in the same way as low-resistivity cells do under concentrated sunlight. The respective values of J_{sc}, V_{oc} and FF are different and they vary with concentration in different ways. However there is a trade-off between these three factors and the resulting efficiency is very similar for both kind of cells in the range of concentrations of practical interest.

For engineering purposes, this conclusion can be stated saying that the resistivity of the base region is not a critical parameter to design BSF solar cells for concentration.

This theoretical result has also been demonstrated qualitatively with experimental cells.

Regarding the theoretical analysis of solar cells under concentration, two more important conclusions are drawn from this work: a) The conventional low-injection theory underestimates the efficiency at moderate concentrations; even for a 0.3 Ω.cm cell it is only applicable up to 30 AM1.5 suns. b) The new analytical high-injection model is correct starting from the concentration at which the base enters in high-injection; this happens at 5AM1.5 suns for a 10 Ω.cm cell.

REFERENCES

1 A. Luque, J. Eguren, Solid St. electron. In press.

2 B.C. Chambers, C.E. Backus, Proc. 3rd E.C. Photov. Solar Energy Conf., p. 418 (1980).

3 R.D. Nasby, C.M. Garner, H.T. Weaver, F.W. Sexton, J.L. Rodríguez, Record of the 15th IEEE Photov. Spec. Conf. p 132 (1981).

4 R.J. Schwartz, M.K. Lundstrom, R.D. Nasby, IEEE Trans. on Electron Devices, ED-28, p. 264 (1981).

5 J. Eguren, J. del Alamo, A. Cuevas, A. Luque, Record of the 15th IEEE Photov. Spec. Conf. p. 1343 (1981).

6 J.C. Plunkett, J.L. Stone, A. Leu, Solid St. Electron. 20, p. 447 (1977).

7 J.G. Fossum, D.S. Lee, Record of the 15th IEEE Photov. Spec. Conf. p. 120 (1981).

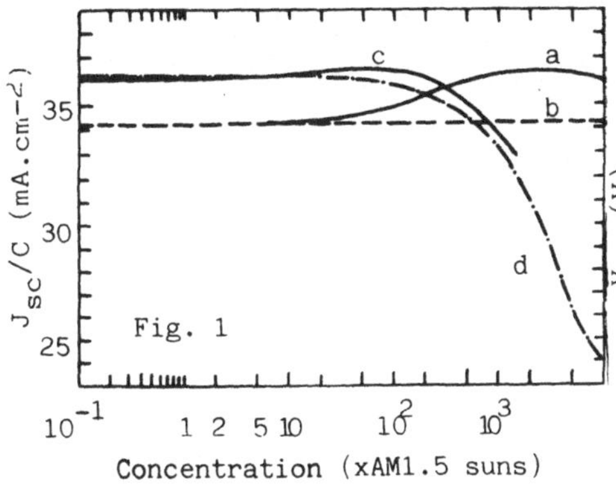

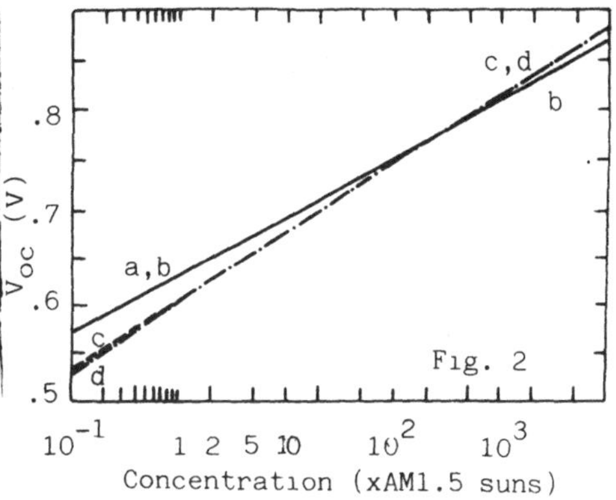

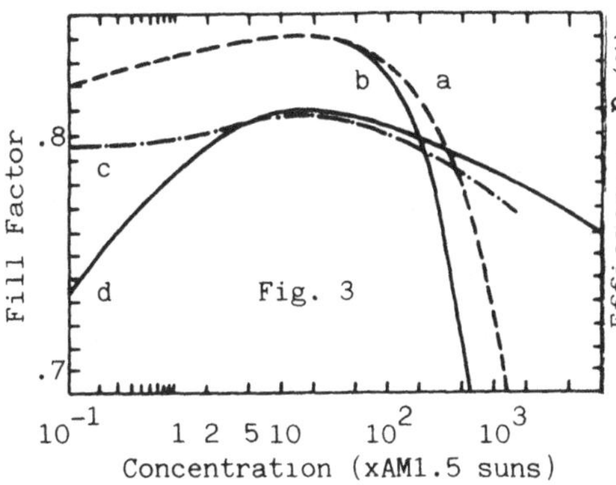

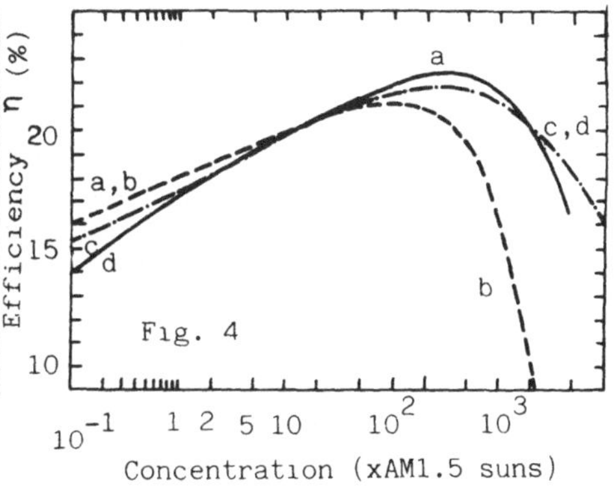

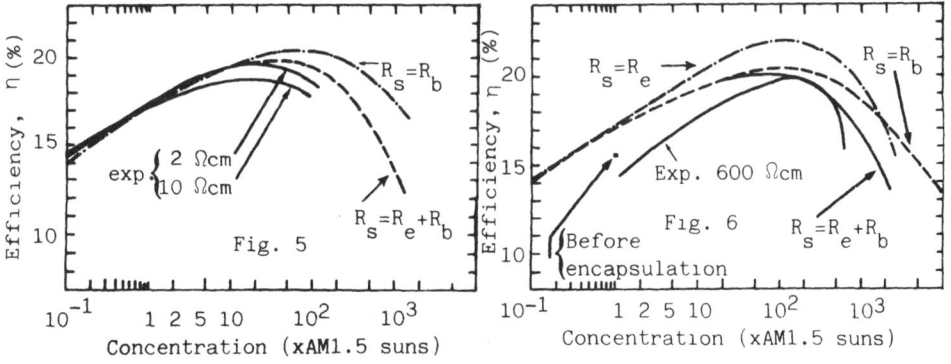

Results of the theoretical models for short-circuit current (fig. 1), open-circuit voltage (fig. 2) fill-factor (fig. 3) and efficiency (fig. 4) versus concentration.
  a: variable injection, $\rho_b = 0.3\ \Omega.cm$; b: low injection $\rho_b = 0.3\ \Omega.cm$.
  c: variable injection, $\rho_b = 10\ \Omega.cm$; d: high injection p^+-i-n^+.

Measured and calculated efficiencies for 2 $\Omega.cm$ and 10 $\Omega.cm$ (fig. 5) and 600 $\Omega.cm$ (fig. 6) base resistivity cells. The calculations include series resistance effects as indicated on the figures.

- 515 -

THE INFLUENCE OF GRAIN-BOUNDARY RECOMBINATION AND GRAIN SIZE ON
THE I(V)-CHARACTERISTICS OF POLYCRYSTALLINE SILICON SOLAR CELLS

M. Böhm, R. Kern, H.G. Wagemann
Institut für Werkstoffe
der Elektrotechnik
Technische Universität Berlin
Jebensstr.1, D-1000 Berlin 12
West-Germany

SUMMARY

The grain-boundary recombination velocity and the grain size determine the I(V)-characteristics and the energy-conversion efficiency of a polycrystalline cell in addition to those parameters known from the monocrystalline cell. The influence of these two additional parameters is analysed quantitatively by means of a model calculation. The calculation is based on an idealized model structure with grain boundaries perpendicular to the pn-junction. The analysis of the carrier-flow problem requires the solution of the 2-dimensional diffusion equation which becomes inhomogeneous under optical illumination. It is shown by means of the balance of charge and currents that the physical properties of the grain boundaries may be expressed by one single parameter, the grain-boundary recombination velocity. Thus the boundary conditions of the carrier-flow problem consider the grain boundaries as recombination planes. The theoretical results for the I(V)-characteristics and the spectral sensitivity are compared to experimental results obtained at commercial polysilicon cells. Calculations and experiments show that the maximum value of the spectral sensitivity shifts to shorter wavelengths for increased recombination and reduced grain size. Furthermore we find the reciprocal slope factor of the I(V)-characteristics to be 1 for mono and polysilicon cells except for small current densities.

1. MODEL STRUCTURE / BOUNDARY-VALUE PROBLEM

Fig. 1 shows the proposed model structure of a polysilicon cell which describes a large central area of technical solar cells by means of the mutual perpendicular orientation of the pn-junction and the grain boundaries [1]. The charge within the grain boundary causes space-charge regions (SCR) adjacent to the metallurgical grain boundary. Furthermore the grain boundary is regarded as

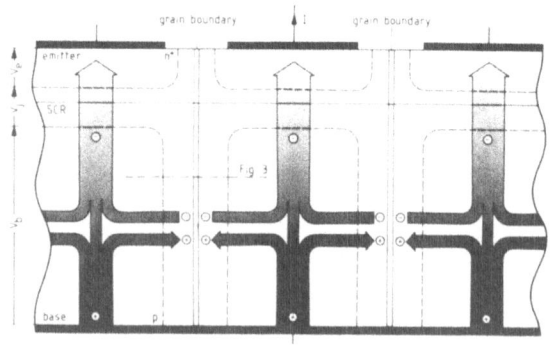

Fig.1 Idealized microstructure of a polycrystalline silicon solar cell

a recombination plane, and as a consequence excess carriers flow to the grain boundary from both sides in order to recombine there. Across the grain boundary no current flow is assumed as long as potential and concentration of carriers are symmetrical within the microcrystallites adjacent to the grain-boundary.

Referring to the geometry of fig. 2 the boundary-value problem to be solved will be defined. Assuming the validity of the Shockley conditions, the carrier concentration profile within the base region of the microcrystallite is determined by the 2-dimensional differential equation for diffusion. With an exponentially decaying rate g of optical generation corresponding to the absorption coefficient α and considering the emitter as a dead layer of the width y_e an inhomogeneous linear differential equation of the elliptic type has to be solved.

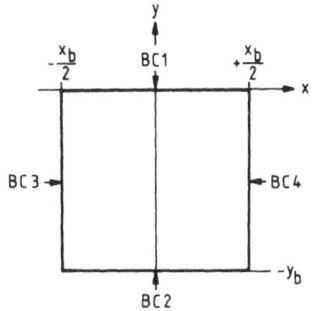

Fig 2 Geometry of the boundary-value problem

(1) $\dfrac{\partial^2 \Delta n}{\partial x^2} + \dfrac{\partial^2 \Delta n}{\partial y^2} - \dfrac{1}{L^2} \Delta n = \dfrac{g(\alpha)}{D} \exp[\alpha(y - y_e)]$

L diffusion length
D diffusion constant

The general solution will be subjected to the boundary conditions 1 to 4.

Boundary Condition 1

Within the plane y=0 the SCR-margin of the pn-junction is located where the minority carrier concentration is altered by the Boltzmann factor. From the solution of the 1-dimensional inhomogeneous diffusion equation in x the following expression is used.

(2) $\Delta n(y = 0) = [1 - C \cosh(x/L)] \cdot n_o \exp(V_j/V_T)$

(2) also satisfies the boundary conditions 3 and 4. The concentration Δn (y=0) decayes into the direction of the grain boundaries and as a consequence the junction voltage V_j as well as the ohmic voltage drops V_e and V_b depend on x. This effect, generally described by an offset voltage [2], is negligeable for realistic recombination velocities in the grain boundaries and realistic grain sizes.

Boundary Condition 2

Within the plane $y = -y_b$ an ohmic contact is assumed.

(3) $\Delta n(y = -y_b) = 0$

Boundary Conditions 3 and 4

The diffusion current approaching the grain boundary equals the recombination current which is assumed to depend linearly on the grain-boundary recombination velocity s_{gb}. A factor 2 makes allowance for diffusion from both sides of the grain boundary.

(4) $\quad 2qD\left(\dfrac{\partial \Delta n}{\partial x}\right)_{x=\mp x_b/2} = \pm qs_{gb}\Delta n(x=\mp x_b/2)$

2. THE GRAIN BOUNDARY AS RECOMBINATION PLANE

Before we continue in dealing with the boundary-value problem a correlation has to be established between the recombination velocity s_{gb} and the inherent physical properties of the grain boundaries, hence the energy distribution of grain-boundary states and their recharge character.

Without restrictions of generality we assume donor-type states within the metallurgical grain boundary (width $2d_{gb}$ corresponds to some lattice constants). From the local behavior of space charge density ρ, field E_o, and potential Ψ_o the energy-band diagram and the distribution of mobile carriers will be obtained (fig. 3). We make up the balance of currents for $x=\pm d_{gb}$. For thermal equilibrium as well as for steady-state non-equilibrium conditions with symmetrical potential the total current is zero.

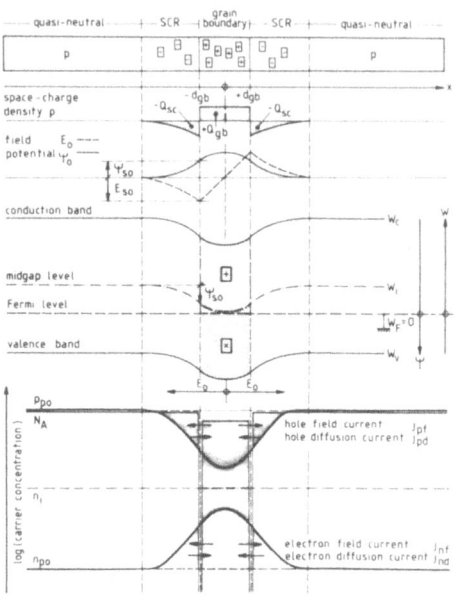

Fig 3 Model of a depletion-type grain boundary (thermal equilibrium)

(5) $\quad j_{pf} + j_{pd} + j_{nf} + j_{nd} = 0$

In addition for thermal equilibrium the field and diffusion currents of each type of carriers compensate each other.

(6) $\quad j_{pf} + j_{pd} = 0$ and $j_{nf} + j_{nd} = 0$

For the case of injection excess recombination occurs.

(7) $\quad j_{pf} + j_{pd} = j_{prec} \neq 0$ and $j_{nf} + j_{nd} = j_{nrec} \neq 0$

Hole and electron recombination currents are equal but show opposite sign.

(8) $\quad j_{prec} = - j_{nrec}$

With the above mentioned linear dependence of recombination current we get for s_{gb} equation (9).

(9) $\quad s_{gb} = \dfrac{j_{nf} + j_{nd}}{q\Delta n} = \dfrac{j_{pf} + j_{pd}}{q\Delta n}$

Assuming that s_{gb} does not depend upon excitation and taking this into account by

(10) $\frac{\partial s_{gb}}{\partial \Delta n} = 0$ and $s_{gb} = \lim\limits_{\Delta n \to 0} \frac{j_{nf} + j_{nd}}{q \Delta n}$

we succeed in evaluating formula (9) without any assumption upon the level of excitation. We obtain (11).

(11) $s_{gb} = \mu E_{so}$

E_{so} is the field strength for thermal equilibrium and μ is the effective mobility in the vicinity of the metallurgical grain boundary. E_{so} may readily be correlated with the energy distribution of grain boundary states N_{gb} by means of the Gaussian theorem and the balance of charges.

(12) $Q_{gb} = -2Q_{sc}$

Fig. 4 shows a calculated $s_{gb}(N_{gb})$-diagram for donor-type states constantly distributed in energy and for the case of constant effective mobility.

3. MINORITY CARRIER PROFILE IN THE BASE REGION

The above defined boundary-value problem can be solved analytically. We obtain the general solution by superposition of the solution of the homogeneous problem with inhomogeneous boundary conditions and the solution of the inhomogeneous problem with homogeneous boundary conditions. The solution can be written as:

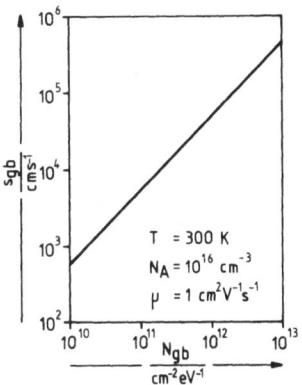

Fig 4 Calculated grain-boundary recombination velocity s_{gb} as a function of grain-boundary state density N_{gb} for donor-type states constantly distributed in energy
N_A impurity concentration
μ effective mobility

(13) $\Delta n = \left\{ \sum\limits_{k} f_{gbk}(y) \cos c_k x \right\} \cdot \left(\exp \frac{V_j}{V_T} - 1 \right) +$

$+ g(\alpha) \cdot \left\{ \sum\limits_{k} g_{gbk}(y,\alpha) \cos c_k x \right\}$

The eigenvalues c_k are solutions of the transcendental equation:

(14) $2 \frac{D}{s_{gb}} c_k = \mathrm{ctg} \frac{c_k x_b}{2}$

The inhomogeneous solution may also be developped in orthogonal functions $\mathrm{sinc}_i y$ with $c_i = i\Pi/y_b$, $i \in N$. This variant, however, converges extremely slowly for high values of the absorption coefficient α.

Fig. 5 presents two typical profiles for monochromatic illumination and for short-circuit conditions.

4. SPECTRAL SENSITIVITY

On the basis of the minority-carrier profiles we expect that the photocurrent losses due to the grain-boundary recombination might be more severe

in the red part of
the spectrum than in
the blue one. For
blue light the concentration gradient
pointing to the
junction is much
steeper than the one
pointing to the grain
boundaries. Only a
relative small fraction of the total
photogenerated
carriers will diffuse
into the grain boundaries and recombine
there. For red illumination, however,

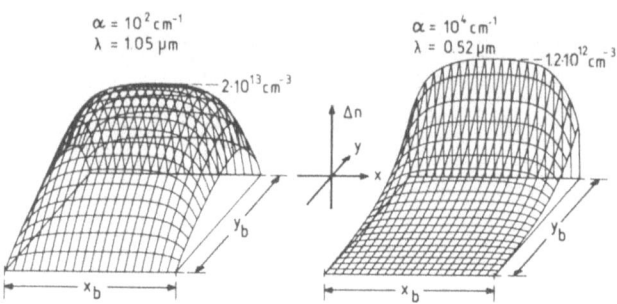

Fig 5 Short-circuit minority-carrier profiles in the base region for monochromatic irradiance of 100 mWcm^{-2}

$s_{gb} = 10^4$ cms^{-1}, $L = 50$ μm, $x_b = 500$ μm, $y_b = 200$ μm, $y_e = 0.2$ μm

the two concentration gradients are comparable and a relative high fraction
of excess carriers will recombine in the grain boundaries.

The photocurrent is obtained by integrating the
components of the diffusion
current vector perpendicular
to the transition region for
$V_j = 0$. From that expression
the absolute spectral sensitivity $S_{abs}(\lambda)$ plotted in
fig. 6 may easily be derived.
For a first approximation the
typical behavior of S_{abs} may
be described phenomenologically by a reduction of the
effective diffusion length
due to surface recombination.
The blue shift of the maximum
of the spectral sensitivity
was validated by locally resolved S_{abs}-measurements at

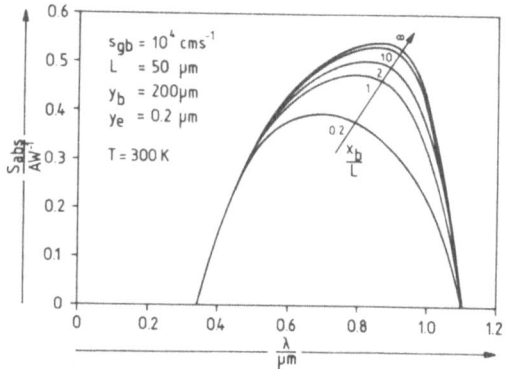

Fig 6 Spectral sensitivity S_{abs} versus wavelength λ for a grain-boundary recombination velocity of 10^4 cms^{-1} with the ratio of grain size x_b and diffusion length L as parameter

commercial 100 cm^2-polysilicon-cells. Variations of the S_{abs}-peak up to 40
nm depending on the grain size were observed for one and the same cell.

5. PHOTOVOLTAIC OUTPUT CHARACTERISTIC

We obtain the reverse saturation current from the non-illuminated
carrier profile by means of the above mentioned integration in x. For the
computation of the white light photocurrent another integration in λ has to
be executed taking into account the spectral distribution of irradiance.

Fig. 7 shows that the photocurrent decayes only about 4 % below the
monocrystalline value for an x_b/L-ratio of 10 even for highest recombination rates. The increase of the open-circuit voltage which is obvious for
$x_b/L = 1$ is due to the reduction of the minority carrier concentration at
the SCR-margin in presence of grain-boundary recombination. For the same
current a higher voltage V_j has to be applied in presence of substantial
grain-boundary recombination than is necessary for the case of negligeable
grain-boundary recombination.

Our calculation which is based on Shockley's model results in a value N=1 for the reciprocal slope factor of the I(V)-characteristic. Fossum and Lindholm [3] obtain N=2 for the grain-boundary recombination current as a result of their model calculation and they refer to experiments done by Mazer, Neugroschel and Feldman.

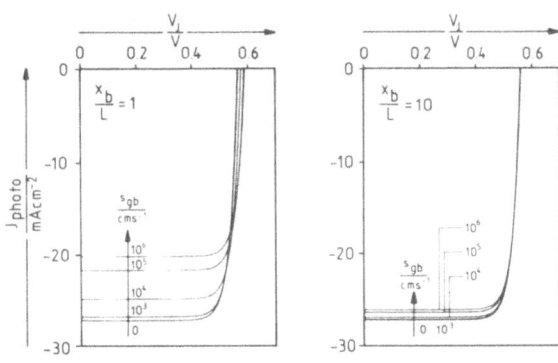

Fig 7 Calculated photovoltaic output characteristics for AM1 spectrum (Thekaekara [8]) for 2 different x_b/L-ratios and s_{gb} as parameter $N_A = 10^{16} cm^{-3}$, $L = 50 \mu m$, $y_b = 200 \mu m$, $y_e = 0.2 \mu m$, $T = 300 K$

For that reason we examined N as a function of current density at several commercially available SILSO and SEMIX cells and at monocrystalline cells. In order to eliminate the influence of the excitation dependent series resistance N(j) was evaluated from the $I_{sc}(V_{oc})$-characteristics [4,5,6]. The parallel resistance, which also affects the N(j)-evaluation, was assumed to be constant and determined from the slope of the I(V)-characteristic for small bias values. Besides extreme care was taken on temperature control. Generally we obtained equal N(j)-characteristics for mono and polysilicon cells. Fig. 8 shows a typical plot for two equally manufactured 25 cm^2-cells which differ only in the original material (mono/poly-Si). Except for small current densities both cells show N≈1.

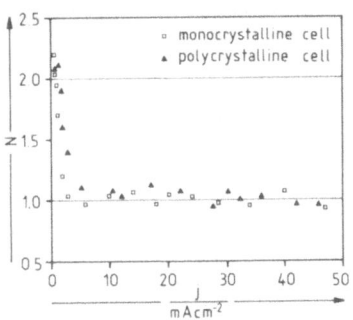

Fig 8 Experimental values of the reciprocal slope factor N as a function of the current density j

As furthermore both cells show a similar increase of N from N≈1 to N≈2 for small current densities, corresponding to small bias values, we conclude that the recombination in the transition region according to the Sah, Noyce, Shockley model [7] is essential for the N(j)-behavior of relatively coarse crystalline cells.

7. LITERATURE

[1] W. Schmidt, G. Friedrich, K.D. Rasch, Proceedings of the 3rd E.C. Photovoltaic Solar Energy Conference, Cannes, 664 (1980)
[2] J.M. Aitchison, F. Berz, Solid-St. Electron. 24, 795 (1981)
[3] J.G. Fossum, F.A. Lindholm, Trans. IEEE, Ed-27, 692 (1980)
[4] M. Wolf, H. Rauschenbach, Advanced Energy Conversion, vol. 3, 455 (1963)
[5] J.A. Mazer, A. Neugroschel, F.A. Lindholm, Trans. IEEE, ED-28, 1530 (1981)
[6] H. Scheer, dissertation TU-Berlin 1982, to be published
[7] C.T. Sah, R.N. Noyce, W. Shockley, Proc. IRE 45, 1228 (1957)
[8] M.P. Thekaekara, Institute of Environmental Sciences, (1974)

ACKNOWLEDGEMENT

The authors wish to thank Dipl.-Ing. H. Scheer for helpful discussions. Financial support of the Deutsche Forschungsgemeinschaft is gratefully acknowledged.

DIFFUSION LENGTH OF MINORITY CARRIERS IN SCANNING ELECTRON BEAM ANNEALED SILICON

H.J. SMITH [a] A. BONTEMPS [b] and R. CILLIERS [a]

[a] CSP, Council for Scientific and Industrial Research, P O Box 395 Pretoria, 0001, South Africa

[b] DRF, Centre d'Etude Nucléaire de Grenoble, 85X 38041, Grenoble Cedex, France

Summary

Ion implantation has advantages for solar cell production, but necessitates an annealing step. Various new transitory annealing methods have appeared recently. A particularly attractive method is multi-scan electron beam annealing of thermally isolated wafers. Energy is applied homogeneously over the whole target surface and the temperature rises throughout the thickness. Backscattering analysis shows good recrystallization in seconds. However the effect of this total heating on the diffusion length (L_D) must be investigated particularly in view of the degradation of L_D due to high temperature oven annealing. The semiconductor-electrolyte diode method was set up to measure the current generated in the cell due to the creation and diffusion of carriers in the silicon under photon irradiation. Comparison with a theoretical model yields L_D. It appears that 3mA.cm^{-2} of 15keV electrons recrystallizes damage in 2.5 seconds and does not decrease L_D in the bulk. In 4 seconds the L_D decreases and dopant diffusion occurs. On technical grounds this method can thus be applied for solar cell production.

1. INTRODUCTION

In research about photovoltaic solar cells two goals are pursued: low cost of materials and processing and high conversion efficiency. High efficiency requires a high value of the minority carrier diffusion length (L_D), which is thus a key parameter.

Ion implantation as a means of doping has definite advantages such as a shallow, well controlled concentration distribution of the dopant impurities, and the fact that it is essentially a low temperature process. High temperature process steps can adversely affect the L_D.

However, implantation necessitates an annealing step to remove the radiation damage. This should be done while retaining the advantages of implantation. New transient annealing techniques have been developed in recent years, such as laser and electron pulse annealing. This melts a surface layer of the material which recrystallizes from the melt. While the temperature of the bulk is not increased, rearrangement of dopants in the molten layer can take place, changing the doping profile. This is avoided by scanned laser or electron beam annealing which operates in the solid phase. One mode of electron beam annealing, which could be called multi-scan electron beam (mseb) annealing of thermally isolated samples is particularly attractive (1,2). It is easier and quicker to use thermally isolated wafers, than to assure good thermal contact. In this variation of mseb the isolated wafer is raster scanned with an electron beam continuously for a duration of the order of seconds. The x- and y-scan frequencies is such that the application of energy can be considered as homogeneous over the target surface. This method combines the advantages of an electron beam (insensitive to reflection and surface inhomogeneity like lasers) and solid phase recrystallization, with simplicity, speed, easy and continuous control of the irradiated area and is amenable to automation. Previous work (2, 3) has shown that a beam intensity of $3mA.cm^{-2}$ of 15keV electrons gives complete recrystallization of a 120keV 3×10^{15} ions.cm^{-2} arsenic implant in silicon in 2 seconds.

However the temperature of the silicon sample rises throughout the thickness to around 1000 °C (4). This raises serious questions about the effect of mseb on the L_D and, accordingly, the applicability of this method for solar cell production. This paper reports first results of an investigation of mseb on the L_D of single crystal silicon.

2. EXPERIMENTAL

For the investigation three different single crystal wafers (A, B and C) of Czochralski-grown p-type 10 Ω.cm $\langle 100 \rangle$- orientation and thickness of 300μm were used. Wafer A was mseb treated without any implantation, while B and C received front surface implantations of 2×10^{15} ions.cm^{-2} of 30keV arsenic ions.

From each wafer samples of 9×9 mm^2 were cut for mseb treatment in a manner and a system that has been described elsewhere (2, 3). An electron beam 15keV, 6mA and focussed to 200μm is raster scanned with x- and y- frequencies of 10.000 and 200Hz respectively to yield a current density of $3mA.cm^{-2}$. Samples are mounted thermally isolated and irradiated for a specified duration. In preliminary experiments it was shown by Rutherford backscattering analysis that irradiation on either the front or back surface at these energy densities gives the same result. This shows that the temperature increases throughout the thickness of the sample.

3. MEASURING THE DIFFUSION LENGTH

To investigate L_D a system was constructed based on the idea of work by Fabre and Mautref (5). A semiconductor-electrolyte diode is formed between the silicon and a hydrofluoric acid solution which contains an auxiliary platinum electrode.

The silicon is irradiated through the electrolyte with monochromatic light of variable wavelength and known intensity, taking into account absorption and reflection losses. The absorption of the photons in the silicon creates electrons and holes which diffuse independently until trapped. Those diffusing electrons that survive trapping until they reach the silicon-electrolyte interface give rise to a current in the suitably polarised external circuit.

The equations for creation, diffusion and recombination of the minority carriers is solved numerically by means of a computer programme for different values of L_D as parameter (6). This leads to a family of curves of the current generated as a function of wavelength (spectral response). Comparison of an experimentally determined curve with the set of calculated curves yields the effective diffusion length in the silicon. The result is in fact an effective value of L_D because it assumes a constant value for L_D throughout the thickness of the sample. It is this effective L_D which is important for the solar cell efficiency.

4. RESULTS

Figure 1 shows the result of a Rutherford backscattering cum channelling analysis with 1.2 MeV He^4 particles on an arsenic implanted sample before and after 2.5 seconds mseb treatment. The damaged surface layer disappears and the arsenic passes into substitutional sites.

The spectral response curves of the unimplanted samples (A) are shown in Figure 2. A_1 (not heated at all) and A_2 (mseb annealed for 2.5 seconds) give practically indistinguishable curves yielding L_D values of 240μm and 230μm respectively in detailed analysis. After 4 seconds of mseb treatment the spectral response is lower (A_3) corresponding to an L_D of 180μm. This shows that longer heating leads to a degradation. These L_D values are summarized with others in the left hand part of Table I.

In Figure 3 two results for B samples (arsenic implanted) are shown. Curve B_1 in the spectral response curve as-implanted. This gives an effective L_D of 24μm. This low value is the result of the damaged implanted layer. After 2.5 seconds mseb heating of this sample the L_D value improves to 150μm (Table I). (This is the case analyzed with Rutherford backscattering in Figure 1). An annealing of 4 seconds results in a degradation to 100μm. The other implanted sample (C) shows the same behaviour.

Table I: Diffusion length in microns after various mseb annealing times for unetched wafers and for wafers with the doped layer stripped.

Wafer \ Annealing time	wafer unetched			doped layer stripped		
	0(sec)	2.5(sec)	4(sec)	0(sec)	2.5(sec)	4(sec)
A: non-implanted	240	230	180	240	230	180
B: arsenic implanted	24	150	100	190	195	160
C: arsenic implanted	22	145	100	175	175	150

The effect of the mseb on the bulk properties is important. To investigate this the influence of the implanted, recrystallized layer on the measurement must be avoided. Removing the implanted unannealed layer on the B sample by means of an HF-etch changes the spectral response curve from B_1 to B_2 (Figure 3). The corresponding L_D value is 190μm.

The evolution of L_D with etching time is shown in Figure 4. In the case of a sample annealed for 2.5 seconds the value increases and after about 25 minutes a constant value is reached. For a sample having been annealed for 4 seconds it takes considerably longer to etch away the doped layer. This corroborates the result previously reported (2) that with a 4 seconds anneal diffusion takes place which broadens the doped layer.

The effect of the mseb annealing on the bulk properties is summarized on the right hand side of Table I where the L_D values after removal of the doped layer is given.

It is important to note that the different bulk values of L_D for A, B and C at zero annealing is not related to having been implanted or not. It is the result of inherent differences in the three wafers at the outset. In a study of 6 untreated, supposedly identical wafers the L_D values varied from 125μm to 240μm (7).

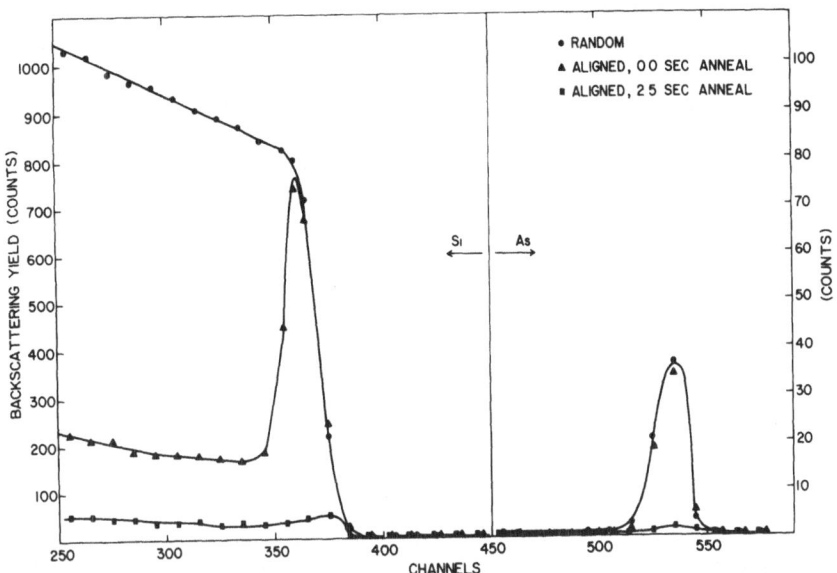

Fig.1: Backscattering spectre for 1.2 MeV He^4 ions incident aligned and random on $\langle 100 \rangle$ silicon 30keV arsenic implanted to 2×10^{15} ions.cm^{-2} before and after 2.5 seconds mseb anneal.

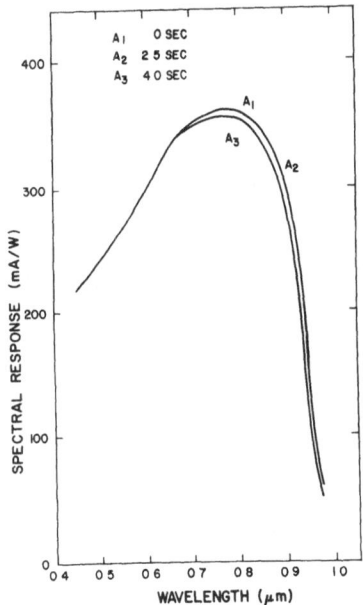

Fig.2: Spectral response curves for un-implanted silicon samples after 0, 2.5 and 4 seconds of mseb anneal with 3mA.cm^{-2} 15keV electrons.

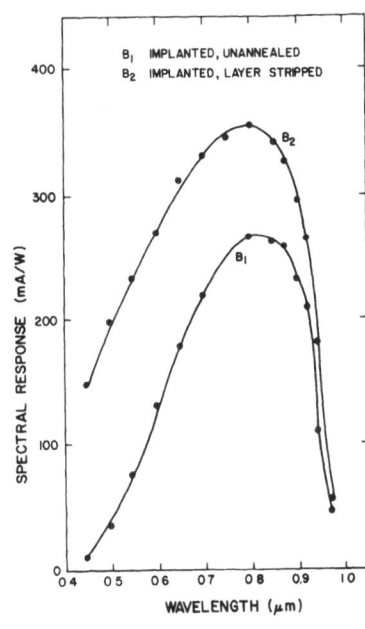

Fig.3: Spectral response curves for silicon implanted to 2×10^{15} ions.cm^{-2} of 30keV arsenic ions. B_1 is as-implanted and B_2 is after stripping of implanted layer (no annealing)

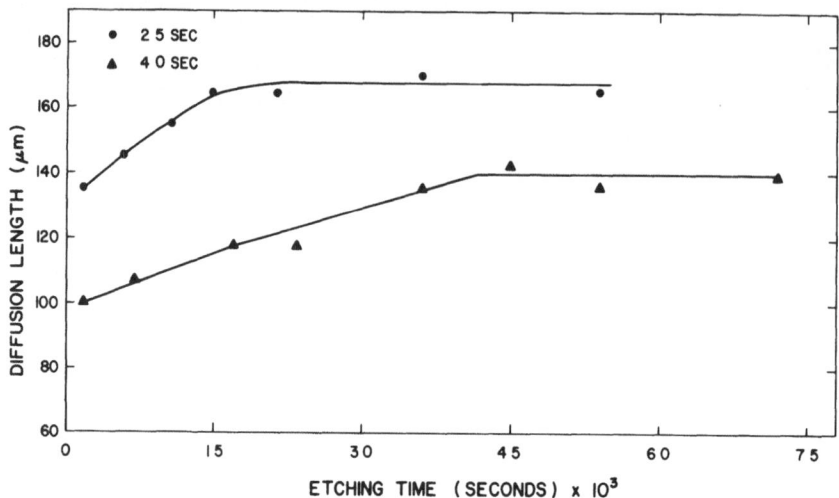

Fig.4: Effective diffusion length at various etching times for 2.5 and 4 seconds mseb annealed silicon samples with 2×10^{15} ions.cm^{-2} of 30keV arsenic implanted.

5. CONCLUSION

All three cases, A, B and C, support the conclusion that this method of multi-scan electron beam annealing does not alter the bulk diffusion length values after a 2.5 seconds anneal (which suffices to completely recrystallize the implantation damage). This is in spite of the fact that the sample is heated throughout its thickness to about 1000 °C. For 4 seconds annealing the diffusion length deteriorates.

On technical grounds this method can thus be considered for solar cell fabrication. Its acceptance will depend on how it will integrate with other process steps and on the economy of the whole procedure.

This method of determining diffusion lengths gives interesting information about the effect of the doped layer. This should be further exploited.

ACKNOWLEDGEMENTS

Thanks are due to Mr M Mautref and to Dr E Ligeon for fruitful discussions.

REFERENCES

1. R.A. McMahon and H. Ahmed, Electron. Lett. 15, 47 (1979)
2. H.J. Smith, E. Ligeon, and A. Bontemps, Appl. Phys. Lett. 37, 1036 (1980)
3. A. Bontemps and H.J. Smith, J. Appl. Phys. accepted for publication 1982
4. A. Bontemps, H.J. Smith and E. Ligeon (to be published)
5. E. Fabre and M. Mautref, Acta Electronica 18 (1975) 331
6. R. Cilliers, MSc Thesis (1982), University of the Orange Free State, Bloemfontein, South Africa
7. R. Cilliers and H.J. Smith, to be submitted to the South African Journal of Physics

PATHOLOGY OF SOLAR CELL CONTACTS

B. ROSS
Bernd Ross Associates
San Diego, CA 92109 U.S.A.

Summary

Energy level diagrams of metal contacts to silicon solar cells are examined, as well as equations governing the tunneling conductance. Symptoms of some solar cell contact problems are mentioned and potential diagnoses are suggested. These include the "flat spot" phenomenon and "S" shaped curves along with the more common series and shunt resistance problems, which all result in poor fill factor and consequently lower efficiencies. Examination of recent solar cells provided with screened thick film copper contacts along with controls, indicate that this contact method does not impose an efficiency limitation.

1. INTRODUCTION

Solar cell contacts provide the electrical interface between the energy conversion element and the load circuit. In order to function efficiently they must provide a low loss path for the generated current. Simultaneously they should be economical in material and fabrication process, and function reliably over long time periods.

Several techniques are available to apply solar cell contacts, including plating,[1] alloying,[2] vacuum evaporation,[3] and thick film screening.[4,5] Despite this variety the final electrical configuration is relatively simple, and is illustrated below.

2. FRONT CONTACT

The front contact of modern solar cells is applied to nearly degenerate n type semiconductor surfaces with impurity concentrations from 10^{19} $\frac{atoms}{cm^3}$ to 10^{21} $\frac{atoms}{cm^3}$ with phosphorus being the donor conventionally in use. The high surface concentration allows better front contacts since the electron tunneling probability is a function of the carrier concentration on both sides of the interface. When a metal coats the surface of a semiconductor, the depletion width within the semiconductor shrinks as a function of carrier concentration, enhancing the tunneling probability. This improves carrier flow through the barrier constituted by the contact potential between contact metal and semiconductor surface.

A portion of this work is the result of one phase of research conducted by the Jet Propulsion Laboratory of the California Institute of Technology for the U.S. Department of Energy through an agreement with the National Aeronautics and Space Administration.

The depletion width, W, can be calculated from

(1) $$W = \sqrt{\frac{2\varepsilon_s}{qN_{D,A}}(V_R + V_D - \frac{kT}{q})} \quad \text{cm} \qquad \text{Ref.6}$$

where ε_s = permittivity of the semiconductor $1.06 \cdot 10^{-12}$ F/cm
q = electronic charge $1.6 \cdot 10^{-19}$ Coulomb
N_D = donor concentrations $1 \cdot 10^{20}$ cm
N_A = acceptor concentration $5 \cdot 10^{18}$ cm^3
V_R = applied bias voltage, $V_R = 0$

(2) $V_D = \phi_B - \phi_n$ = contact potential
ϕ_{BN} = copper - semiconductor barrier = 0.69eV   Ref.7
ϕ_{BP} = 0.46eV   Ref.7

(3) $\phi_n = E_c - E_F$ = 0.0326eV
$\phi_P = E_F - E_V$ = 0.0452eV

(4) E_c = Conduction band edge potential, eV
E_F = Fermi energy, eV
k = Boltzmann constant J/°K
T = Absolute Temperature °K

The depletion width of the front contact is

$$W = 28.9 \text{ Å}$$

in the silicon.
The maximum field ε_m is given by

(5) $$\varepsilon_m = \sqrt{\frac{2qN_{D,A}}{\varepsilon_s}(V_R + V_D - \frac{kT}{q})} = 4.38 \cdot 10^6 \text{ V/cm} \qquad \text{Ref.6}$$

The contact resistance can be calculated from the tunneling conductance

(6) $$R_c = G_o^{-1} = \frac{W}{\sqrt{2m^*\phi}} (\frac{h}{q})^2 \exp \frac{4\pi W}{h}\sqrt{2m^*\phi} \qquad \text{Ref.8}$$

(7) $$= 3.16 \cdot 10^{-11} \frac{W_{\text{Å}}^o}{\sqrt{m^*/m \; \phi}} \exp 1.025 W_{\text{Å}}^o \sqrt{m^*/m \; \phi}$$

where h = Planck's constant
$(m^*/m)_e$ = 0.1905 the effective electronic mass in the ⟨100⟩ direction

(8) $(m^*/m)_h$ = 0.16 the effective hole mass in the ⟨100⟩ direction
$\phi = E_g - \phi_B$ = 0.42eV
E_g = Energy gap of silicon = 1.11eV
Germanium = 0.65eV

The calculated front contact tunneling resistance is

$$R_c = 7.2 \cdot 10^{-6} \Omega \text{ cm}^2$$

Since the barrier voltage values vary among workers, and their magnitude affects the argument of the exponent in equation 6 and 7, the calculated contact resistance is subject to wide variation. The energy band relationships for a copper contact on a silicon front surface doped with $1 \cdot 10^{20}$ phosphorous atoms are shown in Figure 1.

3. BACK CONTACT

When the semiconductor surface has a higher resistivity, as on the back contact, it is necessary to dope the surface of the semiconductor under the electrode metal. In case of high firing temperatures ($\sim 800^\circ C$) an elemental donor impurity may be included in the metal to diffuse into the semiconductor during the firing step. Since only the surface needs to be doped, the donor atom must become a substitutional impurity in the silicon lattice within a few lattice spaces of the surface. A brief firing period will usually suffice.

At very low temperatures ($\lesssim 600^\circ C$) it is difficult to obtain a sufficient surface concentration of dopant atoms by diffusion and a regrowth layer may be produced from a eutectic alloy of a dopant metal (aluminum) and a semiconductor. If a lower energy gap material (germanium) is chosen the contact potential V is reduced, providing an improved contact. Figure 2 shows an energy level diagram for a copper contact including an aluminum eutectic alloy (Al-Si or Al-Ge) and forming a regrowth layer with $N = 5 \cdot 10^{18}$ Al/cm^3.

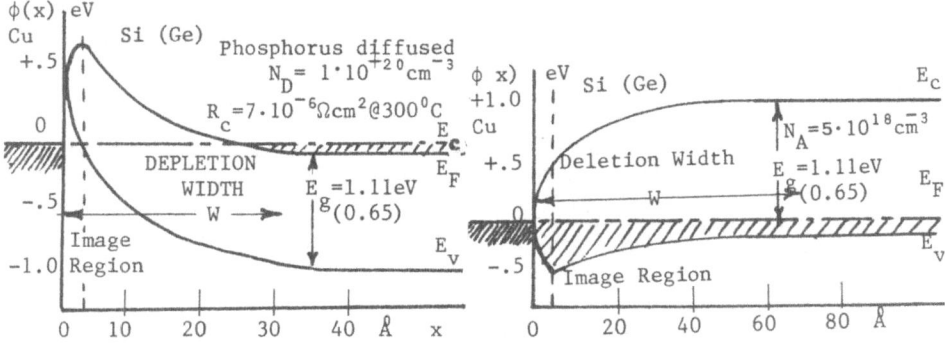

Figure 1. Front Contact   Figure 2. Back Contact

4. FLAT SPOT PHENOMENON

V.G. Weizer et al.[9] have recently published a hypothesis and supporting evidence to explain an IV characteristic anomaly observed on solar cells operated at low temperatures and occassionally at room temperature. The flat spot is a truncation of the maximum power knee of the curve as shown in Figure 3, and is believed[9] to be due to a resistive metal-semiconductor-like junction (MSL) shunting the PN junction. The MSL junction results from the dissolution of silicon by the metalization of the front contact in a low temperature solid state reaction. The authors have shown the phenomenon to be almost independant of the metal system tried for 9 different systems. The silicon surface pitting observed by Weizer et al. is reminiscent of an earlier similar phenomenon observed on germanium surface barrier transistors with indium electrodes[10]. In this case the germanium base dissolved in the indium electrodes, eventually causing device failure. This problem was solved by dimensioning and heat treating the device so that a small germanium-indium alloy region was formed[12] substantially reducing the dissolution rate. Similarly, it can be presumed that a silicon constituent in the solar cell electrodes would suppress the dissolution process.

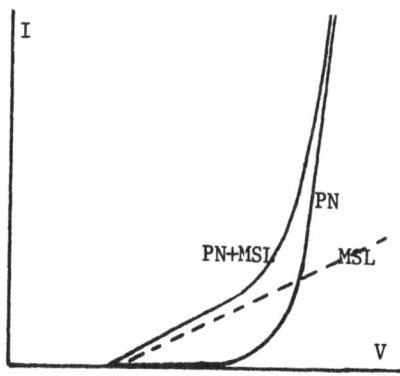

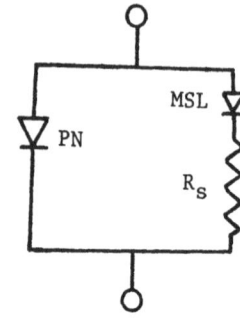

Figure 3. Flat Spot Phenomenon

5. S SHAPED CURVES

During work with screened thick film copper electrodes we have observed S shaped IV curves similar to Figure 4. Such curves were usually seen when contacts were fabricated at relatively low temperatures (near 500°C). The shape of the curve suggested a PN junction or Schottky barrier in series with the current generator of the solar cell. To provide an experimental analog of this situation, a germanium power transistor 2N1136 was connected as shown in the schematic diagram of Figure 5. A normal silicon solar cell was then curve traced by direct connection (curve 1) and through the emitter base diode of the transistor (curve 2). The resulting curve is at least qualitatively similar to Figure 4.

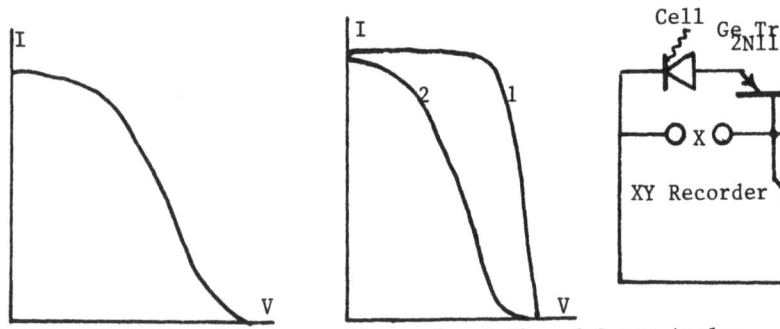

Figure 4. S Shaped Curve   Figure 5. S Shaped Curve Analog

6. PRESENT STATUS OF COPPER THICK FILM CELLS

Figure 6 shows a typical screened copper back contact cell fabricated at 650°C (outer curve) plotted with a typical control cell with titanium-palladium-silver contacts (inside curve). It is apparent that in this system the screened copper contact does not represent an efficiency limitation. Further, the diffusion of copper through the cell has not appeared to either shunt the PN junction or severely degrade the lifetime.

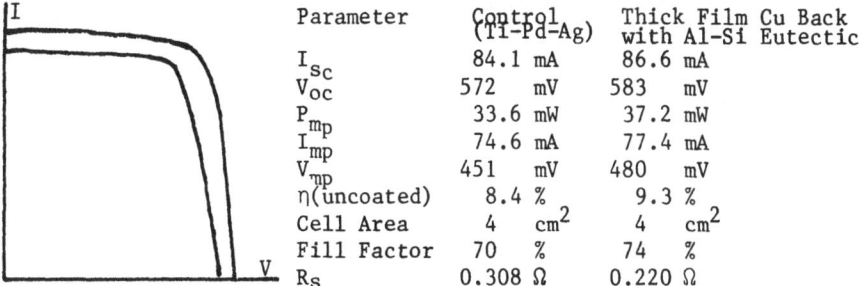

Parameter	Control (Ti-Pd-Ag)	Thick Film Cu Back with Al-Si Eutectic
I_{sc}	84.1 mA	86.6 mA
V_{oc}	572 mV	583 mV
P_{mp}	33.6 mW	37.2 mW
I_{mp}	74.6 mA	77.4 mA
V_{mp}	451 mV	480 mV
η(uncoated)	8.4 %	9.3 %
Cell Area	4 cm^2	4 cm^2
Fill Factor	70 %	74 %
R_s	0.308 Ω	0.220 Ω

Figure 6. IV Curves of Typical Screened Cell and Control

7. CONCLUSION

We have presented two cases of severe contact problems seen in solar cells. While this by no means exhausts the catalog of contact maladies, the examples were thought to be of interest. Energy level diagrams for normal solar cell contacts were also reported.

The author wishes to express his gratitude to Dr. I. Weinberg, NASA Lewis Research Center for permission to reproduce Figure 3.

REFERENCES

1. P. Rappaport, RCA Rev. 20, 373 (Sep 1959)
2. S. Matlow, US Patent #2, 984, 775 (5-1961)
3. P.A. Iles, Conf. Rec. on IEEE Photovolt. Spec.Conf. p.1 (May 1972)
4. E.L. Ralph, Proc. 11th IEEE Photovolt Spec.Conf., p.315, (1975)
5. B. Ross and D.B. Bickler, Proc. 3rd E.C. Photovolt, Solar Energy Conf. p.674 Cannes, France (Oct. 1980)
6. M.P. Lepselter and J.M. Andrews in "Ohmic Contacts to Semiconductors" Ed.B. Schwartz, Electrochem.Soc. p.159 New York (1969)
7. E.H. Rhoderick, "Metal Semiconductor Contacts, Clarendon Press, p.59, Oxford (1978)
8. J.G. Simmons, Jour.Appl.Phys. 34, 1793 (Jun 1973)
9. V.G. Weizer and J.D. Broder, Proc. 15th IEEE Photovolt. Spec.Conf. p.235 (May 1981)
10. J.W. Tiley and C.G. Thornton, Proc.IRE, 41,1706 (Dec 1953)
11. J. Roschen and C.G. Thornton, Jour.Appl.Phys. 29,923 (Jun 1958)
12. A.D. Rittmann, G.C. Messenger, R.W.Williams and E.Zimmerman, IRE Trans. on El.Dev. 5,49 (Apr 1958)

DESIGN OF STABLE METAL-INSULATOR-SEMICONDUCTOR (MIS)
SOLAR CELLS BY OXIDE THICKNESS COMPENSATION

G. Rajeswaran and W. A. Anderson
Department of Electrical and Computer Engineering
State University of New York at Buffalo
Amherst, New York 14226, U.S.A.

Summary

The MIS solar cell structure is an exciting prospect for low-cost, moderately efficient photovoltaic conversion. Like other surface barrier devices, they exhibit degradation of solar cell performance. This degradation emanates from the metal/oxide interface and results in an oxide thickness reduction with time. In a $Cr/SiO_x/Si$ device, for instance, the oxide thickness has to be overcompensated to counter degradation. The performance of such compensated devices shows a higher degree of stability than otherwise. An interface approach to solving stability problems has also resulted in novel ideas for stable MIS structures. A new class of novel Yb-MIS devices, that utilize indirect "oxide thickness reduction" compensation, are reported. These devices have been found to be very stable.

1. INTRODUCTION

The MIS solar cell structure has the potential to ultimately produce low-cost, > 10% efficient solar cells on thin film polycrystalline silicon. The degradation of photovoltaic response of this class of devices stands as a major deterrent in the way of its commercialization. In a recent work on $Cr/SiO_x/Si$ structures [1], we have indicated that an effective oxide thickness reduction at the interface could bring about the observed solar cell deterioration. This degradation occurs at the same rate whether stored in the dark or kept under illumination. It is also independent of the type of substrates used. The origin of degradation lies at the Cr/SiO_x interface.

The dark current characteristics serve as good barometers of the state of degradation. An upward shift of the current density curves is usually observed in degraded devices. This has been observed by other groups as well [2,3] for other types of surface barrier devices. We associate the increase in dark current with a physical occurrence of oxide thickness reduction in our class of devices.

Finally, as a logical step, oxides have been overgrown beyond their optimum thicknesses, to allow for degradation. In this paper, the results of such an oxide thickness compensation technique applied to MIS solar cells are reported.

2. MIS SOLAR CELL FABRICATION

A large volume of Cr-MIS solar cells were fabricated using a revised processing scheme [4]. The substrates used were Hamco-Kayex single crystal silicon or Wacker polycrystalline silicon.

The most crucial processing step, from a stability viewpoint, is the growth of the thin insulating oxide. Oxide growth on our devices was performed in a muffle furnace by ramp heating in air to a maximum temperature of between 580 °C and 650 °C. Ellipsometer measurements have shown [5] that a 620 °C oxidation on similar $Cr/SiO_x/Si$ structures yields an oxide thickness of 23 Å. In this paper, any Cr-MIS solar cell fabricated with an oxidation temperature of and above 620 °C will be referred to as a compensated oxide ($\gtrsim$ 23 Å) device. In contrast, uncompensated oxide (< 23 Å) devices have undergone a heat treatment of less than 620 °C. Based on the above definitions, a Cr-MIS solar cell designed for maximum conversion efficiency, with an oxide thickness of between 18 and 20 Å, belongs to the uncompensated class of devices.

3. ORIGINS OF DEGRADATION

The instabilities that result in solar cell performance degradation originate from the Cr/SiO_x interface. Concurrent Auger, ellipsometry and ESCA measurements were performed on the $Cr/SiO_2/Si$ interface [1]. Figure 1 shows the results of such a study. In both thick and thin oxide devices,

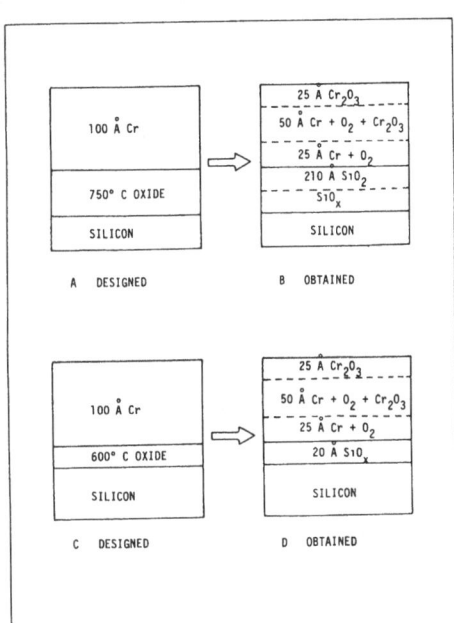

Figure 1

Results of concurrent ESCA, Auger and ellipsometry measurements indicate that chromium at the interface is oxidized.

at the metal/oxide interface, chromium was found to exist partly in the metallic state and partly in the oxidized state. The V_{oc} degradation of Cr-MIS cells is known to proceed rapidly in the days immediately following fabrication. The device then stabilizes and reaches an equilibrium. In addition, the upward shift of dark current is well known. Armed with a knowledge of device behavior, the following mechanisms are suggested.
a) An oxidation-reduction (OR) reaction at the Cr/SiO_x interface may

cause a gradual degradation of the device performance. Standard free energy data [6] support such a speculation, since Cr_2O_3 would be the most stable compound in the neighborhood of this interface. The OR reaction would occur until all the Cr would have converted to Cr_2O_3. The ensuing thin layer of Cr_2O_3 would inhibit further reaction from occurring, leading to device equilibrium. The OR reaction would have occurred at the expense of the SiO_x layer. The reduction of this layer would result in an effective decrease of its thickness.

b) A simple diffusion of metallic chromium into the dips and dents existing on the SiO_x layer would cause an effective oxide thickness reduction.

In practice, both mechanisms could occur concurrently. In any case, the degradation resembles the effect of an oxide thickness reduction.

4. STABILITY CONSIDERATIONS

To understand the role of oxide thickness reduction on Cr-MIS solar cell performance, a comprehensive charge balance analysis of an MIS interface was performed [7]. This includes the effects of an inversion layer, variations in work function (ϕ_M) and oxide thickness (δ). The results are shown in Figure 2. The ϕ_M of Cr has been estimated to be 4.2 for our class of devices [8]. The barrier height (ϕ_{BO}) of a Cr-MIS junction is seen to increase with increasing oxide thickness. This occurs up to a δ of 24 Å beyond which a plateau region is reached. This plateau region is reached for a smaller δ when ϕ_M is decreased (e.g. for ϕ_M = 2.5, δ = 18 Å).

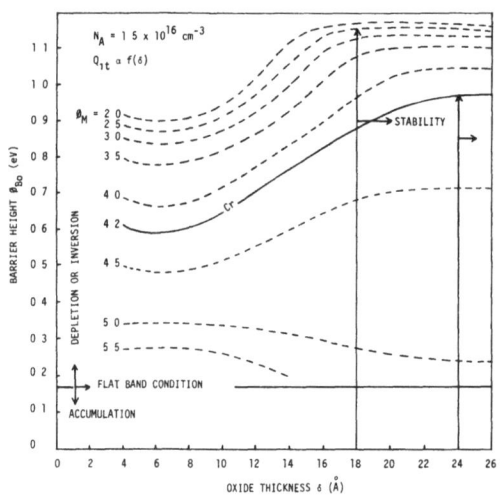

Figure 2.

General barrier height curves obtained from a comprehensive charge balance analysis.

Figure 3 shows some experimentally measured ϕ_{BO} values of Cr-MIS devices as a function of oxidation temperatures (T). There is no direct correspondence between Figures 2 and 3 since oxidation temperatures do not linearly correlate to δ. On the contrary, δ is proportional to T. Ellipsometer measurements of oxide thickness have to be made to correlate δ with T. In fact, at 600 °C oxidation (corresponding to 20 Å of oxide) there is a perfect fit between theoretical and experimental data.

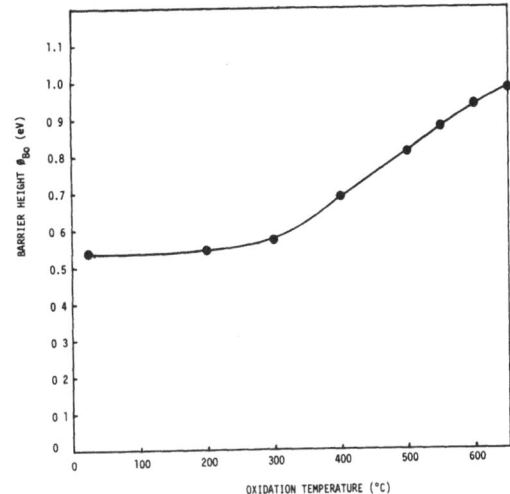

Figure 3.

Experimental Cr-MIS solar cell barrier height data lend support to the analysis in Figure 2.

Nevertheless, some important conclusions may be drawn from these figures.

a) For a Cr-MIS solar cell ($\phi_M = 4.2$), an oxide thickness reduction brings about barrier height relaxation if $\delta < 24$ Å. The combined decrease of δ and ϕ_{BO} would produce an increase of dark current and V_{oc} but not of fill factor or short circuit current (J_{sc}). This is certainly true of all our devices.

b) An oxidation temperature of 620 °C is estimated to produce 23 Å of oxide. Above 23 Å, the rate of decrease of ϕ_{BO} with oxide thickness reduction is negligible. This has, therefore, decided our "critical" temperature of $\sim$ 620 °C as the basis for selecting compensated or uncompensated devices.

c) The "critical" temperature is smaller when low work function metals are used.

d) Oxide growth at 600 °C (for optimum solar cell performance), combined with low ϕ_M (of the Schottky barrier metal) should result in a barrier height that does not reduce even in the event of an oxide thickness reduction due to degradation.

Some stability data for oxide thickness compensated devices are presented in the next section.

5. COMPENSATED Cr-MIS SOLAR CELLS

In all Cr-MIS solar cells major degradation occurs within 20 days of fabrication. The shelf life degradation of V_{oc} and efficiency (η) of a compensated solar cell is compared with that of an uncompensated device in Figure 4. This represents typical performance curves. In the uncompensated case, the drop in V_{oc} (ΔV_{oc}) is close to 30 mV and the decrease in η ($\Delta \eta$) about 10% of its initial value (η_o). For the compensated solar cell, $\Delta V_{oc} \simeq 5$ mV and $\Delta \eta$ is practically negligible. Once the solar cells reach post-degradation equilibrium they continue to deliver power at their degraded minimum.

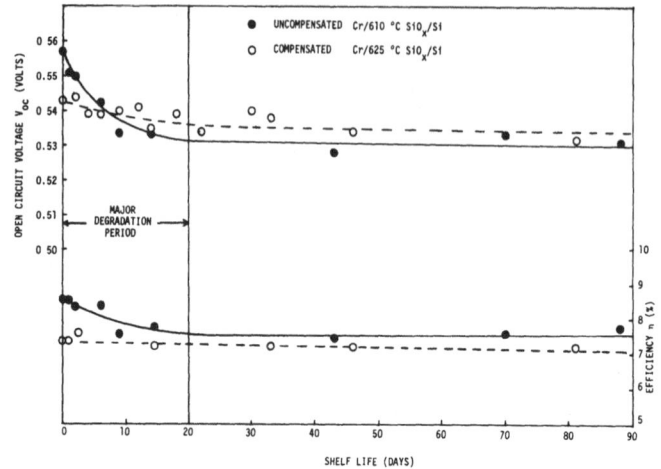

Figure 4.

Compensated Cr-MIS solar cells exhibit stable performance.

6. CONCLUSIONS

Oxide thickness compensation moves the Cr-MIS solar cell into a stable region of operation. But the associated offset from the optimum oxide thickness reduces the photovoltaic conversion efficiency. All Cr-MIS cells, compensated or uncompensated, reach the same operating efficiency during post-degradation equilibrium. This can be seen from Figure 4. The final operating efficiency is governed by the choice of substrates. For instance, a crystalline Cr-MIS solar cell operates close to 10% AM1 efficiency, and a Wacker polycrystalline Cr-MIS solar cell delivers reliable power at 7%.

The insight gained into instability mechanisms of Cr-MIS solar cells suggests that low work function metals cause indirect oxide thickness compensation and will result in highly stable MIS solar cells. Recently, we have developed Ytterbium-MIS solar cells that deliver stable, 11.5% AM1 conversion efficiencies.

REFERENCES

1. G. Rajeswaran, W. A. Anderson, M. Thayer and B. W. Lee, (accepted for publication) IEEE Trans. Rel., 1982.
2. J. K. Kleta and D. L. Pulfrey, IEEE Elec. Dev. Lett., Vol. EDL-1, No. 6, 1980, 107.
3. H. P. Maruska, T. Feng, A. K. Ghosh and D. J. Eustace, 15th Photo. Spec. Conf., 1981, 1412.
4. W. A. Anderson, G. Rajeswaran, K. Rajkanan and G. Hoeft, IEEE Elec. Dev. Lett., Vol. EDL-1, No. 7, 1980, 128.
5. J. K. Kim, W. A. Anderson and A. E. Delahoy, J. Elec. Mat., Vol. 7, No. 3, 1978, 403.
6. R. C. Weast and M. J. Astle, CRC Handbook of Chemistry and Physics, 60th Ed., CRC Press, Inc., 1980, D45.
7. G. Rajeswaran, V. J. Rao, M. A. Jackson, W. A. Anderson, M. Thayer and B. Bhasker Rao (to be published).
8. W. A. Anderson, J. K. Kim and A. E. Delahoy, IEEE Trans. Elec. Dev., Vol. ED-24, No. 4, 1977, 453.

STUDY OF GAP STATES IN a-Si:H BY TRANSIENT CURRENT SPECTROSCOPY

J. BEICHLER and H. MELL
Fachbereich Physik, Universität Marburg, F.R. Germany

Summary

Schottky barrier structures have been made with n-type hydrogenated amorphous silicon (a-Si:H) prepared by the glow discharge decomposition of SiH_4. The density of gap states $N(E)$ in the depletion layer of these diodes has been determined by analysing the current transients caused by the emission of electrons after removal of a small forward bias. Depending on the preparation conditions, the dopant concentration and the history of the structures (exposure to light, electron bombardment) $N(E)$ near midgap ranges from $3 \cdot 10^{15}$ to $10^{18} cm^{-3}$. The distribution of states is more similar to that deduced from field effect and space charge limited current data than to DLTS results.

1. INTRODUCTION

The electronic properties of a-Si:H films and their quality in a solar cell structure depend critically on the density and distribution of the gap states, $N(E)$ (1,2). It is, therefore, of considerable importance to develop techniques which can routinely be used to determine $N(E)$. For many years the measurement of the field effect was the only method applied to a-Si:H films (3-5). Recently, however, several other methods have been employed (6-11) and, in part, have yielded widely different distributions. One of the interesting new techniques is the transient current technique (9,10). In this paper we present a modification of it, which is superior by the simple measurement of the current and by the extended energy range accessible.

2. BASIC CONCEPTS OF MEASUREMENT AND ANALYSIS

The measurement sequence in a conventional CT experiment is as follows (9,10). First, a reverse bias V_r is applied to a Schottky barrier or p-i-n diode. Then V_r is reduced by a voltage pulse thus filling up part of the gap states in the depletion layer with electrons. When this pulse is terminated the trapped electrons are emitted to the conduction band and swept out of the depletion layer by the high electric field therein. The resulting current transient $J(t)$ can be analysed to determine the density and energy of the trapped electrons. Unfortunately the measurement of $J(t)$ is impeded by the reverse current and can, therefore, only be done at short times ($t < 10^{-3}$s) using a complicated technique (9,10). To avoid this difficulty we do not apply a reverse bias in the present study but monitor the emission current at zero bias. By that, it is possible to measure the small decay current flowing at long times ($t \sim 1s$) by conventional techniques. To fill up gap states in the depletion layer we apply a small forward bias V_f to the diode.

Consider the states with a fixed energy E_ν. They are empty from $x = 0$ at the interface of a Schottky barrier diode up to $x = x_\nu$ defined by the intersection of E_ν and the Fermi level E_F (12). When the diode is switched from $V = V_f$ to $V = 0$ those states which are located in the range $\Delta x_\nu = x_\nu(0) - x_\nu(V_f)$ lose their electrons. This occurs by thermal emission to the mobility edge

E_c with a rate (8,13)

$$\nu = \nu_0 \exp[-(E_c - E_\nu)/kT] \qquad (1)$$

where $\nu_0 \sim 10^{13} s^{-1}$ (8) is the attempt to escape frequency. It has been shown (13) that the current in the external circuit, at time t, is mainly determined by those states with $\nu = 1/t$, i.e. with energy

$$E_c - E_\nu = kT \ln(\nu_0 t) \qquad (2)$$

Therefore, if N(E) is a slow function of E, the current density J(t) can be well approximated by

$$J(t) = S_\nu \, e \, \frac{d}{dt} \{N(E_\nu) \, \Delta x_\nu \, dE_\nu\} = S_\nu \, \frac{ekT}{t} \, N(E_\nu) \, \Delta x_\nu \qquad (3)$$

The quantity S_ν in eq. (3) accounts for the fact that the emission occurs in a non-equilibrium state. For this reason the number of electrons with energy E_ν leaving the space charge layer may differ from $N(E_\nu) \, \Delta x_\nu \, dE_\nu$. S_ν can be calculated by solving numerically Poisson's equation. We did this for several model distributions and found that, except for a narrow energy range at E_F, the value of S_ν is close to 1. For the present preliminary analysis we have, therefore, used $S = 1$ in the whole energy range investigated. For small forward bias, $V_f \ll V_D$, Δx_ν is given by

$$\Delta x_\nu = \frac{dx_\nu}{dV_S} V_f = \frac{V_f}{F_S} \qquad (4)$$

where $V_S = V_D - V_f$ and F_S is the electric field strength at the interface. The latter quantity can be calculated from the space charge capacitance C_0 measured at very low frequency where all or most of the states can respond to the change in the applied voltage. We have used the formula $F_S = C_0 V_D/\varepsilon$ valid for $N(E) = N_0$, ε being the dielectric constant. Eq. (3) may then be expressed in the form

$$N(E_\nu) \sim \frac{tJ(t)}{ekT} \frac{C_0 V_D}{\varepsilon V_f} \qquad (5)$$

3. EXPERIMENTAL DETAILS

The n-type a-Si:H films used in this study were deposited onto stainless steel substrates mounted on the grounded anode of one of two capacitively coupled glow discharge systems. The deposition conditions were as follows:

System I: 20 % SiH_4 in Ar, pressure $p \sim 2$ mbar, flow rate $f \sim 90$ sccm, rf-power $P \sim 80$ mW/cm^2 and deposition rate $r \sim 3$ Å/s.
System II: 5 % SiH_4 in He, $p \sim 0.5$ mbar, $f \sim 70$ sccm, $P \sim 50$ mW/cm^2 and $r \sim 2$ Å/s.

The substrate temperature T_S war 280°C and the film thickness $d \sim 0.6$ μm. The barrier metal, usually Pt, was evaporated on top of the freshly prepared a-Si:H film. Its thickness was ~ 10 nm and the area 4 to 30 mm^2. The current transients were monitored in the time interval 0.1 to 10 s using an automated equipment which had been designed and used for the measurement of current-voltage characteristics. It consists of a programmable voltage source, a digital pA-meter with a time resolution of 40 ms and a temperature controller. All three instruments are controlled by a microcomputer. The capacitance C_0 was determined by measuring the total charge ΔQ emitted from

$t = 0$ to $t = 100$ s after removal of $V_f = 50$ mW and putting $C_0 = \Delta Q / V_f$.

4. RESULTS AND DISCUSSION

In Fig. 1a the current density $J(t)$ for $t = 0.5$ s is plotted as a function of temperature T for 6 different diodes (1-3 from system I and 4-6 from system II). Using the formulas (2) and (5), $\nu_0 = 10^{13} s^{-1}$ (8) and the C_0 values listed in Fig. 1b, these $J(T)$ curves were transformed to the $N(E)$ curves in Fig. 1b. All $J(T)$ curves steeply decay towards low T suggesting that in this range $E_\nu > E_F$. We have determined E_F from the series resistance of the diodes near the maxima of the $J(T)$ curves using for the conductivity σ the relation $\sigma = 10^3 \Omega^{-1} cm^{-1} \exp[-(E_C - E_F)/kT]$. These values of E_F are indicated in Fig. 1b by the arrows. Though, in principle, the energy scale of emission can be different from the thermal equilibrium scale determining σ we believe that the arrows indicate with a reasonable accuracy the energies up to which the $N(E)$ curves represent the density of gap states (solid lines). Fig. 1b reveals two interesting results: i) $N(E)$ increases appreciably with the dopant content and ii) $N(E)$ is an order of magnitude larger for the films grown in system I. It is not yet clear whether the latter result is solely due to the different gases used for diluting the SiH_4 (Ar and He, respectively) or also to other differences in the deposition process.

Fig. 2 shows what changes of $N(E)$ can be produced when a diode is subjected to various treatments. After a bombardment with 3 MeV-electrons, $N(E) \sim 10^{18} cm^{-3} eV^{-1}$ near midgap for an undoped a-Si:H film (curve 1 in Fig. 2a). Annealing at $T_A = 125°C$ decreases $N(E)$ by more than one order of magnitude (curve 2) and heating to 180°C returns the film to the original state

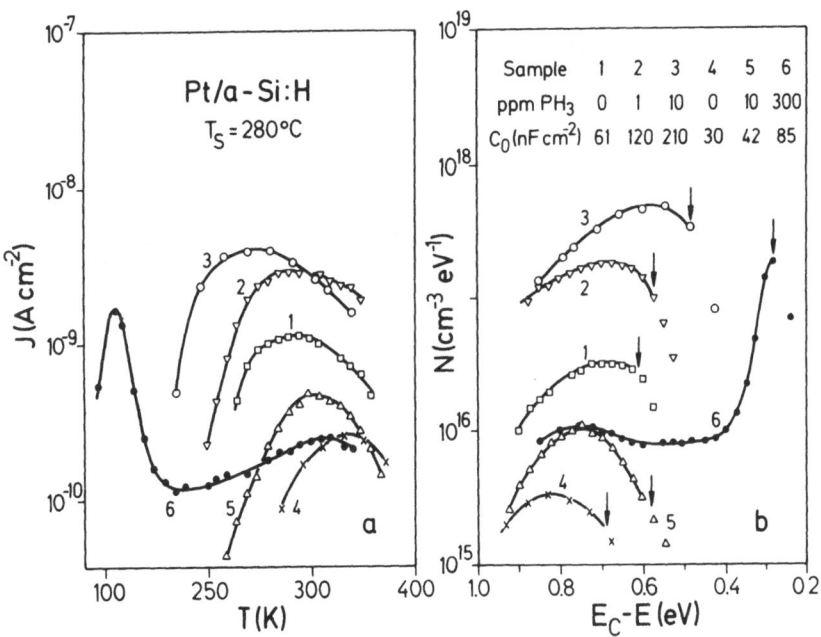

Fig. 1: a) Temperature dependence of the transient current density $J(t)$ measured at $t = 0.5$ s for 6 diodes made with differently prepared a-Si:H-films. b) Density of states distribution $N(E)$ deduced from the data in Fig. 1a.

with $N(E) \sim 3 \cdot 10^{15} \text{cm}^{-3} \text{eV}^{-1}$ (curve 3). When the same film is exposed to white light from a tungsten-iodine lamp (60 hours at 500 mW/cm²) $N(E)$ is increased to $\sim 10^{17} \text{cm}^{-3} \text{eV}^{-1}$ (curve B). The influence of optical exposure was studied in more detail for a film doped with 100 ppm PH_3 (Fig. 2b). Curve A shows $N(E)$ before exposure and after annealing out the photo-induced changes at $T_A = 180°C$. After a 15-hour exposure to uniformly absorbed light ($h\nu < 1.7$ eV) while a forward bias ($V_f > V_{oc}$) was applied curve B1 was obtained. When the diode was reverse biased ($V_r = -2$ V) the same exposure led to curve B2. In both cases $N(E)$ was increased, however, in different energy ranges: in case B1 most strongly near $E_c - E = 0.7$ eV and in case B2 in the tail state region.

The strong increase of $N(E)$ shown by curve B3 in Fig. 2b was produced by a 60-hour exposure to white light of 500 mW/cm² while the diode was forward biased. Here E_F is markedly lowered which is typical for the Staebler-Wronski effect (14). Cohen et al. (15) concluded from DLTS data that this effect is associated with increase of $N(E)$ below midgap. The present results, however, suggest that $N(E)$ is also increased above midgap, in agreement with field effect data (16,17).

Fig. 3 compares three $N(E)$ curves from the present study (curves 6-8) with data from the literature obtained by other techniques. In view of the strong dependence of $N(E)$ on the preparation conditions (Fig. 1) and on the history of the a-Si:H films (Fig. 2) the large spread in $N(E)$ values is not surprising. It could be due to a widely different quality of the films. There is, however, a striking difference in the distribution of states deduced from DLTS data (curve 3) and those obtained by the other techniques. Only the DLTS data predict a deep minimum of $N(E)$ near $E_c - E = 0.45$ eV (7,8). Further experimental and theoretical studies are necessary to solve this discrepancy.

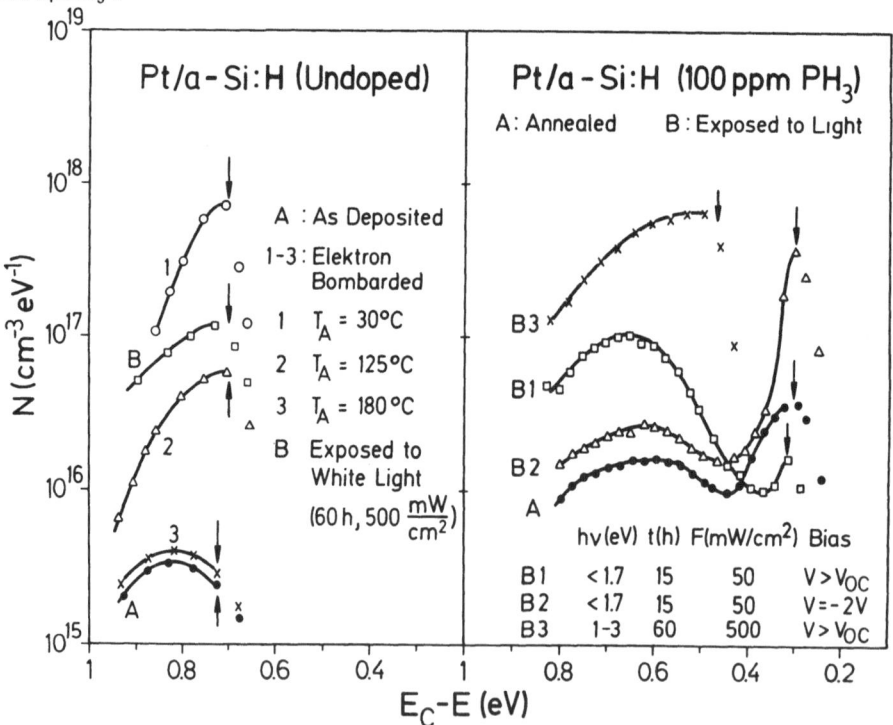

Fig. 2: Dependence of $N(E)$ on the history of the diode made with a) undoped and b) phosphorus doped a-Si:H

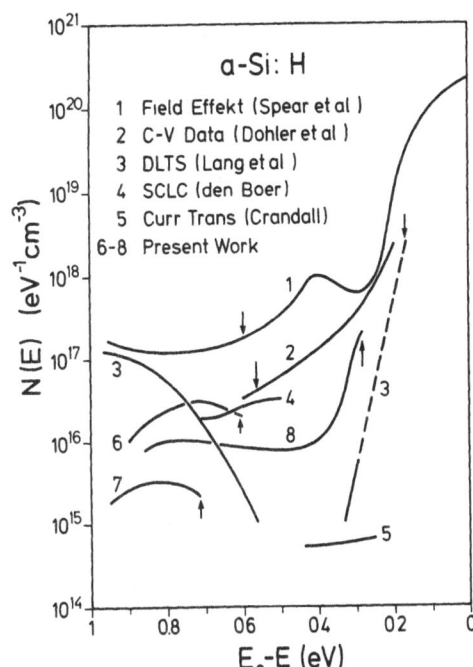

Fig. 3: Density of states distribution N(E) in a-Si:H as determined by various techniques. The curves 1-5 were taken from ref. (1), (6), (8), (11) and (9), respectively.

ACKNOWLEDGEMENT

The authors would like to thank the Bundesminister für Forschung und Technologie for financial support.

REFERENCES

(1) W.E. Spear, Advances in Physics 26, 811 (1977)
(2) D.E. Carlson, J. Non-Cryst. Solids 25/26, 625 (1980)
(3) A. Madan, P.G. LeComber and W.E. Spear, J. Non-Cryst. Solids 20, 239 (1976)
(4) N.B. Goodman and H. Fritzsche, Phil.Mag. B 42, 149 (1980)
(5) K. Weber, M. Grünewald, W. Fuhs and P. Thomas, phys.stat.sol. (b) 110 (1982)
(6) M. Hirose, T. Suzuki and G.H. Döhler, Appl.Phys.Lett. 34, 234 (1979)
(7) J.D. Cohen, D.V. Lang and J.P. Harbison, Phys.Rev.Lett. 45, 197 (1980)
(8) D.V. Lang, J.D. Cohen and J.P. Harbison, Phys.Rev. B, in press
(9) R.S. Crandall, J. Electron. Mat. 9, 713 (1980)
(10) M.J. Thompson, N.M. Johnson and R.A. Street, J. de Physique 10, C4-617 (1981)
(11) W. de Boer, J. de Physique 10, C4-451 (1981)
(12) J. Beichler, W. Fuhs, H. Mell and H.M. Welsch, J. Non-Cryst. Solids 35/36, 587 (1980)
(13) J.G. Simmons and L.S. Wei, Solid State Electr. 17, 117 (1974)
(14) D.L. Staebler and C.R. Wronski, J. Appl.Phys. 51, 3262 (1980)
(15) J.D. Cohen, D.V. Lang, J.P..Harbison, and A.M. Sergent, J. de Physique 10, C4-371 (1981)
(16) M.H. Tanielian, N.B. Goodman and H. Fritzsche, J. de Physique 10, C4-375 (1981)
(17) D. Hauschildt, W. Fuhs, H. Mell and K. Weber, this conference

HIGHLY CONDUCTIVE BORON DOPED Si-LAYERS PREPARED BY PLASMA DECOMPOSITION OF SiH_4

H.Simon, G.Winterling and G.Müller
MESSERSCHMITT BÖLKOW BLOHM GMBH, AE 331, D-8012 OTTOBRUNN, GERMANY

Summary

Highly conductive Si:B films were prepared by plasma decomposition of B_2H_6/SiH_4 mixtures strongly diluted with H_2. The main properties of these films are: i) dark conductivities up to 30 $(\Omega cm)^{-1}$; ii) conductivity activation energies as low as 0.01 eV; iii) an optical gap increased by at least 0.2 eV in comparison to a-Si:H,B although the H-contents are almost the same.

1. PREPARATION AND ELECTRICAL PROPERTIES

The performance of present a-Si:H solar cells of the p-i-n type was shown to be severely limited by the bad quality of the doped layers (1). For example the boron-doped a-Si:H has conductivities of at most 10^{-2} $(\Omega cm)^{-1}$ at a doping ratio of 1% B_2H_6 in SiH_4 and its optical gap is reduced by about 0.3 eV in comparison to intrinsic a-Si:H (2).

For device applications it is of great interest to obtain improved p^+ Si-layers. We, therefore, investigated the preparation of p^+ polycrystalline Si-layers by plasma-decomposition of SiH_4 in a capacitively coupled reactor. By increasing the rf-power density to 0.5 Watt/cm² and by strongly diluting SiH_4 with hydrogen (3) we were able to produce p^+ Si-layers with dark conductivities higher by 3 orders of magnitude (4/5). Values at 300 K are given in the table below and refer to films prepared at substrate temperatures between 300 and 340°C.

	a-Si:H,B	µx-Si:H,B	
rf power density (Watt/cm²)	≅ 0.04	≅ 0.4	
Substrate temperature T_s	300°C	300°C < T_s < 350°C	
Gas	$B_2H_6+SiH_4$	$B_2H_6+SiH_4+H_2$ with SiH_4/H_2 ≅ 0.03	
Doping ratio %B_2H_6 in SiH_4	1%	1%	0.3%
σ_{Dark} $(\Omega cm)^{-1}$	10^{-2}	12 to 20*	2.5**
Activation energy E_a	0.22 eV (near 300 K)	0.010 to 0.014 eV	0.021 eV
Optical gap E_o	< 1.3 eV	≳ 1.5 eV	1.55 eV**
H-content (atomic %)	1.6%	1 to 2%	2.8%

* Samples 294, 295, 298, 299   ** Sample 303

The activation energy, E_a, of the dark conductivity σ_D was determined in the temperature range 500 K<T<300 K: σ_D was measured during cooling down. The strong increase of σ_{Dark} is correlated with a pronounced decrease of E_a as is evident in the right half of the table.

2. STRUCTURE

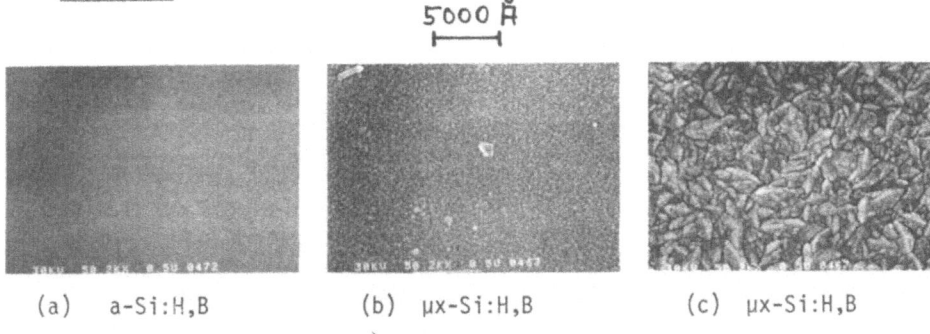

(a)   a-Si:H,B   (b)   µx-Si:H,B   (c)   µx-Si:H,B

Fig.1: Scanning electron micrographs of Si:H,B films

While no structure is detectable on the micrograph of the purely amorphous a-Si:H,B sample (Fig. 1a), the highly conductive films (Fig. 1b) show a granular structure with typical dimensions of ~ 200 Å, which is superimposed by a much coarser structure in many cases, see Fig. 2c. The appearance of the coarse structure, however, was not significantly correlated with the electrical properties of the µx-Si:H,B films.

The coarse structure is similar to the columnar structure observed earlier by Knights on high-power intrinsic Si:H films deposited from a SiH_4/Ar mixture (6). Raman studies have proved our highly conductive films to be partially crystallized. Similar films are often referred to as microcrystalline Si(µx-Si). We have not yet made any attempt to achieve complete crystallisation. Polycrystalline but P-doped films have been prepared earlier by a H_2-plasma assisted transport process (7,8).

3. OPTICAL PROPERTIES

The optical absorption coefficient α was determined from transmission and reflection measurements. α was found to be considerably smaller in µx-Si:H,B films than in a-Si:H,B films (Fig. 2). For comparison, still smaller α-values are observed in µx-Si:H,P films.

Replotting α in Fig. 3, we notice that the quantity $(\alpha h\nu)^{1/2}$ is proportional to $h\nu$ over a relatively large photon energy range. Hence, the Tauc-relation $(\alpha h\nu)^{1/2} \alpha (h\nu - E_o)$ can be applied to determine the optical gap E_o. Evidently E_o is higher in µx-Si:H,B than in a-Si:H,B (2) by at least 0.2 eV. The increase in E_o is accompanied by a remarkable decrease of the index of refraction.

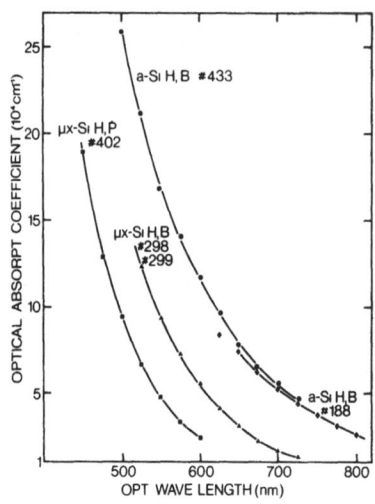

Fig.2: Optical absorption coefficient of Si:H films

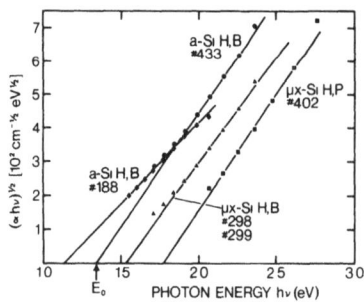

Fig.3: Tauc-plots of the optical absorption to define the optical energy gap E_0

In the case of a-Si:H it was demonstrated that E_0 increases with increasing H-content, C_H, the increase amounting to roughly 0.15 eV for a change in C_H by about 10 at %. In this connection the question arises whether the increased optical gap of µx-Si:H,B is caused by additional H-incorporation.

C_H in our films, as deduced by the ^{15}N nuclear reaction technique, was found to decrease with increasing B_2H_6 doping in a way similar to that observed previously in a-Si:H,B films (10). On the whole we do not find significant differences in C_H going from the amorphous to the highly conductive microcrystalline state as is obvious from the table. Obviously the higher optical gap in µx-Si;H,B cannot be related to the H-content, rather it might be more related to a change in the chemical bonding.

Acknowledgements

We are grateful to Mr.Breitschwerdt, Mr.König, Mr.Richter, MPI für Festkörperforschung, Stuttgart, for providing transmission,reflection,Raman data and to Dr.Kalbitzer, MPI für Kernphysik, Heidelberg, for providing the nuclear reaction data.

References:

(1) D.E.Carlson, Solar Energy Mat. **3**, 503 (1980)

(2) H.Okamoto, Y.Nitta, T.Yamaguchi and Y.Hamakawa, Solar Energy Mat. **2**, 313 (1980)

(3) A.Matsuda, S.Yamasaki, K.Nakagawa, H.Okushi, K.Tanaka et al. Jap.J.Appl.Phys. **19**, L305 (1980)

(4) Previously similar deposition conditions provided $\sigma_{Dark} \cong 10^{-1}(\Omega cm)^{-1}$, see A. Matsuda et al., in AIP Conf. Proceed. No. 73, Tetrahedrally Bonded Amorphous Semiconductors, Am.Inst.Phys., 1981, p.192

(5) T.Hamasaki, H.Kurata, M.Hirose and Y.Osaka, Jap.J.Appl.Phys. **20**, L84 (1981)

(6) J.C.Knights, J.Non-Cryst.Solids Vol. **35&36**, 159 (1980)

(7) Z.Igbal, A.P.Webb and S.Veprek, Appl.Phys.Lett. **36**, 163 (1980)

(8) H.Richter and L.Ley, J.Appl.Phys. 52, 7281 (1981)

(9) G.D.Cody, C.R.Wronski, B.Abeles, R.B.Stephens and B.Brooks, Solar Cells **2**, 227 (1980)

(10) F.J.Demond, G.Müller, H.Damjantschitsch, H.Mannsperger, S.Kalbitzer, P.G.LeComber and W.Spear, J.Physique **42**, Suppl. Nr. 10, C4-779 (1981)

A MODEL FOR ANALYSIS OF OPTICAL MEASUREMENTS CARRIED ON a-Si:H FILMS FOR
PHOTOVOLTAIC APPLICATIONS

L. GUIMARÃES, R.MARTINS, A.G. DIAS and F. BARRADAS

Centro de Física Molecular das Universidades de Lisboa (UNL;INIC)
Complexo I (I.S.T.) Av. Rovisco Pais 1000 LISBOA - PORTUGAL

SUMMARY - A set of absorptance measurements have been performed on
doped and undoped a-Si:H films prepared by r.f. glow discharge at
different deposition conditions. In previous work[1,2,3] we observed the
role of the bias on the electrical properties of the films. We are now
interested to show that the optical properties are also influenced by
deposition conditions. In order to interpret the experimental results
a simple electromagnetic model has been used. This theoretical model
includes the main optical phenomena taking place in the film such as
absorption, multiple reflections and interferences. From this we are
able to determine the film thickness, the optical gap[4] and the
position of the Fermi level at a temperature T. In a general way it is
observed a good agreement between the theoretical model and the
experimental data and also a clear dependence of optical parameters
with deposition conditions.

INTRODUCTION - There is an increasing interest on the study of the photo-
electric properties of a-Si:H films concerning its applications in optoelectronic devices. One of the most important parameters wich caracterizes
the material is its optical behaviour in the wavelenght range from the
infrared up to the ultraviolets. As far as we know the experimental data
related to the high absorption region of absorptance measurements have been
used in order to infer the optical gap of the films and to determine their
absorption coefficients in the spectral range of visible wavelenghts.
Nevertheless, little is known about what happens in the low absorption
region, aside the existence of interference fringes ascribed with the
thickness of the material. In order to correlate such data with intrinsic
properties of the material, we considered the electromagnetic behaviour of
the optical signal crossing two different media (air / a-Si:H / air),
assuming a substrate with refraction index similar to that of the air.
   The derived theoretical model concerning the high and low absorption
regions, taking into account the multireflections and interferences,
allowed us to interpret allmost all optical properties of the material such
as absorption and optical gap.

EXPERIMENTAL DETAILS - Films of a-Si:H were produced by r.f. glow discharge
in 3% SiH_4/Ar with both inductive and capacitive coupling and deposited
onto a glass substrate. Undoped films were produced under different crossed
static electromagnetic fields with the deposition procedure reported in
previous works[1,2,3]. Phosfine and Diborane doped films were produced using
the same deposition procedure as undoped ones, without using crossed static
electromagnetic fields. Doping levels were inferred from the gas pressure
ratio at constant volume between the dopant gas and silane.
   Absorptance measurements were carried out with a Cary 17E spectrophotometer over the spectral range 350 - 2500 nm at room temperature.

MODEL AND RESULTS - In order to interpret our absorptance measurements we used a simple optical model wich consists in neglecting the substrate influence and considering the film as a thin blade of uniform material (see fig. 1). By doing this we deduced the theoretical expression concerning the transmission coefficient for normal incidence of a monocromatic beam. Thus, the transmission coefficient corresponds to the ratio between the electric field $\hat{E}_z$ at each interface:

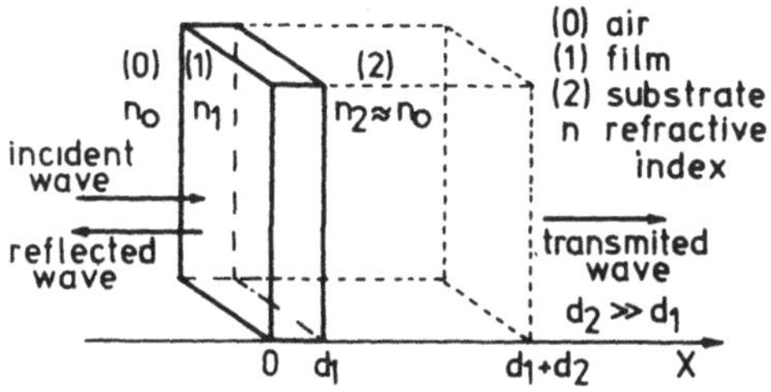

$$\hat{T}' = \frac{\hat{E}_z (d_1^+)}{\hat{E}_z (0^-)} \quad (1)$$

Figure 1. The simple optical model

where $0^- \lesssim 0$ and $d_1^+ \gtrsim d_1$.
Expressing $\hat{E}_z (0^-)$ and $\hat{E}_z (d^+)$ as functions of the optical properties of the interfaces and propagating media, we get

$$\hat{T}' = (1-\Gamma^2) \delta_1 / (1-\Gamma^2 \delta_1^2) \quad (2)$$

where $\delta_1 = e^{-(\alpha_1' d_1 + j \beta_1 d_1)}$ is ascribed to the phase shift $(- j \beta_1 d_1)$ and atenuation $(-\alpha_1' d_1)$ of the optical signal crossing the film, $\Gamma = (1-n_1) / (1+n_1)$ is the reflection coeficient at the air - film interface $x=0$ and n_1 is the film refraction index. Under these conditions, transmittance T corresponds to the square of the transmission coefficient model, i.e.,

$$T = |\hat{T}'|^2 \quad (3)$$

In a range of wavelenghts where the absorption coefficient $\alpha_1 = 2 \alpha_1'$ and the refraction index n_1 are allmost constants we can determine maxima and minima of T by solving $dT / d\lambda = 0$. The solutions are of the form $\lambda_m = 4 n_1 d_1 / m$ where $m=0,1,2,3,\ldots$ is the order of the fringes. Thus, we get the line of the maxima of T for m even:

$$T_{max} = ((1-\Gamma^2) e^{-\alpha_1' d_1} / (1-\Gamma^2 e^{-2\alpha_1' d_1}))^2 \quad (4)$$

and the line of the minima for m odd:

$$T_{min} = ((1-\Gamma^2) e^{-\alpha_1' d_1} / (1+\Gamma^2 e^{-2\alpha_1' d_1}))^2 \quad (5)$$

These are the envelope lines of the fringes in the low absorption region.
In order to get a simple estimation of the refraction index n_1 in the low absorption region we defined $\Delta T = T_{max} - T_{min}$. If we assume $\Gamma^4 e^{-2\alpha_1 d_1} \ll 1$, we get

$$- \ln \Delta T = (8\P \chi_1 d_1 / (h c)) E - 2 \ln 4 \Gamma^2 (1-\Gamma^2) \quad (6)$$

where E, h, c and χ_1 are respectively the photon energy, Planck's constant, speed of light and extinction coefficient. In particular domains of the low absorption region, we verified aproximately $- \ln \Delta T$ as a linear function of E. From that we get the values for $\Gamma^2 (1- \Gamma^2)$ from the intercept with the $- \ln \Delta T$ axis. The results are shown in table I and fig. 2, for two samples. We conclude that these samples have refraction indexes as those published for similar conditions[6].

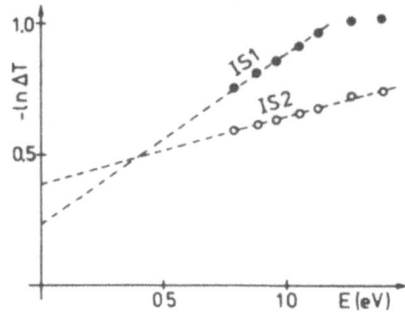

Figure 2. - ln ΔT versus E for undoped inductive samples IS1 and IS2

TABLE I. Some estimated optical parameters

TABLE 1

SAMPLE	m	x_1	Γ_A^2
IS1	3.75	0.107	0.335
IS2	3.35	0.068	0.292
DS1	3.02	0.112	0.253

We know that for an even order m of the fringes the difference between the transmitance values of an uncoated substrate and a coated one are due to film absorption[7]. So using equation (4) we are able to interpret quantitatively absorptance measurements through the simple optical model. Theoreticaly reproduced experimental data are shown in fig. 3 for absorptance A measurements, using the usual definition

$$A = -\log_{10} T \qquad (7)$$

From these results we get the order m of the fringes and the film thickness. Also, from equations (4) and (7) we get equation (8) shown below. By solving equation (8) we obtain the representation of $\sqrt{\alpha E}$ versus E ilustrated in figs. 4a and 4b for several samples. In these figures, B_o is the slope of the high absorption region assimptote and B_1 the low absorption one. E_{to} is defined by the interception of the two assimptotes B_o and B_1. The interception of the high absorption assimptote with the E axis gives the optical gap E_o.

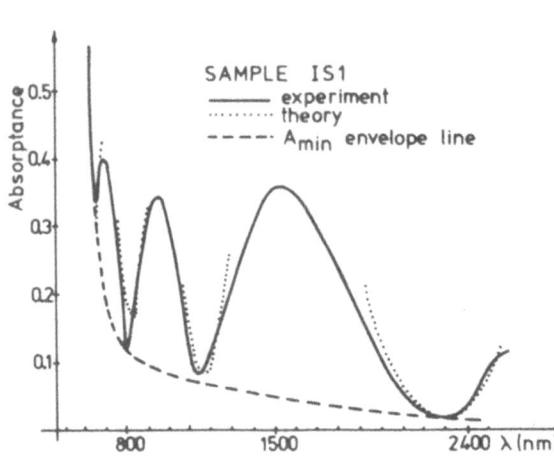

Figure 3. Absorptance spectra for intrinsic sample IS1

$$10^{-A_{min}} = (1-\Gamma^2)\, e^{-\alpha_1' d_1} / (1-\Gamma^2\, e^{-2\alpha_1' d_1}) \qquad (8)$$

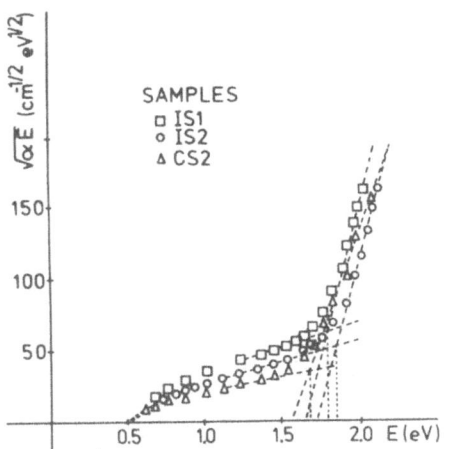

Figure 4a. $\sqrt{\alpha} \, E$ versus E for IS1 (intrinsic inductive), IS2 (idem) and CS2 (intrinsic capacitive)

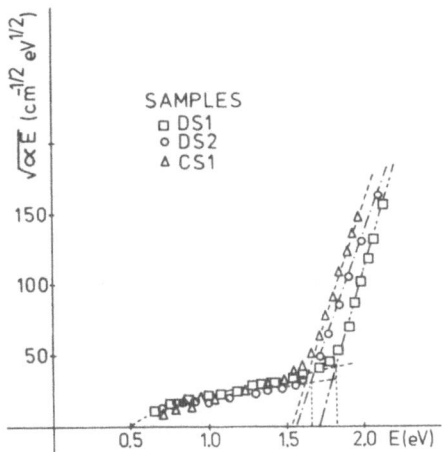

Figure 4b. $\sqrt{\alpha} \, E$ versus E for DS1 (PH_3 doped inductive), DS2 (B_2H_6 idem) and CS1 (intrinsic capacitive)

The extrapolated values for the low absorption region seems to be at least a good approximation of $(E_C - E_F)_T$, the position of the Fermi level at absolute temperature T[5].

From all these results we constructed table II.

TABLE 2

SAMPLE	DC volt (V)	DC (kG) magnetic field	E_G (eV)	E_1 (eV)	$B_1^2 \times 10^{-3}$ (cm^{-1} eV^{-1})	E_0 (eV)	$B_0^2 \times 10^{-5}$ (cm^{-1} eV^{-1})	E_{to} (eV)	$\alpha_{to} \times 10^{-3}$ (cm^{-1})	d_1 (μm)
IS1	+125	0	0.32	~0.50	1.24	1.66	1.77	1.81	2.06	0.31
IS2	0	1	0.40	~0.55	0.96	1.74	1.63	1.86	1.45	0.19
DS1	0	0	0.43	~0.50	0.57	1.71	1.41	1.82	0.87	0.81
DS2	0	0	0.78	~0.50	0.25	1.56	0.92	1.66	0.54	1.15
CS1	-50	0	0.40	~0.45	0.90	1.53	1.08	1.64	0.79	0.83
CS2	+150	1	0.58	~0.45	0.48	1.56	0.84	1.67	0.65	0.72

$E_G = (E_C - E_F)_O$ $E_1 = (E_C - E_F)_T$

TABLE II. List of results for different deposition conditions.

CONCLUSIONS - In our results we observed a fairly good agreement between the simple model and experimental points. Such agreement allowed us to infer the refractive index (see table I) in the low absorption region through equation (6). Deviations of $-\ln \Delta T$ versus E from linearity are ascribed with changes in n. Equation (8) allowed us to plot $\sqrt{\alpha} \, E$ versus E. From such plots we determined the optical gap E_o, the on set of absorption associated with $(E_C - E_F)_T$, the turn over energy E_{to} and the slopes B_1 and B_o as shown in table II. From our values we observed that films are doping sensitized[6] and their properties are biasing dependent[1,2,3]. For photovoltaic purposes we need low E_o (<1.6 eV) and high B_o^2 (~10^5 cm^{-1} eV^{-1})[8,9]. According to our results the capacitive films produced under negative electric static bias (-50 V to -100 V)[1,2,3] seems to be the best way to get films for photovoltaic applications.

REFERENCES

1. - Martins R., Dias A.G. and Guimarães L., A.I.P. Conference Proceedings 73 (1981) 36.
2. - Guimarães L., Martins R. and Dias A.G., Journal de Physique, Colloque C4, Supplement au n^2 10, Tome 42 (1981) C4 - 609
3. - Dias A.G., Guimarães L. and Martins R., Thin Solid Films 89 (1982) 307.
4. - Tauc J., "Optical Properties of Solids" ed. by F. Abeles (North-Holland Amst. 1970).
5. - Anderson D. and Spear W.E., Phil. Mag. B 36 (1977) 695.
6. - Brodsky M.H. and Leary P.A., J. Non - Cryst. Sol. 35 & 36 (1980) 487.
7. - Valeev A.S., Opt.Spectrosc., 15 (1963) 269.
8. - Tsai C.C.and Fritzsche H., Solar Energy Materials 1 (1979) 29.
9. - Fritzsche H., Tsai C.C. and Pearsons P., Solid State Tech. 21 (1978) 55.

This work was partially supported by Junta Nacional de Investigação Científica e Tecnológica under research contract nº 321.81/71.

THE ASSESSMENT OF THIN FILM Cu_xS-CdS SOLAR CELLS
USING CATHODOLUMINESCENCE TECHNIQUES

T.J. Cumberbatch and I.D. McInally,
THORN EMI plc, Central Research Laboratories, Trevor Road, Hayes,
Middlesex, U.K. UB3 1HH

W.K. Ke* and B. Hamilton,
Solid State Electronics, U.M.I.S.T., P.O. Box 88, Manchester,
M60 1QD, U.K.

*Peking Institute of Semiconductors, Chinese Academy of Sciences,
Peking, China

Summary

The object of this work has been to develop a rapid contactless technique for the routine assessment of thin film ($< 5\mu m$) polycrystalline Cu_xS-CdS solar cells prepared by electrophoretic deposition. Cathodoluminescence (CL) is an ideal tool for this task since the data acquired from its three modes of operation (imaging, spectral analysis, decay analysis) can be correlated to provide detailed information about the chemical composition and electrical activity of the constituent materials.

1. INTRODUCTION

Cadmium sulphide films prepared by using electrophoretic deposition techniques typically consist of a 1.5μm thick layer of particles whose average diameter is ∼30 nm (1). Prior to fabrication of the absorber layer this film must be recrystallised whilst preserving its pinhole free nature; a number of different approaches are currently under investigation with pulsed laser radiation a prime contender. After such treatment it is important to be able to assess the uniformity of the layer from electrical, structural and chemical viewpoints. An ideal technique presents all this information quickly and without special sample preparation - analysis of the CL provides useful data about each of these aspects in the first instance. For a more detailed examination, complementary techniques are required although, after initial correlations and with more experience, the CL data can be extrapolated to characterise the properties of a recrystallised layer as will be shown.

Of critical importance to the operation and efficiency of this type of cell is the uniformity of the copper sulphide layer. The complexity of the phase diagram and limited thickness of this material present a problem with regard to its assessment. However, since each Cu_xS phase has an associated band structure and thus characteristic luminescence, the only way to determine unambiguously the type and distribution of phases present on a microscopic scale, is to image the CL through a series of narrow bandpass filters.

This paper discusses both these aspects and illustrates how CL may be used to assess the properties of very thin semiconducting layers on conductive substrates.

2. CATHODOLUMINESCENCE TECHNIQUES AND INSTRUMENTATION

The use of these techniques has, until recently (2), been limited to those materials with high radiative recombination rates and luminescence efficiencies; system development and interpretation grew largely in association with the expanding application of the III-V compounds and their related ternary families (3).

A system with CL facilities may be operated in three different modes, each of which provides unique information.

2.1 Imaging

Integrated and monochromatic micrographs of the luminescence reveal: electrically active extended defects i.e. dislocations, grain boundaries; uniformity of composition, phase and electrical activity (i.e. variations in impurity and defect concentration) - all with high spatial resolution.

2.2 Spectral Analysis

The structure and amplitudes of the peaks in the spectra yield estimations for: the chemical composition and phase; the concentrations of free carriers, point defects and impurities; the energy levels of defects. Obviously complementary measurements are required for correlation or detailed information.

2.3 Decay Analysis

Time resolved measurements of the CL decay are capable of providing values for: the radiative and non-radiative lifetimes; the diffusion length and surface recombination velocity; selected deep level parameters for essentially field free regions. In thin film polycrystalline layers where the size of the grains may control the effective diffusion length/lifetime, the data must be interpreted with extreme caution.

In principle any SEM can be converted for CL work but the operating conditions required are often quite different to those required for conventional topography. Furthermore, when examining films with low CL efficiency such that high excitation levels are necessary, special techniques are often needed to acquire data rapidly lest the condition of the material be changed by the beam. In the course of this work, two systems have been used: a specially designed CL microprobe system (4) and an extensively modified Cambridge Stereoscan S180.

3. CADMIUM SULPHIDE

The CL spectrum from a single crystal of CdS is shown in Fig. 1(a) together with spectra from some evaporated and electrophoretically deposited material*. The differences are striking and perplexing but not dissimilar to those reported for other types of CdS layers for solar cell fabrication (5,6).

---

*(All the spectra presented were acquired with a cooled EMI 9808 tube (S1 photocathode), taken out to 1000 nm and are corrected for the response of the tube).

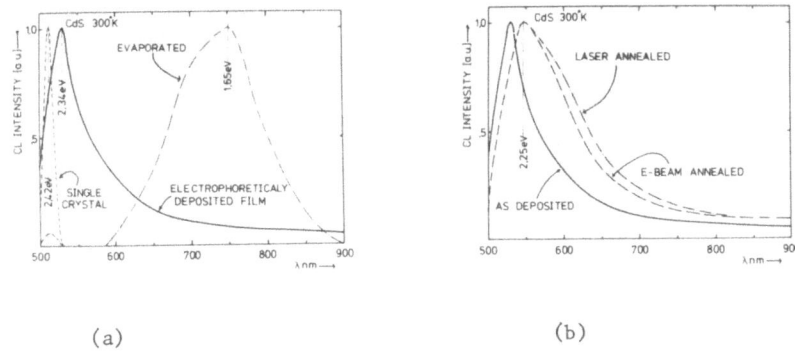

(a)   (b)

Fig. 1   CL spectra from CdS: (a) prepared by different techniques and (b) electrophoreted CdS after transient annealing.

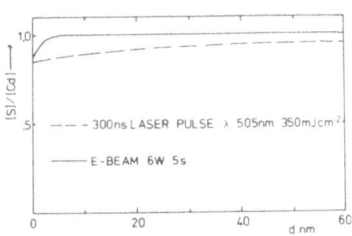

Fig. 2   Film composition after transient annealing

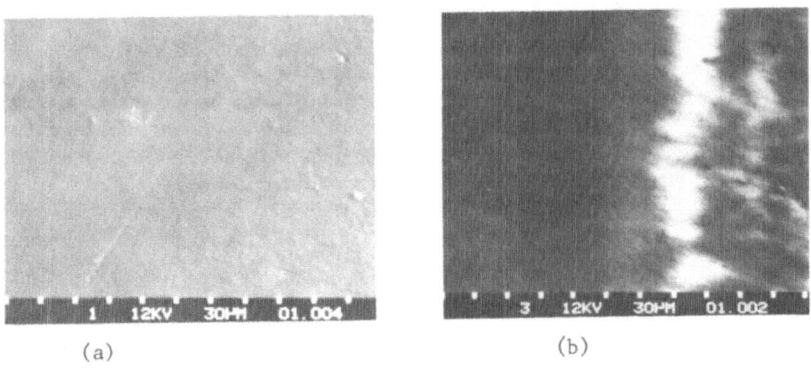

(a)   (b)

Fig. 3   Interface between as deposited film and laser annealed region (a) Emissive   (b) CL

A detailed explanation for the broad peaks commonly observed in thin films of II-VI compounds is still sought; part of this particular project is a study to correlate, if possible, CL spectra with DLTS spectra (preliminary experiments with GaAs:Cr are very promising). The problem is compounded by the electrical activity of antisite defects and the broadness of the peaks down to $\sim 10^\circ K$!

Spectra from electrophoretically deposited CdS films after exposure to a scanned electron beam (306 kW.cm^{-2} for 5s) and a 300 ns 175 mJ laser pulse are shown in Fig. 1(b) with their respective Auger Depth Profiles in Fig. 2. The information gained from all these spectra may be summarised:

1) The as deposited material is cubic (RHEED); the CL peak at 2.34 eV agrees well with reported values.
2) A laser pulse initiates the cubic → hexagonal phase transition; e-beam heating does not. As a result it might be expected that the CL peak of the laser annealed material would move to lower wavelengths whilst that for the e-beam material remains stationary. However, the Auger data reveal a loss of sulphur near the surface suggesting a large sulphur vacancy in the bulk which may give rise to an impurity band. If the activation energy for V_s is $\sim 0.2 eV$ (as reported in the literature) then the peak shift for the laser exposed material is explained, that for the e-beam not, unless the RHEED data are misrepresentative of the bulk.
3) Transient recrystallisation procedures do not introduce defects of the type observed in evaporated material.

The micrographs in Fig. 3, taken at the interface between the as deposited films and a laser exposed region, reveal how the recrystallisation process increases the CL intensity as is expected when the size of the grains is increased substantially (1). The CL image also illustrates how any non uniformities in the recrystallisation may be observed on a microscopic scale.

4. COPPER SULPHIDE

As is well known, the chemistry and electrical properties of this material are still very much open to debate if at all resolved. The properties of this semiconductor have been studied with little enthusiasm, the majority of the work having been performed by Mulder (7), Nakayama (8), Shiozawa et al (9) and more recently Leong (10). There is a lot of disagreement amongst the data with the CL experiments of Loferski et al (11) doing little to resolve the dilemma.

The thickness of this material (a few tenths of a micron) when used as an absorber layer in heterojunction solar cells makes it difficult to excite directly. The data presented here were obtained with an accelerating voltage of 20 kV for films on single crystal substrates; it is therefore possible that the luminescence observed was PL optically pumped by the CL from the underlying CdS - experiments at <10 kV are currently in progress.

The series of micrographs shown in Fig. 4 reveal the multiphase composition of this particular topotaxial film (the CL was imaged through a Hilger and Watts 600 mm Monospex Monochromator with a bandpass of 75Å at the central wavelengths shown). Spectra corresponding to bright spots in the micrographs are shown in Fig. 5. The absence of a chalcocite peak at $\sim 1\mu m$ is difficult to understand since RHEED patterns indicate that this phase is predominant. Furthermore, earlier CL work (11) found that only the chalcocite phase gave rise to appreciable emission; weak luminescence was observed from a $Cu_{1.9}$ tetragonal structure (supposedly unstable at room temperature) with a peak at 1.36 eV.

The broadness of the peaks observed at 1.38 eV and 1.6 eV in this work

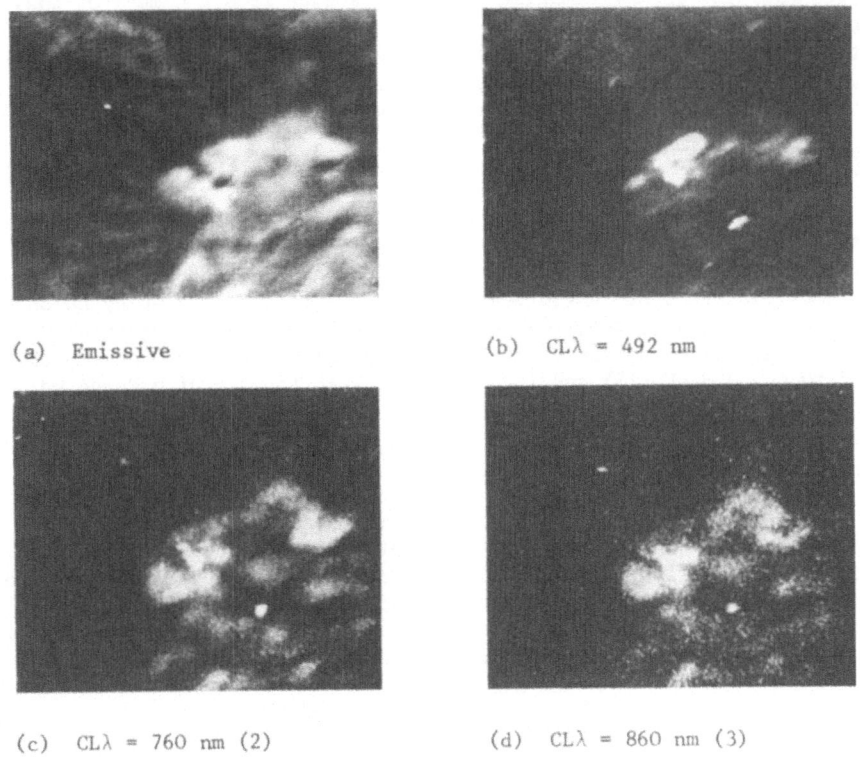

(a) Emissive (b) CLλ = 492 nm

(c) CLλ = 760 nm (2) (d) CLλ = 860 nm (3)

Fig. 4 Copper sulphide layer on single crsytal CdS (77K - Micrographs Δλ = 75Å).

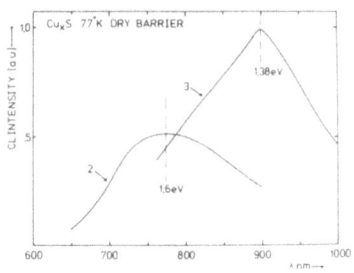

Fig. 5 CL spectra from Cu_xS regions as indicated

suggests that the CL may comprise the luminescence from a number of phases although the energies of the peaks can be related to other work. The 1.38eV spectrum may be the $Cu_{1.9}$ phase (11) or the direct transition in chalcocite (10); the 1.6 eV spectrum may be djurleite (8) or a calculated value for diginite (11). Clearly much more work is needed to understand this result.

5. HETEROJUNCTIONS

Yet to be explored is the nature of the CdS film, after heterojunction formation and subsequent removal of the Cu_xS layer, with regard to the Cu distribution and concentration. Extensive changes in the EBIC response of such junctions after prolonged heat treatment were accompanied by no observable changes in the CL from the CdS (12). Perhaps this was because the strong luminescence associated with Cu in CdS gives rise to bands in the NIR at $\sim 0.6 \rightarrow 0.8$ eV (13) and 1.2 eV (14). With the aid of a North Coast Ge Detector the concentration of copper is to be investigated.

6. CONCLUSION

Although this work is still in its early stages, these results illustrate clearly how CL may be used to provide much sought after information about thin films of CdS and Cu_xS. Techniques for the observation of fast decays have been tested for solar grade silicon (4) and are currently under development for these materials. Preliminary experiments have shown that measurable decays can be obtained from these layers.

7. ACKNOWLEDGEMENT

The authors gratefully acknowledge financial support from the EEC, UK DoI and SERC (UK). We would also like to thank S.P. Gibbons for his work in adding the CL facility to the S180.

8. REFERENCES

1. T.J. Cumberbatch et al (1982) This Conference
2. T.J. Cumberbatch, S.M. Davidson and S. Myhajlenko (1981) I.O.P. Conf. Ser. 60, 197.
3. S.M. Davidson, (1977) J. Microscopy 110, 117.
4. S.M. Davidson, T.J. Cumberbatch, E. Huang and S. Myhajlenko (1981) I.O.P. Conf. Ser. 60, 191.
5. B.J. Feldman and J.A. Duisman (1980) Appl. Phys. Lett. 37, 1092.
6. J.D. Meakin - Private Communication.
7. B.J. Mulder (1973) Phys. Stat. Sol. (a) 18, 633.
8. N. Nakayama (1969) Jpn. J. Appl. Phys. 8, 450.
9. L.R. Shiozawa et al. Final Report Contract AF33 (615)-5224 (1969).
10. J. Leong (1980) Ph.D. Thesis, University of California (LLL).
11. J. Loferski et al. (1979) Solar Energy Materials 1, 157.
12. S.P. Shea (1981) Ph.D. Thesis, University of Delaware.
13. I. Broser, H. Major and U.J. Schulze (1965) Phys. Rev. 140, A2135.
14. A. Suzuki and S. Shionoya (1971) J. Phys. Soc. Jpn 31, 1455.

PHOTOVOLTAIC EFFECT IN SnTe/CdTe JUNCTIONS

M. KANE, G.W. COHEN-SOLAL, D. LAPLAZE
Laboratoire des semi-conducteurs et énergie solaire
Faculté des Sciences Dakar
G. COHEN-SOLAL
Laboratoire de physique des solides C.N.R.S.
Meudon-Bellevue France

Summary

Studies of p SnTe/ n CdTe junctions we have prepared show that band diagrams of these structures are closed to Schottky diode model, with interfacial dead layer. In some cases, structures exhibit surface states. Using such model, theoretical calculations well agree with experimental spectral responses in the range 0.3 - 2 µm. By optimization of preparation parameters, we can hope to get higher quantum efficiencies.

I. INTRODUCTION

Cadmium telluride, which is a semiconducting material with an energy gap around 1.45 eV, is particularly appropriate to photovoltaic conversion of solar energy. Moreover, CdTe allows preparation of different alloys or solid solutions with gradual composition. Such structures are active in spectral range more extended than CdTe p-n homojunctions and can also be used as infrared detectors.

Tin telluride (SnTe), which is a cubic compound with lattice parameter closed to CdTe one (lattice mismatch around 2.5%), can grow by epitaxial technics on CdTe substrates giving gradual gap structures, as Cd Hg Te or other similarly ternary alloys (1), (2), (3).

We have prepared and studied pSnTe/nCdTe junctions and isotype diodes pSnTe/pCdTe. In this paper, we only report results about the p-n junctions ; the other ones have been used merely to determine the electron affinity of SnTe, which is absolutely necessary to build junction band diagrams.

In this work, we have first interpreted electrical properties of our structures using a like type Schottky model. In this way, we obtain also quite good agreement between theoretical spectral responses and experimental curves, in the range 0.3 - 2 µm. (Fig I)

2. SAMPLES PREPARATION

Epitaxial growth of SnTe layers on CdTe monocrystals is carried out by the well known E.D.R.I. (evaporation and diffusion in isothermal system) or C.S.V.T. (close spaced vapour transport) technics (2), (4), (5) both characterized by high deposition efficiencies.

Films obtained by E.D.R.I. technics are made of large crystals grown perpendicularly to substrate surface, showing sharply definite grain boundaries. Using C.S.V.T. method, layers are very thin and are made of small cubic crystals with random orientations.

In both cases, cathodoluminescence and E.B.I.C. observations show that space charge region is inside CdTe material. Concentration profiles measurements exhibit a local zone constituted by Cd Sn Te solid solution, more extended for E.D.R.I. samples than for C.S.V.T. ones. Moreover, it appears that doping impurities (In) in base material and Cd atoms diffuse until the surface of epitaxial layers. In comparison with E.D.R.I. samples, C.S.V.T. junctions are more abrupt, with higher surface states concentration.

3. ELECTRICAL PROPERTIES. I-V , C-V CHARACTERISTICS

In our junctions, SnTe material is strongly degenerate and looks like a metallic compound . This metallic behaviour of SnTe proves availability of the chosen Schottky model.

Detailed studies (8) of dark I-V and C-V characteristics lead to accurate interpretation of experimental results when the later model is used. It has been possible, so, to reach SnTe electron affinity value which is inside the range 4.68 - 4.92 eV.

For each species of diodes, conduction phenomena are mainly due, for $T>300$ K to thermionic emission. In the same way, between 140 and 300 K conduction processes arise from thermionic field emission.

Difference is observed, between E.D.R.I. and C.S.V.T. prepared samples, at low temperatures ($T<140$ K). For the first ones, it was found that the dominant mechanism of carrier transmission through the junction is tunnel effect; for the others, it is always thermionic field emission which is prevailing. Nevertheless, weak tunnel effect contribution is observed.

This analysis has allowed us to build the band diagram drown on figure 2 which is consistent with experimental results for C-V characteristics. For E.D.R.I. structures, linear variation of $1/C^3$ versus reverse bias V displays a gradual interface, in conformity with concentration profiles measurements. On the other hand, C-V characteristics for C.S.V.T. prepared samples can be interpretated with two capacitances in parallel. The first one is due to space charge and varies as in the case of an abrupt junction; the second one is attributed to interfacial layer according to Goodman model (9).

4. SPECTRAL RESPONSES

The spectral dependance of photovoltaic response of pSnTe/nCdTe junction shows two regions: first a large response for radiations of wave-lengths smaller than 0.86 μm; a second, a tail extending to lower energies until wave-length around 2 μm.

When the structure is illuminated on SnTe side, one can

observe all the photocurrent spectra. On other case, when light comes into the structure by CdTe side, the second band is only observed ($\lambda > 0.86$ µm).

Interpretation of this experiment can be achieved using previous model in which one can point out regions, each of them having specific properties:

-The first region is superficial SnTe layer acting as optical filter.

-The second region is the local solid solution Cd Sn Te. Its contribution to photocurrent is negligible for the first part of the response. However carrier generation corresponding to the tail band occurs in this region.

- The third region is the CdTe base in which takes place across-the-gap pair generation for wave-length smaller than 0.86 um.

To take into account interface states and band bending effects unsuitable to separation of the carriers created near the interface, we introduce a dead layer close to CdTe material

We have calculated spectral response for 0.86 µm using a modified Schottky model according to Green calculations (8) to take into account the lowering of collection efficiencies in the space charge region.

The whole processes causing the fall off in spectral responses i.e tunnel effect, trapping effect in side interfacial region, thermionic emission over the barrier, can be represented as a superficial recombination velocity. Moreover, one must take into account the recombination effects due to superficial states. Theses effects are acting on the whole photocarriers whatever may be the generation point.

Let F_s be the electric field on the interface; all the first above-mentioned effects are important in a L_s thickness region lying in CdTe material from interface with $L_s = K_B T / q |F_s|$. In this layer where occurs the absorption of the highest energy photons, the collection efficiency is very less than I. That is the explanation of spectral response vanishing for shortest wave-lengths. The dead region depth is taken equal to L_s and its optical properties are chosen to be those of CdTe.

Taking as a basis these assumptions, photocurrent can be written as :

$$I_{ph.} = q \Phi_0 \frac{(1-R)}{Z} e^{-(\alpha_0 d + \alpha L_s)} \left[(1-e^{-\alpha W}) - \frac{\alpha L_s (1-e^{-(\alpha + \frac{1}{L_s})W})}{(1+\frac{\mu_n |F_s|}{v_{mr}})(\alpha L_d + 1)} + \frac{\alpha L_p e^{-\alpha W}}{\alpha L_p + 1} \right] \quad (1)$$

The needed parameters are only the donor density N_d in CdTe and built in potential V_d, we can deduced from capacitance measurements.

In relation (1), first factor $q \Phi_0 [(1-R)/Z] e^{-(\alpha_0 d + \alpha L_s)}$ is related to reflection and absorption in SnTe and takes into account surface states effects. It can be taken as a constant for a given structure.

We have obtain good agreement between experimental results and theoretical calculations as shown on figure 3.

The second part of spectral response arises from two processes : photocarriers generation in solid solution and electrons escape over the barrier toward semiconductor. A detailled analysis including electron transition probability and escape probability leads to the next expression for photocurrent :

$$I_{ph.} = A F(\hbar\omega) e^{-\alpha d'} (\hbar\omega - \phi)^{3/2} \qquad (2)$$

In this expression ϕ is the barrier height and can be deduced from linear dependance of $(I_{ph.})^{2/3}$ versus $\hbar\omega$.
Figure 4 proves the good accordance between theoretical calculation and experimental data.

5. CONCLUSION

We have prepared p SnTe/ n CdTe structures. Experimental measurements of their photoelectrical characteristics are consistent with a Schottky diode model including interfacial dead layer. Effects of this dead layer can be minimized when growing rate is low (E.D.R.I. method). One can then obtain gradual junctions with interfacial solid solution. When the epitaxial growth is fast (C.S.V.T. method) junctions are abrupt, exhibiting interface states.

Advantages of these structures lie in the extension of the spectral range response (0.3 - 2 µm) overlapping solar spectrum.

Possible enhancement of quantum efficiencies can be reached by optimizing optical properties of SnTe layer.

BIBLIOGRAPHY

1. Y. Marfaing, G. Cohen-Solal, F. Bailly; Conf. int. Phys. Second. Dunod Paris 1964
2. G. Cohen-Solal, Y. Marfaing, F. Bailly; Rev.Phys.app. I - II - 1966
3. F. Bailly, Y. Marfaing, G. Cohen-Solal, J. Melngalis journal de Physique 28 - 573 - 1967
4. J.E. May; J. Electrochem. Soc; 112 - 710 - 1965
5. A.L. Farhenbruch, V. Vasilchenko, F. Buch, K. Mitchell, R.H. Bube; Appl. Phys. Letters 25 - 605 - 1974
6. M. Kane; Thesis Dakar 1982
7. A.M. Goodman; J. Appl. Phys. 34 - 329 - 1963
8. M.A. Green; J. Appl. Phys. 50 - 1116 - 1979

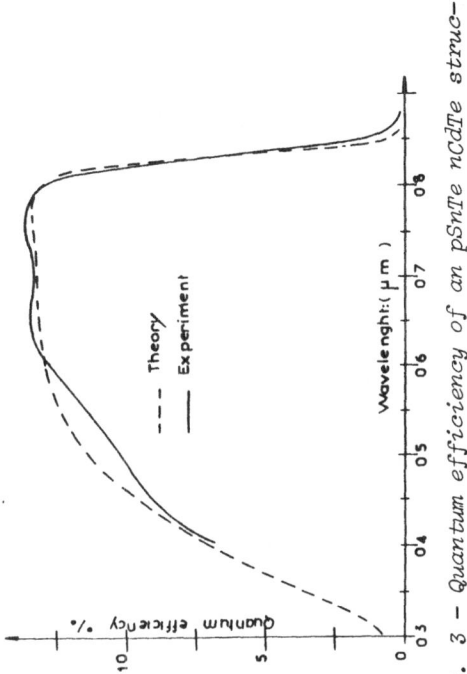

Fig. 1 - Spectral response of an pSnTe nCdTe structure

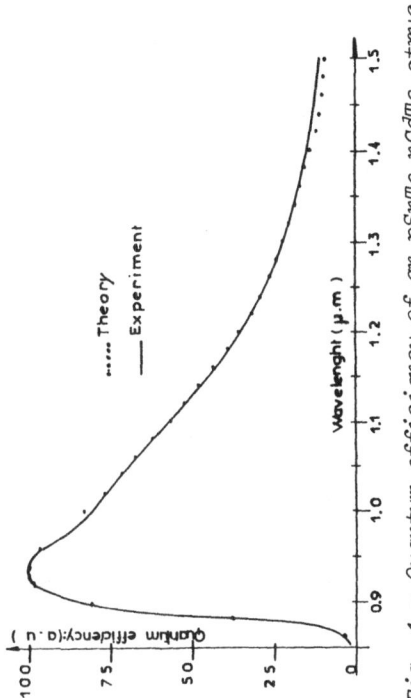

Fig. 2 - Energy band scheme of the pSnTe nCdTe heterojunction

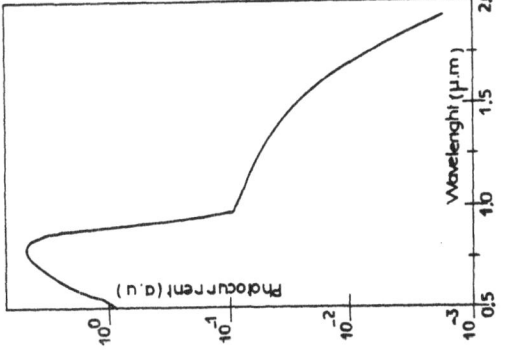

Fig. 3 - Quantum efficiency of an pSnTe nCdTe structure - Illumination from SnTe side

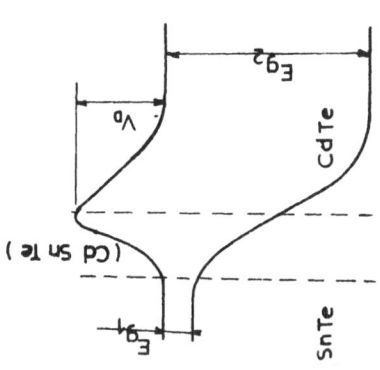

Fig. 4 - Quantum efficiency of an pSnTe nCdTe structure - Illumination from CdTe side

ORGANIC PHOTOVOLTAIC MATERIALS: POLYACETYLENE.

J. KANICKI, P. FEDORKO[#], S. BOUÉ and E. VANDER DONCKT
Chimie Organique Physique, Univeristé Libre de Bruxelles,
50, Av. F.D. Roosevelt, B-1050 Bruxelles (Belgium).

[#] Department of Physics, Electrotechnical Faculty,
Slovak Technical University, Gottwaldovo nám.19,
812 19 Bratislava, Czechoslovakia.

Summary

The electrical properties of contacts between various metals and trans-polyacetylene (hereafter trans-$(CH)_x$) have been examined in the dark and under illumination. The current-voltage, capacitance-voltage and photocurrent-voltage characteristics were measured. Photovoltaic action spectra of blocking contacts were monitored and compared to the absorption spectra. The carrier collection efficiency is approximately 1% at 1.93 eV. The dependence of I_{SC} and V_{OC} on light intensity was determined. From the characteristics recorded in the photovoltaic mode under white light illumination it appears that the light into electrical energy conversion efficiency is $\simeq 0.1\%$.

1. EXPERIMENTAL

Polyacetylene films were synthesized at room temperature from acetylene at the liquid interface of a Ziegler-Natta-type catalyst. The polymer is a mixture of cis and trans isomers. The more stable trans species was obtained by heating up the samples to 175°C during about 25 minutes. All materials were characterized by elemental analysis, X-ray diffraction, infrared and electronic absorption spectra (1,3). A standard technique was used for studying the electrical and photovoltaic characteristics of this system. All measurements were performed between -77° and +110°C under reduced pressure ($\sim 10^{-5}$ Torr). The cell area was varied between 0.1 and 1 cm², and the thickness of the $(CH)_x$ films was between 20 and 75 μm. More details on the fabrication of the cell have been described (1,4,6).

2. RESULTS AND DISCUSSION.

Junction properties between undoped and doped trans-polyacetylene and various metals have been investigated. Rectifying contacts were obtained by direct evaporation of In, Al, Pb, Sn, Ni and Cu on the polymer film(5). Ohmic contacts were made with Electrodag +502. Rectification properties and photovoltaic characteristics of the solar cells involving $(CH)_x$ differed considerably from each other depending on the history of the polymer sample and on the way the contacts were prepared (4,5,6). From these experiments and from the temperature effect on the J-V and C-V curves it appears that polyacetylene surface states and/or the presence of an interfacial layer of oxidized polymer have a strong effect on the current-voltage relationship. Typical rectification characteristics of the Schottky barrier cell for In, Al, Pb and Sn are shown in figure 1. In the dark, the device exhibits a strong rectifying behaviour, with a forward bias corresponding to a negative voltage at metal electrode. Rectification ratios up to 500 were observed. The series resistance of the device (R_S) is dominated by the bulk resistance of the polymer and is thus rather large for the undoped polymer (table I).

TABLE I: Schottky barrier characteristics at 22°C (V_C: contact potential obtained by extrapolation of linear part of J-V characteristics).

trans-$(CH)_x$	$R_S(k\Omega)$	$V_C(V)$	n	$J_o(A/cm^2)$
In	27	0.7	2.4	1.6×10^{-8}
Al	23	0.7	3.6	1.9×10^{-8}
Pb	1.2	0.7	2.1	1.6×10^{-9}
Sn	23	0.7	3.4	1.0×10^{-8}
Ni	52	0.26	1.3	4.6×10^{-8}
trans-$(CHI_y)_x$				
Al	4.7	0.9	8.8	3.8×10^{-6}
trans-$[CH(AsF_5)_y]_x$				
Al	12	1.18	11.6	8.0×10^{-7}
Pb	4.7	0.78	3.2	7.0×10^{-7}

Current density-voltage (J-V) characteristics in the forward direction follow the well known J-V relationship for a Schottky barrier (1):

$$J \simeq J_o \left[\exp q(V-R_S I)/nkT \right]$$

(for $(V-R_S I) > 3kT/q$)

where the symbols have their usual meaning. n, the perfection factor is almost temperature independent. Its departure from unity arises from image-force lowering and/or from interface effects. The reverse characteristics are better explained in terms of image force barrier lowering (1). Furthermore, the J-V curves obtained in the dark and under illumination cross each other as it can be expected with a photoconductive polymer. Figure 3 shows the results of the capacitance-voltage (C-V) measurements carried out at various frequencies. C^{-2} is plotted as a function of V in order to assess N_A and V_C according to the equation:

$$C^{-2} = (2/qN_A\epsilon_S)(V_C+V_R)$$

The average values at room temperature of V_C, N_A, $C(V_R=0)$ and the width of the depletion region (W) are given in table II.

TABLE II: Material parameters derived from junction properties.

trans-$(CH)_x$	$V_C(V)$	$N_A(cm^{-3})$	$C(V_R=0)(nF)$	$W(\overset{\circ}{A})$	f(Hz)
In	0.8	2.3×10^{17}	9.3	380	500
Al	0.8	2.7×10^{17}	54	400	360
Pb	0.5	9.0×10^{16}	81	440	750

The shape of the short-circuit photocurrent action spectra is identical with the absorption spectra of polyacetylene between 450 and 1000 nm (figure 4). This indicates that only light absorbed in the vicinity of the blocking contact leads to charge carriers that will be collected by the external circuit (1). The collection efficiency is of the order of 1% at 1.93 eV. Moreover, the blocking contact becomes negative under irradiation, which is qualitatively consistent with a Schottky junction between the active metal and p-type polyacetylene. The photocurrent-photovoltage characteristics in the photovoltaic mode under constant light intensity are usually linear, figure 5. This is due to the low probability of carrier production or collection and not to high series resistance (1). Doping of $(CH)_x$ by I_2 or AsF_5 decreases I_{SC} and V_{OC} (table III). The open-circuit photovoltage (V_{OC}), the short-circuit current (I_{SC}), the fill factor (FF) and the energy conversion efficiency (η) under white light and monochromatic ($\sim 10^{16}$ photons/cm^{2}sec) illumination are given in table III.

TABLE III: Junction parameters.

trans-$(CH)_x$	Source	P_{in}(mW/cm^2)	FF	J_{SC}(μA/cm^2)	V_{OC}(mV)	η(%)
In	poly	100	0.25	190	88	4.1x10^{-3}
	mono	7	0.25	10	46	1.7x10^{-3}
Al	poly	50	0.15	67	250	5.0x10^{-3}
	mono	1	0.32	13	86	3.6x10^{-2}
Pb	poly	50	0.25	1.23	103	6.0x10^{-5}
trans-$(CHI_y)_x$						
Al	poly	50	-	0.34	25	-
trans-$[CH(AsF_5)_y]_x$						
Al	poly	50	-	1.26	6.4	-
Pb	poly	50	0.25	0.3	0.4	6 x10^{-8}

The dependence of the short-circuit currents and the open-circuit voltages on incident light intensity is shown in figure 6. The short-circuit current varies as $I_{h\nu}^m$, where $I_{h\nu}$ is the incident light intensity and m is a factor which ranges from 0.5 to 1 depending on the metal contact. The photovoltage increases logarithmically with light intensity as shown in figure 6. In any case η is found to be rather low.

3. REFERENCES.

1. J. Kanicki, Propriétés optiques, électriques et photovoltaïques du trans-polyacétylène, Ph.D. Thesis, 1981, Université Libre de Bruxelles, Belgium.

2. E. Vander Donckt, B. Noirhomme and J. Kanicki, R. Deltour and G. Gusman, J. Appl. Pol. Sci., 27, 1 (1982).

3. J. Kanicki, S. Boué and E. Vander Donckt,
   (a) J.C.S. Faraday Trans. II, 77, 2157 (1981).
   (b) Thin Solid Films (1981, in press).

4. J. Kanicki, P. Fedorko, S. Boué and E. Vander Donckt, 2nd General Conference of the Condensed Matter Division of EPS, 22-25 March 1982, Manchester, U.K., 6A, 10.X.59 (1982).
   Ed. Prof. V. Heine, Published by: The European Physical Society.

5. E. Vander Donckt and J. Kanicki, Europ. Pol. J., 16, 677 (1980).

6. J. Kanicki, S. Boué and E. Vander Donckt,
   (a) Mol. Cryst. Liq. Cryst. - Proceeding of the "International Conference on Low-Dimensional Conductors", Boulder, Colorado, U.S.A., 10-14 August 1981 - Part D.
   (b) Solar Cells (to be published).
   (c) J. Phys. D: Appl. Phys. (to be published).

7. J. Kanicki, A.C.S. Polymer Preprints, 23(1), 138 (1982) - Proceedings of the "Symposium on Conducting Polymers at the American Chemical Meeting in Las Vegas, March 28-April 2, 1982, U.S.A.

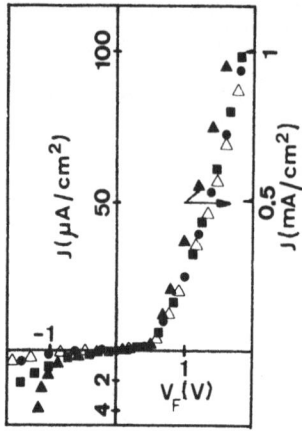

Figure 1. Rectification characteristics of (●)In, (■)Al, (▲)Pb and (△)Sn/trans-$(CH)_x$/Electrodag +502 sandwich cells.

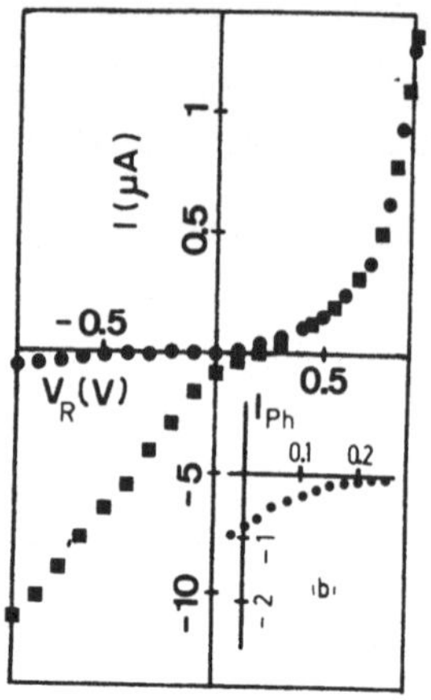

Figure 2. (a) I-V characteristics of a Al/trans-$(CH)_x$/Electrodag +502 cell in the dark (●) and under white irradiation (■). (b) Difference between the dark current and the photocurrent.

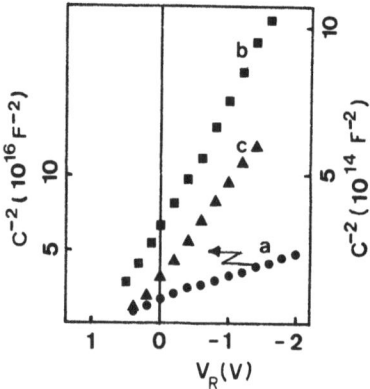

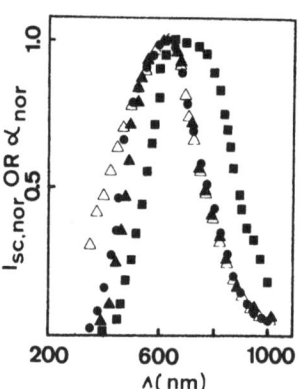

Figure 3. C^{-2}-V plot of the sandwich cell at frequencies of 500 Hz (a), 360 Hz (b) and 750 Hz (c); (●)In, (■)Al and (▲)Pb/trans-$(CH)_x$/Electrodag +502. For each sample a proper choice of modulation frequency has been made.

Figure 4. Electronic absorption spectra (△) of trans-$(CH)_x$. Experimental (●)In, (■)Al and (▲)Pb short-circuit photovoltaic current action spectra with light incident on the blocking contact. α and I_{SC} are normalized to their maximum.

- 566 -

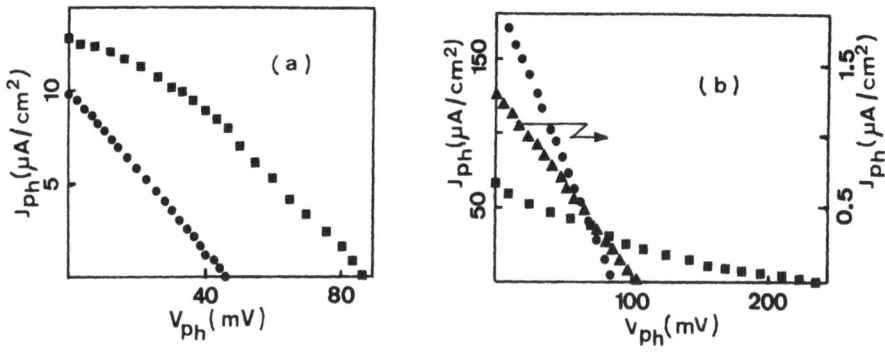

Figure 5. The J_{ph}-V_{ph} characteristics of the (●)In, (■)Al and (▲)Pb sandwich cell under polychromatic (b) and monochromatic (a) illumination.

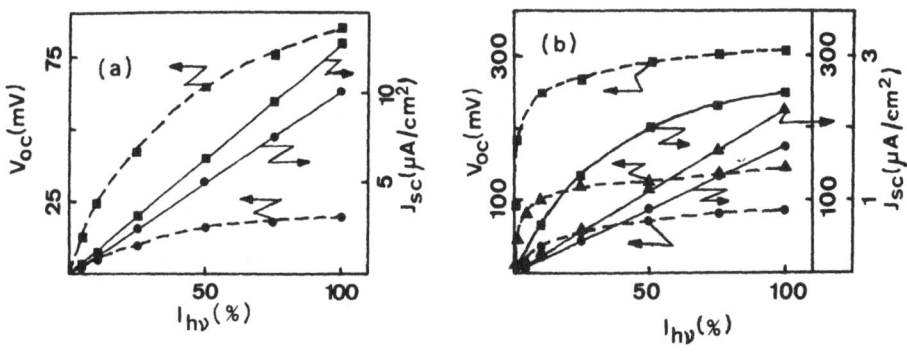

Figure 6. Dependence of the short-circuit current and the open-circuit voltage on incident light intensity ($I_{h\nu}$) under (a) monochromatic and (b) polychromatic illumination.

COMPOSITIONAL ANALYSIS OF CuInS$_2$ CHALCOPYRITE SEMICONDUCTOR

H. L. Hwang, L. M. Liu, M. H. Yang, T. F. Hung, P. Y. Chen, J. R. Chen
National Tsing Hua University
Hsin-chu, Taiwan, R.O.C.

C. Y. Sun
Industrial Technology Research Institute
Hsin-chu, Taiwan, R.O.C.

SUMMARY

A chemical method has been developed for precise determination of elemental compositions of CuInS$_2$ in the form of both synthesized charges and single crystals. The results were used to monitor the processes of material synthesis and crystal growth. Other analytical techniques and prospective methods for thin film analysis are also discussed.

1. INTRODUCTION

CuInS$_2$ has evoked technological interest as a potentially high efficiency solar cell (1-2). In the growth of CuInS$_2$ single crystals, different methods such as melt-growth (3-5) and chemical vapor transport (CVT) (6-8) have been used. CuInS$_2$ was synthesized from its constituent elements (6-8). It is generally agreed that the electrical conducting phenomena are mainly affected by the intrinsic defects of the material (9-10) and the crystals are heavily compensated (2-3). Since material properties are closely related to its compositions and very limited work has been done for I-III-VI$_2$ compounds as for precise identification of stoichiometry, it is, therefore, the purpose of this work to develop a chemical method in precise determination of CuInS$_2$ compositions.

The compositional deviations from the ideal chemical formula can be plotted in a compositional triangle and expressed by two parameters x and y as Cu$_{1-x}$In$_{1+x}$S$_{2+y}$, or by parameters δ and ε as Cu$_{1-\delta}$In$_{1+\delta}$S$_{2+\delta+\varepsilon}$(11).

2. A COMMENT ON ANALYSIS TECHNIQUES

The major factors affecting the accuracy in compositional determination are: (1) the interference of one major constituent of the compound with the determination of another one, (2) the sensitivity and accuracy of the instrumentation used. This explains why the general analytical techniques such as EDAX (Energy Dispersive Analysis of X-ray), RBS (Rutherford Backscattering Spectrometry), PIXE (Particle-Induced X-ray Emission) and X-ray Fluorescence can measure elemental composition and concentration to typically 5% without recourse to secondary standard (12-13).

The improvement in accuracy of EDAX largely depends on the treatment of the X-ray absorption of the constituent elements, and it also depends on the spectrum background treatment and the energy and incidence angle of the impinging electrons as well.

RBS should have the least interference effect among those physical techniques due to elastic scattering of α-particles. Also due to its well-defined cross-section, RBS should possess an accuracy to the extent of about 1%, which nevertheless needs exact fitting of the RBS spectrum with the theoretical curve.

PIXE is equivalent to that of EDAX, but its cross-section is hard to define.

Compositional determination by X-ray fluorescence needs good standards to ensure accuracy. The interference effect is a severe problem.

Atomic absorption (AA) should be immune from the interference effect due to its detection of elements in solution state. However, the accuracy is still limited to a few percent due to its being a trace-element analysis technique.

Experimental results of inter-comparison of the above techniques on some chalcopyrite compounds will be discussed elsewhere.

Since accuracy and precision are the primary goals of stoichiometry analysis, only conventional, well-established methods are considered to be useful. It is commonly believed that the analytical precision attainable for some techniques like titrimetry, gravimetry and electrochemical techniques, especially that of coulometric analysis, can be in the range of 0.01 percent in favorable cases (14). The development for stoichiometry determination will, therefore, emphasize on the way to optimal combination of the above techniques.

3. A CHEMICAL ANALYSIS METHOD

The developed method for stoichiometry analysis is as follows:

Samples of $CuInS_2$ were digested in an oxidizing acid to ensure their complete transformation to Cu^{2+}, In^{3+} and SO_4^{2-} ions. The stoichiometry determination was done sequentially from the same solution by controlled potential electrodeposition of Cu^{2+} ions, then by titrimetric determination of SO_4^{2-} ions. The detailed experimental procedures are described in the flow-chart of Figure 1.

The relative errors associated with the determination of Cu and In were found to be -0.08% and +0.11% respectively, indicating fulfillment of the requirement for accurate stoichiometry assessment; while that of S is -0.66% which, though relatively higher is acceptable.

Uniformity of the synthesized charge was first examined. Table 1 lists the analytical results from different portions from a synthesized charge. It is clear that the composition deviates from sample to sample.

A three zone Lindberg furnace with a long flat zone was used to eliminate the effects of temperature gradient, and this furnace can cool down to room temperature in about one hour. Water quenched from the highest temperature using a "conventional" furnace (in which a temperature gradient may exist) was another method chosen to improve the homogeneity of the synthesized charge. Both results are listed in Table 2. It is shown that the deviations between different samples are much reduced. We can, therefore, conclude that by employing flat zone furnace or a quick quenching method, the homogeneity problem of $CuInS_2$ synthesized charges can be largely improved.

Some stoichiometry data of the CVT crystals and their corresponding sources are listed in Table 3.

It is difficult to correlate the stoichiometry between SB5 and L-B5-1, since the charges used for the CVT growth may have different compositions, although they were taken from the same chunk. Meaningful comparison must be based on the uniformity of the synthesized chunk.

Since water-quenching from high temperature produces a homogeneous chunk, again comparisons were done by using the quenched chunk as the CVT sources (NC2 for L-NC2-2 and N2 for L-N2-1). Consistency of the data between the crystals and the sources are indicated in both cases. If this is true, then the CVT growth could be the proper method to preserve compositions between sources and crystals.

Other analytical results are plotted on the triangle composition diagram as shown in Figure 2. Most data were found to exist in the region: $|y| \leq 8 \times 10^{-2}$, $0 < x < 2 \times 10^{-2}$. Here all the powder used for the above experiments have been identified to be chalcopyrite phase by X-ray diffraction measurements.

4. PROSPECTS FOR THIN FILM ANALYSIS

Thin films are most desirable for low-cost solar cells, and it is essential to acquire an accuracy in film composition determination to be able to correlate with its electrical characteristics. None of the prescribed physical techniques are qualified in this respect.

The chemical method we developed has the accuracy but it is a rather material (≥ 50-100 mg) and time-consuming process. A modified method should therefore be developed which has the advantages of simplicity but still possesses the same accuracy or better.

A method currently under development in our laboratory for a sister material, $CuInSe_2$, by coulometry may be a solution to this problem, in which the sample is dissolved in solution, and by sequentially changing its pH value, by applying different bias, and using fresh Hg electrodes at different stages, the quantities of the constituent elements can be accurately determined. Results of this technique will appear in a future publication.

REFERENCES

(1) J. J. Loferski, J. Appl. Phys. 27 (1956) 777.
(2) S. Mora, N. Romeo and L. Tarricone, Solid State Commun. 29 (1979) 155.
(3) D. C. Look and J. C. Manthuruthil, J. Phys. Chem. Solids 37 (1976) 173.
(4) B. Tell, J. L. Shay and H. M. Kasper, Phys. Rev. B4 (1971) 2463.
(5) H. L. Hwang, C. Y. Lin, C. Y. Sun and H. R. Maa, Thin Solid Films, 81, 2 (1981) 144.
(6) H. L. Hwang, C. Y. Sun, C. Y. Leu, C. L. Cheng and C. C. Tu, Rev. de Phys. Appl. 13 (1978) 745.
(7) C. Y. Sun, H. L. Hwang, C. Y. Leu, L. M. Liu and B. H. Tseng, Jap. J. Appl. Phys. 19 (Suppl. 19-3) (1980) 81.
(8) H. L. Hwang, B. H. Tseng, C. Y. Sun and J. J. Loferski, Solar Energy Mat. 4 (1980) 67.
(9) B. Tell and J. L. Shay, J. Appl. Phys 43 (1972) 2469.
(10) C. Schwab, Jap. J. Appl. Phys. 19 (Suppl. 19-3) (1980) 55.
(11) B. R. Pamplin, Prog. Cryst. Gorw. Charact., 1 (1979) 331.
(12) J. I. Goldstein and H. Yakowitz, Practical Scanning Electron Microscopy (Plenum, New York, 1975) p. 337.

(13) MeV Ion Beam Analysis Manual of General Ionex Co. 1980.
(14) R. A. Landise, Analytical Chemistry, NBS Special Publication 351 (1972) p. 19-74.

TABLE 1. Analytical results of stoichiometry for samples taken from different portion of a synthesized $CuInS_2$ ingot, SA7.

Sample No.	Measured Amount ($\times 10^{-6}$ mole)			Translated Analytical Result			
				$Cu_\alpha In_1 S_\beta$		$Cu_{1-\delta} In_{1+\delta} S_{2+\delta+\epsilon}$	
	Cu	In	S	$\alpha(\pm 0.002)$	$\beta(\pm 0.03)$	$\delta(\pm 0.002)$	$\epsilon(\pm 0.03)$
#1	3.18	3.22	6.38	0.988	1.98	+0.006	-0.01
#2	3.92	4.37	8.56	0.897	1.96	+0.054	+0.01
	3.90	4.32	8.66	0.903	2.00	+0.051	+0.06
#3	3.97	4.30	8.22	0.923	1.91	+0.040	-0.05
#4	4.06	4.23	8.10	0.960	1.92	+0.020	-0.07
	3.98	4.21	8.27	0.945	1.96	+0.028	-0.01
#5	3.94	4.30	8.12	0.916	1.89	+0.044	-0.07

TABLE 2. Analytical results of stoichiometry for samples taken from different portions of $CuInS_2$ ingots.

	Synthesis Conditions	Sample No.	Analytical Result	
			$\delta(\pm 0.002)$	$\epsilon(\pm 0.006)$
SX1	Synthesized charge re-melt in flat zone furnace for 1 day, furnace cooling	#1	+0.012	-0.035
		#2	+0.011	-0.042
		#3	+0.007	+0.025
NC1	600°C 1 day, 1100°C 1½ day in flat zone furnace, furnace cooling	#1	+0.002	+0.001
		#2	0.000	+0.041
		#3	+0.001	+0.010
N2	600°C 1 day 1060°C 1½ day water quenched	#1	+0.006	-0.012
		#2	+0.005	-0.012
		#3	+0.006	-0.019

TABLE 3. Analytical results of stoichiometry of CVT crystals and the corresponding sources.

Sample No.	Sample Preparation Conditions		$Cu_\alpha In_1 S_\beta$		$Cu_{1-\delta} In_{1+\delta} S_{2+\delta+\epsilon}$	
	Source	Temperature	$\alpha(\pm 0.001)$	$\beta(\pm 0.006)$	$\delta(\pm 0.001)$	$\epsilon(\pm 0.006)$
*SB5	$CuInS_{2.1}$	furnace cooling from 1140°C	0.993	2.149	+0.004	+0.153
			0.983	1.956	+0.008	-0.036
L-B5-1	SB5 I_2:65 mg/cc	800 - 750°C furnace cooling	0.969	1.960	+0.016	-0.025
			0.971	1.926	+0.014	-0.061
*NC2	$CuInS_{1.99}$	water quenched from 1100°C	0.995	1.957	+0.002	-0.041
L-NC2-2	NC2, I_2:10 mg/cc	780 - 750°C furnace cooling	0.999	1.978	+0.001	-0.022
*N2	$CuInS_2$	water quenched from 1100°C	0.988	1.975	+0.006	-0.019
L-N2-1	N2, I_2:10 mg/cc	780 - 720°C furnace cooling	0.993	1.974	+0.004	-0.022
L-N2-3	N2 I_2:8 mg/cc ,S added	760 - 720°C furnace cooling	0.983	2.000	+0.008	+0.008

* The synthesized charges are indicated by "*", all the crystals were grown by CVT.

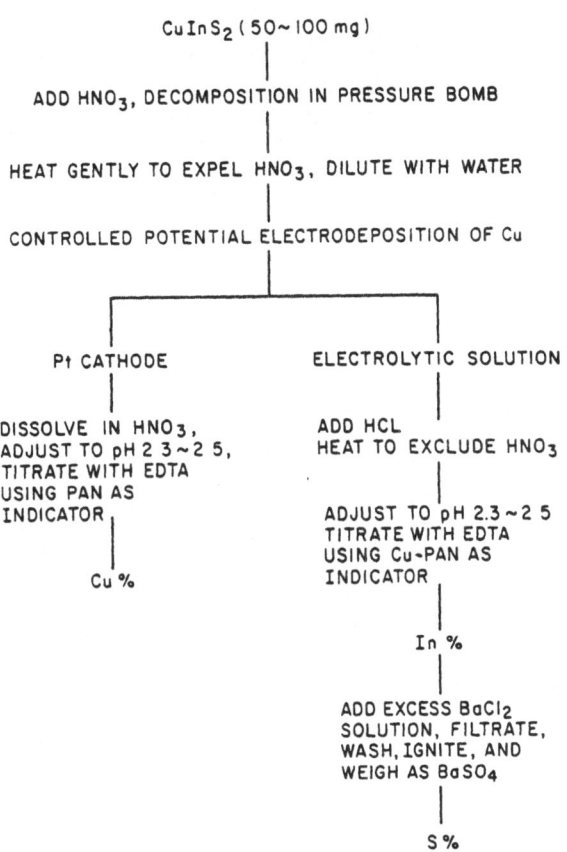

Fig. 1 - Determination scheme for $CuInS_2$ Stoichiometry.

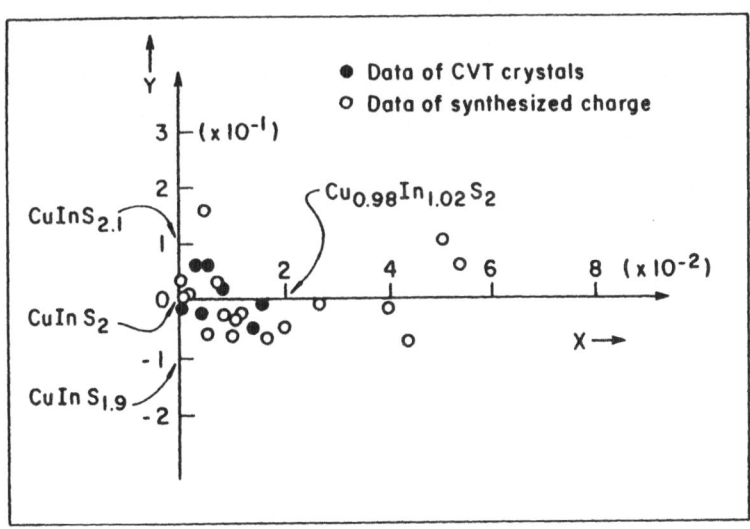

Fig. 2 - Analytical Data of Stoichiometry in the Composition Triangle of $CuInS_2$.

STUDIES ON CdS/n-InP PEC SOLAR CELLS

Y. Ramprakash, S. Basu and D.N. Bose
Materials Science Centre
Indian Institute of Technology
Kharagpur, India.

Summary

The preparation and performance of CdS/n-InP photoelectrochemical solar cells in 0.1M S^{2-}/S_n^{2-} redox electrolyte at pH=10 is reported. The CdS layer 25 μm thick was deposited by spray pyrolysis to act as a wide band-gap window and thus stabilise the InP photoanode. Dark J-V, C-V, spectral response, intensity dependence of V_{OC} and J_{SC} have been studied. V_{fb} was thus found to be 0.48 Volts while the spectral response extended from λ = 0.5-0.8 μm indicating carrier generation mainly in InP. At 75 mW/cm^2 illumination V_{OC} = 0.35 V and J_{SC} = 1.37 mA/cm^2 were obtained with a cell efficiency of 0.1%. Stable operation was obtained upto 72 hours. The reasons for low efficiencies are discussed.

1. INTRODUCTION

InP, a III-V compound semiconductor with E_g = 1.34 eV at 300°K is a suitable material for solar energy conversion. The main problem concerning the use of this semiconductor in photoelectrochemical solar cells is the photoanodic dissolution in highly alkaline electrolyte. Ellis et al (1) reported stabilisation of n-InP in Te^{2-}/Te_2^{2-} redox in 1M NaOH with 1% conversion efficiency of 632.8 nm light to electricity. Recently Heller, Miller and Thiel (2) have reported 11.5% solar conversion efficiency using photocathodically protected p-InP/V^{3+} - V^{2+}-HCl/C system.

Yet another approach for obtaining a stably operated photoanode lies in the choice of suitable semiconducting material deposited on InP which can act as window to the passage of light thus resulting in good stability. Heterojunction solar cells comprise a front layer of a wide band-gap material and a base that has preferably a direct band-gap of 1.5 ± 0.2 eV maximizing the power conversion efficiency. n-CdS/p-InP heterojunctions have shown high conversion efficiency due to excellent energy band and lattice matching at the interface. The band-gaps of CdS and InP are 2.42 eV and 1.34 eV respectively while the electron affinities are 4.50 eV and 4.38 eV respectively. Shay, Wagner et al (3) reported 15% and 5.7% conversion efficiencies respectively with single crystal and polycrystalline InP. Wagner and Shay (4) also reported a photoelectrochemical solar cell incorporating the n-CdS/n-GaAs heterojunction in an electrolyte. Enhancement of photovoltage and photocurrent and the stability was the main feature of such a PEC cell.

In this communication, we report studies on dark J-V, C-V characteristics, spectral response, intensity dependence of open circuit voltage (V_{OC}) and short circuit current density (J_{SC}) and power output characteristics of CdS/n-InP heterojunction in 0.1M S^{2-}/S_n^{2-} redox electrolyte at pH=10.

2. EXPERIMENTAL

n-InP single crystal electrodes used for these experiments were of the following specifications: n-type (100) orientation, $\rho = 0.027$ ohm-cm, $N_D = 2.10^{17}/cm^3$, $\mu_e = 1150$ cm^2/V-sec.

The crystals were slightly lapped on one side and then etched in 3:1:1 mixture of H_2SO_4 (conc):H_2O_2(30%):H_2O for 10 sec followed by a 1% Bromine-Methanol etch for 2 minutes resulting in a smooth shiny surface. The crystals were then ultrasonically cleaned sequentially in baths of Xylene, Acetone and Methanol, and finally in triply distilled water and dried. Ohmic contacts were made by fusing indium on the unpolished side at 400°C in H_2 atmosphere for 1 minute.

CdS was then deposited onto these specimens by spraying a solution of 0.5M $CdCl_2$ and 0.5M thiourea at 350°C in air. The deposition was simultaneously carried out on glass substrates and the CdS films thus obtained were used for evaluating the sheet resistance and structure by X-ray studies. The thickness of the films was found to be 25 μm as determined by weighing. Annealing of the films was done in H_2 atmosphere at 250°C for 10 hours. Such a treatment has been shown to decrease film resistivity and is also expected to reduce porosity.

The photoelectrochemical experiments were performed in a perspex cell with a glass window. Tungsten-halogen illumination through a 1 cm water filter with intensity 75 mW/cm^2 was used. The supporting electrolyte was 5M KOH (pH=10) with 0.1M sulfide/polysulfide redox system. Initially high purity Argon was bubbled through the electrolyte for one hour to remove dissolved oxygen. A saturated calomel electrode (SCE) was used as the reference and a Pt wire kept in the dark served as the counter electrode.

J_{SC} was measured by a Keithley 160-B digital multimeter and V_{OC} by a high impedance voltmeter. The light intensity was varied using wire mesh neutral density filters and in each case the intensity was measured by a calibrated solar intensity meter. The spectral response was studied using interference filters. Corrections were made for the source intensity variation by measuring the output intensity from the filters with a silicon photodiode of known spectral response. The transmission spectra of the redox solution and of the CdS film on glass were obtained by a Beckmann (Model UV 5270) spectrophotometer.

3. RESULTS AND DISCUSSION

The CdS films deposited on glass exhibited a sheet resistance of about 6.5 MΩ in dark and 0.2 MΩ under illumination. X-ray diffraction studies of the annealed films deposited on glass confirmed hexagonal structure.

The dark J-V plots of the heterojunction with and without an electrolyte at the CdS side are shown in fig.1 from which $J_0 = 2.6 \times 10^{-5}$ and 1.8×10^{-3} A/cm^2 respectively were obtained. The ideality factors in the two cases were found to be 3.38 and 6.8. The high values of J_0 and n indicate that generation-recombination processes dominate at these two junctions.

The Mott-Schottky plot for CdS/n-InP heterojunction in indifferent electrolyte at pH = 9.2 (fig.2) gives the flat-band potential V_{fb} = 0.48 V. For n-CdS and n-InP separately at this pH V_{fb} was found to be 0.775 V and 0.7 V respectively. These values are in close agreement with the experimental data of Gomes and Cardon (5). The carrier concentrations calculated from the inverse slope were found to be 1.8×10^{14}/cm^3 and 3.56×10^{17}/cm^3 respectively for CdS and InP.

The normalised spectral response for the CdS/n-InP heterojunction is shown in fig. 3. The response extended from λ=0.5 -0.8 μm with a peak at 0.55 μm. Fig. 4 gives the transmission spectra of the CdS film and the redox-electrolyte from which it is seen that wavelengths shorter than 0.51 μm are absorbed in CdS while the redox has transmission above 0.45 μm. Thus the spectral response is characteristic of InP with the CdS acting mainly as a window. Peak response is observed when radiation is absorbed in a high-field region just near the CdS-InP interface. The weak response at 0.5 μm may be due to absorption in CdS and hole transfer at the CdS-electrolyte interface. The exact nature of the response e.g., dip at 0.65 μm is partly due to the sub-linear current vs intensity relation and varying intensities at different wavelengths. With variation of light intensity J_{sc} was found to increase proportionally but with a slope of 0.3 while the variation of V_{oc} was logarithmic in nature as expected.

A typical power-output characteristic is shown in fig. 5 for the PEC cell. V_{oc} = 0.35 V and J_{sc} = 1.37 mA/cm^2 were obtained under 75 mW/cm^2 illumination, thus giving a conversion efficiency of 0.1%. The cell was stable for 72 hours with no visible etching or weight loss observed during this period.

The low efficiencies obtained in the experiments can be explained with the help of CdS/n-InP band diagram shown in fig. 6. The band diagram reveals a potential discontinuity of 0.96 eV in the valence band. As photons of short wavelength (0.4-0.5 μm) will be absorbed in CdS, these wavelengths are not effective in creating electron-hole pairs at the CdS-InP junction. However, holes created within the CdS at λ= 0.50 μm may diffuse to the surface and be captured by the reducing agents in the electrolyte. In the wavelength region λ= 0.52 to 0.8 μm absorption occurs in InP leading to electron-hole pair generation. Due to the barrier in the valence band, holes are prevented from transfer from InP into CdS but electrons can contribute to current in the external circuit being replenished by flow across the junction from CdS.

4. CONCLUSIONS

Although the V_{oc} of these CdS/n-InP heterojunctions was 0.35 V the overall efficiency was low due to poor J_{sc} values.

This is due to the high values of J_o and ideality factor n indicating carrier recombination at the junction. The spectral response is mainly determined by absorption and carrier generation in InP. Improvement in heterojunction fabrication and doping of the CdS film may result in better device properties.

ACKNOWLEDGEMENTS

The authors acknowledge a grant from C.S.I.R., Govt. of India for this work.

5. REFERENCES

1. A. B. Ellis, J. M. Bolts and M. S. Wrighton, J. Electrochem. Soc., 124, 1603 (1977).
2. A. Heller, B. Miller and F. A. Thiel, Appl. Phys. Lett., 38(4), 282 (1981).
3. J. L. Shay, S. Wagner, K. J. Bachmann and E. Buehler, IEEE Trans. Electron. Dev., ED-24, 483 (1977).
4. S. Wagner and J. L. Shay, "Semiconductor-Liquid Junction solar cells", Ed. A. Heller, 231 (1977).
5. W. P. Gomes and F. Cardon, ibid., 120 (1977).

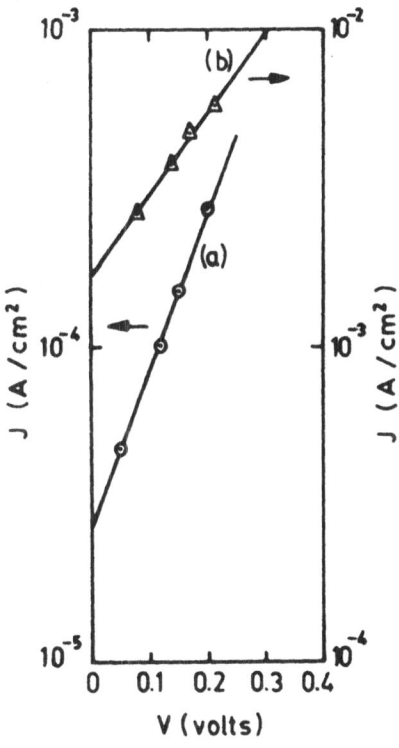

Fig. 1 - Dark J-V characteristics; (a).CdS/n-InP in 5 M KOH-0.1M S^{2-}/Sn^{2-} REDOX; (b) CdS/n-InP heterojunction

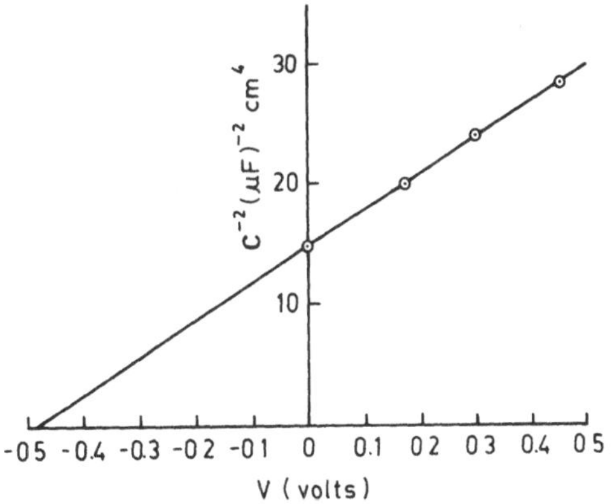

Fig. 2 - Mott-Schottky plot in indifferent electrolyte pH=9.2

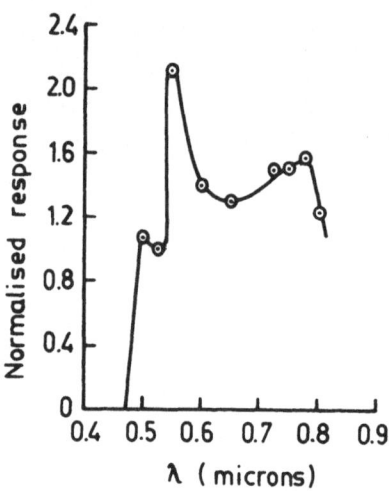

Fig. 3 - Spectral response

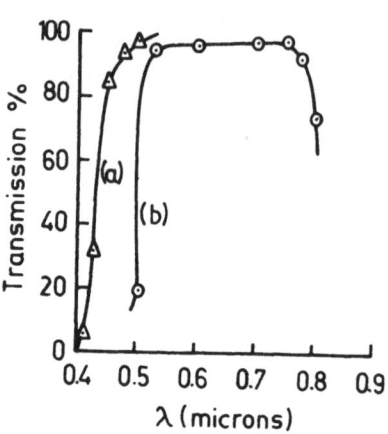

Fig. 4 - Transmission spectra; (a) 5M KOH-0.1M S^{2-}/Sn^{2-} REDOX; (b) n-CdS film on glass

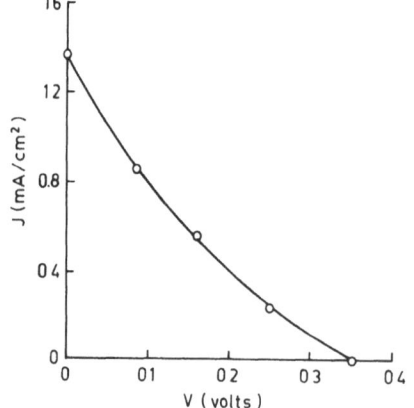

Fig. 5 - Power output characteristics under 75mW/cm^2 illumination

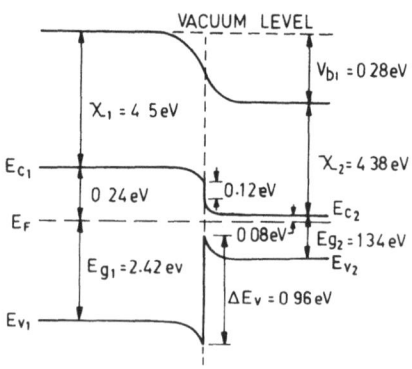

Fig. 6 - CdS/n-InP energy band diagram

SESSION 6

ADVANCED DEVICES AND CONCENTRATION

Chairmen : Prof. V. MAKIOS, University of Patras, Greece

Prof. A. GOETZBERGER, Fraunhofer-Institut für
Solare Energiesysteme, Federal Republic of Germany

Advanced photovoltaic devices

High efficiency GaAs solar cells for concentrator and flat plate arrays

Photovoltaic concentrator technology in the USA

Luminescent solar concentrators (LSC) : technical and economic requirements for a residential system

Recent progress in a residential solar energy system development

High efficiency tandem type solar cells consisting of a-Si:H and a-SiGe:H

High efficiency shallow p^+nn^+ cadmium telluride solar cells

ADVANCED PHOTOVOLTAIC DEVICES

DONALD L. FEUCHT
Solar Energy Research Institute
1617 Cole Boulevard
Golden, Colorado 80401
U.S.A

Summary

A large number of advanced photovoltaic materials, devices and concepts which have potential for efficient low-cost solar electric conversion are being investigated. Photovoltaic cells from sheet polycrystalline silicon, amorphous silicon and other thin-film materials, such as [CdZn]S/CuInSe$_2$, CdTe and Zn$_3$P$_2$, have all increased in their performance during the past few years. Many cells have exceeded or are rapidly approaching a solar-to-electric conversion efficiency of 10%. Advanced concentrator concepts for achieving greater than 30% conversion efficiencies using multijunction cells are showing considerable promise. The flat-plate luminescent concentrator, while relatively low in efficiency, has many features which makes it attractive. Electrochemical photovoltaic cells which can generate electricity directly in a regenerative device or, with the addition of a third electrode, can store energy electrochemically, have exhibited increased lifetimes. The overall research progress on these and other advanced materials, devices and concepts has been outstanding over the past few years.

1. INTRODUCTION

Research on advanced photovoltaic materials and devices has greatly expanded during the past few years. This has resulted in substantial progress in materials preparation, understanding of materials properties and device physics, and cell performance. The research activities can be divided into four major areas of investigation: polycrystalline silicon; a variety of thin-film amorphous and polycrystalline materials; advanced concentrator concepts and devices; and photovoltaic electrochemical systems. In each of these areas, the research is directed toward achieving a thorough understanding of the materials and physics problems which would allow improved device performance and lower cost processing. The research and development status of each of the major areas will be discussed in the following sections. For many of the polycrystalline and thin-film materials, the increase in cell performance over the past seven years is shown in Figure 1.

2. POLYCRYSTALLINE SILICON

The advanced research in polycrystalline silicon is concerned with the development of low-cost polycrystalline sheet silicon, non-conventional cell fabrication and grain boundary research. The vertical growth of silicon, using edge-supported pulling (ESP), and the horizontal growth, using the low-angle silicon sheet (LASS) technique, continue to show excellent promise. In the ESP process, which has yielded 13.8% efficient silicon cells, sheet silicon has been grown using ten different filament materials (1,2). Cells fabricated from polycrystalline sheets using carbon felt, SiC-coated graphite, quartz and mullite are 12.7-13.1% efficient (90-93% of the Cz control wafer). Although quartz is poorly matched in expansion coefficient to silicon, the use of thin quartz capillaries allows for clean thermal shear-stress separation from the sheet. Sheets of 4 cm width and 25 cm length have been grown at speeds of 3 cm/min for sheet thicknesses greater than 100 µm.

Edge-supported pulling has also been applied to sheet growth of metallurgical-grade and direct-arc reactor silicon (3). Because ESP growth is characterized by a 7 mm high meniscus and very good thermal stability is provided by the filament, the growth is relatively insensitive to disruption by interface instability. Impurity segregation was very effective when growth was conducted in the smooth non-constitutionally supercooled mode. It was also found that there is a higher concentration of impurities at the boundary than in the adjacent grains (1.4-12.5 times as high, depending on the impurity). The grain boundary/grain impurity ratio was greater than 5 for Ti, Fe, Cs and B. Cell efficiencies were greater than 50% of the efficiencies of control Cz wafers for diffused cells and nearly equal for epitaxial cells on this material.

The LASS technique is capable of a faster growth rate than vertical pulling techniques. Recent work has demonstrated the growth of a 2 cm wide ribbon, 500 µm thick by 33 meters long at a rate of 55 cm/sec (4). Cell efficiencies of approximately 11% have been achieved on this material.

Although MIS/SIS structures have achieved cell efficiencies of 13-14% on single-crystal and 10-13% on polycrystalline material, stability, reproducibility and scalability of MIS/SIS-type devices continue to be unresolved issues (5-7). The anamalous behavior of ITO/Si junctions, depending on whether the silicon is n or p type, has been attributed to surface damage of the silicon when the ITO is ion-beam deposited rather than sprayed. This damage causes the band edges to bend downward at the surface. For this reason, ion-beam sputtered ITO leads to an ohmic contact on nSi and a rectifying barrier on pSi. Spray or vacuum deposited ITO is ohmic on pSi and rectifying on nSi (8).

The development of low-cost, high-throughput polycrystalline sheet technologies requires that the types and distribution of grain boundaries are such that they do not substantially alter the electrical performance of the cells. Grain boundaries are highly complex structural defects which may contain dangling and distorted bonds and a high density of impurities. These may cause the grain boundaries to act as traps and recombination centers for minority carriers, as potential barriers for majority carrier transport, or as a shunt path and thus lead to significant degradation of the performance of silicon devices. It appears from the research to date that the electrical properties of a grain boundary result from the array of dislocations that form the grain boundary and that dangling bonds are probably responsible for its activity (9).

3. THIN-FILM MATERIALS AND CELLS

There are a variety of thin-film material systems which are receiving considerable research attention and have shown significant progress during the past few years. These include amorphous silicon, cadmium sulfide/copper indium selenide, gallium arsenide and other materials (CdTe, InP, Zn_3P_2, etc). Each of these material systems has a potential cost or performance advantage compared to crystalline silicon but requires further improvements in materials preparation, device performance and basic understanding before the potential will be reached.

Amorphous Silicon. The development of amorphous silicon solar cells has been accelerated by extensive research efforts in the U.S. and Japan. In 1977, the introduction of hydrogen during the glow discharge deposition of amorphous silicon permitted the fabrication of small-area cells with a conversion efficiency approaching 6% (10).

Since then, research has provided improved understanding of materials, processing and device physics, and has resulted in increased efficiencies over larger areas. The preferred cell structure has become pin with amorphous silicon carbide often used for the p region to improve conductivity and provide a wider energy gap window when illuminated from the p side. Recent reports from both the U.S. and Japan suggest that 8% efficient cells have been achieved in single-junction cells (11-13). Most of this improvement can be attributed to an increase of short circuit current density from approximately 12 to 14 mA/cm^2. Series-connected and stacked cell configurations are also being investigated. A 7.7% efficient small-area stacked cell has been reported as well as a 6.2%, 100 cm^2 series-connected cell (14,13).

If the best observed cell parameters could be obtained in a single device, the conversion efficiency for amorphous silicon material would achieve the 10% goal. In spite of these impressive advances, problems still exist in the development and production of these materials for commercial photovoltaic devices. For example, many of the high-efficiency devices have shown a decrease in efficiency upon illumination. Although this performance degradation has been linked to basic property changes in the intrinsic material, the mechanism causing this property change has not been identified. Similar problems occur when looking at doped materials and alloys. The material performance generally degrades, and the reasons for the degradation are not understood. Clearly, more basic work is needed on such materials to remedy these important problems.

CdS/CuInSe$_2$. Intensive research in polycrystalline thin-film CdS-based photovoltaic devices has been performed over the past two decades. Efficiencies exceeding 10% have been reported for 1 cm^2 [Cd,Zn]S/Cu$_2$S wet-processed cells (15). Most of these high-efficiency devices, however, have exhibited significant degradation in their performance when illuminated due to electrochemical decomposition of the Cu$_2$S. In recent years much of this research effort has turned to the investigation of CdS/CuInSe$_2$ cells. Thin-film vacuum evaporated [Cd,Zn]S/CuInSe$_2$ devices with efficiencies of greater than 9.9% have been reported for a 5 μm thick 1 cm^2 cell (16). Presently the CuInSe$_2$ is deposited from individual elemental sources. Initially, a 2.5 μm high conductivity layer is deposited on the substrate, then an 0.8 μm high resistivity layer which forms the absorber layer is deposited. Two [Cd,Zn]S layers are then deposited in order to form the heterojunction; an 0.8 μm thick high resistivity layer is followed by a low resistivity layer. The cell parameters with a $Cd_{0.9}Zn_{0.1}S$ layer are V_{oc} = 0.42V, J_{sc} = 36.3 mA/cm^2, FF = 0.65 for a 1 cm^2 cell.

Investigations of the cell cross-section, using EBIC, have identified the Mo/pCuInSe$_2$ back contact as a Schottky barrier (17). The presence of this blocking barrier, which has been confirmed by C-V measurement, limits the open-circuit voltage of the cell. EBIC measurements have also shown that an annealing process, 200°C for ≈30 min in the ambient, is required following deposition to activate the heterojunction. Oxygen is believed to be responsible for the cell improvement during annealing.

Stability measurements have been conducted for over 8400 h at temperatures as high as 80°C, in room ambient, with no measureable degradation in cell performance. It appears that the CdS/CuInSe$_2$ interface is electrically and compositionally stable up to 100°C. Interdiffusion of sulfur and selenium occurs over the temperature range of 150°C<T<300°C and rapid diffusion of cadmium into the ternary occurs for T>350°C, leading to device degradation.

Further research is necessary to eliminate the Mo/CuInSe$_2$ Schottky barrier and to find a less critical and, hence, less expensive method of depositing the stoichiometric CuInSe$_2$. Although the stability measurements to date are very encouraging, further work is required to understand the interface interaction and to verify the long-term stability.

Gallium Arsenide. Single-junction gallium arsenide homojunction cells have demonstrated efficiencies of nearly 22% under one-sun illuminaton. However, if cost-effective cells for terrestrial applications are going to be produced, a low-cost technique must be found to fabricate thin cells (<5 μm) which retain the high efficiency of single-crystal cells. Three approaches are being investigated: thin-film polycrystalline GaAs; high-efficiency heteroepitaxial GaAs; and high-efficiency separated films.

Improvements in the quality of polycrystalline GaAs films deposited on W-coated graphite have resulted in 8.8%, 1 cm^2 and 8.1%, 7.5 cm^2 p$^+$GaAs/nGaAs/n$^+$GaAs/W/graphite cells (18). Optimization of the device parameters and the addition of an AlGaAs window layer to reduce surface recombination losses could significantly increase the efficiency. Heteroepitaxial GaAs deposited on germanium-coated single-crystal silicon has resulted in 11.7% small area n$^+$GaAs/pGaAs/p$^+$GaAs/Ge/Si cells with V_{oc} = 0.76V (19). Recent results indicate a V_{oc} = 0.85 V and an η = 14% are possible (20). The highest efficiency thin-film GaAs homojunction cell reported is 17% for an 0.5 μm thick-separated film of 0.5 cm^2 area (21).

If high-efficiency cells are going to be obtained using thin-film polycrystalline GaAs, then a method must be found to drastically reduce the effects of grain boundaries. Columnar growth does not readily occur in GaAs. To a large degree, new grains nucleate upon existing areas, producing a large number of transverse grain boundaries which can trap carriers. Although the average grain size obtained can be comparable to the thickness, the grains are non-uniform in size. The fill factor and open-circuit voltage in all thin-film polycrystalline GaAs cells are much lower than in single-crystal cells. It is postulated that J_o is substantially increased as a result of recombination at grain boundaries in the junction space charge region (22).

The growth of columnar- and uniform-sized grains may help reduce J_o but a method is needed to reduce the trapped charge at grain boundaries. Several measures such as capping the surface boundaries with an oxide, treating the material with tin, or diffusing an impurity down the grain boundary can result in some improvement in device performance. A much better technique must be found, however, to passivate the grain boundaries if thin-film polycrystalline GaAs cells are going to exhibit efficiencies approaching that of single-crystal devices.

Other Materials. In addition to the thin-film materials previously discussed, there are a number of other materials that are being investigated. These include InP, CdTe, and Zn$_3$P$_2$. All of these have a near optimum bandgap (1.35-1.5 eV) and a high absorption coefficient ($>10^4$cm^{-1}).

Small area CdS/InP cells, fabricated by growing CdS films on single crystal InP, have exhibited conversion efficiencies as high as 15% (23). Attempts to fabricate thin-film polycrystalline cells by the deposition of CdS and InP have not yielded acceptable performance as the cells exhibit a low open-circuit voltage due to a large J_o or contain shunt paths which degrade performance. ITO has been deposited on InP as an alternate means to obtain a high efficiency InP polycrystalline cell. A maximum efficiency of 14.4% has been reported for devices prepared by the ion-beam deposition of ITO on pInP (24). Recently, a 10.6% efficient ITO/pInP cell (A = 0.8 cm^2) was obtained for ion-beam deposited ITO on bulk polycrystalline InP (25). This compares to an efficiency of 8.7% (A = 2.2 cm^2) when the ITO was r.f. sputter-deposited on the InP. Characterizations of the junction have shown it to be SIS type with P_2O_5 present at the ITO/InP interface. It appears from the research to date, that grain boundaries are extremely important and that InP does not provide any inherent advantage over GaAs for thin-film cells.

CdTe can be doped n- and p-type and reasonably efficient ($\approx$10.5%) heteroface CdS/n-pCdTe cells have been obtained (26). Thin films have also been produced by a variety of methods: vacuum evaporation, CVD, sputtering, electro-deposition, chemical spray, and screen printing. The later technique has been used to fabricate an nCdS/n-pCdTe cell with an overall efficiency of 8.1% (27).

Recently, a small area thin-film Au/CdTe Schottky barrier cell having an efficiency of 8.6% was obtained (28). The CdTe was electro-deposited to a thickness of 4 μm on a cadmium-plated metallic substrate. Following deposition, the sample was annealed in air at 300°C for one hour to produce a surface oxide layer and reduce the density of defects in the CdTe. A 1 cm^2 cell made from the same material had an efficiency of 6.2% with reduced performance in V_{oc}, J_{sc} and FF. Hot wall evaporation, which should result in large grain CdTe films, is being pursued and could result in higher efficiency devices.

Zn_3P_2, because of its bandgap, high-absorption coefficient, long minority-carrier diffusion length, and abundance of its elements, is an interesting new photovoltaic material. The best thin-film cell fabricated to date is a 1 cm^2 Mg/Zn_3P_2 Schottky barrier cell which exhibited an efficiency of 4.3% (29). Interface measurements have shown that, depending on the doping of the Zn_3P_2, two types of junctions can be formed (30). For higher doped (p>10^{17}cm^{-3}) Zn_3P_2, the Mg can interact to form interfacial nMg_3P_2 providing a heterojunction. For the lower doped (p<10^{17}cm^{-3}) Zn_3P_2, a buried homojunction is formed. Junction control is a problem, as Mg readily diffuses (D_o = 4 x 10^{-9} cm^2 at 100°C) into the Zn_3P_2. The resulting junction efficiencies have been poor. Recently a $ZnSe/Zn_3P_2$ heterojunction has been fabricated with an open circuit voltage of 0.75 V.

4. ADVANCED CONCENTRATOR CONCEPTS AND CELLS

Research on advanced photovoltaic concentrators is concerned with the investigation of multijunction devices which have the potential of conversion efficiencies in excess of 30% and low-cost flat-plate luminescent collectors. Research on silicon and gallium arsenide concentrator cells has shown that efficiencies of greater than 20% and 23% are possible (31). Other concepts being explored are the practicability of mechanically stacking different bandgap cells, the feasibility of using the back contact of the large-band gap cell as a mirror to reflect lower energy photons and the possibility of developing a monolithic multijunction cell.

In the multiple-junction concentrator cell, the individiual photovoltaic cells are optimized for a portion of the incident solar illumination and their output is connected in series. Modeling of the various proposed structures indicates that if the bandgaps are optimically selected to maximize the combined photovoltaic output, efficiencies in excess of 32% may be obtained at near ambient temperatures (32). In addition, photovoltaic conversion efficiencies of 25% or greater may be viable at temperatures of 180°C, compatible with low temperature, hybrid photovoltaic/ thermal systems (33). The first demonstration of the concept, using a beam splitter and individual Si and AlGaAs cells, yielded conversion efficiencies of 26.3% (34). Most of the current emphasis is directed toward the development of a monolithic two-cell cascade device. The material problems for this type of structure are particularly stringent as the two homojunction cells must be coupled electrically and optically through a low-resistance interface.

This structure poses combined requirements which include: the top and bottom cells must have bandgaps appropriate for the high and low energy ranges, approximately 1.6 eV and 1.0 eV respectively; the materials for the cells must have a low defect density in order to achieve good cell performance; doping must be controllable up to the high 10^{18} cm^{-3} range with low dopant diffusion coefficients; and the metallization technology must carry high current densities for operation at 500 to 1000 suns. Two materials systems which are capable of meeting these requirements are being investigated. They include alloys of AlGaAsSb on GaAsSb and AlGaInAs on GaInAs. Proof-of-concept cells of more than 16% efficiency have been fabricated using GaAlAs on GaAs with both interconnecting GaAlAs tunnel junctions and a newly developed metal interconnect (35,36). For the latter, a conversion efficiency of 22% has been obtained at AM3 at a concentration of 130x and a junction temperature of 50°C.

In order to reduce stresses and misfit dislocations in the high-energy gap material, which results from the lattice mismatch of the different materials, a super-lattice interconnect between the high- and low-bandgap cell is being investigated (37). The super-lattice system should not only increase the performance of compound-based monolithic cells, but should allow the use of an inexpensive substrate such as silicon.

Luminescent flat-plate solar collectors use luminescent dyes incorporated in a glass or polymer matrix to convert the solar spectrum to a narrow-band luminescense. Approximately 75% of the isotropic emission is trapped by total internal reflection in a collector plate having an index of refraction of 1.5. This trapped light propogates to the plate edges where it is concentrated onto a photovoltaic cell. Since the output is in a narrow spectral band, the photovoltaic cell can be optimized to convert the output light with more than 40% efficiency. Early experimental systems using thin-polymeric films containing Coumarin 6 and Rhodamine 101 dyes on a 200 cm^2 PMMA plate have yielded an overall system conversion of 3.2% using Si cells and 4.5% using GaAs cells (38). Dyes with improved performance and stability need to be developed to make this system cost effective. The present dyes degrade to 1/e of their initial performance in less than one year. The concept can be extended to stacked plates with different dyes and cells to increase the overall conversion efficiency.

The collector can also be used as the cover plate of a module with cells that are mounted to the back of the plate. This can increase the module efficiency by collecting light lost due to the module packing factor and also by red-shifting the ultraviolet and blue regions of the solar spectrum to wavelengths where the cells have better response. This coverplate approach has resulted in a 10% improvement in the efficiency of a GaAlAs/GaAs heteroface cell, solely as a result of wavelength shifting (39).

5. ELECTROCHEMICAL PHOTOVOLTAIC CELLS

Research in photoelectrochemistry, as applied to solar energy conversion, has expanded rapidly during the past five years. The photoelectrochemical cell, consisting of a semiconductor, electrolyte, and counter electrode arranged in a battery configuration, besides having promise as a photovoltaic cell is intrinsically suited for **in-situ** storage. In this system, as compared to other photochemical energy systems, the light is converted directly into electrical energy in a regenerative system in which the overall composition of the electrolyte does not change with time.

High conversion efficiencies have been demonstrated for single-crystal GaAs (14%), CdSe (12.4%), InP (11.5%), WSe_2 (10.4%) and $MoSe_2$ (9.4%) cell configurations (40-43). Thin-film electrodes of CdSe (<5 μm) deposited by chemical bath, electrodeposition, and co-evaporation, have all been shown to produce conversion efficiencies of greater than 6% (44-45).

Significant electrode stability enhancements have been demonstrated for n-type GaAs, CdSe and Si when the surfaces have been coated with an electrodeposited conducting polymer film, such as polypyrrole. The electrodeposited film acts as a barrier to ion/solvent transport, inhibiting photodegradation of the electrode surface, while permitting electron exchange with the electrolyte. The photocurrent remains nearly constant for over 100 h for nGaAs coated with polypyrrole compared to less than 1 min in the absence of the film (46). For silicon, the reduction in photocurrent after 120 h is 30%, compared to complete degradation in seconds without the film (47).

Electrochemical photovoltaic cells may also provide **in-situ** storage. A three-electrode configuration using a two-redox electrolyte separated by a semi-permeable membrane, provides the option to either generate electrical energy or store energy **in-situ**. Such an experimental cell consists of a thin $nCdSe_{1-x}Te_{1-x}$ photoelectrode and a cobalt sulfide covered stainless steel foil counter electrode immersed in a polysulfide electrolyte. A semi-permeable membrane separates this region from the storage compartment, which contains a tin storage electrode immersed in a sulfide solution. System efficiencies of 2% have been achieved and charge/discharge cycling has been performed for two months with no observable degradation in cell performance (48).

A photoelectrochemical energy conversion and storage system based on an HBr system is being developed (49). The system contains photovoltaic arrays constructed of metallized p/n and n/p silicon spheres in contact with an HBr electrolyte. Illumination of the photovoltaic arrays effects the electrolysis of HBr to H_2, which is subsequently stored in a metal-hydride storage cell, and Br_2, which is subsequently stored in a separate compartment. The conversion of stored chemical energy to electrical energy is accomplished using a hydrogen-bromine fuel cell which simultaneously replenishes the HBr electrolyte in the photoelectrolysis cell in a regenerative mode of operation. Thermal energy generated by electrolyte absorption can also be extracted from the system with the use of a heat exchanger. The goal of the four-year program, which began in 1978, is to produce prototype 300 W, 3 m^2 panels and a system with a 1500 Wh storage capacity and a 6% overall efficiency.

To date, 13% solar to electric conversion efficiencies have been demonstrated using single-crystal etched cube arrays and also a corrosion rate consistent with 20 year life has been demonstrated in accelerated tests. The hydrogen storage alloy has been cycled for greater than 5000 charge/discharge cycles without loss in storage capacity and the fuel cell has operated for more than 5300 h at 300 mA/cm^2.

6. CONCLUSION

As shown in Figure 1, the conversion efficiency of most of the polycrystalline thin-film photovoltaic materials has increased dramatically during the past seven years. Specific data on the present performance of representative thin film, advanced concentrator and electrochemical photovoltaic cells is given in Table I. For most materials, several organizations or processes have yielded similar cell performance and thus the technology is not investigation dependent. Although progress on improving the understanding and performance of advanced photovoltaic devices has been outstanding during the past few years, further research is required before these advanced materials are ready to compete with the present silicon technology.

7. ACKNOWLEDGEMENTS

The author wishes to thank many of the staff of the Photovoltaics Program at SERI for their input.

REFERENCES

1. Ciszek, T.F.; Hurd, J.L. 1980 (May). Proc. of Symposium on Novel Silicon Growth Methods. The Electrochemical Society.
2. Ciszek, T.F.; Hurd, J.L.; Schietzelt, M. To be published--J. of the Electrochem. Soc. Also, SERI/TP-212-1536.
3. Ciszek, T.F.; et al. 1981 (May). Proc. 15th IEEE Photovoltaic Specialists Conference. pp. 581-588.
4. Jewett, D.; et al. 1981 (Mar.). Second Quarterly Report. (SERI) Contract XS-9-8041-3).
5. Maruska, H.P.; et al. 1981 (May). Proc. 15th IEEE Photovoltaic Specialists Conference. pp. 1412-1417.
6. Genis, A.P.; et al. 1980 (Aug.). IEEE Electron Device Letters. Vol. EDL-1 (No. 8).
7. Green, M.A.; et al. 1980 (Jan.). Proc. 14th IEEE Photovoltaics Specialists Conference. pp. 684-687.
8. Ashok, S.; et al. To be published--IEEE Electron Device Letters.
9. Tsuo, Y.S.; Milstein, J.B.; Surek, T. Materials Research Society Meeting; Boston, MA; Nov. 16-19, 1981. To be published--Proceedings of the Symposium on Grain Boundaries in Semiconductors. New York, NY: Elsevier North Holland Inc., Pub.
10. Carlson, D.E.; 1977 (Apr.). IEEE Trans. Electron Devices. Ed-24. p. 449.
11. Carlson, D.E. (RCA). Personal Communication.
12. Hamakawa, Y. (Osaka University). Annual Meeting of J.A.A.P., Oct. 9, 1981.
13. Uchida, Y. (Fuji Electric Company). Annual Meeting of J.A.A.P., Oct. 9, 1981.
14. Nakamura, M. (Mitsubishi Electric Company). Annual Meeting of J.A.A.P., Oct. 9, 1981.
15. Hall, R.B.; et al. 1981 (May). Proc. 15th IEEE Photovoltaic Specialists Conference. pp. 777-779.
16. Mickelsen, R.A.; Chen, W.; Bulhaupt, L. (Boeing Aerospace Corporation). Eighth Quarterly Report. (SERI Contract XJ-9-8121-1).
17. Russell, P.E.; et al. To be published--Appl. Phys. Lett.
18. Chu, S.S. 1981 (Mar.). SERI Subcontractor Review Meeting.

19. Gale, R.P.; et al. 1981 (May). Proc. 15th IEEE Photovoltaic Specialists Conference. pp. 1051-1055.
20. Fann, J.C.C. Personal Communication.
21. Fann, J.C.C.; Bozler, C.O.; McClelland, R.W. 1981 (May). Proc. 15th IEEE Photovoltaic Specialists Conference. pp. 666-672.
22. Benner, J.P.; Blakeslee, A.E. 1981 (Nov.). Materials Research Society Meeting; Boston, MA. Also, SERI/TP-212-1493.
23. Wagner, S.; et al. 1977. J. Crystal Growth. Vol. 39; p. 128.
24. Sree Harsha, K.S.; et al. 1977. Appl. Phys. Lett. Vol. 30; p. 645.
25. Coutts, T.J.; et al. 1981 (May). Proc. 15th IEEE Photovoltaic Specialists Conference. pp. 1077-1082.
26. Yamaguchi, K.; et al. 1977. J. Appl. Phys. Vol. 16; p. 1203.
27. Nakayama, N.; et al. 1976. J. Appl. Phys. Vol. 15; p. 2281.
28. Fulop, G.; et al. 1982 (Feb.). Appl. Phys. Lett. Vol. 40; pp. 327-328.
29. Bhushan, M. To be published--Appl. Phys. Lett.
30. Kazmerski, L.L.; Ireland, P.J.; Catalano, A. 1981 (May). Proc. 15th IEEE Photovoltaic Specialists Conference. pp. 1083-1086.
31. Boes, E.C.; Shafer, B.D. 1981 (May). Proc. 15th IEEE Photovoltaic Specialists Conference. pp. 305-310.
32. Moon, R.L.; et al. 1978 (Jun.). Proc. 13th IEEE Photovoltaic Specialists Conference. pp. 859-867.
33. Mitchell, K.W. 1978 (Dec.). Proc. of the International Electron Device Meeting. Washington, DC. pp. 254-257.
34. Borden, P.G.; et al. 1981 (May). Proc. 15th IEEE Photovoltaic Specialists Conference. pp. 311-316.
35. Bedair, S.M.; et al. 1980 (Sep.) SERI High Efficiency Concentrator Review Meeting; Palo Alto, CA.
36. Ludowise, M. (Varian). Personal Communication.
37. Blakeslee, A.E.; Mitchell, K.W. 1980. NASA Conference Publication. No. 2169; pg. 131.
38. Friedman, P.S. Owens-Illinois Report, July 1979-March 1980. (SERI Contract XS-9-8216-1).
39. Friedman, P.S. (Owens-Illinois). Personal Communication.
40. Noufi, R.; Tench, D. 1980. J. Electrochem. Soc. Vol. 127; p. 188.
41. Frese, Jr., K. To be published--Appl. Phys. Lett.
42. Heller, A.; Miller, B.; Thiel, F.A. 1981 (Feb.). Appl. Phys. Lett. Vol. 38; pp. 282-284.
43. Parkinson, B.A. 1980 (Oct.). Final Report. (SERI Contract XP-9-8198-1).
44. Tomkiewicz, M.; et al. 1980 (Sep.). Final Report. (SERI Contract XS-9-8312-1).
45. Russak, M.A.; et al. 1980 (Dec.). Quarterly Technical Progress Report No. 2. (SERI Contract XP-9-8002-8).
46. Noufi, R.; Tench, D.; Warren, L.F. 1980. J. Electrochem. Soc. Vol. 127; p. 2310.
47. Noufi, R.; Frank, A.J.; Nozik, A.J. 1981. J. Am. Chem. Soc. Vol. 103; p. 1849.
48. Manassen, Y.; (Weizmann Inst.). Personal Communication.
49. Johnson, E.L. 1981. The TI Solar Energy System Development. Proceedings IEDM Meeting. Washington, DC.

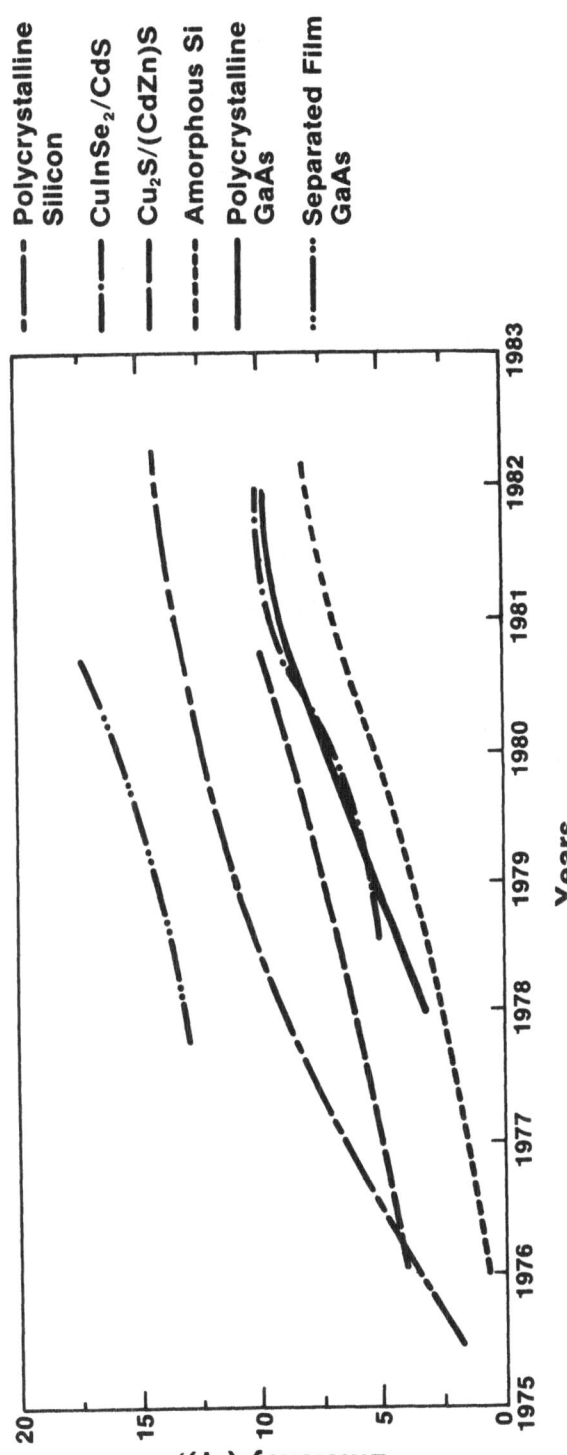

Fig. 1 - Advanced Cell Efficiency vs. Time

TABLE I - PERFORMANCE OF SELECTED PHOTOVOLTAIC RESEARCH CELLS

Material	V_{oc} (Volts)	J_{sc} (mA/cm^2)	FF	η %	Area (cm^2)	Remarks	Ref.
Silicon	0.58	31.9	0.74	13.8	4.0	p-n, Edge-Supported Growth	1
Silicon	0.56	27.0	0.72	11.0	3.8	p-n, Low-Angle Si Sheet	4
Silicon	0.56	29.8	0.67	11.2	1 0	SnO$_2$ or ITO/Wacker n-Si, Spray	5
Silicon	0.52	28.1	0.79	11.5	11.4	ITO/Wacker p-Si, Ion Beam Sputter	6
Silicon	0.54	32.7	0.76	13.3	2.8	Al/Wacker p-Si, Active Area, Grating Cell	7
Amorphous Silicon	0.87	14.0	0.68	7.9	1.1	SnO$_2$·F/p$^+$a-SiC:H/in$^+$a-Si:H	11
Amorphous Silicon	0.88	14.1	0.62	7.7	1.0	ITO/SnO$_2$/pa-SiC/ina-Si:H	12
Amorphous Silicon	0.86	13.9	0.66	7.8	1.2	ITO/n-Si/ipa-Si:H/Al	13
Amorphous Silicon	1.41	9.6	0.57	7.7	0.25	ITO/nipna-Si:H/ia-SiGe:H/pa-Si.H/S.S.	14
Amorphous Silicon	8.50	12.3	0.59	6 2	100.0	Glass/ITO/pa-Si C/ina-Si·H/Al (series connected - 10 cells)	13
Cu$_2$S/Cd$_x$Zn$_{1-x}$S	0.60	22.8	0.75	10.2	1.0	x = 0 16, Solution Ion Exchange	15
CuInSe$_2$/Cd$_x$Zn$_{1-x}$S	0.42	36.3	0.65	9 9	1.0	x = 0.10, Evaporation	16
GaAs	0.57	24.5	0.63	8 8	1.0	p-n, GaAs on W-Coated Graphite	18
GaAs	0.76	24.2	0.63	11 7	0.09	p-n, GaAs on Ge-Coated Si	19
GaAs	0.95	23.0	0.78	17.0	0.5	Separated S.C. Thin Film	21
InP	0.76	21.6	0.65	14.4	0.2	MgF$_2$/nITO/pInP	24
InP	0.62	26.9	0.64	10.6	0 8	ITO/P$_2$O$_5$/InP	25
CdTe	0.70	22.0	0.53	8.1	--	In$_2$O$_3$/nCdS/npCdTe/p$^+$Cu$_2$Te	27
CdTe	0.58	17.3	0.62	6.2	1.0	Electrodeposited Au/nCdTe	28
CdTe	0.72	18 7	0.64	8.6	0 02	Electrodeposited Au/nCdTe	28
Zn$_3$P$_2$	0.48	18.0	0.55	4 5	1.0	Mg/Zn$_3$P$_2$ Thin Film	29
GaAs - Electrochemical	0.70	29.0	0.80	14.0	0.3	n-n$^+$ LPF SC Se^{-2} Se$_2^{-2}$ OH$^-$ 115 mW/cm^2 (Xenon)	40
CdSe - Electrochemical	1.05	14 5	0.72	12.4	--	S C Fe(Cn)$_6^{-4/3}$ OH$^-$ Pt 75 mW/cm^2	41
InP - Electrochemical	0.66	24.8	0.64	11.5	--	S.C pInP/VCl$_3$, VCl$_2$, HCl/C	42
WSe$_2$ - Electrochemical	0.67	22.6	0.57	10.4	0.02	S C , Aqueous I$_2$ I$^-$ 82 6 mW/cm^2	43
MoSe$_2$ - Electrochemical	0.65	25.0	0.56	9.4	0.03	S C , Aqueous, I$_2$ I$^-$ 84 7 mW/cm^2	43
CdSe - Electrochemical	0.56	23.8	0.48	6 4	1.0	Electrodeposited, Aqueous S S^{-2} OH$^-$	44
CdSe - Electrochemical	0 57	17.0	0 51	6.6	0.5	Co-evaporated, Aqueous S S^{-2} OH$^-$	45

HIGH EFFICIENCY GaAs SOLAR CELLS FOR CONCENTRATOR AND FLAT PLATE ARRAYS.

G. Guarini
CISE S.p.A. PO Box 12081 - 20100 Milan (Italy)

Summary

In the present paper some significant events in the recent history of GaAs thin film and concentrator solar cells are first reviewd and discussed.
The activity developed in CISE on GaAs solar cells and modules is then described.
This activity includes the fabrication of high efficiency-high concentration GaAs and AlGaAs cells obtained by LPE and MO-CVD techniques and also a three terminal, 15.5% efficient, AlGaAs - GaAs stacked cell.
Finally, the characteristics of a 750 suns, 130 W spectrum splitting concentrator module employing AlGaAs and GaAs cells, are summarized and commented.

1. INTRODUCTION

High conversion efficiency represents the most precious property for a solar cell, expecially for those photovoltaic applications which imply high Balance of System costs.
In addition to its intrinsically high conversion efficiency, gallium arsenide offers a number of favourable characteristics like a high absorption coefficient, a high resistance to radiations and to elevated temperatures, and also a "tunable" band-gap energy, function of the composition of alloys like AlGaAsSb, InGaAsP or AlGaInAs. Although all these properties seem to indicate gallium arsenide as a highly qualified candidate both for space and terrestrial applications, it has to be recognised that this material is not generally popular in the photovoltaic world.
The principal imputations concern scarse availability, high cost and a cell fabrication technology not suitable for large production volumes.
At the present time, current technology, much more then availability of gallium or arsenic, represents a limitation to large scale production of GaAs solar cells.
Assuming, for example, that 5 μm thick, 10% efficient flat plate GaAs cells had to be fabricated, then gallium and arsenic from identified world sources would allow the production of a few tens of GW, or several orders of magnitude more if also gallium contained in bauxite ores could be utilized (1) (2).
The cost of GaAs single crystals, for large production volumes, has been estimated between 1250 $/Kg and 1600 $/Kg for 1982 with a projection of 1000-1100 $/Kg for 1990 (3) (4). For poly-GaAs, a cost of 600-750 $/Kg. (1982) and 500-600 $/Kg. (1990), has been predicted.
These figures indicate that no significant cost reduction must be expected for GaAs and no possibility exists of employing GaAs substrates

for flat plate arrays even if more economical solutions like ultrathin substrates or ribbons were adopted. In fact the cost contribution of a 300 μm thick GaAs substrate of a 20% efficient cell would result of about 60 $/Wp.

Only material saving solutions like thin film cells or high concentration can be therefore considered. In the present paper, some promising examples of high efficiency GaAs thin film cells for flat plate applications will be first briefly discussed.

Some significant activities and results on high efficiency concentrator multicolor cells will be then reviewed and, finally, the activity developed at CISE on concentrator GaAs and AlGaAs, single-color and bicolor solar cells will be described and preliminary results obtained on a 130 Wp spectrum splitting module will be discussed.

2. GaAs THIN FILM SOLAR CELLS

Two types of thin film structures show promise for conversion efficiencies between 15% and 20%.

The first type of cell is composed of a few μm thick, shallow homojunction GaAs single crystal layer deposited on top of a Ge coated low cost Si substrate. A main limitation to conversion efficiency derives from the high dislocation density generated in the GaAs layer by the lattice mismatch between Ge and Si.

Vapor phase epitaxial techniques, both from chloride ($AsCl_3$ - VPE) and organometallic source (MO-CVD) have been employed so far for the fabrication of these cells (5),(6), (7).

The second type of cell (8), (9) consists of a shallow homojunction GaAs single crystal layer grown by $AsCl_3$ - VPE technique on top of a GaAs substrate, successively separated and reused for a number of other growths (CLEFT method). After the separation, the thin film cell remains attached to a low cost, light weight, transparent support. Due to its higher efficiency, light weight, resistance to radiations and to the possibility of being employed in high efficiency transmission tandem cell configurations, the CLEFT cell is particularly attractive both for space and terrestrial applications.

Some significant results obtained on high efficiency GaAs thin film cells are reported in table I.

Efficiency higher than 15% for GaAs/Ge/Si cells and close to 20% for CLEFT cells seem therefore to represent a realistic short term goal.

TABLE I: SOME RECENT RESULTS ON GaAs THIN FILM CELLS.

YEAR	EVENT	LABORATORY
1980	15% AM1 8 μm CLEFT CELL	LINCOLN LAB.
1981	17% AM1 10 μm .51 cm^2 CLEFT CELL	LINCOLN LAB.
1981	17% AM1 2.36 mm^2 GaAs/Ge CELL	JPL
1981	GOOD MORPHOLOGY GaAs/Ge/Si STRUCTURE	JPL
1981	12% AM1 8μm 9.3 mm^2 GaAs/Ge/Si CELL	LINCOLN LAB.

In addition to the above mentioned single crystal cells, polycrystalline MOS GaAs structures, deposited on tungsten coated graphite substrates have also been obtained with 8.5% AM1 efficiency (10).

Finally, a new type of thin film solar cell, based on photoemission of electrons from the surface of a semiconductor layer, has been proposed by G.J. Williams (Cornell University) (11) and developed by STAR INC.

The basic structure consists of a modular sawtooth shaped glass substrate covered with an ultrathin ($\sim 2\mu m$) layer of polycrystalline GaAs and cesium oxide. One side of each tooth is illuminated emitting electrons which, under low pressure conditions, are collected by the dark side of the next tooth. A constant potential difference is so established between any two adjacent ridges.

In order to guarantee the required low pressure ambient, a tightly sealed glass cover plate must be used for terrestrial applications. The development of the emissive cell has not been ultimated and no full spectrum data are available yet but very high conversion efficiencies (up to 26%) and low cost fabrication processes are predicted for this new type of cell.

3. SOME ECONOMICAL EVALUATIONS

If high efficiency may be considered a short term goal for GaAs thin film cells, the development of a low cost, safe and large scale fabrication technology will certainly require much more research effort. The attention should be focused on the deposition process of thin GaAs layers by epitaxial technique, which represent the main effective extra charge. The rest of the fabrication process, in fact, will not be essentially different from Si cells. Low cost electroless and electrolytic deposition of metal contacts, for example, has already been experimented by Lincoln Lab. (9) and CISE (25) on GaAs cells.

A realistic estimate of the major direct costs for epitaxial growth of small quantities (10.000) of AlGaAs-GaAs layers, was performed by D.W. Almgren and K.I. Csigi (12) on the base of current LPE, MO-CVD and MBE technology.

The results indicate that, even for 20% efficient 11 μm, 4 cm^2 large GaAs solar cells, present epitaxial processes are more than one order of magnitude too expensive (between 20 and 50$/Wp for infinite melt LPE and MO-CVD and about 255$/Wp for MBE, excluding the substrate cost of 60$/Wp).

Molecular beam epitaxy (MBE) is much more expensive than the other techniques due to higher labor and capital equipment costs while for organometallic chemical vapor deposition (MO-CVD) the material cost and gas utilization factor are determinant. In any case MO-CVD, basically similar to Si epitaxial deposition from SiH_4, seems to be the most promising technique for large scale production, even if the high toxicity level of arsine will require the solution of very severe safety problems.

Designs for continous flow MO-CVD reactors have already been proposed and low pressure MO-CVD of GaAs promises very high gas utilization factors (13). Moreover, a severe cost reduction of organometallic compounds re-presents an essential but reasonable goal.

In table II, a preliminary JPL (6) cost projection for epitaxial deposition of 1 μm Ge and 3 μm GaAs on a GaAs/Ge/Si cell structure, is reported. This evaluation was performed by means of a detailed and consistent computer program (SAMICS) which offers a standardized methodology for comparing different photovoltaic approaches.

The results of this calculation, though partially speculative, indicate that low cost flat plate modules can be fabricated employing high efficiency GaAs thin films.

TABLE II: A JPL COST PROJECTION FOR EPITAXIAL DEPOSITION OF 1 μm Ge AND 3 μm GaAs.

CELL STRUCTURE: GaAs/Ge/Si

TECHNOLOGY: LOW PRESSURE MO-CVD
YIELD: 90%
TRIMETYL GALLIUM COST: 5$/g
η MODULE: 18.8%
PLANT CAPACITY : 250MW/YEAR
YEAR: 1986

COST FOR 1 μm Ge AND 3 μm GaAs DEPOSITION AND LASER ANNEALING:

. 11 $/Wp (1980 $)

MODULE COST:

. 56 $/Wp (1980 $)

4. MULTICOLOR CONCENTRATOR SOLAR CELLS

Spectrum splitting approach, which employs two different bandgap concentrator solar cells, coupled with an optical dichroic filter, presently represents the only way of achieving photovoltaic efficiencies higher than for single cells.

A 20.5% AM_2 efficiency has been obtained by VARIAN (14) on a 45 W, 372 suns concentrator module including AlGaAs-Si cell couples and optical filters. A potential 23% efficiency was also predicted for an optimized module design.

The major drawback deriving from the use of filters, is represented by the additional costs of optical components, support structures and installation. For this reason, the development of very economical concentrators supplying concentration and spectrum splitting at the same time, would be definitely welcome (15), (16).

The situation is quite different for monolithic cascade cells which may represent a long term more attractive solution for high efficiency concentrator systems both for space and terrestial applications.

30-35% cell conversion efficiencies, required for cost effective concentrator modules, still seem to be a long term goal. To date a metal interconnected AlGaAs-GaAs stacked cell produced by VARIAN (18) has achieved the best efficiency: 22% AM_3 at 130 suns, while a tunnel junction interconnected AlGaAs-GaAs cell fabricated by LPE at Research Triangle Institute (17) has shown 16.4% AM_1 efficiency without AR coating.

For these cells conversion efficiencies higher than 30% are expected.

A number of different material structures have been investigated in order to optimize interconnect junctions, lattice and current matching and solar spectrum utilization. GaAsP-GaInAs structures grown by vacuum MO-CVD have been investigated by Chevron Research Company (19), (20), while laser bounded AlGaAs-AlGaSbAs cells have been developed by Rockwell International.

AlGaAsSb and AlGaInAs cascade cells have also been fabricated by MO-CVD by Research Triangle Institute and Varian respectively.

The French poject, named "Arc en ciel" and coordinated by CNRS, has the most ambitious goal: a 40% experimental conversion efficiency is projected for a four-cell system including a AlGaAs-GaAs cascade cell, a dichroic filter and a GaInAsP-GaInAs/InP cascade cell. Low resistance tunnel junctions have already been obtained for the intercell contact in the MBE grown AlGaAs-GaAs structure (21).

5. CISE CONCENTRATOR SOLAR CELLS

The photovoltaic activity developed in CISE has been limited so far to the design and fabrication of high concentration solar cells (22) and modules, but the same basic technologies might be extended to the fabrication of thin film GaAs cells.

Table III summarizes the results obtained for different cell structures. Except for the stacked cells which had an exposed area of only $.1 cm^2$, for all the other cases the exposed area was $1.37 cm^2$.

As it can be seen, LPE tecnology approximates its quality goal, while 1 Atm. MO-CVD has still to be improved in terms of reactor geometry and gas contamination.

TABLE III: GaAs CELLS FABRICATED IN CISE.

STRUCTURE	GAP (EV)	TECHNOLOGY	BEST EXP EFF. (AM 1.5)	GOAL
$Al_{.9}Ga_{.1}As/GaAs$	1.43	LPE	22.5% 350x 21% 1000x	23% 350x 21% 1000x
$Al_{.9}Ga_{.1}As/GaAs$	1.43	MO-CVD	18% 635x	23% 350x
$Al_{.9}Ga_{.1}As/Al_{.2}Ga_{.8}As$	1.70	LPE	19% 300x 17.5% 1000x	21% 350x 19% 1000x
GaAs shallow homojunction	1.43	MO-CVD	13% 25x 14% 3x	21% 350x
$Al_{.9}Ga_{.1}As/Al_{.2}Ga_{.8}As/GaAs$	1.7/1.43	LPE	15.5% 23x	25% 350x

The experience on the shallow homojunction GaAs cell is very recent and the result reported in table III, practically represents the first attempt. The efficiency distribution of a lot of LPE GaAlAs/GaAs cells is represented in fig. 1 for two concentration ratios, showing a drop of 3 percentage points from 400 to 750 suns.

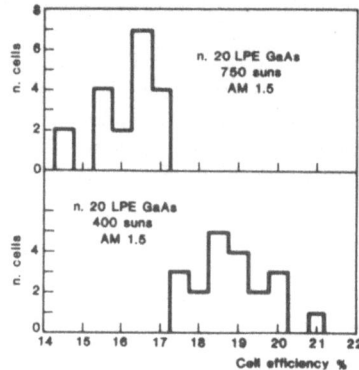

Fig. 1- Efficiency distributions of a lot of GaAs cells for two concentration ratios.

A first simplified version of bicolor stacked cell has also been developed by C. Flores with the principal scope of evaluating the combined photovoltaic performances of two different band-gap cells stacked in the simplest possible way: a p-n-p transistor like structure. (fig. 2).

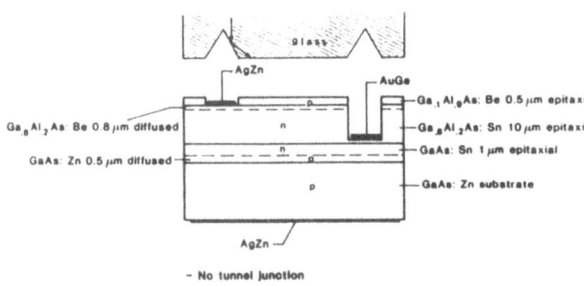

Fig.2-Three-terminal AlGaAs-GaAs

If compared to two terminal cascade cells the advantages of this three terminal structure, are represented by an extremely simplified fabrication process (only three epitaxial layers) and no need for tunnel junction and current matching. Due to the elimination of interconnect junction losses, this type of cell is expected to show better performances at high concentration levels. On the other side, the cell efficiency is reduced by double grid shadowing but the application of the grooved cover glass (fig.2) proposed by P. Borden (23) would be particularly effective. This cover has not been applied yet to the present cell.

Moreover, these cells cannot be series connected directly to each other but simple and effective solutions can be adopted (24).

Quantum efficiency and current voltage curves are reported in fig. 3.

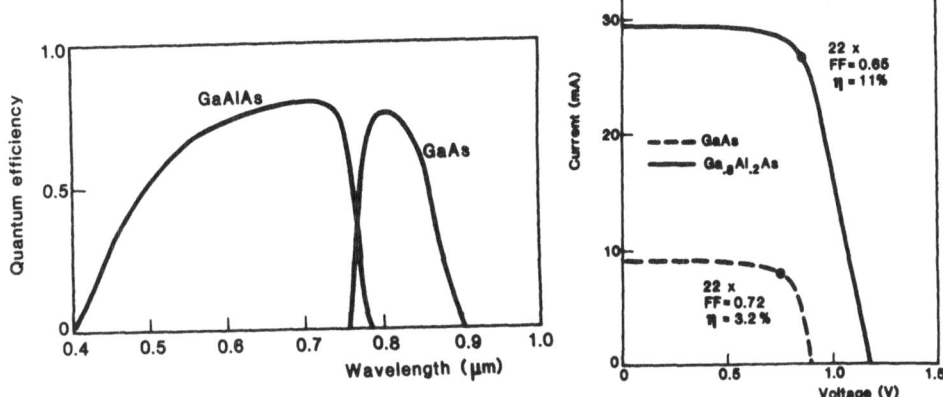

Fig. 3 - Typical experimental quantum efficiencies and I-V curves for the top and the bottom cells of the stacked structure.

Further improvements on this cell are expected to derive from grid design optimization energy gap increase of the top cell, better control of the junction depth in the GaAs cell and use of the grooved cover glass to reduce the double grid shadowing.

6. CISE CONCENTRATOR MODULES

The activity developed on concentration modules was principally directed to make experience on very high concentration as a solution for achieving a drastic reduction of the cost contribution of high efficiency sophisticated cells, without forgetting that some silicon solar cells, like Microwave Associates vertical junction cells, show a conversion efficiency increasing with concentration up to 1000 suns or even more.

Finally, highly concentrated solar energy fluxes may be of particular interest for hybrid applications.

Two 750 suns, Fresnel lens, concentrator modules, have been designed and fabricated. The details are described in another paper of this conference (25) and will not be repeated here.

The first module, approximately 170 Wp with single GaAs cells and 10 large area Fresnel lenses was fabricated with the purpose of testing our concentrator cells in passive cooling conditions.

The second module, containing 6 Fresnel lenses, dichroic filters, and 6 GaAlAs (1.7 eV), GaAs (1.43 eV) actively cooled cell couples, was designed with the scope of combining high concentration and spectrum splitting technique.

The best efficiency measured for a AlGaAs-GaAs cell couple, including filter losses, was 23% AM 1.5, 840 suns.

Preliminary evaluations performed on this module indicate an overall efficiency of 15% AM 1.5 at 750 suns. As reported in table IV, several modifications could be introduced to increase the module efficiency.

TABLE IV: SOME POSSIBLE MODIFICATIONS FOR INCREASING THE EFFICIENCY OF THE SPLITTING MODULE.

MODIFICATION	EFFICIENCY
Presently	15%
400x CONCENTRATION	17%
AR COATING OPTIMIZATION	17.5%
OPTICS OPTIMIZATION	18.2%
SPECIFICALLY DESIGNED Si CELLS	19.7%

As shown in fig. 1, about 3 efficiency percentage points would be gained by decreasing the concentration to 400 suns. More than 1 point would also be gained by employing optimized lenses, filters and a two layer A.R. coating.

Finally, at least 1.5 ponts would be saved if specifically designed Si cells (1.1 eV) were used as low bandgap cells. The combination 1.1 eV-1.7 eV, corresponding to Si-AlGaAs cells, represents in fact a nearly optimal choice. On the other side, the couple GaAs-AlGaAs consisting of two temperature resistant cells, suggest the possibility of a very efficient utilization for spectrum splitting high concentrator modules with recovery of the heat transported by the cooling fluid. Such a hybrid-splitting module may represent the highest efficiency solution compatible with today's photovoltaic technology.

7. CONCLUSION

Recent technical results and middle term cost projection seem to indicate that thin film GaAs solar cells could be conveniently employed in flat-plate array both for space and terrestrial applications. Major efforts should be directed in the future to develop high yield-large scale GaAs deposition process. For high efficiency concentrator systems, the most promising solution seems to be represented by monolithic multicolor stacked cells based on GaAs alloys, but the 30-35% efficiency goal is still far from being achieved and long, intensive research work will be required.

High efficiency spectrum splitting concentrator modules have been fabricated both using AlGaAs-Si (VARIAN) and AlGaAs-GaAs (CISE) cell couples but the success of all these concentration approaches will depend not only on the achievement of very high efficiency cells but also on the development of low cost, reliable concentrators.

ACKNOWLEDGMENTS

The photovoltaic program developed in CISE has been supported under contract by ENEL and, partially, by E.C.

The author wishes to tank all his coworkers for their active contribution to photovoltaic activity and Dr. C. Flores also for helpful discussions.

REFERENCES:

1) J.M. Woodall, H.J. Hovel - J.Vac. Sci. Technol $\underline{12}$, 1000 (1975)
2) P.E. Glaser, D.W. Almgren - Proc. 1st P.V Solar Energy Conf., 836 (1977)
3) M.J. Carr - Crystal Specialities Inc.-private communication (1982)
4) P. Mullen - M/A-COM. Laser Diode Inc.-private communication (1982)
5) R.P. Gale, B.Y. Tsave, J.C. Fan, F.M. Davis, G.W. Turner - Proc. 15th IEEE PVSC, 1051 (1981)
6) R.J. Stirn, K.L. Wang, Y.C. Yeh - Proc. 15th IEEE PVSC, 1045 (1981)
7) M. Garozzo, G. Conte, F. Evangelisti, G. Vitali - Applied Phys. Lett. to be published
8) J.C. Fan, C.O. Borler, R.P. Gale - Proc. IEDM meeting, 534 (1980)
9) J.C. Fan, C.O. Borler, R.W. McClelland - Proc. 15th IEEE PVSC, 666 (1981)
10) S.S. Chu, T.L. Chu, C.L. Jiang, C.W. Loh—Proc. 15th IEEE PVSC, 1310 (1981)
11) G.J. Williams - Patent N°4,094,703 (1978)
12) D.W. Almgren, K.I. Csigi - NASA contractor Report 3361 (1980)
13) L.M. Fraas - J.Appl. Phys $\underline{52}$, 6939 (1981)
14) P.G. Borden, P.E. Gregory, D.E. Moore - Proc. 15th IEEE PVSC, 311 (1981)
15) W.H. Bloss, M. Griesinger, E.R. Reinhardt - Proc. 3rd PV solar Energy Conf. Cannes, 401 (1980)
16) G. Sassi - Proc. 3rd PV. Solar Energy Conf. Cannes, 981 (1980)
17) S.M. Bedair, J.A. Hutchby, J. Chiang, M. Simons, J.R. Hauser - Proc. 15th IEEE PVSC.,21 (1981)
18) P. Borden - Private Communication
19) L.M. Fraas - Proc. 15th IEEE. PVSC., 1353 (1981)
20) L.M. Fraas - Proc. SPIE meeting. L.A. (1982)
21) P. Bouchaib, J.P. Contour, F. Raymond, C. Verie, F. Arnand d'Avitaya, J. Vac. Sci. Technol. $\underline{19}$, 145 (1981)
22) E. Fanetti, C. Flores, G. Guarini, F. Paletta, D. Passoni, - Solar cells $\underline{3}$, 187 (1981)
23) P. Borden - Photovoltaic Concentrator Technology Development Project, 7th Meeting SAND 81-1240 209, (1981)
24) S. Sakai, M. Umeno. J. Appl. Phys. $\underline{51}$, 5018 (1980)
25) E. Fanetti, C. Flores, G. Guarini, F. Paletta - Proc. this conference

PHOTOVOLTAIC CONCENTRATOR TECHNOLOGY IN THE USA*

E.C. BOES and M.W. EDENBURN
Photovoltaic Concentrator Project
Sandia National Laboratories

Summary

One of the alternate electrical energy technologies that has progressed most rapidly over the past several years is photovoltaic concentrator technology. In the US, with funding of approximately $40 million over the past 7 years, this solar electric technology has been transferred from the conceptual stage to the point where several promising designs have been fabricated and tested. The experience of concentrator research and development thus far strongly supports the initial hope that it could be a practical alternative to flat plate photovoltaics. Three major conclusions have emerged from the experience of PV concentrator technology to date:
- PV concentrators are far more efficient than flat plate PV modules
- Designing PV concentrators with 20 year lifetimes appears feasible
- Totally installed PV concentrator array costs are projected to reach $1 to $2 per peak watt.

Fresnel lens concentrators using planar junction Si cells have already achieved measured annual average efficiencies of 13 to 14%, and well understood modifications will increase this to the 15 to 16% range. More advanced module designs have been measured at as high as 20%.

1. INTRODUCTION

This paper presents a summary of photovoltaic concentrator technology in the US.

The two basic motivations for photovoltaic concentrators are to replace expensive solar cells with cheaper optical sunlight concentrators, and to achieve higher conversion efficiencies by taking advantage of the fact that V_{oc} increases logarithmically with intensity causing the efficiency to increase until resistance losses cause fill-factor degradation.

A complete photovoltaic concentrator array consists of the module, which is the basic collection and conversion unit, a tracking support structure on which the modules are mounted, and a control system which keeps the modules pointed at the sun.

2. CONCENTRATOR MODULE DESIGNS AND THEIR PERFORMANCE

A photovoltaic concentrator module consists of an optical concentrator (usually one or more lenses), concentrator cells mounted on a receiver assembly, a cooling mechanism for the cells, and a module housing which combines these into a single package. We shall describe the primary types of PV concentrator modules being developed in the US in four broad categories:
- Point-focus Fresnel (PFF), baseline design
- Linear Fresnel (LF), baseline design
- Reflective parabolic troughs
- Advanced point-focus Fresnel designs

---

*This work supported by US Department of Energy, Division of Photovoltaic Technology.

Baseline point-focus Fresnel modules use flat, acrylic Fresnel lenses and planar junction silicon cells with concentration ratios between 40 and 200. Table 1 lists the primary design features for the principal modules in this category that are currently being developed. All of these designs are passively cooled; that is, waste heat from the cell is conducted into a flat or finned aluminum heat exchanger where it is dispersed and convected to the surrounding air.

The optimal size for the individual, square Fresnel lenses for these point-focus modules is in the range of 200 cm^2 to 700 cm^2 for several reasons: (i) it allows use of a cell in the 1 cm^2 to 10 cm^2 range, where planar junction Si concentrator cells have highest efficiencies; (ii) passive cooling of the cells is relatively easy because only about 35 watts of thermal power per cell needs to be dispersed and dissipated; and (iii) the module housing can be relatively shallow. A disadvantage of larger lenses is that they cause larger cell current, which implies heavier, more costly interconnects.

The more recent modules in this category generally have somewhat higher concentration ratios than the designs of several years ago. This is because more recent planar junction silicon cell designs have peak efficiencies at higher concentration levels.

Table 1 lists NOC* efficiencies above 13% for several recently developed modules. For comparison, most NOC efficiencies for flat plate modules are in the 7 to 9% range. Since most of these recent concentrator modules use cells that have peak efficiencies of about 18%, NOC efficiency increases to the 15 to 16% range are highly likely through the use of 20% Si cells and slightly better lenses.

Two firms, E-Systems and Acurex, are developing actively cooled linear-focus Fresnel modules that have cells mounted on a coolant tube for heat removal. Both use arched, linear Fresnel lenses made by laminating a faceted, 0.55 mm thick acrylic Lensfilm material from 3M Company to clear acrylic sheet. The lenses are mechanically arched and held in that shape by the sheet metal housings. Acurex's module uses a steel coolant tube for cell mounts, while E-Systems' coolant tube is extruded aluminum. Both tubes are porcelain coated to provide electrical isolation for the cells. Both can also provide thermal as well as electric energy.

Electrical efficiencies for linear Fresnel modules thus far are very similar to those for PFF baseline modules. In addition, thermal efficiencies for E-Systems LF modules are about 45 to 60%, depending on the coolant temperature and insolation conditions.

E-Systems is also developing a passively cooled version of their linear Fresnel module.

Table 1 also includes design and performance information for 3 parabolic trough PV concentrators. Our overall experience with PV troughs to date has been that they are generally inferior to lens concentrators because, although possible, it is difficult to obtain an equal illumination on each cell.

Some of the most interesting work in the area of photovoltaic concentrators is advanced PFF modules. These are modules that employ high concentration ratios and exotic cells, such as etched multiple vertical junction (EMVJ) Si cells, GaAlAs cells, or multiple junction devices. Four such modules are described in Table 1. Note that all four acrylic lenses either are domed or have curved facets. Efficiencies for the experimental modules using EMVJ cells are likely to be only in the 14 to 15% range because, to date, these cells have peak efficiencies below 19%.

---

*NOC - Nominal operating condition

The most important conclusion from this experience with advanced PFF modules is that very high solar-to-electric conversion efficiencies are feasible. The 20% measured efficiency for Varian's beam splitter module is a world record over all solar technologies.

3. TRACKING STRUCTURES AND ARRAYS

Line and point focus Fresnel lens modules both require two axis tracking. We have considered five types of tracking structures: turntable (1), pedestal (2), polar axis, frame-with-end-pedestals (3), and tube-with-end-pedestals. We do not recommend the turntable because installation cost is high and we believe that polar axis tracking is only suitable for small arrays. The tube-with-end-pedestals tracker is a new design and is being evaluated.

Concentrating PV arrays have been employed in the experiments listed in Table 2 (4).

Table 2. Array Experiments

Array Fabricator	Location	Installed	Size
Martin Marietta PFF	Sandia PASTF	Sept. 79	2-2.5 kW arrays
	Riyadh	Aug. 81	350 kW
	Phoenix	Apr. 82	225 kW
E-Systems LFF	Dallas, Ft. Worth	Apr. 82	25 kW
BDM (trough)	Albuquerque	Jan. 82	50 kW
Acurex (trough)	Lihue	Dec. 81	60 kW
Solar Kinetics/Solarex (trough)	Blytheville	Apr. 81	240 kW

4. COMPONENTS

Cells—Improvement in concentrator cell efficiency has been the primary reason for the rapid improvement in module efficiency over the last five years. Planar junction silicon cells have achieved 20% efficiencies at concentration levels of 100X (5, 6). GaAlAs cells have achieved 24% at 400X (7). Using both silicon and GaAlAs cells in a spectrum splitting module has attained a combined efficiency of 26% including filter losses (8).

New silicon cell designs e.g. Varian's staircase cell and Microwave Associate's vertical multi-junction cell (9) have the potential of obtaining 20+% efficiencies at concentrations up to 1000X. Research with GaAlAs cells in various spectrum splitting, graded bandgap configurations promises efficiencies of over 30% (10, 11).

A major recent accomplishment in concentrator cell research is the development of a two-dimensional device code at Purdue University (12) which permits the modeling of two-dimensional effects and will play a key role in improving the efficiency of newer cells designed for high concentration levels.

Receivers—Cell efficiency is inversely proportional to cell temperature; therefore, the cell must be attached to its cooling mechanism (active or passive heat exchanger) with a low thermal resistance interface which must be an electrical insulator. It must also accommodate differences in thermal expansion between the cell and heat exchanger, and it must endure temperature cycling and the elements to which it is exposed.

It is also desirable to protect the cell with a transparent cover to which it is optically bonded. This cover must endure the temperatures, radiation and other elements to which it is exposed.

A great deal of experience has been gained with several receiver materials:
- o PVB turns brown and/or bubbles when exposed to UV at temperatures above 90°C
- o Some thermal greases, e.g. ECTC7, squeeze out when heated. Others, e.g. Wakefield 120, work well
- o Some RTV's give good adherence when properly primed e.g. EC 4952. Others are variable, e.g. EC 4954 and 4954W
- o Contact resistance causes higher than expected temperature differences

Passive heat exchangers need not be large. Martin Marietta's second generation heat exchanger weighs only 5.3 kg per m^2 of aperture, and Sandia's baseline module uses the aluminum housing as a heat exchanger without fins (13).

Lenses--Compression molded, flat, point focus, acrylic lenses typically have an optical efficiency of about 85%. Injection molding offers low cost acrylic lenses, but their performance has been disappointing to date. Thin acrylic lens-glass laminates are being developed to take advantage of glass's low cost and resistance to abrasion. Domed point focus lenses are also being considered. They are potentially more efficient than flat ones because refractive losses are lower. They may also be less expensive because they can be injection molded with little performance loss since poorly defined facet tips are not in the light path.

E-Systems uses an arched linear Fresnel lens which has a greater than 85% optical efficiency and is very insensitive to lens deflections.

A great deal of work has been done recently to design lenses using a cell response model so that the illumination profile can be optimized (14).

Control Systems--Most PV concentrator arrays designed thus far employ closed loop control systems with sun sensors for proper array pointing. However, heliostat experience indicates that open loop, ephemeris tracking is also practical. PV concentrator fields use microprocessors to handle all the control functions--sun tracking, daily wake-up and shut-down, and cloudy or foul weather stowing--in an integrated unit. No special development problem is foreseen in this area.

5. DURABILITY

Sandia National Laboratories subjects modules and components to accelerated environmental testing as well as real time testing. Tests include high voltage potential, off axis beam damage, reverse bias heating, temperature-humidity cycling, hail, temperature cycling, and UV exposure. The tests have pointed out many materials and designs that were deficient, allowing the problems to be corrected. Many modules have survived the tests without significant degradation. This does not guarantee that the module will survive for twenty years, but we do believe that the environmental test results are strongly correlated with real time durability.

Martin Marietta's first generation array has been real time tested for over 2-1/2 years at Sandia's test facility. It has experienced a module failure rate of less than 2% per year and the tracking system fails an average of twice a year which is an excellent record for a first generation array.

6. COST ESTIMATES

General Electric and Sandia National Laboratories have performed a projected manufacturing cost study for four current concentrating PV designs at production rates of 10 MW per year (1). The study considered material cost, labor cost, and a markup factor that was typical of similar manufacturing industries. Results are shown in Table 3.

Table 3 - Completely Installed Array Cost ($/m^2)

	Line focus Fresnel	Point Focus Fresnel, Turntable	Trough 1 axis	Point Focus Fresnel, Pedestal
Cells	76	51	72	93
Receiver	44	95	40	56
Lens	45	16	108	16
Housing & Assy.	34	73	3	42
Structure & Tracking	66	23*	77	68
Shipping	17	7	6	10
Installation	76	179	62	75
Total	359	444	368	360
Annual Efficiency	13%	13%	9%	13%
$/W$_{ap}$**	2.76	3.42	4.09	2.77

*The structure cost for this array is included in installation.
**W_{ap} is the annual efficiency times 1000 W/m^2.

Note that, even for <u>current</u> PV concentrator designs with annual efficiencies of about 13%, completely installed field prices below $3 per annualized peak watt are projected. With virtually certain efficiency increases to 15% and design improvements, the price could very easily drop below $2/$W_{ap}$, and advanced PFF module concepts could bring it close to $1/$W_{ap}$.

References
1. Chan, <u>Manufacturing Cost Analysis of 1980 Vintage Photovoltaic Arrays</u>, Sandia Natl. Labs, SAND81-7010, May 1982.
2. Donovan, "Ten Kilowatt Photovoltaic Concentrating Array," IEEE, PVSC, 1978.
3. O'Neill, "The 25 kW Photovoltaic Application Experiment," IEEE, PVSC, 1980.
4. Biringer, "Intermediate Photovoltaic System Experience," IEEE, PVSC, 1981.
5. Nasby, "Characterization of p$^+$nn$^+$ Silicon Concentrator Cells," IEEE, PVSC, 1981.
6. Khemthong, "Improved Silicon Cells for High Concentrator Levels," IEEE, PVSC, 1981.
7. Gregory, "Performance and Durability of AlGaAs/GaAs Cells," IEEE, PVSC, 1981.
8. Borden, "A 10 Unit Dichroic Filter Spectral Splitter Module," IEEE, PVSC, 1981.
9. Frank, "A Low Series Resistance Silicon Cell for High Intensity," IEEE, PVSC, 1980.
10. Borden, "A Monolithic AlGaAs/GaAs Solar Cell Array," IEEE, PVSC, 1980.
11. Mitchell, "High Efficiency Concentrator Cells," IEEE, PVSC, 1981.
12. Lundstrom, "Modeling Solar Cells Containing Heavily Doped Regions," IEEE, PVSC, 1981.
13. Shafer, "Low Cost, High Performance Concentrator Array Design," IEEE, PVSC, 1980.
14. Stillwell, "Compatibility of Fresnel Lenses and Cells," IEEE, PVSC, 1981.

Table 1. Designs and Efficiencies for Concentrating PV Modules

Designer	Module Description	Geom. Conc.	Optics	Cell Size, cm^2	Cooling	Weight, kg/m^2	Efficiency* Peak	Efficiency* NOC	Status
Baseline PFF Modules									
Applied Solar Energy Co.	3x5 lens-cell matrix, Al housing	150	17cmx17cm, acrylic-glass lenses, reflective secondaries	2.9	Passive, cast Al heat exchanger	29			Testing to begin Apr. 82
General Electric	5x6 lens-cell matrix, Al housing	68	17cmx17cm acrylic or acrylic-glass lenses	5.5	Passive, housing back plate	30			Being tested
Martin Marietta, 1st generation	1x4 lens-cell matrix, plastic housing	50	30cmx30cm acrylic lenses	25.7	Passive, Al extrusion heat exchanger	44	10.4	9.5	2-1/2 yrs. testing at Sandia
Martin Marietta, 2nd generation	2x7 lens-cell matrix, plastic housing	84	21cmx21cm acrylic lenses	2.25	Passive, cast Al. heat exchanger	22	14.0	13.2	tests continuing
Sandia Laboratories, baseline I	5x5 lens-cell matrix, Al sheet housing	70	17cmx17cm acrylic lenses	5.5	Passive, housing back plate is heat exchanger	18	13.4	12.1	long term testing
Sandia Laboratories, baseline II	5x6 lens-cell matrix, Al sheet housing	153	17cmx17cm glass laminated and acrylic lenses, refl. & solid secondaries	3.1 with secondary 5.1 with guardband	Passive, housing back plate is heat exchanger	24/18 laminated/ acrylic lens			being fabricated
Baseline LF Modules									
Acurex	Arched linear Fresnel, sheet steel housing	43	3.05x.99m acrylic lens	17.7	Active, steel coolant tube	45			being fabricated
E-Systems, second generation	Arched linear Fresnel, sheet steel housing	40	3.05x.91m 3M lensfilm-acrylic sheet	6.2	Active Al coolant tube	16	12.7	11.5	tests continuing
E-Systems, advanced PVT	Arched linear Fresnel, sheet Al housing	40	3.05x.91m acrylic Fresnel	14.1	Active, Al coolant tube	13			being developed
E-Systems, first gen passive	Arched linear Fresnel, sheet Al housing	40	3.05x.91m 3M lensfilm-acrylic sheet	6.2	Passive Transverse Al fins	16	14.0	12.5	being tested
E-Systems, second gen passive	Arched linear Fresnel, sheet Al housing	40	3.05x.91m acrylic Fresnel	14.1	Passive Transverse Al fins	16			being developed
Trough Modules									
Acurex	Reflective parabolic trough, 90° rim angle	36	3.05x1.83m reflective film	12.5	Active Al coolant tube	21	10.0	8.4	60 kW in Lihue
BDM/Solar Kinetics	Reflective parabolic trough, 90° rim angle	60	6.1x2.15m reflective film	13	Active Al coolant tube	17	9.7	8.2	50 kW in Albuquerque
Sandia Laboratories	Reflective parabolic trough, 90° rim angle	43	2x1m glass	15.1	Active Al coolant tube	22	12.3	11.4	testing complete
Advanced PFF Modules									
Varian	3x4 lens-cell matrix AlGaAs cells, plastic housing	473	29.7cm diag hexagonal lens curved facets	1.6	Active, jet impingement		15.8	15.3	tests continuing
General Electric	3x3 lens-cell matrix VMJ Si cells, plastic housing	1200	17.8cm circular domed acrylic lens	.58	Passive finned heat exchanger				prototype being tested
Thermo Electron	3x4 lens-cell matrix Si cells, plastic housing	590	22.9x22.9cm domed lens TIR secondaries	1.1	Active				cell-lens being tested
Varian	10 lens spectrum splitter with dichroic filter	477	20.4cm diag hexagonal lens curved facets	$.72cm^2$ AlGaAs and $.72cm^2$ Si	Active, jet impingement		20.5	20	tests continuing

*Peak - 800 W/m^2, 28°C cell
NOC - normal operating condition - 800 W/m^2, 20°C ambient, 1 m/s wind or 40°C coolant

LUMINESCENT SOLAR CONCENTRATORS (LSC):

TECHNICAL AND ECONOMIC REQUIREMENTS
FOR A RESIDENTIAL SYSTEM

E. BERMAN and P. D. WILDES
ARCO Solar, Inc.
20554 Plummer Street
Chatsworth, California 91311

Summary

The goal of this work is to develop a luminescent solar concentrator/ solar cell system that will substantially reduce the cost of electricity from the sun. Technical and economic requirements are analyzed using a residential system connected to a utility grid. Based on this model, a fluorescent organic dye dissolved in a solid plastic matrix is a viable candidate for the solar concentrator, and single-crystal silicon is a possible candidate for the light converter. Dye stability is identified as the critical parameter with a quantum efficiency for dye bleaching of less than 10^{-8} the target requirement. Progress with model materials has resulted in a thousand-fold improvement in dye stability and provides mechanistic evidence for applying this work broadly to other dye systems. A new class of dye with almost ideal absorption/emission characteristics is described.

1. INTRODUCTION

Most concentrators either remove or downgrade the advantages of a solar cell system by requiring mechanical tracking and/or high temperature operation. This paper describes a concentrator which offers the promise of reducing the price of solar cell systems while retaining the inherent advantages of solar cells.

A luminescent solar concentrator (LSC) (1) is a planar layer of materials, contained in a transparent matrix, which reemit light after absorption of light. A portion of the reemitted light is trapped within the matrix and can escape only at the layer edges (Figure 1). The fraction trapped can be calculated from Snell's law to be

$$\frac{4(n^2 - 1)^{1/2}}{(n + 1)^2}$$

where n is the refractive index of the matrix (Figure 2). The **maximum** concentration possible is the ratio of the planar area to that of the edges.

The advantages of an LSC are that it collects and concentrates diffuse as well as direct radiation; eliminates heat dissipation problems of solar optical concentrators since excess energy of short wavelength radiation and non-useful long wavelength radiation is not focused on the converter; separates light absorption and light conversion functions; allows more efficient use of narrow spectral response converters.

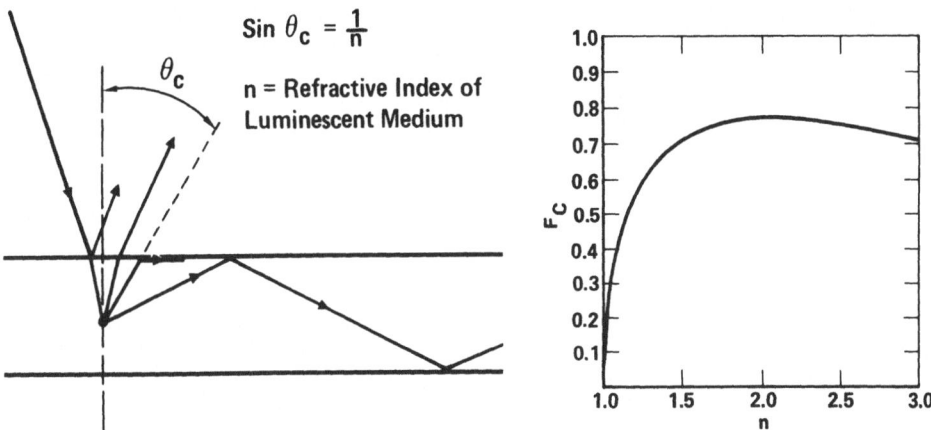

Figure 1. Luminescent Solar Concentrator

Figure 2. LSC Collection Efficiency vs. Refractive Index

2. ECONOMIC POTENTIAL OF AN LSC/SOLAR CELL SYSTEM

In order to evaluate the potential of an LSC/solar cell system, its efficiency, price and value must be known. The model system selected consists of an organic fluorescent dye in a plastic matrix on or in a host matrix with a concentration ratio of 10:1, uses single-crystal silicon solar cells as the light convertor, and is roof-mounted in a utility grid-connected application.

2.1 Potential Efficiency

The current density that can be obtained from a solar cell coupled to an LSC under illumination is given in equation 1:

$$J = \int_{\lambda_0}^{\lambda_{bg}} P_S(\lambda) \, d\lambda \cdot \frac{F}{N_A} \cdot F_A \cdot \phi_F \cdot F_C \cdot F_T \cdot F_{MAT} \cdot \phi_C \qquad (1)$$

where J is the current density, λ_{bg} is wavelength equivalent to the cell bandgap, $P_S(\lambda)$ is the photon flux, F is Faraday's constant, N_A is Avogadro's number, F_A is the fraction of light absorbed by the collector, ϕ_F is fluorescence quantum efficiency, F_C is the fraction of light trapped in the collector, F_T is the fraction of fluoresced light transmitted to the cells, F_{MAT} is the fraction of fluoresced light not lost due to collector material imperfections (e.g., surface scratches, internal scattering centers), and ϕ_C is the solar cell quantum efficiency.

Solar radiation more energetic than 400 nanometers and less energetic than 850 nanometers is excluded for the model system under consideration.

The high energy photons have sufficient energy to break bonds in organic dye molecules. The lower energy photons will be inactive if downshifted further by absorption and reemission. Reasonable values for the LSC efficiency factors in equation 1 are F_A, 0.65; ϕ_F, 0.9; F_C, 0.75; F_T, 0.8; F_{MAT}, 0.9; ϕ_C 0.9. These values result in a 6% efficient device. More ideal values for the factors yield a potential efficiency of 10%.

2.2 Potential Price

Single-crystal silicon solar cell modules presently sell for about $1000/m^2. If an LSC/solar module has a concentration ratio of 10:1, the median price for the solar cell component is $100/m^2. Other components in the system are quite inexpensive; $15-$50/m^2 is taken as their price including fabrication.

2.3 System Cost Effectiveness

The application selected is a roof-mounted LSC/solar cell system connected to a utility grid. It is assumed that the LSC/solar module is the waterproof outer member of the roof. The system cost effectiveness, Y, is defined as the total system cost, P, divided by the value per year for the system, V. Equations 2, 3 and 4 define these parameters:

$$P = (P_A + P_I + IEP_C) A \qquad (2)$$

$$V = (5)(365) AIEV_E/1000 \qquad (3)$$

$$Y = P/V = \frac{P_A + P_I + IEP_C}{1.8\ IEV_E} \qquad (4)$$

where P_A is the LSC/solar module system price, P_I is the installation price, and P_C is the power conditioning price, all in $/m^2; A is the array area in m^2; E is the LSC/solar module efficiency (fraction); V_E is the value of electricity in $/kWh; I is insolation in W/m^2.

Figure 3 presents the relationships among price, efficiency, and cost effectiveness for values of electricity of $0.05 and $0.30 per kilowatt hour.

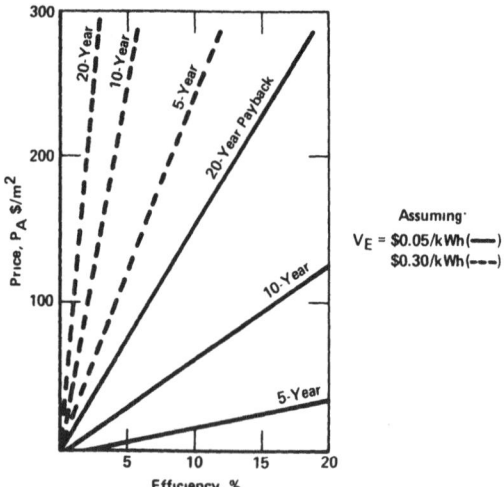

Figure 3. Relationships Among Price, Efficiency, and Cost Effectiveness for Rooftop Systems

An LSC/solar module system with an efficiency of 5-10% and a price of $100-$200/m^2 is clearly cost effective in developing areas of the world (value of electricity $0.30/kWh) and holds commercial promise in developed areas (value of electricity $0.05/kWh).

3. STATUS

LSC/solar module devices using organic fluorescent dyes and yielding an efficiency of about 4% have been reported (1). Dye absorption/emission overlap and dye photostability are major areas of concern with these systems.

3.1 Absorption/Emission Overlap

Figure 4 gives the absorption and fluorescence emission spectra for Pyronin B, a typical organic dye with high fluorescence quantum yield (greater than 0.9). The absorption/emission overlap is substantial in this case with a Stokes shift on the order of 20-30 nm. The major problem here is that each time an emitted photon is reabsorbed and reemitted about 25% of the photons will escape the collector.

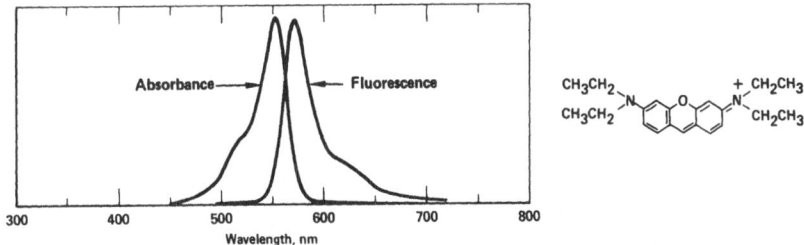

Figure 4. Absorbance and Fluorescence Spectra for Pyronin B in Ethanol

Fortunately, new dyes, like DCM (1), have substantial Stokes shifts (Figure 5).

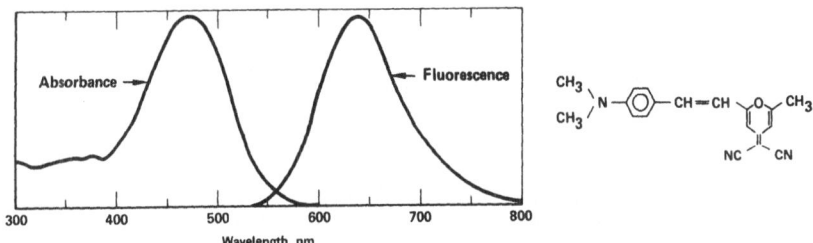

Figure 5. Absorbance and Fluorescence Spectra for DCM in N,N-Dimethylacetamide

3.2 Photostability

A successful product requires an outdoor stability of 10-20 years. Table I compares the quantum efficiency for bleaching, i.e., the number of molecules bleached per absorbed photon, with dye half-life outdoors.

Table I. Quantum Efficiency for Bleaching Compared With Dye Half-Life

ϕB	$\tau 1/2$	ϕB	$\tau 1/2$
10^{-5}	2 days	10^{-8}	4-5 years
10^{-6}	2-3 weeks	10^{-9}	40-50 years
10^{-7}	5-6 months		

Thus the quantum efficiency for bleaching must be in the range of 10^{-8} to 10^{-9}. While organic dyes of exceptional photostability are known, e.g., the phthalocyanines, this requirement is extremely severe. In our laboratory, we have chosen to focus on this issue.

We selected Pyronin B in methanol as a model dye system to investigate in detail. It was selected as being typical of many likely candidates and it allowed us to work with high purity components. Figure 6 shows the

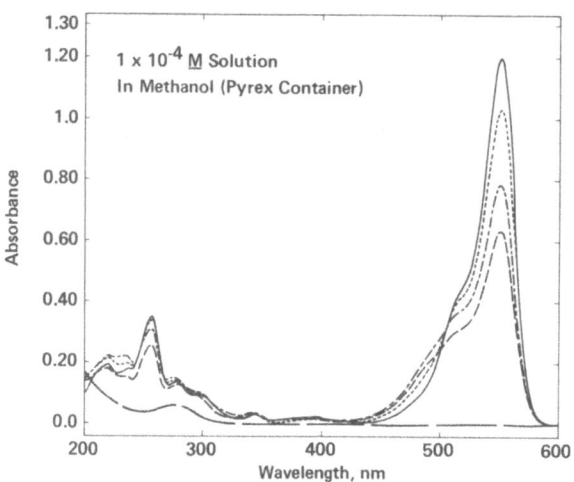

Figure 6. Sunlight Decomposition of Pyronin B

results when Pyronin B in methanol in a Pyrex container undergoes prolonged exposure to sunlight. In time, the dye is completely destroyed. The quantum efficiency for bleaching is 3×10^{-5}, which corresponds to a half-life of only 5 hours! Undaunted, we reduced the quantum efficiency for bleaching to 3×10^{-8}, corresponding to a half-life of about 18 months, by:

    a. Selecting the proper pH
    b. Excluding light below 400 nm and above 850 nm
    c. Excluding oxygen
    d. Adding "triplet state" quenchers

Our research activity continues to emphasize:

    a. Fundamental studies of the primary photoprocess
    b. Synthesis of more stable dyes
    c. Isolation and identification of photodecomposition products

4. CONCLUSION

The promise shown by luminescent solar concentrator systems has reached the point where increased effort to develop the concentrator and solar cell components of the system to meet the requirements defined in this work is fully justified.

REFERENCE

(1) Hermann, A. M. "Luminescent Solar Concentrators -- A Review," <u>Solar Energy</u> (in press).

RECENT PROGRESS IN A RESIDENTIAL SOLAR ENERGY SYSTEM DEVELOPMENT

E. L. Johnson and J. S. Kilby
Texas Instruments Incorporated
Dallas, TX, USA

SUMMARY

The Texas Instruments Solar Energy System (TISES) concept and development status is described. Texas Instruments has had the TISES approach under development for several years. This new system concept combines the energy conversion and storage functions in a novel way. Small silicon solar cells are immersed in an electrolyte, and the current generated by the cells is used directly to electrolyze hydrobromic acid. The hydrogen and bromine produced can be stored separately (hydrogen as a hydride and bromine in solution) until needed and and then recombined in a fuel cell to produce electrical energy on demand. The fuel cell HBr product is returned to the solar chemical convertor to complete the closed loop energy cycle. Component and system feasibilities have been demonstrated.

1. INTRODUCTION

The Texas Instruments Solar Energy System (TISES) under development combines the energy conversion and storage functions in a novel way (1).

Solar energy is an intermittent energy source; therefore, if power is to be supplied continuously or on demand for a residential application as we have chosen, some form of energy storage must be employed. With sufficient storage, the energy produced can command retail price; without storage, its worth is a function of the local utility power mix, peak loading, regulatory climate, solar availability, etc. Rates that have been proposed range from full value to only slightly greater than the utility's fuel cost, ranging from a few tenths of a cent per kWh for a coal facility at today's prices.

The cost estimates made to date indicate that products of the TI system type have a good chance of being able to serve the residential market. The TI system concept is briefly described in the next section. A more detailed description of the system, components, and processes can found in earlier reports (2,3).

2. SYSTEM DESCRIPTION

In the TISES approach arrays of spherical silicon cells are used to electrolyze hydrobromic acid (HBr) for direct storage of the available electrical output as chemical energy. The sphere arrays are the primary component of the Solar Chemical Convertor (SCC) shown in the block diagram of the TISES in Figure 1. In the SCC, HBr is electrolyzed to produce hydrogen gas and bromine which remains in solution as Br_3^- ions. The hydrogen is stored as a metal hydride in a calcium-nickel alloy, the bromine is stored in solution. The stored chemical energy can be used on demand by reacting hydrogen and bromine in the fuel cell to produce electrical energy and HBr which is again routed to the SCC to complete the loop. Excess thermal energy can also be extracted from the electrolyte as low grade heat, useful for domestic hot water and space heating.

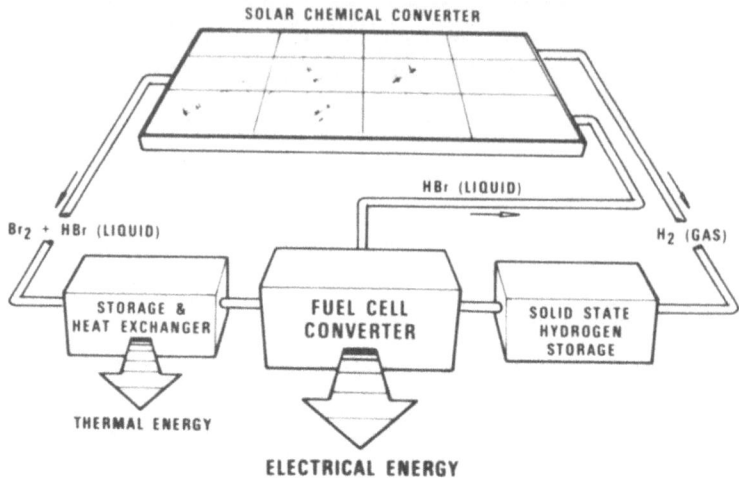

FIGURE 1. Block diagram of Texas Instruments Solar Energy System (TISES)

To provide the voltage required to electrolyze HBr, two silicon solar cells must be connected in series. This can be accomplished in a simple way as shown in Figure 2. A thin glass sheet containing P^+/N and N^+/P spherical solar cells is first etched on one side to expose a portion of the spheres. The diffused junction region is then etched away to expose the core of the sphere. After passivating and then burying the junction perimeter with another layer of glass, ohmic contact is made to the sphere which are connected by a continuous metal layer on the back side of the array. The glass is then etched away on the front side of the array to expose a small portion of the sphere surface. A front side electrode metal is then selectively applied to those exposed sphere surfaces forming an ohmic contact to the diffused layer. Approximately 1 volt is available for the electrolysis reaction with this design.

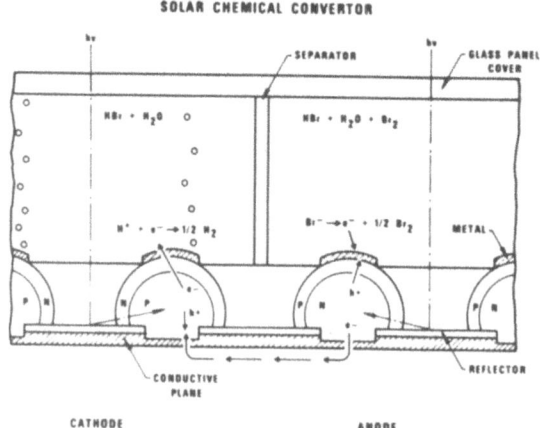

FIGURE 2. Schematic cross section of the TISES Solar Chemical Convertor (SCC)

In the SCC, hydrogen (H^+) ions are reduced to hydrogen gas at the cathodes (N^+/P cells), and Bromide (Br^-) ions are oxidized to bromine at the anode (P^+/N cells). The bromine remains in solution as Br_3^- ions. The anode and cathode cells are segregated to prevent the reduction of bromine back to bromide ions at the cathodes. This back-reaction is a chemical short circuit that would seriously degrade the SCC performance if the cells were mixed. In addition, a chemical separator is placed between the anode and cathode regions to minimize bromine transport to the cathodes.

The spherical cell design offers other advantages that are unique to a photoelectrochemical system such as the TISES. N^+/P and P^+/N cells are easily combined into single glass sheets yielding a voltage output that is twice that of a single cell. Planar cells would require interconnections within the module to achieve this. Most importantly, a small percentage of defective cells have little effect on array performance because the cells are independently connected to the electrochemical load. Each cell independently achieves its own optimum operating point.

3. COMPONENT AND SYSTEM DEVELOPMENT

The SCC array process development has emphasized producibility (automatable processes), simplicity and high material utilization. Small single crystal silicon cells can be made at a fraction of the cost of an equivalent area prepared by conventional or advanced crystal growing technique. The spherical shape is easily produced from the molten state due to surface tension. The high surface area to volume ratio of small spheres results in maximum heat transfer and high material throughput. Arrays of spheres require less silicon per unit area than planar devices. The largest array cast to date is more than an order of magnitude larger than the largest single crystal grown to date, and no barriers to even larger sizes are apparent.

Theoretical modelling of the spherical silicon solar cell shows that the optimum diameter of the sphere is in the range of 150 to 200 μm(4). This diameter intuitively seems correct because this value is roughly twice the diffusion length easily obtained in silicon. The optimum diameter would be larger for material with very long diffusion lengths. It may be more practical to produce sphere arrays with somewhat larger diameter spheres, and the theoretical model shows that the efficiency does not decrease rapidly as the diameter is increased. Most of the arrays produced to date have included spheres in the 250 to 400 μm size range.

The electrical efficiency of individual spheres is difficult to determine due to the light concentration effect of the glass matrix. The efficiency can be accurately determined only by averaging the electrical output of a large number of cells over a large area. Most of the cell arrays produced so far are about 40 cm^2 in area and contain about 25,000 spheres each. Efficiencies are determined by averaging the output of small numbers of cells selected at random over the entire area of the array.

It has proved useful to have a baseline material to which the spheres can be compared directly. Cubes of silicon have been cut from (100) oriented CZ silicon slices so that each face of the cube is a (100) plane. These cubes are then etched in an orientation dependent etch to yield particles of the proper size that approximate a sphere in shape. A small quantity of this material is processed with each sphere lot.

The electrical efficiency goal for sphere arrays at AM1 is 13 percent. This efficiency has been demonstrated with etched cubes indicating that the spherical cell array structure can achieve the required efficiency. This level of efficiency is not routinely obtained at this point, but further optimization of the diffusion, passivation, and front side metallization processes should yield significant improvement. The best efficiency achieved with spheres produced by a potentially low cost process (atomized/remelted spheres) is 10.4 percent.

A hydrogen-bromine fuel cell stack (30-cell) has been developed with excellent electrical performance. The power density has exceeded the original design goal (70 W/SF @ 70% voltage efficiency) by a factor of four.

The hydrogen generated in the SCC is stored as a metal hydride. This is a safe and inexpensive form of storage for this application. The charge and discharge characteristics of the alloy of choice, $CaNi_5$, has low pressure and temperature characteristics well suited to the system and the available thermal energy from the panel. The bromine is stored in solution in the HBr electrolyte.

One concept of a prototype production module is depicted in Figure 3. The modular construction is expected to offer a number of advantages, including manufacturability, system sizing, complete factory assembly and testing, easier installation and maintenance, and redundancy for improved reliability.

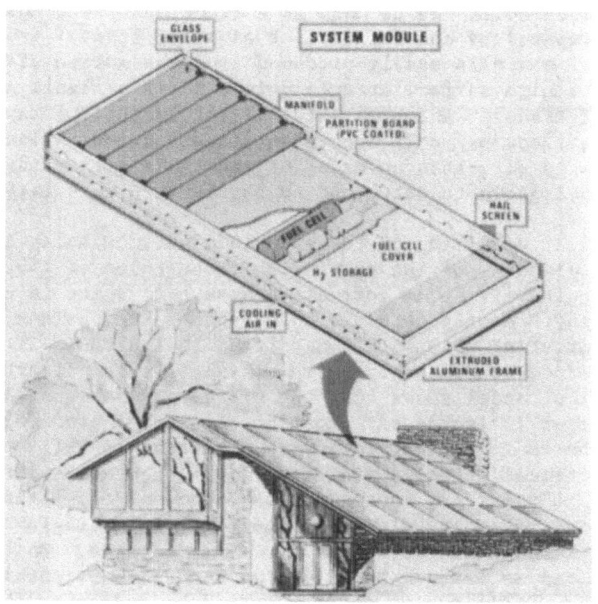

FIGURE 3. A TISES Modular System Concept

4. CONCLUSIONS

Component and system feasibility has been demonstrated. Although a substantial amount of effort has been expended on the proposed system, major efforts will still be required to make it a commercial reality. In the long term, the system is expected to supply approximately 90% of the total energy needs of typical residence at cost to the homeowner competitive with the existing electric utility rates.

All the major milestones have been met and the outlook for achieving the final goals of the program continues to be good.

Finally, although the program is still considered a high risk proposition because of the uncertainty of reliability and economic factors, the continued ontime progress toward program goals supports an encouraging outlook.

ACKNOWLEDGEMENTS

This work is supported by a Cooperative Agreement between the U. S. Department of Energy and Texas Instruments Incorporated under Agreement DE-AC01-79ER10000.

5. REFERENCES

1. J. S. Kilby, J. W. Lathrop, W. A. Porter, "Solar Energy Conversion", U.S. Patent No. 4,021,323, May 3, 1977.

2. E. L. Johnson, "Development of a Solar Energy System", A Special Technical Summary Report, DOE Cooperative Agreement No. FC01-79ER-10000, April 3, 1981.

3. E. L. Johnson, "The TI Solar Energy System Development", IEEE Proceedings of the International Electron Devices Meeting, pp. 2 - 5, December 7-10, 1981.

4. W. R. McKee, "The Analysis and Optimization of a Spherical Silicon Solar Cell," Masters Degree Thesis, Texas A&M University, August, 1976.

HIGH EFFICIENCY TANDEM TYPE SOLAR CELLS
CONSISTING OF a-Si:H AND a-SiGe:H

G.Nakamura,K.Sato,H.Kondo,Y.Yukimoto
and K.Shirahata

LSI Research and Development Lab.,Mitsubishi Elecrtic Corp.
4-1 Mizuhara, Itami 664,Hyogo,JAPAN.

Summary

Electrical properties of a-SiGe:H deposited by RF glow discharge decomposition of SiH_4/GeH_4 mixture and application of a-SiGe:H to tandem type solar cells were investigated. Decrease of residual gas content in discharge atmosphere caused to shift Fermi level to mid-band gap and increased a ratio of photo-conductivity(σph) to dark-conductivity(σd) in a-SiGe:H. Pre-exponential factor of σd in a-SiGe:H varied with conductivity activation energy(Ea) as predicted by Meyer-Neldel rule. The highest conversion efficiency of 8.5% was accomplished in the 3 stacked tandem cell composed of two inverted p-i-n type a-Si:H cells and one inverted p-i-n cell with an intrinsic a-$Si_{0.6}Ge_{0.4}$:H.

1. INTRODUCTION

Although amorphous solar cells have been attaining much attention for their low manufacturing cost, low conversion efficiency of the amorphous solar cells is one of the majour problems. For achieving high conversion efficiency,tandem type amorphous solar cells with a-Si:H and a-SiGe:H have been proposed(1). However,the actual efficiency of the tandem type solar cells developed so far was limited to low value because photo-electric properties of a-SiGe:H, such as σph,degraded with increase of Ge content and also because optimization of cell structure was not sufficient(2).
  Here,we study to improve photo-electric properties of a-SiGe:H prepared by RF glow discharge decomposition of SiH_4/GeH_4 mixture and estimate conversion efficiency of both 2 stacked and 3 stacked tandem cell with various structures. Application of a-SiGe:H to tandem cells are also described.

2. EXPERIMENTAL

A glow discharge equipment with capacitive coupled parallel plate electrodes was used to prepare a-$Si_{1-x}Ge_x$:H($0\leq x \leq 1$) in the mixed gas of helium diluted SiH_4 and hydrogen diluted GeH_4 with various mixing ratio. Table I shows a typical growth condition. Under this condition,decomposition efficiency of SiH_4 and GeH_4 were nearly equal and growth rate was 1-1.5A/sec. which did not vary with GeH_4 content in discharge atmosphere.
  σd and σph of a-SiGe:H were measured using parallel electrodes configuration with 5 mm in width and with 1 mm in

separation. Optical band-gap energy(Egopt) was evaluated from transmission and reflection spectra of the film in visible ligth region.

3. PROPERTIES OF a-SiGe:H

Table I Deposition condition of a-Si$_{1-x}$Ge$_x$:H

Pressure	0.5 Torr.
Flow Rate(Total)	100 sccm
SiH4/He	10 %
GeH4/H2	10 %
RF Power(13.56 MHz)	50-100 W
DC bias	+70 V
Substrate Temperature	230-250°C

Fig.1 shows an experimental relation between Ea and Egopt measured in two groups of a-SiGe a-SiGe:H. a-SiGe:H in group"A"(● in Fig.1) was prepared under the condition of low residual gas content in discharge atmosphere. a-SiGe:H in group"B" (○ in Fig.1) was prepared under the condition of relatively high residual gas content in discharge atmosphere. The former showed higher Ea than the latter at the same Egopt. This showed that Fermi level of a-SiGe:H in group"A" lay closer to mid-band gap than that in group"B". This implied that residual gas in discharge atmosphere caused to increase acceptor like gap states.

Fig.2 shows an experimental relation between pre-exponential factor(σ_o) in σ_d and Ea. σ_o varied with Ea as predicted by Meyer-Neldel rule. The meaning of Meyer-Neldel rule in a-Si a-SiGe:H was not understood yet.

σ ph under simulated AM-1 illumination and electron drift mobility(μn) estimated from current-voltage characteristics of thin film transistor are plotted as a function of Ge content in a-SiGe:H of group"A" in Fig.3. Remarkable improvement in σ_{ph} was observed comparing with previous data(2). The ratio of σph to μn was almost constant in the region where Ge content varied from 0.1 to 0.8. This implied that carrier lifetime was not changed in that region.

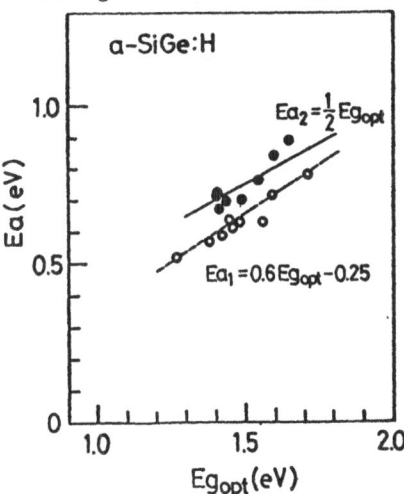

Fig.1 Experimental relation between Ea and Egopt in a-SiGe:H

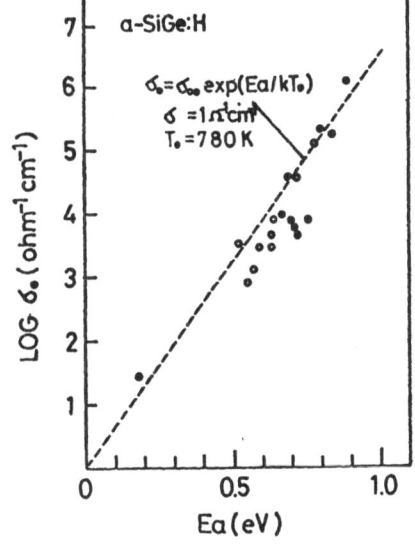

Fig.2 Experimental relation between σo and Ea in a-SiGe:H

4. DESIGN CONSIDERATION OF TANDEM TYPE SOLAR CELLS

Simplified model calculation of tandem type solar cells was carried out to estimate conversion efficiency(η) and for appropreate design of tandem type solar cells.

In calculation, each cell in the tandem structure was treated as single inverted p-i-n cell with exception that photo-current in the tandem type cell was limited by the smallest one of photo-current in each cell.

Our detailed theoretical calculation for single inverted p-i-n cell revealed that short circuit current(Jsc) resulted from drift of photo-generated carriers because intrinsic layer was completely depleted and electric field strength in the depletion layer was larger than 1×10^4 V/cm, in the case that intrinsic layer thickness is less than 1um and minimum gap state density near mid-band gap was $1 \times 10^{15} cm^{-3} \cdot eV^{-1}$. Our calculation also revealed that terminal voltage caused to shrink the depletion layer thickness and degraded both fill factor to the value of 0.7 and open circuit voltage(Voc) to the value of 0.5xEgopt. These results were applied for model calculation.

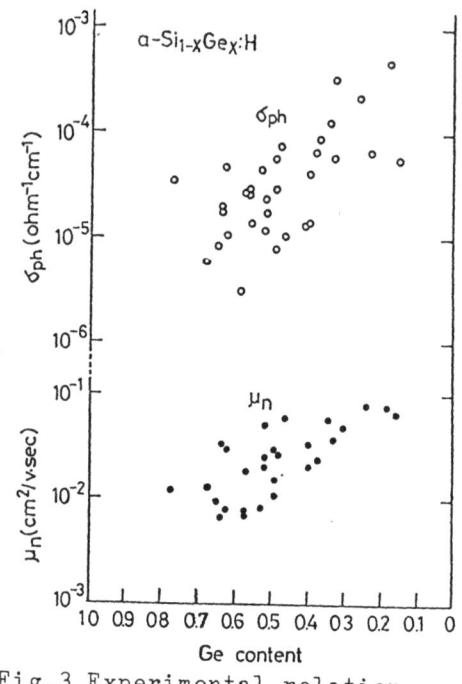

Fig.3 Experimental relation between σph, μn and Ge content in a-SiGe:H

Fig.4 shows models used for calculation. In this figure, Eg, α and W show optical band gap energy, absorption coefficient and width of each layer. α of each layer was given by next empirical equation.

$$\alpha(\lambda) = 8.1 \times 10^5 \cdot (1.24/\lambda - Eg)^2 / (1.24/\lambda) \quad (1)$$

for intrinsic layer and

$$\alpha n(\lambda) = 8.1 \times 10^5 \cdot (1.24/\lambda - Egn)^2 / (1.24/\lambda) \quad (2)$$

$$\alpha p(\lambda) = 1.2 \times 10^6 \cdot (1.24/\lambda - Egp)^2 / (1.24/\lambda) \quad (3)$$

for doped layer. Where λ is wavelength of light. Short circuit current of each inverted p-i-n cell is given by next equation.

$$Jsc1 = q \int_{0.29}^{1.24/Eg_1} \left[\{1 - R(\lambda)\} \cdot Nph(\lambda) \exp(-\alpha n(\lambda) \cdot Wn) \cdot \{1 - \exp(-\alpha_1(\lambda) \cdot W1)\} \right] d\lambda \quad (4)$$

for first cell and

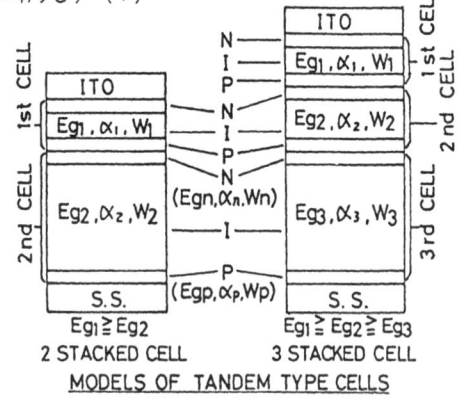

MODELS OF TANDEM TYPE CELLS

Fig.4 Models used for calculation

$$Jsc2 = q \int_{0.29}^{1.24/Eg2} [\{1-R(\lambda)\} \cdot Nph \cdot \exp(-2\alpha_n(\lambda) \cdot Wn - \alpha_1(\lambda) \cdot W1 - \alpha_p(\lambda) \cdot Wp) \cdot \{1-\exp(-\alpha_2(\lambda) \cdot W2)\}] d\lambda \quad (5)$$

for second cell and

$$Jsc3 = q \int_{0.29}^{1.24/Eg3} [\{1-R(\lambda)\} Nph \cdot \exp(-3\alpha_n(\lambda) \cdot Wn - 2\alpha_p(\lambda) \cdot Wp - \alpha_1(\lambda) \cdot W1 - \alpha_2(\lambda) \cdot W2) \cdot \{1-\exp(-\alpha_3 \cdot W3)\}] d\lambda \quad (6)$$

for third cell. Where, $R(\lambda)$ is reflection coefficient at front surface and Nph is photon flux in AM-1.

Voc of the tandem type cell is given by the sum of Voc in each inverted p-i-n cell. Fill factor in tandem type cell was assumed to be equal to that in the single inverted p-i-n cell.
Jsc in 2 stacked tandem type cell(Jsct2) was smaller one of Jsc1 and Jsc2.
Jsc in 3 stacked tandem type cell(Jsct3) was the smallest one of Jsc1, Jsc2 and Jsc3.
Maximum out put power of 2 stacked(Po2) and 3 stacked(Po3) tandem type cells were given by next equations.

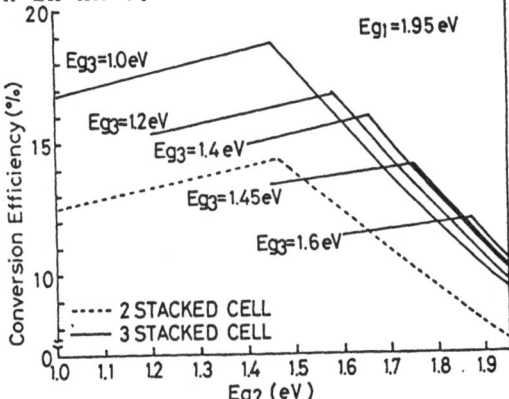

Fig.5 Calculated relation between conversion efficiency and Eg2

Po2 = 0.7 × Jsct2 × 0.5 × (Eg1+Eg2)   (7) (for 2 stacked tandem cell)

Po3 = 0.7 × Jsc3 × 0.5 × (Eg1+Eg2+Eg3)   (8) (for 3 stacked tandem cell)
Conversion efficiency of tandem type cell was given by Pout/Pin ×100.

At first, intrinsic layer thickness in each cell was determined in the region of less than 1μm so as to maximize short circuit current in tandem type cell. Then conversion efficiency of tandem cell was estimated.

Fig.5 shows a calculated relation between η and Eg2 in 2 stacked and 3 stacked tandem type cells. Parameters used for calculation are depicted in the figure. This figure showed that 3 stacked tandem type cell achieved higher conversion efficiency than 2 stacked tandem type cell in the region where Eg2 was less than 1.45eV.

5. TANDEM TYPE SOLAR CELLS

By combining inverted p-i-n type a-Si:H cell and inverted p-i-n type cell with an intrinsic a-SiGe:H, two kinds of tandem type solar cells were fabricated and evaluated.
The first was 2 stacked tandem type cell composed of an inverted p-i-n type a-Si:H cell and an inverted p-i-n cell with an intrinsic a-$Si_{0.6}Ge_{0.4}$:H of about 5000A thick.
The second was 3 stacked tandem type cell composed of two inverted p-i-n type a-Si:H cells and an inverted p-i-n type cell with a-SiGe:H mentioned above.
Thicknesses of a-Si:H cells were determined experimentally

so as to maximize short circuit current. In the former cell thickness of 1st cell was about 1800A and in the latter cell, thickness of 1st cell was about 600A and that of 2nd cell was about 4000A. These values coincided fairly good with the values predicted by the argument in section 4.
Photovoltaic characteristics of tandem type cells are shown in Fig.6 with a characteritics of an inverted p-i-n type a-Si:H cell for comparison.

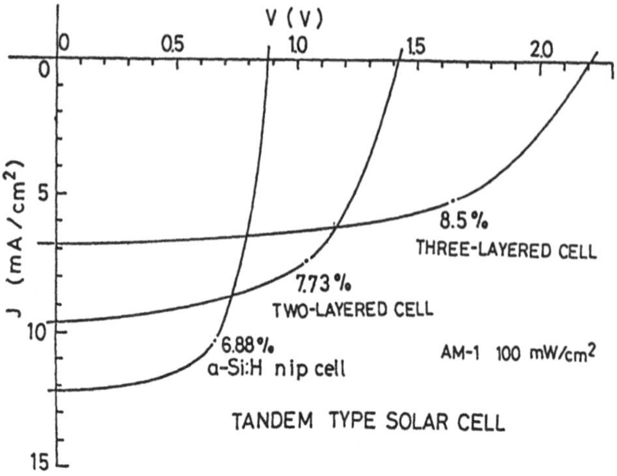

Fig.6 Photo-voltaic characteristics of 2 stacked and 3 stacked tandem type cells comparing with a single type cell

Increase in number of stacked cells caused to decrease short circuit current, while open circuit voltage increased with the number of stacked cells. Tandem type solar cells showed higher conversion efficiency than the single type cell because a-SiGe:H had higher spectral response in long wave length region than a-Si:H. The highest conversion efficiency of 8.5% was accomplished in the 3 stacked tandem type solar cell with area of 9mm^2. Open circuit voltage of 2.2V, short circuit current of 6.74mA/cm^2 and fill factor of 0.57 were obtained under simulated AM-1 sun light of 100mW/cm^2.

6. CONCLUSION

3 stacked tandem type cells with conversion efficiency of 8.5% were fabricated. This tandem type cell was composed of two inverted p-i-n type a-Si:H cells and one inverted p-i-n cell with an intrinsic a-Si$_{0.6}$Ge$_{0.4}$:H. Broadening of efficient spectral response region by improving photo-electric properties of a-SiGe:H and optimization in structure of 3 stacked tandem type cell were main reasons for realizing this high conversion efficiency.

Acknowledgement
     The authors wish to express their sincere appreciation to Prof.Hamakawa for his helpfull discussion. They also thank to Dr.K.Fujikawa and Dr.H.Oka for their interest and encouragement during this work.
     This work was supported by Agency of Industrial Science and Technology under the contract of the Sunshine Project.

References
(1) Y.Marfaing:2nd E.C.Photovoltaic Solar Energy Conf.,(1979) p.287.
(2) G.Nakamura,K.Sato,H.Kondo,Y.Yukimoto and K.Shirahata: Proc. 9th Int.Conf.on Amorphous and Liquid Semi.,(1981) C4-483.

HIGH EFFICIENCY SHALLOW p^+nn^+ CADMIUM TELLURIDE SOLAR CELLS

G. COHEN-SOLAL, D. LINCOT and M. BARBÉ

C.N.R.S. Laboratoire de Physique des Solides
Bellevue-Meudon 92190 - France

Summary

Significant improvements have been achieved in preparing CdTe shallow homojunctions with conversion efficiencies over 11% AM1 (higher than 13% with proper A.R. coating). The cells are fabricated by using a close spaced vapor transport (C.S.V.T.) technique to form a p-type CdTe epitaxial layer upon a single crystal substrate (n-type CdTe). The combined use of a doping gradient in the n-base material, a heavily p doped layer on the front and an optimization of the junction deepness lead to shallow p^+nn^+ structures in a reproducible way. The best AM1 performance parameters of cells are $V_{oc} \sim 0,85$ Volt, $I_{sc} \sim 20$ mA/cm^2 and F.F. $\sim 70\%$. The cells have shown no degradation in conversion efficiencies after 18 months.

1. INTRODUCTION

Le tellurure de cadmium est-il un bon matériau pour photopiles solaires ? Une réponse positive portant sur les potentialités de ce candidat a depuis longtemps été proposée, mais les recherches expérimentales de confirmation ont marqué le pas. Parmi les raisons qui expliquent ce semi-échec figure en premier lieu le caractère marqué, dans ce matériau, du mécanisme d'autocompensation qui a entre autres, pour effets i) de rendre malaisée la réalisation de jonctions pn, de bonne qualité et peu profondes (compte tenu de la valeur élevée du coefficient d'absorption optique $\alpha > 10^4$ cm^{-1} et des faibles longueurs de diffusion des porteurs minoritaires la jonction doit être située à moins de un micron de la surface éclairée), ii) de gêner considérablement l'obtention de contacts ohmiques surtout dans le matériau de type p. C'est pourquoi nous avons choisi de réaliser les jonctions en déposant par épitaxie une couche de type p sur une base monocristalline de type n. Un appareillage spécial de transport en phase vapeur à courte distance en atmosphère d'hydrogène a été développé au laboratoire |1| permettant la réalisation de structures de type p-n$^+$ de 8% de rendement |2| ; les travaux ultérieurs portant sur l'amélioration du taux de collecte des photocourants ont mis en évidence les limitations associées à la faible valeur du dopage des couches épitaxiées |3|. Nous présentons dans cet article les résultats obtenus avec des structures du type p^+nn^+, mieux adaptées à ce type de matériau.

2. REALISATION DES PHOTOPILES

Les structures p^+nn^+ sont préparées en déposant une couche mince épitaxiée (épaisseur 0,2 à 0,8 µm) de CdTe-p, dopé à l'arsenic ou au phosphore ($\sim 10^{17}$ at/cm^3), par la technique C.S.V.T. en pression atmosphérique d'H$_2$, déjà décrite en |4|. Les substrats sont des plaquettes (e ~ 1 mm ; Ø ~ 15 mm) découpées dans des lingots de CdTe de type n, dopés indium (n $\sim 5 \cdot 10^{16}$ at/cm^3) préparés par la méthode Bridgman, par R. Triboulet ; polies mécaniquement et décapées chimiquement, les plaquettes subissent un traitement thermique dans l'enceinte d'épitaxie qui a pour double effet i) de révéler la structure cristalline superficielle par décapage des couches perturbées, favorisant de ce fait le mode épitaxique de la croissance ultérieure et ii) de modifier la distribution des impuretés au voisinage de la surface. La zone p^+ est formée in situ, après le dépôt de la couche, par diffusion gazeuse d'arsenic ou de phosphore, en condition d'espace clos dans l'enceinte toujours sous pression atmosphérique d'hydrogène. Un exemple des profils obtenus d'indium (T = 500 °C, t = 130 secondes) et d'arsenic (T = 500°C, t = 120 secondes) relevés en sonde ionique est présenté sur la figure (1) ; l'étude des différents profils de diffusion qui se poursuit permet déjà d'évaluer les coefficients de diffusion de l'indium $D_{In} \sim 2.10^{-11} cm^2/s$ (500°C) et de l'arsenic $D_{As} \sim 2.10^{-12}\ cm^2/s$ (500°C), et de fixer la concentration d'arsenic à la surface de la couche p^+ à environ 10^{17} at/cm³. Le contact sur la surface avant est réalisé par dépôt sous ultra vide d'une grille d'or évaporé dont le pas est calculé (20 µm-40 µm) pour minimiser les effets de résistance latérale de la couche. On procède ensuite à un léger traitement thermique sous hydrogène (180°C - 120s) pour réduire la résistance de contact. Le contact arrière est pris par fusion et diffusion d'indium sous hydrogène.

3. CARACTERISTIQUES ELECTRIQUES

L'étude des caractéristiques capacité-tension des jonctions ainsi réalisées montre que les zones de charge d'espace à polarisation nulle ont une largeur comprise entre 0,1 et 0,3 µm et que les jonctions sont de type graduel avec un gradient de dopage de l'ordre de $4.10^{20}/cm^4$ au voisinage de la zone de charge d'espace. L'allure des courbes en polarisation directe met en évidence l'absence de couche incluant des états d'interface, ce qui est en bon accord avec les faibles valeurs de courants d'obscurité. Les hauteurs de barrière obtenues par extrapolation sont de l'ordre de 1,1 volt.

L'analyse des caractéristiques courant-tension à l'obscurité (dont un exemple est présenté fig. (2)) permet d'évaluer le courant de saturation des différentes structures I_s compris entre 1,5 à 5.10^{-10} A/cm², avec des facteurs d'idéalité n~2. Ces valeurs sont proches de celles que l'on peut calculer théoriquement dans le cas de courants de génération-recombinaison dans des homojonctions CdTe |5|. A noter des résistances séries généralement faibles ($R_s \sim 0,1\ \Omega.cm^2$) ainsi que des courants inverses de l'ordre de 10 à 20 µA/cm² à polarisation inverse V_R= -4 volts, attestant de la valeur élevée des résistances shunt.

4. CARACTERISTIQUES PHOTOVOLTAIQUES

4.1. Réponses spectrales.

Un exemple de réponse spectrale est donné sur la figure (3) . Le rendement quantique "interne" est obtenu en corrigeant les pertes dues à la réflexion ; il atteint une valeur de 95% entre 0,55 µm et 0,81 µm. Seuls le caractère très superficiel de la jonction et la présence d'un champ accélérateur au voisinage de la surface (associé à la zone p^+ diffusée) peuvent expliquer la réduction des pertes dues à la recombinaison en surface dont on sait qu'elle est aussi élevée que pour le GaAs ($S \sim 10^6$ cm/s), et l'obtention de rendements quantiques supérieurs à 90%. Cet effet, prédit par Ellis et Moss |6| , a été mis en évidence dans les jonctions n^+pp^+ de GaAs |7| et p^+nn^+ de InP |8| .Compte tenu d'une part que la profondeur de la jonction est réduite à environ 0,2 µm de la surface et d'autre part que l'épaisseur de la zone de charge d'espace est de l'ordre de 0,1 µm, une partie non négligeable des photons est absorbée dans la base où l'existence d'un champ électrique associé au gradient de dopage facilite, par entraînement, la diffusion des porteurs ; cet effet, surtout marqué pour les énergies de photons voisines de celle de la largeur de bande interdite, et de même nature que celui adopté dans les cellules au silicium dites B.S.F. (Back Surface Field) |9| , conduit tous calculs faits, à des longueurs de diffusion de trous deux à trois fois supérieures à 0,5 µm (mesuré en absence de champ), et explique la forme de la réponse spectrale vers 0,8 µm.

4.2 Caractéristiques courant-tension sous éclairement

Sur le tableau ci-dessous nous avons porté les caractéristiques d'une des meilleures jonctions, en fonction de la puissance incidente.

Energie incidente	I_{cc} mA/cm^2	V_{oc} volt	Facteur de forme %	Rendement %	Rendement avec couche A.R. (corrigé 20 %)
70 (solaire) *	13$\pm$0,1	0,82	69$\pm$1	10,5$\pm$0,3	13,1
82 (solaire)	15,4	0,82	68	10,5	13,1
107 (simu.AM1)	20,1	0,83	67	10,5	13,1

Le courant de court-circuit I_{cc} = 20,1 mA/cm^2 mesuré passe après correction de 20% des pertes par réflexion à la valeur I_{cc} = 25,1 mA/cm^2, en bon accord avec les réponses spectrales. Par ailleurs en réduisant la surface couverte par la grille, à surface active constante, on pourrait faire passer la tension de circuit ouvert de 0.83 à 0.85 volt, et de ce fait le rendement corrigé de 13,1% à 13,4%.

4.3. Les rendements de conversion

Nous avons vérifié que les pertes associées à l'état de la technologie de nos photopiles (résistances série, shunt et résistances de contact) sont négligeables ; bien que sachant qu'un traitement antireflet adéquat conduira à multiplier les résultats obtenus par un facteur supérieur à 1,3, nous avons décidé de fournir les rendements en n'effectuant que la correction d'ombrage de la grille, ce qui a l'avantage de ne prendre en compte que les effets de recombinaison superficielle, de résistance carrée de la couche p$^+$, de la profondeur de jonction et des profils de dopage. Les mesures ont été effectuées comparativement sous simulateur et au soleil (dans les conditions classiques). La figure (4) présente les caractéristiques d'une cellule typique sous éclairement solaire (AM1~107 mW/cm^2); le rendement de conversion de la surface active est de 10,7 %. Une étude en fonction de l'énergie incidente montre que ce rendement varie peu, comparé à celui des photopiles au silicium |10| , typiquement 2,5 points par deux décades de flux alors que l'on observe une variation de cinq points par deux décades dans le silicium.

En ce qui concerne la stabilité dans le temps des photopiles, conservées sans précautions spéciales, à l'air ambiant, on peut noter sur la figure (5) qu'après une légère baisse dans les trois premiers mois, le rendement de conversion se stabilise et reste pratiquement constant sur dix-huit mois.

5. CONCLUSIONS

Il a été prouvé qu'il était possible d'atteindre et de dépasser 13% de rendement de conversion d'énergie solaire en utilisant comme base du CdTe monocristallin dans une configuration p$^+$nn$^+$, où la zone frontale p$^+$ est constituée par une couche isoépitaxiée de CdTe. L'obtention de rendements supérieurs est liée à la solution de deux types de problèmes : vitesse de recombinaison superficielle et résistance latérale de la zone frontale élevées. La remarquable stabilité au cours du temps de ces dispositifs et le fait qu'il s'agit d'un matériau à absorption "directe" (a priori plus résistant aux irradiations de hautes énergies que le silicium) ouvrent des perspectives d'application spatiale nouvelles. Sur le plan des applications terrestres seules des structures de même type mais réalisées entièrement en couches minces, sur substrats bon marché, et sans réduction notable de la valeur du rendement ($\eta \geqslant 10\%$) présentent un intérêt potentiel certain.

Ce travail a bénéficié de l'aide précieuse de R. Triboulet et F. Bailly que les auteurs remercient en cette occasion.

* Chiffres obtenus par cellule EGG étalon et confirmés par le C.M.N. de Trappes.

REFERENCES

|1| LAROCHE J.-M. et COHEN-SOLAL G., à paraître dans Journal de Physique Appliquée

|2| J. MIMILA-ARROYO, Y. MARFAING, G. COHEN-SOLAL and R. TRIBOULET, Solar Energy Materials, 1, 171 (1979)

|3| LINCOT D., BARBÉ M., COHEN-SOLAL G. and MARFAING Y., Conf. Rec. 3rd Europ. Comm. Photov. Solar En. Conf. Cannes, p. 882 (1980)

|4| LINCOT D., MIMILA-ARROYO J., TRIBOULET R., MARFAING Y., COHEN-SOLAL G. and BARBÉ M., Conf. Rec. 2nd Europ. Comm. Photv. Solar En. Conf. Berlin p 424 (1979)

|5| SAH C.T., NOYCE R.N. and SCHOCKLEY W., Proc. I.R.E., 45, 1228 (1957)

|6| ELLIS B. and MOSS T.S., Solid State Electron. 13, 1 (1970)

|7| FAN J.C.C., BOZLER C.O. and PALM B.J., Appl. Phys. Lett. 35, 875 (1979)

|8| TURNER G.W., FANN J.C.C. and HSIEH J.J., Conf. Rec. IEEE Photo. Spec. Conf. 14th San Diego, p 351 (1980)

|9| MANDELKORN J. and LAMNECK J.H., Jr, Conf. Rec. IEEE Photo. Spec. Conf. 9th Silver Spring, p. 83 (1972)

|10| YASUI R.K. and SCHMIDT L.W., Conf. Rec IEEE Photo. Spec. Conf. 8th Seattle, p. 110 (1970).

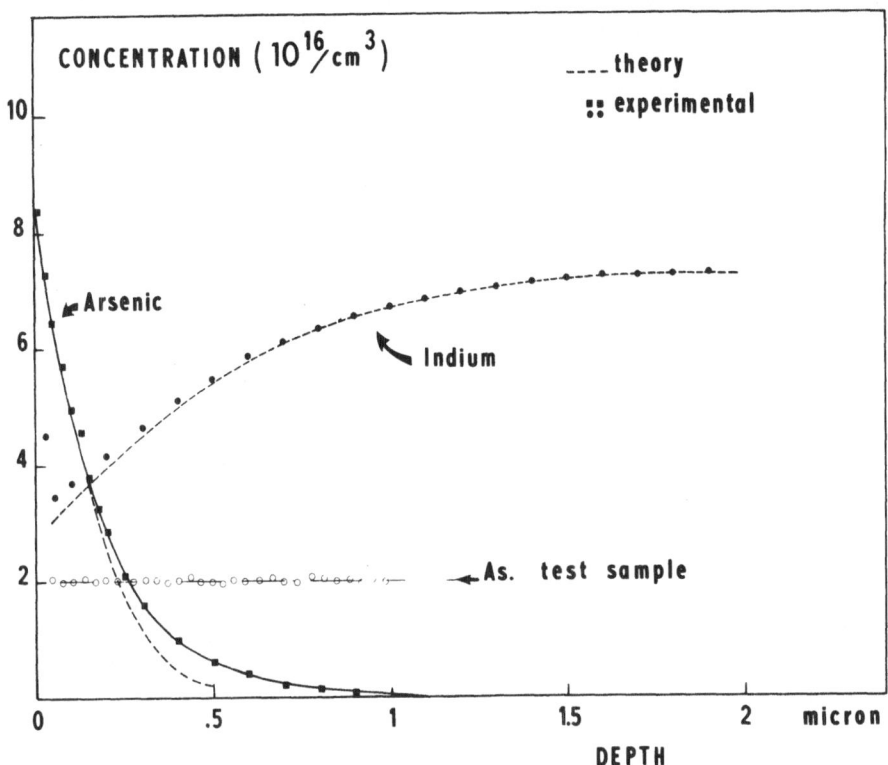

Fig. 1 - Indium and Arsenic impurities profiles from SIMS measurements.

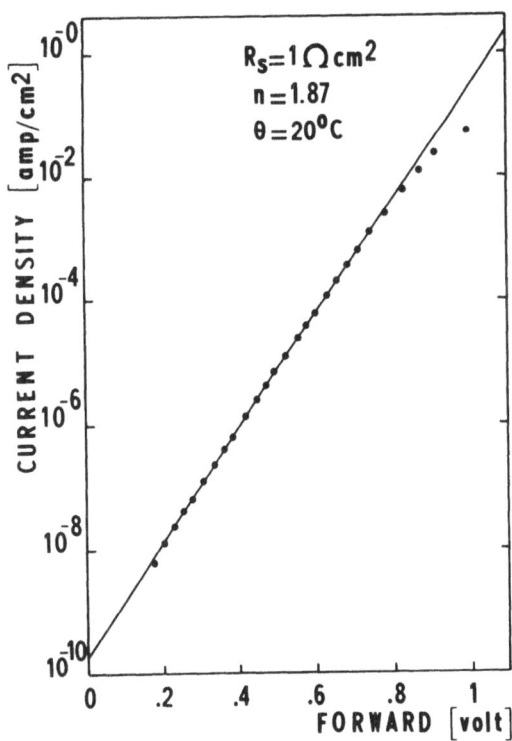

Fig. 2 - Voltage and current output from an illuminated CdTe solar cell.

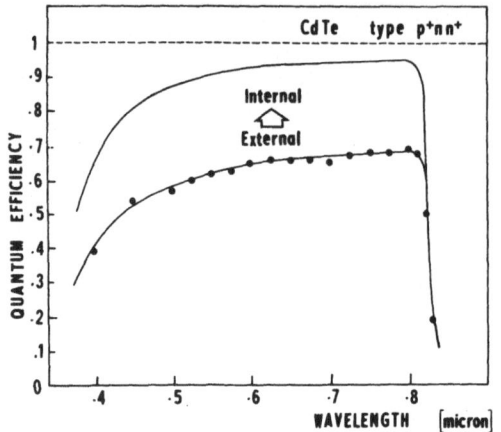

Fig. 3 - Spectral response of a p^+nn^+ CdTe solar cell.

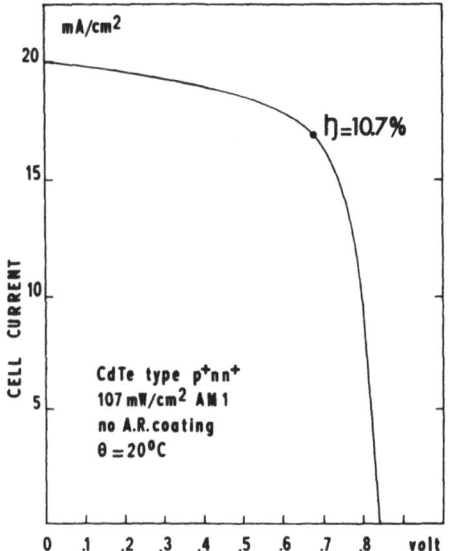

Fig. 4 - The I-V characteristic and performance for a p^+nn^+ CdTe solar cell. The value $\eta = 10.7\ \%$ is obtained at 107 mW/cm^2 AM1.

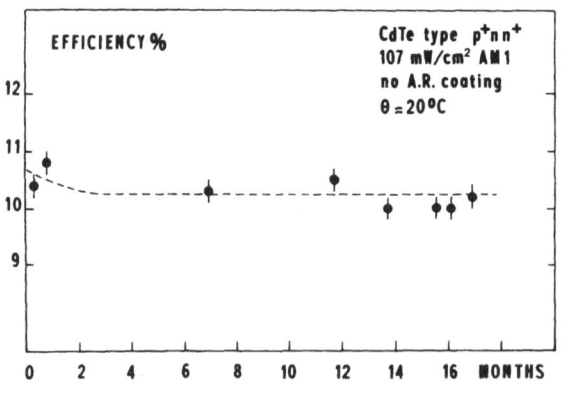

Fig. 5 - Efficiency stability versus time of a p^+nn^+ CdTe solar cell.

POSTER GROUP P6

ADVANCED DEVICES AND CONCENTRATION

- Advanced Solar Cells

The photovoltaic advanced research and development program in the United States

Influence of plasma Si-nitride deposition on the dark I-V curves of MIS contacts for inversion layer solar cells

Limitations of the open circuit voltage of induced junction silicon solar cells due to surface recombination

Some comments on sprayed ITO/semiconductor solar cells

Role of photoluminescence in the efficiency of a $Ga_{1-x}Al_xAs$-GaAs solar cell

Operating characteristics of thin thermophotovoltaic cells with minority carrier mirrors and optical mirrors using selective radiators of erbium and ytterbium oxides

- Concentrators

Lambertian analysis of mirrors and Fresnel lenses for solar concentration

750 suns concentrator modules using GaAs solar cells

A 500 W_{pk} photovoltaic concentrator using a glass laminated metal membrane reflector

Fluorescent planar concentrator (FPC) : Monte-Carlo computer model - Limit efficiency and latest experimental results

THE PHOTOVOLTAIC ADVANCED RESEARCH AND DEVELOPMENT PROGRAM IN THE UNITED STATES[*]

J. L. Stone, D. W. Ritchie, T. Surek, and C. E. Witt
Solar Electric Conversion Research Division
THE SOLAR ENERGY RESEARCH INSTITUTE
1617 Cole Boulevard
Golden, Colorado 80401
USA

SUMMARY

Implementation of the Advanced Research and Development (AR&D) Subprogram element of the National Photovoltaic Program of the United States, along with associated planning, assessment, and integration activities are the responsibility of the Photovoltaic Lead Center for AR&D located at the Solar Energy Research Institute (SERI) in Golden, Colorado. The major thrust of this program is to achieve technical feasibility for various advanced material technologies through long term, high risk, and potentially high payoff research and development.

1. INTRODUCTION

The mission of the PV AR&D Program is implemented through coordinated subcontracting activities with universities and private industries and through the use of SERI scientific staff for the independent assessment and verification of solar photovoltaic technologies. Two approaches are being pursued in order to achieve the cost and efficiency goals. One is to research various polycrystalline and amorphous thin film semiconducting materials on low cost substrates. The other approach is to use solar concentrators in the form of either high efficiency multibandgap cells having 20-40% efficiency potential or luminescent collectors. Each of the approaches is further subdivided by material type and separately statused. Programmatic consideration such as program objectives, problems, and budgets are detailed as well as the technical issues and future directions. Only programs directly sponsored by SERI are discussed. The reader is referred to reference 1 for information on programs outside of the SERI program.

2. POLYCRYSTALLINE THIN FILMS

The research in polycrystalline thin films is the most advanced of the material areas under SERI sponsorship. Several of the polycrystalline silicon and cadmium sulfide approaches from past years activities (1978 - 1981) are currently under development by industry.

2.1 Polycrystalline Silicon

The research in this area addresses new and novel silicon sheet growth techniques such as edge supported pulling (ESP) (2,3) and low angle silicon sheet (LASS) (4), grain boundary effects on polysilicon solar cell performance (5), developing understanding of chemistry/mechanisms of grain boundary passivation, exploring limits of thin film polysilicon solar cells, experimentally categorizing and theoretically modelling selected electrically active defects, exploring various directed energy beam (laser, electron, microwave, flash lamp) processing approaches, and developing thin film ($\sim$10 μm) small grain polysilicon solar cells. Background details have been published (6). Figures 1 and 2 trace the history and show current status of polysilicon cell efficiencies for a variety of p-n junction and MIS/SIS cell types. The SERI confirmed data have resulted from the SERI standards measurements of i-v and

---

[*] Under Contract from the U.S. Department of Energy

spectral response using an AM1 simulator traceable to the NASA-Lewis system. For the two SERI priority programs, ESP solar cells have achieved 13.8% efficiency over approximately 4 cm^2 and LASS has 12% over 4 cm^2. The ESP approach, which uses two filaments (typically quartz or graphite strings) for edge definition of the silicon ribbons, has been grown at speeds up to 9 cm/min and as thin as 90 μm. The thermal controls of this process, innovated at SERI, are at least one order of magnitude more relaxed than that required of the similar WEB approach (7). ESP has also produced direct growth from arc reactor silicon yielding 7% conventional and 10.5% epi cell efficiencies. This process has now been contracted to A. D. Little Co. to further develop its production worthiness. LASS growth has produced, in a 1 hour continuous run, a 5 cm wide ribbon, 33 meters long (55 cm/min) resulting in a 12% solar cell. The development of this process is continuing at Energy Materials Corporation to control the growth from random dendritic to a planar topology. Other high speed techniques such as melt spinning (roller quenching) and plasma spraying have been pursued at a low level without significant success.

The detrimental effects of grain boundaries on solar cell performance, particularly as related to impurity segregation and preferential diffusion, have been recognized for some time. The effects on small grained material are particularly significant. Recently it has been found that passivation of grain boundary effects can occur if the material is suitably treated with atomic hydrogen (8). Most recent results from Sandia Labs on hydrogen passivated small grain (typically about 10 μm) polysilicon solar cells made by Oak Ridge National Lab, using lithium diffusion, glow discharge ion implantation of boron, and laser annealing, treated with a Kaufman ion beam of hydrogen for four minutes resulted in an efficiency increase from 2.2 to 4.3% (J_{sc} from 12.1 to 16.0 mA/cm^2, V_{oc} from 340 to 444 mV, FF from 0.54 to 0.61). Other results include large grain silicon on ceramic (SOC) improved from 2.7% to 6.5% and large grain, 20 μm thick epi, on upgraded MG-silicon improved by approximately 10%. A new cell structure, termed the quasi-grain boundary-free cell (QGBF), conceptualized at the University of Florida (9) avoids loss mechanisms at the intersection of grain boundaries with the space charge region and the emitter and suppresses or completely eliminates the recombination losses due to the presence of grain boundaries. The structure is shown in Figure 3. The ribbon to ribbon (RTR) process previously funded at Motorola was successfully transferred to full industry involvement with the announcement of a joint Motorola-Shell venture to commercialize the process. Motorola had demonstrated 11.2% efficiency on 32 cm^2 using laser beam RTR material.

At SERI, a record of invention was submitted for a one step electro-refining process for extracting PV grade silicon directly from silica. A molten salt of $KF:LiF:K_2SiF_6$ using a semipermeable anode of Si:Cu has produced 99.9996% silicon from 98% silicon, 100-300 μm grain size with 15 μsec lifetime and 25 μm diffusion length. The process has potential for producing polysilicon feedstock at $7/kg.

The fiscal year 1982 budget authority for polysilicon is $1.9M in operating funds, $100K of capital equipment, and $1.544M in carry over from FY1981. Two new programs will be initiated in FY82 covering selected electrically active defects and high efficiency cells.

2.2 II-VI Compounds

This task has as its objectives research on thin-film CdS/Cu-binary and ternary heterojunction photovoltaic devices with potential for high efficiency and long term stability, and research on polycrystalline thin-film devices fabricated from other promising materials including CdTe and Zn_3P_2. Progress has been very positive over the last six years as shown in Figures 4 and 5. Much of the earlier work on CdS/Cu$_2$S has been taken over by industry both in the United States, Europe, and Japan. Principal research activity has been directed at the stability of the device due to both intrinsic (electrochemical) and extrinsic degradation. The Institute for Energy Conversion has recently demonstrated intrinsically stable CdS/Cu$_2$S devices prepared on rigid lattice matched Ni/Fe substrates with laser burn out of shunt defects. Such devices have operated for 586 hours under constant 83 mW/cm^2 illumination without measurable degradation. Regardless, however, the devices must be protected from atmospheric contamination with a hermetic shield or immersed in an inert atmosphere. Efficiencies greater than 10% have been shown for (Cd,Zn)S/Cu$_2$S wet processed cells; the addition of zinc being necessary to increase the open circuit voltage (10). Sputtered devices prepared at Telic Corporation have resulted in 7.1% AM1 efficiency (V_{oc} = 0.53V, J_{sc} = 18.5 mA/cm^2, FF = 0.72) for reactively sputtered Cu$_2$S on evaporated CdS (1 cm^2) and 3.94% AM1 (V_{oc} = 0.527V, J_{sc} = 12.2 mA/cm^2, FF = 0.62, no AR) for an all sputtered cell, 3.6 μm in thickness, 1 cm^2 area. The sputtered devices are being investigated in order to show technical feasibility for a potentially important production worthy process.

In order to circumvent the stability problem associated with Cu$_2$S, research was initiated on CuInSe$_2$ as a replacement for the CdS hetero-partner. CuInSe$_2$ is a direct bandgap material (Eg = 1.04 eV) with a strong absorption coefficient over most of the solar spectrum (α = 10^5 cm^{-1}). The early results have been very encouraging with Boeing having achieved 9.9% (V_{oc} = 0.42V, J_{sc} = 36.3 mA/cm^2, FF = 0.65) for a 1 cm^2, 5 μm thick (Cd,Zn)S/CuInSe$_2$ cell. This is the highest J_{sc} achieved in any photovoltaic cell, single crystal or polycrystalline. The low voltage (expect 700 - 800 mV) is the subject of intense research and if solved could well produce a 15-17% all thin film cell. Even more importantly, these devices have undergone 8400 hours of continuous AM1 illumination (the last 2800 hours at 80°C), unencapsulated, with no measurable degradation. A partial solution to the low V_{oc} may have recently been found at SERI where EBIC measurements indicate a Schottky barrier at the CuInSe$_2$/Mo contact interface. Other back contact metals are currently being tried to obviate this problem. A typical cross section of the cell is shown in Figure 6. A small effort has gone into a CdS/Cu$_{2-x}$Se in an attempt to replace the indium component. Boeing has demonstrated a 5.38% cell (V_{oc} = 0.457V, J_{sc} = 18.7 mA/cm^2, FF = 0.63) heat treated in hydrogen for 5 minutes at 150°C. An all spray process by SRI International has produced 2.5% CdS/CuInSe$_2$ (V_{oc} = 0.26V, J_{sc} = 32 mA/cm^2, FF = 0.3, A = 3 mm^2) cells. Such spray pyrolysis deposition processes may well be the key to truely low cost thin films.

Cadmium telluride is another PV material of interest being direct bandgap (1.47 eV) and having a comparable absorption coefficient to that of CuInSe$_2$. Single crystal CdTe with an evaporated CdS heterojunction have had reported 12% efficiencies (11). The SERI program is directed towards low cost potential thin films produced by spray pyrolysis, sputter deposition, screen printing, CVD, and hot wall vacuum evaporation. Southern Methodist University has produced a 5.3%, 8 cm^2 CVD Au/n-CdTe Schottky barrier cell (V_{oc} = 0.4 V, J_{sc} = 22 mA/cm^2, FF = 0.60) and Radiation Monitoring produced a 4%, 1 cm^2 chemically sprayed cell. Electrodeposited cells from AMETEK (not a SERI contractor) have reported electrodeposited CdTe cells of 6.2% (1 cm^2) and 8.6% (2 mm^2), demonstrating the potential of this inexpensive process approach. The hot wall vacuum evaporation at SERI and Stanford has produced 10 μm or larger grain size In-doped, n-type CdTe films. Zn$_3$P$_2$, another direct bandgap material (Eg = 1.5 eV) with strong optical absorption (α = 10^4 cm^{-1}) has given the highest open circuit voltage for an all thin film heterojunction device (Zn$_3$P$_2$/ZnSe, V_{oc} = 0.75 V) reported by the Institute for Energy Conversion and a 4.3%, 1 cm^2 Mg/Zn$_3$P$_2$ cell. If the cell can be improved, it has excellent potential because of its near optimum bandgap and it being fabricated from inexpensive, readily available material. ZnSiAs$_2$ is also being considered as a monolithic cascade device structure at RTI. Results are, however, only very preliminary. Work on WSe$_2$ was terminated because of the unsatisfactory results. An excellent review of this entire area is available (12).

The fiscal year 1982 budget authority for the II-VI area was $3.21M in operating funds, $190K of capital equipment, and $2.013M in carry over from FY81. FY82 program will emphasize CuInSe$_2$ with new areas of investigation to include single crystal PV devices, thermodynamic studies, and alternate techniques for thin film deposition.

2.3 III-V Compounds

Gallium arsenide bulk devices are the highest efficiency PV available (13). However, cost and potential gallium availability problems will most probably prevent large scale deployment of these devices. Rather, a truly thin film approach (< 5 μm) is sought with properties similar to the bulk counterpart. Various polycrystalline GaAs structures are being pursued as shown in Figure 7. A variety of low cost, sacrificial, or reuseable substrates such as tungsten coated graphite, germanium coated silicon, sodium chloride, and gallium arsenide are being pursued. The objective of this part of the program is to achieve 17% or greater thin film GaAs cells for flat plate modules. Progress to date includes an 8.5%, 9 cm^2 MIS cell of Au/nGaAs/n$^+$GaAs/W/graphite (V_{oc} = 0.63V, J_{sc} = 20.6 mA/cm^2, FF = .656) by Southern Methodist University (14), a pn junction p$^+$GaAs/nGaAs/n$^+$GaAs/W/graphite 8.8%, 1 cm^2 (V_{oc} = 0.57V, J_{sc} = 24.5 mA/cm^2, FF = .63) by Southern Methodist University, a shallow homojunction n$^+$GaAs/pGaAs/ p$^+$GaAs/Ge/Si single crystal of 11.7%, 9.3 mm^2 (V_{oc} = 0.76V, J_{sc} = 24.2 mA/cm^2, FF = .63) by MIT Lincoln Laboratory (15) (recently reported a 14% small area device, V_{oc} = 0.85V), and a GaAs on monocrystalline Ge films separated from NaCl of 4%, .02 cm^2 (V_{oc} = 0.425V, J_{sc} = 14.1 mA/cm^2, FF = .66) by United Technologies (n$^+$/p/p$^+$GaAs/p-Ge). MIT Lincoln Lab has innovated the CLEFT process (Cleavage of Lateral Epitaxial Films for Transfer) whereby thin films of single crystal GaAs (< 5 μm thick) are cleaved from a reuseable GaAs substrate. The film is grown by lateral epitaxy through stripe openings in a carbonized photoresist mask (16). Using this approach, this group has achieved a 17%, 0.5 cm^2 (V_{oc} = 0.95V, J_{sc} = 23 mA/cm^2, FF = 0.78) solar cell. In addition to the device work, basic research in material growth (especially by MOCVD), improved understanding of electrical properties, and improved device processing (by laser and electron beam) are also carried out. The theoretical group at SERI has developed a new first-principle method for band structure calculations, and has shown that Fermi level pinning in Al/GaAs (110) is not controlled by surface states.

The fiscal year 1982 budget for the III-V area was $583K in operating funds, $40K of capital equipment, and $631K in carry over from FY81.

2.4 Photoelectrochemical Cells (PEC)

The major activities of this program include research on PEC cells which convert light energy directly into electrical energy in a regenerative mode of operation, research on PEC cells which first convert light energy into storable chemical products and then convert chemical energy into electrical energy in a redox cell or fuel cell in an overall regenerative cyclic system, and responsibility for monitoring the technical progress of the Texas Instruments Solar Energy System (described in another session of this conference). As shown in Figure 8, the basic PEC cell consists of a semiconductor photoanode (n or p-type) immersed in an electrolyte to form a solid-liquid Schottky barrier. The electrolyte consists of a solvent and redox couple. The circuit is completed through the counter electrode and the load. The junction is spontaneously formed upon immersion and is very uniform even on rough surfaces. The major advantage of the PEC cell is its ability to store energy in situ. One such storage system is the three electrode storage cell shown in Figure 9. Here a third storage electrode is inserted, separated from the conversion cell by a selective semipermeable membrane. The conversion of chemical to electrical energy occurs in the dark discharge between the storage electrode and counterelectrode. The Weizmann Institute has demonstrated a working 3-electrode system using a n-CdSe$_x$Te$_{1-x}$ photoanode, CoS counterelectrode, and a Sn/SnS storage electrode in a basic sulfide electrolyte. They have obtained a 2-3% overall system efficiency with storage efficiencies up to 85%.

By properly chosing the photoanode material and the redox couples, V_{oc} can be controlled to a limited degree. The majority of the research has concentrated on CdSe, CdSe$_x$CdTe$_{1-x}$, GaAs, and the transition metal dichalcogenides MoSe$_2$ and WSe$_2$. Results of polycrystalline and crystalline PEC cell efficiencies are shown in

Table 1. A comprehensive review of this program is published (17). The principal focus of the materials research is to understand and eliminate the photoanode corrosion in the electrolyte. New materials, post deposition treatments, and surface modification are being pursued. Approaches to surface modification include ion adsorption, deposition of submonolayer metal films, thin oxide layers, and conducing polymer films such as polypyrrole. The use of nonaqueous and molten salt electrolytes is being studied to shift the semiconductor lattice decomposition potential.

The fiscal year 1982 budget includes $3.21M operating expenses (including $2.5M for the Texas Instruments program), $40K for capital equipment and $1.249M carry over funds from FY1981. Future activities include studies on new materials such as $CuInSe_2$, basic research on PEC storage, thin film deposition, and innovative approaches to improving stability.

3. AMORPHOUS THIN FILMS

The international interest in amorphous silicon hydrogen alloys, particularly in Japan (18), has resulted from rather spectacular improvements in material quality and device behavior. By the time this paper is published, efficiencies will most likely be in the 8.5 - 9.0% range for small area devices and 6-7% for the large area devices (> 100 cm^2). While much progress has been made towards commercialization, this program has focused on basic understanding of the behavior of amorphous thin films (primarily a-Si:H) and a better understanding of the deposition parameters and their effects on material quality. Of particular interest is the cause for light induced changes in materials and devices. Preliminary evidence attributes the effect to hydrogen or possibly impurities. A workshop, sponsored by SERI, on this subject will precede the 16th IEEE PV Specialists Conference in San Diego on September 24-25, 1982. Other technical problem areas include barriers limiting higher efficiency, relationship between device efficiency and material deposition rate, interface and surface effects between transparent conducting substrates and amorphous material, correlation of optical gap with device performance, research on large area devices (defect density, impurities, uniformity), evaluating the potential of CVD deposition, especially from the higher order silanes, evaluating the potential of Si alloys (SiGe, SiSn) for higher efficiency stacked cells, surface and interface effects between amorphous layers, detailed understanding of plasma conditions on film quality, and the establishment of a theoretical base from which to predict future research direction. A compilation of various a-Si device efficiencies is shown in Figure 10. Various reviews of this program have been published (19, 20). Recent accomplishments of the SERI program include a 7.88%, 1.09 cm^2 glass/SnO_2:F/ p^+SiC:H/i/ n^+a-Si:H cell (V_{oc} = 0.832V, J_{sc} = 14 mA/cm^2, FF = 0.676) by RCA, a 4.66% all sputtered cell by Exxon (V_{oc} = 0.865V, J_{sc} = 11.0 mA/cm^2, FF = 0.49, A = 0.25 cm^2), and the development of good quality transparent conducting substrates (fluorine doped SnO_2, 1 μm thick, 4 Ω/□ on window glass) by Harvard University.

Funding for this task includes $2.8M of operating funds for FY82, $200K for capital equipment, and $3.165M of carry over funds from FY81. Future program focus includes alternate device structures (stacked cells), materials and device research for large area cells using CVD, studies on high deposition rates, improved p-type material, increased effort on light induced changes, and materials research to identify amorphous materials with more optimum bandgaps.

4. HIGH EFFICIENCY CONCENTRATORS

Recent studies by the Electric Power Research Institute (EPRI) detail the premium to be placed on PV efficiency for widespread deployment as electricity producers for utilities (21). Their studies indicate a need for 25% or higher efficiency devices. Since the common flate plate, single cells are limited in theoretical efficiency to about 20-25%, new and innovative device structures and approaches will be required to reach the high efficiency goals. The next section addresses such concepts and reviews the successes to date.

4.1 Multibandgap Structures

The ultimate objective of this program is to achieve greater than 30% efficient multiple junction cells for concentrated sunlight (500 - 1000 x). The basic cell configuration utilizes a high bandgap top cell in electrical and optical series with a low bandgap bottom cell. The means of interconnecting the cells differentiates the various types, including tunnel junction connected, superlattice (a SERI innovation for reduced defect density), and a metal interconnected, hard wired connection. The best performance reported to date is a metal interconnected cascade cell of GaAlAs/GaAs of 22%, 0.5 cm^2 measured at 130x under AM2 conditions. The parameters include V_{oc} = 2.41V, I_{sc} = 420 mA, and FF = 0.758 by Varian Associates. The best reported tunnel junction connected cell was GaAlAs/GaAs with 14.5% efficiency (V_{oc} = 2.05V, J_{sc} = 10.8 mA/cm^2, FF = 0.74) by Research Triangle Institute. The metal interconnected cascade cell (MIC^2) is shown in Figure 11. An excellent review of this work is contained in reference (12). Another cell structure, the series connected high voltage cell has achieved 19.6% for an eight junction cell measured at 4 x under AM2 (V_{oc} = 7.6V, I_{sc} = 3.6 mA, FF = 0.75) by Varian. The technical problems that are addressed in this program include growth of smooth, low defect density III-V ternary and quarternary alloys, incorporation of dopants in III-V's, evaluation of tunnel junction, metal and superlattice interconnects, and grid contact technology for 500-1000 suns.

The funding for FY82 in this area includes $550K operating funds, $40K of capital equipment, and $604K of carry over funds from FY81. This program will take on a higher priority for the future as the need for high efficiencies is stressed.

4.2 Luminescent Collectors

The luminescent collector is a flat plate, optical down shift converter which compresses the solar spectrum into a narrow band so as to more efficiently operate conventional Si or GaAs solar cells. Light is trapped by internal reflection in the host material in which the emitting dyes are placed. The PV cells typically line the edge of the panel. This concept can also be extended to include an application as a flat plate module cover plate. The research in this area is concentrated on dye stability, dye-host interactions, and scaleability to large plate areas. Solar to electric conversion efficiencies of 3.2% over 200 cm^2 using Si cells and 4.5% over 200 cm^2 using GaAs cells have been demonstrated by Owens-Illinois. Testing of 3000 cm^2 plates demonstrates significantly better than predicted performance over long optical path lengths. The FY82 funding for this area includes $590K of operating expenses, $40K for capital equipment, and $233K of carry over funds from FY81.

5. CONCLUSION

The multi-faceted AR&D program in photovoltaics at SERI has been described. Space limitations necessitate the omission of many details that may be of interest to the reader. Questions will be gladly accepted and responded to by the authors. Further details will be presented at the 16th IEEE PV Specialists Conference in San Diego on September 27-30, 1982. The authors greatfully acknowledge the invaluable aid of the SERI Task Managers, John Benner, Allen Hermann, Joe Milstein, Ed Sabisky, and Bill Wallace for their help in compiling the technical information.

6.0 REFERENCES

(1) D. L. Feucht, this conference, session 6, "Advanced Photovoltaic Devices" (Invited).

(2) T. F. Ciszek, J. L. Hurd, "Melt Growth of Silicon Sheets by Edge Supported Pulling", Proceedings of the Symposia on Electronic and Optical Properties of Polycrystalline or Impure Semiconductors and Novel Silicon Growth Methods, St. Louis, MO, May 11-16, 1980.

(3) T. F. Ciszek, M. Schietzelt, L. L. Kazmerski, "Growth of Silicon Sheets from Metallurgical-Grade Silicon", Conf. Record of 15th Photovoltaic Specialists Conf., 1981, p. 581.
(4) D. N. Jewett, H. E. Bates, "Low Angle Crystal Growth of Silicon Ribbon", Conf. Record of 14th Photovoltaic Specialists Conf., 1980, p. 1404.
(5) Y. S. Tsuo, J. B. Milstein, T. Surek, "Grain Boundary Structures and Properties in Polycrystalline Silicon", Proceedings of the Symposia on Grain Boundaries in Semiconductors, Materials Research Society Annual Meeting, Boston, MA, November 16-19, 1981.
(6) T. Surek, A. P. Ariotedjo, G. C. Cheek, R. W. Hardy, J. B. Milstein, Y. S. Tsuo, "Thin-Film Polycrystalline Silicon Solar Cells: Progress and Problems", Conf. Record of 15th Photovoltaic Specialists Conf., 1981, p. 1251.
(7) T. Surek, "The Growth of Silicon Sheets for Photovoltaic Applications", in Reference 2.
(8) C. H. Seager, D. S. Ginley, "Studies of the Hydrogen Passivation of Silicon Grain Boundaries", J. Appl. Phys., $\underline{52}$, 1050 (1981).
(9) F. N. Gonzalez, A. Neugroschel, "Design of Quasi-Grain-Boundary-Free (QGBF) Polycrystalline Solar Cells", IEEE Electron Devices Letts., $\underline{ED-2}$, 141 (1981).
(10) J. Leong, S. Deb, "Advances in the SERI/DOE Program on CdS/Cu_2S and CdS/Cu-Ternary Photovoltaic Cells", Conf. Record of 15th Photovoltaic Specialists Conf., 1981, p. 1016.
(11) K. Yamoguchi, N. Nakayama, H. Matsumoto, S. Ikegami, "CdS-CdTe Solar Cell Prepared by Vapor Phase Epitaxy" Jap. J. Appl. Plys., $\underline{16}$, 1203 (1977).
(12) A. Hermann, L. Fabick, "Research on Polycrystalline Thin-Film Photovoltaic Devices", International Conf. on II-VI Compounds, Durham, England, April, 1982.
(13) H. Hovel, Semiconductor and Semimetals, Volume 11, Solar Cells, Academic Press (1975).
(14) S. Chu, T. Chu, C. Jiang, C. Loh, E. Stokes, J. Yu, "Thin Film Gallium Arsenide Solar Cells With Reduced Film Thickness", Conf. Record of 15th Photovoltaic Specialists Conf., 1981, p. 1310.
(15) R. Gale, B. Tsaur, J. Fan, F. Davis, G. Turner, "GaAs Shallow-Homojunction Solar Cells on Epitaxial Ge Grown on Si Substrates", Conf. Record of 15th Photovoltaic Specialist Conf., 1981, p. 1051.
(16) J. Fan, C. Bozler, R. McClelland, "Thin-Film GaAs Solar Cells", Conf. Record of 15th Photovoltaic Specialists Conf., 1981, p. 666.
(17) W. Wallace, R. Noufi, S. Deb, "Advances in the SERI-DOE Electrochemical Photovoltaic Cell Program", Conf. Record of 15th Photovoltaic Specialists Conf., 1981, p. 1363.
(18) Y. Hamakawa, Amorphous Semiconductor Technologies and Devices, Ohm/North-Holland, 1982.
(19) J. Stone, "Recent Progress in Amorphous Silicon Research and Photovoltaic Applications in the U.S.A.", Japanese Journal of Applied Physics, $\underline{20}$, (1981) 1.
(20) J. Stone, E. Sabisky, H. Mahan, T. McMahon, F. Jeffrey, "The National Photovoltaic Program in Amorphous Material", Conf. Record of 15th Photovoltaic Specialists Conf., 1981, p. 898.
(21) R. Whitaker, "Photovoltaics: A Question of Efficiency", EPRI Journal, December, 1981, p. 45.
(22) K. Mitchell, "High Efficiency Concentrator Cells", Conf. Record of 15th Photovoltaic Specialists Conf., 1981, p. 142.

Table I. Current Status of Major Single Crystal and Polycrystalline Thin Film Efficiency Results in the Photoelectrochemical Cell Program

Cell Structure	Morphology	V_{oc} (V)	J_{sc} (mA/cm^2)	FF	η (%)	Contractor
A. Polycrystalline						
n-CdSe/1.5 M S^{-2}, 0.75 M S, 5 M OH$^-$/Ni	P (ED)	0.56	23.8	0.48	6.4	Brooklyn College
n-CdSe/2.5 F S^{-2}, 1.0 F S, 1.0 F OH$^-$/Pt	P (CE)	0.57	17	0.51	6.6	Grumman
n-CdSe/2.5 FS^{-2}, 1.0 F S, 1.0 F OH$^-$/Pt	P (CE)	0.48	18	0.56	6.2	Grumman
n-CdSe$_{0.80}$Te$_{0.20}$/ 1 F S^{-2}, 1 F S, 1 F OH$^-$/Pt	P (CE)	0.65	16.5	0.49	7.1	Grumman
n-CdSe/1 M S^{-2}, 1 M S, 1 M OH$^-$/Pt	P (CBD)	0.68	19.5	0.41	5.7	EIC
n-CdSe/1 MS^{-2}, 1MS, 1 M OH$^-$/Pt	P (CSD)	0.74	11.5	0.54	6.7	IGT
B. Single Crystal						
n-WSe$_2$/1.0 M I$^-$, 0.05 M I$_2$	SC (CSVT)	0.67	22.6	0.57	10.4	Ames
n-MoSe$_2$/1.0 M I$^-$, 0.05 M I$_2$	SC (CSVT)	0.57	25.5	0.55	9.4	Ames
n$^+$GaAs/n-GaAs/ 1 M Se^{-2}, 0.1 M Se$_2^{-2}$ 1 M OH$^-$/Pt	SC (LPE)	0.7	29	0.80	14	Rockwell
n$^+$GaAs/n-GaAs/ 1 M SE^{-2}, 0.1 M Se$_2^{-2}$, 1 M OH$^-$/Pt	P (MBE)	0.7	29	0.67	12	Rockwell
n-GaAs/Polypyrrole/ 0.2 M Fe (CN)$_6^{-4/-3}$, 0.1 M CN$^-$, 0.1 M OH$^-$/Pt	SC	1.37	18.6	0.70	10.5	Rockwell
n-CdSe/0.5 M FE (CN)$_6^{-4/-3}$, 0.1 M OH$^-$/Pt	SC	1.05	14.5	0.72	12.4	SRI

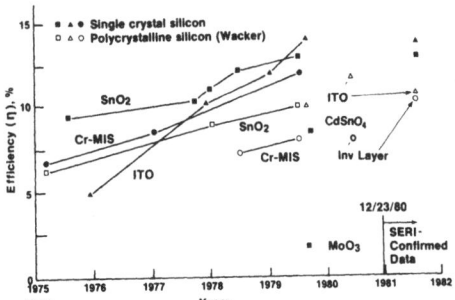

Figure 1. Efficiencies of polycrystalline silicon p-n junction solar cells.

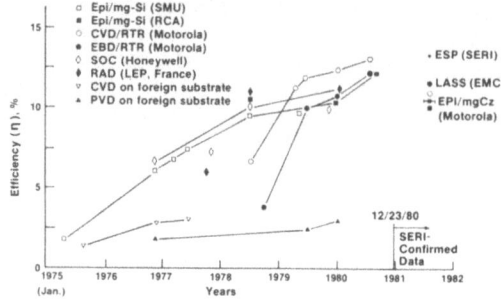

Figure 2. Efficiencies of MIS/SIS junction silicon solar cells.

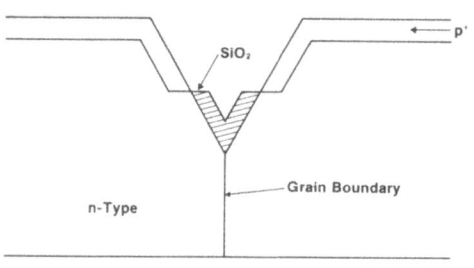

Figure 3. Cross-sectional view of the quasi-grain-boundary-free (QGBF) solar cell structure after the emitter diffusion. The bottom of the V-groove is covered by SiO_2 to prevent diffusion into the grain boundary.

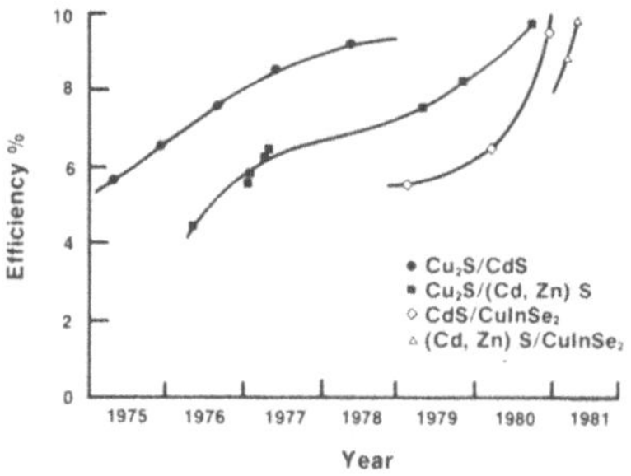

Figure 4. Efficiencies of CdS based solar cells.

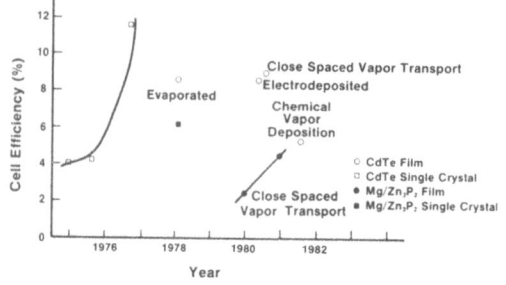

Figure 5. Development of CdTe and Zn_3P_2 solar cells.

Figure 6. Cross section of $CdS/CuInSe_2$ cell.

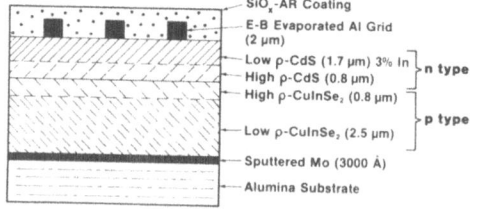

- 636 -

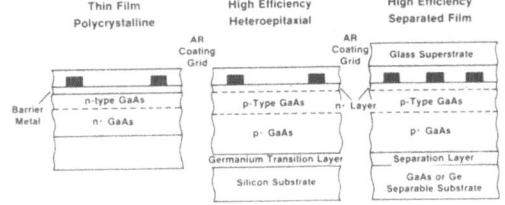

Figure 7. Thin film gallium arsenide solar cells, technical approaches for fabrication.

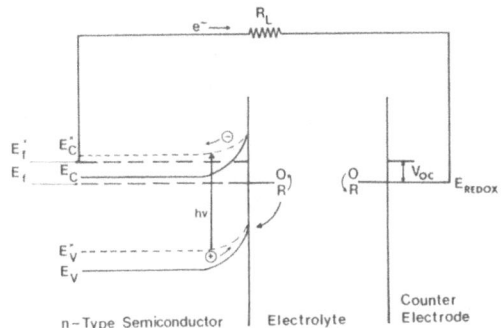

Figure 8. Schematic diagram of an Electrochemical Photovoltaic Cell.

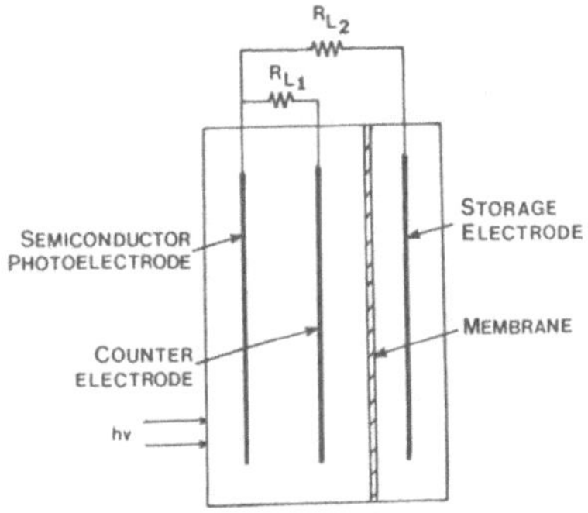

Figure 9. Schematic diagram of a three electrode _in situ_ photoelectrochemical storage cell.

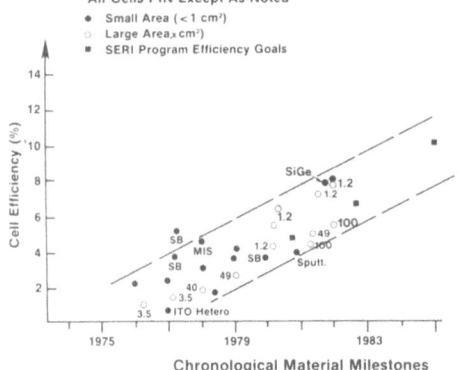

Figure 10. a-Si:H solar cells efficiencies.

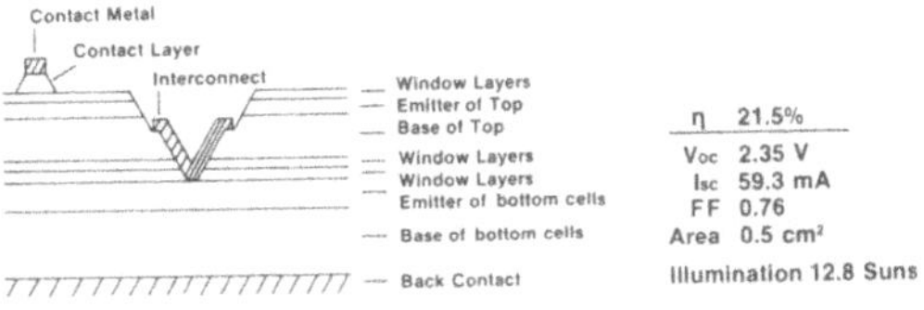

Figure 11. Metal Interconnected Cascade Cell (MIC^2).

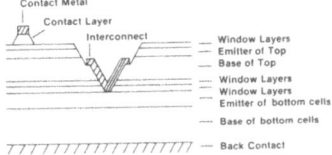

η	21.5%
V_{oc}	2.35 V
I_{sc}	59.3 mA
FF	0.76
Area	0.5 cm^2
Illumination	12.8 Suns

-637-

INFLUENCE OF PLASMA Si-NITRIDE DEPOSITION ON THE DARK I-V CURVES OF MIS CONTACTS FOR INVERSION LAYER SOLAR CELLS

R. SCHÖRNER and R. HEZEL
Institut für Werkstoffwissenschaften VI
Universität Erlangen-Nürnberg, 8520 Erlangen, W-Germany

Summary

The influence of the plasma Si-nitride deposition process on electrical properties of Al/Si-oxide/p-Si MIS contacts was investigated. The Si-nitride was deposited onto the MIS diodes in a rf glow discharge by the reaction of SiH_4 and NH_3 using parameters optimized for application to MIS/inversion layer solar cells. The dark I-V curves before and after the plasma treatment were theoretically fitted with respect to the possible current mechanisms for MIS diodes. It was shown that the reverse saturation current of the diffusion and thermoionic emission component is reduced by about one order of magnitude by the temperature treatment occuring during the plasma deposition. This low reverse saturation current is in good agreement with the theoretical value of a minority carrier MIS diode.

1. INTRODUCTION

In the recently introduced metal-insulator-semiconductor (MIS)/inversion layer solar cells on p-type silicon with Si-nitride as dielectric film, a highly conductive inversion layer at the silicon surface is induced by the fixed positive charges in the Si-nitride in the area between the MIS grid system (1,2,3). A schematic cross section of this cell type is shown in Fig.1.

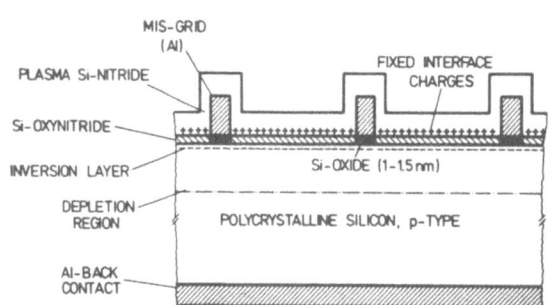

Fig.1: Schematic cross section of the MIS/inversion layer solar cell with plasma Si-nitride as dielectric.

The last but crucial fabrication step for the MIS/inversion layer solar cell using the plasma enhanced chemical vapor deposition process (PECVD) for the Si-nitride preparation is the deposition of Si-nitride on the front side after the Al/Si-oxide/p-Si grid pattern has been defined. During this process the MIS contact is exposed to substrate temperatures between 200 °C and 300 °C and to the bombardment of electrons and ions of the plasma. Under the deposition conditions optimized for obtaining high and stable fixed positive charge densities in the Si-nitride, the electrical characteristics of the MIS contacts are changed.

In this paper the influence of the plasma Si-nitride deposition on the electrical properties of the MIS contacts will be investigated quantitatively on the basis of the dark I-V characteristics. However, these studies could not be performed on the solar cell structure itself, since before the Si-nitride deposition the dark I-V curves of the MIS grid system are dominated by currents due to recombination-generation processes in the vicinity of the metal grid fingers and thus do not reflect the electrical properties of the MIS contact itself. In order to minimize this influence of the bare silicon surface adjacent to the MIS contact the extremely large peripheral length of the metal finger system (up to 400 cm for our 2x2 cm^2 solar cell) had to be drastically reduced. For this purpose the experiments were performed on square shaped MIS contacts. The dark I-V curves of these structures were analyzed before and after the Si-nitride deposition and the reverse saturation currents were compared to the theoretical values of minority carrier MIS diodes (4).

2. EXPERIMENTAL

The MIS contact patterns were prepared on 0.7Ω.cm p/Si (100) substrates, which were cleaned in boiling aceton, trichlorethylene, $NH_4OH-H_2O_2$ and $HCl-H_2O_2$ solutions. After thermal evaporation of the Al back-contact in a vacuum $<10^{-5}$ torr, the thin oxide for the MIS contact was grown at 560 °C in a dry O_2-N_2 atmosphere simultaneously with the formation of the ohmic back contact. The oxide thickness, measured by ellipsometry, was less than 1.5 nm. Subsequently the Al top contact was evaporated and the different patterns were defined by photolithography. The Si-nitride films were deposited in a rf glow discharge (rf power 15W, pressure 0.8 torr) at a substrate temperature of 250 °C by the reaction of SiH_4 and NH_3 with Ar as carrier gas (1).

3. RESULTS AND DISCUSSION

In Fig.2 the dark I-V curves of a square shaped MIS contact (8.6x10^{-2}cm^2 area) before and after plasma Si-nitride deposition are shown. The cross section of the MIS structure after Si-nitride deposition is depicted in the insert.

In order to avoid a possible current path to the back contact of the structure via the inversion layer at the silicon surface, the Si-nitride was removed by etching in buffered HF except for a relatively small area of 0.49cm^2.

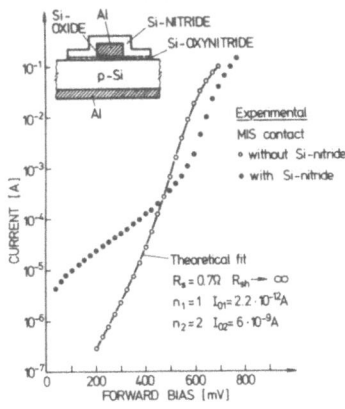

Fig.2: Measured dark I-V curves of a 8.6x10^{-2}cm^2 forward biased MIS diode before and after Si-nitride deposition together with a theoretical fit to the experimental data of the MIS diode before Si-nitride deposition.

MIS contact before plasma Si-nitride deposition. The dark I-V curve of the forward biased MIS contact before the Si-nitride deposition was fitted by using the following diode equation, in which R_s represents the series resistance and R_{sh} the shunt resistance of the diode:

$$I - \frac{V-IR_s}{R_{sh}} = I_{01}\left[\exp\frac{q(V-IR_s)}{n_1 kT} -1\right] + I_{02}\left[\exp\frac{q(V-IR_s)}{n_2 kT} -1\right] \qquad |1|$$

In this equation the following current mechanisms possible for MIS diodes are taken into account. The first two, dominating in the intermediate bias region of the dark I-V curve, are the diffusion of minority carriers in the bulk of the semiconductor (4) and the thermoionic emission of majority carriers (5), both with an ideality factor $n_1=1$. The sum of their reverse saturation currents is represented by I_{01}. The two further current mechanisms, dominating in the low bias range of the I-V curve, are the recombination-generation current in the depletion region beneath the metal contact and the recombination-generation current in the vicinity of the MIS contact (4,6,7). Both can be characterized by the ideality factor $n_2=2$ (4,6,8). The sum of their reverse saturation currents is represented by I_{02}.

For curve fitting I_{02} was taken to be approximately the reverse current measured for the MIS contact. The value for I_{01} was determined by fitting the experimental I-V characteristic with the curve calculated according to Eq.|1| for $R_s=0$ and $R_{sh} \rightarrow \infty$. Finally, by curve fitting in the high bias region, which is mainly affected by series resistance, R_s was obtained.

As can be seen from Fig.2 the dark I-V characteristic of the MIS contact could be approximated in the whole bias range by the curve calculated from Eq.|1| using the following data:

$I_{01} = 2.2 \times 10^{-12}$A; $I_{02} = 6 \times 10^{-9}$A; $R_s = 0.7 \Omega$.

MIS contact after plasma Si-nitride deposition. After the Si-nitride deposition the dark I-V curve of the forward biased MIS contact is drastically changed, as demonstrated in Fig.2. Two characteristic features are observed:(i) a displacement of the I-V curve in the intermediate and high bias region towards higher voltages and (ii) an increase of the current in the low bias regime. This experimental curve cannot be fitted by Eq.|1| since the additional current mechanisms contributing in the low bias region are not included in this relation. As to the origin of this additional current, it is attributed to an inversion layer at the silicon surface beneath the Si-nitride film, resulting in surface channel currents (6). The presence of the inversion layer was demonstrated by high frequency capacitance-voltage measurements. The assumption made above was confirmed by the fact that the currents in the low bias region decreased to their initial value after the plasma Si-nitride layer had been etched off, while in the intermediate and high bias region the I-V curve was not changed. This behavior is demonstrated in Fig.3, where the measured dark I-V curves of the forward biased MIS contact are presented before and after Si-nitride deposition as well as after removal of the Si-nitride film. For the latter case, the experimental data can now be excellently fitted by using Eq.|1| with the following data:

$I_{01} = 1.5 \times 10^{-13}$A; $I_{02} = 2.5 \times 10^{-9}$A and $R_s = 0.4 \Omega$.

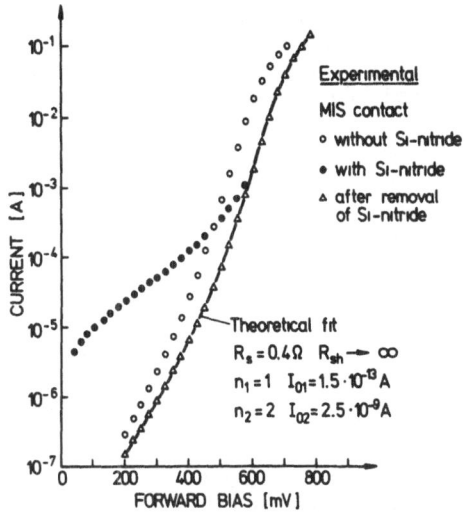

Fig.3: Measured dark I-V curves of a 8.6×10^{-2} cm^2 forward biased MIS diode before and after Si-nitride deposition and after removal of the Si-nitride together with a theoretical fit to the experimental data of the MIS diode after removal of Si-nitride.

For a characterization of the influence of the plasma Si-nitride deposition on the electrical properties of the MIS contact, the results of the theoretical fit of the dark I-V curves before plasma Si-nitride deposition and after removal of the deposited nitride film are compared. As can be seen from Fig.2 and Fig.3, the saturation current I_{02}, which is associated with the recombination-generation processes in the depletion region and in the vicinity of the MIS contact, is not significantly changed by the nitride deposition process.

The reverse saturation current I_{01}, however, decreases from 2.2×10^{-12}A to 1.5×10^{-13}A. The corresponding current densities are: $J_{01} = 2.6 \times 10^{-11}$A/cm^2 and $J_{01} = 1.7 \times 10^{-12}$A/cm^2. This means that the electrical properties of the MIS contact were improved by the Si-nitride deposition treatment. Detailed experiments indicated that this improvement is only due to the temperature treatment occurring during the nitride film deposition (8). Thus for the nitride deposition parameters optimized for inversion layer solar cells any significant influence of the glow discharge on the MIS contact properties can be excluded.

Comparison with theory. In order to get information whether the dark current in these MIS diodes after the Si-nitride deposition is dominated by minority carrier flow as is predicted for Al/Si-oxide/p-Si contacts (4), the experimental value for J_{01} is compared with theory. As outlined in Ref.(4) these MIS structures exhibit in the so called semiconductor limited regime the same electrical characteristics as a one-sided abrupt pn-junction diode. Thus the reverse saturation current density J_{od} for the diffusion component of the minority carrier MIS diode is given by (5)

$$J_{od} = \frac{q\, n_i^2}{N_A} \sqrt{\frac{D_n}{\tau_n}} \qquad |2|$$

With an acceptor impurity density $N_A = 2.3 \times 10^{16}$cm^{-3}, determined from capacitance-voltage characteristics of the MIS diode, a diffusion coefficient $D_n = 25,3$cm^2/s and the reasonable values for the minority-carrier lifetime τ_n in the range of 2µs to 60µs, values of J_{od} from 5.2×10^{-12}A/cm^2 to 9.5×10^{-13}A/cm^2 are resulting. By comparison of these values with the experimental values of $J_{01} = 2.6 \times 10^{-11}$A/cm^2 obtained for the MIS contact before plasma Si-nitride deposition and $J_{01} = 1.7 \times 10^{-12}$A/cm^2 after removal of the deposited Si-nitride the following conclusion can be drawn. After the Si-nitride deposition the reverse saturation current density agrees fairly well with the theoretical current densities J_{od}. This indicates that the dark current in the Al/Si-oxide/p-Si contact after Si-nitride deposition is dominated by minority carrier flow and thus the properties of an ideal one-sided abrupt pn-junction are achieved. Before the Si-nitride deposition the experimental saturation current density J_{o1} exceeds the theoretical value J_{od} by approximately one order of magnitude. This leads to the assumption that in this case thermoionic emission of majority carriers is contributing to the dark current.

4. CONCLUSION

It has been shown, that the deposition of Si-nitride by the reaction of silane and ammonia in a rf glow discharge onto the Al/Si-oxide/p-Si MIS diodes does not degrade the electrical properties of these contacts. By the temperature treatment occuring during the plasma deposition process the MIS contacts are improved in so far as the reverse saturation current is decreased to values characteristic for minority carrier MIS diodes. This fact is a prerequisite for the achievement of high efficiency MIS/inversion layer solar cells with plasma Si-nitride as dielectric layer.

REFERENCES

(1) R. Hezel and R. Schörner, J. Appl. Phys. 52, 3076 (1981)
(2) R. Hezel, Solid St. Electron. 24, 863 (1981)
(3) R. Schörner and R. Hezel, IEEE Trans. El. Dev. ED-28, 1466 (1981)
(4) M.A. Green, F.D. King and J. Shewchun, Solid St. Electron. 17, 551 (1974)
(5) S.M. Sze, "Physics of Semiconductor Devices", Wiley, New York (1969)
(6) C.T. Sah, IRE Trans. El. Dev. ED-9, 94 (1962)
(7) J. Shewchun, M.A. Green and F.D. King, Solid. St. Electron. 17, 563 (1974)
(8) R. Schörner, thesis, University of Erlangen-Nürnberg F.R.G. 1982

LIMITATIONS OF THE OPEN CIRCUIT VOLTAGE OF INDUCED JUNCTION
SILICON SOLAR CELLS DUE TO SURFACE RECOMBINATION.

R. GIRISCH, R.P. MERTENS AND R. VAN OVERSTRAETEN
ESAT Laboratory-Katholieke Universiteit Leuven
Kardinaal Mercierlaan 94, 3030 Heverlee, Belgium

Summary
  In this paper we consider surface recombination under the positively charged insulator between the grating fingers in an induced-junction grating-type solar cell. We investigate this surface recombination both theoretically and experimentally. The numerical solution for the recombination current as a function of internal voltage in the air-insulator-silicon system has been obtained using generalized equations. The impact of fixed insulator charge, the distribution of interface states, nature of interface states (donor or acceptor type) and cross sections on the recombination current was calculated. In order to verify the calculations we developed a grating-type cell with a semi-transparant CVD poly-silicon electrode surrounding the grating emitter. In this structure the amount of inversion layer charge can be controlled by the voltage applied to the electrode. We show that (1) surface recombination may become a limiting mechanism in reaching high open circuit voltages; (2) high open circuit voltages are possible provided a proper combination is chosen of insulator charge and interface state density; (3) even when the surface state density increases linearly with the fixed insulator charge, the device performance improves with increasing insulator charge.

1. Introduction
  An alternative to p-n junction solar cells is the induced junction (IJ) solar cell (1,2) of which the inversion layer cell with MIS-contacts has received much attention during the past few years (3,4,5,6,7,8).
  Very shallow junctions were predicted theoretically with high electric fields in a direction to aid collection of carriers generated by light of short wavelengths. Collection efficiency calculations indicated the IJ cell to be relatively insensitive to lifetime and surface recombination velocity (2). Also complete collection of minority carriers, generated in the inversion layer, has been demonstrated experimentally in IJ photodiodes (9).
  To what extent the open circuit voltage (V_{oc}) may be limited by recombination at the inverted surface has not been discussed throughly in the recent literature. Norman and Thomas (7) assumed that front surface recombination has little influence on the operation of IJ cells, whereas Godfrey and Green (5,6) showed experimental evidence that surface recombination produces an excess dark saturation current that limits the V_{oc} for low resistivity substrates. The last authors described the influence of surface recombination on the V_{oc} using simple analytical expressions. Alam and Yeow (10) numerically calculated the surface photovoltage of IJ solar cells. However, the surface photovoltage is not equal to the V_{oc} as measured at the locally diffused junction or local MIS junction. Also the effects of interface states were neglected.

We worked out a model for recombination at an inverted surface using generalized equations and investigated the impact of fixed insulator charge, the distribution of interface states, the nature of interface states (donor or acceptor-type) and cross sections on the recombination current (J_{rec}) and the resulting V_{oc} of the cell. Experimental evidence for the model was obtained by the results of a grating-type cell with a semi-transparant electrode surrounding the grating emitter.

2. Theoretical model

The model we have used for surface states is that a surface state is either donor or acceptor type and that it is either neutral or singly-ionized. We do not consider surface states with the generalized amphoteric nature as discussed by Sah (11).

With respect to the capture cross sections we have investigated two models. Let σ_{nA}, σ_{pA}, σ_{nD}, σ_{pD} be the cross sections for electrons and holes for acceptor-like and donor-like states respectively; these cross sections are taken to be independent of energy and/or electric field. The first model is the model proposed by Panayotatos and Card who studied the recombination velocity at grain boundaries in polycrystalline silicon (12). Equal "neutral" capture cross sections (σ_N) are assumed for unionized acceptor-like states and unionized donor-like states. Equal "Coulombic" capture cross sections (σ_C) are assumed for ionized acceptor-like states (occupied by an electron/ empty of a hole) and ionized donor-like states (empty of an electron/occupied by a hole), that is

$\sigma_{nA} = \sigma_{pD} = \sigma_N$ and $\sigma_{pA} = \sigma_{nD} = \sigma_C$

The second model used is the more conventional one. We define a cross section σ_n for electrons and a cross section σ_p for holes irrespectively to the nature of the surface state, that is

$\sigma_{nA} = \sigma_{nD} = \sigma_n$ and $\sigma_{pA} = \sigma_{pD} = \sigma_p$

A major assumption in our calculations is that the electron and hole quasi-Fermi levels Φ_n and Φ_p are essentially flat in the space charge region (Figure 1). With this assumption, which is always checked for self-consistency at the end of the calculations, the problem to be solved reduces to a set of six equations in six unknowns. The unknown quantities are the quasi-Fermi levels Φ_n and Φ_p, the electrostatic potential at the surface Ψ_s (Figure 1), the charge induced in the silicon Q_{si}, the charge trapped in the interface states Q_{it} and the charge induced at the insulator-air interface Q_g (Figure 2). Details about the set of equations and the technique to solve this set numerically will be published elsewhere. Once Φ_n, Φ_p and Ψ_s are known for a given internal voltage V, J_{rec} is calculated by integrating the surface recombination rate over the entire energy gap, which is given in an approximated form:

$$U_s (p_s n_s - n_i^2) V_{th} \sigma_o^2 [(\sigma_{nA} n_s + \sigma_{pA} p_s)^{-1} \int_{\Delta E_A} D_A dE + (\sigma_{nD} n_s + \sigma_{pD} p_s)^{-1} \int_{\Delta E_D} D_D dE]$$

where σ_o is defined by: $\sigma_o^2 = \sigma_{nA} \cdot \sigma_{pA} = \sigma_{nD} \cdot \sigma_{pD}$ (1)

and ΔE_A is defined by: $kT [\ln(\sigma_{nA} n_s / \sigma_{pA} n_i + p_s / n_i) + \ln(n_s / n_i + \sigma_{pA} p_s / \sigma_{nA} n_i)]$

and ΔE_D is defined by: $kT [\ln(\sigma_{nD} n_s / \sigma_{pD} n_i + p_s / n_i) + \ln(n_s / n_i + \sigma_{pD} p_s / \sigma_{nD} n_i)]$

3. Results and discussion

First we performed numerous JV-calculations for several types of donor and acceptor state distributions and found that, once the surface is in strong inversion, the total amount of interface states plays a major role whereas the distribution is relatively unimportant. This can be understood easily with the definitions of ΔE_A and ΔE_D of equation (1) and by noting that the electron concentration at the surface n_s is very high: interface states lying in a broad range around midgap are "active" as recombination center.

For the calculations presented in Figures 3, 4 and 5, we used the interface state distribution, normalized on the interface state density at midgap $D_{it}(E_i)$, as shown in the inset of Figure 3; interface states below E_i and above E_i are donor and acceptor states respectively.

Figures 3 and 4 display the recombination current J_{rec} as a function of the fixed insulator charge Q_f for $V=-600mV$, $N_A=10^{16}cm^{-3}$, $D_{it}(E_i)=5 \cdot 10^{10}cm^{-2}eV^{-1}$ and $\sigma_o=10^{-15}cm^2$. We also investigated the influence of the capture cross sections. Figure 3 refers to the first model for cross sections; three values for the ratio σ_N/σ_C were used. Note that J_{rec} splits up into two peaks when $\sigma_C \gg \sigma_N$, because the acceptor term and the donor term in equation (1) reach their maximum value at different values of p_s/n_s. Figure 4 refers to the second model for cross sections; three values for the ratio σ_n/σ_p were used. This model always results in one maximum because the acceptor term and the donor term reach their maximum value at the same value of p_s/n_s. These calculations reveal that, in the inversion regime, a large cross section for holes rather than a large cross section for electrons results in high surface recombination: the capture of holes predominantly determines the recombination rate. In terms of the first model for cross sections this means that acceptor states rather than donor states degrade cell performance.

Until now D_{it} was kept constant while varying Q_f. In Figure 5, which displays J_{rec} as a function of Q_f for $N_A=5 \cdot 10^{16}cm^{-3}$ and all cross sections equal to $10^{-15}cm^2$, the interface state density D_{it} increases linearly with Q_f; the factor of proportionality F is defined by:

$$\int D_{it} \, dE = F \cdot Q_f / q \qquad (2)$$

The line at 36 mA.cm^{-2} represents a common short circuit current density value at AM1 conditions. This figure reveals that generally surface recombination may degrade the V_{oc} of IJ solar cells. However, provided a proper combination is chosen of Q_f and D_{it} (for example $F<.50$ and $Q_f>3 \cdot 10^{-7}$ C.cm^{-2} for $N_A=5 \cdot 10^{16}cm^{-3}$), a high V_{oc} is possible. Another interesting result is that, once in inversion, the device performance improves with increasing Q_f even when D_{it} increases linearly with Q_f.

In order to verify the theoretical model we developed a grating- type cell with a semi-transparant CVD poly-silicon electrode surrounding the grating emitter (inset of Figure 6). In this structure the amount of charge induced in the silicon surface region can be controlled by the voltage applied to the electrode. In Figure 6 we show the experimental V_{oc}-Vg dependence for a structure with a high quality C33 oxide as gate isolator (solid line) together with the theoretical curve (dashed line); the values of Q_f, $D_{it}(E_i)$ and D_{ox} are $3 \cdot 10^{-8}$C.cm^{-2}, $5 \cdot 10^9$eV^{-1}.cm^{-2} and 121 nm respectively and were determined independently. We used a U-shaped distribution of interface states. The first model for cross sections did not yield a good curve fit unless we assumed the interface states to be donor states predominantly. The second model with $\sigma_n=1.8 \cdot 10^{-15}cm^2$ and $\sigma_p=1.8 \cdot 10^{-17}cm^2$ fits the experimental data rather closely. Discrepancies between the experimental and theoretical curves, particularly the decrease in V_{oc} with increasing positive gate voltage, are under study.

Finally we comment on the assumption of essentially flat quasi-Fermi levels in the space charge region. The reason for the validity is due to the relatively low surface recombination currents at an inverted surface. Panayotatos and Card, who studied the recombination velocity at grain boundaries in n-type silicon (12), found the minority carrier quasi-Fermi level not to be constant in the space charge layer adjacent to the grain boundary. However, the high grain boundary state density and the absence of fixed charge result in much higher grain boundary recombination currents.

4. Conclusions

(1) Generally surface recombination may degrade the V_{oc} of IJ solar cells. However (2), provided a proper combination is chosen of Q_f and D_{it}, a high V_{oc} is possible. (3) Even when D_{it} increases linearly with Q_f, the device performance improves with increasing Q_f. (4) The cross sections for holes rather than the cross sections for electrons determine the surface recombination current.

Aknowledgement.

This paper is part of a report of the Belgian National R-D Program (Prime Ministers Office of Scientific Research, Wetenschapstraat 8, 1040 Brussels, Belgium). R. Girisch has been supported by the European Community.

References.
1. G.C.Salter and R.E.Thomas, Proc. 11th IEEE Photovoltaic Specialists Conf., Phoenix, 364-370 (1975).
2. G.C.Salter and R.E.Thomas, Solid State Electron. 20, 95-104 (1977).
3. P.Van Halen et al., IEEE, ED-25(5), 507-511 (1978).
4. R.B.Godfrey and M.A.Green, Appl. Phys. Lett. 33(7), 637-639 (1978).
5. R.B.Godfrey and M.A.Green, Appl. Phys. Lett. 34(11), 790-793 (1979).
6. R.B.Godfrey and M.A.Green, IEEE, ED-27(4), 737-745 (1980).
7. C.E.Norman and R.E.Thomas, IEEE, ED-27(4), 731-737 (1980).
8. R.Hezel, Solid State Electron. 24(9), 863-868 (1981).
9. J.Geist, E.Liang and A.Russell Schaefer, J. Appl. Phys. 52(7), 4879-4881 (1981).
10. M.K.Alam and Y.T.Yeow, Appl. Phys. Lett. 37(5), 469-470 (1980).
11. C.T.Sah, IEEE(NS), NS-23, 1563 (1976).
12. P.Panayotatos and H.C.Card, IEEE Electron Dev. Lett., ED1-1(12), 263-266 (1980).

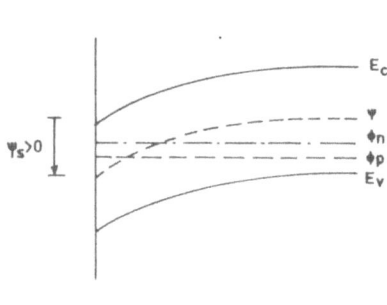

Fig. 1 : Band diagram at the silicon surface.

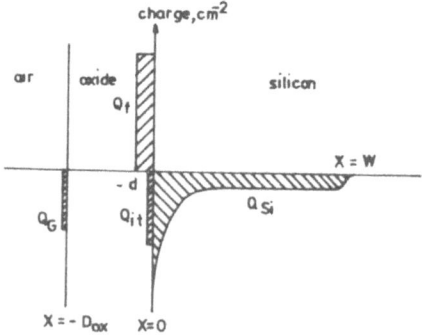

Fig. 2 : Charge representation of the Si-insulator-air system.

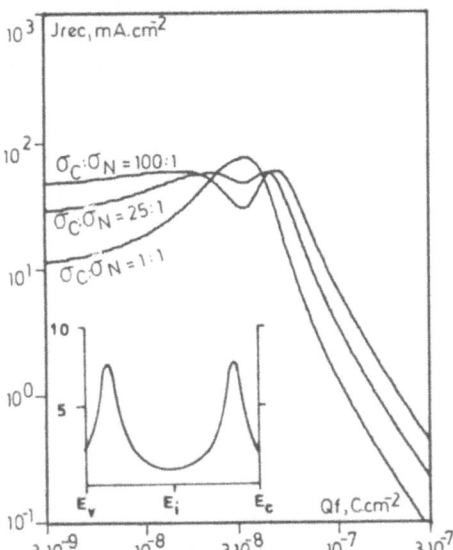

Fig. 3 : J_{rec} as a function of Q_f for V = -600 mV for the first model of cross sections.

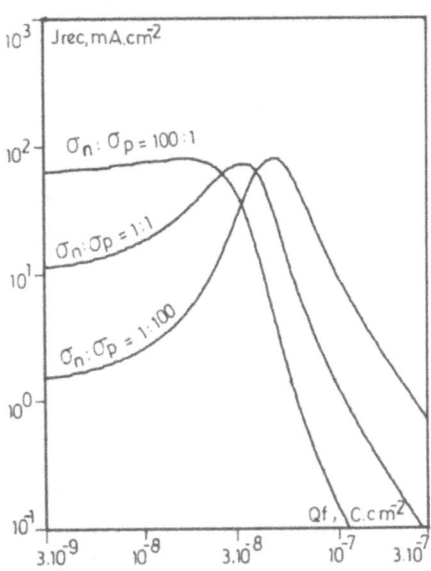

Fig. 4 : J_{rec} as a function of Q_f for V = -600 mV for the second model of cross sections.

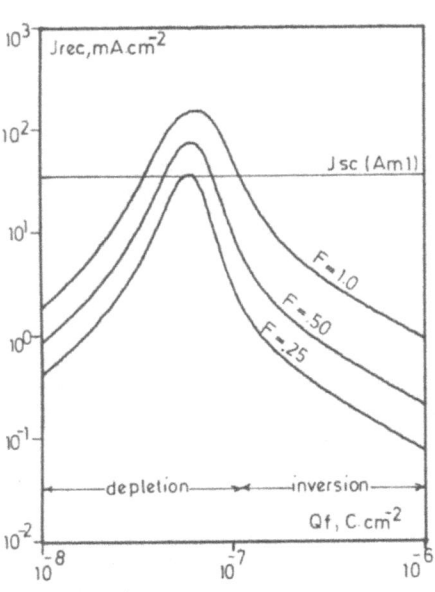

Fig. 5 : J_{rec} as a function of Q_f for V = -650 mV; D_{it} increases linearly with Q_f.

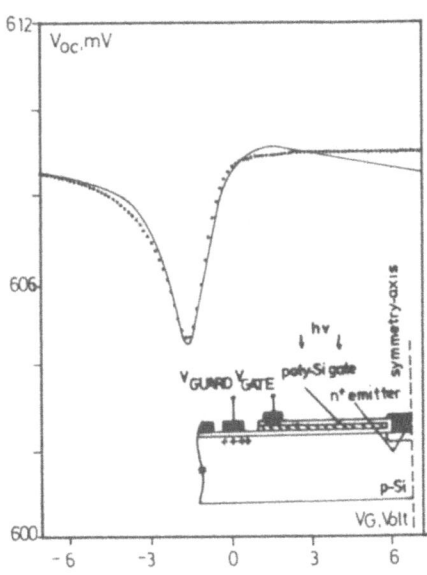

Fig. 6 : Experimental (solid line) and numerical (dashed line) $V_{oc} - V_g$ dependence.

SOME COMMENTS ON SPRAYED ITO/SEMICONDUCTOR SOLAR CELLS

J.C. MANIFACIER, H. LUQUET, L. GOUSKOV, C. GRIL, A. OEMRY, A. CHAOUI

Centre d'Etudes d'Electronique des Solides, associé au C.N.R.S.,(LA 21), Université des Sciences et Techniques du Languedoc, 34060 Montpellier cédex, France.

Summary

The realization of efficient photocells involving a semiconductor and I.T.O. window has recently received increased interest. We present here the properties of three types of such photocells obtained with three different semiconductors: n Si, p InP, n or p CdTe and a front ITO layer deposited by spray. The optical and electrical properties of our sprayed ITO layers are comparable to the best ITO films fabricated by more sophisticated methods (transmission $\simeq$ 90%, square resistance = 10 Ω for 0.5 µm thick film). Electrical and optical measurements such as dark I-V characteristics versus temperature, C-V characteristics, photocurrent spectral response, I-V characteristics under AM1 illumination are compared for the three devices.

Results obtained on ITO/Si and SnO_2/n CdTe devices are well explained by a Schottky model involving an interfacial layer. On ITO/p CdTe devices the existence of an interfacial layer determines the behaviour of the devices. A buried homojunction model allows to account for the main properties of ITO/InP diodes. These examples show that ITO deposition taking place at relatively high temperatures and under oxidizing atmosphere, leads to structures having an interfacial layer where interdiffusion processes may occur between the substrate and the ITO layer, affecting the photocell performances. The conversion efficiencies for ITO/CdTe, ITO/Si, ITO/InP are 2.1%, 10%, 14% respectively.

1. INTRODUCTION

Efficiencies comparable to those of homojunctions have been obtained on various ITO/semiconductor devices. The first structure investigated was ITO/Si (1-2). Solar cells involving III-V compounds (3-4) and II-VI compounds (5) were subsequently studied. ITO deposition has been obtained by a variety of methods like C.V.D., RF sputtering, magnetron sputtering and ion beam sputtering. Various models have been proposed to describe the behaviours of such ITO devices (Schottky, MIS, buried homojunction) taking into account the effect of an interfacial layer. We describe here the properties of ITO/semiconductor photovoltaic devices in which ITO deposition was obtained by the spray method.

2. SPRAYED ITO

Figure 1 describes the method of fabrication. A mixture of tin and indium chloride is dissolved in methanol. This solution is sprayed onto the heated substrate under a steady nitrogen flux. The chemical reactions leading to ITO deposition can be written:

$$SnCl_4 + 2H_2O \rightarrow SnO_2 + 4HCl$$

$$2InCl_3 + 3H_2O \rightarrow In_2O_3 + 6HCl.$$

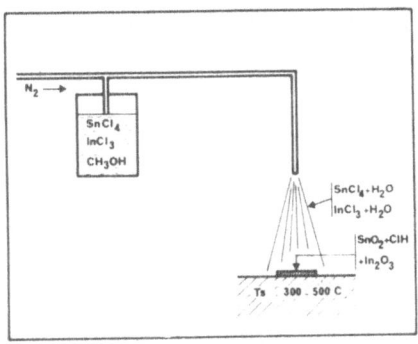

Fig.1- Principle of ITO spray deposition

The main fabrication parameters are the substrate temperature T_s ($300 < T_s < 500°C$). The atomic ratio $C_{Sn} = Sn/In$ in the solution, the growth rate (about 0.15 $\mu m/min$). the thickness t of the layer was chosen around 0.4 μm. With such a thickness, the mean transparency is about 90% in the visible range. The square resistance of the best ITO is 10Ω for this 0.4 μm thick layer. A grid must be deposited for efficient carrier collection on large area devices involving such ITO layers.

Figure 2 gives the reflection R and transmission T spectra of an

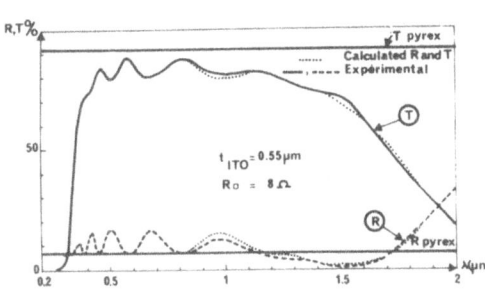

Fig. 2- R and T spectra of a sprayed ITO film on pyrex, thickness 0.55 μm.

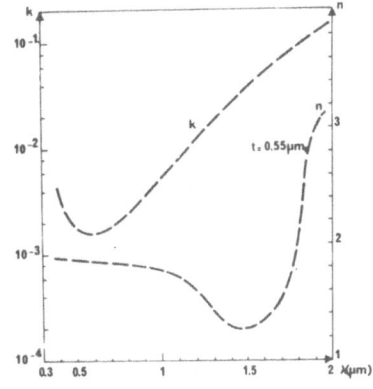

Fig. 3- Refractive index n and extinction coefficient k deduced from R and T of Fig. 2

ITO film deposited on pyrex. The refractive index n and extinction coefficient k deduced from these experimental curves are given on Fig. 3.

3. DIODE FABRICATION

Table I describes the technologies involved in the various fabrication steps.

	Si (1-6)	InP (7-8)	CdTe (9)
Substrate Free carrier density (300K) Orientation	n-type single crystal $n \approx 10^{15}$ cm^{-3} (111)	p-type single crystal $p = 1.2 \times 10^{16}$ cm^{-3} (111) or (100)	p-type single crystal $p = 3.3 \times 10^{16}$ cm^{-3}
Back contact fabrication	Au-Sb (0.9%) evaporation + diffusion at 430°C under H_2 ambient	Au-Zn (4% Zn) alloy evaporation. Diffusion during 2 min. at 450°C under H_2 ambient	Gold deposition from gold salts. H_2 annealing.
Surface treatment before spray	Chemical polishing with CP_4 rinsed for 1 min. in HF.	Mechanical polishing. Chemical etching (1% Br in methanol) for about 1 min.	Mechanical polishing. Chemical etching (1% Br in methanol) for about 1 min.
Optimized spray parameters	$T_s \approx 380-420°C$ $C_{Sn} \approx 15-25\%$	$T_s = 450°C$ $C_{Sn} = 1\%$	$T_s = 450°C$ $C_{Sn} = 10\%$

Table I - Sprayed ITO photodiode fabrication process.

The main difficulty we found in diode fabrication was the realization of good ohmic contacts on InP and CdTe. On InP the contact are still not reproducible, the best ones have a sheet resistance of 0.65 Ωcm^2.

4. EXPERIMENTAL RESULTS

- Dark I-V and C-V characteristics

Table II presents the main results of the dark I-V and C-V measurements. The best junction model deduced from these results is given.

ITO/Si	ITO/InP	ITO/CdTe
$J = J_o(\exp\frac{q(V-JR_S)}{n(T)kT} - 1)$	$J = J_o(\exp\frac{qV}{nkT} - 1)$	$J = J_o(\exp aV - 1)$
$J_o = A^* T^2 \exp^{-\frac{\Phi_{Bn}}{kT}}$	$J_o = J_{oo} \exp^{-\frac{\Delta E}{kT}}$	$a_{300K} = 12.1$
$n \approx 2.3$ to $n = 5.2$	$n \approx 2$	$J_{o_{300K}} = 1.2 \times 10^{-5} A/cm^2$
$A^* < 10^{-4} A.cm^{-2} K^{-2}$	$J_{o_{300K}} \approx 10^{-8} A/cm^2$	
$\Phi_{Bn} \approx 0.35 eV$ (I-V)	$\Delta E \approx 760$ meV	
$\Phi_{Bn} > 0.95 eV$ (C-V)	($\Delta E \approx E_G/2$)	
Important discrepancy observed for the barrier height between I-V and C-V measurements.	Rather good agreement with G-R transport in the space charge of an homojunction.	For SnO_2/n CdTe: good agreement with Schottky diode model. For ITO/p CdTe: the mechanism of interface recombination is preponderant.
The $C^{-2}(V)$ relation is perfectly linear from 77K to 300K	The $C^{-2}(V)$ relation is linear	The $C^{-2}(V)$ relation shows a poor linear variation
$V_D = 700 mV$ (300K)	$V_D = 960$ mV,	$V(C^{-2}=0) > V_D$
$N_d = n$ (Si)	$N_a > p$ (InP)	$N_a < p$ (CdTe)

Table II - Dark I-V and C-V results and diode models for the three ITO devices.

- Photoelectrical properties

Figure 4 compares the I-V characteristics under AM1 illumination. In table III the photovoltaic parameters of each ITO device is compared to those of the homojunction. These results show that ITO/Si and ITO/InP devices give efficiencies which can compete with those of the homojunctions. The low efficiency of ITO/CdTe diodes is probably due to a large interface states density.

The photocurrent spectrum measured for a constant photon flux is given on Figure 5 for the three devices. The slopes of the high wavelength edges is characteristic of the differences between direct gaps (CdTe-InP) and indirect gap (Si) materials. The photoresponse of ITO/Si diode is enhanced in the short wavelength range (shallow junction). For ITO/InP the photoresponse is rather flat. The model of a 0.5 µm deep buried n^+/p InP homojunction accounts very well for the whole spectrum of this device. The ITO/CdTe photoresponse is flat. The interfacial recombination diminishes the photocurrent independently of the wavelength.

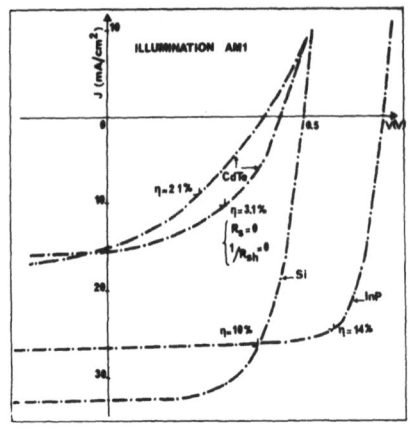

Fig. 4
I-V characteristics under AM 1 illuminations for the three ITO/devices.

		Si	InP	CdTe
ITO	V_{co} (mV)	500	700	424
	I_{sc} (mA/cm^2)	32	26.6	15.8
	η % AM 1	10	14	3.1
homo junct.	V_{co} (mV)	540	780	820
	I_{sc} (mA/cm^2)	40	26.5	20
	η % AM 1	14	14.8	13

Table III
Comparison between photovoltaic properties of ITO and homojunction devices (10, 11, 12).

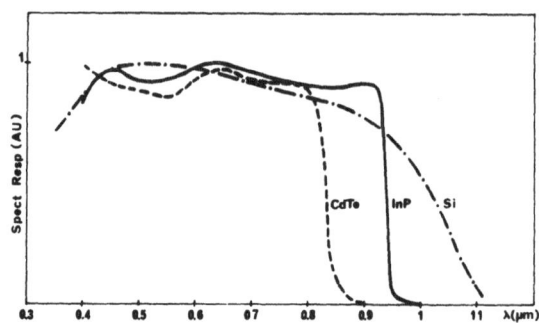

Fig.5 - Relative photoresponse spectra for the three devices.

5. CONCLUSION

Using a very simple spray technique photovoltaic devices have been obtained on Si, InP and CdTe. For Si and InP substrates, the photovoltaic performances are comparable to those of homojunction. The ITO layer adds to ease of preparation, low cost, low temperature processing an interesting combination of properties: it is an integral part of the barrier a protective coating and can act as an antireflecting layer. It is then clear that in the field of solar cells, such a spraying technique is a promising alternative to the standard diffusion technology.

REFERENCES

(1) J.C. Manifacier and L. Szepessy, Appl. Phys. Lett. 31, 1979, 459

(2) J. Calderer, J.C. Manifacier, L. Szepessy, J.M. Darolles et M. Perotin, Rev. Phys. Appl. 14, 1979, 485
(3) L. Hsu, E.Y. Wang, 13th Photovoltaic Specialists Conf., Washington 1978, p.536
(4) K.J. Bachmann, H. Schreiber, W.R. Sinclair, P.H. Schmidt, F.A. Thiel, E.G. Spencer, G. Pasteur, W.L. Feldmann, K. Sree Harsha, J. Appl. Phys. 50-5, 1979, 3441
(5) A.L. Fahrenbruch, J. Aranovich, F. Courreges, T. Chinoweth and R.H. Bube, 13th IEEE Photovoltaic Specialists Conf., Washington 1978
(6) J.C. Manifacier, Thin Solid Films 90, 1982
(7) L. Gouskov, H. Luquet, J. Esta, C. Gril, Solar Cells 5, 1981-82, 51
(8) L. Gouskov, H. Luquet, C. Gril, A. Oemry, M. Savelli, Rev. Phys. Appl. 17, 1982, 125
(9) J. Calderer, J. Esta, H. Luquet, M. Savelli, Solar Energy Materials 5, 1981, 337-347
(10) H.J. Hovel, Solar Cells in R.K. Willardson and A.C. Beers, Eds., Semiconductors and Semimetals, Vol.II, Academic Press, 1975
(11) G.W. Turner, J.C.C. Fan, J.J. Hsieh, Appl. Phys. Lett. 37-4, 1980, 400
(12) G. Cohen Solal et al. (this issue)

ROLE OF PHOTOLUMINESCENCE IN THE EFFICIENCY OF A $Ga_{1-x}Al_xAs$-GaAs SOLAR CELL

P. BARUCH and M. CUNIOT
Groupe de Physique des Solides de l'Ecole Normale Supérieure
Laboratoire associé au C.N.R.S. n°17
Université Paris 7
2, Place Jussieu - 75251 Paris Cédex 05

Summary

Following a suggestion by Alferov, we have studied, in a $Ga_{1-x}Al_xAs$-GaAs cell the contribution to the photocurrent of the photoluminescence induced in the $Ga_{1-x}Al_xAs$ front layer, of gap E_2 (1.4 eV < E_2 < 2.8 eV). The incident photons of energy larger than E_2 are absorbed in this layer and create electron-hole pairs which recombine, emitting band gap photons with $h\nu = E_2$, with a high efficiency if the gap is direct (x < 0.4). This radiation is not absorbed in the front layer and, almost entirely trapped by internal reflection, travels to the GaAs base, where it can create pairs in a process very similar to fluorescence conversion. We have modelled the different processes, as a function of the thickness of the front and base layer, of diffusion lengths, front surface recombination velocity, and radiative efficiency. We have calculated separately the contribution of the base and front to the photocurrent, from generation by direct incoming photons and by recombination generated photons and we will present the main results. It will be shown that this design allows a thick (> 3 μm) front layer, while retaining a high collection efficiency. This will lead to a low series resistance for concentration cells.

1. INTRODUCTION

One of the problems to solve in photovoltaïc cells for solar concentrators is the reduction of the series resistance which in view of the high currents flowing through the cell, must be kept very low. However, in the GaAs cell, the high optical absorption and the low lifetime restrict the thickness of the front layer and give rise to a high resistance. Even the addition of a GaAlAs window, to improve the interface properties, does not reduce much this resistance.

A way out of this dilemma has been suggested by Alferov (1), by using energy transport through photon recycling : the front layer is made up of a material showing a high radiative efficiency so that the incoming photons, absorbed in this layer, create electron-hole pairs which recombine into band gap photons. These photons travel in the front material without much absorption, and, arriving in the lower band gap base material, are absorbed photoelectrically, to yield electron hole pairs which can be collected. This process is, in fact, identical to the well known "fluorescent converter", except that here, the fluorescent material is an integral part of the cell. Another analogy can be drawn with a tandem cell, where the first cell is open circuited.

As we will show below, the front layer can be thicker than in the usual cell, which allows the reduction of the series resistance. We have attempted the modelization of the collection efficiency of such photovoltaïc

2. DESCRIPTION OF THE MODEL

As a first approximation, we will compute here the collection efficiency of a heterojunction with a GaAs base and $Ga_{1-x}Al_xAs$ front layer

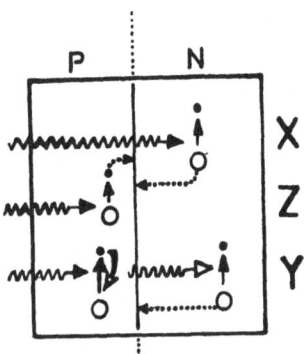

- Energy gap in the base: E_1
  the front: E_2
- Diffusion length in the base : L_1
  the front : L_2
- Recombination velocity on the front face : S
- Absorption coefficient of the base : $\alpha_1(E)$
  the front : $\alpha_2(E)$
  where E is the energy of the incident photon
- Diffusion constant in the front : D_2
- Radiative efficiency in the front : R
- Thickness of the base : d_1
  the front : d_2

In this model no recombination takes places at the heterojunction interface : all minority carriers arriving at the interface are collected. The carrier transport takes place through diffusion.

With these assumptions, the diffusion equations are easily solved; the spectral response (number of carriers collected per incident photon) can be decomposed into three terms :
$Q(E) = X(E) + Y(E) + Z(E)$

. $X(E)$ is the base (GaAs) contribution, due to pairs created by incident photons absorbed in the base ($E > E_1$) and diffusing to the junction.

$$X(E) = \frac{\alpha_1 L_1 e^{-\alpha_2 d_2}}{\alpha_1 L_1 + 1}$$

. $Z(E)$ is the front ($Ga_{1-x}Al_xAs$) contribution due to the same process in the front ($E > E_2$).

$$Z(E) = \frac{\alpha_2 L_2}{(\alpha_2 L_2)^2 - 1} e^{-\alpha_2 d_2} \left[\frac{e^{+\alpha_2 d_2}\left(1 + \frac{\alpha_2 D_2}{S}\right) - ch\frac{d_2}{L_2} - \frac{D_2}{L_2 S} sh\frac{d_2}{L_2}}{sh\frac{d_2}{L_2} + \frac{D_2}{L_2 S} ch\frac{d_2}{L_2}} - \alpha_2 L_2 \right]$$

. $Y(E)$ is the photoluminescence contribution due to the pairs created in the front but not collected, recombining radiatively in the bulk, generating photons of energy E_2. These photons are absorbed in the base, where they create pairs, the number of these photons is proportional, with the factor R, to the number of bulk recombinations in the front. Surface recombination is assumed to be non radiative. The escape of photons through the front surface is neglected, because of trapping through internal reflections : this loss (around 1,8%) can formally be incorporated in R.

$$Y(E) = R \frac{\alpha_1 L_1}{\alpha_1 L_1 + 1} \left[1 - e^{-\alpha_2 d_2} \frac{\alpha_2 L_2}{\alpha_2^2 L_2^2 - 1} \left[\frac{(1 + \frac{\alpha_2 D_2}{S} + e^{-\alpha_2 d_2})(1 - ch\, d_2/L_2)}{sh\, d_2/L_2 + (D_2/L_2 S) ch\, d_2/L_2} - \frac{(D_2/L_2 S) e^{-\alpha_2 d_2} sh\, d_2/L_2}{sh\, d_2/L_2 + (D_2/L_2 S) ch\, d_2/L_2} \right] + \frac{(\alpha_2 L_2)^2}{(\alpha_2 L_2)^2 - 1} (e^{-\alpha_2 d_2} - 1) \right]$$

where α'_1 is the value of α_1, for recombination photons ($\alpha'_1 = \alpha_1 (E_2)$).

Integrating over the standard solar spectrum at AM 1 gives the short circuit current per unit area I_{cc}.

As a comparison, similar calculations have been made for a conventional GaAs homojunction.

No attempt are made here to calculate the open circuit voltage and the fill factor.

GaAs and $Ga_{1-x}Al_xAs$ properties have been obtained from the literature (2, 3, 4, 5), with $x = 0,3$ ($E_2 \sim 1,8$ eV) giving a minimum band gap, and yet a direct band structure. Then $\alpha'_1 = 3.7 \times 10^4$ cm^{-1}.

3. RESULTS

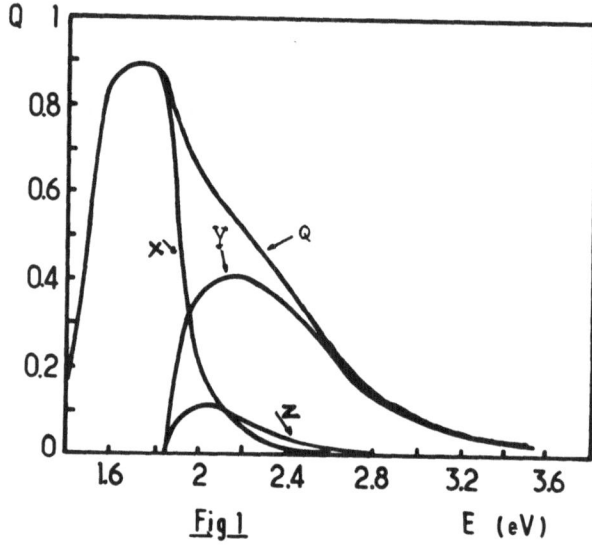

Fig 1

Fig. 1. shows an example of the three contributions for the following set of parameters :
$d_2 = 1.1$ μm $L_2 = 0.27$ μm $L_1 = 2.5$ μm $d_1 \gg L_1$; $s = +\infty$ $R = 1$

This is a worst case, since the surface recombination velocity is the highest. Yet, it can be seen that the photoluminescence contribution is important for $h\nu > 2,1$ eV, and enhances considerably the short wave length response.

Fig. 2. shows the effect of improving the surface for $R = 1$, $d_2 = 6$ μm, $L_1 = 2.5$ μm, $L_2 = 0.27$ μm, $D_2 = 4$ cm^2/s. A value of s as large as 10^4 cm/s will still show a high photoluminescence contribution, since the direct front contribution is negligible here ($L_2 \ll d_2$). This gives $I_{cc} = 22.7$ mA/cm^2, where 12.8 mA/cm^2 are due to photoluminescence. The collection efficiency in this case is 70%. Even for s infinite, $I_{cc} = 15.9$ mA/cm^2, with 5.9 mA/cm^2 from photoluminescence.

Fig. 3 shows the influence of d_2, in a worst case (s =∞) : as expected for a thin front layer (0.1 μm), the photoluminescence contribution is negligible, since direct pair diffusion to the junction will be effective in the front. On the other hand, with a thick front layer (6 μm), the photoluminescence term is predominant above 1.9 eV. The total efficiency is lower than with a thin front layer, but is still relatively important, even if about half of the pairs are lost by non-radiative surface recombination.

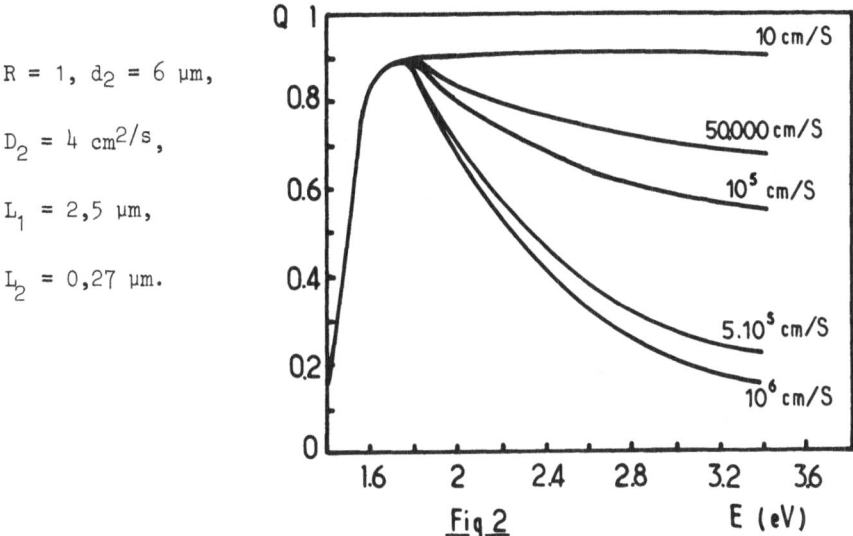

$R = 1$, $d_2 = 6$ μm,

$D_2 = 4$ cm^2/s,

$L_1 = 2,5$ μm,

$L_2 = 0,27$ μm.

Fig 2

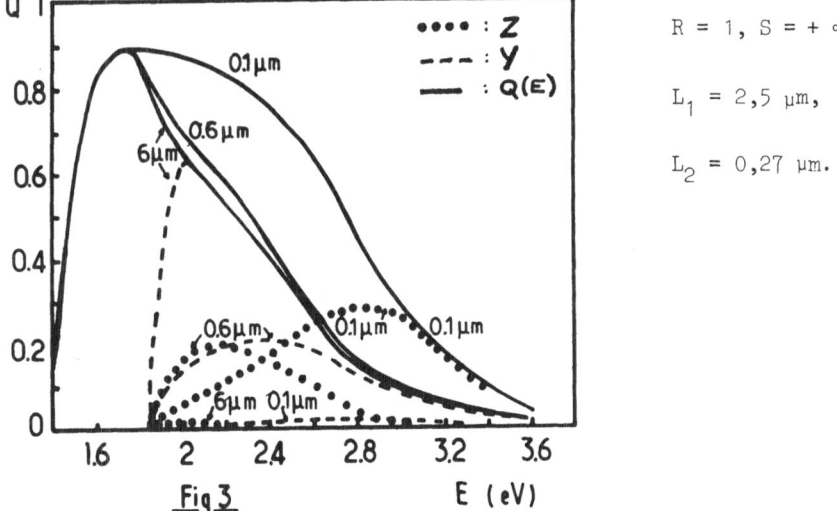

$R = 1$, $S = + \infty$,

$L_1 = 2,5$ μm,

$L_2 = 0,27$ μm.

Fig 3

A comparison can be made with the conventional GaAs homojunction. As we have seen in the case illustrated in fig. 2, the thick ($d_2 = 6$ μm) heterojunction has a good (50%) collection efficiency, even with s infinite : about one third of the current comes from the photoluminescence. On the other hand, a homojunction with the same dimensions would have a low (10%) collection efficiency : not only the front contribution is very small, since $L_2 \ll d_2$, but also the high optical absorption limits severely the number of photons reaching the junction and the base. Evidently a thin homojunction, (d_2 = 0.5 m) would be much better, specially if the surface recombination is limited by the addition of a thin GaAlAs window, with a high Al concentration (5), but the cell fabrication is much more critical.

4. DISCUSSION

The case shown on Fig. 2, with s = 10^4 cm/s appears to give a cell with interesting performances, as far as collection efficiency is concerned. However, one should discuss the possibility of obtaining these parameters. First, a p^+-n heterojunction should be used ; because of the influence of the factor D_2/L_2s in the expression of Y, a large D_2 should be retained, corresponding to a p front layer. As in Alferov's papers (1), we take a moderate doping (5.10^{17} cm^{-3}), which combines a relatively low sheet resistance and the possibility of a good value of R. However, this point is very critical, and will depend much on crystal perfection. Also, the low s value will require a thin GaAlAs window, with a high Al concentration.

The model we have studied is very crude. No attempt has been made to account for the heterojunction properties and to predict open-circuit voltage or fill factor. Experimental data should be obtained soon, on some samples which have been prepared according to this design.

Nevertheless, the overall performance will probably not be comparable to the best GaAs cells or to tandem cells, but the possibility of a low series resistance and the relative simplicity of fabrication make this type of cells attractive.

References

1) Zh. I. Alferov et al., Sov. Phys. Semicond. 11 1034 (1977), 14 403 (1980).
2) M.D. Sturge, Phys. Rev. 127. 768 (1968),
3) H.C. Casey, D.D. Sell, J. Appl. Phys. 46 250 (1975),
4) A. Laugier, J.A. Roger, Les photopiles solaires (Techn. et Doc. Paris 1981),
5) H.J. HOVEL, IBM Journal 22 112 (1978).

OPERATING CHARACTERISTICS OF THIN THERMOPHOTOVOLTAIC CELLS WITH MINORITY CARRIER MIRRORS AND OPTICAL MIRRORS USING SELECTIVE RADIATORS OF ERBIUM AND YTTERBIUM OXIDES

E.S. Vera*, J.J. Loferski** and M. Spitzer*†
Department of Physics*
Division of Engineering**
Brown University

J. Severns
Naval Research Laboratory

Summary

The performance of thermophotovoltaic (TPV) germanium based systems in which the radiation emitters are coated with selective absorber-emitters of erbium (Er_2O_3) and ytterbium (Yb_2O_3) oxides is studied. The principal characteristic of these high efficiency TPV cells is that they are thin optimized p-n junctions which incorporate minority carrier mirrors and optical mirrors at the end surfaces of the photovoltaically active volume of the cells. The efficiency of a single germanium cell (in excess of 20%) and of a two-junction tandem silicon-germanium cell (in excess of 22%) having an optimized opto-electronic structure is found to be higher for an Er_2O_3 coated emitter than with a conventional blackbody, this advantage being more pronounced for the lower temperatures in the range considered (1300-2000°K).

1. INTRODUCTION

Solar thermophotovoltaic (TPV) conversion systems are being considered as power sources for radiation hardened satellites. In this mode of operation the incident radiation is first absorbed by an intermediate absorber and then reemitted onto a conventional solar cell. Germanium has been selected (1) as the most suitable semiconductor for this kind of application since its energy bandgap ($E_G = 0.66$ eV) provides a better spectral match with the lower temperature (approximately 1300-2000°C) thermal radiation sources contemplated for use in TPV systems than those of other materials (2).

In earlier papers (3-5) we have calculated the maximum conversion efficiency from TPV systems utilizing optimized p-n junction structures which incorporate minority carrier mirrors (MCM) and optical mirrors (OM) at the front and back surfaces of the cell. When different thermal radiation sources emitting blackbody radiation within the temperature range of 1500 to 2000°C were considered, the upper limit to the conversion efficiency was found to be about 22% for an optimum design germanium cell 90 microns thick and about 26% for a two-junction silicon-germanium tandem arrangement 50 and 90 microns thick, respectively, both systems receiving an overall input power density of 25 W/cm^2 from a 2000°C blackbody radiation source.

High purity rare earth oxides are promising candidates as selective spectral radiation sources (6-8). In this paper we explore further improvements in the performance of the previous MCM and OM cell structures using erbium (Er_2O_3) and ytterbium (Yb_2O_3) oxides as intermediate radiation absorber-emitter sources. This work describes the results of two studies. The first is a comparison of germanium operating parameters, including efficiency, when alternately illuminated by a blackbody and an Er_2O_3 selective radiator at the same temperature, for radiator temperatures from 1600 to 2000°K. The second study involves a split spectrum system or a tandem cell combination comprised of silicon and germanium units. The silicon cell here is taken to be of similar design to the MCM-OM germanium structure. The operating parameters of the combination are computed for illumination by a blackbody, a pure Er_2O_3 radiator, or an Er_2O_3-Yb_2O_3 mixture, as would be obtained if Er_2O_3 and Yb_2O_3 were chemically mixed and a radiator then fabricated. Cell performance is predicted for the same temperatures used in the first study, and as a function of Er_2O_3-Yb_2O_3 mixture ratio.

In the first study, the germaniun cell efficiency is greater with Er_2O_3 than with the blackbody, due to the better spectral match of the Er_2O_3 to the germanium characteristic. The second study shows an advantage of the dual cell combination with the spectra studied, especially the selective radiators, this advantage being more pronounced as the radiator temperature decreases. The inclusion of Yb_2O_3 with the Er_2O_3 appears to decrease dual efficiency considerably.

2. CALCULATIONS FOR AN ERBIUM OXIDE SELECTIVE RADIATOR

High purity ceramic oxides are among the most promising candidates as selective spectral radiation sources. A specific feature of interest with the rare earth oxides is their strong band emission which ranges from the visible to the near-infrared wavelength region. These bands permit strong thermal excitation at temperatures compatible with the high temperature stability of these materials. Therefore, their potential use in TPV energy conversion systems has become a subject of increasing interest (6-8). Figure 2 shows the spectral distribution of photons of a blackbody and an erbium oxide radiator. Using the Er_2O_3 emittance data measured by G.E. Guazzoni (8) for different temperatures, we were able to evaluate the performance of an optimized germanium p-n junction structure, as previously described, under illumination from this selective radiator and compare it to its performance under illumination from a blackbody. In all the calculations we have used the set of parameters given in Table I under the assumptions discussed in earlier work (3-5).

Figure 4 shows the conversion efficiency as a function of thickness for a germanium cell with ideal MCM (surface recombination velocity s = 0) and nearly-ideal OM (internal reflection coefficients R_F = 0.95, R_B = 0.98) under illumination from a blackbody and from an erbium oxide radiator, both at the same temperature (T = 2000°K). We notice that an optimal structure 90 microns thick has a slightly higher efficiency with an Er_2O_3 selective radiator (η_{SR} = 21.3%) than with a blackbody radiator (η_{BR} = 20.9%) despite its lower intensity emission. However, the difference ($\Delta\eta = \eta_{SR} - \eta_{BR}$) is small because, for solar cells located at the same distance from the source (fixed geometry), the input power density is an order of magnitude less with the erbium oxide radiator (H_{INC} = 2.4 W/cm^2) than with the blackbody (H_{INC} = 25 W/cm^2); the cell efficiency increases with increasing input power.

Figures 5 and 6 show the same calculation for Er_2O_3 sources having different temperatures (T = 1800 and 1600°K, respectively) but delivering the same 25 W/cm^2 power density to the cell as the sources in our blackbody calculations. The power density is controlled by the spacing between the source and the solar cell. If we consider a spherically symmetric TPV system, like the one depicted in Figure 1, the radiation density (H_{INC}) incident on the solar cells is related to the geometry of the arrangement through the simple expression,

$$H_{INC} = \mu H_{EMI} \text{ with } \mu = \left(\frac{R_{EMI}}{R_{INC}}\right)^2 \tag{1}$$

where R_{EMI} and R_{INC} are the distances from the origin to the surfaces of the blackbody and solar cell array respectively. For a fixed temperature, the emitted radiation power density (H_{EMI}) is constant. Thus, a variation of incident power density means a change in geometry, since,

$$H_{INC} = \mu H_{EMI} = \mu \sigma T^4 \tag{2}$$

with

$$\mu = \left(\frac{R_{EMI}}{R_{EMI} + d}\right)^2, \quad \sigma = 5.67 \cdot 10^{-12} \text{ W/cm}^2 \text{ }^\circ K^4$$

To give a quantitative idea, in a TPV system with a blackbody of radius R_{EMI} = 10 cm, solar cells placed at a distance d=9 cm away from the surface of a blackbody radiator at T = 2000 K would receive an input density H_{INC} = 25 W/cm^2; to receive the same level of radiation intensity from blackbody sources at T = 1800 and 1600°K, the solar cells would need to be placed at distances d=5.4 and 2 cm respectively. We observe that the efficiency difference between the blackbody and Er_2O_3 sources increases with decreasing temperature because the illumination level ratio between the Er_2O_3 and the blackbody increases with decreasing temperature. The result is important because it shows that a selective radiator becomes more advantageous with respect to a blackbody radiator as the temperature of the emitter source decreases (see Figure 7).

3. CALCULATIONS FOR AN ERBIUM-YTTERBIUM OXIDE MIXTURE SELECTIVE RADIATOR

Now we extend these calculations to the case of a two-junction silicon-germanium tandem solar cell arrangement intended for TPV application, using erbium and ytterbium oxides as intermediate absorber-emitter sources. The study compares the overall efficiency of the system under illumination by a blackbody, a pure Er_2O_3, a pure Yb_2O_3 and an Er_2O_3-Yb_2O_3 mixture, as would be obtained if Er_2O_3 and Yb_2O_3 were chemically mixed and a radiator fabricated. The system performance is calculated for radiator temperatures in the range of 1600 to 2000°K which is usual to TPV energy conversion. Figures 2 and 3 show the spectral distribution of photons of a blackbody and of different oxide mixtures emitting electromagnetic radiation at T = 2000°K. We notice that while Er_2O_3 has a peak distribution of photons with energies just above the germanium bandgap, Yb_2O_3 emits strongly above the silicon band edge.

Following the theoretical framework described in earlier work (5), we investigate the performance of a system consisting of a silicon and a germanium p-n junction, both of which have ideal MCM (s=0) and nearly-ideal OM (R_F = 0.95, R_B = 0.98). First, we calculate the conversion efficiency for these structures under illumination from a blackbody and from a pure Er_2O_3 radiator at T = 2000°K. The upper limit to the conversion efficiency is 5.7 and 5.4% for an optimal silicon solar cell 50 microns thick illuminated by the Er_2O_3 selective radiator and the blackbody respectively (Figure 8a); for an optimal germanium structure 90 microns thick, these values are 17.2 and 16.8% respectively (Figure 8b). Thus, the two-junction system has a slightly higher overall efficiency ($\eta_T = \eta_{Si} + \eta_{Ge}$) when illuminated with an Er_2O_3 selective radiator (η_{TSR} = 22.9%) than with a blackbody radiator (η_{TBR} = 22.5%). Although the overall efficiency of the system decreases with increasing temperature, the efficiency difference ($\Delta\eta_T = \eta_{TSR} - \eta_{TBR}$) is greater for lower temperature sources (Figure 9). The reason for this is that the ratio between the illumination level of the selective radiator and the blackbody is higher for lower temperatures, because of the strong temperature dependence of the blackbody power density described by Equation (2).

Next, we vary the composition of the oxide coating in the radiator reproducing the previous calculations for three other cases: i) 60% Er_2O_3 - 40% Yb_2O_3, ii) 40% Er_2O_3 - 60% Yb_2O_3, and iii) pure Yb_2O_3. The results of these calculations are shown in Figure 10, where the overall conversion efficiency of the two-junction silicon-germanium PV cell system is plotted as a function of oxide composition for the radiation levels at three differenc temperatures (T = 2000, 1800 and 1600°K). We notice that the conversion efficiency of the system continuously deteriorates when the photovoltaically active semiconductors are illuminated by absorber-emitter sources containing increasing proportion of Yb_2O_3. In fact, for the case of pure Yb_2O_3 the system performs better under illumination from the blackbody then illuminated by the ytterbium oxide selective radiator at all temperatures in the range considered. The higher efficiency performance of the erbium oxide dominant mixtures can be explained because the lower energy peak of its spectral distribution of photons has a better match with the blackbody spectrum at the temperature range for the sources used in TPV energy conversion. Therefore, we find out that the good spectral match between the silicon and the ytterbium oxide plays a less important role than the match between the germanium and the erbium oxide photon emission distribution.

4. CONCLUSIONS

The study of the performance of optimized TPV cell structures using erbium and ytterbium oxides, as intermediate absorber-emitter radiation sources, indicates that for TPV systems based on germanium PV structures only the use of Er_2O_3 as selective radiator presents an advantage over the choice of a blackbody radiator emitting at the same temperature. We observe that the advantage or efficiency difference between the blackbody and Er_2O_3 is greater for the lower temperatures in the range considered. The efficiency difference is small for the higher temperatures in the range considered because the illumination level ratio between the Er_2O_3 and the blackbody radiators decreases with increasing temperature. Thus, if the potential advantage of using selective radiators at higher temperatures is to be fully realized, ways of concentrating its electromagnetic emission to higher levels of radiation intensity should be conceived. Construction of these high efficiency structures appears to be possible using present available technology, as discussed in earlier work (3-5).

5. ACKNOWLEDGEMENT

Part of the work reported here (at Brown University) was supported by the Naval Research Laboratory.

6. REFERENCES

1. E. Kittl and G. Guazzoni, Proc. of the 25th Power Sources Symposium, (1972), p. 106.
2. E. Kittl, Proc. of the 10th IEEE Photovoltaic Specialists Conference, Palo Alto, California, (1973), p. 103.

3. E.S. Vera, J.J. Loferski, M. Spitzer and J. Shewchun, Proc. of the 3rd European Communities Photovoltaic Solar Energy Conferency, Cannes, France, (1980), p. 911.
4. E.S. Vera, J.J. Loferski and M. Spitzer, Proc. of the Second Chilean Symposium of Physics, University of Santiago Press, Santiago, Chile, (1980).
5. E.S. Vera, J.J. Loferski and M. Spitzer, Proc. of the 15th IEEE Photovoltaic Specialists Conference, Orlando, Florida, (1981), p. 877.
6. W.R. McMahon and D.R. Wilder, "High Temperature Spectral Emissitivity of Yttrium, Samarium, Gadolinium, Erbium and Lutetium Oxides", Report IS-578, Ames Laboratory, Iowa State University of Science and Technology, Ames, Iowa, (January 1963).
7. G.E. Guazzoni and S.J. Shapiro, R&D Technical Report ECOM-3281, (May 1980).
8. G.E. Guazzoni, Applied Spectroscopy, 26-1, 60 (1972).

Table I Parameters used in the calculations of cell thickness dependence

Parameter	Si Value	Ge Value	Units
Temperature	300	300	°K
Intrinsic Concentration	$2 \cdot 10^{10}$	$2.4 \cdot 10^{13}$	cm^{-3}
Donor Concentration	10^{19}	10^{19}	cm^{-3}
Acceptor Concentration	$2 \cdot 10^{17}$	$2 \cdot 10^{17}$	cm^{-3}
Electron Mobility	600	2500	$cm^2 v^{-1} sec^{-1}$
Hole Mobility	90	115	$cm^2 v^{-1} sec^{-1}$
Electron Lifetime	20	300	μsec
Hole Lifetime	0.1	0.03	μsec
Junction Depth	0.5	0.1	μsec
Illumination	25 Wcm^{-2} from a 2000°C blackbody radiation source		

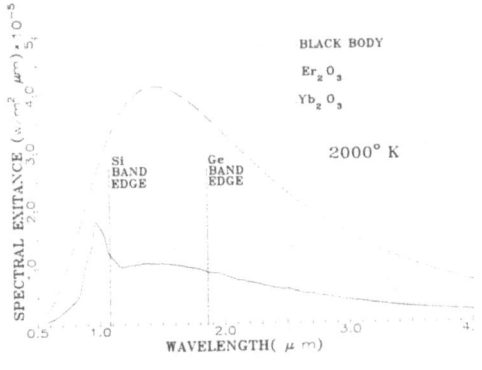

Fig.2: Spectral distribution for a blackbody and two selective radiators (pure Er_2O_3 and pure Yb_2O_3) emitting at T=2000°K.

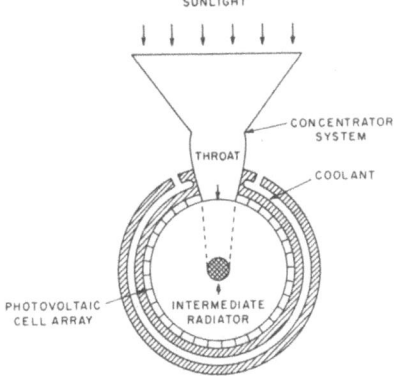

Fig.1: Geometrical arrangement for a TPV conversion system.

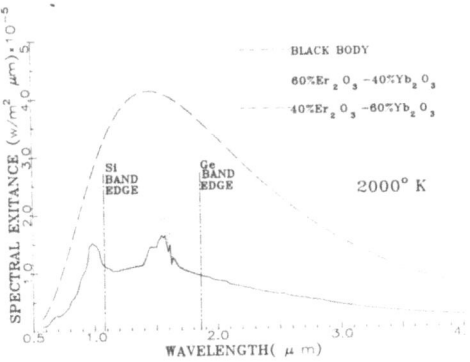

Fig.3: Spectral distribution for a blackbody and two selective radiators (40% Er_2O_3-60% Yb_2O_3 and 60% Er_2O_3-40% Yb_2O_3) emitting at T=2000°K.

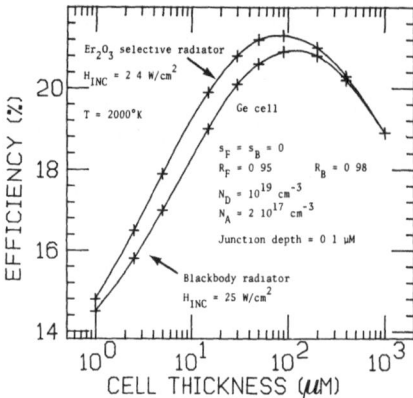

Fig.4: Calculated conversion efficiency as a function of cell thickness for a Ge p-n junction TPV system with two different intermediate radiators at T=2000°K.

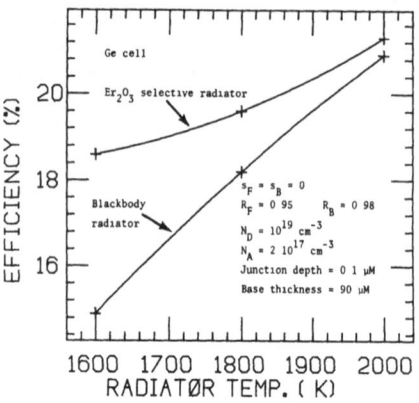

Fig.7: Calculated conversion efficiency as a function of radiator temperature for an optimal Ge structure 90µm thick with Er_2O_3 and blackbody (H_{inc}=25W/cm²) radiators.

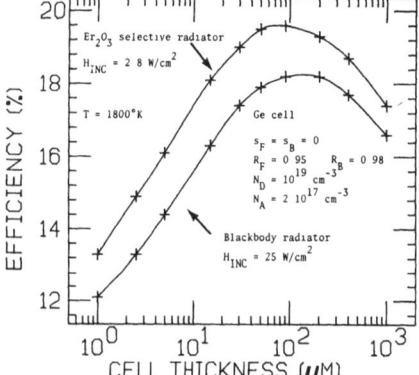

Fig.5 Calculated conversion efficiency as a function of cell thickness for a Ge p-n junction TPV system with two different intermediate radiators at T=1800°K

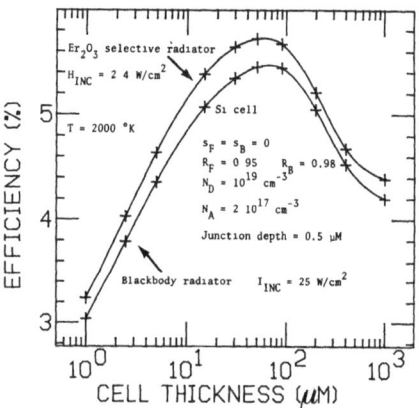

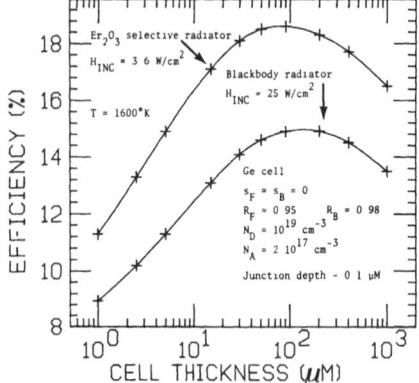

Fig.6: Calculated conversion efficiency as a function of cell thickness for a Ge p-n junction TPV system with two different intermediate radiators at T=1600°K.

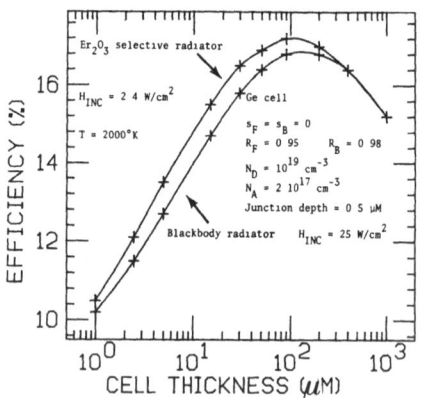

Fig.8: Calculated conversion efficiency as a function of cell thickness with two different intermediate radiators at T=2000°K for optimal structures of: (a) 50µm thick Si and (b) 90µm thick Ge.

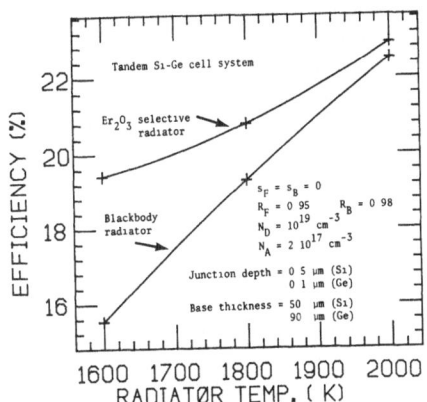

Fig.9: Calculated total conversion efficiency ($\eta_T=\eta_{Si}+\eta_{Ge}$) as a function of radiator temperature for optimal Si and Ge structures (50 and 90µm thick respectively) with pure Er_2O_3 and blackbody ($H_{inc}=25W/cm^2$) radiators.

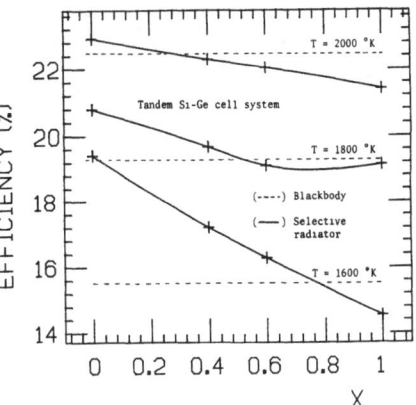

Fig.10: Calculated total conversion efficiency ($\eta_T=\eta_{Si}+\eta_{Ge}$) as a function of oxide composition, $(Yb_2O_3)_x(Er_2O_3)_{1-x}$, for optimal Si and Ge structures for three radiator temperatures (T=1600,1800,2000°K).

† Present address: Spire Corporation, Patriots Park, Bedford, Massachusetts 01730, U.S.A.

LAMBERTIAN ANALYSIS OF MIRRORS AND FRESNEL LENSES FOR SOLAR CONCENTRATION

A. Luque, E. Lorenzo
Instituto de Energía Solar - E.T.S.I.T. - Universidad Politécnica Madrid

Summary

Concentrators for extended sources usefull for tracking and quasi-static applications are analysed, based on the idea that maximum energy on the collector is cast if the concentrator illuminated by the source becomes a Lambertian source. It is demonstrated that the achievement of this condition is not compatible with the condition of casting on the collector all the energy entering the concentrator. Different analysis of concentrators are presented based in these ideas.

1. INTRODUCTION

The solar light source must be usually considered as an extended source. Intrinsically it is so because of the finite sun's diameter, the circumsolar radiation and the diffuse radiation. Static concentrators which constitute increasingly interesting devices must concentrate the sun rays into the collector no matter what the sun's position is. A way for their study is to consider the sun as an extended source covering the sky regions which are actually occupied by the sun at some moment during the year, thus benefiting the studies on extended light sources. This study has been performed by us and presented elsewhere (1).

Tracking concentrators must also be designed for extended sources. The main source of angular extension of tracking concentrators is the error-step inherent to the tracking mechanism. In many cases the tracking is driven by an on-off system and in these cases the source can be considered as an homogeneous extended source, particularly if the concentrator is linear. The area of the source (the area in the space p-q of direction cosines of incoming rays) is in this case considerably smaller than in the case of static concentrators. Since we have shown (1) that there is an upper bound for the optical concentration (i.e. the ratio of power in a cell in a loseless concentrator to the power in a cell outside of it), and that this upper bound is inversely proportional to the source area, we conclude that much higher values of the optical concentration can be used with tracking concentrators than with static ones. The same can be said if we design a concentrator for quasi-static ones which only requires seasonal adjustment of elevation angle and because of it constitutes an appropiate solution for high-concentration low-maintenance systems for rural use.

The concentrators designed for extended sources as analysed for the static cases (1), which are related to CPC shapes, become extremely deep when the sourcehas little angular extension, therefore becoming non-practical. Besides, in such deep concentrators many reflections are produced, every one of them implying reflection losses.

To reduce the active reflecting or refracting surface its normal in any point has to be oriented, approximately, towards the source. Under this condition, in most cases the collector cannot be surrounded by the concentrator. Since the production of isotropic illumination on the cell is a condition for achieving the upper bound of optical concentration a lower value of optical concentration than that of deep concentrators is to be expected. A possible solution for this drawback is to employ a second stage concentrator in the place of the collector. In Fig. 1 we represent a typical tracking concentrator using a Fresnel lens or a mirror as first stage and possibly a second stage. Such second stage can be designed considering the first one as a finite Lambertian source according to the theory of Rabl and Winston (2).

In this work we restrict our attention to the first stage of either a lens or a mirror and for simplicity' sake we consider only the case of linear concentrators where the rays are defined for one position coordinate X and a direction coordinate p which are the cosines of the angle of the ray with the concentrator's entry aperture axis OX (or AA' in Fig. 1).

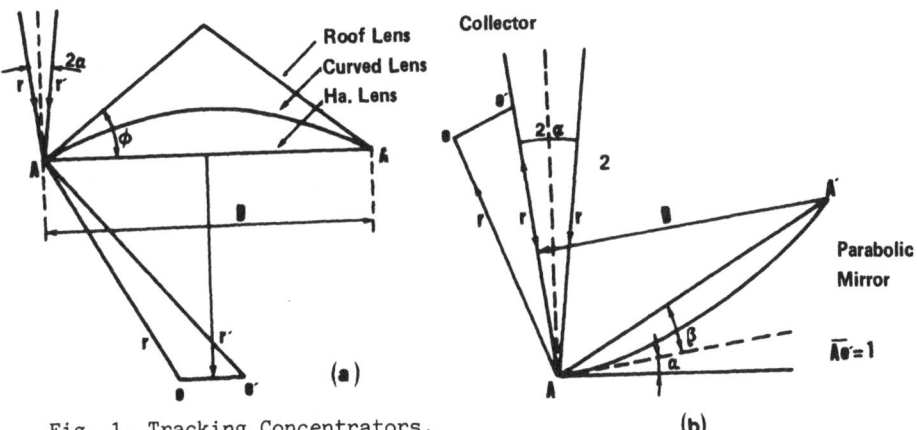

Fig. 1. Tracking Concentrators.

2. LAMBERTIAN ANALYSIS

A basic parameter of a concentrator is the directional intercept factor (1) $\alpha(p)$ which is defined as the ratio of the power cast into the collector (or second stage aperture) or a loseless concentrator from the direction p to the power entering its entry aperture from the same direction. Its average value, for a source between p_a and $p_o + \Delta p$ is

$$I = \frac{1}{\Delta p} \int_{p_o}^{p_o + \Delta p} \alpha(p) \, dp \quad (1)$$

is the intercept factor or ratio of power in the cell to power entering the loseless concentrator from an extended source. Obviously $I \leq 1$. If $I = 1$ the concentrator is called ideal. It has been shown (2) that the intercept factor relates the optical gain C_o with the geometrical gain or ratio of entry aperture area to collector area C_g through the equation

$$C_o = I \, C_g \quad (2)$$

Therefore ideal concentrators are those with smallest entry aperture for a given optical concentrator.

The ideality condition insures that no power entering the concentrator is lost but it does not insure that the highest power has been cast on the collector (or second stage aperture). In fact we have demonstrated that the maximum power on the collector is achieved when the first stage is illuminated in such way that it is seen by the collector as if it were a Lambertian source of the same radiance R than the infinite source. Many concentrators cannot achieve this limit and a shape quality factor Q is defined as the ratio of the actual power in the collector to the power of the Lambertian source of the same size. We have demonstrated that this factor can be calculated as

$$Q = \frac{1}{E_L} \int_{p_o}^{p_o+\Delta p} \alpha(p) \, dp \qquad (3)$$

where according to an elegant demonstration by Winston and Weldford (4) E_L, which is the power cast on the collector OO by the Lambertian source of radiance unity leaning on the points AA', is given by

$$E_L = \overline{OA'} + \overline{O'A} - \overline{OA} - \overline{O'A'} \qquad (4)$$

The factor Q is also the ratio of the optical gain C_o to its maximum usue value C_{om} for concentrators leaning in AA'.

In the case of mirrors equation (2) can also be written in the form

$$Q = \int_{p_o}^{p_o+\Delta p} \alpha(p) \, dp \Big/ \int_{-1}^{1} \alpha(p) \, dp \qquad (5)$$

Equations (1) and (5) give an interesting pictorial representation of Q and I valid for mirrors. For that let us represent $\alpha(p)$ as in Fig. 5. The concentrator will be Lambertian (Q = 1) and ideal (I = 1) if $\alpha(p)$ has a square shape. This is usually not achieved and real concentrators have a more complex shape. The dotted area within the square and outside $\alpha(p)$ is proportional to the departure from the ideal condition (I < 1) and the dashed area outside the square and inside $\alpha(p)$ is proportional to the departure from Lambertian quality (Q < 1).

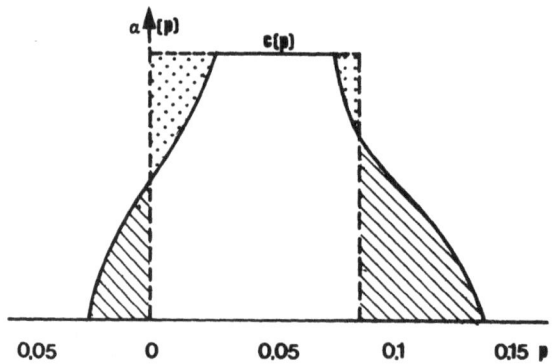

Fig. 2. The directional intercept factor of a typical parabolic mirror with some shape error and 5 degrees of acceptance angle.

In general Lambertian concentrators can be achieved with mirrors but not with lenses. The specific way for obtaining Lambertian concentrators is outlined elsewhere (3) but a complete development of these concepts in still pending.

Of course a Lambertian first stage is the one casting maximum power on the collector (second stage) among these lying in the same points AA'. If AA' is bigger, so that more isotropic illumination is achieved in the collector, the optical concentration becomes higher provided that the concentrator will remain Lambertian. When a two stage concentrator is used and the first stage is Lambertian, the combined optical gain of both stages achieves the upper limit of concentration as it does in classical Winston's CPC's but possibly these two stage concentrators require less optical surface than CPC's, so becoming more practical for concentrators of high concentration adapted to sources of small area.

Concentrators cannot simoultaneosly be Ideal and Lambertian. For that, let us restrict our analysis to symmetric concentrators like those of Fig. 1(a) and let us consider the locus of the points AA' such that the geometrical gain G_g of such concentrators equals their optical gain when considered Lambertian G_L. This locus is well known and can be drawn. It divides the plane into two regions. All the concentrators built leaning in the outer region have an optical gain that at most reaches maximum value wich corresponds to the Lambertian case and therefore they have $G_g > G_o$ so that $I < 1$ and they are not ideal. We have been able to found Lambertian concentrators in this outer region (3).

On the other hand, concentrators in the inner region have a geometrical gain and, therefore, an optical gain below the Lambertian value, i.e. they are not Lambertian and $Q < 1$. Ideal concentrators have been found (5)(6)(7) in this inner region.

3. MIRRORS AND LENSES FOR HIGH CONCENTRATION

Ideal concentrators can be designed approaching as much as possible the Lambertian conditions. We call this type of concentrators ideal quasi-Lambertian concentrators. We have developed a quite complete study of mirrors and Fresnel lenses of these characteristics (5)(6)(7). Among the lenses, three types have been studied: flat lenses, roof shaped lenses and curved lenses, whose outline give highest Q (the details of construction of these lenses are given in the corresponding references). We represent in Fig. 3 the value of the optical gain (since they are ideal that is also the geometrical gain) compared to the thermodynamical limit of the gain (see Eq. 8 for q = 1) for the corresponding source. The shape quality factor is also represented. The shape quality factor is also the ratio of optical gain to its themodynamical limit for a two-stage concentrator. The abscissa is in all cases the luminosity of the concentrator (1/f-number) i.e., the ratio of entry aperture to distance concentrator-to-collector. We included that the curve shaped Fresnel lens is the best concentrator which almost does not require a second stage to achieve its best performances, and this in spite of the theoretical superiority of the reflecting concentrators. The behaviour of curved shaped Fresnel lenses could only be improved by Fresnel mirrors not yet studied.

Most of the work we have already done refers to ideal quasi-Lambertian concentrators. Another approach is to try to cast highest energy on the collector. This is the most appropriate if the objective of the concentrator is to reduce the area of costly solar cells. For that, Lambertian concentrators are to be used. The feasibility of Lambertian reflecting concentrators has been shown by us (3). It is important to realize that such

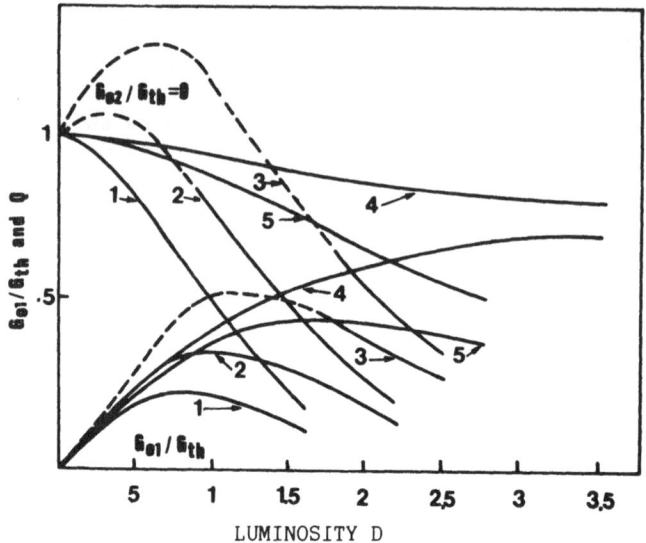

Fig. 3. Ratio of the one stage (G_{o1}/G_{th}) and the two stage (G_{o2}/G_{th}) concentrations to the thermodynamical limit vs. the luminosity factor. G_{o2}/G_{th} is also the shape quality factor Q. (1) for flat lenses, (2) for roof lenses of 30°, (3) for roof lenses of 50°, (4) for curve shaped lenses and (5) for mirrors.

concentrators cast the same energy than today's classical Compound Parabolic Concentrators. Unlike these, our concentrators do not cast all the energy entering into them, but still they cast the thermodynamical limit of concentration if a two-stage concentrator is used. Our concentrator, in some cases, requires less mirror surface than the corresponding CPC and therefore, they can be cheaper. Not much work has been done yet in this direction.

4. CONCLUSIONS

The Lambertian analysis of concentrators allow for the design of practical quasi-static concentrators and tracking concentrators that either cast all the energy from an extended source reaching the concentrator into the collector (ideal concentrators) or produces maximum energy casting into the collector (Lambertian concentrators); both conditions cannot be simultaneously reached. For ideal concentrators the characteristics of several types of concentrators are compared.

REFERENCES

1 A. Luque, Solar cells, 3, 355, (1981).
2 A. Rabl and R. Winston, Applied Optics, 15, (1976), 2880.
3 A. Luque and J.M. Gómez, Applied Optics, 20, (1981), 4193.
4 R. Winston and W.T. Welford, J. Opt. Soc. Am., 68, 289, (1978).
5 A. Luque, Appl. Opt., 19, 2398, 1980.
6 E. Lorenzo and A. Luque, Appl. Opt. 20, (17), 2941-2945 (1981).
7 E. Lorenzo and A. Luque, Appl. Opt. in press (May 1982).

750 SUNS CONCENTRATOR MODULES USING GaAs SOLAR CELLS

E. Fanetti, C. Flores, G. Guarini, F. Paletta
CISE S.p.A. - PO Box 12081, 20100 Milan (Italy)

Summary

In order to make experience on high concentration and check the maturity of GaAs technology, two small size modules have been fabricated. In the first module, single GaAs cells, Fresnel lens 750 suns concentrators and passive cooling are used while in the second one a spectrum split approach and active cooling are applied in order to increase the conversion efficiency. In this last case, GaAs (1.43 eV bandgap) cells are combined with GaAlAs (1.7 eV bandgap) cells and optical filters transmitting the high energy part of sunlight to the high bandgap cell and reflecting the low energy part to the other one. The characteristics of the principal components such as GaAs and GaAlAs solar cells, Fresnel lenses, spectrum splitting optical filters and heat dissipators, will be described in the present paper as well as the performances of the two photovoltaic modules.

1. INTRODUCTION

In photovoltaic power stations the contribution of Balance of System costs, including ground preparation, materials, support structures, wiring etc., will be determinant and hardly reducible. This evidences the importance of using very high efficiency cells which, in turn, may be expected to require highly sophisticated and expensive technology.

A very high concentration of solar energy may be therefore requested both for a drastic cost reduction of high efficiency cells and an efficient recovery of the thermal energy dissipated in hybrid systems. GaAs and GaAlAs solar cells have been therefore designed and fabricated for operation at very high solar fluxes and characterized up to 1000 suns.

Presently, the combined use of different band-gap cells with spectrum splitting optical filters is the only available method capable of achieving cell conversion efficiencies higher than 25% (1,2).

In this paper two small size 750 suns concentrator modules, employing GaAs cells in one case and combined GaAlAs and GaAs cells in a spectrum splitting configuration in the second case, are described. Even if the couple Si-GaAlAs has to be preferred because of higher potential efficiency, the module has been fabricated using GaAs-GaAlAs couples because both GaAs and GaAlAs cell are well suitable for operation at temperatures higher than 100°C and could therefore been used in a hybrid system producing hot water as well as electrical energy.

2. GaAs AND GaAlAs SOLAR CELLS

1.37 cm^2 GaAs or $Ga_{.8}Al_{.2}As$ LPE grown solar cells have been used for both modules. In each case the contact grid touches the p-type active layer through grooves selectively etched in the $Ga_{.1}Al_{.9}As$ window layer and p-n junction is obtained by diffusion of beryllium during the epitaxial growth of the p-type window layer (3). The thickness of the window layer is less than 0.5 μm in order to improve the cell response in the UV spectral region

An n-type tin doped buffer layer is grown on top of a commercial GaAs substrate in order to improve the cell characteristics. The cross section of the 1.7 eV gap $Ga_{.8}Al_{.2}As$ cell is shown in fig. 1 while the 1.43 eV gap GaAs cell had a similar structure with GaAs in place of $Ga_{.8}Al_{.2}As$ layers.

For both types of cells, the grid configuration shown in fig. 2, was designed by a computer modelling program taking into account the specific contact resistance, the sheet resistance, the grid thickness as well as the bell shaped light intensity distribution on the cell.

Ohmic contact were obtained by vacuum evaporation of AuGeNi on the substrate and AgZnAg on the top p side. A low cost process for ohmic contacts on GaAs cells, that does not require any vapour deposition cycle, was also developed. The process is based on the electroless deposition of palladium as p-type contact, while electrolytic gold is used as n-type contact.

Indoor I-V measurements were performed by means of a solar simulator up to 100 suns and by flash technique (4) up to 800 suns. Outdoor measurements were extended up to 1000 suns or to higher concentration by means of Fresnel lens concentrators. The best conversion efficiencies measured for GaAs and GaAlAs cells are shown in fig. 3 as a function of the concentration ratio.

3. OPTICAL COMPONENTS

For both modules, the same type of Fresnel lenses, supplied by Tokyo Plastic Lens Co., were used. The external dimension were 40x40 cm^2, the focal length 56 cm., the thickness 3 mm and transmission 80%.

The spatial distribution of the light transmitted by the lens was also measured by means of a small cell, cooled and mounted on a translation unit. A satisfactory uniformity was found on the cell plane.

The optical filters, used in the spectrum split module, consisted of a sequence of 20-33 dielectric layers with high and low refractive index (cryolite and zinc sulphide) alternately deposited on a transparent substrate by vacuum evaporation or sputtering. The diameter was 8 cm, the cutoff wavelength 0.73 μm for an incidence angle of 45°. The high energy part of the solar spectrum was transmitted and the low energy part reflected to the low bandgap cell. Filter optical losses could be evaluated of about 10-15%.

The best efficiency measured on small areas GaAlAs and GaAs cell combined with a dicroic filter has been 23% at 840 suns.

4. THE CONCENTRATOR MODULES

The characteristics of the two concentrator modules are summarized in table I.

GaAs	MODULE CHARACTERISTICS	GaAs + GaAlAs
Fresnel lens	Concentrator	Fresnel lens
Square	Lens geometry	Square
80%	Lens efficiency	80%
10	Number of lenses	6
1.46 m^2	Lenses area	0.87 m^2
LPE	Cell technology	LPE
Circular	Cell geometry	Circular
1.37 cm^2	Cell exposed area	1.37 cm^2
750	Concentration ratio	750
Passive	Cooling	Active
12%	Module efficiency	15%
75°C	Cell temperature	45°C
175 W	Module peak power	130W

Table I Characteristics of the two concentrator modules.

Both modules include a support structure with a two axis tracking system. The GaAs module (fig. 4) contains 10 Fresnel lenses, each focusing the sunlight on a GaAs solar cell mounted by expoxy resin on a passive dissipator. The thermal resistance was 0.63°C/W for each cell operating at a temperature of 75°C.

The advantage of avoiding water circulation, pumping and energy consumption (as for an active cooling) was payed in terms of 1 percentage point lost in cell conversion efficiency. All GaAs cells were connected in series, giving a total Voc of 10.6 V and I_{SC} of 17.2 A with a fill factor of 0.66 for an illumination of 72 mW/cm^2.

In the spectrum split module, containing only 6 Fresnel lenses, the light concentrated by each lens was directed on to a spectrum splitting optical filter positioned at 45° with respect to the lens plane (fig. 5)

The high energy part of the solar spectrum is transmitted, through the filter, to the high band-gap (1.7 eV) GaAlAs cell while the low energy part is reflected to the GaAs low band-gap (1.4 eV) cell. Both cell are symmetrically placed around the filter on adjustable supports, so that the cell position could be corrected in order to maximize the current output.

Each cell is mounted on a molybdenum base, having a good thermal expansion match with the cell, and electrically isolated by a thin alumina layer. The active cooling is supplied by a closed loop hydraulic circuit consisting of three parallel branches and a heat exchanger dissipating to the air the thermal energy transferred by the cooling fluid.

A circulation pump is series-connected to the cooling loop, together with a flow-switch driving the tracking system. The module will be misoriented and sunlight will not be focused on the cells, each time an incorrect operation of the cooling circuit will be detected.

The electrical output of the module is connected to a maximum power track (MPT) circuit in order to maintain the maximum power output down to sun power levels of about 1/10 of the peak power ($1KW/m^2$). The cells are connected in series and protected by a diode for each cell. Photograph of the module is shown in fig. 6.

Under sun illumination of 74 mW/cm^2, preliminary test performed at 750 suns indicated an average current of 9.35 A for the GaAs cells and of 8.1 A for the GaAlAs cells in spectrum splitting configuration. GaAs cells were series connected together as well as GaAlAs cells forming two indipendent raws with an overall efficiency of 15%. In a higher power version of this module, the two raws could be connected in parallel compensating the difference in voltage output by the insertion of one additional GaAs cell every seven GaAlAs cells.

By a proper adjustment of relative cell and filter positions an acceptable current matching between GaAs and GaAlAs cells could also be obtained and all the cells could be series connected with minimum efficiency reduction (from 15% to 14.2%)

5. CONCLUSIONS

The feasibility of very high concentrator modules using GaAs and GaAlAs solar cells has been demonstrated both with a single cell and tandem cell configurations.

Two or three additional percentage efficiency parts would have been gained if the concentration ratio were reduced to 400 suns, but the experience achieved on these very high concentration modules is thougth to be worth for several possible high efficiency tandem or hybrid photovoltaic systems.

REFERENCES

1) H.A. Vander Plas et al. Proc. P.V. Solar Energy Conf. Berlin. April 1979 p. 507
2) P.G. Borden et al. Proc. 15th IEEE P.V. Spec. Conf. May 1981 - p.311
3) E. Fanetti et al. Solar Cells 3 (1981) 187.
4) E. Fanetti - Electronic Letters 17 (1981) 469.

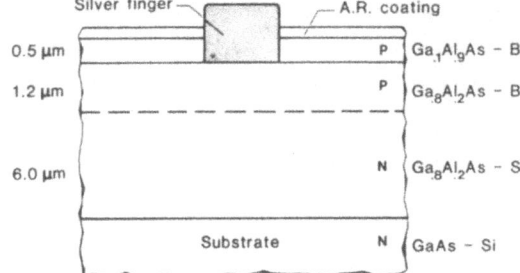

Fig. 1 - Cross section of the GaAlAs cell.

Fig. 2 - Top view of 1.37 cm^2 concentrator solar cell.

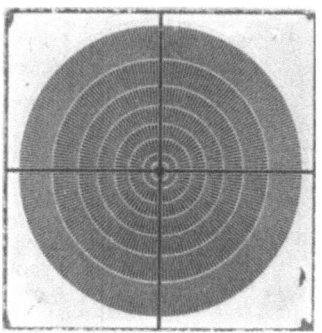

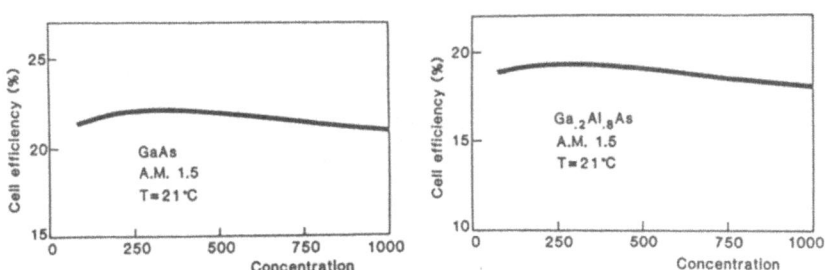

Fig. 3 - Experimental effciencies of GaAs and GaAlAs cells.

Fig. 4 - 175 Wp GaAs concentrator module.

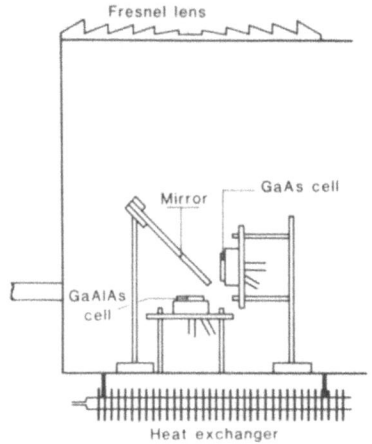

Fig. 5 - Spectrum splitting set-up with optical filter and two mounted cells.

Fig. 6 - 130 Wp spectrum splitting module.

A 500 W_{PK} PHOTOVOLTAIC CONCENTRATOR
USING A GLASS LAMINATED METAL MEMBRANE REFLECTOR

W. HAAF, Schlaich + Partner, Stuttgart
K. HAGENLOCHER, Zeppelin Metallwerke, Friedrichshafen

Summary

A 3 m diameter parabolic dish has been constructed. The reflector shape has been chosen such as to give a most homogeneous intensity distribution for a geometric concentration ratio of C = 50. As converter a C50 generator using new monocrystalline Si concentrator cells has been developed. The system is placed in polar position and tracked by driving one axis. The generator drives a consumer water pump and feeds two batteries as a buffer storage for tracking, closed water circuit cooling, and electronic control. Outdoor tests have been started.

THE SYSTEM

The system (Fig. 1) is the prototype of a solar concentrator for decentral electric power generation below 1 kW. This prototype has been designed to match the requirements of a Si C50 photovoltaic generator. The time axis is placed in polar position (depending on latitude), tracking is done by electronic control. The cooling operates by a closed water circuit; cooling, tracking and electronic control are fed by a battery buffer, which, in turn, gets energy from one of the four generator strings when a certain lower load level is reached. At maximum power output, the consumer pump has a flow rate of 20 m³ per hour at a lift level of 3.5 m. Some technical data are listed on the following page.

Fig. 1: Operating system

Table 1.): Technical data
(for I_o = 1000 W/m² and T_{Gen} = 40° C)

Reflector diameter	300 cm
Mirror reflectivity	0.91
Tracking accuracy	0.15°
Intercept factor	0.96
Losses by covering:	
a) glass with antireflex treatment	0.04
b) pexiglass	0.08
Max. diameter of active area	42 cm
Active absorbing area	1100 cm²
Geometric concentration ratio	50
Shading factor	0.79
Total number of cells	176
Number of strings	4
Generator operation temperature	40-45° C
Si cell efficiency	0.11
Matching losses	
a) 1. reflector version	0.97
b) 2. reflector version	0.80
Generator voltage	26.5 V
System efficiency a)	0.07
b)	0.06
Power output a)	495 W
b)	420 W
Power supply for tracking ⎫ discontinuous	12 W
Power supply for cooling ⎭ operation	70 W
Consumption per day - tracking	40 Wh
- cooling	140 Wh
Battery capacity	2 x 88 Ah
Power consumption of consumer pump (at 1500 rpm)	430 W
Consumer pump flow rate (h = 3.5 m)	20 m³/h

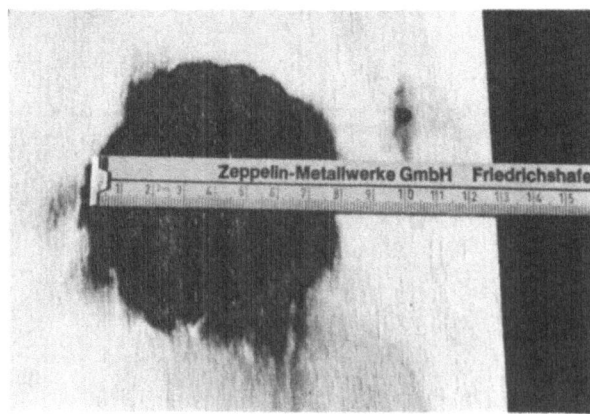

Fig. 2:
Focus image at
f = 292 cm
burned into wood,
corresponding to a
geometric concentration ratio of
C = 1400

THE REFLECTOR

The shape of the metal membrane is roughly a parabolic one rectified in the sun regions in order to yield a most homogeneous intensity distribution at the receiver plane which, for C = 50, is placed at 256 cm in front of the mirror center. The metal membranes are laminated with mirror glass segments in a compound of high durability. The surface slope of the membrane is of high accuracy (surface errors: 0.2 mm RMS) and may be chosen within certain limits in order to match other receiver bodies and higher concentration ratios. The geometric concentration ratio of the burning spot shown in Fig. 2 may be calculated to be C = 1400.

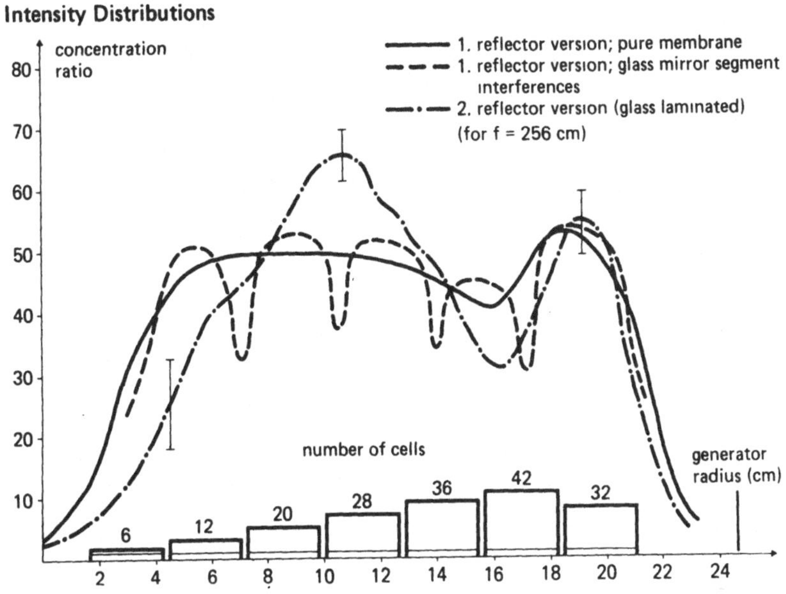

Intensity Distributions

——— 1. reflector version; pure membrane
– – – 1. reflector version; glass mirror segment interferences
—·— 2. reflector version (glass laminated)
(for f = 256 cm)

Fig. 3: Intensity distribution at f = 256 cm

In a first membrane version, the intensity distribution turned out to be as wanted for the membrane itself, however, when laminating the membrane with mirrors, a disturbing periodically changing pattern was superimposed, the origin of which were periodic vaultings in the 25 x 25 cm^2 mirror segments. This problem could be overcome in a second version with 10 x 10 cm^2 segments (this time, however, including an unfortunate mistake in the membrane shape itself leading to a somewhat more inhomogeneous distribution). The corresponding intensity distributions are shown in Fig. 3*). The experiments show that, in future versions, the reflector performance will allow for matching losses smaller than 2 % using photovoltaic converters. In Fig. 4, the intensity pattern is shown on a PV panel hold onto the generator covering. The images of the shadows of both, the panel and the operating person, may be recognized.

*) All measurements carried through by the Institut für Theorie der Elektrotechnik der Universität Stuttgart.

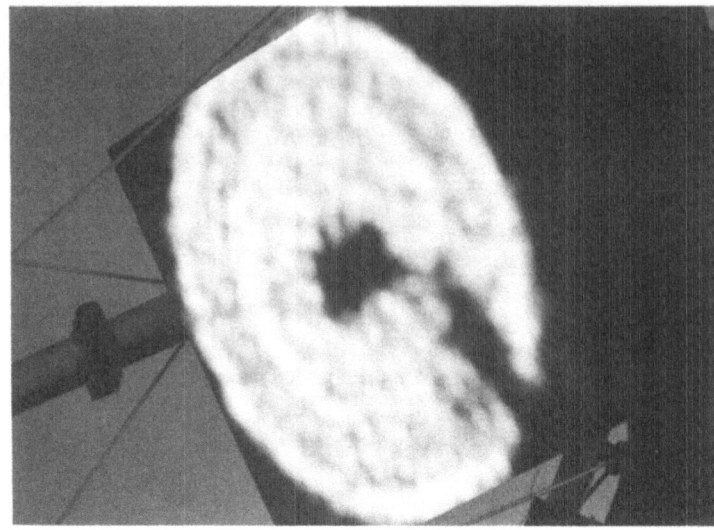

Fig. 4:
Intensity pattern on a PV panel. The dark spots are the images of the shadows of the panel and of the operator

THE GENERATOR

A new C50 photovoltaic generator has been developed by AEG TELEFUNKEN using monocrystalline Si concentrator cells with high grid density. 176 cells 2.5 x 2.5 cm² are elastically attached onto the copper cooling body; a rather good heat conductivity could be achieved. The cells are welded in series to give 4 strings, each yielding 26.5 V at the operating temperature T = 40° C of the generator. The generator is protected with a air-tight glass covering with antireflex treatment (4 % losses). Fig. 5 shows a picture of the generator in operation. Fig. 6 shows some J-V curves measured during operation.

THE COOLING SYSTEM

The generator is cooled by a closed water circuit system at a flow rate of 800 l per hour. The cooling water enters the back of the copper cooling plate at its center and is distributed all over the surface in radially placed channels. Homogeneous water flow in all directions independent from the generator position is reached by a pressure screen at the outlet of these channels. Cooling back to pumped (consumer) water temperature or ambient temperature is done by a standard truck radiator. Two different operation modes are possible: 1) permanent operation resulting in a generator temperature of + 9° C above the consumer water temperature, 2) automatic interval operation at a selected and optimized upper generator temperature level. During first outdoor test, this upper level was chosen to T_{Gen} = 40° C resulting in a daily cooling pump power consumption of 140 Wh.

Fig. 5: AEG C50 generator in operation

SYSTEM OPERATION AND ELECTRONIC CONTROL

The time axis is tracked by electronic control in steps of $\pm$ 0.12°. The declination has to be adjusted by means of an integrated pointing device. At sunset, the reflector goes back to sunrise position within 25 min and automatically starts again in the next morning. In the emergency case (e.g. if the generator temperature exceeds 85° C) the generator turns out of the focal plane within 20 s. The consumer pump has been selected such as to match optimally the J-V characteristics of the generator. At maximum power the flow rate is about 20 m³ per hour at a lift height of 3.5 m. The charging control connects one of the four strings to the batteries when the charging level becomes lower than 80 %. Operation is possible even in cloudy periods for at least ten days.

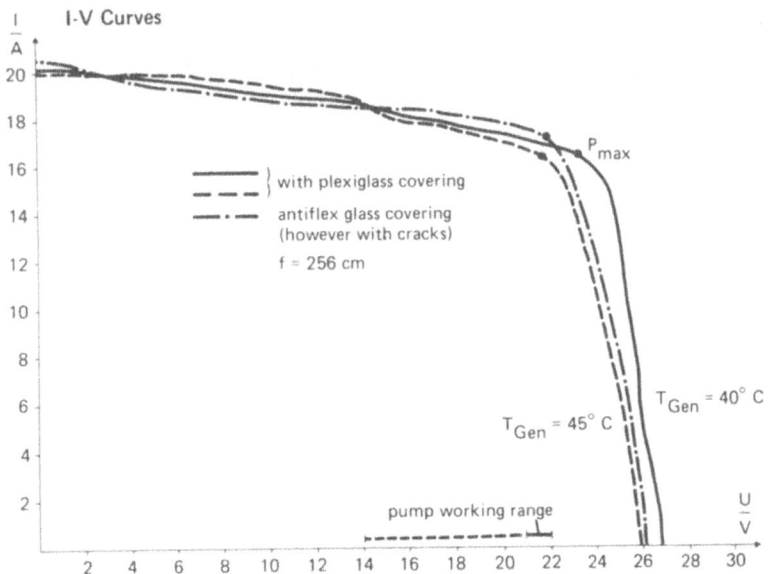

FLUORESCENT PLANAR CONCENTRATOR (FPC)
MONTE-CARLO COMPUTER MODEL
LIMIT EFFICIENCY AND LATEST EXPERIMENTAL RESULTS

K. Heidler, A. Goetzberger and V. Wittwer
Fraunhofer-Institut für Solare Energiesysteme
Oltmannsstr. 22, D-7800 Freiburg, W-Germany

Summary

A Monte-Carlo computer model for simulation and optimization of fluorescent concentrators is described. The program calculates absolute efficiencies "ab initio"; agreement between simulated and experimental data is good. Power flow diagrams and optimization characteristics for single sheet collectors and collector stacks are discussed. It is shown that stacking of up to three sheets does not degrade the concentration ratio. For a "realistic-ideal" three-sheet collector stack, 9 % electrical efficiency at an optical concentration ratio of 10 is derived as a conservative upper limit for the conversion of solar radiation into electrical energy by a photovoltaic FPC-system. Presently the best experimental values for a right angle triangle with 400 cm² area and 3 mm thickness are 9 % optical and 2 % total electrical efficiency with Si-solar cells. The concentration ratio at the output edge is 2.9. For a three sheet stack an electrical efficiency of 3.1 % could be reached.

1. INTRODUCTION

Development of a highly efficient fluorescent concentrator is mainly a problem of dye-chemistry. As a consequence many dyes have to be tested. Most of them are ruled out during pre-examinations but a certain number of candidates must still be tested in collectors. In addition dye concentration and collector geometry have to be optimized for each individual dye. Performing these investigations experimentally is not only extremely time consuming but is also unsatisfactory from the point of view of a thorough physical understanding. Hence the need for a theoretical collector model arises. Many analytical approaches have been set up to describe the collector (1), (2), (3), (4), (5) but either they use semi-empirical approximations or accuracy is poor. The same applies for (6) where the statistical random walk method is used. The main reason for these difficulties is the interdependence of statistical processes in the collector. Therefore the pure statistical Monte-Carlo method was chosen to build a computer model of the FPC. Its structure, comparison with experiment and some applications will be described in this paper, followed by the latest experimental results.

2. MONTE-CARLO COMPUTER MODEL (*CARLOS*)

A simplified flow chart is shown in Fig. 1. Incorporated physical processes are reflection, refraction, absorption, emission, reabsorption, total reflection, back-reflection and loss cone recycling. Not included are matrix extinction, total reflection losses and losses due to imperfect coupling of the solar cell. The fate of each photon is calculated individually, whereby the statistical nature of the processes is simulated by a "random generator" subroutine.

The program *CARLOS* calculates absolute efficiencies "ab initio". Input are the experimental parameters solar spectrum, dye absorption and emission spectra, dye quantum yield, dye concentration, collector length and thickness, refractive index and back reflector characteristics. Output is a detailed analysis of collector processes including absolute values of internal and external efficiencies and concentration ratios, information about reabsorption and total reflection, spectral and angular distributions of fluorescence at the collector output and in the loss cones. The program simulates and optimizes single-sheet collectors as well as collector stacks of up to three sheets.

Agreement between experiment and simulation is good: absolute experimental efficiencies are reproduced within typically 10 - 15 %, spectral data within 1 % (Fig. 2).

3. MONTE-CARLO RESULTS

The following definitions are used throughout the paper:

electrical efficiency $\eta_{el} = \frac{\text{electrical output power}}{\text{power of total input light}}$

optical efficiency $\eta^p_{opt} = \frac{\text{power of output light}}{\text{power of total input light}}$

internal efficiency $\eta^p_{int} = \frac{\text{power of output light}}{\text{power of absorbed light}}$

optical concentration ratio $C^p_{opt} = \frac{\text{power flux density of output light}}{\text{power flux density of input light}}$

3.1 Single-sheet collector

For comparison and analysis of collectors the following standard conditions apply:
(i) input is an AM1.5 solar spectrum of a clear cool spring day (8)
(ii) optical concentration ratio is 3
(iii) dye concentrations and collector geometry are optimized for $C^p_{opt} = 3$.

As an example two of the best single-sheet collectors with similar optical efficiency are compared. Simplified power flow diagrams are given in Fig. 3. While absorption of LPERO is low and internal efficiency is higher, BA291 has a high absorption and a poor internal efficiency mainly due to radiationless transitions. A typical variation of efficiency with relative molar dye concentration K is shown for several collector thicknesses D in Fig. 4. An efficiency maximum results because high K not only increases solar absorption but also the self-absorption losses. Decreasing D demands increasing K for constant solar absorption, hence self-absorption losses increase with decreasing D. Efficiency levels off for high K because red-shift of transmitted fluorescence is so high that reabsorption probability drops significantly.

3.2 Collector stack

Stacking of collectors combines high solar absorption and separate light collection of different spectral regions. Co-doping of dyes in a single-sheet collector is less efficient because Stokes losses are higher and only one type of solar cell may be used at the output.

The enhanced reflection loss of a stack is partly offset by loss cone recycling. In general stacks are thicker than single-sheet collectors thus reducing the ratio of input area to output area. For the dyes investigated (absorption range 350 to 675 nm) it could be shown that <u>stacking of up to three sheets does not degrade the concentration ratio</u>. For equal concentration ratio, stack efficiencies are higher than the best single-sheet efficiency and decrease of efficiencies with concentration ratio is less pronounced for the stack (Fig. 7). For the cited absorption range 3 sheets seem to be the maximum number. Efficiency gain of additional sheets most probably cannot balance increased reflection losses.

In the following the analysis of the best "real" 3-sheet stack (LPERO// K27//BA291; first dye on top etc) is combined with the investigation of a "realistic-ideal" system (ID2I//ID1H//ID4C). The real system has been optimized for C^p_{opt} = 3, the ideal system for C^p_{opt} = 10. Input for the real system were experimental dye data for which optimum collector parameters have been calculated. For the ideal system, it is only the absorption range, Stokes shift and dye quantum yield of experimental dyes which have been theoretically "improved" within realistic physical limits. In Fig. 6 the power flow diagrams of the two stacks are given. Outstanding features of the ideal system are:
(i) higher solar absorption
(ii) good fitting of individual absorption bands leads to high loss cone recycling rates
(iii) high Stokes shift and hence low reabsorption losses allow for high concentration ratios.

A major problem is poor dye quantum yield in the infrared which affects internal efficiency seriously. With solar cell data from (9) (GaAlAs, GaAs) and (10) (Si) 9 % electrical efficiency and a concentration ratio of 10 were derived as a conservative upper limit for the conversion of solar radiation into electrical energy by a photovoltaic FPC-system.

3.3 Summary of Monte-Carlo results

(i) The Monte-Carlo computer model *CARLOS* simulates and optimizes fluorescent concentrator systems of up to three sheets by means of an absolute "ab initio" calculation. Agreement between experimental and simulated data is good.
(ii) The optimization of collector geometry and dye concentration for collectors with high optical efficiency is not very critical.
(iii) Stacking of up to three plates in the absorption range 350 to 675 nm does not degrade concentration ratio and enhances optical efficiency compared to the best single-sheet collector.
(iv) Improvement of collector performance is mainly a problem of dye chemistry: extension of the absorption range into the near IR, improvement of dye quantum yield and Stokes shift.
(v) 9 % electrical efficiency at an optical concentration ratio of 10 seems to be attainable with a realistic-ideal 3-sheet collector stack.
(vi) Significant reduction of self-absorption losses will have dramatic impact on the concentration ratio while the efficiency gain will be moderate (thermodynamic limit (11)).
(vii) Selective transparency of the FPC makes hybrid systems (windows, greenhouses, transparent roofs, etc.) the most attractive application.

4. EXPERIMENTAL RESULTS

Most experimental measurements were made with right angle triangles (area 400 cm², thickness 3 mm, reflectors on the two shorter edges) and silicon solar cells as active elements for conversion. The best results were obtained with cells specially optimized for concentrated red light by Sandia Laboratories Albuquerque. Table 1 shows the results for two different single collectors and one stack with three stages. The overall concentration ratio for the stack is smaller than for the single plates because the third stage (a 5 mm thick neodymium doped glass plate) was not optimized.

Fig. 11 shows the results of a long-term stability test of one collector of standard size on the roof of our institute. The dye stability has been improved in the last few years. Nevertheless the collector efficiency falls to about 70 % in the first months of weathering due to the production of absorbing but non-fluorescent photoproducts generated during the decomposition of the dye.

Table 1

	Efficiency		Concentration ratio
	optical	electrical	
single plate (1 dye, red)	7.0 %	1.5 %	2.4
single plate (2 dyes, green and yellow)	8.7 %	2.1 %	2.9
stack (3 plates)	13.9 %	3.2 %	1.4

REFERENCES

(1) A. Goetzberger, W. Greubel, Appl.Phys. 14, 123 (1977)
(2) J.A. Levitt, W.H. Weber, Appl.Opt. 16, 10 (1977)
(3) J.S. Batchelder, A.H. Zewail, T. Cole, Appl.Opt. 18, 18 (1979)
(4) A. Goetzberger, V. Wittwer, Festkörperprobleme XIX, Vieweg Braunschweig (1979)
(5) J.S. Batchelder, H.A. Zewail, T. Cole, Appl.Opt. 20, 21 (1981)
(6) R.W. Olson, R.F. Loring, M.D. Fayer, Appl.Opt. 20, 17 (1981)
(7) J.M. Hammersky, D.C. Handscomb, "Monte Carlo methods", Chapman and Hall London, 5th Edition (1979)
(8) K.W. Boer, Phys.Stat.Sol. (a) 40, 355 (1977)
(9) E. Fanetti, C. Flores, G. Guarini, F. Paletta, D. Passoni, Solar Cells 3, 187 (1981)
(10) C.M. Garner, F.W. Sexton, R.O. Nasby, Solar Cells 4, 37 (1981)
(11) E. Yablonovitch, J.Opt.Soc.Am. 70, 1362 (1980)

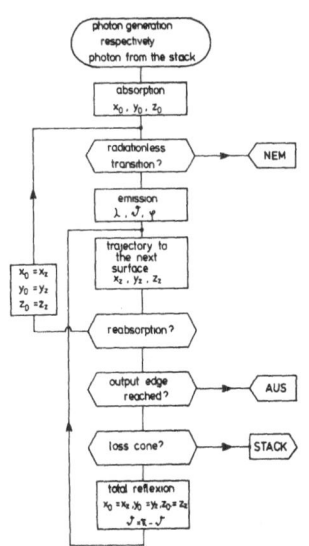
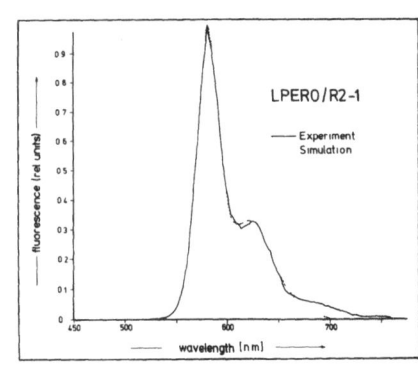

Fig.2: Output spectrum of a typical single sheet collector. Comparison of experiment and simulation

Fig.1: Simplified flow chart of the collector simulation subroutine

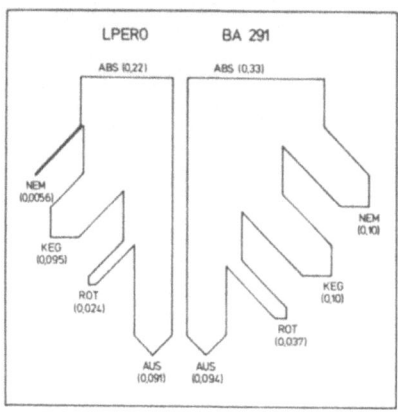

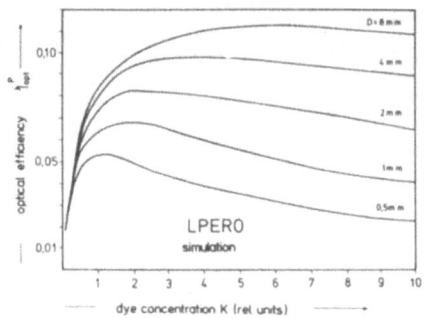

Fig.3: Simplified power flow diagrams of two dyes $c_{opt}^p = 3$ (optimized simulation)
ABS: absorbed
NEM: not emitted
KEG: lost through loss cone
ROT: power loss due to redshift
AUS: output power

Fig. 4: Optical efficiency versus dye concentration for several collector thickness. (simulation)

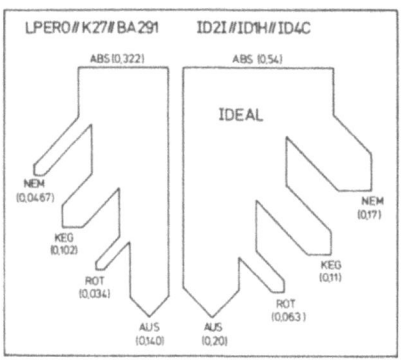

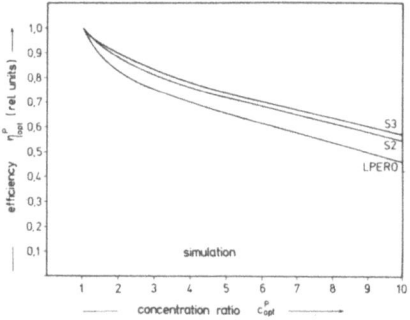

Fig.4: Optical efficiency versus optical concentration ratio for single, two and three sheet stack

Fig.6: Simplified power flow diagrams for simulation of experimental (left, $c_{opt}^p = 3$) and idealized (right, $c_{opt}^p = 10$) stacks
ABS: absorbed
NEM: not emitted
KEG: lost through loss cone
ROT: power loss due to redshift
AUS: output power

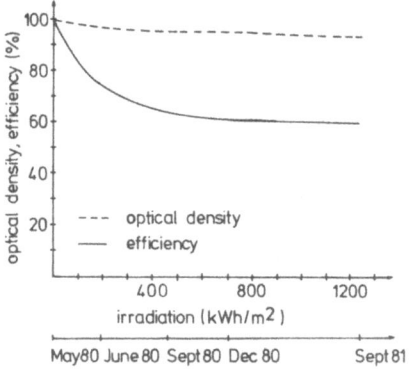

Fig.7: Outdoor stability test (Collector dimension: triangle, 400cm², 3mm thickness)

- 686 -

SESSION 7

THIN FILM SOLAR CELL TECHNOLOGY

Chairmen : Prof. A. FROVA, Istituto di Fisica "G. Marconi", Rome, Italy

Prof. M. SAVELLI, Université des Sciences et Techniques du Languedoc, Montpellier, France

Physical limitations of present thin film solar cells

8% efficiency a-SiC:H/a-Si:H heterojunction solar cells

Amorphous silicon solar cells produced by a consecutive, separated reaction chambre method

Charge collection in a-Si:H solar cells

The effect of glow discharge excitation frequency on the performance of microcrystalline Si:H thin films and devices

Electrodeposited CdS/CdTe heterojunction solar cells

Thin film heterojunction CdS/Cu ternary alloys - solar cells with minority carrier mirrors

Large area CdS/Cu$_x$S thin film solar cells produced by electrophoretic deposition

Sprayed zinc-cadmium sulfide films for backwall $Cu_2S/(ZnCd)S$ cells

PHYSICAL LIMITATIONS OF PRESENT THIN FILM SOLAR CELLS

Y. MARFAING
Laboratoire de Physique des Solides, CNRS 92190 MEUDON - FRANCE

Summary

A comparison is made between the various thin film solar cells by stressing the physical factors which limit their present performances. A general diagram is drawn up on which each cell is localized by two coordinates : one is relative to photocurrent generation, the other is indicative of diode rectifying properties. These aspects are detailed by considering the diode saturation current, the electron-hole pair collection length and the curve filling factor in an unified way which makes use of reference values. The efficiency limitations are thus precisely pointed out and the possibilities of future improvements are discussed.

1. INTRODUCTION

Thin film solar cells are being actively studied in many laboratories as an alternative to crystalline Si cells for reaching the low specific cost required in terrestrial applications.

Since these efforts are expected to come off in 10-20 years all possible technological ways deserve to be now considered. As a matter of fact a number of materials, photovoltaïc structures are being explored, mainly polycrystalline Cu_2S, $CuInSe_2$, $CdTe$, Zn_3P_2, $GaAs$ and amorphous Si : H. The purpose of this review is to make a critical comparison between these different cells considered at their respective stage of technological development.

In a previous EC Conference an overview was given based on economic aspects (1). Here I rather want to make an analysis of present performances in terms of simple and meaningful physical quantities. This leads to a unified description and discussion of the various systems.

2. A COMPARATIVE TWO-DIMENSIONAL DIAGRAM

It is well known that a solar cell is primarily a current generator whose the voltage ouput depends on the rectifying diode properties. Current and voltage are thus the basic quantities representative of two physical aspects : current generation, diode properties. A comparison analysis requires the use of reduced values. We choose the following :

- short circuit current i_{sc} is refered to the maximum AM1 current i_g corresponding to the material band gap E_g. i_g was calculated by several workers [e.g (2)]. We define

$$G = \frac{i_{sc}}{i_g} \qquad (1)$$

G is indicative of the current generation performance.

- product fill factor by open circuit voltage $f \cdot V_{oc}$ is referred to a reference quantity $(fV_{oc})_o$ defined as :

$$i_g (fV_{oc})_o = 10 \text{ mWcm}^{-2} \qquad (2)$$

Thus $(fV_{oc})_o$ appears to represent the minimum voltage

- because i_g is a maximum current-needed to give a somewhat arbitrary 10 % conversion efficiency. We put :

$$D = \frac{fV_{oc}}{(fV_{oc})_o} \qquad (3)$$

D is a relative indication of the diode properties, despite the fact that $(fV_{oc})_o$ is not the maximum achievable value. Indeed the usefulness of the two parameters G and D lies in that the product GD is proportional to the real conversion efficiency of the considered cell.

Every cell can now be represented by a point in a G-D plot on which iso-efficiency curves are hyperbolas (Fig. 1). The data used for this figure are taken from the recent literature and are collected in Table I with the corresponding point numbers. It immediately appears that some cells are efficient current generators close to the fundamental limit : GaAs, CuInSe$_2$ and, to a lesser extent, Zn$_3$P$_2$. Conversely the more recent Cu$_2$S cells have the best relative diodes properties. CdTe is in the middle of the plot and a-Si is presently qualified by both modest current generation performances and diode properties. With regard to the latter Zn$_3$P$_2$ is in the worst situation.

Besides this general picture it is instructive to analyse in more detail the underlying physical limitations. This will be done below in terms of diode saturation current, electron-hole pair collection length and fill factor.

3. DIODE SATURATION CURRENT

The variations in open circuit voltage mainly reflect the variations in diode saturation current i_o according to the formula

$$V_{oc} = n \frac{kT}{q} \ln \left(\frac{i_{sc}}{i_o} \right) \qquad (4)$$

i_o is determined by the mechanisms of injection-recombination in the diode. One can show that the various cases can be covered by a unique relationship of the type :

$$i_o = q N_c S \exp \left(- \frac{E_B}{kT} \right) \qquad (5)$$

S is the recombination velocity of carriers (supposed here to be electrons) injected form the n-side over the barrier E_B (Fig. 2). This situation strictly applies to interface recombination current in Cu$_2$S-CdS heterojunctions. There S is related to the density of interface states and the factor n depends on the doping ratio between the n and p type sides (3).

For the Shockley diffusion-recombination case S is the equivalent diffusion velocity of minority carriers $S = L_n/\tau_n$ and E_B is the height of the minority carrier band with respect to the equilibrium Fermi level

For recombination at the centre of the space charge region n = 2, $E_B = E_g/2$ and S is given as usual by :

$$S = \frac{w}{2\tau_o}$$

where w is the space charge zone width and τ_o is an equivalent lifetime. More generally space charge recombination can occur over a barrier $E_B = (qV_D + \varphi_n)/n$ where V_D is the diffusion voltage, φ_n the electron chemical potential and n is the same factor which enters expression (4).

The saturation current values of the various diodes investigated were analysed along these lines. The input data and the extracted parameters are gathered in Table II. When i_o had not been directly measured it was computed from the photovoltaic parameters using expression (4) and introducing an appropriate n value. n value for a-Si p-i-n junctions was obtained from (4).

For the purpose of comparison a reference saturation current i_o is brought in. It corresponds to the Shockley diffusion current in a strongly doped junction ($E_B \approx E_g$) with an equivalent recombination velocity $S_o = 10^4$ cm/sec. This value is attained for instance in good GaAs p-n junctions where

$$\frac{L_n}{\tau_n} = \frac{3.10^{-4} \text{ cm}}{3.10^{-8} \text{ sec}} = 10^4 \text{ cm/sec.}$$

In that way a reduced saturation current is defined as

$$\frac{i_o}{i_{oR}} = S/S_o \exp\frac{E_g - E_B}{kT} \qquad (6)$$

This relationship is illustrated in Fig. 3 by straight lines corresponding to different values of the ratio i_o/i_{oR}. Each cell investigated is represented by a point on this plot. Several aspects are worth noticing. The Schottky or MIS structures are characterized by high recombination velocity values as expected. Volumic or interface space charge recombination leads to about the same S value of 10^5 cm/sec. This corresponds to an effective space charge lifetime of $10^{-10} - 10^{-9}$ sec. The differences lie in the injection barrier height. The best performance is obtained in the $Cu_2S/CdZnS$ structure due to a better affinity matching compared to Cu_2S/CdS. Note that the high resistivity of thin film $CuInSe_2$ limits the barrier height E_B. Similarly volumic space charge recombination is inherently associated with a low barrier height especially when grain boundaries are present (5). It is clear that the saturation currents in CdTe, GaAs, Zn_3P_2, a-Si are far from the reference values. Any progress would first require an increase in the space charge lifetime. In the case of a-Si this lifetime is related to the band gap state density. Recent studies have shown that the band gap states act as recombination centres with a capture cross section around 10^{-15} cm^2 [6]. Thus a lifetime of 10^{-9} sec corresponds to gap state concentration of 10^{17} cm^{-3} which is indeed the value quoted for present technology a-Si films.

4. ELECTRON-HOLE PAIR COLLECTION LENGTH

The short circuit current attainable in a solar cell depends on the relative values of the optical absorption depth and the electron-hole pair collection length. In materials of interest for thin film devices the optical absorption depth is between 1 and 3 microns to which values the collection length has to be comparable.

The total collection length is the sum of the minority carrier diffusion lengths in the neutral regions and of the field-assisted drift length in the space charge region

$$L_c = L_n + L_p + L_E (1 - \exp - \frac{w}{L_E}) \qquad (7)$$

The last term obviously amounts to the space charge width w when $w/L_E < 1$. Similarly L_n (or L_p) is to be replaced by the junction depth for a thin enough frontal region. The field-assisted drift length has the following form for holes

$$L_E = \frac{L_p}{\sqrt{1 + \frac{q^2 E^2 L_p^2}{4k^2 T^2}} - \frac{q\, E L_p}{2kT}} \qquad (8)$$

where E is the electric field. A limiting expression at high electric field is

$$L_E = \frac{q\, L_p^2\, E}{kT} = \mu_p \tau_p E \qquad (9)$$

The preceding expressions are able to cover all the particular thin film cell cases. A graphical representation is shown in Fig. 4.

The left-hand side of the diagram gives the collection length contribution of the space charge region. In the case of a-Si introducing a hole diffusion length $L_p = 0.1$ μm and an electric field $E = 10^4$ V/cm we obtain a drift length $L_E = 0.4$ μm. For CdTe (and similar materials) the starting value $L_p = 0.3$ μm is already larger than the usual space charge width of 0.2 μm. Therefore the latter is the space charge contribution. The right-hand side of the diagram alllows one to express the collection length as the sum of the field and neutral region contributions, respectively L_E and $L_n + L_p$.

In a-Si Schottky and p-i-n diodes L_c is reduced to $L_E = 0.4$ μm because the contribution of the doped regions is negligible.

In CdTe, Zn_3P_2, GaAs cells L_c is the sum of the space charge width w and of L_n and/or L_p depending on the exact cell structure. The total collection length is close to or over 1 micron.

In CdS-type heterojunctions the collection volume is limited to the low band gap material - either Cu_2S or $CuInSe_2$. In Cu_2S cells the collection length is only L_n as the space charge region is in CdS. In high resistivity $CuInSe_2$ cells there is in addition some space charge contribution. Nevertheless we have seen that a high resistivity state is detrimental to the diode properties because of the low barrier height.

This analysis points up an inherent limitation of CdS heterojuncton cells for current generation. Only one neutral region contributes to the latter which reinforces the requirement for a large diffusion length. It appears that the present electronic properties of Cu_2S which set the diffusion length at about 0.5 μm are a real limiting factor.

5. CURVE FILLING FACTOR

In an ideal photovoltaïc cell the curve factor f is directly related to the dimensionless quantity (qV_{oc}/nkT) as first shown by Shockley and Queisser (7). This relationship is shown in Fig. 5 with the related curves 0.9 f and 0.8 f. The representative points for the various thin film cells are plotted using the data of Tables I and II. The actual f factors are reduced with respect to the theoretical values by about 10 to 20 %.

The possible origines fo this loss are two fold : series resistance and voltage-dependent photocurrent. The last effect is present in the small diffusion length materials (a-Si, CdTe) where the space charge contribution is significant but also in CdS heterojunctions where interface recombination varies with the interface field (3) (8).

The effet of series resistance is important in Schottky barriers (Zn_3P_2, CdTe), high resistivity materials (a-Si) and more generally in cells of non mature technology (such as the polycristalline GaAs junctions).

6. CONCLUSIONS

The conversion efficiency of present thin film solar cells lies in the range 6-10 %. Further progress require a technological effort but also a good understanding of the physical limitations. From this point of view the various thin film cells call for different particular appreciations which suggest corresponding ways of improvement :

- Cu_2S/CdS and $CuInSe_2$/CdS heterojunctions have basically good diode properties, because space charge recombination is eliminated in this type of structure. The junction barrier height increases when the resistivity of the low band gap material decreases.

But in the same time the electron diffusion length decreases what reduces the photogenerated current ; thus an optimum doping is to exist and has to be searched for $CuInSe_2$. As Cu_2S is degenerate p type the question arises whether this optimum could be ever obtained in that material.

Indeed the diffusion length seems to be too short to give a high collection efficiency.

- all n/p and n-i-p homojunctions are presently dominated by space charge recombination currents which limit the attainable photovoltage. That kind of recombination is sensitive to all types of defects including killing impurities, grain boundaries, dangling bonds and disorder associated gap states.

To reduce the concentration of those defects demands strong efforts both in technology and fundamental understanding. Covalent materials such as GaAs, InP, Zn_3P_2 appears to be more affected by the polycrystalline state. Also it is not clear at present whether the gap state density in a-Si : H could be decreased below the observed value of $10^{16} - 10^{17} cm^{-3}$ which corresponds to a recombination lifetime around 10^{-9}sec. By comparison longer lifetimes are encountered in some polycrystalline materials such as Zn_3P_2 and GaAs. An advantage of amorphous materials is the ease with which complex structures can be realized : window heterojunction [25] , tandem cells [9].

- small diffusion length materials can be used in structures where carrier collection occurs partly or totally by drift in a field region. However such arrangement leads to a voltage dependent photocurrent and consequently to a reduced filling factor.

This aspect would deserve more attention and special experiments could be conducted to evaluate this loss as done for Cu_2S/CdS [10]

Finally while all the present systems have potentialities of improvement it is difficult to predict which one will first reach a decisive 12 % conversion efficiency.

REFERENCES.

[1] A.M. BARNETT, 2nd EC Photovoltaïc Solar Energy Conference D. Reidel, Dordrecht (1979) p. 37.
[2] M. WOLF, Proc. IRE, 48, 1246 (1960).
[3] A.ROTHWARF, J. PHILLIPS and N. CONVERS WYETH, 13th IEEE Photovoltaïc Specialists Conf. (1978) p. 419.
[4] R.A. GIBSON, P.G. Le COMBER and W.E. SPEAR, Appl. Phys. 21, 307 (1980).
[5] J.G. FOSSUM, R.D. NASBY and S.C. PAO, IEEE Trans. Electron. Dev. ED-27 (1980) 785.
[6] T.D. MOUSTAKAS, C.R. WRONSKI and T. TIEDJE, Appl. Phys. Lett. 39, 721 (1981).
[7] W. SHOCKLEY and H.J. QUEISSER, J. Appl. Phys. 32, 510 (1961).
[8] G. BORDURE, M.O. HENRY, J.L. JACQUEMIN and M. SAVELLI, 2nd EC Photovoltaïc Solar Energy Conference D. Reidel, Dordrecht (1979) p. 868.
[9] G. NAKAMURA, K. SATO, H. KONDO, Y. YUKIMOTO and K. SHIRANATA, 9th Amorphous, Liquid Semicond. Conf. J. Physique 42, C4-433 (1981).
[10] L.M. KILGREN, 2nd EC Photovoltaïc Solar Energy Conference D. Reidel Dordrecht (1979)p. 344.
[11] J.D. MEAKIN, Proc. CdS/Cu_2S and CdS/Cu-Ternary Photovoltaïc Cells Subcontractors In-depth Review Meeting, Alexandria V.A. SERI (1980) p. 147-173.
[12] A.M. BARNETT, J.A. BRAGAGNOLO, R.B. HALL, J.E. PHILLIPS and J.D. MEAKIN 13th IEEE Photovoltaïc Specialists Conf. (1978) p. 419.
[13] R.B. HALL, P.W. BIRKMIRE, J.E. PHILLIPS and J.D. MEAKIN, 3rd EC Photovoltaïc Solar Energy Conf. D. Reidel, Dordrecht (1981) p. 1094.
[14] R.A. MICKELSEN and W.S. CHEN, 15th IEEE Photovoltaïc Specialists Conf. (1981) p. 800.
[15] R. MICKELSEN, Cited by D.L. FEUCHT, 15th IEEE Photovoltaïc Specialists Conf. (1981) p. 648.
[16] H. UDA, H. TANIGUCHI, M. YOSHIDA and T.YAMASHITA , Japan J. Appl. Phys. 17, 585 (1978).
[17] G. FULOP, M. DOTY, P. MEYERS, J. BETZ and C.H. LIU, Appl. Phys. Lett. 40, 327 (1982).
[18] G. COHEN-SOLAL, D. LINCOT, M. BARBE, 4th EC Photovoltaïc Solar Energy Conference Stresa (1982).
[19] R.P. GALE, B.Y. TSAUR, J.CC. FAN, F.M. DAVIS and G.W. TURNER, 15th IEEE Photovoltaïc Specialists Conf. (1981) p. 1051.
[20] S. CHU, T. CHU, C. JIANG, C. LOH, E. STOKES and J. YU, 15th IEEE Photovoltaïc Specialists Conf. (1981) p. 1310.
[21] S. CHU, T. CHU, F.S. ZHANG, L. BOOK and J.M. YU, Appl. Phys. Lett. 39 803 (1981).
[22] A. MADAN, W. CZUBATYJ, J. YANG, J. Mc GILL and S.R. OVSHINSKY, 9th Amorphous Liquid Semicond. Conf. J. Physique 42 C4-463 (1981).
[23] D.E. CARLSON, 3rd E.C. Photovoltaïc Solar Energy Conf. D. Reidel, Dordrecht (1981) p. 294.

[24] Reported by J.S. STONE, 15th IEEE Photovoltaïc Specialists Conf. (1981) p. 898.
[25] Y. TAWADA, M. KONDO, H. OKAMOTO and Y. HAMAKAWA, 15th IEEE Photovoltaïc Specialists Conf. (1981) p. 245.
[26] A. CATALANO, V. DALAL, W.E. DEVANEY, E.A. FAGEN, R.B. HALL, J.V. MASI, J.D. MEAKIN, G. WARFIELD, N.C. WYETH and A.M. BARNETT, 13th IEEE Photovoltaïc Specialists Conf. (1978) p. 288.
[27] M. BUSHAN, Appl. Phys. Lett. 40, 51 (1982).
[28] M. BUSHAN, J. Appl. Phys. 53, 514 (1982).
[29] S. WAGNER, J.L. SHAY, K.J. BACHMANN, E. BUEHLER and M. BETTINI, J. Crystal Growth 39, 128 (1977).

n°	Cell	E_g (eV)	i_g mAcm^{-2}	i_{sc} mAcm^{-2}	f	V_{oc} (V)	η(%)	Ref
1	Cu_2S/CdS	1.2	37	22	0.72	0.56	8.8	11
2	Cu_2S/CdS	"	"	24.8	0.71	0.52	9.1	12
3	Cu_2S/CdZnS	"	"	22.8	0.75	0.6	10.2	13
4	$CuInSe_2$/CdS	1.05	44	38	0.63	0.4	9.5	14
5	$CuInSe_2$/CdZnS	"	"	36.2	0.66	0.418	9.9	15
6	CdTe/CdS	1.5	25	20	0.58	0.75	8.7	16
7	CdTe-MIS	"	"	18.7	0.64	0.723	8.6	17
8	CdTe p/n	"	"	18.8	0.68	0.82	10.5	18
9	GaAs/Ge/Si	1.43	28	24.4	0.63	0.76	11.7	19
10	GaAs-MIS	"	"	22.7	0.67	0.56	8.5	20
11	GaAs p/n	"	"	23.7	0.52	0.575	7.1	21
12	a-Si MIS	1.7	20	13.1	0.57	0.88	6.6	22
13	a-Si p i n	"	"	12.5	0.6	0.822	6.2	23
14	a-Si p i n	"	"	11.1	0.66	0.878	6.4	24
15	a-SiC/a-Si	"	"	12.3	0.65	0.887	7.1	25
16	Zn_3P_2 MIS	1.5	25	19	0.64	0.5	6.1	26
17	Zn_3P_2 p/n	"	"	18	0.55	0.48	4.7	27
18	Zn_3P_2 p/n	"	"	13.2	0.46	0.54	3.3	28
19	InP/CdS	1.34	32	13.5	0.68	0.46	4.2	29

TABLE I - Photovoltaic Parameters of thin film solar cells (8 and 16 are single crystal cells).

n°	Cell	i_o Acm^{-2}	n	E_B (eV)	S cm/sec
2	Cu_2S/CdS	4×10^{-11}	1	0.9	3×10^5
3	Cu_2S/CdS	4×10^{-12}	1	0.96	3×10^5
4	$Cu_2S/CuInSe_2$	1.8×10^{-7}	1.3	0.65	10^5
6	CdTe/CdS	1.3×10^{-9}	1.7	0.7	10^5
7	CdTe MIS	1.1×10^{-10}	1.3	0.92	6×10^7
8	CdTe p/n	3×10^{-10}	1.8	0.72	6×10^4
9	GaAs/Ge/Si	3×10^{-9}	1.7	0.71	4×10^5
10	GaAs - MIS	2×10^{-7}	2.2	0.6	3×10^5
11	GaAs p/n	1.5×10^{-5}	3.6	0.43	6×10^4
12	a-Si MIS	4×10^{-11}	1.2	0.82	2×10^6
15	a-SiC/a-Si	7×10^{-13}	1.5	1	10^5
17	Zn_3P_2 MIS	10^{-7}	1.8	0.75	6×10^7
18	Zn_3P_2 p/n	6.4×10^{-8}	1.7	0.76	6×10^7

TABLE II - Dark diode current parameters.

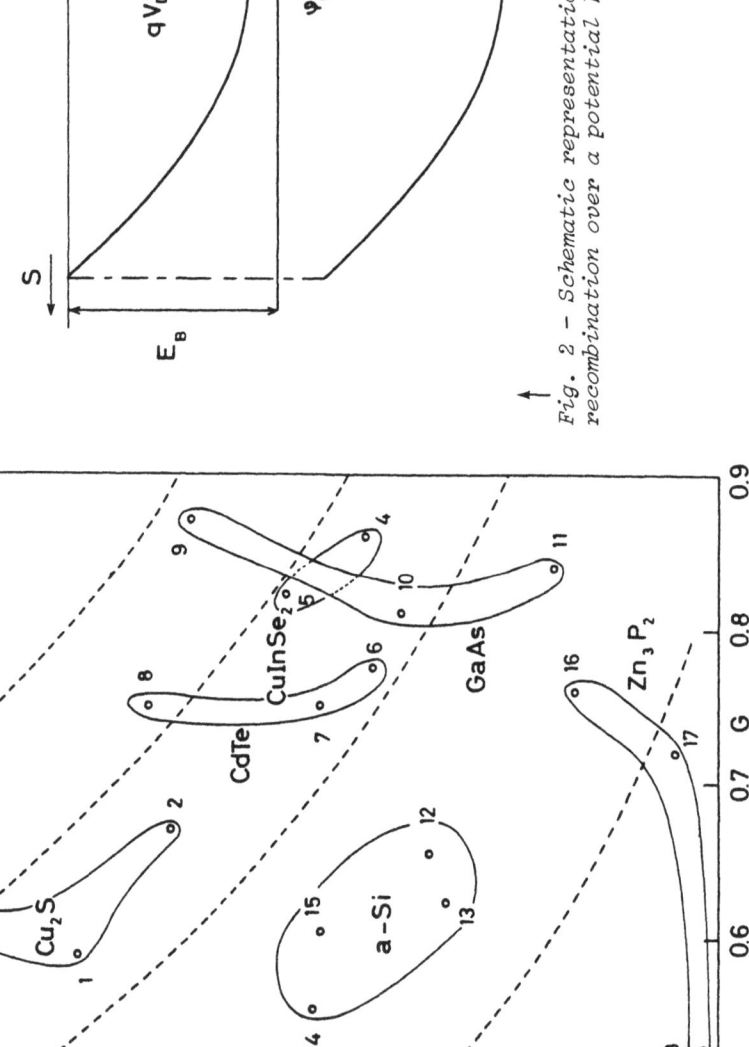

Fig. 1 - Decomposition of conversion efficiency in the reduced quantities G, related to short circuit current and D related to fill factor and open circuit voltage.

Fig. 2 - Schematic representation of injection-recombination over a potential barrier.

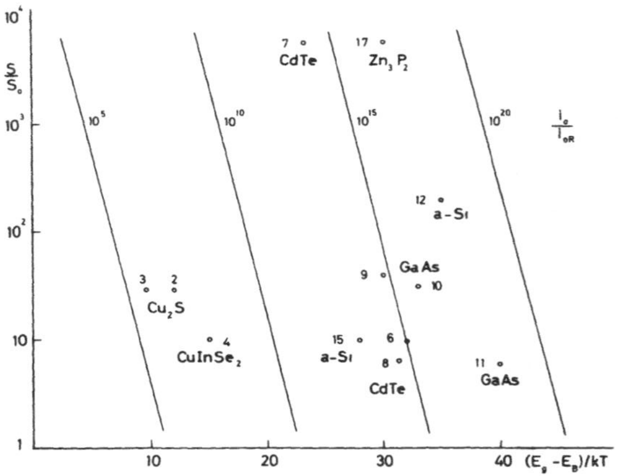

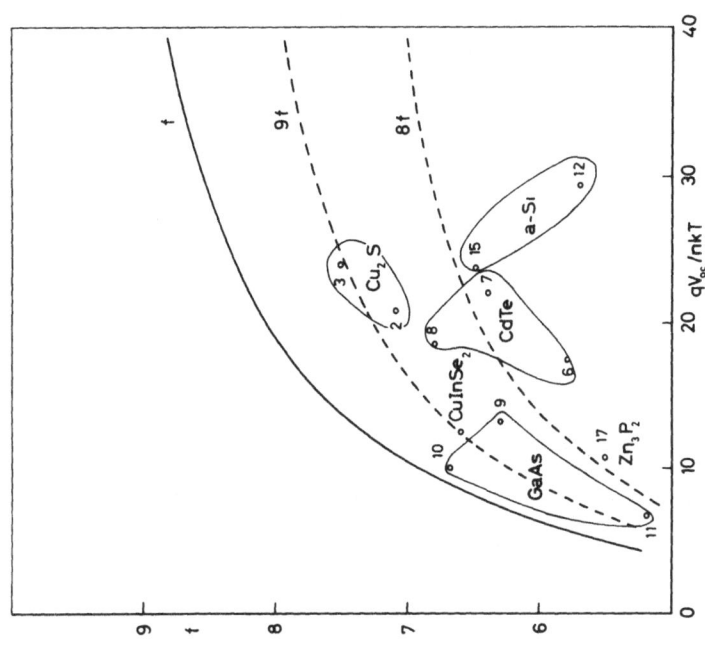

Fig. 3 — Decomposition of the diode saturation current in recombination velocity S and injection barrier height E_B.

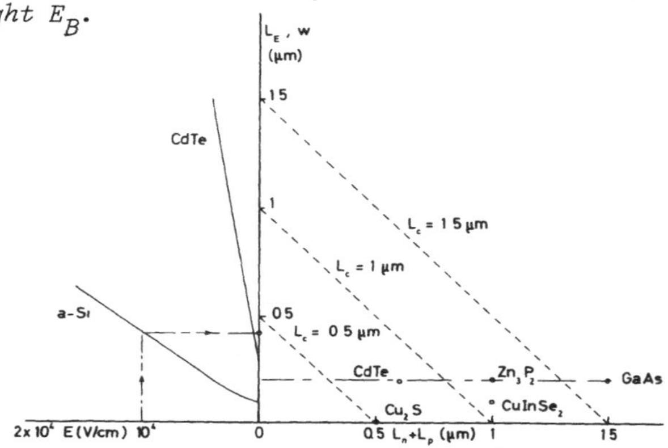

Fig. 4 — Contributions of the neutral regions (L_n, L_p) and of the field zone (L_E) to the collection length.

Fig. 5 — Plot of the ideal and actual fill factors.

8% EFFICIENCY a-SiC:H/a-Si:H HETEROJUNCTION SOLAR CELLS

Y. TAWADA,[*] K. TSUGE,[*] M. KONDO, K. NISHIMURA,[*]
H. OKAMOTO and Y. HAMAKAWA

Faculty of Engineering Science, Osaka University
Toyonaka, Osaka 560, Japan

Summary

An experimental investigation for the wide gap window material in the amorphous silicon solar cell is shown on methane based and ethylene based a-SiC:H. The methane based a-SiC:H/a-Si:H heterojunction solar cell shows a larger short-circuit current density than the ethylene based one. From IR absorption analysis, methane based a-SiC:H film is recognized a rather ideal amorphous SiC alloy as compared with ethylene based one. It has been found through these investigations that the chemical bonding structure is an important factor for a window material. 8% efficiency barrier has been firstly broken through with this methane based a-SiC:H/a-Si:H heterojunction solar cell.

1. INTRODUCTION

A hydrogenated amorphous silicon carbide (a-SiC:H) film fabricated by the plasma decomposition of silane and hydrocarbon gas mixture has a good optical transparency with higher mechanical strength and chemical stability than a hydrogenated amorphous silicon[1]. While, the optical band gap of a-SiC:H can be controlled in a range from 1.8 to 2.8 eV. We have conducted a series of experimantal verifications of the optoelectronic properties in these materials to utilize as a window side junction for p-i-n a-Si solar cell[2] and also examined the relationship between the deposition conditions, the electrical and optical properties of a-SiC:H films[3]. It has been found through these systematic investigations that there exists a good valency electron controllability in an a-SiC:H prepared by the plasma decomposition of $[SiH_4{(1-x)} + CH_4{(x)}]$ gas mixture[4]. Utilizing this valency controlled a-SiC:H as a window side p-layer, a-SiC:H/a-Si:H heterojunction solar cells have been developed[5, 6]. As a result, we have succeeded to break through an 8% efficiency barrier with this material.

In this talk, a series of technical data on the material synthesis, and the optical and optoelectronic properties of a-SiC:H is presented as a promised a-Si solar cell material for the efficiency improvement. The photovoltaic performances of a-SiC:H/a-Si:H heterojunction solar cells are also demonstrated and discussed.

2. OPTICAL AND OPTOELECTRONIC PROPERTIES OF METHANE BASED AND ETHYLENE BASED a-SiC:H FILMS

Properties and chemical bonding structure of a-SiC:H films might be influenced by not only deposition conditions but also carbon sources. Figure 1 shows the AM-1 photoconductivity σ_{ph} and the optical band gap $E_g(opt)$ of undoped and boron doped a-SiC:H films prepared by the decomposition of $[SiH_4{(1-x)} + CH_4{(x)}]$. AM-1 photoconductivity of undoped a-SiC:H

---
[*] on leave of absence from Kanegafuchi Chemical Industry Co. Ltd.
1-2-80 Yoshida-cho, Hyogo, Kobe 652, Japan.

films significantly decreases with increasing the methane fraction. On the other hand, boron doped one shows one or three orders of magnitude larger photoconductivity as compared with undoped one. These photoconductivity recovery effects have been similarly seen in phosphorus doped a-SiC:H films. These effects are accompanied by the decrease of ESR spin density by doping. In other word, boron or phosphorus atoms would compensate the dangling bond in amorphous SiC network and thus enhance the carrier life time.

Figure 2 shows the optical band gap $E_{g(opt)}$ and AM-1 photoconductivty σ_{ph} of undoped and boron doped a-SiC:H films prepared by [$SiH_4(1-x)$ + $1/2C_2H_4(x)$] gas mixture. The optical band gap of these ethylene based a-SiC:H films increases from 1.76 to 2.8 eV with increasing ethylene fraction. The reason why the ethylene based a-SiC:H film shows a larger optical band gap in the same gas phase carbon fraction is that the carbon content in the ethylene based a-SiC:H film is larger than that of the methane based a-SiC:H film. It is interesting in this case that $E_{g(opt)}$ of boron doped a-SiC:H films is larger in the gas fraction less than 0.5 and smaller in the gas fraction more than 0.5 than that of undoped a-SiC:H films. The hydrogen content attached to silicon (Si) is not affected with boron doping but the hydrogen content attached to carbon (C) is affected with boron doping, that is, these narrowing and widening effects are caused by the hydrogen attached to carbon.

In the range where the optical band gap widening occures, the photoconductivity of boron doped a-SiC:H based on ethylene is only one order larger than that of undoped a-SiC:H. Comparing the ethylene based a-SiC:H and the methane based a-SiC:H, it is recognized that the methane based a-SiC:H shows one or two orders larger magnitude of photoconductivity recovery effect of boron doping than the ethylene based a-SiC:H. The amount of hydrogen attached to carbon is a factor responsible for these photoconductivity recovery effect.

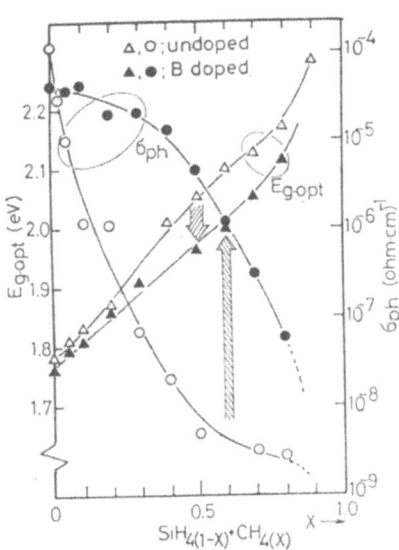

Fig. 1. Photoconductivity and optical band gap of undoped and boron doped a-SiC:H films prepared by the decomposition of [$SiH_{4(1-x)}$+ $CH_{4(x)}$].

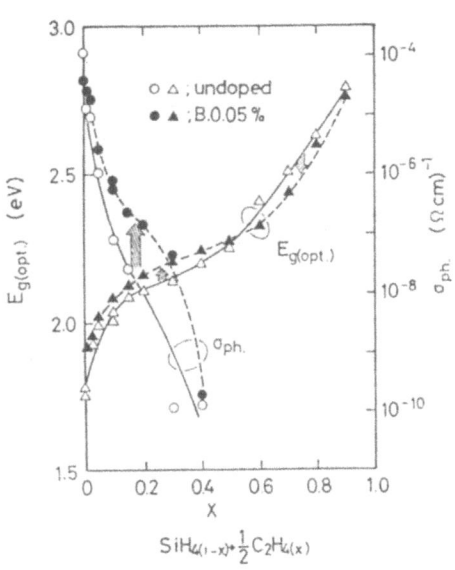

Fig. 2. Photoconductivity and optical band gap of undoped and boron doped a-SiC:H films prepared by the decomposition of [$SiH_{4(1-x)}$+ $1/2C_2H_4$].

2. IR SPECTRA AND CHEMICAL BONDING STRUCTURE OF a-SiC:H FILMS

Methane based a-SiC:H films show one or two orders larger magnitude of photoconductivity recovery effect of boron doping than ethylene based a-SiC:H films. To understand the difference between the two, the chemical bonding structure of the methane based and the ethylene based a-SiC:H has been investigated by means of IR measurement. Figure 3 shows the typical IR spectra of boron doped a-SiC:H films. The carbon content of these films might be estimated to be about 12 atm% in terms of Auger electron spectroscopy (AES peak ratio C/Si=0.06, and crystalline SiC standard). a-SiC:H films exhibit four main absorption regions. One of the strong features can be seen in the C-H stretching mode (2800-3000 cm^{-1}). There are four peaks in the ethylene based a-SiC:H, that is, 2940, 2910, 2890 and 2870 cm^{-1}. On the other hand, the methane based a-SiC:H film shows only two peaks at 2940 and 2890 cm^{-1}. From the detail analysis[7], we concluded that carbons are almost incorporated as ethyl group ($-C_2H_5$) in the ethylene based a-SiC:H and are incorporated as methyle group ($-CH_3$) in the methane based a-SiC:H. However, carbon content as $Si-CH_3$ bond in the methane based one is only 1 or 2%.

Another noticeable feature in these spectra can be seen at 860 - 890 cm^{-1} band which is assigned to SiH_2 of $(SiH_2)_n$ bending mode by Brodsky[8] and Fritzsche[9]. In amorphous silicon films deposited in the same chamber in the absence of carbon sources this band is too weak to detect. While, this absorption band can be seen in a-SiC:H films, especially in the ethylene based a-SiC:H film. This result shows that carbons promote silicon dihydride bonds in amorphous network. The 890 cm^{-1} peak absorption coefficient of the methane based and the ethylene based a-SiC:H films are 350/cm and 1100/cm, respectively, so that the ethylene based a-SiC:H film contains about three times larger content of silicon dihydride than the methane based one.

The hydrogen content of these a-SiC:H films was estimated by IR streching absorption and $^1H(^{15}N, \alpha\gamma)^{12}C$ nuclear reaction [10]. The results are summarized in Fig. 3. The content of hydrogen attached to Si is 20 atm% and 16 atm% for the ethylene based and the methane based a-SiC:H, respectively, and the content of hydrogen attached to C is 23 atm% and 5.5 atm% for the ethylene based and the methane based a-SiC:H films, respectively. The most improtant feature can be seen in the carbon attached hydrogen which is four times larger amount in the ethylene based a-SiC:H than in the methane based one. Assuming that carbons are incorporated as $-C_2H_5$ group in the ethylene based a-SiC:H and as $-CH_3$ group in the methane based one, one can estimate amount of $Si-C_2H_5$ and $Si-CH_3$ bond: 9.2 atm%

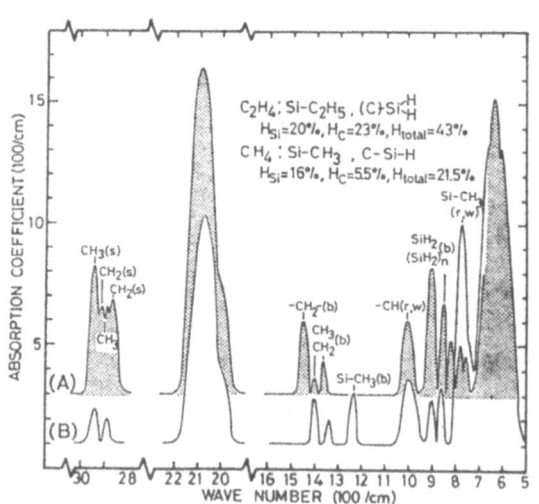

Fig. 3. Infra-red spectra of boron doped ethylene based a-SiC:H film(A) and boron doped methane based a-SiC:H film(B).

carbons are incorporated as $-C_2H_5$ group in the ethylene based a-SiC:H film and 1.8 amt% carbons are incorporated as $-CH_3$ group in the methane based a-SiC:H film, and other carbons (about 10 %) are incorporated as terahedrally bonded carbons or carbons attached with one hydrogen (C-H) in the methane based a-SiC:H. From these investigations, we propose the structual models of ethylene based and methane based a-SiC:H as shown in Fig. 4.

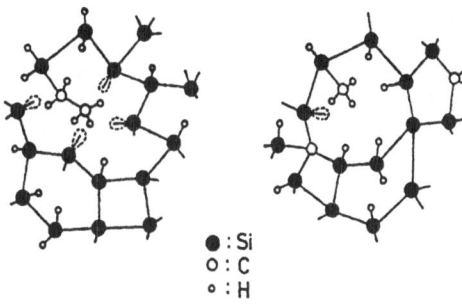

● : Si
◐ : C
○ : H

C_2H_4 based a-SiC:H CH_4 based a-SiC:H

Fig. 4. Models of the chemical bonding structure in hydrogenated amorphous SiC network. (A)=ethylene based a-SiC:H and (B)=methane based a-SiC:H.

It is recognized from these investigations that the methane based a-SiC:H is a rather ideal amorphous SiC alloy in contrast to the ethylene based a-SiC:H which has an organosilane like structure. It might be basically caused by the structural difference between the two that the methane based a-SiC:H film shows one or two orders larger magnitude of photoconductivity recovery effect of doping than the ethylene based one. Furthermore, this structural difference affects the photovoltaic performances of a-SiC:H/a-Si:H heterojunction solar cells.

4. PHOTOVOLTAIC PERFORMANCES OF METHANE BASED AND ETHYLENE BASED a-SiC:H/a-Si:H HETEROJUNCTION SOLAR CELLS

The methane based a-SiC:H films show one or two orders larger magnitude of photoconductivity recovery effect by boron doping than the ethylene based one. To investigate the optoelectronic difference between the two, a series of experimental examination has been carried out on the photovoltaic performances. Figure 5 shows the photovoltaic performances of a-SiC:H/a-Si:H heterojunction solar cells as a function of the optical band gap $E_{g(opt)}$ of p-type methane based and ethylene based a-SiC:H. As is seen in this figure, not only the short-circuit current density J_{sc} but also the open-circuit voltage V_{oc} of methane based a-SiC:H/a-Si:H heterojunction cells increase as the increase of the optical band gap of p-type methane based a-SiC:H. On the other hand, in the range from 1.9 to 2.13 eV the short-circuit current density J_{sc} of ethylene based a-SiC:H/a-Si:H heterojunction cells is smaller than that of the p-i-n a-Si:H homojunction solar cell ($E_{g(opt)}$=1.76 eV). The open-circuit voltage of the cells increases with increasing $E_{g(opt)}$ of p-type ethylene based a-SiC:H layer as well as methane based a-SiC:H/a-Si:H heterojunction cells. An essential matter required to increase the short-circuit

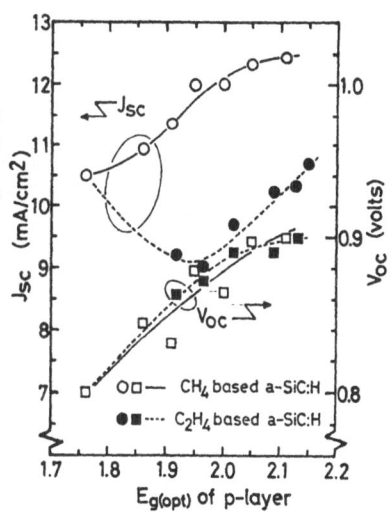

Fig. 5. Effects of p-type methane based and ethylene based a-SiC:H as a window material in p-i-n a-Si solar cells on the short-circuit current density J_{sc} and the open-circuit voltage V_{oc}.

current density is to effectively introduce the incident photons into i-layer where the photocurrent is mainly produced. The increase of J_{sc} in the methane based a-SiC:H/a-Si:H heterojunction cells is first of all due to the wide gap window of a-SiC:H. Another effect of the increased band gap in p-layer is a corresponding enhancement of the blocking barrier for electrons at p-i interface. The alignment of conduction band in methane based a-SiC:H/a-Si:H heterojunction structure was chosen so that the discontinuity appears in the band edge at p-i interface. Therefore, the back diffusion of electrons from the i-layer into p-layer might be practically blocked. This blocking barrier effect is remarkable, especially in the short wave length region of the incident photons. In the ethylene based a-SiC:H/a-Si:H heterojunction cells, it seems that an energy spike might exist at the p-i interface in valence band edge. Because holes could not traverse freely into p-layer, the photocurrent might be limited in the ethylene based a-SiC:H/a-Si:H heterostructure. The experimental evidence can be seen in the collection efficiency spectra. Figure 6 shows the collection efficiency data of an ordinary p-i-n a-Si:H homojunction solar cell, and methane based and ethylene based a-SiC:H/a-Si:H heterojunction solar cells. The collection efficiency of this methane based heterojunction solar cell is improved by more than two times at the short wave length region as compared with the homojunction solar cell, while this improvement is only 20 % at the wave length of 550 nm. It is concluded that the increase of collection efficiency at the short wave length region is mainly caused by the blocking barrier at the p-i interface of methane based a-SiC:H/a-Si:H heterojunction solar cell. The collection efficiency of ethylene based a-SiC:H/a-Si:H heterojunction cell is relatively smaller than methane based one. Normalizing the peak efficiency of ethylene based a-SiC:H/a-Si:H cell to that of methane based a-SiC:H/a-Si:H cell, the former spectrum perfectly coincides with the latter one. The result suggests an existence of energy spike at p-i interface of ethylene based a-SiC:H/a-Si:H heterojunction cell. Accordingly, the distinction of the photocurrent between the two may be attributed to the structural difference as is discussed above, and it cones to the conclusion that wide gap amorphous materials are not always useful as a window side junction material and that the structure of these materials is an important factors for the junction formation.

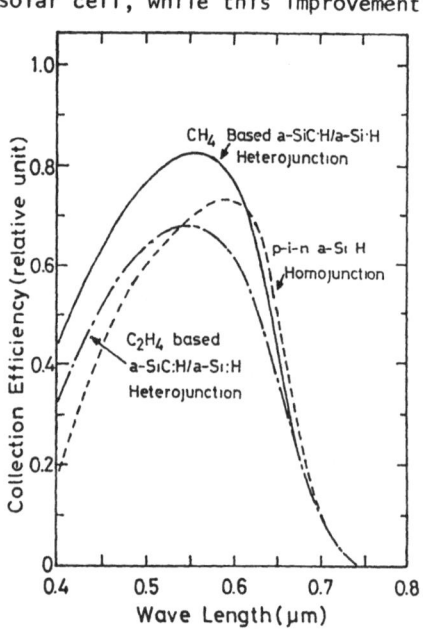

Fig. 6. Collection efficiency spectra of methane based and ethylene based a-SiC:H/a-Si:H heterojunction cells, and p-i-n a-Si:H homojunction cell.

5. TYPICAL J-V CHARACTERISTICS OF a-SiC:H/a-Si:H HETEROJUNCTION SOLAR CELLS

It has been found through these systematic investigations that the methane based a-SiC:H film is a rather ideal amorphous SiC alloy and is superior to the ethylene based one as a window material. Utilizing the methane based a-SiC:H, we have succeeded to break through an 8% efficiency barrier with a-SiC:H/a-Si:H heterojunction cells. Typical performances of this cell are J_{sc}=15.2 mA/cm^2, V_{oc}=0.88 volts, FF=60.1 % and η=8.04 % with the sensitive area of 3.3 mm^2, as shown in Fig. 7. Comparing an ordinary p-i-n a-Si:H homojunction solar cell, a-SiC:H/a-Si:H heterojunction solar cell is improved by 38 % in J_{sc}, 9.9 % in V_{oc} and 41% in η. The interface between the transparent electrode and p-layer is important to improve the efficiency of large area solar cells. We have developed a glass/ITO-SnO$_2$ (100A) substrate to improve the interface. Employing this substrate, 7.7 % efficiency has been obtained with J_{sc} of 14.04 mA/cm^2, V_{oc} of 0.88 volts and FF of 62.4 %.

Furthermore, the top data of J_{sc}, V_{oc} and FF separately obtained in a-SiC:H/a-Si:H heterojunction solar cells are 16.4 mA/cm^2, 0.91 volts and 71 %, respectively. If we assume these realistic top data will become a routine performances on one cell, the efficiency of 10.6 % might be obtained in a near future with this a-SiC:H/a-Si:H heterojunction solar cell.

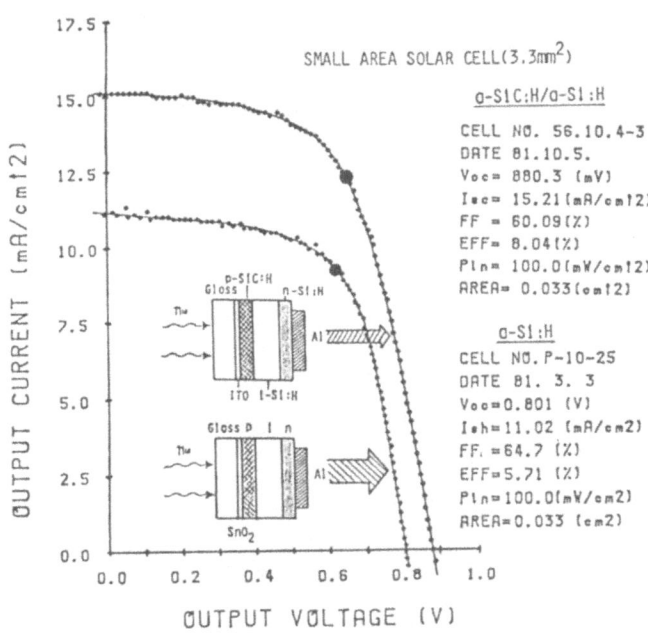

Fig. 6. Typical J-V characteristics of methane based a-SiC:H/a-Si:H heterojunction solar cell having a sensitive

REFERENCES
[1] D.A. Anderson and W.E. Spear: Phil. Mag., 35, 1(1977).
[2] Y. Tawada, H. Okamoto and Y. Hamakawa: Appl. Phys. Lett., 39, 237(1981).
[3] Y. Tawada, M. Kondo, H. Okamoto and Y. Hamakawa: J. du Physique, 42 C-4, suppl. 10, 471(1981).
[4] Y. Tawada, M. Kondo, H. Okamoto and Y. Hamakawa: Solar Energy Mat.(1982) in press.
[5] Y. Tawada, M. Kondo, H. Okamoto and Y. Hamakawa: Proc. *15th IEEE Photovoltaic Specialists Conf.*, Florida (1981), p 245.
[6] Y. Tawada, M. Kondo, H. Okamoto and Y. Hamakawa: Jpn. J. Appl. Phys., 21, suppl. 21-1, 297(1982).
[7] Y. Tawada, K. Tsuge, M. Kondo, H. Okamoto and Y. Hamakawa: J. Appl. Phys., 53(1982) in press.
[8] M.H. Brodsky, M. Cardona and J.J. Cuomo: Phys. Rev., B16, 3556(1977).
[9] C.C. Tsai and H. Fritzsche: Solar Energy Mat., 1, 29(1979).
[10] Y. Tawada, K. Tsuge, K. Nishimura, M. Kondo, H. Okamoto and Y. Hamakawa: Proc. *3rd Photovoltaic Science & Engineering Conf.*, Kyoto, Japan (1982) to be published.

AMORPHOUS SILICON SOLAR CELLS PRODUCED BY A CONSECUTIVE, SEPARATED REACTION CHAMBER METHOD

Y. Kuwano, M. Ohnishi, S. Nakano,
T. Fukatsu, H. Nishiwaki, and S. Tsuda
Research Center, SANYO Electric Co., Ltd.
1-18-13, Hashiridani, Hirakata, Osaka, Japan

Summary

A consecutive, separated reaction chamber method has been developed for the fabrication of a-Si solar cells. In this method, p. i, and n layers are deposited in different reaction chambers, and the undesirable doping caused by residual dopant gases which remain in the reaction chamber can be avoided. Following various kinds of fundamental experiments, Glass/SnO$_2$/p(a-SiC)-i-n/Me type a-Si solar cells were fabricated by this method, and the best conversion efficiency was 8.15% with a size of 2mm x 2mm in sunlight of AM-1. The best conversion efficiency for integrated type a-Si solar cells fabricated by this method was 6.35% with a size of 10cm x 10cm. The integrated type a-Si solar cells are being experimentally used in battery chargers, radios, TV receivers, and a 2kW demonstration plant.

1. INTRODUCTION

The conventional glow discharge method for the fabrication of a-Si solar cells uses a single reaction chamber.[1] In this method, the residual dopant gases which remain in the reaction chamber cause an undesirable doping of the film.[2] To avoid this, the authors have developed a consecutive, separated reaction chamber method in whic p, i, and n-layers are deposited in different reaction chambers.

This paper presents the preparation of a-Si:H film and a-SiC:H film[3] by this new method, and describes the photovoltaic performance of the fabricated a-Si solar cells. Some applications of integrated type a-Si solar cells prepared by this method are also described.

2. THE CONSECUTIVE, SEPARATED REACTION CHAMBER METHOD

In order to obtain high quality[4] and reproducibility in a-Si, we have developed a new fabrication process for a-Si films. This new process is shown in Fig. 1.

In this method, the p- and n-type layers are deposited in different reaction chambers, as shown in Fig. 1. The substrates are set on substrate holders in the preparation chamber, and carried to the first reaction chamber, where p-type a-Si layers are

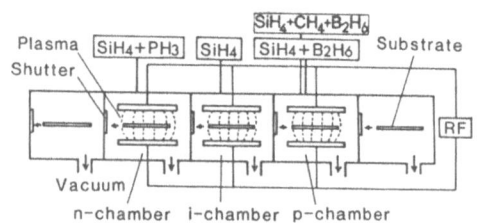

Fig.1 Consecutive, separated reaction chamber apparatus for the fabrication of a-Si films

deposited. Next, they are carried to the second chamber, where i-type a-Si layers are deposited, and then to the third chamber, where n-type a-Si layers are deposited. Finally, they are carried to the fourth chamber to be removed. The features of this method are as follows.

It is possible to (1) avoid the influence of residual gaseous dopants which remain in the reaction chamber, (2) sufficiently control the reaction conditions and amount of dopant in each layer, (3) produce a-Si films consecutively, and (4) prevent the inside of the reaction chamber from being exposed to the open air.

These features present a number of merits in comparison with the single reaction chamber method.

3. PROPERTIES OF THE I-LAYER

In order to investigate the difference between a-Si film deposited by the conventional single reaction chamber method and a-Si film deposited by the consecutive, separated reaction chamber method, i-type a-Si films were prepared in the p-chamber, i-chamber, and n-chamber of the consecutive, separated reaction chambers, as shown in Fig. 1. Before preparing the i-layers in the p-chamber and n-chamber, glow discharge reactions were performed in $SiH_4 + B_2H_6$ (B_2H_6/SiH_4 = 1%) in the p-chamber, and in $SiH_4 + PH_3$ (PH_3/SiH_4 = 1%) in the n-chamber, in order to duplicate the conventional preparation conditions for p-i-n a-Si solar cells, and then the chambers were exhausted to a vacuum. The substrate temperature (T_S), RF power, and gas pressure were 250°C, 22 mW/cm^2 and 0.4 Torr, respectively.

The dark conductivity (σ_d) and its temperature dependence for the i-layers prepared in the p-chamber, i-chamber, and n-chamber are shown in Fig. 2. These results indicate that the residual gaseous dopants which remained in the p-chamber and n-chamber were additionally doped to the i-layers which were prepared in each chamber.

The doping efficiency in the consecutive, separated reaction chamber method is superior to that in the single reaction chamber method, because there is no additional of the opposite type of residual gaseous dopant.

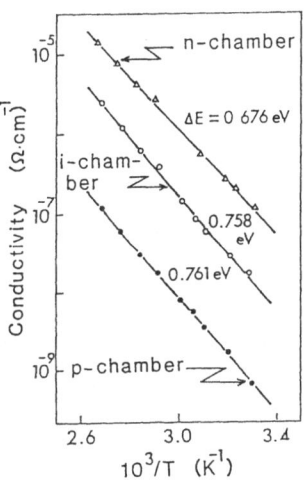

Fig.2 Temperature dependence of the dark conductivity (σ_d) of i-layers prepared in the p-chamber (·), i-chamber (o) and n-chamber (Δ) of the consecutive, separated reaction chamber method.

4. IMPURITY DISTRIBUTION OF THE p-n AND p-i-n a-Si DIODES

In order to investigate the impurity profile of a-Si device fabricated by this method, p-n a-Si diodes were prepared. The same diodes were also prepared by the single reaction chamber method. The impurity distribution was measured by ion microanalyzer (IMA). The results for p-n diodes are shown in Fig. 3.

As this figure shows, the undesirable mixing of boron from the p-layer to the n-layer [Fig. 3] and that of boron from the p-layer to the n-layer deposited by the separated reaction chamber method decrease sharply and are very small compared with those of the single reaction chamber method. This result indicates that the residual gaseous dopant was undesirably added to the subsequently deposited a-Si film in the case of the single reaction chamber method.

This indicates that the separated reaction chamber method is superior to the single reaction chamber method for prevention of the undesirable mixing of dopants.

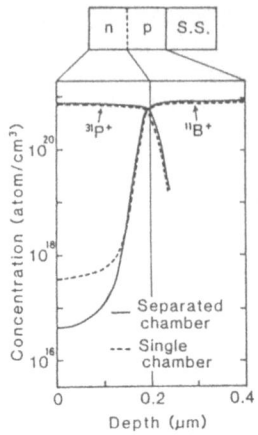

Fig. 3 IMA (ion microanalysis) depth profile of impurity concentration in p-n a-Si diodes

5. COLLECTION EFFICIENCY SPECTRUM AND ILLUMINATED I-V CHARACTERISTICS OF a-Si SOLAR CELLS

In order to investigate the influence of the residual gaseous dopants on a-Si p-i-n solar cells, three kinds of Glass/ITO/p(SiC)-i-n/Me a-Si solar cells, in which i-layers were prepared in the p-chamber, i-chamber, and n-chamber of the consecutive, separated reaction chambers, were fabricated.

The collection efficiency of these a-Si solar cells was measured. Their collection efficiency spectra are shown in Fig. 4, where the notations i(p), i(i), and i(n) show the i-layers prepared in the p-chamber, i-chamber, and n-chamber, respectively.

The collection efficiency of the a-Si p-i(p)-n cell decreases compared with that of the a-Si p-i(i)-n cell in the short wave length region, as shown in Fig.4. On the other hand, the collection efficiency of the a-Si p-i(n)-n cell decreases compared with that of the of the a-Si p-i(i)-n cell in the long wavelength region, as shown in Fig. 4. This decrease in the collection efficiency is due to the additional impurity doping which is causes the change of potential energy distribution in a-Si solar cells.

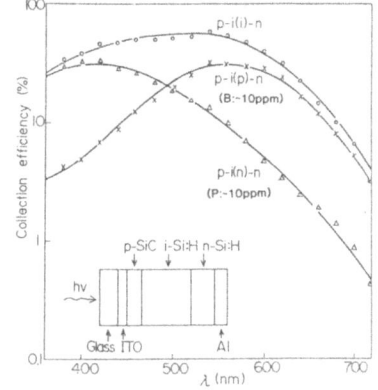

Fig. 4 Collection efficiency spectra of p-i-n a-Si solar cells whose i-layers are prepared in p-chamber, i-chamber, and n-chamber in the consecutive, separated reaction chamber method.

6. PHOTOVOLTAIC PERFORMANCE OF a-Si SOLAR CELLS

We fabricated a-Si solar cells by the consecutive, separated reaction chamber method. For comparison, we fabricated the same kind of cells by the single reaction chamber method. The same reaction conditions were used for the two methods. The illuminated I-V characteristics of Glass/SnO_2/p(SiC)-i-n/Me a-Si solar cells are shown in Fig. 5.
The best conversion efficiencies of Glass/SnO_2/p(SiC)-i-n/Me a-Si solar cells prepared by the single reaction chamber method and the separated reaction chamber method were 6.10% and 8.15%, respectively, with a size of 4 mm^2 in sunlight of 100mW/cm^2 (AM-1).

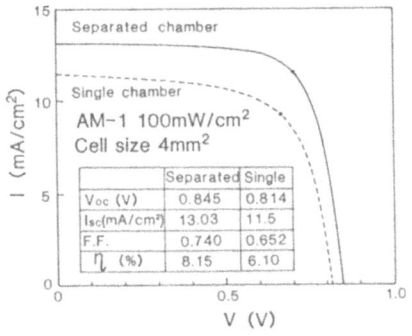

Fig. 5 Illuminated I-V characteristics of SnO_2/p-SiC/-i-n/Me. a-Si solar cells deposited by the consecutive, separated reaction chamber method (solid line) and the single reaction chamber method (broken line).

7. APPLICATION

We have reported on integrated type a-Si solar cells for enlarging the size of a-Si solar cells.[5)] Integrated type a-Si solar cells, in which nine cells were connected in series, were fabricated by this method. The cell structure was Glass/SnO_2/p(a-SiC)-i-n/Me, and the best conversion efficiency was 6.35% with a size of 10cm x 10cm in sunlight of AM-1 (100mW/cm^2). The V_{OC}, I_{SC}, and F.F. were 7.11 (V), 12.41 (mA/cm^2), and 0.648, respectively.

Some examples of these applications are shown in

Fig. 6 Various consumer product applications and a-Si solar cell panel (right rear)

Fig. 7 Residence with a-Si solar panels which generate 2 kW

Some examples of these applications are shown in Fig. 6. In this figure, pocket calculators and watches powered by integrated type a-Si solar cells, which were marketed by SANYO, are the world's first practical use of a-Si solar cells. Other instruments powered by the a-Si solar cells in Fig. 6 are being experimentally produced.

The a-Si solar cell panel, which is also shwon behind the instruments in Fig. 6, was designed for electric power generating systems. A 2kW electric power generating system using this panel has been constructed as shown in Fig. 7.

8. CONCLUSION

A new preparation method for a-Si films, in which each of the p, i, and n layers is deposited in a separated reaction chamber, was developed. This method, called the consecutive, separated reaction chamber method, can avoid the intermixing of doped impurities. The best conversion efficiencies of the p-i-n a-Si solar cells prepared by this method were 8.15% and 6.35% in sunlight of AM-1 (100mW/cm^2), where cell size was 2mm x 2mm and 10cm x 10cm, respectively. The a-Si solar cells were applied to consumer electronic products and power generating systems.

ACKNOWLEDGEMENTS

The authors wish to express their sincere appreciation to Prof. Y. Hamakawa of Osaka University for his kind guidance.
This work was supported by the Agency of Industrial Science and Technology, Ministry of Industrial Trade and Industry (MITT), under the Sunshine Project.

References

1) Y. Kuwano, T. Imai, M. Ohnishi, S. Nakano, and T. Fukatsu: Proc. 1st Photovoltaic Science & Engineering Conf., Tokyo, 1981, Jpn. J. Appl. Phys. 19 (1980) Suppl. 19-2, p137
2) H. Haruki, Y. Uchida, H. Sakai, M. Nishiura, and M. Kamiyama: Proc. 13th Conf. Solid State Devices, Tokyo, 1981, Jpn. J. Appl. Phys. 21 (1982) Suppl. 21-2
3) Y. Tawada, T. Yamaguchi, S. Nonomura, M. Kondo, H. Okamoto and Y. Hamakawa: Proc. 2nd Photovoltaic Science & Engineering Conf., Tokyo, 1980, Jpn. J. Appl. Phys. 20 (1981) Suppl. 20-2, p219
4) W. Schockley: Electrons and Holes (Van Nostrand, New York, 1950) p258
5) Y. Kuwano, S. Tsuda, M. Ohnishi, H. Nishiwaki, H. Shibuya, S. Nakano and T. Imai: Jpn. J. Appl. Phys., 20 (1981) Suppl. 20-2 p213.

CHARGE COLLECTION IN a-Si:H SOLAR CELLS

G. Müller, G. Mück, M. Simon and G. Winterling
MESSERSCHMITT BÖLKOW BLOHM GMBH, AE 331, D-8012 OTTOBRUNN, GERMANY

Summary

The performance of p-i-n and Schottky barrier cells is investigated as a function of intrinsic layer thickness, d_i, up to $d_i = 2.5\mu m$. The main result is that the efficiency of p-i-n/ITO cells is limited by the extraction of photo-generated holes. The hole $(\mu\tau)$-product is determined from the saturation of the photocurrent with increasing reverse bias. The d_i-dependence of the hole $(\mu\tau)$-product indicates that charge collection in a-Si:H is severely limited by surface - and interfacial effects.

1. INTRODUCTION

Work on a-Si:H cells has shown that charge collection in such devices is limited by the relatively small hole range in this material (1). In general $(\mu\tau)_p$ is found to be of the order of 10^{-8} cm^2V^{-1}. Values of this order allow efficient extraction of charge only from relatively thin and weakly absorbing films and thus present a serious limitation to the maximum achievable conversion efficiency. Work on the xerographic applications of a-Si:H, on the other hand, has indicated that $(\mu\tau)_p$-values one or two orders of magnitude higher can be achieved in relatively thick films (2,3). Such values would not impose any limitation to the efficiency of a-Si:H cells. Attempting to settle this controversy we have investigated the performance of a-Si:H cells as a function of d_i and in particular determined the variation of $(\mu\tau)_p$ with d_i.

2. JUNCTION PREPARATION AND MATERIALS ASSESSMENT

We prepared two series of cells with d_i varying in the range 0.5 µm $\leq d_i \leq$ 2.5 µm. The first series consisted of p-i-n/ITO cells illuminated through the n^+-layer and the second of Schottky barrier cells illuminated through a semi-transparent Pt-contact. All cells were prepared in a capacitively coupled glow discharge reactor by decomposition of silane onto metallised Corning glass substrates heated to about 300°C during Si-deposition.

In order to assess the quality of our material, the junction capacitance, C_o, was measured in the dark as a function of external bias in the range of frequencies, 10 Hz $\leq f \leq 10^6$ Hz. The general finding was that C_o was almost independent on frequency in the range of V_{ext} extending from 0.3 V forward to 2 V reverse bias. We take this as evidence that our devices are fully depleted under the action of the built-in potential, V_{bi}, and that charge in deep traps does not significantly contribute to C_o. Fig. 1 moreover reveals that $C_o \propto 1/d_i$, i.e. that our devices are fully depleted in the dark up to $d_i^o = 2.5$ µm.

This finding suggests that the density of states in our a-Si:H is very low, i.e., in the order of 10^{14} cm^{-3} eV^{-1}. A discussion of this aspect will be published elsewhere.

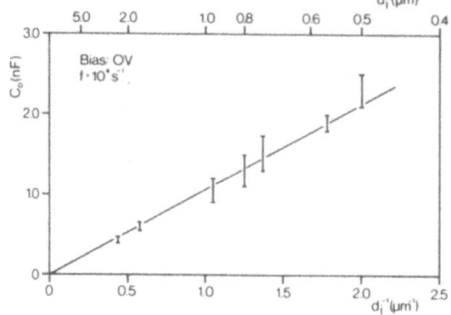

Fig. 1: Junction capacitance as a function of intrinsic layer thickness d_i

3. SOLAR CELL PROPERTIES

In fig. 2 we have plotted the variation of the short circuit current density, J_{sc}, with d_i for the two series of solar cells. For comparison we have also included theoretical estimates of the maximum J_{sc} (4) that can be drawn from a device of thickness d_i assuming no reflection losses, complete charge collection and no reflection from the rear contact of the cell. For relatively thin p-i-n cells the actually observed values of J_{sc} approach the theoretical limit quite closely. In this range of thicknesses efficiencies $\approx$ 7% could be obtained (F.F. $\approx$ 0.60, $V_{oc} \approx$ 0.80 V). Increasing d_i, however, does not increase J_{sc} any further and finally leads to a severe reduction in J_{sc}. Quite obviously, no use can be made of the increased optical absorption in thicker films. The form J_{sc} depends on d_i moreover was found to be critically dependent on the temperature of the cells during operation. Heating the thicker cells to 60 or 70°C resulted in a complete loss of the d_i-dependence of J_{sc} as indicated by the dashed line in fig. 2. Most likely this effect reflects the temperature dependence of the hole drift mobility. In the case of the Schottky barrier cells J_{sc}, although smaller due to the absence of anti-reflection coating, did not depend on d_i at all. Keeping in mind that the variation of d_i probes the range of the photo-generated holes in the case of p-i-n/ITO cells and the range of the photo-generated electrons in the case of Schottky barrier cells the data of fig. 2 suggests that charge collection in a-Si:H is limited by the extraction of the hole charge.

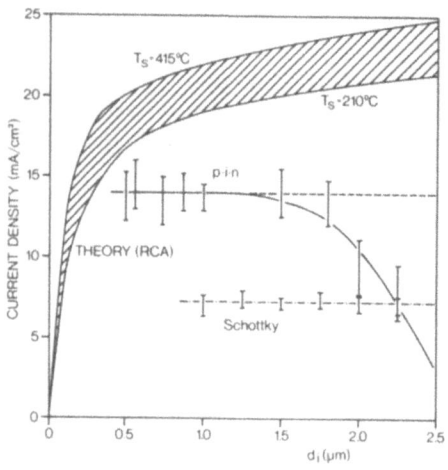

Fig.2: Theoretically expected and experimentally observed variation of the short circuit current density with d_i

Fig.3 brings out the effect of the limited hole range on the spectral response and the fill factor of a-Si:H cells. The maximum spectral response of p-i-n cells decreases with increasing d_i (fig.3B) and at the same time shifts from the green

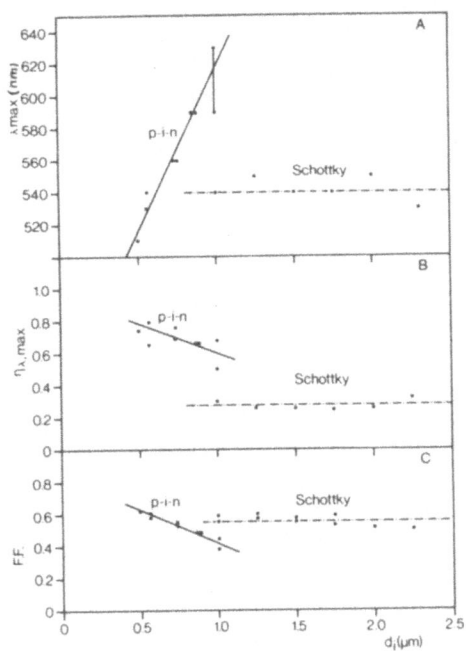

to the red (fig.3A). Moreover, the data of fig.3C suggests that the fill factor FF is affected by the small hole range as well. As increasing values of d_i tend to reduce the average field in the device the data of fig.3C supports recent findings of Spear and coworkers (1) that FF is field limited. In the case of Schottky barrier cells the quantities λ_{max}, $n_{\lambda max}$ and FF all do not depend on d_i as there the electron range does not limit the extraction of the photo-generated charge.

Fig.3: Variation of spectral response and fill factor with d_i (λ_{max} = wave length of maximum spectral response, $n_{\lambda max}$ = quantum efficiency at λ_{max}, FF = fill factor)

4. THE HOLE (μτ)-PRODUCT IN a-Si:H CELLS

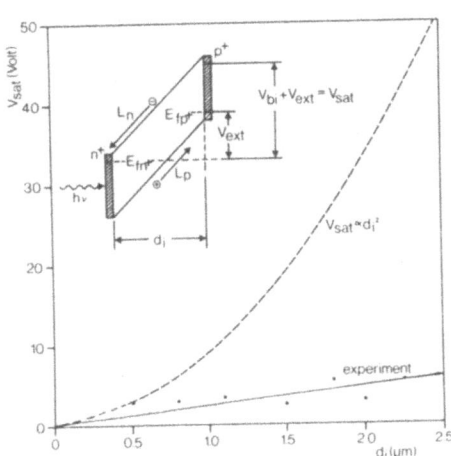

Fig.4: Comparison of experimental (full line) and theoretical (dashed line) values of V_{sat} as a function of d_i. V_{sat} is defined in the inset

As charge collection in p-i-n/ITO cells is limited by the transport of holes across the intrinsic layer a measure of $(\mu\tau)_p$ can be obtained by increasing the reverse bias voltage until saturation of the photocurrent is observed. Capacitance measurements showed that at this bias level the cells are fully depleted under illumination. In this situation the average internal field, $E_{av} = (V_{bi}+V_{ext})/d_i$, has become sufficient to make the hole drift length, $L_p=(\mu\tau)_p E_{av}$, larger than d_i by a certain factor. Accordingly, the d_i-dependence of V_{sat} should be

$$V_{sat}=(V_{bi}+V_{ext}) \alpha\ d_i^2/(\mu\tau)_p \quad (1)$$

The experimental values of V_{sat}, which were independent of light intensity, are plotted in fig.4 showing little dependence on d_i in contrast to the theoretical expetations based on equation 1. Obviously hole transport cannot be

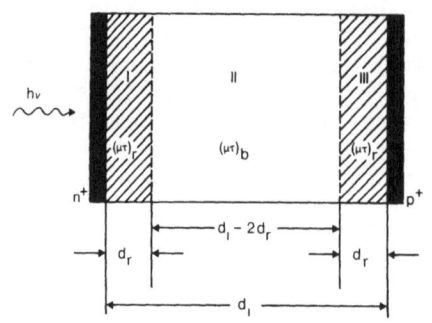

Fig.5: Simplified model to account for surface- and interfacial effects in the charge collection process in a-Si:H cells

described by the simple picture outlined above.

Observing 'wrong' d_i-dependences is clear indication that surface- and/or interfacial effects play a significant role. The existence of such effects has been established by investigations of the thickness dependence of both the dangling bond ESR-signal (5), and of the dark- and photoconductivity (6) of a-Si:H films and by hydrogen profiling measurements (7) on a-Si:H films and devices. These effects can be modelled, as sketched in fig. 5, by subdividing the intrinsic layer into a true bulk region with a large $(\mu\tau)$-product, $(\mu\tau)_b$, and into two boundary layers with small $(\mu\tau)$-products, $(\mu\tau)_r$. As will be shown elsewhere the saturation voltage in this case is given by:

$$V_{sat} = \frac{10 d_i^2}{(\mu\tau)_b} \times \left[1 + \left(\frac{2 d_r}{d_i}\right) \frac{(\mu\tau)_b}{(\mu\tau)_r} \right] \qquad (2)$$

assuming $d_i \gg d_r$. In the case boundary effects dominate $V_{sat}(d_i)$ reduces to:

$$V_{sat} = \frac{20 d_r}{(\mu\tau)_r} \times d_i \qquad (3)$$

Comparing the d_i-dependences of equations 2 and 3 it is suggested that the latter case applies to our a-Si:H cells (fig. 4). Using equation 3 $(\mu\tau)_r$ can be estimated to be about 10^{-8} cm^{2}V^{-1} assuming $d_r = 0.1$ μm. As equation 3 implies that the boundary layer correction in equation 2 is large, say of the order of 10, $(\mu\tau)_b$ can be estimated to be of the order of 5×10^{-7} cm^{2}V^{-1} assuming $2 d_r / d_i = 0.1^b$. Such high values have indeed been observed in thick films normally used in xerographic applications (2,3).

5. OUTLOOK

In conclusion our results indicate that the extraction of holes from a-Si:H cells is severely limited by surface- and interfacial effects. Whether these limitations are of principle nature or whether they merely reflect our presently limited technological skills cannot be answered at the moment. In the latter case it may be expected that improvement of the surface- and interfacial properties will increase $(\mu\tau)_r$ towards $(\mu\tau)_b$ with the consequence that the hole range would become sufficient to allow the collection of charge even from relatively thick and strongly absorbing films. In this case it should be possible to reach values of J_{sc} exceeding 20 mA/cm^2. Moreover the reduction of space charge in the boundary layers would increase the effective field in the interior of the junctions which should help increasing the fill factor.

REFERENCES:

(1) W.E.Spear, R.A.Gibson, D.Yang, P.G.LeComber, G.Müller and S.Kalbitzer; J.de Physique, $\underline{42}$, (1981), 1143

(2) J.Mort; J.de Physique, $\underline{42}$, (1981), 433

(3) I.Shimizu, S.Oda, K.Saito, and E.Inone; J.Appl.Phys., (1980), 6422

(4) D.E.Carlson; J.Non-Cryst.Solids, $\underline{35,36}$, (1980), 707

(5) I.Solomon, "Amorphous Semiconductors", M.H.Brodsky ed., Springer Verlag, Berlin, Heidelberg, New York, (1979), 189

(6) M.H.Brodsky, F.Evangelisti, R.Fischer, R.W.Johnson, W.Reuther, and I.Solomon; Solar Cells, $\underline{2}$, (1980), 401

(7) G.Müller, F.Demond, S.Kalbitzer, H.Damjantschitsch, H.Mannsperger, W.E.Spear, P.G.LeComber, R.A.Gibson; Phil.Mag.B, $\underline{41}$, (1980), 571

THE EFFECT OF GLOW DISCHARGE EXCITATION FREQUENCY
ON THE PERFORMANCE OF MICROCRYSTALLINE
Si:H THIN FILMS AND DEVICES

R. R. GAY, D. L. MOREL, D. P. TANNER,
D. KANANI and H. S. ULLAL
ARCO Solar, Inc.
20554 Plummer Street
Chatsworth, California 91311

Summary

Alloys of Si and H prepared by the glow discharge decomposition of SiH_4 are being developed world-wide for use in thin-film solar cells. The present study focuses on two important aspects of this technology. The first is the effect of excitation frequency on intrinsic material properties, and the second is the influence of processing methodology on device performance. Measurements made on intrinsic materials indicate that electron mobility increases with decreasing excitation frequency. Additional measurements made on devices indicate no systematic change in hole properties with frequency, and thus the conduction band is the main material property affected by excitation frequency. Device performance is found to be very sensitive to exposing the incomplete structure to the atmosphere between layers. Further, the n-i interface is found to be sensitive to the frequency at which the intrinsic layer is deposited. The implications of these observations to device fabrication technology are discussed.

1. INTRODUCTION

The pervasive interest in so-called "amorphous silicon" has led to a myriad of system designs and preparation procedures to study and optimize its performance. Consequently there is no single material involved, but actually a family of materials whose properties are as much a part of the method of fabrication as they are of the components themselves. In order to find commonalities in these efforts several groups have tried to study the effects of variations in material preparation procedures within a given laboratory so that direct comparisons could be made. In this paper we report the results of such an effort. The main parameter of interest is discharge frequency. The purpose is to determine what differences exist between materials fabricated at frequencies ranging from DC to 13 MHz, particularly those differences which affect device performance.

2. EXPERIMENTAL

The deposition system is a conventional capacitively coupled glow discharge unit having a 32-cm diameter cathode. Different power supplies and matching networks were used to run at the different frequencies. Because of the interrelationships of the various parameters, it was difficult to vary only the frequency while keeping other parameters constant. Growth rate was the main parameter of interest and it was maintained within the 1-2 Å/sec range for all frequencies. In order to control growth rate, it was necessary to allow power densities to vary 0.01-0.1 watts/cm^2 and self-bias potential to vary 6-100 volts.

Two different preparation procedures were used for device fabrication. In one, each of the p, i and n layers was deposited in a separate pumpdown, so that the respective layers were exposed to ambient conditions before completing the device. In the second procedure, complete devices were made in a single pumpdown. The glow was turned off between depositions to allow for the necessary change in frequency. In each procedure all p and n layers were deposited under the same conditions using 13.56 MHz. Only the frequency for the i layer depositions was changed. The specific preparation conditions are shown in Table I and the resulting cell structures in Figure 1.

Table I. Cell Preparation Conditions

Hydrogen Pretreatment		n Layers	
H$_2$ Flow Rate:	30 sccm	All conditions identical to intrinsic with addition of 12.9 sccm of 1% PH$_3$ in He (0.43% PH$_3$ to SiH$_4$)	
Power:	200 Watts (13.56 MHz)		
Pressure:	0.4 Torr	Deposition Rate:	4 Å/sec
Substrate Temperature:	280°C		
Time	5 Minutes	p Layers	
Intrinsic Layers		All conditions identical to intrinsic with addition of 13.4 sccm of 10% B$_2$H$_6$ in He (4.5% B$_2$H$_6$ to SiH$_4$)	
SiH$_4$ Flow Rate:	30 sccm	Deposition Rate:	4 Å/sec
Pressure:	0.7 Torr		
Power:	10-100 Watts		
Substrate Temperature:	280°C		
Deposition Rate:	1 – 2 Å/sec		

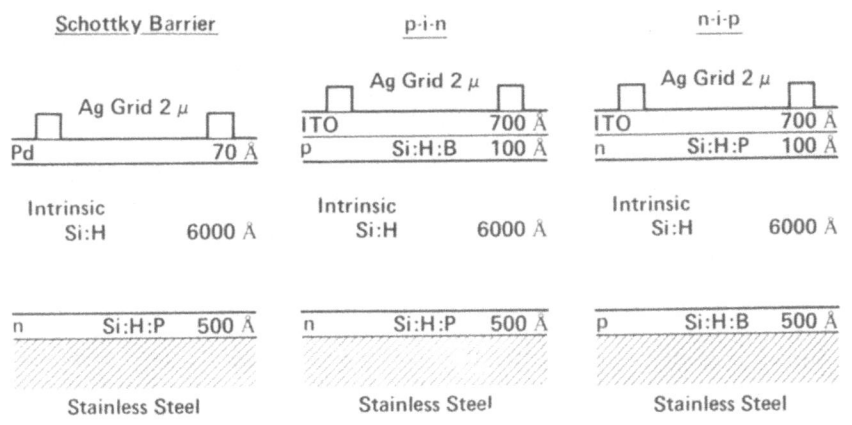

Figure 1. Cell Structures

Standard measurement procedures were used for all electronic and optical properties. IR measurements were taken on crystalline silicon witness wafers for each run. Those results are not presented here because all of the Si-H stretch absorptions were primarily at 2000 cm^{-1} with little 2100 cm^{-1} contribution, and there was no significant variation of these signals with plasma frequency.

Each data point in the following graphs for device properties represents the mean for about 20 cells. The error bars shown represent the standard deviation.

3. RESULTS AND DISCUSSION

3.1 Films

Table II lists the optical and electrical properties of the individual layers. Optical gaps for the i and n layers are in the typical 1.7-1.8 eV range, while the p layers show characteristically lower gaps of around 1.5 eV. σ_D and σ_L values are again within the range normally observed (σ_L for the doped layers is actually thermal modulation of σ_D). Optical and electrical properties of intrinsic material prepared with a DC excitation are not included in Table II since the measurements require that the material be deposited on glass, an insulator which would distort a DC plasma.

Table II. Electrical Properties

Material	Frequency	E_{opt} eV	σ_D $(\Omega cm)^{-1}$	σ_L $(\Omega cm)^{-1}$	σ_L/σ_D	E_A eV
p	13 MHz	1.51	3.9×10^{-3}	4.5×10^{-3}	1.2	0.26
n	13 MHz	1.77	4.9×10^{-3}	6.0×10^{-3}	1.2	0.23
i	13 MHz	1.79	2.0×10^{-10}	3.2×10^{-5}	1.6×10^5	0.80
i	100 kHz	1.74	8.7×10^{-10}	7.6×10^{-5}	8.7×10^4	0.75
i	10 kHz	1.73	1.8×10^{-9}	1.4×10^{-4}	7.8×10^4	0.79

One of the more interesting results is the dependence of σ_D and σ_L on frequency as shown in Figure 2. In most cases, for so-called "i layers" majority carriers are electrons, so these properties are attributed to electrons, although no direct verification of this has been made. Since σ_D and σ_L track each other with frequency, mobility must be the affected property since it is the common denominator. This hypothesis is supported by the fact that the band gap and the dark conductivity activation energy are nearly independent of frequency (Figure 3). The increasing electron mobility with decreasing frequency might manifest itself as a sharpening of the conduction band edge, which will be determined by closer examination of the optical properties.

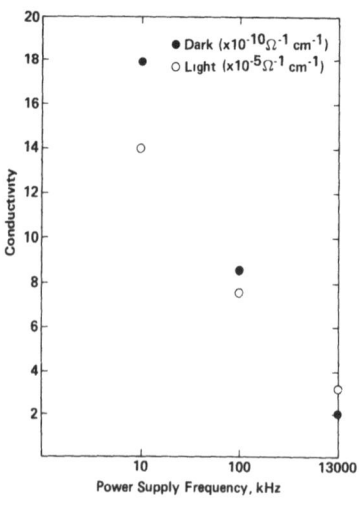

Figure 2. Conductivity vs Frequency

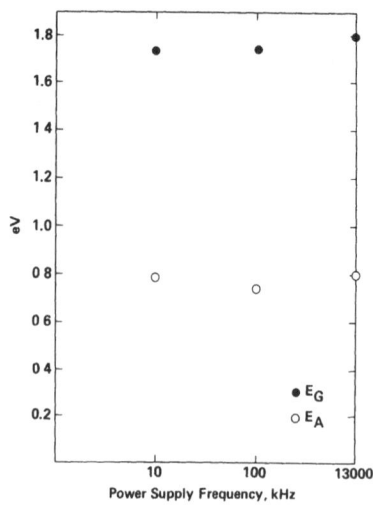

Figure 3. Band Gap and Activation Energy vs Frequency

3.2 Devices

Results for devices made by breaking vacuum between layers are shown in Figures 4 and 5. I_{sc} for Schottky barriers is independent of frequency as is the case for p-i-n devices (data have not been included to reduce clutter). I_{sc} for n-i-p structures on the other hand shows a strong decrease with decreasing frequency. As can be seen by examining the corresponding data for devices made without breaking vacuum (Figure 6), this dependence is peculiar to this particular series. In fact, the time order in which these devices were made is that of decreasing frequency. It is hypothesized therefore that the decrease with frequency observed here is in reality due to increasing degradation of the bare p layers as they awaited processing. The flat performance of the other two structures in which n is put down first indicates that these structures are less susceptible to such degradation. This also applies to the i layers for these structures, which also had varying exposure times before p layers were applied. As seen in Figure 5, V_{oc} is less susceptible to such temporal factors for the n-i-p structures, though the magnitude of V_{oc} in each case is suppressed by about 0.1 volt from its counterpart made without breaking vacuum.

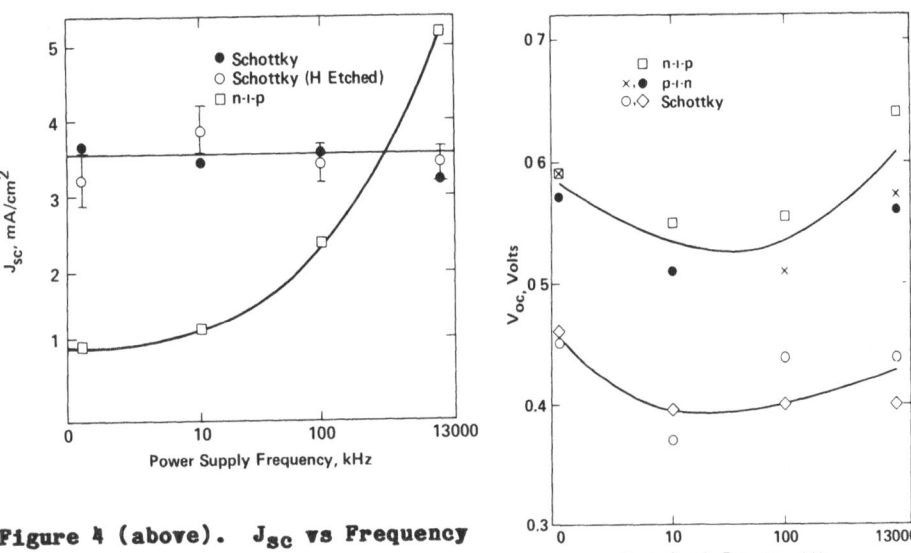

Figure 4 (above). J_{sc} vs Frequency

Figure 5 (right). V_{oc} vs Frequency

The results in Figures 6 and 7 show additional correlations of interest. The intermediate frequency performance in both device structures is inferior to the performance of the two extreme frequencies, and this is true for all three parameters: I_{sc}, V_{oc}, and FF. The origin of this problem is the n-i interface. This is illustrated in Figure 8 which shows quantum efficiency versus wavelength for 130 kHz i layers in n-i-p and p-i-n structures compared with a 13 MHz i layer device. In the case of the n-i-p 130 kHz device, light is incident on the n side. The device exhibits poorer blue response, indicating poor performance of the n-i interface region. Conversely the p-i-n device which has light incident from the p side exhibits poor response in the red which is weakly absorbed and thus again implies poor n-i performance. Devices made with interrupted vacuum show flat I_{sc} response with varying frequency for p-i-n and Schottky barrier structures, and decreasing I_{sc} with decreasing frequency for n-i-p structures. For these devices however, overall performance is much poorer than the performance of their counterparts made without breaking vacuum. This is because of the effects at the interfaces due to exposure and contamination, although spectral details also indicate poor n-i performance, which is in agreement with the above.

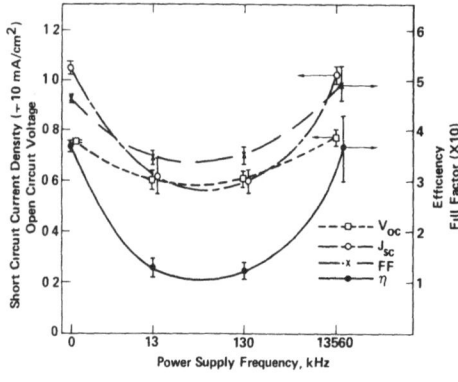

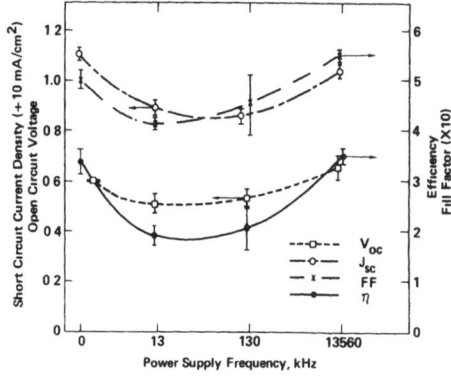

Figure 6 (above). n-i-p Device Performance

Figure 7 (above, right). p-i-n Device Performance

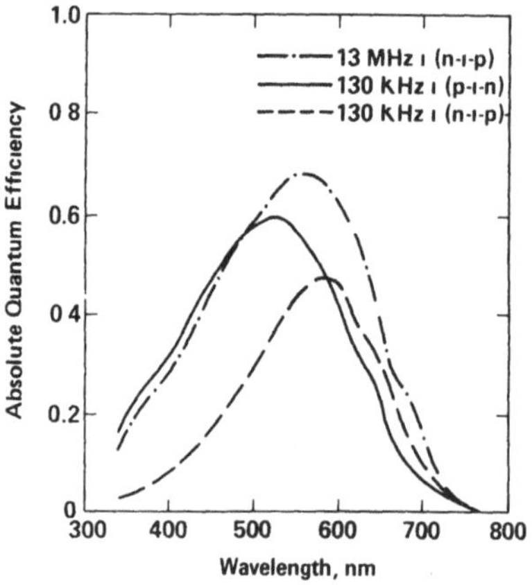

Figure 8 (right). Quantum Efficiency for i Layers at 13 MHz and 130 kHz

Based upon the findings for devices prepared without breaking vacuum, it is apparent that the intermediate frequency i layers result in inferior devices. Spectral response of these devices clearly shows that the reduction in performance comes from the n-i interface, regardless of which layer is put down first. Additional properties of the material grown at intermediate frequencies are that it grows slowly and has a tendency to peel. The tendency to peel is indicative of high stress, and this stress occurs whether the substrate is stainless steel, glass, or doped Si:H. One possibility is that a 13 kHz or 130 kHz plasma is not clearly "on" at all times, but instead is continually going on and off as the excitation changes polarity. At 13 MHz this oscillation occurs faster than the decay time of the plasma, hence 13 MHz and DC plasmas are continuous. But since plasma behavior is a function of many factors, including system geometry, pressure, power, and DC bias, a close study of all of these parameters is in order before definitive conclusions can be drawn. Such experimentation is being pursued so that these phenomena can be unraveled and their effects on films and device performance understood.

ELECTRODEPOSITED CdS/CdTe HETEROJUNCTION SOLAR CELLS

BULENT M. BASOL, ROBERT L. ROD, and ERIC S. TSENG

Monosolar, Inc., a unit of Monogram Industries, Inc.
1299 Ocean Avenue - Santa Monica, California 90401, USA

Summary

The latest in a series of solar cells made using electrochemical deposition is the CdS/p-CdTe heterojunction described here. The two ultrathin polycrystalline active layers were electrodeposited to yield cells with an efficiency of 7% over 0.2 cm^2 area. Best observed solar cell parameters were V_{OC} = 0.79 V, J_{SC} = 18.8 mA/cm^2 and F.F. = 0.56 under illumination of 100 mW/cm^2. An improved window material should increase the efficiency of this potentially low-cost cell.

1. INTRODUCTION

Thin film CdS/p-CdTe heterojunction devices are among the most promising candidates for high efficiency, low-cost, terrestrial solar cells. The theoretical efficiency value is about 17% for this structure (1), and efficiencies close to and over 10% have been achieved experimentally. The highest reported efficiency was 10.5% for a device fabricated by growing a CdS epitaxial layer on a p-CdTe single crystal (2). Polycrystalline cells having efficiencies of 5.4% (3), 8.7% (4), 8.1% (5), and 6.3% (6) have been reported, highest efficiencies being for screen printed cells (4,5,6). Careful studies of these devices showed that most of them were buried homojunctions rather than genuine CdS/p-CdTe heterojunctions.

Electrodeposition is a simple, low-cost method of preparing uniform, ultrathin layers of semiconductors. This technique was shown to be useful in preparing II-VI compound semiconductor films such as CdTe (7) and CdSe (8). Our early work on Schottky barrier devices made on electrodeposited n-CdTe films showed the presence of a large density of traps in these 'as-deposited' films giving rise to small minority carrier diffusion lengths (9,10). As a result, such devices exhibited relatively low solar cell efficiencies (11,12). Annealing the films yielded improved diffusion lengths, and surface etching techniques that were developed (KOH etching of Au/CdTe junctions) gave better solar cell parameters (12).

Our most recent efforts have concentrated on producing heterojunction structures because of the known advantages of these structures over Schottky barriers on n-CdTe. The only previous report of a heterojunction solar cell was for ITO/CdTe:As devices which yielded efficiencies below 1% (13). Here we report on the first efficient, all-electroplated cell which has a near-term potential for 10% efficiency.

2. EXPERIMENTAL

The present solar cell has a backwall structure as shown in Figure 1. The five layers comprising the device are: glass, ITO, CdS, CdTe and Au where the two materials forming the active junction (CdS and CdTe) were electrodeposited. In Figure 2 is shown the plating system used in preparing the active films. ITO-coated glass with a nominal sheet resistance of 10 Ω per square served as the substrate (cathode). CdS film was then electrodeposited on the ITO using the cathodic deposition technique described by Panicker et al (7) and by Panicker (14). Thickness was kept small (under 1,000 Å) to maximize light transmission through the glass/ITO/CdS window of the solar cell. SEM analysis of the CdS films indicated a fine grain structure where the grain size was comparable to the film thickness.

A layer of CdTe then was electrodeposited on the CdS film to a thickness of 1-2 μm. The details of this process were described previously in several papers by us and our colleagues (7,11,14,15) and will not be repeated here. The resulting CdTe films were polycrystalline with columnar grains around 0.5-1 μm. Material had a strong <111> preferred orientation as indicated by X-ray diffraction. Cd-to-Te ratio through the film was determined by Auger depth profile and found to be very close to 1. Such a profile is shown in Figure 3 where the underlying CdS layer also can be observed. The type of these 'as-deposited' films was high resistivity n-CdTe.

Formation of the heterojunction was achieved by converting the originally high resistivity n-CdTe into a p-type layer at the same time forming the junction barrier between the CdS and the converted film. The details of this process are the subject of a patent pending and will be published at a later date. Figure 4 clearly demonstrates how the process works. Curve A of this figure is the spectral response of a Au/n-CdTe/CdS structure where the CdTe thickness was approximately 1 μm. These data were taken by shining light through a 100 Å thick Au layer. The peak at 0.5 μm is indicative of a Schottky barrier device as expected. Curve B of the same figure is the spectral response of the same structure after the type-conversion of the CdTe was achieved. The new peak at 0.8 μm clearly shows the shift of the barrier from the Au/n-CdTe junction to the CdS/p-CdTe junction.

3. RESULTS AND DISCUSSION

Figure 5 shows the temperature dependence of the dark current for a device of 0.2 cm^2 area. At high temperatures the current follows the regular diode equation

$$I = I_o \left[\exp(\frac{qV}{nkT}) - 1 \right]$$

with a room temperature saturation current of J_o = 2.7 x 10^{-9} A and a diode factor of 2. The variations of these parameters among a large number of cells we made were J_o = 7.5 x 10^{-10} - 1.5 x 10^{-8} and n = 1.5 - 2.2. The dark I-V at room temperature for a device with the lowest value of J_o we obtained to date is shown in Figure 6. Also shown in the same figure is the I_{sc} vs. V_{oc} data for the same device. The temperature dependence of the J_o was obtained from Figure 5 and is plotted in Figure 7. The high temperature portion of this curve gives a thermal activation energy of 0.57 eV and can be represented by a recombination-dominated current in the form of

$$I_o = I_{\infty} \exp\left[-\frac{qV_b}{nkT}\right]$$

Using the ionization energy of 0.57 eV and the diode factor of 2, the value of the built-in potential V_b can be found to be 1.14 V which is close to the value expected for a CdS/p-CdTe junction.

It is interesting that our results defining the current mechanism are in very close agreement with the previously reported analysis of single crystal p-CdTe/CdS junctions (16,17). The ionization energy for J_o for junctions prepared by evaporating CdS on single crystal CdTe was measured to be 0.59 eV (16). Devices obtained by CVD deposition of CdS on a single crystal CdTe gave a value of 0.6 eV (17). Both devices had n values comparable to ours. This close agreement between three different device preparation techniques, with various surface preparation procedures and different CdTe materials, strengthens the argument that the observed recombination centers are inherent to the CdS/CdTe interface rather than being introduced as a result of fabrication procedures. It should be noted that our process is a continuous one as opposed to the others mentioned above. There are no surface preparation steps needed, as is the case for single crystal cells, and junction formation takes place *in situ* away from possible outside effects.

The spectral response of our devices were typical of CdS/p-CdTe heterojunctions except that the low wavelength cutoff extended deeper into the ultraviolet. This is because of the thinness of the CdS layer. Higher transmission in the $\lambda > 0.4$ μm region, as well as the collection of photogenerated carriers within the CdS, gives rise to such behavior. Figure 8 compares the quantum efficiency curves of a thick CdS/single crystal CdTe cell (16) and our 7% efficient device (Curve B). Absolute measurements were made using a calibrated Si detector and no correction for reflection was made. The light I-V characteristic of our cell is shown in Figure 9. The intensity of the dichroic filtered tungsten sunlight simulator was set to 100 mW/cm^2. Measurements of the same cell made outside in sunlight gave higher efficiency values of 7.3%. This is due to the fact the simulator is poor in blue where the response of our cells is good. It should also be noted that all measurements were made without any A/R coating on the surface of the glass substrates. The use of an A/R film would cancel out a measured reflection loss averaging about 7% and increase the cell efficiency accordingly.

4. CONCLUSIONS

We have demonstrated the first efficient, all-electrodeposited heterojunction solar cell, and we have obtained, to the best of our knowledge, the highest open-circuit voltage (0.79 V) in an efficient CdS/p-CdTe solar cell. Best solar cell parameters we observed (V_{oc} = 0.79 V, J_{sc} = 18.6 mA/cm^2, and F.F. = 0.56 under 100 mW/cm^2 illumination without an A/R coating) demonstrate the potential for higher cell efficiencies using our process. We expect an improved window material will assist us in reaching toward the 10% efficiency goal in the near future.

5. ACKNOWLEDGEMENTS

The authors acknowledge contributions of Monosolar lab personnel of Dr. O. Stafsudd and S. Ou of UCLA, and of M. Schwartz, W. Kemmel, Jr. and L. Yoder.

6. REFERENCES

1. A.L. Fahrenbruch, V. Vasilchenko, F. Buch, K. Mitchell and R.H. Bube, Appl. Phys. Lett., 25, 605 (1974)
2. K. Yamaguchi, N. Nakayama, H. Matsumoto and S. Ikegami, Japan. J. Appl. Phys., 16, 1203 (1977)
3. D. Bonnet and H. Rabenhorst, 9th IEEE Photovoltaic Specialists Conf., p.129 (1972)
4. H. Uda, H. Taniguchi, M. Yoshida and T. Yamashita, Japan. J. Appl. Phys., 17, 585 (1978)
5. N. Nakayama, H, Matsumoto, K. Yamaguchi, S. Ikegami and Y. Hioki, Japan. J. Appl. Phys., 15, 2281 (1976)
6. N. Nakayama, H. Matsumoto, A. Nakano, S. Ikegami, H. Uda, and T. Yamashita, Japan. J. Appl. Phys., 19, 703 (1980)
7. M.P.R. Panicker, M. Knaster and F.A. Kröger, J. Electrochem. Soc., 125, 556 (1978)
8. M. Kazacos and B. Miller, J. Electrochem. Soc., 127, 2378 (1980)
9. B.M. Basol and O.M. Stafsudd, Solid State Electron., 24, 121 (1981)
10. B.M. Basol and O.M. Stafsudd, Thin Solid Films, 78, 217 (1981)
11. Z. Shkedi and R.L. Rod, 14th IEEE Photovoltaic Specialists Conf., p.472 (1980)
12. B.M. Basol, O.M. Stafsudd, R.L. Rod and E.S. Tseng, Proc. 3rd EC Photovoltaic Solar Energy Conf., p.878 (1980)
13. "Development of Large Area Solar Cells Based on CdTe," Monosolar, Inc. Final Report, DOE Contract No. EX-76-C-01-2457 (1978)
14. M.P.R. Panicker, Ph.D. dissertation, USC (1978)
15. U.K. Patent 1,532,616. Other patents issued and pending.
16. K. Mitchell, Ph.D. dissertation, Stanford (1976)
17. M. Arienzo and J.J. Loferski, J. Appl. Phys. 51, 3393 (1980)

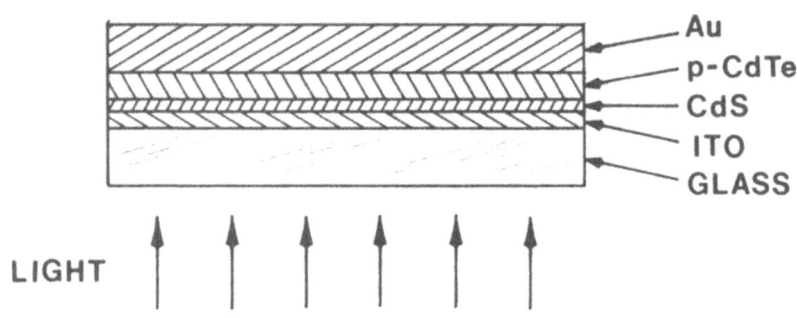

Fig. 1 - *Solar cell structure*

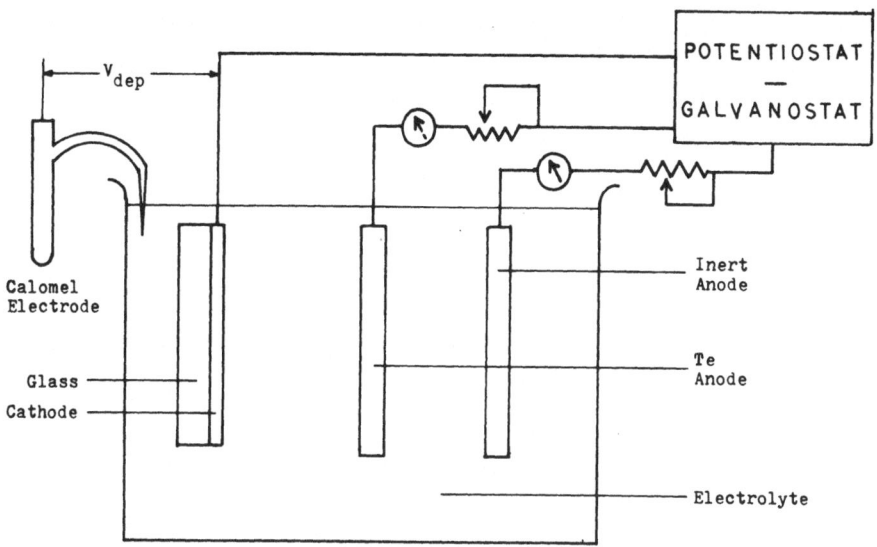

Fig. 2 - *Electrodeposition set-up*

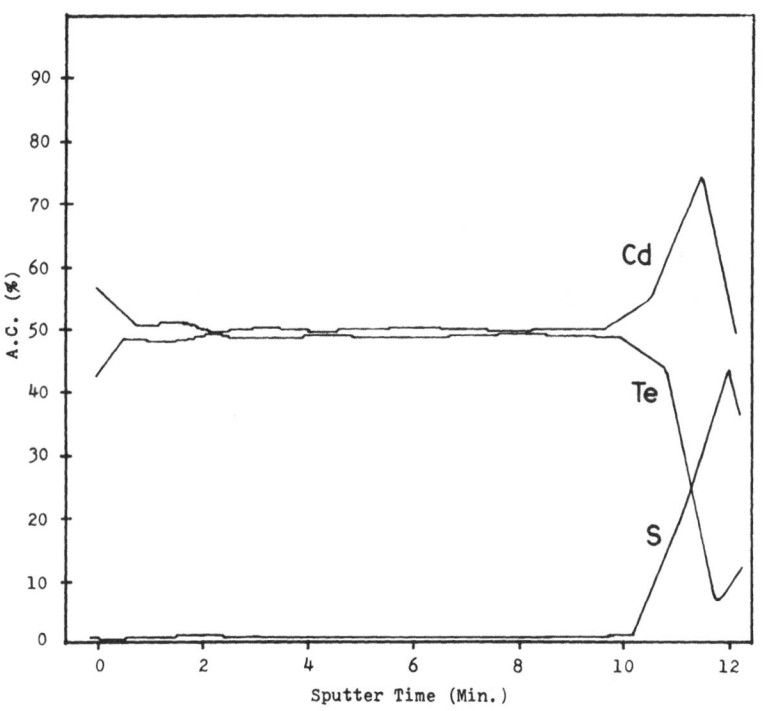

Fig. 3 - *Auger depth profile of the CdTe/CdS structure*

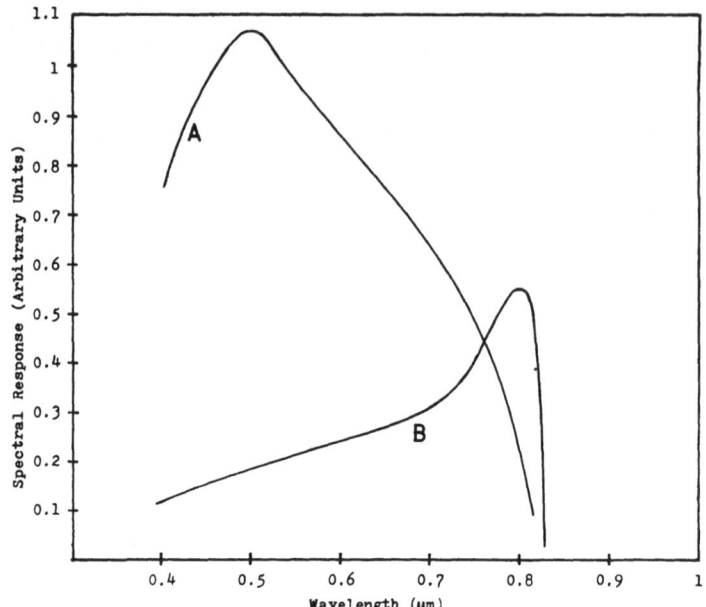

Fig. 4 - Spectral response of a Au/CdTe/CdS device
  A - for as-deposited n-CdTe
  B - for converted p-CdTe

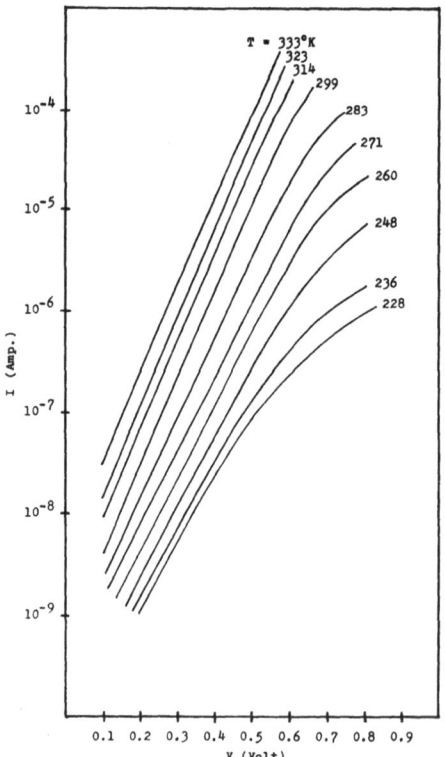

Fig. 5 - Dark I-V of the CdS/p-CdTe heterojunction

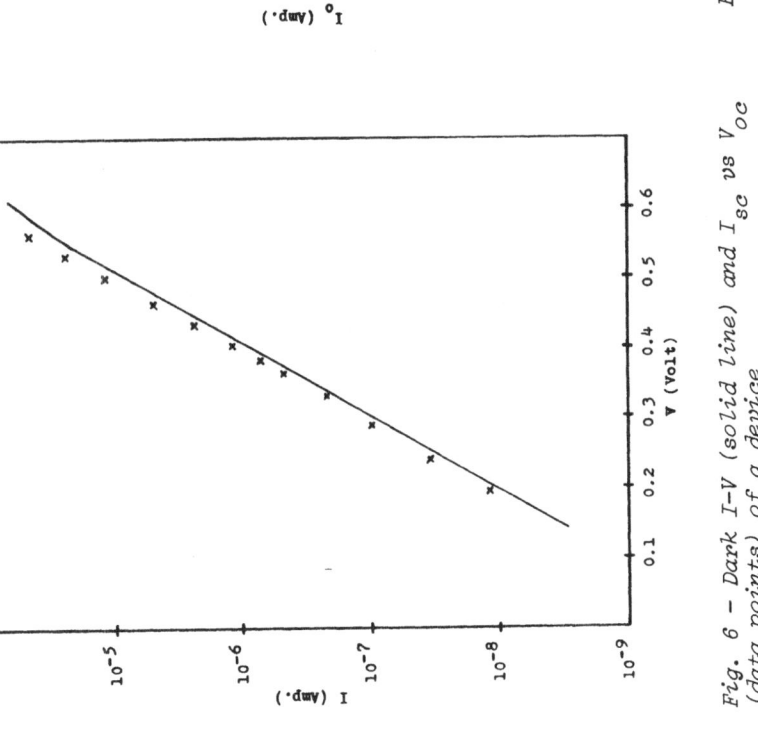

Fig. 6 - Dark I-V (solid line) and I_{sc} vs V_{oc} (data points) of a device

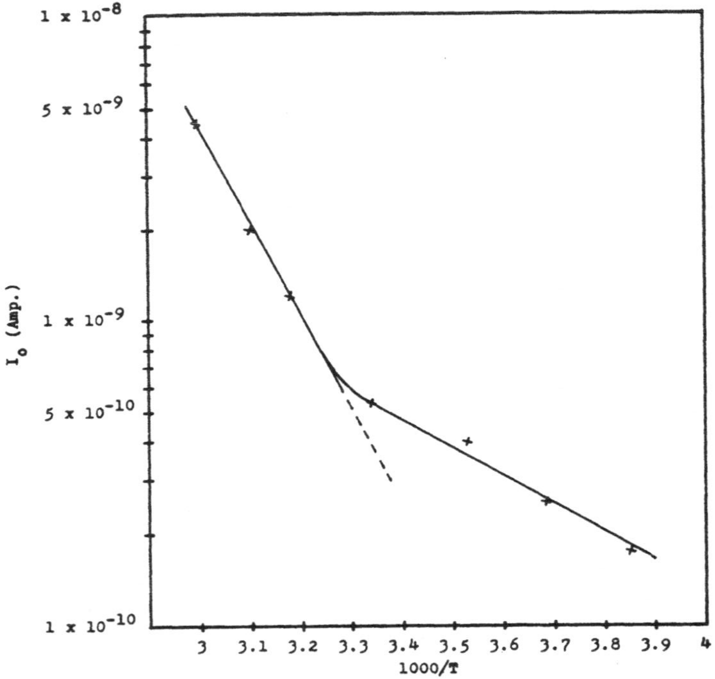

Fig. 7 - Temperature dependence of reverse current

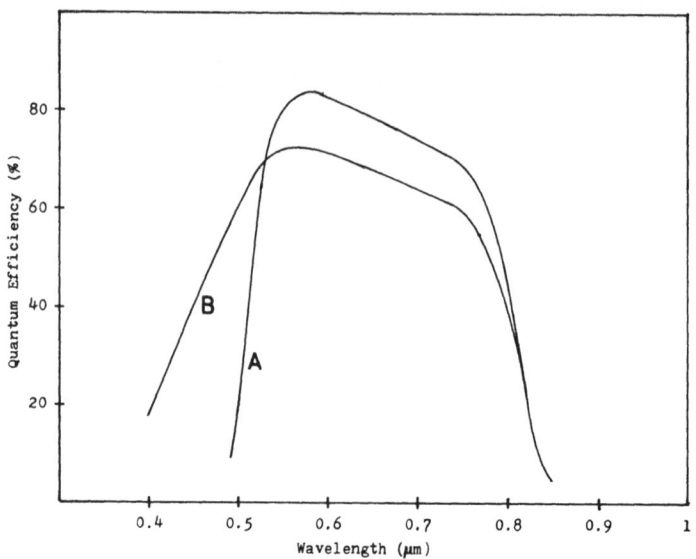

Fig. 8 — Quantum efficiency of our cell (curve B) vs the cell of ref. 16 (curve A).

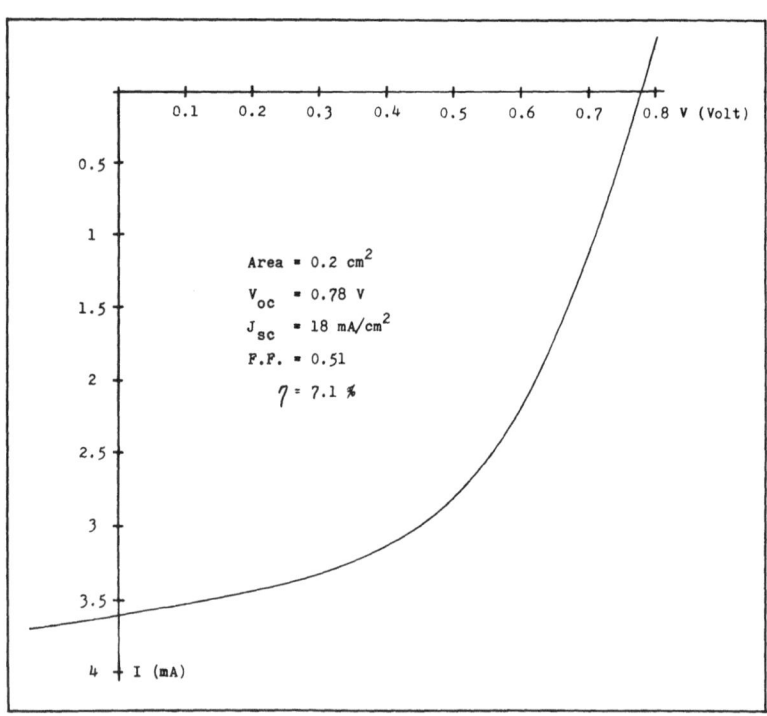

Fig. 9 — Light I-V of a CdS/p-CdTe solar cell.

THIN FILM HETEROJUNCTION CdS/Cu TERNARY ALLOYS SOLAR CELLS WITH MINORITY CARRIER MIRRORS

M. KWIETNIAK, J.J. LOFERSKI, R. BEAULIEU, R.R. ARYA, E. VERA
Division of Engineering
Brown University

L. KAZMERSKI
Solar Energy Research Institute

Summary

A new concept in the fabrication of thin film solar cells with a multilayer structure in which the base region contains a minority carrier mirror (MCM) is reported. The theory of heterojunctions employing CdS as a wide bandgap window and layers of $CuInSe_2$ and $CuGaSe_{0.9}Te_{1.1}$ with MCM as the photovoltaically active semiconductor is presented. A first cell of this type was made by rf-sputtering the successive layers; its AM1 efficiency was about 4%.

1. INTRODUCTION

In previous papers published by the Brown University group, we analyzed a p/n homojunction cell made from $CuInSe_2$ in which the active volume of the PV cell is terminated by minority carrier mirrors, MCM; i.e., electrostatic barriers to the flow of minority carriers (1,2,3) at which the interface recombination velocity is zero. This work extended the concept first introduced by M. Wolf (4) for the indirect gap semiconductor silicon. Essentially, cells of this structure contain both front-surface and back-surface fields.
Theory shows that if d, the distance between the edge of the space charge region and the MCM, is small compared to the magnitude of the minority carrier diffusion length, L, in each of the two pertinent regions of the cell, then the contribution to reverse saturation current from each region, is decreased from the value it would have had if d >> L by the factor $\tanh(d/L)$. Since $\tanh(d/L)$ approaches the value d/L as d/L approaches zero, the reduction in I_o can be large. Decreased I_o means increased open circuit voltage V_{oc}. In addition, the MCM increases the short circuit current I_{sc} since carriers that would have been lost by recombination on the terminating surfaces of the PV cell are reflected back toward the SCR and may contribute to I_{sc}.
In what follows, we describe the extension of these ideas to heterojunction cells of the type shown in Fig. 1. In these cells, CdS is a wide bandgap window which does not absorb any significant fraction of photons from the solar system, but rather transmits them to the photovoltaically active semiconductor (PVAS) which, in Figure 1, is $CuInSe_2$. The cells contained two layers of CdS. The first, or an inner layer, had a resistivity of about 10 ohm cm and the second, or light receiving layer, a resistivity of about 0.1 ohm cm. The resistivity of the $CuInSe_2$ is about 10 ohm cm; this insures that the SCR is principally in the $CuInSe_2$. The resistivity of the wide bandgap (1.43 eV) pseudobinary alloy semiconductor $CuGaSe_{0.9}Te_{1.1}$ is about 0.1 ohm cm. In contrast to p/n homojunction cells like those discussed in Ref. 1, this heterojunction requires only one MCM since

the wide bandgap window (CdS) does not contribute to either I_{sc} or I_o. Essentially, the cell contains a back surface consisting of two reinforcing features: 1) a p/p^+ junction and 2) a conduction band discontinuity arising from the differences between the bandgaps of $CuInSe_2$ ($\sim$1.02 eV) and $CuGaSe_{0.9}Te_{1.1}$ ($\sim$1.43 eV). Note that this particular ternary alloy has the same lattice structure and lattice constant as $CuInSe_2$.

We have also calculated upper limit efficiencies for cells in which the $CuGaSe_{0.9}Te_{1.1}$ is the PVAS and contains a back surface field produced either by a conductivity step (p/p^+) or a p^+ ternary alloy having a larger bandgap than $CuGaSe_{0.9}Te_{1.1}$ (e.g., $CuGaSe_2$).

2. THEORY

To calculate the upper limit efficiency for the structures described above, we used the formalism developed in our previous publications (1,2). The ambipolar diffusion equation containing an optical generation term is assumed to govern the transport of minority carriers in the PVAS. Solutions of this equation, coupled with the known absorption constant vs photon energy curves of the PVAS, ultimately yield values of I_{sc}, V_{oc} and P_{max}; these parameters were calculated for the AM1 solar spectrum. The diffusion length and other parameters of the PVAS had the values used and justified in our previous calculations.

We considered the following cases: a) the "ideal" case of no absorption of photons in the CdS and an MCM with interface recombination velocity equal to zero terminating the photovoltaically active volume; b) photon absorption in a 5 μm thick CdS layer of the type we have been preparing in our laboratory (about 80% transmission over the solar spectrum) and an MCM with s = 0; and c) photon absorption in the CdS layer as in (b), but now the back surface of the cell is covered by a conventional ohmic contact (s infinite). Figure 2 shows the efficiency η vs base thickness for the $CdS/CuInSe_2$ cells while Figure 3 presents the results for the $CdS/CuGaSe_{0.9}Te_{1.1}$ case. In both cases, the highest efficiency occurs for a PVAS thickness of the order of 1 μm. The calculations show how important it is to minimize optical absorption in (and reflection from) the wide bandgap window layer. They also show that, for the set of parameter values used in this calculation, the MCM has an advantage over an ohmic contact for PVAS thicknesses less than 2 μm.

3. CELL FABRICATION

As previously reported, the Brown group has had considerable experience with rf-sputtering of ternary and ternary alloy films (5). We, therefore, set out to fabricate a structure like that of Figure 1, by rf-sputtering the films of $CuInSe_2$ and $CuGaSe_{0.9}Te_{1.1}$ and the first layer of CdS and evaporating a second, low resistivity layer of CdS over sputtered CdS.

The preparation of targets used for sputtering ternary films and CdS was described in previous publications (2,4). Briefly the materials were synthesized in powder form and cold pressed to form a compact layer 2 mm thick and about 5 cm in diameter. The CdS target was made from Cd rich CdS powder produced by heating as received CdS in pure hydrogen; the same kind of CdS powder was used as the charge for evaporating the top layer of CdS. Analysis of this powder (by EDAX) after heat treatment indicated that the Cd content was about 50.8% and the sulfur content, 49.2%.

The heterojunction thin cell was fabricated by first sputtering a thin film of $CuGaSe_{0.9}Te_{1.1}$ over gold coated alumina substrates held at a temperature of 490°C and a thin film of $CuInSe_2$ was sputtered over the $CuGaSe_{0.9}Te_{1.1}$ layer for about 25 min. which resulted in a layer of $CuInSe_2$ having a thickness of about 1.5 μm. The substrate was then cooled down to 150°C and a very thin layer of CdS was sputtered over the $CuInSe_2$ for about 15 minutes (thickness about 1 μm). All three layers were sputtered without breaking vacuum. The system was back filled with pure argon and the sample was cooled to room temperature.

The sputtered sample was taken out of the sputtering system, cleaned in acetone and loaded into the CdS deposition system. The experimental set-up used for CdS deposition is described elsewhere (6). During the deposition of CdS the source and the substrate temperatures were kept at 1100°C and 170°C respectively. A 3 micron thick film of CdS was deposited over the sputtered CdS layer. The resistivity of the evaporated CdS film, as measured on glass was 0.04 ohm cm.

4. DEVICE CHARACTERIZATION

The cell was heated in air at 200°C for 20 minutes before the i-v measurements were made.

The illuminated i-v measurements of this cell are shown in Figure 4. The cell was illuminated by an ELH lamp with an intensity of 100 mW/cm^2 as calibrated by a standard silicon cell.

The PV device had the following parameters: V_{oc} = 310 mV; I_{sc} = 28.9 mA/cm^2; Fill Factor = 0.45; $\eta \sim 4.0\%$ for a device with an active area of 0.045 cm^2.

An Auger analysis (conducted at SERI in Golden, Colorado) of a device having a similar structure is shown in Figure 5. The depth-profile analysis clearly shows the junctions between $CdS/CuInSe_2$ and $CuInSe_2/CuGaSe_{0.9}Te_{1.1}$. It further shows that the 4-element compound and CdS are more homogeneous in composition that the $CuInSe_2$ layer. This lack of homogeneity in the $CuInSe_2$ which is the PVAS may account for lower values of the photovoltaic parameters than those we had expected. For comparison, a $CdS/CuInSe_2$ without an MCM; i.e., with an ohmic contact covering its back surface, fabricated in our laboratory by the technique described above, had the following photovoltaic characteristics: V_{oc} = 280 mV; I_{sc} = 24.5 mA/cm^2; Fill Factor = 0.58; $\eta \sim 4.0\%$. The cell containing the MCM has a 10% higher V_{oc} and a larger I_{sc}; it is the fill factor which prevents it from significantly exceeding the performance of the conventional cell.

5. SUMMARY AND CONCLUSIONS

1. Theory shows that the incorporation of minority carrier mirrors in thin film $CuInSe_2$ type solar cells should result in higher efficiencies for thinner layers.

2. Such cells can be realized by deposition of a multi-layer structure containing a highly conducting wide bandgap semiconductor over which a less conducting PVAS is deposited.

3. A first structure of this type was produced by rf-sputtering $CuInSe_2$ (1.0 eV) over an rf-sputtered film of $CuGaSe_{0.9}Te_{1.1}$ ($E_G \sim 1.4$ eV). The cell was made from a thinner layer of $CuInSe_2$ than in conventional cells. Its V_{oc} and I_{sc} were somewhat higher than the values of these parameters in a conventional cell with an ohmic contact covering the back surface of a thicker $CuInSe_2$ layer. These results, while encouraging, are not a definitive demonstration that the effect of the MCM has been observed.

6. ACKNOWLEDGEMENT

This work was supported by subcontracts SERI-XI-9-8012-1 between the U.S. Solar Energy Research Institute and Brown University and by the U.S. NSF supported Materials Research Laboratory at Brown University.

7. REFERENCES

1. M. Spitzer, J.J. Loferski and J. Shewchun, "Proc. of the Fourteenth IEEE PV Specialists Conference, San Diego, Jan. 1980", pp. 585-590.
2. E.S. Vera, J.J. Loferski, M. Spitzer and J. Shewchun, "Proc. of the Third European PV Solar Energy Conference, Cannes, France, October 1980", pp. 911-919.
3. J.J. Loferski, "Conversion of Solar Energy Using Tandem PV Cells Made from Multi-Element Semiconductors", in the book Photovoltaic and Photoelectrochemical Solar Energy Conversion, F. Cardon, W.P. Gomes and W. Dekeyser, Editors, p. 157-198, Plenum Press, New York, 1981.
4. M. Wolf, "Proc. of the Fourteenth IEEE PV Specialists Conference, San Diego, Jan. 1980", pp. 563-568.
5. J.J. Loferski, M. Kwietniak, J. Piekoszewski, M. Spitzer, R. Arya, B. Roessler, R. Beaulieu, E. Vera, J. Shewchun, and L. Kazmerski, "Proceedings of the Fifteenth IEEE Photovoltaic Specialists Conference, Orlando, Florida, June 1981", pp. 1277-1282.
6. R.R. Arya, R. Beaulieu, M. Kwietniak, J.J. Loferski and L. Kazmerski, J. Vac. Sci, Technol. 20(3), 306 (1982).

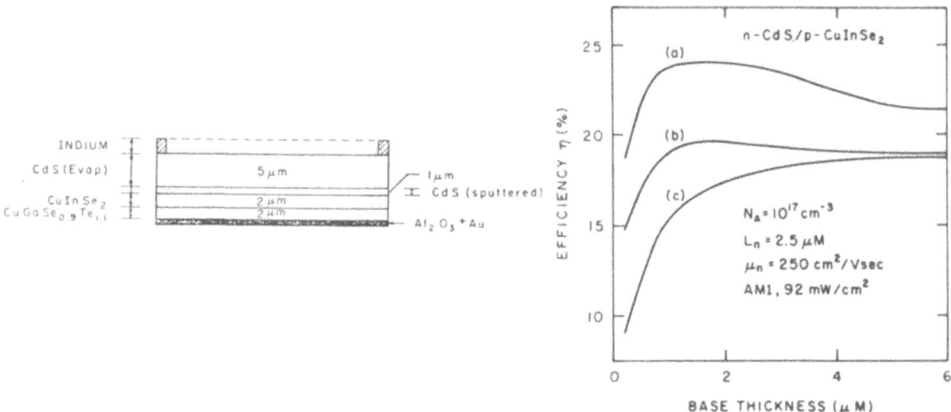

Fig. 1 - Structure of the heterojunction solar cell with MCM

Fig. 2 - Efficiency η vs base thickness for CdS/CuInSe$_2$ cell

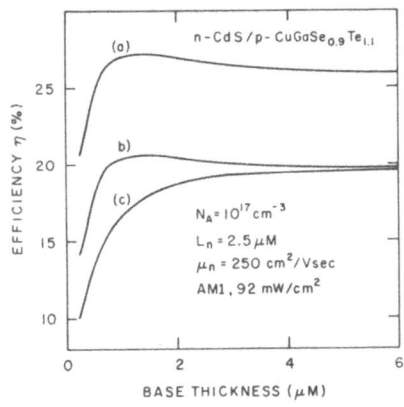
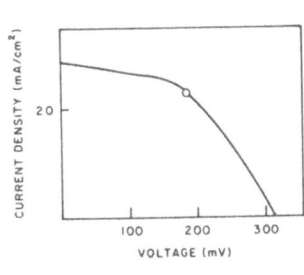

Fig. 3 - Efficiency η vs base thickness for $CdS/CuGaSe_{0.9}Te_{1.1}$

Fig. 4 - I-V characteristics under illumination of $CdS/CuInSe_2/CuGaSe_{0.9}Te_{1.1}$ solar cell

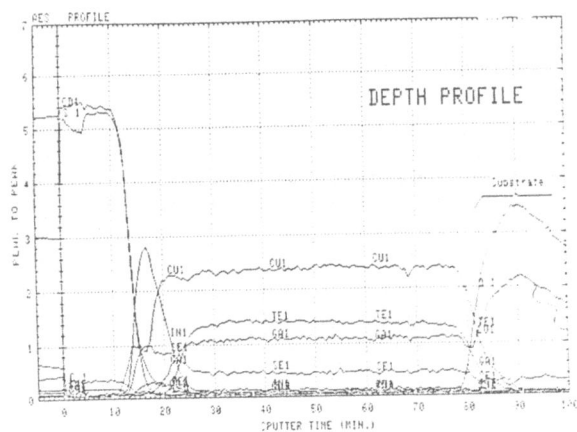

Fig. 5 - Auger analysis of $CdS/CuInSe_2/CuGaSe_{0.9}Te_{1.2}$ solar cell

LARGE AREA CdS/Cu$_x$S THIN FILM SOLAR CELLS PRODUCED BY ELECTROPHORETIC DEPOSITION

T.J. Cumberbatch, I.D. McInally, E.W. Williams, D.J. Gibbons,
M. Claybourn, H. Clow and P.M.G. Dickinson
THORN EMI plc, Central Research Laboratories, Trevor Road,
Hayes, Middlesex, U.K. UB3 1HH

R. Hill and N.M. Pearsall,
Newcastle Polytechnic, Ellison Building, Newcastle upon Tyne

J. Woods, G. Russell and P.C. Pande,
Science Laboratories, Durham University, South Road, Durham

Summary

Electrophoretic deposition can be used to deposit large area thin films of CdS or ZnCdS: Cu$_x$S thin films have also been prepared. This technique has significant advantages over more conventional procedures with deposition rates of up to 3μm min^{-1} achieved.

Recrystallisation of the as deposited powder layer is necessary and has been achieved using both laser and thermal treatments. A pulsed dye laser has also been used to aid the formation of heterojunctions prepared by a number of vacuum and non vacuum techniques. Preliminary results from these heterojunctions are outlined.

1. INTRODUCTION

Electrophoretic deposition involves the movement of colloidal particles of CdS (or ZnCdS or Cu$_x$S) by the application of an electric field and their deposition onto a SnO$_x$ coated glass substrate acting as an electrode (Fig.1). The hydrophobic colloidal suspensions are prepared by the precipitation of CdS from aqueous solutions of cadmium salts under rigorously controlled conditions.

The advantages of depositing cadmium sulphide in this way are:

- fast deposition rates (>3μmin^{-1}) has been achieved)
- large areas can be coated (limited only by size of bath)
- pinhole free dense films are produced (therefore the film can be very thin)
- low material wastage (only the conductive substrate is coated)
- low equipment costs (non vacuum, suitable for third world production)
- ZnCdS can easily be prepared
- Continuous coating is feasible

The disadvantages of the process are:

- the CdS deposited is cubic phase
- the grain size is very small (20 - 30 nm) therefore the as-deposited films have high resistivity
- electrolysis of the solution causes problems with gas evolution
- cracking of the layers after deposition limits thickness.

2. RECRYSTALLISATION

Recrystallisation is required to transform the as-deposited cubic phase powder layer (Fig. 2) into a hexagonal phase polycrystalline layer. Recrystallisation of CdS has been widely investigated for powder layers (as used in electroluminescence and phosphor displays) and in general it is well established that grain growth in such layers is dependent on many factors, most importantly

- stoichiometry (which is usually related to the method of preparation)
- type and concentration of impurities
- the nature of the glass substrate (1).

For pure CdS powders grain growth is significant above $700°C$ with the cubic → hexagonal phase change also occurring at these temperatures (2,3). For ZnCdS this temperature drops to about $600°C$ (3,4).

Our present work on electrophoretically deposited CdS layers has shown that temperatures in excess of $700°C$ are required for homogenisation as is shown in the micrograph (Fig. 3). The application of a thin layer of $CdCl_2$ onto a preheated film (by spraying an aqueous solution) reduces this to $600°C$. However, these temperatures are still too high for standard glasses (or for plastic films which could be used in a continuous process) therefore, pulsed laser and scanned electron beam (as a prelude to cw laser) annealing have been investigated (Table I). Whilst the electron beam annealing results are not fully understood at present, the laser annealing has produced more interesting and very promising results. Experiments using pulsed lasers have suggested that pulse duration is the most crucial factor.

- a 30 ns pulse from a Q switched ruby laser removes small sections of CdS to form a pitted surface on a macroscopic scale
- a 300 ns pulse from a coaxial excited flashlamp pumped dye laser produces the effect shown in figure 4. Pulse lengths of this order appear to melt a thin layer of CdS adjacent to the top surface (Fig. 5). Auger analysis indicates that this layer is slightly deficient in sulphur. With a low temperature preheat prior to laser annealing the original topography is unaffected.
- a pulse width of $2.3 \mu S$ from a linearly excited flashlamp pumped dye laser produces a recrystallised layer as shown in figure 6
- a pulse width of $1.5 \mu s$ onto a preheated layer which had been sprayed with $CuCl_2$ solution gave rise to the structure illustrated in figure 6. Since this has only just been achieved its structure and composition are unknown. However, the electrical and optical properties suggest that it is a heterojunction.

We ascribe the general behaviour of these films under pulsed laser irradiation to the poor thermal coupling between the grains i.e. as the power of the laser pulse is reduced so are the thermal gradients; preheating the film improves the thermal coupling.

3. TECHNIQUES

A number of techniques are being used to assess the $CdS-Cu_xS$ films including RHEED, DLTS, Auger and C L. For a further description of C L. reference should be made to the poster paper included in section P5 of these proceedings.

4. HETEROJUNCTION FORMATION

Standard vacuum evaporation techniques have been used to fabricate heterojunctions on electrophoretically deposited CdS (Table II); the thickness of the films is currently felt to preclude the Clevite process.

However, since the object of the process is to avoid the use of vacuum processing, alternative wet chemical methods have been explored (Table II). We stress that many of these techniques are in the early stages of investigation and will require a detailed study to ascertain what is happening. Experience to date supports the following conclusions.

- the method of recrystallisation of the CdS is important in determining which of the techniques can be used to form a junction
- a laser pulse has a marked beneficial effect on the junction formation
- the junctions formed so far are not very stable
- very uniform layers of Cu_xS can be deposited on electrophoreted CdS by electrophoresis

5. CONCLUSION

Although the heterojunctions produced to date have not been very efficient, considering that the process is far from optimised the results are very encouraging for the future production of a non vacuum CdS/Cu_xS solar cell on a rapid time scale.

6. ACKNOWLEDGEMENT

This work was supported in part by both the Commission of the European Communities and the United Kingdom Department of Industry whose assistance is gratefully acknowledged.

7. REFERENCES

1. A. Vecht and A. Apling, Phys. Stat. Sol. 3, 1238, 1963.
2. N. Nakayama, Jap. J. Appl. Phys. 8, 450, 1969.
3. E.S. Rittner and J.H. Schulmann, J. Phys. Chem. 47, 537, 1943.
4. A.O. Dmitrienko et al, Inorganic Materials, 13, 1578, 1977.
5. M. Sakaguchi et al, J. Electrochem. Soc. 124, 550, 1977.

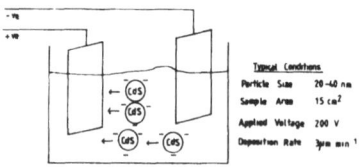

Fig. 1 Electrophoretic Deposition

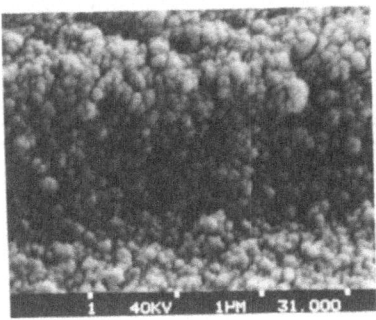

Fig. 2 As deposited CdS

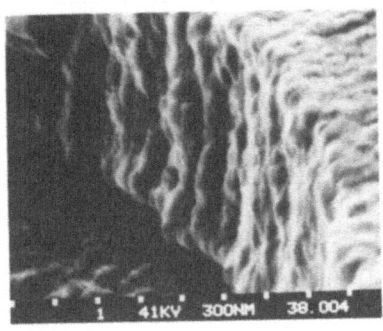

Fig. 3 750°C for 10 mins in N_2

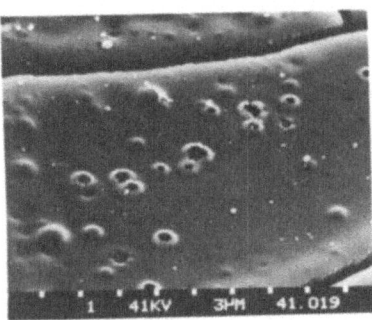

Fig. 4 300 ns laser pulse at 350 mJ cm^{-2}

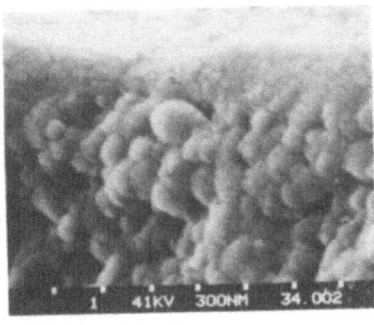

Fig. 5 300 ns laser pulse at 350 mJ cm^{-2} onto pre-heated film (650°C)

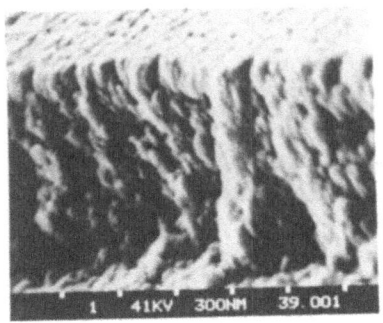

Fig. 6 2.3µs laser pulse at 220 mJ cm^{-2}

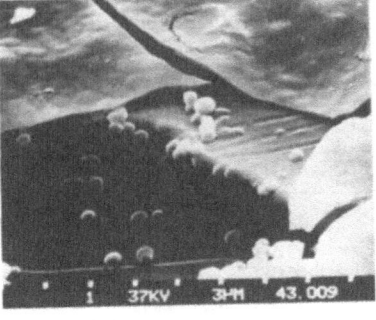

Fig. 7 1.5µs laser pulse 350 mJ cm^{-2} after $CuCl_2$ solution sprayed on pre-heated film

THERMAL HEATING:	
Temperature:	200 - 800C
Duration:	~10s - 120 Hrs
Ambient Atm:	N_2, Ar, H_2/N_2, Air, Cd Vapour
Flux:	In, Ag: $CdCl_2$ (sprayed and evaporated)
SCANNED ELECTRON BEAM:	50μM DIAMETER SPOT
Dwell Time:	500 ps - 5 ns
Power Density:	250 $kW.cm^{-2}$ - 15 $MW.cm^{-2}$
Exposure Time:	0.1s - 20s
Total Energy:	1.2 $kJ.cm^{-2}$ - 3 $MJ.cm^{-2}$
Temperature:	500 - 1000C (in vacuum)
PULSED LASER RADIATION:	
Pulse Width:	30 ns - 2.3μs (FWHM)
Pulse Energy:	300 $mJ.cm^{-2}$ - 2.8 $J\ cm^{-2}$
Wavelength:	347 nm; 480, 505 (broadband) Dye
Ambient:	Air
Flux, etc.	$CdCl_2$ (sprayed); 300 - 400C 30 min prebake

RESULTS:					
Technique:	Conditions:	Cubic → Hex:	Grain Growth:	Notes:	
Thermal	N_2; T ≥ 540C	Yes	Slight		
Thermal	N_2; T > 700C	Yes	Significant	Badly Cracked	
E-Beam	All	No	No	Sintering	
Laser	2.3μs; 200mJ	Yes	Significant	λ = 505 nm	

TABLE I - RECRYSTALLISATION PROCEDURE

COPPER COMPOUND:	DEPOSITION TECHNIQUE:	HEAT SOURCE:	ION EXCHANGE:	HETEROJUNCTION RESPONSE:
CuCl	Evaporation	Thermal	Yes	Yes (Best)
CuCl	Evaporation	Laser	Yes	Yes
Cu	Evaporation	Thermal	Yes	Yes
Cu	Evaporation	Laser	Yes	Yes
Cu	Electroplate	Laser	Yes?	No
$CuCl_2$	Spray	Laser	Yes	No
$CuCl_2$	Spray (CdZnS)	Laser	Yes	No
$CuSO_4$	Spray	Laser	Yes	Yes
Cu_xS	Electrophoresis	Laser	-	-

Comments:
1. Phase of Cu_xS layers not yet determined
2. Heterojunction response defined as measured V_{oc}, I_{sc} under 632 nm illumination
3. ESCA shows Cu^+ surface
4. Layers recrystallised with $CdCl_2$ yield best result

TABLE II - HETEROJUNCTION FORMATION

SPRAYED ZINC-CADMIUM SULFIDE FILMS FOR

BACKWALL $Cu_2S/(ZnCd)S$ CELLS

V. P. Singh, M. C. Bost
J. F. Jordan and D. M. Spitzer, Jr.

PHOTON POWER, INC.
EL PASO, TX. 79906 USA

SUMMARY

$Zn_xCd_{1-x}S$ films ($0 < x < .5$) were sprayed on tin oxide coated glass. Film thickness showed little variation with x; resistivity increased exponentially with x. $Cu_2S/Zn_xCd_{1-x}S$ cells formed on these films were found to have much lower J_o, higher Voc and α, and lower Jsc than their CdS counterparts. Also, the spectral response of their short-circuit current was more sensitive to the white light bias than that of the $Cu_2S/$CdS cell. Photocapacitance showed a quenching band around 900 nm which accompanied an enhancement band for the Voc.

I. INTRODUCTION

Backwall Cu_2S/CdS cells made on sprayed CdS films have considerable advantage over traditional frontwall cells in terms of cost and adaptability to large scale production (1, 2). However, they have not yet reached the relatively high efficiency levels obtained in the frontwall cells made on evaporated CdS (3). Earlier work in our laboratory on the optimization of various material parameters, such as the crystallinity and composition of CdS and thickness of Cu_xS film, resulted in sprayed CdS/Cu_2S cells of up to 8.6% efficiency. Further advances in cell performance were sought in terms of increased open-circuit voltage. Because of a better match with Cu_2S in terms of lattice constant and electron affinity, zinc-cadmium sulfide has been shown to yield higher Voc than cadmium sulfide (4, 5, 6). The technique of CdS film deposition by spray pyrolysis (7) can easily be extended to $Zn_xCd_{1-x}S$ by the addition of a suitable amount of zinc salt in the spray solution. $Cu_2S/Zn_xCd_{1-x}S$ cells made on our zinc-cadmium sulfide film exhibited Voc as high as 0.784 V (8); short-circuit currents, however, were drastically reduced. A further study of the film and junction characteristics was therefore undertaken.

II. EXPERIMENTAL

Zinc-cadmium sulfide films were prepared by spraying a solution containing cadmium chloride and zinc chloride in suitable proportions onto a heated substrate consisting of glass coated with tin oxide. The percentage of zinc content in the film was changed by varying the relative molarity of zinc and cadmium in the spray solution and was measured by atomic absorption spectroscopy. In the following sections, x will denote the atomic percentage of zinc in the film. The total amount of zinc and cadmium content in the solution as well as the substrate temperature during the spray was held fixed for all values of x while x was varied between zero and 0.5. However, because of very high resistivity and low current values yielded by high zinc films fabrication of $Cu_2S/(ZnCd)S$ cells for x greater than 0.22 was deemed impractical.

The films were analyzed for their resistivity and optical absorption. Optical absorption was measured by placing the samples in an integrating sphere and measuring transmission and reflection as a function of wavelength. The films were also examined under a scanning electron microscope for morphology while thickness was obtained with a Tencor Alpha-Step thickness profiler.

Copper sulfide was formed by a solid state reaction process consisting of vacuum depositing cuprous chloride on zinc-cadmium sulfide followed by heat treatment and a rinse. The copper sulfide was then covered with a conductive paste which constituted the positive electrode of the device. The cell area in all cases was 0.785 cm^2.

The junction capacitance was measured at a frequency of 1 MHz with a test signal of 20 mV, r.m.s. Data on the spectral response of photocapacitance and open-circuit voltage were taken with a white light bias along with the secondary illumination from a Bausch and Lomb monochromator; short-circuit current data were taken with and without the bias light.

III. RESULTS

The thickness of the films as a function of zinc content is plotted in Figure 1 and shows little variation with x. Deposition rate of CdS films by spray pyrolysis is a much more sensitive function of the substrate temperature (9) than it is of the added impurities.

X-ray diffraction data on (ZnCd)S films revealed a hexagonal wurtzite structure with a preferred c-axis orientation. Scanning electron micrographs revealed crystallite sizes in the range of 1-3 microns; average crystallite size changed little with zinc content except at higher values of x where a reduction was observed.

Resistivity of the films as a function of x is plotted in figure 2 and shows an exponential increase with zinc content. Earlier studies on single crystal (10), vacuum evaporated (5) and sprayed (11, 12, 9) films also reported that resistivity of this material is a sensitive function of zinc content.

The optical bandgap, as determined from the measurement of spectral absorption, is plotted in figure 3 and shows an approximately linear increase with x indicating the formation of zinc-cadmium solid state alloy in these films. Earlier studies have reported slightly superlinear variation of bandgap with x (9).

Table 1 lists several cell parameters as functions of zinc content. Open-circuit voltage, Voc, increased with x, while the short-circuit current density, Jsc, showed a decrease. The resultant efficiency, η, exhibited a maximum at x = 0.058. We note that these cells were made under a fixed set of process parameters without any attempt at optimization for a given amount of zinc. The diode parameters, α and J_O were computed for each x from the plot of Voc against Jsc under variable intensities of illumination (obtained with a series of neutral density filters). Jsc varied linearly with the intensity of illumination. Current-voltage characteristic was taken to be of the form,

$$J + J_L = J_O \exp(\alpha V) \quad\quad\quad\quad\quad\quad\quad\quad\quad\quad\quad\quad\quad\quad (1)$$

where J_L is the light generated current.

It is interesting to note the large difference in the values of α and J_O between CdS and (ZnCd)S. J_O decreased by four orders of magnitude between x = 0 and x = .058 while α increased by approximately 50%. These step changes in α and J_O may reflect a basic change in the current flow mechanism at the junction or they might be caused by a reduced space charge density in the depletion region while the basic electron transport mechanisms remain the same. More experimental data, especially the dependence of α and J_0 on temperature, is necessary to ascertain this. Measurements in the temperature range -65°C to 20°C show very little change in α for Cu_2S/CdS as well as $Cu_2S/(ZnCd)S$ cells. This indicates that tunneling plays a significant role in carrier transport in these cells. Furthermore, capacitance measurements point to a reduced space charge density and wider depletion layer width (d) with increasing x; d increased from 0.63 microns at x = 0 to 1.01 microns at x = .058.

Normalized spectral response of the short-circuit current with and without white light bias is shown in figure 4 for x = 0 and x = 0.21. In the blue region, $Cu_2S/(ZnCd)S$ cell begins to respond at a shorter wavelength than Cu_2S/CdS because of the wider optical bandgap of (ZnCd)S. In the long wavelength region, bias light strongly affects the relative response of the two cells. With the white light bias, cell responses are comparable, $Cu_2S/(ZnCd)S$ being slightly better. In the absence of bias light, the long wavelength response is substantially reduced, $Cu_2S/(ZnCd)S$ cutting off even earlier than Cu_2S/CdS. The quantum efficiency of these cells depends upon the absorption co-efficient, thickness and diffusion length of copper sulfide and on the junction characteristics like the interface states and the electric field, E. A spike in the energy band diagram at the $Cu_2S/(ZnCd)S$ interface (13) would also affect the quantum efficiency and make it strongly dependent upon E. Capacitance-voltage measurements on these cells yield a space charge density (and hence E) in $Zn_{.21}Cd_{.79}S$ an order of magnitude lower than CdS.

Spectral response of change in capacitance and open-circuit voltage is shown in figure 5 for x = 0. These measurements were taken in the presence of the white light bias. Enhancement of capacitance is seen at 500 nm which corresponds to the bandgap of CdS; a quenching band centered around 900 nm is also present. The quenching of photocapacitance is accompanied by an increase in Voc, indicating that a wider depletion layer corresponds to a higher Voc. These phenomena, similar to the ones observed earlier in Cu_2S/CdS cells by Haines and Bube (14), can be explained in terms of deep levels in CdS near the interface; these levels can act as tunneling centers between Cu_2S and CdS.

IV. CONCLUSION

Since $Zn_xCd_{1-x}S$ provides a better window to the solar spectrum than CdS, backwall $Cu_2S/Zn_xCd_{1-x}S$ cells can potentially yield higher Jsc; this was confirmed with the normalized spectral response data. The reduced magnitude of Jsc may partially be due to the reduced junction field observed in these cells; however, other factors such as a potential spike at the interface between Cu_2S and (ZnCd)S along some or all parts of the film are likely to be playing the dominant role. Further investigations are necessary to ascertain this and also to determine the role played by tunneling in the carrier transport across these junctions.

REFERENCES

1. J. F. Jordan, Proceedings of the 11th. IEEE Photovoltaic Specialist Conference, p. 508, (1975).
2. A. M. Barnett and A. Rothwarf, IEEE Trans. on Electron Devices, ED-17, pp. 615-30 (1980).
3. R. B. Hall, R. W. Birkmire, J. E. Phillips and J. D. Meakin, Proceedings of the 15th. IEEE Photovoltaic Specialists Conf., p. 777 (1981).

4. A. Rothwarf, Proc. International Conf. on Solar Electricity, Toulouse, CNES and CNRS, p. 273, (1976).
5. L. C. Burton, Appl Phys. Letters, 29, p. 126 (1976).
6. L. C. Burton, Solar Cells, 1, p. 519-174 (1979-80).
7. R. R. Chamberlin and J. S. Skarman, Journal of the Electrochemical Society, Vol., 113, p. 86 (1966).
8. V. P. Singh and J. F. Jordan, IEEE Electron Device Letters, Vol. EDL-2, No. 6, June 1981, p. 137.
9. N. A. Reshamwala, W. B. Hsu, L. C. Burton and V. P. Singh, Proc. Third E. C. Photovoltaic Solar Energy Conference, Oct. 1980, Cannes France, p. 787.
10. E. A. Davis and E. L. Lind, J. Phys., Chem. Solids, Vol. 29, 1968.
11. R. S. Feigelson, A. N. Diaye, S. Y. Yin and R. H. Bube, J. Appl. Phys., 48, No. 7, July 1977, p. 3162.
12. A. Banerjee, P. Nath, V. D. Vankar, and K. L. Chopra, Phys. Status Solidi A 46, p. 723 (1978).
13. K. W. Boer, Phys. Status Solidi A. 49, p. 255 (1978).
14. W. G. Haines and R. H. Bube, IEEE Trans., on Electron Devices, ED-27, No. 11, p. 2133, Nov. 1980.

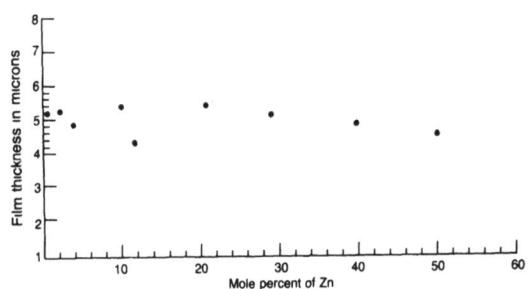

Fig. 1 - Mole percent zinc versus film thickness

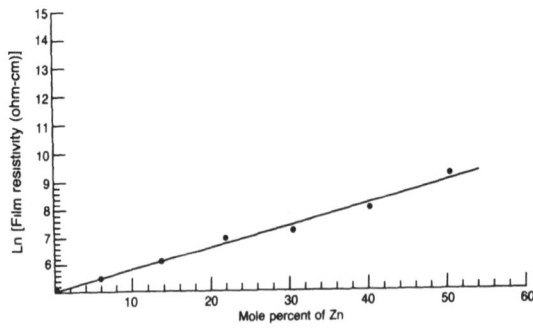

Fig. 2 - Mole percent zinc versus Ln(resistivity)

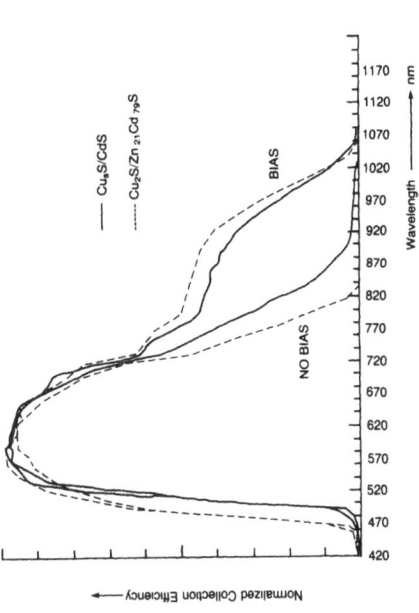

Fig. 4 — Normalized collection efficiency of Cu_2S/CdS and $Cu_2S/Zn_{.21}Cd_{.79}S$ cells with and without bias light.

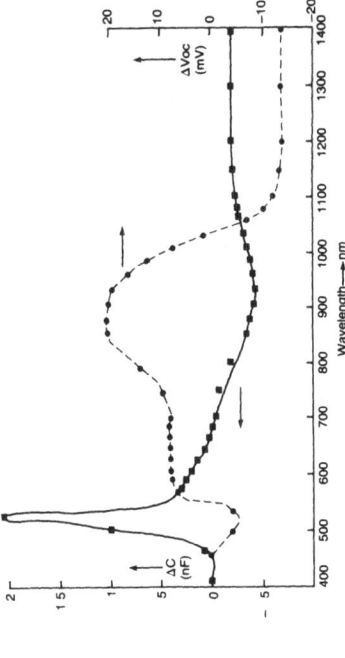

Fig. 5 — Spectral response of photocapacitance and open-circuit voltage for the Cu_2S/CdS cell (white light bias) $C_0 = 3.26\ nF$ $V_0 = 394\ mV$

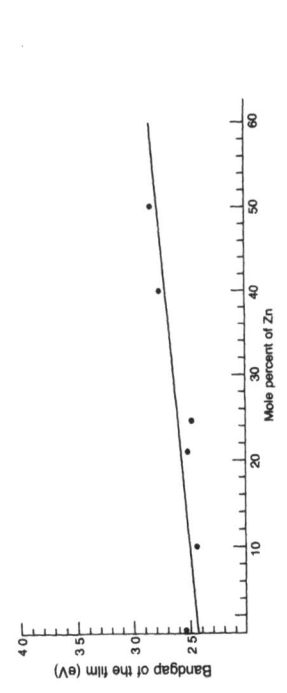

Fig. 3 — Mole percent zinc versus optical bandgap of the film

PARAMETERS	x = 0	0.058	106	119	217
Voc (Volts)	0.501	0.572	623	650	683
Jsc (mA/cm²)	20	17.8	10.9	9.4	2.7
η (%)	6	6.6	4.8	4.0	0.8
x (Volts⁻¹)	19.7	30.9	31.8	28.4	27.9
Jo (mA/cm²)	10⁻³	2.6×10⁻⁷	1.6×10⁻⁷	1.7×10⁻⁸	1.9×10⁻⁸

TABLE I — Cell parameters as functions of zinc content (illumination intensity = $100\ mW/cm^2$)

POSTER GROUP P7

THIN FILM SOLAR CELL TECHNOLOGY

- Amorphous Semiconductors

Large area and high efficiency a-Si:H solar cell

Post-hydrogenated CVD amorphous silicon p-i-n diodes for photovoltaic applications

Stability of amorphous silicon solar cells with pin structure

Carrier conduction in a-Si:H solar cells

Optical optimization of amorphous silicon solar cells

Large area hydrogenated amorphous silicon for photovoltaic application

Novel plasma chemical methods for doping a-Si:H

Electronic properties of doped amorphous SiO_x

- Crystalline Silicon Based Cells

Antimony doping in vacuum deposited thin film silicon photovoltaic cells

Photovoltaic performance of CdS heterojunctions on poly-crystalline silicon

- II-IV Compound Solar Cells

Temperature dependence of the IV-characteristic of Cu_2S-CdS thin film solar cells and related phenomena

Continuous deposition of photovoltaic grade CdS sheet at the unit operations scale

$Cu_xS(p)$ - $CdZnS(n)$ - $CdS(n^+)$ evaporated thin film solar cells

Thin film Cu_2S/CdS junctions produced by evaporation and sputtering: effect of thermal treatments in vaccum

Airless sprayed CdS solar cells

Electrochemical preparation and conditioning of Cu_2S for Cu_2S-CdS solar cells

Physico-chemical properties of Cu_xS

Photovoltaic behaviour of CdSe thin film solar cells

All thin film n-CdTe/ITO Solar Cell

Preparation of high purity II-VI compounds by laser annealing

- Other Thin Film Systems

Zn_3P_2 thin-film solar cells

Cadmium sulfide polyacetylene photovoltaic hererojunction

Antimony sulphide thin films

LARGE AREA AND HIGH EFFICIENCY A-SI:H SOLAR CELL

Y. HIGAKI, M. KATO, M. AIGA, and Y. YUKIMOTO
LSI R & D Lab., Mitsubishi Electric Corp.

Summary

A conversion efficiency of 5.4% for a-Si:H solar cell on 10x10 cm^2 stainless steel substrate has been obtained. Large area a-Si:H solar cells were fabricated by the C-coupled (60x60 cm^2 parallel plates) glow discharge method. The stainless steel substrate were polished down to 0.2 µm maximum roughness. Boron concentration profile was measured by IMA technique. The inverted (nip/s.s.) configuration is superior to normal (pin/s.s) type for a-Si:H solar cell, since boron atoms likely pile up at the interface between the substrate and the a-Si:H even though B_2H_6 gas is not introduced. The undesirable boron atoms seem to come from the susceptor or the side wall of the reaction chamber, which cannot be etched out easily. Power dissipation at the transparent conducting film, at the comb electrode, and at the bus electrode on this large area solar cell were calculated respectively. It was concluded that the loss at the bus electrode is the largest and with the reduction of its resistance, the cell performance would be improved. As a result, Voc of 0.85V, Jsc of 11.5mA/cm^2, F.F. of 0.554 and efficiency of 5.4% have been obtained.

1. Introduction

In these few years there has been considerable advancies in the technology of amorphous silicon (a-Si:H) produced by the glow discharge decomposition of silane. These developments and researches are proceeded vigorously especially in Japan. Nowadays, the conversion efficiency of a-Si:H solar cells reaches up to 8%.

However, it can be said that the properties of amorphous films are not understood enough, and the difference between the idealistic film and the realistic one is not fully recognized. It is true that a slight difference in processing may leads exceedingly different results.

We investigated the impurity atom distribution profile in real a-Si:H films, which is very important for the p-i-n arrangement.

The degradation of the performance of the large area solar cell is inevitable compared with small area one. We analyzed that the power loss on the large area S.C. mainly occurs at the surface electrode. And with the decrease of electrical resistance of surface metal electrode, fill factor (F.F.) and conversion efficiency have been remarkably improved.

2. a-Si:H deposition

A-Si:H films were prepared in the G.D. plasma deposition system which had C-coupled 60x60 cm^2 parallel plates. The upper plate was an anode and used as a susceptor. 9 substrates in the size of 10x10 cm^2 can be inserted in it upside down at one charge.

SiH_4 gas diluted with Ar was introduced for the undoped a-Si:H deposition and PH_3 diluted with Ar, B_2H_6 with Ar was added for n-type a-Si:H film and for p-type, respectively. The doping ratio was 1 mole per cent. All the undoped and doped a-Si:H were deposited in a same single chamber, and this fact is one of the most important point which will be described next section.

The stainless steel substrates were chemical-mechanically polished down to 0.2 μm maximum roughness for fear of the generation of defects on a-Si:H film throughout the 10x10 cm^2 area.

3. IMA measurements

The impurity atom distribution in a-Si:H film were investigated with IMA technique.

Fig.1 shows boron profile in undoped a-Si:H film. Boron concentration was calibrated by the ion implanted boron in single crystal silicon.

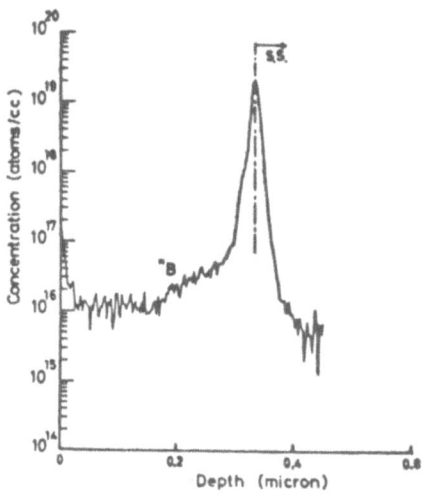

Fig.1 Boron concentration profile in non doped a-Si:H

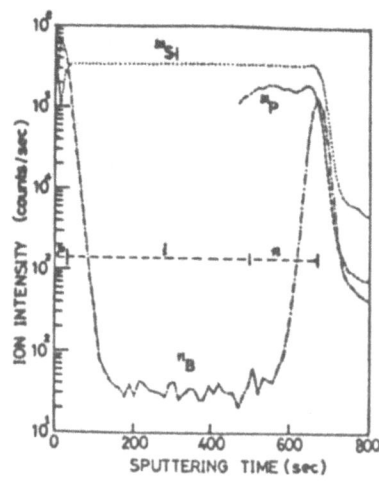

Fig.2 Depth profile of a-Si:H pin cell

The concentration of about 10^{16} atoms/cc inside of the film and over 10^{19} atoms/cc at the interface of the film and the stainless steel substrate are detected, nevertheless it is an undoped sample. No B_2H_6 gas was introduced. This pile-up of boron atoms at the interface seems to be inevitable as long as the doped and undoped films are deposited in a same reaction chamber. Boron atoms cannot be cleaned or etched out easily. The reason of this pile-up phenomenon is not clear at present. But it may be caused by some electrical interactions because it does not appear at the a-Si:H and a-Si:H interface. A-Si:H has a very high resistivity.

The example of p-i-n cell is shown in Fig.2 . As expected, boron atoms pile up at the interface of n-layer and substrate. Hence, it is easy to suppose that the phosphorus doped n-layer would be compensated in some part and cannot act ideally.

Therefore, we concluded that the inverted n-i-p cell is more advantageous than p-i-n cell on the stainless steel substrate. The pile-uped boron atoms are masked by doped boron atoms when the plasma deposition are performed in the sequence of p, i, n.

4. Power dissipation at surface electrode

As the area of solar cell becomes large, the problem of internal power dissipation becomes an issue.

Dissipation occurs mainly at the transparent conducting film (ITO) the comb electrode (metal), and bus electrode (metal) as illustrated in Fig.3 . We analyzed the power disspation at these three electrodes with regard to 10x10 cm^2 cell. Contact resistances between a-Si:H and ITO, and between ITO and metal electrode were ignored in this analysis.

The comb spacing dependences of power loss for three electrodes are calculated and shown in Fig. 4 . Here, resistivity of ITO is $3x10^{-4}$ Ω-cm, thickness of ITO is 700Å, resistivity of Ag electrode is

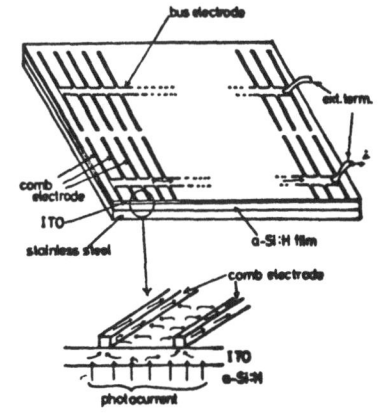

Fig.3 Illustration of surface electrode

2×10^{-6} Ω-cm, thickness of Ag electrode is 1 μm. As can be seen, the most large power loss occurs at the bus electrode in any comb spacing. And Fig.4 indicates that one should make comb spacing narrower and decrease the resistance of bus electrode somhow.

We put kovar lead which is 2 mm wide and 0.1 mm thick on the bus electrode with solder in order to decrease the resistivity of it. The improvement of cell performance with this process is shown in Fig.5 .

Voc, Jsc and especially fill factor are improved and hence conversion efficiency is improved more than 15%. The most excellent data which is achieved are Voc of 0.85V, Jsc of 11.5mA/cm^2, F.F. of 0.554 and η of 5.4% under AM1 100mW/cm^2 illumination.

This work was supported by the Agency of Industrial Science and Technology under the contract of the Sunshine Project.

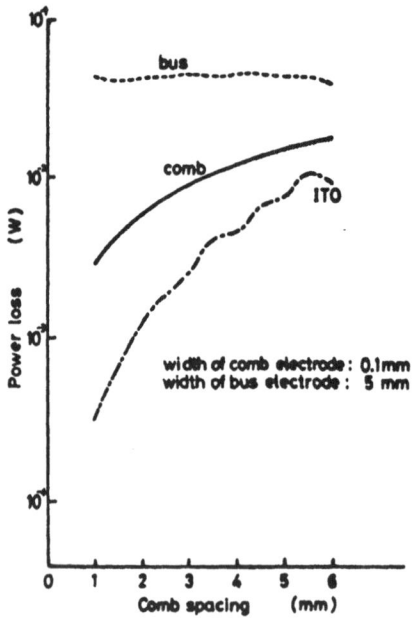

Fig.4 Power loss vs comb spacing

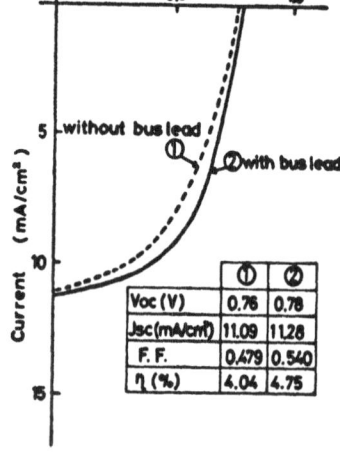

Fig.5 Improved I-V characteristics of nip cell with bus lead

POST-HYDROGENATED CVD AMORPHOUS SILICON p-i-n DIODES
FOR PHOTOVOLTAIC APPLICATIONS *

N. SZYDLO, E. CHARTIER, N. PROUST, J. MAGARIÑO and D. KAPLAN
Thomson-CSF, Laboratoire Central de Recherches,
B.P. N° 10, 91401 Orsay (France)

Summary

We present the first report on a-Si p-i-n photovoltaic
structures prepared by thermal decomposition of silane
(CVD) followed by a plasma post-hydrogenation treatment.
The technical and economical advantages of this process
are discussed. The main technical difficulty is stability
of the doped thin top layer under the post-hydrogenation
treatment. Preliminary results presented will concern
structures with relatively thick top layers (50-200nm).
These yield open-circuit voltages exceeding 0.7 volts.

1. INTRODUCTION

The majority of amorphous silicon solar cells have used
to date thin films obtained by plasma decomposition of
silane (glow discharge or plasma CVD). Thermal chemical vapor
deposition (CVD) is an alternative method to obtain amorphous
silicon films on large areas. As deposited films contain a
low concentration of hydrogen : about 0.3% of the silicon
concentration. These films must be post-hydrogenated in a
hydrogen plasma in order to passivate dangling bond defects
and to improve the electronic transport properties. By this
method one can adjust the quantity of hydrogen typically
between 1 and 10%[1]. The incentives for exploring this
method are both technical and economic. The technical arguments stem from the fact that by adjusting the hydrogen
content one can optimize the optical absorption and also the
quality of the n^+ and p^+ type layers, which can be made an
order of magnitude more conductive than for the glow discharge process. The economic arguments, beside the simplicity of
the CVD process, come from its ability, using low pressure
deposition, to deposit films on many substrates at the same
time, stacked parallel to each other. This is common practice
in the electronic industry and is not practically feasible
in plasma CVD systems, which inherently deposit one layer at
a time. Since much of the economic competitiveness of amorphous silicon depends on obtaining large production rates
for each deposition machine to decrease investment costs,
this aspect will be of great importance when large scale
production of solar panels is envisaged. Schottky diodes with
photovoltaic performances comparable to those obtained by

the glow discharge method have been reported previously in post-hydrogenated CVD amorphous silicon[2,3,4]. We report here preliminary results on p-i-n diodes prepared by this method. The electrical characteristics and photovoltaic properties of these diodes and the related physical properties of the undoped and doped films which constitute the structure of the junction are discussed.

2. PREPARATION DETAILS AND PHYSICAL PROPERTIES

Amorphous silicon films have been deposited from hydrogen diluted silane in a conventional atmospheric CVD system, at temperatures ranging between 500 and 600°C[1]. Boron and phosphorus doping were achieved by adding B_2H_6 and PH_3. The a-Si layers were deposited on either molybdenum or p-type crystalline silicon substrates, the latter being in some cases covered with a thin evaporated molybdenum film. After deposition of the three layers corresponding to the p-i-n structure, the samples are treated around 400°C in a low pressure hydrogen plasma excited by microwaves during typically 2 hours.

To achieve good photovoltaic performances, the absorption of light in the doped front layer must be minimized. The optimized thickness is around 10nm for a-Si glow discharge p-i-n photovoltaic devices. From this point of view, we have encountered major difficulties with the present technique : thin top layers (<30nm) did not generally survive the post-hydrogenation treatment which produces either etching of this layer and/or concentrations of hydrogen at the interfaces resulting in defects in the form of bubbles. Thicker top layers, of thickness ranging between 50-200nm proved to be more resistent to the hydrogenation process and the results presented here will be confined to such structures, which are clearly not optimal for photovoltaic application.

Figure 1 shows the phosphorus, boron and deuterium concentrations as a function of depth, obtained by secondary ion mass spectrometry (SIMS). The doping ratios in the solid phase are $\sim$ 5 and $\sim$ 200 times the ratios in the gas phase respectively for P and B. The junction is not graded, as clearly shown in the figure. The deuterium concentration at the surface is $\sim$ 6% and in half of the undoped layer it is $\sim$ 2%, these concentrations being a factor two to five lower than those obtained in glow discharge films[5]. One can note in the figure a rapid decrease of the deuterium profile in the undoped layer near the i/p^+ interface and an accumulation of deuterium in the boron doped layer related to a segregation of the diffused deuterium from the undoped layer to the doped layer. The ratio between the deuterium concentrations at the two sides is of the order of three. This result is also observed by deuterium diffusion in the p-i-n structure inverted with respect to the one described here.

The electrical and optical properties of CVD undoped and phosphorus doped films and the implications to the photovoltaic performances have been previously reported[2]. In the case of high boron doping, as deposited films show higher room temperature conductivities ($10^{-1}\Omega^{-1}cm^{-1}$) and lower activation energies ($\sim$0.1eV) as compared with glow discharge

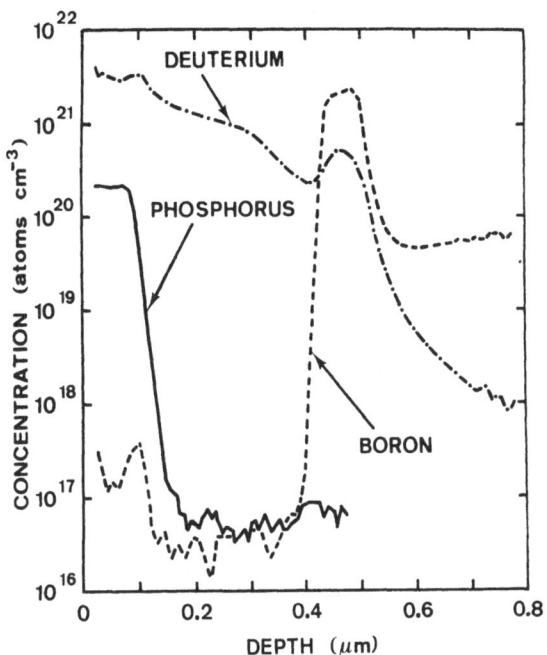

Fig. 1 - Phophorus, Boron and Deuterium concentration profiles. Phosphine and diborane concentrations in the gas phase are respectively 0.1% and 0.03% for the two doped layers. Deuterium diffusion is made at 400°C during 90 mn.

boron doped amorphous silicon[1]. A mild post-hydrogenation does not change very much the measured conductivities. On the contrary for comparable post-hydrogenation treatments as those used in this work, the conductivity decreases by one or two orders of magnitude and the activation energy increases. This effect is associated with a reduction of the doping efficiency produced by the introduction of hydrogen[1,6]. Because of the shape of the hydrogen profile the concentration of hydrogen in the p layer is lower when it is deposited first than when it is deposited last. This is why we chose the first configuration. Note however that there is a trade-off between efficient hydrogenation of the undoped layer and excessive hydrogen of the boron doped layer.

3. CURRENT-VOLTAGE CHARACTERISTICS

The J-V characteristics of two diodes, after mesa etching to avoid edge contributions to the current, are shown in figure 2. At low forward voltages (<0.5 volts) the current increases exponentially with the voltage, and an ideality factor $n \sim 2$ is obtained by fitting these characteristics. The saturation currents are respectively 2×10^{-9} and 10^{-8} A/cm^2 for these two diodes.

Rectification ratios of 10^7 at 2 volts are obtained. In both cases the reverse current increases between one and two orders of magnitude from 0 to 2 volts.

The large values of n and J_0 can be explained by the importance of the recombination in these diodes[7]. A possible explanation of the rather large recombination currents observed is the ineffective passivation of defects, in particular at the i/p$^+$ junction where the hydrogen concentration is low.

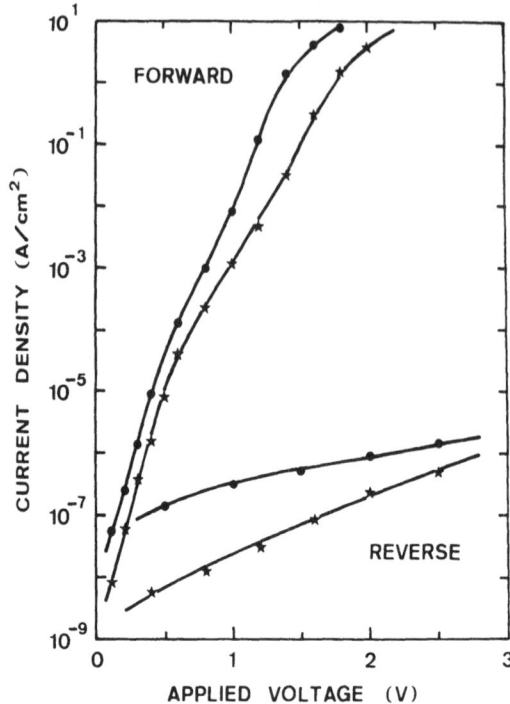

Fig. 2 - Dark current-voltage characteristics of two diodes (Substrate/(1000nm)p$^+$/(500nm)i/(100nm)n$^+$/Au),
● <Si>/Mo substrate,
★ Mo substrate.

4. PHOTOCURRENT MEASUREMENTS AND PHOTOVOLTAIC PERFORMANCES

The results presented here concern p-i-n structures with thicker top films of 50-200nm. In this case, the photovoltaic performances are mainly reduced by the strong absorption in the phosphorus doped layer where photogenerated carriers are ineffective. For example, at λ=600nm the calculated transmitted light in the undoped layer through a front layer of 100nm thickness is only $\sim$29% compared with $\sim$54% for a top layer of 10nm.

The photocurrent normalized at constant incident photon density as a function of light is shown in figure 3. The strong absorption of light in the front layer is manifested by the rapid decrease of the photocurrent at short wavelengths. The photovoltaic performances on these diodes (area=2.35mm^2) have been tested under AM1 solar simulator conditions (100mW/cm^2). Open-circuit voltages of 0.72 volts have been obtained. Short-circuit currents of 2-4.5 mA/cm^2 are a factor $\sim$ 2 lower than those obtained in Pt-Schottky diodes by the same method[2,4]. The fill factor of $\sim$ 0.5 should be improved by an optimized passivation of defects by hydrogenation.

In conclusion, p-i-n jucntions have for the first time been obtained by CVD deposition and post-hydrogenation. Good quality diodes are obtained, the limitation of the actual photovoltaic performances being the excessive thickness of the top doped layer. We are currently working on improvements of deposition and hydrogenation conditions to achieve thin top layers.

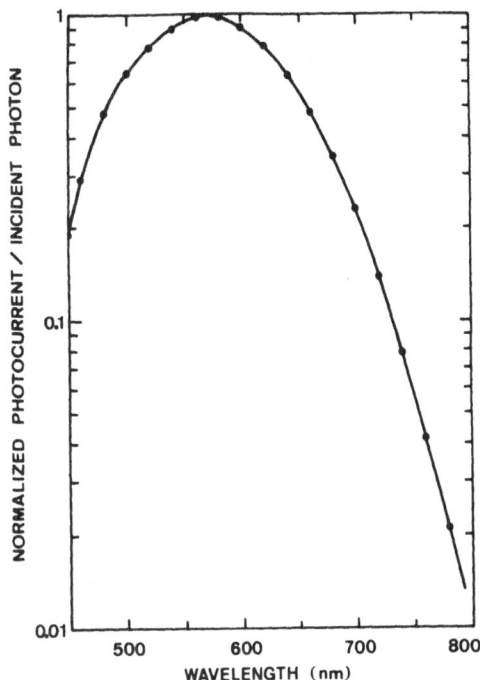

Fig. 3 - Normalized photocurrent versus wavelength for the same structure as in Fig. 2.

Acknowledgments : The authors wish to acknowledge cooperation with A. Huber, G. Morillot and R. Poirier, and technical assistance provided by P. Landouar, E. Criton and R. Kasprzak.

REFERENCES

(1) J. MAGARINO, D. KAPLAN, A. FRIEDERICH and A. DENEUVILLE, Phil. Mag.B, 45, 285 (1982).
(2) N. SZYDLO, D. KAPLAN and R. POIRIER, Proceedings of Third EC Photovoltaic Solar Conference, Ed. by W. Palz, Reidel, Dordreeht, Holland, 317 (1980).
(3) M. HIROSE, J. Phys. (Paris) 42, Suppl. N°10, C4-705 (1981)
(4) N. SZYDLO, J. MAGARINO and D. KAPLAN, to be published in J. Appl. Phys. (1982).
(5) D.E. CARLSON and C.R. WRONSKI, Amorphous Semiconductors in "Topics in Applied Physics", Ed. M.H. Brodsky, Springer-Verlag Berlin, 36, 287 (1979).
(6) J. MAGARINO, A. FRIEDERICH, D. KAPLAN and A. DENEUVILLE, J. Phys. (Paris) 42, Suppl. N°10, C4- 737 (1981).
(7) I. CHEN and S. LEE, J. Phys. (Paris) 42, Suppl. N°10, C4 - 499 (1981).

* This work was partially supported by Commissariat à l'Energie Solaire.

STABILITY OF AMORPHOUS SILICON SOLAR CELLS WITH pin STRUCTURE *)

W. KRÜHLER, M. MÖLLER, H. PFLEIDERER, R. PLÄTTNER
and B. RAUSCHER
Research Laboratories of Siemens AG
D-8000 Munich 83

Summary

Solar cells made in our laboratory from hydrogenated amorphous Silicon with pin structure show an efficiency of up to 7,4 % on 6 mm^2 and up to 5 % on 2" x 2". Degradation experiments with illumination and under forward bias in the dark, were performed. These measurements as well as capacitance and spectral response measurements show, that optically or electrically injected carriers induce recombination enhanced defect reactions, which lead to an increase of the density of states and, respectively, to a decrease of the $\mu\tau$-product, mainly of that of the holes. Compensation of the normally n-type i-layer by doping with Boron leads to higher cell stability due to a shifting of the Fermi level closer to midgap. The effect of different concentrations of Boron in the i-layer on its dark and photoconductivity was investigated. A maximum of photoconductivity is seen for 5 ppm. Preliminary tests on cells suggest, that the degradation can be minimized by a controlled Boron concentration of the i-layer.

1. INTRODUCTION

During the past years solar cells made from hydrogenated amorphous Silicon (a-Si:H) have made good progress with respect to their conversion efficiency and cell area /1/. Cells of the type glass/metal/pin/ITO produced by an rf glow discharge from SiH_4 in our laboratory have been described in /2/. Recently, we obtained an AM1-efficiency η of 7,4 % (V_{oc} = 840 mV, I_{sc} = 13.8 mA/cm^2, FF = 64 %, area 6 mm^2). The red response and therefore I_{sc} of this cell were improved by using Ag as a bottom contact which is more reflecting /3/ than the Cr or steel normally used. On large area cells (2 x 2 inch) deposited on steel and with tin soldered grid contacts an efficiency of 5,0 % (V_{oc} = 800 mV, I_{sc} = 10,9 mA/cm^2 on the active area, FF = 57 %) was reached. Cell stability which is important for future application of pin cells was the subject of the following degradation experiments with 4 - 6 % efficiency cells.

2. CELL STABILITY

In air at room temperature pin cells have a good shelf live. Within a period of about two years no measureable

---

*) This work has been supported under the technological program of the Federal Department of Research and Technology of the FRG. The authors alone are responsible for the contents.

change of their electrical properties was found.

On the other hand, pin cells can change their properties during exposure to light /4/. Our normally prepared cells (without intentionally doping of the n-type i-layer) exposed to AM1 illumination show an instability of η in a typical range between curve A und B (fig. 1). The cell stability varies with the deposition conditions, the different concentrations of residual Boron in the reactor chamber being most likely the main parameter. Within approx. 20 hrs the efficiency can decrease to 70 % of the initial values, while cells whose i-layer has been compensated by Boron (see chapt 5) are more stable (curve C).

The degradation of η is mainly due to a decrease of I_{sc} and FF, while V_{oc} remains stable in general. This is shown in fig. 2, in which a heavily degrading cell is taken for demonstration. It is well known that the initial value of η can be reestablished within half an hour by annealing the cell at a temperature of about 180°C. In a-Si:H layers the decrease of the dark- and photoconductivity during exposure to light (Staebler-Wronski-Effect, SWE) is generally explained by an increase of the density of defect states and an associated formation of recombination centers /5,6/ accompanied by a shift of the Fermi level E_F to midgap. To be sure, that in the case of pin solar cells the SWE is responsible for the degradation rather than some interface effects, degradation experiments were carried out on solar cells with monochromatic blue light (λ = 476 nm) using its low penetration depth (40 nm). Beside the pin/ITO cell deposited on metal a structure glass/ITO/pin/ITO was prepared which allows to separately illuminate each interface (fig. 3). For comparison, measurements were carried out with red light (λ = 647 nm) leading to homogeneous illumination throughout the i-layer of 0,5 μm thickness.

The experiments indeed showed that in all cases a to d (fig. 3) and under AM1 illumination (open circuit conditions) degradation occurs with similar results. Therefore we conclude that the illumination degrades the bulk of the i-layer, according to the SWE. The results can be interpreted in the way that the carriers generated at some interface under blue light are injected into the i-layer where they recombine and cause a degradation homogeneously distributed over the entire i-layer which is what obviously happens when the sample is homogeneously illuminated with red light. The assumption that it is not the light itself but the recombination of charge carriers which causes defects and therefore the degradation is also affirmed by the fact that our unstable pin cells degrade similarly when carriers are injected by a forward bias of $\approx$ 1 V in the dark as was first proposed in /4/. On the other hand they do not degrade when exposed to illumination and to a reverse bias of 2 V simultaneously whereby the recombination current is heavily reduced in favor of the external current.

3. SPECTRAL RESPONSE

To gain the $\mu\tau$-product of both carriers, electrons and holes, in the annealed or normal (N) case and in the light-soaked (LS) case, we measured the spectral response with chopped monochromatic light and constant background illumination (AM1) by the lock-in technique /2/. The values for the $\mu\tau$-products were calculated using a charge collection model described also in /2/. Fig. 4 shows a typical spectral response in the N-case and in the LS-case at zero bias. A decrease of the response mainly in the blue range of the spectrum can be observed. The analysis of the experimental data yields the following values for the $\mu\tau$-product of electrons and holes in the N- and LS-case for a heavily degrading cell:

$\mu_e \cdot \tau_e$ (N) = 5.6 $10^{-9} cm^2/V$; $\mu_p \cdot \tau_p$ (N) = 9.9 $10^{-9} cm^2/V$
$\mu_e \cdot \tau_e$ (LS) = 4.5 $10^{-9} cm^2/V$; $\mu_p \cdot \tau_p$ (LS) = 1.5 $10^{-9} cm^2/V$.

In less degrading cells the change of $\mu\tau$-values show a similar behavior but with less extend. In both cases (LS or N) the electric field was assumed to be constant over the active i-layer. Therefore the possibility of field distortion being the reason for enhanced recombination /4/ is not included in our model. Actually the electric field, which is derived from the voltage where the low signal currents /2/ change sign, does not change significantly when going from state N to LS. However, the results show reduced $\mu\tau$-products for both carriers in the LS-state. The greater reduction for the hole $\mu\tau$-product accounts for the drop in the blue response. The fact that the carrier lifetime τ is reduced by enhanced recombination due to new recombination centers in the LS-state was already seen for a-Si:H layers in photoconductivity studies /6/.

4. IMPEDANCE LOCUS OF pin CELLS

To obtain information on the density of states in the light soaked cells we measured the impedance locus of a heavily degrading pin cell and calculated the frequency dependence of the capacitance $C(\nu)$ as described earlier /2/ with the bias voltage U_B as parameter for the N- and LS-case. The impedance curves are well described by semicircles. Therefore the equivalent circuit used for the $C(\nu)$-analysis /2/ is an RC-circuit with R parallel to C. R and C are interpreted as the resistance and capacitance of the i- layer. Fig. 5 shows the $C(\nu)$ curves at U_B = 0.3 V for the N-case (a), for a LS-case with 19 h AM1 (b) and for a cell degraded 16 h under forward bias of + 1.2 V (c). The capacitance value beyond 1000 Hz can be attributed to the geometric capacitance of the i-layer whereas for 100 Hz $< \nu <$ 1000 Hz the shallow states in the gap follow the modulation of the applied alternating voltage, which therefore shows a slight increase. Only the deep gap states are affected by the degradation experiments since the curves for (b) and (c) deviate essentially from curve (a) only below 10 Hz. This corresponds to an increase of the density of states for (c) compared to (a) of $\approx$ 20 % (at 1 Hz).

5. STABILITY OF B-COMPENSATED pin-CELLS

From measurements of the conductivity Tanielian et al. /7/ found, that the SWE does not occur in a-Si:H when E_F is in midgap. Assuming that this holds also for the i-layer in the pin cell, those cells whose i-layer has slightly been compensated with Boron, should be more stable. Fig. 6 shows the activation energy E_A of the dark conductivity σ_D, as well as the photoconductivity σ_{ph} and σ_{ph}/σ_D, as a function of a slight B-doping. B-doping of about 2 to 5 ppm should lead to a material with E_F in midgap and with a high photoconductivity. Therefore we made pin cells with B-compensated i-layer and tested them for optical stability. Indeed, we found less degradation for these cells (fig. 1, curve C). Because of the residual Boron in the reaction chamber following the deposition of the p-layer the reproducible doping of the i-layer by small amounts of Boron is still a problem and reflects the variation in stability behavior given in fig. 1 (A-B range). We expect, to eliminate the degradation completely by better control of the deposition in separated chamber reactors.

6. CONCLUSION

Our results confirm the interpretation of the Staebler-Wronski-Effect of pin cells that the density of states in the i-layer increases by the recombination of charge carriers produced optically or by electrical injection. It can be concluded from the capacitance measurements that only the deep gap states are affected. From our current collection model and the spectral response measurements it follows that the $\mu\tau$-products decrease for holes especially which leads us to the conclusion that the reduced lifetime τ is due to increased recombination at activated recombination centers. These centers which may be active corresponding to their charge state are neutralized when the Fermi level is near midgap. Therefore the degradation of cells can be minimized through compensation of the normally n-type i-layer by doping slightly with Boron.

7. REFERENCES

/1/ Y.Kuwano, M.Ohnishi: Proc. of the 3rd E.C. Photovoltaic Solar Energy Conf., Cannes, France (1980) 309-316

/2/ R.D.Plättner, H.Pfleiderer, B.Rauscher, W.Krühler and M.Möller: Proc. of the 15th IEEE Photovoltaic Specialists Conf., Orlando, USA (1981) 917-921

/3/ M.Ondris and W. den Boer: Proc. of the 3rd E.C. Photovoltaic Solar Energy Conf., Cannes, France (1980) 809-814

/4/ D.L.Staebler, R.S.Crandall and R.Williams: Appl. Phys. Lett. 39 (1981) 733-735

/5/ D.L.Staebler and C.R.Wronski: J. Appl. Phys. 51 (1980) 3262-3268

/6/ C.R.Wronski and R.E.Daniel: Phys. Rev. B, 23 (1981) 794-804

/7/ M.H.Tanielian, N.B.Goodman and H.Fritzsche: Proc. of the 9th Intern. Conf. on Amorphous and Liquid Semiconductors, Grenoble, France (1981) C4-375--C4-378

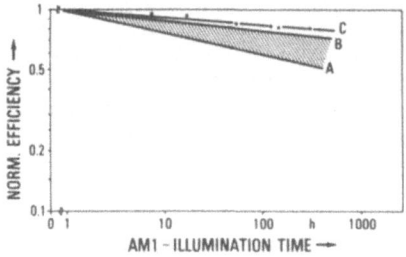

Fig. 1. pin cell stability range
A-B: range for uncompensated cells
C: for compensated cells

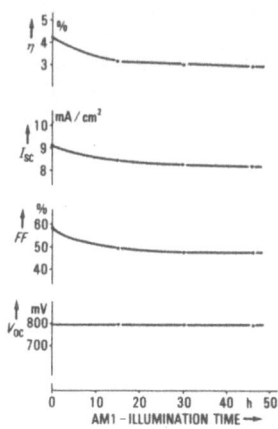

Fig. 2. Dependence of cell parameter on the AM1 illum.

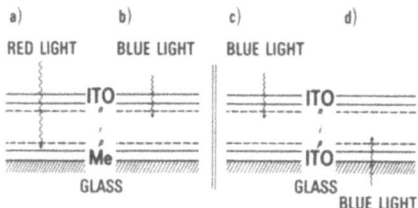

Fig. 3. Stability test after monochromatic light exposure

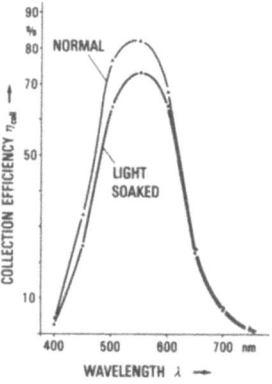

Fig. 4. Collection efficiency of a pin cell before and after illuminat.

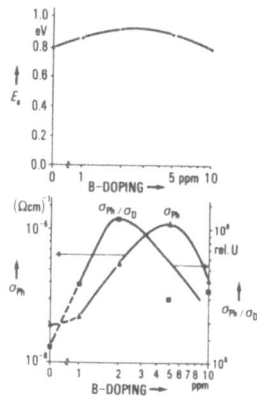

Fig. 6. Activation energy E_A, dark and photoconductivity (σ_D, σ_{ph}) in B-compensated a-Si:H

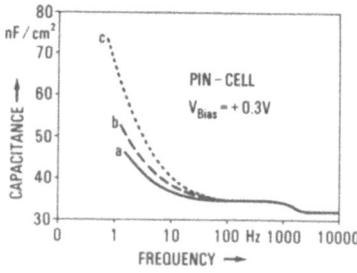

Fig. 5. C(ν)-curves of a pin cell ($U_B = 0.3$ V); a) in the annealed state, b) in the light-soaked state, c) in a degraded state (16 h, 1.2 V, dark)

- 758 -

CARRIER CONDUCTION IN a-Si:H SOLAR CELLS

M. K. Han, P. Sung, R. Lahri and W. A. Anderson
Department of Electrical and Computer Engineering
State University of New York at Buffalo
4232 Ridge Lea Road
Amherst, New York 14226

Summary

The conduction process in a-Si:H Schottky structures ($I-N^+/SS$) is found to be barrier controlled at low forward bias (< 0.2 V) and bulk controlled at large forward bias (> 0.6 V). The I-layer shows evidence of space charge limited current conduction (SCLC) in the dark. A model incorporating these considerations has been presented to develop equations for the terminal I-V characteristics of these structures. Dark I-V characteristics of the P-I-N structure are also explained on the basis of the above model. Dark C-V-f characteristics of the Schottky structures also support the existence of SCLC conduction in these structures.

1. INTRODUCTION

Dark I-V-T measurements showing evidence of space charge limited current (SCLC) conduction in a-Si:H Schottky structures (metal-I-N^+/substrate) have been reported [1,2]. Large dark resistivity (10^6-10^{10} Ω-cm) combined with the amorphous nature of the I-layer makes it amenable to SCLC conduction in the dark. The dark I-V characteristics of these structures are found to have an exponential I-V relationship ($I \alpha \exp qV/nkT$) at low forward bias (< 0.2 V) and a power law relationship ($I \alpha V^m$) at high forward bias (> 0.6 V). This corresponds to a barrier controlled and bulk controlled conduction process, respectively. In the present work, a model is proposed to distinguish the two regimes and to calculate the voltage drops across the depletion and neutral bulk regions under an external bias. Dark C-V-f characteristics of the Schottky structures, supporting this model, have also been presented. The model is later extended to explain the dark I-V characteristics of P-I-N a-Si:H structures.

2. THEORETICAL ANAYLSIS

a-Si:H Schottky structures ($I-N^+/SS$) can be divided into two regions: the depletion region and the neutral bulk region. The bottom N^+-region can be treated as a part of the ohmic contact. Any external bias (V) would be divided across the two regions according to their dynamic resistance. The voltage drop across the depletion region (V_1) and the neutral bulk region (V_2) can be obtained from the theoretical I-V characteristics of the corresponding region [3]. The depletion region I-V characteristics are given by the diffusion theory of metal-semiconductor rectification [4]. On the other hand, the neutral bulk region shows a power law relationship,

$I \propto V^m$. Trap distribution in this region is found to be exponential [2] and is characterized by

$$N_t(E) = N_o \exp\left(\frac{E-E_c}{kT_t}\right) \qquad (1)$$

where $N_t(E)$ is the concentration of traps per unit energy E, E_c is the conduction band minimum and $T_t (> T)$ is a temperature parameter characterizing the trap distribution. In that case, the I-V characteristics are given by [5]

$$I_2 = q\, \mu_n N_c A \left[\frac{\varepsilon}{qN_o kT_t}\right]^\ell \frac{V_2^{\ell+1}}{[d-W(V_1)]^{2\ell+1}} \qquad (2)$$

In equation (2), A is the cell area, μ_n the free electron mobility, N_c the density of states in the conduction band, d the total thickness of the cell $W(V_1)$ the width of the depletion region and $\ell = T_t/T$.

The depletion and the neutral bulk regions are in series and, therefore, at any external bias (V), current (I) passing through the two regions should be the same i.e. $I = I_1 = I_2$ and $V = V_1 + V_2$. These conditions inconjunction with the I-V characteristics of the individual regions would yield V_1 and V_2 for any value of V. Figure 1 shows the results of computations for V_1 and V_2 for two Pd-I-N$^+$/SS structures with different I-layer thickness but otherwise identical. As expected, V_2 becomes more important for thicker I-layers. A higher ℓ-value would yield the same result. It is also evident from the figure that V_1 is much higher than V_2 in the low forward bias region (< 0.2 V) and approaches a saturation value V_{bi} for large forward bias (> 0.6 V). This suggests that in the low forward bias region, the characteristics are barrier controlled while in the large forward bias region then are controlled by the neutral bulk region. A knowledge of V_1 and V_2 in conjunction with equations (1) and (4) can be used to find the terminal dark I-V characteristics.

3. EXPERIMENTAL RESULTS AND DISCUSSION

A. Experimental Setup

The a-Si:H films have been deposited on stainless steel (SS) substrates utilizing dc glow discharge decomposition of silane. The doping is done using phosphine and diborane for N$^+$- and P$^+$-layers, respectively. Substrates are maintained at 250 °C with a deposition rate of 0.5 μm/hr. Typical Schottky and P-I-N structures fabricated for the present study are shown in Figure 2. Dark I-V measurements at room temperature have been done using a Keithley Microvolt Digital Multimeter that can read to the order of pico-amperes. Dark C-V measurements are carried out using a lock-in-analyzer (PAR Model 5204). Most of the measurements are done below 50 Hz with a modulation voltage less than 15 mV rms.

B. Results and Discussion

Figure 3 shows the theoretically obtained dark I-V characteristics of a Cr-I-N$^+$/SS and Pd-I-N$^+$/SS structure with I-layer thickness around 0.45 μm.

The corresponding experimental data is also shown. The agreement between the experimental data and theoretical curves is evident. Figure 4 shows the corresponding characteristics of two similar Schottky structures with I-layer thickness around 1.30 μm. The parameters chosen for the calculations are similar to those used for the Schottky barriers with I-layer thickness of 0.45 μm. This accounts for slight deviation in the theoretical and experimental curves. The properties of the thicker I-layer may be slightly different due to the longer deposition times and associated annealing effects. Figure 5 shows the dark I-V characteristics of a Cr-N-I-P/SS structure. These structures show low shunt resistance at low forward bias (< 0.2 V) and yield diode quality factor (n) around 1.5 - 2.3 indicating junction recombination. Evidence of SCLC conduction could not be observed in these structures. This is due to the fact that these structures have high built-in potential and, therefore, barrier controlled current conduction can be observed up to much higher forward bias levels as compared to that for the Schottky structures.

Figure 6 shows the dark C-V-f characteristics of a Schottky structure. These characteristics show a monotonic increase in capacitance with bias as expected from the trap controlled conduction in the I-layer [6]. The ohmic behavior of the I-layer predicts a peak in capacitance around 0.3 - 0.4 V [7]. This further confirms that SCLC conduction takes place, in dark, in the I-layer.

4. CONCLUSIONS

Dark I-V characteristics of a-Si:H Schottky structures show barrier controlled conduction in low forward bias (< 0.2 V) regime and bulk controlled conduction in the large forward bias regime. The neutral bulk region shows evidence of SCLC conduction in the dark. Model calculations have been done using the above considerations and a general agreement between the theory and experiments is found. P-I-N structures show barrier controlled conduction even at high forward bias due to their larger built-in potential. C-V-f data for Schottky structures also confirms SCLC conduction, in dark, in the I-layer.

REFERENCES

1. M. K. Han, W. A. Anderson, Y. Onuma, P. Sung, R. Lahri and J. Coleman, IEEE Electron Device Letters, Vol. EDL-2, No. 8, 198 (1981)

2. S. Ashok, A. Lester and S. J. Fonash, IEEE Electron Device Letters, Vol. EDL-1, No. 10, 200 (1980)

3. R. Lahri, M. K. Han, P. Sung and W. A. Anderson, Journ. of Appl. Phys., submitted.

4. S. M. Sze, Physics of Semiconductor Devices, Wiley-Interscience, New York, 1969.

5. M. A. Lampert and P. Mark, Current Injection in Solids, Academic Press, New York, 1970.

6. R. Lahri, M. K. Han and W. A. Anderson, IEEE Trans. on Electron Devices, to be published.

7. A. J. Snell, K. D. Mackenzie, P. G. LeComber and W. E. Spear, Phil. Mag. B, $\underline{40}$(1), 1 (1979)

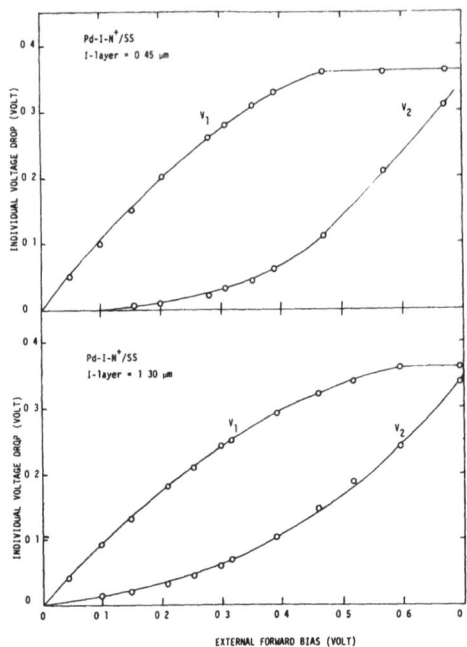

Fig. 1. Variation of voltage drops across the depletion and the neutral bulk regions under an external bias.

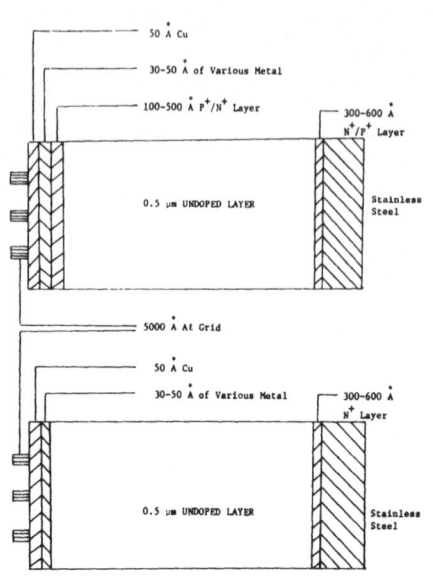

Fig. 2. Typical Schottky and P-I-N structures fabricated for the present work.

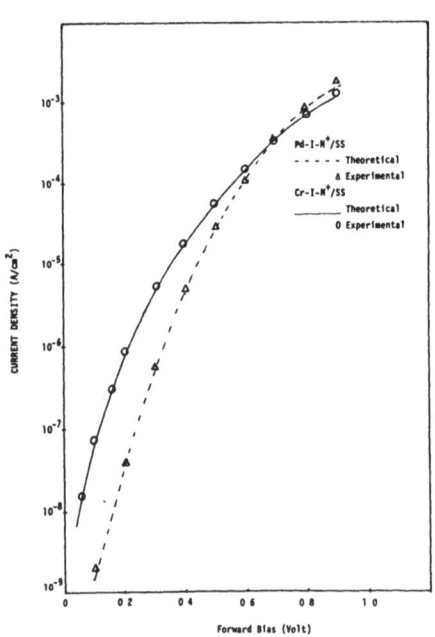

Fig. 3. Dark I-V characteristics of Pd and Cr Schottky structures (I-layer thickness = 0.45 μm).

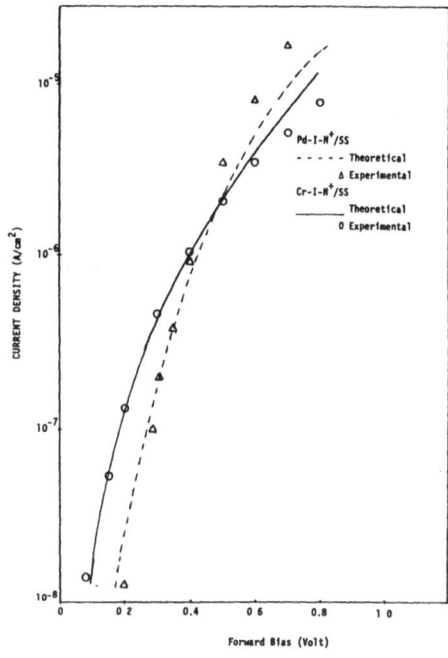

Fig. 4. Dark I-V characteristics of Pd and Cr Schottky structures (I-layer thickness = 1.3 μm).

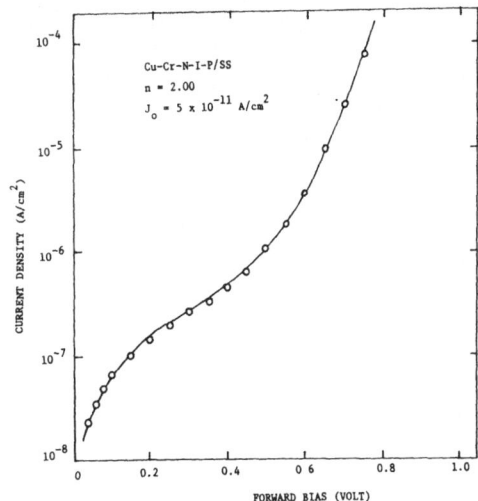

Fig. 5. Dark I-V characteristics of a CR-N^+-I-P^+/SS structures.

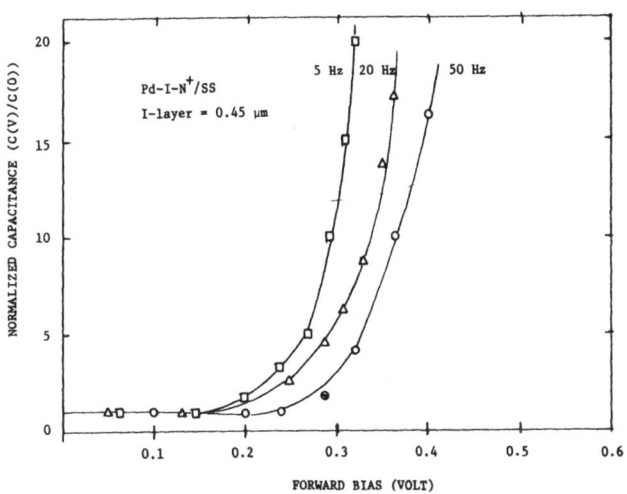

Fig. 6. Dark C-V-f characteristics of a Pd-I-N^+/SS structures.

OPTICAL OPTIMIZATION OF AMORPHOUS SILICON SOLAR CELLS

W. DEN BOER and R.M. VAN STRIJP
Department of Electrical Engineering
Delft University of Technology

Summary

A computer simulation study is presented to calculate the optical behaviour of thin film solar cells. Application to Schottky-barrier a-Si:H solar cells with highly reflective back contacts indicates that multiple reflections are strong for long wavelength light and can be used to improve the efficiency. Relative maxima in the AM1 integrated absorption occur at certain a-Si:H film thicknesses. For a given cell configuration the optimum antireflection coating is calculated. For weakly absorbed monochromatic radiation the electron-hole pair generation rate varies periodically with distance from the front contact due to interference. Integrated over the solar spectrum interference peaks in the generation rate are reduced, but remain present near the back contact. The computer program can be easily adapted for calculations on p-i-n-cells, tandem cells and multi-junction cells.

1. INTRODUCTION

The recent improvement of amorphous silicon solar cells to efficiencies beyond 7% has been mainly due to the optimization of the undoped material and the p^+ and n^+ contact layers.

Relatively little attention has been paid to the optical optimization of the device. This optimization can be achieved by using optimized antireflection coatings and highly reflective back contacts, so that a maximal number of solar photons is absorbed in the active semiconductor layer. Since the active layer is only a few wavelengths thick (0.3 - 1 μm), interference effects will occur in cells with a reflective back contact and in combination with multiple reflections these can be used to improve the efficiency.

Several authors have described interference peaks in the spectral response of a-Si:H solar cells (1,2,3,4), but no systematic study has been published yet on the various interference effects.

Since analytical treatment of the absorption in multilayer systems results in extremely complicated formulae, we have simulated the cells on computer using the refractive indices of real materials.

2. SIMULATION PROCEDURE

A thin film solar cell is optically simulated by a multilayer configuration (fig. 1), in which each layer is characterized by complex refractive indices $N_i = n_i - jk_i$. n_i and $-k_i$ are the real and imaginary part of N_i respectively, and are in general both a function of wavelength λ. k_i is related to the absorption coefficient α_i by

$$k_i = \frac{\alpha_i \lambda}{4\pi}$$

Monochromatic light is incident from the top in fig. 1 perpendicular to the interfaces, which are assumed to be perfectly planparallel, so that incoherent reflections are absent.

We use a procedure similar to that outlined by Heavens (5). The wave amplitudes ψ^+ and ψ^- for waves propagating in the positive and negative z-direction respectively can be calculated for arbitrary z by putting $\psi^+ = 1$ and $\psi^- = 0$ at the interface in the last medium. ψ^+ and ψ^- can be either the magnetic field or electric field amplitudes. For the calculations the matrices for the interface transitions and transitions in each medium are used. By substituting ψ^+ and ψ^- in the Poynting vector the net energy flux density $I(z)$, which we normalize to the incident energy flux density, can be obtained.

The fraction of the incident energy absorbed between two arbitrary planes at z_1 and z_2 is then equal to

$$A_{1,2} = I(z_1) - I(z_2)$$

The electron-hole pair generation rate $G(z)$ is given by

$$G(z) \sim \frac{dI}{dz}$$

assuming a quantum efficiency of unity. For AM1 illumination the fractional absorption of photons between z_1 and z_2 is given by

$$A_{1,2}^{total} = \int A_{1,2}(\lambda) \frac{S(\lambda)}{S^{total}} d\lambda$$

where $S(\lambda)$ is the AM1 photon flux spectrum and S^{total} is the total photon flux of the relevant part of the spectrum over which we integrate. The total generation rate $G^{total}(z)$ is written as

$$G^{total}(z) = \int G(z,\lambda) \frac{S(\lambda)}{S^{total}} d\lambda$$

3. RESULTS AND DISCUSSION

The computer model has been applied to Schottky diode a-Si:H solar cells as shown in the inset of fig. 2. The back contact is either molybdenum (low reflective) or silver (highly reflective). The transparent Schottky metal contact is 7.5 nm gold and the anti-reflection coating has a refractive index n=2 and a thickness of 50 nm. Data for the refractive indices of the metals were taken from ref.(6), and of the a-Si:H from ref.(3).

Fig. 2 shows the calculated absorption vs. λ in the a-Si:H film for a cell of 0.35 μm thickness and either a silver or a molybdenum back contact. The cell with the silver back contact exhibits an interference peak. This peak has been experimentally verified in the spectral response of a cell with this configuration (3) and improves the efficiency with 10 to 20%.

Fig. 3 shows the AM1 integrated absorption A^{total} as a function of a-Si:H film thickness t. The integration is over the wavelength region 400-800 nm, since the contribution of longer wavelength photons to the photocurrent is negligible. Relative maxima occur at t = 0.18, 0.27, 0.35, 0.43, 0.51, 0.60 and 0.69 μm. For a molybdenum back contact the increase in AM1 integrated absorption with thickness is monotonous. The absorption in a 0.35 μm silver back contact cell is considerably larger than in a 0.4 μm cell and is almost equal to the AM1 absorption in a 0.55 μm cell. From the electronic point of view a 0.35 μm cell is preferable to a 0.55 μm cell, because the ratio of drift length to cell thickness ($\mu\tau E/t$) is higher, so that recombination is reduced.

For a given solar cell configuration the optimum refractive index n and thickness d for an a.r. coating are found by calculating the AM1 integrated absorption in the a-Si:H film for a range of values of n and d. This is shown in Table 1. A^{total} is in this case normalized to the optimum value. For the cell with t = 0.35 µm the optimum values are n = 2.3 and d = 40 nm. It is also seen that a broad range of refractive indices from n = 2 to n = 2.6 are suitable. The combinations for which the AM1 integrated absorption is less than 1% below the maximum are underlined. They occupy the diagonal from the bottom left to top right in Table 1, for which the product of n and d is approximately constant. The result for the optimized cell corresponds with a maximal short-circuit current density of 15.5 mA/cm^2.

Table 1.

$d_{(nm)}$ \ n	2.0	2.1	2.2	2.3	2.4	2.5	2.6
32	0.886					0.985	0.989
35	0.913			0.981		0.995	0.990
37		0.956	0.977	0.992	0.997		0.969
40		0.976	0.992	[1]	0.998		
43		0.989	0.998	0.998	0.989		
45		0.993	0.998	0.993	0.979		
47	0.984	0.995	0.995	0.985	0.965	0.937	0.904
50	0.988	0.992	0.985	0.966	0.941	0.907	0.868

We are not only interested in the total absorption in each layer, but also in the absorption profile, i.e. the electron-hole pair generation rate G(z) as a function of z, the distance from the front contact. Fig. 4 shows G(z) for monochromatic illumination with λ = 650 nm on a 0.35 µm cell with either a silver or a molybdenum back contact. G(z) varies periodically and the variation amounts to more than one order of magnitude in the silver back contact cell.

When integrating over the solar spectrum the interference effects in $G^{total}(z)$ are reduced by compensation for the different wavelengths. Fig. 5 shows that they remain present near the silver back contact and are virtually absent in a cell with a low reflectance back contact.

Metals with a high reflectance at the a-Si:H - metal interface include copper, silver, gold and aluminium (7). Unfortunately none of these metals is suitable as a substrate for a-Si:H because of interdiffusion, silicide formation or insufficient adhesion. However, when the back contact is deposited after a-Si:H film growth, these metals can be applied. The results are not significantly different for weakly absorbed light incident under an angle < 60° with the normal to the interfaces, since this light is refracted towards the normal in the a-Si:H film within 15°. The resulting increase in optical path length is less than 3%. A different spectral distribution of the sunlight does not modify the principal results either. For AM2 illumination interference effects are even more important, since the relative contribution of long wavelength light is larger.

The computer program can be easily adapted for calculations on more complicated configurations, such as multi-junction cells and tandem cells. Thin film tandem cells are likely to attract many researchers in the future, since they promise to combine cost-effectiveness with higher efficiencies. For the proper matching of the two semiconductor layer thicknesses in a tandem cell computer calculations might be very useful.

In summary we have optically optimized a-Si:H Schottky diode solar cells using computer simulations. Multiple reflections can be exploited to improve their efficiency.

This work was financially supported by the Foundation for Fundamental Research on Matter (FOM).

REFERENCES

(1) V.L. Dalal, Solar Cells 2 (1980) 261.
(2) G.D. Cody, C.R. Wronski, B. Abeles, R.B. Stephens, B. Brooks, Solar Cells 2 (1980) 227.
(3) W. den Boer, M. Ondris, Solar Cells 3 (1981) 209.
(4) O. von Roos, J. Appl. Phys. 51 (1980) 6426.
(5) O.S. Heavens, Thin Film Physics, Methuen, London (1970).
(6) American Institute of Physics Handbook, McGraw-Hill, New York, 3rd edn., 1972, p. 6-124.
(7) M. Ondris, W. den Boer, Proc. 3rd E.C. Photovoltaic Solar Energy Conference, Cannes, 1980, Reidel, Dordrecht, p. 809.

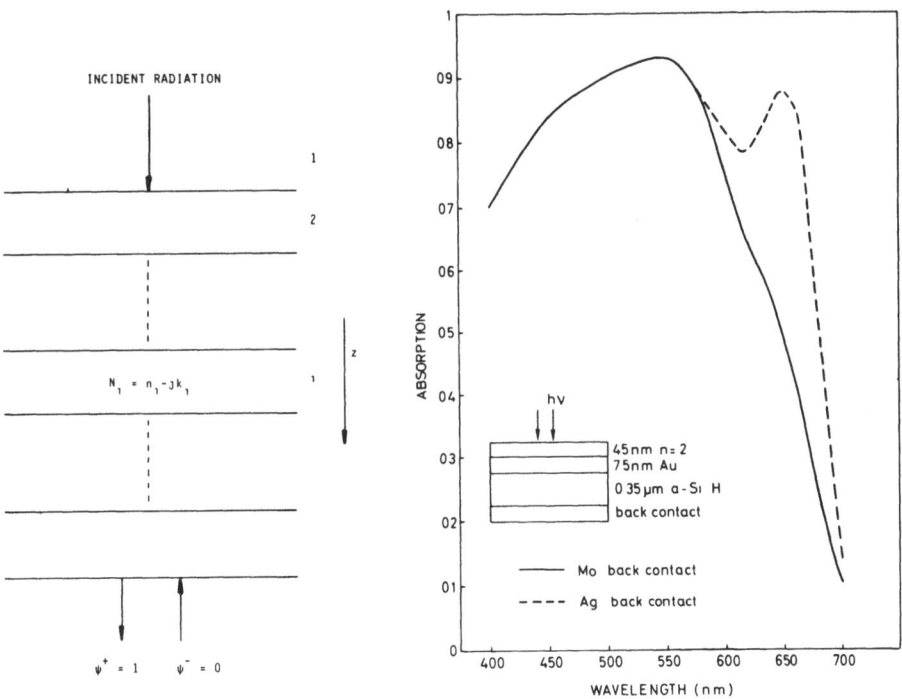

Fig. 1: Multilayer configuration for optical simulation of amorphous silicon solar cells

Fig. 2: Spectral absorption of cell shown in the inset

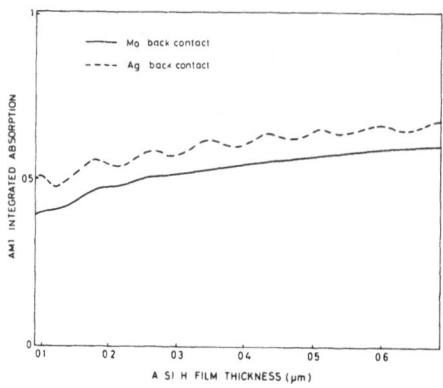

Fig. 3: AM1 integrated absorption as a function of a-Si:H film thickness

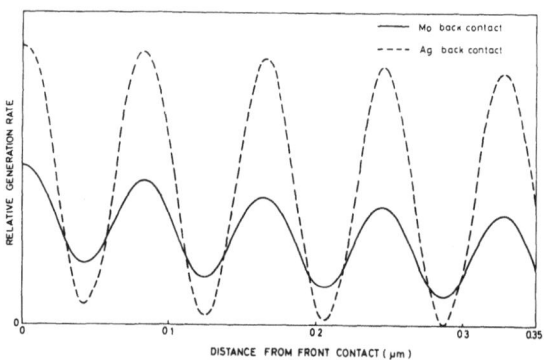

Fig. 4: Electron-hole pair generation rate vs. distance from front contact for monochromatic illumination with = 650 nm

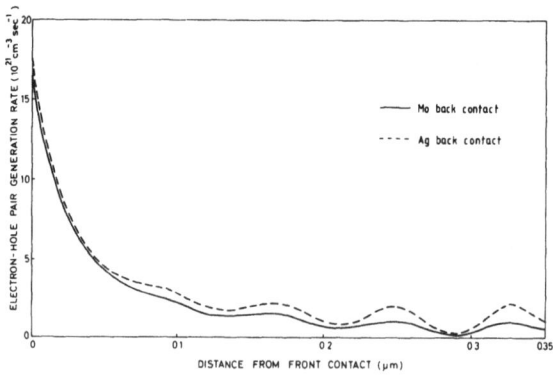

Fig. 5: Electron-hole pair generation rate vs. distance from front contact for AM1 illumination

LARGE AREA HYDROGENATED AMORPHOUS SILICON FOR PHOTOVOLTAIC APPLICATION

G.J. Smith   Cambridge University
W.I. Milne   Engineering Department (GB)

P. Blackborow   Electrotech (ET Associates) *

ABSTRACT

Satisfactory production techniques for depositing hydrogenated amorphous silicon (a-Si:H) over large areas remain to be developed. This paper reports the manufacture of uniform films of a-Si:H by glow discharge of silane over an area 60 cm in diameter. Electrical and optical properties and film quality are discussed.

INTRODUCTION

Hydrogenated amorphous silicon (a-Si:H) is widely regarded as a promising material for the fabrication of low cost terrestrial photovoltaic conversion systems. Many laboratories are working on this material with small scale plasma deposition systems, and for large scale deposition a two stage process of CVD followed by post-hydrogenation [1] has been proposed as a possible production process. We present details of a technique for glow discharge deposition of uniform films of a-Si:H up to 60 cm in diameter, using an existing commercial plasma deposition machine. The advantages of this machine are that it uses a single stage process, and is already proven in similar industrial use depositing silicon nitride films.

DEPOSITION PROCESS

The deposition system is a commercial automated plasma depositer using a capacitative plasma discharge process. The constructional details have been given in a previous paper [2]. The normal use of such machines is to provide thin films of silicon nitride as insulation and passivation layers in integrated circuit manufacture. This deposition process is similar to the conventional glow discharge technique [3] of producing a-Si:H. Simply by altering the plasma and deposition conditions the machine can be made to deposit a-Si:H. As well as smaller areas for characterisation experiments, we have produced a single film 60 cm in diameter on window glass as a graphic illustration of the large deposition area available in the machine. When large area high efficiency a-Si:H solar cells are eventually constructed, deposition systems like this will be needed for low cost mass production.

To deposit a-Si:H a simple silane plasma has been used, although dopant and dilutant gases may be included. The properties of large area a-Si:H made with argon as a dilutant have been reported previously [2]. The films discussed here were made under the following deposition conditions:

TABLE OF DEPOSITION CONDITIONS

Silane pressure	110	millitorr
Silane flow rate	250	scc/min
Plasma frequency	380	kilohertz
Plasma power	1200	Watts
Deposition area	2800	square centimetres
Substrate temperature	275	C
Resultant deposition rate	2.5	micrometres/hour

The advantage of using a sub-megahertz plasma frequency is that power matching problems are removed. This is especially important for large area deposition, as the large plasma volume requires a large power input. The low frequency enables the construction of a simple and efficient production machine.

MATERIAL PROPERTIES

During the deposition process the substrate table is rotated and produces excellent circumferential uniformity. There is a slight decrease in deposition rate towards the edge of the table, but overall a deviation of less than five percent from the mean thickness is found. In a typical characterisation sample, deposited onto a carefully cleaned 25 mm diameter quartz disc there are no pinholes, and no visible microstructure.

Infrared analysis was used to estimate the hydrogen concentration in this a-Si:H, and gave a figure of 19 atomic percent. Calibrated analysis enables not only the total hydrogen content to be established, but also indicates the distribution of hydrogen atoms between the different hydride structural units [4]. Films deposited as above show a predominance of monohydride groups (2000 cm^{-1}) with only extremely small traces of the isolated dihydride (870 cm^{-1}) and the multiple dihydride (840 cm^{-1}) sites. The dihydride peak at 2090 cm^{-1} is completely masked by the adjacent monohydride peak [Fig 1]. This distribution is believed to depend principally on substrate temperature and plasma pressure [4]. Films manufactured under identical plasma conditions but onto a substrate held at 325 C were deposited at the same rate, but contained only ten atomic percent of hydrogen and showed no trace of any dihydride groups under infrared analysis. This agrees with the generally observed trend that a higher substrate temperature will give a film with a lower hydrogen content, and a larger fraction bound in the monohydride structure.

The optical bandgap of a-Si:H lies in the visible region, and the material is strongly absorbing [Fig 2] over the solar spectrum. From a Tauc plot the optical bandgap of films deposited as above has been determined to be 1.74 eV. The higher temperature films with less hydrogen have an optical bandgap of 1.65 eV. The value for films deposited under similar conditions but from a plasma also containing argon is also approximately 1.65 eV. These films are found to contain less hydrogen, typically 8-10 atomic percent.

This a-Si:H shows Staebler - Wronski photoinduced effects [5], but these are small. The properties of the films also show very little substrate dependence. All results here refer to material in the light-soaked state B which results from continuous prolonged illumination. At 300 K the dark conductivity is 2×10^{-10} S/cm and the activation energy is 0.7 eV. The AM1 photoconductivity is 3×10^{-6} S/cm. These conductivities are lower than the best glow discharge material, and about half the values found in similar films made in a plasma containing argon. We would expect films containing no argon to have improved electrical properties [6], but these are heavily dependent on the deposition conditions and insufficient films have been produced as yet to allow optimisation of deposition conditions.

TABLE OF MATERIAL PROPERTIES

a-Si:H films grown in :	silane plasma	silane-in-argon plasma	
Hydrogen content	19	8-10	atomic percent
Optical bandgap	1.74	1.65	eV
Activation energy	0.7	0.7	eV
Dark conductivity	2×10^{-10}	4×10^{-9}	S/cm
AM1 Photoconductivity	3×10^{-6}	1×10^{-5}	S/cm

DISCUSSION

Before a-Si:H can be widely utilised in the electronics industry in photovoltaics, TFTs, or any of its other potential applications, reliable techniques have to be developed to deposit it in large area uniform films. We have shown that a conventional commercial plasma depositer can deposit such films from plasmas of pure silane or silane-in-argon. Insufficient material has been produced yet to enable optimisation of the conductivity properties with respect to deposition conditions, and these properties are not yet as good as those of other glow discharge a-Si:H. However the films are of good physical quality, having excellent uniformity and very low pinhole densities. The optical absorption spectrum closely resembles that of other glow discharge and also sputtered a-Si:H. Infrared analysis shows that it is possible to produce these films with the hydrogen content almost exclusively incoporated as monohydride.

REFERENCES

1. N.Szydlo, D.Kaplan, R.Poirier
   3rd EC Photovoltaic Conference (1980) pp 317-321

2. G.J.Smith, W.I.Milne, P.Blackborow
   Elec. Lett 18-5 (1982) pp 211-213

3. H.Fritzsche   Solar Energy Mats. 3 (1980) pp 447-501

4. C.C.Tsai, H.Fritzsche   Solar Energy Mats. 1 (1979) pp 29-42

5. D.L.Staebler, C.R.Wronski   Appl. Phys. Lett. 31 (1977) pp 292-294

6. K.Tanaka et al   J. Non-Cryst. Solids 35 (1980) pp 475-480

* P.Blackborow is now with Plasmatech, Banwell, Bristol (GB)

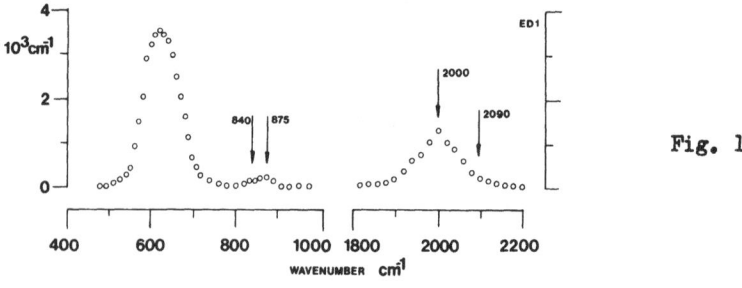

Fig. 1

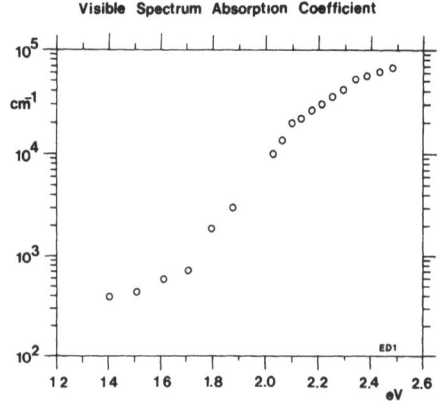

Fig. 2

NOVEL PLASMA CHEMICAL METHODS FOR DOPING a-Si:H

G.H. BAUER and G.BILGER
Institut fuer Physikalische Elektronik
Universitaet Stuttgart
F.R. Germany

Summary

Hydrogenated amorphous silicon prepared by decomposition of silane in a dc-glow discharge or by rf-sputtering in Ar/H_2 mixtures has been doped by methods of dopands injection into the plasma which base on interactions of plasma or the neutral gas with surfaces at different temperatures and varying electric potentials (co-sputtering, dc-sputtering, thermally activated injection to the neutral gas, plasma assisted transport). By these techniques various kinds of dopands (In, B, P, Sb and N) have been injected into the plasma and by this means incorporated into the film partially acting as electronically active donors and acceptors resp.. Analyses of doping effects and efficiencies by evaluation of temperature dependent conductivity and thermoelectric power data yield increases in room temperature conductivity up to a factor 10^8 (to $\approx 10^{-2}(\Omega cm)^{-1}$) and decreases of activation energies for extended states conduction which indicate the position of the Fermi level to < 0.2 eV.

1. INTRODUCTION

Hydrogenated amorphous silicon promises an excellent suitability for thin film photovoltaic applications. Since the amorphous structure of this material strongly affects diffusion length and life time of carriers only barrier types effecting an internal electric field - usually p-i-n-structures - are used /1,2/. The p- and n-doped layer typically having a thickness of some tens of nanometers, should show high electrical conductivity $\gtrsim 10^{-2}(\Omega cm)^{-1}$ in comparison to the intrinsic a-Si:H layer ($10^{-10}(\Omega cm)^{-1}$) and the Fermi level near by (0.2 eV) the conduction or the valence band.

Currently substitutional doping of amorphous silicon films is achieved during deposition processes by an addition of gases containing dopands (PH_3 or B_2H_6) to the neutral gas e.g. to silane which is decomposed in rf- or dc-glow discharges /3/ or by CVD-methods /4/ and to mixtures of hydrogen and nobel gases in rf-sputtering discharges /5/.

As a consequence of the relatively high density of gap states in a-Si:H ($\approx 10^{17} cm^{-3} eV^{-1}$) the concentration of electronically active dopands in the film (e.g. B- or P-atoms) and thereby their concentration in the plasma and in the neutral gas have to amount 10^{-3}-10^{-2} related to the number of silicon atoms /6/. Additional components of gases in those ranges of some 10^{-3} in the neutral silane gas might dramatically change the parameters of the plasma which strongly influence the electronic and optical properties of the amorphous films (absorption, H-content, density of states) as previously has been shown in detail /7,8/.

Substitutional doping of amorphous silicon by ion implantation after film preparation /9/ seems not to be of such importance for technical applications and might cause some more defects in the films though this technique provides a chance for the incorporation of potentially doping atoms of any kind into the film.

2. INJECTION OF DOPANDS BY PLASMACHEMICAL METHODS

Two plasmachemical methods for film deposition have been investigated (rf-sputtering in H_2/nobel gas mixtures and dc-glow discharge in SiH_4), which both have turned out to be suitable for reproducible preparation of intrinsic a-Si:H layers and which both satisfy the main condition for efficient substitutional doping namely a low density of states ($\lesssim 10^{17} cm^{-3} eV^{-1}$) in the mobility gap. Evaluation of field effect data have been employed as a direct criterion for quality of films, whereas temperature dependent electrical conductivity, $\mu n \tau$ products from photoconductivity and yields and spectral functions of photoluminescence have been taken as indirect criteria /10/.

For doping of these films during plasmachemical preparation in both types of discharges, methods for injection of dopands into the plasma have been investigated which base on interactions of the plasma or the neutral gas with surfaces which contain dopands at different temperatures and electric potentials. These injection techniques combine two advantages of the above mentioned methods:
  in situ doping during preparation of the film and
  the possibility for injection of doping material of various kinds (no restriction to stable gaseous hydrides).
Furthermore a smaller variation of plasma parameters enforced by an application of these methods, indicated in matrix I, than by an addition of PH_3 or B_2H_6 resp. to the neutral gas might be expected.

	dc-glow discharge	rf-sputtering
methods for injection of dopands into the plasma	dc-sputter effect from separated cathode containing dopands/8/	sputtering from highly doped targets
	generation of gaseous hydrides by Plasma Assisted Transport/11/	co-sputtering from locally separated cathode with adjustable electric potential
	Thermally Activated Injection (evaporation into neutral gas flow)	
	Thermally Activated Diffusion from thin substrate layers containing dopands	

Matrix I: Summary of methods for plasmachemical injection of dopands.

3. EXPERIMENTAL RESULTS

The effects and efficiencies of injection of dopands by the above mentioned methods for rf-sputtering and dc-glow discharge film preparation are analysed by electronic properties of the layers like temperature dependent conductivity for evaluation of σ_{RT} and activation energy mainly for extended states conduction and temperature dependent thermoelectric power for determination of p- and n-type conduction resp., and calculation of the position of the Fermi level.

Figures (1,2,3,4) show temperature dependent electrical conductivities for layers prepared by dc-glow discharge and rf-sputtering for various dopands, different rates of injections by diverse injection methods.

Injections of In, B, Sb, P and even N into the plasma by these methods have hitherto effected an increase in room temperature conductivity by up to eight orders of magnitude and a shift of the activation energy which has been interpreted with the help of thermoelectric power analyses as a shift of the Fermi level up to ≤ 0,2 eV to the valence (In, B) and to the conduction band (Sb, P, N) respectively.

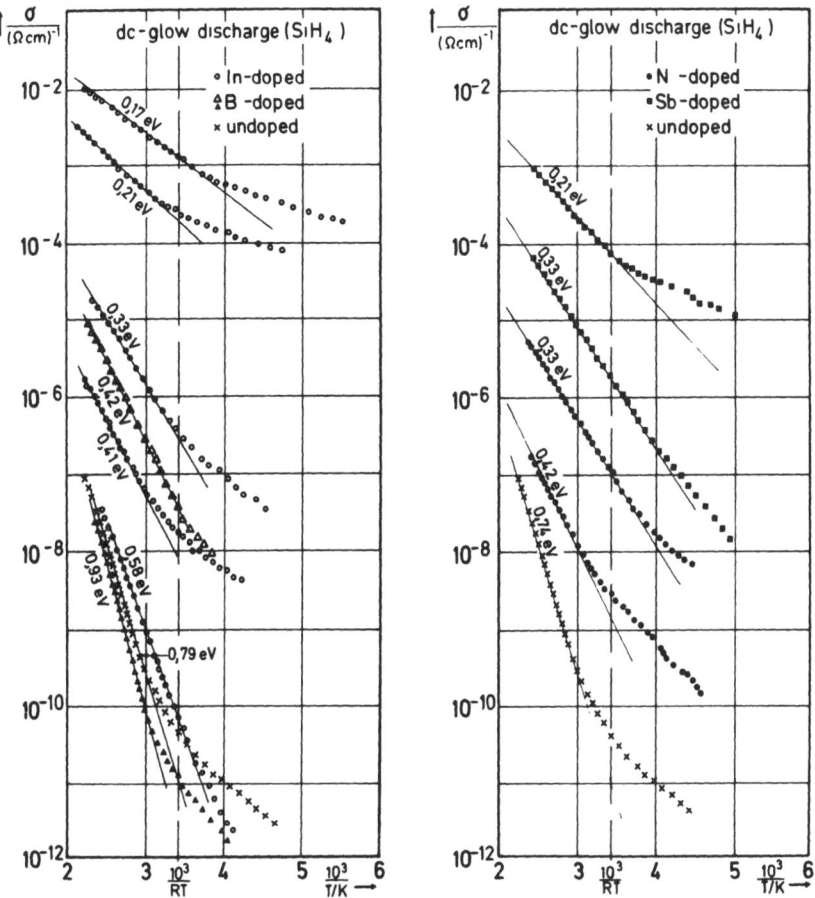

Figures 1,2: Temperature dependent conductivities for undoped and doped a-Si:H films prepared by dc-glow discharge.

4. INTERPRETATION

The variation of electrical conductivity and of the position of the Fermi level are caused by injection of dopands into the plasma/neutral gas which in a fraction of some percent can be assumed to be incorporated into the amorphous network acting as electronically active donors or aceptors /6/.

Increasing acceptor material injection firstly effects a decrease in conductivity and an increase in activation energy since the mechanism of conduction is governed by electron transport. A further increase in p-type doping increases conductivity and decreases activation energy for extended states conduction as soon as p-type transport process predominates (fig. 5,6).

Injection of donor material continuously increases conductivity and decreases activation energy (shifts the Fermi level to the conduction band) of the originally undoped and slightly n-type specimens.

With the single exception of N injection the differing methods and the different materials show a linear relation of $\ln(\sigma_{RT})$ and activation

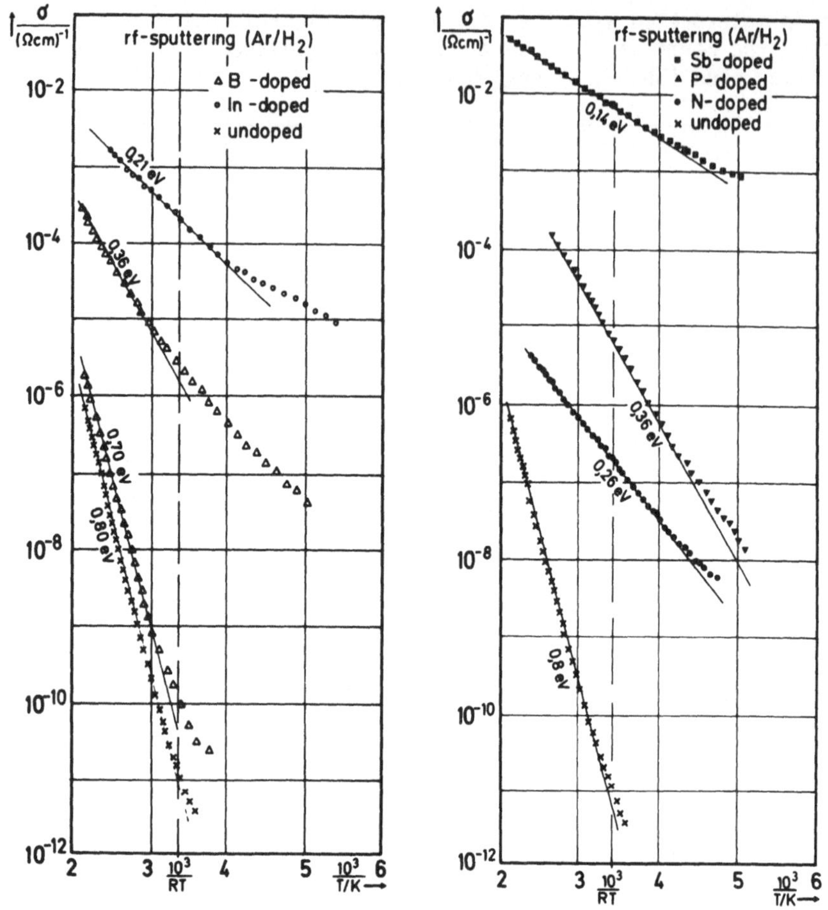

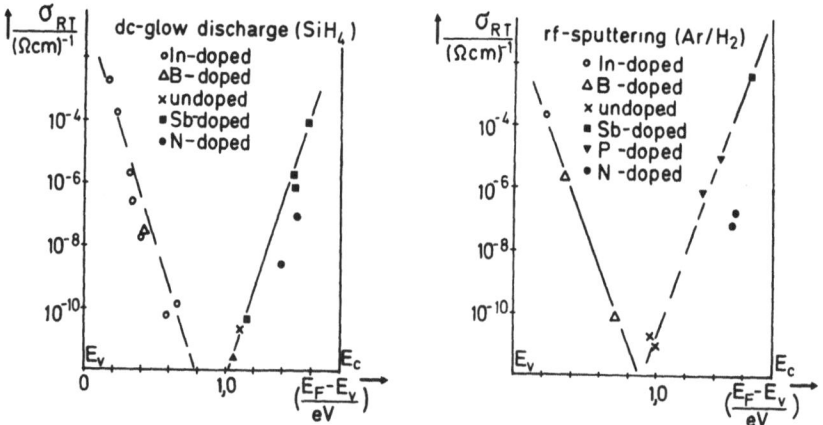

Figures 3,4: Temperature dependent conductivities for undoped and doped a-Si:H films prepared by rf-sputtering.

Figures 5,6: Room temperature conductivities as function of position of the Fermi level for a-Si:H films prepared by dc-glow discharge and rf-sputtering.

energy which indicates the position of the Fermi level in good agreement with the theory of extended states conduction. The minimum metallic conductivity σ_0 for p-doped (In, B), undoped and n-doped (P, Sb) films amounts to about 10^0 $(\Omega cm)^{-1}$. Nitrogen doped specimens do not correspond to this linear function; there is a reduction of σ_0 which could be explained by a remarkable amount of defects enforced by nitrogen incorporation.

Additionally, further increase in N_2 injection into the plasma of either rf-sputtering or dc-glow discharge increases the band gap of the amorphous semiconductor up to >2.5 eV indicating the transition to amorphous hydrogenated silicon nitride /12/.

5. CONCLUSION

By injecting dopands of various kinds into the plasma by means of some novel plasmachemical methods, conductivity and position of Fermi level in amorphous silicon films can be shifted over a wide range. Evaluations of analyses of conductivity and thermoelectric power show p- and n-type charge transport processes and indicate the suitability of these methods for substitutional doping.

These methods, showing advantageous facts as incorporation of any kind of potentially doping atoms, incorporation during film deposition and a chance for less variations of the plasma state in comparison to the addition of diborane and phosphine resp. provide means for an optimization of plasmachemical doping methods concerning injection techniques and materials/combination of materials with respect to photovoltaic applications.

6. ACKNOWLEDGEMENT

This work was supported by the Bundesministerium für Forschung und Technologie contract no. ET 4045 B.

7. REFERENCES

/1/ D.E. Carlson, Proc. 3rd E.C. Photovolt. Solar Energy Conf., Cannes, Oct. 1980, p. 294, D. Reidel Publ. Comp.,Dordrecht (Holland)
/2/ Y. Hamakawa, Proc. 9th Int. Conf. Amorph. Liqu. Semicond., Grenoble, July 1981, p. C4-1131, les éditions de physique, Les Ulis (France)
/3/ W.E. Spear, P.G. LeComber, Phil. Mag., 33, 935 (1976)
/4/ S.C. Gau et al., Appl. Phys. Lett., 39, 436 (1981)
/5/ E.C. Freemann, W. Paul, Phys. Rev. B, 20, 716 (1979)
/6/ W.E. Spear, P.G. LeComber, Sol. Stat. Comm., 17, 1193 (1975)
/7/ G.H. Bauer, G. Bilger, Thin Sol. Films, 83/84, 223, (1981)
/8/ G.H. Bauer, G. Bilger, Proc. 5th Int. Sympos. Plasma Chem., Edinburgh, Aug. 1981, p. 639
/9/ S. Kalbitzer et al., Phil. Mag. B, 41, 439 (1980)
/10/ BMFT-Forschungsbericht, Proj. ET 4045 B, Techn. Nutz. Sol. Energie/ Dünnschichtsolarzellen, accepted for publication
/11/ A.P. Webb, S. Veprek, Chem. Phys. Lett., 62, 173 (1979)
/12/ G.H. Bauer, G. Bilger, to be published

ELECTRONIC PROPERTIES OF DOPED AMORPHOUS SiO_x

E. HOLZENKÄMPFER, J. STUKE and R. FISCHER*

Fachbereich Physik, Universität Marburg, F.R. Germany
*AEG-Telefunken, Frankfurt, F.R. Germany

Summary

Films of a-SiO_x:H ($0 \leq x \lesssim 1$) were prepared in a glow discharge of SiH_4-N_2O-mixtures. It is found that the band gap widens at a rate $dE_{04}/dx = 1$ eV. For fixed oxygen content, the band gap shrinks upon doping. This effect sets in at about 10^3 ppm B; on the P side this effect is much weaker. - For $x = 0.23$, the maximum values for the dark conductivity are $\approx 3 \cdot 10^{-5} \Omega^{-1} cm^{-1}$ (B-doping) and $\approx 10^{-6} \Omega^{-1} cm^{-1}$ (P-doping).

1. INTRODUCTION

One of the critical components of p-i-n or n-i-p thin film amorphous silicon solar cells is the top layer of doped a-Si, the "window". On the one hand, the doping level has to be high because the conductivity and the Fermi level shift of this top layer have to be as high as possible. On the other hand it is always found that doping a-Si:H generates network defects (1) and enhances the optical absorption in the band edge region (2). The competition between these two contrary effects can be neutralized by the use of admixtures to a-Si which widen the band gap. For example, the application of doped a-$Si_{1-x}C_x$:H as window material has permitted the fabrication of a-Si:H based solar cells with efficiencies larger than 7 % (3).

There are, however, yet other possibilities to enlarge the band gap of a-Si:H, e.g. the admixture of oxygen (4). The electronic properties of doped a-SiO_x:H are the subject of this contribution. Previous luminescence and absorption measurements (5) have shown that a-SiO_x:H might, for small x, be a suitable material for the doped window layer in an a-Si solar cell.

2. PREPARATION

The SiO_x:H films were prepared by reactive plasma deposition from suitable mixtures of the gases SiH_4 and N_2O. The oxygen content x in the alloy system SiO_x:H was varied by changing the ratio of the partial pressures $p(N_2O)/p(SiH_4)$. This gas mixture was decomposed in a capacitive HF discharge at the HF power level of roughly 0.07 W/cm^2. The substrates were held at $T_s = 280°C$, and the deposition rate was about 3 Å/s. The films used had thicknesses of around 2 μm. The substrate materials were amorphous silica, sapphire, and thin beryllium sheets; the latter ones were needed for the proton backscattering experiments for the determination of the oxygen content.

For the doping of the films, diborane or phosphine were admitted to the plasma discharge in addition to the silane-nitrous oxide mixture.

3. RESULTS

The dependence of the absorption edges $\alpha(h\nu)$ on the oxygen content is shown in Fig. 1 for the case of undoped samples. As these edges are feature-

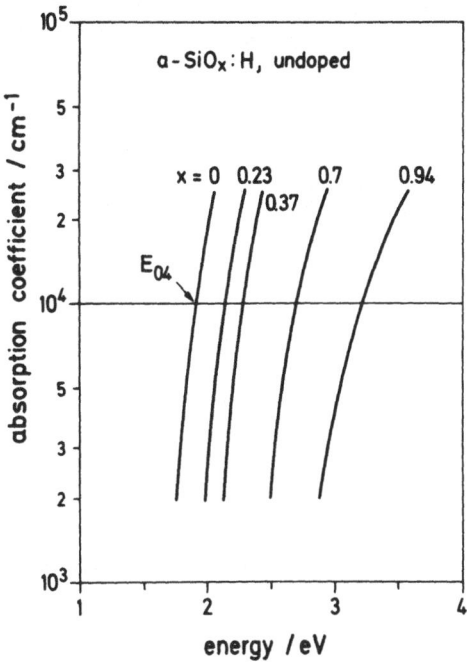

Fig. 1: The absorption edges of undoped a-SiO$_x$:H films for various values of the oxygen content x.

less, the energy E_{04} is used to characterize the position of the edges, where $\alpha = 10^4$ cm^{-1}, i.e. at $h\nu = E_{04}$ the penetration depth of the light is 1 μm which is in the range of typical sample thicknesses.

There is a clear increase of E_{04} with x, shown in Fig. 2 in some more detail. Fig. 2 also contains the activation energies E_a of the dark conductivity and E_{ph} of the photoconductivity, for undoped as well as for doped samples with a fixed doping ratio of 1 % diborane.

The dependences of the characteristic energy E_{04} on oxygen content for undoped and 1 % B-doped SiO$_x$ were very similar. The rise with x is first practically linear ($dE_{04}/dx = 1$ eV); it becomes steeper above $x \sim 1$. The band gap E_{04} of 1 % B-doped SiO$_x$:H is lower by the amount of 0.3 eV which is practically constant in the whole x range investigated.

The variation of the activation energy E_a with x is also linear in the undoped case; the slope is somewhat smaller (0.6 eV), as is expected when the Fermi level lies roughly in the middle of the gap. The behaviour of E_a for B-doped SiO$_x$:H is markedly different. There is a clear decrease of E_a which is largest for $x > 0.5$ and for $x = 0$. For oxygen contents around $x = 0.2$ the change of E_a by doping is relatively small. We conjecture that in this range part of the oxygen atoms are incorporated into the network as deep centers, thus attenuating the influence of the doping atoms.

The activation energies E_{ph} of the photoconductivity are again different for the doped and undoped cases. In the doped case, E_{ph} begins at a rather high value at $x = 0$ but further on varies less rapidly than for undoped SiO$_x$:H.

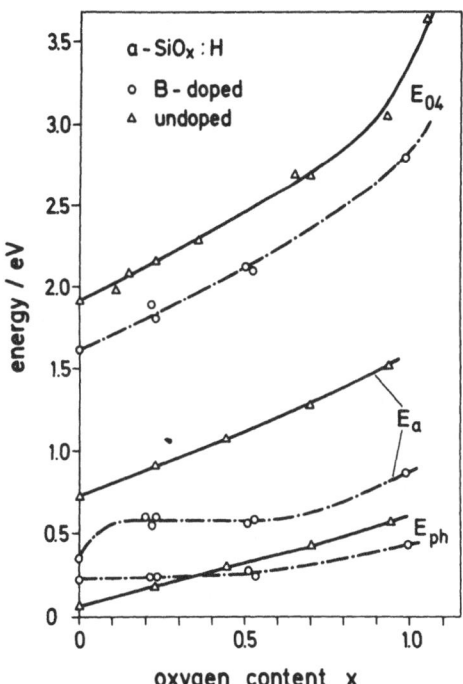

Fig. 2: E_{04}, E_a and E_{ph} vs. oxygen content x.

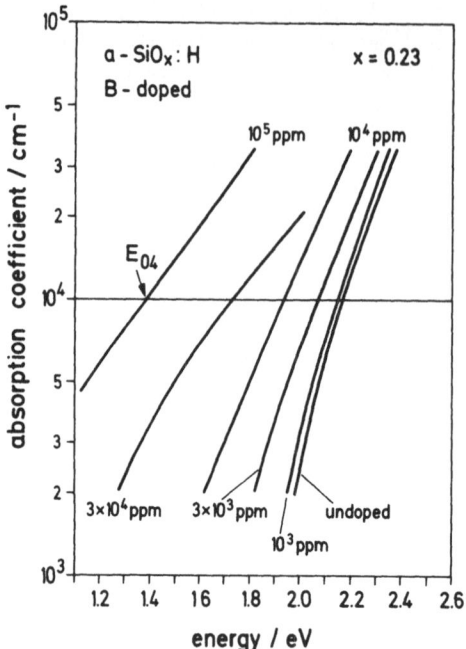

Fig. 3: The absorption edges of a-SiO$_{0.23}$:H films for various B-doping levels.

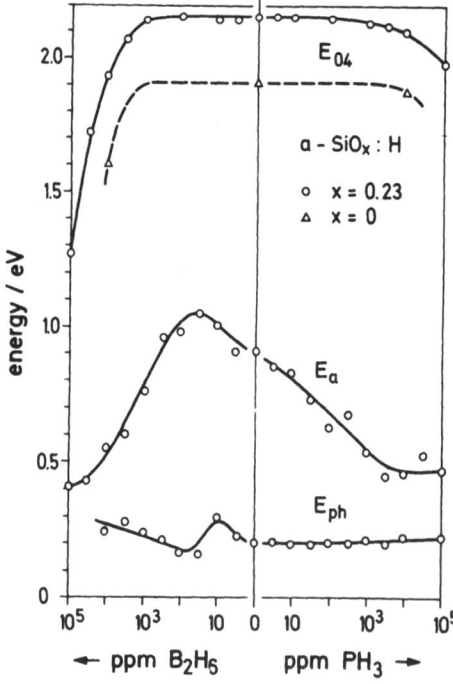

Fig. 4: E_{04}, E_a, and E_{ph} vs. doping level.

The influence of boron doping on the optical gap of SiO$_x$:H, with x fixed at 0.23, is shown in Fig. 3 with the relative B$_2$H$_6$ partial pressure as parameter. For undoped SiO$_{0.23}$:H we obtain E_{04} = 2.15 eV, which value changes only little by doping up to about 10^3 ppm diborane. We find, however, a strong decrease of E_{04} for higher B-doping: at 1 % diborane we find E_{04} = 1.9 eV which is the same value as for undoped a-Si:H (Figs. 1 and 2).

Not shown here are results from phosphorus doped SiO$_x$:H. The influence on the absorption edge is qualitatively similar as for B-doping, however much weaker: The E_{04} shift of 10 % P-doped samples is about equal to that of 1 % B-doped samples. Fig. 4 gives an overview over the influence of doping on the characteristic energies E_{04}, E_a and E_{ph} of SiO$_{0.23}$:H. The uppermost curve exhibits the optical edge E_{04}, the difference between the effects of boron and phosphorus doping are clearly demonstrated. The important result is that E_{04} remains practically unchanged in the doping range between 10^3 ppm B and 10^4 ppm P, the influence of doping on E_{04} of a-Si:H is shown for comparison.

The activation energy E_a of the dark conductivity can be varied by doping from about 1 eV (intrinsic case) to 0.4 eV (doped case). As in a-Si:H the maximum of E_a is found for slight boron doping. This indicates that for x = 0.23, as for x = 0, E_F lies somewhat above its position for intrinsic conduction. - E_{ph} of the photoconductivity changes relatively little with doping and, again as in a-Si:H, there is a maximum for about 10 ppm B-doping.

The key properties of doped SiO$_x$:H are summarized in the next two figures. Fig. 5 shows the dependence of σ_d and σ_{ph} on doping for x = 0.23 and, for comparison, σ_{ph} of doped a-Si:H (6) (dashed line). σ_d has a deep minimum at 20 ppm B-doping and rises by about eight orders of magnitude with increasing B content to some 10^{-5}Ω^{-1}cm^{-1}. For

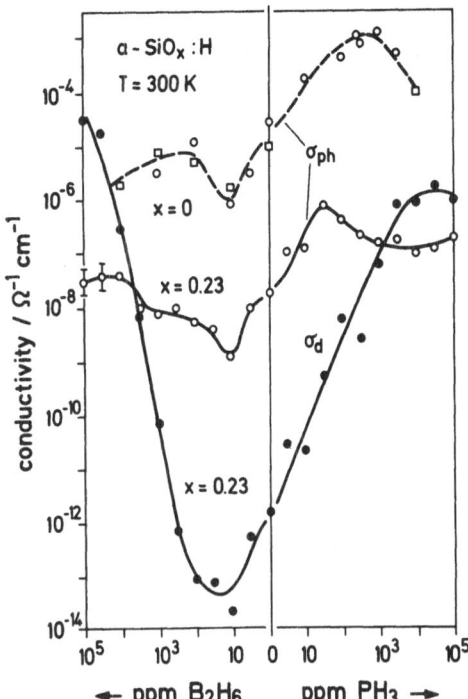

Fig. 5: The dark conductivity of a-SiO$_{0.23}$:H and the photoconductivity of a-Si:H (6) and a-SiO$_{0.23}$:H vs. doping level.

P-doping, σ_d saturates at about 1 % P, at $\sigma_d \sim 10^{-6} \Omega^{-1} cm^{-1}$ which is lower than on the B side. At present we explain this unsymmetry of the optimally attainable conductivities by the shrinkage of the band gap (E_{04}) with doping which is more pronounced in the boron-doped case as is shown in Fig. 4.

The effect of doping on the room temperature photoconductivity is qualitatively similar for a-Si:H and a-SiO$_{0.23}$:H; in both cases there is a minimum in the intrinsic case (weak B-doping). σ_{ph} of a-SiO$_{0.23}$:H is, however, smaller by a factor of about 10^3. This apparently indicates that part of the oxygen atoms are incorporated not as bridging atoms but instead as deep impurity centers, e.g. when saturating dangling bonds or in connection with the other atoms still present in the films (hydrogen, and to a small degree, nitrogen) as OH or NO groups. Final understanding of this phenomenon needs, however, further study.

In Fig. 6 the present results on a-SiO$_{0.23}$:H are summarized (dashed-dotted lines) and compared with the analog results on a-Si:H (solid lines) and a-Si$_{1-x}$C$_x$:H (dashed lines) (3). It is surprising that phosphorus doping yields very similar σ_d values for Si$_{1-x}$C$_x$:H and for a-Si:H, whereas σ_d of a-SiO$_{0.23}$:H is lower than those by roughly 10^{-5}. On the boron side, however, σ_d of a-Si$_{1-x}$C$_x$:H and a-SiO$_{0.23}$:H are similar, both are lower than σ_d of a-Si:H. This means that at high B-doping SiO$_x$ could also be used as a window layer of p-i-n solar cells.

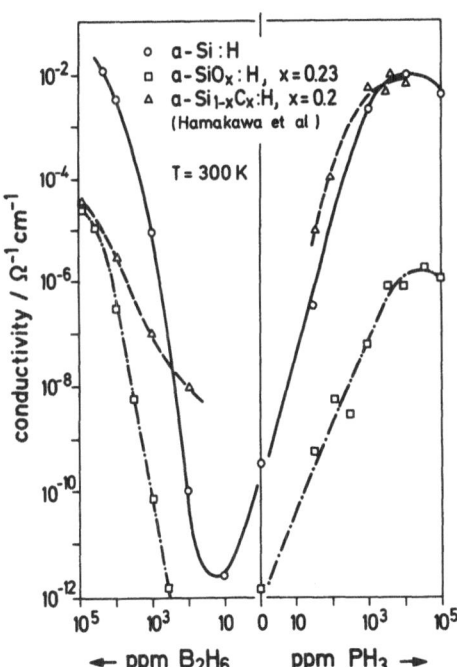

Fig. 6: The dark conductivity of a-Si:H, a-SiO$_x$:H and a-Si$_{1-x}$C$_x$:H (3) vs. doping level.

ACKNOWLEDGEMENT

We should like to thank the Deutsche Forschungsgemeinschaft for financial support.

REFERENCES

(1) H. Dersch, J. Stuke and J. Beichler, phys.stat.sol. (b) 105, 265 (1981)
(2) J.C. Knights, AIP Conf. Proc. 31, 296 (1976)
(3) Y. Tawada, M. Kondo, H. Okamoto and Y. Hamakawa, J. de Physique 10, C4-471 (1981)
(4) E. Holzenkämpfer, J. Stuke, Verhandl. DPG (VI) 15, HL 63 (1980)
(5) R. Carius, R. Fischer and E. Holzenkämpfer, J. of Luminescence 24/25, 47 (1981)*
(6) Circels for a-Si:H: D. Weller, Marburg, unpublished

*The sharp drop of the luminescence intensity at $x \sim 0.2$ as well as the double-peak structure of the luminescence spectra reported in reference (5) were found to be most certainly caused by a sharp extinction band in our detection system. We thank B. Rauscher of Bundeswehr-Hochschule, München, for complementary measurements.

ANTIMONY DOPING IN VACUUM DEPOSITED THIN FILM SILICON PHOTOVOLTAIC CELLS

C. Feldman, F.G. Satkiewicz, N. A. Blum, and K. G. Hoggarth
The Johns Hopkins University, Applied Physics Laboratory

SUMMARY

A method for antimony doping silicon polycrystalline thin films and single crystals has been investigated. The method is compatible with the concept of forming photovoltaic cells completely in a vacuum system. Layers of Sb_2O_3 and Si were deposited either simultaneously or sequentially through masks onto the silicon surface. Heating (e.g. 1100° C, 1 hr) in either an inert atmosphere or oxygen brought about the formation of an SiO_x -Sb glassy layer and caused Sb to diffuse into the base silicon surface. The oxide complex is then etched off leaving n-type regions on the surface. Reactions of the layers were examined by secondary ion mass spectrometry and X-ray diffraction.

1. INTRODUCTION

Research is being conducted on forming thin film silicon photovoltaic cells entirely by physical vapor deposition and vacuum processing. It is anticipated that this research will lead to a method for producing large areas of inexpensive cells. Previous studies have led to a cell structure consisting of the following layers on an Al_2O_3 substrate: TiB_2 bottom electrode (1-2 μm thick), p-type Si polycrystalline layer (10-20 μm thick), n-type diffused layer, top metal electrode (1). In order to isolate research problems, the phosphorous doped n-type layer was formed by standard gaseous diffusion, and was thus the only portion of the cell not formed in the vacuum.

The usual n-type dopants (phosphorous, arsenic, and antimony) have high vapor pressures and are far more difficult to apply in a vacuum process than the p-type dopant, boron. Low atomic sticking, outdiffusion, and desorption occur at the high substrate temperatures needed to form large grain Si layers. Recent studies on low temperature molecular beam epitaxy have, however, resulted in successful doping by codeposition (Si and Sb), providing the substrate temperature was carefully controlled and vacuum conditions were very good (2). Vacuum deposited antimony thin films followed by laser (3), or electron beam heating (4) and ion implantation (5) have also resulted in doping but these techniques lead to surface damage, require annealing, and may not be suitable for the extremely large areas envisioned for low cost production of solar cells. Earlier work in this laboratory on depositing Si and Sb at low temperatures, followed by heating, met with some success; however, the resulting film consisted of a fine grained layer of silicon separated from the sample surface by a thin oxide. It was necessary to carry out a high temperature diffusion step in order to overcome the oxide barrier and a severe of loss of antimony occurred.

In order to eliminate some of the problems in the Sb doping techniques discussed above, and perhaps develop an inexpensive, workable approach, research has been conducted on the Sb_2O_3-Si system. Two configurations were examined: (i) a bilayered structure in which an Sb_2O_3 layer was deposited first and then covered with a silicon layer, and (ii) a codeposited layer of Sb_2O_3 and Si forming essentially a single uniform layer. The results of these studies will be described here. Other aspects of the thin film solar cell, such as purity and grain size, will not be discussed. The Si grain size in the films used in this study was 10 to 20 μm.

2. DEPOSITION AND FORMATION PROCEDURES

Depositions of Sb_2O_3 and Si were carried out on single crystal silicon wafers and polycystalline thin film, p-type silicon samples in the low 10^{-6} torr range. The silicon was vaporized by electron beam heating of single crystal silicon by an arrangement previously described (1). Sb_2O_3(99.999% pure) was sublimed from a baffled tantalum crucible. The substrates were not deliberately heated, but generally rose to a temperature of approximately 150° C during deposition. Six 1" x 1/3" samples were coated simultaneously. Kovar masks were used to define variously sized n-type regions on the sample surface(1).

Following deposition, the samples were heated in a tube furnace in either oxygen or an inert atmosphere. Heating in an inert atmosphere was intended to simulate heating in the vacuum, since a clean vacuum chamber was not available. After heating, the samples were etched in buffered HF followed, when necessary, by dilute CP-4.

3. RESULTS

Pertinent parameters and the results of heating of representative samples are listed in Table I. The samples are listed in pairs, e.g. 6D,E for the single crystal (D) and the polycrystalline (E) sample. Samples 6-11 have the bilayered configuration while samples 12 and 13 have codeposited structure. A bilayered sample coated with SiO instead of Si seemed to behave similarly to those coated with Si, however, the low purity of the available SiO made it advisable to use only Si in these experiments. In the samples marked "no junction", doping was not achieved; this occurred in polycrystalline films and was probably related to incomplete coverage due to surface roughness. The samples were tested for junction formation by means of point contact probes placed on the surface of the n-type and p-type regions. These measurements give a reasonably reliable indication of the resulting open circuit voltage of completed devices. The voltages listed for the polycrystalline and single crystal samples in Table I are similar to the voltages observed for phosphorous diffusion-doped samples reported previously (1). The open circuit voltage (V_{oc}) is low in the polycrystalline films due to both small grain size and the presence of impurities. A completed single crystal sample (12D), AR coated, yielded V_{oc} = 0.53V, ff=0.55 and η= 8% under AM-1 conditions.

The deposited layers were analyzed before and after heating by means of secondary ion mass spectrometry (SIMS), X-ray diffraction, and scanning electron microscope. In the SIMS analysis, both atomic and polyatomic spectra were used to determine composition and compound formation in the layers. In addition to antimony, oxygen, and silicon shown in the SIMS profiles, boron and carbon were also present. As might be expected, coatings heated in inert atmosphere contained less oxygen than heated in O_2.

3.1 BILAYERED STRUCTURE

Prior to heating, the deposited layers of antimony oxide covered with silicon were uniform and smooth in appearance. Cracks and wrinkles were observed on the surface following heating. The cracks appear to occur on cooling, and the wrinkles appear to be due to the high vapor pressure of the underlying antimony oxide layer and may be formed before the glassy phase is created. Before heating, the X-ray analysis indicated that cubic Sb_2O_3 was present and that the silicon layer was amorphous. After heating, the X-ray showed in most cases, no distinct lines indicating that only a glassy phase remains. In samples with very thick layers of silicon (e.g. 8A), a separate silicon phase was detected.

The SIMS analysis prior to heating showed the two separate layers as expected. The SIMS determined profiles after heating generally indicated a mixed glassy phase, with silicon oxide and then Si on the surface in some cases. The profile of Sample 8A which contained a relatively thick silicon layer, is shown in Fig. 1. Due to the presence of various phases with different and unknown sputtering rates, sputtering time rather than depth is given on the horizontal scale. The following layers can be identified in Fig. 1: (i) silicon dioxide (ii) silicon with reduced oxygen content, (iii) mixed glassy complex (SiO_x and Sb). Near the surface of the silicon substrate excess silicon appears, and the oxygen level falls off. The antimony level in the various layers can be seen in the figure and ranges from 0.8-3 at%. The total amount of antimony in the sample prior to heating, determined from the film thickness, was approximately 8at%. Thus antimony was lost during the heating process.

3.2 CODEPOSITED STRUCTURES

The codeposited samples were uniform in appearance both before and after heating. Various oxide interference colors were present in oxygen heated samples. X-ray analysis before heating revealed no diffraction lines, indicating only a non-crystalline phase. Following heating, the only X-ray lines that could be identified were attributed to hexagonal antimony.

The SIMS profiles of codeposited samples before and after heating are shown in Fig. 2 and 3. The initial glassy phase Fig. 2 consists of silicon, oxygen, and antimony (15at%), rather uniformly mixed. After heating, the amount of antimony at the Si surface was 2.4at% for the crystal and 1.2% for the polycrystalline film. In all samples the loss of antimony was greater in the polycrystalline sample than in the single crystal. In Fig. 3 the composition of the oxide layer can be interpreted as silicon dioxide with antimony partly in solid solution and partly in elemental form.

The antimony profiles in Samples 12D and 12C, following etching of the glassy phase, are shown in Fig. 4. The boron profiles are also shown to indicate junction depths. An erfc curve was fitted to the experimental data using the surface concentrations determined from the experimental curve. The diffusion constants at 1100°C (70 min heating) are D (xtal) = $3.9 \times 10^{-14} cm^2/sec$ and D (poly) = $8.4 \times 10^{-14} cm^2/sec$. The single crystal value compares favorably with the reported value of D = $3.2 \times 10^{-14} cm^2/sec$ (6). The diffusion coefficient for the polycrystalline film is greater than for the single crystal due to the presence of grain boundaries and dislocations.

4. DISCUSSION

Vacuum deposited films of Sb_2O_3-Si in both the bilayered and codeposited structures can serve as diffusion sources. Some Sb is, however, lost during heating. Codeposited layers appear to produce more uniform sources. The observation that SiO_x and Sb are formed following heating in an inert atmosphere as well as in O_2 indicates that the following type of reaction takes place:

$$xSb_2O_3 + 3Si \rightarrow 3SiO_x + 2xSb \qquad (0 < x \leq 2)$$

The reaction product of samples heated in O_2 contained SiO_2, while those heated in He or Ar contained a composition close to SiO. Heating in a reducing atmosphere (vacuum) would probably result in compositions with even less oxygen; in this case SiO or SiO_2 could be used in place of Si. It is probable that the above reaction and Sb diffusion into silicon take place at the same time; however, the relative rates of the two processes have not been determined. The formation of SiO_x allows the layer to be selectively etched following diffusion. However, when there is insufficient oxygen, $x < 1$, etchants other than HF are required. Etching could be carried out by a reactive plasma process and thus the sample could be prepared for electrode deposition (and antireflection coating) without removal from the system.

REFERENCES

1. C. Feldman, N. A. Blum, and F. G. Satkiewicz, 14th IEEE Photovoltaic Specialists Conference, San Diego, CA., Jan. 7, 1980, IEEE New York 1980, p 391. Thin Solid Films, 90, 1982.
2. Y. Ota, J. Electrochem Soc. 126, 1761 (1979).
3. E. Fogarassy, R. Stuck, J. J. Grob, and P. Siffert, J. Appl. Phys. 52, 1076, (1981).
4. M. Maenpaa, S.S. Lau, M. Von Allmen, L. Golecki, M. A. Nicolet, J. Minnucci, Thin Solid Films, 67, pp 293, (1980).
5. See,for example, B. L. Crowder and J. M. Fairfield, J. Electrochem. Soc, 117, 363 (1970).
6. J. J. Rohan, N. E. Pickering and J. Kennedy, J. Electrochem Soc., 106, 705 (1959).

Table I

Antimony doping parameters

Sample	Thickness (μm) Sb_2O_3	Thickness (μm) Si	Heating Conditions 1100°C	V_{oc} (v) (point contacts) X-tal	V_{oc} (v) (point contacts) Poly
6 D,E	0.16	0.08	O_2 2 hrs.	0.50	no junct
8 A	0.16	0 73	O_2 2 hrs.	–	0.23
8 E,C	0.16	0.43	O_2 2 hrs.	0.55	no junct
9 B,E	0.32	0.32	O_2 2 hrs.	0.42	0.28
9 D	0 32	0.32	He 2 hrs	0.25*	–
10 B,C+	0.16	0.88	O_2 ¾ hrs.	0.33	0.23
12 D,C	0 76		O_2 1 hr.	0.50	0.12
12 F,E	0.76		Ar 1 hr.	0.41	0.19
13 A	1.2		O_2 1 hr.	0 53	–
13 D,E	1.2		He 1 hr.	0.48*	0.13

*Epitaxial layer on single crystal +Heated 1150°C

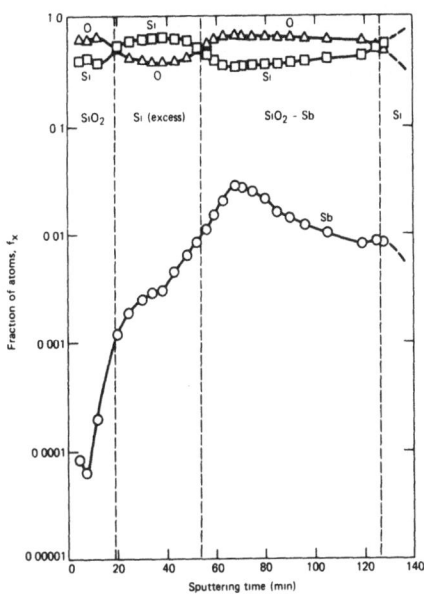

Fig 1. SIMS profiles of glassy coating on Si (8A) after heating.

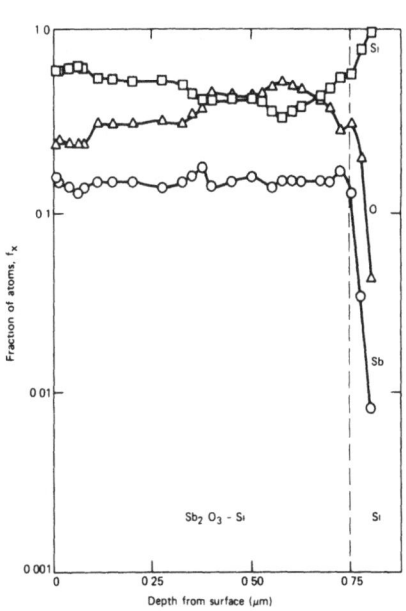

Fig 2. Profiles of glassy coating on Si (12D) before heating.

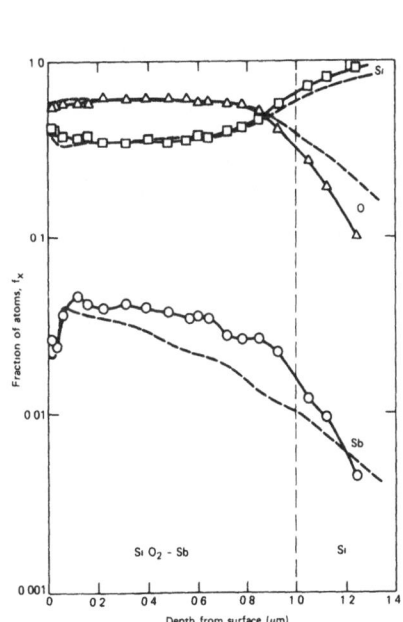

Fig 3. Profiles of coating on 12D (———) and 12C (---) after heating.

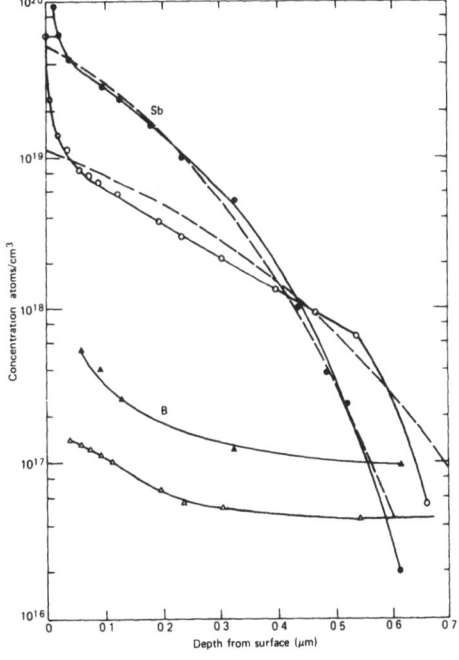

Fig 4. Sb profiles in Si xtal. 12D (●,▲), 12C (o,△) after heating and etching, erfc plots (---).

PHOTOVOLTAIC PERFORMANCE OF CdS HETEROJUNCTIONS ON POLYCRYSTALLINE SILICON

E. SCAFE', G. MALETTA, R. TOMACIELLO, P. ALESSANDRINI,
A. CAMANZI, L. DE ANGELIS and F. GALLUZZI
Laboratori Ricerche ASSORENI, Monterotondo (Roma), Italy

Summary

New results on photovoltaic performance of n-CdS/p-Si heterojunctions are reported. Conversion efficiencies up to 11.1% for single crystal Si and 9.2% for semi-crystal Si have been obtained (without ARC and BSF) by a systematic study of CdS film doping, Silicon substrate preparation and interfacial oxide thickness.

1. INTRODUCTION

Solar cells based on heterojunctions are expected to show good performance when strict conditions on structural and electronic parameters are fulfilled. However good conversion efficiencies (> 10%) have been reported for junctions between poorly matched materials, such as conducting wide-gap oxides (SnO_2, ITO) and Silicon (1). Recently very promising photovoltaic characteristics have been also observed in our laboratory (2, 3) for a quite non-ideal heterojunction, formed by deposition of a conducting CdS "window" on a p-Si substrate. Such results have been optimized by a systematic study on the properties of In doped CdS films, on the influence of Si substrate structure and on the role of interfacial layers. An account of this work is reported here.

2. PROPERTIES OF In DOPED CdS FILMS

Fig. 1 shows the Indium content (measured by A.A.) dependence of electrical properties of CdS:In films grown on quartz substrates. Undoped CdS films show high resistivity (> $10^5 \Omega$ cm) due to the grain boundary barriers (affecting the mobility μ), and to the relatively low carrier concentration n. Addition of Indium causes the films resistivities to drop because of the increase of both n and μ. The increase of mobility depends essentially on the lowering of intergrain barrier height and the increased tunnelling probability.

At Indium concentration higher than about 5% there is a decrease of resistivity due to an increase of both n and μ. Such a behaviour has been also observed on tin oxide films (doped Sb) and on Indium oxide films (doped Sn) (4) and it has been explained in terms of crystal lattice distortion (decrease of carrier density) and of disorder in the crystal lattice which enhances scattering mechanisms (decrease of mobility).

In undoped CdS films, scanning electron micrographs show the characteristic columnar structure with the Z axis perpendicular to the substrate surface. When Indium is added up to 5%, structure is essentially maintained but grain dimension decreases (fig. 2). Further amount of In up to 15% causes the appearance of slightly larger grains. At Indium concentration higher than 20% CdS films became practically amorphous (fig. 2) and the formation of other chemical phases is suggested by electron microanalysis.

3. PROPERTIES OF PHOTOVOLTAIC DEVICE n-CdS/p-SI

a) Photovoltaic characteristics. Photovoltaic parameters (V_{oc}, J_{sc}, F.F.) have been also investigated as function of Indium concentration in CdS films (fig. 3). There is a general increase of these parameters for Indium content up to about 10-13% (a similar optimum condition is observed in ITO). Further increase of Indium amount causes all the parameters to decrease dramatically.
Devices with mono and semicrystalline Si substrates and with 10% In doped windows exhibit under simulated AM1 illumination short-circuit currents of 25-28 mA/cm^2, open-circuit voltages about 530-560 mV, fill factors of .65-.70. Best results give 11.1% efficiency on Si single crystal and 9.2% on cast semicrystal (SILSO), without using ARC and BSF (Fig. 4).

b) Photoresponses. External peak quantum yields exceeding 80% - theoretically expected - are actually observed in devices with Si single crystals and semicrystals. Such low losses can be essentially attributed to optical reflection, so that recombination at interface traps - so important for dark current (3) - seems to be negligible for photogenerated minority carriers.
At short wavelengths (around 500 nm) photoresponses reflect optical properties of CdS "window". In this region lateral photocurrent profiles measured by scanning light-spot technique (fig. 5) - as well as EBIC profiles at low electron acceleration voltage - show a remarkable uniformity, suggesting a high omogeneity in the composition and in the thickness of evaporated films. On the other hand, spectral responses in the near infra-red region (800-1000 nm) reflect transport properties of Silicon. In accordance, near IR photocurrent profiles (fig. 5) and local spectral responses (fig. 6) depend on spatial distribution of Si diffusion lengths and give information on grain boundaries and extended defects. Similar information on Si substrates - with much greater spatial resolution - have been obtained by EBIC maps.

c) Interfacial layer. Best performance of CdS/Si cells have been obtained without any particular treatment after de-oxidation. In this condition, as shown by ESCA and X fluorescence measurements, SiO_x layers are thinner than 10 A.

On the other hand, when on the Si surface a thicker SiO_x layer is thermally grown, photovoltaic parameters show a dramatic reduction and dark I-V characteristics are limited by the transport across the oxide layers (fig. 7). It is noteworthy that such effects are already observed for oxide thickness of about 15 A, when MIS structure (in particular Cr/Si), prepared as reference, still shows a good electrical and photovoltaic behaviour.

These results seem to suggest that "well behaved" n-CdS/p-Si junctions represent an intermediate case between ideal abrupt heterojunctions and proper semiconductor-insulator-semiconductor (SIS) structures (7).

REFERENCES

(1) For a review see R. Singh, M.A. Green and K. Rajkenen, Solar Cells 3, 95 (1981).
(2) C. Coluzza, M. Garozzo, G. Maletta, D. Margadonna, R. Tomaciello and P. Migliorato, Appl. Phys. Lett 37, 569 (1980).
(3) L. De Angelis, F. Galluzzi, G. Maletta, E. Scafè and R. Tomaciello, Proceedings Third E.C. Photovoltaic Solar Energy Conf., W. Palz ed, Reidel, Dordrecht (1981) p. 906.
(4) M. Mizuhashi, Thin Solid Films 70, 91 (1980).
(5) J. Scewchun, J. Dubow, A. Myszkowsti and R. Singh, J. Appl. Phys. 49, 855 (1978).

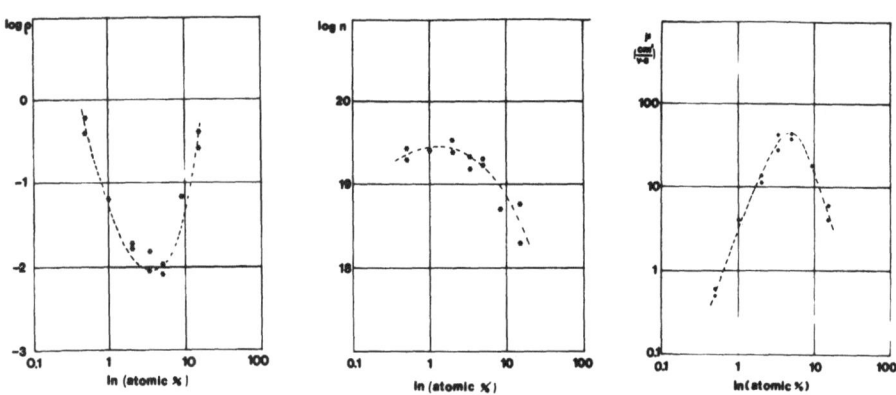

Fig.1: Electrical properties of In doped CdS films on quartz. The ratio In/(In+Cd) was measured by AA.

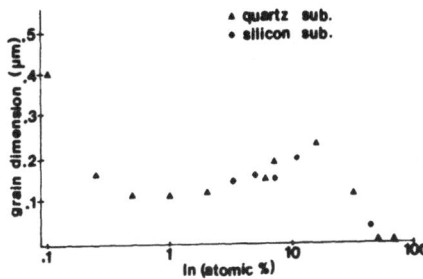

Fig.2: Grain size vs Indium content. Film thickness is fixed at about 2 μm.

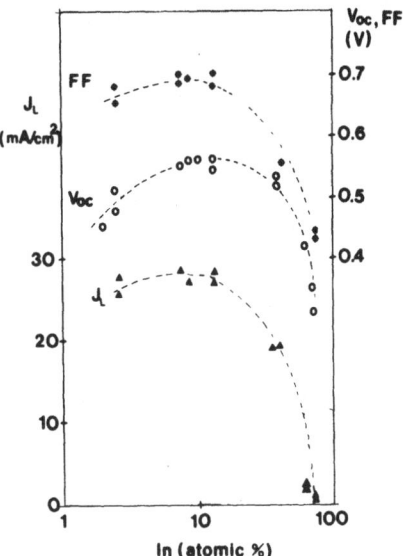

Fig.3: Photovoltaic parameters vs Indium content in n-CdS/p-Si mono junctions.

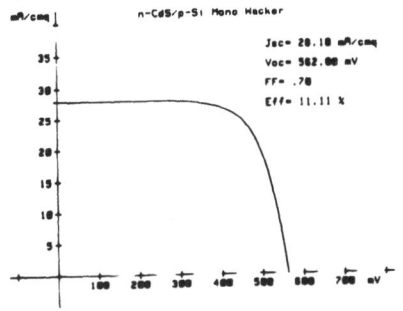

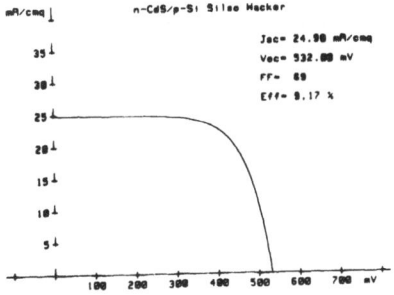

Fig.4: AM1 characteristics of n-CdS/p-Si. Active area is 1.5 cm^2 with non optimized collection grid.

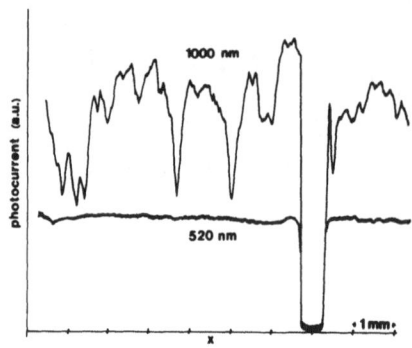

Fig.5: Photocurrent profiles of a n-CdS/p-Si (SILSO) cell at 520 and 1000 nm.

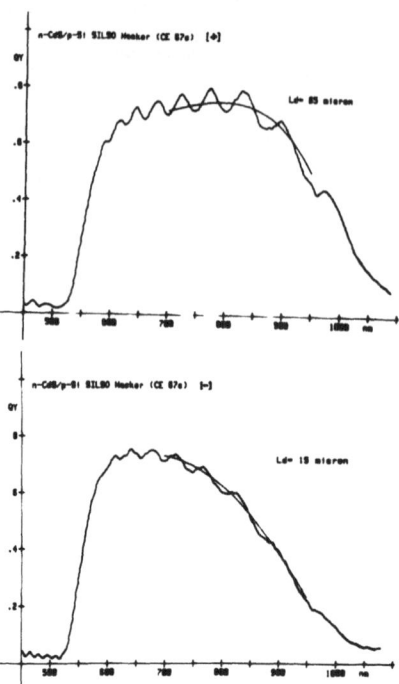

Fig.6: Local spectral responses at two different grains of the same polycrystalline device. Spot size: 90 µm.

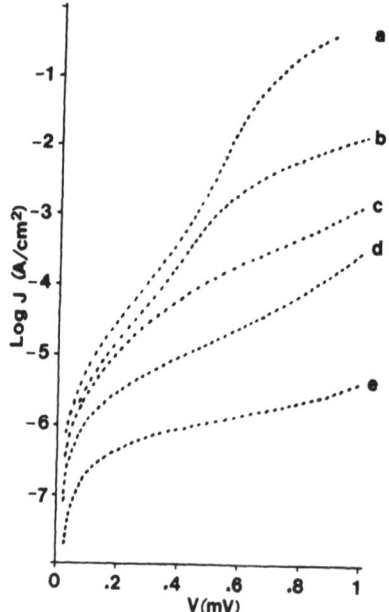

Fig.7: Dark I-V characteristics of Si-SiO$_x$-CdS:In devices. The thickness of SiO$_x$ layer is respectively:
a $\leqslant$ 10 A
b = 13 A
c = 16 A
d = 20 A
e = 50 A

TEMPERATURE DEPENDENCE OF THE IV-CHARACTERISTIC OF Cu_2S-CdS THIN FILM SOLAR CELLS AND RELATED PHENOMENA

G.H. HEWIG, F. PFISTERER, and H.W. SCHOCK
Institut fuer Physikalische Elektronik
Universitaet Stuttgart
F.R. Germany

SUMMARY

The temperature dependence of the IV-characteristic of Cu_2S-CdS solar cells in the dark and under various illumination conditions has been measured in the range -100 to $+100$ C. The major findings are:
1.) Cells processed in different ways exhibit different temperature dependences of the IV-characteristics.
2.) The temperature dependence is influenced in different ways by illuminating the cells with varying monochromatic light.
3.) Cells showing a strong light-effect also show a strong temperature dependence of the IV-characteristics.

The population density of copper levels in the CdS region adjacent to junction is a function of temperature and of illumination and causes variation of the field strength and of the interface recombination velocity at the metallurgical junction. These mechanisms are responsible for the observed effects.

1. INTRODUCTION

The temperature dependence of the IV-characteristics of crystalline Si-solar cells can be explained by classical models where a linear decrease of the open circuit voltage and a slight increase of the short circuit current with increasing temperature can be expected /1/. Cu_2S-CdS heterojunction cells show a behaviour which is much more complex and which depends on the fabrication parameters.

For investigating the respective mechanisms the temperature dependences of various cell parameters (V_{OC}, I_{SC}, FF, η) have been measured as a function of spectral illumination. The measurements were carried out mainly with two types of cells: Type A which were heat treated at 180 °C in air for 15 minutes and Type B with 150 minutes heat treatment under the same conditions. The results of these experiments have been considered in the frame of earlier investigations and are used for supplementing the theoretical model of the Cu_2S-CdS solar cell.

2. EXPERIMENTAL

Cu_2S-CdS thin film solar cells have been fabricated by vacuum deposition of the CdS and by applying the wet topotaxial process for the Cu_2S formation as it is described in detail e.g. in /1,2/. After dipping the post treatment as described and analyzed in /3,4/ is applied. The two types of cells differ only in the duration of the final heat treatment: The type-A-cells-15 min. heat treatment corresponds to the average heat treatment time for obtaining maximum efficiency, the type-B-cells-150 min. heat treatment have been carried out in view to experimentally emphasize the effects of a prolonged heat treatment. The cells were gridded and encapsulated using vacuum deposited grid-lines and laminated cross-bars as described in /5/.

For the measurements of the electrical cell parameters in the temperature range 150 - 370 K they are mounted onto a temperature controlled Cu-block and kept in vacuum. A thin film thermocouple attached to the surface of the cell indicated a maximum deviation of the surface temperature from the temperature of the copper block of - 7 to + 3 °C.

Three different kinds of illumination have been applied:
1. White light (ELH-lamp).
2. Red light (ELH-lamp + corning 4308-filter transmitting $\lambda > 600$ nm).
3. Blue light (ELH-lamp + corning 2408-filter, transmitting 350 nm - 600 nm).

The intensity of the illumination was adjusted in such a way that equal short circuit currents at 300 K had been obtained.

3. RESULTS

The experimental data for the temperature dependence of the open circuit voltage V_{OC} and the short circuit current I_{SC} are plotted in the Figs. 1 and 2 for type A and type B cells respectively. Fig. 3 shows the variations of η with temperature.

3.1 Temperature dependence of V_{OC}

Both cells show a linear decrease of the open circuit voltage with temperature, type B cell exhibits a sublinear increase towards low temperatures. By extrapolating the linear part of the curve with "red" illumination to 0 K a value of 1 V is found, corresponding to the expectet height of the emission barrier for the reverse current. "Blue" illumination reduces this value by 20 mV for type A and by 10 mV for type B cells. White light illumination yields an intermediate value.

3.2 Temperature dependence of I_{SC}

The short circuit current is much less affected by temperature changes. However, type B cells show a strong decrease of I_{SC} at elevated temperatures especially at "red" illumination.

3.3 Temperature dependence of the efficiency

The efficiency of type B cell is much more temperature dependent than that of type A cells. An almost linear decrease according to the decrease of V_{OC} is observed in type A cells, whereas in type B cells an additional reduction of I_{SC} and the fill factor has to be taken into account. Furthermore these cells are much more sensitive to spectral changes of the illumination. (The curves for "blue" and "red" illumination can not be taken as absolute values because the short circuit current is normalized to white illumination. The deviation from the white light curve is thus due to changes of V_{OC} and the fill factor).

3.4 Other observations

Especially type-B-cells show a strong cross-over of the light and dark IV-characteristics. The dark IV-characteristic shows a hysteresis. At Increasing forward bias a lower forward current is observed compared to the value when the forward voltage is reduced again. This effect is more pronounced at elevated temperature. Furthermore the log/log plot of the IV-characteristic indicates the presence of space charge limited currents /6/.

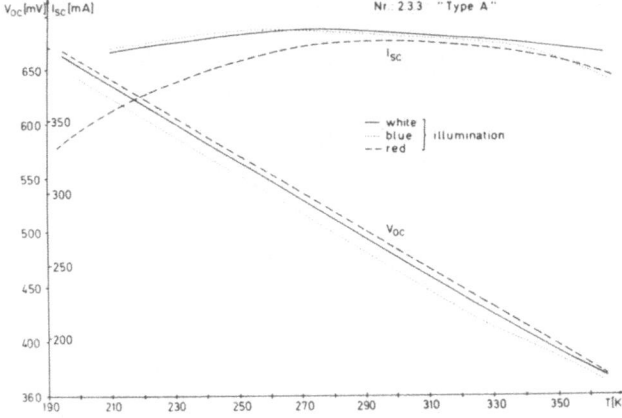

Fig. 1 Temperature dependence of V_{OC} and I_{SC} of type-A-cells.

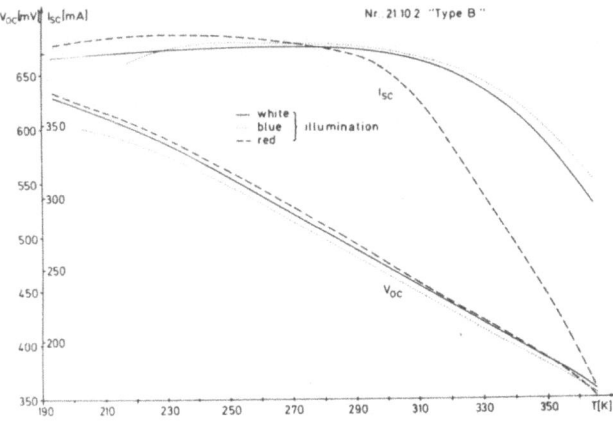

Fig. 2 Temperature dependence of V_{OC} and I_{SC} of type-B-cells.

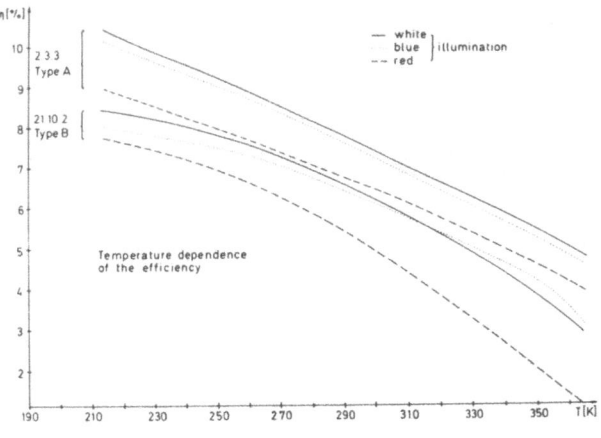

Fig. 3 Temperature dependence of η for type-A and type-B-cells respectively.

The width of the space charge region in the dark of type-B-cells is about 3 times larger than of type-B-cells, as it was determined by means of capacitance measurements.

4. DISCUSSION

A model which describes the junction in the enhanced state (solid line) and in the quenched state (dashed line) an the corresponding space charge distribution is shown in Fig. 4.

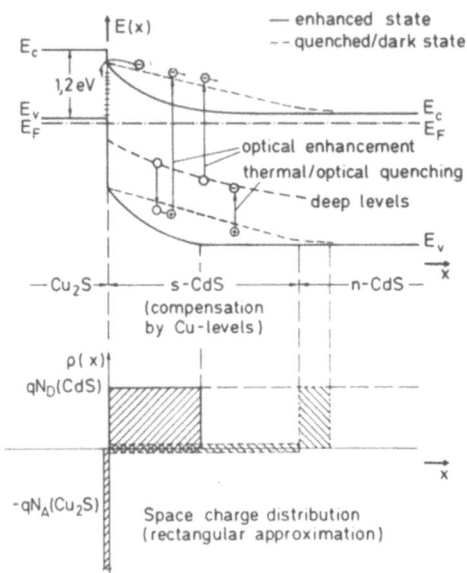

Band diagram of the $Cu_{2-x}S$-CdS heterojunction

Fig. 4 Band diagram and space charge distribution of the Cu_2S-CdS heterojunction.

The deep acceptor levels on the CdS side of the junction can be charged or neutralized either by light or temperature, resulting in a change of the space charge and hence the electrical field at the junction. The optical enhancement is most effective at photon energies greater than the bandgap energy (2.4 eV) of CdS, because this process provides free holes and therefore a neutralization of the deep hole traps. Less effective is the process which leads to the emission of electrons from the hole traps to the conduction band at photon energies $h\nu > 1.6$ eV.

4.1 Open circuit voltage

The temperature dependence of V_{oc} has to be related to emission mechanisms, because diffusion components of the reverse current from both sides of the junction are very small. Thus the open circuit voltage can be expressed by the equation which includes the emission of electrons from the conduction band of the CdS into the Cu_2S via interface states /7/:

$$V_{oc} = \phi_B - \frac{kT}{q} \ln \frac{q \cdot N_C \cdot S_I}{J_{ph}}$$

with potential barrier $\phi_B = 1$ V, effective density of states in the conduction band of CdS $N_C = 3 \times 10^{18}$ cm^{-3}, $S_I = 10^6 - 5 \times 10^6$ cm/s.

The increase or decrease of the open circuit voltage for "blue" and "red" illumination respectively can be explained by image force effects at the interface (lowering of the barrier height by the electric field at the interface). A change of the effective doping concentration of the CdS at the junction as it is described by the model in Fig. 4 from $N_D-N_A=10^{17}$ to $N_D-N_A=10^{16}$ for type A and from $N_D-N_A=10^{16}$ to $N_D-N_A=10^{15}$ for type-B-cells can explain observed effects.

4.2 Short circuit current

The pronounced decrease of I_{sc} of type-B-cell can be explained by thermal quenching due to the ionization of the deep acceptor levels according to the model in Fig. 4. Especially at "red" illumination the thermal quenching is the dominating process.

4.3 Efficiency

The efficiency of type-A-cells is only effected by the reduction of the open circuit voltage wheras type-B-cells also suffer from the reduction of I_{sc} and the fill factor.

4.4 Other observations

The hysteresis effects during measuring the IV-characteristics indicate the precense of field dependent space charges caused by field ionization of deep hole traps (Frenkel-Poole effect) /8/. These effects together with space charge limited forward current transport occurs predominantly in cells with a very wide compensated layer (type B).

5. CONCLUSIONS

The measurements showed that the IV-characteristics are governed by a number of different mechanisms where the dominance of one mechanism is related to the fabrication process. Extensive copper diffusion and thus strongly compensated CdS at the interface favour temperature and light-bias effects. Further investigations should help to clarify all observed phenomena in order to optimize the cell with respect to an optimum operation also at elevated temperature.

6. ACKNOWLEDGEMENT

This work has been supported by the Bundesministerium für Forschung und Technologie, F.R. Germany, Contract No ET 4045 B.

7. REFERENCES

/1/ H.J. Hovel, in: Semiconductors and Semimetals, Vol. 11, Academic Press New York, 1975.
/2/ W.H. Bloss, F. Pfisterer, and H.W. Schock, Proc. 3rd EC Photovoltaic Solar Energy Conf., Cannes, 1980, (D. Reidel Publ. Comp. Dordrecht, 1980), p. 340.
/3/ F. Pfisterer, H.W. Schock, and W.H. Bloss, Conf. Rec. XIIth IEEE Photov. Spec. Conf., Baton Rouge, 1976, (IEEE New York, 1976), p. 502.
/4/ F. Pfisterer, H.W. Schock, and J. Woerner, ref. 2, p. 762.
/5/ W. Arndt and H.W. Schock, this issue.
/6/ L.D. Partain et al, J. Electron. Mater. 9, 3, 1980.
/7/ A. Rothwarf, Proc. Internat. Workshop on CdS Solar Cells and other Abrupt Heterojunction, Univ. of Deleware, Apr. 30 - May 2, 1975, NSF-RANN AER 75 15858, p. 9.
/8/ K.W. Boer, ref. 2, p. 949.

CONTINUOUS DEPOSITION OF PHOTOVOLTAIC GRADE CdS SHEET
AT THE UNIT OPERATIONS SCALE

R. E. Rocheleau, P. J. Lutz, D. F. Brestovansky, B. N. Baron
and T. W. F. Russell
Institute of Energy Conversion
Department of Chemical Engineering
University of Delaware
Newark, Delaware 19711

Summary

Uniform photovoltaic grade CdS sheet has been reproducibly deposited on a continuously moving flexible substrate in a reel to reel vacuum coater. Materials characterization by scanning electron microscopy, photoluminescence, and resistivity revealed that continuously deposited CdS is essentially equivalent to material that was deposited for making high efficiency Cu_2S/CdS cells in the laboratory scale process. Cells made using continuously deposited CdS sheet had efficiencies as high as 7.85%.

1. INTRODUCTION

The promise of thin-film solar cells rests on low materials cost and applicability of continuous processing. An analysis (1) of manufacturing costs shows that CdS/Cu_2S solar cells can be produced at a cost of less than $0.70/Wp in a continuous process using thermal evaporation of the active semiconductor layers onto a continuous moving substrate. High efficiency thin film cells utilizing evaporated CdS in conjunction with Cu_2S and $CuInSe_2$ have been developed at the laboratory scale. Vacuum deposition of the CdS semiconductor layer is, therefore, a key unit operation in the manufacture of these thin film cells.

The logic for development of a commercial scale process involving continuous deposition of the semiconductor layers was described at the previous E. C. Photovoltaic Solar Energy Conference in October 1980 (2). At that time, we described a reel-to-reel vacuum coater which was built to our specifications for continuous deposition of CdS. The CdS evaporant source and web temperature control systems have been modified to meet requirements for high materials utilization and extended operation under controlled conditions. In this paper, we report recent results and cell data obtained from continuously deposited CdS sheet.

2. EQUIPMENT

The reel-to-reel vacuum coater is shown schematically in Figure 1 and is described more fully in Reference 1. The evaporation source which makes it possible to uniformly deposit CdS on a moving substrate is also described elsewhere (3).

3. MATERIALS CHARACTERIZATION

CdS was deposited on rolls of zinc-plated copper foil. Depositions were carried out with substrate throughputs ranging from 180 to 600 cm^2 per hour at film growth rates ranging from 0.5 to 2 µm/min. The substrate temperature was controlled within $10^{\circ}C$ of the set point between $200^{\circ}C$ and $250^{\circ}C$. Runs producing up to 3000 cm^2 of 25 µm thick CdS have been carried out at throughputs of approximately 400 cm^2 per hour. Materials utilization was typically 25% at a substrate temperature of $230^{\circ}C$. Ambient pressure during deposition was less than 10^{-4} mbar. Measured deposition rate, thickness uniformity, and utilization were found to agree with predictions based on a model for CdS evaporation (4).

Figure 2 shows scanning electron micrographs of CdS deposited in the continuous coating unit and in a batch process. The batch deposition of CdS was carried out in a bell jar evaporator that is used in producing CdS cells over 8% efficiency for IEC's concurrent research on durability and device physics. The micrographs show the top surface growth morphology, grain structure, and etching behavior in 25% HCl, a processing step used in cell fabrication (5).

Cross-sectional micrographs of continuously deposited CdS (Figure 3) show characteristic columnar growth. X-ray diffraction confirms predominantly c-axis orientation for continuously deposited CdS.

The resistivities of the CdS films were determined by measuring the resistance between a 4×10^{-3} cm^2 indium dot pressed onto the film and the copper substrate. Resistivities of continuously deposited CdS ranged from 1 to 100 ohm-cm. High efficiency cells produced in the laboratory are usually fabricated from CdS films having resistivities in the range of 1 to 10 ohm-cm (5).

Photoluminescence emission spectra from 450 to 850 nm were made with a Perkin Elmer MPR-44A Flourescence Spectrophotometer at $77^{\circ}K$ using 340nm excitation. Hall et al (5) have used this measurement to evaluate CdS deposited at the laboratory scale. Continuously deposited samples of CdS exhibit both the 490nm peak and the critical green edge emission peak at 520nm, although the 490nm peak is sometimes absent.

4. CELL RESULTS

Although materials characterizations of CdS are routinely carried out, the principle means of evaluating the CdS is the photovoltaic response of cells fabricated on this material using a laboratory scale (batch) procedure (6).

Reverse bias laser scans (7) following junction formation were used to survey the uniformity of current response and the density of macrodefects such as pinholes and shunts. One cm^2 specimens have been made which exhibit uniform response and the absence of macrodefects. Laser scans for both laboratory and sheet CdS are shown in Figure 4. One cm^2 cells with evaporated grids (6) were made from specimens which were cut out randomly along the CdS sheet. Figure (5) is a histogram of efficiencies of cells made with continuously deposited CdS taken from two sheets which were 2.2 and 2.8 meters long respectively. The highest efficiency cells, 7.68 and 7.85%, were made using CdS taken from positions separated by 100 cm along the continuously coated sheet.

5. CONCLUSIONS

Photovoltaic grade CdS thin film sheet has been deposited on a continuously moving flexible substrate. Experiments now underway are directed to formation of the Cu_2S layer by the evaporation of CuCl and solid state reaction in the reel-to-reel vacuum coater. The results of this and other unit operations research are necessary for a successful commercial scale process design.

6. ACKNOWLEDGEMENT

This work was carried out under a contract with Chevron Research Company whose support is gratefully acknowledged. We also wish to thank the Device Design & Analysis and Solar Cell Development groups at IEC for their efforts in support of this work.

7. REFERENCES

(1) T. W. F. Russell et al. in Fundamentals and Applications of Solar Energy II, I. H. Farag and S. S. Melsheimer, Eds. AIChE Symposium Series S-210 (1981).
(2) T. W. F. Russell, B. N. Baron and R. E. Rocheleau, Proc. Third European Communities Photovoltaic Solar Energy Conference, Cannes, October 1980, D. Reidel, Boston (1980), pp. 348-352.
(3) B. N. Baron, et al., United States Patent No. 4,325,986; April 20, 1982.
(4) R. E. Rocheleau et al., accepted for publications in AIChE Journal.
(5) R. B. Hall et al., Proc. 14th IEEE Photovoltaic Specialists Conference, San Diego (1980) pp. 706-711.
(6) R. B. Hall and J. D. Meakin, Thin Solid Films, 63, 203-211 (1979).
(7) B. C. Plunkett and P. G. Lasswell, Proc. SPIE 24th Annual Technical Symposium, Vol. 24B, 142-147 (1980).

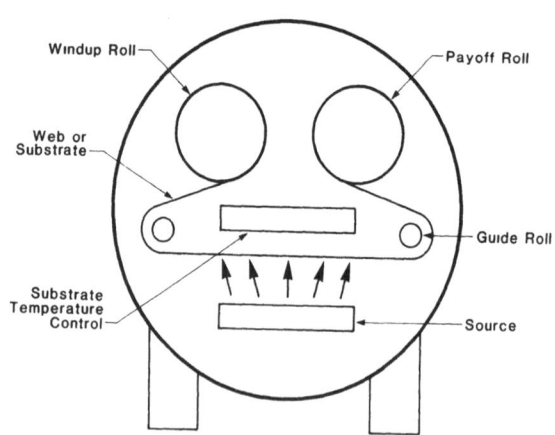

FIG. 1: SCHEMATIC OF REEL-TO-REEL VACUUM COATING UNIT

a. Continuous-As Grown

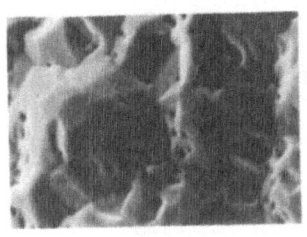

b. Batch-As Grown

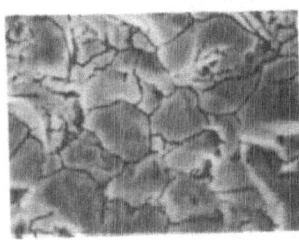

5 μr

c. Continuous-Grain Boundaries

d. Batch-Grain Boundaries

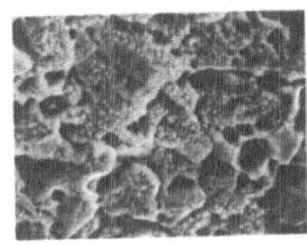

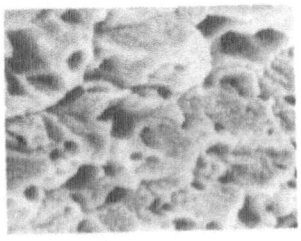

e. Continuous-Etch Behavior

f. Batch-Etch Behavior

<u>FIG. 2.</u> Scanning Electron Micrographs of Batch and Continuously Deposited CdS (magn. = 2000 x)

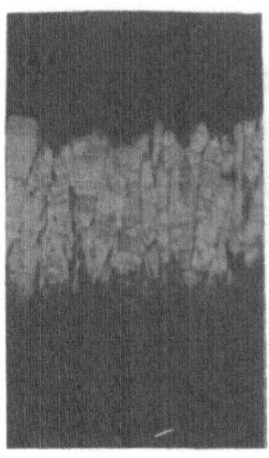 

FIG. 3a: Cross Section of Continuously Deposited CdS Film (Grain boundaries enhanced by etch).

FIG. 3b: Cross Section of Batch Evaporated CdS Film (Grain boundaries enhanced by etch).

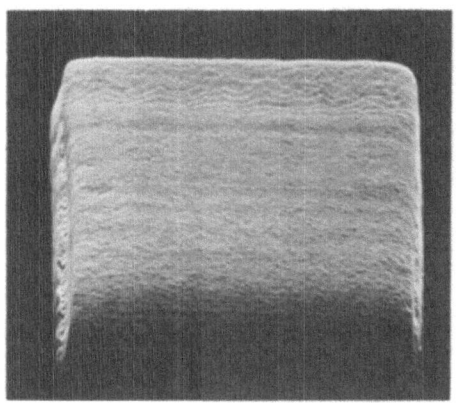 

FIG. 4a: Laser Scan of Continuously Deposited CdS Showing Uniformity of Current Response.

FIG. 4b: Laser Scan of Batch Evaporated CdS Showing Uniformity of Current Response.

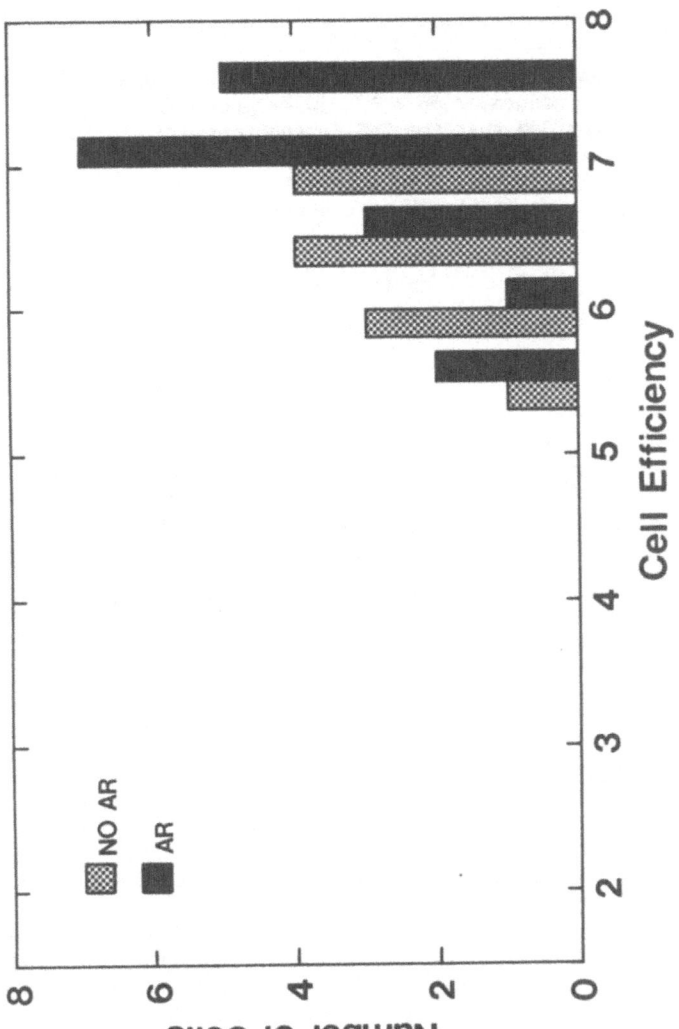

FIG. 5 Histogram of Cell Efficiencies: Samples from Two Continuously-deposited CdS Sheets

$Cu_xS(p)$ - $CdZnS(n)$ - $CdS(n^+)$ EVAPORATED THIN FILM SOLAR CELLS

B. BOUCHIKHI, S. CHANDRASEKHAR, F. ZAPIEN NATAREN and S. MARTINUZZI
Laboratoire de Photoélectricité des Semi-conducteurs
University of Marseille - F.13397 MARSEILLE Cedex 13

Summary

The orientation and size of crystallites may be improved in $Cd_{1-y}Zn_yS$ films when they are condensed on a CdS layer previously deposited in the same vacuum provided that the ZnS concentration y does not exceed 12 %. Backwall cells were made by means of the solid state reaction for different ZnS concentrations. The photovoltage increases with ZnS concentration y while very little changes occur for the photocurrent when y values are in the range 0-0.12.
The current transport mechanism remains dominated by recombination tunneling currents. L.B.I.C. scannings indicate that the photoresponse of the cells are quite homogenous.

1. INTRODUCTION

The conversion efficiency of Cu_xS-CdS cells is limited by relatively low values of the photovoltage, which are at least the consequence of lattice and electron affinity mismatches. These mismatches can be reduced by the use of a $Cd_{1-y}Zn_yS$ films in place of the CdS layer, and photovoltages as high as 0.7V were reported (1). However there were generally a fall off in the photocurrent when the ZnS concentration increased.
Several origins were proposed to explain the degradation of the photocurrent. Misorientations in the columnar crystallite growth and crystallographic defects which appears in these ternary layers (2) can be extended in the topotaxially grown Cu_xS film reducing the electron mobility and diffusion length. Inhomogeneities of ZnS concentration can impede the growth of the Cu_xS layer. Zn retention in the Cu_xS layer has a detrimental effect on the electron properties of the copper sulfide film (3). The interface dislocation field existing at the junction may be enhanced (4).
The goal of the present work is to show that the crystallite defects of the CdZnS layers contribute to degrade the photocurrent for low ZnS concentrations.

2. EXPERIMENTAL

Backwall cells (area 0.5 cm^2) were made using bifilm structure :
a $Cd_{1-y}Zn_yS$ film (2 to 3 µm) was deposited on a thicker layer of CdS (10 to 12 µm - 5 Ω.cm) previously condensed on SnO_2 or In_2O_3 substrates heated to 200°C (figure 1).
Conventionnal thermal evaporation of large grain size CdS powders and $Cd_{1-y}Zn_yS$ solid solutions was used. The solid solutions were previously annealed in vacuum and were rapidly heated to the evaporation temperature (1100°C - 5.10^{-6} torr). The two evaporations were made successively in the same vacuum, and after cooling, a CuCℓ film (0.35 - 0.4 µm) is condensed. The structure is heated to 200°C during 3 mn in an H_2 flux. A thin film of

copper (150 Å) was deposited on the Cu_xS (after the removal of the $CdCl_2$) to improve the stoichiometry and to realize a suitable ohmic contact with the top electrode. Endly a mylar foil applied in vacuum (10^{-2} torr - 60°C) protected the cells.

Several series of ten backwall cells were realized with different ZnS concentrations : (0.6 ; 0.8 ; 0.10 ; 0.12 ; 0.15). Some cells were also prepared by means of a four crucible electron gun, which can evaporate successively layers of CdS+In, CdS, CdZnS and CuCl, in the same vacuum.

The methods of analysis of the layers (X rays, molecular absorption spectroscopy, ESCA) and the photoelectric measurement technology have been described previously (2,6,7).

3. RESULTS

As shown by figure 2 , the top layer of $Cd_{1-y}Zn_yS$ has a tendency to grow on the base layer. There is a continuity in the crystallite growth from one side to the other of the interface. The columnar growths are well defined, the crystallite sizes are in the range 1 to 3 µm and the misorientations are reduced compared to monolayers of $Cd_{1-y}Zn_yS$ (2) thus confirming the X rays analysis. These improvements occur for y values up to .12.

The X-rays and ESCA analyses indicate that the composition of the $Cd_{1-y}Zn_yS$ layers are transversally homogeneous and that the ZnS concentration in the evaporated powders and in the layers are similar. ESCA reveal also that Zn atoms are chemically bound to sulphur atoms.

Typical I-V curves (80 mW.cm^{-2}) are given by the figure 3 for different y values. The mean value of the photovoltage increases from 0.52 V to 0.65 V when y varies between 0 an 0.12, while the photocurrent remains practically unchanged ($\simeq$ 16 to 17 mA.cm^{-2} AM1). Consequently the mean efficiencies increase from 4.5 to 6 % with y.

The variations with T of the dark forward I-V curves indicate that tunnel recombination currents dominated the transport mechanism through the heterojunction. For temperatures below 20°C and applied voltages in the range 0.3 to 0.7 V, these dark currents may be expressed by :

$$I = I_{oo} \exp(-\frac{q\phi}{kT}) \exp \beta V$$

where the activation energy ϕ increases with ZnS concentration y ($\phi \simeq 0.52$ V and $\beta \simeq 18$ V^{-1} for y = 0.08 ; $\phi \simeq 0.65$ V for y $\simeq 0.12$).

To verify the homogeneity of the cell response, we have made light beam induced current scannings and the figure 4 shows a typical picture (y $\simeq 0.10$). It is clear that the homogeneity is acceptable.

The figure 5 shows a typical spectral variation of the photocurrent (y $\simeq 0.10$) and the inset represents the determination of the electron diffusion length L_n which were found around 0.3 µm ± 0.1 µm for y < 0.12.

For y > 0.12 in the bifilm cells and for y > 0.6 in CdZnS monolayer cells, L_n and so the photocurrents tend to decrease.

The $C^{-2}(V)$ characteristics indicate that the densities of donors in the CdZnS layers decrease in the 10^{16} cm^{-3} range when y increases. The intercepts of the graphs with the voltage axis at $C^{-2} = 0$ occur between 1.5 and 2.5 V, suggesting that interfaces states are present in all the cells studied and that their density is not reduced by the use of a ternary layer.

4. CONCLUSION

The bifilm structure improves the $Cd_{1-y}Zn_yS$ layers. The photocurrents remain constant while V_{OC}, and the conversion efficiency increase, for $y < 0.12$. The results suggest that for $y < 0.12$, the degradation of the photocurrent may be the consequence of a large density of crystallographic defects.

This work was partly supported by PIRDES-CNRS France and by the Commission of the European Communities.

5. REFERENCES

(1) L.C. BURTON - Appl.Phys.Lett. 35 (1979) 780.

(2) P. DOMENS et al - Phys.Stat.Sol.(a) 59, 201 (1980) 201.

(3) L.C. BURTON - Solar Cells 1 (1979/80) 159.

(4) K.W. BOER - Phys.Stat.Sol.(a) 49, (1978) 456.

(5) R.B. HALL et al. - Appl.Phys.Lett. 38 (11), 1, (1981) 925.

(6) S. MARTINUZZI et al. - Proc. of 3rd European Communities Photovoltaïc Solar Energy Conf. D. REIDEL Pub.Comp. 1980 p.782.

(7) J. OUALID et al. - Rev. Phys.Appliquée 17 (1982), 119.

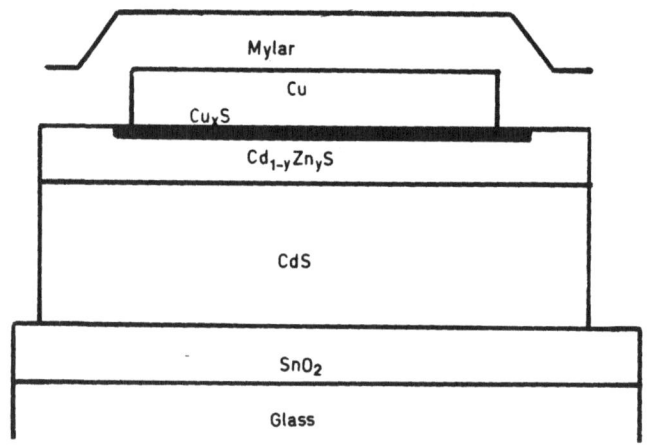

Fig.1. Structure of the studied photocells.

CdZnS

CdS

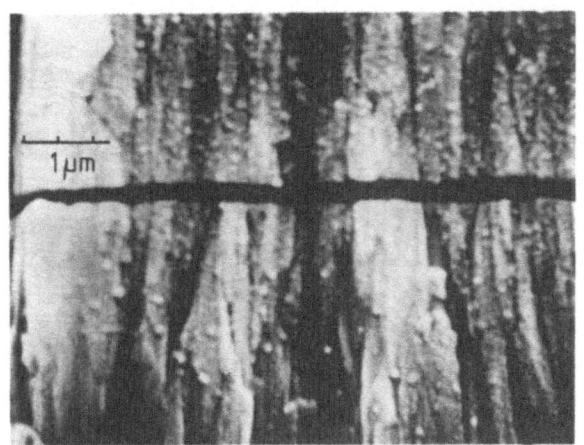

Fig.2. Microphotograph of a bifilm edge.

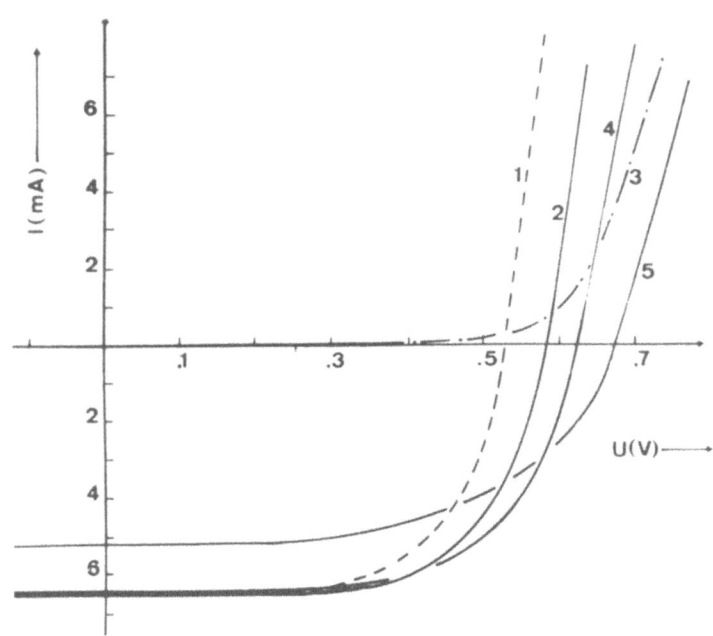

Fig.3. I-V curves (80 mW/cm^{-2})
(1) CdS ; (2) 8 % ZnS ;
(4) 12 % ZnS ;
(5) 15 % ZnS.

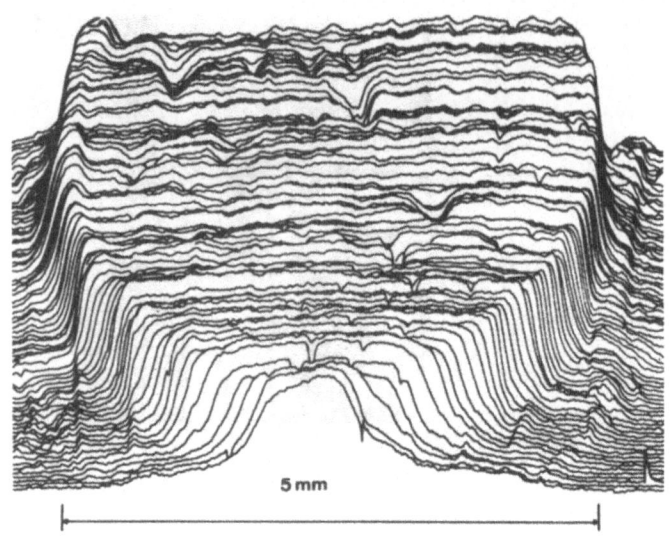

Fig.4. LBIC scanning of a cell.

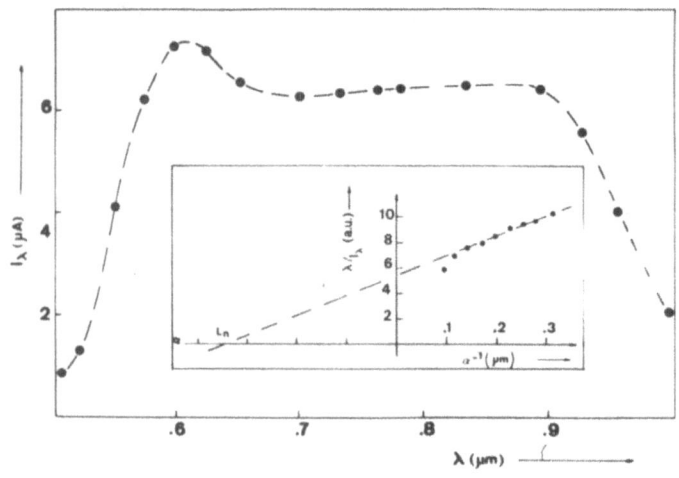

Fig.5. Spectral response vs wavelength and determination of L_n.

THIN FILM Cu_2S/CdS JUNCTIONS PRODUCED BY EVAPORATION AND
SPUTTERING: EFFECT OF THERMAL TREATMENTS IN
VACUUM.

E. ELIZALDE, M. LEON, F. RUEDA and F. ARJONA
Departamento de Física Aplicada, Universidad Autónoma de Madrid
Cantoblanco, Madrid, Spain.

Summary

Heterojunctions Cu_xS-CdS produced by evaporation and sputtering of chalcocite on evaporated CdS have been produced. The effect of vacuum annealing on the spectral quantum efficiency has been used as a means of deducing the chalcocite and djurleite presence. A peak at 500nm is tentatively assigned to djurleite. Open circuit voltages are similar to those reported for junctions produced by the "Clevite" method.

1. INTRODUCTION

The use of evaporation or sputtering of cuprous sulphide as an alternative technique to the "Clevite method"(1) has been attempted by several authors (2-5). However, the best solar cells have been obtained by the Clevite method. One of the potential advantages of these techniques is an independent control of stechiometry and doping of the sulphide layer that would result in different phase composition and optical and electrical properties of the heterojunction. Chalcocite has been identified as the phase producing the best characteristics in solar cells (6). The spectral response of heterojunctions varies largely from one author to another. Stanley (7) and Savelli and Bougnot (8) have tabulated the results of about 30 workers and a large diversity of peaks. In this respect, the work of Loferski et al.(19) is illustrative of the diversity of results; these authors use cathodo-luminiscence as a test of the presence of chalcocite and djurleite. Even when they find by cathodo luminescence chalcocite as the only detectable phase the spectral response of the junctions present a diversity of results where a peak at ~500nm seems common to all the different modes of fabrication. Recently, Robertson and Woods (10) have shown that Cu_xS-CdS junctions obtained in single crystals by the "Clevite" method may have chalcocite and djurleite depending on the temperature of the CuCl solution, respectively at 96-92°C and 80-85°C. The spectral responses present peaks at 500, 740 and 920nm in either case, being predominant those at 920 for chalcocite and 740nm for djurleite. The peak at ~500nm has been interpreted by these and other authors as due to the copper ions diffused into CdS during the junction forming annealing.

Following these lines the present work intends to asign the different peaks to the presence of chalcocite and djurleite, producing the sulphide films by sputtering (non reactive and reactive) and evaporation, and following the evolution of the spectral response as a function of vacuum thermal treatments.

2. EXPERIMENTAL

CdS films of about 20μm thickness evaporated on a Ni-Cr plated glass substrate[+] have been used as substrates, after etching in HCl solution for some seconds. The cuprous sulfide layers have been produced by sputtering in a triode type system built in our laboratory. Djurleite and chalcocite targets have been made, by sintering in an evacuated ampoule the powder, whose X-ray diffractograms show the lines of these phases assigned by Potter and Evans (11).

The sputtering was carried out at 10^{-2} torr in pure Ar; in several sputtering runs of chalcocite addition of 5.10^{-3} torr of H_2 was also used in order to favor some sulphur loss, with the hope to attain chalcocite as the film phase. The substrate of CdS was kept at 20°C by cooling the holder or allowed to heat by the plasma in the 100-150°C range. The sputtering conditions were 1KV-1mA and produced a deposition rate of ~15Å/min The film thickness was determined with a "Talystep-Rank precision" instrument, from 300 to 2000Å. In some samples a film of pure copper of ~75Å thickness was deposited on the sulphide film by evaporation or sputtering.

Some heterojunctions have been produced by evaporating chalcocite from a tungsten crucible in high vacuum ($<1.10^{-5}$ torr with the CdS substrate at 100-140°C or at 20°C. Sulphide thickness ranged from 600 to 2500Å.

Electron diffraction of films, deposited simultaneouly over acetate butyl cellulose a that was later dissolved, was carried out and the main phases present were determined in the "as grown" condition. Optical measurements in transmission and reflexion were also done and the optical gaps and index of refraction of the films were determined. Spectral quantum efficiency measurements were done on the heterojunctions in the "as grown" condition and after subsequent annealings at temperatures in the 160-200°C range for 30min in high vacuum. A grating monochromator with tungsten lamp and HP-8330 Fluxmeter were used in these measurements.

The heterojunctions after taking them to the air, were measured pressing by suction a gold plated grid over the cuprous sulphide layer and illuminating in the "front wall" mode. The contact of the grid was not as good as in laminated cells and therefore the short circuit currents cannot be high and are to be taken as indicative. However, this method allows to perform succesive thermal treatments with reduced contamination as well as further deposition of a copper film on cuprous sulphide in a single sample.

[+]Kindly supplied by Prof. W. Bloss, Stuttgart University, BRD.

3. RESULTS AND DISCUSSION

The quantum efficiencies of four heterojunctions were obtained by sputtering pure djurleite in Ar atmosphere. Fig.1 shows the spectral quantum efficiency in the "as grown" condition with cold substrate and after succesive annealings in vacuum at 165 and 200°C for 30min. The "as grown" film presents amorphous structure with several diffuse reflexions, correponding to high-chalcocite, when heating the samples by the beam of electrons during observation in the E.M.

This is compatible with djurleite phase at room temperature after recrystallisation. The heterojunctions obtained by sputtering chalcocite in pure Ar, with the substrate reaching ~110°C during deposition, present a quantum efficiency as shown in Fig. 2. The effect of vacuum annealing is to increase the response up to 170°C, decreasing at higher temperatures and dessapearing for treatments above 200°C. The deposition of a new sulphide layer in similar conditions on top of the previous film produces a response (Fig. 2,e) that presents a peak at 500nm. This would indicate that the peak at ~500nm is not associated to copper diffused into CdS but to the photoresponse of the sulphide phase of the heterojunction. Electron diffraction on films produced in the same run indicate djurleite as the main phase.

In order to reduce the sulphur content of the sputtered film, a mixture of Ar-H_2 was used and a film of 50Å of copper was deposited on top of it. Fig. 3 shows the spectral quantum efficiency of a heterojunction grown on cold substrate for different annealing treatments. One can see that the peak at 500nm has been greatly reduced in the "as grown" condition and extends its response in the red with a peak at ~920nm. On annealing the spectral response increases in particular in the IR region, resulting an increase in I_{SC} in sun light.

The evaporated cells show a behaviour similar to the sputtered ones, presenting peak at ~500nm in same cases or extended red response with peak at ~920nm. The thin sulphide films show peak at 500nm that increases with thermal treatments this effect being more pronounced when cold substrates are used. Thicker samples show peak at ~920nm but fail to do so at 500nm and this is essentially unchanged with annealing.

Fig. 4 shows the spectral quantum efficiency of a heterojunction with evaporated cuprous sulphide of 1000A (420A/min) on substrate at 130°C in the "as grown" condition and after annealing at 170°C. The electron difraction of the policristalline sample indicates chalcocite as the major component, djurleite is not specifically observed but can not be excluded.

Fig. 5 shows similar results with a sulphide film of 1000A (420A/min) deposited on CdS substrate at 20°C, "as grown" and after annealing at 175°C. Electron diffraction indicates the presence of chalcocite and djurleite, the latter in small proportion.

Fig. 6 refers to a junction with sulphide film of 2000A, (1680A/min) deposited on CdS substrate at 140°C, "as grown" and after annealing at 175°C. Electron diffraction could not be done but thicker samples obtained in similar conditions (~6000A) show chalcocite as the only observed phase and we may assume that this film has a higher chalcocite content than the previous ones.

As general observations on sixty samples, it should be mentioned that in photoresponse the peak at 500nm appears in cells where djurleite is probably present and this peak increases in a normalized response in films with less than 1000A after thermal treatment up to 170°C. Junctions with thicker films do not show this increase so clearly. The peak at 920nm appears in junctions with sulphide films rich in chalcocite and the peak a 500nm does not appear. The junctions presenting peak at 500nm, after deposition of 75A of Cu and annealing, show a decrease of this peak and an increase of the peaks at about 660 and 740nm.

The intensity and FF always increase with the annealings up to 180°C-30min. Two samples produced by sputtering have shown $V_{OC} \sim 520mV$ after annealing at 170°C, however, the current was low in these cells. The best cells results are indicated in Table I and Fig. 7.

Figs 1-5 show results obtained in the "as grown" condition on substrate at room temperature during deposition and the peak at 500nm appears. In this conditions it is improbable that copper diffused into CdS and therefore this peak should be due to the sulphide layer and is temptatively assigned to djurleite. The increase of this peak with vacuum annealing is compatible with an increase in the djurleite content of the film. This assumption is consistent with the results shown in Fig. 3, where a film deposited on a cold substrate does not show the 500nm peak in photoresponse in "as grown" condition and neither does it, after annealing. Fig. 2,d and e where a new sulphide layer is deposited after prolonged annealing at 200°C indicates that copper diffussion is not sufficient to produce the 500nm peak.

Films deposited in hot substrate, Fig 6, do not show peak at 500nm nor after annealing. As in these samples the expected phase in the film is chalcocite, the 500nm peak seems to be characteristic of djurleite. The diffusion of copper into CdS improves the junction characteristic but does not seem to cause the 500nm peak. This conclusion is in contrast with that of Robertson and Woods (10) and other authors[1,12].

4. CONCLUSIONS

The spectral response of a Cu_xS-CdS heterojunction appears to be correlated with the cuprous sulphide phase content, the peak at ~500nm being temptatively assigned to djurleite. The peaks at ~740 and 920nm are consistent with the presence of chalcocite, as indicated in (10).

The values of V_{OC} in junctions with sulphide layers both evaporated and sputtered are comparable to those obtained by the "Clevite" method. The deposition conditions should assure the presence of chalcocite: hot substrate, high deposition rate and thickness above 1500A for evaporated sulphide films and mixture of $Ar-H_2$ in the sputtered ones.

The shape of the spectral response does not change with vacuum annealings up to 180°C when the sulphur layer is chalco cite, indicating that only a small fraction of Cu ions is diffused into CdS. However, when djurleite is present this diffusion produces an increase in the 500nm peak. If a thin film of copper is deposited on the outer surface the 500nm peak decreases on annealing and the supply of copper overcomes the loss by diffusion.

REFERENCES

1. L.R. SHIOZAWA, F. AUGUSTINE, G.A. SULLIVAN, J.M. SMITH and W.R. COOK Jr, Clevite Final Report, Contract AF33(615)-5224 (1969).

2. A.E. CARLSON, L.R. SHIOZAWA and J.D. FINEGAN, US Patent 2.820.841; S. YU. PAVELETS and G.A. FODORUS, Ukr. Fiz.Zh.$\underline{11}$ 686 (1966).

3. B. SELLE, W. LUDWING and R. MACH, Phys.Stat.Sol.$\underline{24}$,K149(1966)

4. G.A. ARMANTROUT, J.H. YEE, E. FISCHER-COLBRIE, J. LEONG, D.E. MILLER, E.J. HSIEH, K.E. VINDELOV and T.G. BROWN, Procs 13th IEEE Photov.Spec.Conf. 383 (1978); N.K. ANNAMALAI,Procs 12th IEEE Photov.Spec.Conf. 547 (1976).

5. W.W. ANDERSON, A.D. JONATH and J.A. THORNTON, 2nd E.C. Photov. Solar Energy Conf., 890 (1979).

6. J. BRAGAGNOLO, A. BARNETT, Procs. XIII IEEE Photovoltaic Specialist Conference, Washington, 1978.

7. A.G. STANLEY, Applied Solid State Science, vol.5, p.251, R. WOLFE Ed. Academic Press, 1975.

8. M. SAVELLI and J. BOUGNOT, Topics in Applied Physics, $\underline{31}$, p. 213, B.O. Seraphin Ed. Springer Ver., 1979.

9. J.J. LOFERSKI, S. SHEWCHUN, E.A. DEMEO, R. ARNOW, E.E.CRISMAN H.L. HEVANS, R. BEAULIEU and C.C. WIC, Procs. Coll. Int. d'Electricité Solaire, Toulouse, p.317 (1975).

10. M.J. ROBERTSON and J. WOODS, Procs.2nd E.C. Photovo. Solar Energy Conf. 909 (1979).

11. R.W. POTTER and H.T. EVANS, J. Res. US Geol.Survey $\underline{4}$(2),203 (1976).

12. A.L. FAHRENBRUCH and R.H. BUBE, J. App. Phys. $\underline{45}$ 1264 (1974).

Table 1

	Best Sputtered Cell (Ar+H$_2$)		Best Evaporated Cell	
	"As grown"	After Annealing (170°C, 30min, 75ÅCu)	"as grown"	After annealing (110°C, 30min, 75ÅCu)
V_{OC} : Open Circuit Voltage	350mV	480mV	484mV	372mV
I_{SC} : Short Circuit Current	1.1mA	3.5mA	4.2mA	4.7mA
V_{mp} : Voltage at Maximum Power	256mV	312mV	346mV	288mV
I_{mp} : Current at Maximum Power	0.64mA	2.2mA	2.6 mA	4.2mA
Fill Factor	42%	41%	45%	65%
Area of the Cell	0.55cm^2	0.55cm^2	0.55cm^2	0.55cm^2
CuXS thickness	1500Å	1500Å	1000Å	1000Å

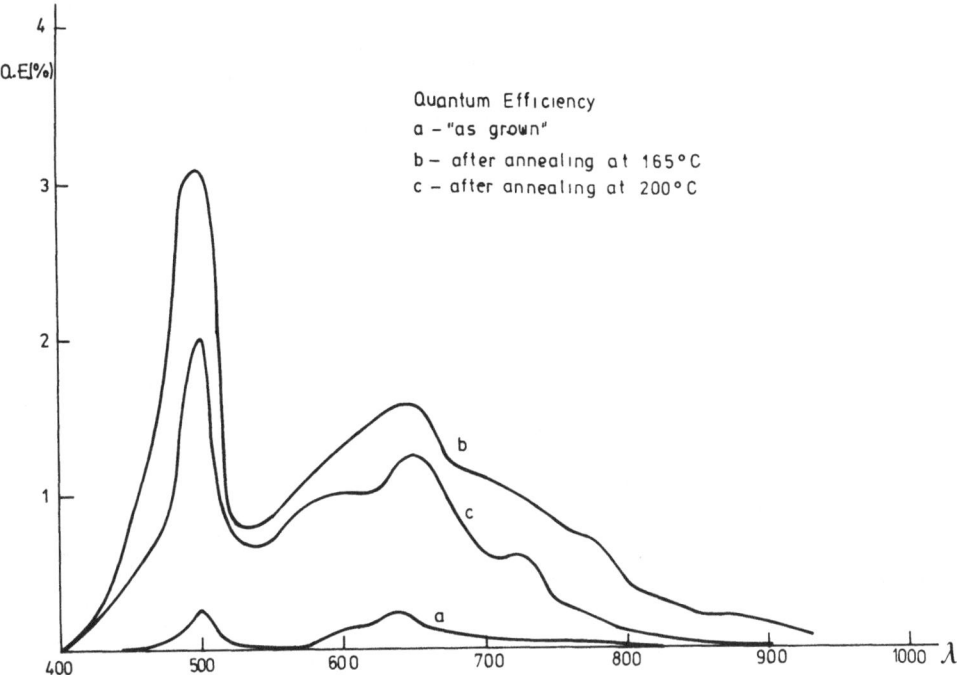

Fig. 1 - *Quantum efficiency of a CdS-Cu_xS heterojunction obtained by sputtering pure djurleite in Ar atmosphere*

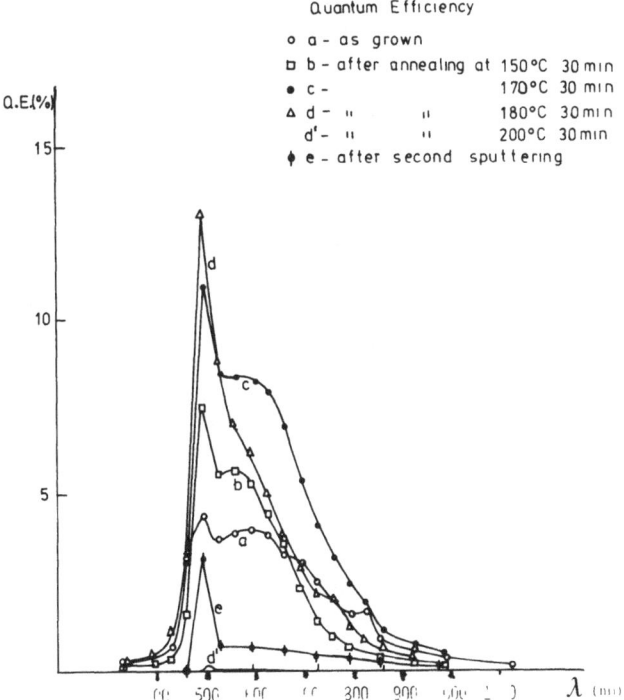

Fig. 2 - *Spectrul quantum efficiency of a CdS-Cu_xS heterojunction obtained by sputtering pure chalcocite in Ar atmosphere*

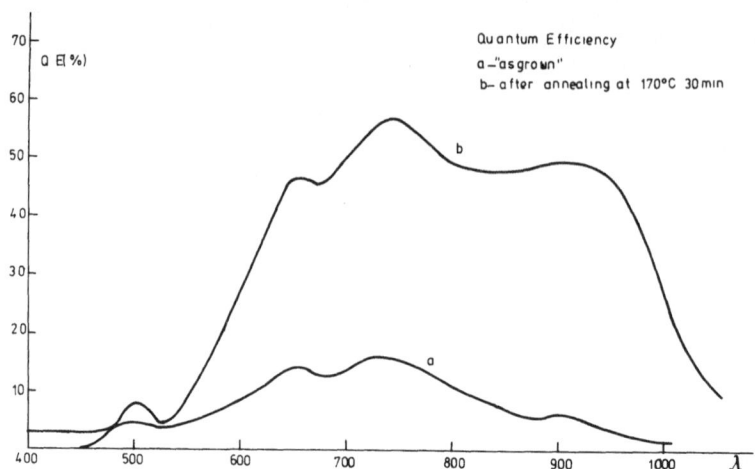

Fig. 3 - Spectral quantum efficiency of a CdS-Cu_xS heterojunction obtained by sputtering pure chalcocite in Ar_H_2 mixture with 50 A of copper deposited on top of it.

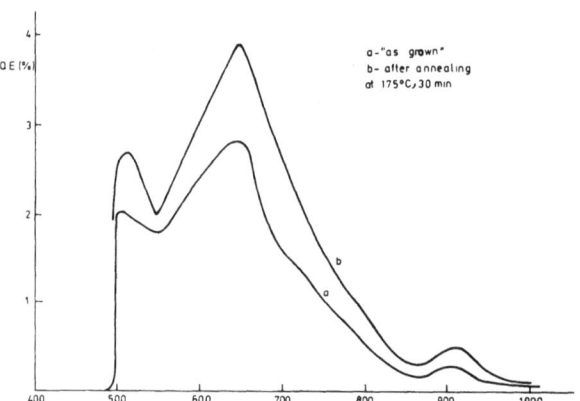

Fig. 4 - Spectral quantum efficiency of a evaporated Cu_xS film of 1000 A on CdS at 130°C.

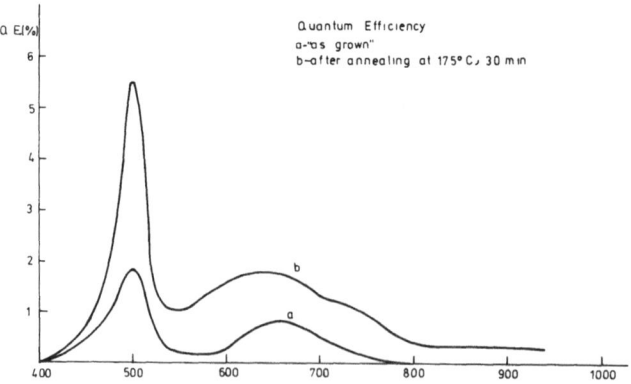

Fig. 5 - Spectral quantum efficiency of a heterojunction with evaporated cuprous sulphide of 1000 A deposited on CdS at 20°C.

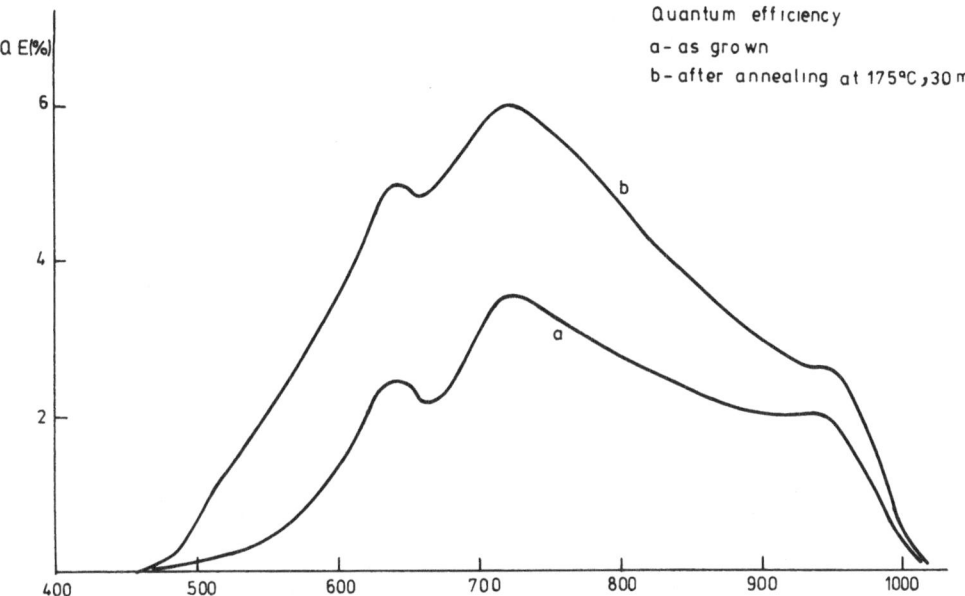

Fig. 6 - Spectral quantum efficiency of a heterojunction with evaporated Cu_xS of 2000 A deposited on CdS at 140°C.

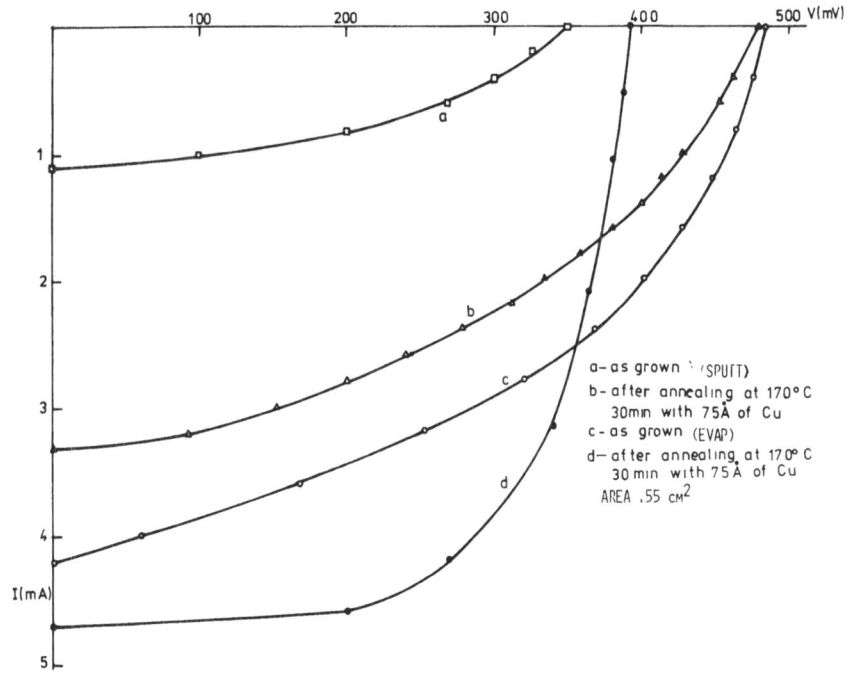

Fig. 7 - The best cells results of sputtered and evaporated Cu_xS films, before and after vacuum annealing.

AIRLESS SPRAYED CdS SOLAR CELLS

J. VEDEL, B. THIEBAUT and M. LEVART
Laboratoire d'Electrochimie Analytique et Appliquée de l'ENSCP
Laboratoire associé au CNRS (L.A. 216)
Ecole Nationale Supérieure de Chimie
11, rue Pierre et Marie Curie, 75231-PARIS CEDEX 05, France

Summary

This communication presents the results obtained using a new technique for making CdS solar cells. The technique employs an airless spray gun which allows the use of more concentrated solution, as well as a decrease in the heating energy. Laboratory scale experiments have been achieved and compared with those obtained using the classical way. Photovoltaic cells were prepared using the dipping process. The resulting cuprous sulfide being far from stoichiometry, a copper layer was vacuum deposited at the end of dipping. After heat treatment the efficiency of the cell was about 5.2 % (V_{OC} = 0.44 V ; I_{SC} = 18 mA cm^{-2}).

1. INTRODUCTION

Pneumatical spray[1], usually employed for making CdS solar cells shows two drawbacks, connected with the gaseous flow needed for the nebulization :
- the evaporation of the solution causes the precipitation of the solute (the complex $CdCl_2$-thiourea) in the nozzle and blocks it up ;
- the surface temperature of the substrate is decreased.

Airless spray pyrolysis (ASP) consists in strongly compressing the solution to be aerosolized in a cylinder with a small hole in the end, through which it is sprayed in a multitude of droplets, which are ejected against the substrate. There is no carrier gas and a higher growth rate can be achieved by increasing concentration of the solute and flow rate of the solution with a minimal temperature loss.

2. EXPERIMENTAL

A slightly modified COPROM, Minicop type, airless paint spray gun was used. The liquid flow was 20 ml mn^{-1}. The spray gun was manually driven.

The heater was a 10x10x2 cm^2 stainless steel block. Differences in surface temperature were not greater than 5°C. The relation between the temperature of regulation (θ_R) and the surface temperature was θ_s/°C = 130 + 0.55 θ_R.

Substrate were 2mm-thick Pyrex sheets, covered (by ASP) with conductive tin oxide (solution : anh.$SnCl_4$: 20 ml l^{-1}, NH_4F : 5.5g l^{-1} ; solvent : methanol ; θ_R = 600° C) CdS was sprayed on SnO_x (solution : $CdCl_2$: 73.3 gl^{-1} ; thiourea : 30.5 gl^{-1}, solvent : water (60°C)).

Photovoltaic cells were made upon a layer of CdS, heavily doped with alumina(2). The CdS layer was sprayed on top of the aluminated layer.

Cu_2S was prepared by the dipping process and equilibrated by vacuum evaporating a copper layer. The cell was annealed on air, at 155°C for 120mn.

3. RESULTS

Tin oxide. The growth rate was similar to that obtained with pneumatical spray (0.1 µm mn^{-1}). Figure 1 and 2 show the variation of transmission vs λ and vs square resistance, resp. layers with R = 10Ω were used as a substrate for CdS cells.

Cadmium sulfide. Figure 3 shows the thickness variation with time and temperature. A 9 µm thick CdS layer was prepared in 20 mn by spraying 400 ml of solution.

Cristallinity. Figure 4 shows the diffractograms obtained with various CdS layers (vacuum deposited, sprayed and airless sprayed at θ_R = 450 and θ_R = 550°C resp.). The increase in θ_R favors the preferential orientation of CdS.

Photovoltaic cells. IV characteristic is shown on fig. 5. The thicknesses of the various layers were : SnO_x : 0.5 ; CdS + Al_2O_3 : 2 µm; CdS : 8 µm ; Cu_2S : 0.4 µm ; Cu : 300 A°.

4. DISCUSSION AND CONCLUSION

The comparison between airless and classical spray is made using our own results and summarized by Maris (3). The data signalized by an asterisk were determined in our lab, to get comparable values. The photovoltaic cell suffers a too high series resistance, which decreases its efficiency, but there is an important gain in both time of spray and energy needed for the fabrication.

TABLE 1. Comparison of spray and airless spray processes

	Spray	Airless spray
Transparent Oxide	ITO	SnO_x
R	5 Ω	10 Ω
T %	90 %	85 %
Solution concentration	$CdCl_2$ + $SC(NH_2)_2$ 5 10^{-2} mol l^{-1}	$CdCl_2$ + $SC(NH_2)_2$ 0.4 mol l^{-1}
flow	9 cm^3 mn^{-1}	20 cm^3 mn^{-1}
volume (for 9µm)	2750 cm^3	400 cm^3
Spray of CdS Time (for 9µm)	5 hours	20 mn
Efficiency*	2 %	2 %
Thermal balance*	5,25 kWh	0.25 kWh
Growth rate	2 µm h^{-1}	30 µm h^{-1}
Substract parameter	10-40	20-40
Physical properties I(101)/I(002)	0.66	0.64
V_{OC}	0.420 V	0.440 V
I_{SC}	23 mA cm^{-2}	18 mA/cm^2
η	7.1%	5.2%
FF	0.73	0.66

5. REFERENCES

(1) Rh. Chamberlin and J. Skarman, J. Electrochem Soc., <u>113</u>, 86, (1966).
(2) J.F. Jordan, Solar Electricity, Internat. Conf., Toulouse (1976), p.57
(3) O. Maris, Thèse, Montpellier, 1980.

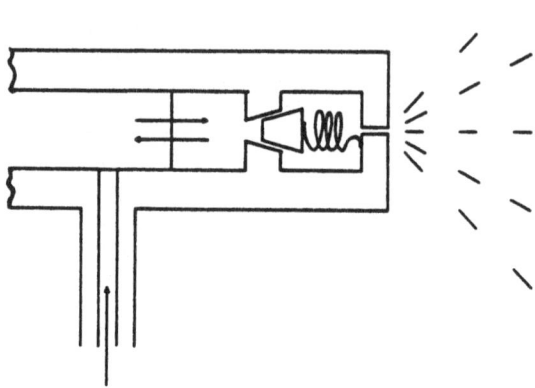

Figure 1 : Principle of airless spray.

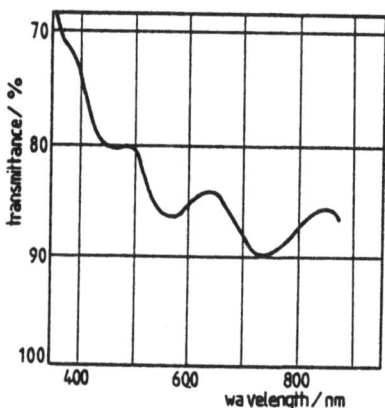

Figure 2 : transmittance of tin oxide layer vs wavelength.

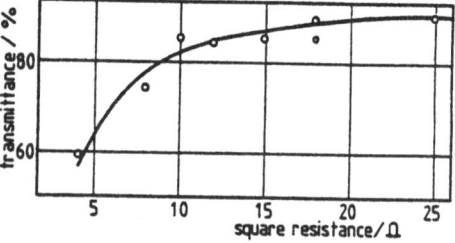

Figure 3 : transmittance of tin oxide layer vs square resistance at 550 nm.

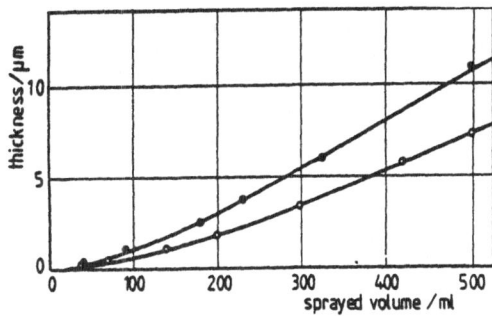

Figure 4 : thickness of CdS layer vs sprayed volume.

Figure 5 :

X-ray diffractogram :
(a) vacuum deposited CdS
(b) pneumatically sprayed CdS
(c) airless sprayed CdS ($\theta_R = 550°C$)
(d) airless sprayed CdS ($\theta_R = 450°C$)

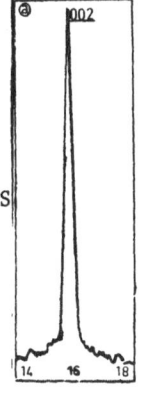 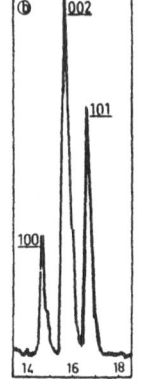 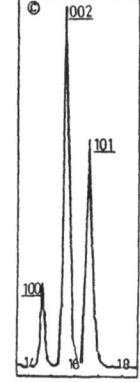 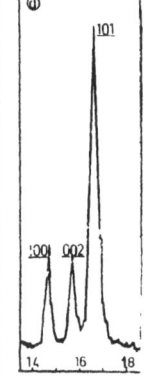

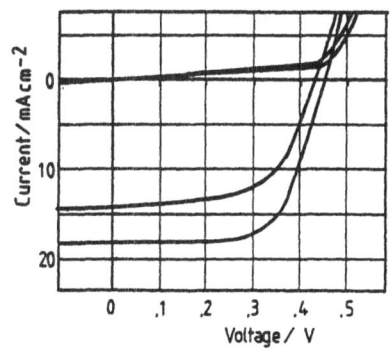

Figure 6 : I.V. characteristics :

With copper treatment
$V_{oc} = 0.42$ V
$I_{SC} = 14$ mA.cm^{-2}
$\eta = 3.8\%$
FF = 0.65

without copper treatment
$V_{oc} = 0.44$ V
$I_{SC} = 18$ mA.cm^{-2}
$\eta = 5.2\%$
FF = 0.66

ELECTROCHEMICAL PREPARATION AND CONDITIONING
OF Cu_2S FOR Cu_2S-CdS SOLAR CELLS

J. VEDEL, P. COWACHE, D. LINCOT

Laboratoire d'Electrochimie Analytique et Appliquée de l'ENSCP
Laboratoire associé au CNRS (LA 216)
Ecole Nationale Supérieure de Chimie
11, rue Pierre et Marie Curie, 75005 PARIS, France

Summary

Cuprous sulfide is electroplated from Cu(II) solution on thermally evaporated CdS. A solution composition is given, which suppresses the direct reaction between Cu^{2+} ions and CdS. Stoichiometric Cu_2S is formed. Photocells are realized, with a front grid directly electroplated on the cell. I-V curves, spectral responses and C^{-2}-V plots are given. The main difference with dipped cells lies in the C^{-2}-V plot which shows a less pronounced variation for electroplated cells. Efficiency of 7% were obtained.

I. INTRODUCTION

Electroplating of cuprous sulfide on cadmium sulfide is potentially more easily controlled than the dipping process[1] : the amount of formed Cu_2S depends only on the quantity of electricity consumed but not on the time of contact with the solution, which can be very short for high temperature formation. The formation reaction is very simple :

$$CdS + 2 Cu^{2+} + 2e \rightarrow Cu_2S + Cd^{2+} \qquad (1)$$

but is disturbed by two side reactions, the direct reaction of cupric ions into copper :

$$Cu^{2+} + 2e \rightarrow Cu \qquad (2)$$

which occurs preferentially to (1) at low temperature (25°C), and the spontaneous reaction of Cu^{2+} ions with CdS :

$$CdS + Cu^{2+} \rightarrow CuS + Cd^{2+} \qquad (3)$$

which leads to non stoichiometric cuprous sulfide :

$$(2-x)CuS + (x-1)Cu_2S \rightarrow Cu_xS \qquad (4)$$

These reactions, which decrease the photocell quality occurs preferently at high temperature.

Reaction (3) can be entirely masked by adding to the electrolytical bath, a convenient amine, like ethylenediamine, tetraethylenepentamine or triethylenediamine (TRIEN). Here is described the elaboration of photocells by electroplating Cu_2S on evaporated CdS, the front grid also being electroplated.

2. EXPERIMENTAL

Thermally evaporated CdS substrates were supplied by the Institut für Physikalische Elektronik (Stuttgart). The carrier concentration, determined by a Mott-Schottky plot in aqueous NaOH solution was $6.10^{17} cm^{-3}$. The CdS layers were cleaned and etched following standard techniques.

The composition of a typical electroplating solution was :

sodium acetate	0.1 mol l^{-1}
TRIEN	0.07 mol l^{-1}
cupric perchlorate	0.002 mol l^{-1}
solvent water, temperature : 80°C	

Before use, the solution must be anodically oxidized at controlled potential (-0.1 V vs Ag-AgCl) to remove a reducing impurity.

The amount and stoichiometric ratio of deposited copper sulfide were determined by electrochemical analysis[2]. The total copper deposited (as Cu and Cu_xS) was determined after dissolution by anodic stripping.

Cuprous sulfide was potentiostatically deposited at various deposition potential. The reference electrode was an Ag-AgCl one but the potentials are referred to the copper electrode (-0.48 V vs Ag/AgCl) owing to its thermodynamical significance.

The front grid was electroplated before the photovoltaic Cu_2S, i.e., before etching : the CdS was cleaned and covered with photoresist ; a grid was exposed and developed and 0.5 µm thick layer of cuprous sulfide was electroplated in the grid design. A gold layer was then electroplated, using a cyanide free bath[3]. The remaining photoresist was then removed, the CdS etched and the photovoltaic Cu_2S electroplated.

Capacity determination were carried out at 100 kHz, in the darkness.

3. FORMATION OF THE JUNCTION

At 90°C, for 5 hours, there is no direct reaction between the electroplating solution and the evaporated CdS, provided the purification has been completely done. Stoichiometric ratio are close to 2 : substrates of about 10 cm^2 were transformed and analyzed from place to place (analyzed area : 5 mm^2). For non etched and for etched CdS, x values were respectively 1.989 $\pm$ 0.005 and 1.980 $\pm$ 0.007, for mean thickness of 0.55 $\pm$ 0.05 µm (deposition potential - 0.2V). There was no evidence of copper oxides. A comparison between the proposed solution and a 2.10^{-3}M $Cu(ClO_4)_2$ solution was achieved. With TRIEN, the x value was 1.987 and without TRIEN, it was 1.90.

Figure 1 shows the variation of the mean thickness of deposited Cu_2S (obtained by integrating the electrolysis current) vs time. After an induction period, the increase becomes linear showing the reaction is governed by an interfacial process and not by diffusion through Cu_2S. In fact, the log of the slope linearly varies with the imposed potential (electrochemically controlled reaction). The copper deposited as Cu_2S was compared with the total copper present on the substrate after electroplating: the ratio Cu/Cu_2S increases with decreasing the deposition potential. At E_{dep} =-0.2 V vs Cu, there is practically no free copper in the substrate.

4. ANALYSIS OF THE JUNCTION

Immediately after formation, the cells generally show high photocurrents (20 mA cm^{-2} at AM1). However, low shunt resistance limits the cell properties. The properties are improved by heating, without changing the photocurrent. On figure 2 are shown the I.V of an annealed junction (180°C, 10 mn, air). Dark and light curves intersect, due to the formation of a Cu compensated layer[4]. This effect also appears on the I-λ curves (figure 3). Before annealing, a good quantum efficiency is observed. In particular, it is approximately equal to 0.7 between 0.7 and 1μm. Assuming a mean value of 0.6 over the wavelength range, the integration of the current for the AM1 spectrum (96.3 mWcm^{-2}) leads to a I_{sc} value of 24 mAcm^{-2}|5|, equal to the effectively observed value.

After annealing, a decrease of the quantum efficiency is observed for wavelength higher than 0.5 μm, with response times larger than before annealing. This effect is observed for spectra obtained in the darkness. It is caused by the compensated region and does not appreciably affect the photocurrent (21.5 mAcm^{-2}, AM1). It depends on the illumination state of the junction immediately before the determination (figure 4).

On figure 5 are plotted the C^{-2} vs V variations for non etched CdS devices. The initial doping of CdS was determined by analysis of a Mott-Schottky plot obtained in aqueous 1M NaOH. The obtained value is $N_D \simeq 6.3 \ 10^{17}$ cm^{-3}. A more precise value could be obtained using the exact value of the area factor (or roughness factor). After formation of the junction (curve B), the C^{-2} vs V plot is typical of an abrupt junction, of space charge region of which extending in the CdS part, as shown by the observed N_D value (6.10^{17} cm^{-3}). The barried height is $\sim$ 1,1 eV, close to the Cu$_2$S gap, showing good properties of the sulfide and no significative migration of copper in the CdS during the plating. The narrowness of the space charge region is presumably the cause of the observed low shunt resistance, due to tunneling (W = 4.10^{-2} μm).

After annealing, the compensated layer causes a low relative variation of C^{-2} vs V(Curve C). For V < 0.5 V, the slope shows that the S.C. regions lies in the non-compensated CdS. For V > 0.5, the linearity of the plot supports the hypothesis of uniform doping of the compensated layer ($N_D' = 10^{-15}$ cm^{-3})|6| as well as the lowering of the barrier height.

For etched substrates, the observed phenomenas are qualitatively the same. However, the real area of the junction is increased : typically at zerobias the capacity is $\sim$ 83 instead of 13 nF cm^{-2}. This gives an area factor of 6.4. A difference between electroplated and "dipped" Cu$_2$S lies in the variation of C^{-2} vs V (curve D, figure 5). In the first case, the plot remains practically constant with the bias contrary to the pronounced variation observed in the later case|7|. This could be due to a more regular growth of the electroplated cuprous sulfide (fig. 6).

5. CONCLUSION

Figure 2 shows one of the best IV characteristics, for a 1 cm^2 cell. The efficiency is 7% at AM1. The use of specially designed electroplating solutions allow the obtention of good Cu$_2$S-CdS photocells in simpler technological conditions.

This work has been realized under CEE contract n° ESC/040/F.

We gratefully acknowledge the Stuttgart group, particularly Pr Bloss and Dr Pfisterer for supplying CdS substrates.

REFERENCES

|1| N. NAKAYAMA, Jap. J. Appl. Phys., __8__, 450 (1969).

|2| J. VEDEL and M. SOUBEYRAND, J. Electrochem. Soc., __127__, 1730, (1980).

|3| F. MOHRNHEIM, Plating, (1961), 1104.

|4| W.D. GILL and R.H. BUBE, J. Appl. Phys., __41__, 3731 (1970).

|5| A.M'BAYE, Données solaires pour différentes caractéristiques atmosphériques, CNRS (1980).

|6| P.F. LINDQUIST and R.H. BUBE, J. Appl. Phys., __43__, 2838 (1972).

|7| F. PFISTERER, Workshop on the II-VI solar cells and similar compounds, Montpellier (1979).

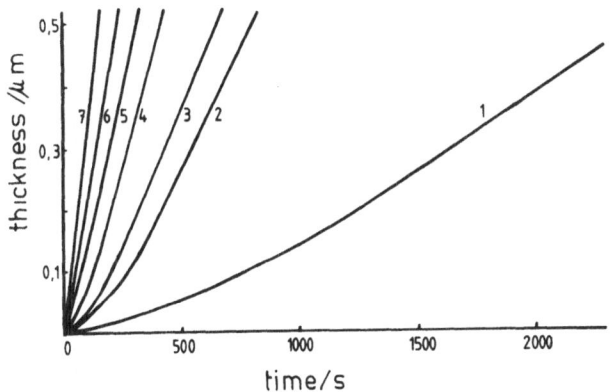

Fig. 1. Variation of the mean thickness of plated Cu_2S at various potential(vs Cu) 1:-0.05 - 2:-0.1 - 3:-0.15 - 4:-0.2 - 5:-0.25 - 6:-0.3 7:-0.35.

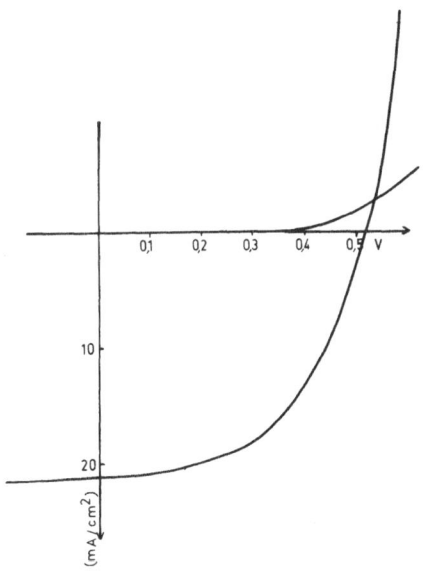

Fig.2. I.V. characteristic of an annealed cell (V_{oc}=0.510, I_{sc}=22mA/cm^2, η=7%).

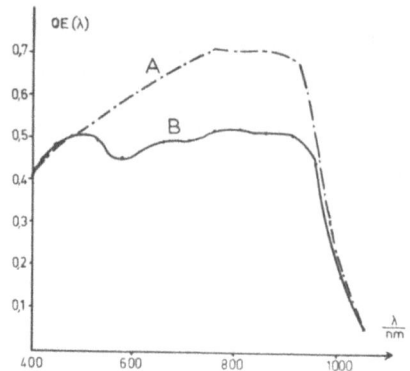

Fig.3. Influence of annealing on spectral response.

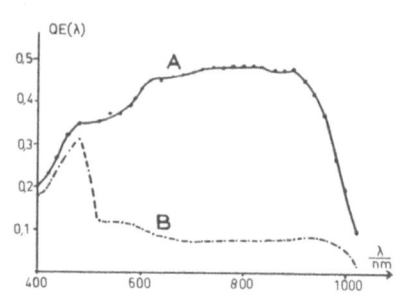

Fig.4. Spectral response of an annealed cell. A: just after illumination at 0.5 µm; B: after a long stay in the dark (or a quenching illumination at 0.9 µm).

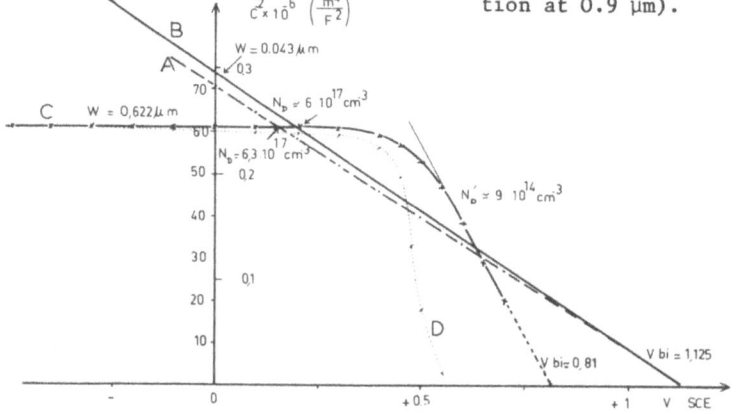

Fig.5. C^{-2} vs V plots for non etched CdS layers. A: non plated CdS, in a 1 M NaOH solution; B: just after plating; C: after heat treatment (180°C, 10mn,air) D: Etched CdS after plating and heat treatment-area factor $\simeq$ 6.

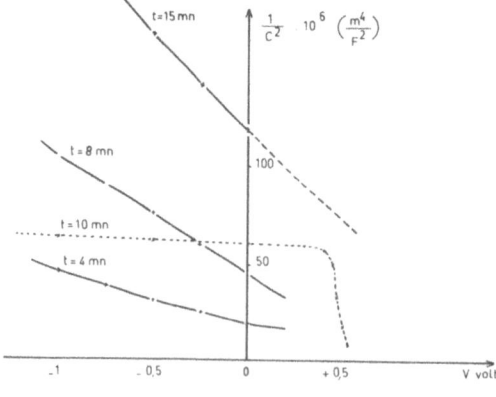

Fig.6. Comparison between dipped[7] Cu_2S (full lines) and electroplated Cu_2S (broken line).

PHYSICO-CHEMICAL PROPERTIES OF Cu_xS

H. RICKERT, H.-D. WIEMHÖFER, I.E. SCHMIDT, R. WAGNER
University of Dortmund
Department of Physical Chemistry I

Summary

Transport properties of Cu_xS, i.e. the electrical conductivity and the chemical diffusion coefficient of copper, describing the rate of stoichiometric changes within Cu_xS have been measured for monoclinic chalcocite and djurleite at 20-80 °C. At 20 °C the electrical conductivity ranges between $3 \; (\Omega \; cm)^{-1}$ for $Cu_{2.000}S$ and $500 \; (\Omega \; cm)^{-1}$ for $Cu_{1.93}S$. The chemical diffusion coefficient is $D_{Cu} = 1.1 \; 10^{-6} \; cm^2/s$ in the case of $Cu_{2.000}S$ and changes to a value of $D_{Cu} = 2 \cdot 10^{-8} \; cm^2/s$ for $Cu_{1.93}S$.

The current voltage characteristics of some metal/Cu_xS contacts have been studied with Me = Au, Ag, Cu, Ni, Pt, Zn.

The current voltage curves with Au, Ni or Pt contacts are mainly influenced by the stoichiometry of Cu_xS and not by the type of metal. Usually one observes an increasing resistance of Cu_xS if the metal contact has a negative polarity against Cu_xS; with positive metal contact the resistance of Cu_xS decreases. The results are explained by reversible stoichiometry changes within the Cu_xS. Stationary gradients of stoichiometry are generated within Cu_xS by diffusion of copper in the presence of applied voltages.

Within a second group of contact materials, Ag, Cu, Zn, one observes irreversible changes of resistance which indicate chemical effects at these contacts on Cu_xS.

Introduction

The properties of the CdS-Cu_xS thin film solar cell are to a large degree characterized by the cuprous sulfide Cu_xS, a substance which is also very interesting for the solid state electrochemist.
There are several phases of Cu_xS with different molar ratios x of Cu to S. (1) The most important phases in this context are the monoclinic low temperature form of chalcocite with compositions $2.000 > x > 1.995$ and the monoclinic djurleite with approximate compositions $1.96 > x > 1.93$. These two p-conducting Cu_xS phases show a slight ionic conductivity. Applying a voltage to Cu_xS results in an additional transport of Cu^+-ions besides the prevailing electron current so that a gradient of the composition x in Cu_xS is built up ("gradient of stoichiometry") which is connected with a change in the local equilibrium concentrations of electron defects.
Our aim was to get quantitative data of the transport properties of electron defects and of copper ions in Cu_xS, in particular we were interested in the dependences of these data on the composition of Cu_xS. The results involve the behaviour of different contact metals on Cu_xS, too. The obtained data allow to calculate the influence of time-dependent transport phenomena within the Cu_xS of thin film solar cells.

Electrical conductivity of electron defects

Fig. 1 and Fig. 2 sum up results of conductivity measurements on thin films and on bulk material of Cu_xS. (5) Special emphasize has been put on the range of stoichiometries near x=2,000. The conductivity of pure stoichiometric $Cu_{2,000}S$ is about $\sigma = 3 (\Omega\ cm)^{-1}$.
The conductivity always increases with decreasing Cu content, the different slopes of the two curves for thin Cu_xS films and bulk Cu_xS respectively are caused by lower electrical mobilities of the electron defects in thin films.

Ionic transport

Applying a voltage to Cu_xS between two metal electrodes leads to a gradient of stoichiometry within Cu_xS because of the transport of Cu. (2,3,4) If the experimental arrangement is given by

$$Me\ ''/\ Cu_xS\ /'\ Me$$

the voltage U determines the difference of the electrochemical potential $\tilde{\eta}_e$ of the electrons identical with the difference of the Fermi energy (6) between the two metal contacts by

$$eU = \tilde{\eta}_e'' - \tilde{\eta}_e' = E_F'' - E_F'$$

We can describe the thermodynamic equilibrium of copper, copper ions and electrons by ("$\sim$" means per particle)

$$\tilde{\mu}_{Cu} = \tilde{\eta}_{Cu^+} + \tilde{\eta}_e$$

where $\tilde{\mu}_{Cu}$ and $\tilde{\eta}_{Cu^+}$ denote the chemical potential of copper and the electrochemical potential of copper ions respectively. So the voltage is also expressed by

$$eU = (\tilde{\mu}_{Cu}'' - \tilde{\mu}_{Cu}') - (\tilde{\eta}_{Cu^+}'' - \tilde{\eta}_{Cu^+}')$$

For a homogeneous sample we have $\tilde{\mu}_{Cu}'' = \tilde{\mu}_{Cu}'$. Applying a voltage U to such a sample gives $eU = \tilde{\eta}_{Cu^+}'' - \tilde{\eta}_{Cu^+}'$ at $t = 0$, this means that an ionic current occurs.
The sample becomes inhomogeneous and the difference $\tilde{\mu}_{Cu}'' - \tilde{\mu}_{Cu}'$ increases as a function of time with $\tilde{\eta}_{Cu^+}'' - \tilde{\eta}_{Cu^+}'$ decreases. If the chemical potential does not exceed certain limits a steady state is reached with vanishing, ionic current and a steady state gradient of composition of Cu_xS described by a difference of the chemical potential $\tilde{\mu}_{Cu}$

$$eU = \tilde{\mu}_{Cu}'' - \tilde{\mu}_{Cu}'$$

Turning off the external voltage U leads to the reverse process: back diffusion of Cu in form of Cu^+ ions and electron defects h until the sample is homogeneous.
The transport of copper in Cu_xS is termed chemical diffusion if the building-up or the decay of the composition or chemical potential gradient are involved. The time dependence is characterized by the chemical diffusion coefficient D_{Cu}.
In the case of Cu_xS the local electrical conductivity changes because of the above mentioned dependence of the conductivity σ on the composition x. The change in conductivity or resistance can be used to measure the kinetics of Cu transport (= chemical diffusion) in Cu_xS although the partial current of Cu^+ ions itself is too low to be detected. (4). Transient measurements of the resistance as a function of time after a voltage step have been

carried out with point electrodes on Cu_xS according to the scheme in Fig.3. Some results for the chemical diffusion coefficient of copper as a function of temperature are shown in Fig. 4.
In the case of $Cu_{2.000}S$ we have $D_{Cu} = 1.1 \cdot 10^{-6}$ cm^2/s which gives $\sigma_{Cu^+} = 6.0 \cdot 10^{-5}$ Ω^{-1} cm^{-1} for the partial electrical conductivity of the Cu^+-ions.

Contact materials

The measurement of the steady state current as a function of the applied voltage at point contacts on Cu_xS leads to the current voltage curves in Fig. 5 and Fig. 6. The curves in Fig. 5 with the contact metals Me = Au, Ni, Pt show the expected results in the light of the above considerations if one only takes into account the reversible transport of copper in Cu_xS. (3) For negative polarity at the point contacts the copper content in the vicinity of the point contact and with it the resistance increases and we get a limiting current for higher voltages $|U| > 100$ mV. The inverse is true for positive point contacts.

The second group of contact materials (Fig. 6) produces irreversible effects, the current voltage curves do not have such an uniform appearance like those in Fig. 5. In addition to it there is no exact steady state obtained. These effects become stronger the lower the copper content. In the case of Cu and Ag point contacts a dissolution of the contact metal as Cu^+ and Ag^+ respectively seems very probable, in the case of Zn we assume a dissolution of zinc in Cu_xS or a chemical reaction with Cu_xS to give ZnS.

Acknowledgement

This work is supported by the "Bundesministerium für Forschung und Technologie", contract No. ET 4272 A.

References

1. H.J. Mathieu, H. Rickert, Z. physikal. Chem. NF **79** (1972) 315-330

2. H.J. Mathieu, K.K. Reinhartz, H. Rickert, 10th IEEE Photovoltaic Specialists Conference, (IEEE, New York, 1973), p. 93

3. H. Rickert, H.-D. Wiemhöfer, in: Statusbericht Sonnenenergie, Vol. II; Ed. ISES, VDI Verlag, Düsseldorf 1980

4. H.-D. Wiemhöfer, Dissertation, Dortmund 1982

5. R. Wagner, H.-D. Wiemhöfer, to be published

6. H. Rickert, Einführung in die Elektrochemie fester Stoffe, Springer Verlag, Berlin, Heidelberg, New York, 1973

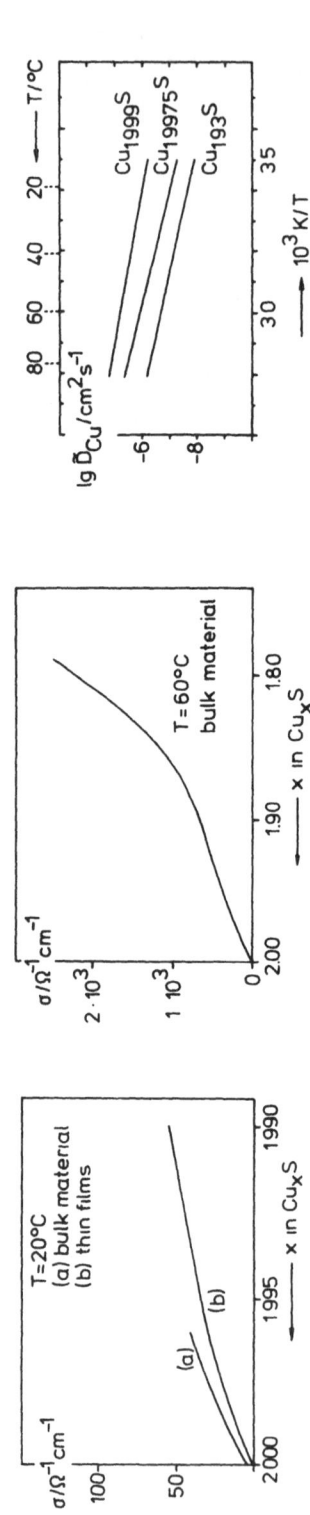

Fig. 1: Electrical conductivity of Cu_xS: $2000 > x > 1.990$, $T = 20°C$

Fig. 2: Electrical conductivity of Cu_xS: $2.00 > x > 1.75$, $T = 60°C$

Fig. 4: Chemical diffusion coefficient $\tilde{D}_{Cu}$ in Cu_xS

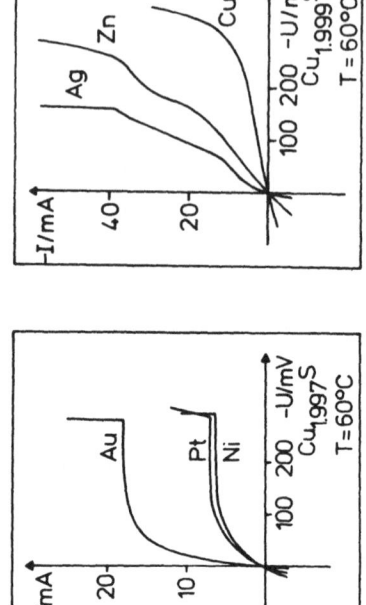

Fig. 5 and Fig. 6:
Steady state current-voltage curves with Ag, Au, Cu, Ni, Pt, Zn as contact materials on Cu_xS

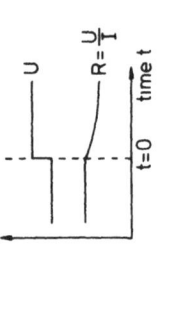

Fig. 3: Point contacts on Cu_xS

- 830 -

PHOTOVOLTAIC BEHAVIOUR OF CdSe THIN FILM SOLAR CELLS[*]

E. Rickus
Am Roemerhof 35
6000 Frankfurt am Main 90
Germany

Summary

CdSe is a semiconductor with a direct band gap of 1.7 eV. Its high absorption coefficient for visible light allows the fabrication of very thin solar cells (1 to 2 µm thick) with low material consumption. Currently the cells, representing MIS structures, are produced solely by standard vacuum evaporation processes. They show efficiencies above 6 percent.
The surface barrier in CdSe can be positively influenced by the I-layer, by the contact material and by irradiating the cells with photons of an energy greater than 3.5 eV.
Stability tests have established that the structures are very insensitive to humidity. Storage on the shelf for more than one year and even immersion into water does not affect the unsealed cells seriously.

Introduction

MIS structures have proved their potential as alternative, efficient solar cells. Compared to p-n-junctions they have a significant potential for lower costs because of their inherent simplicity. The application of this technology to polycrystalline, thin semiconductor films fulfils all major conditions for the production of low cost solar cells.

The purpose of the work reported here is the development of such a cell for terrestrial applications which finally should reach an efficiency around 10 percent.

Structure of the cells

In order to produce an efficient thin film solar cell of 1 µm to 2 µm thickness one has to use a semiconductor with a high absorption coefficient for visible light and with sufficiently long diffusion length for light-generated carriers. CdSe with a direct band gap of 1.7 eV can satisfy these requirements and therefore is used as the base material of our cells. At substrate temperatures of 425 °C it is deposited onto a chromium-covered glass-substrate by vacuum evaporation and has a final thickness of about 2 µm. The deposition conditions, i.e. substrate temperature, evaporation rate and oxygen partial pressure as well as the substrate material can influence the crystallographic and photovoltaic properties of the cells /1, 2/.

Before depositing the transparent Schottky contact, wich presently consists of a 6 nm thick gold layer, a thin ZnSe film serving as "I-layer" is evaporated onto the CdSe (Fig. 1).

---

[*] This work is being sponsored by the Commission of the European Communities

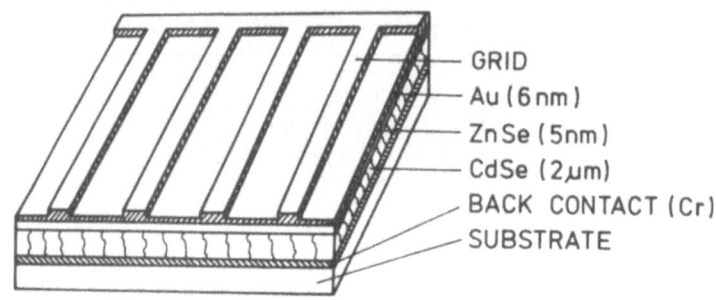

Fig. 1: Structure of the CdSe MIS cell

ZnSe is a semiconductor with an energy gap of 2.7 eV which represents an additional barrier for majority carriers (electrons) in CdSe resulting in an enhancement of the open circuit voltage.

Compared to common insulators such as SiO or SiO_2 the energy gap of 2.7 eV is relatively small and thus the matching of the Fermi levels at the interface can only produce a low potential barrier for light generated minority carriers in CdSe. This means, that they have a higher tunneling probability and that the "I-layer" therefore can be made thicker than in common MIS devices.

As has been shown experimentally, the optimum thickness of the ZnSe-film in our structures therefore is in the range of 4 to 5 nm. This and the fact that the photovoltaic parameters are relatively insensitive to thickness variations of the "I-layer" represents a distinct fabrication-related advantage over native oxide MIS-structures.

The current collection grid on our cells (Fig. 1) has not been optimized yet. In order to reduce the reflection from the Schottky contact we evaporate a ZnS antireflection coating on top of the structure.

After deposition the structures have to be annealed in a nitrogen atmosphere at temperatures up to 200 °C to improve the photovoltaic properties /1/.

The potential barrier in CdSe

One of the most important parameters of MIS solar cells is the barrier height in the photosensitive semiconductor determining the open circuit voltage of the cells under illumination.

As deduced from Hall effect measurements our CdSe films have a native doping level of 10^{13} to 10^{14} electrons per cm³ which is enhanced to 10^{15} to 10^{16} per cm³ by illuminating the films with white light of 100 mW/cm². The deposition of metals with a high work function therefore produce a surface barrier by matching the Fermi levels at the interface /3/. Calculating the depth of the depletion region results in values around 1 µm, a value far greater than the absorption depth of incident photons. As has been confirmed by the analysis of spectral response measurements /4/, this causes a relatively high (near unity) internal collection efficiency of the photogenerated carriers.

In order to study the influence of the metal work function on the open circuit voltage or the barrier height in CdSe, we deposited different metals onto the I-layer. The resulting open circuit voltage (U_{oc}) of the cells can be seen from figure 2. The standard annealing step is omitted in order to show the direct influence of the metals. As predicted by the Schottky theory, the open circuit voltage varies significantly with the work function of the contact material. This behavior is in contrast to studies on MS diodes on single crystalline CdSe resulting in surface-state-controlled barriers which are practically independent of the metal /5, 6/. Obviously the different behavior of our structures (Fig. 2) is caused by the ZnSe film. With respect to pure MS diodes this layer drastically reduces the number of surface states so that the Fermi level no longer is pinned at a fixed position.

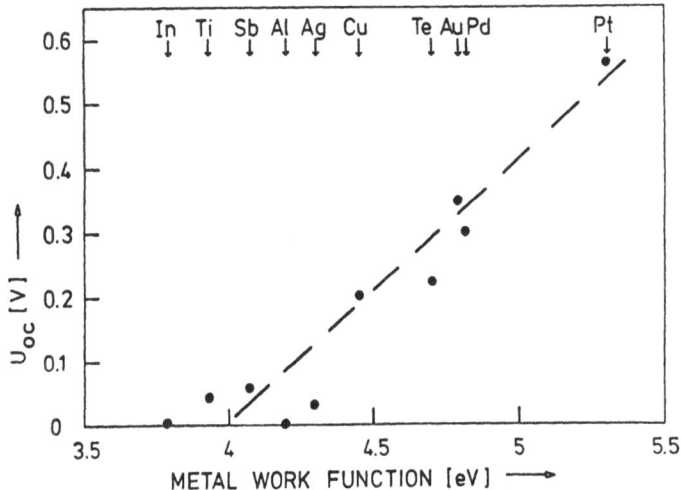

Fig. 2: Influence of the contact material on the open circuit voltage of unannealed MIS cells.

Because of interactions between contact material and I-layer which can change the electronic properties of ZnSe, annealing the cells causes deviations from the behavior shown in Fig. 2 and opens up possibilities to influence the photovoltaic properties of the cells. The greatest increase in U_{oc} after annealing occurs with gold contacts. Open circuit voltages up to 850 mV have been observed.

Another method of influencing the potential barrier in CdSe is irradiating the active area of the solar cells with UV light of 250 to 350 nm wavelength. This procedure probably creates negatively charged centers in the ZnSe-film which enhance the surface barrier in CdSe. The improvements of the photovoltaic parameters of the CdSe MIS cells induced by this process saturates after an irradiation dose of about 1.5 Ws/cm². It results in a final enhancement of the open circuit voltage up to 15 % and in a slight improvement of the fill factor and the short circuit current density (Fig. 3). The electronic centers in the ZnSe-layer are stable up to temperatures around 150 °C. Under normal operation conditions they therefore cause no stability problems so that UV irradiation represents a very simple method to improve the efficiency of the CdSe MIS solar cells.

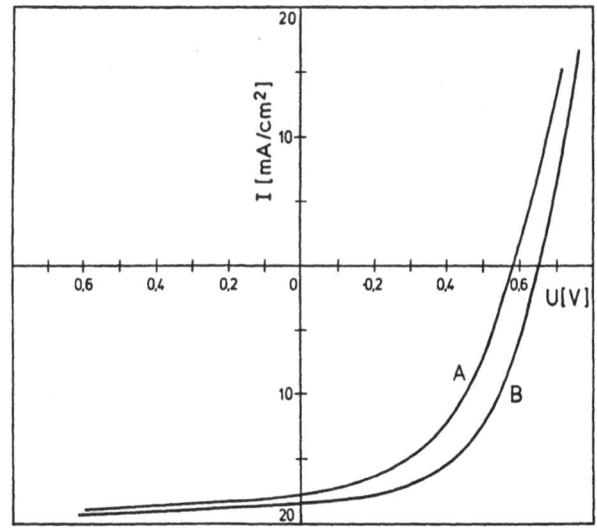

Fig. 3:

Standard I-V-characteristic of CdSe MIS cells.
A: after annealing,
B: after annealing and UV-irradiation.

Stability tests

The use of ZnSe as "I-layer" in CdSe MIS structures results in a very low sensitivity of the cells to oxidative ambients because oxygen cannot cause the "I-layer" to grow further with time. This and the fact that CdSe itself is a very stable material probably are the main reasons for the high stability of the CdSe MIS solar cells. Tests have established that the structures are very insensitive to humidity. Even immersion into water or unsealed storage on the shelf for more than one year does not affect the cells seriously.

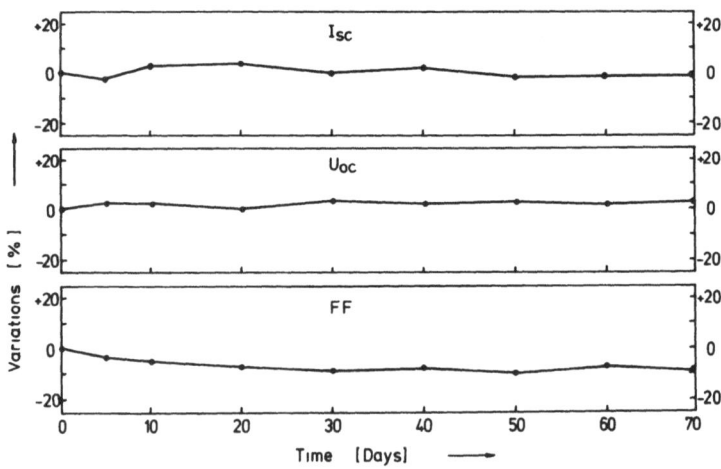

Fig. 4: Stability tests in moist air (100 % humidity) of 75 °C.

In order to get a careful measure of the stability of our MIS structures, we stored some unsealed cells in moist air (100 % humidity) of 75 °C. The resulting modifications of the photovoltaic parameters are shown in figure 4. Even after 70 days under this stress the short circuit current density (I_{sc}) and the open circuit voltage (U_{oc}) pracitcally are not affected (all structures even showed a somewhat enhanced U_{oc}). Solely the fill factor (FF) decreases slightly. Its reduction is lower than 10 percent (Fig. 4).

Cell area and efficiency

The work on the CdSe MIS cells has lead to structures of 1 cm² active area showing efficiencies above 6 percent under simulated solar light. Because of the good orientation of the crystallites which have their x-axis perpendicular to the substrate /2/, the short circuit current density is above 18 mA/cm² representing about 75 percent of the theoretical value. The open circuit voltage around 0.65 V and the fill factor slightly over 50 % have to be enhanced by further optimizations.

Increasing the active area of the cells to 30 cm² poses no fundamental problems. While in this case a lowering of I_{sc} and fill factor is caused by the series resistance of the not adapted contacts, the U_{oc} even is somewhat improved because of the relative reduction of the edge lenght.

Conclusion

Investigations on CdSe MIS diodes have demonstrated that these structures may have a significant potential for terrestrial use as stable, low-cost thin film solar cells. Further work will have to concentrate on doping of the CdSe- and ZnSe-layer.

Efficiencies above 6 percent achieved with a relatively little effort compared to other solar cell developments give reasons to expect further progress. Individual values of I_{sc}, U_{oc} and fill factor show that CdSe MIS solar cells can yield in efficiencies around 10 percent.

References

/1/ D. Bonnet, E. Rickus: Proc. 14th Photovoltaic Specialists' Conf., San Diego, 1980, p. 629

/2/ E. Rickus, D. Bonnet: Proc. 3rd E.C. Photovolt. Solar Energy Conf., Cannes, 1980, p. 871

/3/ D.L. Pulfrey: IEEE Transactions on Electron Devices, No. 11, Nov. 1978, p. 1308

/4/ E. Rickus: Proc. 15th Photovolt. Specialists Conf., Orlando, 1981, p. 1073

/5/ R.L. Consigny III, J.R. Madigan: Solid State Communications, Vol. 7, (1969) p. 189

/6/ C.A. Mead: Appl. Phys. Lett., Vol. 6, No. 6 (1965) p. 103

"ALL THIN FILM n-CdTe/ITO SOLAR CELL".

C. MENEZES, F. SANCHEZ-SINENCIO, C. VAZQUEZ-LOPEZ, A. SOUZA E.
Depto. Física, CINVESTAV, Apdo. Postal 14-740
México 14, D. F.

R. Bube
Dept. of Materials Science
Stanford University
Stanford, Cal. 94305

Summary

This paper describes a simple all thin film solar cell consisting of a layer of n-CdTe deposited by the GREG* technique onto tin oxide coated glass slides. The CdTe layers are polycrystalline but have a grain size between 10-30 μm. I_{sc} is between 22-24 ma/cm^2 and the spectral response extends from 350 μm to 850 μm.

THE GREG SYSTEM

We describe here a rather novel technique adapted to the growth of large grain polycrystalline and single crystal semiconductor layers, in this case, CdTe layers. The technique involves the use of a gradient recrystallization and growth system hence the acronym GREG.
The system is in fact suitable for two main roles: growth of CdTe layers using a variety of CdTe starting materials (platelets, slices, granules, compressed powder) and secondly the recrystallization of already deposited films under close to equilibrium conditions, utilising a sharp temperature gradient to aid the recrystallization along prefered orientations.
Essentially the GREG system is a modified C.S.V.T. deposition system where the source and substrate are separated by a small gap and are surrounded by a gas (H_2 typically) at athmospheric pressure.
The important high lights of the GREG process are
1) Thermal etching of the source is the principal process involved in transport.
2) Gradient cycling helps to choose larger nuclei and thus obtain large grain layers.
3) On single crystal substrates single crystal homoepitaxial layers are obtained. (Fig. 1).
4) The incorporation of a hot wall chamber between the source block and substrate block, which helps to contain the source and its vapours at a high temperature during deposition. (Fig. 2).
5) Preparation of free standing CdTe layers with large grain size (30-50μ).
6) On suitable substrates, it is possible to separate the grown layer

from the substrate on which it was grown. This raises interesting possibilities for producing free standing layers with good electrical and structural (large grain size) properties.

We now describe an all-thin-film solar cell consisting of n-CdTe deposited on tin-oxide coated glass slides.

n-CdTe/SnO$_2$ Solar Cell

n-CdTe/SnO$_2$ solar cell is remarkable for its simplicity of preparation and fabrication. A layer of n-CdTe is deposited onto tin-oxide conducting glass. The deposition is done by the GREG technique (see above). Following deposition, contacts are made to the conducting glass and the free CdTe surface by dots of In-Ga alloy. The photosensitive surface lies between the glass and the CdTe layer and hence is self protected.

The I_{SC} is quite large; we have obtained 22-25 ma/cm^2 at AMI. However the V_{OC} is low, 0.37 volts. Efforts are now in progress to improve this value.

The spectral response (Fig.4) is very good, with a range extending from 350 nm to 850 nm.

In summary, the cell we have developed shows the following features:

1) I_{SC} of 22-25 ma/cm^2 (AMI)
2) Non-saturation of I_{SC}, indicating possibility of improving I_{SC} further.
3) Very wide spectral response.
4) Very simple construction, virtually a one-step process.

However, the efficiency at present is low 3.3% because of the low open circuit voltage.

Fig. 1: Edge-on view of a CdTe layer grown epitaxially on CdTe single crystal. Shows smooth epitaxial layer free from voids or grain boundaries. Magnification 4.0K. Taken on AMR scanning electron microscope at CMR Stanford University.

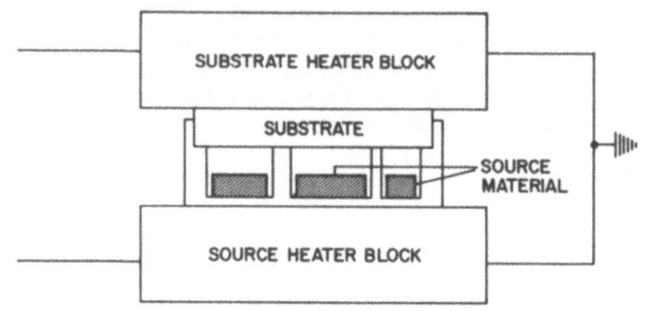

Fig. 2: Schematic diagram of hot wall-C.S.V.T. (GREG) system for deposition of CdTe

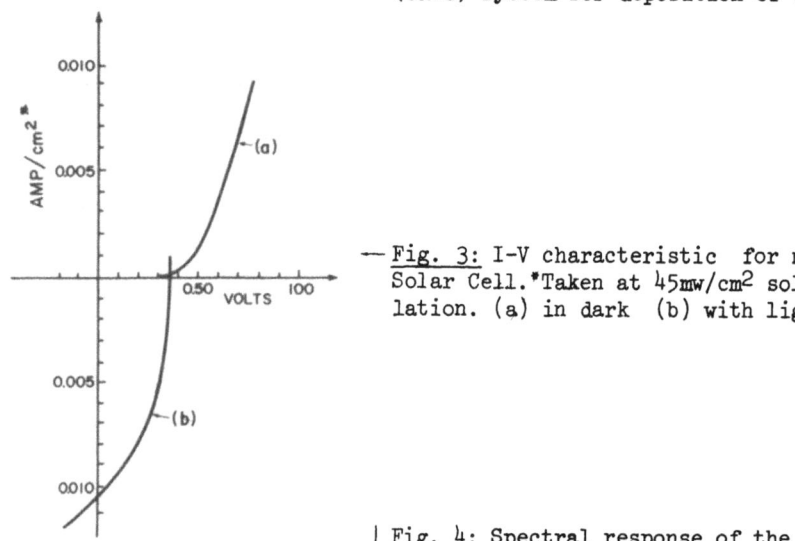

Fig. 3: I-V characteristic for n-CdTe-ITO Solar Cell. *Taken at 45mw/cm^2 solar insolation. (a) in dark (b) with light

Fig. 4: Spectral response of the n-CdTe-ITO solar cell

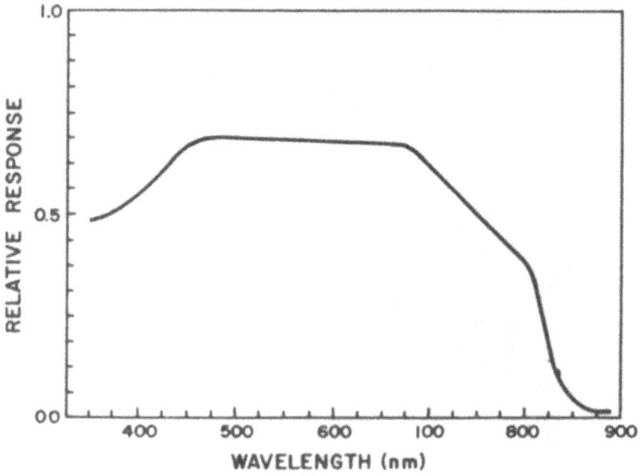

PREPARATION OF HIGH PURITY II-VI COMPOUNDS BY LASER ANNEALING

L. BAUFAY, D. DISPA, A. PIGEOLET, M.-C. JOLIET and L.D. LAUDE
IRIS- Faculté des Sciences- Université de l'Etat à Mons,
Avenue Maistriau, 23, B-7000 Mons.

Summary

Films of semiconducting compounds are of great interest for photovoltaic applications. Nevertheless, high quality materials have to be produced by a versatile and low cost processing. In this paper, a laser is found to be useful and powerful tool in II-VI compounds formation. We describe some laser annealing experiments which are performed on multilayer films of alternately Cd and Te and/or Se in order to obtain the II-VI semiconducting compounds CdTe, CdSe and $CdTe_xSe_{1-x}$. The high purity of the material is stressed by optical absorption and electrical transport measurements. The characterization is also achieved by electron diffraction and transmission electron microscopy. High quality semiconductors are obtained under preparation conditions which may be considered to be approximate.

1. INTRODUCTION

It has been demonstrated that laser annealing is a powerful technique in material processing for instance in recrystallizing ion implanted regions of Si single crystals but also in preparing III-V semiconducting compound (1,2,3).
The method used previously for AlSb formation is extended to the II-VI compounds preparation.
In our experiments, the irradiation is performed on multilayer films of alternately Cd and Te and/or Se in order to obtain the II-VI binary compounds CdTe and CdSe and their ad-mixture. The choice of the materials is suggested by their potential application as photovoltaic materials and their relative difficulty and high cost of their preparation by other means. The laser initiated formation of III-V and II-VI compounds from multilayer metallic films has been discussed elsewhere (4,5,6). This paper is restricted to the characterisation of such films by a combination of measurements: TEM (transmission electron microscopy), HRED (high resolution electron diffraction), optical absorption in the range 1.4-4.0 eV, electrical conductivity and photocourant measurements.

2. PREPARATION

Cd and Te and/or Se are successively and alternately condensed onto either cleaved salt or sputter-cleaned glass substrate using a multiple crucible e-gun source and a quartz thickness monitor at a base pressure of 10^{-7}-10^{-6} torr. The goal is to deposit equal numbers of Cd and Te/Se atoms (accuracy of the initial stoechiometry $\leq 2\%$) in the overall film. Moreover, the relative proportion between the numbers of Te and Se atoms determine the composition of the ternary system $CdTe_xSe_{1-x}$. The total thickness ranges from 1000 Å to 1 μm. The more volatile components are buried in the film, away from the outer surfaces. Sometimes, 250 Å thick

SiO layer(s) is (are) deposited on these outer surface to avoid film degradation in air.

Portions of the films deposited on NaCl single crystal are floated off onto copper grid for TEM work, they will be referred to in the following as free standing samples.

The irradiation necessary to achieve the phase transformation is always performed in air, at room temperature. Three types of laser beam are used:
1) a dye laser pulse of 2 eV photon energy and about 1 µs pulse duration. The threshold energy is either 8 or 13 mJ/cm^2 for 1000 Å thick Cd-Se or Cd-Te samples respectively.
2) a cw argon ion laser beam scanned on the film at a speed ranging from 0.5 to 1.5 cm/s while the beam power ranges from 50 to 100 W/cm^2. (The beam diameter is about 1.6 mm - and all green lines are used).
3) a chopped argon ion laser beam (all green lines): the irradiation time is then 36 ± 4 ms, the beam power is of the order of 40-45 W/cm^2.

3. IDENTIFICATION

Table I summarizes the measurements made as a function of preparation and irradiation conditions.

	CW laser supported sample	chopped CW laser free standing film	Dye laser free standing film
CdTe	Ph,R,OA,HRED,TEM	HRED,TEM	HRED,TEM
CdSe	R,OA,HRED,TEM	HRED,TEM	HRED,TEM
$CdTe_xSe_{1-x}$	OA,HRED,TEM	-	-

Table I :   R: electrical conductivity measurements
              OA: optical absorption measurements
          HRED: high resolution electron diffraction measurements
           TEM: transmission electron microscopy measurements
            Ph: photocourant measurement as a function of wavelength.

A. Binary system
A-1) Optical absorption

Supported samples are irradiated in CW regime and their optical transmission measured in the visible and near infra-red ranges. Figs 1 and 2 show the (1-T) spectra (T: transmission) and their first derivative obtained for CdTe and CdSe, respectively.

Both materials are direct gap materials with a fundamental absorption edge (E_o) , associated with transitions occuring between Γ_{15} and Γ_1. Our measurements yield $E_o = 1,47 \pm 0.05$ eV and $E_o = 1,63 \pm 0,05$ eV respectively for CdTe and CdSe. Other transitions $E_o + \Delta_o$, occuring between the spin-orbit split off valence band and the conduction band at Γ , is also evidenced. We have measured $E_o + \Delta_o = 2,38 \pm 0,10$ eV and $E_o + \Delta_o = 2,04 \pm 0,10$ eV respectively for CdTe and CdSe. These results agree with the published date [7,8] which indicate $E_o \simeq 1,49$ eV, $E_o + \Delta_o \simeq 2,41$ eV for CdTe and $E_o \simeq 1,8$ eV; $E_o + \Delta_o = 2,22$ eV for CdSe.

A-2) Electrical transport measurements

After CW irradiation, supported samples present at room temperature a high dark resistivity of the order of 10^6 Ωcm for CdTe and 10^5 Ωcm for CdSe. Fig. 3 (Log.ρ = f(1/T) indicates that the width of the mobility gap is 1,62 $\pm$ 0,10 eV for CdTe and 1,85 $\pm$ 0,10 eV for CdSe. The intrinsic behaviour is evidenced down to ~100°c and ~150°c respectively for CdTe and CdSe. Below these temperatures, extrinsic conduction is controlled by impurity activation. For instance, in the CdTe case, one single activation energy at 0,16 $\pm$ 0,10 eV is detected down to -100°c, which might be associated with a singly ionized Cd vacancy (9) acting as a hole trap.

From these informations, the Fermi level position is determined (10), which seems to indicate that the Cd vacancy concentration would be less than $10^{16} - 10^{17}$ cm^{-3} for CdTe.

A-3) Photocurrent measurements

Fig. 4 shows the photocurrent as a function of wavelength for a 1000 Å thick CdTe film. These data are in excellent agreement with published values (11) . Plotting the square root of the photoresponse as a function of the photon energy indicates also that the width of the energy gap is 1,43 $\pm$ 0,05 eV;

A-4) Electron diffraction and electron microscopy measurements

CdTe and CdSe are known to exist under two stable crystallographic configurations: the blende structure and the wurtzite structure. It has been found in this work. that for both materials, the type of structure obtained upon laser irradiation depends on the irradiation characteristics as summarized in the table II.

Table II. Crystallographic structure as a function of the preparation conditions.

	CW laser either chopped or not	Dye laser
CdTe	cubic	hexagonal (+cubic)
CdSe	cubic	hexagonal

Additional traces of free components (Te,Se) are sometimes detected.

B. Ternary system

Optical absorption and electron diffraction measurements have been performed for a variety of composition. They clearly indicate that there are no other structure than the well-established binary ones, which might be related to a ternary compound formation.

These results are in agreement with previous indications about the CdTe-CdSe system formation by other means (12).

CONCLUSION

The possibility of preparing well-crystallized films of the semiconducting binary compounds CdTe and CdSe and of their ad-mixtures using various laser processing methods has been demonstrated. It is shown that laser annealing proves to be a very effective, versatile and low cost technique for the preparation of such materials, correlating results already obtained for the III-V compound AlSb. Moreover, optical and transport data presented here show the good quality of the materials obtained by laser annealing. A remarkable feature is that such a quality (in particular the absence of deep levels) is obtained from films which were only approximately stoechiometric (to ~1-2%) before irradiation.

Finally, this work shows the possibility to choose ab-initio the crystallographic structure of the processed material by tuning adequately the actual irradiation parameters: energy and power.

Acknowledgement

The authors would like to thank R. Andrew and M. Wautelet for many useful discussions and M.C. Joliet for the TEM work. This work is supported by projects "Energy" and "IRIS" of the Belgian Ministry for Science Policy.

References

[1] R. Andrew, M. Ledezma, M. Lovato, M. Wautelet and L.D. Laude
Appl. Phys. Lett. 35 (1979) 418.

[2] R. Andrew, L. Baufay, L.D. Laude, M. Lovato and M. Wautelet
J. de Physique 41, C4-71 (1980).

[3] L. Baufay, M. Failly-Lovato, R. Andrew, M.C. Joliet, L.D. Laude, A. Pigeolet, M. Wautelet
Insulating Films on Semiconductors in Proceedings Inter Conf. on (ed. M. Schulz and G. Pensl) Springer Series in Electrophysics 7, 242 (1981).

[4] R. Andrew, L. Baufay, A. Pigeolet and L.D. Laude
to be published

[5] L. Baufay, R. Andrew, A. Pigeolet and L.D. Laude
to be published in Thin Solid Films, Proceeding of 5th Thin Film Congress (Israel) 1981.

[6] R. Andrew, L. Baufay, A. Pigeolet and L.D. Laude
to be published (Laser and Electron Beam Interactions with Solids-Boston, 1981).

[7] T.H. Myers, S.W. Edwards and J.F. Schetzina
J. Appl. Phys. 52 (6) (1981) 4231.

[8] J.O. Dimmock in Proceedings Inter Conf. on II-VI Semiconducting Compounds (1967) p. 227 ed. D.G. Thomas (W.A. Benjamin, N-Y).

[9] A.M. Mancini and D. Manfredotti, and C. de Blasi, G. Miccocci and A. Tefore
Rev. Phys. Appl. 12 (1977) 255-261.

[10] Physics of Semiconductor Devices p. 25-38
S.M. Sze (1979) ed. Wiley-Interscience

[11] G. Falop, M. Doty, P. Meyers, J. Betz, and C.H. Liu
Appl. Phys. Lett. 40 (4), 327, (1982).

[12] Semiconducting II-VI, IV-VI and V-VI compounds
N. Kh. Abrikosov, V.F. Banknia, L.V. Poretskaya, L.E. Shelimova and E.V. Skudnova (1969) Plenum Press, p. 31.

Figure captions

Fig. 1 and 2: Absorbance spectra of laser annealed CdTe and CdSe films. First derivatives of these spectra are shown at the bottom. Structure location are indicated. These measurements are performed on ~ 1000 Å thick samples.

Fig. 3 : Logarithm of the resistivity against the temperature inverse, for CdTe. Activation energies, films composition before irradiation, resolution are indicated.

Fig. 4 : The photocurrent is plotted against the wavelength of the incident light. (The photon flux is normalized and equal to 4.10^{13} photons/cm^2.s). The width of the energy gap is deduced from the plott showing the square root of the photoresponse as a function of the photon energy (in insert).

Fig.1 ▼

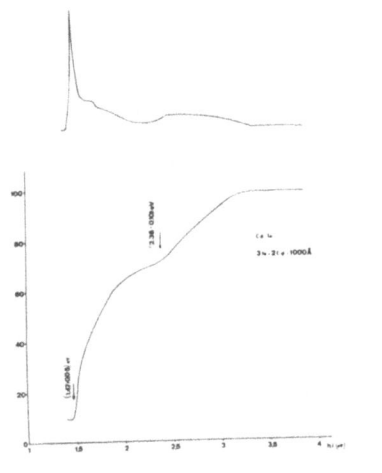

Fig.2 ▼

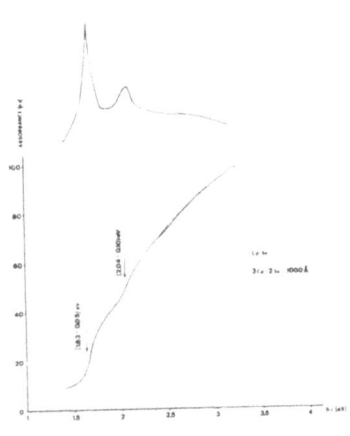

Fig.3 ▼

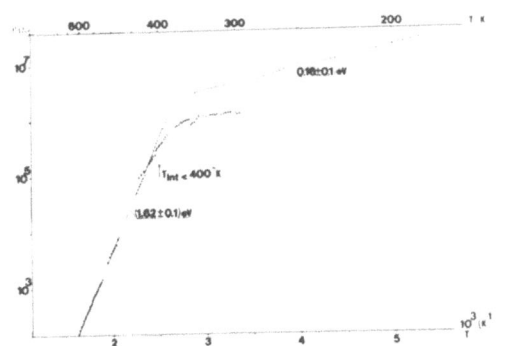

Fig.4 ▼

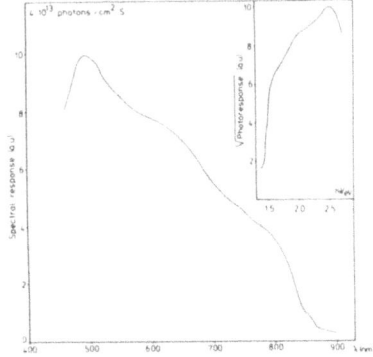

Zn_3P_2 THIN-FILM SOLAR CELLS

M. Bhushan
Institute of Energy Conversion
University of Delaware
Newark, Delaware 19711

Summary

Thin polycrystalline films of Zn_3P_2 were deposited on metallized mica substrates by close-spaced vapor transport. Mg-Zn_3P_2 Schottky barrier solar cells were prepared by depositing a thin semi-transparent film of Mg by d.c. sputtering. Comparison of thin-film cells with those made on single crystal Zn_3P_2 showed that the presence of grain-boundaries had no significant influence on the light-generated current and the open-circuit voltage. Heating the cells in air at temperatures > 50°C resulted in diffusion of Mg in Zn_3P_2 and creation of an n/p junction. The problem of rapid diffusion of Mg at low temperatures was overcome by using a Mg/p/p$^+$ Zn_3P_2 cell structure. A total area conversion efficiency of 4.3% was achieved under simulated AM-1 illumination on a Mg-Zn_3P_2 thin-film solar cell, 1 cm^2 in area.

1. INTRODUCTION

Zinc phosphide, Zn_3P_2 is presently being developed as a material for low-cost solar cells. It has an appropriate energy band gap, high absorption coefficient and relatively long minority carrier diffusion length (1,2). In addition, it has been shown that the presence of grain boundaries in thin polycrystalline films of Zn_3P_2 has no significant influence on the light-generated current and open-circuit voltage (3,4). Thin films deposited by close-spaced vapor transport (CSVT) are found to be more suitable for solar cell applications than by vacuum evaporation (5). In this paper, results on Mg-Zn_3P_2 thin film solar cells on p-type Zn_3P_2 deposited by the CSVT method are described.

2. THIN FILM DEPOSITION

A polycrystalline Zn_3P_2 wafer, heated to a temperature of 675°C was used as the source in the CSVT system. The substrate was placed at a distance of 0.1 cm from the source and the film deposition took place in an open-tube system in argon atmosphere. The source-to-substrate temperature difference was ~ 100°C and a growth rate of 0.75 μm/min was obtained.

A high substrate temperature during the film growth and a relatively large thermal expansion coefficient of Zn_3P_2 (1.4 x 10^{-5} K^{-1}) put severe restrictions on the substrate material. The substrate should also be chemically inert and make a low resistance ohmic contact to Zn_3P_2. Although low-cost metal substrates (e.g. silicon steel) have been used, the best results to date have been achieved on metallized mica substrates

(6). A 100 nm thick film of Fe followed by a 10 nm thick film of Si were deposited in a multiple source electron beam evaporator on freshly cleaved muscovite mica. Silicon was found to be more effective in retarding the interdiffusion of iron and phosphorus atoms during the film deposition than carbon used previously (4,6).

The average grain size in the films was $\simeq 10\,\mu m$. The electron diffusion length (0.4 to 4 μm), Hall mobility ($\sim 10\,cm^2\,V^{-1}\,sec^{-1}$) and carrier concentration ($10^{15} - 10^{16}\,cm^{-3}$) in these thin films were comparable to the values obtained on crystals grown by vapor transport (7,8). Recently, the properties of thin films grown by chemical vapor transport, by reacting zinc vapor with phosphine, (9) and quasi-rheotaxy (10), have been reported to be similar to the CSVT-grown films.

3. Mg-Zn$_3$P$_2$ SOLAR CELLS

Barrier heights of Schottky-type devices on p-type Zn$_3$P$_2$ have been measured using several low-work function metals. The highest barrier height of ~ 0.8 eV has been found on a Mg-Zn$_3$P$_2$ diode (11).

Mg-Zn$_3$P$_2$ devices were prepared on Zn$_3$P$_2$ thin films by depositing a semi-transparent film of Mg after chemically polishing the Zn$_3$P$_2$ in a jet of 0.25% solution of Br in methanol. Although Mg oxidizes rapidly in air, thin Mg films (~ 100 Å) deposited by d.c. sputtering, on an rf sputter cleaned Zn$_3$P$_2$ surface were found to be very stable. The sheet resistivity of the Mg films was 30 $\Omega/\square$. On a 1 cm^2 area cell, a 0.05 cm wide and 500 nm thick bar of thermally evaporated Mg was used as a contact to the transparent Mg film. A 70 nm thick film of vacuum evaporated zinc sulphide was used as an anti-reflection coating, reducing the total reflection losses over the AM-1 spectrum to $\lesssim 10\%$.

Although this device structure leads to relatively high optical losses, as the cell is illuminated through the metal film, it has been the most successful way to date to prepare large area Mg-Zn$_3$P$_2$ devices. One alternative, a metal grid type structure, requires a grid spacing of the order of the electron diffusion length ($\sim 1\,\mu m$) and high resolution photolithographic techniques must be used for cell fabrication. The optical losses remain high due to grid shading and reflection from the Zn$_3$P$_2$ surface. Illuminating the cell through the substrate requires deposition of a very thin ($< 2\,\mu m$) pinhole free film of Zn$_3$P$_2$ on a suitable transparent substrate with a low sheet resistance. In both these alternative designs the light-generated current is strongly restricted by the relatively short diffusion length.

The J-V characteristics of the cells on mica/Fe-Si substrates, were measured in the dark and under simulated AM-1 illumination at 26°C. The reverse saturation current, J_0 and the diode ideality factor were determined from the plots of log J vs V in the dark and from log J_{sc} vs V_{oc} under illumination. The barrier height, Φ_B, was measured by (1) temperature dependence of V_{oc} (2) log J_0 vs 1/T plots and (3) capacitance variation with voltage.

A maximum total area conversion efficiency of 4.3% was achieved on a 1 cm^2 area cell as shown in Figure 1. The light-generated current was only 60% of the maximum theoretical estimate, mainly due to high optical losses in the thin Mg film (reflection and absorption). The diode factors were between 1.5 and 2.0 and J_0 was $\sim 10^{-7}$ Amp/cm^2 both in the light and the dark. The values were comparable to Mg-Zn$_3$P$_2$ devices on bulk Zn$_3$P$_2$ wafers and suggest recombination in the space-charge region to be dominant. A value of $\Phi_B \gtrsim 1$ eV was obtained from all the three methods of measurements, higher than the value reported on bulk Zn$_3$P$_2$ material.

Mg-Zn$_3$P$_2$ cells were made on a wafer cut from a large grain size boule. On heating in air at temperatures > 50°C, Mg atoms diffuse into Zn$_3$P$_2$ creating an n/p junction. As a result, an increase in the barrier height and V$_{OC}$ was observed (12) but the short-circuit current, J$_{SC}$, decreased with prolonged heating and progressive growth of the Mg diffused layer. Similar changes in V$_{OC}$, J$_{SC}$, and the spectral response (Figure 2) were observed on Mg-Zn$_3$P$_2$ thin film devices. However, no preferential diffusion of Mg atoms along the grain-boundaries was observed and in contrast to the behavior in bulk Zn$_3$P$_2$ the depth of the n-type layer was always limited to less than the film thickness. In the thin films prepared on mica substrates coated with Fe-Si, it was suspected that a p$^+$ Fe doped layer existed near the substrate due to out diffusion of Fe during the film deposition. This limited the junction depth in the Mg diffused cells.

If cells are prepared on heavily doped p$^+$ Zn$_3$P$_2$ with a carrier concentration, N$_p$, exceeding the solid solubility limit of Mg ($\simeq$ 1 x 10^{17} cm^{-3}), there is no change in the J-V parameters even after prolonged heating. However, these devices suffer from higher leakage currents and a lower V$_{OC}$ ($\simeq$ 0.3 V). To maintain the enhanced V$_{OC}$ created by the n/p junction while preventing a rapid decrease in J$_{SC}$, it is necessary to stabilize a shallow junction ($\sim$ 1μm). This was achieved by depositing a 1 m thick lightly doped Zn$_3$P$_2$ film (N$_p \simeq$ 10^{16} cm^{-3}) by CSVT on a more heavily Ag doped p$^+$ substrate (N$_p \geq$ 10^{17} cm^{-3}).

In summary, Mg-Zn$_3$P$_2$ solar cells, 1 cm^2 in area have been prepared on thin polycrystalline films and a maximum total area conversion efficiency of 4.3% has been achieved. The junction depth created by diffusion of Mg atoms in Zn$_3$P$_2$ is limited to < 1μm by using a p/p$^+$ Zn$_3$P$_2$ substrate.

4. ACKNOWLEDGEMENTS

The author would like to thank Dr. A. W. Catalano and Dr. J. M. Pawlikowski for many helpful suggestions. This work was supported by DOE/SERI.

5. REFERENCES

(1) A. Catalano, V. Dalal, W. E. Devaney, E. A. Fagen, R. B. Hall, J. V. Masi, G. Warfield, Proc. of 13th IEEE PVSC. 288 (1978).
(2) J. M. Pawlikowski, Infrared Physics, 21, 181 (1981).
(3) M. Bhushan and A. Catalano, Appl. Phys. Lett. 38, 39 (1981).
(4) M. Bhushan and A. Catalano, Proc. of 15th IEEE PVSC, 1261 (1981).
(5) A. Catalano, M. Bhushan and N. C. Wyeth, Proc. of 14th IEEE PVSC, 641 (1980).
(6) M. Bhushan, Appl. Phys. Lett. 40, 51 (1982).
(7) A. Catalano and R. B. Hall, J. Phys. Chem. Sol., 41, 635 (1980).
(8) N. C. Wyeth and A. Catalano, J. Appl. Phys. 50, 1403 (1979).
(9) T. L. Chu, Southern Methodist University, Private Communications.
(10) G. Sberveglieri and N. Romeo, Thin Solid Films, 83, L133 (1981).
(11) N. C. Wyeth and A. Catalano, J. Appl. Phys., 51, 2286 (1980).
(12) M. Bhushan, J. Appl. Phys, 53, 514 (1982).

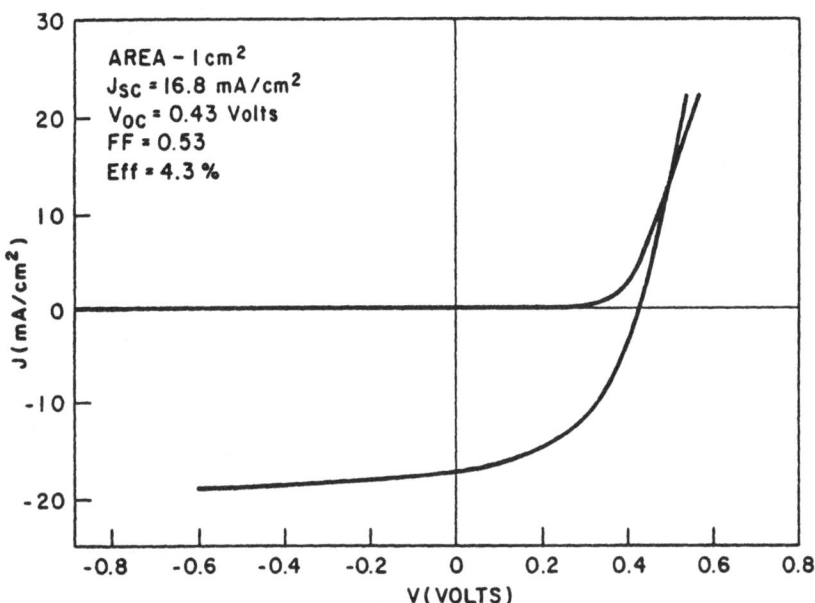

Fig. 1 – Dark and illuminated (simulated AM-1) I-V characteristics of a Mg-Zn_3P_2 thin film solar cell

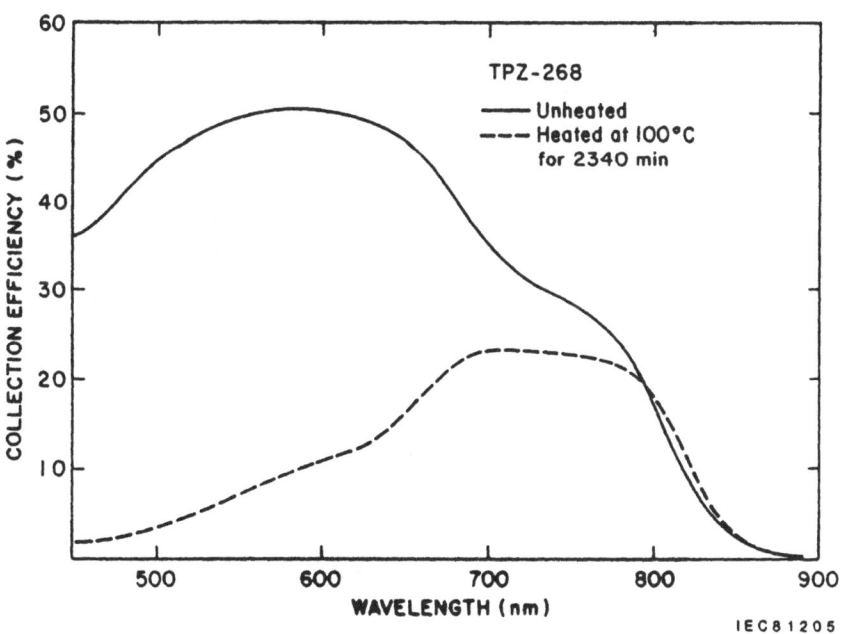

Fig. 2 – Spectral response characteristics of Mg-Zn_3P_2 thin film solar cell before and after heating in air at 100°C

CADMIUM SULFIDE POLYACETYLENE PHOTOVOLTAIC HEREROJUNCTION

M. CADENE, M. ROLLAND, M. ALDISSI*, M. ABADIE*

Groupe de Dynamique des Phases Condensées (L.A. 233)
* Laboratoire de Chimie Macromoléculaire
Université des Sciences et Techniques du Languedoc
Place Eugène Bataillon - F. 34060 MONTPELLIER Cédex

Summary

Since few years a strong interest is devoted to the organic semiconductors, specially to doped polyacetylene films with an insulating, semiconducting and even metallic behavior. Its optical and transport properties, its low cost and easeness to polymerise very large surfaces, leads to study its potential application in the field of future photovoltaic solar cells.

The aim of this communication is to present results obtained with thin film heterojunction nCdS - $\underline{p}(CHA_y)_x$. Dark and illuminated I-V characteristics at various temperatures as well as C-V characteristic are given. Typically we obtained (at AM1) $V_{OC} \simeq 300$ mV, and $I_{cc} \simeq 3$ mA/cm^2. and 0.5 percent efficiency.

- INTRODUCTION

The polyacetylene simplest conjugated polymer looks like as use of the most interesting material in the field of organic semiconductors. Synthetised as a flexible film of large surfaces and with a low cost, its electrical conductivity ranges from 10^{-9} Ω.cm if undoped and reaches 10^{+3} when AsF$_5$ doped. At this time its physical properties have been extensively studied as well in the semiconducting range as in the metallic one.

Consequently this hopefull material is now used in order to develope a number of applications : at this time it allows to built batteries of various types |1| |2|. On the other hand, many workers try to realise elctronic devices such as p-n junctions |3| and specially photovoltaic devices of various types. When undoped or lightly doped, trans (CH)$_x$ appears to be a direct gap semiconductor, the optical band gap lies from 1,4 to 1,8 eV, well matched to the solar spectrum as well as the absorptivity ($\alpha > 10^5$ cm^{-1}).

Tani obtained p-n junctions using undoped (CH)$_x$ as p-type material and n-type silicon or gallium arsenide |4|. When highly doping the polyacetylene of these junctions one obtain Schottky diodes in which the polymer acts as the backside metallic contact |5| but such devices are not low cost because they use single crystals. In order to lower the price of the device we realized cells using (CH)$_x$ with cadmium sulfide thin films and aluminium sheets.

In this paper we report the first results on thin film heterojunction between undoped or I$_2$-doped trans (CH)$_x$ and n-CdS evaporated layers. The different junctions were caracterized by measurements of current-voltage, impedance variation versus frequency and capacitance voltage versus temperature.

- SAMPLES PREPARATION :

The cadmium sulfide thin film (10 µm, 30 µm thick) were prepared by thermal evaporation onto conducting coated glass slides around 200°C. The CdS crystallites diameter are about 2 microns and the classical orientation of c-axis perpendicular to the substrate as shown by S.E.M. view graph (figure I and II, cross section, fig. III surface)

We directly polymerised acetylene on the CdS substrate by including them in a glass vessel. The polymerisation was performed by the ITO et al. technique |6| in the following experimental conditions: catalyser ratio Al/Ti = 4, AlEt$_3$ concentration 0,4 mole/l of solvent (toluene), acetylene pressure 1 atmosphere, reaction temperature -78°C. In all cases the (CH)$_x$ thickness range from 1 µm up to 5 µm

In both cases we obtained the trans stable isomer by a thermal treatment under vacuum at 140°C during a few minuts.

Various dopant species were used to obtained the metallic polymer, for example, SbF$_5$, MoCl$_5$ and Iodine and we have already pointed out that the (CH)$_x$ morphology strongly depends of the dopant specie and of the doping process |7|.

- DARK J-V CHARACTERISTICS.

J-V curve always displayed good rectifying behavior.

a)- Underlined (CH)$_x$

At room temperature, the relationship $J = J_s(\exp \frac{qV}{nkT} - 1)$ is well verified. The saturation current density ranges between 0.1 to 1 µA/cm^2. The coefficient n, about 3 to 4 greatly deviates from ideality.

At high direct voltage (V > 1 volt) the curves have a departure from the above law, due to the series resistance, around 1500 Ω/mm^2 providing to the high resistivity of the undoped (CH)$_x$ film.

The figure (V) report a set of curves for various temperatures. We must notice the slope is pratically temperature independant and the J_s dependence is weaker than would be predicted by the expression :

$$J_s = A^* T^2 \exp\left(-\frac{q \Phi_B}{kT}\right)$$

The simple thermoionic emission model of current transport is not sufficient. However the same slope suggest an alone conducting process between 77 K and 300 K. From the expression

$$J_s = A^* T^2 \exp\left(-\frac{q \Phi_B}{kT}\right) \text{ and } A^* = 232.10^3 \text{ A/m}^2 \text{ for CdS (8)}$$

we obtain a approximative value Φ_B = 0,85 eV at room temperature.

b)- <u>Doped</u> $(CH)_x$

With iodine, for example, the dark current density were more important, around 1 mA.cm^{-2}, but the serie resistance decreases to 100 Ω/mm^2.

Fig. VI shows a set of forward biased J-V characteristics of an n-CdS metallic $(CHI_y)_x$ Schottky diode, for various temperatures. The linear dependence of Log J versus voltage is well verified for all temperatures but the coefficient n $\simeq$ 3 deviate greatly from ideality. With the same value for A^*, we have found $\Phi_B \simeq$ 0,68 eV at room temperature. We must notice an identical slope of the curves and also low $J_s(T)$ dependence.

- IMPEDANCE AND CAPACITANCE MEASUREMENTS.

Preliminar studies of impedance-frequency variation are important to determine in the classical model of the solar cell, the serie and shunt resistance and the high frequency capacity value.

For undoped $(CH)_x$ device, these parameters are R_s = 5000 Ω R_{sh} = 35 000 Ω and C = 3.10^{-10}F for a 1 cm^2 area. These serie resistance is consistent with a resistivity about 10^5 Ω.cm for undoped trans $(CH)_x$.

From C-V data over the frequency range 1-150 kHzs (fig. IV) we determine a built-in potential $V_D \simeq$ 1 volt much larger than the observed V_{OC}. The frequency dependence arises certainly from the relaxation times associated with the population of the deep levels lying in the forbidden gap.

- J-V CHARACTERISTICS UNDER ILLUMINATION.

Under simulated solar irradiation, the I-V characteristics are displayed in fig. VIIa for undoped, and, VIIb for doped cells.

In the both cases, under AM1 illumination at 25°C, the open circuit voltage ranges around 300 mV.

Concerning the short circuit current density, its weak value ($J_{cc} \simeq$ 20 µA/cm^2) for undoped samples is due to the high serie resistance of the polymer. On the other hand the figure VIIb clearly exhibits that J_{cc} strongly increases when doping and typically reachs $J_{cc} \simeq$ 3 mA/cm^2 that is to say about two orders of magnitude higher than in the previous case.

- CONCLUSION

The actually highest solar photovoltaic efficiency, η = 0.5 percent, we have obtained for a thin film organic device based on doped $(CH)_x$, can be improved by the optimization of the fill factor ($\simeq$ 0,4) of the cells, and by using $(CH\,A_y)_x$ material with improved morphology.

However such result exhibits the potentiality of $(CH)_x$ as an photovoltaic sensitive material.

- REFERENCES.

(1) S.N. Chen, A.J. Heeger et al., Appl. Phys. Lett. 36 (1980), 96.
(2) P.J. Nigrey, D. McInnes et al., J. Electroch. Soc. (to be published).
(3) M. Ozaki, D.L. Peebles, B.R. Weinberger, A.J. Heeger and A.G. Mc Diarmid, J. Appl., Phys. 51 (1980) 4252.
(4) T. Tani, P.M. Grant, W.D. Gill et al., Sol. St. Comm. 33 (1980), 499.
(5) M. Ozaki, D.L. Peebles et al., Appl. Phys. Lett., 35 (1979), 83.
(6) T. Ito, H. Shirakawa and S. Ikeda, J. Polym. Sci. Chem. Educ. 12 (1974) 11.
(7) M. Rolland, M. Cadène et al., Mat. Res. Bull. 16 (1981), 1045.
(8) M. Luquet, Thèse, Montpellier (1977).

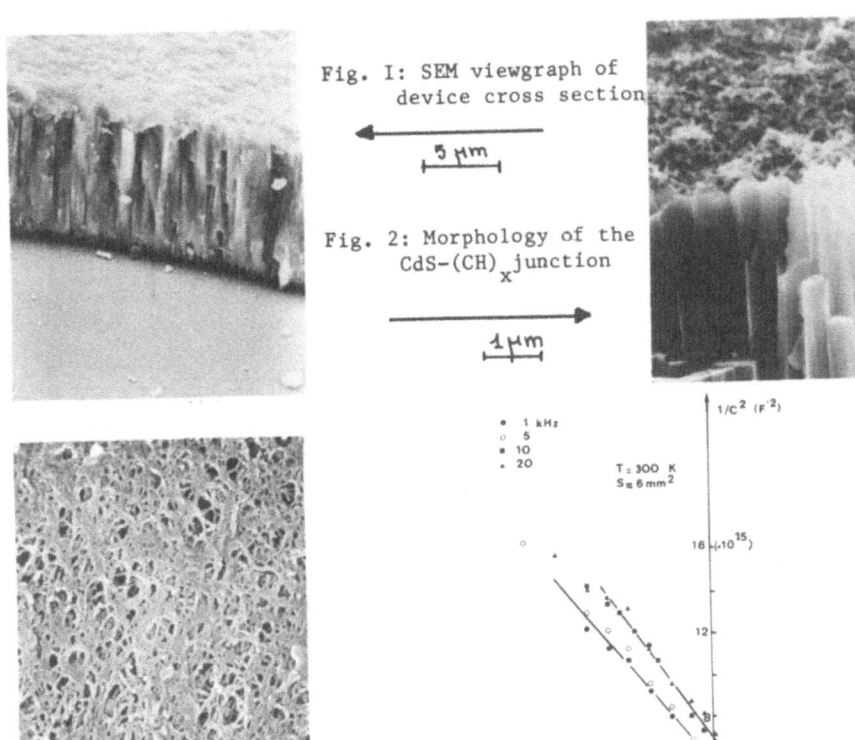

Fig. 1: SEM viewgraph of device cross section

Fig. 2: Morphology of the CdS-$(CH)_x$ junction

Fig. 3: SEM viewgraph of the $(CH)_x$ surface

Fig. 4: Typical C-vs V characteristic of the heterojunction

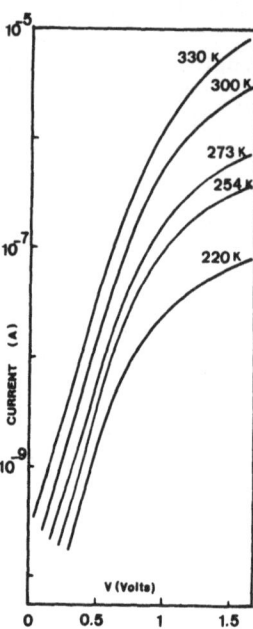

Fig.5 : Dark Log I=f(V) curves at various temperatures (undoped sample)

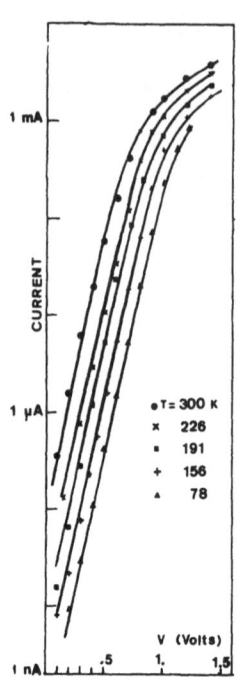

Fig.6 : Dark Log I=f(V) curves at various temperatures (**doped** sample)

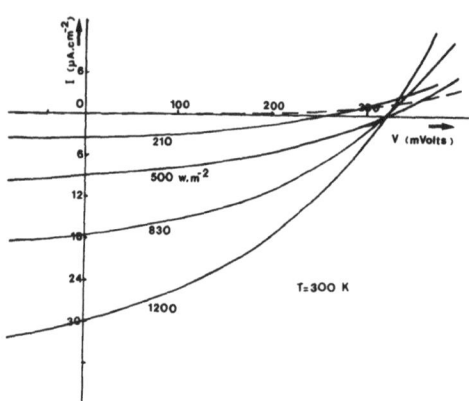

Fig.7a : Illuminated I-V characteristics of an undoped cell (surface 0.1 cm^2)

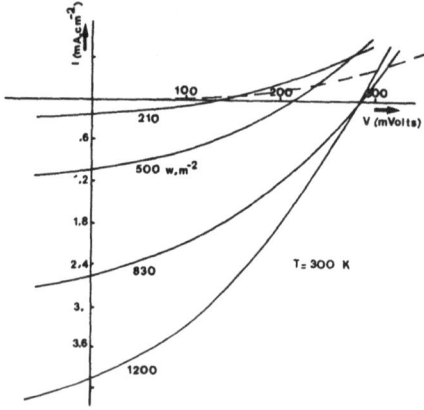

Fig. 7b : Illuminated I-V characteristics of a doped cell (surface 0.1 cm^2)

ANTIMONY SULPHIDE THIN FILMS

J.S.CURRAN and R.PHILIPPE
Laboratoire de Physicochimie des Interfaces
Ecole Centrale de Lyon

Summary

Antimony trisulphide has been prepared with n-type conductivity, in the form of thin polycrystalline films. These films were obtained by a tarnishing, or gas-solid reaction between metallic antimony and sulphur vapour, and their properties studied by photoelectrochemical techniques. An analysis of photocurrent-voltage curves and photocurrent spectra allows an estimation of the following parameters: bandgap 1.62 eV, flat band potential -0.45 V (SCE) and donor density approximately 10^{14} cm^{-3}. Spectroellipsometric measurements confirmed the high optical absorption coefficients previously reported for this material. Photovoltages of up to 400 mV have been observed, and monochromatic quantum yieds of over 60%. Preliminary experiments have indicated that photocorrosion may be suppressed in organic electrolytes.

INTRODUCTION

Antimony trisulphide is a semiconductor with reported bandgap of 1.6-1.7 eV (1,2,3,4) which has been used in photoconductive light detectors. It possesses an orthorhombic chain structure, and a rather low melting point (5). All literature reports known to the authors report p-type conductivity. Optical absorption coefficients, determined by several groups (1,2) are high. A crystalline-amorphous transition takes place at around 220°C (6). Hall effect measurements indicate a low hole mobility in the 10 cm^2 V^{-1} s^{-1} range (3,4), but the strong optical absorption allows at least in principle the exploitation of this material in a device relying on minority carrier collection by drift.

We have found that preparing Sb_2S_3 films by a "tarnishing" reaction:

$$4\ Sb\ (solid) + 3\ S_2 (gas) \xrightarrow{300°C} 2\ Sb_2S_3\ (solid)$$

with sulphur vapour at it's saturated value of 50 mm Hg at this temperature leads to n-type Sb_2S_3 with a very low donor concentration. Thus we have obtained high quantum yields from semiconductor-electrolyte junctions made with these films. A disadvantage of the preparation method is the very small grain size obtained.

EXPERIMENTAL

Electrodes were prepared in a tube furnace; to which sulphur vapour could be admitted from a sulphur "boiler" at the same temperature, the vapour being carried by a slow stream of argon. The antimony metal used was 99.999% (Bracconot et Cie) and the sulphur reported free of As, Se and Te (Elf-Aquitaine). Reactions were carried byt for one hour, and then the electrodes cooled rapidly.

Electrodes were mounted in a 3-electrode cell and electrochemical measurements made with standard commercial instruments. The light source was a 500 W Oriel xenon lamp, used with a Jobin-Yvon H 20 IR monochromator. Light intensity was measured with a calibrated Photodyne GGXŁA photodiode.

Photocurrent measurements were made in concentrated solutions of potassium ferrocyanide, in which photocorrosion is partially suppressed.

Some spectroellipsometric measurements were made with an instrument constructed in this laboratory using a Jobin-Yvon H 10 D double grating monochromator (250-760 nm) and a rotating analyser controlled by a Digital Corporation MINC computer.

RESULTS AND DISCUSSION

X-ray diffraction showed the films, about 3 µm thick as measured by dissolution and chemical analysis of the solution, to be crystalline. S.E.M. shows the presence of many short needles on the surface (about 1 x 5 µm).

Spectroellipsometric measurements confirm the very high values of the optical absorption coefficient (α) reported in the literature, reaching 6×10^5 cm^{-1} at 600 nm. At long wavelength however, these measurements showed large disagreements with single crystal values (2) due to grain boundary effects (7).

Figure 1 illustrates the photoaction spectrum of Sb_2S_3. A notable feature is the fall in quantum yield (η) at short wavelength typical of surface recombination. Figure 2 shows the photoaction spectrum at long wavelength replotted as $\eta^{1/2}$ versus $h\nu$. The straight line may be extrapolated to give an estimate of the bandgap, 1.62 ± 0.01 eV.

The square of the quantum yield plotted against the electrode potential yields straight lines such as shown in Fig.3 The extrapolation of these plots allows an estimate of the flat band potential, $V_{fb} \simeq -0.45$ v (S.C.E.). Such plots may be interpreted by an approximation (8) neglecting diffusion of holes from outside the space charge layer, width ω :

$$\eta = 1 - \exp(-\alpha \omega)$$

$$\simeq \alpha \omega \quad \text{for} \quad \alpha \omega \ll 1 \quad \text{(long wavelength)}$$

$$\simeq \alpha \left(\frac{2\varepsilon\varepsilon_0}{qN_D}\right)^{1/2} (V - V_{fb})^{1/2}$$

In addition, if the optical excitation is via an indirect transition, and the bandgap value is E_g :

$$\eta \simeq A (h\nu - E_g)^2 \omega \quad \text{(A being a constant).}$$

N_D may be calculated using single crystal data for α (2).

We find N_D values in the region of 10^{14} cm^{-3}. It must be admitted however that such a model may be inappropriate since if the donor concentration is so low the needle-like structures are likely to be fully depleted, all the mobile charges being trapped in surface states (9). In such a case a treatment analogous to that used for p - i - n photocells (10) would be correct.

Photocorrosion takes place under illumination in all aqueous electrolyte systems tried. However practically stable photocurrents have been observed in ethanolic solutions of the reducing agent ferrocene, when the illumination level was sufficiently low to prevent depletion of the reducing agent at the electrode surface, i.e. with photocurrents of less than about 0.1 mA cm^{-2}. This photocell, with a carbon counterelectrode, shows photovoltages up to 400 mV. Sb_2S_3 might be entirely stabilized in even lower donicity solvents, especially since the triply charged Sb^{3+} ion must have a high solvent-solvent free energy of transfer (11). We are currently investigating other such solvents.

References-

1. A. Audzijonis and A. Karpus, Leit. Fiz. Rinkinys. 18 (1978) 127.
2. A.A. Agasiev et al. Sov. Phys. Semicond. 6 (1972) 560.
3. B. Roy et al., Solid State Comm, 25 (1978) 937.
4. P. Bohac and P. Kaufmann, Mat. Res. Bull. 10 (1975) 613.
5. M. Schoijet, Solar Energy Materials 1 (1979) 43.
6. A. Gaumann, A. Orliukas and P. Bohac, Helv. Phys. Acta, 50 (1977) 773.
7. D.E. Aspnes, E. Kinsbron and D.D. Bacon, Phys. Rev. B 21 (1980) 3920.
8. M.A. Butler, J. Appl. Phys. 48 (1977) 1914.
9. G. Baccarini, B. Ricco and G. Spadini, J. Appl. Phys. 49 (1978) 5565.
10. T.I. Chappell, IEEE Trans. Electron Devices, ED-27 (1980) 760.
11. J.S. Curran, J. Electrochem. Soc. 127 (1980) 2063.

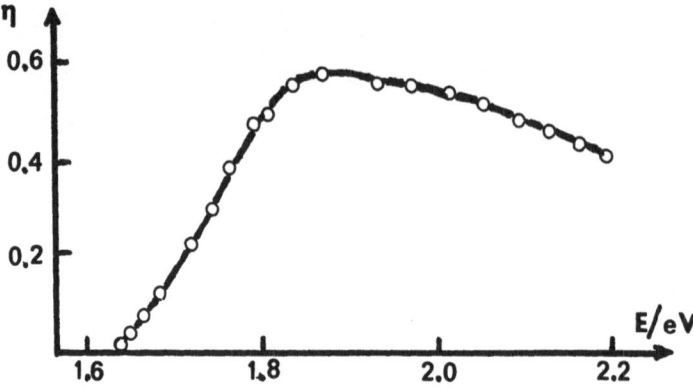

Figure 1 - The quantum yield versus photon energy for a polycrystalline Sb_2S_3 electrode held potentiostatically at -0.15 V (S.C.E.) in a saturated solution of potassium ferrocyanate. Optical bandpass 0.006 eV.

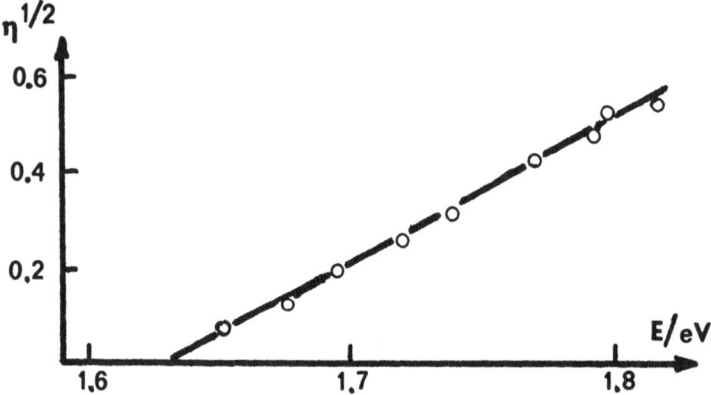

Figure 2 - A plot of the square root of the quantum yield versus photon energy at low energies under the same conditions as in 1.

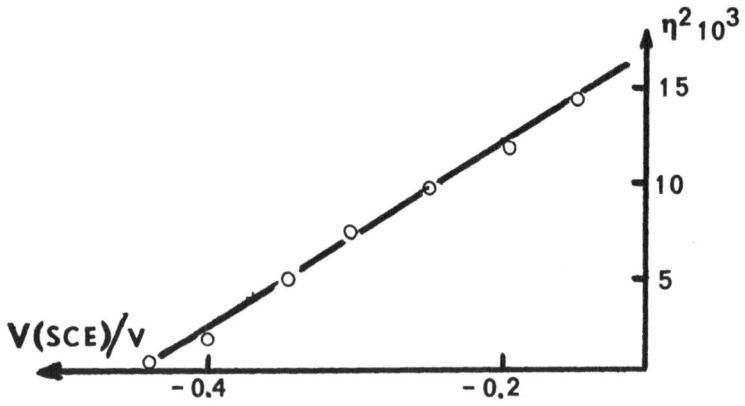

Figure 3 - The square of the quantum yield under low intensity illumination at 710 nm (110 µW) versus the applied potential with respect to the saturated calomel electrode. Solution saturated with potassium ferrocyanide.

SESSION 8

CRYSTALLINE SILICON SOLAR CELL TECHNOLOGY

Chairmen : Dr. E. FABRE, Photowatt International, Rueil
Malmaison, France

Prof. S. PIZZINI, Heliosil, Milan, Italy

Progress in unconventional crystallization of silicon

Production of solar grade silicon in an arc furnace using high purity starting materials

Segregation of impurities at grain boundaries and other compositional inhomogeneities in chill-casted silicon ingots

Advanced slicing techniques

Continuous growth of thin polysilicon sheets on a temporary carbon shaper by the R.A.D. process

Possibilities of ion implantation in silicon solar cell manufacturing

Low cost processes for cast silicon solar cells

Polycrystalline silicon solar cells utilizing an integral screen printing technique

Photovoltaic solar cell comparison methodology

PROGRESS IN UNCONVENTIONAL CRYSTALLIZATION
OF SILICON

E. Sirtl
Heliotronic GmbH
D-8263 Burghausen

Summary

A survey of the present status in advanced crystallization methods for silicon in terrestrial photovoltaics is given on the basis of their potential in large-scale technology and reproducible material quality. Additional emphasis is put on the question of compatibility between refining and crystallization concepts. For its partial illustration particular accomplishments in the program of Wacker are used.

1. INTRODUCTION

Recently, different work of comparative or comprehensive character has been published on the subject of advanced silicon crystal growth [1-4]. In contrast to the more detailed studies referred, this paper is intended to deal much more with the future aspects of those techniques which have experienced in the meantime technical testing on a larger scale or have clearly shown their high potential in long-range conceptualization. Besides some comparative compilations, the quantitative interaction of correlated technologies (refining, crystallization, cutting) in terms of charge sizes and output per equipment are areas of increasing interest and will be given special attention.

2. METHODS

Ingot technology

"Pure" bulk crystallization

In this particular field, we have to distinguish between two basic technologies mainly. In the one case, improvements have been made in economizing the leading processes for making monocrystalline rods, namely Czochralski pulling (CZ) and vertical float zoning (FZ), toward semiconventional techniques (products). The second case, directional solidification of the melt in containers in form of polycrystalline structures of any kind, has been increasingly developing since attractive PV properties of such materials have been corroborated at most different places.

In the first case, continuous or semicontinuous melt (feed) replenishment has been the major issue and further development still continues. Reduced specifications already have led to cost savings through modifications in crystal pulling conditions, quality of starting material (both influencing the defect situation), and slicing operation. For the next five years this kind of semiconventional solar material undoubtedly will have its importance in PV technology, as any of the unconventional processes needs more developmental work before reaching appropriate maturity in terms of large-scale production /̲ 5 /̲.

Ingot growth from a (square-shaped) melt container is utilized at several places. Wacker has been studying from the beginning a die casting process, with the target of multiple reuse of the cooled graphite mould /̲ see e.g. 6/̲ and recent investigations in Japan are aimed at protecting a carbon container by silicon nitride powder /̲ 7 /̲. The Heat Exchanger Method of Crystal Systems is utilizing quartz container material that will crack during the cooling process. The same supposedly happens during the SEMIX process. The basic idea of some inexpensive one-way mould already had been pursued in a process studied at Union Carbide /̲ 8 /̲. While the only process leading here to a monocrystal is the HEM technique /̲ 9 /̲, all others result in a polycrystalline ingot of at least partial columnar structure. To avoid mix-ups with the term "poly" being used for the starting material in conventional silicon crystal growth, such materials have been named "multicrystalline"(IBM, Heliotronic) or "semicrystalline" (Solarex), respectively, or characterized by a trademark (e.g. "SILSO"). All products of this type have shown maximum conversion efficiency data well over 10 % on a $10 \times 10 \text{ cm}^2$ basis.

Solar-grade "synthesis" through bulk segregation

Various methods are presently studied to obtain intermediate qualities of silicon aimed at further refining to reach "solar-grade" quality. Nearly all techniques being mentioned in the foregoing section have been claimed to be useful for this additional refining action. Segregation certainly shows distinct effects in all bulk crystallization techniques. It has already been shown for slow solidification rates of polycrystalline ingots that impurity distribution phenomena are even more pronounced there and very close to theoretical limits /̲ 10 /̲. The chart in Fig. 1 shows the different situations possible. Some combinations will, however, have their limited application. By using semiconventional solidification processes - after all - one has to end up with semiconventional economy. Considering the quality of partially refined starting material it remains to be seen which quality criteria may guarantee sufficiently high values of conversion efficiency. Certainly boron and phosphorus have to exist here at concentration levels which are acceptable to the skilled device maker. Some of the quality ranges are demonstrated in Fig. 2.

Slicing

Surprisingly, the increasing efforts in ingot growth technology are not matched by adequate activities in the development of unconventional slicing processes. Such techniques are absolutely linked with any low-cost ingot-based crystallization system. Problems and progresses in this area are separately treated in an extensive manner in another paper by G. Werner /̲ 11 /̲. What has been mentioned about advanced slicing, to a certain extent also applies to contouring and portioning of as-grown square-shaped ingots.

Ribbon technology

Ribbon pulling

Well organized literature already exists on the subject of comparative studies of ribbon growth methods /⎯1,4,12⎯7. Depending on width, thickness, crystal perfection and purity, all "typical" pulling methods range in their linear growth rates between 1 and 10 cm/min. Solidification processes based on much faster growth rates are rare and either already terminated or still in an early stage of development. In terms of pulling operation the HRG and LASS method /⎯13⎯7 and a modified RAD process /⎯14⎯7 seem to be the most promising candidates left.

Ribbon (foil) casting

Starting from the extreme case of roller quenching /⎯15⎯7 or splash cooling /⎯16⎯7, these methods certainly have their special application in manufacturing amorphous metal sheets. In the case of silicon, very low values of average grain size and extreme material brittleness forbid harnessing of those extremely high throughput rates being correlated with such procedures. The Wacker ICC method /⎯17⎯7, although started as a pulling method (Fig. 3), has the potential of very fast growth as a casting process (Fig. 4). Its speciality is the liquid encapsulation of the silicon melt by a thin slag film which markedly reduces its surface tension (ribbon width!), prevents random silicon nucleation from the supporting material (large grains!) and provokes preferred (111) grain orientation parallel to the bottom surface (rheotaxy).

In all cases of continuous ribbon growth segregation is neither wanted nor possible. The quality of the starting material, consequently, and principally, has to be of higher quality than that being used in ingot techniques.

3. Criteria for future potential

Process and equipment austerity studies

In Helmreich's paper /⎯3⎯7 a compilation of "parameters" is given that hardly has any parallel in the pertinent literature. For the subject of ingot technologies a chart has been put together for the best-investigated methods in this field (Fig. 5). Key words like practicability or productivity stand for the making and maintenance of the equipment, the stability of a given process, and the easiness of process control or production costs, throughput, chances for upscaling and continuous operation, respectively. The last two columns deal with the problems of integrability in terms of differently refined starting materials and their technology and the character of the slicing technology to be offered subsequently. This chart is completed by another one (Fig. 6) in which the different criteria in terms of material properties are compiled.

The curves which are shown in Fig. 7 illustrate the necessity to concurrently improve our knowledge about the peculiarities and necessities of large-scale operation for all relevant sub-technologies. This applies to the means of process engineering and their impact on material quality as well. As a typical example, open-hearth casting at Wacker (see Fig. 8) is one of those parallel studied objects to accelerate - among other aspects - our experience in preparing continuous block casting.

Throughput rates per production (equipment) unit are one of the stronger parameters allowing judgement of the future potential of some of the most-discussed groups in crystallization technology. The compilation of data in Table 1, although prepared in a crude manner, reveals some striking results. There are two ways of preparing crystalline silicon being far ahead of others. Whereas continuous block casting, however, has to suffer from the bottleneck of an even advanced slicing technology, continuous foil casting appears to be the least-limited candidate for the future.

Table 1. Estimates for future large-scale production conditions

ingot technology [1]	throughput per equipment unit (MW_p/year) [2]
growth methods with limited heat extraction	3
continuous block casting	300
slicing [3]	10
sheet technology	
multiple ribbon pulling ("standard")	3
continuous foil casting	300

1) kerf loss etc. considered
2) PV performance 10 %; 100 $cm^2 \triangleq$ 1 W_p
3) Wacker

Material quality aspects

Comparative estimates of that type are of limited value if the properties (geometrical considerations included) of a selected corresponding product are in a very early stage of investigation and qualification. Most of the presently discussed products are of multicrystalline character. There is a strong need for deeper insight into the nature of the various of (large-angle) grain boundaries and their interaction with different kinds of extrinsic and intrinsic crystal defects. An additional problem is the optimization of device processing based on such material. To start or continue true research work along this line is indebatable.

What is required in the meantime are numerous experiments in the sense Fig. 6. Topographically demonstrated performance can be of great help to identify certain types of crystallite patterns as detrimental for cell performance (Fig. 9). The study of precipitate formation in grain boundaries and bulk crystals as well (see Fig. 10) is another invaluable aid in reaching acceptable cell performance as soon as possible.

Different ways of empirical adaptation and optimization in given "integrated systems" will and have to be found with certainty before the "fine tuning" on the basis of the type of research work already mentioned will be able to take over.

ACKNOWLEDGEMENT

I would like to thank my co-workers J. Dietl and D. Helmreich for extensive discussions and their special assistance in preparing this paper. This study has been supported by the Bundesministerium für Forschung und Technologie under contract Nr. NT 0845/0846.

LITERATURE REFERENCES

[1] G. Cullen, T. Surek (eds.): Shaped Crystal Growth (Suppl.Issue), J. Cryst. Growth 50 (1980)

[2] J. Dietl, D. Helmreich, and E. Sirtl in "Crystals: Growth Properties and Applications", Vol. 5, J. Grabmaier (ed.), Springer Verlag, Berlin (1981), p. 43

[3] D. Helmreich, Proc. of Symp. Electronic and Optical Properties of Polycrystalline or Impure Semiconductors and Novel Silicon Growth Methods, K.V. Ravi and B. O'Mara (eds.), The Electrochemical Soc. Inc., Pennington (1980), p. 184

[4] T. Ciszek, ECS Spring Meeting, Montreal, May 1982, Symp. Materials and New Processing Technologies for Photovoltaics

[5] W. Freiesleben, Proc. 3rd E.C. Photovoltaic Solar Energy Conf., Cannes (D. Reidel Publ. Comp., Dordrecht, 1981), p. 166

[6] D. Helmreich, ref. [4]

[7] T. Saito, A. Shimura, and S. Ichikawa, Proc. 15th IEEE Photovoltaic Specialists Conf., Orlando (IEEE, New York, 1981), p. 576

[8] J.I. Hanoka, H.B. Strock, and P.S. Kotval, Proc. 13th IEEE Photovoltaics Specialists Conf., Washington, D.C. (IEEE, New York 1978), p. 485

[9] C.P. Khattak and F. Schmid, ref. [8], p. 137

[10] D. Helmreich and G. Ast, ref. [3], p. 233

[11] P. G. Werner, this conference

[12] T. Surek, ref. [3], p. 173

[13] a) B. Kudo, ref. [1], p. 247
     b) D.N. Jewett and H.E. Bates, Proc. 14th IEEE Photovoltaic Specialists Conf., San Diego, CA (IEEE, New York 1980), p.1404

[14] C. Texier-Hervo, M. Mantref and C. Belouet, this conference

[15] K.I. Arai, N. Tsuya, and T. Takeuchi, ref. [13 b], p. 31

[16] D.B. Bickler, L.E. Sanchez and W.J. Sampson, ref. [13 b], p.36

[17] German Patent (DOS) No. 29 03 061

[18] D. Helmreich and F. Suk, ref. [4]
[19] D. Huber, R. Wahlich, and D. Helmreich, ref. [13 b], p. 316

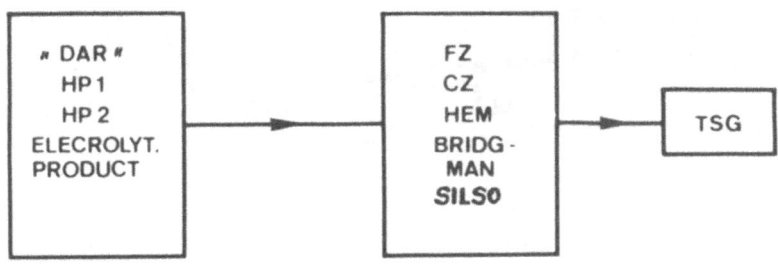

Fig.1 "Solar-grade synthesis" from partially refined silicon by directional solidification

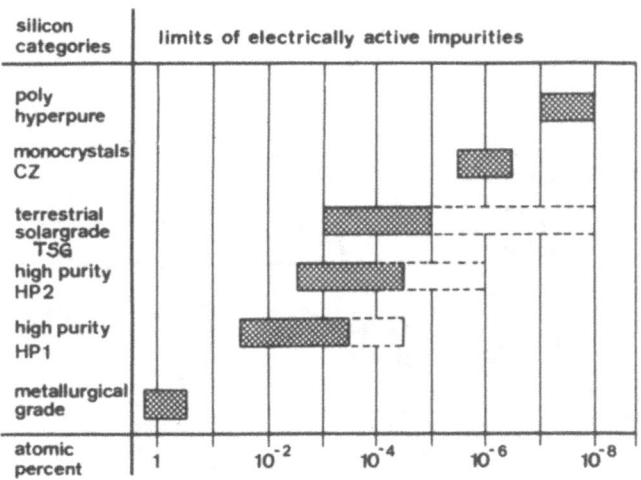

Fig.2 Different categories of silicon quality (Dashed lines indicate concentrations of fast diffusing metals)

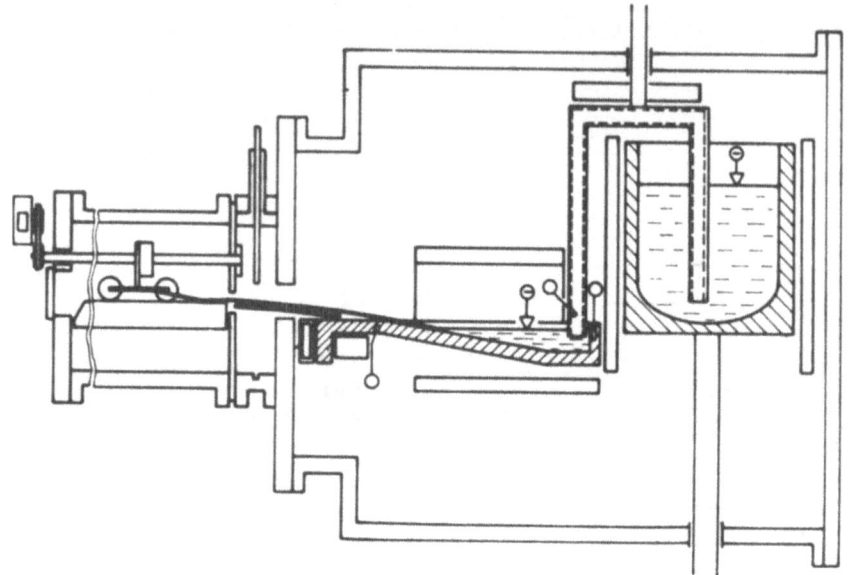

Fig. 3 Set-up of horizontal ICC pulling unit
  ○—● temperature control system
  ⓪—▷ melt level control system

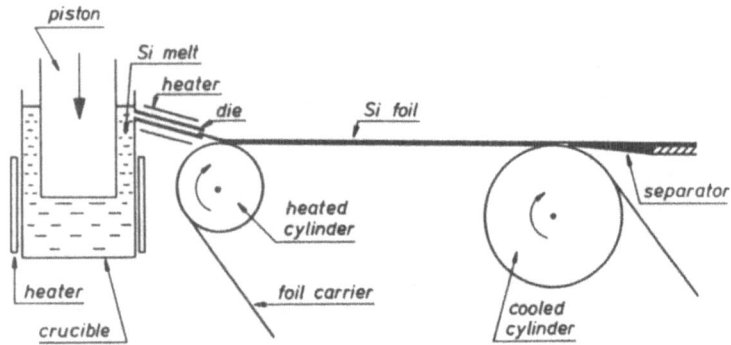

Fig. 4 Schematic drawing of continuous foil casting system after /2/

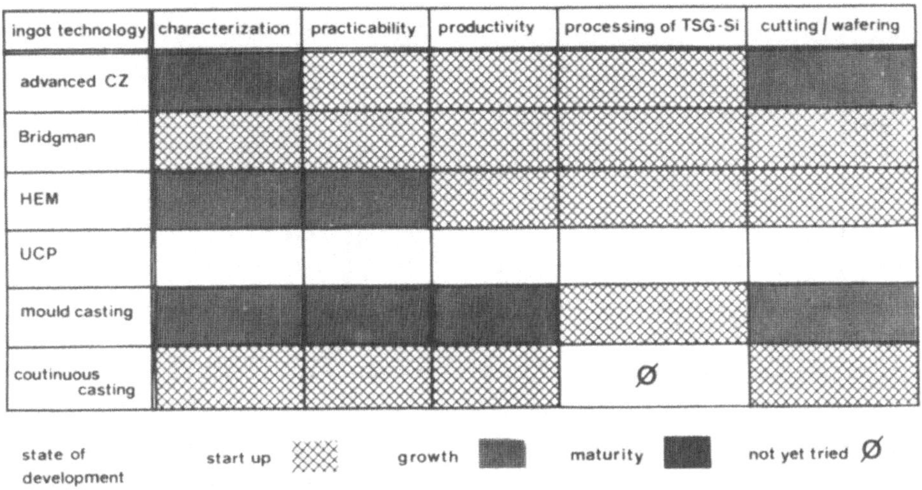

ingot technology	characterization	practicability	productivity	processing of TSG-Si	cutting / wafering
advanced CZ	maturity	start up	start up	start up	maturity
Bridgman	start up	start up	start up	start up	start up
HEM	maturity	start up	growth	start up	start up
UCP					
mould casting	maturity	maturity	maturity	start up	maturity
coutinuous casting	start up	start up	start up	∅	start up

state of development | start up ▩ | growth ■ | maturity ■ | not yet tried ∅

Fig. 5 Classification of ingot technologies - after Helmreich /3/ (reliable data missing in the case of UCP)

ingot technology	soundness	crystal perfection	impurities	PV performance
advanced CZ	+	+	+ +	+ +
Bridgman	O	O	O	O
HEM	O	+	+ +	+
UCP				
mould casting	+	O	+	+
continuous casting	O	O	−	−

+ + excellent   + good   O fair   − insufficient

Fig. 6 Characterization of ingots - after Helmreich /3/

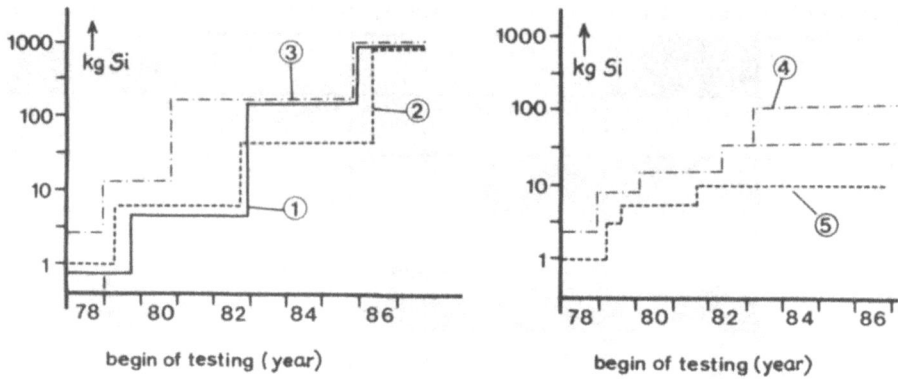

Fig. 7 Development of charge dimensions in the Wacker project
Curve 1: Liquid-liquid extraction (pyrometallurgy)
Curve 2: Solid-liquid extraction (hydrometallurgy)
Curve 3: Melt stock for casting
Curve 4: Ingots, as-cast
Curve 5: Ingot charging per slicing unit

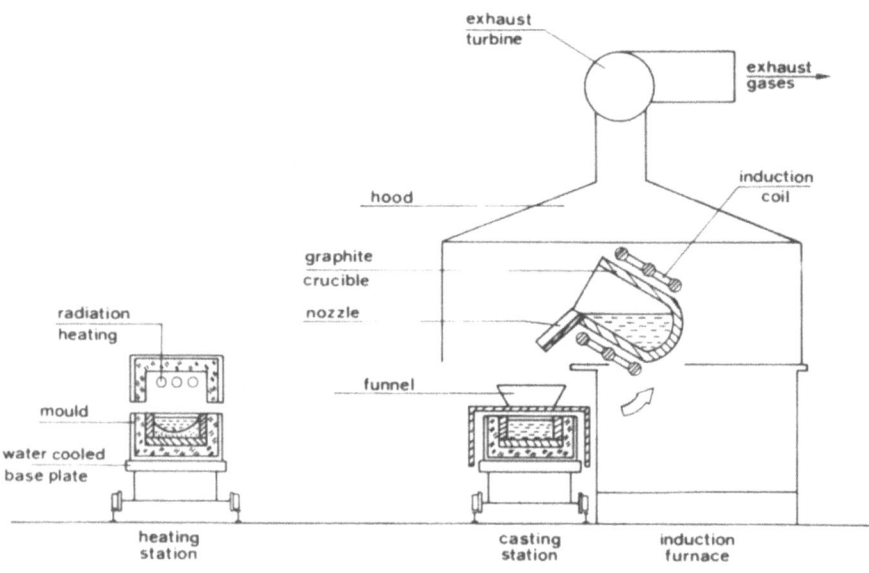

Fig. 8 Schematic view of open hearth casting system - after Helmreich and Suk [18]

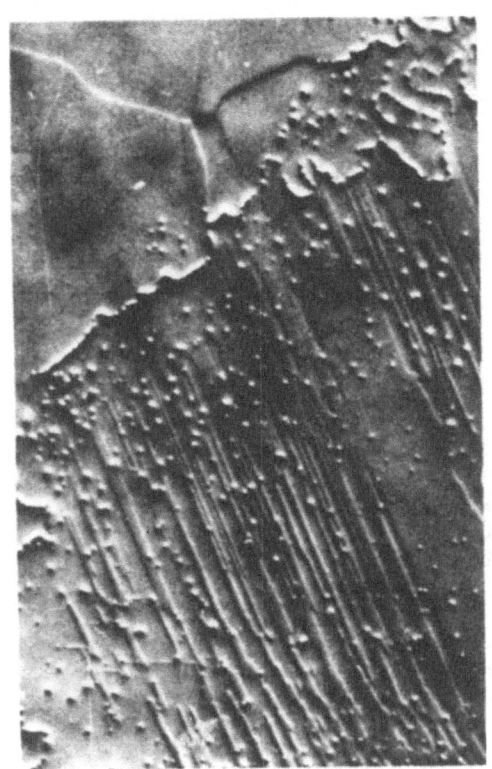

Fig. 10 Optical micrograph of a mechanically polished SILSO sample showing SiC precipitation (30x) - after /‾2‾/

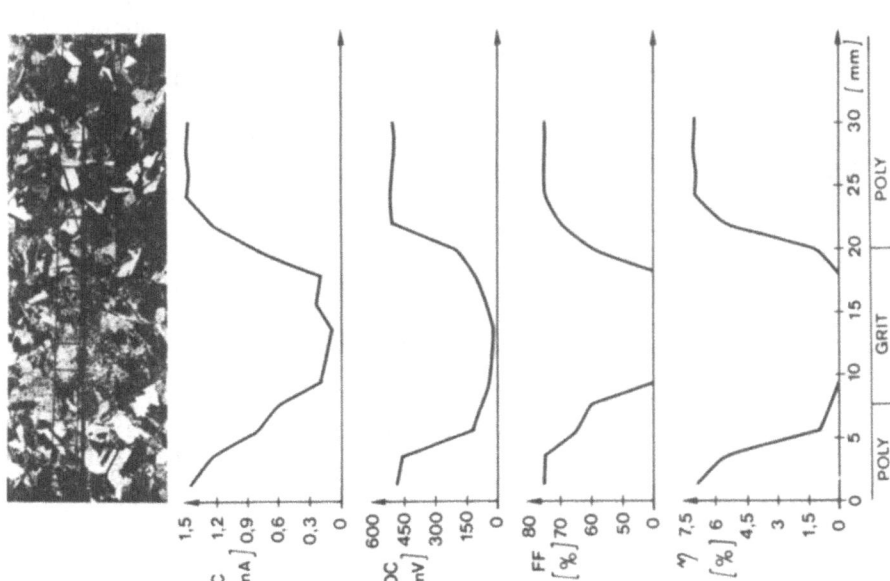

Fig. 9 Photovoltaic parameters as topographic indicators for unfavorable crystalline structure - after /‾19‾/

PRODUCTION OF SOLAR GRADE SILICON IN AN ARC FURNACE USING HIGH PURITY STARTING MATERIALS

H.A. AULICH[1)2)], W. DIETZE[2)], K.-H. EISENRITH[1)], J. SCHÄFER[2)]
F.W. SCHULZE[2)], H.-P. URBACH[2)]

[1)] SIEMENS AG
Research Laboratories
[2)] Components Group, Discrete Semiconductors
Otto-Hahn-Ring 6, 8000 Muenchen 83

Summary

The carbothermic reduction of silica has been adopted to produce solar-grade silicon using high-purity raw materials. Inexpensive and abundant quartz sand is purified by fusing it with glass-forming oxides to form a melt from which glass fibers are drawn. Subsequent treatment of the fibers with hot HCl leaches out all impurities, resulting in high-purity SiO_2 analysed to have B, P and transition metal concentrations of less than 1 ppmw. High-purity carbon (B, P and transition metal concentration < 1 ppm) is prepared by treating carbon black with hot HCl. When reacting these purified materials in a small arc furnace, the impurity concentration of the silicon obtained corresponded to the impurities present in the starting materials. A three-phase, 550 kVA-arc furnace was constructed to prepare silicon on a larger scale. Quartz and charcoal were used as raw materials to gain experience in operating the furnace and to study the process parameters. The silicon produced from these impure materials was further purified employing the Czochralski method.

1. INTRODUCTION

One of the prerequisites for large scale photovoltaic energy conversion using crystalline silicon solar cells is an economical process for the production of high purity silicon. The commercial solar cells presently available are produced almost exclusively from high-purity semiconductor grade silicon, a material too expensive for large scale terrestrial applications.

A feasible process for producing large quantities of inexpensive silicon is the conventional carbothermic reduction of silica. For solar cells, however, the impurity concentration of the resulting metallurgical-grade (MG) Si is too high by a factor of $10^4 - 10^5$. Since the impurity level of this silicon is largely determined by the impurities present in the raw materials, quartzite and coke, it was natural to use raw materials of higher purity to produce high-purity silicon. This approach, first employed by L.P. Hunt et al. /1/, yielded silicon with an impurity concentration of approximately 200 ppm. A further reduction of this impurity level requires the use of SiO_2 and carbon of even higher purity and consistent quality. To achieve this goal, we developed a process for preparing high-purity silica and carbon and designed a 550 kVA-three-phase submerged arc furnace which permits the carbothermic reduction under high-purity conditions. This paper describes the preparation of high-purity starting materials and provides preliminary results on the reduction of these materials in a small reactor. In addition, results on the reduction of natural quartz with charcoal using the 550 kVA arc furnace are presented.

2. PREPARATION OF HIGH-PURITY SILICA

Pure natural quartz mined from various deposits located in Brazil, the U.S.A., Africa etc. has an impurity concentration which requires further purification of the silicon produced by carbothermic reduction /1/. Even in the purest materials the main impurities, Al, Fe, Ti are typically in the 10 to 100 ppm region, while the B and P concentrations typically vary between 1 and 10 ppm. Furthermore, these naturally occurring materials are expensive, and their impurity concentration may vary significantly. In an attempt to decrease the impurity level of natural quartz, the quartz lumps are milled to a grain size corresponding to the average size in the bulk material and then acid-treated to dissolve the impurities present at the grain boundaries. Although this treatment decreases the concentration of some impurities, it fails to remove impurities present inside the grains. To overcome these problems, we developed a process in which abundant and inexpensive quartz sand is purified using a combined glass melting/fiber leaching(FL) process: By mixing sand with other metal oxides such as Na_2O, CaO, Al_2O_3 etc. in a crucible, a glass is prepared at approx. $1360^\circ C$ (Fig. 1). After the glass has been homogenized and fined, thin fibers are drawn from the melt through a hole in the bottom of the crucible. During the drawing process the fiber diameter is continuously monitored to insure uniform thickness.

The glass fibers are subsequently treated with hot HCl solution, and as a result the metal oxides and impurities present in the glass matrix are leached out by an ion exchange mechanism resulting in high-purity silica. The basic idea behind this FL-process is to first provide a solvent for the impurities in the quartz using glass forming oxides and then to remove the solvent from the SiO_2-network by acid leaching treatment. When glass fibers of suitable composition are immersed in hot HCl, destruction of the glass network by ion exchange takes place. Ions present in the glass such as Na^+, Ca^{++}, Fe^{+++} etc. are exchanged for H^+ ions in the acid, resulting first in a porous layer of silica on the fiber surface. Ion exchange may proceed through this layer by diffusion of H^+ ions, so that the underlying glass layer is continuously attacked by the acid until all impurities are removed, leaving a structure of pure silica. This leaching process is facilitated by the large ratio of surface area to glass volume obtained with fibers, thus providing a short diffusion length. Furthermore, the exchange of alkali and earthalkali ions by H^+ ions leads to considerable tensile forces at the fiber surface, causing cracking and peeling of the SiO_2 surface layer. This in turn produces new surfaces which aid the leaching process so that a completely leached and homogeneous silica product is readily obtained. Fig. 2 shows a scanning electron micrograph of a fiber treated with HCl for 4 minutes; the cracked and partly peeled silica surface is clearly visible.

One important parameter in the FL-process that determines the purity of the silica produced is the chemical composition of the glass. The prerequisites for glasses used in the FL process include the use of inexpensive and abundant raw materials, good glassforming capability, ready conversion into fibers and low corrosion resistance to acid. The first experiments were carried out using phase separable glasses from the sodium-borosilicate-system known to be leachable with mineral acids /2/. Although the transition metal concentration of the SiO_2 obtained in this way was below 1 ppm, the B-concentration still varied between 10 - 100 ppm. To obviate the problems encountered in further reducing the B content, fibers were prepared using boron-free glasses from the $Na_2O-MgO-CaO-Al_2O_3-SiO_2$-system. To demonstrate the feasibility of the FL-process in preparing high purity SiO_2, sea sand with a high impurity level was used as starting material. Table 1 shows the impurities present in the sea sand and the high purity level of the SiO_2 achieved using the glass system mentioned above. The results show that the low concentration of boron and transition metals in particular should qualify the SiO_2 as an excellent starting material for the production of solar grade silicon.

3. PREPARATION OF HIGH PURITY CARBON

The production of solar grade silicon by carbothermic reduction requires the use of carbon comparable in purity to the SiO_2 produced by the FL process. In addition, the carbon must exhibit high chemical reactivity during the reduction process. Carbon blacks were shown to be sufficiently reactive /1/, but the impurity concentrations of the materials available at acceptable cost do not meet our requirements. To overcome this problem, carbon black was purified by treating it with hot HCl for several hours in order to leach out harmful impurities, in particular iron. Table 2 shows that this treatment resulted in a significant improvement in the purity of carbon black.

4. CARBOTHERMIC REDUCTION IN AN EXPERIMENTAL ARC FURNACE

Both SiO_2 and C are finely powdered after purification

and must therefore be pelletized before being introduced into the arc furnace. The pellets must exhibit sufficient reactivity and mechanical strength to prevent premature disintegration into fine particles during reduction. A phenol resin of high-purity was used as a binder to prepare pellets with a disc pelletizer. The pellets obtained varied in size from 5 - 20 mm. On account of the small quantities of high purity SiO_2 and C presently prepared, the suitability of the pellets for carbothermic reduction was tested in a small high-purity carbon reactor (Ø = 120 mm) equipped with one electrode (40 mm). Since such a small reactor does not allow the removal of silicon during reaction, the Si is always contaminated with unreacted starting materials and reaction products such as SiC. Our results have shown, however, that the pellets have sufficient mechanical strength and that the purity of the silicon obtained corresponds to the impurities present in the starting materials.

To carry out experiments with high purity SiO_2 and C under more realistic conditions, a single electrode (Ø = 150 mm), one-phase 90 kVA arc furnace with an inside diameter of 400 mm was constructed that permits periodic tapping of the silicon produced. Preliminary experiments were carried out using natural quartz and charcoal as starting materials. Silicon was produced at a rate of 1 - 2 kg/h and removed from the reaction zone every 2 - 3 hours. The material was of p-type and had a resistivity of 0.1 to 0.05 $\Omega \cdot cm$. This reactor will determine the experimental parameters for the preparation of solar-grade silicon using charges of 100 - 200 kg of high-purity starting materials.

5. CARBOTHERMIC REDUCTION IN THE 550 kVA ARC FURNACE

On account of the encouraging results achieved in the preparation of high-purity starting materials and their reduction without further contamination, a three-electrode, three-phase 550 kVA arc furnace was constructed. This furnace has an inside diameter of 1200 mm, a capacity of 1200 kg and an electrode arrangement similar to those commonly used in large arc furnaces. All parts of the reactor that come into contact with raw materials or liquid silicon are made of high-purity carbon to prevent contamination of the silicon produced. To determine the optimal process parameters for silicon production, the furnace was first operated using quartz and charcoal, starting materials with a relatively high impurity level. Silicon was produced at a rate of $\approx$ 15 kg/h, collected in graphite moulds and then transferred into a Czochralski crystal puller for further purification. The resulting polycrystalline ingot (Ø = 78 mm) was remelted, and a single-crystal ingot was pulled for solar cell preparation. Table 3 shows the impurity concentration of the silicon before CZ-purification as determined by optical emission spectroscopy. Solar cells (Ø 44 cm^2, resistivity 3 $\Omega \cdot cm$, p-type) fabricated from the single-crystalline ingot (Ø = 76 mm) using conventional cell technology had an AM1 conversion efficiency of 6,8 %. Best results were 11,2 % obtained on 4 cm^2 cells. These results show the feasibility of the process; it is expected that solar grade silicon can be produced directly with our purified starting materials.

CONCLUSION

Silicon dioxide and carbon of high purity were prepared on a laboratory scale and reduced to silicon in a small arc furnace. A 550 kVA arc furnace was constructed for larger quantities and employed to produce silicon from quartz and charcoal. Further work is aimed at producing high-purity starting materials in sufficient quantity for the large arc furnace in order to determine the process parameters, the economy of the process and the purity of the silicon obtained.

ACKNOWLEDGMENT

The authors are indebted to Dr. H. Schroth, A. Hartl and K. Seppi for chemical analysis, H. Fenzl for crystal pulling, H.-D. Hecht and G. Glasow for solar cell fabrication, L. Bernewitz for metallographic evaluation and K. Geim for electrical measurements on Si-samples. This work was supported by the Bundesministerium für Forschung und Technologie.

REFERENCES

/1/ Hunt, L.P.; Dosaj, V.D.: Proc. 2nd Photovoltaic Solar Energy Conf., April 23-26, 1979, p. 98
/2/ Nordberg, M.E.; Hood, H.P.: U.S. Patent Nr. 2315329, 30.3.1934

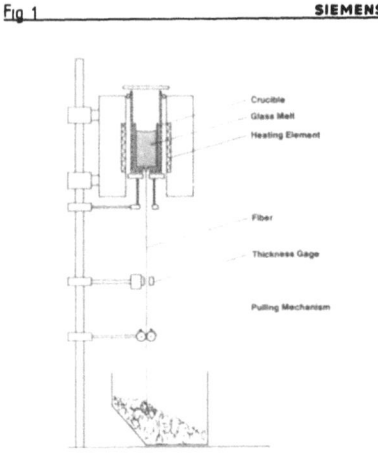

Fig 1 — Glass melting- and fiber drawing apparatus   FL FKE 333

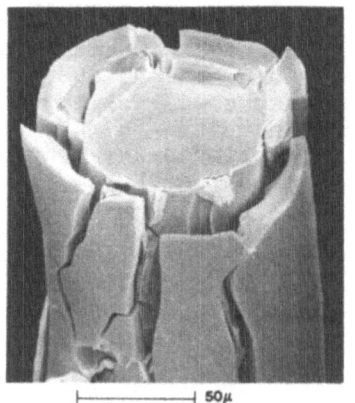

Fig. 2 — Glass fiber treated 4' with hot HCl

Table 1   SIEMENS

Impurity	Starting Material Sea Sand Concentration in ppmw	SiO$_2$ after FL-process Concentration in ppmw
Na	10^4–10^5	<5
Mg	10^3–10^4	0.05–0.1
Ca	10^3–10^4	<5
Al	10^4–10^5	0.5–1
B	1–10	<1
Fe	10^3–10^4	<0.5
Ni	1–10	<0.05
Mn	10–100	<0.05
Cu	1–10	<0.05
Ti	100–1000	0.05–0.1

Impurity Concentration in SiO$_2$ Purified with the Fiber Leaching (FL)-Process   FL FKE 333

Table 2   SIEMENS

Impurity	Starting material Concentration in ppmw	After purification Concentration in ppmw
Al	<0.062	<1
Cr	<0.62	<10
Cu	0.25	<0.1
Fe	62	<1
Mn	0.62	<0.1
Mg	6.2	0.1–1
Mo	0.062	<1
Ni	3.1	<0.1
Ti	<0.124	<0.1
Ca	100–1000	<10

Purification of Carbon Black with Hot HCL   FL FKE 333

Table 3   SIEMENS

Impurity	Concentration [ppmw]
Mg	1–10
Ca	100–1000
Al	36
B	8.5
P	47
Cr	1,7
Cu	1,7
Fe	170
Mn	10–100
Ni	2,7
Ti	19
V	<10
Mo	<10
W, Zr, Ta	<100

Analysis of Silicon produced from Quartz and Charcoal   FL FKE 333

SEGREGATION OF IMPURITIES AT GRAIN BOUNDARIES AND OTHER COMPOSITIONAL INHOMOGENEITIES IN CHILL-CASTED SILICON INGOTS

S. PIZZINI^{-x}, L. BRAICOVICH^{o+}, L. CALLIARI^{+}, M. GASPARINI^{-}, C.M. MARI^{x}, F. REDAELLI^{x}, M. SANCROTTI^{+}

x Istituto di Elettrochimica, Via Venezian 21, Milano
- Heliosil S.p.A., Via Franchetti 1/A, Milano
° Istituto di Fisica, Politecnico di Milano, Piazza Leonardo da Vinci, Milano

+ IRST, Trento

Summary

Segregation of residual impurities within a matrix of solar grade silicon is a process which could take place when a crystallization process is carried out far from equilibrium conditions or residual impurities content exceedes solubility limits. As information on this topic is however very scarce, a systematic investigation has been carried out using MG-silicon or heavily doped EG-silicon as model material and chemical analysis, SEM, x-ray micro analysis and Auger spectroscopy as diagnostic methods. Results of this investigation show that residual impurities segregate at the grain boundaries or within the single grains as silicides and that, therefore, chemical bonding is the driving force for excess impurity removal. Grain-boundaries could act as sinks for impurities but it appears as well that infra-granular precipitates tend to coalesce in a network of sub-grain boundaries. Eventually, the fact that the precipitate composition is constant while the distribution of impurities is inhomogeneous within the crystals shows that the local density of precipitates varies within the crystals, depending on the thermal history of the crystal.

1. INTRODUCTION

In the due of a research which aims obtaining solar grade silicon from MG-silicon, prepared from pure raw materials and/or suitably upgraded, a knowledge of the influence of residual impurities on the ultimate efficiency of polycrystalline silicon solar cells is compulsory.

The solar silicon definition which could be adopted for crystalline silicon is infact unusable for polycristalline materials and therefore the final crystallization method that will be chosen in order to obtain ingots directly usable for the fabrication of diffuse junction solar cells depends on its capability to confine impurities at the grain boundaries (provided it will be definitly ascertained that impurities at grain boundaries are electrically inactive or could be rendered inactive) and/or to concentrate the impurities exclusively in a macroscopic region of the as grown-ingot which could be removed by cutting without an excessive loss of material.

To this scope, one should investigate the features of the macroscopic distribution of the different impurities and then look for a model which could be used for predetermining, on the base of the actual impurity content of a certain feedstock, the crystallization technique which shall be useful to get from it usable solar grade ingots.

As the sensitivity and resolution of the trace analysis method which should permit a mapping of the microscopic distribution of impurities in a solid matrix, is too low to permit an investigation to be carried out on a material which is already solar grade, MG-Si have been choosen as a model material .

This paper reports therefore results dealing with behaviour of impurities in a material which, although not necessarily representative of their behaviour in S.G. silicon, gives more insight on the features of the micro-and-macro heterogeneties observed in chill-casted and unidirectionally solidified polycristalline ingots.

2. EXPERIMENTAL RESULTS

Preparation of the samples

MG-silicon has been used throughout as reference material. For the preparation of the polycristalline ingots, a directional solidification furnace has been used, which has a 300 mm. diameter hot zone and could be operated under vacuum (up to 10^{-3} mbar) or under argor-purge at selected total pressures.

Dense graphite crucibles were used for the solidification runs which were carried out by melting the silicon charge under reduced pressure of argon, after a backing-up step under vacuum up to 1000°C. Solidification has been conducted either leaving the samples cooling down with the natural cooling rate of the furnace, with the sample in the hot zone of the furnace, or subjecting it to a directional solidification cycle, pulling down the crucible from the hot zone at a rate of 5 cm/hr.

In either case the ingots are macrocrystalline, with grains having a size ranging from some mm. to cm. while only in the case of directionally solidified materials the grains are strictly columnar.

Distribution of the impurities within the samples.

In order to determine the distribution of impurities among the samples, slices parallel or normal to the axis of the ingots were cut and macroscopic portions used for impurity determination via dissolution of the sample and atomic absorption analysis. Typical results of these determinations are illustrated in fig. 1a for a sample cut from a naturally-cooled ingot and fig. 1b for a sample cut from a directionally solified ingot, which show that in both cases impurities are non-homogeneously distributed and tend to concentrate in the region of the ingot which is the last to solidify.

The fact that also in the directionally solidified ingot the distribution is inhomogeneous indicates that the heat flux from the sample to the cold sink was far to be optimized.

Improvements in the thermal geometry of the crucible-heat sink system did however succeed in confining the impurities on the top of the directionally solidified ingot (see fig. 1c which refers to a 100 ppmw iron doped E.G. silicon) which appears also to be columnar, as it is shown in fig. 2.

Microscopic distribution of impurities.

SEM and XR microanalysis have been used for the determination of microscopic distribution of impurities. To this scope a Cambridge Stereoscan 150 has been used, equipped with a K860 energy dispersion system.

XR microanalysis has been carried out on mechanically lapped and polished slices. SEM and XR microanalysis runs on all the samples investigated showed that impurities segregate in relatively large ($\sim 1\ \mu m$)^x size globular or elongated precipitates. For this reason the analysis of the impurities was confined to the determination of the composition of the precipitates, scanning the samples along lines running parallel or normal to the axis of the ingot.

Results of this analysis, reported in terms of a single impurity content normalized to the total content of all the impurities plus silicon within the precipitates are shown in fig. 3a, b, c, d for iron, aluminium, calcium. It appears that the composition of the precipitates remains essentially constant from the border of the ingot to the centre or from the bottom of the ingot to the top.

Influence of grain boundaries on the local distribution of microsegregations.

As prints of microsegregations remain after Sirtl-etch (20" at room temperature) of the polished samples, an analysis of the local distribution of microsegregation is possible.

---

x Smaller size precipitates are present in W-25 samples: this will be discussed in a later paper.

Fig. 4 shows that microsegregations are constituted by precipitates at grain boundaries and by globular precipitates.

Fig. 5 shows however that not all grain boundaries act as a sink for precipitates and that some precipitates could even cross a boundary constituted possibly by a twin.

Fig. 6, eventually, shows a region constituted by a single grain, which is intersecated by a (closed) network of subgrain boundaries constituted by precipitates.

In all these microphotographs one observes that the surface of the samples is decorated by microetches, which could as well be the prints of microsegregations outside the grain boundaries or zones of local stress released in microcracks, as it shows in fig. 7.

Auger-spectroscopy of Si-Fe bond.

By comparing the results of XR-microanalysis (fig. 3) in the precipitates with the macroscopic chemical analysis of the samples (fig. 1) one observes the remarkable constancy of the composition of the precipitates which does not reflect the variation of macroscopic composition.

This calls for the importance of the chemistry of the interaction between Si and the impurities; in the case of Fe this conclusion was strongly supported by the analysis of Auger lineshapes carried out with a scanning instrument, i.e. with a method of valence spectroscopy able to give chemical information with spatial resolution. We have studied the Si ($L_{2,3}VV$) lineshapes in a model system obtained by adding 1% Fe to high purity Si and by producing ingots by chill-casting. The lineshapes (not the derivatives of the spectra as usually done in conventional Auger spectroscopy) have been measured for Si within the precipitates and in pure Si region. The spatial resolution was 0,3 um and the measurements were taken on wafers as cut; the surfaces were cleaned by gentle argon sputtering in situ. The results are given in fig. 8 where we compare the ($L_{2,3}VV$) lineshapes of pure Si and of Si in the precipitates; the shaded area points out the main difference between the two cases. The clean Si ($L_{2,3}VV$) lineshape is in excellent agreement with the literature (1) and is basically determined by the selfconvolution of Si p density of states (1-3); thus its modification is due to a change of Si p-states connected with the presence of Fe in the precipitates. The results of fig. 8 are very typical of the formation of silicides in analogy with the results found in Pd-Si by Ho et al. (4); we note that this is a general trend in consideration of the nature of the chemical bond in silicides (4,5).

This behaviour of Auger lineshape resulted to be basically independent of the local variation of composition in the precipitates.

It is thus possible to state that the precipitates are always formed by silicides in the sense that the bond between Si and Fe is always silicide-like. We notice that this could not have been established on the mere basis of the average composition, which is around $FeSi_2$, obtained with conventional Auger spectroscopy on the same samples and Fe_2Si_3 on MG-silicon samples measured with X-Ray microprobe (see fig. 9).

The present results have also another important implication; in fact no evidence of silicon oxides has been found in the precipitates after sputtering (i.e. in the bulk of the precipitates) as it is shown by Auger line energies (6). This indicates that the oxygen present in the cast does not influence significantly the precipitate growth.

3. DISCUSSION AND CONCLUSION

The fact that precipitates have an almost constant composition within the samples while the chemical composition of the sample itself is non-homogeneous calls for an inhomogeneous distribution of the density of the precipitates, which tend to concentrate in the regions of the ingots where impurities are more abundant.

Moreover, the fact that the composition of the precipitates is almost constant, is a proof that chemical bonding between silicon and impurities is the driving force for the formation of the precipitates, and that grain-boundary segregation is a second-order effect at least in the impurity concentration range investigated.

This conclusion does not drop out possible grain-boundary sinking effects in less concentrated solutions of impurities, but puts in evidence that local silicide clusters could well exists in polycristalline matrixes grown from solar grade silicon (which is a typical very diluted solution of impurities) unless by careful directional solidification procedures the solid matrix is freed from impurities to such a low level to remain substantially unaffected by residual impurities. Eventually, the fact that only directional solidification confines the impurities to the top-region of the ingots drops out conventional chill-casting methods for the production of solar grade ingots, that appear only suited for the use of off-grade EG silicon, as originally pursued by Wacker.

AKNOWLEDGEMENTS

The authors are greatly endebted to Giorgio Terzaghi, who carried out the SEM and microanalysis determinations and to Dr. Cristina Calligarich for extended discussion.

This work has been partially carried out under the sponsorship of C.N.R., Programma finalizzato energetica, contratto n. 80.02583.92.

REFERENCES

(1) R.H. Brockman and G.J. Russell
Phys. Rev. B 22, 6302 (1980)

(2) J.E. Houston, M.G. Lagally and G. Moore,
Solid State Commun. 21, 879 (1977)

(3) P.J. Feibelman, E.J. McGuire and K.C. Pandey,
Phys. Rev. B15, 2202 (1977)

P.J. Feibelman, and E.J. McGuire,
Phys. Rev. B17, 690 (1978)

(4) P.S. Ho, G.W. Rubloff, J.E. Lewis, V.L. Moruzzi and
A.R. Williams
Phys. Rev. B 22, 4784 (1980)

(5) G. Rossi, I. Abbati, L. Braicovich, I. Lindau and
W.E. Spicer
Solid State Commun. 39, 195 (1981)

(6) J.A. Tagle, M.C. Munoz and J.L. Sacedow
Surface Sci. 81, 519 (1979)

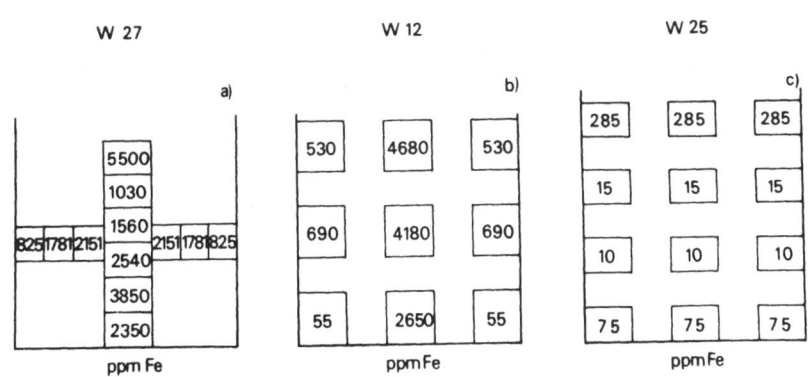

Fig. 1 - Macroscopic distribution of iron in
  a) chill-cast (W 27)
  b) directionally solidified MG-silicon ingots containing
     initially 2200 ppmw of iron (W27) (W12)
  c) directionally solidified ingot containing initially 100 ppmw
     of iron (W 25)

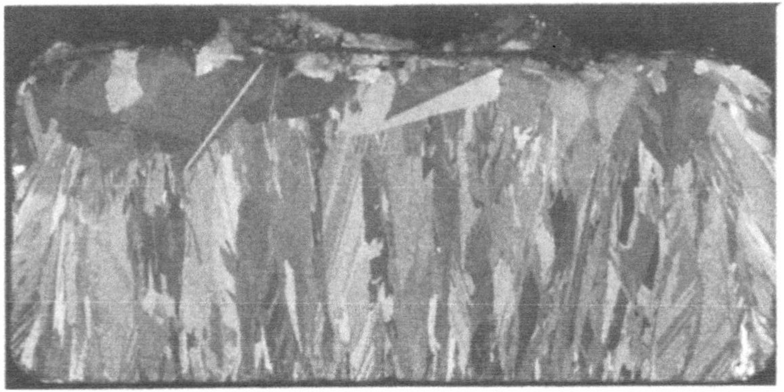

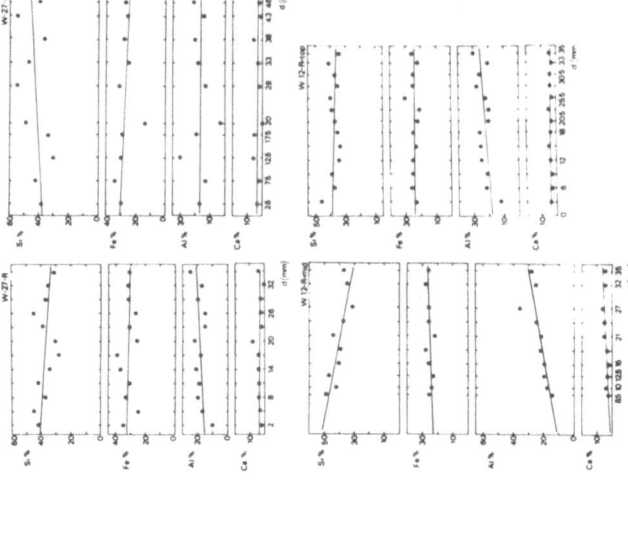

Fig. 3 - XR microanalysis results of axial (AX) and radial (R) sections of W27 and W12 ingots.

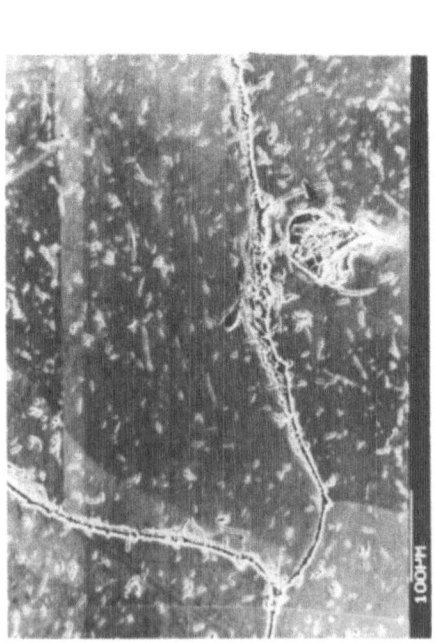

Fig. 4 - SEM photomicrograph of linear and globular precipitates in slices coming from W27 ingots

Fig. 2 - Vertical section of the directionally solidified ingot W25

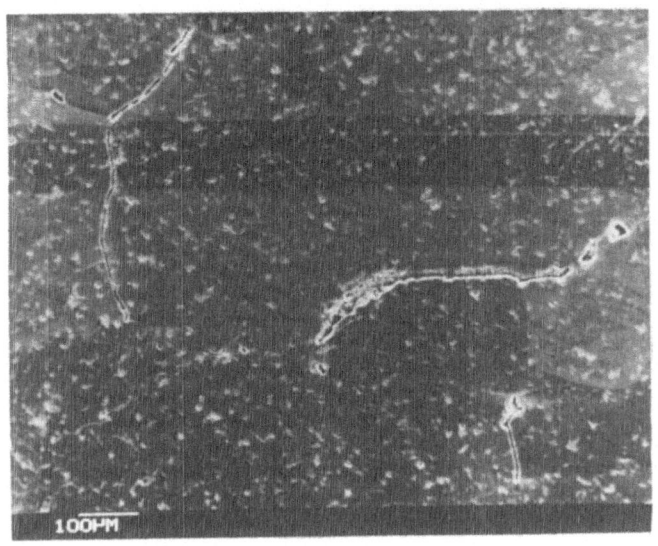

Fig. 5 - SEM photomicrograph of a precipitate crossing a twin.

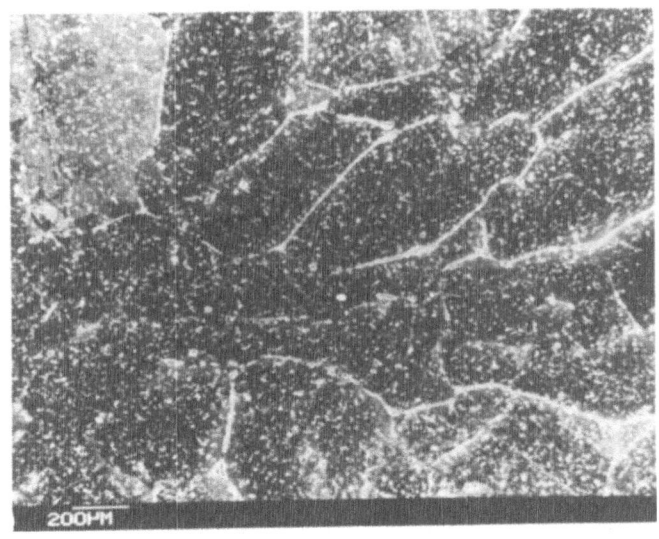

Fig. 6 - Network of sub-grain boundaries generated by clustering of precipitates

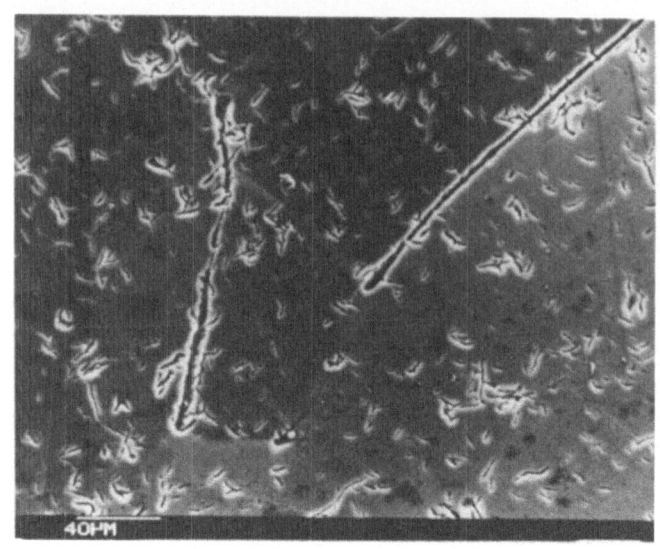

Fig. 7 - Microetches and microcracks in a sample coming from the W27 ingot.

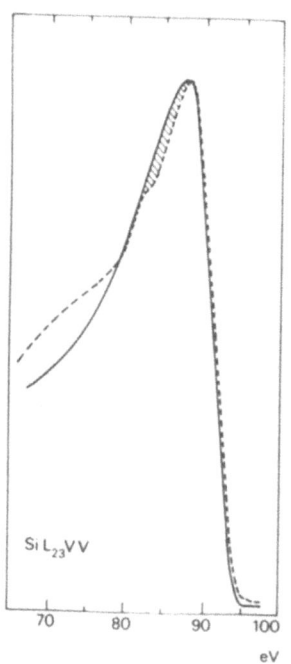

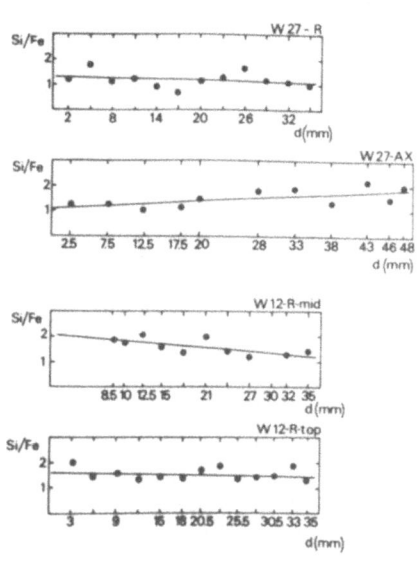

Fig. 8 - Si ($L_{2,3}$VV) Auger lineshapes measured in an iron-rich precipitate (dashed) and in an iron-free region.

Fig. 9 - Si/Fe ratios from XR microanalysis of the different samples examined.

ADVANCED SLICING TECHNIQUES

P. G. WERNER
Laboratory for Industrial Production Techniques
University of Bremen

Summary

Only slicing techniques with a definite potential for increased removal rates, high yields, small kerf losses, good work surface quality, as well as sufficient feasibility for automatic operation and/or in-process control capacity will be applied in future on a larger scale in the mass production of silicon wafers for solar power applications. By means of a comparative investigation, covering fundamentals of machines and tooling, mechanics of material removal, applied and projected removal rates, as well as tool wear and related production cost, alternative slicing techniques will be discussed and evaluated with regard to their feasibility for high-efficiency silicon wafer slicing. Beside of actual advances in ID-sawing, the paper stresses high-efficiency slurry sawing in detail, covering advances recently achieved in Europe. Improved operational concepts and pertinent slicing results will be presented proving that cutting rates of up to $2\,cm^2$/min/blade are obtainable. As a result of this, an assessment will be executed on future advances in silicon slicing by means of high-efficiency techniques, based on possible reductions of add-on cost per square meter of wafer surface cut.

1. INTRODUCTION

Silicon is a material of considerable hardness and rigidity, and as a consequence machining and slicing by conventional methods are difficult tasks. Therefor special processes have been developed, such as internal-diameter (ID) sawing and multi-blade slurry (MBS) sawing. For low cost mass production of solar silicon wafer, these manufacturing problems are even greater, because of high demands regarding minimum wafer thickness and kerf loss, high production rates, and yields near 100%. Up till now, no practical solutions are generally available yet, and therefore slicing of silicon is still to be regarded one of the main abstacles on the way to low-cost photovoltaic energy.

In this urgent situation, many attemps have been made to increase cutting rates and yields, mainly by modification of existing slicing processes. From these investments, two processes will most probably come through showing a high potential for the required increase of productivity. One is ID sawing with larger saw blades and in-process control capabilities, and the other one is high-efficiency slurry sawing (HESS) with increased cutting speed and tool load.

Because of the high technological and operational reliability, conventional ID sawing is still the dominant slicing method for solar silicon wafer production. As long as no other process with better productivity and equivalent reliability is available, industry is bound to live with the relatively low production rates and the rather labor-intensive technology of ID-sawing.

However, alternative silicon slicing processes with high-efficiency and low-cost capabilities are in the state of practical testing and will most probably be introduced in near future.

2. SURVEY ON SLICING METHODS FOR SILICON MATERIAL

The technical state of the art in silicon slicing is represented by only five different methods. Three of them are used in practice for solar wafer production, while the remaining two are still in the state of laboratory testing:

- Internal diameter sawing (ID),
- Conventional multi-blade slurry sawing (MBS),
- Multi-wire slurry sawing (MWS),
- High-efficiency multi-blade slurry sawing (HESS),
- Fixed abrasive multi-wire cutting (FAST).

Figure 1 shows a conventional ID-saw (type STC-16) with a maximum saw blade size of 8" internal diameter, a horizontal work feed mechanism and a vertically rotating saw blade. The silicon ingot is fed into the saw blade according to wafer thickness and kerf loss, and with each plunge cut a single wafer is produced. Modern types of ID sawing machines feature integrated in-process control systems to check cutting conditions, tool wear, and tool dynamics. This and other features, like multi-ingot cutting with larger blades, increased infeed rates, and reduction of labor by automatic operation, will be utilized in near future to increase production rates /1/.

In contrast to the former illustration, Figure 2 shows a conventional Meyer + Burger multi-blade slurry (MBS) saw, type GS-1. By means of a lapping suspension, which is brought into action in the contact zone between saw blades and work surface, several hundred wafers are cut at the same

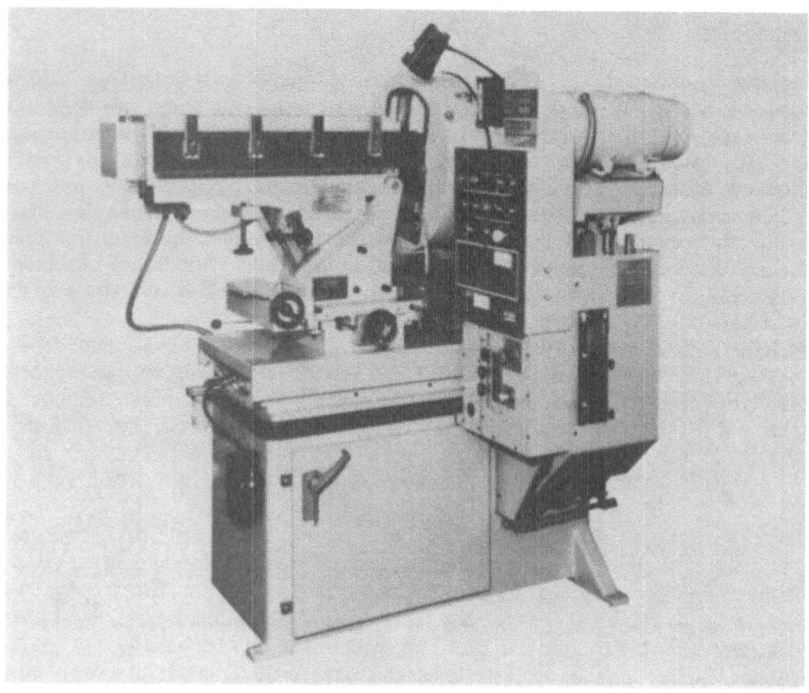

Figure 1: ID Sawing Machine for Silicon Slicing; Model STC-16

Figure 2: Conventional Multi-Blade Slurry Saw, Model Meyer + Burger, GS-1

time. However, the infeed rate is a few hundred times lower than in the case of ID sawing. Consequently, the total production of wafers per unit of time is of the same order of magnitude for both these conventional slicing processes with a slight advantage for ID sawing. In high-efficiency slurry sawing (HESS) on the other hand, infeed rates have been increased in laboratory tests to an extend that up to 270 wafers of $10 \times 10\ cm^2$ can be produced per hour. This is 8 to 10 times more than in conventional ID sawing.

A recently published survey /1/ provides detailed information on the state of technology and machine tool design for both, ID and MBS sawing,

Slicing Methods	Mode of Applic.	Production Rate $\left[\frac{Wafer}{h}\right]$	Yield [%]	Surface Quality -	Technical Reliability -
ID-Sawing	convent. advanced	average high	high very high	average good	high very high
Multi-Blade Slurry Sawing	convent. advanced	low average	average high	average good	good fair
Wire Slurry Sawing	convent.	low	average	good	fair
Fixed Abrasive Multi-Wire Cutting	testing state	average	average	good	low

Table I: Comperative Evaluation of Operational Conditions of Alternative Silicon Slicing Methods

as well as multi-wire slurry sawing (MWS). From here and from other pertinent articles /2, 3, 4/, a qualitative evaluation on the operational and economic feasibility of alternative slicing methods can be derived, as presented in Table I. It indicates why the advanced version of the conventional ID sawing process as available today for immediate practical application are superior to all other conventional methods. Conventional multi-blade slurry sawing also shows acceptable results, while wire slurry processes are assigned with the lowest ratings of all practically applied conventional slicing methods. Cutting of silicon with fixed abrasive multi-wire packs is hampered by similar disadvantages, mainly caused by the high flexibility of the wire tools and the related low engagement forces between tool and work piece. So far, fixed abrasive wire slicing has only been carried out under laboratory test conditions and is referred to in publications as fixed abrasive slicing technique (FAST). Its chances to become a competive silicon slicing method are not regarded to be high /4/.

3. TECHNOLOGICAL AND ECONOMIC EVALUATION OF PRACTICALLY APPLIED SLICING TECHNIQUES

For a better understanding and a more reliable quantitave evaluation of the economics of alternative slicing methods, first the fundamentals of the respective removal processes will be described, and from here criteria will be derived to evaluate tool wear, tool life, and work surface quality. Secondly, economic considerations will be applied to evaluate actual activities regarding the development and practical implementation of high-efficiency silicon slicing methods.

3.1. MATERIAL REMOVAL PROCESS AND TOOL WEAR

In the case of ID sawing, silicon stock is removed by diamond grains which are firmly bonded in a nickle base applied on the internal tool rim by electroplating. Thus, the cutting action is in essence a plunge grinding

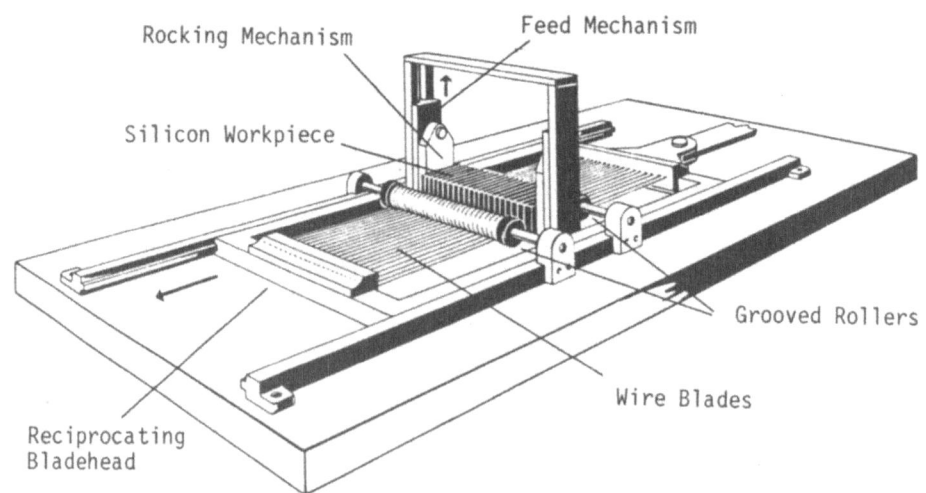

Figure 3: Principle of Fixed-Abrasive Wire Cutting (FAST) /5/

operation, characterized by high cutting speeds of v_S = 15 - 20 m/s and a rather large contact zone between tool and workpiece. This in turn results in comparatively high total contact forces and single cutting edge forces as well which, on the other side, represent the most important prerequisite for an efficient material removal process with diamond abrasives. Another vital condition is a smoothly running blade with an appropriate state of pretensioning, enabling the very thin tool body to cope with the high cutting forces. This interactive combination of technological and operational conditions in ID-sawing results in specific removal rates per tool of Z' = 50 - 80 cm^2/min, which is up to 1000 times higher than in wire sawing with diamond-clad tools (FAST). Also with regard to the so-called wear ratio G (i.e. work material removed / abrasive layer lost by wear), ID sawing with G $\simeq$ 500 000 cm^2/cm^2 is superior to the FAST-process with G $\simeq$ 7 500 cm^2/cm^2. In spite of high specific removal rates, ID sawing as well as other slicing processes require a moderate application of coolant, only, when cutting silicon. This is due to the low level of thermal effects when silicon material is removed by breaking-off small irregular particles at a high frequency by the compressive engagement of a larger number of diamond cutting edges.

As mentioned before, a similar removal process also using electroplated diamond abrasives bonded to the surface of a wire-type of tool is ef-

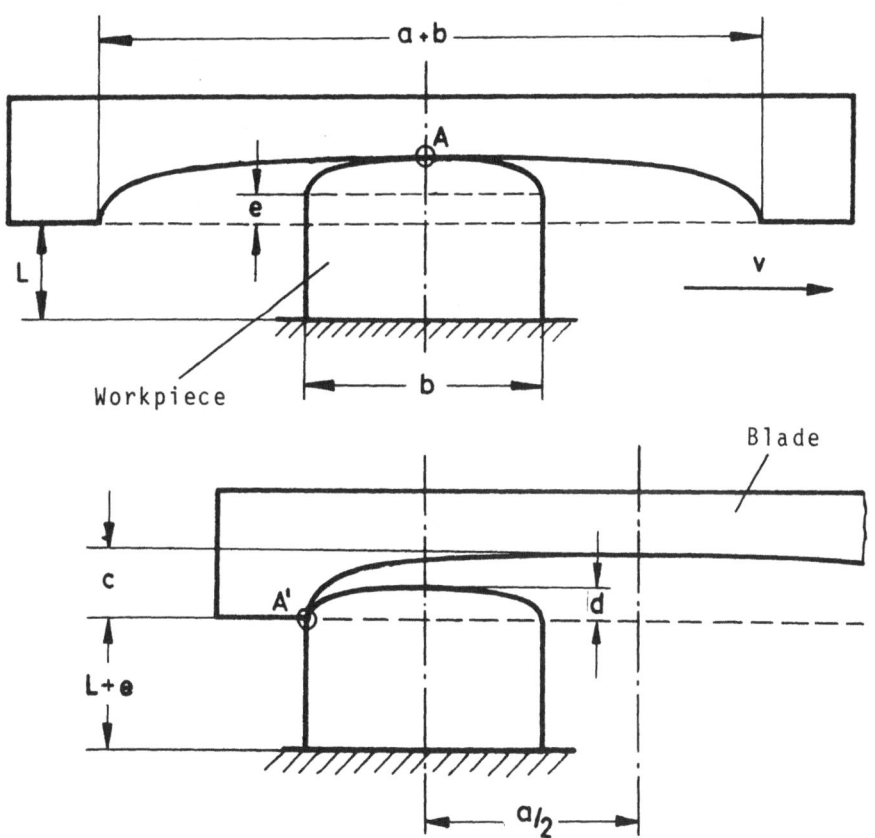

Figure 4: Kinematic and Geometrical Interrelation between Tool and Workpiece in Slurry Sawing /4, 6/

fective in the fixed abrasive multi-wire sawing process (FAST). **Figure 3** shows the operational principle of this method, the main advantage of which is the relative low mass of the moving tool frame and the relative high cutting speed of v_s = 100 - 150 m/min. However, in spite of the pretensioning of the wires and the two grooved guidance rollers positioned as near as possible to the work piece, the effective tool load is rather limited due to the elastic deformation of the wires perpendicular to the work surface. The resulting specific normal forces do hardly exceed 5 g per 1 cm of contact length, while in the case of ID sawing the respective values are 50 to 100 times higher. Thus, the necessary grooving action with a brittle type of fracture to seperate small particles from the work surface cannot take place in fixed abrasive wire sawing with suitable efficiency.

A very different removal process is effective in the so-called slurry sawing processes. Again, these methods do not have very much in common with a real sawing process, especially with regard to their specific stock removal process. Actually, they represent specially modified lapping processes, characterized by loose abrasive particles which are conveyed into the contact zone between tool and work piece by means of a lapping suspension. As effective in all lapping processes /10/ and supported by the specific kinematical and mechanical conditions of slurry sawing, the lapping grains are forced to roll through the contact zone, effecting the work surface with an extremely high impact frequency, which can reach a few hundredthousand impacts per second and per square millimeter of the work surface.

In contrast to the conventional operational model, slurry sawing with a pack of reciprocating blades is characterized by a point-type of contact between tool and work surface (**Figure 4**), caused by the formation of a quasi-elliptic wear contour in the tool blade /6/. This contour is automatically transduced into the workpiece, forming a congruent profile there with a greater overall curvature. As a consequence, the actual contact pressure in the limited vicinity of the contact point A is amplified considerably, and the related removal process becomes more effective at the same time. While the tool reciprocates over the workpiece with a stroke length "a", the contact point reciprocates too, but, in the opposite direction,

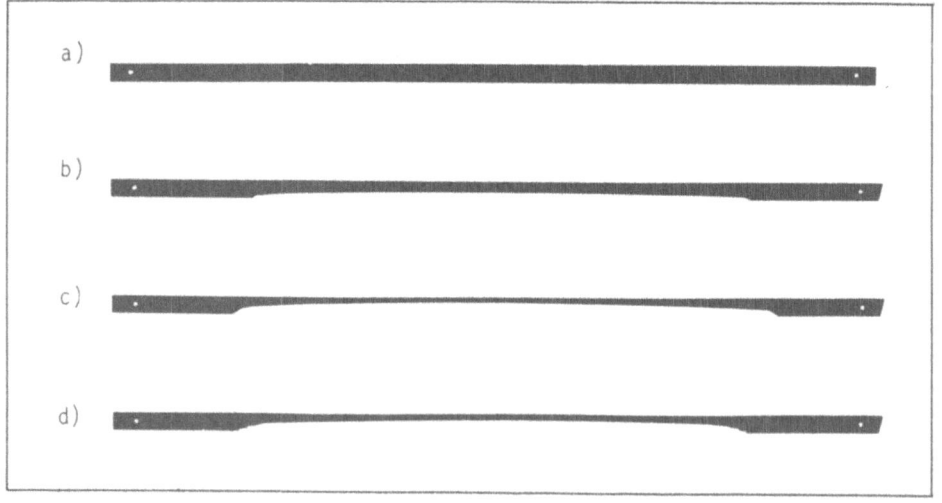

Figure 5: Representation of Wear Contours on Slurry Sawing Blades

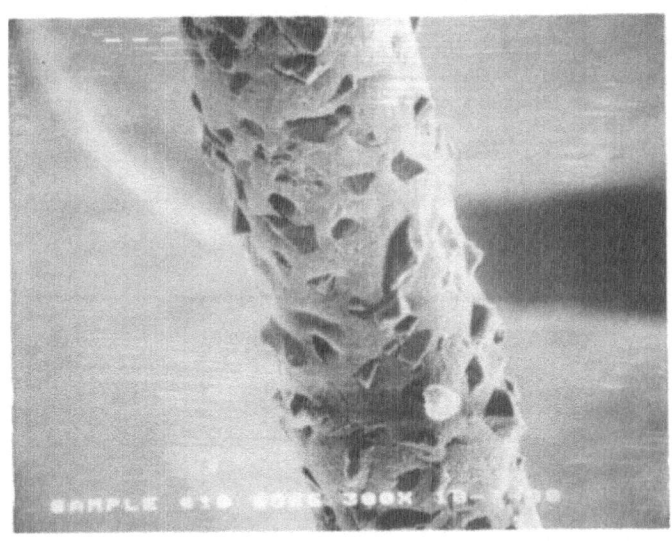

Figure 6: View on Diamond-Clad Wire Tool Used in FAST-Slicing a Total Silicon Wafer Surface of 280 cm²

thus covering the total work surface with each individual stroke. This unique interaction is also responsible for a major disturbancy caused by an enforced vertical movement of the blades when sliding over the work profile. The maximum 'e' is reached at the end of each stroke. It grows in magnitude while the tool wear proceeds and causes severe dynamic shocks which can result in wafer breakage.

Beside of the material removal from the work surface, the rolling grits also have a removal effect on the tool blades made of conventional steel. Due to its elastic and/or ductile behaviour, the wear volume is 15 to 20 times smaller than the stock removed from the workpiece, thus enabling a 6 mm blade to cut through a 10 x 10 cm² ingot once. In Figure 5 respective contours of worn slurry saw blades are shown, generated by slicing two 5 x 5 cm² ingots simultaneously. The basic elliptical wear contour is clearly visible, while at both ends a short slope is noticed (b) and (c), related to an oblique positioning of the two silicon blocks. This is done to shorten the effective length of the tool profile continuously versus its infeed, thus counteracting the vertical stroke impacts. Example (d) shows little steps at both the profile ends, caused by intermediate reductions of the nominal stroke length 'a' (Figure 4).

In the case of wire slurry sawing, this specific work-tool interaction does not occur, as the elastic wires are always in full contact with the generated work surface, which shows a sligth convex curvature. In order to compensate in part for the ineffectively low contact pressure, the workpiece is rocked, thus reducing the work-tool contact length and increasing the local cutting forces. Figure 6 shows a magnified section of a diamond-clad wire as used in the FAST process. After cutting three times through a 10^{ϕ} cm silicon ingot, there is no trace of any wear on the imbedded diamond grits. Nontheless the wire has reached the end of its tool life after a total cut of 880 cm², because it began to deviate laterally from the required straight cut, probably caused by uneven wear of a smaller fraction of the engaged cutting edges.

3.2. ECONOMIC ASSESSMENT

The total production cost for solar silicon wafers are composed of different cost elements, related to material, tool, machine and labor, and other minor expenses. This is demonstrated in Table II by the example of a conventional ID sawing process /2/, showing that the dominant cost factor is the material (83.1%), based on a price of 400 DM/kg. From the remaining add-on costs, machine and labor has the strongest influence with 13.7%, while the tool cost are of secondary importance.

Cost Elements	Cost per Wafer [sfr]	Relative Cost Factor [%]
Material	4.67	83.1
Tool	0.12	2.1
Machine and Labor	0.77	13.7
Other	0.06	1.1
Total	5.62	100.0

Table II: Cost Elements in Silicon Wafer Production by ID-Sawing /2/

These relations are valid in principle for other slicing processes, too. Therefore minimal production cost are likely to be achieved with slicing methods which provide low values for wafer thickness and kerf loss, and which at the same time produce a maximum number of wafers per unit of time.

Table III indicates that ID sawing meets these criteria quite well, while a slightly lower performance is stated for conventional multi-blade slurry sawing (MBS). For multi-wire slurry sawing (MWS) very high values for the total wafer surface produced per unit of ingot volume are recorded, but at the same time the tooling cost and the technical process reliability are rather unfavourable compared to competitive silicing processes /7/.

Very attractive performance values can, on the other hand, be attributed to high-efficiency multi-blade slurry sawing (HESS). Pertinent machines with increased cutting speeds and blade loads have been built and tested with very good results, as demonstrated in detail in the following chapter.

A seperate evaluation of the add-on production cost /2, 8, 9, 10/ also shows a significant advantage of the novel HESS process (Figure 7). Optimized test runs carried out under production conditions with an improved HESS-machine resulted in add-on costs of less than 15 \$/m^2, unmatched so far by any other slicing technique. Normal add-on production cost for ID and MBS slicing are in the range of 40 - 90 \$/m^2. Even higher values are reported for mulitiple-wire slurry sawing processes.

		Surface Speed [m/min]	Cutting Force [N]		Surface Cut [m²]		Infeed Rate [cm/min]	Cutting Rate [cm²/min]		Wafers per h [1/h]	Wafers per cm [1/cm]	Yield [%]	Surface Damage [µm]
			per tool	total	per tool	compl. pack		per tool	compl. pack				
I D (1 blade)	aver.	1 000	28	–	30	–	5	50	–	30	14	90	10
	max.	1 200	56	–	40	–	8	80	–	48	18	95	10
MBS (300 blades)	aver.	30	0.8	240	0.01	3.0	0.008	0.08	24	15	16	85	10
	max.	50	1.5	450	0.01	3.0	0.015	0.15	45	27	21	90	8
MBS/HESS (300 blades)	aver.	80	1.5	450	0.01	3.0	0.120	1.20	360	216	16	90	10
	max.	120	2.0	600	0.01	3.0	0.180	1.50	450	270	21	95	8
MWS (150 wires)	aver.	75	1.0	150	–	1.2	0.007	–	10	6	22	95	5
	max.	85	1.0	150	–	1.2	0.020	–	30	18	25	98	3
MWF/FAST (250 wires)	aver.	100	0.4	100	0.03	7.5	0.012	0.12	30	18	22	90	5
	max.	140	0.4	100	0.05	12.5	0.016	0.16	40	24	25	96	3

Table III: Comparative Survey on Operational Criteria of Different Silicon Slicing Techniques

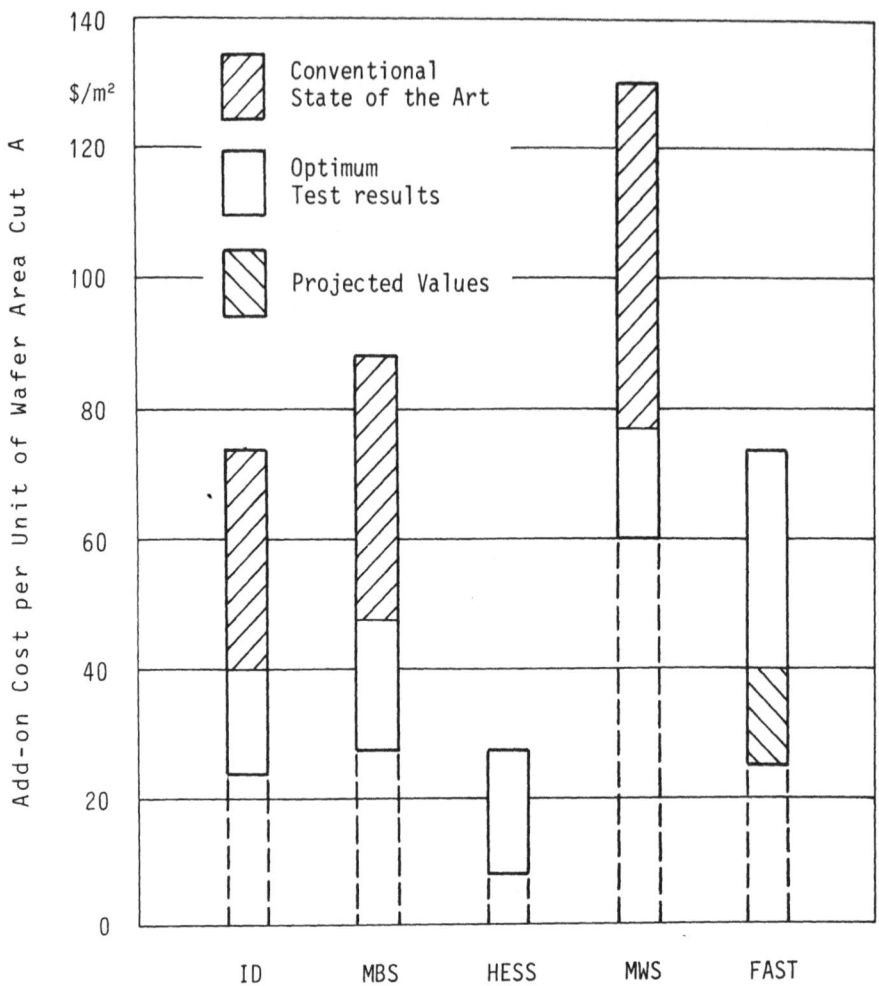

Figure 7: Applied and Projected Ranges of Add-on Cost for Different Silicon Slicing Processes

4. TECHNOLOGY OF HIGH EFFICIENCY SLURRY SAWING (HESS)

Basically there are two ways to increase the production rate of multi-blade slurry saws:

a) Increase of cutting blade number,

b) Increase of cutting rate.

Both these measures have been investigated intensively during the last few years. One definite result of these studies is that blade packs with more than 300 blades become inpractical. This is due to the enormous tensioning forces and the intensified problems with tool handling an tool accuracy, as experieced in a major development task /11/. Actually, a real productivity increase can hardly be achieved by building the volumetric capacity of three machines into one and paying for it by an overproportional increase of costs for machine, labor and tooling. A more resonable way to boost productivity and to lower production cost at the same time is to increase the infeed rate of conventionally sized blade packs by means of higher cutting speeds, reduced stroke and blade length and accordingly increased blade loads, as recently demonstrated with a new type of high-efficiency slurry sawing machines /12, 13/.

In **Figure 9** the prototype of such a machine is shown. In spite of reciprocating cutting speeds of up to $\bar{v}$ = 120 m/min and increased cutting forces, the machine shows an extreme dynamic stability. The cutting blades are packend and tensioned in a rather conventional frame and the work feed is actuated by a force controlled stepping motor. The machine shown was originally built by Meyer + Burger AG in a joint development task together with Heliotronic GmbH. In the meantime both companies are continuing to develop their own proprietary systems.

Reported slicing results achieved in production-oriented laboratory tests prove that the HESS-system provides a cutting potential which is much higher than experienced in conventional MBS-processes /14/ (**Figure 8**). With

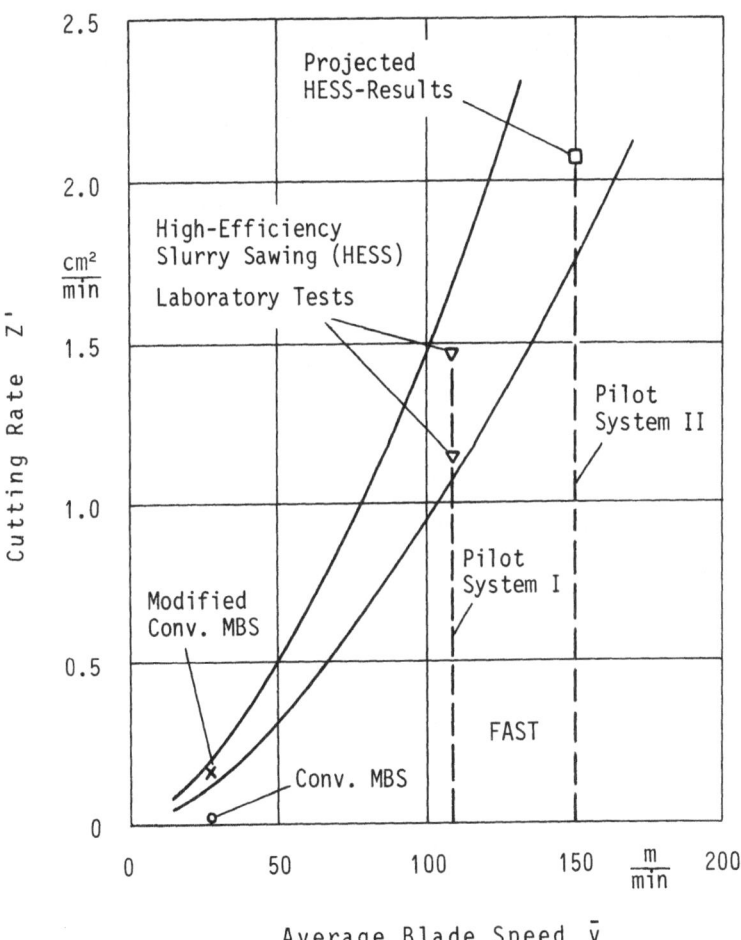

Figure 8: Cutting Rates versus Blade Speed in Conventional and High-Efficiency Sawing of Silicon /14/

Figure 9: High-Efficiency Slurry Saw for Solar Silicon Wafer Production /12/

a removal rate of 1,5 cm²/min for example, a 10 x 10 cm² silicon ingot can be cut in 66 minutes, producing 270 wafers or 2,7 m² in an hour. The progressive character of the removal rate curve versus blade speed results from the superimposition of two effects, both having a proportionally increasing influence on the cutting rate Z': The first one is the blade speed $\bar{v}$ itself, the second one is assigned to the incrasing blade load. Under optimal working conditions, these rates result in total add-on cost of 10 - 15 \$/m².

5. CONCLUSIONS

From the technological and economic comparison of alternative silicon slicing methods, two trends can be extracted which most probably will determine the future state of application:

a) Regarding future mass production of solar silicon wafers, ID sawing with diamond-plated saw blades has an effective advantage against all other processes based on its well introduced and accepted machine and tooling technology. With increasing blade diameters (up to 27 inches and eventually more) the output will be increased by cutting larger ingots or cutting more than one ingot at the same time. In-process control systems, as already available with the most advanced ID-systems, will increase yield and cutting quality considerably and will also reduce labor cost. All that will result in add-on costs of about 20 \$/m² and will certainly secure a high frequency of application for the ID-process in future mass production of solar silicon wafers.

b) High-efficiency slurry sawing (HESS) is the second slicing technique, which in near future will have a definite chance to play a major roll in the production of solar silicon wafers. From all slicing techniques it provides the highest production rates and the lowest add-on costs. Before this method, which actually is undergoing an intensive phase of technological testing, will be introduced into practice at a significant scale, major efforts have to be spent to guarantee the required technical reliability. However, because of the striking economic advantages, and the proven technological feasibility, high-efficiency slurry sawing (HESS) is expected to become an attractive alternative for ID-sawing.

6. REFERENCES

/1/ Iscoff, R., Crystal Slicing Equipment Directions; Semiconductor International, February 1982

/2/ Stauffer, A., Wirtschaftlichkeitsberechnungen zum Sägen von Silizium auf Innentrennsägen; Industrie Diamanten Rundschau (IDR), Nr. 3, 1981

/3/ Wolf, M., Comparison of Various Silicon Sawing Methods; Proceeding of the Low-Cost Solar Array Wafering Workshop, Jet Propulsion Laboratory, Pasadena, California, June 8-10, 1981

/4/ Werner, G., Minke E., Burkhardt, W., Grundlagen der Prozeßtechnologie des Trennläppens; Abschlußbericht zu Forschungsvorhaben; Bremer Institut für Angewandte Strahltechnik (BIAS), Bremen, Febr. 1982

/5/ Schmidt, F., Smith, M.B., Khattak, C.P., Overview of a New Slicing Method - Fixed Abrasive Slicing Technique (FAST). Proceedings of the Low-Cost Solar Array Wafering Workshop, Jet Propulsion Laboratory, Pasadena, California, June 8-10, 1981

/6/ Werner, G., Kinematical and Mechanical Aspects of Wafer Slicing; Proceedings of the Low-Cost Solar Arry Wafering Workshop, Jet Propulsion Laboratory, Pasadena, California, June 8-10, 1981

/7/ Goldmann, H., Wolf, M., Quarterly Report DOE/JPL - 954796 - 78/4, University of Pensylvania, Nov. 1978

/8/ Rand, T., Lin, J.K., Fiegl, G., Economics of Ingot Slicing with an ID-Saw for Low-Cost Solar Cells; 15th IEEE Photovoltaic Specialists Conference, Kissimmee, Florida, May 12, 1981

/9/ Mokashi, A.R., Sensitivity Analysis of Add-On Price Estimate for Selected Silicon Wafering Techniques; Proceedings of the Low-Cost Solar Array Wafering Workshop, Jet Propulsion Laboratory, Pasadena, California, June 8-10, 1981

/10/ Martin, K., Neuere Erkenntnisse über den Werkstoffabtrag beim Läppen; Fachberichte für Oberflächentechnik, Bd. 10, Nr. 6, 1972

/11/ Holden, S.C., Fleming, J.R., Slicing of Silicon into Sheet Material; Series of Quarterly Reports, ERDA/JPL Contract No. 954374-77/5, 1977-79

/12/ Sirtl, E., Regler, D., Moritz, A., Process for Multiple Lap Cutting of Solid Materials; USA Patent, No. 4 187 827, Febr. 1980

/13/ Stauffer, A., Personal Information, Zürich, Oct. 1980

/14/ Dietl, J., Helmreich, D., Sirtl, E., Solar Silicon; in Crystals - Growth, Properties and Applications, Vol. 5, Springer-Verlag, Berlin, 1981

CONTINUOUS GROWTH OF THIN POLYSILICON SHEETS

ON A TEMPORARY CARBON SHAPER

BY THE R.A.D. PROCESS*

C. TEXIER-HERVO, M. MAUTREF, C. BELOUET, E. KERRAND

LABORATOIRES DE MARCOUSSIS, Centre de Recherches de la C.G.E.
Route de Nozay, 91460 MARCOUSSIS - FRANCE

SUMMARY

The current status of the RAD process is presented. Growth results obtained with a new puller equipped with a continuous melt replenishment are described. The definite progress due to both the chlorine purification and the burn-off of the carbon shaper is discussed in relation to the process capabilities. It is illustrated by the achievement of AM1 conversion efficiencies in the 9 to 11 % range (A.R. coated cells) with the best results around 12 %.

1. INTRODUCTION

The RAD process has been developed for the continuous growth of silicon sheets in view of achieving low-cost solar cells. The use of a carbon ribbon as a shaper of the freezing meniscus is a unique feature of the process, which potentially results in reliable growth conditions well suited for a continuous operation, a minimal silicon consumption and a high throughput rate [1]. However, major drawbacks were also reported to be associated with the carbon ribbon [2]; they relate mainly to i) the contamination of the melt, ii) the thermo-elastic strains in the silicon layers during cooling and iii) the limitation of the solar cell performances due to the poor electrical contact at the substrate side. The thermo-elastic strains were minimized by adjusting the carbon ribbon relevant characteristics ; a definite progress regarding the two other items came along with the chlorine purification of the carbon ribbon and its burn-off after growth.

This paper reports on the current status of the RAD process, the separation of the silicon layers from the carbon shaper with their consequences and the performances of the solar cells made from the thin, self-supporting sheets thus obtained.

2. RIBBON GROWTH

The results below refer to layers grown with a new puller equipped with a melt replenishment line for the continuous growth of ribbons 5 cm in width, figs.1,2. Its basic concept is derived from that of a previous puller [1]. Heat is supplied to the melt by RF induction (150 KHz) on a graphite susceptor ; the crucible rests on a pile of pyrolytic graphite plates which ensure

---

* This work was supported by the Commission of European Communities (CEC) and the Commissariat à l'Energie Solaire (COMES).

good heat conduction in the horizontal planes and efficient insulation in the vertical direction. The growth chamber is protected against air contamination by an argon stream. The melt replenishment line in this experimental puller is based on a controllable rate of injection of cubic silicon pellets 5 mm in edge. The pellets are progressively moved on a silica ramp where they are heated up to about 1000 °C prior to their fall into the melt. The injection area is restricted to one end of the rectangular crucible [3] and is partly isolated from the growth area by a quartz barrier which also damps oscillations of the melt surface. The pellet feeding rate is driven by a feed-back loop which maintains the melt height constant.

The best achievements obtained with this equipment in its early operation stage are given in table 1. The total thickness of the opposite silicon layers grown in the vicinity of the smooth-to-dendritic transition region was about 200 µm for a pulling rate of 8 cm/mn ; this finding closely corroborates theoretical predictions in fig.3 based on past experiments using the previous puller. Fig.3 also shows that the goal of a pulling rate of 10 cm/mn with a silicon thickness of 50 µm for each layer (noted 1 and 3) may be achieved with a carbon shaper 200 µm thick (curve B). This preliminary phase is continued with the investigation of i) higher values of the pulling rate and ii) longer growth periods of time.

3. CARBON SHAPER

The carbon shaper consists of a graphite ribbon (Papyex)* tens of meters in length, coated by a thin layer of lamellar pyrocarbon. Previous studies established that the erosion of the pyrocarbon coating by molten silicon in the actual experimental conditions could be ascribed to the presence of the competing catalytic pyrocarbon variety, generated by metallic impurities coming from the graphite ribbon [4]. The purification of the "Papyex" material at 2800 °C in an ambient containing chlorine prior to the deposition of pyrocarbon drastically reduced the concentration of these impurities and concomittantly, the catalytic variety formation. Consequently, the reactivity with molten silicon was almost suppressed and the contamination of the silicon melt and layers was considerably reduced [5]. Residual imperfections in the form of impurity clusters in which Mo, Ca, Ti, Fe and S were found, were identified by SEM observation of the surface of the pyrocarbon coating, fig.4. The characteristics of the carbon shaper are given in table 2.

4. CARBON SHAPER BURN-OFF

The separation of the opposite silicon layers relies upon the burn-off of the carbon ribbon at high temperatures in an oxygen ambient, followed by a mechanical (or laser) cutting of the ribbon edges. The experiments were conducted with ribbons 20 cm in length with carbon exposed to the ambient at the two ribbon ends only. A detailed investigation of the influence of the parameters in the burn-off step, e.g. the composition of the ambient, the temperature and duration of burn-off, the heating and cooling rate of the Si layers, on the electrical performances of the final devices has been initiated. The results in this paper refer to a burn-off at 1020 °C in dry oxygen for 1 hour with fast heating and cooling rates (duration, about 5 mn). Si layers 60 to 150 µm thick were separated in this way with a high degree of success, thus establishing the mechanical feasibility of the process.

The potential consequences of the burn-off process are listed in table 3.

---

* Papyex is the trade name of the graphite ribbon manufactured by LE CARBONE LORRAINE (France).

As compared with the conventional RAD process, it directly results in a reduction of the effective cost of the carbon shaper and a higher throughput rate. Indirectly, the growth with opposite Si layers of equal thicknesses (symmetrical growth) allows thicknesses as low as 50 µm or less [5], which results in i) a faster pulling rate (or higher throughput rate), fig.3, ii) a minimal silicon consumption and iii) a thin film solar cell technology with eventually improved conversion efficiencies [6]. Finally, this process delivers self-sustained layers, thus obviating the processing of a multi-layered structure during the cell fabrication.

5. SOLAR CELL PERFORMANCES

Solar cells were made from the silicon layers (1 to 2 Ω.cm, p type) obtained as described above (see fabrication chart in table 4) ; the layers were first submitted to an HF etch to remove the oxide of the front face and a chemical lapping by an HF, HNO_3, CH_3COOH mixture to eliminate the thin and porous silicon carbide layer at the back side.

n^+/p homojunctions were made by $POCl_3$ diffusion at 850 °C (junction depth about 0,5 µm) ; a p^+ contact was formed at the back side by the annealing of evaporated aluminium and the front contact was a Ti, Pd, Ag evaporated grid.

Typical performances of solar cells 4 cm^2 in size thus prepared are presented in table 5. Most of the cells exhibit AM1 conversion efficiencies (η) in the 9-11 % range (active area), the best η value obtained was 12.4 % with the following characteristics : photocurrent density 30.9 mA/cm^2, open-circuit voltage 552 mV and fill-factor 0.73. These results show a considerable improvement as compared with those of solar cells made on layers not separated from the carbon shaper [3]. The relatively large spread in the cell characteristics appears to be related mainly to large variations of the density of trapping centers ; temperature annealings not described here were found to be very effective in reducing this spread.

6. CONCLUSIONS

The continuous growth of RAD ribbons 5 cm in width at pulling rates up to 8 cm/mn was demonstrated using a new puller equipped with a melt replenishment line. Two major drawbacks of the RAD process originating from the carbon shaper, namely the contamination of the melt and the degradation of the solar cell performances, were overcome. The contamination of the melt was considerably reduced by the use of carbon shapers made from graphite ribbons purified in chlorine at high temperatures, and the solar cell back-contact problem was obviated by the burn-off of the carbon shaper after growth. Solar cells ($n^+/p/p^+$ structures) made from the self-supporting layers thus obtained exhibit AM1 conversion efficiencies η in the 9-11 % range with the best η values above 12 % (A.R. coated, active area, 4 cm^2 in size). The optimization of the burn-off step in relation to its influence on the solar cell performances is now in progress.

REFERENCES

[1] C.Belouet, Novel silicon growth methods, ESC Meeting 80-5, 195-212.
[2] C.Belouet, J.Schneider, C.Belin, C.Texier, R.Martres, Proc.XIVth IEEE PVSC (1980) 49.
[3] C.Belouet, Submitted for publication to J. of Crystal Growth.
[4] J.Goma, M.Oberlin, A.Oberlin, J.Schneider & C.Belouet, High temperatures - High pressures 13 (1981) 263.
[5] J.Revel, N.Deschamps, J.Deville, C.Texier-Hervo, C.Belouet, this Conf.

[6] H.J.Hovel, Semiconductors and semimetals, Vol.II, Academic Press, p.100.

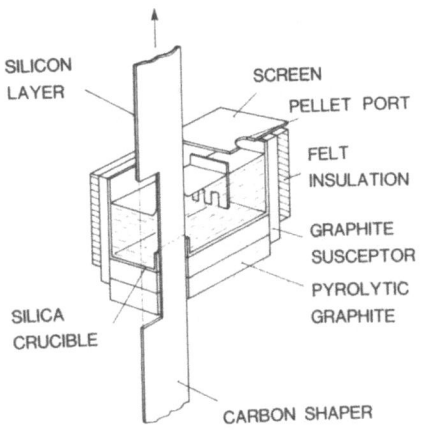

Fig.1 - Schematic of the RAD growth process.

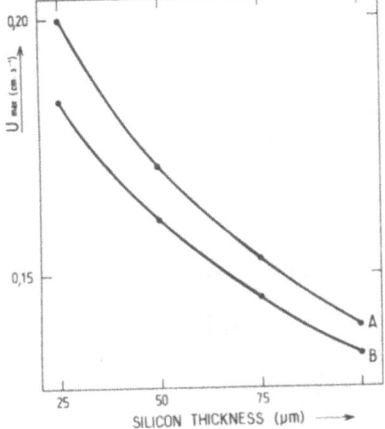

Fig.3 - Pulling rate Umax vs. silicon thickness. Shaper thickness : 150 μm (a) and 200 μm (b), effective ambient : 500 K, heat convection coefficient h = 0.01 W.cm^{-2}.°C^{-1}.

Fig. 2 - Full view of the puller with continuous melt replenishment.

Fig.4 - SEM micrograph of pyrocarbon surface, a precipitate is indicated by an arrow.

Table 1 : Growth status (new puller).

SLOT CLEARANCE	06	MM
MELT HEIGHT	2	CM
RIBBON WIDTH	5	CM
PULLING RATE	8	CM/MN
SILICON THICKNESS (Total)	200	μM
PULLING LENGTH	15	M

Table 2 : Carbon shaper characteristics.

PAPYEX BASE		
LENGTH	50	M
WIDTH	5	CM
THICKNESS	200	μM
DENSITY	07 G/CM^3	
YOUNG'S MODULUS	< 30 daN/MM^2	
PYROCARBON COATING		
VARIETY	LAMELLAR TYPE	
THICKNESS	2 TO 4	μM

Table 3 : Consequences of the burn-off step on the RAD process capabilities.

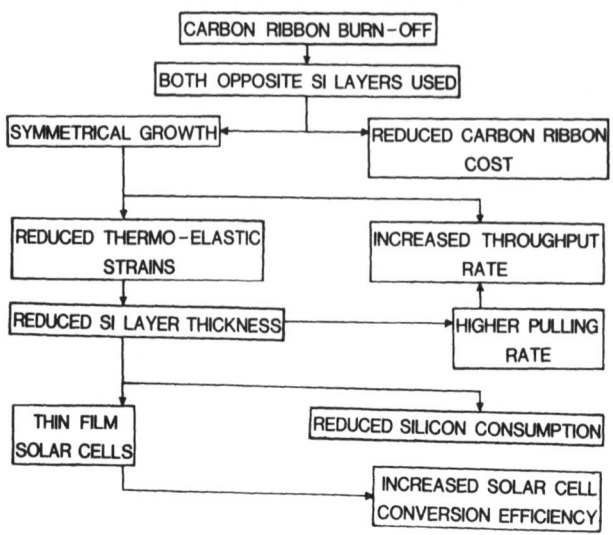

Table 4 : Solar cell fabrication chart.

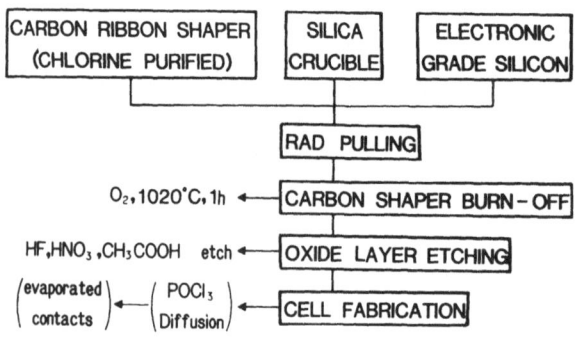

Table 5 : Typical solar cell performances.

SOLAR CELL : TYPICAL RESULTS	
(AM$_1$ ILLUMINATION A.R COATED)	
SIZE	4 CM^2
PHOTOCURRENT DENSITY	24 - 29 mA/cm^2
OPEN-CIRCUIT VOLTAGE	525 - 555 mV
FILL FACTOR	0.69 - 0.73
CONVERSION EFFICIENCY	9 - 11 %
BEST CELL	12.4 %

POSSIBILITIES OF ION IMPLANTATION IN SILICON
SOLAR CELL MANUFACTURING

P. SIFFERT

Laboratoire de Physique et Applications des Semiconducteurs
(PHASE)

CENTRE DE RECHERCHES NUCLEAIRES
67037 STRASBOURG-Cedex (France)

Abstract

In this review paper, the various methods which have been developped for introducing the dopants for making the P-N junction of silicon solar cells using ion implantation. The lattice annealing and dopant incorporation has also be considered. The current possibilities of these various methods are critically considered and the most significant results on cells presented.

1. INTRODUCTION : HISTORICAL SURVEY

Since the earlier days of semiconductors, it was recognized that these materials degrade under irradiation. Despite that, some authors attempted to achieve chemical doping by using ion bombardment. The beams existed at that time essentially on magnetic isotope separators. A historical description of these earlier work, starting around 1950 at Bell and Purdue, is given in ref.(1). However, at this time, the doping effect was masked by the radiation damage occuring during the slowing down of the ion. But, in a patent issued 1957 by SHOCKLEY (2) on ion doping, thermal annealing was proposed to restore the crystallinity of the semiconductor.

The possibility to prepare shallow P-N junctions was first mentionned in the literature by ROURKE (3) but without any details concerning manufacturing process and performance. But, this work probably initiated efforts in many nuclear laboratories to prepare particle detectors, the first results being published by ALVAGER (4) and the optimal conditions being soon worked out (5,6) for achieving very low current diodes. In particular, the advantage of using low energy projectiles was recognized.

Starting 1962 in USA (7) solar cells began to be investigated, the first results being presented at the 4th Photovoltaïc Specialists Conference by KING and BURRIL (8), as well as on other meetings (9). In an extensive paper, the same authors (10) proposed quite sophisticated cells by implanting $^{31}P^+$ ions at energies up to 100 keV into 10 Ω.cm Si. Efficiencies up to 11 % (AM0) have been measured. Even some efforts have been done on dendritic webb from Westinghouse, the efficiency reached 9.3 %. A particular doping profile was achieved (Fig. 1) either by tilting the sample during implantation or by changing the projectiles energy. A drift field of 6,000 V/cm resulted from this structure insuring the charge collection from the near surface zone. Even the degradation of the cells under electron bombardment (simulation of space) was considered.

In USSR, GUSEV (11,12) implanted 30 keV $^{31}P^+$ ions into 4 Ω. cm material and reached in 1965 efficiencies of 6-8 %, with poor U.V. response due to incomplete damage annealing.

Despite these promising results and the number of investigations which have been done since that time, no industrial production of cells was initiated and all of them are made by thermal diffusion. This is just the opposite to the situation in integrated circuit manufacturing, where hundreds of machines are in use, handled by non-specialists, a development never expected by the older researchers in the field. There, ion implantation shows decisive advantages over thermal diffusion, which are summarized in Table I. We feel that many of them are still valid here.

This paper is restricted to implanted silicon cells, essentially single or polycrystalline. First, we consider the slowing down of the projectiles in the target, then the various implant procedures which have been used, as well as the annealing procedures. A comparative study of the most significant results constitutes the next chapter. Finally, a discussion over the real possibilities of this doping technique, including the economical point of view is considered.

2. ION SLOWING -DOWN IN SILICON : A RESUME

While in diffusion, the atomic dopants penetrate towards the bulk through an increase of thermal vibrations under the influence of a gradient driving force (energy involved in the eV renge), here the dopants are ionized (except for recoil implantation) and accelerated typically between 5-100 keV, they penetrate the target through their kinetic energy.

The slowing down of these projectiles has been extensively investigated in the literature for both amorphous and crystalline targets. Since this subject is not really in the scope of this paper, we restrict ourself to a few generalities :
- when impinging the solid, the projectiles loose their energy be electronic and nuclear collisions. The latter have a strong influence on the projectile's trajectory, which does no longer remain linear. Since only the junction depth is of interest here, the projected range R_p is the most important parameter (Fig. 2). It has been computed for many situations in the literature (12-14) and the values compared to experience (15). Fig. 3 (16) reports some values for a few dopants.
- nuclear collisions are responsible of the damage which is created, their importance increases with projectile mass and with lower speeds.

3. IMPLANT PROCEDURES

Essentially four different methods have been considered to perform the dopant implantation :

A. CONVENTIONAL ION IMPLANTATION

Fig. 4a gives a schematic view of the equipment : dopant ions are produced in a source starting from a gas or vapour compound containing the dopant atom.

These positive ions are extracted, handled in a correct optics and accelerated before being analyzed either by a magnet or a speed filter in order to get an isotopically pure ion beam. This beam can eventually be further accelerated or reduced in speed. In recent years, such machines have become commercially available ; they are fully automated and the beam currents progressively increased from 1 µA to 1 mA. The typical implant dose is around $1-5 \times 10^{15}$ cm^{-2}, therefore, about 1 m^2 can be produced per hour, a value close to that currently achieved by diffusion (200 wafers/h, 4" in diameter) if handlings are excluded. But, by contrast to diffusion, a gaussian distribution is obtained (see above), its maximum being located at R_p ; however, some deep penetrating low concentration tail is present, extending typically to 1 µm in depth, which is due to channelling (15, 17). But, this tail plays no role for the cell performance, if thermal annealing is used. To achieve low sheet resistance with such a profile it is just possible to increase the dose ; best results are obtained when using low energy projectiles, but commercial machines operate, in general, best above 25 keV. Another possibility is to achieve a flat distribution starting from surface, followed by a sharp cut-off, giving rise to an abrupt P-N junction, just be changing the projectile's energy.

B. ION INCRUSTATION

The equipment needed for mass analysis and beam handling in conventional ion implantation is quite large and expensive. After several years of research on interaction of heavy ions with silicon, we get convinced that mass separation is not really necessary for realizing solar cells and we proposed 1975 (18) the direct use of glow discharge beams issued from a gas including dopant atoms, in which both atomic and molecular ions are implanted simultaneously without any mass selection. At the same conference WICKNER (19), in a "post deadline" paper, proposed a similar approach, but working in the Corona discharge regime. Our set-up was quickly improved (Fig. 4b (20)) to get much more flexibility, in particular, to vary the energy between 2 and 50 keV.

A similar set-up has been recently realized by NIELSEN (21). ITOH (22) uses the same principle but the ions are produced in a microwave source and magnetic fields separate the dopant compounds (PH^+, PH^+_2, PH^+_3) from H^+ and H^+_2. It seems to me that in polycrystalline substrates the presence of H_2 can improve the grain boundaries zones.

The principle having been demonstrated, sources of nearly any size can be realized by using multi-holes extraction structures, as for ion milling systems. Presently, our source has 300 holes and delivers a total ion current of 30 mA, uniformly distributed over more than 30 cm^2. Recently, SIRKIS (23) modified a commercial ion miller to implant large area cells at 2 keV.

The dopant distribution in ion incrustation is no longer gaussian, as shown on Fig. 5, but is a superposition of the various contributions.

C. RECOIL IMPLANTATION

In this procedure, dopants are first deposited at the silicon surface at thicknesses typically 50-200 Å by any technique and then, this film is bombarded by heavy ions, generally rare gases, for convenience reasons. They push the dopants into the target, with a yield which is larger then unity (Fig. 6) i.e. the number of recoiling dopant atoms is larger than the incident beam intensity. The penetration is very shallow (typically 1000 Å) (Fig. 7) before annealing ; however severe annealing problems exist, the small gain in yield is not sufficient to make this procedure attractive for cell manufacturing.

D. IONIZED CLUSTER BEAM DEPOSITION

This procedure has been developped by TAKAGI (26) : material to be deposited is vaporized in a crucible (Fig. 4 c) and is adiabatically expanded when going through the exit nozzle into about $N \simeq 1,000$ loosely coupled atoms. By bombarding this cluster either with a beam of electrons or ions, in the keV, range it becomes single positively charged and can be accelerated (voltage V). When impinging the silicon surface the ionized clusters are broken into individual atoms, each of them having a mean energy qV/N (q : electron charge). No beam transport limitations due to space charge effects exist, but the mean energy in presently available systems is low (1-1000 eV), therefore, the dopants are just deposited on surface, a situation intermediate between ion plating and implantation.

4. ANNEAL PROCEDURES

To achieve the low sheet resistance which is needed for solar cells, all authors implant doses in excess of 10^{15} cm^{-2}. At this dosis the bombarded layer is fully amorphized for all common dopants except boron. In order to anneal the damage and put the dopants in substitutional sites annealing is necessary. Essentially three heat treatment regimes are available today :

A. STEADY STATE UNIFORM WAFER HEATING

Conventional thermal annealing is, in fact, a rather complicated problem, since the speed of regrowth depends strongly on nature of the implanted ion species, their energy, dose and dose rate, temperature reached during implantation...

Quite often, authors refer to the data of CSEPREGI (27) which indicate a regrowth speed for Si$^+$ implantation given by $V = V_o \exp(-\Delta E/kT)$ with $V_o = 1.5 \times 10^{10} \mu$/s. and $\Delta E = 2.3 - 2.5$ eV, depending on crystalline orientation (best with <100> (Fig. 8). However, these data have no real signification here, for the reasons already mentionned above : in particular, for high dopant doses, a strong reduction of regrowth, even an inhibition is possible and even several temperatures steps may not fully restore the crystallinity (28, 29).

In practice, for preparing implanted cells, various kind of annealing procedures have been proposed :

a) CONVENTIONAL FURNACE at least three procedures are used :

- single temperature step under N$_2$ atmosphere. Typically, temperatures between 750-1000°C are used for 30-60', followed by cooling at a slow rate (2°/min). Efficiencies in excess of 15 % (AM1) have been reached for implanted cells (30) and 14 % for incrustation (31).

- three temperature steps, under N$_2$ atmosphere. The initial 550°C (2 h) causes, following the authors (32), an effective epitaxial regrowth of the damage lattice, the second at 850°C (15') activates the dopants, the last step at 550°C (2h) increases the minority carrier lifetime (32, 33). Efficiencies as high as 16.5 % (AM1), with mean values centered at 15 % for 3" cells have been measured on single crystalline material.

- four temperature steps performed under O$_2$ atmosphere. The two first steps act as above, while heating at 700°C (1h) and then at 550°C (2h) promotes oxide growth on surface to control recombination and restore lifetime. For low resistivity materials, V_{oc} of 0.645 V (AM0) have been measured, but the efficiency was limited to 14.5 % (34).

I am not fully convinced that procedures running over 4 hours just for annealing constitue the best conditions, even if the results are spectacular, some compromise may be necessary.

b) UNCONVENTIOANL APPROACHES

Heating surface zones with lamps is a well know technique, which has only recently been applied to anneal damage due to implantation (35). Xenon (36) or Argon (37) arc lamps have been used to anneal implantation damage. For example, for one geometry proposed (Fig. 9a) the temperature increases quickly with time (Fig. 9b).

Moveable strip heaters made from graphite have also been employed for silicon recrystallization (38). Microwave thermal processing are also under investigation.

To my knowledge, no cells have been made up to now. It seems to me, that even if low sheet resistance (100 $\Omega/\square$) have been reached it has still to be demonstrated that the bulk lifetime is maintained.

B. LONG PULSES OR SCANNING METHODS

These techniques are under fast development and a series of conferences has been devoted to pulsed annealing (39). We reviewed recently the use of lasers in solar cell manufacturing (40). The principle is quite simple : we have reported on fig. 8 the time needed theoretically to regrowth a certain film as a function of temperature. If sufficiently high power densities are available, pulsed annealing becomes possible, since the time needed to reach a high temperature is short. Solid phase epitaxial regrowth of the damaged film constitutes the mechanism of annealing. Three kind of techniques are employed :

- <u>incoherent light spots</u> : by focusing a rather low power (500 W electric) arc lamp, recrystallization of P^+ implantation has been achieved (40). Scanned light sources, or flash lamps have also been used with some success (41-43). Generally, the samples are preheated to about 400°C. Up to now, the recrystallization is probably not complete, but good electrical characteristics, with implanted profiles undisturbed by the heat treatment, have been reported.

- <u>scanned lasers</u> : instead of using incoherent focused light, laser emission scanned over the sample can be used. Visible light (argon, krypton lasers) or I.R. (CO_2) have been employed, in the energy range 10-20 W (optical) focused on 50-100 μm^2, the scanning speed being 1-100 cm/s and the displacement typically 50-100 $\mu m/s$. Efficiencies around 12 % (AM1) have been reported (45).

- scanned electron beams can be employed to regrow the damage induced by implantation, as shown by several authors (39). However, <u>less</u> attention has been given to this procedure when compared to lasers.

C. SHORT DURATION PULSES

Both laser and electron pulses are intensively investigated since a couple of years. The first commercial set-ups are even available.

-<u>Lasers</u> : essentially ruby and YAG lasers have been used, typically of 1-2 J/cm² and 10-100 ns pulse duration. Above a certain threshold, melting is reached, several models have been described to explain the mechanisms of liquid phase regrowth (40). As an example, Fig. 10 gives some indication about the time evolution vs time and depth of the molten zone. For P-N junction manufacturing, this procedure has some advantages ; especially, the implanted dopant distribution is fully modified, depending on effective segregation coefficient (49) as shown on Fig. 11 (50) ; furthermore, solubility well in excess the solubility limit can be achieved as shown on Table II (49, 51, 52).

Efficiencies of 16.6 % (AM1) have been achieved (46) by performing the laser illumination on the sample maintained at 600°C (V_o = 605 mV, FF = 0.80, I_{SC} = 34,3 mA/cm²). Generally, a conventional heating at 600°C is necessary in order to suppress the deeply penetrating tail of defects resulting from channelled projectiles (47), this heat treatment is, automatically done when back contact sintering is performed. However, if necessary, it can be omitted by mixing with a YAG laser the 1.06 and 0.53 µm wavelength (53). To achieve large area, an x-y scanning with overlapping pulses is used. Some problems may arise, which have been considered in literature (48). In general, very good diffraction patterns are obtained even in polycristalline material (Fig. 12) (54).

Electrons (PEBA)

Fast surface zone heating can also been achieved by large pulses of electrons (55-57). Here, the penetration is larger than for laser, therefore, zones in excess of 1 µm can be annealed. The calculated time evolution of temperature and molten depth is given in Fig. 13 (58), several other authors calculated it (55, 59). Despite the greater depth which is annealed, some microscopic defects (60) due to the fast regrowth are present, which must be thermally annealed, as for the laser. Cell efficiencies in excess of 15 % have been found (30, 61).

5. OPTIMAL MANUFACTURING CONDITIONS

In my opinion, ion beam doping has no definitive advantage over conventional diffusion procedures for preparing cells on single crystalline silicon, since the technology on electronic grade material is well adapted. But, these new procedures are of greatest interest for the tomorrow mass production of polycristalline cells. Here, we restrict our considerations to technologies which are sufficiently developped today.

A. IMPLANTATION OR INCRUSTATION : we have shown recently (62) that on single crystalline materials quite similar results can be achieved by both techniques. The cost of an incrustation set-up is much lower than of a high current implanter and low energy beams are much easier to achieve on incrustation systems. However, other parameters have to be taken into account : availability of commercial automatic equipments, service on systems which have to work 24 hours a day with very high ion currents, probably in the 50-100 mA range.

B. NATURE OF DOPANTS AND ENERGY
- For implantation, $^{31}P^+$ and $^{11}B^+$ ions on P - and N-type substrates gave identical results, As^+ and a combination of $As^+ + P^+$ was not better (63). Best results are achieved with low energy projectiles,

in the 5-10 keV range, as shown on Fig. 14 (63). It should be mentionned that for such low velocity projectiles channelling is important; therefore, on polycrystalline substrates of various orientations, small range deviations are possible from grain to grain.

- The same is also true for PF_5 or BF_3 incrustation, where the best results have also been achieved in the low energy range (62).

C. DOSE

- On single crystalline materials, the optimum dosis is around $2-5 \times 10^{15}$ cm^{-2} for implantation (63), depending on energy, and 1×10^{16} cm^{-2} for incrustation (62).

- For polycrystalline substrates, slightly higher values, about a factor of two, are the best choice, depending on the annealing procedure.

D. FURNACE ANNEALING

Here, no general solution can be given, since the procedure is very sensitive on beam flux, nature of the material, heat exchange during implantation...

On single crystalline cells, values around 900 - 1000°C for 30-60' have been most reported, with eventually several temperature steps, as already mentionned. Fig. 15 compares a cell performance of implanted and diffused diodes prepared on the same material (65) ; it appears that the results are rather close together.

In our opinion, heating of less pure polycrystalline material should be avoided, since impurity migration may degrade the cell's performance. Today, on electronic grade polycrystalline silicon, the temperature effect depends strongly on the nature of the material : on some, diffusion length degradation was observed after heating to 600°C, whereas on Mobil Tyco ribbons an improvement is found.

E. PULSED ANNEALING

Today, it seems not possible to tell which kind of annealing will be the best, the decision will probably result from equipment handling costs : lasers have not reached their final capabilities and quite simple electron pulse machines are under development. Quantum efficiency curves for cells made on the same single crystalline Si by this procedure are quite similar to that of diffused cells (Fig. 16) (30) and on polycrystalline Silso , the I-V characteristics under illumination are not so far away from conventional procedures (65) as shown on fig. 17. Annealing after the pulsed procedure at the same time as ohmic contact sintering, or heating during the pulsed process is always necessary (Fig. 18)(46).

6. DISCUSSION

- Surface doping by ion beams seems to be today the most promising procedure for preparing the junction of polycrystalline silicon cells. However, it should be kept in mind that other technologies, like dopant spin-on followed by conventional or pulsed annealing have also good chances to succeed if the grain boundary migrations can be controlled.

- Conventional ion implanters, as those used today cannot fullfill the high producting speeds over days, which are necessary to achieve acceptable costs. Simplified machines, which, probably, will also be used for ion doping in metallurgy, seem much better suited for a large scale automatic production. Therefore, the wet chemistry will be strongly reduced or even suppressed.

- The damage anneal procedure has not yet reached its final stage ; higher production rates, high fiability are necessary for the major technologies under investigation.

- The cost reduction, which is expected through these methods is quite significant, as calculated by GOLDMAN (Fig. 19) (66).

7. DOPING OF AMORPHOUS SILICON

This paper has been devoted to crystalline silicon, but it should be mentionned that ion implantation can lead to the same range of control of conductivity of amorphous silicon as gas phase doping. However, some problems on spectral response in the blue part are still present. Today. It appears that P, B and Te are the most efficient ions for N - and P -type substitutional doping of a-Si and stable and efficient interstitial N-type doping is possible with alkali ions (Na, K, Pb, Cs). Efficiencies up to 5.4 % have been achieved (Fig. 20) (67-70).

8. CCNCLUSION

The semiconductor device industry is entered since a couple of years in directions in which ions, electrons and photons enter for a large part in the technology. Solar cells have great chances to follow the same evolution. However, due to the high doping and very large areas involved here, new approaches have to be invented. Most of these which are under investigation today have been reviewed. But, it is still too soon to tell today which set-up will be the best choice, even if the various possibilities show progressively their capabilities. The selection must probably be done in the next two years, when solar grade polycrystalline silicon will be really available.

9. REFERENCES

1. G. DEARNALEY, J.H. FREEMAN, R.S. NELSON, J. STEPHEN, Ion Implantation North Holland (1973) 1.

2. W. SHOCKLEY, US Patent n° 2 787 564 (1957).

3. F.M. ROURKE, J.C. SHEFFIELD, F.A. WHITE, Rev. Sci. Instr. 32 (1961) 455.

4. T. AVAGER, N.J. HANSEN, Rev. Sci. Instr. 33 (1962) 567.

5. F.W. MARTIN, W.J. KING, S. HARRISON, IEEE Trans. Nucl. Sci. NS 11 (1964) 280.

6. J.W. MAYER in Semiconductor Nuclear Particle Detectors. Publication 1593 of National Academy of Sciences Washington (1969) p. 284.

7. Ion Physics Co. Burlington Mass. Air Force Contracts AF 33 (657) 10505, AF 33 (615) 1097, AF 33 (615) 2292 (1962-1965).
8. W.J. KING, J.T. BURRIL, Proceed. 4th IEEE Photovoltaïc Specialists Conference Cleveland, Ohio Pic sect. B2.
9. W.J. KING, J.T. BURRIL, S. HARRISON, F.W. MARTIN, C.M. KELLETT. Conf. on Electromagnetic Isotope Separator Aarhus (1965). Nucl. Instr. Meth. 38 (1965) 178.
10. J.T. BURRIL, W.J. KING, S. HARRISON, P. Mc NALLY, IEEE Trans. on Electron Devices ED 14 (1967) 10.
11. V.M. GUSEV, V.V. TITOV, M.I. GUSEVA, V.I. KURINNYI, FTT 7 (1965) 2077 (Sov. Phys. Semicond. Solid State 7 (1965) 1673).
12. J.F. GIBBONS, W.S. JOHNSON, S.W. MYLROIE "Projected Range Statistics" Ed. Dowten Hutchinson and Ross, Stroudsburg (1975).
13. D.K. BRICE, Ion Implantation Range and Energy Deposition Distributions vol. 1 Plenum Press (1975).
14. K.B. WINTERBON, Ion Implantation Range and Energy Deposition Distribution vol. 2, Plenum Press (1975).
15. H. MAES, W. VANDERVORST, R. VAN OVERSTRAETEN, Material Processing Theory and Practices Series Vol 2 "Impurity Doping processes in Silicon" North Holland (1981) 443.
16. B. SMITH, Ion Implantation Range Data for Silicon and Gaussian Device Technology. Learned Information Oxford (1977).
17. P. BLOOD, G. DEARNALEY, M.A. WILKINS, J. Appl. Phys. 45 (1974) 5123.
18. J.P. PONPON, P. SIFFERT, Proc. 11th IEEE Photovoltaïc Specialists Conference (1975) 342.
19. R. WICHNER, Proc. 11th IEEE Photovoltaïc Specialists Conference (1975) 243.
20. French patent 78.02.828, US 88 70 76, Germany P 28.114.149...
21. L.D. NIELSEN, P. BALSLEV, Proc. Third E.C. Photovoltaïc Solar Energy Conference (1980) D. Reidel, Dordrecht, p. 698.
22. H. ITOH, K. TOKIGUCHI, T. WARABISAKO, T. SAITOH, T. TOHUYAMA, Proc. 2nd Photovoltaïc Science and Energy Conf. in Japan (1980).
23. D. SIRKIS, D. SALTZMANN, Proc. 15th IEEE Photovoltaïc Specialists Conference (1981) 981.
24. O. CHRISTENSEN, H. BAY, Appl. Phys. Lett. 28 (1976) 491.
25. A. GROB, J.J. GROB, N. MESLI, D. SALLES, P. SIFFERT, Nucl. Instr. Meth. 182 (1981) 85.
26. T. TAKAGI, I. YAMADA, A. SASAKI, J. Vac. Sci. Tech. 12 (1975)1128 ; id. Instr. Phys. Conf. Ser. 38 (1978) 229.
27. L. CSEPREGI, E.F. KENNEDY, T.J. GALLACHER, J.W. MAYER, T.W. SIGMON, J. Appl. Phys. 48 (1977) 4234.
28. KEICHIRDHODA, MOTOTAKA, KAMOSHIDA, J. Appl. Phys. 40 (1977) 18.
29. V.C. KANNAN, D.D. CASEY, Appl. Phys. Lett. 31 (1977) 74.
30. G. MICCLLI, P. OSTOJA, S. SOLMI, Third E.C. Photovoltaïc Solar Energy Conference (1980) D. Reidel, Dordrecht, p. 719.
31. J.C. MULLER, A. and J.J. GROB, R. STUCK, P. SIFFERT, 13th IEEE Photovoltaïc Specialists Conf. (1978) 711.
32. A.R. KIRKPATRICK, J.A. MINNUCI, A.C. GREENWALD, R.H. JOSEPHS Proceed. 13th IEEE Photovoltaïc Specialists Conference (1978) 707, idem 14th (1980) 820, idem Proceed. 9th Project Integration Meeting LSA Low Cost Solar Array Project,

J.P.L. (1978) 4.
33. E. C. D'AIELCC, 14th. IEEE Photovoltaïc Specialists Conference (1980) 825.
34. J.A. MINNUCCI, K.W. MATTHEI, A.R. KIRKPATRICK, A. Mc. CROSKY, Proc. 14th IEEE Photovoltaïc Specialists Conf. (1980) 93 id. IEEE Trans. on Electron Devices ED-27 (1980) 802.
35. K. NISHIYAMA, M. ARAI, N. WATANABE, Japan J. Appl. Phys. 19 (1980) L 563.
36. R.A. POWELL, T.O. YEP, R.T. FULKS, Appl. Phys. Lett. 32 (1981) 150.
37. EOTON Semiconductor Equipment, Beverly MA (USA).
38. B.Y. TSAU, J.C.C. FAN, N.W. GEIS, D.J. SILVERSMITH, R.W. MOUNTAIN, Appl. Phys. Lett. (in press).
39. Material Research Society meetings, Boston.
40. J.C. MULLER, P. SIFFERT, Rev. Phys. Lett. 15 (1980) 611.
41. L. CORRERA, private communication.
42. R.L. COHEN, J.S. WILLIAMS, L.C. FELDMAN, K.W. WEST, Appl. Phys. Lett. 33 (1978) 751.
43. L. CORRERA, L. PEDULLI, Appl. Phys. Lett. 37 (1980) 55.
44. A. GAT, IEEE Electron Device Letters EDL-2 (1981) 85.
45. J.L. REGOLINI, J.F. GIBBONS, T.W. SIGMON, R.F. PEASE, T.J. MAGEE, J. PENG, Appl. Phys. Lett. 34 (1979) 410.
46. R.T. YOUNG, R.F. WOOD, E. JELLISON, W.H. CHRISTIE, Third E.C. Photovoltaïc Solar Energy Conference (1980) Reidel p. 703.
47. A. MESLI, J.C. MULLER, D. SALLES, P. SIFFERT, Appl. Phys. Lett. 39 (1981) 159.
48. G.K. CELLER, Workshop on laser effects in ion implanted semiconductors. Catania (1978) ed. Rimini.
49. R. STUCK, E. FOGARASSY, J.J. GROB, P. SIFFERT, Applied Physics 23 (1980) 15.
50. C.W. WHITE, W.H. CHRISTIE, B.R. APPLETON Appl. Phys. Lett. 33(1978)662
51. E. FOGARASSY, R. STUCK, J.J. GROB, P. SIFFERT, Laser and Electron Beam Processing of Materials Academic Press (1980) Boston p. 117.
52. C.W. WHITE, S.R. WILSON, B.R. APPLETON, F.W. YOUNG, J. NARAYAN, idem ref. 51 p. 111.
53. J.S. KATZEFF, M. LOPEZ, R.H. JOSEPHS, Third EC Photovoltaïc Solar Energy Conference (1980) Reidel p. 708.
54. F. WALD, J. HO, J. HANDKA, Mobil Tyco Private communication.
55. A.R. KIRPATRICK, J.A. MINUCCI, A.C. GREENWALD, R.H. JOSEPHS, Proc. 13th IEEE Photovoltaïc Specialists Conf. (1978) 706.
56. A.R. KIRPATRICK, Laser and Electron Beam Processing of Electronic Materials, Electrochemical Society ed. (1980) 108.
57. G.A. LANDIS, A.J. ARMINI, A.C. GREENWALD, R.A. KIESLING, 15th Proc. IEEE Photovoltaïc Specialists Conf. (1981) 976.
58. M. TOULEMONDE, P. SIFFERT, Appl. Phys. 25(1981) 159.
59. J. MICHEL, D. BARBIER, A. LAUGIER, 15th Proc. IEEE Photovoltaïc Specialists Conf. (1981) 1007.
60. A. MESLI, J.C. MULLER, P. SIFFERT, Appl. Phys. Lett. (in press).
61. A.R. KIRPATRICK, J.A. MINNUCCI, A.C. GREENWALD, 15th Proc. IEEE Photovoltaïc Specialists Conf. (1980) 821.

62. J.C. MULLER, P. SIFFERT, Intern. Workshop on Ion Implantation, Laser Treatment and Ion Beam Analysis of Materials, Bombay (1981) in press, Rad. Effects.
63. E.C. DOUGLAS, R.V. D'AIELLO, Proc. 14th IEEE Photovoltaïc Specialists Conf. (1980) 825 ; idem IEEE Electron Devices ED 27 (1980) 792.
64. E. COURCELLE, E. FOGARASSY, J.C. MULLER, P. SIFFERT, this conference, paper P 8.17.
65. W. SCHMIDT, K.D. RASCH, this conference, paper P 8.13.
66. H. GOLDMAN, M. WOLF, Proc. 14th IEEE Photovoltaïc Specialists Conf. (1980) 923.
67. G. MULLER, P.G. LECOMBER, Phil. Mag. B, 43 (1981) 429.
68. P.G. LE COMBER, W.E. SPEAR, S. KALBITZER, G. MULLER, F. DEMOND, J. Phys. Soc. Japan 49 (1980) Suppl. A, 1221.
69. W.E. SPEAR, R.A. GIBSON, D. YANG, P.G. LECOMBER, G. MULLER, S. KALBITZER, J. de Phys. Colloque C4 42 (1981) C4 - 1143.
70. S. KALBITZER, G. MULLER, P.G. LE COMBER, W.E. SPEAR Phil. Mag. B 41 (1980) 439.

ACKNOWLEDGMENTS

Many thanks are due to the french and european administrations supporting this work on ion implantation and incrustation in our laboratory. The help of J.C. MULLER and E. FOGARASSY is greatly appreciated for preparing this paper. Furthermore, I wish to thank Drs W. SCHMIDT and K.D. RASCH from AEG TELEFUNKEN for allowing me to make use of results presented at this conference. Finally, the contribution of F. WALD from MOBIL TYCO is greatly acknowledged.

TABLE I

MAIN ADVANTAGES	MAIN DRAWBACKS
- nearly all elements can be implanted - process independent from temperature - easy control of dopant concentration through dose (current measurement) - easy modification of dopant profile (by changing energy) - high degree of reproducibility and uniformity even over very large area - high degree of dopant purity - distribution of dopants independent on the presence of other dopants or defects - any doping level can be achieved and controlled - no effect on the substrate behind the doped zone, especially no heating	- each projectile produces several defects which must be annealed, even surface amorphization is reached - annealing is very sensitive to many parameters (dosis, dose rate, temperature) - difficulty to anneal high dose implantations by conventional techniques - samples have to be handled individually, but automatic loading systems exist - process done under high vacuum

TABLE II

Impurity	Segregation coefficient at equilibrium	Solubility at equilibrium (cm^{-3})	Solubility after ruby laser annealing (cm^{-3})
B	8×10^{-1}	5×10^{20}	
Al	2×10^{-3}	2×10^{19}	
Ga	8×10^{-3}	4×10^{19}	9×10^{20}
In	4×10^{-4}	8×10^{17}	2×10^{20}
C		5×10^{17}	2×10^{20}
Ge			4×10^{20}
Sn	16×10^{-3}	5×10^{19}	
P	3.5×10^{-1}	1×10^{21}	5×10^{21}
As	3×10^{-1}	1×10^{21}	6×10^{21}
Sb	2×10^{-2}	7×10^{19}	2×10^{21}
Bi	7×10^{-4}	8×10^{17}	4×10^{20}
Se	10^{-6}	1×10^{16}	$\sim 10^{20}$
Te	8×10^{-6}		$\sim 10^{20}$
Zn	1×10^{-5}	8×10^{16}	

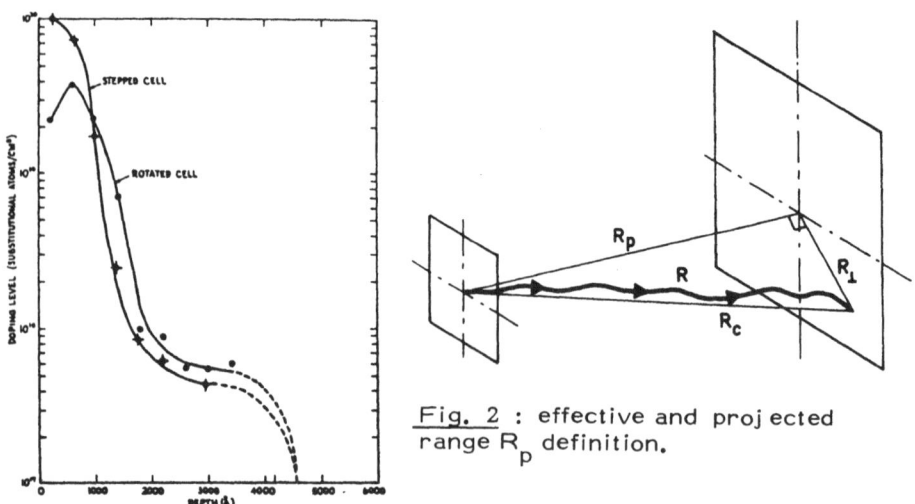

Fig. 2 : effective and projected range R_p definition.

Fig. 1 : Dopant profile obtained either by tilting the silicon wafer during implantation or by changing the projectile's energy (8).

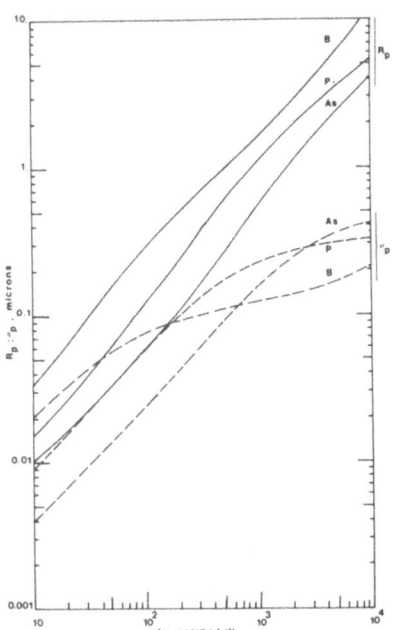

Fig. 3: Projected range R_p and range straggling σ_p for a few dopants in Si as a function of energy.

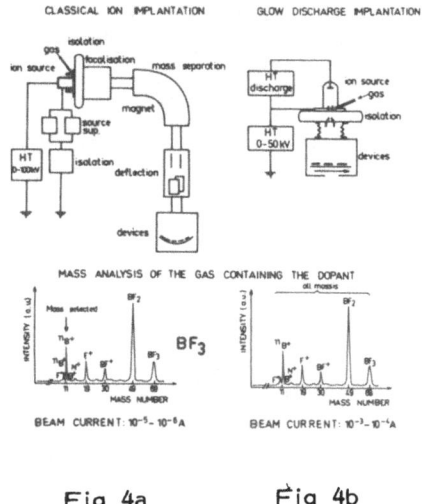

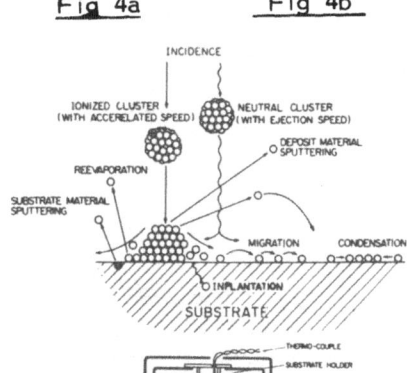

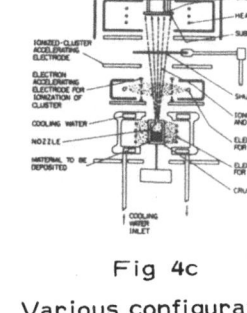

Fig 4c

Various configurations used for: conventional ion implantation (a), ion incrustation (b) and cluster beam deposition (c).

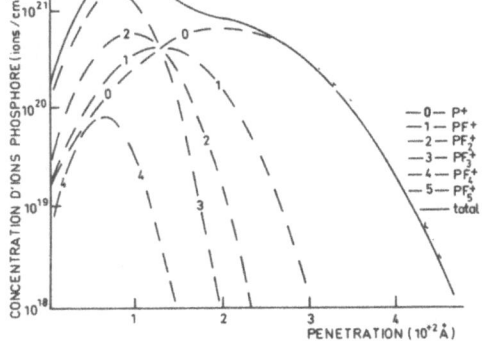

Fig. 5: calculated distribution of phosphorus for an ion incrustation from PF_5 gas performed under the following conditions: discharge 8kV, acceleration 15 kV, dosis $5.10 \text{ E } 15 \text{ cmE} -2$.

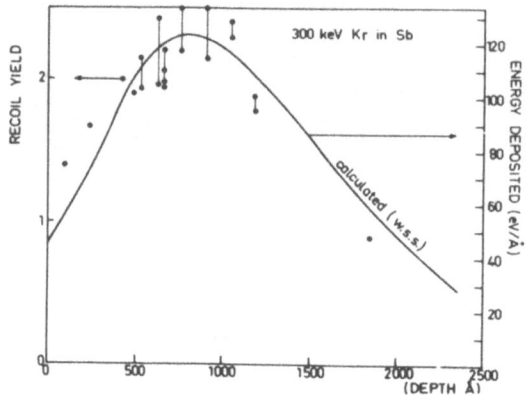

Fig. 6: recoil yield obtained for a 300 keV Kr^+ beam ($10E14\ cmE-2$) vs the deposited dopant layer thickness of Sb. The solid line is calculated from Winterbon's model (25).

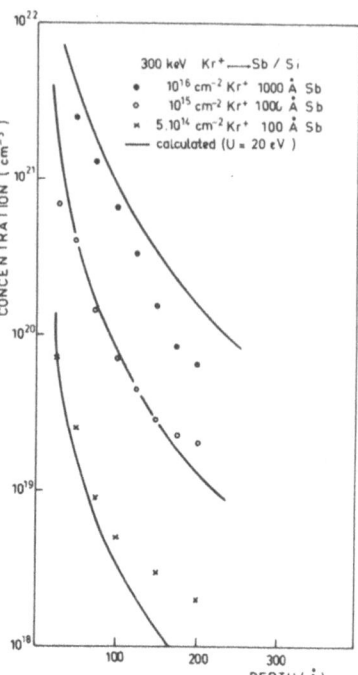

Fig 7: Spatial distribution of Sb in Si for different implant conditions (25).

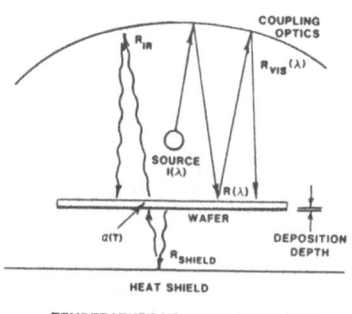

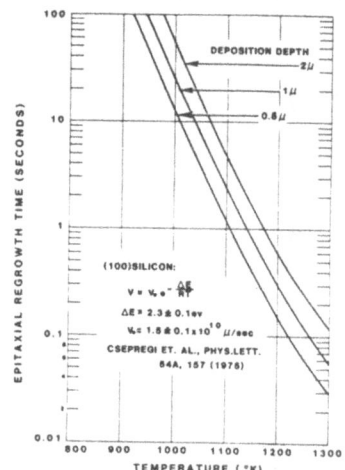

Fig. 8: Calculated epitaxial regrow time vs temperature for an ideal situation (27, 37)

Fig 9a: arc lamp annealing geometry for solid phase annealing (37)
Fig 9b: Calculated temperature vs time for the above geometry (37)

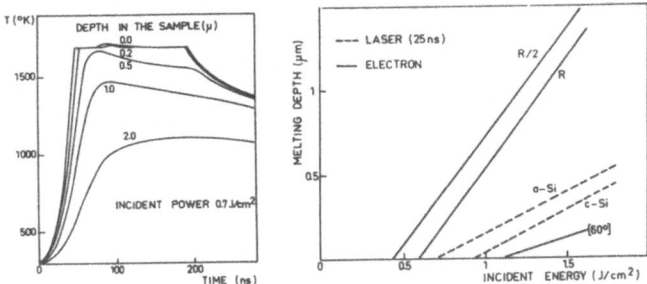

Fig. 10: Calculated evolution of temperature vs time for ruby laser annealing ($0.7 J/cm^2$) and depth of melting vs energy for ruby laser and pulsed electrons (40).

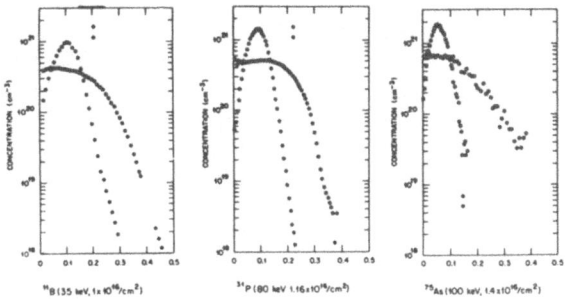

Fig. 11: Effect of ruby laser annealing on some dopant distribution (50)

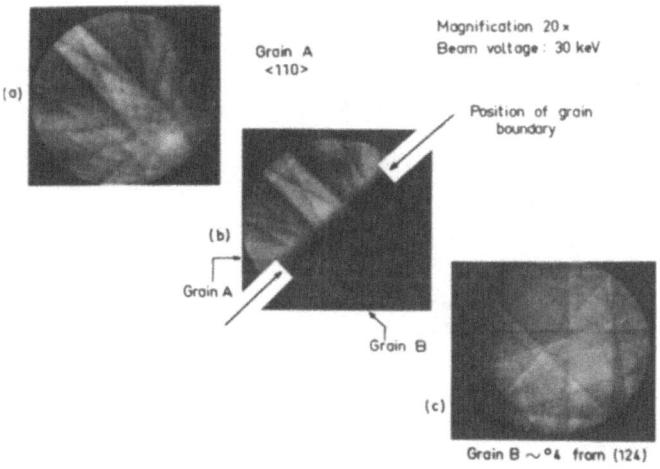

Fig. 12: Diffraction pattern observed on Mobil-Tyco ribbon after Yag laser annealing on two grains A and B and at their boundary

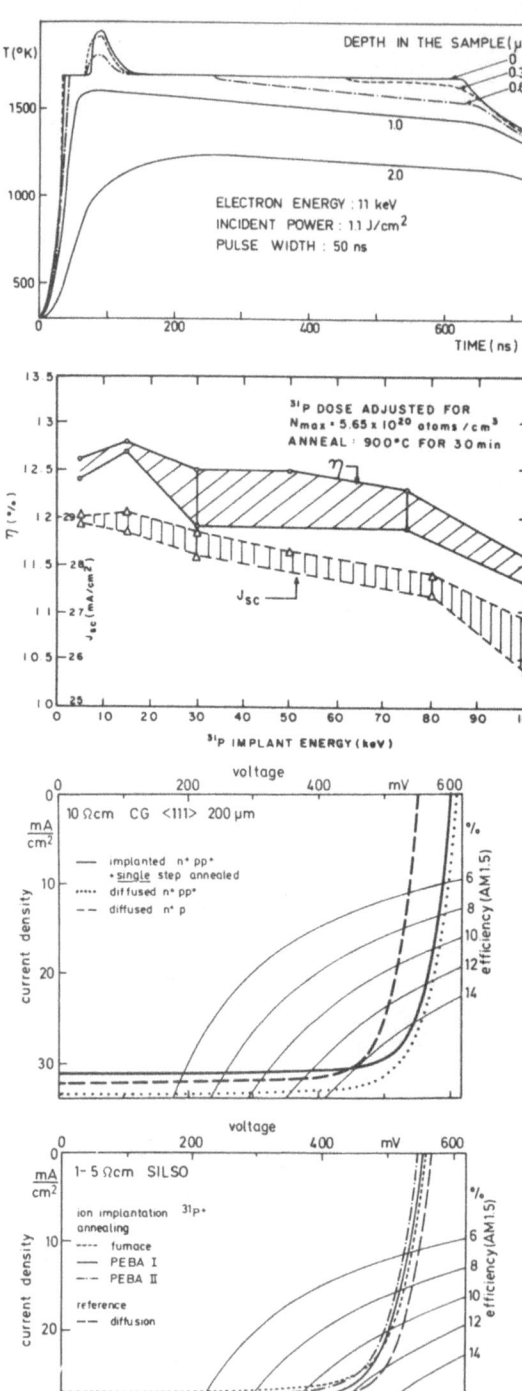

Fig. 13: Calculated evolution of temperature for pulsed electron annealing; for conditions see ref (58).

Fig. 14: Effect on cell efficiency and current of ion implant energy (63).

Fig 15a: Comparison of implanted and diffused cell's efficiency under AM 1 illumination (65).

Single crystalline Si

Fig. 15 b: same as above but on SILSO poly (65).

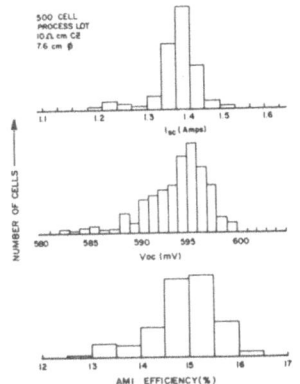 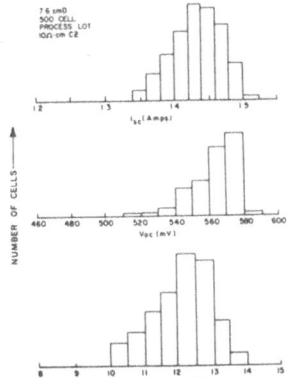

Fig 15c:

Statistical distribution on cell performance for implanted and annealed structures:
left : furnace
right : pulsed electrons (32)

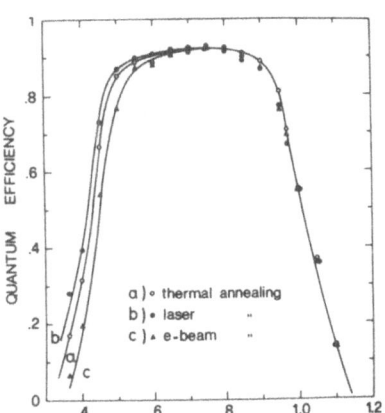 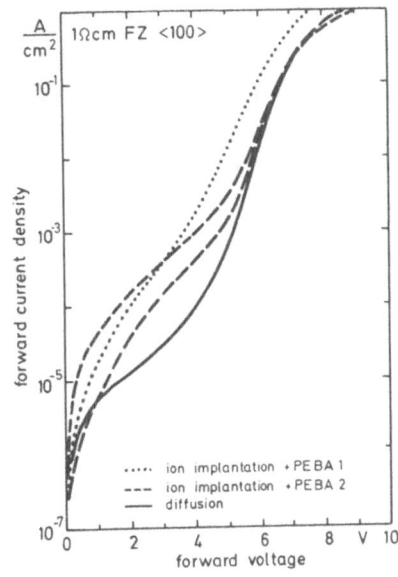

Fig. 16: Comparison of spectral response for implanted cells annealed by furnace, pulsed ruby laser and electrons (30)

Fig. 17: I-V characteristics comparison between implanted and electron annealed cells with conventional diffused structures (65)

Fig. 18: Effect of thermal annealing during the ruby laser treatment of implanted cells ((46)

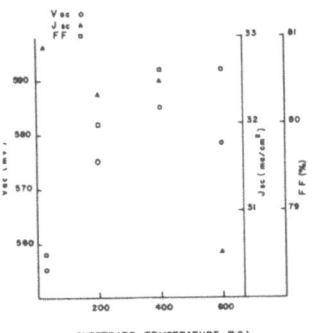

-917-

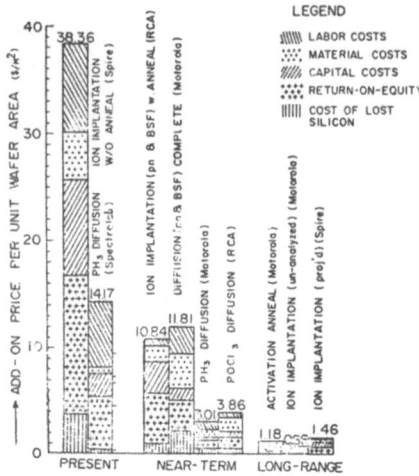

Fig. 19: Calculated cost reduction in short and long term range for various junction manufacturing techniques (66).

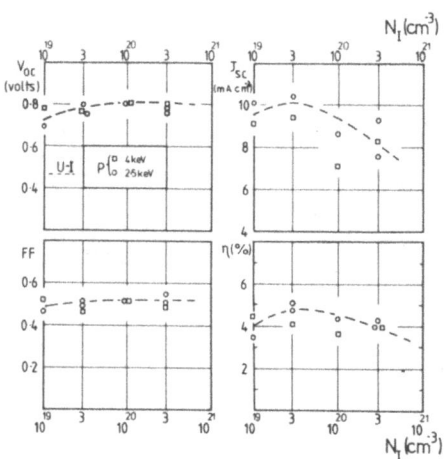

Fig. 20: Implanted amorphous silicon cell performance under illumination, for 4 and 25 keV P^+ implantations. (67-70)

LOW COST PROCESSES FOR CAST SILICON SOLAR CELLS

K.-D. Rasch   K. Roy   W. Schmidt   G. Wahl

AEG-TELEFUNKEN
Electronic Components Division
Heilbronn, Germany

Summary

The possibilities of a low cost process sequence for silicon solar cells using thick film technology have been investigated. The feasibility and problems of screen-printing technology for junction formation, metallization and antireflective layer coating will be discussed in respect to special cast silicon demand. Results from a development line will be presented.

Another possibility for cost reduction is the efficiency improvement. Hydrogen plasma treatment can reduce the influence of grain boundaries in cast silicon solar cells. Therefore comparable efficiencies of multigrain and single crystalline solar cells are attainable.

1. INTRODUCTION

There are three main efforts in terrestrial silicon solar cell development:

- reducing the cost of the starting silicon
- reducing the cost of fabricating cells
- improving the conversion efficiency as high as possible under economic aspects.

Large scale production for terrestrial solar cells using unconventional multigrain cast silicon has been demonstrated with conventional technology (1). About 1 MW has been produced with this process with an overall efficiency of 10 % AM 1.5. A typical process sequence and the capacity of the relevant steps for a 100 kW/yr and a 1 MW/yr pilotline are given in Figure 1. The sequence is characterized by open-tube diffusion and high vacuum deposition of contact metallizations and antireflective coating. The TiPdAg contact system is replaced by a non-noble metal. The capacity of the vacuum deposition - the bottleneck of the process - is enhanced to 8 square meters/run with new advanced evaporation chambers. However, current solar cell processing involves considerable individual device handling.

TYPICAL PROCESS SEQUENCE

Pilotline 100 kW/a	Si-WAFER (SILSO) 5x5 and 10x10 (cm²)	Pilotline 1 MW/a
0.5 m²/Run	ETCH DAMAGE	2 m²/Run
Silica	SPIN ON	
1 m²/Run	DIFFUSION	3 m²/Run
1 m²/Run Ti Pd Ag	EVAPORAT. BACK CONTACT	8 m²/Run
	HEAT TREATMENT	
1 m²/Run Ti Pd Ag	EVAPORAT. FRONT CONTACT	8 m²/Run
1 m²/Run Ti Ox	EVAPORAT AR-LAYER	8 m²/Run
	Test	

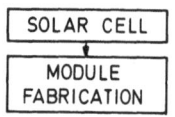

```
SOLAR CELL
   ↓
MODULE
FABRICATION
```

For long-term prospects in the multi MW-range a low cost process should be developed. One possibility is the use of thick film technology conventionally used in production of microcircuits. The feasibility of this concept for an automated silicon solar cell fabrication will be discussed in this work.

Another possibility for cost reduction is the efficiency improvement of solar cells from multigrain silicon. These cells show a reduced efficiency in comparison to the efficiency of single crystalline solar cells. Recombination losses at the grain boundaries are responsible for this difference. The potential of hydrogen plasma treatment to reduce the influence of grain boundaries will be discussed.

Figure 1   Conventional process sequence and capacity of relevant steps

2. THICK FILM TECHNOLOGY

The advantages of thick film technology commonly used in hybrid circuits and microcircuits are

- automated fabrication
- continuous processing with high throughput
- non vacuum processing
- low equipment costs.

Therefore soon screen printed contact metallizations as an alternative method are in usage for the manufacture of terrestrial silicon cells. However the purpose of this paper is the concept of a complete use of thick film technology for the relevant process steps in an automated solar cell fabrication.

This means a continuous procedure for junction formation, contact metallizations and antireflective coating layer. Suitable pastes or gels are applied by a screen printing technique, dried in a IR dryer and driven-in or fired in a belt furnace. All the processes can be realized with a modular system which is shown in principle in Figure 2.

The feasibility of this concept was tested in respect to multigrain cast p-type silicon (SILSO), but the process sequence can be used without restriction for low cost single crystalline silicon.

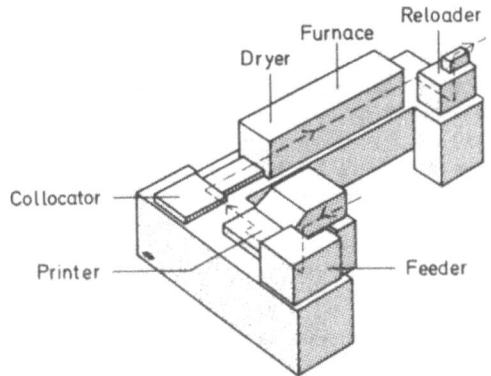

Figure 2
Automated solar cell fabrication with a modular system using thick film technology

2.1 Junction formation

Commercially available liquid phosphorous dopants for solar cells normally deposited either by spinning or spraying are modified by organic additives to obtain the desired viscosity for screen printing applications. It is also important to mix appropriate fluid properties to ensure a suitable surface dopant concentration. One side of the wafer is then printed with the n-type dopant paste. The diffusion can be carried out in a modified belt furnace in nitrogen atmosphere with about 3 % oxygen. Typical sheet resistance for 15 min and 60 min diffusion as a function of temperature is given in Figure 3. The results realized with the diffusion paste are comparable with values from open-tube diffusion with a gaseous source.

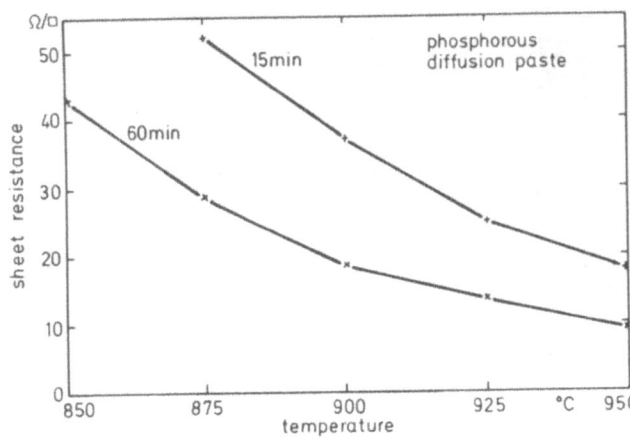

Figure 3
Sheet resistance of diffusion paste as a function of temperature and time

2.2 Contact metallization

A process for screen printed metal front and back contacts as a substitute for evaporated contacts is in development. Paste composition, screens, printing conditions and firing conditions were optimized. Best results are obtained with silver conductive pastes. The firing temperature in a nitrogen/oxygen atmosphere is between 650 and 750 °C. Critical is the front contact metallization. It is necessary to use relatively deep junctions to avoid shunting effects brought about by firing in the screen printed contacts. A trade-off must be found between junction depth and firing temperature which determines besides others the reliability properties. A silver conductive paste with a modified frit content results in a definite improvement of the paste performance in both avoiding junction shunting and decreasing the sensitivity to changes in the firing time and temperature. Typical firing temperatures of the front metallization are in the range of 650 °C. In Figure 4 a screen printed front contact metallization on a 10 cm x 10 cm multigrain solar cell is shown. The contact coverage of the front surface area is 11.5 %. The gridline width is between 250 and 300 μm.

Because of the relatively high cost of silver and because of the uncertain availability of silver for long term prospects some attempts were made to use non-noble metal pastes. Aluminium conductor pastes can be a replacement for the silver on the rear side, but for the front side until now a non-silver metallization could not be found which results in comparable cell efficiencies and reliability. A typical quantity of silver paste required to achieve a front contact of a 10 cm x 10 cm solar cell is approximately 0.14 g. That means the front side can be metallized for about 0.14 DM, if one assumes 1 DM/g silver paste.

Figure 4
Screen printed front contact metallization on a 10 cm x 10 cm multicrystalline solar cell

2.3 Antireflective layer coating

An organic titanium complex is mixed in-house with suitable organic additives. Typical viscosity of the paste is 1770 mPas. The solar cell is coated with the titanium complex paste by screen printing. The coated cells are heated in a belt furnace up to 500 °C to drive off solvent and organics and to form a one-fourth wavelength thick TiOx coating. The screen printed antireflective coating layer is relatively uniform and shows comparable optical properties as the evaporated TiOx films. In Figure 5 a comparison of evaporated and screen printed TiOx coatings is given. An increase in short-circuit current of more than 32 % is found after application of screen printed TiOx coating on multigrain silicon solar cells.

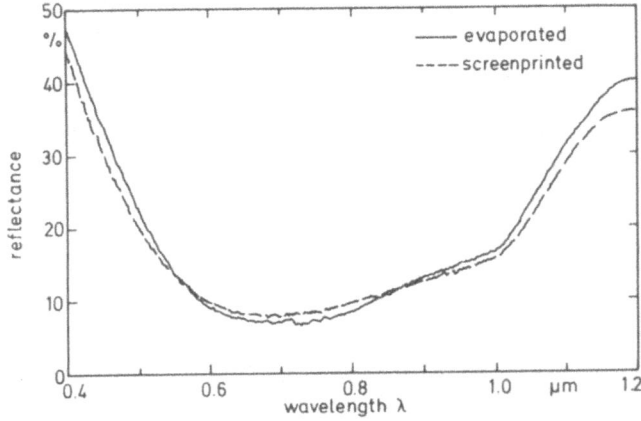

Figure 5
Reflectance of 10 cm x 10 cm multicrystalline solar cells with evaporated or screen-printed TiOx antireflective coating

2.4 Cell properties

Multigrain silicon solar cells 10 cm x 10 cm processed with thick film technology show average conversion efficiencies of 9.5 % AM 1.5 (Figure 6). The solar cells with n^+p structure have screen printed silver contact metallizations and screen printed antireflective coating layers. The open-circuit voltage is typically 550 mV, the short-circuit current density 23,6 mA/cm^2, the curve factor greater than 72 %. The lower efficiency compared to conventional processed multigrain silicon solar cells can be due to the deep junction or the high contact coverage.

But it has been demonstrated that a low cost continuous process sequence with thick film technology is feasible.

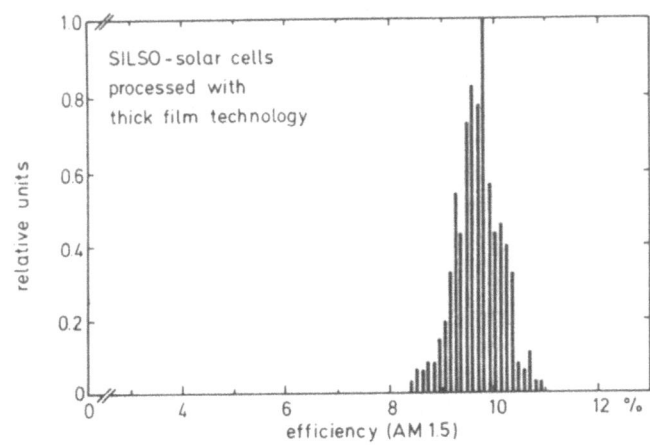

Figure 6
Efficiency distribution of 10 cm x 10 cm multicrystalline (SILSO) solar cells processed with thick film technology

3. IMPROVED EFFICIENCY

The efficiencies of cast silicon solar cells are 5 - 25 % lower than those of single crystalline silicon. This difference exhibits the influence of the grain boundaries in cast silicon. Figure 7 shows the local photoresponse of a multigrain solar cell. Hydrogen plasma treatment can reduce the influence of grain boundaries (2, 3), also in multigrain cast silicon (SILSO) as demonstrated in Figure 8. The recombination loss at most grain boundaries is reduced. The hydrogenated multigrain solar cells show a remarkable increase in open-circuit voltage and short-circuit current. This results in a significant efficiency improvement of 1 - 15 % for the cast silicon solar cells, depending on the efficiency before treatment - "good" cells show no or minor increase, "bad" cells a significant increase. The long term stability of the hydrogenation is tested since 18 month at 200 °C. No relevant change in cell performance is observed. The hydrogen treatment therefore can be a possibility for effiency improvement of multigrain solar cells. However the feasibility and the economy of this treatment in a production line is not yet demonstrated.

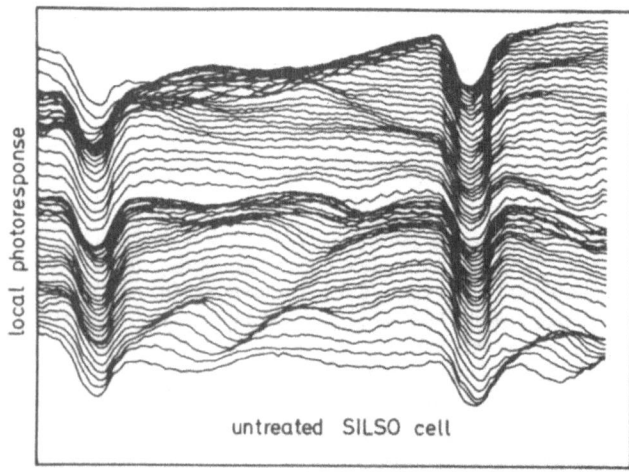

Figure 7
Local photoresponse of untreated multigrain cast (SILSO) solar cell

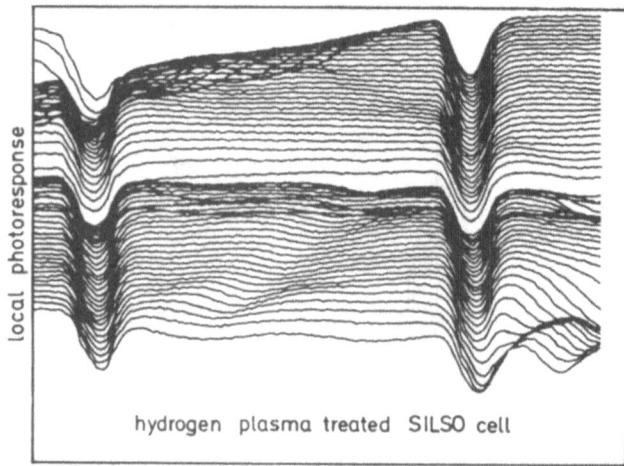

Figure 8
Local photoresponse of hydrogen plasma treated multigrain cast (SILSO) solar cell

4. CONCLUSIONS

The feasibility of a low cost process based on thick film technology has been demonstrated. Besides the chemical damage removal all other relevant process steps can be done by screen printing of suitable pastes for the junction formation, the contact metallizations and the antireflective coating layer followed by firing in a belt furnace. Efficiencies of nearly 10 % can be obtained with 10 cm x 10 cm multigrain silicon solar cells.

Hydrogen treatment of cast silicon solar cells has shown the potential for improved efficiencies due to reduced recombination at grain boundaries.

5. ACKNOWLEDGEMENT

The authors wish to thank Dr. K. Lutz, Demetron, for support and helpful discussions during metal pastes development.

They also want to thank W. Notz for TiOx paste preparation, and K. Niemann for performing the hydrogen plasma treatments. This work is supported by the Bundesministerium für Forschung und Technologie.

6. REFERENCES

(1) K. Roy, W. Pschunder, Comparison of solar cells from nonsingle and single crystalline silicon in a pilot production line, 3rd EC Photovoltaic Solar Energy Conf, Cannes, 263 (1980)

(2) P. H. Robinson, R. V. D'Aiello, The effect of atomic hydrogen passivation on poycrystalline silicon epitaxial solar cells, Appl. Phys. Lett. 39 (1) 63 (1981)

(3) C. H. Seager, D. S. Ginley, J. D. Zook, Appl. Phys. Lett. 36, 831 (1980)

POLYCRYSTALLINE SILICON SOLAR CELLS UTILIZING AN INTEGRAL
SCREEN PRINTING TECHNIQUE

G. Cheek, R. Janssens, M. Leempoels, L. Frisson(+), R. Mertens
and R. Van Overstraeten

Katholieke Universiteit Leuven E.S.A.T.
Kardinaal Mercierlaan 94, B-3030 Heverlee, Belgium

Summary

The integral screen printing process for the fabrication of silicon solar cells offers a low-cost, continuous technique. Printing has been used for the deposition of active layers, antireflection coatings, front and back metallizations and a back surface field. Processing parameters such as initial surface etching, pre-gettering, types of phosphorous diffusion and back surface field formation are discussed in this paper. A comparison is made between three potentially low-cost silicon materials using the integral screen printing process. These materials include Wacker Silso, HEM polycrystalline silicon, and highly dislocated single crystal silcon.

1.0 Introduction

Early reports indicate that a total screen printing process for the fabrication of silicon solar cells is an attractive technique.(1-2) Since all processing is done in atmosphere and there is no requirement for ultra-clean working conditions, the printing process is easily scalable for continuous, large scale production. Several thousand single crystal silicon cells have been made in our laboratory using the printing process. A typical batch of 400, 3 inch, (111) oriented wafers results in an average efficiency of 11.2% with a 98% yield: defined as cells higher than 10% efficiency. The implementation of this process on potentially low-cost silicon materials has begun with an emphasis on process optimization and material evaluation. The evaluation of low-cost silicon materials using conventional processing technology has been done earlier.(3) The work reported here, using some similar silicon materials, is for non-conventional, low-cost cell processing. Solar cell results, processing optimization and material evaluation are reported in this paper.

2.0 Cell Processing

A typical processing sequence consists of planar acid etching the silicon surface, the printing of a home-made phosphorous paste and the junction diffusion in a conveyor belt furnace. After re-cleaning the wafer, an antireflection coating (ARC) of either TiO_2 or Ta_2O_5 is spun-on or printed. The printing and firing of the silver (Ag) (Electro-Science Laboratory) front contact is followed by the printing and firing of the aluminum (Al) (Englehard) back contact. Finally, a solderable pad of silver-palladium (Englehard) is printed and fired on the backside.

Earlier reports indicate that a deep phosphorous diffusion with a subsequent etching can "getter" impurities and/or defects from the bulk and grain boundaries in Wacker Silso material.(4) Our experience with a similar "pre-gettering" step at either 900°C or 1100°C for 60 minutes, with subsequent etching of the junction region, has resulted in a small average increase in Jsc of about 0.2 mA/cm^2. The improvement is consistent but not significant enough for implementation into our normal processing.

The initial surface etching is an important step in the processing. Wafers etched in NaOH are not suitable for device processing because the surface is texturized and the ARC is not uniform. A non-uniform ARC results in a poor Jsc and an increased second diode factor. Also, semicrystalline silicon etched in NaOH has a highly irregular surface due to the orientation dependent etch rates. Our best results are from a CP4 etchant which yields a smooth surface.

Both a conventional ($POCl_3$) vapor phase diffusion (VPD) and a screen printed diffusion (SPD), using a Phosphorus dopant paste developed in our laboratory, have been investigated. Efficiency distributions for both VPD and SPD, with TiO_2 spin-on ARC, are given in Figure 1. The results indicate there is no penalty for using a SPD rather than a VPD since the difference in average efficiency is marginally small. The average efficiency for the SPD cells is slightly lower but the distribution of cell characteristics is more narrow compared with VPD cells. Peak efficiencies on Wacker Silso (3.24 cm^2) exceed 10% for both VPD and SPD. The optimized sheet resistance (Rsht) for this process is 22 ohms/□. For the same Rsht, the SPD cells have a deeper junction which implies a lower surface concentration. As a result, the Jsc is slightly reduced which accounts for the lower average efficiency in Figure 1. The Voc and ff for the SPD cells is typically higher than the VPD cells. The higher values are a result of the bulk diffusion length (Ld), measured from long wavelength spectral response, being longer for the SPD cells. The Ld is preserved in the printing process due to the conveyor belt furnace's temperature profile that reduces excessive thermal shock.

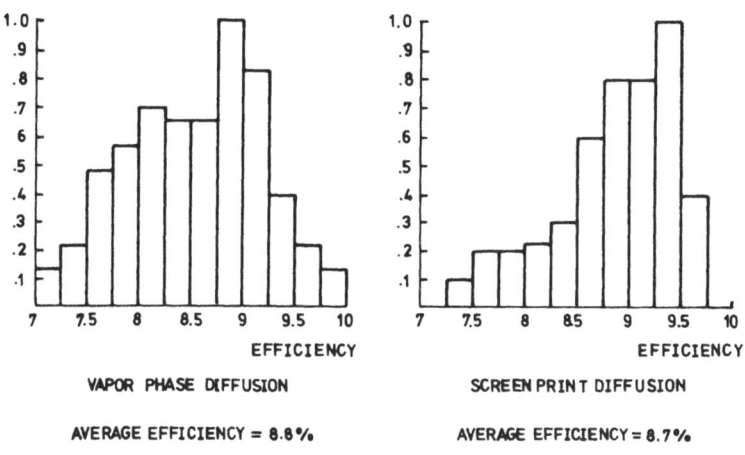

FIGURE 1

Figure 1: Normalized distribution of 150 cell efficiencies AM1, total area, 3.24 cm^2, for both VPD and SPD sources on Wacker Silso material, both with printed metallization. Similar distributions are obtained for batches of monocrystalline cells (400 wafers) only with higher average efficiencies.

Initial results using screen printed ARC (Englehard) are encouraging. Cells on monocrystalline silicon made with Ta_2O_5 printed and fired before the printing of the Ag ink, have an average efficiency of 11.3% compared with 11.8% for the control cells with a spin-on TiO_2 (Emulsitone C). The difference in efficiency is due to a reduced Jsc resulting from an increased surface reflection. The printed ARC's refractive index of 1.85, compared to 2.1 for the spin-on ARC, accounts for the reflection loss. Also, the ff is slightly reduced because of a decreased shunt resistance. Peak efficiencies for the true, integral printed cells exceeds 12% (AM1, 3.24 cm^2).

The Ag ink firing temperature is important because these inks are fired through the ARC.(1) In this process, the ARC can act as a diffusion barrier for metallic impurities sometimes found in the Ag inks. SIMS analysis indicate that Ag,Cr and Mg are reduced from diffusing into the silicon surface region. The high temperatures associated with the fired through contacting technique eliminate the need for special hydrofluoric acid (HF) etching steps. The HF treatment is commonly used to reduce contact resistance that is caused by the glass frit in the Ag inks. The HF etching is critical and known to damage the mechanical and electrical integrity of the printed Ag contacts. (5) Series resistances of 0.6 ohm-cm^2 can easily be achieved using the fired through contacting technique without HF treatment.

A Back Surface Field (BSF) can be formed by an aluminum-silicon alloying during the firing of the printed aluminum back contact. A screen mesh number of 165 is sufficient to produce an Al layer of about 15 μm which is required for a deep p^+ layer. The depth of the Al doped p^+ layer, measured with a Spreading Resistance Probe, exceeds 5 μm. Cells on monocrystalline silicon with Voc greater than 600 mV have been made with this process.

3.0 <u>Low Cost Materials</u>

A comparison of 3 potentially low-cost materials for use with the integral screen printing process has begun. Our aim is to utilize baseline processing established for single crystal silicon and through cell characterization and material evaluation, optimize the printing process for non-conventional materials. The materials being studied include;
1) Silso Blanks, 10 X 10 cm, average thickness 450 μm, 5 ohm-cm, p-type, supplied by Wacker;
2) HEM: Heat Exchanger Method grown large grain polycrystalline silicon, 10 X 10 cm, average thickness 400 μm, 1 ohm-cm, p-type, supplied by Crystal Systems;
3) MHO: Dislocated single crystal silicon, CZ grown <111>, 94 cm^2 cut semi-square from a 4.5 inch crystal, average thickness 460 μm, 2 ohm-cm, p-type, 5000 and 50000 dislocations / cm^2, supplied by Metallurgie Hoboken Overpelt.

Our initial work was testing the material uniformity and process sensitivity by fabricating a matrix of 9 cells (3.24 cm^2) on each wafer. The average value of these 9 cells illuminated characteristics and diffusion length is given in Table I.

Table I: Average values of nine cells from same wafer.

	Wacker	HEM	MHO (50000)
Voc (mV)	533 (14.3)	551 (5.2)	558 (4.3)
Jsc (mA/cm^2)	24.7 (1.7)	25.8 (0.5)	27.2 (0.5)
FF (%)	64.8 (6.1)	68.6 (3.2)	73.4 (2.9)
η (%)	8.6 (1.4)	9.8 (0.6)	11.2 (0.6)
Ld (μm)	49.4	64.0	70.1

The standard deviation data, numbers in parentheses in Table I, indicate that the cells on Wacker Silso material are more sensitive to the position on the wafer, ie) edge vs. center, than the HEM or MHO materials. The lower Voc and fill factor in the Wacker Silso cells is most likely a result of the higher resistivity material. The fill factor losses increase significantly in order of decreasing grain size. The factors contributing to the reduced fill factors are detailed in Table II. The loss components are given as a percentage of the total fill factor reduction.

Table II: Fill Factor Losses

Material	Ideal ff	Actual ff	% reduction	Loss Components (%)		
				2nd Diode	Rs	Rsh
Wacker	81.1	64.8	20.9	82.8	15.8	1.4
HEM	81.6	68.6	16.5	80	18.1	1.9
MHO 50000	81.8	73.4	10.6	68	30	2
MHO 5000 *	82.0	73.8	10.0	33	64	3.0

(*) = Large Area Cell 86.2 cm^2

It is clear that most of the fill factor losses in the small cells are associated with 2nd diode effects. This may be accounted for from impurities from the Ag ink diffusing into the junction space charge region. Measurements of Voc-Isc for the three materials indicate that the 2nd diode factor, m_2, is often greater than 2 as indicated in Table III.

Table III: Voc - Isc Data

	m_2	Io_1	Io_2
Wacker	2.34	1.4×10^{-11}	5.3×10^{-6}
HEM	2.65	1.5×10^{-11}	2.4×10^{-6}
MHO 50000	2.85	1.1×10^{-11}	1.4×10^{-6}

The low fill factor in the HEM material may be due to inclusions or precipitates in the bulk material (3). Also some shunting occurs resulting from a dislocation density as high as $1 \times 10^6/cm^2$ (6). It is interesting to note that the loss in fill factor on a large cell (86.2 cm^2) is dominated by the series resistance component and the second diode loss becomes less significant due to the large current of 2.3 amps present in these cells. Several large and small area cells have been made using these materials and the best results, using a spin-on ARC, are give in Table IV.

Table IV: Best Results on Low Cost Materials

Material	Area (cm^2)	Voc (mV)	Jsc (mA/cm^2)	FF (%)	Efficiency (%)
Wacker	3.2	545	26.1	70.7	10.1
	90.3	537	24.9	62.6	8.4
HEM	3.2	554	26.4	72.5	10.6
	90.3	555	24.5	60.7	8.3
MHO	3.2	561	27.8	75.7	11.8
(50000)	86.3	551	25.4	70.1	9.9
(5000)	86.3	569	26.6	73.8	11.2
Monocrystal	78.5	595	29.2	72.5	12.7

Based on total cell area, 28°C, AM1

The absolute spectral response (A/W) measured with bandpass filters indicates a poor IR response for the HEM and the Wacker cells. The blue response is similar for all materials. The average diffusion lengths of 25 cells, made from different wafers and calculated from the long wavelength spectral response, are 51 μm, 71 μm, and 72 μm for the Wacker, HEM and MHO (50000) materials respectively. The difference with the values of Table I results from the large number of cells used here; in Table I, the numbers were quoted for an average efficiency of 9 cells from 1 wafer only. Significant non-linearities in the spectral response of the HEM and Wacker materials, with and without AM1 white light bias, indicate the diffusion potential at the grain boundaries is reduced under illumination.

4.0 Conclusion

The true, integral printing process is being optimized for use with low cost, semicrystalline materials. Reasonable efficiencies have been achieved in small batch quantities. This process is readily automatable with existing equipment and good throughput rates are practicable on cells of 100 cm^2 in area. Efficiency distributions are similar to the single crystal cells but with a lower efficiency.

5.0 Acknowledgement

This paper is part of a report of the Belgian National R-D Energy Program (Prime Ministers Office of Scientific Research, Wetenschapstraat 8, 1040 Brussels, Belgium). It has also been supported by the Solar Energy program of the European Community under Contract No. ESC-R-018-B-(G). Dr. Paul Van Halen is gratefully acknowledged for supplying some materials used in this work. R.Mertens is supported by the N.F.W.O. (+) Author now with Fabricable in Huizingen, Belgium.

6.0 References

1) A.D.Haigh 12th IEEE Photovoltaic Specialists Conference p.360 (1976)
2) L.Frisson, P.Lauwers, L.DeSmet, R.Mertens, R.Van Overstraeten, P.Bulteel and R.Govaerts 13th IEEE Photovoltaic Specialists Conference p. 590 (1978)
3) H.I.Yoo, P.A.Iles, D.C.Leung and S.Hyland, 15th IEEE Photovoltaic Specialists Conference p.598 (1981)
4) S.Chitre and J.Donon 3rd Commission of European Communities Photovoltaic Solar Energy Conference p.608 (1980)
5) K.Firor and S.Hogan Solar Cells, 5 (1981-82) 87-100
6) K.A.Dumas, C.P.Khattak and F.Schmid 15th IEEE Photovoltaic Specialists Conference p. 954 (1981)

PHOTOVOLTAIC SOLAR CELL COMPARISON METHODOLOGY

ALLEN M. BARNETT
Electrical Engineering Department
University of Delaware
Newark, Delaware 19711, USA

Summary

The lack of a consistent analytical basis for different solar cell materials, configurations and manufacturing processes has led to a tendency to optimistically estimate the potential of any individual approach and to force this estimate to conform to previously established target values. The consistent analytical technique presented here permits the quantitative comparison of all options, using identical criteria.

There are more than thirty-two solar cell configurations based on fourteen materials systems under active investigation. The potential of each of these systems can be evaluated quantitatively using the same bases. This evaluation considers the solar cell performance based on first principles using the materials parameters. The most promising designs are then evaluated for cost, based on standard cost estimating techniques and using demonstrated manufacturing processes. The results from this detailed analysis are shown for the three primary crystalline silicon approaches: ingots, ribbons and epitaxy, and for two thin film approaches: amorphous silicon and CdS-based heterojunctions.

1. PERFORMANCE MAXIMIZATION

The maximum performance for each of the more than thirty-two solar cell configurations can be determined using basic materials and device parameters. A set of consistent equations which quantitatively describe solar cell performance can be applied to each design. Since the solar cell is a monochromatic energy converter, the theoretical maximum efficiency can be determined in its simplest form from the energy gap, assuming an ideal diode (1,2). Deviations from this theoretical maximum can be described as losses. The losses can be analyzed from the basic materials and device parameters and the engineering design. Each design can be analyzed in terms of minimizing these losses. This "loss minimization" (3,4) leads to a quantitative analysis of the maximum performance for any solar cell configuration. This minimum loss is described in terms of current, voltage and power.

A. Power Generation and Loss Minimization

The <u>maximum generated current</u> for any solar cell configuration can be calculated from the absorption coefficient (α), the photon density (F), the semiconductor reflection (R), the minority carrier diffusion length (L) and the surface recombination (S). The detailed derivation is described elsewhere (5,6,7).

The maximum light generated current occurs when the thickness of the absorber layer allows virtually all of the light at energies equal to or greater than the energy gap to be absorbed, when the surface recombination velocity ≈ 0 and the diffusion length is long compared to the distance that the carriers must travel to be collected. This condition would occur in GaAs with surface passivation and 8 micron diffusion lengths. Under these conditions, and considering a two-layer anti-reflection (AR) coating, 93% of maximum theoretical current can be collected in GaAs. For silicon a practical optimized current occurs with a 300 micron thick absorber, a 600 micron diffusion length, back surface recombination = 0, two-layer AR, and a shallow junction. This condition allows 83% of the maximum theoretical current to be collected.

The <u>maximum voltage</u> can be determined from the energy gap (E_G), diode ideality (n), the nature of the diode leakage current (J_o) and the diffusion length (L). The magnitude of the voltage is determined by the diode equation.

<u>Maximum power</u> can be determined from the diode equation and is primarily influenced by the ideality factor, series and shunt resistances. The maximum power occurs when the product of J and V from the diode equation is maximized, leading to J_{mp} and V_{mp} at the maximum power point. The maximum value occurs when the shape of the diode curve approaches a square, leading to n = 1.

The major power losses, in addition to those intrinsic to the diode, are related to the structure designed to remove the power from the solar cell. The contact facing the sun reflects (or absorbs) some of the light. The other major losses are heating losses, I^2R, where R is the resistance of the semiconductor layers, the metal-semiconductor contact and the design of the contacts.

Other geometric losses which can be based on the solar cell design or its implementation are shunt losses which are the inadvertent formation of parallel diodes or electric shunts that form a lower resistance path in parallel with the diode.

When solar cells are connected into modules, additional losses occur due to the mis-match of solar cell characteristics and the use of an encapsulant to protect the module from extrinsic degradation.

The <u>maximum performance design</u> would include a thickness that led to the absorption of virtually all (99%) of the available sunlight, a diffusion length long enough to permit collection of virtually all of these generated carriers, minimum reflection, minimum surface recombination, diode factor = 1, maximum voltage (J_o minimized), and minimum power losses.

Most of these design criteria can be implemented for the 32 solar cell configurations. Crystalline silicon will only absorb 91% of the available sunlight in a 600 micron thick layer and diffusion lengths are rarely longer than 300 microns. This problem can be reduced by adding a back surface reflector (BSR) to a 300 micron or thinner device.

Back surface recombination can be reduced by adding a more heavily doped p-region at the back surface. This region creates a back surface field (BSF) which assists the drift of minority carriers toward the collector. This BSF can also serve to reduce J_o, thereby increasing the voltage and power. Since the generated minority carrier density is much higher on the surface facing the sun, it is usually advisable to "collect" the current from this top surface. The shallow junction design minimizes light absorption (which will not lead to collected minority carriers) in the collector region. Finally, although single layer anti-reflection AR coatings can reduce the reflection from approximately 30% to 6%, two layers can further reduce the reflection to 2%.

Therefore a solar cell <u>design optimized</u> for <u>performance</u> will have a two-layer AR coating, a shallow junction, back surface field (BSF), and back surface reflector (BSR). The maximum practical thickness is determined by the diffusion length and the absorption coefficient. The absorption coefficient is a basic material parameter.

An optimized device design for each configuration can be made based on the best consistent materials data. This design will be based on the fundamental materials parameters energy gap (E_G), absorption coefficient (α), thickness (H), minority carrier diffusion length (L_n), semiconductor reflection (R), and surface recombination velocity (S). The small area device design can then be engineered to a larger area solar cell based on an analysis of the area related losses and non-uniformities and also to a module.

B. Performance of Crystalline Silicon and Thin Film Solar Cells

This analysis has been applied to a number of solar cell designs, including three silicon configurations and two thin films.

Cast silicon solar cell ingots were analyzed with a thickness of 300 microns. BSF, BSR, shallow junction and two-layer AR were included. The most common N on P structure was analyzed. Based on the available data, the best small area diffusion length was estimated to be 200 microns, for this N on P configuration, with the large area diffusion length at 100 microns. The predicted and measured small area results plus the extrapolated large area and module efficiencies were analyzed.

Ribbon solar cells with a thickness of 150 microns were analyzed using the same design with small area diffusion lengths of 200 microns and the large area diffusion length at 100 microns.

For epitaxial growth on metallurgical grade silicon substrates, a back surface field was included. A back surface reflector provides no advantage. The epi thickness was 20 microns. The best small area diffusion length was estimated at 70 microns with the average large area diffusion length estimated at 35 microns.

The estimated optimum small area efficiency for each of these configurations, the best reported experimental results, large area efficiency estimates and module efficiencies are shown for these silicon solar cell configurations in Table I.

This performance analysis has been applied to two of the most popular thin film solar cell configurations, amorphous silicon and a cadmium sulfide-based heterojunction. The amorphous silicon P-I-N structure was analyzed using the best reported data.

There is active development work on three CdS-based heterojunctions. Efficiencies over 10% have been reported for copper sulfide and CdS (with approximately 10% of the Cd replaced by Zn, (ZnCd)S, and copper indium diselenide with (CdZn)S. The addition of Zn increases the voltage due to an improved electron affinity match. Efficiencies of 8.7% have been reported for thin film CdTe/CdS heterojunctions. $CuInSe_2$ has been selected for this analysis since greater intrinsic stability when compared to Cu_2S has been demonstrated. The CdTe structure promises similar efficiencies at higher voltage which would lead to reduced I^2R losses.

An analysis of the optimum performance for these two thin film configurations based on the best materials data leads to an efficiency estimate of 11.1% for amorphous silicon and 14.3% for $CuInSe_2$/(CdZn)S. This difference of 29% is mirrored by the difference in experimentally reported efficiencies of 25%. This comparison is surprising since considerably more effort and resources have been directed towards amorphous silicon. These results plus the large area and module analysis are shown in Table II.

C. Performance Potential

An application of the performance analysis leads to all of the most promising solar cell materials systems in order of potential efficiency:
1. Single Crystal Gallium Arsenide, 2. Single Crystal Silicon, 3. Gallium Arsenide Film (2 to 10 microns thick), 4. Silicon Ribbon (150 microns thick), 5. Crystalline Silicon Film (10 to 30 microns thick), 6. Cadmium Sulfide Film-Based (with CdTe, $CuInSe_2$, Cu_2S or InP), 7. Cadmium Telleride Film, 8. Indium Phosphide Film, 9. Amorphous Silicon Film, and 10. Zinc Phosphide Film.

New materials and new device configurations can be analyzed in terms of their fundamental parameters and assessed in comparison to the preceding list. Since the only new promising material for photovoltaic applications, Zn_3P_2, was discovered in 1976 and since there has been considerable effort expended in this search for new photovoltaic materials, there is a low probability that a significant new photovoltaic material will appear. On the other hand, there is the potential for improved device configurations. Areas for new, important configurations include thin GaAs, thin Si and tandem (stacked junction) solar cells.

2. COST MINIMIZATION

The total solar cell cost for each of the 32 configurations can be estimated using standard cost estimating techniques (8). Since the solar cell market is very price elastic at selling prices below $2 per watt, the total product cost for large scale manufacturing plants (approximately 100 megawatts annual capacity) is required. A consistent analytical methodology based on current manufacturing experience and demonstrated laboratory results has been developed. This technique avoids excessive speculation on the nature of proposed manufacturing processes by firmly basing the cost analysis on present experience and existing manufacturing equipment.

Costs were estimated for three manufacturing plant annual capacities, 1, 10 and 100 megawatts. The product cost for the one megawatt plant was based on an existing pilot-scale manufacturing plant. The 10 megawatt design included demonstrated laboratory processes plus improved capital utilization and labor productivity. A scale factor for the product cost difference for these two designs was calculated and used to extrapolate the cost for the 100 megawatt plant design. The module efficiencies estimated in the previous section were used with the assumptions that 80% of the estimated module energy conversion efficiency would be achieved with the one megawatt plant and 90% of the estimated efficiency for the 10 megawatt plant. This analysis remains firmly grounded in demonstrated results and reduces the required speculation. Of greatest importance, this approach allows all the 32 solar cell configurations to be analyzed on a similar basis.

The <u>total product cost</u> was estimated based on the relationships between raw materials, direct labor and capital costs shown in Table III. The primary differences between this analytical technique and the SAMICS technique (9) are the relative transparency of the assumptions and the inclusion of General Administrative (G&A) costs, rather than only manufacturing costs. These G&A costs increase the total product cost by approximately 30%. The relationships were based on the 100 megawatt plant and some of them, particularly G&A, may be too optimistic for the smallest plant sizes. Nonetheless, this methodology can be applied uniformly to the 32 solar cell configurations and is relatively easy to use and adapt.

The <u>product cost</u> for the <u>three silicon processes</u> described previously, cast ingot, ribbon and epitaxial growth on metallurgical grade substrates were calculated. Silicon costs of $70 and $14 per kilogram were considered. The cost estimates ranged from a high of $4.04 per watt for the one megawatt plant to a low of $1.16 for 100 megawatts. The 100 megawatt plant costs ranged from $1.16 for the ribbon ($14/Kg) to $2.35 with the cast ingot ($70/Kg).

The product cost for the two thin film solar cells analyzed previously, amorphous silicon P-I-N and $CuInSe_2$/CdS heterojunctions, was calculated. Glass substrates and modified vacuum deposition systems were considered for both configurations. The cost estimates ranged from a high of $1.43 for the one megawatt plant to a low of $0.31 for the 100 megawatt plant. The primary difference between the costs for these two solar cell configurations was due to the difference in estimated efficiency. These costs are summarized in Table IV.

The time for completion of these manufacturing plants has been estimated. A time interval of five years has been used to separate the major events of laboratory demonstration, one megawatt and each of the larger plants. Variations in this timing of plus or minus two years are possible.

Cast ingot, advanced process plants were completed in 1981. An epi on MG substrates plant could be operational by 1984. An amorphous silicon plant at the one megawatt scale could be completed by 1985, a ribbon plant by 1986 and a CdS-based heterojunction by 1987. The timing of these plant completions is shown graphically in Figure 1.

3. SUMMARY AND CONCLUSIONS

A consistent basis for the analysis of cost and performance using current manufacturing information and measured materials parameters for any solar cell has been described. It has been applied to three crystalline silicon configurations and two thin film configurations. The advanced ingot technology should continue to dominate the crystalline silicon approaches and the market for the rest of this decade. Provided long-term durability can be demonstrated, a thin film approach could begin to dominate by the end of the decade. The reliability of the thin film product will be more important than ultimate efficiency in the short term. Accordingly, amorphous silicon is favored, provided its durability can be demonstrated. In the longer term the CdS-based thin films show the promise of higher efficiencies.

New approaches which show significant potential include thin film crystalline silicon based on its demonstrated durability, thin film gallium arsenide based on its potential efficiency and tandem (multi-junction) solar cells for enhanced efficiency.

The application of this analytical technique can lead to the focusing of resources on the critical materials and device parameters. This analysis can be used to rank all the different solar cell materials systems in order of potential performance, cost and timing in the marketplace. This focus can reduce the solar cell product development time and cost while increasing the overall probability of success.

References

1. M. Prince, J. Appl. Phys. 26, 534 (1955).
2. J. J. Loferski, J. Appl. Phys. 27, 777 (1956).
3. A. M. Barnett, Proc. 2nd European Community Photovoltaic Solar Energy Conference, Berlin: D. Reidel, 328 (1979).

4. A. M. Barnett and A. Rothwarf, IEEE Trans. Electron Devices, ED-27, 615 (1980).
5. M. Wolf, Proc. IRE 48, 1426 (1960).
6. M. Wolf, Proc. IEEE 51, 674 (1963).
7. H. J. Hovel, "Semiconductors and Semimetals" (R. K. Willardson and A. C. Beer, eds.), Vol. 11, Solar Cells, Academic Press, New York (1966).
8. M. S. Peters and K. D. Timmerhaus, Plant Design and Economics for Chemical Engineers, New York, McGraw-Hill (1982).
9. R. G. Chamberlain, Jet Propulsion Laboratory, Pasadena, CA, JPL publication 78-98, JPL document S101-94 (1979).

Table I
Silicon Solar Cell Efficiency Comparison

Configuration		J_{sc} (mA/cm^2)	V_{oc} (volts)	FF	Efficiency (%)	Large Area Solar Cell Efficiency	Solar Cell Efficiency Within Module/ Module Area Efficiency
N on P Cast Ingot	Estimated	34.9	.594	.802	16.6	14.7	13.4/12.7
	Reported	34.0	.597	.78	15.8		
N on P Ribbon	Estimated	33.8	.594	.802	16.1	14.1	12.9/12.0
	Reported	30.2	.567	.782	13.4		
Epi on MG-Silicon	Estimated	28.6	.614	.822	14.4	12.9	11.8/11.2
	Reported	25.3	.600	.81	12.4		

Table II
Thin Film Solar Cell Efficiency Comparison

Configuration		J_{sc} (mA/cm^2)	V_{oc} (volts)	FF	Efficiency (%)	Large Area Solar Cell Efficiency	Solar Cell Efficiency Within Module/ Module Area Efficiency
Amorphous Silicon	Estimated	16	.92	.75	11.1	9.9	9.0/ 8.7
	Reported	15.2	.88	.601	8.04		
CuInSe$_2$/ CdS	Estimated	39	.54	.68	14.3	12.8	11.7/11.2
	Reported	37.2	.476	.649	10.04		

Table III
Estimation of Total Product Cost

MANUFACTURING COST

Materials Costs
  Raw materials + Utilities
Production Labor
  Operating labor
  Direct supervisory and clerical labor (18% of operating labor)
  Laboratory charge (15% of operating labor)
  Maintenance and repairs (4% of fixed capital investment)
  Plant overhead (40% of operating labor, supervision and maintenance)
Capital Costs
  Fixed capital investment (259% of purchased equipment)
  Depreciation (8 years equipment, etc.; 15 years building, etc.)
  Local taxes (2% of fixed capital investment)
  Insurance (0.7% of fixed capital investment)
  Total capital investment (305% of purchased equipment)

GENERAL EXPENSES

General and Administrative Expenses
  Administrative costs (12% of operating labor, supervision and maintenance)
  Distribution and selling (12% of total product cost)
  Research and development (8% of total product cost)
TOTAL PRODUCT COST = MANUFACTURING COST + GENERAL EXPENSES

Table IV
Solar Cell Cost Comparison

Configuration		Module Efficiency (%)	Total Product Cost ($/watt)		
			One Megawatt	10 Megawatts	100 Megawatts
Silicon Cast Ingot	($70/Kg)	12.7	4.04	3.04	2.35
Cast Ingot	($14/Kg)		2.85	1.98	1.40
Ribbon	($70/Kg)	12.0	3.45	2.21	1.42
Ribbon	($14/Kg)		3.11	1.90	1.16
Silicon Epi on MG Substrate		11.2	3.44	2.17	1.37
Thin Film Amorphous Silicon		8.7	1.43	.74	.38
Thin Film $CuInSe_2$/CdS		11.2	1.12	.59	.31

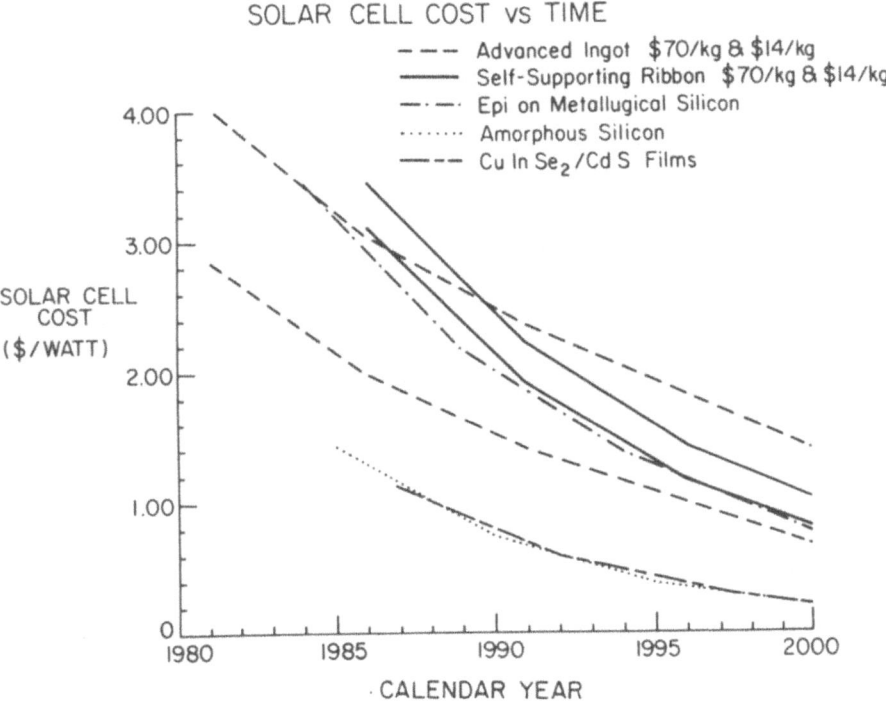

Fig. 1 – Estimate of Five Different Solar Cell Product Costs as a Function of Calendar Year for 1, 10 and 100 Megawatt Plants

POSTER GROUP P8

CRYSTALLINE SILICON SOLAR CELL TECHNOLOGY

- Silicon Material

Aluminothermic reduction of quartz sand

Solar grade floating-zone silicon

Current aspects of the CGE semicrystalline silicon ingots elaboration method

An approach to solargrade silicon layers epitaxially grown on mg silicon substrates

Method of raw material continuous feeding on silicon ribbon growth

Impurity incorporation in R.A.D. polysilicon layers and consequences on their electrical properties

Fast silicon-sheet growth with the supported-web method

Recent developments in multi-wire fixed abrasive slicing technique (FAST)

Critical technology limits to silicon material and sheet production

Economic viability of the UCP semicrystalline silicon sheet technology

- Ion Implantation, Cell Processing

Comparison between various ion beam doping procedures and anneal techniques used in manufacturing silicon solar cells

Status of ion-implanted silicon solar cells

Optimization of pulsed electron beam annealing process for silicon solar cells

Silicon solar cells by ion implantation : E-beam and self annealing

An automated ion implant/pulse anneal machine for low cost silicon cell production

Laser processing in the preparation of high efficiency polycristalline silicon solar cells

The influence of surface texture and thermal treatment on the performance of laser-annealed silicon solar cells

Implantation of boron and boron fluoride compounds into silicon for production of solar cells

Grain boundary photocurrent enhancement in solar cells made by laser diffusion

Dry process for economic cell manufacturing

ALUMINOTHERMIC REDUCTION OF QUARTZ SAND

J. Dietl, C. Holm, E. Sirtl

Heliotronic GmbH

D-8263 Burghausen

Summary

The exothermic reaction of quartz with aluminum is a promising way to make high purity silicon on a large scale - according to

$$3 \ SiO_2 + 4 \ Al \longrightarrow 2 \ Al_2O_3 + 3 \ Si$$

A suitable technical process on this basis requires liquid phases of both oxidic and metallic nature acting as solvents for the reaction products. The two liquid and immiscible systems are represented by an Al-Si alloy and an Al_2S_3-Al_2O_3-SiO_2 slag system. The reaction between them is controlled by an adjustable stirring system which is placed at or close to the phase boundary. Silicon is obtained in form of platelets. Their special structure is described in detail. Typical impurity levels are lower than 10 ppma total (except Al). Prior to further refining the product is milled and leached.

The production of silicon by aluminothermic reduction of SiO_2 must not utilize chunks of quartz rock, as they are required in the electric arc process, but is based on particulate matter, like powder or sand, the latter being available in large quantities and high purity. The semicontinuous character of the process is one of the preconditions for guaranteeing low cost production. Thus recycling of aluminum by recrystallizing the by-product Al_2O_3 from the slag system and reducing it by electrolysis is a rather important part of the process (Fig.1). To initiate the process, quartz sand is introduced into the reaction vessel while stirring the liquefied slag to mix it and aluminum with the slag. Droplets of Al being dispersed there then starts to reduce SiO_2. The final products Si and Al_2O_3 are kept in the liquid state by dissolving in excess Al and Al_2S_3, respectively. This reaction step is followed by a phase in which the Al droplets containing dissolved Si separate from the slag as a liquid metal pool. The reaction takes place in a temperature range of 1100-1200°C. Whereas the carbothermic reduction of SiO_2 is inefficient below 1800°C, the reduction with Al is a pronounced exothermic reaction over the whole temperature range (Fig. 2). At temperatures above 1200°C the production of volatile silicon-sulfur compounds resulting from Al_2S_3 and SiO_2 is enhanced. At 1200°C the solubility of Al_2O_3 in the slag is in the order of 25% by weight. To gain the slag for multiple use, most of the aluminum oxide is getting crystallized by lowering the temperature of the slag thus enabling its removal. To characterize the ternary system Al_2S_3-Al_2O_3-SiO_2, DTA measurements were performed to reveal the melting and solidification temperatures of different mixtures in the concentration field of interest. Al_2S_3 has a melting point of $1100^\circ C$. By adding low concentrations of Al_2O_3 to the Al_2S_3 slag a decrease of the melting point of about $100^\circ C$ is attainable (Tab.1). Higher Al_2O_3 amounts raise the melting temperature of the mixture markedly. The addition of SiO_2 to Al_2S_3 results in a more complex phase system, accompanied by the formation of gaseous silicon-sulfur compounds. The addition of both SiO_2 and Al_2O_3 (about 10 % by weight each) causes a decrease of the melting point of more than $200^\circ C$. The viscosity of the ternary slag melt increases and approximates to the character of a glass melt.

At the temperature of the controlled reduction process the aluminum melt is able to dissolve up to about 65 % by weight according to the Al/Si phase diagram (Fig. 3). When the slag phase and the Al/Si phase have separated, the Al/Si phase is transferred into a crystallization chamber. There it is slowly cooled causing the silicon to recrystallize in form of intergrown platelets (Fig. 4). The platelets are separated from the liquid phase at a temperature slightly above the solidification of the Al/Si eutectic. Finally the Al melt containing residual Si is returned to the reaction vessel.

The silicon crystals have the habitus of tri- or hexagonal platelets. The surface of the platelets shows typical relief patterns as demonstrated in 4 b. The micrograph of a solidified Al/Si mass (5 a) reveals inclusions of Al/Si eutectic encapsulating different impurity precipitates (5 b,c). After removing adherent Al the platelets are comminuted and leached with diluted hydrochloric acid. In table 2 silicon samples are analytically characterized as produced by the aluminothermic reduction of different quartz sands.

Most of the impurity concentrations are below the detection limits of our analytical tools (AAS, ICP, SSMS). The remaining Al level results from temperature dependent solid solubility in silicon when recrystallizing from aluminum melt. To reduce this Al level an additional purification step is required. A sufficiently low B content is achievable by selecting adequate quartz sand qualities.

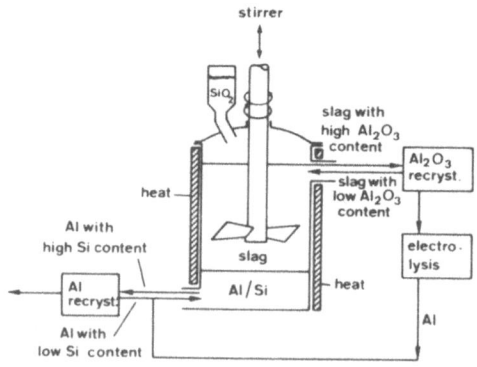

Fig. 1: Processing sequence of the aluminothermic reduction

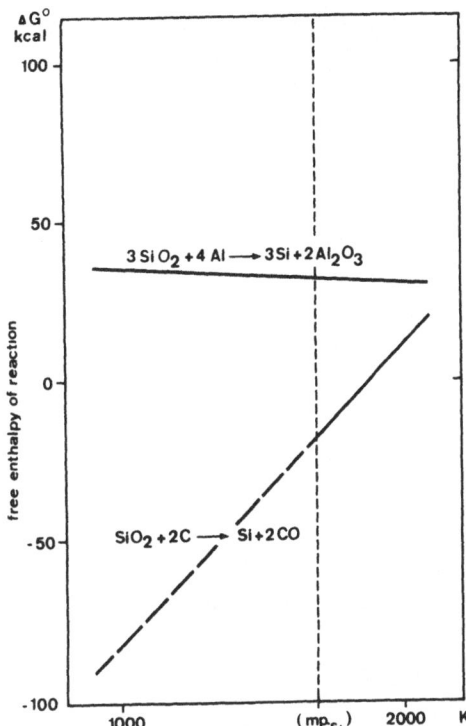

Fig. 2: Gibbs Free energy of aluminothermic and carbothermic reduction of quartz

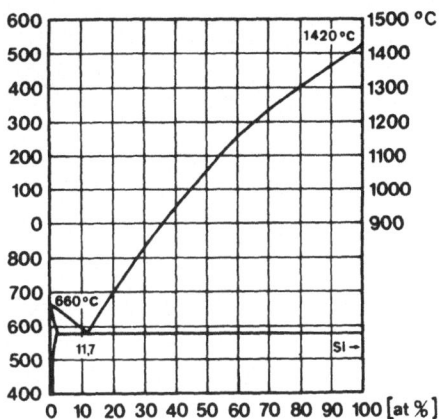

Fig. 3: Phase diagram of the Al/Si system (from Hansen: Constitution of Binary Alloys)

Tab. 1: Melting temperatures of Al_2S_3 with various contents of Al_2O_3 and/or SiO_2

composition [weight %]	100% Al_2S_3	90% Al_2S_3 10% Al_2O_3	80% Al_2S_3 20% Al_2O_3	90% Al_2S_3 10% SiO_2	80% Al_2S_3 20% SiO_2	80% Al_2S_3 10% Al_2O_3 10% SiO_2
melt. temperature [°C]	1109	1046	1060	978	918	902

Fig.4a

4b

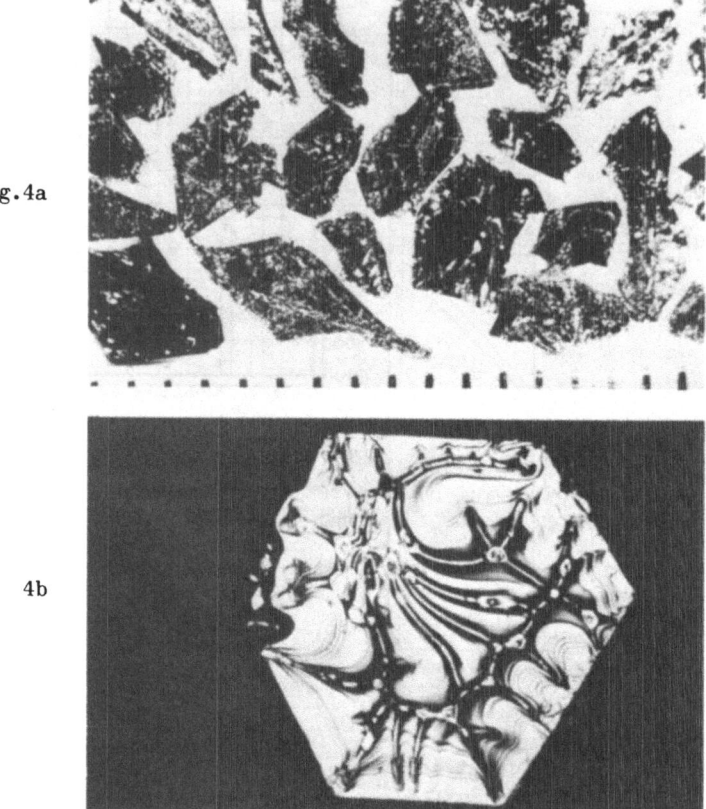

Fig. 4: Silicon recrystallized from Al-solution: a) platelets (6x),
b) topographic structure of platelet (50x)

Tab. 2: impurity concentrations in silicon, made from different quartz sand qualities

impurity-concentration [ppma]	Mn	Cr	Cu	Ni	Fe	Al	Ca	Mg	Ti	B	P
sample 1	<2	<2	<2	<2	<2	327	4,2	<2	0,1	9,1	<1
sample 2	<2	<2	<2	<2	2,5	285	5,7	<2	0,2	15,6	<0,5
sample 3	<1	<1	<1	0,7	0,6	296	<1	<1	0,1	3,9	<0,5

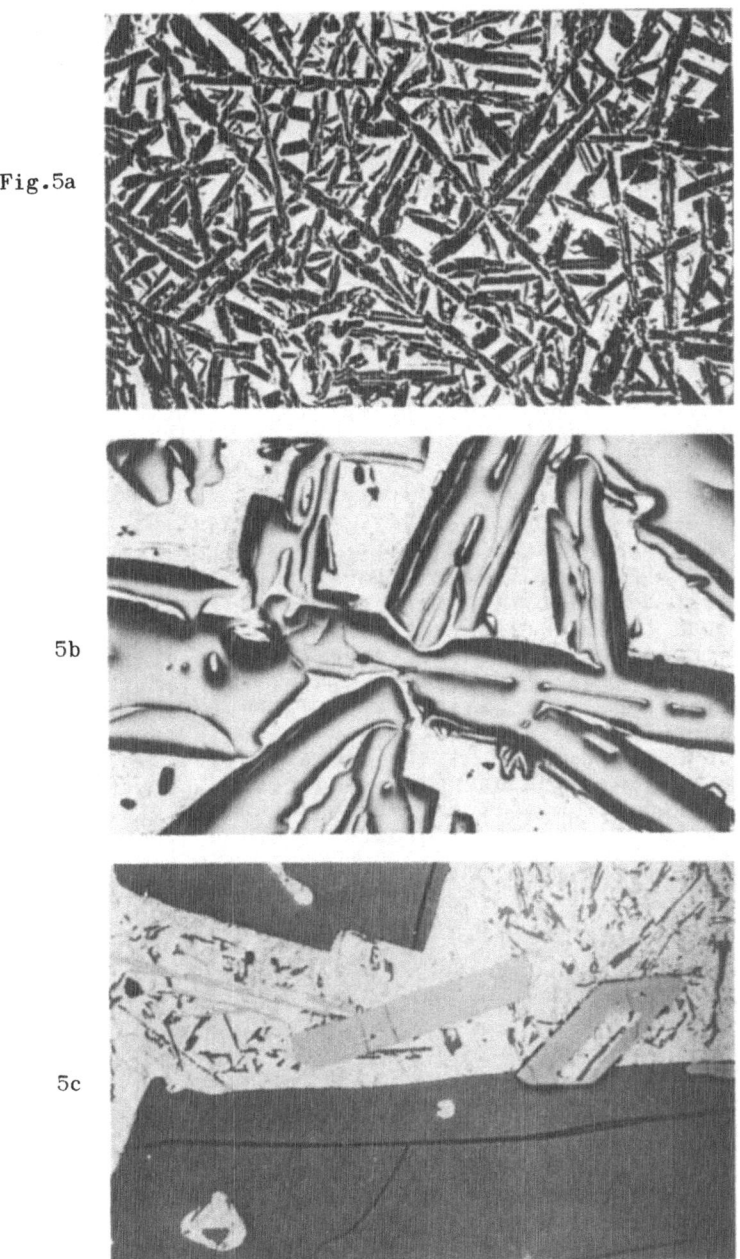

Fig. 5: Silicon recrystallized from Al-solution: a) optical micrograph (3x), inclusion of eutectic phase (35x), c) impurity precipitates (75)

SOLAR GRADE FLOATING-ZONE SILICON

A. LUDSTECK and H.J. FENZL

Siemens AG, Frankfurter Ring 152
D-8000 München 46

Summary

The economic use of photovoltaic solar energy on a large scale is based on the availability of low-cost solar silicon qualified for the production of solar cells with a high conversion efficiency. A proper process for the manufacture of high quality solar silicon is float-zone growth. Starting from a low-cost polycrystalline silicon, i.e. up-graded metallurgical grade silicon, FZ silicon can be produced at lower cost than CZ material because of the higher crystallization rate and the lack of quartz and graphite parts. Furthermore there is a high purification of the silicon by float-zone growth. In particular the metallic impurities which act as lifetime killers can be reduced drastically by one single zone pass. The concentration of metals are at least reduced to less than 1% by one pass. A detailed comparison of the reduction of the concentration of metallic impurities by CZ and FZ growth is given. The impurity level of more than 20 elements in the starting metallurgical silicon as well as in the grown ingots have been determined by neutron activation analysis.

1. INTRODUCTION

The economic use of photovoltaic solar energy on a large scale is based on two conditions: the availability of low-cost solar silicon and a high efficiency of the conversion of sunlight into electrical energy. This solar grade silicon will never be a second-rate material but a high quality product, the properties of which will differ from those of conventional semiconductor grade silicon. Parameters important to the performance of solar cells will be specified as precisely as in the case of semiconductor silicon.
The cost goals can be achieved by the use of metallurgical grade silicon, a material which however has an impurity level in the 1 % range, a value much higher than tolerable for solar cell production. A proper purification of the metallurgical silicon therefore has to be done prior to the production of high

efficiency solar cells. The conversion efficiency is strongly related to the diffusion length of the minority carriers which again demands silicon with a low level of recombination centers.

2. THEORY

Grain boundaries and metallic impurities are the most important defects reducing the carrier lifetime. Single crystals are free of grain boundaries. Additionally the concentration of the metallic impurities can be effectively reduced by a crystal pulling process because the distribution coefficients of metals are much lower than unity, resulting in an accumulation of these impurities in the melt. The equilibrium distribution coefficients of some important elements are tabulated in Fig.1, showing values between 2×10^{-3} for Al and 1×10^{-7} for Ta. Other coefficients are expected to be in the same order of magnitude. Crystals are usually grown under conditions which do not allow an equilibrium at the liquid-solid interface. The effective distribution coefficient which depends on the growth conditions is higher, leading to a purification effect of the crystal growth which is less than that expected from the equilibrium value.

In principle each kind of pulling process, e.g. Czochralski pulling, float zone or sheet growth, results in a purification, the amount of which differs for the different growth processes. The float zone growth seems to be more advantageous than the others for some reasons: It is an economical process. Furthermore, it is the only process which does not contaminate the silicon melt by additional impurities like oxygen and carbon. Finally, the conversion efficiency of solar cells produced from float zoned silicon is 1 to 2 % higher than that of CZ material, because the carrier lifetime of float zoned silicon is higher by a factor of about 10, a value which is well supported by experience.

3. ECONOMIC ANALYSIS

The most important processes for the manufacture of single-crystalline silicon are float-zone growth and Czochralski-pulling. Both methods are equivalent concerning equipment cost, labor, floor area, maintenance and spare parts, but they differ strongly in the crystallization rate and the production volume, resulting in different production cost.

The crystallization rate for a 100 mm diameter rod is 3.8 kg/h for FZ growth and 1.65 kg/h for CZ pulling, leading to production rates of 1.45 kg/h and 0.8 kg/h, respectively. A comparison of the direct production cost of both techniques is given in Fig. 2. The high value of the cost of expendable material in the case of CZ-growth is due to the consumption of crucibles. On the other hand the polysilicon used in a float-zoner has to be rod-shaped and crack-free which demands an additional production step like casting. Therefore the polycrystalline silicon for FZ growth is more expensive than that for Czochralski pulling. Nevertheless the total direct production cost of silicon with 100 mm diameter via the

FZ growth for one pass are only about 75 % of those of the
CZ pulling.

4. EXPERIMENTAL

The samples were prepared from different qualities of
metallurgical grade silicon. For purification by float zone
technique the sample has to be rod shaped. The rods were
produced by casting. The silicon rods were then purified in
a medium size Siemens float zoner VZA6 by one crucible-free
zone pass in a vacuum.

The evaluation of the sample was done by means of
emission spectrography and neutron activation analysis. The
first one gives only qualitative results and is not suitable
for the evaluation of impurity distributions.

The neutron activation analysis gives exact and reliable
quantitative results by means of multi-channel x·ray analysis,
the accuracy of which is roughly 25 % at present. Unfortunately,
two very important elements, B and P, cannot be detected by
neutron activation analysis.

5. RESULTS

The concentration of more than 20 elements have been
determined quantitatively in this way. In Fig. 3 - 5 the
concentrations of those metals have been plotted which result
in a degradation of the conversion efficiency if the treshold
concentrations are exceeded.

Fig. 3 shows the purification of metallurgical grade
silicon by zone refining. The hatched columns show the impurity
concentrations of the starting material which was a rod cast
from normal metallurgical silicon. The metal contents range
from 3×10^{15} atoms/cm^3 for Ta to more than 1×10^{20}
atoms/cm^3 for Al. On the left hand side of the figure the
concentration of those elements is shown which could be determined exactly after one zone pass in a vacuum. The light
columns indicate the remaining impurity levels. For the other
elements - shown on right - the concentrations of the impurities
are below the detection limit after the refining step. The
detection limit is marked by a dash for each element. The real
concentrations are less than indicated by the dashes. They
could be determined exactly by an additional long time activation which was not done in this case. After one single zone
pass in a vacuum the impurity levels were reduced drastically
to less than 1 % of the original concentration for each element.

For comparison the purification effect of the Czochralski
process was also investigated. The result for the same impurities as before is given in figure 4.   Rods were pulled
from normal metallurgical grade silicon in a modern cable
puller. The impurity levels of the starting metallurgical
silicon are shown by the hatched columns as before. The
impurity values are similar to that of the cast rod shown in
Fig. 4. The light columns indicate the concentrations of
metallic impurities after the Czochralski growth. The purification effect of the CZ growth is very small here, the

remaining impurities are nearly as high as in the starting silicon. No change of impurity levels along the axis of the Czochralski grown rod is detectable.

In Fig. 5 the reduction of the impurity level by FZ and Czochralski growth is compared for each individual impurity. Here the relative remaining impurity concentrations which are found in the silicon rods after one pulling process are plotted on a logarithmic scale. In this figure the relative remaining impurity levels after FZ growth and CZ pulling are shown by the hatched and light columns, respectively. Whereas the remaining impurity concentration is below 1 % in the case of float zone growth, the corresponding value is more than 50 % for CZ pulling.

6. DISCUSSION

This result is very surprising because in the case of CZ growth it contradicts the known values of the segregation coefficients. While the purification effect of the float zone process is in agreement with the distribution coefficients within accurately measurable limits, the lacking purification in the case of Czochralski growth cannot be explained in this way.

We do not have an exact explanation of this effect at this stage of the investigations. Further efforts are necessary to find out which of the differences between CZ and FZ growth are responsible for this unexpected result.

Coming back to the refining by float zone technique it is necessary to say that the impurity contents of normal metallurgical grade silicon is too high to bring it down by one single zone pass to a level where the metallic impurities have no influence on the conversion efficiency of solar cells. This is especially true for those elements like Ti or Ta for which the treshold concentrations are in the region or below 1×10^{12} atoms/cm^3. Today more than one zone pass is necessary for a proper crystallisation of metallurgical grade silicon, resulting in the growth of a dislocation free single crystal with a high conversion efficiency even if the resistivity of the silicon is very low. For example, the conversion efficiency of p-type silicon solar cells made from dislocation-free silicon with a resistivity of only 0.04 Ωcm were in the same range as that of cells made from semiconductor grade silicon of 1.5 Ωcm using the same cell technology.

The authors would like to thank Mr. E.W. Haas from KWU Erlangen for the neutron activation analysis.

Element	Distribution Coefficient k_o
Lithium	1×10^{-2}
Copper	4×10^{-4}
Silver	1×10^{-6}
Gold	2.5×10^{-5}
Zinc	1×10^{-5}
Cadmium	1×10^{-6}
Boron	8×10^{-1}
Aluminium	2×10^{-3}
Gallium	8×10^{-3}
Indium	4×10^{-4}
Thallium	1.7×10^{-4}
Carbon	6×10^{-2}
Germanium	3.3×10^{-1}
Tin	1.6×10^{-2}
Nitrogen	7×10^{-4}
Phosphorus	3.5×10^{-1}
Arsenic	3×10^{-1}
Antimony	2.3×10^{-2}
Bismuth	7×10^{-4}
Oxygen	1.25
Sulfur	1×10^{-5}
Manganese	1×10^{-5}
Iron	8×10^{-6}
Cobalt	8×10^{-6}
Nickel	3×10^{-5}
Tantalum	1×10^{-7}

Fig. 1: Distribution Coefficients of Various Impurities

	FZ-Growth	CZ-Growth
Growth Parameters		
Diameter	100 mm	100 mm
Crystallization rate	3,80 kg/h	1,65 kg/h
Production rate	1,45 kg/h	0,80 kg/h
<u>Cell Efficiency</u> (AM 1) [*] (active area)	17,6 %	16,8 %
<u>Direct Production Cost of</u> <u>Monocrystalline Silicon ($/kg)</u>		
Capital cost	3,20	6,90
Laboratory cost	1,20	2,40
Energy cost	2,00	4,10
Cost of expendable materials	2,70	13,50
Cost of solar grade polycrystalline silicon	18,90	11,70
	28,10	38,60

[*] R.B. Godfrey and M.A. Green
App. Phys. Lett. <u>34</u> (11), 1979, 790

Fig. 2: Parameters and Direct Production Cost of Solar Grade Monocrystalline Silicon

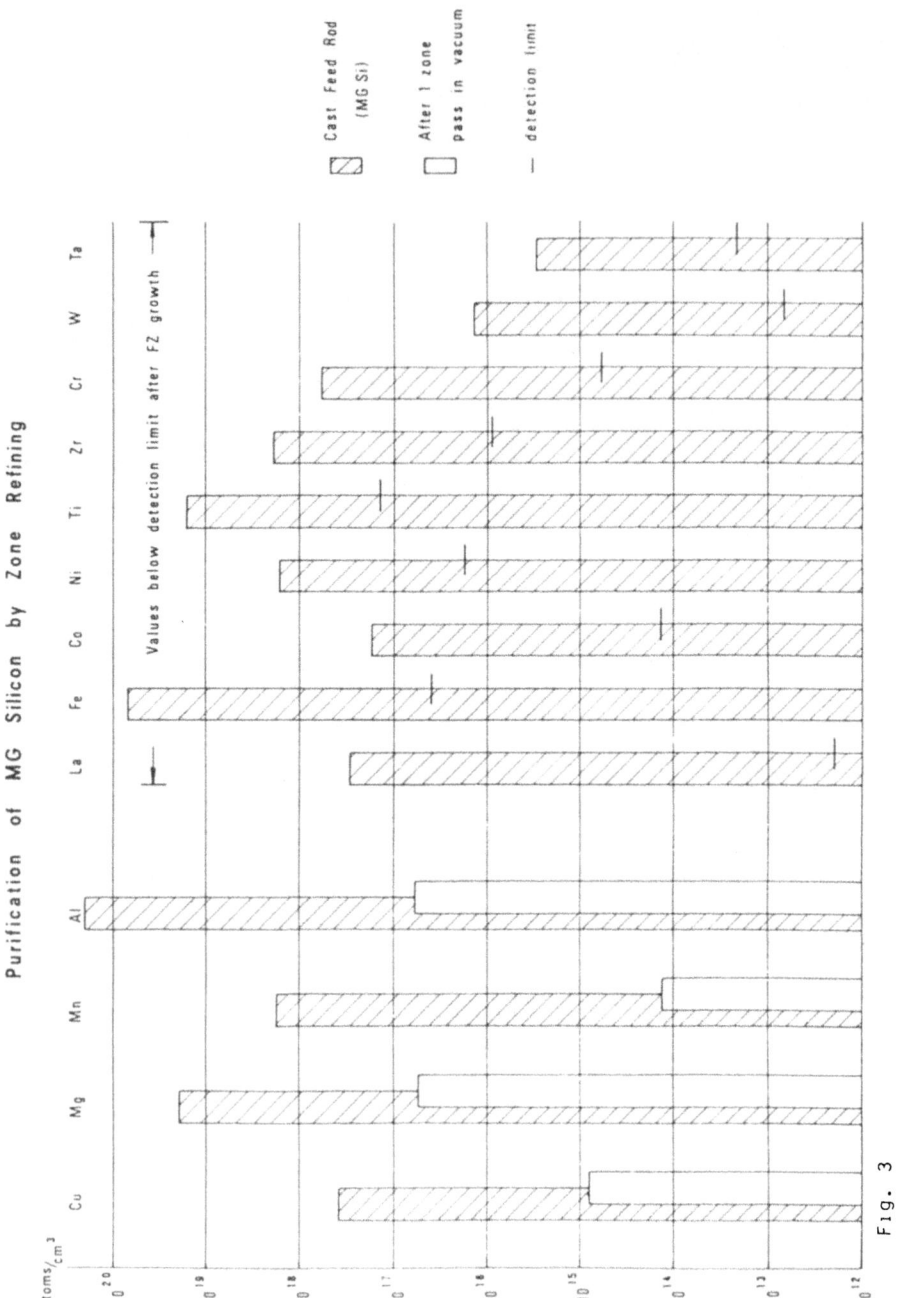

Fig. 3

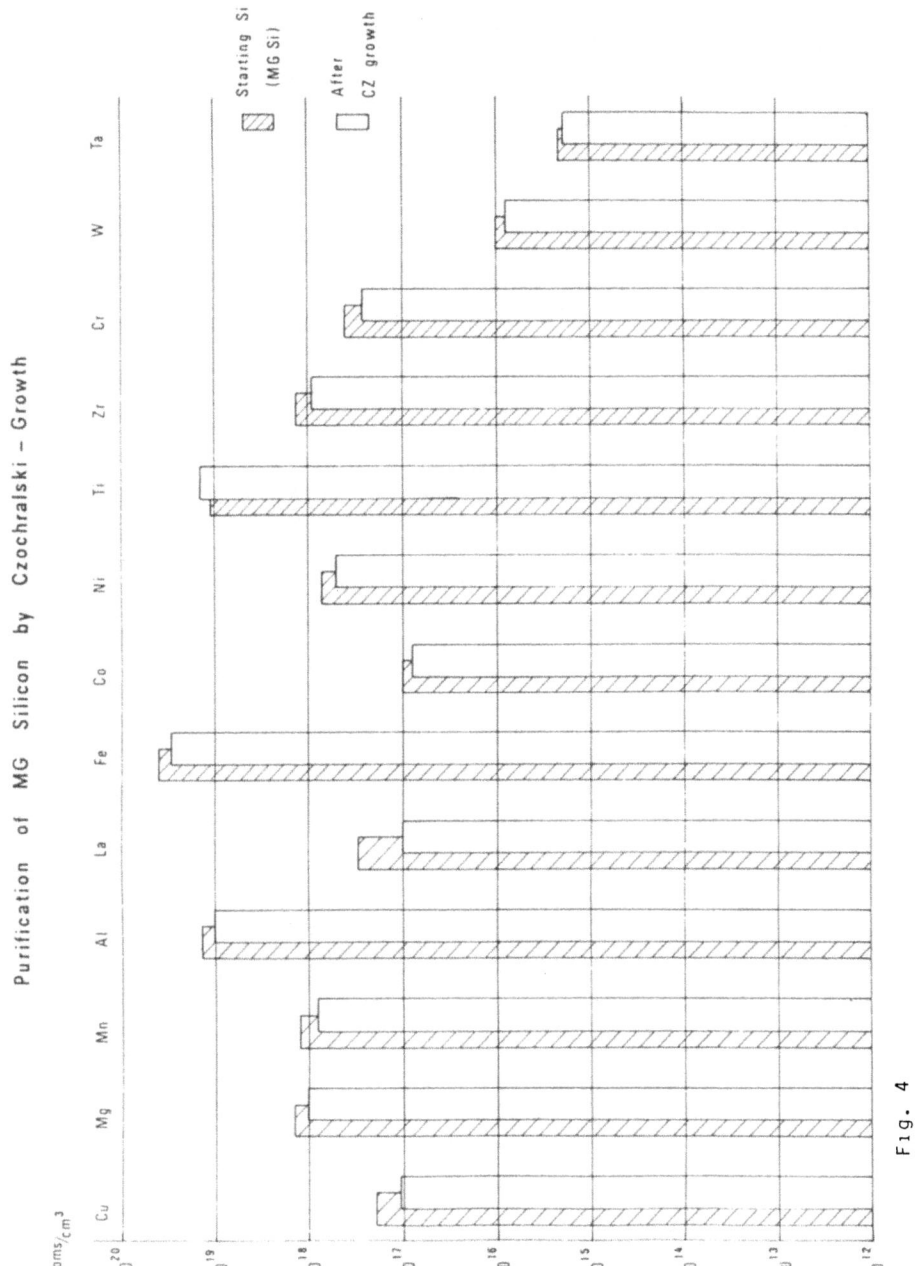

Fig. 4

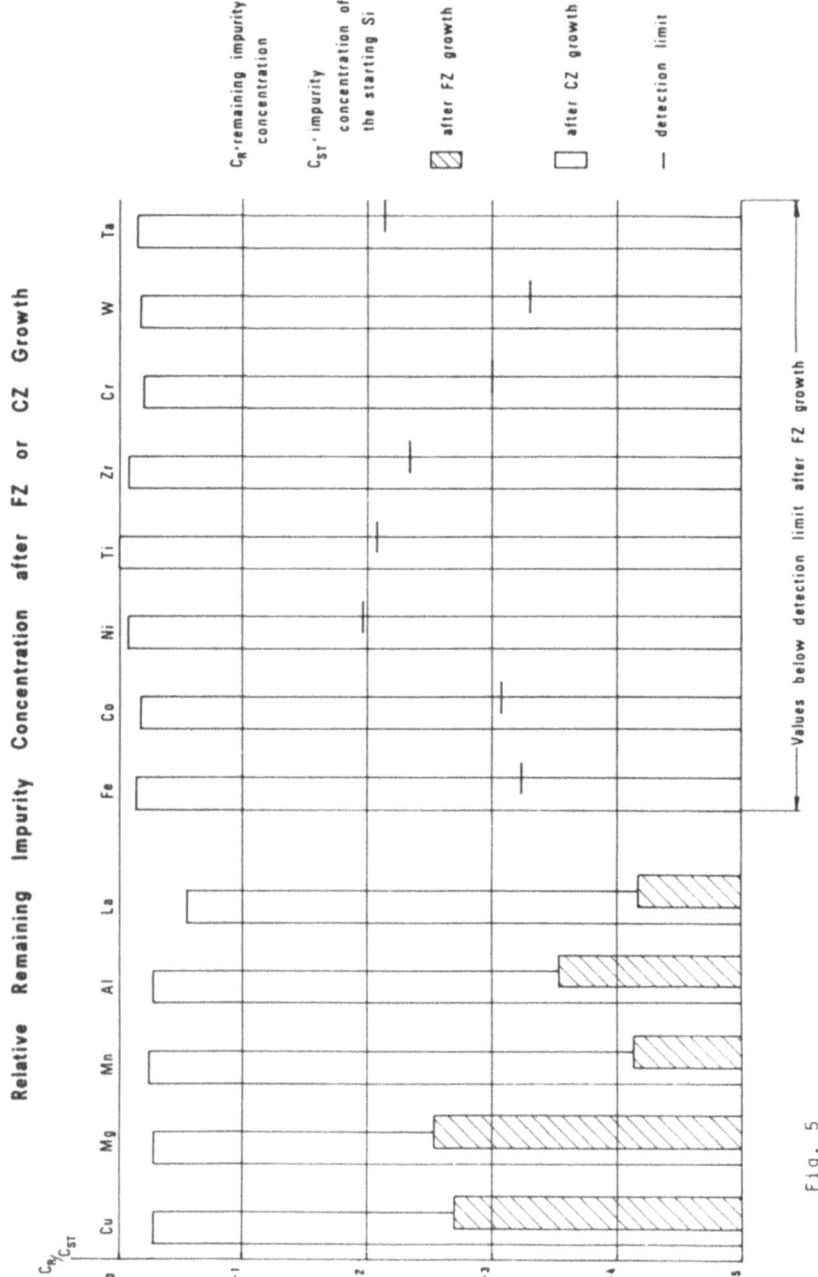

Fig. 5

CURRENT ASPECTS OF THE C.G.E. SEMICRYSTALLINE SILICON INGOTS ELABORATION METHOD

J. FALLY and C. GUENEL
Laboratoires de Marcoussis, Centre de Recherches de la C.G.E.
Route de Nozay - 91460 MARCOUSSIS - FRANCE -

SUMMARY

The feasibility of the elaboration of semicrystalline silicon ingots by the process of unidirectionnal crystallization in the crucible has been demonstrated on the scale of a laboratory furnace in a previous step of our study.
The aim of the present work concerns i) the scale-up : design and construction of a crystallization furnace to the industrial prototype level for the elaboration of 25 kg maximum unitary weight ingots, and ii) the optimization of the properties of the semicrystalline material thus obtained, starting from electronic grade silicon charges.
The main results obtained during this study are :
. basaltic semicrystalline silicon ingots 15 kg in weight were currently produced by means of the scaled-up furnace.
. photovoltaic efficiency of the elaborated semicrystalline cells (100 cm^2) using this material reach the 10 % range.

1. INTRODUCTION

In order to sufficiently lower the production cost of silicon solar cells, we have developped a process intended to obtain silicon semicrystalline blocks to be wafered afterwards. The first part of our program has consisted in a feasibility study of the elaboration of such ingots by a unidirectionnal crystallization method using a fast variant of the so-called Bridgman process. The study led to a material quality sufficient to make solar cells of reasonably good photovoltaic efficiency ($\eta_{AM1} \approx 8$ %) ; the weight of the silicon blocks was about 1 kg [1].
The main limitations of this material were related to crystallographic imperfections, combined with the relatively high bulk-grain impurity content coming from the crucible walls, both limiting the short circuit density current.
Beside the necessary correction of these quality parameters, and in order to thoroughly work out the economic potential of the chosen process, it was needed to increase the capacity of the equipment. So the work here presented concerns the design and construction of an industrial prototype furnace, able to permit the adaptation of the directionnal solidification process to the elaboration of semicrystalline silicon 25 kg ingots and to optimize the characteristics of basaltic, electronic grade semicrystalline silicon blocks.
The use of a heating mode different of that used in the laboratory furnace (i.e. induction vs. resistance), combined with the adaptation to the important increase in weigth of the ingots led us to modify consistently the crystallization method.

2. ELABORATION OF SEMICRYSTALLINE INGOTS

A view of the scaled-up crystallization furnace is given in Fig. 1. The main features of this installation appears on Table I, which gives the current specifications.

As said above, we have introduced a modification of the crystallization method, which is described in Fig. 2. This figure indicates that the seedless crystallization step is operated in a crucible stationnary with respect to the working coil (direct induction in the crucible wall).

During the whole solidification step the heating power is held constant. Heat extraction occuring by radiation directly from the bare bottom of the graphite crucible is sufficiently active to control the heat balance at the freezing front.

In comparison to the straight Bridgman process in which the crucible moves down the thermal gradient, the directionnal crystallization using the scheme described in Fig. 2 leds to a steadier solidification rate, and the shape of the solid/liquid interface is more regular, i.e. nearly flat over a substantial extent of its surface.

These two characteristics greatly improve the crystallographic properties [2] of the basaltic semicrystalline material (grain dimensions, EPD, lineages concentration). On the technological point of view, this evolution of the solidification technique permit a simplification of the furnace construction.

The original procedure we have experienced in order to extract the ingot out of the crucible during the first part of the program [1], was reconducted with the 25 kg furnace.

The current status of the CGE ingots growth method is summarized in Table II. The main attributes of the basaltic ingots elaboration process are : high troughput rate, reusable crucible, moderate furnace investment, and possible use of Solar Grade Silicon.

3. PROPERTIES OF THE SILICON INGOTS

The current material quality criteria of the obtained silicon ingots or wafers are summarized on Table II. We call "finished" ingot the state of the material just prior to the I.D. wafering.

4. SOLAR CELLS PERFORMANCES

Solar cells are realized using square 100 x 100 mm wafers, obtained by I.D. slicing industrial-sized ingots (15 kg unitary weight). The standard process used by PHOTOWATT S.A. includes : NaOH etching, $POCl_3$ diffusion, silver screen-printing metallization and TiO_2 A.R. coating. Table II shows the photovoltaic results obtained on wafers cut parallel to the growth direction ("vertically sliced").

The statistic showed is related to about 70 cells taken in one ingot. The dispersion of results, as indicated by σ_{N-1} is relatively narrow, indicating a good homogeneity of properties in planes parallel to the growth direction.

In order to study the variation of properties in the growth direction, we have laser-scribed the square 100 cm^2 cells in four 25 cm^2 cells, distributed vertically in four sections extending from the bottom to the top-regions of the ingot. The best results obtained on a "bottom" representative 25 cm^2 cell are given on Table II ; AM1 efficiency of 10.25 % was reached.

The best photovoltaic efficiency of 100 cm^2 cells is 9,45 % (Table II, Fig. 3).

The study of the vertical distribution of the photovoltaic parameters in the 15 kg ingot led to the results summarized in Fig. 4. We have observed an almost general trend on our basaltic material of decreasing photovoltaic characteristics, namely JSC, from the botton to the top of ingots. This is ascribed to impurity segregation effect due to the directionnal crystallization (see also [3]).

In Fig. 5, we show an experimental relationship between JSC and L_D (diffusion length of minority carriers), these results being obtained on another basaltic ingot. L_D measurements were derived from the spectral response curves of solar cells, the values of the absorption coefficient of silicon used are those published in [4] .

The study in progress of the influence of slicing direction, i.e. "vertical" vs "horizontal" cells, in connection with the orientation of metallization pattern, was initiated following the hypothesis of a possible improvement of the photocurrent extraction [5] , if the metallization fine lines are normal to the greatest length of the grains.

5. CONCLUSIONS

The main results obtained during this study are :
. basaltic semicrystalline ingots 15 kg unitary weight were produced.
. AM1 photovoltaic efficiency of the elaborated 100 cm^2 semicrystalline cells, made from these ingots, reach the 10 % range.

AKNOWLEDGEMENT

This work was supported jointly by COMES and CEC.

REFERENCES

[1] J. FALLY, C. GUENEL : "Study of the elaboration of semicrystalline silicon ingots" - Proc. 3rd. E.C. Photovoltaic Solar Cell Energy Conf. CANNES, 27-31 oct. 1980 - p. 598.

[2] B. CHALMERS : "Principles of Solidification" John Wiley (1964) p. 306

[3] J. OUALID et al., this Conference.

[4] K.A. DUMAS, R.T. SWIMM "Minority carrier diffusion length and absorption coefficients in silicon sheet material" SPIE vol. 248 - Role of Electro-Optics in Photovoltaic Energy Conversion - July 31, Aug. 1, 1980 - SAN DIEGO - p. 16.

[5] H. MATARE "Defect Electronics in semiconductors" Wiley Interscience (1971) - p. 282.

TABLE I

FEATURES

FURNACE

. DESIGNED CURRENTLY FOR A 25 KG SILICON CAPACITY, POSSIBLE SCALE-UP TO 60 KG
. HEATING IS PROVIDED BY MEDIUM FREQUENCY CURRENTS ($\sim$ 10 kHz).
. THE SQUARE SHAPED COIL SURROUNDS LATERALLY THE CRUCIBLE, A PANCAKE COIL HEATS THE TOP
. POWER ON COIL DURING MELTING IS 45 KW

WORKING ATMOSPHERE IS VACUUM

CRUCIBLE

SQUARE SECTION REUSABLE CRUCIBLE, INTERNAL DIMENSIONS   230x230x360 mm
(MAX CAPACITY = 25 KG SILICON , TYPICAL CAPACITY = 20 KG)

CRUCIBLE IS ESSENTIALLY MADE OF SPECIAL GRADE HIGH DENSITY GRAPHITE.

DURING CRYSTALLIZATION, HEAT FLOW IS CONTROLLED BY MEANS OF

- LATERAL AND TOP GRAPHITE FELT THERMAL INSULATION
- HEAT EXTRACTION BY RADIATION FROM THE BOTTOM OF THE CRUCIBLE

Fig. 1 - Crystallization Furnace

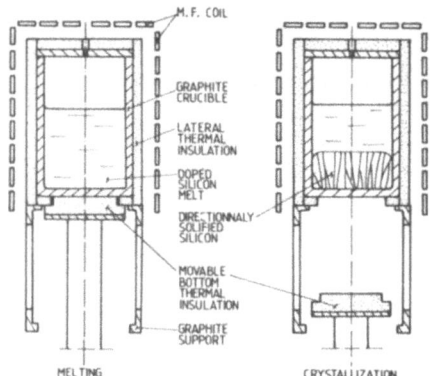

Fig. 2 - Crystallization method

TABLE II

STATUS

MATERIAL
200 x 200 x 160 mm FINISHED SEMICRYSTALLINE BASALTIC BLOCK
SOLIDIFICATION RATE = 6 kg/HR
TYPE P, RESISTIVITY 0.5 - 1 Ω cm
TYPICAL DIFFUSION LENGHT LD = 35 μm
TYPICAL EPD 20% WAFER AREA $\geq 10^5$ EPD cm^{-2}
TYPICAL GRAIN WIDTH 2 to 10 mm
TYPICAL GRAIN LENGTH 20 to 80 mm
ELECTRONIC GRADE SILICON (SEG)

PHOTOVOLTAIC RESULTS

(15 kg FINISHED INGOT - STANDARD PHOTOWATT S A PROCESS)

• ON 100 cm^2 "VERTICALLY SLICED" CELLS

PARAMETER	ηAM1(%)	JSC(mA cm^{-2})	VOC (mV)	FF(%)
MEAN VALUE	8.83	21.76	573	71.25
STD DEVIAT σN-1	0.30	0.40	2.90	1.42
MAX VALUE *	9.45	22.45	577	73
MIN. VALUE *	8.3	20.85	567	70.2

• ON A 24.75 cm^2 "VERTICAL" CELL, BEST RESULTS

PARAMETER	ηAM1(%)	JSC (mA cm^{-2})	VOC (mV)	FF(%)
MAX VALUE *	10.25	24.2	580	73

* simultaneously obtained on a cell

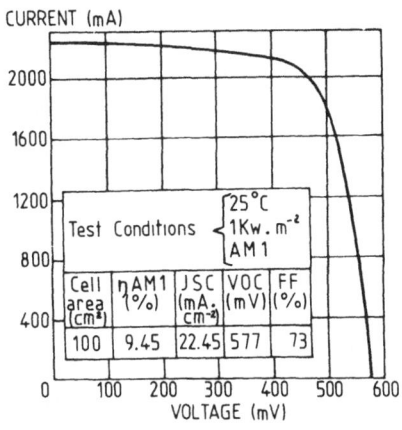

Fig. 3 - Electrical performance of a 100 cm^2 basaltic cell

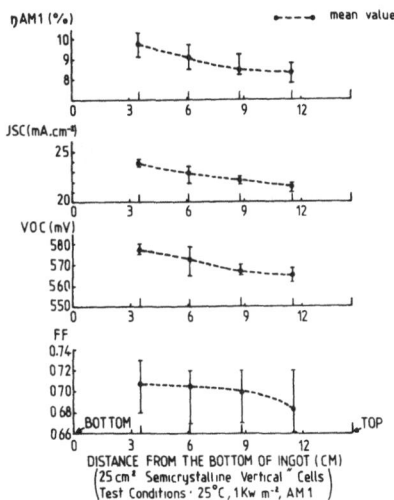

Fig. 4 - Vertical distribution of the photovoltaic parameters in a 15 kg-ingot

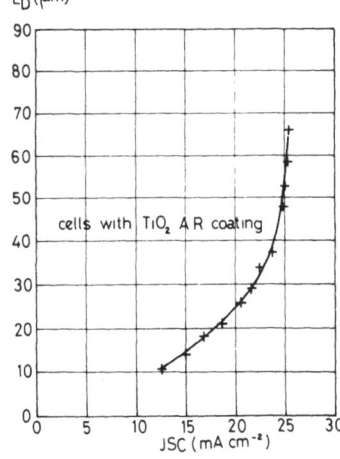

Fig. 5 - Observed relationship between diffusion length LD and short-circuit current density

AN APPROACH TO SOLARGRADE SILICON LAYERS
EPITAXIALLY GROWN ON MG SILICON SUBSTRATES

V. SCHLOSSER, F. KUCHAR, K. SEEGER
Ludwig Boltzmann Institut für Festkörperphysik, Wien and
Institut für Festkörperphysik der Universität Wien, Austria[x]

Summary

We report about a study on the preparation of mg-Si substrates for growing silicon solar cells by CVD. Two processes for preparing mg-Si substrates are investigated: (1) The substrate is etched to a thickness which is in the range of the grown layer thickness. (2) Prior to deposition, a gettering process with a-Si is applied to the substrates. The applicability of a-Si gettering to silicon of varying purity has been demonstrated by results from I-V measurements and photoresponse measurements of pn-junctions prepared on these substrates.

1. INTRODUCTION

A comparison of results obtained from diffused and epitaxial solar cells made on different mg-Si substrates show significantly better results for epitaxial cells (1). In the latter case the requirements for the substrate to be met are the following: (A) good crystallinity (B) the number of impurities diffusing from the substrate into the epitaxial layer must be reduced to the SoG level (2). Especially metals which are present at high concentrations in the substrate are diffusing fast in the growing layer. We suggest two processes to meet the requirements stated in (B): (i) Use of very thin substrates (thickness in the same range like the one of the deposited layer). In this case the doping of the epitaxial layer by outdiffusion is limited by the total impurity content of the substrate rather than by the impurity concentration. (ii) Removal of a large amount of fast diffusing impurities. Taking into account the applicability to a large scale production, it is desirable to integrate the gettering process in the CVD process itself. Gettering by heavily damaged silicon surfaces is well known (3-5). The method suggested here is to deposit an a-Si layer on the substrate instead of inducing a damaged surface on the substrate. This is done in situ prior to deposition of the epitaxial layer.

2. EXPERIMENTAL

Substrates from single crystal silicon (CZ ⟨100⟩ oriented 6-10Ωcm p-type), polycrystalline silicon (SILSO Wacker Chemitronic, p-type), and mg-Si (pulled from the melt without significant purification) were used. Substrate size was either 1.5 x 1.5 cm^2 or 1 cm diameter. Chemical vapor deposition was carried out in a conventionally operated CVD system (rec-

---

[x] Work supported by Shell-Austria, the Forschungsförderungsfonds der Gewerblichen Wirtschaft, Austria, and the Bundesministerium für Wissenschaft und Forschung, Austria.

tangular reactor and RF-heated susceptor). pn-junctions were either formed by depositing n-type layers on the substrates using PH_3 as the dopant gas or by P diffusion from a P_2O_5 source. For gettering with a-Si the same CVD system was used. In the case of epitaxial layers, the gettering was carried out either in situ prior to deposition or after the layers were grown. If P diffusion was made the substrate was gettered before the diffusion. Additional doping of single crystals with Au was carried out by evaporating Au onto the substrate surface and subsequent solid state diffusion.

pn-junctions on gettered and ungettered substrates were examined by I-V measurements in the dark and under illumination as well as by spectral response measurements. The minority carrier diffusion length in the base region was obtained with a short circuit current method using published values for the absorption coefficient (6,7).

On mg-Si substrates etched to thicknesses ranging from 30-100µm n-Si, layers, 7-50µm thick, were grown. For van der Pauw measurements of the epitaxial layers, the substrates were removed by wet etching.

3. RESULTS AND DISCUSSION

3.1. a-Si gettering of mg-Si substrates

Considerations with simple assumptions (not taking into account grain boundary effects or high impurity concentrations) are made. As a result a-Si gettering should be effective on fast diffusing impurities since the gettering rate G_I should be proportional to $D_I \times t_a$ where D_I is the diffusion coefficent in the crystal and t_a the annealing time. For Cu and Ni experimentally determined values of G_I for ion damaged silicon gettering are published (3). For Ti G_I was calculated from the following equation (3):

$$G_I = C_I D_I t_a / d \qquad (3.1)$$

where C_I is the solid solubility in the crystal, d is the substrate thickness and the other expressions are the same as above. Although the use of C_I in the equation does not seem to be a good assumption when applying a-Si the calculated G_I could give a rough idea of the gettering effect to be expected. Table I shows calculated values of the impurity concentration N_I in the layers grown on mg-Si substrates taking into account (i) the effect of a thin substate (ii) gettering with a-Si using gettering rates described above. The values are evaluated from the diffusion equation for typical growth conditions. In Table II measured values of the Hall mobility μ_H and the resitivity ρ of epitaxial layers on gettered substrates are compared to μ_H and ρ obtained for mg-Si substrates. ρ of the layers corresponds to the dopant concentration expected from the results on pure substrates. μ_H is very close to values of μ_H in pure single crystals with the same dopant concentration (7).

3.2. Gettering of Au by a-Si in single crystal substrates

The effect of a-Si gettering is demonstrated on Au doped substrates. Au is one of the best known fast diffusing interstitial impurities in silicon. Pure single crystal substrates are used in order to exclude any additional effect from grain boundaries or other impurities. The solid state diffusion was made under conditions where the effect of substitutional doping is negligible (7). a-Si was deposited on two samples of a series of three. One sample with a-Si and one without a-Si were annealed for

30 min. The second sample with a-Si was annealed for 60 min at the same temperature. After the a-Si was removed by wet etching, P was diffused simultanously in the three samples. Results of the measurements of the minority carrier diffusion length are shown in Fig.1. Fig.2 shows the influence of the remaining Au on the reverse biased diode in the dark.

3.3. Effect of a-Si gettering on solar cell parameters

The influence of gettering on the solar cell performance is demonstrated with pn-junctions on polycrystalline SILSO material with different cell configurations. Two junctions were prepared simultanously for each experiment. One junction was gettered with a-Si. AM1 effeciencies of the ungettered control cells and the gettered cells are compared in Table III. Fig.3 shows the photoresponse $s(\lambda)$ for the gettered cells relative to $s(\lambda)$ of the control cells. From the improved values in the short wavelength range for gettered cells, the effect of the gettering process as a function of the distance from the surface can be clearly seen.

4. CONCLUSIONS

The applicability of a-Si gettering of impurities in silicon of varying purity has been demonstrated by several experiments. Epitaxial layers grown on gettered mg-Si substrates show properties comparable to those on pure substrates. Experiments on Au doped single crystal demonstrated the effectiveness of the process for fast diffusing impurities. From these results and results on gettered solar cells produced on SILSO substrates, the conclusion is drawn that the a-Si gettering can be a useful process for improving cell parameters when using impure substrates.

REFERENCES

(1) F.SCHMID, M.BASARAN, C.P.KHATTAK, 3rd Photovoltaic Solarenergy Conference 1980, W.Palz, ed. (D.Reidel, 1980) p253

(2) G.F.WAKEFIELD, Solar Silicon Definition
(NSF/RANN/SE/AER 75-03972/FR/75/1, Final Report 1975)

(3) T.M.BUCK et al., Appl.Phys.Lett. 21 (1972) p485

(4) C.M.HSIEM et al., Appl.Phys.Lett. 22 (1973) p238

(5) T.SAITOH et al., 14th IEEE PV Specialists Conf.Rec.(New York 1980) p912

(6) E.D.STOKES, T.L.CHU, Appl.Phys.Lett. 30 (1977) p425

(7) W.R.RUNYAN, Silicon Semiconductor Technology (McGraw-Hill 1965)

TABLE I. Cu, Ni AND Ti IMPURITY CONCENTRATIONS IN THE SUBSTRATE AND THE EPITAXIAL LAYERS

ELEMENT	CONCENTRATION IN MG-SI [PPMA]	$N_I(X) \times 10^{17}$ [cm^{-3}] IN EPITAXIAL LAYERS 10μM AND 25μM FROM SUBSTRATE SURFACE AS CALCULATED FOR TYPICAL GROWTH CONDITIONS:					
		(1) THICK SUBSTRATE		(2) THIN SUBSTRATE		(3) THIN, GETTERED SUBSTRATE	
		$N_I(10)$	$N_I(25)$	$N_I(10)$	$N_I(25)$	$N_I(10)$	$N_I(25)$
Cu	30	15	15	7.4	7.4	-	-
Ni	300	140	120	74	71	-	-
Ti	60	10	0.48	9.3	0.42	9.3	0.42

TABLE II. RESULTS FROM VAN DER PAUW MEASUREMENTS ON MG-SI SUBSTRATES AND EPITAXIAL LAYERS

	SUBSTRATES	LAYERS
	P - TYPE	N - TYPE
ς [Ωcm]	0.03	0.02 - 0.03
μ_H [$cm^2 V^{-1} s^{-1}$]	50	400 - 450
$\|N_D - N_A\|$ [cm^{-3}]	4×10^{18}	$(5 - 7) \times 10^{17}$

TABLE III. COMPARISON OF AM1 EFFICIENCIES FOR GETTERED AND UNGETTERED CELLS

RUN NO.	CELL CONFIGURATION	$\langle L_N \rangle$ [μM]		$\eta(AM1)^*$ [%]
1	5μM N-SI (CVD)	2	C.C	4.1
			G.C	4.8
2	2.5μM N-SI (CVD)	10	C.C	5.6
			G.C	6.3
3	< 1μM N-SI (P-DIFFUSION)	35	C.C	5.9
			G.C	7.0

*WITHOUT AR COATING UNDER SIMULATED AM1 CONDITIONS

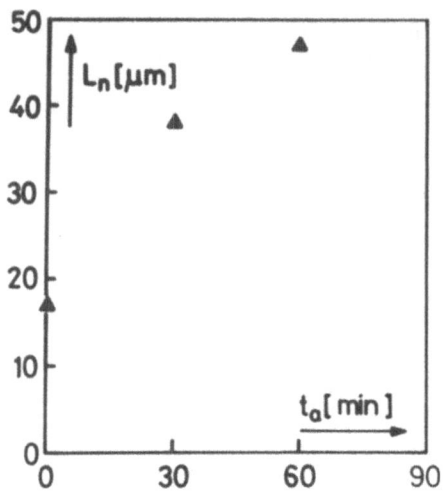

FIG.1. MINORITY CARRIER DIFFUSION LENGTH OF Au DOPED PN-JUNCTIONS AS A FUNCTION OF THE ANNEALING TIME

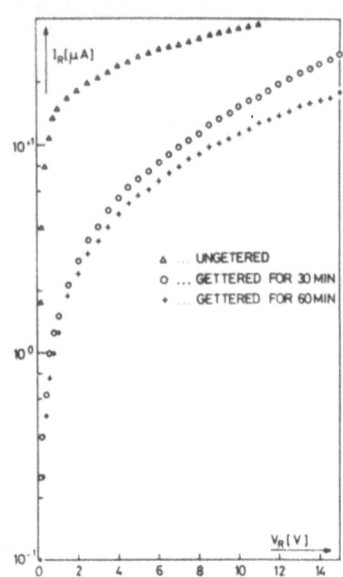

FIG.2 THE INFLUENCE OF a-Si GETTERING ON THE REVERSE BIAS DIODE CHARACTE- RISTIC OF Au DOPED PN-JUNCTIONS

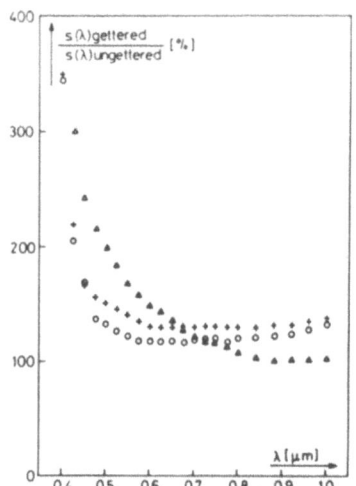

FIG.3 RELATIVE SPECTRAL RESPONSE OF GETTERED CELLS COMPARED TO CONTROL CELLS
- △ CORRESPONDS TO RUN 1
- + CORRESPONDS TO RUN 2
- ○ CORRESPONDS TO RUN 3
  IN TABLE III

METHOD OF RAW MATERIAL CONTINUOUS FEEDING ON SILICON RIBBON GROWTH

N. MAKI, T. SAWADA, M. IIDA, T. MATSUI, K. TAMAI and M. NAKAGAWA
Electron Device Engineering Lab,
Electron Tube & Device Div, Toshiba Corporation
1, Komukai Toshiba-cho, Saiwaiku, Kawasaki, 210, Japan

SUMMARY

   To get low cost solar cell substrates, it is needed to cut down both material cost and process cost. Ribbon crystal pulling method has a potential to satisfy this requirement.
   On the material cost, ribbon crystal pulling method has an advantage for cutting down the cost. It is well known that there is no slicing loss in the method when to get the substrate. However, on the process cost, in spite of having a strong potential to decrease the cost, not so much concern has been paid on this essential feature of the method.
   Process cost would be decreased when solar cell substrates are produced continuously from raw material. In this paper, we report an approach to make all the processes of solar cell substrate production be combined in one continuous prosess. Ribbon crystal pulling method is useful to construct this process.

1. INTRODUCTION

   Ribbon crystal pulling method has an ability to decrease solar cell price drastically.
   One feature of the method is widely known, that it needs no slicing process to get solar cell substrate. Although, another feature is less known, or less said often, even if this will contribute much for the cost reduction of solar cells.
   Ribbon crystal pulling method has a potential to become a complete continuous process to get the substrates. Continuous process is meaningful in the following points, that first to pull ribbon crystals for a long duration, and second, to pull them from a smaller crucible which needs less heating power than used commonly. These points are potentially effective to cut down process cost of the products. In different ways such as Czochralski method, there is no possibility to get these effectiveness.
   Silicon ribbon crystals can be grown directly from the molten silicon by several ways.[1~5] Some of these ways can be classified into the methods which use shaping die on which ribbon crystals are grown. In these methods of category, it is possible to feed raw material directly into the crucible. The temperature condition of the die top (where the ribbon is grown) can be controlled almost separately from that of the die bottom (where the raw material is fed). Then temperature at which silicon ribbon is grown does not change much even when low temperature raw material is fed. It is said that, in ribbon crystal pulling method, raw material can be fed easier than in the other method. Then we can make this raw

material feeding stage as a continuous feeding process.

Adding to this advantage, ribbon crystal has some more preferable points to be developed as a cotinuous process. One point is that the shape of ribbon crystal is adequate to be pulled continuously. Rollers touched on the ribbon surface can give an continuous drive motion to it. Then we can make this pulling stage as a continuous pulling process.

Another point is that ribbon crystal can be cut away easily, so that previously grown products do not prevent succesive crystal pulling. It is done, for example, by laser beam. Then we can make this removal stage as a continuous removal process.

Combining above mentioned continuous feeding process of raw material, contineous pulling process, and continious removal process of the products, we can arrange continuous process of ribbon crystal pulling method. We describe the results of our development.

2. APPARATUS OF CONTINUOUS PROCESS AND EXPERIMENTS

2.1 Continuous feeding process

Feeding apparatus is perspectively illustrated in Fig.1. This apparatus has three components: feeding device, feed rate controller and liquid level sensor. The feeding device supplies granular silicon into the crucible, and the feed rate controller regulates supply volume of granular silicon. The liquid level sensor plays a role of setting upper and lower limits of molten silicon level height.

The first problem on this process was that when granular silicon was fed, the temperature of molten silicon in the crucible decreased and consequently that of molten silicon at the die top also decreased. This lowered the position of solid-liquid interface toward the die top, and ribbon growth could not continue successively.

In order to keep ribbon growth, next procedure was required. Feed rate was gradually increased in the range of 0.5g/min to 2g/min by the feed rate controller, and at the same time, growth speed and heater power was adjusted to keep the position of solid-liquid interface slightly above the die top.

The second problem was that it was not sure to hold liquid level constantly by using only feed rate controller and pulling speed controller. This was solved by using an on-off type liquid level sensor. This consists of three carbon electrodes. The first electrode sets the upper level of molten silicon, the second electrode sets the lower level, and the third is common. In the case of ribbon growth using die, the allowance of liquid level height is rather large, requirement is only to prevent overflow and exhaustion of molten silicon.

Liquid level sensor works as follows. For example, when the first electrode touches on the molten silicon, the current flows between the first and third. The electric signal is fed back to the feed rate controller and then the feed rate is decreased to the optimum value.

The last problem was whether the resistivity of ribbon crystal was kept homogeneously. Our method of impurity doping is simple, that only granular silicon boron alloy is mixed into the raw material as dopant. It is not sure to keep homogeneity in this way. Adequate ratio of boron to silicon of the dopant was calculated taking account of molten silicon volume, the size of impurity alloy. By above method, resistivity of the ribbon was maintained in the range of $0.4\,\Omega$-cm to $1.5\,\Omega$-cm.

2.2 Continuous pulling process

The continuous pulling is to grow a straight and long ribbon crystal stably. The perspective of the apparatus is shown in Fig.2. This is functionally divided into a roller drive mechanism and a limiter mechanism. In the roller drive mechanism there are two pairs of rollers and the face of rollers are in contact with ribbon surface to give the ribbon upward pulling motion.

The limiter mechanism which is set between the die top and the rollers, plays a role of preventing ribbon crystal from swinging, for a swinging of the ribbon crystal during the growth often resulted in stop of the continuous growth. The limiter mechanism consists of a pair of carbon plates which are set to make enough slit separation that the ribbons can pass through. This slit separation is changeable by replacing four carbon spacers which are placed at the corners of the carbon plate.

Experiments proved that the slit separation was suitable when selected about twice as large as ribbon thickness. It is important that the die top, the limiter slit and the tangent of two rollers are in a nearly flat hypothetical plane even at high temperature. The die top should be able to be viewed down through the center of the limiter slit from an upper side. At the case of good adjustment was attained, ribbon crystal which had the roughness of only in the range of 20μm to 30μm was obtained.

2.3 Continious removal process

Mechanism of continuous removal consists of two devices: laser cutting device and a ribbon transfer device. Main part of laser cutting device is shown in Fig. 3, where a ribbon crystal is being cut and a lot of spark scatterings can be seen. Ribbon crystals could be cut into any size by scanning YAG laser beam along the ribbon width direction simultaneously when they were growing. Focussing of laser beam was performed by using a TV camera focussing mechanism.

During cutting procedure, ribbon crystals were often cracked mainly along the pulling direction. This was considered that there remained thermal stress in wide ribbon crystals. To reduce the residual stress, a pair of resistance heaters was set at the outside of the both surface of limiter carbon plates and ribbon crystals were annealed when growth was continued. Maximum temperature in the limiter slit was about 1000 °C measured by a Pt-13%RhPt thermocouple. After annealing, silicon ribbons could be cut off almost without cracking.

One problem was how to prevent air leakage from atmosphare to the furnace. Several ways were adopted and one of these were gas curtain mechanism. This was placed on the top of the furnace from which ribbons were pulled out. This was a device which made a lamination gas flow to prevent air leakage. About 20 l/min argon was blown from a small slit. Ribbon quality was improved by using gus curtain.

2.4 Ribbon growth

Schematic view of the furnace for wide ribbon growth is illustrated in Fig. 4. Heating was done by using a pair of resistance heaters in an argon atmosphere. Each heater was made of a carbon plate. The furnace contained a fused silica crucible which was a little bit shorter than the heater. The maximum volume of silicon raw material was about 600g. Heat shields were set around the heater, crucible and other components. A die was stood upright on the crucible through the uppermost heat shield and the die top was slightly higher than the top of the shield. The die was made of purified carbon and its width was 105mm. The temperature was controled by a signal of thermocouple which was set below the heat shields. Die top temperature uniformity was kept attainable by operating two carbon blocks placed movable toward ribbon width direction.

Ribbon crystals grown in this furnace were 94mm in width and 0.5mm in thickness and pulling speed was kept about 20mm/min. 94mm square and 0.5mm thick silicon substrates were obtained after the laser beam cutting, as shown in Fig. 5.

3. RESULTS

A method of continuous process on ribbon growth was developed. This consisted continuous feeding, continuous pulling and continuous removal process. Continuous feeding process was establisbed by feeding granular silicon directly into the molten silicon in the crucible. Three problems in this process, die top temperature fluctuation, deviation of liquid level, and inhomogeneity of resistivity were solved.

Continuous pulling was achieved by using roller drive mechanism. Tendency to pull wavy shaped ribbon was eliminated by using a pair of carbon plates set parallel configuration above the die top.

Continuous removal was achieved by using YAG laser cutting mechanism and transfer mechanism. The problem of air leakage into the furnace was solved by using gas curtain mechanism.

By above continuous process, we successfully pulled ribbon crystals of 94mm in width and 0.5mm in thickness continuously and obtained 94mm x 94mm x 0.5mm silicon substrates.

The conductivity was p type, the resistivity was kept in the range of $0.4\,\Omega$-cm to $1.5\,\Omega$-cm and, the roughness of ribbon surface was within about 20 30um.

4. CONCLUSION

As described above, most of the problems of continuous pulling of ribbon crystals have been solved. 94mm wide ribbon is continuously pulled when glanular silicon is fed continuously. And then ribbon crystal is cut by YAG laser continually into adequate size for solar cells.

We believe this success promisses to contribute the cost reduction of solar cell substrates. We have a plan to confirm the effectiveness of the results of this work, on a pilot plant level.

ACKNOWLEDGEMENT

The authors are greatly grateful to Dr. H. Hirano for his encouragement and useful discussion. We would also like to thank our colleagues for providing the mechanical system. This work is financially supported by the SUNSHINE PROJECT, MITI through a contract with New Energy Development Organization.

REFERENCES

1. S. O'hara and A.I. Bennett: J. Appl. phys. 35 (1964) 686.
2. A. V. Stepanov: Bull, Acad, Sci.USSR 33 (1969) 1775.
3. T. Surek and B. Chalmers: J. Cryst. Growth, 29 (1975) 1.
4. T. F. Ciszek and G. H. Schwuttke: Phys. Status Solidi (a), 27 (1975) 231.
5. T. F. Ciszek, M. Schietzelt, L. L. Kazmersky, J. L. Hurd and B. Bernelius: Proc. of the 14th IEEE Photovoltaic Specialists Conference, p.581 (1981).

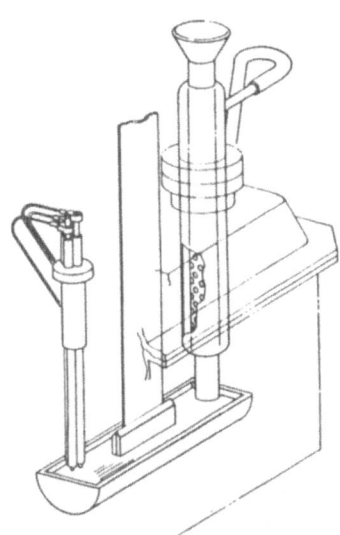

Fig. 1. Apparatus for fecaing granular silicon into the crucible.

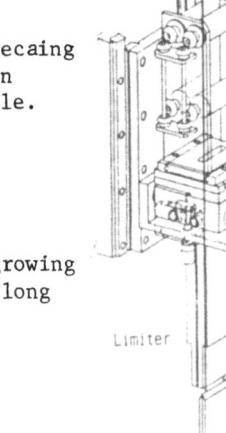

Fig. 2. Apparatus for growing a straight and long ribbon crystal staloly.

Fig. 3. Laser cutting of a ribbon crystal. A lot of sparkes are scattering.

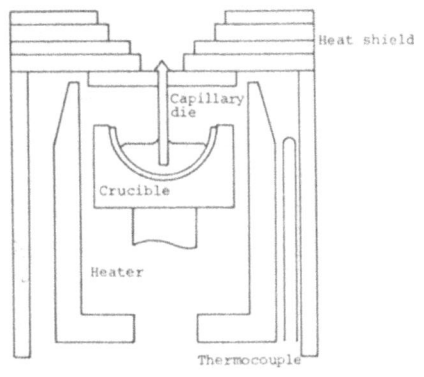

Fig. 4. Schematic view of the furnace for wide ribbon growth.

Fig. 5. 94mm square and 0.5mm thick silicon substrate.

IMPURITY INCORPORATION IN R.A.D. POLYSILICON LAYERS
AND CONSEQUENCES ON THEIR ELECTRICAL PROPERTIES*

G. REVEL, N. DESCHAMPS
Laboratoire Pierre Süe, CEN/SACLAY - 91191 GIF sur YVETTE CEDEX (FRANCE)

J.P. DEVILLE
Laboratoire d'Etude des Surfaces, Université LOUIS PASTEUR
4, rue Blaise Pascal - 67070 STRASBOURG CEDEX (FRANCE)

C. TEXIER-HERVO, C. BELOUET
Laboratoire de MARCOUSSIS, Centre de Recherches de la C.G.E.
Route de Nozay - 91460 MARCOUSSIS (FRANCE)

Summary

The growth of polysilicon layers by direct freezing of a film on a carbon ribbon by the RAD process goes along with a contamination of the silicon melt by carbon and its compositional impurities. This paper reports on this contamination effect studied mainly by means of neutron activation analyses (NAA) and its consequences on the electrical properties of the layers. The purification of the carbon ribbons in chlorine at high temperatures results in low contamination levels of the melt ; NAA evidenced a significant segregation at the growth front, the effective partition coefficients determined being in the 10^{-1} to 10^{-3} range. Even though impurities are shown to impair the device performances, it is concluded that they are not necessarily incorporated in an electrically active form. Finally, AM_1 conversion efficiencies of 12% are reported for RAD cells 4 cm^2 in size.

1. INTRODUCTION

The growth of polysilicon layers by direct freezing of a silicon film on a carbon ribbon by the RAD process in view of achieving low-cost solar cells goes along with a contamination of the layers by carbon and its compositional impurities. This contamination may be a severe limitation to the performances of the solar cells made from this material.

The deleterious effects of impurities, including carbon and oxygen, on the silicon solar cell performances have been discussed in a voluminous literature, e.g. (2, 3). In a series of papers, R.H. HOPKINS et al. have evaluated the role of a number of impurities, among which the transition metals, on solar cells made from Czochralski and float-zone single crystals grown with controlled additions of impurities (2). These authors established a quantitative classification of the impurities for their effectiveness in reducing the overall cell efficiency ; in particular, they reported an efficiency degradation at concentrations as low as 10^{11} atoms.cm^{-3} for titanium and molybdenum. However, it was also shown in this work that high temperature treatments of the silicon wafers in an ambient containing $POCl_3$ could neutralize, or "getter", contaminants incorporated during crystal growth.

Similar conclusions were drawn concerning the complex roles of carbon and oxygen (3). Thus, the extent to which impurities may impair silicon

* This work was supported by the Commission of European Communities (CEC) and the Commissariat à l'Energie Solaire (COMES).

solar cell performances appear to depend first on crystal growth and second, on the subsequent processing steps leading to the final device.

This paper reports on a comprehensive study of the contamination effects in the RAD process ; the impurity incorporation during growth and its consequences on the performances of solar cells are discussed.

2. EXPERIMENTAL

These studies relate to growth runs conducted with equipment described elsewhere (1, 4). Runs noted B and C were made using the puller equipped with a continuous replenishment of the melt.

The silicon starting material was a p-type, 1 to 2 Ω.cm electronic grade material. The carbon ribbons differed considerably for the various runs in relation to the progress of their fabrication process ; they consisted of a Papyex* ribbon purified (in some cases only) in an ambient containing chlorine at 2000 or 2800°C and then coated by a thin pyrocarbon layer in an open furnace (5).

Impurity bulk concentration measurements were performed on the carbon ribbon at its main fabrication steps, the silica crucible and the silicon materials-starting product, silicon left in the crucible and polycrystalline layers separated from the carbon substrate. They were conducted using a neutron activation analysis (NAA) technique ; irradiations were made on samples about 100 mg in weight in a neutron flux of 1×10^{14} n.cm^{-2}.s^{-1} in the OSIRIS reactor at CEN/Saclay, France. In all cases, the layer samples were chemically lapped after irradiation in order to eliminate surface contaminants. Concentrations were derived using non-destructive γ-spectrometry. Thus, about 20 elements were systematically followed with the important exceptions of Ti and Zr for which a chemical separation process should be applied to improve their detection limits.

Impurity surface analyses were restricted to the as-grown silicon layers. X-ray photo-electron spectroscopy (XPS) and Auger electron spectroscopy (AES) were used to analyse the surface (areas and depth explored of 1 cm^2 - 0,4 nm and 10^{-3} cm^2 - 2 nm respectively) and in-depth concentration profiles of the major impurities detected. In the latter case, argon milling was operated at a rate of 0,3 nm. $\times \mu A^{-1} \times mn^{-1}$ with 2 KeV argon ions. It is worth noting that the chemical shifts of the emitted photo-electrons and the line structure of the Auger spectrum provide information on the nature of the impurity bonding to the lattice.

Spectral photoresponse measurements were made on the RAD layers at each step of fabrication using the electrolyte -semiconductor junction. Current - voltage curves of solar cells 2 and 4 cm^2 in size were drawn under AM$_1$ illumination. The solar cells were fabricated on layers separated from the carbon ribbon by a burn-off at 1020°C for 1 hour (6) ; the junction was made by a 90 minute phosphorus diffusion at 850°C (POCl$_3$ process) and the p$^+$ back-contact was formed by the annealing at 660°C of evaporated aluminium. The front-contact was a typical Ti, Pd, Ag evaporated grid.

3. RESULTS

Bulk analyses

NAA results are given in tables 1 and 2. Table 1 shows the concentration of impurities in the carbon ribbon at three stages of its fabrication.

* Papyex is the trade name of a carbon ribbon obtained by rolling of expended graphite crystallites and manufactured by LE CARBONE LORRAINE.

namely, the Papyex raw material, Papyex after purification in a chlorine ambient at 2800°C and the final ribbon coated with pyrocarbon. The last column is typical of a ribbon generation without chlorine purification ; such a ribbon was used in the run noted A below.

The results in each column were averaged out except for Na because of the large concentration spread of this element along the ribbons ; a number of elements, e.g. As, Au, Ba, Br, Cd, Ga, La and Sb, present at concentrations below 0.1 µg/g in all cases were not listed in table 1.

The Papyex raw material appears to be heavily contaminated by metallic impurities eventually harmful to solar cell performances, among which Fe, Mo, Ti and Zr. The chlorine purification is very effective for most impurities with the exception of Cl and Na (compare with the ribbon noted run A). However, the Mo content remains fairly large. The results regarging Ti and Zr contents are not significant within the scope of this study because of their poor detection limits.

Local analyses conducted in a scanning electron microscope strongly suggested that most impurities were concentrated in clusters located at the surface of the pyrocarbon coating for the case of Mo rather than homogeneously distributed across the ribbons.

The results in table 2 relate to the impurity content in the silicon materials ; runs A, B and C were conducted with the same silicon starting material (column 1) and the same crucible quality. A comparison between the contents of contaminants in the melts of runs A and B suggests that the carbon ribbon is directly responsible for the contamination (cf. table 1). The melt of run B conducted with a purified carbon ribbon closely matches that of the starting material with the exception of Mo and Na. This result illustrates a direct consequence of the reduction of contaminants by chlorine purification, namely, the achievement of quasi-pure lamellar pyrocarbon coatings free of catalytic growth and not eroded by the silicon melt in the experimental conditions (7). Mo and Na may originate from the carbon ribbon surface ; Na is also a major impurity of the silica crucible.

Table 2 evidences a substantial purification effect at the freezing front, the effective partition coefficients K of most impurities being in the range 10^{-1} to 10^{-3} with the exceptions of Na and Cr. This property of the process may be readily ascribed to the growth from a non-confined freezing meniscus, which allows a redistribution of impurities towards the melt. NAA of layer samples taken 2.5 m apart on the ribbon of run A (no melt replenishment) evidenced an increase of the contamination in the growth direction never exceeding a factor of two for all impurities.

Surface analyses

Surface analyses are of primary importance in ribbon growth because of the large segregation effects expected in the vicinity of the surface. Some typical results of XPS analyses are given in Fig. 1. The as-grown surface was found to be partly oxidized down to a depth presumably smaller than it appears in the figure (because of the surface roughness). The Si-O bond was identified as that in silica. In the same region, accumulations of Ca, Na and C with a Si-C type bond were always observed (profiles of the latter elements are not produced in the figure for sake of clarity) ; it was also unambiguously established that Na did segregate rapidly to the surface at room temperature.

Underneath this layer, O (not linked to silicon) and C (as SiC and graphite) were found at concentrations estimated around their solubility limits ; Fe was the only metallic impurity detected besides Ca, with a profile ending up at the transition between bulk silicon and the oxidized layer (despite its large bulk content Cr was not observed in the ribbon of run A - table 2).

Electrical measurements

It was previously reported that the effective minority-carrier diffusion length Leff in as-grown layers exhibited a bell-shaped dependence against the pulling rate, the maximum of Leff occurring at the transition from dentritic to smooth growth. Thus, Leff values up to 65 µm were measured on dip-coated samples grown at 8 cm/mn (also see interpretation in Ref. (8)). Similar findings were made with RAD layers and Leff values of 90 µm were obtained.

Other experiments evidenced large decreases of Leff with rapid recovery, whereas the mean value was constant throughout the ribbon length, table 3. It is suggested that abrupt Leff decreases are due to local releases of contaminants from the substrate into the melt which are drawn along with the freezing film. In the case of the purified ribbons the release of contaminants may occur at impurity clusters (e.g. Mo) on the pyrocarbon surface or at direct melt Papyex contacts through the pyrocarbon coating ; such microscopic and isolated contacts were actually observed even on purified ribbons. A detailed understanding of the effects of impurities on the layer electrical performances would require additional analytical data (now in progress) especially regarding Ti and Zr contents. However, present results indicate that all the impurities determined by NAA may not be electrically active in as-grown layers and solar cells as well. For instance, Leff values of 30 to 35 µm were obtained in layer of run A despite a Mo content of $2.6.10^{14}$ at.cm^{-3} (2). Nevertheless, the results in table 4 and fig. 2 may be interpreted in terms of a lower contamination in the layer of run C.

4. CONCLUSIONS

The carbon shaper is a potential source of contaminants in the RAD process ; the high temperature chlorine purification treatment has permitted the achievement of pyrocarbon coatings resistant to molten silicon and overcome the contamination problem with the possible exceptions of Mo and Na. The RAD process goes along with a significant purification at the freezing front, the effective partition coefficients of the impurities analysed being in the 10^{-1} to 10^{-3} range. The results of electrical measurements suggest that the impurities may not be necessarily incorporated in the layers in an electrically active form.

Indeed, RAD solar cells exhibit AM_1 conversion efficiencies (AR-coated 4 cm^2 in size) in the 9 to 11% range with the best values around 12%. This study would indicate the possibility of using a slightly degraded silicon starting material as compared with the conventional electronic grade silicon.

REFERENCES

1 - C. BELOUET, Novel Silicon growth methods, ESC meeting 80-5, 195-212

2 - R.H. HOPKINS, R.G. SEIDENSTICKER, J.R. DAVIS, P. RAI-CHOUDHURY, P.D. BLAIS, J. Cryst. Growth 42 (1977), 493

3 - Y. MATSUSHITA, J. Cryst. Growth 56 (1982), 516

4 - C. BELOUET, Paper submitted for publication to J. Cryst. Growth

5 - J. MAIRE, J.P. SLONINA and J. GELLON, Proc. 3rd Int. Carbon Conf. Baden-Baden (1980), 667

6 - C. TEXIER-HERVO, M. MAUTREF, C. BELOUET, E. KERRAND, this conference

7 - J. GOMA, M. OBERLIN, A. OBERLIN, J. SCHNEIDER and C. BELOUET
High temperatures - High pressures 13 (1981), 263

8 - C. TEXIER-HERVO, M. MAUTREF, S. MAKRAM-EBEID, C. BELOUET
Proc. 2nd Photovoltaic Solar Energy Conf., Cannes (1980), 563

TABLE I
NEUTRON ACTIVATION ANALYSIS OF THE CARBON RIBBONS
Concentrations : $\mu g.g^{-1}$

CARBON RIBBONS ELEMENTS	RAW MATERIAL	CHLORINE PURIFIED	CHLORINE PURIFIED PYROCARBON COATED (RUNS B & C)	PYROCARBON COATED (RUN A)
Cl	25	6	7	5
Co	1.5	4×10^{-3}	4×10^{-3}	3×10^{-1}
Cr	40	5×10^{-2}	6×10^{-2}	1
Cu	20	2×10^{-1}	2×10^{-1}	1×10^{-1}
Fe	700	1	1	65
Mn	8.5	5×10^{-3}	1×10^{-2}	1×10^{-2}
Mo	100	2	1.5	35
Na	16	5 to 7	4×10^{-1} to 5	1
Ni	61	$< 5 \times 10^{-1}$	4×10^{-1}	2.5
Ti	10	< 3	< 3	< 5
W	5	1×10^{-1}	1×10^{-1}	2
Zn	170	5×10^{-1}	3×10^{-1}	$< 2 \times 10^{-1}$
Zr	2	$< 4 \times 10^{-1}$	$< 5 \times 10^{-1}$	< 1

TABLE II
NEUTRON ACTIVATION ANALYSIS OF THE SILICON MATERIALS
Concentrations : $atom.cm^{-3}$

SILICON MATERIALS ELEMENTS	STARTING MATERIALS	RUN A			RUN B	RUN C
		MELT	LAYER	K	MELT	LAYER
As	$< 7.5 \times 10^{12}$	2.0×10^{12}	$< 7.5 \times 10^{11}$	< 0.4	2×10^{12}	1.0×10^{12}
Au	2.0×10^{10}	7.0×10^{12}	7.0×10^{11}	0.1	4×10^{10}	2.0×10^{11}
Co	$< 1.0 \times 10^{13}$	5.0×10^{14}	2.4×10^{13}	0.05	1×10^{13}	1.7×10^{13}
Cr	$< 5.0 \times 10^{13}$	5.0×10^{15}	2.0×10^{15}	0.4	$< 1 \times 10^{13}$	$< 8.0 \times 10^{12}$
Cu	$< 4.0 \times 10^{13}$	3.0×10^{15}	4.0×10^{13}	0.01	$< 4 \times 10^{13}$	6.0×10^{14}
Fe	$< 1.5 \times 10^{15}$	2.8×10^{17}	$< 5.0 \times 10^{15}$	< 0.02	$< 2 \times 10^{15}$	$< 2.0 \times 10^{15}$
Mo	$< 9.0 \times 10^{13}$	2.2×10^{15}	2.6×10^{14}	0.1	1×10^{15}	1.5×10^{13}
Na	-	6.0×10^{14}	1.2×10^{14}	0.2	2×10^{14}	1.0×10^{15}
Ni	$< 1.0 \times 10^{15}$	2.0×10^{16}	$< 5.0 \times 10^{14}$	< 0.3	$< 2 \times 10^{15}$	$< 2.0 \times 10^{14}$
Sb	$< 1.0 \times 10^{12}$	9.0×10^{14}	1.2×10^{12}	0.001	1×10^{13}	3.5×10^{12}
W	$< 5.0 \times 10^{10}$	3.8×10^{14}	$< 5.0 \times 10^{11}$	< 0.001	$< 1 \times 10^{12}$	$< 1.5 \times 10^{12}$
Zn	$< 1.0 \times 10^{14}$	1.0×10^{15}	$< 2.0 \times 10^{14}$	< 0.2	$< 1 \times 10^{14}$	$< 1.0 \times 10^{14}$

POSITION (m)	L_{eff} (μm)
1.47	28
1.28	40
1.31	40
1.34	80
1.53	42
1.54	77

Table 3 : Example of variation of the minority-carrier diffusion length in as-grown layers along the ribbon.

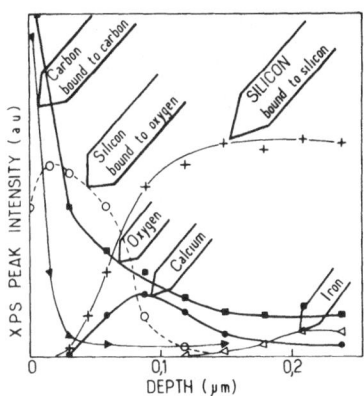

Fig. 1 : XPS peak intensity in-depth profiles for a series of elements ; chemical bondings of C and Si are indicated.

DATA	RUN A	RUN C
Leff (as-grown)	30 - 35	45 - 60
Ribbon width (cm)	2	5
Pulling rate (cm/mn)	4	4

Table 4 : Electrical and growth data for runs A and C.

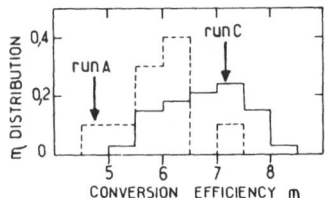

Fig. 2 : Normalized distribution of conversion efficiency for runs A and C (cells not A.R. coated).

FAST SILICON-SHEET GROWTH WITH THE SUPPORTED-WEB METHOD

J.G. GRABMAIER[1)2)], H. FÖLL[1)], B. FREIENSTEIN[2)] and K. GEIM[1)]

SIEMENS AG
1) Components Group, Discrete Semiconductors
2) Research Laboratories
Otto-Hahn-Ring 6, D-8000 München 83

Summary

Directly grown Si ribbons or sheets are attractive for the manufacture of low-cost Si solar cells. Intensive research efforts have led to the development of ribbon growth methods that yield products suitable for the manufacture of efficient solar cells. However, the areal growth rates of most methods are rather small and this constitutes a serious obstacle to their large-scale implementation. The newly developed supported-web (S-Web) technique attempts to overcome this problem. It envisions the use of a carbon-fibre net which is pulled through a melt of liquid Si at a high pulling speed. Liquid films or webs of Si are spread out within the meshes of the net and crystallize some time after leaving the melt. In this way the formation and the crystallization of the Si ribbon is decoupled and very large areal growth rates should be possible. First experiments demonstrated the feasibility of this concept and pulling speeds of $\gtrsim 2$ m/min have been achieved. Problems exist with respect to the crystalline quality and the topography of the specimens obtained.

1. INTRODUCTION

Si crystals directly grown as sheets or ribbons are considered to provide an attractive alternative to single-crystalline Si wafers for the manufacture of low-cost solar cells. In recent years, several techniques for growing Si ribbons have been established and efficient solar cells have been made from these ribbons. A serious disadvantage of ribbon growth techniques, however, lies in the rather small areal growth rates achievable in most cases. The areal growth rates, e.g., of the EFG, WEB, SOC, and RAD techniques (for the meaning of the abbreviations cf. (1)) are usually below ≈ 100 cm^2/min as compared to ≈ 500 cm^2/min in the conventional CZ-technique.

The growth-rate limiting factor in all cases is the removal of the heat of crystallization from the liquid-solid interface. Ribbon growth methods with improved heat-removal capabilities, as, e.g., roller-quenching (heat removal by intimate contact to a cooled metal wheel (2)) or horizontal ribbon growth (HRG; heat removal through very large liquid-solid interface (3)), therefore find increasing attention.

A completely new technique with the potential of very high areal growth rates is the so-called supported-web (S-Web) technique. This method is presently being explored in the Siemens laboratories and preliminary results are given in this paper.

2. PRINCIPLES OF THE S-WEB TECHNIQUE

The intention of the S-Web technique is to pull a net of carbon fibres (or carbon-coated quartz fibres) through a crucible containing liquid Si (Fig. 1). Because carbon is wetted by liquid Si, a liquid web of Si is spread out within the meshes of the net and kept stable by the high surface tension of liquid Si. The web crystallizes some time after leaving the melt, depending on the thermal environment (Fig. 2). In consequence, the forming (or "pulling") of a Si ribbon and its crystallization are decoupled and the extraction of the heat of crystallization is no longer a speed-limiting factor. Very high growth rates should be possible and first experiments support this view.

3. EXPERIMENTAL RESULTS

Graphite sheets (4 cm x 6 cm; 0.5 mm thick) have been machined into grids (typical mesh size 2 mm x 2 mm) by laser-cutting. These grids were immersed into liquid Si at a temperature close to the melting point and then withdrawn with pulling speeds ranging from 2 cm/min to 200 cm/min either vertically ($\alpha = 90°$) or inclined ($\alpha = 70° - 80°$) with respect to the Si surface. At low pulling speeds ($v \lesssim 5$ cm/min) continuous crystallization occurs and the behaviour is similar to the SOC (4), RAD (5) or "contiguous coating" (6) processes. For pulling angles $\alpha \lesssim 80°$, the grids are coated on one side only (Fig. 3) in a way reminiscent of the RAD process. The quality of the Si sheets obtained in this way is promising; solar cells fabricated in a non-optimized way had efficiencies up to 8 %, indicating that the presence of the graphite grid is not very critical to solar cell performance.

At high pulling rates liquid webs are drawn out as was expected. Crystallization occurs some time after the formation of the webs; this could be directly observed. The webs within the meshes of the grid crystallize independently from each other and this process has been called "mesh crystallization". Without a thermal control of the crystallization process, solidification occurs rapidly and often in a dendritic fashion, and the quality of the specimen is rather poor. Fig. 4 shows a cross-section through a specimen pulled at $v = 1.8$ m/min. The protrusions visible in this figure are a result of meniscus effects during the crystallization, cf. (7). Specimens akin to the one shown in Fig. 4 have been obtained at pulling speeds $\gtrsim 2$ m/min.

The experiments clearly demonstrated the validity of the basic S-Web concept. The objective of present research is to upgrade the topography and the crystalline perfection of rapidly pulled specimens. At the same time the technical necessities for continuous pulling of a carbon-fibre net through a melt of liquid Si as shown in Fig. 1 are explored. Very recently, nets have been made by a weaving process and first results indicate that, given proper slot design, no serious problems are encountered in the continuous pulling process.

ACKNOWLEDGMENT

The able technical assistance of Ms. L. Bernewitz is gratefully acknowledged. Thanks are also due to P. Glasow and coworkers who made the solar cells and to Dr. Aulich for numerous discussions. This work was supported by the Bundesministerium für Forschung und Technologie.

REFERENCES

(1) T.Surek, in: Proc. Symp. Electronic and Optical Prop. of Polycryst. or Impure Semiconductors and Novel Si Growth Methods; eds.: K.V.Ravi and B.O'Hara (The Electrochem. Soc.,) Pennington 1980, p. 173
(2) N.Tsuya, K.I.Arai, T.Takeudi, T.Ojima and A.Kuriowa, Jap. J. Appl. Phys. 19, Suppl. 19-2 (1980) 13
(3) B.Kudo, J. Cryst. Growth, 50 (1980) 247
(4) J.D.Zook, B.G.Koepke, B.L.Grung and M.H.Leipold, J. Cryst. Growth, 50 (1980) 260
(5) C.Belouet, J. Cryst. Growth, 50 (1980) 275
(6) T.F.Ciszek and L.J.Hurd, Proc. 14th IEEE Photovoltaic Specialists Conf., San Diego (1980) p. 397
(7) T.Surek and B.Chalmers, J. Cryst. Growth, 29 (1975) 1

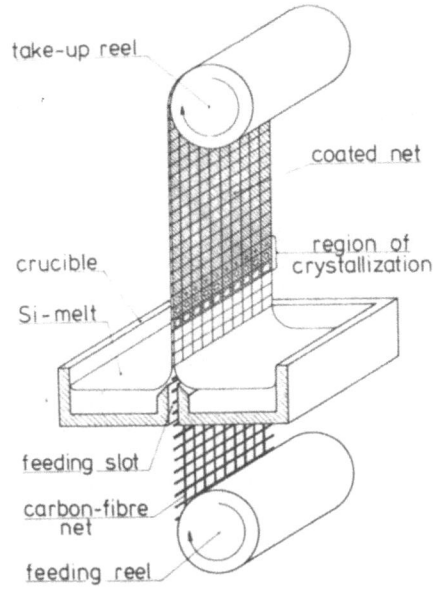

Fig. 1
Principles of S-Web technique

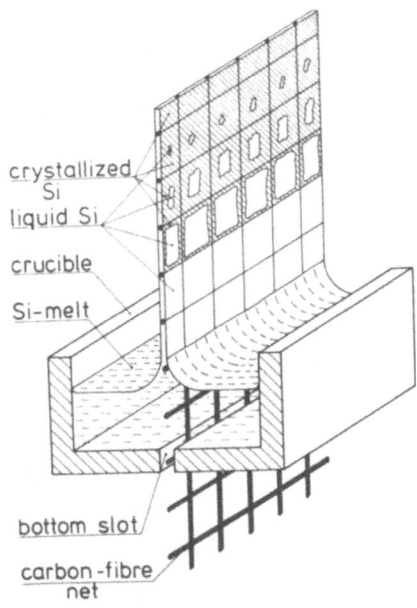

Fig. 2
Mesh-crystallization at high pulling speeds

Fig. 3
Cross-section through S-Web pulled at low pulling speeds

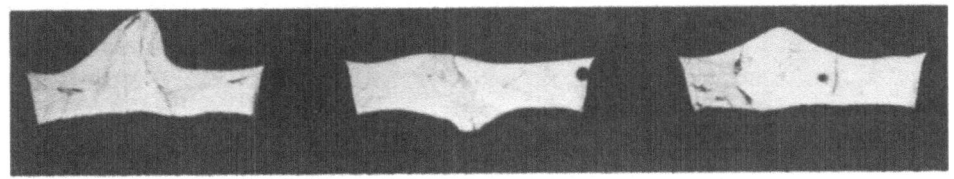

Fig. 4
Cross-section through S-Web pulled at high pulling speeds

RECENT DEVELOPMENTS IN MULTI-WIRE
FIXED ABRASIVE SLICING TECHNIQUE (FAST)

F. Schmid, C. P. Khattak, M. B. Smith and L. D. Lynch
Crystal Systems, Inc.
35 Congress Street, Salem, MA 01970 (USA)

Summary

The FAST process has shown effective slicing of 10 cm and 15 cm diameter ingots at 25 and 19 wafers/cm respectively. Correlation of area generation rate with slicing parameters has shown a strong relationship to the pressure at the diamond tips. This pressure is increased by rocking the workpiece and thereby decreasing the contact length. The area generation rate varies linearly with the surface speed of the FAST slicer. Novel techniques of electroforming have been developed in fixing diamonds whereby a controlled, predetermined kerf can be achieved.

1. INTRODUCTION

Slicing is the limiting process step for all ingot technologies. Considerable effort has been directed toward improving existing slicing techniques (1) and developing new techniques (2,3). At present ingot processes combine high performance and throughput as compared to ribbons (4). Slicing must be a low-cost, high-throughput process that combines minimum kerf and slice thickness to maximize material utilization. The high cost of polysilicon and ingot growth makes material utilization one of the most important parameters to make photovoltaics cost effective for terrestrial applications.

An economic comparison (5) of three slicing techniques, *viz.*, internal diameter (ID), multiblade slurry (MBS), and fixed abrasive slicing technique (FAST), has shown that FAST has the lowest price estimate and a better potential of achieving price goals.

FAST is a process designed to achieve the high material utilization characteristic of wire, but with high reliability and with low expendable cost. To accomplish this, diamonds are fixed to the wires to prevent wear. The wires are reciprocated in a straight line to prevent fatigue from winding around pulleys. To accomplish this, a technique was developed to stretch wires in a frame so all the wires are equally spaced, tensioned, and clamped in the frame. This bladepack of wires is reciprocated in a machine similar to a multiblade machine shown in Figure 1. Two of the novel features developed have been grooved guide rollers on either side of the workpiece and rocking of the crystal. Rocking the workpiece generates a radial cut profile and minimizes the contact length. These features increase contact pressure and minimize wire wander.

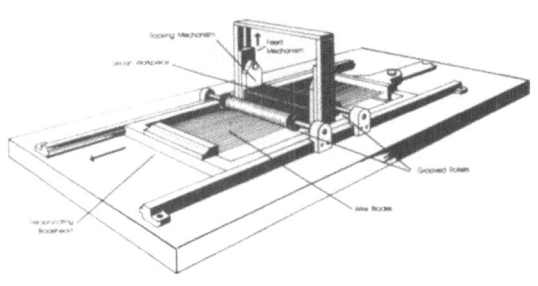

Figure 1. Schematic of FAST slicer

Various factors such as rocking angle, wire deflection and workpiece size have a significant effect on the contact length. These factors will be discussed and test results will be presented for slicing 25 wafers from 5 x 5 x 9 cm workpieces as a function of rocking angle and surface speed. An electroforming technique will be discussed which allowed the kerf to be reduced from 0.2 to 0.16 mm.

2. FACTORS AFFECTING CONTACT LENGTH

High pressure between the diamond tip and the workpiece is important for effectively slicing with fixed diamond. High feed forces increase cutting rate, but cause wire wander. It was found that high pressure between the diamond tip and workpiece could be achieved without increasing the feed force by rocking the workpiece to decrease the contact length.

The contact length between the wires and workpiece is determined by the deflection of the wires and shape of the cut profile. The deflection of wires is related to properties of wires, feed force and length of unsupported section. The effect of length of wires is minimized by placing the grooved support roller as close to the workpiece as possible. This would reduce the contact length by reducing the deflection for the same feed force. The shape of the cut profile is an arc and is determined by the rocking angle of the workpiece, α. Assuming the radius of the cut profile is a circle of radius R and the contact length, ℓ, subtends an angle ϕ at the center, then

$$\ell = \frac{\phi}{360} (2\pi R) = \frac{\phi R}{57.296} \tag{1}$$

where ϕ is in degrees. From geometrical considerations, as shown in Figure 2, ϕ is equal to twice the angle of deflection of the wires at the point of contact of the guide roller, θ, or

$$\phi \simeq 2\theta = 2 \tan^{-1}\left(\frac{D}{L}\right) \tag{2}$$

where D is the deflection of the wires and L is the half-distance between the guide rollers. Combining equations (1) and (2)

$$\ell \simeq \frac{2R \tan^{-1}\left(\frac{D}{L}\right)}{57.296} \tag{3}$$

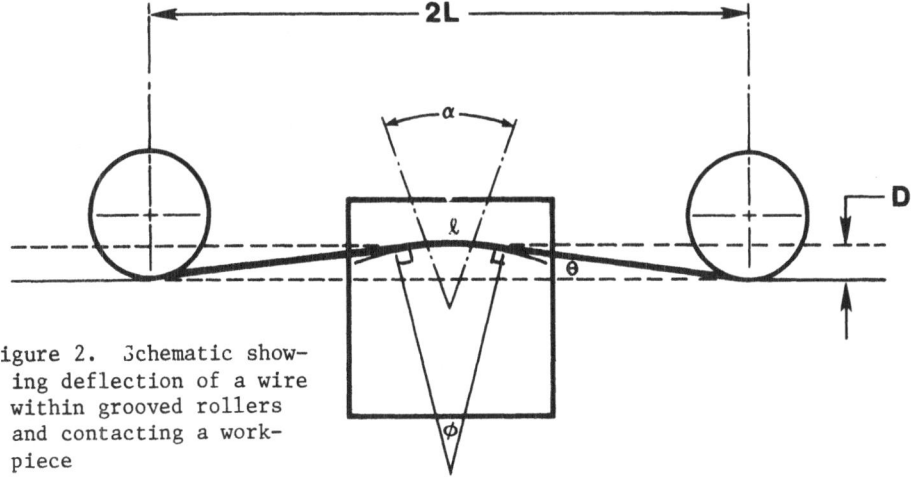

Figure 2. Schematic showing deflection of a wire within grooved rollers and contacting a workpiece

In order to achieve high pressures at the diamond tip ℓ has to be minimized. The radius of the cut profile, R, is controlled by the rocking angle; D/L can be reduced by enlarging L; however, this also increases D.

Slicing experiments were carried out to show the effect of contact length, rocking angle and bladehead speed. All the experiments were carried out using the following standard conditions: Two hundred twenty-five wafers were sliced from 9-cm long, 5 x 5 cm silicon for each experiment at 25 wafers per cm with a feed force of 26 grams per wire. Cutting was performed using 225, 0.125-mm diameter tungsten wires nickel plated with 30 μm natural diamond to achieve a 0.2 mm kerf width.

Kerf Width

Slicing experiments using a stationary step block shown as inset in Figure 3 were carried out to experimentally establish the effect of kerf width on the slicing rate. The width of steps is 5.1, 8.4, 11.7, 15.0, 18.3, 21.6 and 22.4 mm, which at a feed force of 26 gm and a kerf of 0.2 mm corresponds to a pressure of 25.5, 15.5, 11.1, 8.7, 7.1, 6.0 and 5.2 gms/mm^2, respectively. The material removal rate mm^2/min for each step was calculated by dividing the area generated per wire by the time. This is shown in Figure 3 as a function of step width. This plot shows that at a feed force of 26 gms/wire the removal rate is inversely proportional to kerf width or directly proportional to pressure of the wire against the workpiece. At low kerf widths or high pressures the material removal rate increases rapidly. For larger kerf widths the material removal rate reduces significantly.

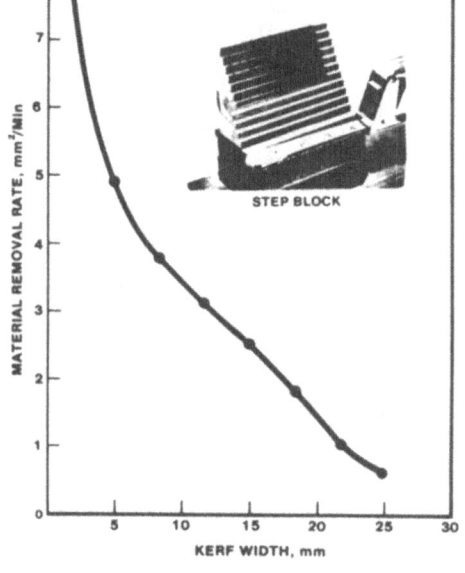

Figure 3. A plot of the area generation rate per wire as a function of kerf width for a step block without rocking

Rocking Angle

Equation 3 shows that the contact length is dependent on the radius of cut profile which in turn is dependent on the rocking angle. Slicing experiments were conducted at a 20° and 40° angle using 5 x 5 x 9 cm workpieces, with all other parameters the same as the step-block experiment to determine the effect of rocking on the removal rate. The average removal rate of the piece rocked at 20° was 5.125 mm^2/min, and 6.5 mm^2/min for a 40° rocking angle; the corresponding cut profiles were 12.7 cm and 7.6 cm respectively. Contact lengths of 2.8 and 4.69 mm for 40° and 20° rocking angles, respectively, were calculated using equation 3, where D and L are 1.524 and 82.6 mm. The material removal rate produced from Figure 3 for the 2.8 and 4.69 mm kerf width is 6.5 and 5.1 mm^2/min respectively. This is the same as the actual removal rate.

Ingot Size

Slicing experiments with 100 mm and 150 mm diameter silicon and 40° rocking angle has shown that the maximum contact lengths are 5.33 and 8.89 mm respectively. In the case of 150 mm slicing, effective slicing occurred during initial stages; however, considerable breakage of wafers was observed as slicing progressed. This data showed that the critical contact length for FAST slicing at a 40° rocking angle and 26 gms feed force is between 5.33 and 8.89 mm.

Surface Speed

Slicing experiments to determine the cutting rate as a function of average surface speed were carried out. The average removal rate was determined by dividing 2500 mm^2 cross-section area of the 50 x 50 mm workpiece by the time to cut through it. The average surface speed was calculated by doubling the product of the 40.64 cm stroke length and RPM. The machine was run at 120, 79 and 37.5 RPM corresponding to 97.5, 64.2 and 30.5 m/min respectively. The removal rate for each of these average speeds was 6.5, 4.375 and 2 mm^2/min, respectively, as shown in Figure 4. This linear plot shows that cutting rate is directly proportional to speed. From the plot approximately 0.067 mm^2 is generated for every meter of fixed diamond abrasive traversed across the workpiece.

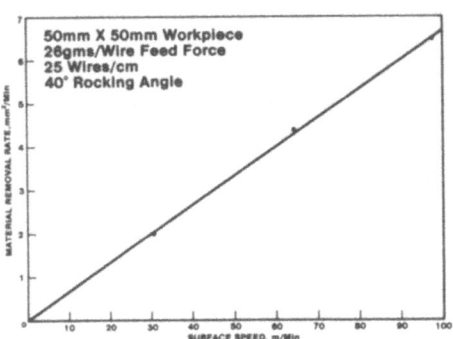

Figure 4. Area generation rate per wire as a function of surface speed of FAST slicer

3. BLADE DEVELOPMENT

High concentration of diamonds on wire has been achieved in wire-packs used for FAST slicing. A life of three 10-cm diameter ingots with the same wirepack has been demonstrated (2). A comparison of wires before use and after slicing three 10-cm diameter ingots shows no diamond degradation under SEM examination (Figure 5). These wires were plated with diamonds over the entire circumference; the kerf achieved was about 0.200 mm.

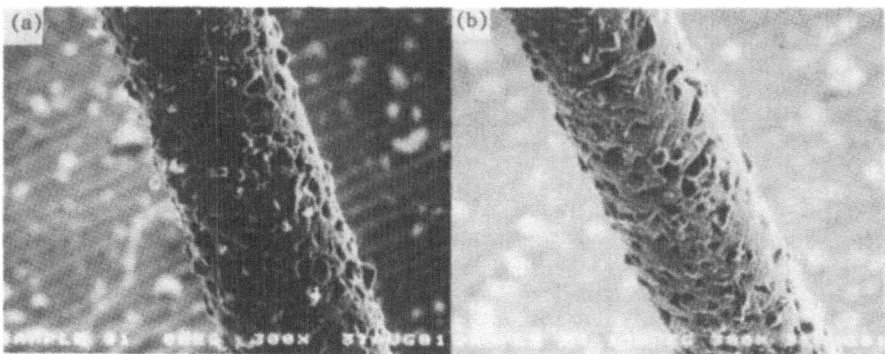

Figure 5. A longitudinal view of a wire before use (a) and after slicing three 100 mm diameter silicon ingots showing no degradation (b).

Besides producing thicker wafers it is desirable to reduce kerf in order to improve slicing effectiveness. An electroforming technique (6) was therefore invented that allows plating of diamonds in a predetermined form over the wires (Figure 6). This procedure allows plating to conform to any desired shape and size. Under these conditions larger diamonds can be used without adding to the kerf. A controlled low kerf will allow higher pressures at the diamond tip without increasing the feed force during slicing.

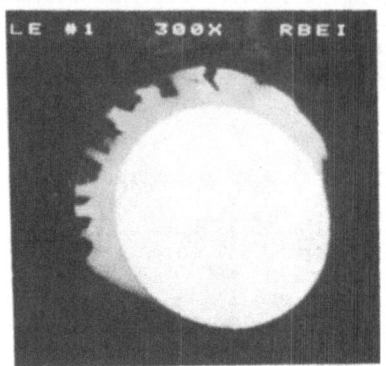

Figure 6. Cross-section of electroformed wire showing preferential plating of diamonds

4. CONCLUSION

The FAST process has been developed to a stage where all the technological aspects have been demonstrated. This has been achieved by slicing 10-cm diameter silicon at 25 wafers/cm, 15-cm diameter silicon at 19/cm, reduction of kerf to as low as 0.15 mm, and production of wafers with a surface damage of 3-5 μm.

It has been seen that the correlation of material removal rate is directly proportional to the pressure at the diamond tips. It is desirable to increase this pressure by reducing the contact length between the wires and the workpiece. For ingots of 100 mm or larger the rocking angle has to be 40° or larger in order to reduce the contact length. The relationship of material removal rate is linear with the surface speed of the fixed diamond wires.

Blade development has progressed whereby consistent wirepacks are produced which show no visible diamond degradation after slicing three 100 mm diameter silicon ingots. Electroforming techniques show potential for even better performance towards reducing kerf and increased slicing effectiveness.

5. ACKNOWLEDGMENT

This paper presents results of research performed in part for the Flat-Plate Solar Array Project, Jet Propulsion Laboratory, California Institute of Technology, sponsored by the U.S. Department of Energy through an interagency agreement with NASA.

6. REFERENCES

1. K.M. Koliwad, "LSA Project Perspective of Wafering Technology," Proc. Low-Cost Solar Array Wafering Workshop," JPL Publication 82-9 (Pasadena, CA, 1982), p. 3.
2. F. Schmid, M.B. Smith, and C.P. Khattak, "Overview of a New Slicing Method--Fixed Abrasive Slicing Technique (FAST)," *ibid.*, p. 233.
3. H. Lauvray, A. Talpied, and J.P. Besselere, "New Wire Silicon Slicing Technology for Solar Cells," Third E.C. Photovoltaic Solar Energy Conf. (D. Reidel Publ. Co., 1981), p. 603.
4. E. Fabre and C. Belouet, "Ribbons and Sheets as an Alternative to Ingots in Solar Cell Technology," *ibid.*, p. 24.
5. A.R. Mokashi, "Sensitivity Analysis of Add-on Price Estimate for Select Silicon Wafering Technologies," Proc. Low-Cost Solar Array Wafering Workshop, JPL Publication 82-9 (Pasadena, CA, 1982), p. 197.
6. C.P. Khattak, F. Schmid, and M.B. Smith, "Wire-Blade Development for Fixed Abrasive Slicing Technique (FAST) Slicing," *ibid.*, p. 111.

CRITICAL TECHNOLOGY LIMITS TO SILICON MATERIAL AND SHEET PRODUCTION

Martin H. Leipold
Deputy Manager, PV Components Research
California Institute of Technology
Jet Propulsion Laboratory

Summary

The reduction in scope of the Flat-Plate Solar Array Project has resulted in sharply focusing the developments within the programs involving silicon material and silicon sheet production. This has been accomplished by limiting the research to those elements which have the most impact on the long-term potential of each process under development.

Silicon material development efforts include two approaches. The first involves the production of high purity silane and conversion to silicon particles in a fluidized-bed reactor. Here the critical element is the performance of the fluidized-bed reactor, specifically minimizing fines production and wall deposits and demonstrating purity. In the second, the dichlorosilane process, wall deposits in the rod reactor are limiting the conversion of dichlorosilane to silicon.

Sheet preparation efforts involving Czochralski growth and wafering have largely been eliminated accompanied by reduction in casting development. Within ribbon development, the most critical elements here are reduction of ribbon stress and maintenance of flatness during high speed growth. Secondary efforts in improving conversion efficiency continue.

1. INTRODUCTION

The preparation of high-purity silicon and its conversion to silicon sheet requires major technology improvements in the Flat-Plate Solar Array (FSA) Project. Earlier studies have indicated that these two elements represent from 40-52% of the cost of the entire panel[1]. This importance, coupled with reduction in Project scope[2] makes the selection of the critical elements within these areas vital to long-range payoff. This paper presents the elements which were selected for investigation and the basis for selection.

2. SILICON MATERIAL PREPARATION

The first of two technologies continuing under development within the FSA Project has been the conversion of metallurgical-grade silicon through a silane purification process to silicon particles under development by the Union Carbide Corp. (UCC). The second involves the conversion of trichlorosilane to dichlorosilane and subsequent production of silicon using modified rod reactors of the Siemens type. This has been developed by the Hemlock Corp.. Additional information on the broad programmatic changes is available[2]. The discussion here will be limited to the critical elements evaluation.

The silane conversion and purification process under development by UCC had previously shown satisfactory technical performance with only total process integration and concomitant economics to be done. The latter is now being assessed by UCC separately. The remaining technology element, conversion of the silane gas to high purity silicon particles, has followed two parallel development paths. These were: 1) a free-space reactor (FSR) followed by conversion of the submicron particles to pellets by means of a shot tower or, 2) growth of silicon seed particles in a fluidized-bed reactor (FBR) by deposition of silicon from the silane. Two factors were considered in the selection, (1) the economics and (2) recent technical progress. Analyses of the economics indicated that the FBR was capable of producing silicon particles at an additional cost of $1/per kg[3]. The FSR process, including the shotting step, had been projected to cost between $4.00 and $6.00 per kg, with the difference and uncertainty being largely associated with the production of shot.

Technical progress in the silane to silicon conversion using FBR has been significant with research proceeding at both UCC and JPL. At UCC, a 6" diameter fluidized-bed reactor has been operating for some time and recent results have been very encouraging. Additional significant results have been obtained as part of an in-house program at JPL in which a similar 2" system is being used[4]. Results have shown that almost complete conversion of silane to silicon can be obtained with temperatures above 650°C. Dense deposits on the seed particles are produced. Previous problems with clogging at the bottom of the reactor have been eliminated. The mechanism of particle growth is envisioned by both direct deposition (CVD) of silane onto the particles and/or by scavenging of submicron particles by the larger seed particles. The JPL system is being scaled now to 6" diameter. Pilot plant systems are expected to be ~18" in diameter. A major concern for the final scale-up is related to heat transfer through the bed as surface to volume ratio decreases.

A parallel program has been underway involving first the conversion of the silane to silicon in a FSR with subsequent conversion of the submicron particles to small spheres by means of a shot tower. FSR design and performance had developed to where 100% conversion of the silane to submicron particles could be obtained; however, the purity of these fine particles has not yet been demonstrated. Efforts with the shot tower have demonstrated shot both from semiconductor grade and FSR product. However, erosion of the fused silica melting vessel and orifice indicated either a materials development program, multiple orifice designs or other approaches would be necessary to make the shotting approach suitable for a long term operation[5]. The technical problems, including submicron particle melting, and the inferior economics have resulted in termination of efforts on the FSR approach.

The program with Hemlock Semiconductor is to produce low-cost, high-purity silicon by the conversion of trichlorosilane to dichlorosilane in a redistribution reactor and subsequent conversion to silicon in a modified Siemens rod reactor. The redistribution process has been working very well. A 5" diameter redistribution reactor has been operated intermittently for four months with no degradation of the catalyst. It is believed that this portion of the process is under control. With the rod deposition reactor, three problems remain: (1) excessive deposition of silicon on the reactor bell jar walls; (2) achieving conversion efficiency and deposition rate simultaneously, and (3) achieving reactor power consumption goal of 60 KWH/kg Si. Presently the wall deposit amounts to 5-10% of the total silicon and several solutions are under investigation. The deposition reactor has been scaled to large production size. Solutions to the other problems appear available.

3. SILICON SHEET PREPARATION

Efforts have focused both on preparation of ingots, followed by wafering, and direct crystallization of molten silicon into a ribbon or film. The ingot preparation methods included advanced Czochralski (CZ) growth and direct casting by the heat exchanger method (HEM) and ubiquitous crystallization process (UCP). Subsequent wafering is by internal diameter (ID), multi-blade slurry sawing (MBS) and a fixed abrasive slicing technique (FAST) using wires. The direct sheet formation methods included edged-defined film fed growth (EFG), web dendrite growth (WD) and silicon on ceramic (SOC).

The advanced CZ methods have focused on multiple ingot growth from a single crucible. One approach (Hamco) involves additions of solid between growth of crystals and a second (Siltec Corp.) involves liquid additions during crystal growth[6]. The Hamco process has been carried further under this program. Multiple growth of ingots totaling 150 kg, 8.5 cm/hr average crystallization rates, uniform conversion efficiency, and automated control have been demonstrated for 15 cm ingots. Only limited single crystallinity (~60%), possibly related to ambient gas phase composition, remains critical. This element is being reviewed for further investigations. Cell performance has been excellent, averaging >15% for single crystal regions and >12% for poly regions.

Ingot casting by HEM has been developed to where technical feasibility has been demonstrated. Analyses of costs indicate that the goals have been reached if economies of large-scale production, optimization and yield control can be achieved. Such demonstrations do not fall within the definition of high-risk, long-range, high-payoff, research and efforts with HEM now fall outside of the FSA Project.

Similarly, the UCP process used by Semix corporation has progressed well toward its project goals. The identification of remaining critical technical elements is under review. Because the process is proprietary, a discussion of those elements cannot be presented here.

The FSA Project has extensively supported the development of improved wafering for silicon ingots. The approach consisted largely of incremental changes in the various technologies designed to reduce costs, increase cutting speeds, and improve silicon ingot utilization. Initially, considerable improvements were observed, but recent technical progress has not been apparent. For this reason, research will now be directed at a more fundamental understanding of the cutting process, rather than an empirical and incremental approach. A critical review of the fundamental and technical studies was held[7]. The specific details of such fundamental approaches are under review and so cannot be presented in detail at this time. Potentially useful critical elements may include abrasive-silicon interaction, silicon material property modification, stress and fracture control in wafers and new novel approaches to wafering. Such high-risk, high-payoff technological developments will likely form a portion of the continuing FSA program.

The SOC process under development by Honeywell Corp. had been transferred to The Solar Energy Research Institute so it will not be considered for critical elements here.

The EFG process being developed by Mobil Tyco Solar Energy Corp. has demonstrated many of the elements necessary for technical feasibility.

However, several critical elements remain. An extensive analysis of the sensitivity of price to performance is available[8] in which a number of elements are shown to be relatively unimportant. These include equipment cost, power consumption, floor area and direct materials. The most sensitive elements are growth rate and number of machines per operator. Specifically, for 10 cm wide ribbon, growth rate must exceed 4.3 cm per minute for cost effective operation. While this growth rate has been achieved for EFG, the ribbon produced suffered from mechanical ripples and buckles and fractures from internal stresses. Effort has been expended in attempting to optimize the growth conditions to maximize growth rates and minimize these mechanical perturbations; however, the empirical approach has not been successful. It has been shown that internal stresses in the ribbon are generated during the cooling from the melting point and emphasis is now placed on this one critical element of growth, e.g., the understanding of stress generation and relaxation during the crystallization and cooling process. By this means, growth costs could be reduced substantially.

The second item critical to low-cost EFG growth, that of number of machines per operator, appears to be tractable while not yet demonstrated. Automatic control systems exist for such ribbon growth and the problem appears to be one of engineering optimization under development by MTSEC. Such developments are not a question of technical feasibility.

EFG conversion efficiency, which was not considered in reference (3), remains important. Cells that are 50 cm square in area have demonstrated performance approaching 12%, the project goal, and it appears that incremental improvement in material and its subsequent processing will result in achievement of that goal.

The web-dendrite process (Westinghouse) continues to progress. An analysis of the price sensitivity[9] is largely consistent with that for EFG. The growth rate again is important (rates of 15-20 cm^2 per minute are cost effective). Experimentally, growth rates as high as 27 cm^2 per minute have been demonstrated occasionally, and 30-35 cm^2 per minute appears to be within reach. Such increases could reduce the cost of the web growth by a factor of 1/2, reducing the total module price below Project goals. Again, the primary problem with high speed growth is ribbon flatness and internal stress. Consequently, the emphasis on this particular problem is warranted. It is reasonable to extrapolate that all ribbon growth processes, particularly those with a crystallization front lying nearly parallel to the growth direction, are limited by this stress problem.

The second critical area for WD again is the number of machines which can be operated by one operator and the duty cycle for growth. This again represents an engineering optimization problem lying outside the scope of technical feasibility demonstrations. Finally, conversion efficiency for WD material has generally been in the range of that required for cost effective material and difficulties are not anticipated.

4. CONCLUSIONS

Critical elements, requiring further study in the silicon material synthesis and the conversion to sheet, have been identified. With the silicon material synthesis gas phase preparation well established, the problem is entirely the conversion of the gas phase to silicon. With the silane process, the fluidized-bed approach has been found to be superior to the FSR-shotting approach. Performance of the FBR has been good and scaling to larger sizes is underway. With the dichlorosilane process, investigations continue in an attempt to reduce or eliminate deposition of silicon on the reactor walls of the Siemens-type rod reactors, achieve the conversion efficiency and deposition rate goals simultaneously, and reduce the reactor power consumption.

The parallel process for conversion of silicon to sheet forms, has been greatly reduced in scope and number. Technical feasibility of replenished Czochralski growth has been demonstrated and only limited work involving ambient gas phase control will be done. Ingot growth by casting has largely been completed, although a few technical elements of a proprietary nature are still under review with the UCP process. Wafering efforts will now focus on understanding the critical aspects of abrasive materials interaction. Empirical incremental improvements will not be sought in the technologies as presently being utilized. The details of such critical studies are still in the formative stage.

The efforts within direct crystallization of ribbons are now focusing on the understanding and control of ribbon stresses, flatness and fracture during cooling. This has been critical to all types of ribbons for which analyses have been conducted. Additionally, the efficient use of operator time is critical for these processes, but improved performance will likely come from engineering design and optimization and thus does not represent a critical element.

5. ACKNOWLEDGEMENT

This work was performed for the Jet Propulsion Laboratory, California Institute of Technology and was sponsored by the U.S. Department of Energy, Division of Solar Energy through an agreement with NASA. The assistance of A. Briglio, G. Hsu, A.H. Kachare, and A.D. Morrison is acknowledged.

6. REFERENCES

(1) R.W. Aster, Price Allocation Guidelines, Report DOE/JPL 1012-47, Jan. 15, 1980.
(2) W.T. Callaghan, "Flat-Plate Solar Array Project", this publication.
(3) Union Carbide Corporation Final Report to JPL/FSA, Contract #954334 for Phase I & II; Feasibility and Process Design Studies, June 1981.
(4) R. Hogle, G. Hsu, N. Rohatgi and R. Lutwack, "Fines Formation and Bed Agglomeration in Fluidized Silicon Deposition", to be submitted to J. of Chem. Eng.
(5) R.L. Lane, E. Roberts and J. Scott Shea, Subcontract to UCC for equipment and Design for a Silicon Powder Consolidation System, Final Report, and Technology Development Report, Union Carbide Contract 825-50106 Kayex Corporate Technology Center, Oct. 16, 1981.
(6) T. Daud and A. Kachare, "Advanced Czochralski Silicon Growth for Photovoltaic Modules", Report DOE/JPL 1012-70 for the U.S. Dept. of Energy (In Print).
(7) A. Morrison, Chairman, "Proceedings of the Low-Cost Solar Array Wafering Workshop", June 8-10, 1981, Report DOE/JPL 1012-66, Feb. 1, 1982.
(8) A.R. Mokashi and A.H. Kachare,"Sensitivity Analysis of the Add-On Price Estimate For Edge-Defined Film Fed Growth Process." Report DOE/JPL 1012-55 for U.S. Dept. of Energy, March 15, 1981.
(9) A.R. Mokashi, "Sensitivity Analysis of the Add-On Price Estimate for Silicon Web Growth Process." Report DOE/JPL 1012-61 for the U.S. Dept. of Energy, Dec. 15, 1981.

ECONOMIC VIABILITY OF THE UCP SEMICRYSTALLINE SILICON SHEET TECHNOLOGY*

Z. Putney, T. Rosenfield, and C. Wrigley
SEMIX INCORPORATED
15801 Gaither Drive
Gaithersburg, Maryland U.S.A.

Summary

Extensive economic analysis performed under DOE sponsorship has shown that photovoltaics can be competitive with central power station grid electricity when the price of photovoltaic modules is reduced to below $0.70/Wp. This in turn requires a reduction in the cost of sheet material from which to fabricate cells.

Starting from work begun in the early 1970's, Semix has developed a new industry, now some three years old, manufacturing semicrystalline silicon for photovoltaic cells. The Ubiquitous Crystallization Process (UCP) so developed is now approaching its second megawatt of output of semicrystalline silicon sheet. In a recent study employing the JPL SAMIS analysis it has been projected that at the 100 MWp annual level this industry can provide photovoltaic panels close to $0.50/Wp, well below the $0.70/Wp level, using the Semix-developed Simultaneously Present Large Impurity Technology (SPLIT) process for silicon feedstock.

In conjunction with the UCP industry, the SPLIT feedstock, and mildly aggressive projections for near-term wafering technology development, this analysis projects reaching $0.60/Wp with silicon solar cell efficiencies as low as 13%. Extension of the large-volume solar cell fabrication technology to maintain a 15% conversion efficiency, based upon recent developments on a smaller scale, results in a module cost projection of $0.52/Wp.

1. INTRODUCTION

Some six years ago the initial results of a new approach to generating silicon sheet for photovoltaic applications were reported (1). The approach was named the Ubiquitous Crystallization Process (UCP) and the resulting material is called semicrystalline silicon. The term semicrystalline is employed to differentiate the middle ground between single-crystal material and fine-grained polycrystalline forms, referring to a structure in which the individual crystallites are multi-millimeters to centimeters in size. Later developments of this approach led to the achievement in 1981 of a 17% AM1 efficiency semicrystalline silicon 4 cm^2 solar cell (2) as shown in Figure 1. Industrialization of the UCP approach has now resulted in the manufacture of semicrystalline silicon by the UCP to an accumulated output approaching its second megawatt for photovoltaic application. Quantities of 10 cm x 10 cm semicrystalline silicon wafers from

*This work was partially supported by U.S. DOE Co-operative Agreement No. DE-FC01-80ET 23197.

this UCP technique were processed into large-area solar cells in 1981 and 1982 with the results shown in Figure 2. The solar cell fabrication technologies applied to the semicrystalline wafers were improved during the period to include a back surface field and multilayer antireflective coating, which is evidenced by the upward trend in conversion efficiencies. A laboratory process for large-area semicrystalline solar cells has produced efficiencies up to 13.5% as shown in the example of Figure 3. This series of achievements provides confidence that future developments will undoubtedly lead to solar cell conversion efficiencies in large volume manufacturing which are in the neighborhood of 15%, an important point in projecting the economic viability of any approach to photovoltaic technology.

Another important aspect of the UCP technology for semicrystalline silicon was found to be a decreased sensitivity of photovoltaic performance to the impurity content of the silicon. In 1981 it was reported (3) that semicrystalline silicon solar cells made by the UCP from metallurgical grade silicon resulted in AM1 conversion efficiencies up to 8.2%. Since that time Semix has been developing a lower-purity silicon approach specifically for the UCP, termed the Simultaneously Present Large Impurity Technology (SPLIT) to provide a low-cost UCP feedstock. Semix has recently announced that it is building the world's first manufacturing plant for production of SPLIT silicon, the first facility for silicon optimized specifically for photovoltaics. Progress in utilizing SPLIT silicon for the UCP is exemplified by the 4 cm^2 solar cell achieving 14% AM1 conversion efficiency depicted in Figure 4.

The relatively rapid progress to very respectable photovoltaic performance with the UCP and the SPLIT feedstock for semicrystalline silicon foretell a very high probability of reaching the requisite low cost of grid-competitive photovoltaics with this approach. The following economic analyses utilizing the SAMIS system of the Jet Propulsion Laboratory support this expectation.

2. ECONOMIC ANALYSES

Projection of the cost for future production of photovoltaic modules with the UCP and SPLIT approaches for silicon material at an annual industry level of 100 MWp was accomplished with the JPL SAMIS computation program (4). Inputs were generated for equipment and processes associated with the UCP, sizing of the crystallized silicon bricks to a 10 cm by 15 cm cross-section, wafering with inner-diameter saws, wafer cleaning, handling, quality control functions and shipping of sawn wafers. Fairly conservative engineering estimates for upscaling to 100 MWp were used. The JPL/DOE guidelines (5) were used for solar cell processing and module add-on costs, since the technologies proposed there are very similar to portions already compatibly utilized on UCP semicrystalline wafers.

Table 1 lists the input assumptions employed in the SAMIS computations. The incoming silicon feedstock cases examined were the JPL projection of 14 $/kg and the Semix expectation of 5 $/kg for its lower-purity SPLIT silicon. The assumptions for I.D. wafering are mildly aggressive for today, but certainly appear to be achievable extrapolations for the moderate future.

Calculations of the projected costs were performed and the results are depicted graphically in Figure 5. The efficiencies of resulting solar cells were left as a variable, ranging from 13% AM1 to 15% AM1. Both the projected $14/kg result for Siemens process derived silicon and Semix' SPLIT silicon at $5/kg were used in the calculations.

TABLE 1
SAMIS INPUTS ASSUMPTIONS

UCP		WAFERING	
Ingot size	41.4 kg	Blade life	2500 cuts/blade
Output rate	2.07 kg/min	Blade cost	$100
Labor	1 man/machine	Ctr.-Ctr. spacing	0.46 mm (0.018")
Machine cost	$100,000	Wafer size	10cm x 15cm x0.23mm
Heating elements	$530/month	Machine cost	$40,000
Electric power	$0.03/kWh	Labor	1 man/12 machines
Power required	470 KW	Electric power	3 KW
Refractory parts	$20/ingot		
Gasses	$0.052/min	STEPS COSTED	YIELDS
Miscellaneous	$0.057/min	UCP	99%
		Sizing to 10x15 cm	99.5%
FINISH SIZING		Brick Q.C.	99%
Output rate	7.5 mm/brick	Wafering	98%
Machine cost	$80,000	Rinse/Clean	99.8%
Labor	1 man/2 machines	Ship	
Material loss	5%		
Electric power	500 W	MEASUREMENTS & QUALITY CONTROL	
Blade life	1000 cuts/blade	Output Rate	1 mm/brick
		Machine Cost	$50,000
		Labor	1 man/4 machines
		Electric Power	1 KW

3. CONCLUSIONS

As can be seen from Figure 5, the projected cost for modules made with UCP silicon can closely approach $0.50/Wp with SPLIT silicon feedstock and $0.60/Wp with $14/kg feedstock. The assumptions made for 100 MWp production levels are reasonable extrapolations of existing technology and the progress to date supports the rather conservative conversion efficiency values. Consequently, we conclude that the UCP semicrystalline silicon sheet technology does indeed have the economic viability for low-cost photovoltaics.

ACKNOWLEDGEMENT

Discussions with JPL personnel are gratefully acknowledged for their contribution in assuring the reasonableness of the inputs and in the utilization of SAMIS.

REFERENCES

(1) J. Lindmayer, "Semi-Crystalline Silicon Solar Cells", 12th IEEE Photovoltaic Specialists Conference (1976), p. 82.
(2) G. Storti, "The Fabrication of a 17% AM1 Efficient Semicrystalline Silicon Solar Cell", 15th IEEE Photovoltaic Specialists Conference (1981), p. 442.
(3) J. Lindmayer & Z. Putney, "Semicrystalline Material from Metallurgical Grade Silicon, 15th IEEE Photovoltaic Specialists Conference, (1981), p. 572.
(4) P. J. Firnett, "Standard Assembly-Line Manufacturing Industry Simulation (SAMIS), Computer Program User's Guide, Release No. 4", 5101-60, Rev. C, Flat Plate Solar Array Project, April 1982.
(5) R. W. Aster, "Price Allocation Guidelines, January 1980", DOE/JPL-1012-47, 5101-68, Rev. A, Low Cost Solar Array Project, JPL Publication 80-51.

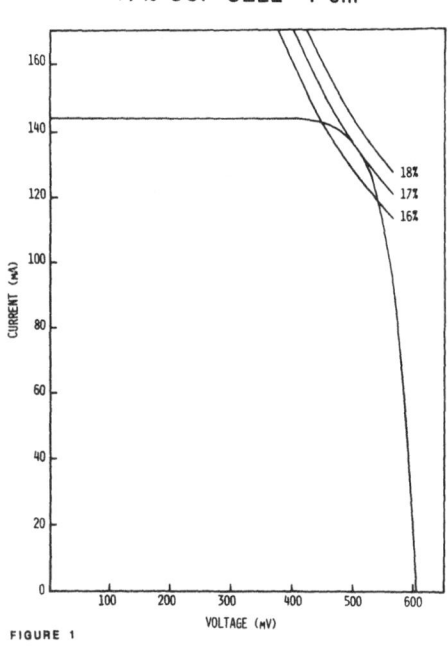

FIGURE 1

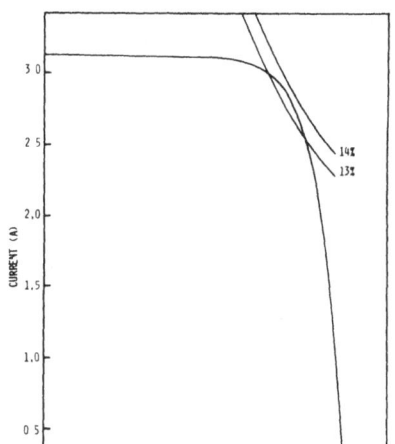

FIGURE 2

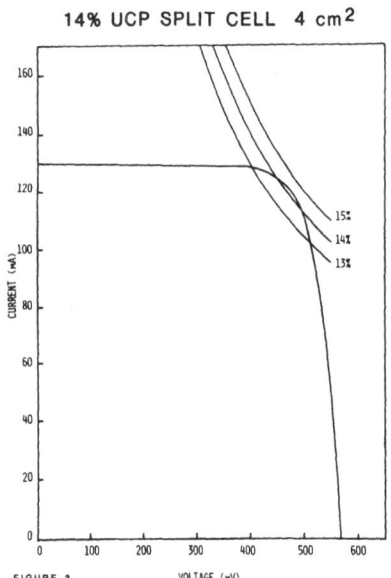

FIGURE 3

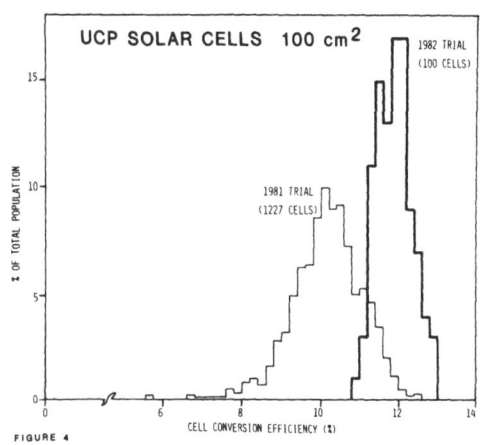

FIGURE 4

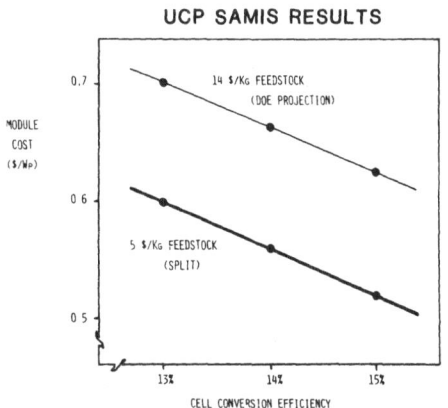

FIGURE 5

COMPARISON BETWEEN VARIOUS ION BEAM DOPING PROCEDURES AND ANNEAL
TECHNIQUES USED IN MANUFACTURING SILICON SOLAR CELLS

JC. MULLER, A. MESLI, P. SIFFERT
Centre de Recherches Nucléaires, Groupe PHASE, 67037 - STRASBOURG Cedex

J. COM-NOUGUÉ, C. TESSARI, JP.DUMAS
Laboratoires de Marcoussis, Centre de Recherches de la COMPAGNIE GENERALE
D'ELECTRICITE, 91460 - MARCOUSSIS (France)

Summary

Conventional ion implantation of phosphorus and ion incrustation
(non mass separated PF_5 molecular ions), have been used to realize
the N^+ layer of single cristalline silicon solar cells. Various
regrowth techniques have been investigated : classical thermal treatment, laser annealing in the liquid and solid phase regimes and
election beam annealing. A systematic investigation by DLTS has allowed
the determination of the optimal conditions of the post thermal treatment. Finally, large scale cells ($\geqslant$ 10 cm^2) have been prepared by
using industrial processes and efficiencies higher than 13% have been
measured for the two doping procedures. These techniques are presently
applied to polycrystalline silicon from various sources. The first
results obtained on solar cells elaborated with these materials, in
particular the silicon under development at Laboratoires de Marcoussis
are given in the paper.

1. INTRODUCTION

Ion beam techniques are, in principle, well adapted to automatic
manufacturing of P-N junctions for solar cells. Strong cost reduction is
expected from these procedures [1]. However, the severe radiation damage
associated with the ion bombardment makes an annealing of the defects of
absolute necessity. Today, this is generally performed by thermal treatment. Various approaches of limited heating, are investigated to regrowth
the damaged surface zone. For laser, the regrowth mechanism is generally
the liquid phase epitaxy, which, furthermore, gives interesting doping
profiles and allows high doping concentration, in excess of solubility
leading to good cells performances [2]. The strong influence of a post-
laser thermal annealing was also noticed by other authors [3].

Here, we present the results obtained essentially on single crystalline silicon cells but also on polycrystalline cells manufactured by
using routine procedures. Two implant techniques have been employed namely
conventionnal phosphorus implantation at low energy and ion "incrustation"
(AMI) which can introduce the dopant at very low energies at high flux
(10 keV, 0.4 mA/cm^2) by using a non-mass separated beam [3,4] . This latter
procedure we developed at CRN is now also under investigation in two other
groups [5,6]. The anneal procedures were the following : thermal treatment,
laser in a pulsed mode (liquid phase epitaxy), scanned CW laser (solid

phase regrowth).

The characteristics of the doped layer, as well as the influence of residual defects on the electrical properties of the cells have been investigated. A study of the residual microscopic defects, by DLTS [7] allowed us to determine the optimal conditions. Finally, the photovoltaic performances of cells prepared by these techniques have been considered. The best results are very close to those obtained on standard diffused cells.

2. CELLS PREPARATION

The steps of the cells fabrication process were the following :

- Material :   Mono Si (p) 1 Ω.cm (111)
                Poly Si (p)$\sim$ 1 Ω.cm

- Doping procedures :
  - 31 p^+ implantation
    typically 5.10^{15} ions/cm^2 - 20 Kev

  - PF_5 molecular ion beam incrustation without mass separation
    typically 5.10^{16}/cm^2 - 10 Kev
         beam current 0.4 mA/cm^2

- Recrystallization

  • by laser - Pulsed YAG (Liquide Phase Epitaxy)
        λ = 0.53 µm - repetition rate 1 to 10 KHz
        spot size   : Ø 0.1 mm
        spot duration: 100 nsec.

     - CW CO_2 - 20 W (Solide Phase Epitaxy)
        focalized beam : Ø 0.6 mm
        overlopping   : 50 %

  • by pulsed electron beam : pulse duration : 300 nsec.

  • by thermal treatment : 850°C - 50 min.

- Metallization of the back contact : Ti - Pd - Ag

- Post thermal annealing :
    - sintering of the back contact
    - annealing of the residual defect subsisting after pulsed annealing.

- Metallization of the front contact : Ti - Pd - Ag.

- AR coating : $Si_3 N_4$.

3. PROPERTIES OF THE DOPED LAYERS

a) <u>Macroscopic damage</u> : The low speed and heavy molecular ions produce a complete amorphization of the bombarded zone [4, 8]. It was shown by RBS that the pulsed laser annealing (Ruby, YAG) results in a better recrystallization compared to a thermal treatment. A surface peak remains, probably due to a precipitation of fluorine in the layer. This point has been confirmed by TEM [4]. No crystallography defects, such as dislocations or stacking faults are detected in the major part of the recrystallized layer.

b) <u>Residual defects after pulsed annealing</u> : Despite the crystalline perfection after pulsed annealing, no structure has been obtained up to now which presents performances in excess of conventional diffused cells due to the presence of residual microscopy defects. Most of them are minority carrier traps, which introduce compensation near the junction [7]. They mainly result from damage generated in the deeply penetrating tail of the dopant distribution (1 micron) which is not reached by the laser annealing. Figure 2 shows two DLTS spectra recorded under injection conditions. The measurements have been performed after equivalent energy annealing using ruby and YAG lasers, as calculated (see fig. 3) and corresponding to the power density needed to melt the same material thickness [9].

It should be noticed that the H_1 and E_3 levels have already disappeared for the YAG laser. The concentration of the other defects is lower for the YAG anneal. The results obtained in the case of the ruby laser annealing suggests that the laser does not play the dominant role in the observed defects, which are rather due to the implantation [7]. The observed levels can be divided into two groups : i) E_1, E_2, E_4 which have a deep penetration ; their thermal annealing occurs at rather high temperature (600°C), indicating that their nature is rather complex ; ii) the levels H_1, H_2, E_3 and E_5 are visible only for implantations performed under high fluxes ; they only anneal thermally at high temperatures.

The differences observed between the two kind of lasers can be explained by considering that i) the pulse duration of the YAG (100 ns) is much longer than that of the Rubis (20 ns), giving rise to a smaller recrystallization speed and a deeper penetration of the heat ; ii) the high repetitive YAG laser, with overlapping pulses can increase the mean temperature of the surface zone and melt deeper-in.

4. APPLICATION TO SOLAR CELL MANUFACTURING

a. Role of the implantation parameters and post annealing

On figure 4, we have reported the variation of the open circuit voltage Voc as a function of the post laser annealing temperature for both kind of pulsed laser annealing (ruby and YAG) technique : i) the best annealing temperature lies between 600 and 660°C for ruby laser treatment and arroun 550-600°C for the YAG laser, ii) Voc increases when the energy of the projectiles is dimished. This may be due to the reduction of the deeply penetrating "tail", we already discussed before [7]. Voc is systematically lower for a ruby laser annealing compared to a YAG laser processing, for the same implantation energy. This may be explained by the fact that a pulse annealing with a high repetition rate overlapping results in a lower concentration of defects.

b. Performances of the solar cells :

The silicon solar cells elaborated by using a routine procedure as indicated in 2) were measured under AM1 illumination at 28°C. Their performances are given in Table I and II.. The values of short circuit currents and efficiencies are given for the active area (10 cm^2) of the cells (junction total surface 11 cm^2).

. ion implantation : in the case of thermal annealing, the best efficiencies higher than 13.5 % AM1 are obtained for an implantation of 5.10^{15} ions/cm^2 at 20 Kev followed by a treatment at 850°C.

For the same conditions of implantation, a pulsed annealing at 2.5 J/cm^2 gives comparable results and efficiencies higher than 13% AM1. The results obtained up to now on cells processed with a CW CO_2 laser annealing indicate slightly lower performances, resulting essentialy from a lower Voc.

. ion incrustation : a pulsed laser annealing at an energy 2.5 J/cm^2 or higher gives better performances than a thermal treatment and efficiencies higher than 13 % AM1 are achieved.
Fig. 4 gives the spectral responses of implanted cells annealed with a pulsed YAG and a CW CO_2 lasers. A first result obtained on PF_5 implanted cell annealed with an electron beam canon [10] is reported in table II.

. the various processes are presently applied to polycrystalline materials, in particular a technique of elaboration of large semicrystalline silicon ingots is under development at LABORATOIRES DE MARCOUSSIS [11] . Implanted and laser annealed cells produced with such a material, not yet optimized present efficiencies of 9,5% AM1. This first result must be compared to efficiencies higher than 11% (active area) obtained on diffused 100 cm^2 cells produced with a more recent ingot [12] .

5. REFERENCES

[1] H. GOLDMAN AND WOLF - 14th IEEE Phot. Spec. Conf. SAN DIEGO (1980) p. 923
[2] J.S. KATZEFF, M. LOPEZ and RH. JOSEPHS, 3rd E.C. Phot. Sol. En. Conf. Cannes (1980) p. 708
[3] G. MICCOLI, P. OSTOJA and S. SOLMI in Ref. 2 p. 719
[4] JC. MULLER and P. SIFFERT, Inter. Workshop on Ion Impl., Laser Treat. and ion beam anal. of Materials, BOMBAY (1981)
[5] L.D. NIELSEN and P. BALSLEV in Ref.2, p. 698.
[6] M.D. SIRKIS and D.L. SALZMAN, 15th IEEE Phot. Spec. Conf. ORLANDO (1981).
[7] A MESLI, A. GOLZENE, JC. MULLER, B. MEYER, C. SCHWAB and P. SIFFERT Meeting on laser and E.B. Inter. with Solids, BOSTON (1981).
[8] J.C. MULLER, A and JJ GROB, R. STUCK and P. SIFFERT Appl. Physics Lett. 34 (1979), 287
[9] M. TOULEMONDE   Unpublished.
[10] J. GEERK, F. RATZEL, Kernforschungs Centrum Karlsruhe.
[11] J. FALLY and C. GUENEL, in ref.2, p. 598
[12] J. FALLY and C. GUENEL, 4th Phot. Sol. Energ. Conf. STRESA (1982)

This study was supported by the COMES and the CEC.

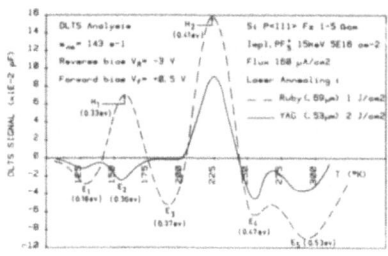

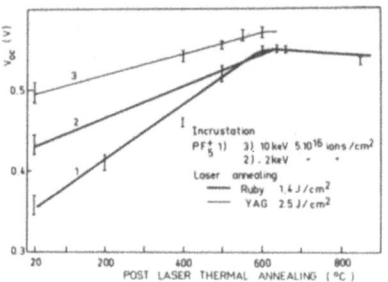

Fig.1 : DLTS Spectra after ruby and YAG laser annealing. (PF$_5$ incrustation)

Fig.3 : Variation of V_{oc} as a function of the post laser annealing temp.

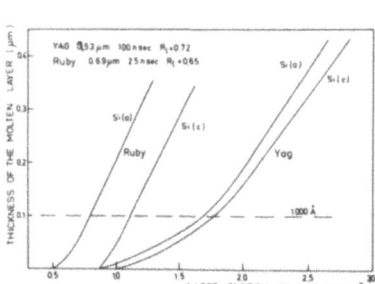

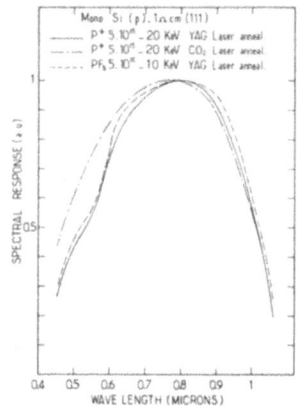

Fig.2 : Calculated thickness of the melt layer as a function of the laser energy.

Fig.4 : Spectral response of implanted cells annealed by YAG or CO_2 laser.

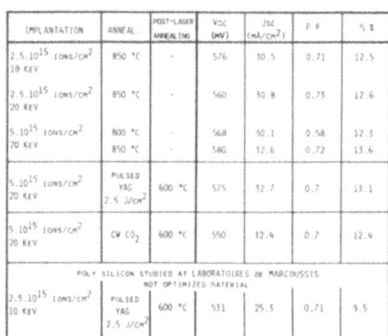

TABLE I
IMPLANTATION $^{31}P^+$

IMPLANTATION	ANNEAL	POST-LASER ANNEALING	VOC (mV)	JSC (mA/cm²)	F F	η %
2.5.10¹⁵ IONS/CM² 10 KEV	850 °C	-	576	30.5	0.71	12.5
2.5.10¹⁵ IONS/CM² 20 KEV	850 °C	-	560	30.8	0.73	12.6
5.10¹⁵ IONS/CM² 20 KEV	800 °C	-	568	30.1	0.58	12.3
	850 °C	-	580	32.6	0.72	13.6
5.10¹⁵ IONS/CM² 20 KEV	PULSED YAG 2.5 J/CM²	600 °C	575	12.7	0.7	13.1
5.10¹⁵ IONS/CM² 20 KEV	CW CO₂	600 °C	550	12.4	0.7	12.4
POLY-SILICON STUDIED AT LABORATOIRES DE MARCOUSSIS NOT OPTIMIZED MATERIAL						
2.5.10¹⁵ IONS/CM² 10 KEV	PULSED YAG 2.5 J/CM²	600 °C	531	25.3	0.71	9.5

TABLE II
INCRUSTATION PF$_5$

ANNEALING	POST-LASER ANNEALING	VOC (mV)	JSC (mA/cm²)	F F	η %
900 °C	-	560	34	0.65	12.3
PULSED YAG					
2 J/CM²	550 °C	564	32.4	0.64	11.6
2.5 J/CM²	550 °C	548	31.1	0.56	9.5
	600 °C	571	33.3	0.7	13.3
3 J/CM²	550 °C	576	30	0.71	12.2
	600 °C	585	32.6	0.7	13.4
PERA 1.5 J/CM² CELL : 1CM²	600 °C	552	33.4	0.6	11.1
WACKER POLY-SILICON					
2.5 J/CM²	600 °C	539	28.1	0.71	10.7

- 998 -

STATUS OF ION-IMPLANTED SILICON SOLAR CELLS

W. Schmidt   K.-D. Rasch

AEG-Telefunken, Heilbronn,
Germany

Summary

Ion implantation is considered as a possible low-cost "cold" process for junction formation in terrestrial solar cell fabrication. But it requires a subsequent implantation damage annealing step. Various annealing techniques as thermal treatment or irradiation with laser or electron beams have been applied. The results indicate for all annealing techniques lower efficiencies compared with standard diffused cells. The problems of annealing are discussed. The influence of a post-anneal heat treatment and of resistivity on the performance of pulsed electron beam annealed cells is evaluated.

1. INTRODUCTION

In conventional terrestrial solar cells production open-tube diffusion to form the p-n junction is a standard process. But this means a high temperature step in the cell processing which can cause a change in material properties resulting in cell performance degradation. For cast silicon (SILSO) for example, an increased carrier recombination at grain boundaries was observed after heat treatment (1 - 3). Future low cost silicon with its probably higher impurity content will be presumable more sensitive to heat treatments.

Therefore an alternative "cold" or at least rapid thermal junction formation process would be desirable for that sensitive silicon. This could be ion implantation in connection with a suitable annealing step. Since some new "non-furnace" annealing techniques were developed during the last years (4), it seems to be useful to test them for solar cell application.

So it is the purpose of this paper to investigate the status of ion implantation-annealing techniques and to compare efficiencies of ion-implanted solar cells with standard diffused cells, using standard single crystalline and multicrystalline silicon. Therefore experiments were carried out to determine the influence of the factors which can affect the performance of ion-implanted solar cells as

- implantation species and machine parameter
- annealing process
- substrate material

2. SOLAR CELL FABRICATION AND MEASUREMENTS

To compare the different junction formation processes, the test solar cells with n^+p-structure were fabricated as follows. The cleaned wafers were implanted and annealed. If necessary, a 400 °C or 600 °C post thermal treatment for 15 min was performed. The contacts were formed by evaporation. The front contact grid is suitable for sheet resistances up to 120 $\Omega/\square$. An antireflective coating of evaporated titaniumoxide is finally added. As process control and for reference, wafers with a diffused junction with 40 $\Omega/\square$ sheet resistance were co-processed.

Implantation was carried out at AEG-Telefunken and at Spire Corp., Bedford, USA. For implantation at AEG-Telefunken, an implanter of Balzers type Scanibal SCI 218 was used. Lowest implantation energy for reasonable process times was 30 keV. The implantation time for 4×10^{15} ions/cm^2 was 50 min for P^+ and 54 min for As^+ at 1.3 mA implanter current.

The different annealing methods and equipments under investigation were

- furnace annealing with three-step procedure (5), activation step optimized
- rapid isothermal annealing (RIA) with thermal radiation source (6) carried out using Varian model IA 200 system
- scanned laser beam annealing (SLBA) (7) carried out using Quantronix model 609 LCP system with Q-switched Nd:YAG laser ($\lambda = 0.53$ μm)
- pulsed laser beam annealing (PLBA) carried out at Kernforschungsanlage Jülich, using ruby laser ($\lambda = 0.694$ μm, 15 ns pulse length, about 1 J/cm^2 pulse energy)
- pulsed electron beam annealing (PEBA), version I (8) and II (9) beam energy about 1 J/cm^2, version I carried out at Kernforschungszentrum Karlsruhe, version II at Spire Corp.

The sheet resistance was measured with a four-point probe equipment. The I-V characteristics were measured at 25 °C under simulated AM 1.5 irradiation of 100 mW/cm^2, adjusted with a solar cell calibrated at ESTEC. Dark diode characteristics were also measured at 25 °C. The diffusion lengths were measured using short-circuit current method (10).

3. RESULTS AND DISCUSSION

3.1 Influence of implantation parameters

At first, the influence of implantation parameters as implantation species and dose on sheet resistance and solar cell performance was investigated. 1 Ωcm <111> silicon was used because of its more critical solid-state regrowth behaviour during annealing (11). Implantation energy was always 30 keV. Implantation damage annealing was carried out by conventional furnace annealing. The influence of implantation species, dose and annealing temperature on sheet resistance is shown in Figure 1 and 2. To obtain comparable sheet resistance values for a given dose, higher temperatures must be used for arsenic. The solar cell efficiency is nearly independent of the applied dose range of $1 - 4 \times 10^{15}$/cm^2. But it strongly depends on the annealing temperature. Especially for arsenic, maximum efficiencies up to 14 % could only be obtained for annealing at 900 °C.

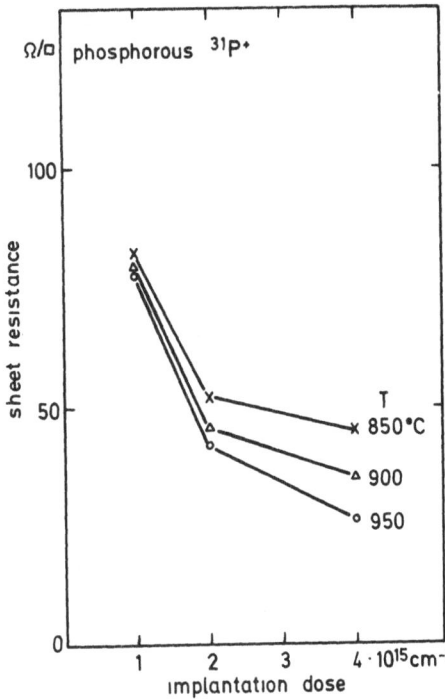

Figure 1
Sheet resistance versus implanted dose of P^+ at 30 keV into 1 Ωcm FZ silicon, furnace annealed for 15 min at temperature T

Figure 2
Sheet resistance versus implanted dose of As^+ at 30 keV into 1 Ωcm FZ silicon, furnace annealed for 30 min at temperature T

With an optimized annealing procedure efficiencies of almost 14 % are achievable for 2×10^{15} P^+/cm^2 implanted cells, too. Since the related sheet resistance of 40 - 50 $\Omega/\square$ corresponds to that of standard diffused junctions, this dose was chosen in the comparison of annealing techniques.

3.2 Comparison of annealing techniques (present status)

1 Ωcm <111> FZ and 1 - 5 Ωcm SILSO silicon wafers were implanted with 2×10^{15} P^+/cm^2 at 30 keV, annealed and processed to solar cells including the 600 °C/15 min post heat treatment. Results of best cells are shown in Figure 3 and 4.

1 Ωcm <100> FZ silicon wafers, implanted with 2.5×10^{15} P^+/cm^2 at 10 keV and PEBA (version II) treated, were provided by Spire Corp. and also processed to solar cells. Performance distribution of this cells (PEBA 2) together with a typical result of a previous experiment (PEBA 1) is shown in Figure 5. Results of diffused solar cells which served as process control are added for comparison.

Only implanted and furnace annealed or PEBA II treated single crystalline solar cells show nearly comparable efficiencies to diffused cells. Implanted and annealed SILSO cells show always a lower efficiency. However, the SILSO results are additionally modulated by material properties (density of grain boundaries etc.). Reproducible results are only obtained until now with implanted and furnace annealed single crystalline solar cells.

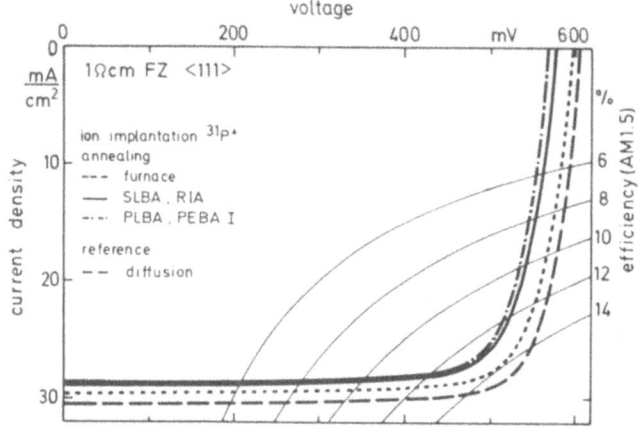

Figure 3
Best results for ion-implanted (30 keV, 2×10^{15} P^+/cm^2), various annealed and diffused single crystalline solar cells (2 cm x 2 cm)

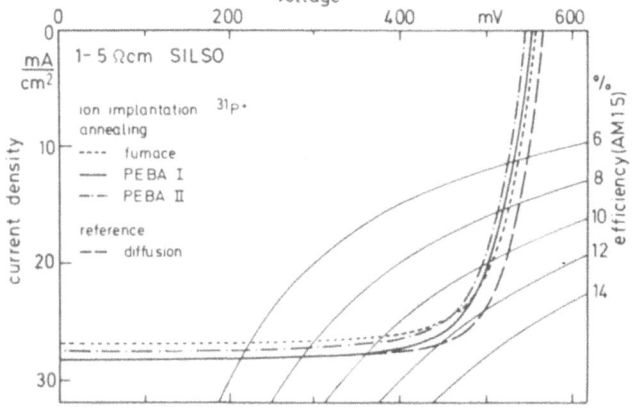

Figure 4
Best results for ion-implanted (30 keV, 2×10^{15} P^+/cm^2), various annealed and diffused SILSO solar cells (2 cm x 2 cm)

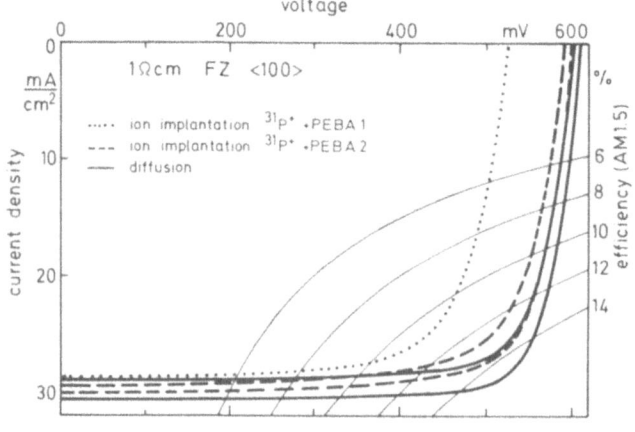

Figure 5
Comparison of ion-implanted (10 keV, 2.5×10^{15} P^+/cm^2), PEBA (II) treated and diffused single crystalline solar cells (2 cm x 2 cm)

The lower efficiencies of ion-implanted cells, as can be seen in Figure 3 - 5, are due to a lower curve factor for PEBA II, lower short-circuit current I_{sc} and in particular lower open-circuit voltage V_{oc}. To investigate the reasons, the blue response, bulk diffusion length L_n (Table 1) and dark I-V characteristics (Figure 6) of the single crystalline cells were measured. From dark I-V curves the parameters in the diode equation (1)

$$I = I_{01}(e^{qV/kT} - 1) + I_{02}(e^{qV/nkT} - 1) + \frac{V}{R_{sh}} \; ; \; n \approx 2 \qquad (1)$$

with

$$I_{01} = qn_i^2 \left(\frac{D_n}{N_A L_n} \coth \frac{W_p}{L_n} + \frac{D_p}{N_D L_p} \coth \frac{W_n}{L_p} \right) \qquad (2)$$

were graphically evaluated (Table 2).

Table 1: Relative blue response (λ = 505 nm) and bulk diffusion length L_n of solar cells fabricated with various annealing techniques.

Annealing	blue response (relative units)	L_n (μm)
furnace	101	115
RIA	111	50
SLBA	102	90
PLBA	101	110
PEBA I	106	80
diffusion (reference)	100	145
starting material (MIS-cell, reference)	–	200

Table 2: Approximate values of diode equation parameters of solar cells, fabricated with various annealing techniques.

Annealing	I_{01} (A/cm^2)	I_{02} (A/cm^2)	R_{sh} (kΩ)	substrate
furnace	2 x 10^{-12}	1 x 10^{-6}	4	
RIA	4 x 10^{-12}	7 x 10^{-7}	1.3	
SLBA	4 x 10^{-12}	3 x 10^{-7}	5	1 Ωcm FZ <111>
PLBA	5 x 10^{-12}	2 x 10^{-7}	12	
PEBA I	5 x 10^{-12}	1 x 10^{-6}	0.6	
diffusion (reference)	1.5 x 10^{-12}	4 x 10^{-7}	10	
PEBA II/1	1.5 x 10^{-11}	1.5 x 10^{-6}	2	
PEBA II/2	1-2 x 10^{-12}	1-5 x 10^{-6}	0.5-5	1 Ωcm FZ <100>
diffusion (reference)	1 x 10^{-12}	$\leq$ 10^{-7}	4	

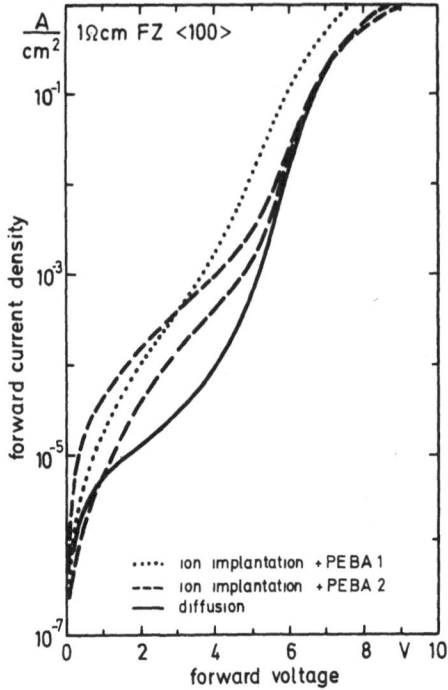

Figure 6
Dark I-V curves of diffused and 30 keV, 2.5×10^{15} P^+/cm^2 implanted, PEBA II treated solar cells (2 cm x 2 cm)

Shunting problems in connection with higher values for I_{O2} can be hold responsible for the reduced curve factor of PEBA II processed cells.

The diminution in the short-circuit current is obviously due to lower diffusion length in the bulk. The blue response of ion-implanted cells is always better than that of diffused cells. The lower values of L_n are caused by the implantation/annealing procedure since L_n in the starting material is high.

Significant in our implantation experiments is also a high blue response and a low L_n and vice versa. This can not be explained until now.

Lower values of V_{oc} are correlated to higher values of I_{O1}. From equation (2), a reason for this increase are the lower values of L_n. But same values of I_{O1} for different L_n are found. Further investigations are necessary to reveal the reasons.

Above results of a lower L_n indicate, that the alternative annealing processes as PEBA, PLBA or SLBA only anneal defects in the n^+-zone and leave defects in the bulk. Therefore the influence of an additional post-anneal heat treatment was investigated.

3.3 Influence of post-anneal heat treatment and of wafer resistivity

The application of a pre-anneal (12) or a post-anneal (13) thermal treatment to improve the performance of ion-implanted and laser annealed samples was already reported. In Figure 7, for ion-implanted and PEBA treated wafers of 1 and 10 Ωcm resistivity, the influence of a post-anneal heat treatment on solar cell parameters is exhibited.

The parameters improve with higher temperature due to removal of residual damage. This removal is accomplished better for the lower (10 Ωcm) than the higher (1 Ωcm) doped silicon material. Similar dependence on sheet resistivity was found for cells prepared with furnace annealing.

Diode characteristic measurements of 1 Ωcm samples indicate unchanged values of I_{O2} and R_{sh} but I_{O1} is about 3×10^{-12} A/cm^2 for T = 400 °C, compared with 1.5×10^{-12} A/cm^2 for 600 °C. As discussed in (12, 13) defects can be created in regions beyond 0.4 µm from the surface by a few deeper penetrating ions and remain after annealing. This can explain the decrease for post-anneal at 400 °C. And as shown in the last chapter, even 600 °C seems to be not sufficient for complete defect removal. Additional work especially with DLTS will be necessary to understand the created defect and complex types.

3.4 Implanted n^+pp^+ structures

As demonstrated before, 10 Ωcm silicon is suitable to fabricate ion-implanted solar cells of satisfactory quality. So it should be possible to fabricate ion-implanted high-efficient n^+pp^+ cells, using single-step simultaneous furnace annealing. For non-optimized implantation parameters, a typical result is shown in Figure 8 together with standard diffused cells. Since ion-implantation and furnace annealing is simply to perform, ion-implanted n^+pp^+-structures could become a near-term application in terrestrial solar cell fabrication.

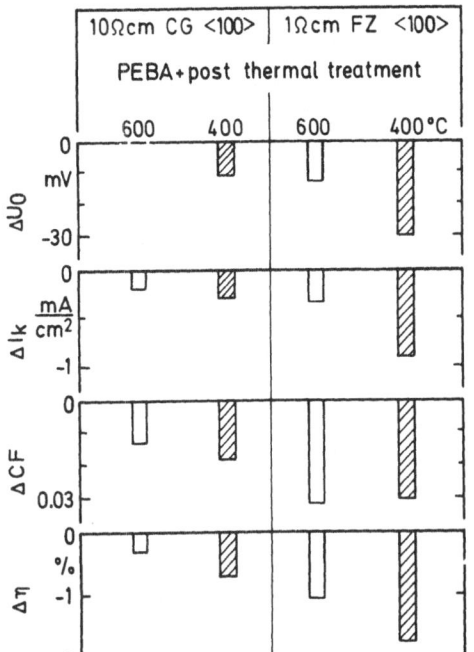

Figure 7
Comparison of ion-implanted (10 keV, 2.5×10^{15} P^+/cm^2), PEBA II treated and diffused solar cells (2 cm x 2 cm)

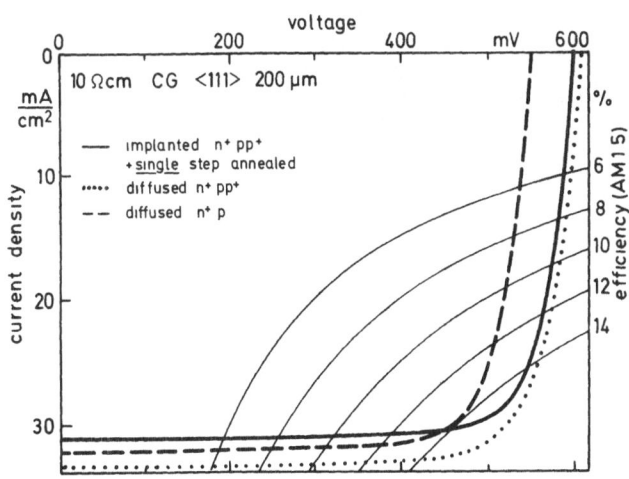

Figure 8
With phosphorus and boron implantation and single-step simultaneous annealing (furnace) fabricated solar cells (2 cm x 4 cm)

4. CONCLUSIONS

The results of the experiments can be summerized as follows:

- Ion-implanted and furnace or pulsed electron beam annealed single crystalline solar cells have nearly the same efficiency as those using diffusion. Ion-implanted SILSO solar cells are always poorer.
- Good reproducibility of solar cell performance was found until now only for furnace annealed cells.
- Implantation/annealing causes lowered bulk diffusion length and increased reverse saturation current. This is more critical for 1 Ωcm than for 10 Ωcm silicon.
- A post-anneal thermal treatment is necessary to attain satisfactory efficiencies for ion-implanted and PEBA or PLBA treated cells.

Further experiments with ion implantation and "alternative" annealing processes should be carried out, using solar-grade silicon of a resistivity below 1 Ωcm.

5. ACKNOWLEDGEMENT

The authors wish to thank F. Ratzel (Kernforschungszentrum Karlsruhe), Dr. B. Stritzker (Kernforschungsanlage Jülich), Dr. E. G. Arthurs (Quantronix Corp.), M. B. Spitzer (Spire Corp.) and J. White (Varian Assoc., Inc.) for carrying out annealing processes. This work was supported by the Bundesministerium für Forschung und Technologie.

6. REFERENCES

(1) B. Legros, G. R. David, M. C. Boissy, J. Lebailly, T. P. Ngo, E. Fabre, Proc. 3rd EC Photovoltaic Solar Energy Conf., Cannes, 593 (1981)
(2) P. E. Russel, C. R. Herrington, D. E. Burke, P. H. Halloway, The effect of heat treatment on grain boundary properties in cast polycrystalline silicon, Proc. of the Materials Research Society Symposium on Grain Boundaries in Semiconductors, Boston, 1981
(3) D. Redfield, Appl. Phys. Lett. 40 (2), 163 (1982)
(4) R. Iscoff, Semicond. Intern. 4 (11), 69 (1981)
(5) A. R. Kirkpatrick, J. A. Minucci, A. C. Geenwald, R. H. Josephs, Proc. 13th IEEE Photovoltaic Spec. Conf., Washington, 706 (1978)
(6) R. T. Fulks, C. J. Russo, P. R. Hanley, T. I. Kamins, Appl. Phys. Lett, 39 (8), 604 (1981)
(7) R. A. Kaplan, M. G. Cohen, K. C. Liu, Laser cold processing, Elec. Chem. Soc. Meeting, Los Angeles, 1979
(8) J. Geerk, F. Ratzel, KFK Report 2912 (in press), cited from A. Turos, J. Geerk, Appl. Phys. 22, 385 (1980)
(10) E.D. Stokes, T. L. Chu, Appl. Phys. Lett. 30, 425 (1977)
(11) J. F. Ready, B. T. Mc Clure, W. L. Larson, Semicond. International 4 (11), 93 (1981)
(12) A. Mesli, J. C. Muller, D. Salles, P. Siffert, Appl. Phys. Lett. 39 (2), 159
(13) H. Itoh, H. Tamura, M. Miyao, T. Warabisako, K. Itoh, Y. Sasaki Jap. Journ. Appl. Phys. 19, Supplement 19 - 2, 55 (1980)

OPTIMIZATION OF PULSED ELECTRON BEAM ANNEALING PROCESS
FOR SILICON SOLAR CELLS 1)

A. LAUGIER, D. BARBIER, G. CHEMISKY
Laboratoire de Physique de la Matière
Institut National des Sciences Appliquées de Lyon
20, Avenue Albert Einstein - 69621 VILLEURBANNE CEDEX (France)

SUMMARY

Ion implanted silicon solar cells characteristics are related to pulsed electron beam annealing parameters (PEBA). Evolution of electron beam energy deposition profile versus charging voltage and plasma field emission diode geometry is studied. By computer simulation of thermal effects it is shown that molten depth and liquid phase duration can be adjusted to yield the best ion implanted junction characteristics. Effect of heating Si wafers above 400°C before pulsing is investigated. SIMS profiling is used to determine junction profiles after PEBA in (100) phosphorous implanted silicon. PEBA solar cells characteristics are presented.

1. INTRODUCTION

Ion implantation followed by a single submicrosecond electron beam pulse is now considered as among the most attractive processes for high throughput solar cells production. Short duration electron beam pulses (less than 100 ns) can be obtained by discharging a capacitive energy store into a vacuum plasma field emission diode (1). With low mean electron energies (10 to 15 keV) a 1 J/cm^2 beam fluence is sufficient to produce fast liquid phase epitaxial regrowth of a few hundred of nanometers thick silicon layer starting from the single crystal substrate. The capability of processing large diameter wafers is submitted to the availability of a sufficiently high energy store capacity. Production rate of 1800 per hour on 100 mm diameter silicon wafers are expected with recently designed electron beam processors (2).

However, solar cells characteristics being related to junction parameters it is necessary to optimize pulsed electron beam annealing (PEBA) in the overall junction process. Transient annealing effects depend on the time-dependent beam energy deposition profile in the material. In the case of electron control of this profile is possible by mean of electron energy and current adjustment.

In this paper silicon solar cells characteristics are related to adjustable parameters of a pulsed electron beam processor (electron energy spectrum, fluence, process diameter). PEBA thermal effects such as molten depths, liquid-solid interface kinetics, cooling rates and thermal gradients have been studied by computer simulation starting from real-time measurement of pulse parameters.

2. PULSED ELECTRON BEAM PROCESSOR ADJUSTABLE PARAMETERS

The schematic diagram of fig. 1 shows the main parts of a coaxial pulsed electron beam processor. Electron spectra depends on impedance matching between diode and storage line. Diode impedance varies with $(d/r)^2$, r being the cathode radius and d the cathode-anode distance. Hence plasma field emission diode geometry, capacity and charging voltage of the energy store determine pulse characteristics (3). Time-resolved spectroscopy of the electron beam pulse has been achieved by mean of analysis of diode current and voltage waveforms. Measurements were made on a SPI-300 processor.

1) Work supporterd by French COMES and CNRS.

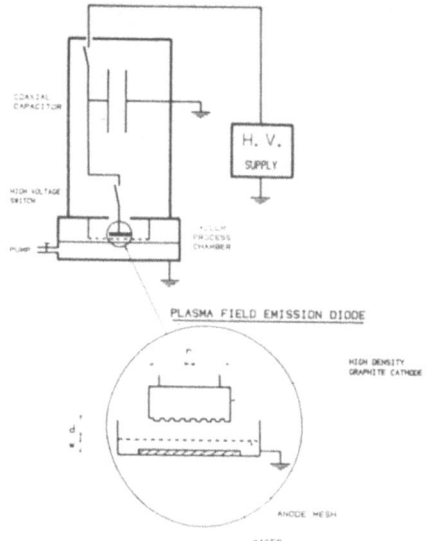

Figure 1 - Schematic diagram of a coaxial pulsed electron beam processor.

Monte-Carlo calculation was then used to simulate the time-dependent energy deposition profile of the pulse. Fig.2 shows typical normalized time-integrated energy deposition profiles in silicon as a function of the r/d ratio. Fluence is keeped near 1 J/cm^2 by adjusting charging voltage The doted line indicates the equilibrium threshold dose to reach the melt temperature of silicon for 1 J/cm^2. Then if an about one micron thick molten depth is desired pulses of 15 keV mean electron energy or less will be convenient. In the case of more penetrating electron pulses beam fluence should be increased to reach a molten layer thickness of one micron or more in silicon.

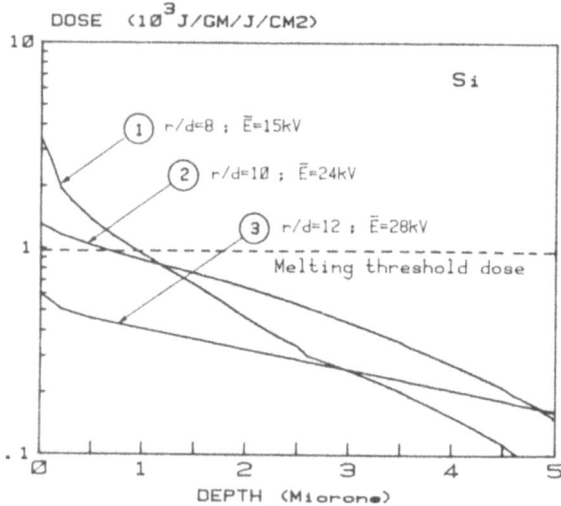

Figure 2 - Normalized energy deposition profiles of a SPI-300 pulsed electron beam processor for various electron pulses of 1 J/cm^2 as a function of the r/d ratio.

3. COMPUTER SIMULATION STUDY OF THERMAL EFFECTS AND JUNCTIONS CHARACTERISTICS

Starting from the heat generation function determined from electron pulses parameters the one dimensional time-dependent heat flow equation has been numerically solved in order to study thermal effects of a typical electron beam pulse as a function of fluence assuming a 100 nm thick amorphous surface layer. Sample temperature before pulsing was fixed at 20°C except for one case in which we simulate thermal effects for a starting temperature of 400°C.

Fig.3 shows melt front motion curves as a function of beam fluence for depth-dose n°1 of fig.2. Because of focusing effects at high current densities beam fluence can be adjusted by varying sample-anode distance w other parameters being unchanged. However energy deposition profiles undergo only minor changes when fluence is adjusted by varying capacitor charging voltage less than 30 % around a mean value. On can see that the maximum molten depth is reached within the pulse duration ($\simeq$ 50 ns). The liquid-solid interface moves then back toward the surface, and melt front velocity during the regrowth phase decreases with fluence. Maximum molten depths are plotted against fluence on fig.4. The saturation at 1.2 micron is related to the depth-limited energy deposition profile which is presently a nearly fixed parameter. Taking a 400°C starting temperature, results in a molten depth higher than 1.2 micron at only 1.2 J/cm^2, because a lower dose is sufficient to reach the melt temperature.

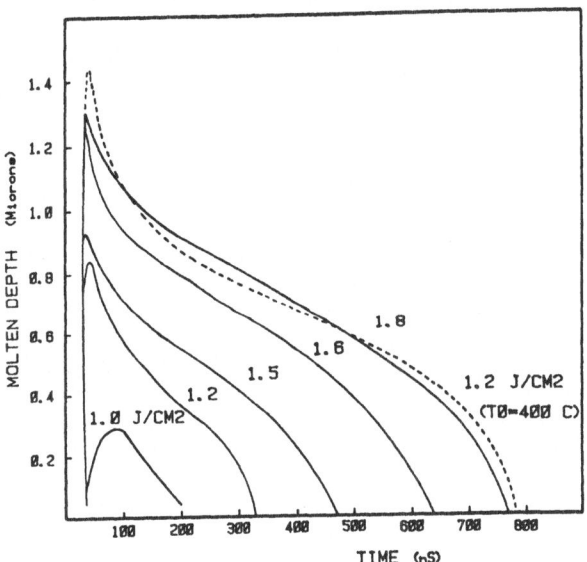

Figure 3 - Time dependence of the melt front position vs fluence

Melt front velocity and liquid phase duration are quasi linear function of fluence as shown on fig.5. These two parameters are not independent and determine the regrowth layer quality and dopants redistribution (4).

With sample at 20°C before pulsing dopants redistribution as a function of fluence can be estimated assuming a diffusivity of 10^{-4} cm^2/s. When fluence is raised from 1.2 to 1.8 J/cm^2 mean diffusion length of dopants varies from 40 to 70 nm. Possible segregation effects depends on melt front velocity for a given specimen orientation. So junction profiles after PEBA will be strongly dependent on melt front kinetics. Comparing the curves obtained when starting with a sample

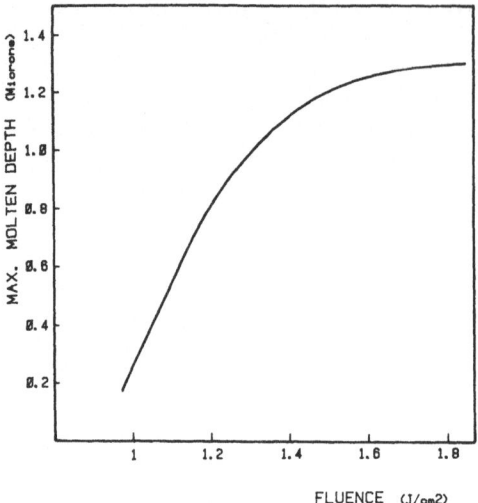
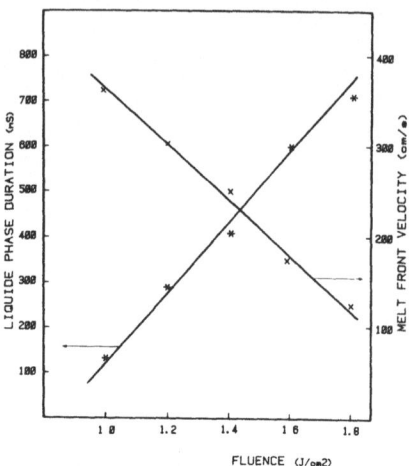

Figure 4 - Maximum molten depth variation vs fluence

Figure 5 - Liquid phase duration variation vs fluence

temperature of 400°C with others one can see that melt front kinetics can be changed by varying either fluence or material temperature before pulsing. Hence a 1.2 J/cm^2 shot on a 400°C heated sample yields nearly the same melt front kinetic as a 1.8 J/cm^2 shot on a sample at 20°C.

Moreover for the same beam fluence cooling rates and thermal gradients are lowered when the specimen is heated before pulsing. So starting temperature should be an important parameter for the control of regrowth layers crystal quality.

4. EXPERIMENTAL PEBA SILICON SOLAR CELLS CHARACTERISTICS

Experimental results were carried out with a SPI-300 pulsed electron beam processor. This machine mostly designed for research applications has a maximum energy store capacity of 15 J. The cathode diameter is then limited to a maximum of 25 mm for fluences higher than 1 J/cm^2.

The starting material is (100) C.Z. boron doped silicon 1.5 Ω.cm in resistivity. After phosphorous implantation at 15 keV with a dose of 2×10^{15} cm^{-2} wafers were annealed at 20°C with a 50 ns duration electron beam pulse. The r/d value was adjusted around 8 in order to have depth dose profiles slightly different from n°1 on fig.2. Fluence has been varied between 1.2 and 1.7 J/cm^2. Annealing diameter is about 15 mm which allows 2x2 cm wafers processing in 4 shots. Experiments made on 1x1 cm wafers with only one shot do not show any significant difference.

PEBA induced doping profile has been studied by SIMS. Fig.6 shows redistribution of phosphorus atoms after a 1.2 or a 1.5 J/cm^2 electron beam pulse. In both cases spreading of the profile is consistent with the computer simulations of section 2 and leads to junction depth less than 200 nm with a N$^+$ layer dopant concentration higher than 10^{20}/cm^3. Assuming a uniform junction model the expected short circuit current Isc and Voc for our starting material are respectively 33 mA/cm^2 and .58 V.

Solar cells were realized with PEBA wafers according to the following process. Ti-Pd-Ag grid contacts were evaporated and a 80 nm thick TiO_2 A.R. coating was deposited at 325°C. The transparency of the grid is 94.5 %. An aluminium layer was evaporated on the back side of the wafers and annealed during 15 mn to one hour at 400°C.

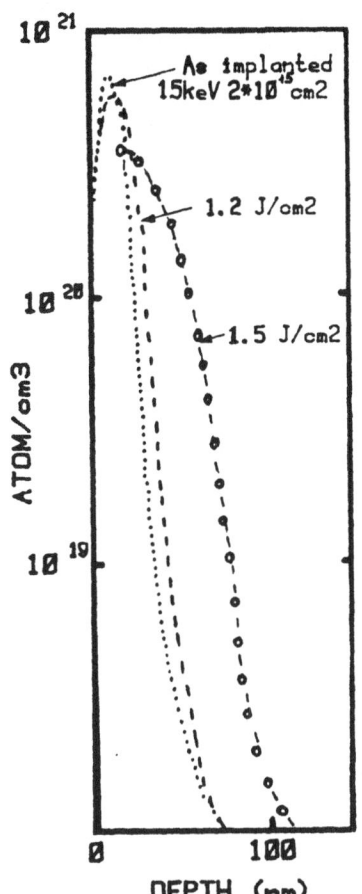

Fluence (J/cm^2)	Icc (mA/cm^2)	Voc (mV)
1.2	23.1	468
1.3	23.6	471
1 5	23	475
1.7	22 7	469

Table 1 - Solar cell parameters vs fluence

Characterization of cells involves spectral response measurements and AM1 I(V) curves determination. A typical spectral response is shown on fig.7 for a PEBA fluence of 1.5 J/cm^2. The fairly good short wavelength response indicates that a shallow and active junction has been achieved. The 880 nm spectral response peak is consistent with a base minority carrier diffusion length of about 100 µm indicating that the overall process does not affect the starting material properties Mean electrical parameters without A.R. coating are presented on table 1. A maximum appears near 1.3 J/cm^2 for Isc and near 1.5 for Voc. One must note that Voc is rather low compared to expected values. This could be related to defects induced by PEBA in the active layer because of the fast cooling rates when wafers are pulsed at room temperature (5). PEBA on samples heated up to 400°C has not yet been tried with the SPI-300 electron beam processor but experiments carried out by SPIRE Inc. have shown a 40 mV increase of Voc when sample starting temperature is 400°C (6).

	Voc (mV)	Isc (mA/cm^2)	FF	Efficiency (%)
Diffusion	575	28.5	0.74	12.1
PEBA	540	32.4	0.70	12.2

Table 2 - Comparison between diffused and PEBA cells. (Same basic material, best results).

However, the high value of Isc confirms the good quality of the junction optimized for fluences between 1.3 and 1.5 J/cm^2.
Table 2 gives the best AM1 solar cells characteristics obtained by mean of classical diffusion technics with the same starting material for comparison with PEBA best solar cells (with A.R. coating).

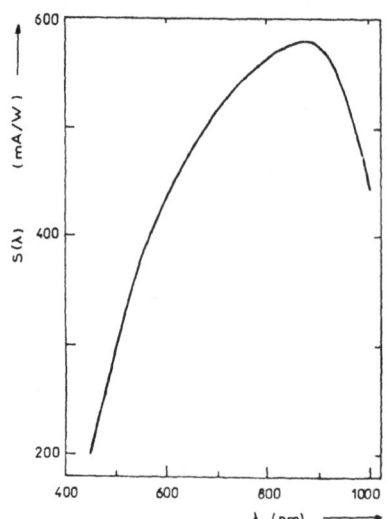

Figure 7 - Typical spectral response of Phosphorus implanted PEBA solar cells

5. CONCLUSION

An optimization study of PEBA has been performed for the silicon solar cells junctions process. It has been shown by computer simulation of thermal effects that the electron beam parameters can be modified to yield a desired doping profile. With shallow electron energy deposition profiles (mean electron energy less than 15 keV) a fluence of the order of 1 J/cm^2 is sufficient to melt an about one micron thick silicon layer. Junction characteristics can be controlled by either fluence or sample starting temperature adjustment. Experiments carried out on phosphorous implanted silicon have shown that PEBA junction profiles are consistent with computed annealing effects. Moreover high short circuit currents have been measured on PEBA solar cells indicating that fairly good quality and shallow junctions have been achieved with an optimized electron beam pulse.

BIBLIOGRAPHY

(1) A.C. GREENWALD, A.R. KIRKPATRICK, R.G. LITTLE and J.A. MINNUCCI : J.Appl.Phys. 50, 2 783, 1979
(2) G.A. LANDIS, A.J. ARMINI, A.C. GREENWALD, R.A. KIESLING : Proceed. of the 15th IEEE Photovoltaic Specialists Conference, p. 976-980, Orlando 1981
(3) D. BARBIER, G. CHEMISKY, A. LAUGIER : to be published
(4) P. BAERI, G. FOTI, J. PONTE : Laser and Electron beam solid interactions and material Processing. Gibbons, Hess and Sigmon eds, 1981
(5) M. THOLOMIER, D. BARBIER. M. PITAVAL, M. AMBRI, A. LAUGIER : J. of Appl.Phys. to be published
(6) A. GREENWALD (private communication)

SILICON SOLAR CELLS BY ION IMPLANTATION: E-BEAM AND SELF ANNEALING

G.F. Cembali, R. Galloni, G. Lulli, A. Mazzone, P.G. Merli, R. Nipoti
CNR - Istituto LAMEL, Via Castagnoli, 1 - 40126 Bologna (Italy)
and F. Zignani
Ist. Chimico - Facoltà di Ingegneria - Università di Bologna.
(Research supported by CEE and CNR-PFE)

Summary

Two different techniques, based on ion implantation, have been experimented to obtain solar cells: a) ion implantation followed by electron beam annealing, b) self-annealed ion implantation. Silicon single crystals have been used to get a better understanding of the processes. In all cases radiation damage is annealed by a solid phase process. Solar cells tested at AM1.5 show efficiencies between 13 and 13.5% which may be compared with a 15% AM1.5 efficiency obtained by ion implantation and furnace annealing. Experimental results are compared with data obtained from a computer simulation to evaluate emitter tailoring effects. A new electron gun especially designed and set up for annealing purpose is presented.

1. INTRODUCTION

Ion Implantation process could become at medium term an interesting technological choice for large scale solar cell manufacturing, to day based on p/n junctions obtained by thermal diffusion. Main advantages offered by this technique are the followings: (a) good control of doping profile which makes possible emitter tailoring for higher conversion efficiencies; (b) reduction of contamination by fast diffusing impurities; (c) reduction of annealing induced bulk lifetime degradation; (d) better automation possibilities for large scale production; (e) high conversion efficiency; (f) compatibility with the use of Solar Grade Silicon.
Two different annealing processes have been recently employed to recover the radiation damage introduced by Ion Implantation: 1) single or interlaced short energy pulses (10-100 ns), (laser or e-beam), by which a liquid-phase epitaxial regrowth of the implanted layer is produced; 2) continuous wave e-beam or laser treatments by which a continuous beam is scanned over the implanted surface for a few seconds, inducing a solid phase regrowth (1). By the first technique heating is limited to the surface and as a consequence bulk lifetime is unaffected, but doping profile is redistributed due to the liquid phase process. By the second technique the doping profile is not modified, but heating is homogeneous over the entire thickness and bulk carriers lifetime may be affected more seriously. Care has to be taken to reduce residual point defects after liquid or solid phase regrowth.
In the present work we employed two different techniques, both involving solid phase processes, to obtain an electrically active dopant distribution for solar cells applications:
i) Implantation followed by irradiation with an homogeneous e-beam of large cross section.
ii) Implantation by an high current density ion beam to obtain heating of the sample during the process and therefore anneal the radiation damage (Self-annealed Implantation).
In case of e-beam annealing the samples have been implanted by phosphorus ions at to successive energies to obtain a good emitter tailoring as suggested by a computer simulation. Self-annealed implantation provides a

theoretically less efficient box shaped dopant distribution. Results are compared with cells obtained by ion implantation plus furnace annealing (2). All the solar cells were realized on (100) Si wafers, p type, 1 cm.

2. ELECTRON BEAM ANNEALING

A specially designed electron gun, which was developed by LAMEL-CNR in collaboration with INTERNOVA, Milan, has been used to perform thermal treatments in semiconductors and in particular annealing of ion implantation damage. The beam energy can be varied between 2 to 60 keV with beam current up to 200 mA. The configuration of the electron column allows to perform homogeneuos annealing of up to about 3 inches wafers without beam scanning. Alternatively the beam can be focalized and scanned at variable raster and frame frequencies. In the annealing experiments here reported we emploied the electron gun in the first mode. A scheme of the experimental arrangement is shown in Fig.1: square shaped samples (2.5x2.5 cm^2) were supported on the target holder by the four corners so that only thermal losses of radiative nature can be assumed and the temperature reached by the samples can be easily evaluated.

Samples to be e-beam annealed, were P^+ implanted by two successive energies, first at 10 keV under (100) channelling conditions, easily obtained with our Lintott III accelerator by simply setting the wafers at right angle to the beam, and then at 40 keV random (8° off axis). In this way the emitter tailoring suggested by the computer model was obtained. The total implanted dose was 2.5×10^{15} P^+/cm^2. Several experimental conditions of irradiation (power, time) were tested; the best results in the solar cells application have been obtained up to now, by irradiating with a power density of 5.8 W/cm^2 for a time of about 12 sec, sufficient to heat the sample at about 730°C for the same time.

3. SELF-ANNEALING IMPLANTATION

As already reported in a previous paper (3), radiation damage recovery can be obtained during the implantation process itself. A computer simulation (4) of the thermal effects produced by an high current density ion beam in a Silicon specimen suggested the best geometrical and beam parameters to carry out the process in solid phase.

Self-annealed implants were performed in a Lintott III implanter with 100 KeV P^+ beam, 2 cm diameter (as defined by a graphite diaphragm) and 2mA total current; sample dimension was 1.5 x 1.5 cm^2. A sample holder similar to the one used in case of e-beam annealing was employed (see Fig. 1); also in this case only thermal losses of radiative nature can be assumed.

With the power density deposition of 50 W/cm^2 used in this experiment the extimated temperature reached by the sample is of about 1300°C. Irradiation of 2.5 sec resulted in a total implanted dose of 4×10^{15} P^+/cm^2. An electronic timer acting on a mechanical shutter controlled the implant time. To improve dose uniformity during the process, implantation was performed with the sample rotating at 3000 cps about its normal axis and tilted of 60° with respect to the ion beam direction.

4. NUMERICAL SIMULATION

A computer model (5,6) has been employed to simulate the effect of emitter tayloring on cell efficiency.

The main points emerging from the simulation can thus be summarized. The features of the cells are similar though electron beam annealed cells have slightly higher short-circuit currents and efficiencies owing to the lower retrograde field near the surface. Representative figures are about 16% for electron beam irradiation, about 15% for self-implanted cells. These numbers must be judiciously read as the available values for some parameters of the simulation span a fairly large range. For instance, the previous figures refer to $n_i = 7.6 \times 10^9$ cm^{-3}. If the standard $n_i = 1.5 \times 10^{10}$ cm^{-3} is used V_{oc} decreases of 25 mV and scales of 1 point.

The simulation indicates that band-gap narrowing effects are negligible as the bulk is lightly doped. As far as the other physical parameters of the emitter are concerned lifetime might be effective only if it reaches values as low as few ps. In this case V_{oc} decrease of about 20 mV and the efficiency of 1 point.

5. EXPERIMENTAL RESULTS AND DISCUSSION

The best results obtained on a set of ten solar cells for each annealing technique used are shown in Table 1.

Solar cells have TiAg, front grid and Al back contact (sintering at 450°C 15 min), TiO_2-SiO_2 double layer antireflecting coating, grid fingers distance 1 mm, coverage ratio about 8%.

Shown in Fig. 2 are the electrical activity profiles evaluated by differential resistivity and Hall effect measurements. Comparison between experimental mobility values and mobilities drawn from the literature relative to samples doped by thermal diffusion, gives evidence of the depth location of scattering centers (residual damage). Particulary evident in the self-annealed samples the region with low mobility values due to a high concentration of defects. Very good mobility values (better than on diffused samples are instead obtained with electron beam annealing. It must be pointed out however that mobility values in the emitter region only influence the series resistance of the device that in case of self-annealed implantation is however very low (0.4) due to the deep junction and high doping.

Quantum efficiencies reported in Fig.3 confirmed that deeper junctions (self-annealing) decrease the short wavelenghts response, however the self annealed cells have a better response to the long wavelenghts so that the total collected current is the same in case of self-annealing and e-beam and cell efficiency is almost the same in the two cases.

Fig.4 shows the dark I-V characteristics of e-beam and self annealed cells. As foreseen, better characteristics are obtained in the tailored emitter e-beam annealed (curve a).

Bulk carriers diffusion lenghts have been measured with the surface photovoltage technique on the finished cells: values similar in case of self-annealing and e-beam (150 um) show that bulk properties have not been affected.

6. CONCLUSIONS

From the experiments reported it is possible to state that:
(a) The Ion Implantation and e-beam annealing technique offers a real possibility of manifacturing Silicon Solar Cell of high conversion efficiency, with a low cost high rate production process. Good chances exist to reach the 15% AM1 efficiency goal, due to the possibility of further optimizing the annealing parameters to reduce residual damage in the device space charge region and further increase bulk carriers lifetime.

(b) The self-annealing Ion Implantation technique shows the same possibility offered by the e-beam annealing, but more work has to be done in the next future; i) to improve emitter tayloring by using higher mass dopant (e.g. As$^+$); ii) to design a new special Implant apparatus with large cross section and high current density Ion beam for omogeneous large area implant.

(c) More experiments are planned to use these techniques on different types of Solar Grade Silicon.

REFERENCES

1. "Laser and Electron beam solid interactions and Material Processing", J.F.Gibbons, L.D. Hess and T.W. Sigmon Ed., North Holland (1981).

2. G. Miccoli, P. Ostoja and S. Solmi: Third E.C. Photovoltaic Solar Energy Conference, by W. Palz Ed., D. Reidel Publishing Company (1981), pag. 719-723.

3. G.F. Cembali, P.G. Merli and F. Zignani, Appl. Phys. Lett. $\underline{38}$, 808 (1981).

4. P.G. Merli and F. Zignani, Rad. Effects Lett., $\underline{50}$, 115 (1980).

5. P.U. Calzolari and A.M. Mazzone, Proc. of photovoltaic Solar Energy Conf., Luxembourg 1977.

6. J.W. Slotboom and H.C. de Graaff, Solid St.Electron $\underline{19}$, 857 (1976).

EXPERIMENTAL ARRANGEMENT FOR E-BEAM ANNEALING

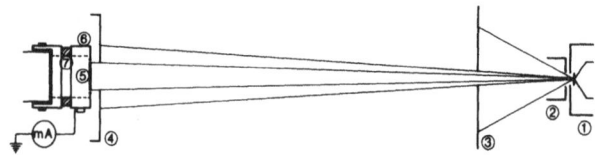

FRONT VIEW OF
THE SPECIMEN HOLDER

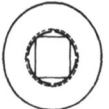

① ELECTRON GUN
② ACCELERATING ELECTRODE
③ INTERCHANGEABLE MOLY APERTURE
④ MOLY WINDOW
⑤ SPECIMEN
⑥ SPECIMEN HOLDER
⑦ CERAMIC INSULATOR

Fig. 1

	J_{SC} (mA/cm²)	V_{OC} (mV)	FF	$\eta\%$ (AM 1.5)
SELF ANNEALED I.I.	33.4	560	.72	13.5
I.I. + E-BEAM	32.8	555	.73	13.3
I.I. + FURNACE	35.8	580	.73	15.2

TABLE 1. Solar Cells parameters.

	a) e-beam	b) self anneal
implanted dose	2.5×10^{15} P⁺/cm²	$\sim 4\times10^{15}$ P⁺/cm²
active dose	96%	95%
sheet resistivity	50 Ω/□	27 Ω/□

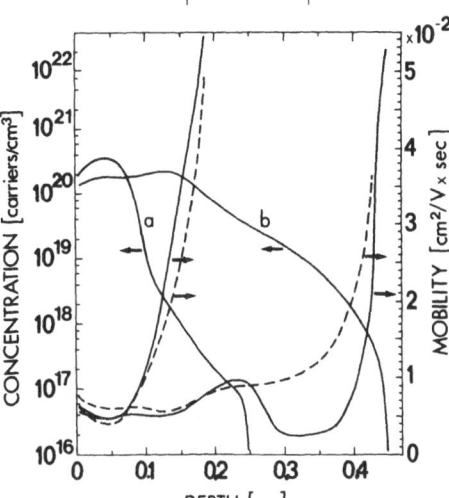

Fig. 2. Electrical activity profiles of: a) tailored implantation of 10 keV 6×10^{14} P⁺/cm² into (100) Si sample at right angles to the incident beam (channeled implant) plus 40 keV 1.86×10^{15} P⁺/cm² random (8° off axis); b) self annealed implantation at 100 keV, $\sim 4 \times 10^{15}$ P⁺/cm², ~ 50 W/cm². Dashed curves are best mobility values available (thermal diffused samples)

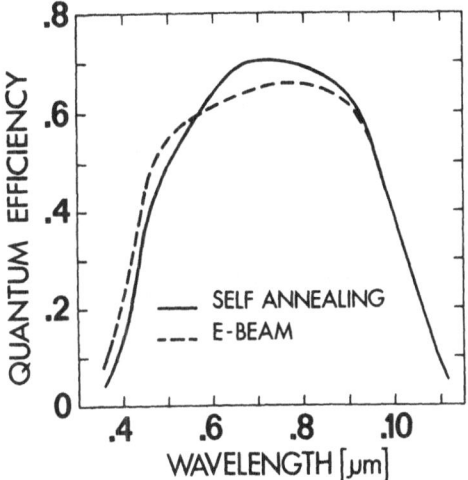

Fig. 3. Quantum efficiency of electron-beam annealed (a) and self-annealed (b) Solar Cells, without ARC.

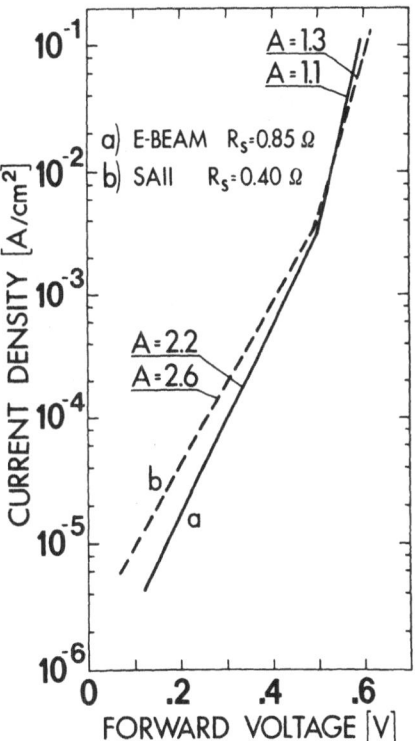

Fig. 4. Dark I-V characteristics of electron-beam annealed (a) and self-annealed (b) Solar Cells.

AN AUTOMATED ION IMPLANT/PULSE ANNEAL MACHINE
FOR LOW COST SILICON CELL PRODUCTION

A.J. Armini, S.N. Bunker and M.B. Spitzer
Spire Corporation
Bedford, Massachusetts USA

Summary

The continuing development of a high throughput ion implanter and a pulsed electron beam annealer designed for dedicated silicon solar cell manufacture is reviewed. This equipment is intended for production of junctions in 10 cm wide wafers at a throughput up to 10 MW_p per year. The principal features of the implanter are the lack of mass analysis and defocusing utilizing electrostatic deflection. The implanted surface is annealed by liquid phase epitaxy resulting from a single burst of a large area electron beam. Cells with non-mass analyzed ion implantation have yielded AM1 cell efficiencies in excess of 15%. Pulse annealed Czochralski cells have been made with AM1 efficiencies of 13% vs. 15% for a furnace annealed group. J_{sc} is 3-4% higher for pulse annealing; V_{oc} is usually lower. Results of pulse annealing of polycrystalline materials indicate that cell performance comparable to diffusion can be obtained.

This work was supported by the Jet Propulsion Laboratory under NASA Contract NAS7-100.

1. INTRODUCTION

Ion implantation and pulsed electron beam annealing for formation of n^+/p junctions in silicon solar cells have been studied for the past 5 years(1-5). Industrial semiconductor process equipment has been used which has been adequate for proof of scientific feasibility and low volume production. Equipment designed for large volume processing of cells is required to verify process yield and commercial feasibility.

Based on several generations of research equipment previously developed(5), we have designed a dedicated solar cell implanter and a pulsed electron beam annealer. This design is based on the processing of 10 cm (round or square) silicon wafers at a rate of 1200 per hour. The current status of the construction of prototype machinery, as well as solar cell studies using this technology will be dicussed.

2. NON-MASS ANALYZED ION IMPLANTER

The ion implanter under development is designed for solar cell production. Cost constraints, which demand a simple and inexpensive machine, make the use of versatile semiconductor equipment undesirable. Our approach to cost reduction is the elimination of mass analysis and fine-line beam scanning. In our machine the wafers, which pass in single file under a broad non-mass analyzed (NMA) phosphorus beam, are carried by a "walking beam" transport system. Wafer speeds up to 5 cm/sec are possible. Figures 1 and 2 show the overall configuration and beam path. The ion beam column is vertical, pointing down at the wafers held by gravity on the track. The system is inherently simple, consisting only of the ion source, the defocus plates, and the walking beam.

This design is based on studies using the NMA test facility shown in Figure 3. Phosphorus ions were produced from a conventional Freeman-type ion source(6) utilizing commercial grade elemental phosphorus vaporized in an oven. A spectrum of the mass components found in the NMA beam indicates the absence of impurities which affect solar cell lifetime(7).

The test chamber has an electrostatic beam deflection unit which spreads the narrow beam spot to the width of a 10 cm wafer. Defocusing results when an electric field disrupts the naturally occurring space-charge neutralization, producing charge-induced beam expansion. Deflection also separates the ions from neutral phosphorus gas and charge-neutralized beam. A large area Faraday cup measures total current and a carbon/sodium target defines the beam spot area.

Three values of electrostatic bend were tested corresponding to $0°$, $17°$, and $25°$. The ions subtended a wide range of incident angles of up to $43°$ for the $25°$ bend. Beam currents up to 6 milliamperes have been readily produced at 10 keV, and currents above 10 milliamperes have been obtained with modifications to the ion source. Operation in the range between 5 and 20 keV has been tested with adequate current.

Results of the cell testing program have indicated that neither large angle deflection nor precision dose is required to obtain high cell efficiency. A modified unit consisting of opposing right and left deflections has been tested for in-line defocusing, and implant uniformity was measured. The standard deviation of the dose for the defocused beam was reduced to 2.6% from 12.5% for the natural beam.

3. ANNEALER DESCRIPTION

The pulsed electron beam annealer (PEBA) shown in Figure 4 is designed to process 10 cm diameter silicon wafers at a rate of up to 1200 per hour. This annealer has been described at length in a previous paper(4) and will be briefly reviewed here. The annealer consists of a high voltage power supply, energy store, diode, and walking beam transport system. The power supply charges the energy store, which is then discharged across the diode. This discharge is triggered when a wafer is directly beneath the anode. The electron pulse of about 80 nanoseconds duration delivers approximately 1 joule/cm^2, which melts the implanted surface to a depth of approximately 0.3 micrometers. The surface regrows epitaxially from the liquid-solid interface. Further details of the anneal mechanism can be found in reference 5.

4. PROCESSING RESULTS

Solar cell processing using NMA phosphorus implantation was compared to mass-analyzed phosphorus implantation using p-type 10 ohm-cm (100) CZ silicon. Wafer backs were implanted with mass-analyzed boron to form a BSF(3) and the fronts were implanted at 10 keV and three implant angles. Estimated dose was 1×10^{15} ions/cm^2. All wafers received a three-step furnace anneal(3) and were metallized with Ti-Pd-Ag contacts. Final cell area was 4 cm^2 with no anti-reflection coating.

Table I shows the measured solar cell performance tested under simulated AM0 insolation. Cell temperature was 25°C. AM1 performance is obtained by multiplying the AM0 efficiency by a correction factor of 1.18 obtained from the cell spectral response and by 1.4, which is the gain typically realized by application of an antireflection coating. The cells implanted with NMA phosphorus perform as well as cells implanted utilizing mass analysis (15.3%). The results also indicate that the implant angle is relatively unimportant; the 0-degree implants performed as well as the bent-beam implants. An important feature of the NMA-implanted cells is the uniformity of the performance. Including all cells not shown in the table, the entire group had AM1 efficiency of 15.4 ± 0.11 percent, despite an order of magnitude variation in dose. The complete results are reported in reference 7.

The application of pulsed electron beams to implantation annealing in single crystal silicon has been discussed elsewhere(5); here we report results on polycrystalline wafers. Cells were fabricated from HEM (semiconductor grade with grain size > 1 cm), SEMIX (grain size >1 cm), SILSO (grain size > .5 cm) and 10 ohm-cm CZ silicon. Implantation consisted of 10 keV $^{31}P^+$ at a dose of 2.5×10^{15} ions/cm^2 with no back surface implantation. The implantation damage was annealed by a single electron pulse with a fluence of approximately 1 joule/cm^2, followed by a one hour thermal treatment at 550°C. This treatment was used because it has been observed to increase V_{oc} in single crystal silicon cells, but it is not believed to be necessary for some types of sheet material(8). Cell metallization consisted of patterned Ti-Pd-Ag on the front and full area Al-Ti-Pd-Ag on the back. Cells of area 4 cm^2 were cut from the wafers and a TiO_2 antireflection coating was applied.

Table II shows the AM1 performance of selected cells as measured by the Solar Energy Research Institute (SERI). The efficiency of the Czochralski control cell is only 12.6% without a BSF. The HEM cell performs almost as well as the control. The performance of the SILSO cell is comparable to diffused cells made from SILSO(9).

5. ECONOMIC ANALYSIS

An analysis of cell production economics has been performed on each of the machines (implanter and annealer) operated separately to produce junctions in silicon wafers. The procedure used is the IPEG II (Interim Price Estimating Guidelines) developed by the U.S. Jet Propulsion Laboratory(10). A production rate of 10 MW and 8280 hours per year are assumed for Table III. The junction production costs for both implantation and annealing are 4.5¢/W_p. With larger versions of these machines, (30 MW/yr), the total junction cost is reduced to 1.8¢/W_p.

6. CONCLUSIONS

We have designed and are constructing a pair of machines which will be capable of implanting and electron beam annealing 10 cm diameter solar cells at a 10 MW per year rate. The process techniques have been used to form n^+-p junctions on CZ wafers as well as other low cost sheet material. NMA implantation has yielded cells equivalent to mass-analyzed controls. AM1 efficiencies of cells formed by pulse annealing, as measured by SERI, were 11.3% for HEM and 10.5% for SILSO solar cells. Further process development specifically for sheet materials is expected to lead to better cell performance than can be obtained by diffusion. Production costs when operated at full capacity are estimated at 0.026\$/W_p for ion implantation and 0.018\$/W_p for annealing.

REFERENCES

1. A.R. Kirkpatrick, et al., 12th IEEE Photovoltaic Specialists Conference (1976), page 299.
2. A.R. Kirkpatrick, et al., 13th IEEE Photovoltaic Specialists Conference (1978), page 706.
3. A.R. Kirkpatrick, et al., 14th IEEE Photovoltaic Specialists Conference (1980), page 820.
4. G.A. Landis, et al., 15th IEEE Photovoltaic Specialists Conference (1981), page 976.
5. A.C. Greenwald, et al., J. Appl. Phys. 50, 783 (1979).
6. J.H. Freeman and G. Sidenius, Nucl. Instrum. Methods 107, 477 (1973).
7. M.B. Spitzer et al., accepted by Appl. Phys. Lett. (June 1982).
8. J. Ho, private communication.
9. K. Roy, et al., 14th IEEE Photovoltaic Specialists Conference (1980), page 897.
10. R.W. Aster, et al., Improved Price Estimation Guidelines (IPEG), Rev. A, JPL Report 5101-158, July, 1980.

Table I. 25°C AM0 Performance of NMA-Implanted CZ Cells (see Text)

Implant Angle	V_{OC} (mV)	J_{SC} (mA/cm^2)	FF (%)	AM0 Eff (%)	AM1 Eff (%)
0	579	29.0	75.1	9.3 $\pm$ .03	15.4 $\pm$ .1
17	580	29.0	75.3	9.36 $\pm$ .06	15.5 $\pm$ .1
25	579	29.0	75.5	9.35 $\pm$ .03	15.4 $\pm$ .1

Table II. 28°C AM1 Performance of PEBA-Processed Cells

Material Type	V_{OC} (mV)	J_{SC} (mA/cm^2)	FF (%)	Eff (%)
HEM	540	27.6	76.1	11.3
SEMIX	519	23.8	71.1	8.8
SILSO	518	27.2	74.2	10.5
Cz	541	30.5	76.6	12.6

Table III. Operating Costs at 10 MW/yr Throughput

		Ion Implanter	Pulse Annealer
Equipment Cost	(X.489)	$166,000	$138,000
Floor Space	(X 123)	$ 2,950	2,950
Direct Labor	(X 2.1)	39,000	39,000
Materials	(X 1.3)	52,000	1,560
Utilities	(X 1.3)	1,800	1,400
		$261,750	$182,910
Cost per W_p		0.0261 $/W_p$	0.0183$/W_p$

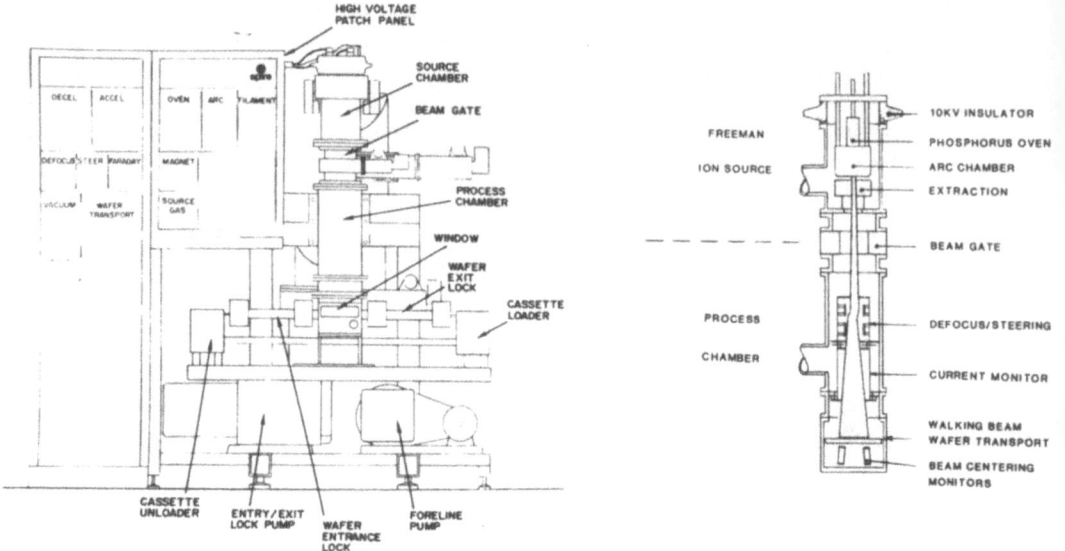

Fig. 1 - NMA implanter design

Fig. 2 - NMA implanter beam path

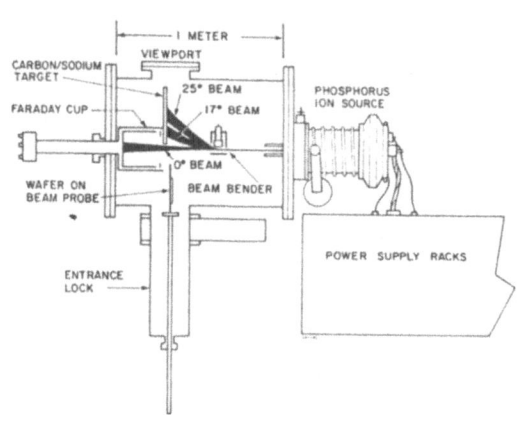

Fig. 3 - NMA development system

Fig. 4 - Electron beam annealer

LASER PROCESSING IN THE PREPARATION OF HIGH EFFICIENCY POLYCRISTALLINE SILICON SOLAR CELLS

E. COURCELLE, E. FOGARASSY, J.C. MULLER, P. SIFFERT
CENTRE DE RECHERCHES NUCLEAIRES
Groupe de Physique et Applications des Semiconducteurs (Phase)
67037 STRASBOURG-CEDEX (France)

Summary

Two doping processes, based on the use of high power lasers as surface localized energy sources have been employed to realize the junction on P-type polycristalline silicon, prepared by a cast technique (Wacker).
Doping - Non mass separated ion implantation, the ions being generated from PF_5
- Laser induced diffusion of Sb deposited on surface.
Annealing
- A Q-switched pulsed Ruby laser ($\lambda = 0.69 \mu m$, $E = 1.5$ J/cm^2 providing a large area beam ($\emptyset$ 8 mm).
- A repetitively Q-switched, CW pumped Yag ($\lambda = 0.53 \mu m$, $E = 2.5$ J/cm^2) laser focused on 100 microns diameter spots scanned on large area. The properties of the doped layers have been investigated and compared to that of single cristalline silicon handled in the same way. The characteristics of solar cells were studied as a function of various parameters, such as ion implantation conditions, thickness of the dopant layers and effect of post laser thermal treatments. Cell performances under AM1 conditions are rather promising, since efficiencies up to 11 % have been achieved on large grained cast silicons for large area cells.

1. INTRODUCTION

Development of solar energy on a large scale is conditionned by lowering of the costs of material and of manufacturing the junction. New techniques for elaborating the latter have been recently proposed and developped. They require, in principle, less handlings than thermal diffusion and should, therefore, be cheaper, the more so as they can be easily automated. Furthermore, more of the polycristalline materials proposed today degrade in performance when heated at high temperature, as a consequence, cold processes would be prefered. Here, we describe two techniques that we developped for manufacturing cells on poly-silicon.

2. EXPERIMENTAL CONDITIONS

Samples used in this study were 300 μm thick polycristalline wafers of Wacker's Silso material, having 1-5 Ω.cm resistivity. After chemical etching with a white etch, the junction was realized by two techniques.

A. GLOW DISCHARGE INCRUSTATION

This method (1) differs from conventional ion implantation in that no mass separation is used for selecting the impinging projectiles. The ion source contains a gas mixture containing the dopant, here PF_5 at a pressure of 10^{-2} to 10^{-3} torr. A stable plasma is generated under a constant D.C. voltage of 8 KV. The ions are extracted from the plasma and accelerated directly towards the samples at an energy of 10 keV, with a current density of 0.4 mA/cm^2. The implanted doses varied between 5.10^{16} and 2.10^{17} cm^{-2}.

B. SURFACE DEPOSITION OF THE DOPANT

In this method (2), a film of dopant, here Antimony, is deposited by evaporation in vacuum (p = 10^{-6} torr) on the etched silicon wafer. The film thickness is monitored by a quartz monitor and the rate of deposition is maintained below 2 Å/s, in order to avoid heating of the substrate. Thicknesses ranging between 50 and 150 Å have been used.

C. LASER IRRADIATION

The implanted as well as deposited samples were irradiated using two different lasers. Laser annealing is necessary for the incrustated cells to remove the damage from the cells. For the deposited cells, laser heating makes the dopant to diffuse in the lattice. We used successively :

- a pulse ruby laser (λ = 0.69 μm) with energy densities ranging between 0.8 and 2 J/cm^2, a pulse duration of 20 ns, and a beam size of 8 mm.
- a high repetitive rate (5 KHz) Q-switched YAG laser (λ = 0.53 μm) which delivers pulses of about 100 μm in diameter of 100 ns duration and an energy density of 1-2.5 J/cm^2. The pulse overlapping can be adjusted and the full set-up is micro-processor controlled.
Post thermal annealing (600°C 1/2 h) of the laser annealed implanted junctions must be performed to remove the tail implantation defects.

D. SOLAR CELL PREPARATION

Non optimized technologies have been used for manufacturing the cells after laser annealing : a 3000Å thick back contact (Au) was evaporated and the front grid was obtained by evaporation of 3000 Å silver through a mask. The antireflecting coating was, finally, deposited by evaporating a 700 Å thick SiO film in vacuum.

E. MEASUREMENTS PERFORMED

- Sheet resistance of the doped layers were estimated by a four point probe, before deposition of the grid.
- Electrical dark properties of the junction included I-V as well as C-V measurements.
- Cells performance under irradiation were done at 28°C using an ELH lamp, which produces a reasonable approximation of the AM1 spectrum. An energy level of 100 mW/cm^2 was need, determined by a calibrated cell.

2. RESULTS

A. OPTIMIZATION OF DOPING CONDITIONS

They were determined by using the sheet resistance measurements of the doped layers as well as the I-V characteristics. Despite variations of the sheet resistance from grain to grain, very low values are obtained for the conditions reported on Fig. 1. The I-V characteristics of the junctions follow the following equation :

$$J = J_{or} \exp \frac{qV}{n_r kT} + J_{od} \exp \frac{qV}{n_d kT}$$

where J_{or} is the recombination saturation current and J_{od} is the diffusion saturation current, n_r and n_d being the corresponding quality factors. Fig. 2 a-b shows the variations of J_{or}, n_r, J_{od} and n_d as a function of the dopant dose.

As in the case of sheet resistance, these curves present a minimum corresponding to the optimized doping conditions, which are reported on Table I for both mono and polycristalline silicon after ruby (1.5 J/cm^2) and YAG (2.5 J/cm^2) laser annealing. It appears that for poly-silicon annealed with the ruby laser, film thicknesses and implanted doses must be twice as high as for single crystalline silicon. This can be interpreted in terms of an accumulation of dopants at grain boundaries. With our YAG set-up, the very localized fusion of silicon, typically 100 μm spots, prevents lateral diffusion of dopant in the molten region on longer distances, as for the ruby laser, reducing the segregation of impurities at grain boundaries. Sheet resistance measurements show also less dispersion, which means that the doping is more homogeneous. Besides the found reverse currents are always higher than for thermal diffused cells, as in the case of single crystalline silicon. This can be interpreted in two ways : either by surface effects associated with the very high doping of the front layer, due to the very high solubility of dopants after laser processing, or to defects induced by the fast epitaxial regrowth (3 m/s) of the surface irradiated with the laser.

B. C-V MEASUREMENTS

As shown on Fig. 3, it appears that by plotting $1/C^2$ against V a straight line is obtained for the deposited dopants, except for bias voltages below 1V, where a more complicated behaviour is observed ($\sim 1/C^3$). This effect disappears after thermal annealing at about 600°C for 30 min. On implanted cells, such an effect cannot be observed since they are always thermally annealed. Such effects, well known on single cristalline cells (2) has been explained by a compensation effect in the P-type substrate by a donor defect created in the molten zone during the epitaxial high speed regrowth (3).

C. SOLAR CELL RESULTS

a) SPECTRAL RESPONSE

The spectral response of the cells has been measured for the two kinds of dopants introduction with the two types of lasers. It must be remembered that the results are very sensitive on the poly-cristalline material, since the low temperature annealing is known to induce contamination effects (4, 5) at grain boundaries and introduce deep levels (6). On Fig. 4 we reported the normalized spectral response found for ruby and YAG annealing, for the same material.

It appears that the response of the implanted cells degrades more rapidly in the I.R. This results from the 600°C thermal treatment we always performed on these structures and from thermal sensitivity of this Silso crystal, degrading the bulk minority carrier diffusion length,as shown by the spectral response measured on a post-laser thermal annealed (600°C 1/2 h) deposited cell.

b) DIFFUSION LENGTH

By measuring the charge collection efficiency vs wavelength in the energy gap region a linear function of the reciprocal absorption coefficient was obtained, from which the effective diffusion length was deduced (7). Results are shown on Table II. They show clearly the degradation due to the thermal annealing of the implanted cells. But, as already indicated this behaviour is sensitive to the kind of material used.

c) PERFORMANCE UNDER ILLUMINATION

Table III summarizes the results. The efficiency of the ruby laser deposited cells is somewhat lower than that of the corresponding implanted cells. This difference disappears when the annealing is performed with the YAG laser.

It can be explained by surface effects, probably due to accumulation of dopant at grain boundaries, induced by the ruby laser treatment.

CONCLUSION

These simple methods allow the preparation of high quality solar cells. The YAG laser system "EPITHERM" (8) we used in this work is quite superior to the conventional ruby lasers, probably due to the small light spot diameter, which avoid the lateral diffusions in polycrystalline materials. Optimalization of these techniques for poly Si is under way in the laboratory.

REFERENCES

1. J.C. MULLER, J.P. PONPON, J.J. GROB and P. SIFFERT, in Proc. European Photovoltaïc Solar Energy Conf. D. Reidel (Dordrecht, Holland), pp. 897-909, sept. 1977.
2. E. FOGARASSY, R. STUCK, J.J. GROB and P. SIFFERT, J. of Appl. Phys. 52 (2), 1076 (1980).
3. L.C. KIMERLING and J.L. BENTON, Laser and Electron Beam Processing of Materials, edited by C.W. White and P.S. Percy, p. 385 (1980).
4. P.E. RUSSEL, C.R. HERRINGTON, D.E. BURKE and P.H. NOLLOWAY The Proceedings of a Symposium on Laser and Electron beam Interactions with Solids, Boston 1981.
5. R.T. YOUNG, R.F. WOOD, J. NARAYAN and C.W. WHITE, Laser and Electron Beam Processing of Materials, Edited by C.W. White and P.S. Percy p. 651 (1980).
6. M. MESLI, private communication.
7. E.D. STOKES and T.L.CHU, 102, 30, 425 (1977).
8. QUANTRONIX, Smithtown, N.Y. (USA).

TABLE I
OPTIMAL CONDITIONS FOR JUNCTION MANUFACTURING

	Single crystalline		Polycrystalline	
	deposited thickness Sb (Å)	implant dose (cm^{-2})	deposited thickness Sb (Å)	implant dose (cm^{-2})
RUBY	50	5.10^{16}	130	1.10^{17}
YAG	50	5.10^{16}	50	5.10^{16}

TABLE II
EFFECTIVE DIFFUSION LENGTH

	Deposited Sb	Incrustation + 600°C thermal anneal
RUBY	90 μm	30 μm
RUBY + 600°C thermal anneal	60 μm	-
YAG	100 μm	35 μm
YAG + 600°C thermal anneal	40 μm	-

TABLE III
SOLAR CELL PERFORMANCE

	V_{oc} (mV)	I_{sc} (mA/cm^2)	FF	η %
deposited Ruby	510	23	0.74	8.7
implanted Ruby	515	27	0,75	10.4
deposited Yag	540	27	0.71	10.4
implanted Yag	515	29.4	0.69	10.4

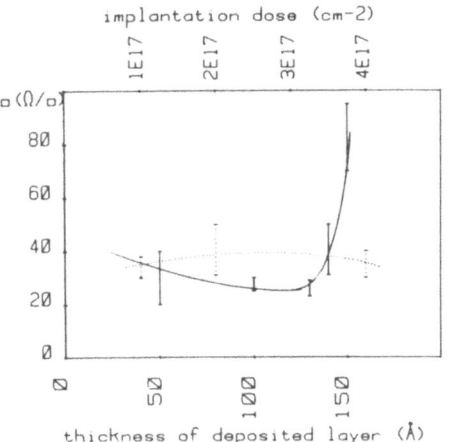

Fig. 1 : Variation of the sheet resistance as a function of the implantation dose (dotted line) or the thickness of the deposited layer (solid curve) (Ruby laser 1.5 J/cm^2).

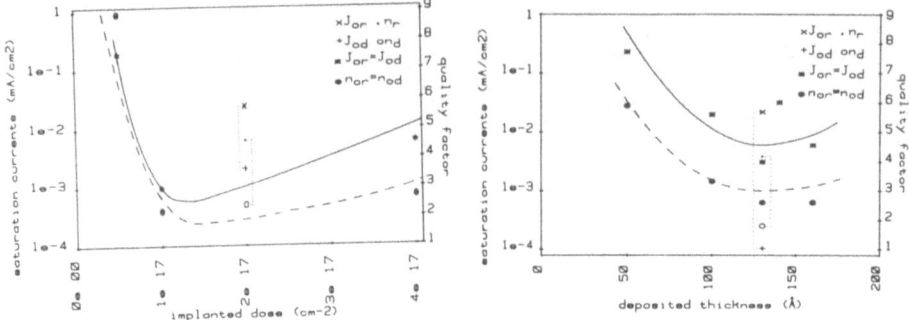

Fig. 2 a-b : Variation of the saturation currents (recombination and diffusion)(solid curve) and the quality factors (dashed line) as a function of the implantation dose (2a) and the thickness of the deposited layer (2b) for ruby laser annealing (1.5 J/cm^2).
Points connected by dotted vertical lines correspond to diffusion and recombination currents or quality factors for the same sample.

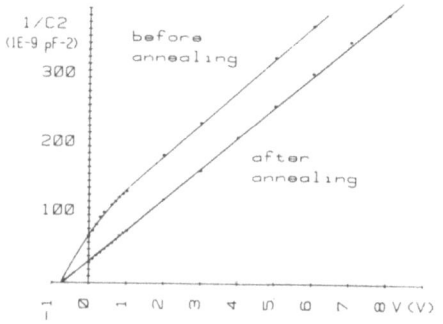

Fig. 3 : Reverse biased C-V measurements of laser treated deposited layers, before and after post-laser thermal annealing (600°C, 1/2 h).

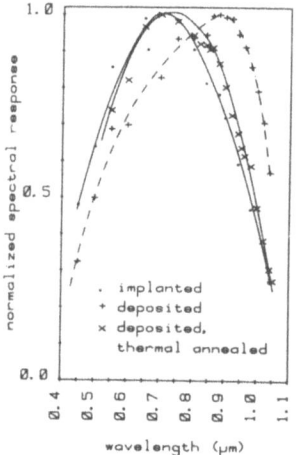

Fig. 4 : Normalized spectral responses for the two types of doping processings. Comparison between a deposited cell, with and without thermal annealing and an implanted cell.

THE INFLUENCE OF SURFACE TEXTURE AND THERMAL TREATMENT
ON THE PERFORMANCE OF LASER-ANNEALED SILICON SOLAR CELLS

W. SINKE, D. HOONHOUT and F.W. SARIS
FOM-Institute for Atomic and Molecular Physics
Kruislaan 407, 1098 SJ Amsterdam

Summary

Single-crystal silicon cells have been made by shallow, mass-analyzed ion-beam implantation followed by Q-switched ruby-laser annealing. The influence of surface texture of the substrate, and of a thermal treatment at $400^{o}C$ or $600^{o}C$ during 40 min. either prior to, during or after pulsed-laser annealing on the performance of the cells has been studied. Structure and composition of implanted and annealed silicon have been investigated by Rutherford Backscattering and channeling. Solar cell performance was characterized by measuring the I-V-curve. It was found that texturized surfaces are very prone to laser-induced damage as compared to non-texturized or polished surfaces. Treating the implanted silicon wafers to $400^{o}C$ during pulsed-laser-annealing results in improved open-circuit-voltage, with respect to laser-annealing without substrate-heating. We found no evidence that prolonged thermal heating prior to or after pulsed-laser annealing can further improve cell performance.

1. INTRODUCTION

The interest in applying ion-implantation plus pulsed-laser annealing to the fabrication of silicon solar cells stems from the possibility to form very shallow and abrupt pn-junctions, while extensive and prolonged heating of the bulk silicon to temperatures in the order of $1000^{o}C$ can be avoided. The latter is especially important in view of the great emphasis which is presently being put on the use of poly-crystalline solar-grade silicon as a substrate, since high temperatures in these materials can be the cause of severe deterioration in carrier-diffusion length. We have shown earlier that with 2x2 cm^2 single-crystalline substrates, mass-analyzed ion-implantation followed by pulsed-laser annealing using overlapping pulses, yields solar cells comparable in performance to those made with standard diffusion techniques on the same substrate (1). Moreover, we have found that heating the substrate to $450^{o}C$ for 10 min. after laser-annealing, which was necessary to sinter evaporated front - and/or back-contacts, had a favourable effect on cell performance. Thermal treatments at moderate temperatures before or during pulsed-laser annealing have also been claimed by various authors (2-4) to yield substantial improvements.
In the present work we compare the influence of thermal heating at $400^{o}C$ during 40 min. either prior to, during or after pulsed-laser annealing, and of thermal heating at $600^{o}C$ after pulsed-laser annealing, on the current-voltage characteristics of single-crystal silicon solar cells made with ion-implantation and laser-annealing. In addition, we compare the structure and electrical characteristics of cells made in this way on the basis of texturized and non-texturized substrates. Structure and composition of the samples have been measured by Rutherford Backscattering (RBS)

plus channeling. Solar cell performance was characterized by measuring the current-voltage characteristics.

2. EXPERIMENT

The substrates used were (100)-oriented, single-crystal, p-type silicon wafers of 3 inch diameter. Before implantation the wafers were treated in 20% NaOH, which resulted in a slightly roughened surface. In addition, the cells labelled as texturized received an additional etch, resulting in a more pronounced roughening of the surface. Then a back-contact was made by screen-printing of Ag, followed by sintering at 680°C for a few minutes. The wafers were ion-implanted with a mass-analyzed 10 keV As^+-beam, up to a dose of 3×10^{15} cm^{-2}.

Laser annealing was carried out with a single stage 5/8"-diameter Q-switched ruby laser with max. energy output of 4J and a pulse length of 20-25 ns. In most cases the laser-light was shone through a converging lense and a glass-plate diffuser. The laser-energy-profile on target, as measured with a 1 mm-pinhole and a volume-absorbing calorimeter, was Gaussian-shaped with a FWHM on target of 12 mm without diffuser and 9 mm with diffuser. The 3 inch wafers were annealed with overlapping pulses in a triangular pattern, with steps of 6 mm in one direction and 5 mm in the other. Energy output of the laser was adjusted such that each spot was at least once irradiated with an energy density in the range 1.0-1.4 J/cm^2, in the case of the non-texturized wafers, and in the range 0.8-1.2 J/cm^2, in the case of the texturized wafers.

For each type of surface-texture 6 sets of samples were made, each under different annealing conditions as follows: pulsed-laser annealing via lense and diffuser, with thermal heating to 400°C during 40 min. prior to, during or after the laser-treatment, or with thermal heating to 600°C for 40 min. after the laser-treatment, or pulsed-laser annealing without any additional thermal treatment (apart from contact sintering) either directly or via lense and diffuser. Laser-annealing as well as thermal heating were carried out in open air.

The implanted and annealed wafers were processed into 3" cells by screen-printing and sintering of a front-contact grid. On some non-texturized cells a TiO_2 anti-reflective coating was made. No chemical cleaning or etching was applied after implantation and annealing.

3. RESULTS AND DISCUSSION

It appeared that surface texture is an important parameter when assessing the possibilities of pulsed-laser annealing. In previous work we found that polished surfaces exhibit no damage visible with the naked eye up to a laser energy density of 2 J/cm^2 (1). It should be noted that with an implantation of 10 keV, the best results were obtained at a laser energy density of 1.2 J/cm^2, when using a Q-switched ruby-laser having an approximately rectangular energy profile (5). In the case of the texturized surface however, visible changes could be observed when the local energy density exceeded 1.3 J/cm^2. Figs. 1-3 show micrographs of texturized surfaces before and after irradiation with densities of 1.2 and 2.0 J/cm^2. For the surface irradiated with 2.0 J/cm^2, RBS and channeling measurements showed a min. channeling yield which was 6-9% of the random yield, indicating that a large fraction of the silicon atoms within a few hundred nm from the surface was not on substitutional sites. The lower damage-threshold for texturized surfaces at least in part is related to more effective laser-absorption due to a lower reflectivity. By the same token, a lower energy density is sufficient to completely anneal implantation damage than in the case of polished surfaces. RBS and channeling measurements provided no evidence for residual damage on texturized wafers annealed with a laser scan such that each spot was irradiated with an energy density between 0.8 and 1.2 J/cm^2. The non-texturized surfaces were less sensitive to surface damage, and were annealed with a local energy density between 1.0 and 1.4 J/cm^2.

The most important parameters of the current-voltage characteristics for all sets of 3-inch solar cells are shown in table I. Although from this table one cannot conclude that the non-texturized wafers yielded the better cells, non-texturized wafers are more easily applied, as the useful energy-window between the annealing-threshold and damage threshold is larger (1). Due to lower reflectivity, the short-circuit current is always higher in texturized cells when compared to non-texturized cells without anti-reflective coating.

Although various authors have reported that the thermal treatments studied result in improved solar cell performance, we could only reproduce these findings for heating of the substrate <u>during</u> laser-annealing. Within the framework of laser-induced melting, two explanations for this favourable effect are possible. On the one hand, when the substrate is already heated less energy is needed to reach the melting point. Therefore, for a given laser pulse with a Gaussian-shaped energy profile, a larger area will reach the melting point and recrystallize. Indeed, for a single pulse the annealed area, as judged from the vanishing of the "milky appearance" typical of amorphous silicon, appeared significantly larger when the substrate was heated during laser-anneling. Consequently, the overlap between neighboring pulses is more ideal. On the other hand, it has been argued, on the basis of heat-flow calculations (3), that during the recrystallization stage, the phase boundary moves at a slower rate when the substrate is hot, such that less structural defects are frozen in than at high freezing rates. In both models, structural quality of the junction is improved, resulting in a higher open-circuit voltage, as observed. Further verification of these models is required.

Contrary to polished silicon cells, made with laser-annealing, the fillfactor appears to be rather low, indicating a high series resistance or a low shunt-resistance. In most cases, the best fit of theoretical and experimetal I-V curves (as measured on 2x2 cm^2-cells) is obtained for a series resistance above 0.75 Ω and a shunt-resistance below 75 Ω. Probably the high series-resistance is related to the low junction depth in combination with the surface texture. This is further substantiated by the fact the R_n appeared to be highest on texturized material.

Pulsed-laser annealing without diffuser gave poor results, also in comparison with previous work on 2x2 cm^2-cells annealed without diffuser. This can be understood, as the probability for hot spots which are deleterious to cell performance increases with cell-size.

RBS measurements revealed the presence of 1-3 monolayers of Ag, if the backcontact is screen-printed before ion-implantation and laser-annealing. The contacting-procedure therefore must be modified accordingly.

Fig. 1

surface of a texturized silicon wafer before laser annealing

Fig. 2

surface of a texturized silicon wafer after laser annealing at 1.2 J/cm^{-2}

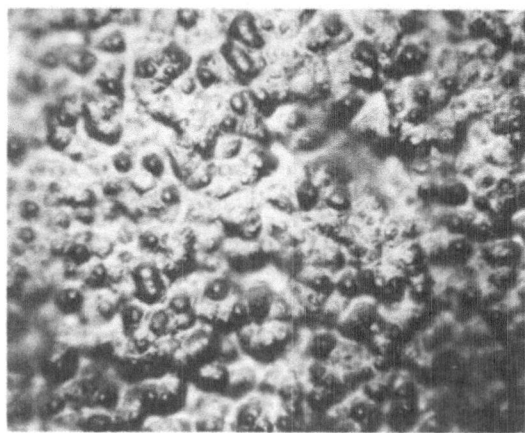

Fig. 3

surface of a texturized silicon wafer after laser annealing at 2.0 J/cm^{-2}

4. CONCLUSION

We have studied the influence of surface texture and additional thermal treatment on the performance of single-crystal silicon solar cells made by ion-implantation and pulsed-laser annealing. Texturized surfaces are found to be prone to laser-induced damage limiting the useful laser-energy window for annealing. Moreover, a high series resistance is found, related to the shallow junction depth in combination with the surface texture. The best results are obtained with pulsed laser annealing while the substrate is heated to $400°C$. We found no evidence for improved cell performance when a thermal treatment is applied before or after laser-annealing.

ACKNOWLEDGEMENT

The authors wish to thank S. Doorn for technical assistance, specifically the ion-implantations; and J. Donon of Photowatt International S.A. for his work on the preparation and measurements of the cells. This work is part of the researchprogram of FOM with financial support from ZWO and the Commission of the EC.

Table I

	Voc (mV)	Jsc (mA/cm^2)	FF	N(%)
Non Text. (NO AR)				
R.T.	530	20.9	0.44	4.9
40' at 400°C before	530	21.5	0.57	6.5
" " during	555	21.1	0.59	6.9
" " " (AR)	560	27.0	0.60	9.1
" " after	515	20.9	0.45	4.8
" 600°C "	490	21.6	0.52	5.6
R.T. no diffuser	485	20.0	0.49	4.8
Text. (NO AR)				
R.T.	535	28.7	0.57	8.7
40' at 400°C before	510	27.6	0.48	6.8
" " during	560	28.2	0.57	9.0
" " after	535	28.9	0.54	8.3
" 600°C "	510	24.4	0.42	5.3
R.T. no diffuser	475	15.0	0.26	1.9

REFERENCES

(1) J. Michel, C. Fages, J.C. Muller, P. Siffert, D. Hoonhout, T. de Jong and F.W. Saris, Proc. Third E.C. Photovoltaic Solar Energy Conf., Cannes, 1980 (Reidel Publ., Dordrecht, Holland, 1981)
(2) A. Mesli, J.C. Muller, D. Salles. P. Siffert, Appl.Phys.Lett. 39, 159 (1981)
(3) R.T. Young, R.F. Wood, W.M. Christie, J.Appl.Phys. 53, 1178 (1982)
(4) Z.K. Fan, V.Q. Ho, T. Sugano, Appl. Phys. Lett. 40, 418 (1982)
(5) D. Hoonhout, F.W. Saris, J. Michel, C. Fages, E. Fabre, Proc. 15th IEEE Photovoltaic Specialists Conf., Orlando, Fl., 1981

IMPLANTATION OF BORON AND BORON FLUORIDE COMPOUNDS INTO SILICON FOR PRODUCTION OF SOLAR CELLS

A.Nylandsted Larsen, F.Nielsen, and G.Sørensen
Institute of Physics, University of Aarhus
DK-8000 Aarhus C, Denmark

SUMMARY

The effects of pulsed laser irradiation on single crystalline silicon implanted with mass-separated B, BF, BF_2, and BF_3 have been studied by channeling, transmission electron microscopy, and sheet resistivity measurements. It is shown that a high implantation dose of fluorine in the surface layer causes a decrease in the threshold energy density for recrystallization. Low sheet resistivities before recrystallization is also found to correlate to the fluorine concentration.

1. INTRODUCTION

Ion implantation of single crystal silicon for the production of solar cells has received a considerable attention, since it is easy to control impurity concentration and profile, and shallow junctions on large areas can be easily obtained[1]. Combined with laser or electron beam annealing to remove the damage caused by the ionic bombardment, the ion implantation method is fast, easily automatized, and practicable at room temperature[2]. For a further cost reduction, ion implantation without mass separation has been utilized by several groups[3,4,5]. Although a high beam current is easily obtainable, the composition of the ionic beam is difficult to control. This may influence both the annealing process and the electrical properties in the near-surface region of the specimen. For instance, when boron implantations are performed by feeding boron trifluoride to the ion source, ionic species such as B^+, F^+, BF^+, BF_2^+, BF_3, SiF_2, and CF_2 are known to be present in the beam in proportions depending on the conditions in the ion source. This will give rise to boron depth profiles which are dependent on the actual conditions in the ion source and to radiation damage which is mainly determined by the F^+-component. The radiation damage has an effect on coupling the laser pulse into the solid and besides, high concentration of fluorine implanted in the near surface layer may influence the electrical activity of the boron impurities.

The present paper deals with simulating boron implantations without magnetic analysis of the ion beam by implanting molecular ions. With an electromagnetic ion accelerator, pure boron as well as mass-separated molecular boron fluoride (BF^+, BF_2^+, and BF_3^+) have been implanted into single crystalline silicon. The effects of pulsed laser irradiation have been studied by channeling, transmission electron microscopy, and sheet resistivity measurements in order to reveal any possible influence of the fluorine concentration on the threshold laser energy density for annealing as well as the sheet resistivity of the surface layers after laser irradiation. Preliminary results from this study are reported.

2. EXPERIMENTAL

Single-crystal, 2 Ωcm, n-type <100> silicon samples were implanted at room temperature with mass-separated B^+, BF^+, BF_2^+, or BF_3^+. The implantation energies of the different ions were chosen so that the projected range of the boron dopants was identical for all implantations. Relevant implantation parameters are collected in table I, together with estimated peak concentrations. Implantations were carried out with the single crystals tilted by 7° relative to the beam axis in order to avoid implantation in a channeled direction. Laser irradiations were performed with light from a Q-switched Nd:glass laser having a pulse length of 40 nsec (λ = 1.06 µm). Each sample was exposed to only a single laser pulse during which a bent silica rod, 10 mm in diameter, was used as a beam homogenizer.

Table I

Implantation parameters. The projected ranges of boron and fluorine are for all implantations 400 Å and 410 Å, respectively. The peak concentrations have been estimated from the relation $N_V = N_d/2.5\, R_p$, where R_p is the range straggling[6].

Implanted Impurity	Implantation Energy (keV)	Implanted Dose, N_d cm^{-2}	Estimated peak concentration, N_V, of	
			Boron cm^{-2}	Fluorine cm^{-2}
B^+	11	5×10^{14}	1.3×10^{20}	–
B^+	11	5×10^{15}	1.3×10^{21}	–
BF^+	30	5×10^{14}	1.3×10^{20}	1.1×10^{20}
BF^+	30	5×10^{15}	1.3×10^{21}	1.1×10^{21}
BF_2^+	50	5×10^{14}	1.3×10^{20}	2.1×10^{20}
BF_2^+	50	5×10^{15}	1.3×10^{21}	2.1×10^{21}
BF_3^+	68	5×10^{14}	1.3×10^{20}	3.2×10^{20}
BF_3^+	68	5×10^{15}	1.3×10^{21}	3.2×10^{21}

Channeling measurements were performed with 2-MeV He^+ particles, with a typical beam current of 5 nA and a beam divergence of less than 0.03°. The backscattered He^+-particles were detected at an angle of 135° relative to the beam with a surface-barrier detector. Samples for transmission electron microscopy (TEM) were prepared by jet etching from the unimplanted side with a mixture of 1 part HF and 9 parts HNO_3. The electron microscope was operated at 100 kV. Sheet resistivity of the implanted crystals were measured by the four point probe method.

3. RESULTS AND DISCUSSION

Channeling measurements along the <100> axis on the as-implanted crystals yielded spectra showing complete amorphization for BF_2 and BF_3 implanted to doses of 5×10^{14} cm^{-2} and 5×10^{15} cm^{-2}, while for BF only the 5×10^{15} cm^{-2} showed complete amorphization. The widths of the surface damage peaks after correction for detector resolution were $\sim$500 Å and $\sim$750 Å for doses of 5×10^{14} cm^{-2} and 5×10^{15} cm^{-2}, respectively. In case of pure boron implantations, both ion doses resulted in only a marginal radiation damage. The normalized yield χ_{min}(Si) from the <100> channeling spectra at a depth just below the disordered layer has been used as a measure of the degree of recrystallization achieved by laser exposure. Figure 1a and 1b display χ_{min}(Si) as a function of laser energy density for all the implanted species as well as for the two ion doses.

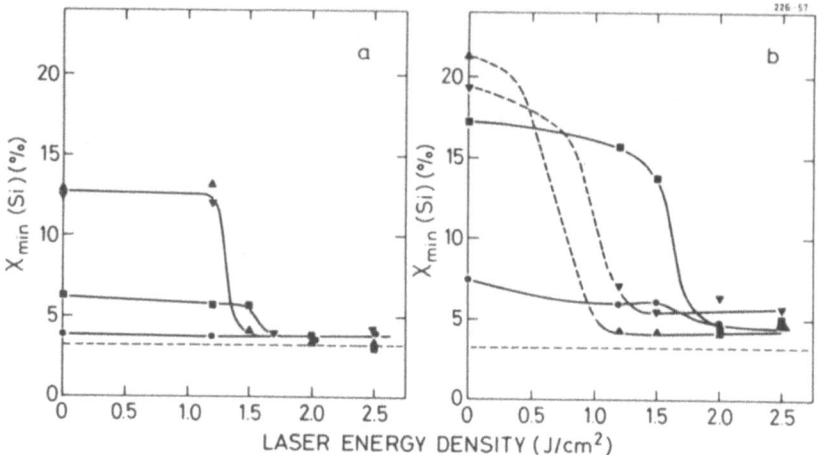

Fig.1 Results from 2-MeV He^+ channeling measurements along the <100> axis for implanted doses of 5×10^{14} cm^{-2} (a) and 5×10^{15} cm^{-2} (b). (·): B, (■): BF, (▲): BF_2, (▼): BF_3. The dashed line gives the χ_{min}-value for a virgin Si crystal.

The results for the 5×10^{14} cm^{-2} implantations (Fig.1a) show a threshold laser energy density for the BF_3, BF_2, and BF implantations at about 1.5 J/cm^2. Above this threshold the χ_{min}(Si) values are, within the uncertainty, equal to the value for a perfect crystal, measured to be χ_{min}^{virgin}(Si) = 3.2%. The χ_{min}(Si) value for the pure boron implantations is not affected by the laser exposure, but remains at a value of about 4% in the investigated laser energy density range. Figure 1b displays the results for an implanted dose of 5×10^{15} cm^{-2}. At this dose only the pure boron and BF implantations showed an annealing behaviour as in fig.1a.

Recrystallization was for the fluorine excess ions, BF_2 and BF_3, at the high dose observable (fig.1b) at low laser energy densities. Even at 1.2 J/cm^2 which is below the usually assumed threshold value for recrystallization, the present data indicate recrystallization.

Transmission electron microscopy of the 2.0 J/cm^2 laser annealed 5×10^{15} cm^{-2} implantations yielded results equivalent to those obtained in the channeling measurements. Figure 2 displays a typical result. The laser annealed samples were defect free for all four implanted species, and diffraction patterns indicated complete recrystallization.

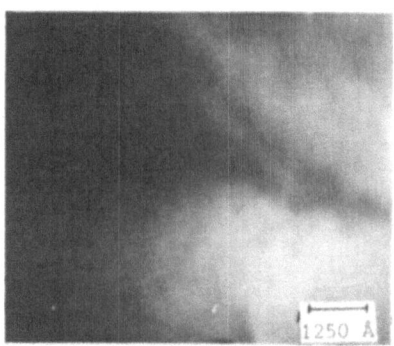

Fig.2 TEM-micrograph of a BF_3 implanted silicon single crystal. Implanted dose: 5×10^{15} cm^{-2}, laser energy density: 2.0 J/cm^2.

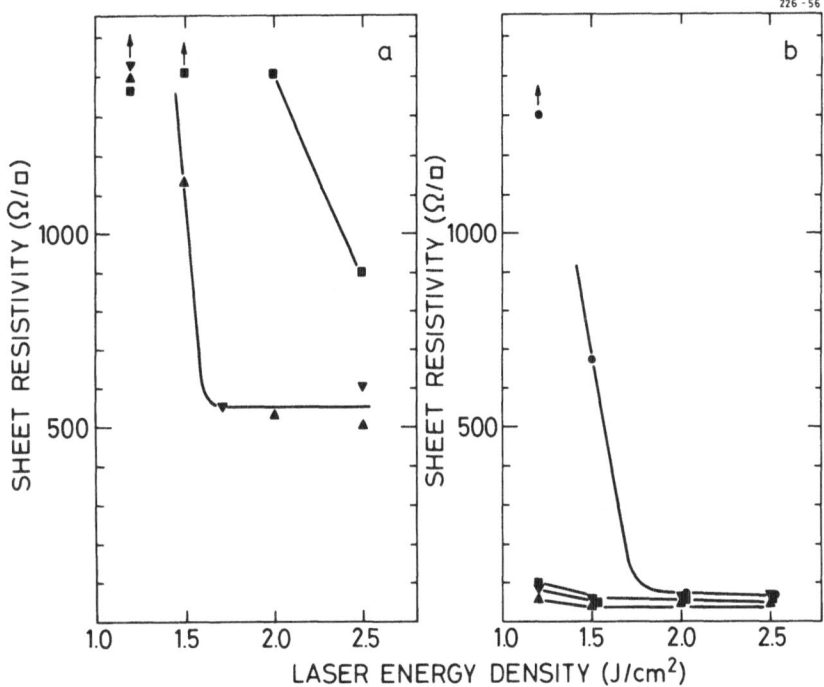

Fig.3 Sheet resistivity measurements for implanted doses of 5×10^{14} cm^{-2} (a) and 5×10^{15} cm^{-2} (b). (•): B, (■): BF, (▲): BF$_2$, (▼): BF$_3$.

The results from the sheet resistivity measurements on the same samples as used in the channeling measurements are displayed in Fig.3. The data from the 5×10^{14} cm^{-2} implantations (Fig.2a) show for BF$_2$ and BF$_3$ a similar trend as observed in the channeling measurements. The sheet resistivity reach a constant value of about 500 Ω/□ for a laser irradiation at about 1.5 J/cm^2. The values for the B and BF implantations were for all energy densities high and did not reach a constant value even after a 2.5 J/cm^2 irradiation.

The sheet resistivity measurements on the 5×10^{15} cm^{-2} implantations (Fig.2b) are almost identical for BF, BF$_2$, and BF$_3$ despite the fact that the channeling analysis gave different results in the three cases. Low sheet resistivities are obtained already at an laser energy density of 1.2 J/cm^2 and a constant value of about 50 Ω/□ is reached at 1.5 J/cm^2. The results from the pure B-implantation resemble those obtained for the low dose BF$_2$ and BF$_3$ implantations.

From the present results it is evident that the fluorine concentration in the surface layer has an influence on the threshold for the laser annealing. A possible explanation may have relation to a modification of the density of states in the energy gap of silicon due to the presence of fluorine atoms. The absorption of light from a Nd:glass laser ($\lambda = 1.06$ μm, $E_\nu = 1.1$ eV) in Si is very sensitive to the density of states in the band gap, as the photon energy is barely sufficient for interband absorption. Thus, a high fluorine concentration would give rise to a low threshold energy density[7], which is also observed. This could also explain that the low dose B-implantation is not affected by the laser exposure. A second conclusion which can be drawn from the data is that low sheet resistivity at low laser energy density is found in samples amorphized by the implantation or with high fluorine concentration in the doped layer.

The data are still not conclusive with respect to which of the two mechanisms is responsible for the observed effect. Previous experimental work[8] indicates that for BF_2 implanted in silicon a substantial fraction of fluorine atoms remain in the surface layer even at 2.0 J/cm^2 laser energy. It is reasonable to compare the observation with the effect of saturation of dangling bonds by fluorine in amorphous silicon[9], and accordingly a low sheet resistivity is feasible even though channeling measurements show disorder.

The preliminary data presented demonstrate that for a production of solar cells with ion implantations without mass separation, it is of crucial importance to get a more detailed understanding of the role of fluorine atoms in the near-surface region, particularly if laser annealing is used to reduce the damage caused by the ionic bombardment.

Thanks are due to J.Chevallier for fruitful discussions and help with the experiment. This work was supported by the Commission of the European Communities under contract ESC-R-020-DK(G)

REFERENCES

1) G.Soncini, in: Proc.3rd Int.E.C.Photovoltaic Solar Energy Conference, Cannes, France, 1980, W.Palz (ed.) (Reidel, Dordrecht, 1981) p. 270.

2) See, for example: Proc.3rd Int.E.C.Photovoltaic Solar Energy Conference, Cannes, France, 1980, W.Palz (ed.) (Reidel, Dordrecht, 1981)

3) I.Michel, C.Fages, J.C.Muller, P.Siffert, D.Hoonhout, T.de Jong, and F.W.Saris, in Ref.2, p.713

4) J.C.Muller, P.Siffert, J.Michel, and E.Fabre, in Laser and Electron Beam Processing of Materials, White and Percy (eds.) (Academic Press, 1980), p.278

5) L.D.Nielsen and P.Balslev, in ref.2, p.698

6) J.W.Mayer, L.Eriksson, and J.A.Davies, Ion Implantation in Semiconductors (Academic Press, New York and London, 1970) p.36

7) D.Hoonhout and F.Saris, in Laser and Electron-Beam Solid Interactions and Materials Processing, J.F.Gibbons, L.D.Hess, and T.W.Sigmon (eds.) (North-Holland, New York and Oxford, 1981) p.31

8) A.Nylandsted Larsen and R.Jarjis, to be published

9) W.Y.Ching, J.Non-Crystalline Solids 35 and 36, 61 (1980)

GRAIN BOUNDARY PHOTOCURRENT ENHANCEMENT IN SOLAR CELLS MADE BY LASER DIFFUSION

G. B. TURNER, D. TARRANT and D. ALDRICH
ARCO Solar, Inc.
20554 Plummer Street
Chatsworth, California 91311

R. Pressley and R. Press
XMR, Inc.
3350 Scott Boulevard #57
Santa Clara, California 95051

Summary

An enhanced photocurrent has been observed at grain boundaries in solar cells made by Gas Immersion Laser Diffusion (GILD). Wafers of Wacker "Silso" polycrystalline material were sliced from the cast block, kept in order, and subjected to different treatments. Many grain boundaries extended through three wafers. One wafer was furnace-diffused with phosphorus (850°C, 1 hr) and the next was made by the GILD method. A third wafer was heat treated, without dopant, the same as the furnace-diffused cell. Laser photocurrent scans show that many grain boundaries degrade photocurrent in cells that have been heat treated but enhance it slightly in GILDed cells. Photocurrent scans on the backs of the cells show the effect strongly amplified. Two models are presented which are consistent with these results.

1. INTRODUCTION

The Gas Immersion Laser Diffusion (GILD) process provides a means of making solar cells without substantially heating the bulk of the wafer. The GILD process is described elsewhere in this Proceedings (1). When cells are made in polycrystalline material by this technique, the grain boundary recombination normally seen in diffused cells is replaced, at many boundaries, by enhanced photocurrent.

2. MATERIALS AND HEAT TREATMENTS

Wafers for these cells were cut from a block of "Silso" p-type polycrystalline Si cast by the Wacker Chemitronic Corp. Many of the grains in this material are several mm long, so many grain boundaries extend through a number of the 330 µm thick slices. Three different processing treatments were applied to three adjacent slices in order to examine the different effects on the same grain boundaries. All wafers were chemically-mechanically polished before processing.

The bottom wafer was diffused at 850°C for 1 hour using $POCl_3$ as the dopant. The middle wafer was GILDed to produce a similar junction without heating the bulk of the wafer significantly. The top wafer was heat treated at 850°C for 1 hour the same as the diffused one, but in Ar with no $POCl_3$ present. The three sister wafers -- diffused, GILD, and HEAT+GILD --

were then cut into 2 x 2 cm cells. Contacts were evaporated Cr-Pd-Ag, annealed for 10 s at < 500 °C. As described in a preliminary report on this work (2) the uncoated GILD cell produced J_{sc} = 19-20 mA/cm^2, about 10% better than their HEAT+GILD sisters. Base doping was 6.5 x 10^{15} cm^{-3}.

The cells were scanned with a low power 633 nm laser to produce photocurrent pictures in which regions of high photocurrent response appear light and those of low response appear dark. In all laser scans the intensity was reduced until all features showed a linear response. Figure 1a shows these pictures for the same area of three sister cells. For the diffused cell at the bottom, and for the HEAT+GILD cell at the top, grain boundaries show up as dark lines. For the GILD cell in the middle, however, the response is almost uniform, with a slight enhancement perceptible at the boundaries. Figure 2 shows the photocurrent variation along the white horizontal lines in Fig. 1a.

Photocurrent scans at 1153 nm show 30-50% enhancement at many grain boundaries in the GILD cell (2), but for even more sensitivity, windows were etched through the cell backs and scans were made from the back with 633 nm light. Since the carriers are generated close to the back surface, the photocurrent is very sensitive to recombination.

These back scans for the same areas are shown in Fig. 1b. The diffused and HEAT+GILD cells at bottom and top respectively show back response similar to the front scans, but with the grain boundaries much darker and broader. The GILD cell, however, shows greatly enhanced collection near grain boundaries. The gain of the middle (GILD) picture in Fig. 1b has been reduced to avoid saturating the light regions. Figure 3 shows photocurrent variation along the horizontal lines of Fig. 1b for the HEAT+GILD and GILD cells. Grain boundary response in the GILD cell is up to 10 times larger than mid-grain response in the heated cells. For peak B' the response falls off to 1/e of its peak value in about 75 μm. Grain boundary response in the two heated cells is drastically degraded.

3. DISCUSSION

The minority carriers in back scans are electrons generated within about 2.5 μm of the back surface and the increased photocurrent at grain boundaries represents a greatly increased ability to reach the emitter without recombination. Two possible models are consistent with this effect: a) increased minority carrier diffusion length, L, and b) grain boundary collection.

Since diffusion of carriers a distance W depends on exp(-W/L), moderate changes in L could have a major effect on photocurrent. The regions near grain boundaries could be denuded of recombination centers by gettering to the boundary during casting, thus increasing L. If so, the boundary itself must be passive (i.e., have low "surface" recombination velocity). The boundary recombination and perhaps diffusion length must then be degraded by heat treatment.

In the grain boundary collection model, depletion regions on each side of the boundary, and extending nearly through the wafer, collect electrons and transport them by drift to the junction. These depletion regions (i.e., double Schottky barriers) might be caused by native or impurity states on the grain boundary surface. Heat treatment must then destroy the barriers. In an attempt to confirm these barriers, four-point probe measurements were made within grains and aross boundaries on the backs of GILD and HEAT+GILD wafers as shown in Fig. 4. The results indicate high boundary resistance, consistent with Schottky barriers in the GILD cells which showed the enhanced photocurrent. However Redfield (3,4) previously

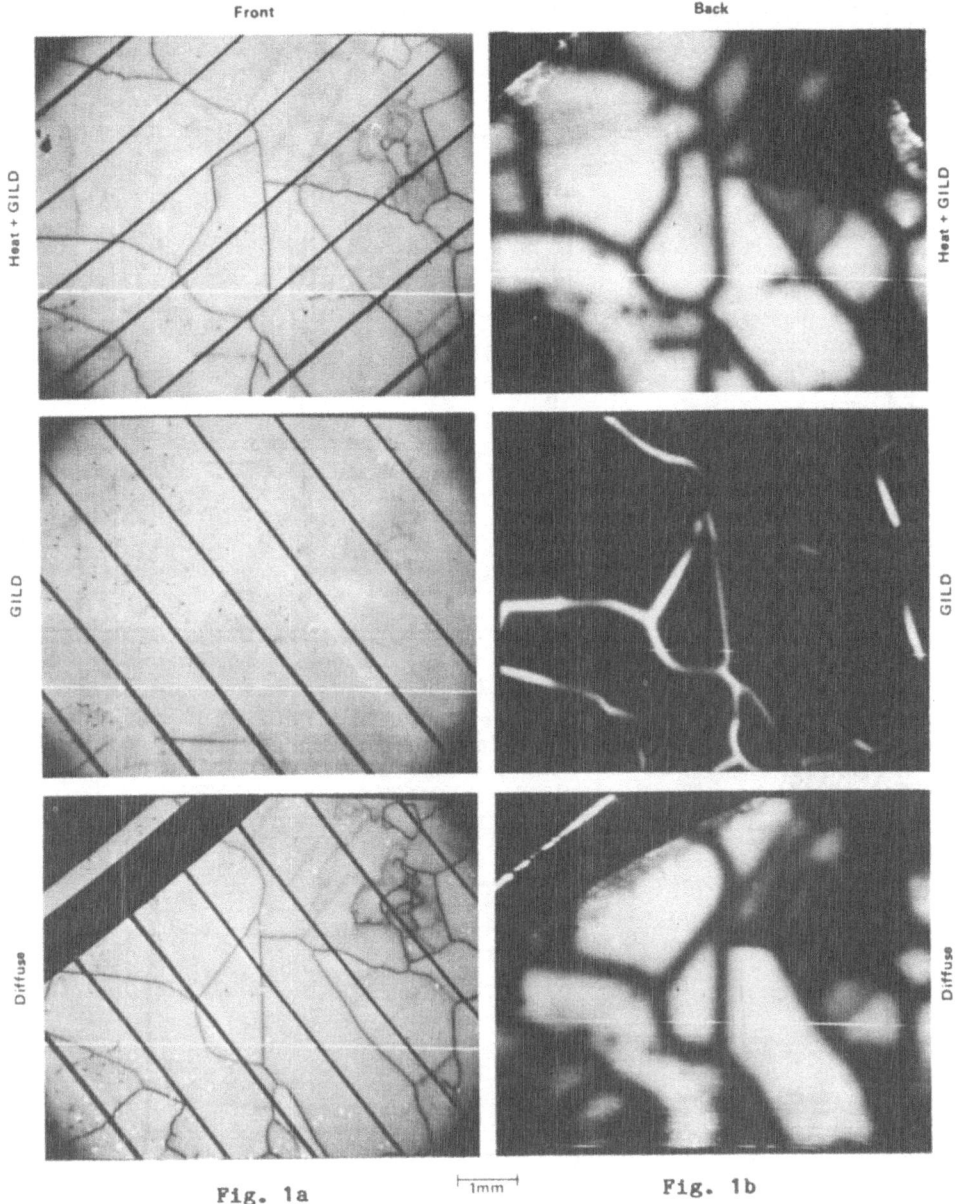

Fig. 1a. Laser photocurrent scans of three cells scanned from the front with 633 nm light. The shape of the grains varies slightly among the sister wafers because the boundaries are not exactly perpendicular to the surface. Dark straight lines are the front grid.

Fig. 1b. Photocurrent scans from the back. The gain is lower for the middle (GILD) picture than for the top and bottom (see text).

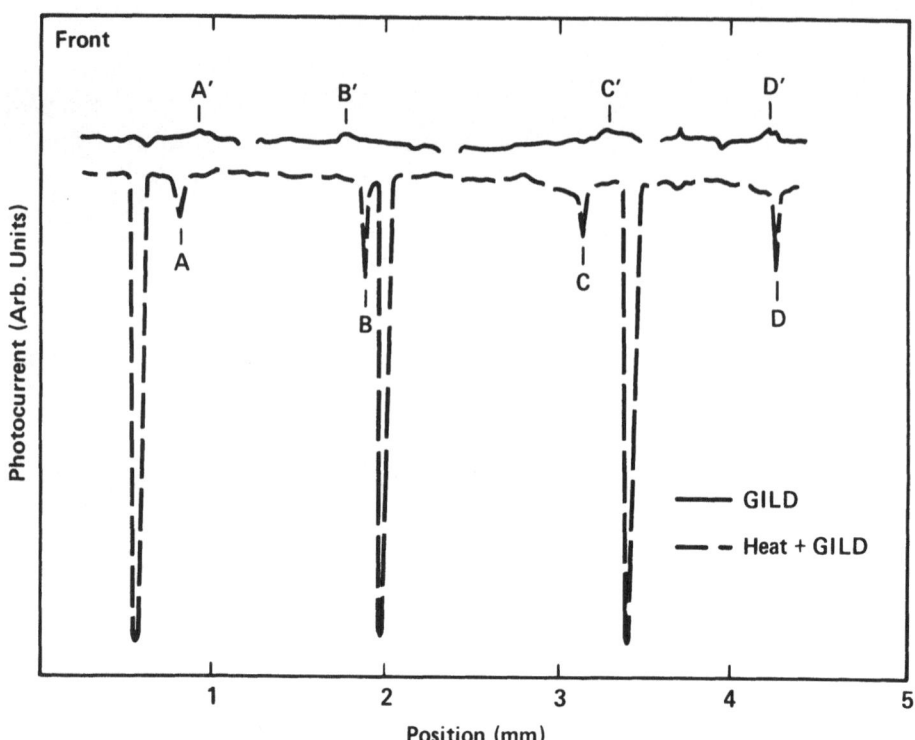

Fig. 2. Front photocurrent for two cells from Fig. 1a, to same vertical scale. The 3 deep grid notches are suppressed on GILD cell for clarity.

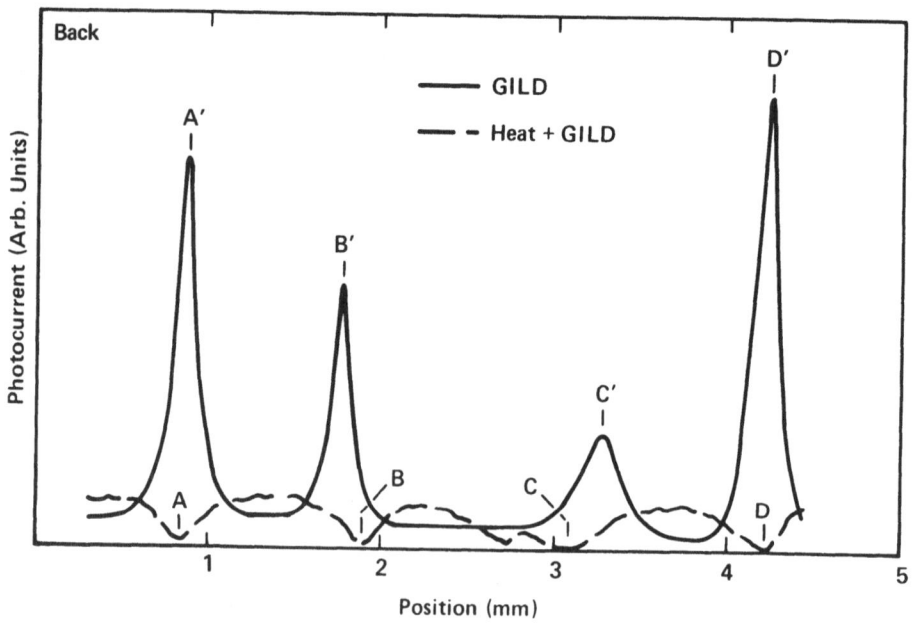

Fig. 3. Back photocurrent from two cells (see Fig. 1b), to the same scale (scale amplified 20x from Fig. 2).

reported opposite 4-point probe results (i.e., heat treatment caused barriers) in Silso and confirmed them by other techniques.

Perhaps not all Silso is alike in this respect. In any case, the enhancement seen here raises the possibility that appropriately prepared fine grain polycrystalline Si could result in superior solar cells when low temperature processes like GILD are used.

We wish to thank Richard Schwartz of Purdue for valuable discussions. Significant experimental contributions were made by Gary Brown, Twyla Byers, Dorothy Houk, Ruth LeMay and Ron Merkord.

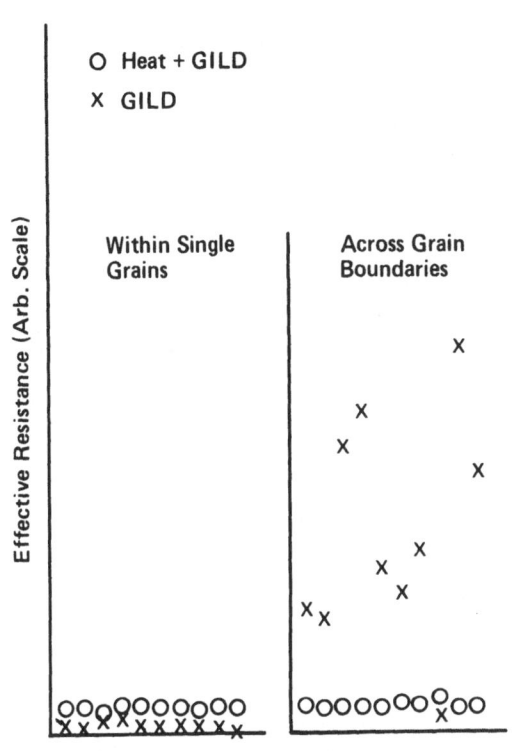

Fig. 4. Four-point resistance measurements on backs of wafers.

4. REFERENCES

(1) G. B. Turner, D. Tarrant, D. Aldrich, R. Pressley, and R. Press, "Gas Immersion Laser Diffusion: A New Method for Making Efficient Si Solar Cells," this conference.

(2) G. B. Turner, D. Tarrant, D. Aldrich, R. Pressley, and R. Press. To be published in Proceedings of the Symposium on Grain Boundaries in Semiconductors, Materials Research Society Meeting, Nov. 16-19, 1981, Elsevier, North Holland.

(3) David Redfield, Record of the Fifteenth IEEE Photovoltaic Specialists Conference (IEEE, New York, 1981), p. 1179.

(4) David Redfield, Appl. Phys. Lett. 40, 163 (1982).

DRY PROCESS FOR ECONOMIC CELL MANUFACTURING

J. DONON, H. LAUVRAY, P. AUBRIL, G. DAVID, P. LOUBLY
PHOTOWATT International S.A.
6, rue de la Girafe
14001 - CAEN

Summary

Different dry processes for cell manufacturing are under development leading to important cost reduction :

- PLASMA ETCHING in order to remove the thin damaged layer induced during sawing with a wire saw, to etch away the oxyde layer after diffusion and to open the junction on the edges,

- GENERAL USE OF SCREEN PRINTING in order to obtain N+/P and N+/P/P+ structures with screen printed doping sources, to print front and back contacts (silver can be used for the front, aluminium and silver for the back with the possibility of diffusing through the oxyde and the junction when needed), to deposit the A.R. coating, either at the end of the process or during the junction formation.

The main advantages of these dry processes using plasma etching and screen printing are :

- low material consumption,
- high throughput versus capital investment ratio,
- little pollution,
- automation capability.

A preliminary economic analysis is made to show the interest of using such a process in production capacity.

1. INTRODUCTION

The objectives of this study were to develop a completely dry process for cell manufacturing with the use of the plasma dry etching technologies and of the screen printing processes for both dopant and contacts. The main advantages of this dry process are the low material consumption, the high throughput versus capital investment ratio , the weak pollution, and the automation capability. The working programme was first the study of various plasma etching steps : surface sawing damage removal, edge leakage current reduction ("ring etching") and the selective oxyde etching, second the study of diffusion from screen printed doping layers.
The above technologies were used on mono and polycrystalline Si wafers.

2. PLASMA ETCHING TECHNOLOGIES

The first step to be investigated was surface plasma etching.
The basic problem is to etch the sawing damage which is at the maximum $10 \mu m$ with a wire saw and $30 \mu m$ with an ID saw. In the classical industrial process this etching is chemical with either acid or basic etching which are expensive and give pollution.
With the use of a plasma the cost and pollution reduction are very significant.

The apparatus used were barrel reactors; an optimization of the etching rate was done by varying all the etching parameters (table I).

TABLE I

Gases	TIME	200 cc/min.			250 cc/min.		
CF$_4$ + 10% O$_2$	15 min.	0,3 μm	0,56 μm	0,35 μm	0,33 μm	0,4 μm	0,43 μm
	25 min.	0,32 μm	0,53 μm	0,35 μm	0,33 μm	0,4 μm	0,43 μm
	35 min.	0,34 μm	0,4 μm	0,35 μm	0,33 μm	0,4 μm	0,43 μm
CF$_4$ + 4% O$_2$	15 min.		0,36 m			0,28 m	
	25 min.		0,32 m			0,28 m	
	35 min.		0,34 m				0,34 m
	POWER	500 W	650 W	750 W	500 W	650 W	750 W

The etch rate is between 0,3 μm/mn and 0,6 μm/mn which is low in comparison with acid etching 13 μm/mn or basic etching 5 μm/mn. The most influential parameters are the time and the percentage of oxygen.

The second step studied was the ring etching. In most of the process the ring etching is done by chemical ways or with masking technologies permitting diffusion only on the front surface. In the case of the plasma, all the wafers are positioned in stack and etched on the edge by the gas. The apparatus used was a barrel reactor and the best results were obtained with the following parameters of the plasma and for two stacks of 200 wafers at the same time
- power .... 850 W
- gas .... CF$_4$ + 8% O$_2$
- gas rate .... 200 cc/min.
- duration .... 15 min.

For example, figures 1 to 4 give the cell electrical parameters versus duration of the etching for small stacks of 20 wafers and the parameters were the following :
- gas .... CF$_4$ + 4% O$_2$
- power .... 150 W
- gas rate .... 200 cc/min.

Reproducibility of the ring etching was demonstrated.

Cells were manufactured with the following process to verify the possibility of use of the plasma etching technologies ;

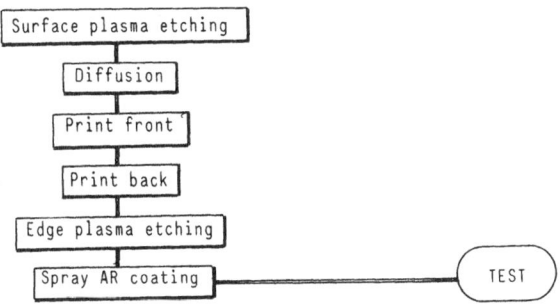

Table II shows the average values obtained :

	Isc (mA)	Voc (mV)	Imax. (mA)	FF	RS (mΩ)	η (%)	I_R-400mV (mA)	I_R+400mV (mA)
mono.	2100	575	1950 at 455 mV	0,735	20	11,4	14	24
Silso	2250	535	2000 at 430 mV	0,700	25	8,6	10	30

TABLE II

The selective oxyde etching was tested but only bad results were obtained due to non uniformity of the etch rate on the surface of the wafer.

3. DIFFUSION FROM SCREEN PRINTED DOPING LAYERS

The starting doping materials were for N+ source the N250 from "EMULSITONE" and for P+ source borosilicate from "EMULSITONE". The added binder elements for optimum deposition conditions were :
- terpinol .... 85 % in weight
- ethyl cellulose .... 15 % in weight

and the final viscosity of the mix adjusted at 600 Cps by mixing 50 % of vehicle and 50 % of doping material.

Good diffusion homogeneity were obtained with this doping paste after a thermal treatment under oxygen at 850°C for 1 hour. The junction depth were measured at 0,5 - 0,6 μm and the sheet resistance at 25 - 30 ohm/□ .

Cells were manufactured with the following process :

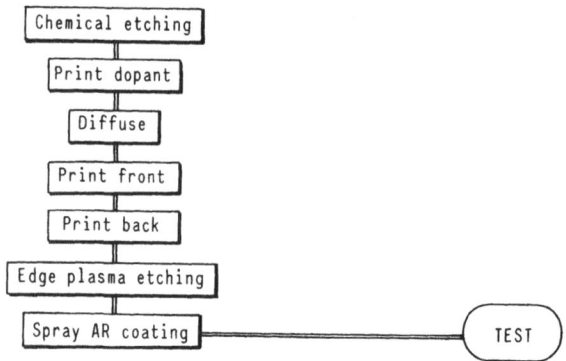

Results obtained with this process on the mono and polycrystalline wafers are reported in table III.

		Chemical acid etching	Chemical basic etching
mono.	η (%)	11,8	12
	FF (%)	72,7	73

TABLE III (part one)

		N250 - Plasma etching	POCl$_3$ - Plasma etching
Silso before AR coating	Isc (mA)	1686	1633
	Voc (mV)	546	546
	FF (%)	72,3	74,5
	η (%)	6,6	6,7
	Ld	44 - 50 µm	34 - 38 µm

<u>TABLE III</u>
(part two)

The results obtained show no significant difference between screen printed doping layers and POCl$_3$ diffusion.

4. PRELIMINARY ECONOMICAL ANALYSIS

For the ring etching which can be used in production with commercially available apparatus the reduction in cost for the operation in comparison to the chemical etching or to the masking operation is the following :

chemical etching (base per watt)	masking operation for ring etching	plasma ring etching
100 ⟶	30 ⟶	5

For the surface etching it is too early to give operation cost. All the commercially available apparatus have too low throughput to be used. Further investigations with new machine are under developpement.

For the printed dopant the best advantage is the possibility of full automation in the line for a total cost which is very similar to the cost of a classical diffusion furnace.

5. CONCLUSION

General use of plasma etching for front surface to remove damaged layer from the sawing operation and to open the junction and general use of screen printed process for dopant deposition were demonstrated during this contract.

The advantage of this process are the low material consumption particularly for the plasma operations, the automation capability particularly for dopant deposition and contact operation, the little pollution by utilizing only dry processes : plasma and printing and the high throughput versus capital investment ratio particularly for the plasma.

This work was partly supported by CEC.

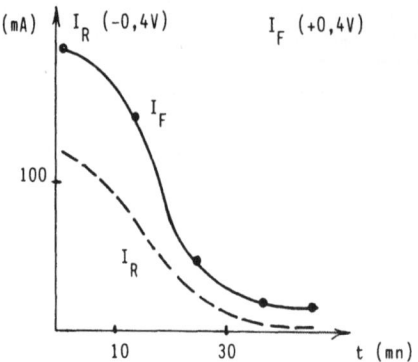

Figure 1 : leakage current

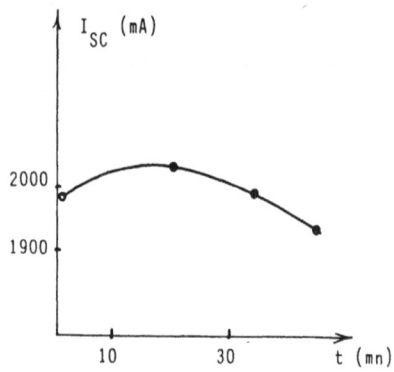

Figure 2 : short circuit current

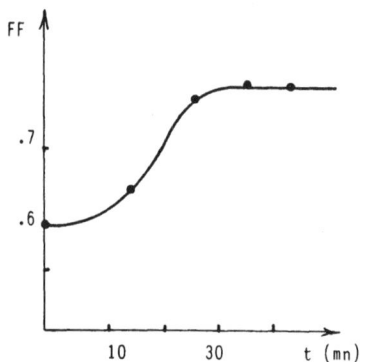

Figure 3 : fill factor

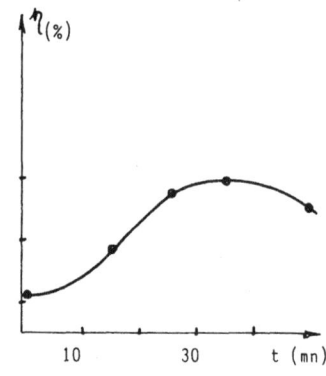

Figure 4 : efficiency

LATE NEWS PAPERS

Low cost structures and optimization of support structures

Screen printed SIS-type solar cells

An improved derivation of solar cell parameters in terms of transition probabilities

Impurity diffusion in amorphous silicon and its implications for solar cells

The world's largest 12 Volt single string photovoltaic module

Light assisted pulsed annealing of photovoltaic silicon by microwave energy

Microprogrammed coupling system for photovoltaic generators with multiple receptors

LOW-COST STRUCTURES AND OPTIMIZATION OF SUPPORT STRUCTURES

J. Glöckl, P. Helm and K. Träder

WIP
Sylvensteinstrasse 2
8000 München 70, West Germany
Tel. (089) 776041   Telex 5-212186

SUMMARY

This paper presents a newly developed low-cost support structure for PV panels based on elements commonly used for fence construction. Thereafter trade-off methods to arrive at an optimization in terms of total costs for support structures have been described.

1. LOW-COST STRUCTURES

A substantial cost reduction for support structures is possible when standard elements usually applied in similar constructions are used. This will be shown in the example of a construction which we have developed.

The suggested support structure (sketched on the following page) uses connection parts which are customary for fences, i.e. connection clips and screws.

The advantage of such a construction lies in the fact that this technology is widespread. The supporting elements (hot galvanized iron pipes) as well as the connection clips are ready-made. They are mass-produced and thus reasonably priced.

Only one part must be made-to-order, i.e. the clips connecting the modules to the horizontal pipes.

The net production costs for this support construction, calculated for a reference PV plant of approx. 2,100 m² panel area, amount to roughly 32 ECU/m². These costs comprise the following items:

	approx. percentage of total construction costs
- foundation	16 %
- pipes for columns and supports	21 %
- connection parts	37 %
- manpower	17 %
- site installation, furnishing and operation of machinery	9 %
	100 %

The costs refer only to the construction of the support structure. Overheads and contingencies must be added thereto.

dimensions in cm

connection clip

ø 6.3
ø 2.0

2. OPTIMIZATION OF SUPPORT STRUCTURES

Since even for small PV power plants many similar support structures have to be built the latter represent a good example for saving construction costs through optimization.

An optimization concerning the total costs of a support structure can only be done for a given structure. Therefore the optimization must comprise two steps.

The first step will address the choice of the adequate construction material and construction design. The second step will concern the optimization of the chosen support structure itself, since altering the dimensions of a structural part (for example the length of a beam) causes in most cases an impact on the balance of the system. Owing to the considerable amount of variables which act differently on the costs it is not possible to say a priori where the minimum cost will be.

Therefore an optimization to that effect is required. An example of the outcome of such an optimization is given in the following graph.

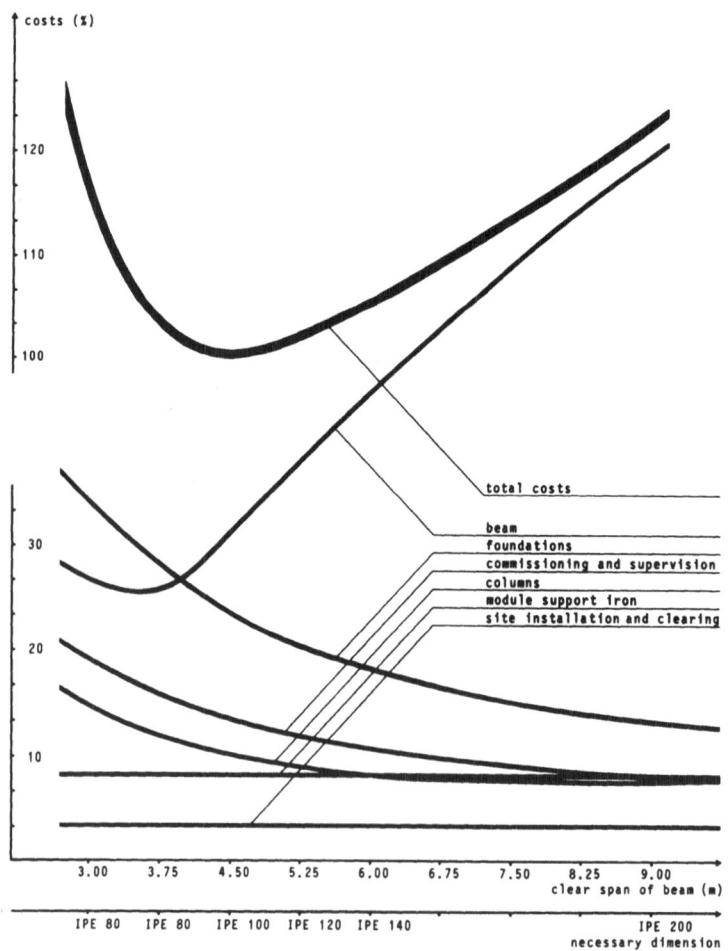

SCREEN PRINTED SIS-TYPE SOLAR CELLS

J.N.Avaritsiotis[+], D.S.Campbell[*], and C. Caroubalos[+]
+University of Athens,Department of Physics,Division
of Electronics,104 Solonos Str.,Athens 144,GREECE.
*University of Technology,Department of Electrical
and Electronic Engineering,Loughborough,Leics LE11
3TU ,U.K.

Summary

The experimental results of a feasibility study that has been undertaken in order to investigate the ponteciality of the thick film technology in the fabrication of ITO-single crystal Si solar cells,are discussed. Newly developed thick film pastes,in use in the optoelectronics industry have been tried.In this structure the ITO thick film acts not only as a conducting surface layer that induces the SIS junction but also as an antireflective coating.Promising cells were fabricated employing conventional thick film apparatus that is widely used in the microelectronics industry, showing a V_{oc} = 310mV and a J_{sc} = 1mA/cm^2,under a total insolation of 613W/m^2.Further work is necessary for the optimisation of the cell paramet.

INTRODUCTION

SIS-type solar cells such as ITO on Si are being extensively investigated[1] because of the relative ease with which high efficiencies are achieved employing simpler fabricational technology,in relation to the one used for conventional p-n Si solar cells.One of the most crucial processing step in SIS junction fabrication is the interfacial Si oxide growth.It has been established[2] both theoretically and experimentally that an oxide thickness of around 16Å favors high solar cell efficiencies (above 12%). If the oxide thickness becomes more than 22Å,the efficiency falls dramatically.

The ITO film,normally a wide-band gap highly conducting oxide,is deposited on the base semiconductor by various rather expensive techniques such as thermal or e-beam evaporation,ion beam or magnetron sputtering,CVD, etc,followed generally by thermal annealing.

On the other hand,thick film techniques are commercially available for the fabrication of the back and front metallizations of solar cells[3,4], and their great advantages for a high volume automated metallization and diffusion process for Si solar cells,are known[5].

Consequently,it was decided to investigate the possibility of inducing an SIS junction on n-type single crystal Si wafers,employing all thick film techniques,using an indium - tin oxide thick film paste (ESL 3050) that has been marketed for the printing of optically transmitting and conducting paths on glass substrates (LCD).

SOLAR CELL PREPARATION

N-type,phos.doped single crystal silicon wafers,(111),optically polished on one side and showing a resistivity of 1-2 Ωcm ,280 μm thick,and having a diameter of 2 in were cut into four segments with a laser to be used as the base semiconductor for the SIS solar cells.The Si wafers were kindly provided by the Lucas Research Centre,U.K.

The Si segments were etched for 20 min in 20% HF to remove the native Si dioxide and possible surface damages, and then were rinsed in deionized water and dried rapidly. After that the Si substrates were loaded on a screen printer and a gold film (paste ESL 8835-1B) was printed, dried and fired according to the manufacturer's recommendations, to provide the back contact With the aid of an ellipsometer and by varying the firing time, we found that a Si oxide with thickness of approximately 20Å was growing on the optically smooth side of the wafer segments, if the substrates were kept in the active region of the furnace for 10 min at $650^\circ C$. This is in accordance with the typical results on the oxidation of Si(6).

The ITO screen printing process was optimized so that films of an average thickness of 2000Å were printed on the oxidized and optically smooth surfaces. The post-fired films were exhibiting good mechanical, electrical ($R_s = 170\Omega/\square$), and optical properties and also a very good adhesion to the Si substrate.

Finally, a silver front thick-film-grid (paste ESL 9990) of a simple and arbitrary geometry was printed, dried, and fired according to the manufacturers spec. A large number of samples was fabricated at different firing temperatures in a first attempt to investigate the possibility of optimizing the structure of the cells and thus to increase the V_{oc} and J_{sc} Also, various other pastes were unsuccessfully tried in order to replace gold with a less expensive paste. Fig.1 shows the structure of the cells and Fig.2 is an SEM picture of the Ag/ITO/Si interface, taken at point A.

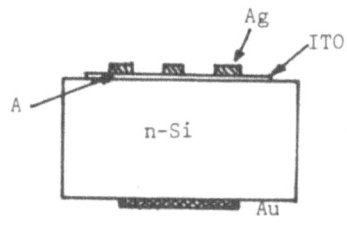

Fig.1: Schematic diagram of the structure of the cell.

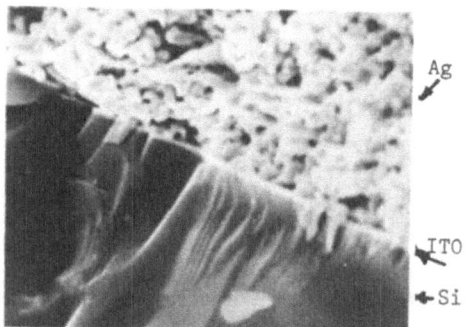

Fig.2: SEM picture of the interface Ag/ITO/Si (x1900).

EXPERIMENTAL RESULTS AND CONCLUSIONS

A preliminary investigation showed (see Fig.3) that the most promising cells should be prepared under the preparation conditions that apply for sample F. The process was repeatable and the V_{oc} and J_{sc} of the samples prepared as the sample F were the same within 10%. The junction behavior shown in Fig.4 is typical of the best cells. The dark forward-bias ITO/Si curve fits the standard equation:

$$J = J(T)\exp(qV/nkT)$$

At higher currents ($\geq 1.1 mA/cm^2$) the curve deviates from the exponential behavior, consistent with a resistance in the ITO layer(7). The open circuit photovoltage of the best samples was 293mV for Φ between 55 and $90 mW/cm^2$ and it was falling to 280mV for $\Phi = 108 mW/cm^2$; Φ was measured with an Eppley black and white pyranometer. Respectively J_{sc} was increasing linearly, with Φ varying between 55 and $80 mW/cm^2$, and it was leveling off for higher values of Φ. A calculated typical value for FF was 0.49 and the efficiency of the best cells was less than 1. Measurements with a grating monochromator and a commercial p-n Si solar cell showed for the ratio J_{ITO}/J_{p-n} a peak at 650 nm. It is worth noting that despite the fact that the cells were not passivated, V_{oc} was reduced by only 10% in a period of six months; the cells were kept in a domestic enviroment. It is believed that further manipulation of the ITO paste and the use of proper conditions for the growth of the Si oxide may give better cells.

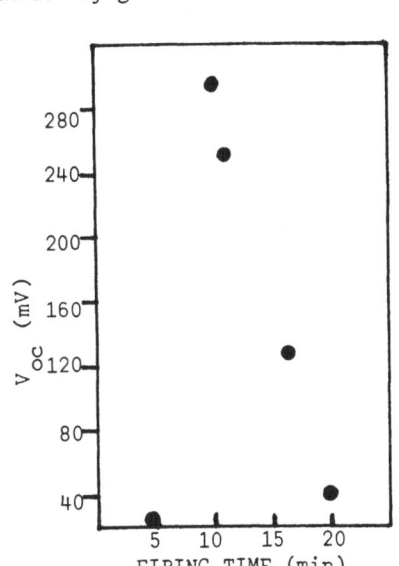

Fig.3: Open circuit photovoltage for various samples fired for different times.

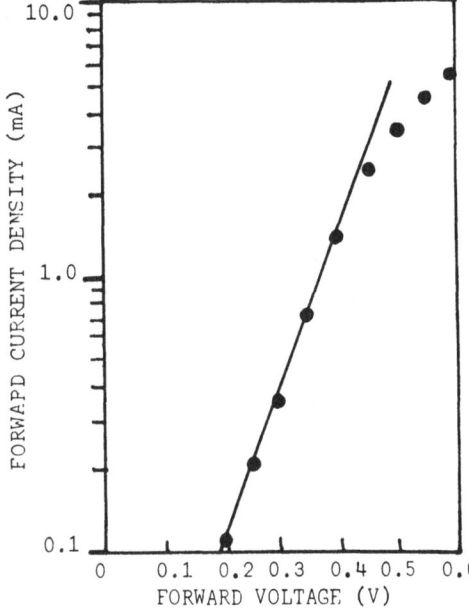

Fig.4: Dark forward characteristic at 300K of sample F.

REFERENCES

1. J.Shewchun,D.Burk,and M.Spitzer,IEEE-ED 27,No4,(1980)pp705-716.
2. J.Shewchun,J.Dubow,C.W.Wilmsen,R.Singh,D.Burk,and J.Wager,J.Appl.Phys. 50,No4,(1979)pp2832-39.
3. B.Ross and D.Bickler,Proceedings of the 3d EC Photovoltaic Solar Energy Conference 1980,D.Reidel Publ.Co.,pp674-8.
4. J.Michel,H.Baudry,D.Diguet,and G.David,Proc.3d EC P/V Solar Energy Conf., pp679-83.
5. L.Frisson,M.Honore,R.Mertens,R.Govaerts,and R.Overstraeten,14th Photovoltaic Specialists Conf.,1980,IEEE,pp941-42.
6. R.A.Colclaser,"Microelectronics:processing and device design",John Wiley and Sons,1980,pp92-93.
7. N.S.Charg and J.R.Sites,J.Appl.Phys.,49,No9,(1978)pp4833-37.

AN IMPROVED DERIVATION OF SOLAR CELL PARAMETERS

IN TERMS OF TRANSITION PROBABILITIES

P.T. LANDSBERG
The University, Southampton, England.

Abstract

Using rate equation and quasi-chemical potentials, the relation $j = j_L - j_{rev}[\exp(qV/kT) - 1]$ is obtained. It yields identifications for j_L and j_{rev} in terms of radiative and non-radiative transition probabilities per unit time.

1. Introduction

At a recent International Conference on Luminescence a simple radiation-driven energy conversion model was presented which yielded the solar cell equation in a special case [1]. The main point of this deduction was to show (a) that the solar equation can be based on a variety of very different models, and (b) that one can obtain an interpretation of j_L and j_{rev} in

$$j = j_L - j_{rev}(\exp \frac{eV}{kT} - 1)$$

which depends only on the transition probabilities per unit time which are involved. This leaves to a further generalisation the question of transporting the current carriers. The deduction was, however, too restrictive in that it had utilised the identification of the "upward" reaction constants in terms of the "downward" ones by detailed balance. This is however only valid if these constants are excitation – independent [2] and this must not be assumed for all constants. The removal of this blemish is, however, possible, and in no way detracts from the elegance of this particular argument, as will be shown here.

2. The Model

Our top group of states $i = 1, 2, \ldots n$ are specified by a quasi-chemical potential μ and lie at energies $e_1, \ldots e_n$. The ground state at energy e_o has a quasi-chemical potential μ_o. Let N_i be the number of photons with energy $e_i - e_o$ and let the downward and upward radiative transition rates be respectively ($\mu_1 = \mu_2 = \cdots = \mu_n \equiv \mu$)

$$B_i(1+N_i)p_i \;,\quad B_i N_i p_o \;,\quad p_i \equiv \exp \frac{\mu_i - e_i}{kT} \quad (i=1,2,\ldots n;0) \;. \quad (1)$$

Let the non-radiative transition rates be L_i, L_i', the latter being upward. Transitions are needed to drive a current through an external load or alternatively to charge an accumulator. Let S_{si}, S_{si}' be the appropriate downward and upward storage reaction constants. We shall write

$$p_i/p_o = \exp(y - x_i), \quad y \equiv \frac{\mu - \mu_o}{kT}, \quad x_i \equiv \frac{e_i - e_o}{kT}. \tag{2}$$

This completes the model and the notation.

We have reaction constants B_i, L_i, L_i', S_{si}, S_{si}' and in thermal equilibrium by detailed balance $B_i(1+N_i)p_i = B_i N_i p_o$ which yields the Bose-Einstein distribution. In addition,

$$\left[L_i' = L_i e^{-x_i}\right]_{eq}, \quad \left[S_{si}' = S_{si} e^{-x_i}\right]_{eq} \tag{3a,b}$$

but we shall not extrapolate these relations into the non-equilibrium situation, thus allowing L_i and S_i to be excitation dependent, in extension of [1].

3. The general two-level argument

The steady state requires

$$\sum_{i=1}^{n}\left[B_i(1+N_i) + L_i + S_{si}\right]\exp(y-x_i) = \sum_{i=1}^{n}\left[B_i N_i + L_i' + S_{si}'\right]. \tag{4}$$

This is an equation for the steady-state value of y, defined in (2). If A is the area of the exposed surface of the converter, and q the charge on the charge carrier, then define the following electric current densities

$$\frac{p_o q}{A}\{\Sigma B_i N_i, \; \Sigma(L_i + S_{si})e^{-x_i}, \; \Sigma(L_i' + S_{si}'), \; \Sigma B_i(1+N_i)e^{-x_i}\} \equiv \{j_a, j_s, j_s', j_r\}\} \tag{5}$$

By (4,5) the steady state condition implies simply

$$e^y j_s - j_s' = j_a - e^y j_r. \tag{6}$$

Turning now to the storage reaction, the rate of energy deposition in the store (per unit area), divided by kT is, if $\varepsilon_{ps} \equiv kT\, y_{ps}$ is the energy stored per particle making the transition,

$$\dot{W}/kT = (y_{ps}/q)\left[e^y(j_s - j_{so}) - (j_s' - j_{so}')\right] \tag{7}$$

where j_{so} is j_s without the storage reaction, and j_{so}' is defined similarly. Using (6) to eliminate j_s, j_s',

$$\dot{W}/kT = (y_{ps}/q)\left[j_a + j_{so}' - j_r - j_{so} - (j_r + j_{so})(e^y - 1)\right]. \tag{8}$$

The theory is simple in that it relies on two-equations only. The storage rate (7), and the steady-state condition (4) or (6).

4. Application to the solar cell concept

The load current density j and the voltage V across the device are introduced by

$$\dot{W} \to jV, \quad ykT \to qV. \tag{9}$$

In addition, put $\varepsilon_{ps} = qV_s$ to find that (8) is

$$\frac{V}{V_s} j = j_L - j_{rev}(\exp\frac{qV}{kT} - 1) \tag{10}$$

where

$$j_L \equiv \frac{P_o q}{A} \sum_i \left[B_i N_i - B_i(1+N_i)e^{-x_i} + L_i' - L_i e^{-x_i} \right] \qquad (11)$$

$$j_{rev} \equiv \frac{P_o q}{A} \sum_i \left[B_i(1+N_i) + L_i \right] e^{-x_i} . \qquad (12)$$

We have to assume that the energy qV_s is in fact equal to qV. The result (12) is the earlier one [1]; the result (11) reduces to the earlier one if (3a) is used also when one is not equilibrium. Note that (11) vanishes for a cell in the dark, and that reduction of non-radiative transitions reduces (12) (which is desirable for a good efficiency).

While (3a) may be used, (3b) cannot be used away from equilibrium (as I did in [1]). If it is so used, one has the problem that (7) yields

$$\frac{jV}{kT} = \frac{\dot{W}}{kT} = \frac{y_{ps} P_o}{q} (e^y - 1)(\sum_i S_{si}') .$$

This would imply that the short-circuit current always vanishes since $y = 0$ in this case. I am indebted to Dr. A. Pimpale for pointing this out.

An extension to three-level systems with additional applications (also to photosynthesis) will appear [3].

REFERENCES

[1] P.T. Landsberg, J. Luminescence 24/25, 861-864 (1981).
[2] P.T. Landsberg, J. Luminescence 18/19, 1-10 (1979).
[3] P.T. Landsberg and G. Tonge, Photochemistry and Photobiology 35, 769-781 (1982).

IMPURITY DIFFUSION IN AMORPHOUS SILICON AND
ITS IMPLICATIONS FOR SOLAR CELLS

S. KALBITZER, M. REINELT and W. STOLZ
Max-Planck-Institut für Kernphysik, Heidelberg

Summary
Diffusion coefficients of a variety of impurities from different columns of the atomic table have been measured by high resolution nuclear spectroscopy. The relevance of the resulting distribution lengths with respect to solar cell properties is discussed for short time deposition processes at elevated temperatures and long time operation at ambient temperatures.

1. INTRODUCTION

The fabrication process of a solar cell based on amorphous silicon (a-Si) will usually involve temperature treatments at one step or other. In the case of glow discharge deposition of a p-i-n structure, such as displayed in Fig. 1, usually a temperature in the vicinity of 300°C is employed for a time of the order of one hour. During this time dopants, especially from the first deposited highly doped part of the junction structure, will diffuse a certain distance into the bulk i layer. Also, impurities from the backing material, usually glass or metal, could penetrate into the layer at the same time and cause unwanted effects, be it by additional doping or counterdoping, or be it by introduction of deep electronic levels impairing the lifetime of photo-generated carriers. Similar considerations apply for other processes at the top layer, where conductive and antireflecting coatings are deposited and annealed, or where ion implantation doping, followed by a temperature treatment, is applied. Also, the finalized structure, the solar cell, is expected to operate for about ten years at ambient conditions, possibly reaching temperatures of 50°C. During this time, especially the junction top layer, forming an opaque window 100 Å thick or better even less, is required not to essentially change its optoelectronic properties. This again could happen by diffusion and also drift processes in the device due to the built-in electric field (Fig. 1).

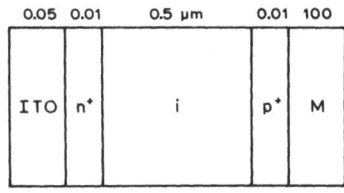

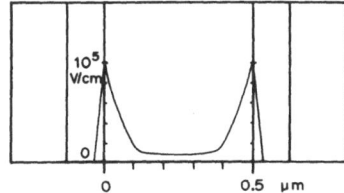

Fig. 1. a) Schematic diagram of a typical solar cell deposited onto a metallic substrate and coated with a conductive and antireflecting layer of indium tin oxide. b) Electrical field distribution in this cell (8).

Therefore, it is desirable to establish the basic transport parameters for certain classes of impurities being used as dopants and others being potential "poisons" for carrier lifetime as known from crystalline silicon, c-Si, as well as for H as a necessary constituent of solar grade a-Si.

2. EXPERIMENTAL

The amorphous samples have been prepared by bombarding c-Si with heavy ions such as Si or Ne. In the case of H effusion measurements, glow discharge samples have been provided by Prof. Spear, Dundee, and by Dr. G. Müller, MBB Ottobrunn. The impurities of interest have been introduced by a second implantation process (1).

The diffused profiles have been measured by Rutherford backscattering in the case of impurities of higher atomic number than that of the host material. Depth resolutions of the order of 10 Å have been achieved by using an electrostatic spectrometer for the energy analysis of the backscattered He ions of about 0.5 MeV incident energy. In the case of H profiling, the well-known ^{15}N analysis has been employed with depth resolutions of 100 Å or better, if required.

Because of this high resolving power, diffusion constants as low as $D \sim 10^{-20}$ cm^2/s can be measured (1).

3. RESULTS

The results, to be published in full detail elsewhere, may be summarized as follows (2). In general, the diffusion of all impurities in a-Si studied so far is either many orders of magnitude slower than in c-Si or below the above detection limit. This statement is based on values extrap-

Table of impurity diffusion in a-Si

Impurity (Group)	T = 400°C, t = 1 h		T = 50°C, t = 10 a	
	D (cm^2/s)	δ (Å)	D (cm^2/s)	δ (Å)
H(I)[a]	~7 × 10^{-14}	~3000	≤3 × 10^{-24}	≤6
Li(I)[a]	~1 × 10^{-14}	~1000	≤10^{-24}	≤3
K(I)[a]	~2 × 10^{-17}	~40	≪10^{-25}	≪1
Rb(I)[b]	≤3 × 10^{-20}	≤2	≪10^{-25}	≪1
Cs(I)[b]	≤1 × 10^{-20}	≤1	≪10^{-25}	≪1
Cu(Ia)[a]	~2 × 10^{-13}	~5000	<10^{-16}[a] <10^{-19}[c]	<30000 <1000
Au(Ia)[a]	~4 × 10^{-15}	~800	≤10^{-25}	≤1
In(III)[b]	≤1 × 10^{-20}	≤1	≪10^{-25}	≪1
Tl(III)[b]	<2 × 10^{-19}	<5	≪10^{-25}	≪1
Sb(V)[b]	<2 × 10^{-19}	<5	≪10^{-25}	≪1
Ar(VIII)[b]	≤1 × 10^{-20}	≤1	≪10^{-25}	≪1

[a] Free diffusion without trapping.
[b] Below detection limit.
[c] Trap controlled diffusion.

olated from published diffusion parameters of c-Si (3,4). The table above lists the relevant data for elevated and ambient temperatures.

Turning first to D (400°C), we note that the group of fast diffusors, H, Cu, Au, known from c-Si to move interstitially, will be redistributed during the deposition over distances $\delta = 2(Dt)^{1/2}$ comparable to the thickness of the solar cell and hence very likely will introduce detrimental effects as known definitely in the case of H. This effect can be greatly reduced for H by using deposition temperatures of 280°C, where D(H) is smaller by more than two orders of magnitude and thus the corresponding δ amounts to some 100 Å. This figure is in good agreement with measurements of H profiles on glow-discharge a-Si layers and devices (10). For Cu, and likely also for Au, the corresponding reduction would be much less, ~0.1 in D and hence ~0.3 in δ.

The heavy alkali impurities, interstitial donors diffusing very likely by an interstitial mechanism in a-Si also, are practically immobile with K at the tolerable limit (5,6,1). Li - D(Li) estimated from the literature (7) - however, cannot be used for device applications as regards stable junctions, and very likely the same is true for Na. Ar, presumably an interstitial diffusor, is practically immobile over the entire accessible temperature regime below the recrystallization temperature at about 600°C, and so the heavier rare gases are expected. The substitutional dopants of groups III and V appear to be slow diffusors, at least as regards the investigated impurities In, Tl, and Sb. Since, however, we expect the relative scaling of the D's not to differ much from that valid in c-Si, similar relative values should be appropriate for the more commonly used dopants B and P in a-Si, i.e., D(B), D(P) ~ 10 × D(Sb) would be an upper limit. The corresponding diffusion lengths would be quite small, of the order of 10 Å. With respect to the diffusion data by the RCA group, we should like to argue that their D's of the substitutional dopants are high by orders of magnitude (9). First, D(Sb) at 400°C amounts to 2×10^{-15} cm^2/s according to this reference, whereas we find 2×10^{-19} cm^2/s as an upper limit. Second, D(B) = 1×10^{-12} cm^2/s, as measured at T = 400°C by the RCA group, would lead to $\delta > 1$ µm, i.e., the total cell volume would be p doped; similar conclusions apply to D(P) = 4×10^{-14} cm^2/s.

At an ambient temperature of about 50°C and a diffusion time of 10 a ~ 10^5 h, the individual D's would have to scale down by a factor of 10^5 to obtain the same δ's as at 400°C in 1 h. This would be the case, if the activation energy of the respective impurity is $Q \geq 0.5$ eV. So far, this appears to be true for all impurities, with the possible exception of Cu. This matter is under further investigation.

One may ask what the conditions for negligible diffusion under these conditions are. In terms of the maximum allowable diffusivity, the condition of $\delta \leq 30$ Å would be fulfilled by $D = \delta^2/4t \leq 10^{-23}$ cm^2/s.

The values of D (50°C) in the table have been conservatively extrapolated by using the lowest activation energies observed in the elevated temperature regime. Clearly, this may overestimate room temperature values, for example for H, Cu, and Au, where trap-controlled diffusion is expected to give rise to considerably reduced mobilities. In all other cases the extrapolated D's are much smaller than 10^{-25} cm^2/s, the equivalent to δ ~ 1 Å, as indicated by the double inequality signs in the table.

Finally, what is the effect of the built-in field on charged impurities in a solar cell? Fig. 1 shows that the electrical fields amount to about 10^5 V/cm in the vicinity of the p$^+$-i and n$^+$-i junctions. Fig. 2 shows how the drift length λ relates to a given diffusion length δ. The proportionality factor p, derived by using the Einstein relation D = µkT/q, is a measure of the drift strength in comparison with random motion. It is seen that for p = 10^6 cm^{-1}, as obtained from V = 10^5 V/cm and kT = 25 meV, the ion drift effect is smaller than the diffusion effect for δ < 100 Å, and thus can be neglected for a tolerated diffusion length of δ ≤ 30 Å. With increasing temperature the relative magnitude of the drift effect will be further reduced, since p ~ 1/T.

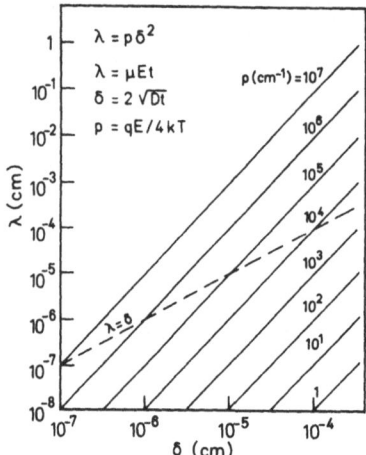

Fig. 2. Relation between drift length λ and diffusion length δ for various drift strength parameters p. The dashed line separates the two regimes of dominant diffusion and drift, respectively.

4. CONCLUSIONS

a-Si solar cells with long term stability against atomic redistribution processes can be made by using suitable group III, V and I dopants. The effusion of H would be problematic even under ambient conditions, unless here a slower diffusion mechanism takes over, which fortunately appears to be the case. The cells should be protected against incorporation of fast diffusing impurities, e.g., Cu, during the deposition process and other elevated temperature treatments; at operation temperatures trapping is expected to prevent penetration into larger depths.

5. ACKNOWLEDGEMENTS

We wish to thank R. Schwan for the sample preparation, W. Hirschel for assisting in the ion implantations, and Th. Dollinger for operating the van de Graaff accelerator.

References

1) M. Reinelt and S. Kalbitzer, J.Physique, suppl.10, **42** (1981) C4-843.
2) M. Reinelt, W. Stolz and S. Kalbitzer, Proc. 16th Int.Conf.Physics of Semiconductors, Montpellier 1982.
3) H.F. Wolf, Silicon Semiconductor Data, Pergamon Press 1969.
4) D.L. Kendall and D.B. De Vries, Semiconductor Silicon, The Electronic Society 1969, p.358.
5) R.A. Gibson, D. Young, P.G. LeComber, W.E. Spear, G. Müller and S. Kalbitzer, J.Physique, suppl.10 **42** (1981) C4-1143.
6) P.G. LeComber, W.E. Spear, G. Müller and S. Kalbitzer, Proc.8th Int. Conf.Amorphous and Liquid Semiconductors, Cambridge 1979, North-Holland Publishing Company 1980, p.327.
7) W. Beyer and R. Fischer, Appl.Phys.Lett. 31 (1977) 850.
8) H. Okamoto, T. Yamaguchi, S. Nomomura and Y. Hamakawa, J.Physique, suppl.10, 42 (1981) C4-507.
9) D.E. Carlson, C.R. Wronski, J.I. Pankove, P.J. Zanzucchi and D.L. Staebler, RCA Review 38 (1977) 211.
10) F.J. Demond, G. Müller, H. Damjantschitsch, H. Mannsperger, S. Kalbitzer, P.G. LeComber and W.E. Spear, J.Physique, suppl.10, **42** (1981) C4-779.

THE WORLD LARGEST 12VOLT SINGLE STRING PHOTOVOLTAIC MODULE

Ph. LAUWERS and G.R. SMEKENS.
Energies Nouvelles et Environnement. S.A.
E.N.E. , Brussels - Belgium.

Summary

This paper presents the world's largest nominal 12Volt single string photovoltaic module ever made until today. It is a 1m2 module consisting of 40 semicristalline cells, each 15cm by 15cm. The module delivers 72Watt under 1000W/m2, AM-1.5 illumination.
The results listed here are those of a first batch of 100 wafers processed from starting material to encapsulated module. The module was completed on May 12, 1982 for presentation as late news at the fourth European Photovoltaic Conference in Stresa.

TECHNICAL DESCRIPTION.

The starting material is the UCP semicristalline S_i material fabricated by SEMIX. A 20cmx20cm ingot was shaped to 15cmx15cm dimensions, and was sliced and etched into 500μ thick wafers by PASAN and INTERSEMIX, Switzerland.
The cell processing, assembling and encapsulation were performed with standard screen printing and related technologies, and with standard encapsulation (tempered glass front and silicone-back) by E.N.E. s.a. Belgium.
No fundamental problems were encountered in the fabrication sequences. The parameters of the front metallisation were adapted to maintain low metal coverage but further optimisation of the processing steps will need to be done to improve yield and efficiency over what has been acheived with this first experimental batch. Nevertheless, all process equipment used for standard 10x10cm2 cells is fully compatible with 15x15cm2 cells.
A large 15x15cm2 cell has shown some technical advantages, as for example reduced perimeter to surface ratio (0,27cm-1 for a 15x15 cell compared to 0.4cm-1 for a 10x10cell). Consequently this results in a lower edge shunt current per cm2, which is a problem often encountered in cell fabrication.
However, the most pronounced advantages of a large cell are cost related. A preliminary cost evaluation on this 100 cell-batch made us conclude that assuming identical yield and efficiency as with 10x10cells, a 20% reduction in cell-manufacturing costs, and a 25% reduction in module-fabrication costs can be acheived.

TECHNICAL SPECIFICATIONS.

Material : UCP Semicristalline Silicon.

Wafers : 500 μ thick 15x15cm2.

Cells   : Short circuit current : 5.5A
          Charging current at 16V: 4.5A
          Efficiency   : 8%
          Metal coverage   :10.4%

Module : Dimensions : 1.26mx0.8m
          Surface   : 1m2
          Weight   : 16Kg

          Maximum power   : 72Watt.
          Total module efficiency : 7.2%.

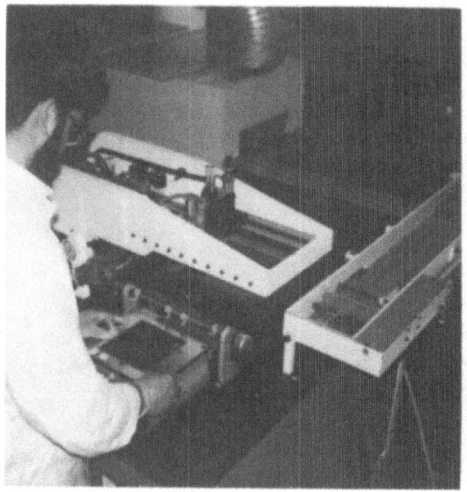 

Fig. 1.

Processing a 15x15cm wafer by screen printing.

Fig. 2.

The world's largest 12Volt single string photovoltaic module was exhibited at the 4th European Photovoltaic Conference in Stresa, Italia.

LIGHT ASSISTED PULSED ANNEALING OF PHOTOVOLTAIC SILICON
BY MICROWAVE ENERGY

P. CHENEVIER, J. COHEN, G. KAMARINOS
Laboratoire de Physique des Composants à Semiconducteurs - ERA CNRS N°659 -
ENSERG, 23, rue des Martyrs - 38031 - GRENOBLE CEDEX - FRANCE

Résumé

Une nouvelle méthode de recuit pulsé de semiconducteurs par énergie microonde est proposée et testée. Elle présente un très bon rendement énergétique global (20 à 30 %) et elle permet des recuits d'une très bonne homogénéité spatiale des surfaces importantes. Le principe de la méthode ainsi que le montage expérimental relatif aux premiers essais sont décrits. Nous montrons la possibilité de recuire des couches implantées de silicium à condition d'éclairer l'échantillon. Elle peut être utilisée pour la fabrication des cellules solaires de faible coût.

1. INTRODUCTION

In this paper, we recommend a new method of pulsed annealing using microwave energy (1). This method allows annealings of sufficiently short duration (on the order of 100 ms) to operate in an ordinary clean room ambience, and to treat relatively large areas. Moreover, the total energy efficiency is far higher (up to 20 to 30 %) than with a laser or electron beam. In addition, microwave generators also allow profiling the power pulse, so that the treatment can be accurately controlled.

2. PRINCIPLE OF THE METHOD

The treatment is carried out in a resonant cavity. The sample is located in a zone with homogeneous electric field, and forms the cavity wall locally (1). At resonance, large currents flow through the sample in a thickness of approximately its skin thickness (10 µm at 10 GHz for Si at 1 000°C). It is in this surface region that the power is dissipated by joule effect. To the extent that the cavity is of good quality, and that it is at the critical coupling, practically all the energy supplied by the generator is dissipated in the sample.
This operation implies that the surface conductivity of the semiconductor is adequate. While this is true when the sample is at high temperature, it is no longer the case at ambient temperature for implanted silicon samples, for example. To initiate power absorption, free carriers are generated at the surface by illuminating the sample (1). Note that apart from any considerations concerning the power that the generator can supply, the sample area that can be treated in one pulse is determined by the zone in which the electric field can be considered homogeneous. Hence it depends essentially on the configuration of the cavity. Fig.1 shows chematically the experimental rig.

3. EXPERIMENTAL RESULTS

The treatments presented here were carried out with relatively low microwave powers (about 500 W), which were nevertheless sufficient for feasibility measurements. We give, in figure 2, as an example, the results of the annealing of a As implanted sample. After annealing under illumination, the impurity profile is not substantially altered, but recrystallization occurs as shown by the collapse of the square resistance of the annealed sample. Measurements of the latter showed that with pulses of 500 W power, or 1.25 kW.cm^{-2}, acting on the sample, practically perfect annealing is obtained within 200 ms (figure 3). This figure shows that over practically 5 mm we obtained a square resistance closely comparable to that obtained by standard thermal annealing. The sudden rise in the square resistance at the ends of the sample is essentially due to its thermal contact with the cavity walls.

The propose annealing process may be used advantageously in the fabrication of cheap solar cells on monocristalline or polycristalline silicon wafers or Si ribbons.

(1) J. COHEN, G. KAMARINOS, P. CHENEVIER, CNRS patent 29/9/1981, N° 81-18656.

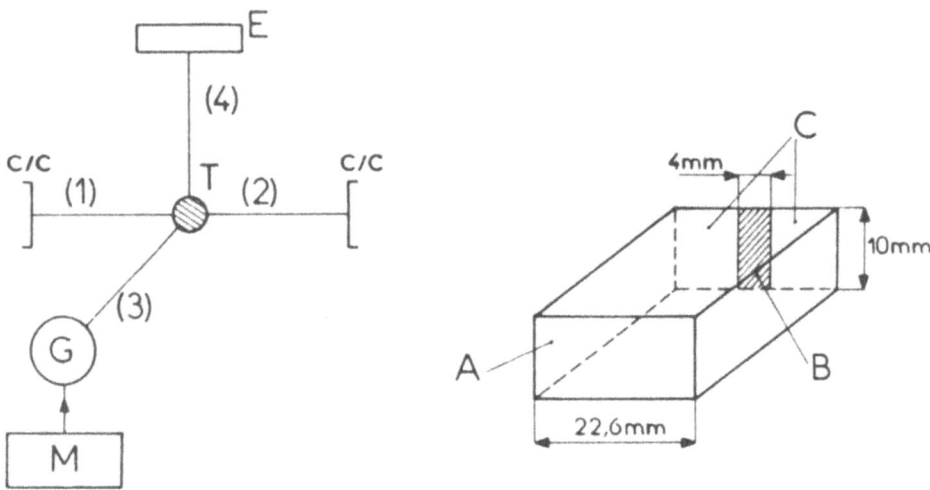

Fig.1 - Experimental set-up. a) Block-diagram ; G : microwave generator ; M : modulator ; T : magic T ; C/C : short-circuit ; E : sample holder. b) Sample holder ; A : guide X-band ; B : sample ; C : metallic sheets closing the guide.

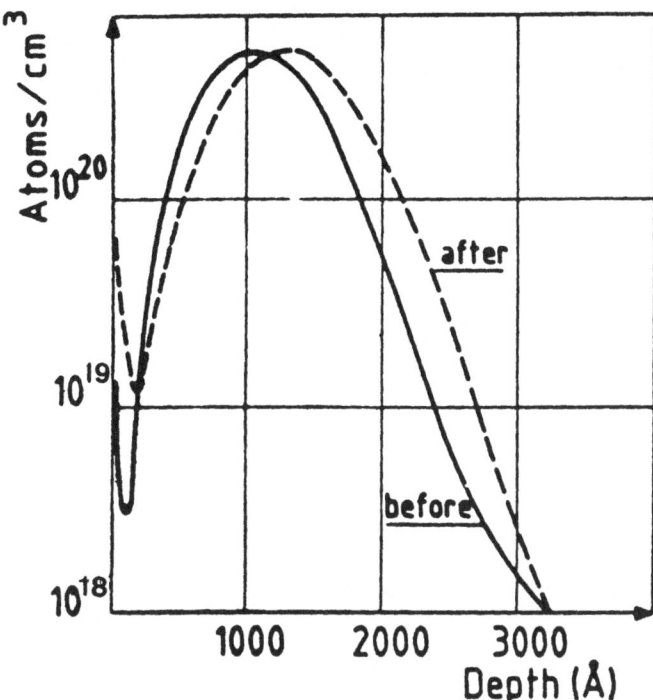

Fig.2 - Arsenic profile (obtained by SIMS) before and after pulsed annealing by microwave energy and with illumination of the sample. Implantation dose : $5 \times 10^{15} cm^{-2}$ under 200 kV.

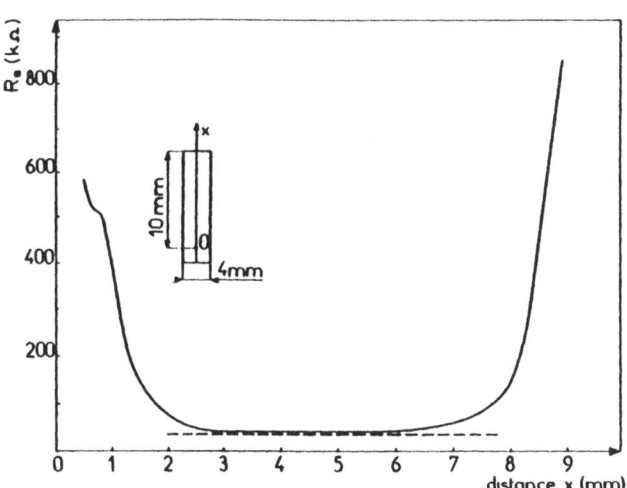

Fig.3 - Square resistance in X direction of the Si sample implanted by As : ——— after microwave annealing ; ---- after thermal (classical) annealing.

MICROPROGRAMMED COUPLING SYSTEM FOR PHOTOVOLTAÏC GENERATORS WITH MULTIPLE RECEPTORS

G. CHAUMAIN, M. BARLAUD, P. ROUAN, JP. REQUIER
Laboratoire de microinformatique et électronique appliquées aux énergies renouvelables - ENSUT de Dakar - BP. 5085 - SENEGAL

SUMMARY

When photovoltaïc generators have a power greater than a few kilowatts, they are often coupled to multiple receptors of diverse characteristics. In such cases it is necessary to optimise their operation and arrange for the proper allocation of energy among them.
Our aim is to present a set-up wich achieves this optimization and distribution.

I. GENERAL REMARKS

Several systems for bringing about the optimum operation of photovoltaïc generators have been put forward (1) (2).
Elsewhere studies have highlighted the harmful effect both of a defective cell on the overall output of photovoltaïc generators, and of the important age-induced modifications of their characteristics.
    To offset these drawbacks, it is desirable to organise the photovoltaic generator into electrically independent modules, each one équipped with an automatic power-tracking device control of the different generator and receptor modules as well as of the overall energy distribution can be carried out with a microprocessor.

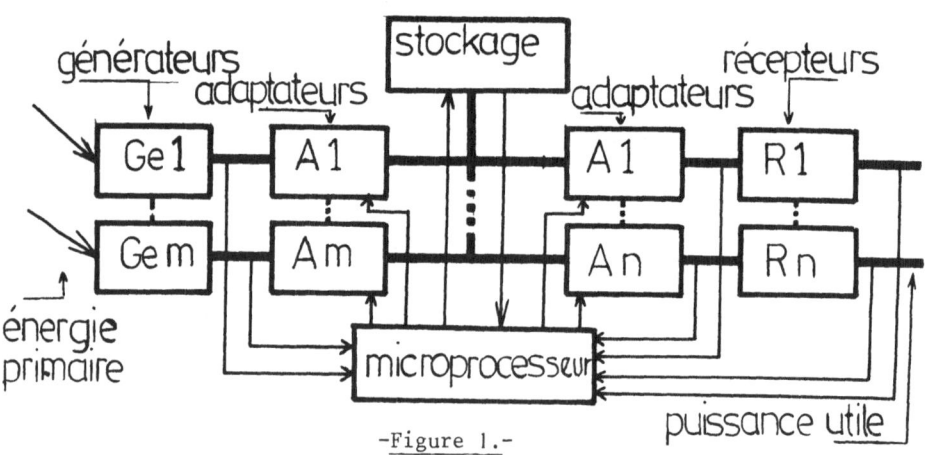

-Figure 1.-

Figure (1) gives the comprehensive organisation of the system, which consists of impedance-adjusting devices coupled to each generator module and each receptor, one buffer storage unit and a microcomputer.
Notice that to this structure one may readily add other generators such as windmills or electric generating sets. Operating principles for the various static impedance adapters are outlined.

For instance, in the case of D.C. sources, a chopper is inserted between the generator and the buffer storage unit in order to control power transferseveral approaches are possible.

   1) by considering the generator a dipole to which maximum power is supplied by an algorithmic procedure activating the chopper ;

   2) by designing a programmable receptor with a static performance adapted to that of the generator.

   The latter approach also enables us to design a programmable generator adapted to any receptor.

II. OPERATING GUIDELINES FOR PROGRAMMABLE RECEPTOR

   We will illustrate the operating guidelines for such adapters by looking at a programmable receptor. Figure (2) presents the static features of a photovoltaïc generator in terms of its solar insulation E_i.

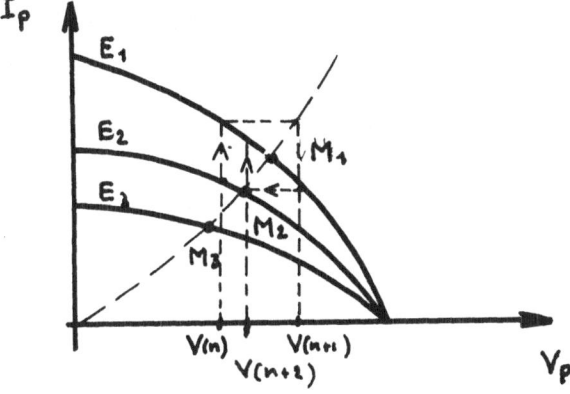

-Figure 2.-

   The points labelled M_i correspond to maximum power the characteristics of the ideal receptor are shown by the curve passing through points M_i. This curve is defined by a table of number pairs stored in the mircoprocessor memory. An algorihm insures that the functionning point converges towards one of these number pairs, thus making it possible to carry out impedance adaptation. Similarly, one can design a programmable generator control of these adapters requires the use of multiplexing to determine power values or (I,V) pairs ; orders in the form of analog voltages must also be issued.

III. STORAGE PROBLEMS

   This structure makes overall energy management possible, and thus control of storage as well. In certain applications, however, it is desirable to minimize storage. Nonetheless, the latter is indispensable for lingking up generators and receptors with different reponse times. Since renewable energy sources are subject to chance, it is an advantage to use more than one source and to manage these resources to the best possible effect.

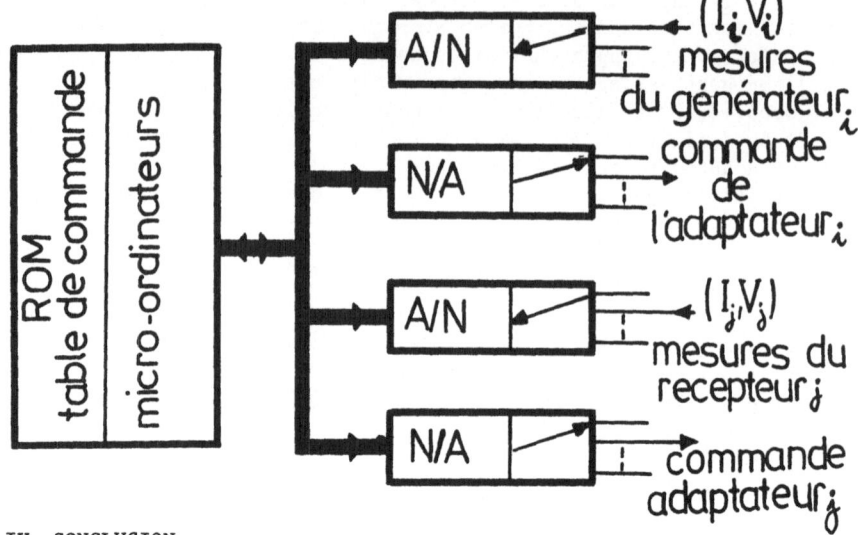

IV. CONCLUSION

Microprocessor management of solar energy resources is a flexible solution which helps in constructing set-up based on modest means. Adding any new generator or receptor component to our structure is quite easy and requires only one change in the management program. Noreover it is possible to install a defect detection system.

The gain in power thus obtained and the modular organization of the set-up are enough in themselves to justify choosing a microprocessor system for management and optimization.

BIBLIOGRAPHY

(1) BARLAUD, PRADAL, REQUIER - Recherche de la puissance optimale délivrée par le générateur photovoltaïque à l'aide d'un dispositif microprogrammé. Photovoltaïc Solar Energy - CANNES 1980.

(2) REQUIER, BARLAUD, ROUAN - Optimization of Photovoltaïc Solar pump by means of static converters driven by microprocessor. 15ème I.E.E.E. Photovoltaïc Specialists Conference - ORLANDO Mai 1981.

LIST OF PARTICIPANTS

ABBOTT D.
 Solar Technology
 Associated Research
 P.O. Box 8
 USA - 13025 AURORA NX

ABERGER N.
 Wacker-Chemietronic GmbH
 Postfach 1140
 D - 8263 BURGHAUSEN

AGHIB PORTA R.
 Porta Electronics
 V. Porta Tenaglia 1/3
 I - 20121 MILANO

ALBERGAMO V.
 ENEA
 S.P. Anguillarese, Km 1.300
 I - 00060 CASACCIA, ROMA

ALBERTAZZI-BOSSI P.
 Corriere del Ticino
 Via Sorengo 22
 CH - 6903 LUGANO

ALLISON J.
 University of Sheffield
 Dept. of Electronic Eng.
 Mappin Street
 GB - SHEFFIELD S13JD

AMALRIC C.
 L'Usine Nouvelle
 rue du Rocher 52
 F - PARIS

ARCIDIACONO V.
 ENEL
 Via Carducci 1/3
 I - 20123 MILANO

ARIENZO M.
 Pragma Solaris
 Via Poli 25
 I - 00198 ROMA

ARMINI A.J.
 Spire Corporation
 Patriots Park
 USA - 01730 BEDFORD, MA

ARNAULT R.
 Arco Solar Inc.
 911 Wilshire boulevard
 USA - LOS ANGELES, CA

ARNE NYLANDSTED L.
 Institute of Physics
 University of Aarhus
 Ny Munkegade
 DK - 8000 AARHUS C

ASCHENBRENNER P.
 Arco Solar Europe SPA
 Via Rasori 15
 I - 20100 MILANO

AUGELLI V.
 Istituto di Fisica
 Via Amendola 173
 I - 70126 BARI

AULICH H.
 Siemens
 Otto-Hahn-Ring 6
 D - 83 MUENCHEN 8

AUTHIER B.
 Heliotronic GmbH
 Johannes-Hess-Str.
 D - 8263 BURGHAUSEN

AVARITSIOTIS J.
 University of Athens
 Solanos Street 104
 GR - 744 ATHENS

BAGLEY E.
 Shell Intern. Petroleum Co Ltd
 Shell Centre
 GB - LONDON SE1 7NA

BANDYOPADHYAY B.
 Centre of Energy Studies
 Indian Institute of Technology
 Houz Khas
 India - 110016 NEW DELHI

BARBE P.
 La Stampa di Torino
 I - TORINO

BARNETT A.
University of Delaware
Dept. of Elec. Eng.
USA - 19711 NEWARK, DE

BARON B.
University of Delaware
Institute of Energy Conversion
USA - 19711 NEWARK, DE

BARUCH P.
Université de Paris VIIe
Groupe de Physique des Solides
2, Place Jussieu
F - 75251 PARIS Cedex 05

BARY A.
Université de Caen
116, rue de Lebisey
F - 14034 CAEN

BASOL B.
Monogram Industries
1299, Ocean Avenue
USA - 90401 SANTA MONICA, CA

BASU P.
University of Kalyani
India - 741235 KALYANI, West Bengal

BAUER G.H.
Inst. für Physikal. Elektron.
Universität Stuttgart
Böblinger Str. 70
D - 7000 STUTTGART 1

BAUFAY L.
Université de l'Etat à Mons
rue Maistriau 23
B - 7000 MONS

BAULAC R.
Pechiney-Ugine-Kuhlmann
23, rue Balzac
F - 75008 PARIS

BEER G.
Phoebus
Via G. Leopardi 148
I - 95100 CATANIA

BEGHI G.
Commission of the European Communities
Joint Research Centre Ispra
I - 21020 ISPRA (Varese)

BEICHLER J.
Universität Marburg
Phys. Institut, AG Halbleiter Renthof
D - 3550 MARBURG

BELLAN D.
BBC London
GB - LONDON

BELLUGUE J.
"Moulin-Neuf" PAUNAT
F - 24510 Ste ALVERE

BELOUET C.
Laboratoires de Marcoussis
route de Nozay
F - 91460 MARCOUSSIS

BELZE A.
Elf Aquitaine
Centre de Recherche
B.P. 22
F - 69360 ST SYMPHORIEN D'OZON

BENEVOLO G.
Agip Nucleare
Viale Brenta 29
I - 20139 MILANO

BEN GHANEM B.
Université de Paris VIIe
2, Place Jussieu
F - 75251 PARIS Cedex 05

BENMALEK M.
Centre des Sciences et
Technologies Nucléaires
B.P. 1017 Alger-Gare
Algérie - ALGER

BENNACEUR R.
Faculté des Sciences de Tunis
Campus Universitaire le Belveder
Tunisie - 1002 TUNIS

BERESOVSKI T.
UNESCO
1, rue Miollis
F - 75015 PARIS

BERMAN E.
ARCO, Solar Incorporation
20554 Plummer Street
USA - 91311 CHATSWORTH, CA

BERNER H.
Universität Konstanz
Postfach 5560
D - 7750 KONSTANZ

BEROTE G.
Univ. Catholique de Louvain
Place Croix du Sud
B - 1348 LOUVAIN LA NEUVE

BEYER U.
RWE
Kruppstrasse 5
D - 4300 ESSEN 1

BHAT R.S.
Institut de Microtechnique
71, rue Maladière
CH - 2007 NEUCHATEL

BHUSHAN M.
University of Delaware
Institut of Energy Conversion
USA - 19711 NEWARK, DE

BIELIKOFF S.
Elf Solaire
Tour Gan
F - 92082 PARIS LA DEFENSE Cedex 13

BIERMANN E.
Deutsche Gesellschaft für
Technische Zusammenarbeit (GTZ)
Dag-Hammarskjöld-Weg 1
D - 6236 ESCHBORN

BITER W.
STD Oil of Ohio
3092 Broadway
USA - 44115 CLEVELAND, OH

BLOSS W.H.
Inst. für Physikal. Elektronik
Universität Stuttgart
Böblinger Strasse 70
D - 7000 STUTTGART 1

BOEHM M.
Institut für Werkstoffe
Technische Universität Berlin
Jebensstrasse 1
D - 1000 BERLIN 12

BOEKE W.
Fraunhofer - ICT
Postfach 1240
D - 7507 PFINZTAL

BOES E.C.
Sandia National Laboratories
USA - ALBUQUERQUE, N. Mexico 87185

BONTE L.
Lab. of Electronics & Metrology
St. Pietersnieuwstraat 41
B - 9000 GENT

BOSE D.N.
Indian Inst. of Techn. Kharagpur
Materials Science Centre
India - 721302 KHARAGPUR, Bengal

BOTTENBERG W.R.
Arco Solar Incorporation
20554 Plummer Street
USA - 91311 CHATSWORTH, CA

BOUAZZI A.S.
ENIT
P.O. Box 37
Tunisia - 1012 TUNIS BELVEDERE

BOUDJANI A.
SONELEC-CEGP
route de Mascava
Algérie - SIDI BEL ABBES

BOUREE J.E.
CNRS
1, Place Aristide Briand
F - 92190 MEUDON

BOUSQUET A.
    Elf Solaire
    Tour Gan
    F - 92082 PARIS LA DEFENSE Cedex 13

BOUTELEUX M.
    SFBP
    CRL 14
    B.P. 1
    F - 13117 LAVERA

BOUZOUITA H.
    Ecole Nationale de Tunis
    Campus Universitaire
    Tunisie - 37 TUNIS

BRAICOVICH L.
    IEFE - Università Bocconi
    Via Sarfatti 25
    I - 20136 MILANO

BRANDHORST H.W.
    NASA - Lewis Research Centre
    21000 Brookpark Road MS 302-1
    USA - 44135 CLEVELAND, Ohio

BRAWLEY T.
    Applied Solar Energy Corporation
    15251 E. Don Julian Road
    USA - 91746 CITY OF INDUSTRY, CA

BRIGANTI G.
    ENEA
    Via Anguillarese
    I - 00060 CASACCIA, Roma

BROSER I.
    Technische Universität Berlin
    Institut für Festkörperphysik II
    Strasse des 17. Juni 112
    D - 1000 BERLIN 12

BROSER R.
    Fritz-Haber-Institut der MPG
    Faraday-Weg 4 - 6
    D - 1000 BERLIN 33

BRUNSTROEM C.
    Swedish State Power Board
    Hydraulics Laboratory
    S - 81071 ALVKARLEBY

BUCHER E.
    Universität Konstanz
    Jakob-Burckhardtstrasse 27
    D - 7750 KONSTANZ

BUCKLEY R.W.
    Frank Wheldon School
    Coningswath Road
    GB - CARLTON NG4 3SH, Nottingham

BULLO P.
    ENEL
    Via Carducci 1/3
    I - 20123 MILANO

BURGEL R.J.
    Holecsol Systems B.V.
    P.O. Box 2300
    NL - 5600 CH EINDHOVEN

BUYSSE J.
    Fabricable Huizingen
    Heideveld 1
    B - 1511 HUIZINGEN

CABESTANY J.
    ETSI Telecomunicacion
    Jordi Girona Salgado, S/N
    E - 34 BARCELONA

CADENE M.
    Univ. Sciences Techn. Languedoc
    Place E. Bataillon
    F - 34060 MONTPELLIER

CALLAGHAN W.T.
    American Jet Propulsion Lab.
    4800 Oak Grove Drive, M/S 502-422
    USA - 91109 PASADENA, CA

CALLIGARICH C.
    Heliosil S.p.A.
    Via Franchetti 1/A
    I - 20124 MILANO

CALZIA J.
    Rhone-Poulenc Specialités Chim.
    18, boucle d'Alsace
    F - 92408 COURBEVOIE

CAMANI M.
　Dipart. dell'ambiente
　Cantone Ticino
　CH - 6500 BELLINZONA

CANAL B.
　Photowatt International S.A.
　131, route de l'Empereur
　F - 92500 RUEIL MALMAISON

CARDINALI G.
　CNR Istituto LAMEL
　Via Castagnoli 1
　I - 40126 BOLOGNA

CASTANER L.
　ETSI Telecomunicacion
　Jordi Girona Salgado S/N
　E - 34 BARCELONA

CAVALLINI M.
　Assoreni
　Ercole Ramarini
　I - 00015 MONTEROTONDO, Roma

CEPPI P.
　Eidgenössische Technische Hochschule
　CH - 8093 ZUERICH

CERESARA S.
　La Metalli Industriale SpA
　Via della Repubblica
　I - 55052 FORNACI DI BARGA

CESARONI C.
　ENEA
　Via Anguillarese
　I - 00060 CASACCIA, Roma

CETINCELIK M.
　Energy World
　P.K. 680 Kizilay
　Turkey - ANKARA

CHABOT B.
　COMES
　route des Lucioles
　F - 06565 VALBONNE

CHAMBERLAIN G.A.
　Shell Research Ltd
　Thornton Research Centre
　P.O. Box 1
　GB - CHESTER

CHAPUIS R.M.
　Université de Lausanne
　Inst. Crystallography
　B.S.P. 1020
　CH - LAUSANNE

CHARTIER E.
　Thomson - CSF
　Domaine de Corbeville
　F - 91401 ORSAY

CHEEK G.
　Katholieke Universiteit Leuven
　Kardinaal Mercierlaan 94
　B - 3030 HEVERLEE

CHEMISKY G.
　INSA Lyon
　20, av. Albert Einstein, Bte 502
　F - 69621 VILLEURBANNE

CHENLO F.
　Instituto Energia Solar
　ETSI Telecomunicacion
　E - MADRID 3

CHEVALIER Y.
　COMES
　26, rue Chambéry
　F - 75015 PARIS

CHIARELLI F.
　Assoreni
　Ercole Ramarini
　I - 00015 MONTEROTONDO, Roma

CHIDESTER L.
　Lockheed
　21438 Krzich Place
　USA - 95014 CUPERTINO, CA

CHITRE S.
　Spire Corporation
　Patriots Park
　USA - 01730 BEDFORD, MA

CID M.
　Instituto de Energia Solar
　ETSI Telecomunicacion
　E - MADRID 3

CLARK C.W.
　Solenergy Corporation
　171 Merrimac Street
　USA - 01801 WOBURN, Mass.

COHEN-SOLAL G.
　CNRS - Bellevue
　1, Place A. Briand
　F - 92190 MEUDON-BELLEVUE

COLLINS R.W.
　Lucas BP Solar Systems
　Windmill Road
　GB - HADDENHAM, ALESBURY HP178JB

COM-NOUGUE J.
　Laboratoires de Marcoussis
　route de Nozay
　F - 91460 MARCOUSSIS

COOK W.
　46, St. Georges Court
　GB - GLOUCESTER, LONDON

COPPA-ZUCCARI G.
　Technical Journalist
　c/o Stampa Estera
　Via della Mercede 55
　I - 00127 ROMA

CORNU J.P.
　SAFT
　119, rue du Président Wilson
　F - 92300 LEVALLOIS PERRET

CORSI S.
　ENEL
　Via Carducci 1/3
　I - 20123 MILANO

COSSEMENT G.
　Univ. Catholique de Louvain
　Place Croix du Sud
　B - 1348 LOUVAIN LA NEUVE

COWACHE P.
　CNRS
　LEAA ENSCP
　11, rue Pierre et Marie Curie
　F - 75231 PARIS

CRISTIANO L.
　16, Viale A. Gramsci
　I - 80122 NAPOLI

CUMBERBATCH T.
　Thorn EMI
　Central Research Laboratories
　Trevor Road
　GB - HAYES UB3 7HH

CURRAN J.
　Ecole Centrale de Lyon
　Lab. Physicochimie des interfac
　F - 69130 ECULLY

CURTINS H.
　Institut de Microtechnique
　71, rue Maladière
　CH - 2007 NEUCHATEL

CZUBATYJ W.
　Energy Conversion Devices
　1675 W. Maple Road
　USA - 48084 TROY

D'AIELLO R.V.
　RCA Laboratories
　USA - PRINCETON, N.J. 08540

DARDINI C.
　ENEA
　Via Anguillarese
　I - 00060 CASACCIA, Roma

DE ANGELIS B.
　Assoreni
　Ercole Ramarini
　I - 00015 MONTEROTONDO, Roma

DE KOUSEMAEKER
　Holecsol
　P.O. Box 2300
　NL - 5600 CH EINDHOVEN

DEMICHELIS F.
    Politecnico di Torino
    Via Duca degli Abruzzi 24
    I - 10129 TORINO

DEMONGEOT F.
    Photowatt International SA
    131, route de l'Empereur
    F - 92500 RUEIL MALMAISON

DEN BOER W.
    Delft University of Technology
    Department of Electr. Engineering
    Mekelweg 4
    NL - 2600 GA DELFT

DESOMBRE A.
    Photowatt International SA
    131, route de l'Empereur
    F - 92500 RUEIL MALMAISON

DE VOS A.
    Rijksuniversiteit te Gent
    St. Pietersnieuwstraat 41
    B - 9000 GENT

D'HARCOURT A.
    Ministry of Industry
    120, rue du Cherche Midi
    F - 75006 PARIS

DICHLER A.
    Solapak Ltd
    Factory Three
    Cock Lane
    GB - HIGH WYCOMBE, Bucks HP1

DIETL J.
    Heliotronic
    Postfach 1129
    D - 8263 BURGHAUSEN

DI MARTILE F.
    SUCCESSO
    Via Turati 6
    I - 20121 MILANO

DINKESPILER J.A.
    Commission des Communautés
    européennes
    120, rue de la Loi
    B - BRUXELLES

DOGNIAUX R.
    Institut Royal Météorologique
    148, rue Groeselenberg
    B - 1180 BRUXELLES

DOMINGUEZ E.
    Torres Quevedo (CSIC)
    C/ Serrano 144
    E - 6 MADRID

DON E.
    Newcastle Polytechnic
    School of Physics
    Ellison Place
    GB - NEWCASTLE NE1 8ST

DONON J.
    Photowatt International SA
    131, route de l'Empereur
    F - 92500 RUEIL MALMAISON

DRAGHETTI G.
    Società ANIC
    Divisione Chimica Secondaria
    Via Mecenate 89
    I - MILANO

DROUOT L.
    COMES
    route des Lucioles
    F - 06565 VALBONNE

DUBAL Léo
    P.O. Box 1324
    CH - 3001 BERNE

DUCHEMIN S.
    Univ. Sciences Techn. Languedoc
    Place E. Bataillon
    F - 34060 MONTPELLIER

DUMAS J.P.
    Laboratoires de Marcoussis
    route de Nozay
    F - 91460 MARCOUSSIS

DUMONT P.
    ACEC Charleroi
    P.O. Box 4
    B - 6000 CHARLEROI

DUNKEL W.
   AEG Telefunken
   Lilienthalstrasse 150
   D - 3500 KASSEL

DURAND H.
   COMES
   Commissariat de l'Energie Solaire
   208, rue Raymond Losserand
   F - PARIS

DURIAN A.G.
   Automot. Mgmt. Serv., Inc.
   base-info-press-serv.
   69, avenue Isola Bella
   F - 06400 CANNES

EGELHAAF P.
   Robert Bosch GmbH
   Abt. FMG 1
   Robert-Bosch-Platz 1,
   Postfach 50
   D - STUTTGART

ELIAS G.
   CNR
   Aera di Ricerca di Roma
   I - ROMA

ELIZALDE E.
   Universidad Autonoma Madrid
   Garcia Molina 1
   E - 35 MADRID

EMANUELE G.
   Phoebus
   Via G. Leopardi 148
   I - 95100 CATANIA

ENGEMANN D.
   Battelle Institut Frankfurt
   Am Römerhof 35
   D - 6000 FRANKFURT

EQUER B.
   PIRSEM / CNRS
   282, boulevard St Germain
   F - 75007 PARIS

ECHARD
   Photowatt International
   131, route de l'Empereur
   F - 92500 RUEIL MALMAISON

EVANS R.J.
   Lucas BP Solar Systems
   Windmill Road
   GB - HADDENHAM, Aylesbury HP17 8JB

FABRE E.
   Photowatt International SA
   131, route de l'Empereur
   F - 92500 RUEIL MALMAISON

FALLY J.
   Laboratoires de Marcoussis
   route de Nozay
   F - 91460 MARCOUSSIS

FARADAY B.J.
   Naval Research Laboratory
   Code 6604
   4555 Overlook Ave., SW
   USA - 0375 WASHINGTON, D.C.

FARINELLI U.
   ENEA
   Via Regina Margherita 125
   I - 00198 ROMA

FELDMAN C.
   The Johns Hopkins University
   Applied Physics Laboratory
   Johns Hopkins Road
   USA - 20707 LAUREL, Maryland

FERBER R.
   Jet Propulsion Lab.
   4800 Oak Grove Drive
   USA - PASADENA, CA 91109

FEUCHT D.
   Solar Energy Research Institute
   1617 Cole Bvd.
   USA - 80228 GOLDEN, Colorado

FINCH D.L.
   Strategies Unlimited
   201 San Antonio Circle, Suite 205
   USA - 94040 MOUNTAIN VIEW, CA

FIRESTER A.H.
   RCA
   route 1
   USA - 08540 PRINCETON, NJ

FISCHER M.
  Deutsche Forschungs- und Versuchs-
  anstalt f. Luft- und Raumfahrt e.V.
  Pfaffenwaldring 38/40
  D - 7000 STUTTGART 80

FISCHER R.
  AEG - Telefunken
  Goldstein Strasse 235
  D - 6000 FRANKFURT

FITTIPALDI F.
  ENEL
  Via Martini 1-3
  I - ROMA

FLAMMIA N.
  ENEA
  Editorial Staff Member
  Viale Regina Margherita 125
  I - 00198 ROMA

FLORES C.
  CISE SpA
  Via Reggio Emilia 39
  I - 20090 SEGRATE (Milano)

FOELL H.
  Siemens AG
  Otto-Hahn-Ring 6
  D - 8000 MUENCHEN 83

FOGARASSY E.
  CRN - Groupe PHASE
  23, rue du Loess
  F - 67037 STRASBOURG

FORESTIERI A.
  NASA, Lewis Research Centre
  21000 Brookpark Road, MS 302-1
  USA - 44135 CLEVELAND, Ohio

FRATTI F.
  Feal S.p.A.
  Via Bernardino Verro 90
  I - 20141 MILANO

FREUDENBERG K.
  Universität München
  Amalienstrasse 54
  D - 8000 MUENCHEN

FRISCH F.
  GEO-Magazine
  Postfach 302040
  D - HAMBURG 36

FRISSON L.
  Fabricable Huizingen
  Heideveld 1
  B - 1511 HUIZINGEN

FROGGATT N.
  Delco Remy
  P.O. Box 242, Delaware Drive
  GB - TONGWELL, Milton Keynes

FROVA A.
  Università di Roma
  Istituto di Fisica G. Marconi
  P. Aldo Moro 2
  I - 00185 ROMA

FUHS W.
  Universität Marburg
  Renthof 5
  D - 355 MARBURG

GALLONI R.
  CNR - LAMEL
  Via Castagnoli 1
  I - 40126 BOLOGNA

GALLUZZI F.
  Assoreni
  Via E. Ramarini 32
  I - 00015 MONTEROTONDO, Roma

GARCIA MENDOZA F.
  INSA Lyon
  Lab. de Physique Matière B-502
  20, av. Albert Einstein
  F - VILLEURBANNE

GARNER I.F.
  British Imperial College
  48, Princes Gardens
  GB - LONDON SW7 1LU

GASBARRA A.M.
  ENEL
  Via Carducci 1/3
  I - 20123 MILANO

GASPARINI M.
 Università di Parma
 Via Agadir 16/B
 I - 20097 SAN DONATO MIL.

GAY C.F.
 Arco Solar Incorporation
 20554 Plummer Street
 USA - 91311 CHATSWORTH, CA

GAY R.R.
 Arco Solar Incorporation
 20554 Plummer Street
 USA - 91311 CHATSWORTH, CA

GAZZOLA G.
 ENEA
 Via Anguillarese, Km 1.300
 I - 00060 CASACCIA, Roma

GEISSLER J.
 Heliotronic GmbH
 Postfach 1129
 D - 8263 BURGHAUSEN

GENTILUCCI R.
 FEAL SpA
 Via Cristoforo Colombo 436
 I - 00145 ROMA

GILING L.J.
 University of Nijmegen
 NL - NIJMEGEN

GILSON P.
 GILSON SA
 rue du Centre 55A
 CH - 1025 ST SULPICE

GIRISCH R.
 Katholieke Universiteit Leuven
 Kardinaal Mercierlaan 94
 B - 3030 LEUVEN

GISLON R.
 ENEA - FARE - Casaccia
 I - S.M. GALERIA, ROMA 00060

GIUFFRIDA M.
 Ansaldo Impianti
 Via G. d'Annunzio 113
 I - 16121 GENOVA

GOCHERMANN H.
 AEG - Telefunken
 Schmidt-Isserstedt-Weg 2
 D - 2081 HOLM

GODDAER M.
 Fabricable - Belgosolar
 Heideveld 1
 B - 1511 HUIZINGEN

GOETZBERGER A.
 Fraunhofer-Institut für
 Solare Energiesysteme
 Oltmannstrasse 22
 D - 7800 FREIBURG

GOLDNER F.
 American Embassy
 US Department of Energy
 D - BONN

GOLDSMITH P.
 TRW
 One Space Park, R5/1281
 USA - 90278 REDONDO BEACH, CA

GOMEZ-COSCOLLAR T.
 Solarex Trading
 Baron de Carcer 28 - 1°
 E - VALENCIA 1

GONZALO A.L.
 Standard Electrica SA
 Josefa Valcarcel 27
 E - 27 MADRID

GORDILLO G.
 Institut für Physik. Elektronik
 Waldburgstrasse 25
 D - 7000 STUTTGART 80

GOUSKOV L.
 Université de Montpellier
 F - 34000 MONTPELLIER

GRASSI G.
Commission des Communautés
européennes
D.G. Science, recherche et
développement
200, rue de Loi
B - 1049 BRUXELLES

GRENIER C.
Photowatt International SA
131, route de l'Empereur
F - 92500 RUEIL MALMAISON

GRUBER E.
Resart-Ihm AG
Gassnerallee 40
D - 6500 MAINZ

GRUETER J.W.
Kernforschungsanlage - STE
Stetternicher Stadtforst
D - 5170 JUELICH

GUARINI G.
CISE SpA
Via Reggio Emilia 39
I - 20090 SEGRATE (Milano)

GUCHER P.
Institut de Chimie et
de Physique Industrielle
31, Place Bellecour
F - 69002 LYON

GUILLOT J.M.
LEPI - ISIN
F - 54500 VANDOEUVRE

GUIMARES L.
Centro de Fisica Molecular
das Univers. de Lisboa
Complexo I, av. Rovisco Paes
P - 1000 LISBOA

HABOECK A.
Siemens AG
Günther-Scharowsky-Strasse 2
D - 8520 ERLANGEN

HAMAKAWA Y.
Faculty of Electrical Engineering
Osaka University
1-1 Machikameyama-cho
Japan - TOYONAKA, Osaka 560

HAMMER H.
Siemens AG, E STE 13
Günther-Scharowsky-Strasse 2
D - 8520 ERLANGEN

HAMOUDA C.
Sonelec
16, avenue de la République
Algérie - BATNA

HANIA D.
CNRS - CECM
15, rue Georges Urbain
F - 94400 VITRY-SUR-SEINE

HEDGES S.
BBC London
GB - LONDON

HEDSTROEM J.
Institute of Microwave Technology
S - 10044 STOCKHOLM

HEIERS W.
Kommission der Europäischen
Gemeinschaft
G.D. Informationsmarkt und
Innovation
P.B. 1907
L - LUXEMBURG

HELM P.
WIP-München
Sylvensteinstrasse 2
D - 8 MUENCHEN

HENN K.
Kernforschungsanlage Jülich
Postfach 1913
D - 5170 JUELICH

HERNANDEZ G.
Heliocastilla
Arco de Ladrillo 29
E - VALLADOLID

HERVO C.
Laboratoires de Marcoussis
route de Nozay
F - 91460 MARCOUSSIS

HEZEL R.
 Universität Erlangen
 Martensstrasse 7, WW6
 D - 8520 ERLANGEN

HILL R.
 Newcastle Polytechnic
 Ellison Place
 GB - NEWCASTLE UPON TYNE NE1 8ST

HINTRINGER O.
 Siemens AG
 Frankfurter Ring 152
 D - 8000 MUENCHEN 46

Ho. C.T.
 Mobil Tyco Solar Energy Corp.
 16, Hickory Drive
 USA - 02254 WALTHAM, MA

HOLM C.
 Heliotronic GmbH
 Postfach 1129
 D - 8263 BURGHAUSEN

HOLZENKAEMPFER E.
 Universität Marburg
 Renthof 5
 D - 3550 MARBURG

HOUSSARX A.
 Scientific Studies and
 Research Centre
 P.O. Box 4470
 Syria - DAMASCUS

HUMENBERGER J.
 Universität Linz
 Altenbergerstrasse 69
 A - 4040 LINZ

HUSCHKA H.
 NUKEM GmbH
 Postfach 110080
 D - 6450 HANAU 11

ILICETO A.
 ENEL
 Via Carducci 1/3
 I - 20123 MILANO

IND T.J.
 Lucas BP Solar Systems Ltd.
 Windmill Road
 GB - HADDENHAM, AYLESBURY HP17 8JF

JACOB J.
 SARTI / GLAVERSEL
 Via G. Compagnoni 1
 I - MILANO

JACOBS H.
 Siemens AG
 Werner-von-Siemens-Str. 50
 D - 8520 ERLANGEN

JAILLET P.
 Photowatt International SA
 131, route de l'Empereur
 F - 92500 RUEIL MALMAISON

JAIN S.C.
 Solid State Physics Lab.
 Lucknow Road
 India - 110007 DELHI

JANSSEN R.
 Katholieke Universiteit Leuven
 Kardinaal Mercierlaan 94
 B - 3030 HEVERLEE

JARGELIUS M.
 Institute of Microwave Technology
 S - 10044 STOCKHOLM

JOHNSON E.L.
 Texas Instruments Incorporation
 P.O. Box 225303, MS 158
 USA - 75265 DALLAS, Texas

JOHNSON S.
 Frank Wheldon School
 Cuningswath Road
 GB - CARLTON, Nottingham NG4 3SH

KAESEN K.
 Siemens AG
 Frankfurterring 152
 D - 8000 MUENCHEN 46

KAESTNER R.
   Consol GmbH
   CH - 6851 CASIMA,TI

KALBITZER S.
   Max-Planck-Institut
   Saupferch Eckweg 1
   D - 6900 HEIDELBERG

KALLBACK B.
   Institute of Microwave Technology
   S - 10044 STOCKHOLM

KAMARINOS G.
   ENSERG
   23, avenue des Martyrs
   F - 38031 GRENOBLE CEDEX

KAMEL F.
   Universität Hannover
   4, Wilhelm-Busch-Strasse
   D - 3000 HANNOVER 1

KANE M.
   Faculté des Sciences Dakar
   Villa 69, Cité des Professeurs
   Sénégal - DAKAR

KANICKI J.
   Université Libre de Bruxelles
   Chimie Organique
   50, avenue F.D. Roosevelt
   B - 1050 BRUXELLES

KASZETA W.
   Solavolt International
   P.O. Box 2934
   USA - 85062 PHOENIX

KERBER M.
   Universität Innsbruck
   Schoepfstrasse 41
   A - 6020 INNSBRUCK

KHATTAK C.
   Crystal Systems Incorporation
   35, Congress Street
   USA - 01970 SALEM, MA

KIESSLING K.P.
   KSB, Klein, Schanzlin & Becker
   Johann-Klein-Strasse 9
   D - 6710 FRANKENTHAL

KINNELL G.H.
   Department of Minerals & Energy
   P.O. Box 5148
   Papua New Guinea - BOROKO

KIPPERMAN A.H.M.
   University of Technology
   den Dolech 2, P.O. Box 513
   NL - 5600 MB EINDHOVEN

KIRKBRIDE B.J.
   Pilkington Brothers PLC
   R + D Laboratories
   Lathom
   GB - ORMSKIRK L40 5OF

KLEIN R.J.
   345 East 47th Street
   USA - 10017 NEW YORK CITY, NY

KLERSY R.
   CCR -EEC
   I - ISPRA (Varese)

KLOCKGETHER J.
   Klein, Schanzlin & Becker AG.
   Johann-Klein-Strasse 6
   D - 6710 FRANKENTHAL

KOCH W.
   Bayer AG
   AC-F/Ue, R 86
   Rheinuferstrasse
   D - 4150 KREFELD

KOEPKE R.
   KFA Jülich
   Postfach 1913
   D - 5170 JUELICH

KORUPP K.H.
   AEG - Telefunken
   Industriestrasse 29
   D - 2000 WEDEL

KREBS K.
 Commission of the European
 Communities
 Joint Research Centre
 P.O. Box 1
 I - 21020 ISPRA (Varese)

KRUEHLER W.
 Siemens AG, FL OPT 31
 Otto-Hahn-Ring 6
 D - 8000 MUENCHEN 83

KRUSE K.
 Grundvos AS
 Revlingvej 7
 DK - 8850 BJERRINGBRO D4

KUMAR S.
 Central Electronics Limited
 4, Industrial Area
 India - SAHIBABAD (U.P.)

KUWANO Y.
 Sanyo Electric Co Limited
 Sanyo Research Center
 13 Hashiridani
 Japan - 570 HIRAKATA CITY, Osaka

LAMBRI L.
 ENEL
 Via Carducci 1/3
 I - 20123 MILANO

LANDSBERG P.T.
 University of Southampton
 GB - SOUTHAMPTON SO9 5NH

LAPLAZE D.
 Faculté des Sciences Dakar
 Sénégal - DAKAR

LAUGIER A.
 INSA - 502
 Laboratoire Physique Matière
 20, av. Albert Einstein
 F - 69621 VILLEURBANNE

LAURENT B.
 LAAS du CNRS
 7, avenue du Colonel Roche
 F - 31400 TOULOUSE

LAUVRAY H.
 Photowatt International SA
 131, route de l'Empereur
 F - 92500 RUEIL MALMAISON

LAUWERS P.
 ENE
 188, av. van der Meerschen
 B - 1150 BRUXELLES

LEBRUN P.
 Laboratoires de Marcoussis
 route de Nozay
 F - 91460 MARCOUSSIS

LEMETAYER M.
 Centre National d'Etudes Spatiale
 18, avenue E. Belin
 F - 31055 TOULOUSE Cedex

LE POULL F.
 SNES
 113, Chemin de Ronde
 F - CROISSY/SEINE

LEPPIHALME M.J.
 Technical Research Centre
 Semiconductor Laboratoies
 Otakaari 5a
 SF - 02150 ESPOO 15

LEQUEUX P.
 Commission des Communautés
 européennes
 200, rue de la Loi
 B - 1049 BRUXELLES

LEVHA L.
 Elf
 Tour Gan
 F - 92082 PARIS Cedex 13

LINCOT D.
 CNRS - ENSCP
 rue Pierre et Marie Curie 11
 F - 75005 PARIS

LINDMAYER J.
 Solarex Trading Office
 CH - 1196 GLAND

LIPPMANN G.
  Dornier-System
  D - 7990 FRIEDRICHSHAFEN

LOBRY P.
  ICAM
  rue Auber 6
  F - 59650 LILLE

LOESCH H.
  DFVLR
  Linderhöhe
  D - 5000 KOELN 90

LOFERSKI J.J.
  Brown University
  Division of Engineering
  USA - 02912 PROVIDENCE, Rhode Island

LUCA L.
  SCR Associati
  Foro Buonaparte 70
  I - 20121 MILANO

LUDSTECK A.
  Siemens AG
  Frankfurter Ring 152
  D - 8000 MUENCHEN 46

LUQUE A.
  Instituto de Energia Solar
  ETSI Telecomunicacion
  E - MADRID 3

LUQUET H.
  CEES - USTL
  Université de Montpellier
  Place E. Bataillon
  F - 34060 MONTPELLIER

LUTHER J.
  Universität Oldenburg
  Ammerländer Heerstrasse 67-99
  D - 2900 OLDENBURG

MAASS K.P.
  AEG - Telefunken
  Industriestrasse 29
  D - 2000 WEDEL

MACFALL K.
  Marubeni America Incorporation
  200 Park Avenue NYC NY
  USA - NEW YOK

MACOMBER H.L.
  Monegon Ltd.
  4, Professional Drive, Suite 130
  USA - GAITHERSBURG, Maryland 20879

MADET D.
  Electricité de France
  6, Quai Watier
  F - 78400 CHATOU

MAGID L.
  Solarex Sheet
  1335 Piccard Drive
  USA - BROCKVILLE, Mass. HD 20850

MAGNI M.
  Commission of the European Communities
  Joint Research Centre - Euratom
  I - 21020 ISPRA

MAKI N.
  Toshiba
  Saiwaiku, Kawasaki
  Japan - Z10 KOMUKAI TOSHIBACHO

MAKIOS V.
  University of Patras
  GR - PATRAS

MALTAGLIATI M.M.
  Solaris SpA
  Via T. Alderotti 26
  I - 50100 FIRENZE

MANFREDI A.
  Arco Solar Europe SpA
  Viale Milanofiori, Fabb. E5
  I - 20094 ASSAGO, Milano

MANGELSDORF M.A.
  Solpower Industries
  19371 Titus CT
  USA - 95070 SARATOGA, CA

MANIFACIER J.C.
    Université des Sciences - CEES
    Place E. Bataillon
    F - 34060 MONTPELLIER Cedex

MANUEL P.
    Electricité de France
    1, avenue du Général de Gaulle
    F - 92141 CLAMART

MARFAING Y.
    Physique des Solides CNRS
    1, Place A. Briand
    F - 92190 MEUDON-BELLEVUE

MARGADONNA D.
    Assoreni
    Via Ramarini
    I - 00015 MONTEROTONDO, Roma

MARGARIA T.
    Nobel Bozel
    Division Bozel Electrométallurgie
    Tour Roussel Nobel,
    3, av. Général de Gaulle
    F - 92800 PUTEAUX

MARIS O.
    Saint-Gobain Recherches
    39, Quai Lucien Lefrancs
    F - 93300 AUBERVILLIERS

MARTINS R.
    Centro de Fisica Molecular
    Complexo I - I.S.T.
    P - 1000 LISBOA

MARTINUZZI S.
    Faculté des Sc. et Techn.
    de Saint-Jérôme
    Laboratoire de Photoélectricité
    F - 13397 MARSEILLE Cedex 4

MAS A.
    ETSI Telecomunicacion B.
    Jordi Girona Salgado S/N
    E - 34 BARCELONA

MASSAAD S.
    Université Claude Bernard Lyon
    Dépt. Physique des Matériaux
    43, bvd du 11 Novembre 1918
    F - 69621 VILLEUBANNE

MATARE H.F.
    ISSEC
    P.O. Box 49177
    USA - 90049 LOS ANGELES, CA

McGINNIS R.W.
    Photowatt International
    2414 W. 14th Street
    USA - 85281 TEMPE, Arizona

McINALLY I.
    Thorn EMI
    Central Research Laboratories
    Trevor Road
    GB - HAYES 483 IMM, Middlesex

McNELIS B.
    Intermediate Technology
    Power Limited
    Mortimer Hill
    GB - MORTIMER RG7 3PG

MEHRMANN A.
    Gesamthochschule Kassel
    Wilhelmshöhe Allee 73
    D - 3500 KASSEL

MEISEL T.
    Universität Erlangen
    Martensstrasse 7
    D - 8520 ERLANGEN

MELEBECK T.
    ENI - Antwerp
    Kontichsesteenweg 25
    B - 2630 AARTSELAAR

MENEZES C.
    Centro de Investigacion IPN
    Departamento de Fisica
    Apdo. Postal 14-740
    Mexico - Z.P. 14 MEXICO, D.F.

MERLI P.G.
    CNR Istituto Lamel
    Via de Castagnoli 1
    I - 40126 BOLOGNA

MERTENS R.
    Katholieke Universiteit Leuven
    Kardinaal Mercierlaan 94
    B - 3030 LEUVEN

MERTIG D.
  AEG - Telefunken
  Industriestrasse 29
  D - 2000 WEDEL

MESSANA C.
  Assoreni
  Ercole Ramarini
  I - 00015 MONTEROTONDO, Roma

MEUNIER B.
  SEMA
  rue Barbès 16
  F - 92128 MONTROUGE Cedex

MICHEL M.C.
  Laboratoires de Marcoussis
  route de Nozay
  F - 91460 MARCOUSSIS

MICKAL H.
  Siemens AG
  Schuhstrasse 48
  D - 8520 ERLANGEN

MILLAR J.H.
  Advanced Solar Electric
  Power Systems Incorporation
  3, Place du Palais
  MONACO

MILNE W.I.
  Cambridge University
  Engineering Department
  Trumpington Street
  GB - CAMBRIDGE

MINOT M.
  SNES
  113 Chemin de Ronde
  F - 78290 CROISSY/SEINE

MISSOMI G.
  Agip Nucleare
  Via Giorgione 59
  I - ROMA

MORETTI M.P.
  Commission of the European
  Communities
  Joint Research Centre
  I - 21020 ISPRA

MOUHOUB A.
  Centre des Sciences et
  Technologie Nucléaires
  B.P. 1017 Alger - Gare
  Algérie - ALGER

MOURIDSEN J.
  Arco Solar
  Lendoevej 125
  DK - 7840 HOEJSLEN

MUELLER E.
  Nukem GmbH
  Hühnerberg 25
  D - 6466 GRUENDAU 2

MUELLER I.M.
  Nukem GmbH
  Rodenbacher Chaussee 6
  D - 6450 HANAU 11

MUIGG L.
  Universität Innsbruck
  Hunoldstrasse 3
  A - INNSBRUCK

MULEO A.
  Solaris SpA
  Via T. Alderotti 26
  I - 50100 FIRENZE

MUELLER J.G.
  MBB
  AE 331, Postfach 801149
  D - 8 MUENCHEN 80

MULLER J.C.
  Centre de Recherches Nucléaires
  Groupe PHASE
  23, rue de Loess
  F - 67037 STRASBOURG

MURAIB F.
  Airi/Soleras
  P.O. Box 5927
  Saudi Arabia - 27 RIYAD

MURRAY W.
  Strategies Unlimited
  201, San Antonio Circle, Suite 205
  USA - 94040 MOUNTAIN VIEW, CA

MUSELLA F.
ENEA
S.P. Anguillarese, Km 1.300
I - 00100 ROMA

NENCINI G.
Assoreni
Via Ramarini 15
I - 00015 MONTEROTONDO, Roma

NEUHAEUSSER G.
AEG - Telefunken
Industriestrasse 29
D - 2000 WEDEL

NEWHAM M.
Financial Times
10, Cannon Street
GB - LONDON ECA

NEWTON K.
AERE Harwell
Energy Technology Support
Building 156
GB - HARWELL, Oxfordshire OX11 ORA

NGUEMA NDONG F.
Ministère des Mines
Etat Gabonais
B.P. 874
Gabon - LIBREVILLE

N'GUYEN DINH H.
CIES - CNRS
15, rue Georges Urbain
F - 94400 VITRY-SUR-SEINE

NGUYEN P.H.
Institut des Sciences
de l'Ingénieur Nancy - LEPI
Parc Robert Bentz
F - 54500 VANDOEUVRE

NGUYEN V.D.
Commissariat à l'Energie Atomique
CEN SACLAY
F - 91191 GIF-SUR-YVETTE

NICOLAY D.
Commission des Communautés
européennes
D.G. Marché de l'information et
Innovation
Bât. Jean Monnet, B.P. 1907
L - LUXEMBOURG

NIEKISCH E.A.
Kernforschungsanlage Jülich
Haubourdinstrasse 6
D - 5170 JUELICH

NIELSEN F
Institute of Physics
University of Aarhus
Ny Munkegade
DK - 8000 AARHUS C

NOETZOLD H.
TUV Rheinland
Am Grauen Stein
D - 5000 KOELN 91

OPPERMANN H.
Sun Power GmbH
Eschenheimer Anlage 31
D - 6000 FRANKFURT/MAIN

ORR R.
International Solar Utilities Corp
41 E. 42nd Street, Suite 1400
USA - 10017 NEW YORK, NY

OSTOIA P.
CNR - Istituto Lamel
Via Castagnoli 1
I - 40126 BOLOGNA

OVSHINSKY B.
Energy Conversion Devices
6, Gayville Road
GB- LONDON SW11

PACE S.
Commission of the European
Communities
Joint Research Center
C.P. 1
I - 21020 ISPRA (Varese)

PALLETT R.G.
Hydraulics Research Station
GB - OX10 8BA WALLINGFORD

PALZ W.
Commission des Communautés
européennes
D.G. Science, recherche et
développement
200, rue de la Loi
B - 1049 BRUXELLES

PAPAVASSILIOU C.
   National Investment Bank for
   Industrial Development S.A.
   14, Amalias Avenue
   GR - TT 118 ATHENS

PARAGGIO V.
   Solaris S.p.A.
   Via T. Alderotti 26
   I - 50100 FIRENZE

PARISI PRESICCE L.
   ENEA
   Via Anguillarese, Km 1.200
   I - 00060 ROMA

PAVIOT J.L.
   INSA Lyon
   Laboratoire Physique de la Matière
   20, avenue A. Einstein, B- 502
   F - 69621 VILLEURBANNE

PEARSALL N.M.
   Newcastle upon Tyne Polytechnic
   Ellison Place
   GB - NEWCASTLE UPON TYNE NE1 8ST

PEETERS G.
   ENE Bruxelles
   188, van der Meerschen
   B - 1150 BRUXELLES

PELLERIN A.
   Photowatt International SA
   131, route de l'Empereur
   F - 92500 RUEIL MALMAISON

PEROTIN M.
   CEES - Univ. Sciences et Techn.
   Place E. Bataillon
   F - 34060 MONTPELLIER

PETER F.
   Total CFP
   5, rue Michel Anee
   F - 75781 PARIS Cedex 16

PETRI S.
   Solarex Trading
   Case postale
   CH - 1196 GLAND

PETRONIO F
   Parlemento Europeo
   Via Felice Casati 20
   I - 20124 MILANO

PFISTERER F.
   Inst. für Physikal. Elektr.
   Universität Stuttgart
   Böblinger Strasse 70
   D - 7000 STUTTGART 1

PIRAZZI R.
   Samim Abrasivi
   Piave 60
   I - 28037 DOMODOSSOLA

PIVOT J.
   Université de Lyon 1
   43, Bd 11 Novembre
   F - 69100 VILLEURBANNE

PIZZINI S.
   Heliosil S.p.A.
   Via Franchetti 1/A
   I - 20124 MILANO

POSBIC J.
   Université Claude Bernard Lyon
   Dépt. Physique des matériaux
   43, bvd. 11. Novembre 1918
   F - 69621 VILLEURBANNE

POTARD C.
   Centre d'Etudes Nucléaires
   DMG/SEM/LES
   85 X
   F - 38041 GRENOBLE Cedex

PRENI A.
   ENEL
   Via Bastioni di Porta Volta 10
   I - 20100 MILANO

PRINCE M.
   US Department of Energy
   1000 Independence Avenue, S.W.
   USA - 20585 WASHINGTON,DC

PUTNEY Z.
   Semix Incorporated
   15801 Gaither Drive
   USA - 20877 GAITHERSBURG, Maryland

RADELAAR S.
  University of Technology
  Rotterdamseweg 137
  NL - 2628 AL DELFT

RAGA F.
  Istituto di Fisica
  Università di Cagliari
  Via Ospedele 72
  I - 09100 CAGLIARI

RAHN O.
  Heliotronic GmbH
  Postfach 1129
  D - 8263 BURGHAUSEN

RAMAMURTHY E.S.
  Bharat Heavy Electricals
  Limited, CED
  Mysore Road, P.B. No 2606
  India - 560026 BANGALORE

RAMOS M.R.
  Solarex Trading Office
  CH - 1196 GLAND

RASCH K.D.
  AEG - Telefunken
  Theresienstrasse 2
  D - 7100 HEILBRONN

RASMUSSEN J.E.
  Danfoss, Att. D11-1
  DK - 6430 NORDBORG

RAU H.
  Am Bruch 20
  D - 2413 MOERS

RAU K.
  Siemens AG
  Werner-von-Siemens-Str. 50
  D - 8520 ERLANGEN

RAVA P.
  MIT
  Via Druento 20
  I - 10040 SAVONERA

RAYNE A.R.
  International Copper Research Ass.
  Brosnan House, Darkes Lane
  GB - EN6 1BW POTTERS BAR, Herts.

REA S.N.
  Arco Solar Incorporation
  20554 Plummer Street
  USA - 91311 CHATSWORTH, CA

REAL M.G.
  Swiss Federal Institut
  for Reactor Research
  CH - 5303 WUERENLINGEN

REMY C.
  Photowatt International SA
  131, route de l'Empereur
  F - 92500 RUEIL MALMAISON

RENDITORE C
  Astor Solar Energy Srl.
  Valtellina 20
  I - 20159 MILANO

RICAUD A.
  France-Photon
  ZI Les Agriers
  F - 16015 ANGOULLEME

RICHMOND E.
  Electricity Supply Board
  27 LWR Fitzwilliam Street
  Ireland - DUBLIN 2

RICKUS
  Battelle Institut E.V.
  Am Römerhof 35
  D - 6000 FRANKFURT/MAIN

RIMMER A.
  Pilkington Bros. P.L.C.
  Prescot Road
  GB - ST. HELENS WA10 3TT, Merseyside

RITTER H.
  Siemens AG
  Werner-von-Siemens-Str. 50
  D - 8520 ERLANGEN

ROCHELEAU R.
 University of Delaware
 Institute of Energy Conversion
 USA - 19711 NEWARK, DE

ROD R.L
 Monogram Industries Incorporation
 1299 Ocean Avenue
 USA - 90401 SANTA MONICA

RODWELL B.
 Corporation of Trinity House
 Savage Gdns, Tower Hill
 GB - LONDON ECJN4DH

ROGER J.A.
 Université Claude Bernard
 Lyon 1
 43, bvd du 11 Novembre 1918
 F - 69622 VILLEURBANNE

ROLL H.
 Zeppelin Metall
 Lilienweg 12
 D - 7770 UEBERLINGEN

ROLLAND M.
 Université de Sciences et
 Techniques du Languedoc
 Place E. Bataillon
 F - 34060 MONTPELLIER

ROMEO N.
 Istituto di Fisica Parma
 Via M. D'Azeglio 85
 I - 43100 PARMA

ROSSI E.
 Commission of the European
 Communities
 Joint Research Center
 C.P. 1
 I - 21020 ISPRA (Varese)

ROY R.B.
 Studsvik Energiteknik AB
 S - 61182 NIKOEPING

ROZSA R.S.
 FONE-MAT S.A. Ind. p. Telecom.
 Al. Santos, 285 Ap. 141
 Brazil - 01419 SAO PAULO

RUEDA SANCHEZ F.
 Univ. Autonoma Madrid
 Garcia Molina
 E - 35 MADRID

RUIZ J.M.
 Instituto Energia Solar
 ETSI Telecomunicacion
 E - MADRID 3

RUMBOLD G.
 National Research Council
 Solar Energy Project
 Building 97
 Canada - OTTAWA K1AORG

SAB-NIFE
 Viale Cembrano 11
 I - 16148 GENOVA

SADEGHI M.
 Experimentalphysik
 Universität Linz
 Altenbergerstrasse 69
 A - 4040 LINZ

SALA G.
 Instituto Energia Solar
 ETSI Telecomunicacion
 E - MADRID 3

SALKALACHEN S.
 Department of Physics
 University of Hull
 GB- HULL HU6 7RX

SALLES D.
 Elf Aquitaine
 Centre de Recherches
 B.P. 22
 F - 69360 ST SYMPHORIEN D'OZON

SAMI G.E.
 Consulting Engineer
 P.O. Box 394
 Egypt - MEADI

SANCHEZ QUESADA F.
 Universidad Complutense
 Ciudad Universitaria F. Fisicas
 E - MADRID 1

SANKIES M.
University of Guyana
Turkeyen Campus
P.O. Box 101110
Guyana - GEORGETOWN

SARIS F.W.
FOM Institute
Kruislaan 407
NL - 1098 SJ AMSTERDAM

SARNO A.
ENEA
S.P. Anguillarese
I - 00100 ROMA

SASSI G.
Università di Milano
Via Celoria 16
I - 20133 MILANO

SAUTREUIL B.
Institut National des
Sciences Appliqués - INSA
20, avenue A. Einstein, bât 502
F - 69621 VILLEURBANNE

SBERVEGLIERI G.
Università di Parma
I - PARMA

SCAFE E.
Assoreni
Via E. Ramarini 32
I - 00015 MONTEROTONDO, Roma

SCERRI E.
University of Malta
Rosendale Old Railway Avenue
Malta - BALZAN

SCHEER H.
Inst. für Werkstoffe der E-Technik
Technische Universität Berlin
Jebenstrasse 1
D - 1000 BERLIN 12

SCHEI A.
Elkem a/s, R & D Center
P.O. Box 40
N - 4620 VAGSBYGD

SCHELLEMAN F.
Ministry of Economic Affairs
Grashof 72
NL - 2403 VD ALPHEN a d. RIJN

SCHLOSSER V.
L. Boltzmanninstitut für
Festkörperphysik
Kopernikusgasse 15
A - 1060 WIEN

SCHMID F.
Crystal Systems Incorporation
35 Congress Street
USA - 01970 SALEM, MA

SCHMID J.
ISE - Freiburg
Oltmannsstrasse 22
D - 7800 FREIBURG

SCHMIDT W.
AEG - Telefunken
Theresienstrasse 2
D - 7100 HEILBRONN

SCHNELL W.
Kommission der Europäischen
Gemeinschaft
rue de la Loi 200
B - 1049 BRUXELLES

SCHNOELLER M.
Siemens AG
Otto-Hahn-Ring 6
D - 8000 MUENCHEN

SCHOCK H.W.
Inst.f. Physikalische Elektronik
Universität Stuttgart
Böblinger Strasse 70
D - 7000 STUTTGART 1

SCHOEFER C.
Commission des Communautés
européennes
D.G. Science, recherche et
développement
200, rue de la Loi
B - 1049 BRUXELLES

SCHOTT T.
DFVLR
Pfaffenwaldring 38
D - 7000 STUTTGART 80

SCHROEDER M.
   VDM Aluminium GmbH
   D - 6000 FRANKFURT

SCHUNCK J.P.
   CRN Laboratoire PREN
   23, rue du Loess
   F - 67200 STRASBOURG

SCHURICH B.
   NUKEM GmbH
   Postfach 110080
   D - 6450 HANAU 11

SCHURINK F.
   Kema NV
   Utrechtse Weg 310
   NL - 6812 AR ARNHEM

SCHUSTER G.
   Commission des Communautés
   européennes
   D.G. Science, recherche et
   développement
   200, rue de la Loi
   B - 1049 BRUXELLES

SCHWARZ R.
   Bellevue 26
   CH - 2000 NEUCHATEL

SCOTT R.D.W.
   Lucas BP Solar Systems Ltd.
   Windmill Road
   GB - HADDENHAM, Aylesbury HP17 8JB

SCOTT P.
   Calif. Export Int.
   17500 Lemarsh Street
   USA - 91325 NORTHRIDGE

SCOTT G.
   BP Research Center
   New Technology Division
   Chertsby Road
   GB - SUNBURY ON THAMES

SEEGER K.
   L. Boltzmann Institut
   für Festkörperphysik
   Kopernikusgasse 15
   A - 1030 WIEN

SELLES L.
   Seri Renault Engineering
   2, avenue du Vieil Etang
   F - 78390 BOIS D'ARCY

SEVERNS J.G.
   Naval Research Laboratory
   Code 7712
   USA - 20375 WASHINGTON

SGARD
   Photowatt International SA
   131, rue de l'Empereur
   F - 92500 RUEIL MALMAISON

SHAW D.
   Frank Wheldon School
   Coningswath Road
   GB - NG4 3SH CARLTON, Nottingham

SIFFERT P.
   Centre de Recherches Nucléaires
   Groupe PHASE
   F - 67037 STRASBOURG Cedex

SIGMUND H.
   Fraunhofer Institut
   Paul-Gerhardt-Allee 42
   D - 8000 MUENCHEN

SIGURD D.E.
   Institute of Microwave Technology
   S - 10044 STOCKHOLM

SIMONI G.
   Pragma S.p.A.
   Via Po 25/A
   I - 00198 ROMA

SINGAL C.M.
   University of Roorkee
   India - 247672 ROORKEE

SINKE W.
   FOM Institute
   Kruislaan 407
   NL - 1098 SJ AMSTERDAM

SIRTL E.
Heliotronic
Johann-Hess-Strasse 24
D - 8263 BURGHAUSEN

SMITH G.J.
Engineering Department
Cambridge University
Trumpington Street
GB - CAMBRIDGE

SMITH H.J.
CSP, Council for Scientific
and Industrial Research
42, Trent Street
South Africa - MURRAYFIELD, PRETORIA

SMITH H.V.
Mississippi Country
Community College
915 Holly Street
USA - 72315, Blythville, Arkansas

SOLMI S.
CNR-LAMEL
Via Castagnoli 1
I - 40125 BOLOGNA

SOMMER E.
Dornröschenstrasse 45
D - 8000 MUENCHEN 83 WALDPERLACH

SONCINI G.
CNR-LAMEL
Via Castagnoli 1
I - 40126 BOLOGNA

SPENCER B.
Acurex Corporation
485 Clyde Avenue
USA - 94042 MOUNTAIN VIEW, Ca

SPIRITO P.
Università di Napoli
Via Claudio 21
I - 80125 NAPOLI

SPITZER Jr. D.M.
Photon Power Incorporation
13, Founders
USA - 79906 EL PASO, Texas

STAHL W.
Fraunhofer Institut für
Solare Energiesysteme
Oltmannsstrasse 22
D - 7800 FREIBURG

STAMPA L.
ENEA
Via Anguillarese
I - ROMA

STARR M.R.
Sir William Halcrow & Partners
Burderop Park
GB - SWINDON SN4 OQD

STAUFFER F.
Meyer & Burger
CH - 3613 STEFFISBURG

STAUGAARD H.
Grundvos AS
Revlingvej 7
DK - 8850 BJERRINGBRO D4

STEEB H.
DFVLR/FRG
Pfaffenwaldring 38-40
D - 7000 STUTTGART

STEVENSON W.C.
Solar Generators Singapore PTE Ltd
151 Lorong Chuan
Singapore - 1955 SINGAPORE

STOCK H.
Wacker-Chemitronic
D - 8263 BURGHAUSEN

STOJANOVIC M.
Institut "Boris Kidrich" Vinca
P.O. Box 522
Yougoslavia - BEOGRAD

STONE J.L.
Solar Energy Research Institut
1617 Cole Boulevard
USA - 80401 GOLDEN, Colorado

STORTI G.
  Solarex Corporation
  1335 Piccard Drive
  USA - 20850 ROCKVILLE

STRAAYER A.
  Eindhoven University of Technology
  den Dolech 2 (NV 01.04)
  NL - 5612 AZ EINDHOVEN

STRUB A.
  Commission des Communautés
  européennes
  D.G. Science, recherche et
  développement
  200, rue de la Loi
  B - 1049 BRUXELLES

SUTTON C.
  Frank Wheldon School
  Coningswath Road
  GB - CARLTON, Nottingham NG4 3SH

SZYDLO N.
  Thomson - CSF
  Domaine de Corbeville
  F - 91401 ORSAY

TALINI M.
  La Metalli Industriae SpA
  Via della Repubblica
  I - 55052 FORNACI-DI-BARGA

TARRICONE L.
  Istituto di Fisico
  I - PARMA

TASCHINI A.
  ENEL
  Via Carducci 1/3
  I - 20123 MILANO

TAWADA Y.
  Kanegafuchi Chemical Industry Co Ltd.
  Central Research Laboratories
  1-2 80 Yoshida-cho
  Japan - HYOGO-KU

TEJEDOR P.
  Torres Quevedo (CSIC)
  C/ Serrano 144
  E - 6 MADRID

TELETTRA S.p.A.
  Via Trento 30
  I - VIMERCATE

THEREZ F.
  LAAS du CNRS
  7, avenue du Colonel Roche
  F - 31400 TOULOUSE

THIEBAUT B.
  CNRS
  LEAA ENSCP
  11, rue Pierre et Marie Curie
  F - 75231 PARIS Cedex 05

THISSEN H.G.
  ITW GmbH
  Theaterstrasse 90
  D - 5100 AACHEN

THOMAS R.
  Tideland Signal Ltd
  25B Trowers Way
  GB - RH1 2LH REDHILL, Surrey

THOMAS R.E.
  Carleton University
  Colonel By Drive
  Canada - OTTAWA K15 5B6

THOMPSON W.
  Applied Solar Energy Corporation
  15251 E. Don Julian Road
  USA - 91746 CITY OF INDUSTRY, Ca

TOMSIN M.
  Fabricable Belgo Solar
  Heideveld 1
  B - 1511 HUIZINGEN

TOPEL B.
  Heliodinamica SA
  C.P. 8085
  Brazil - 01000 SAO PAULO

TOWNSEND W.G.
  Royal Military College of Science
  Shrivenham
  GB - SWINDON SN6 8LA

TRAEDER K.
 WIP München
 Sylvensteinstrasse 2
 D - 8 MUENCHEN

TREBLE F.C.
 Consultant
 43, Pierrefondes Avenue
 GB - GU14 8PA FARNBOROUGH

TSCHARNER R.
 Institut de Microtechnique
 71, rue Maladière
 CH - 2007 NEUCHATEL

TURKENBURG W.C.
 Fysisch Laboratorium
 Riks Universiteit Utrecht
 Princetonplein 5, P.O. 80000
 NL - 3508 TA UTRECHT

TURNER D.P.
 Sheffield University
 Department of Electronic Engineering
 Mappin Street
 GB - SHEFFIELD S1 8JD

TURNER G.B.
 Arco Solar Incorporation
 20554 Plummer Street
 USA - 91311 CHATSWORTH, CA

ULLRICH K.
 Kernforschungsanlage Jülich
 Postfach 1913
 D - 5170 JUELICH

ULMI R.
 Institut für angewandte Physik
 ETH - Hänggerberg
 CH - 8093 ZUERICH

VACHTSEVANOS
 Polytechnic School
 GR - XANTHI

VAN BEEK F.
 University of Technology Eindhoven
 den Dolech 2 (NV 01.09)
 NL - 5612 AZ EINDHOVEN

VANDERMAESEN F.
 Technical University Eindhoven
 den Dolech 2, P.O. Box 513
 NL - 5600 MB EINDHOVEN

VAN DER WEG W.F.
 Utrecht State University
 Department Technical Physics
 Princetonplein 5
 NL - 3508 TA UTRECHT

VANECEK M.
 Institute of Physics
 Czechoslovak Academy of Sciences
 Na Slovance 2
 CZ - 18040 PRAGUE 8

VAN GYSEL M.A.
 IDE
 Parc Industriel
 B - 5430 ROCHEFORT

VANHALEN P.
 Metallurgie Hoboken
 Overpelt NV
 Greinerstraat 14
 B - 2710 HOBOKEN

VANHECKE L.
 Metallurgie Hoboken
 Overpelt NV
 Greinerstraat 14
 B - 2710 HOBOKEN

VAN OVERSTRAETEN R.J.
 ESAT - Katholieke Univ. Leuven
 Kardinaal Mercierlaan 94
 B - 3030 HEVERLEE

VAN ZOLINGEN R.J.C.
 Holecsol
 P.O. Box 30
 NL - 5700 AA HELMOND

VEDEL J.
 CNRS - ENSCP
 11, rue Pierre et Marie Curie
 F - 75005 PARIS

VERA E.S.
 Brown University
 Division of Engineering, Box D
 USA - 02912 PROVIDENCE, Rhode Island

VERDIER J.M.
    Rhone Poulenc Recherches
    12, rue des Gardinoux
    F - 93308 AUBERVILLIERS

VERIE C.
    CNRS
    Laboratoire Physique du Solide
    et Energie Solaire
    F - 06560 VALBONNE

VIALART J.C.
    Photowatt International SA
    131, route de l'Empereur
    F - 92500 RUEIL MALMAISON

VIEUX-ROCHAZ L.
    CEN.G LETI
    avenue des Martyrs 85X
    F - 38041 GRENOBLE

VILLOZ M.
    Solarex Trading Office
    CH - 1196 GLAND

VOIGT C.
    DFVLR
    Pfaffenwaldring 38-40
    D - 7000 STUTTGART 80

VOJDANI S.
    AEG - Telefunken
    Egenbuttel Weg 29
    D - 2 WEDEL

VON BACHHAUS A.
    DARCOM
    Bremerstrasse
    D - 6000 FRANKFURT 1

VON TOBEL G.
    Swiss Federal Institute for
    Reactor Research
    CH - 5303 WUERENLINGEN

VOSS B.
    FH6 -ISE
    Oltmannsstrasse 22
    D - 7800 FREIBURG

VYVERMAN A.
    Programmatie van het
    Wetenschapsbeleid
    Wetenschapsstraat 8
    B - 1040 BRUXELLES

WAGEMANN H.G.
    Inst. f. Werkstoffe der E-Technik
    Technische Universität Berlin
    Jebenstrasse 1
    D - 1000 BERLIN 12

WAGNER A.
    Universität Stuttgart
    Breitscheidstrasse 3
    D - 7000 STUTTGART 1

WAGNER P.
    Heliotronic GmbH
    P.O. Box 1129
    D - 8263 BURGHAUSEN

WAHL G.
    AEG - Telefunken
    Theresienstrasse 2
    D - 7100 HEILBRONNE

WAKEFIELD G.F.
    Arco Solar Incorporation
    20554 Plummer Street
    USA - 91311 CHATSWORTH, Ca

WENZLER M.F.
    Photon Power Incorporation
    13, Founders Blvd
    USA - 79906 EL PASO, Texas

WERNER G.
    Universität Bremen
    P.O. Box 330440
    D - 2800 BREMEN

WHITAKER R.D.
    DSET Laboratories Incorporation
    Box 1850
    Black Canyon Stage I
    USA - 85029 PHOENIX

WIEMHOEFER H.D.
    University of Dortmund
    Physical Chemistry I
    Otto-Hahn-Strasse
    D - 4600 DORTMUND 50

WILLMES H.
    Varta Batterie AG
    Dieckstrasse 42
    D - 5800 HAGEN

WILSON H.R.
    Inst. für Solare Energiesysteme
    der Fraunhofer Gesellschaft
    Oltmannsstrasse 22
    D - 7800 FREIBURG

WINTERUNG G.
    MBB
    AE 331, Postfach 801149
    D - 8 MUENCHEN 80

WITTWER V.
    Fraunhofer Institut für
    Solare Energiesysteme
    Oltmannsstrasse 22
    D - 7800 FREIBURG

WODITSCH P.
    Bayer AG
    Deswatinesstrasse 83
    D - 4150 KREFELD

WOLFE P.R.
    Solapak
    Factory Three
    Cock Lane
    GB - HIGH WYCOMBE HP13 7DE

WOERNER J.
    Nukem GmbH
    Postfach 110080
    D - 6450 HANAU 11

WRIGHT D.E.
    Sir William Halcrow & Partners
    3, Shortlands, Hammersmith
    GB - LONDON W6

WRIXON G.
    University College
    Ireland - CORK

WYERS Jr. L.T.
    2312 Moultrie Road
    USA - 29020 LAMDEN, So Car.

YABLONOVITCH E.
    Exxon Research
    P.O. Box 45
    USA - 07036 LINDEN, N.J.

YERKES B.
    Arco Solar Incorporation
    Plummer Street 20554
    USA - CHATSWORTH, Ca

YUKIMOTO Y.
    Mitsubishi Electric Corporation
    4 - 1 Mizuhara Itami
    Japan - HJOGO

ZAMA W.
    Shell Oil Company
    900 Louisiana Street
    USA - 77002 HOUSTON, Texas

ZANI P.E.
    Ansaldo S.p.A.
    Via N. Lorenzi 8
    I - 16152 GENOVA

ZIGNANI F.
    LAMEL -CNR
    Via Castagnoli 1
    I - BOLOGNA

ZWIBEL H.S.
    N.M. Solar Energy Institute
    P.O. Box 3 SOL
    USA - 88003 LAS CRUCES, New Mexico

INDEX OF AUTHORS

ABADIE, M., 848
AIGA, M., 745
AL-SANI, A., 57
ALBERGAMO, V., 270,296
ALDISSI, M., 848
ALDRICH, D., 427,1039
ALESSANDRINI, P., 788
AMZIL, H., 421
ANDERSON, W.A., 532,759
ANGUET, J., 387
ARCIDIACONO, V., 115,258
ARJONA, F., 809
ARMINI, A.J., 1018
ARNDT, W., 404
ARYA, R.R., 727
AUBERT, B., 139
AUBRIL, P., 1033
AULICH. H.A., 868
AVARITSIOTIS, J.N., 1053

BANERJEE, T., 203
BARBE', M., 621
BARBIER, D., 453,1007
BARLAUD, M., 1068
BARNETT, A.M., 931
BARON, B.N., 798
BARRADAS, R., 546
BARRETT, A.L., jr., 20
BARUCH, P., 654
BASOL, B.M., 719
BASU, P., 203
BASU, S., 574
BAUER, G.H., 773
BAUFAY, L., 839
BEAULIEU, R., 727
BEICHLER, J., 537
BELOUET, C., 896,970
BENMALIK, M., 81
BERMAN, E., 606
BHUSHAN, M., 844
BILGER, G., 773
BLACKBOROW, P., 769
BLOSS, W., v
BLUM, N.A., 783
BOBO, J.C., 377
BOEHM, M., 516

BOER, K.W., 488
BOES, E.C., 600
BONTEMPS, A., 522
BORLENGHI, P.L., 270
BOSE, D.N., 574
BOST, M.C., 737
BOUCHIKHI, B., 804
BOUE', S., 562
BOUTRIT, C., 492
BOZZOLO, D., 97
BRABBEN, T.E., 125
BRAICOVICH, L., 874
BRESTOVANSKY, D.F., 798
BUBE, R., 836
BULLO, P., 286
BUNKER, S.N., 1018
BURGESS, E.L., 238

CABESTANY, J., 498
CADENE, M., 848
CAHARD, A., 453
CALLAGHAN. W.T., 364
CALLIARI, L., 874
CALZOLARI, P.U., 106
CAMANI, M., 97
CAMANZI, A., 788
CAMPBELL, D.S., 1053
CARDINALI, G.C., 106,232
CARMICHAEL, D.C., 275
CAROUBALOS, C., 1053
CASTANER, L., 340
CASTLE, J.A., 275
CELIO, T., 97
CEMBALI, G.F., 1013
CEPPI, P., 320
CHANDRASEKHAR, S., 804
CHAOUI, A., 648
CHARTIER, E., 749
CHAUMAIN, G., 1068
CHEEK, G., 432,926
CHELI, F., 270
CHEMISKY, G., 1007
CHEN, P.Y., 568
CHEN. J.R., 568
CHENEVIER, P., 1065
CHENLO, F., 217

CID, M., 511
CILLIERS, R., 522
CLAYBOURN, M., 732
CLOSAS, L., 340
CLOW, H., 732
COHEN, J., 1065
COHEN-SOLAL, G.W., 557
COHEN-SOLAL, G., 557,621
COM-NOUGUE, J., 994
CORDES, V., 89
CORSI, S., 115
COURCELLE, E., 1023
COUREAU, P., 387
COUTTS, T.J., 459
COWACHE. P., 822
CREST, J.P., 421
CUEVAS, A., 511
CUMBERBATCH, T.J., 551,732
CUNIOT, M., 654
CURRAN, J.S., 853

DALDINI, O., 97
DAS, S., 203
DAVID, G., 1033
DE ANGELIS, L., 788
DE PAUW, Pù., 432
DE ZELICOURT, Y., 377
DEN BOER, W., 764
DESCHAMPS, N., 970
DESOMBRE, A., 377,387
DEVILLE, J.P., 970
DIANA, G., 270
DIAS, A.G., 546
DICKINSON, P.M.G., 732
DIETL, J., 941
DIETZE, W., 868
DISPA, D., 839
DONON, J., 377,387,1033
DUGAS, J., 421
DUMAS, J.P., 994

EDENBURN, M.W., 600
EISENRITH, K.H., 868
ELIAS, G., 17

ELIZALDE, E., 809
EMANUELE, G., 286

FABRE, E., 346
FALCO, M., 270
FALDELLA, E., 232
FALLY, J., 421,455
FANETTI, E., 671
FARINELLI, U., 9
FEDORKO, P., 562
FELDMAN, C., 783
FENZL, H.J., 946
FERBER, R.R., 364
FEUCHT, D.L., 580
FISCHER, R., 778
FITTIPALDI, F., 13
FLORES, C., 671
FOELL, H., 976
FOGARASSY, E., 1023
FRAENKEL, P.L., 71
FREIENSTEIN, B., 976
FREUDENBERG, K., 135
FRISSON, L., 926
FUHS, W., 448
FUKATSU, T., 704

GALLONI, R., 1013
GALUZZI, F., 477,788
GARCIA, J., 340
GARULLI, A., 106
GASBARRA, M., 286
GASPARINI, M., 874
GAY, C.F., 335,353
GAY, R.R., 714
GEIM, K., 976
GERVAIS, J., 421
GERWIN, H.J., 222
GIBBONS, D.J., 732
GIOVANNINI, C., 97
GIRISCH, R., 643
GLOECKL, J., 1050
GOETZBERGER, A., 682
GOUSKOV, L, 648
GRABMAIER, J.G., 976
GRASSI, G., 248
GRIL, C., 648
GRIMBAS, E.J., 325
GUARINI, G., 591,671
GUCHER, P., 330
GUEKOS, G., 320
GUENEL, C., 955

GUIMARAES, L., 546

HA, H.T., 443
HAAF, W., 677
HAGENLOCHER, K., 677
HALDER, N.C., 437
HAMILTON, B., 551
HAN, M.K., 759
HAUSCHILDT, D., 448
HEDSTROEM, J., 227
HEIDLER, K., 682
HELM, P., 1050
HEWIG, G.H., 291,793
HEZEL, R., 638
HIGAKI, Y., 745
HILL, R., 732
HOGGARTH, K.G., 783
HOLM, C., 941
HOLZENKAEMPFER, E., 778
HOCHNHOUT, D., 1029
HUEBNER, H.P., 291
HUNG, T.F., 568
HURAIB, F., 57
HUSCHKA, H., 399
HWANG, H.L., 568

IIDA, M., 965
ILICETO, A., 115
IMAMURA, M.S., 57
INSAUSTI, M., 340

JACKSON, J.L., 101
JAIN, S.C., 432
JANSSENS, R., 432,926
JOHNSON, E.L., 611
JOLIET, M.C., 839
JORDAN, J.F., 737

KAELLBAECK, B., 227
KALAITZAKIS, C.K., 325
KALBITZER, S., 1059
KAMARINOS, G., 1065
KANANI, D., 714
KANE. M., 557
KANICKI, J., 562
KAPLAN, D., 749
KAPUR, V.K., 335

KATO, M., 745
KAZMERSKI, L., 727
KE, K.E., 551
KERN, R., 516
KERRAND, E., 896
KEY, T.S., 238
KHATTAK, C.P., 980
KHOSHAIM, B., 57
KIESSLING, K.P., 130
KILBY, J.S., 611
KINNELL, G.H., 189
KIPPERMAN, A.H.M., 280
KOCKA, J., 443
KONDO, H., 616
KONDO, M., 698
KORUPP. K.H., 89
KREBS, K., 158
KRUEHLER, W., 754
KUCHAR, F., 960
KUWANO, Y., 704
KWIETNIAK, M., 727

LAHRI, R., 759
LANDSBERG, P.T., 485,1056
LAPLAZE, D., 557
LAUDE, L.D., 839
LAUGIER, A., 453,1007
LAURENT, B., 212
LAUVRAY, H., 1033
LAUWERS, P., 1063
LEEMPOELS, M., 926
LEIPOLD, M.H., 985
LEON, M., 809
LEPLEY, B., 492
LEVART, M., 818
LEVHA, L, 120
LIMBOURG, N., 312
LINCOT, D., 621,822
LIU, L.M., 568
LOFERSKI, J.J., 659,727
LORENZO, E., 666
LOUBLY, P., 1033
LUDSTECK, A., 946
LULLI, G., 1013
LUQUE, A., 511,666
LUQUET, H., 648
LUTZ, P.J., 798
LYNCH, L.D., 980

MACOMBER, H.L., 30
MADET, D., 307

MAGARINO, J., 749
MAKI, N., 965
MALETTA, G., 788
MALIK, M.A.S., 71
MANFREDI, A., 335
MANIFACIER, J.C., 648
MARFAING, Y., 688
MARI, C.M., 874
MARTINS, R., 546
MARTINUZZI, S., 421,804
MAS, A., 340
MASSAAD, S., 301,330
MATHIAN, G., 421
MATSUI, T., 965
MAUTREF, M., 896
MAZZONE, A., 1013
McINALLY, I.D., 551,732
McNELIS, B., 144
MELL, H., 448,537
MENEZES, C., 836
MERLI, P.G., 1013
MERTENS, R., 432,643,926
MESLI, A., 994
MICHEL, M.C., 374
MILNE, W.I., 769
MINARI, F., 421
MOELLER, M., 754
MOHN. J., 67
MOREL, D.L., 714
MOUHOUB, A., 81
MUECK, G., 709
MUELLER, G., 542,709
MUELLER, J.C., 994
MUKHOIPADHYAY, K., 203
MULEO, A., 382
MULLER, J.C., 1023

NAKAGAWA, M., 965
NAKAMURA, C., 616
NAKANO, S., 704
NARDI, F., 382
NEUHAEUSSER, 67
NGUYEN, P.H., 492
NIELSEN, F., 1034
NIPOTI, R., 1013
NISHIMURA, K., 698
NISHIWAKI, H., 704
NOBILI, D., 106,410
NYLANDSTED LARSEN, A., 1034

OEMRY, A., 648

OHNISHI, M., 704
OKAMATO, H., 698
OUALID, J., 421

PALETTA, F., 671
PALLETT, R.G., 125
PALZ, W., 3
PAMINI, R.,, 97
PANDE, P.C., 732
PARAGGIO, V., 382
PEARSALL, N.M., 459,732
PETERSEN, G., 67
PFISTERER, F., 793
PFLEIDERER, H., 754
PHAM VAN VUI, 212
PHILIPPE, R., 853
PICHAUD, B., 421
PICMAUS, E., 130
PIGEOLET, A., 839
PIMPALE, A., 485
PIVOT, J., 301,330
PIZZINI, S., 874
PLAETTNER, R. 754
POSBIC, J., 301,330
POST, H.N., 238,275
PRESS. R., 427,1039
PRESSLEY, R., 427,1039
PREVI, A., 115
PRINCE, M.B., 20
PROUST, N., 749
PUTNEY. Z., 990
PYLE, B., 335

RAJESWARAN, G., 532
RAMPRAKASH, Y., 574
RASCH, K.D., 919,999
RAUSCHER, B., 754
RAVELET, S., 492
REAL. M., 51
REDEALLI, F., 874
REINELT, M., 1059
REQUIER, J.P., 1068
REUYL, J.S., 111
REVEL, G., 970
RICAUD, A.M., 392
RICKERT, H., 827
RICKUS, E., 831
RISSER, V.V., 179
RITCHIE, D.W., 628
ROCHELEAU, R.E., 798
ROD, R.L., 719

ROGER, J.A., 301,330
ROLLAND, M., 848
ROSENFIELD, T., 990
ROSS, B., 527
ROSS, R.G., jr., 169
ROUAN, P., 1068
ROY, K., 919
RUEDA, F., 809
RUIZ, J.M., 511
RUMBURG, J., 335
RUSSEL, T.W.F., 798
RUSSELL, G., 732

SAHA, H., 203
SALA, G., 217
SALIM. A.A., 57
SALVADE', G., 97
SANCHEZ-SINENCIO, F., 836
SANCROTTI, M., 874
SARIS, F.W., 1029
SATKIEWICZ, F.G., 783
SATO, K., 616
SAUTREUIL, B., 453
SAWADA, T., 965
SCAFE', E., 477,788
SCHAEFER, J., 868
SCHAETZLE, R., 316
SCHLOSSER, V., 960
SCHMID, F., 980
SCHMID, J., 316
SCHMIDT, I.E., 827
SCHMIDT, W., 919,999
SCHOCK, H.W., 404,793
SCHOERNER, R., 638
SCHULZE, F.W., 868
SCHURICH, B., 399
SCHUTT, R.D., 111
SEEGER, K., 960
SELLES, L, 139
SEVERNS., J., 659
SHIRAHATA, K., 616
SIFFERT, P., 901,994,1023
SIGURD, D., 227
SIMON, H., 542
SIMON, M., 709
SINGAL, C.M., 503
SINGH, R.V., 503
SINGH, V.P., 737
SINKE, W., 1029
SIPEK, E., 443
SIRTL, E., 858,941
SMEKENS, G.R., 1063
SMITH, G.J., 769

SMITH, H.V., 94
SMITH, H.J., 522
SMITH, M.B., 980
SOLCA', F., 97
SØRENSEN, G., 1034
SOUZA E., A., 836
SPENCER, R.M., 184
SPINEDI, C., 97
SPIRITO, P., 296
SPITZER, D.M., jr., 737
SPITZER, M., 659
SPITZER, M.B., 1018
STARR, M.R., 40,144
STIKA, O., 443
STOLZ, W., 1059
STONE, J.L., 628
STRONG, S.J., 151
STUCHLIK, J., 443
STUKE, J., 778
SUN, C.Y., 568
SUNG, P., 759
SUREK, T., 628
SZYDLO, N., 749

VANACEK, M., 443
VANDER DONCKT, E., 562
VAZQUEZ-LOPEZ, C., 836
VEDEL, J., 818,822
VERA, E.S., 659,727
VIALARET, G., 212

WAGEMANN, H.G., 516
WAGNER, R., 827
WAHL, G., 919
WEBER, K., 448
WERNER, P.G., 883
WIEMHOEFER, H.D., 827
WILDES, P.D., 606
WILLIAMS, E.W., 732
WINTERLING, G., 542,709
WITT, C.E., 628
WITTWER, V., 682
WOERNER, J., 399
WOODS. J., 732
WRIGHT, D.E., 71
WRIGLEY, C., 990

TAMAI, K., 965
TANNER, D.P., 714
TARRANT, D., 427,1039
TASCHINI, A., 115,258
TAWADA, Y., 698
TESSARI C., 994
TEXIER-HERVO, C., 896,970
THIEBAUT, B., 818
TOMACIELLO, R., 788
TRAEDER, K., 1050
TREBLE, F.C., vii
TRISKA, A., 443
TSENG, E.S., 719
TSUDA, S., 704
TSUGE, K., 698
TURNER, G.B., 427,1039

YABLONOVITCH, E., 465
YANG, M.H., 568
YUKIMOTO, Y., 616,745

ZAMBONI, F., 97
ZANI, A., 106
ZAPIEN NATEREN, F., 804
ZEHAF, M., 421
ZIGNANI, F., 1013
ZWIBEL, H.S., 179

ULLAL, H.S., 714
ULMI, R., 320
URBACH, H.P., 868

VACHTSEVANOS, G.J., 325
VAN GYSEL, M., 312
VAN OVERSTRAETEN, R., 432,
            643,926
VAN STRIJP, R.M., 764